D1201889

Planetary Sciences

The space age, with lunar missions and interplanetary probes, has revolutionized our understanding of the Solar System. Planets and large moons, which previously appeared in telescopes only as fuzzy disks, have become familiar worlds, with a range of diverse properties. Large numbers of asteroids, comets and small moons have now been discovered, and many of these objects studied in detail. Moreover, the number of planets now known to orbit stars other than our Sun far exceeds those within our own Solar System. As a result, our understanding of the process of star and planet formation is increasing all the time.

Planetary Sciences presents a comprehensive coverage of this fascinating and expanding field at a level appropriate for graduate students and juniors and seniors majoring in the physical sciences. The book explains the wide variety of physical, chemical and geological processes that govern the motions and properties of planets. Observations of the planets, moons, asteroids, comets and planetary rings in our Solar System, as well as extrasolar planets, are described and the process of planetary formation is discussed.

This text will be used by graduate students and researchers in astronomy, planetary science and Earth science.

IMKE DE PATER is a Professor at the University of California, in the departments of Astronomy and Earth and Planetary Science. She was born in the Netherlands and received her Ph.D. cum Laude in 1980 from Leiden University. She held a postdoctoral fellowship at the Lunar and Planetary Laboratory at the University of Arizona from 1980 to 1983, after which she moved to the University of California at Berkeley.

JACK J. LISSAUER is a Space Scientist at NASA's Ames Research Center in Moffett Field, California. He was born in San Francisco in 1957, received his S.B. from the Massachusetts Institute of Technology in 1978 and his Ph.D. in Applied Mathematics from U.C. Berkeley in 1982. He held postdoctoral fellowships at NASA Ames and U.C. Santa Barbara and was on the faculty of the State University of New York at Stony Brook from 1987 to 1996.

Planetary Sciences

Imke de Pater
University of California, Berkeley

and

Jack J. Lissauer
NASA, Ames Research Center

PUBLISHED BY THE PRESS SYNDICATE OF THE UNIVERSITY OF CAMBRIDGE
The Pitt Building, Trumpington Street, Cambridge, United Kingdom

CAMBRIDGE UNIVERSITY PRESS
The Edinburgh Building, Cambridge CB2 2RU, UK
40 West 20th Street, New York, NY 10011-4211, USA
10 Stamford Road, Oakleigh, VIC 3166, Australia
Ruiz de Alarcón 13, 28014, Madrid, Spain
Dock House, The Waterfront, Cape Town 8001, South Africa

http://www.cambridge.org

First published 2001

Printed in the United Kingdom at the University Press, Cambridge

Typeface Times 9/12pt. *System* LATEX 2_ε [DBD]

A catalogue record of this book is available from the British Library

Library of Congress Cataloguing in Publication data

De Pater, Imke, 1952–
Planetary sciences / Imke de Pater and Jack J. Lissauer.
p. cm.
Includes bibliographical references and index.
ISBN 0 521 48219 4
1. Planetology. I. Lissauer, Jack Jonathan. II. Title.
QB601.D38 2001
559.9′2–dc21 00-052938

ISBN 0 521 48219 4 hardback

Contents

Tables

Preface

The study of Solar System objects was the dominant branch of Astronomy from antiquity until the nineteenth century. Analysis of planetary motion by Isaac Newton and others helped reveal the workings of the Universe. While the first astronomical uses of the telescope were primarily to study planetary bodies, improvements in telescope and detector technology in the nineteenth and early twentieth centuries brought the greatest advances in stellar and galactic astrophysics. Our understanding of the Earth and its relationship to the other planets advanced greatly during this period. The advent of the Space Age, with lunar missions and interplanetary probes, has revolutionized our understanding of our Solar System over the past forty years. Dozens of planets in orbit about stars other than our Sun have been discovered since 1995; these massive extrasolar planets have orbits quite different from the giant planets in our Solar System, and their discovery is fueling research into the process of planetary formation.

Planetary Science is now a major interdisciplinary field, combining aspects of Astronomy/Astrophysics with Geology/Geophysics, Meteorology/Atmospheric Sciences and Space Science/Plasma Physics. We are aware of more than ten thousand small bodies in orbit about the Sun and the giant planets. Many objects have been studied as individual worlds rather than merely as points of light. We now realize that the Solar System contains a more dynamic and rapidly evolving group of objects than previously imagined. The cratering record on dozens of imaged bodies shows that impacts have been quite important in the evolution of the Solar System, especially during the epoch of planetary formation. Other evidence, including the compositions of meteorites and asteroids and the high bulk density of the planet Mercury, suggests that even more energetic collisions have disrupted objects. More modest impacts, such as the collision of comet D/Shoemaker–Levy 9 with Jupiter in 1994, continue to occur in the current era. Dynamical investigations have destroyed the regular 'clockwork' image of the Solar System that had held prominence since the time of Newton. Resonances and chaotic orbital variations are now believed to have been important for the evolution of many small and possibly some large planetary bodies.

The renewed importance of the Planetary Sciences as a subfield of Astronomy implies that some exposure to Solar System studies is an important component to the education of astronomers. Planetary Sciences' close relationship to Geophysics, Atmospheric and Space Sciences means that the study of the planets offers the unique opportunity for comparison available to Earth scientists.

The amount of material contained in this book is difficult to cover in a one year graduate-level course. Moreover, many professors will prefer to cover their favorite topics at greater depth using supplemental materials. Most students using this book are likely to be taking one semester classes, and many will be undergraduates. Although many superficially differing aspects of the Planetary Sciences are interconnected, and we have included extensive cross-referencing between chapters, we have also attempted to organize the text in a manner that allows for courses to focus on more limited topics. Chapter 1 and the first sections of Chapters 2 and 3 should be covered by all students. The remainder of Chapter 2 is particularly useful for Chapters 9–13, and is essential for Chapter 11 and portions of Chapter 12. The remainder of Chapter 3 is essential for Chapter 4 and useful for Chapters 5, 6, 9 and 10. Portions of Chapter 5 are needed for Chapter 6. Chapter 7 is probably the most technical. Chapter 8 contains necessary material for Chapters 9 and 12, and parts of Chapters 9 and 10 are closely related. Although details of observing techniques are beyond the scope of this book, we think it is important that the students are familiar with the variety of observational methods. We have therefore included a general summary of observational techniques in Chapter 9.

Various symbols are commonly used to represent variables and constants in both equations and the text. Some variables have a unique correspondence with standard symbols in the literature, whereas other variables are represented by differing symbols by different authors and many symbols have multiple uses. The interdisciplinary nature of the Planetary Sciences exacerbates the problem because standard notation differs between fields. We have endeavored to minimize confusion within the text and provide the student with the greatest access to the literature by using standard symbols, sometimes augmented by nonstandard subscripts or printed using calligraphic fonts in order to avoid duplication of meanings whenever practical. A list of the symbols used in this book is presented as Appendix A.

Inclusion of high-quality color figures within the main text would have added substantially to this book's production costs and consequently to its price. We have thus used monochrome illustrations wherever possible and included color plates in a separate section. However, to facilitate the flow of figures in the book, we have included a monochrome representation and the figure caption within the main text, with the color image and figure number presented in the plates.

We feel that the learning of concepts in the physical sciences, as well as obtaining a feel for Solar System properties, is greatly enhanced when students get their 'hands dirty' by solving problems. Thus, we have included an extensive collection of exercises at the end of each chapter in this text. We rank these problems by degree of conceptual difficulty: The easiest problems, denoted by E, should be accessible to most upper level undergraduate science majors; indeed, some are simply plugging numbers into a given formula. Intermediate (I) problems involve more sophisticated reasoning, and are geared towards graduate students. Some of the difficult (D) problems are quite challenging. Note that these rankings are not related to the number of calculations required, and some E problems take most graduate students longer to solve than some of the I problems.

The breadth of the material covered in the text extends well beyond the area of expertise of the authors. As such, we benefitted greatly from comments by many of our colleagues. Especially useful suggestions were provided by Michael A'Hearn, James Bauer, Alice Berman, Donald DePaolo, John Dickel, Luke Dones, Martin Duncan, Stephen Gramsch, Russell Hemley, Bill Hubbard, Donald Hunten, Andy Ingersoll, Raymond Jeanloz, David Kary, Monika Kress, Janet Luhmann, Geoffrey Marcy, Jay Melosh, Bill Nellis, Typhoon Lee, Eugenia Ruskol, Victor Safronov, Mark Showalter, David Stevenson, John Wood, and Dorothy Woolum. Our special thanks go to Catherine Flack, our initial editor, who helped make the book more readable. We enjoyed discussions and comments by the students who were taught with drafts of book chapters, and who worked through half-baked problem sets. This book, like the rapidly evolving field of Planetary Sciences, is a work in progress; as such, we welcome corrections, updates and other suggestions that we may use to improve future editions. Cambridge University Press has set up a web page for this book on their website: www.cup.cam.ac.uk/scripts/textbook.asp. This page includes errata, various updates, color versions of some of the figures that appear in black and white in this volume, and links to various Solar System information sites.

We dedicate this book to our parents and teachers, and to family, friends and colleagues who provided us encouragement and support over many years, and to Floris van Breugel, now a teenager, who has never known his mother *not* to be working on this book.

Imke de Pater and Jack J. Lissauer
Berkeley, California

FIGURE 1.1a

(a)

(b)

(c)

FIGURE 4.28

FIGURE 4.38a

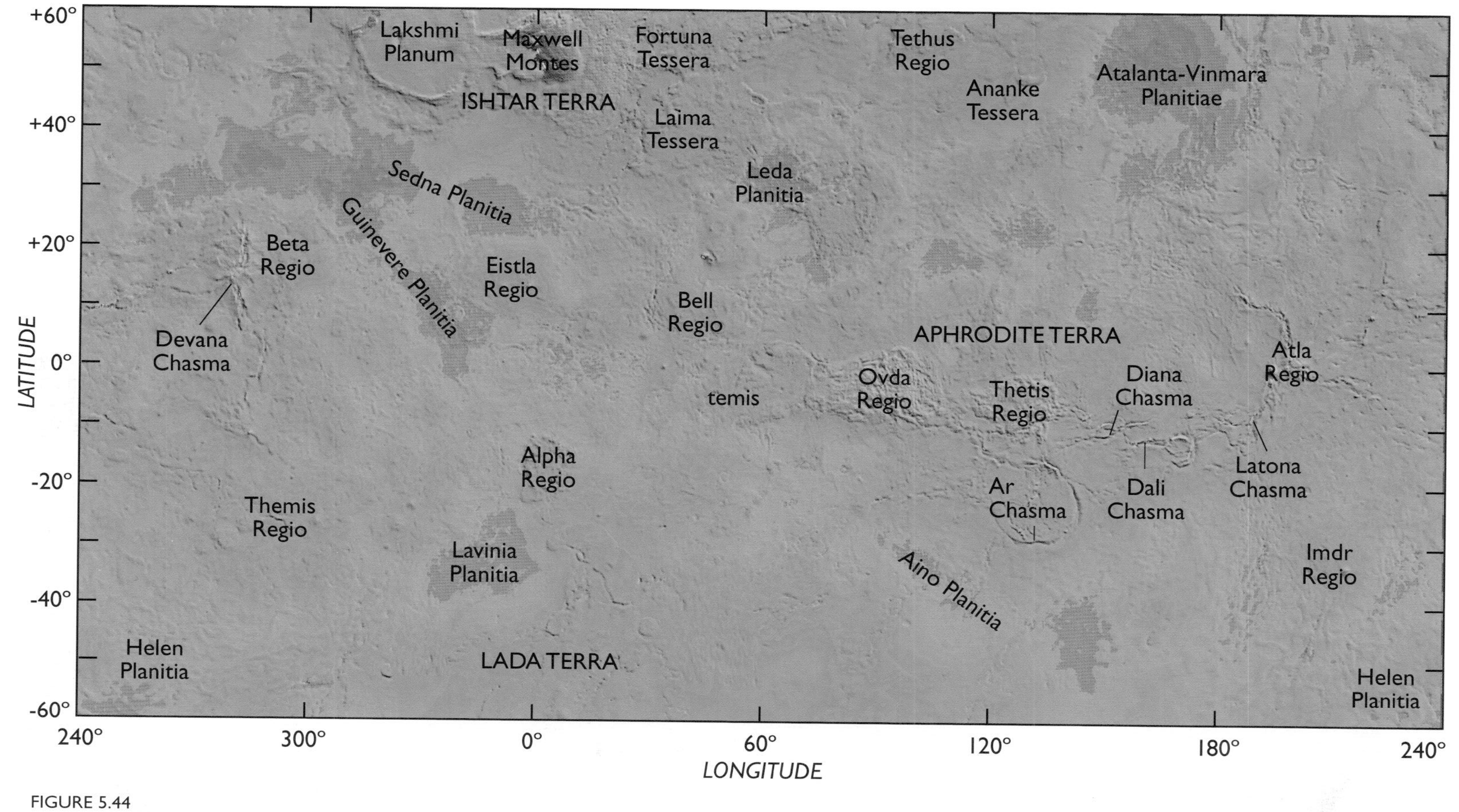

+60°
+40°
+20°
0°
-20°
-40°
-60°
LATITUDE
240°
300°
0°
60°
120°
180°
240°
LONGITUDE
Lakshmi Planum
Maxwell Montes
ISHTAR TERRA
Fortuna Tessera
Laima Tessera
Tethus Regio
Ananke Tessera
Atalanta-Vinmara Planitiae
Sedna Planitia
Guinevere Planitia
Leda Planitia
Beta Regio
Devana Chasma
Eistla Regio
Bell Regio
APHRODITE TERRA
Atla Regio
Ovda Regio
Thetis Regio
Diana Chasma
temis
Alpha Regio
Ar Chasma
Dali Chasma
Latona Chasma
Themis Regio
Lavinia Planitia
Aino Planitia
Imdr Regio
Helen Planitia
LADA TERRA
Helen Planitia

FIGURE 5.44

FIGURE 5.49

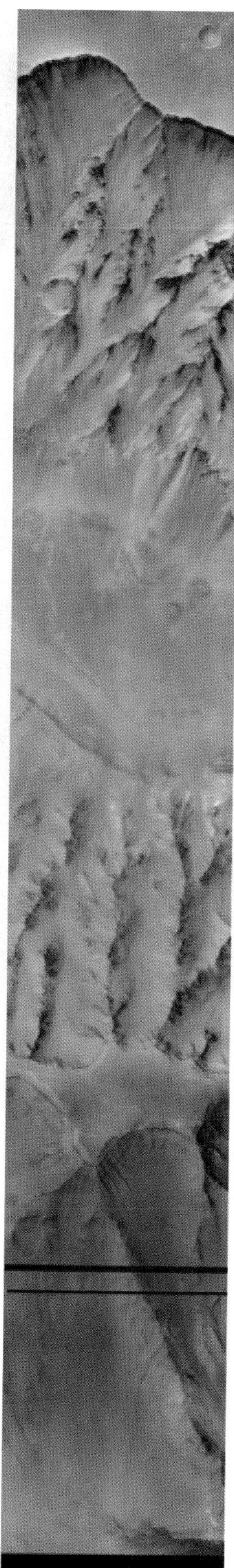

FIGURE 5.50b

FIGURE 5.52a

(a)

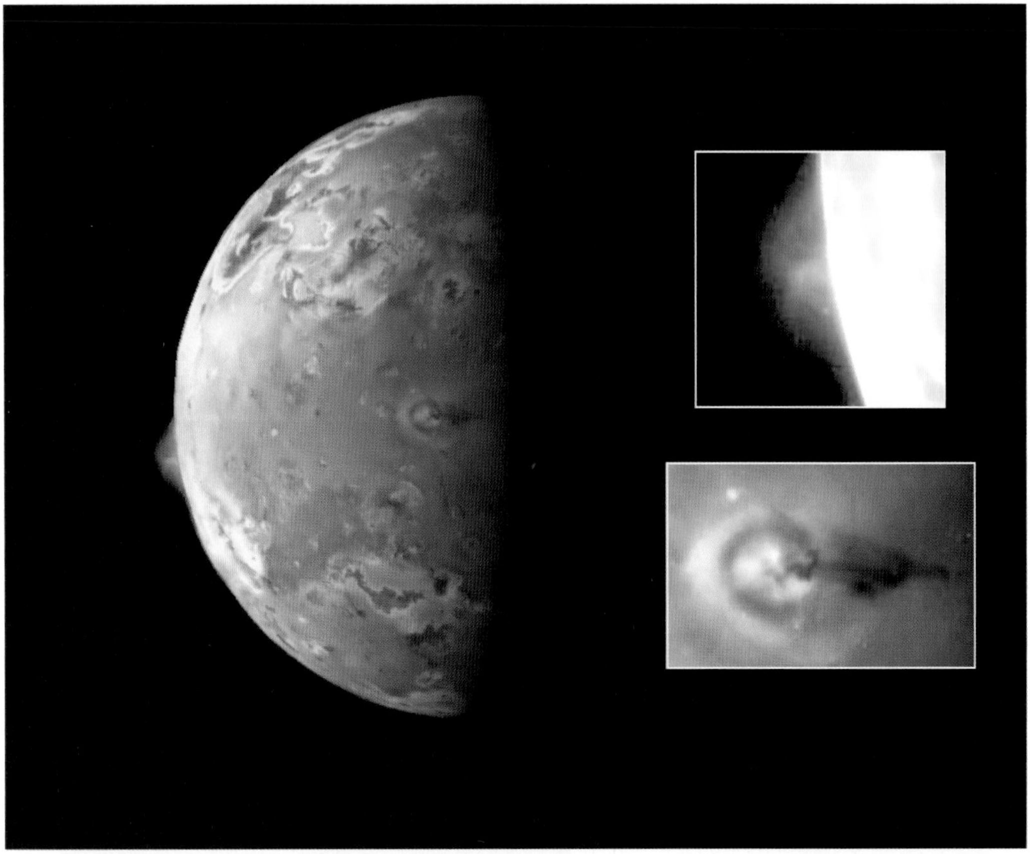

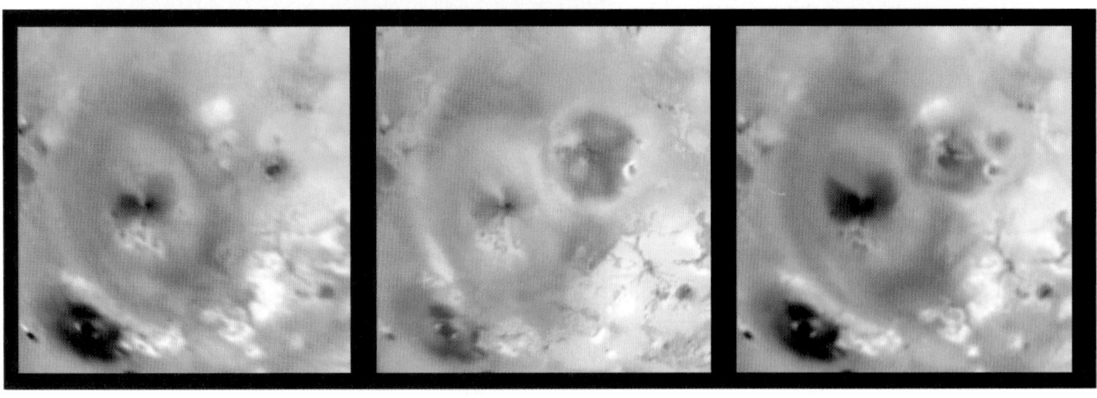

(b) April 1997 September 1997 July 1999

FIGURE 5.53

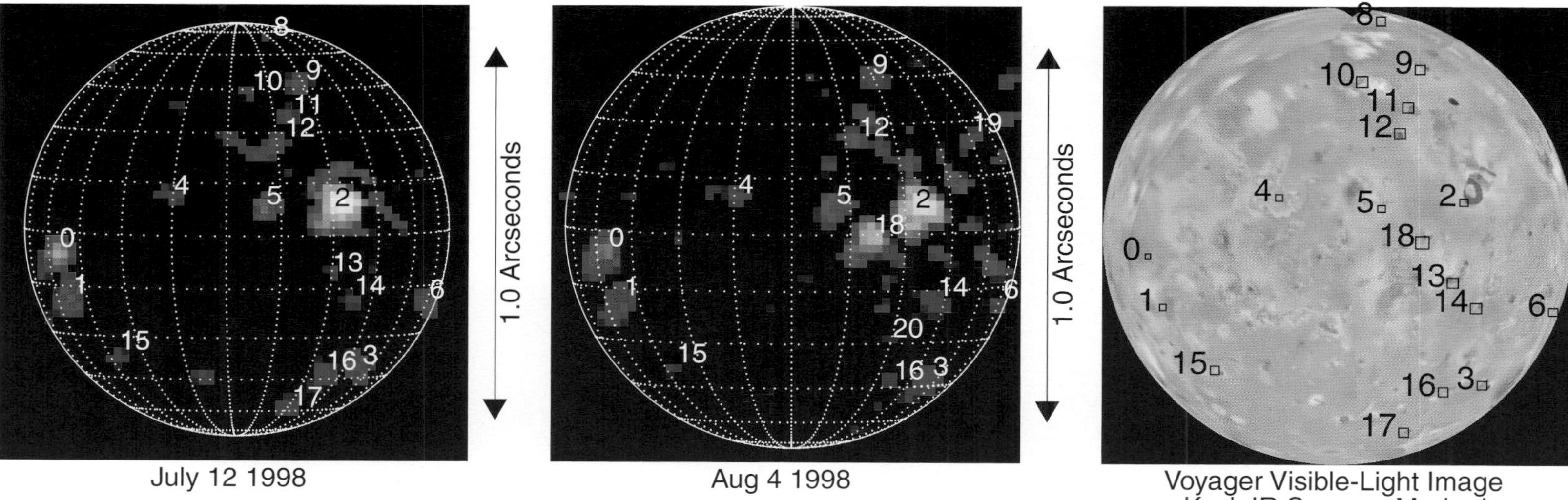
8
10
9
11
12
4
5
2
0
13
1
14
6
15
16
3
17
July 12 1998
1.0 Arcseconds
9
12
19
4
5
2
18
0
1
14
6
20
15
16
3
Aug 4 1998
1.0 Arcseconds
8
10
9
11
12
4
5
2
18
0
13
1
14
6
15
16
3
17
Voyager Visible-Light Image
Keck IR Sources Marked

FIGURE 5.55a

FIGURE 5.56c

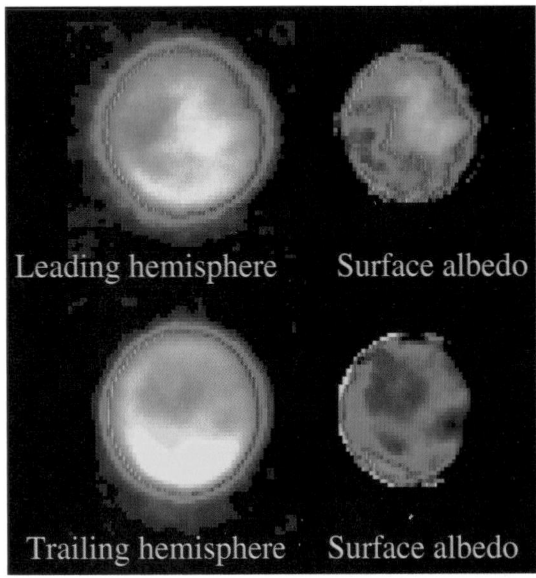

FIGURE 5.61

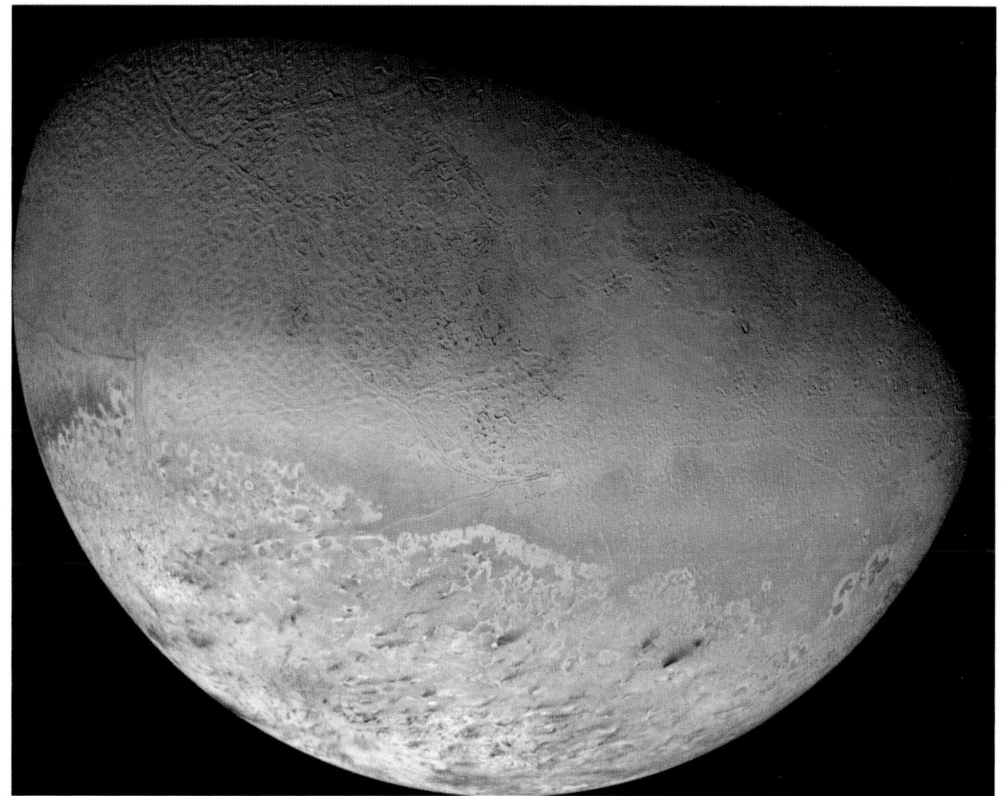

FIGURE 5.64a

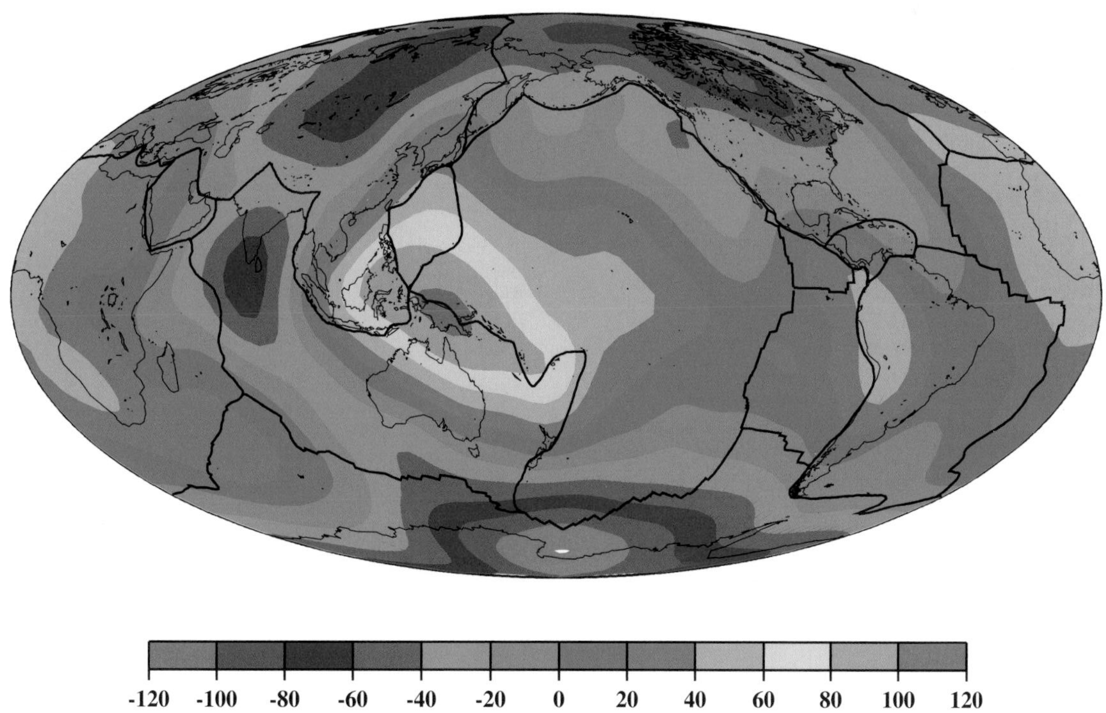

FIGURE 6.7

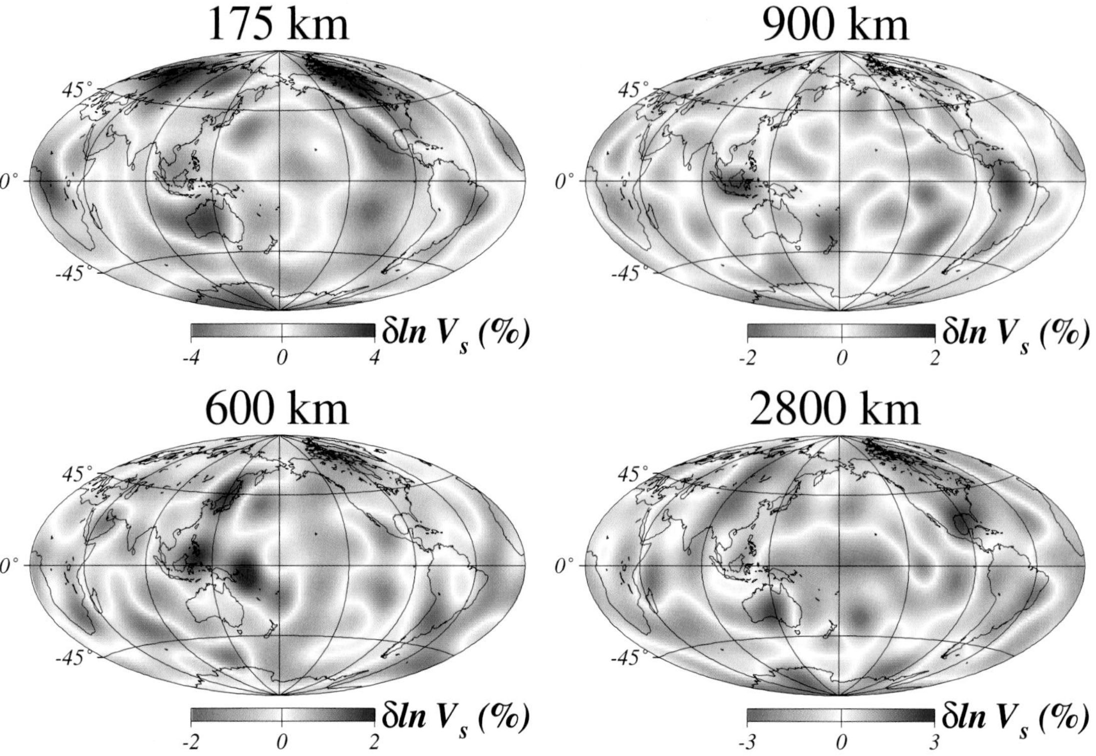

FIGURE 6.15a

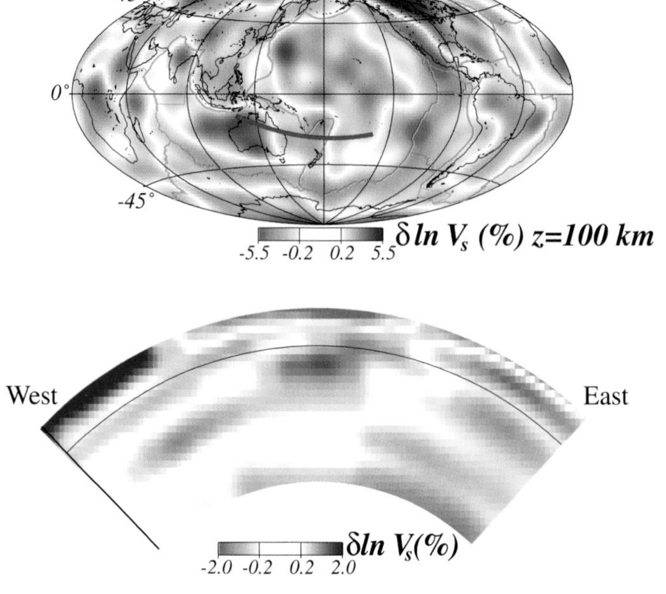

FIGURE 6.15b

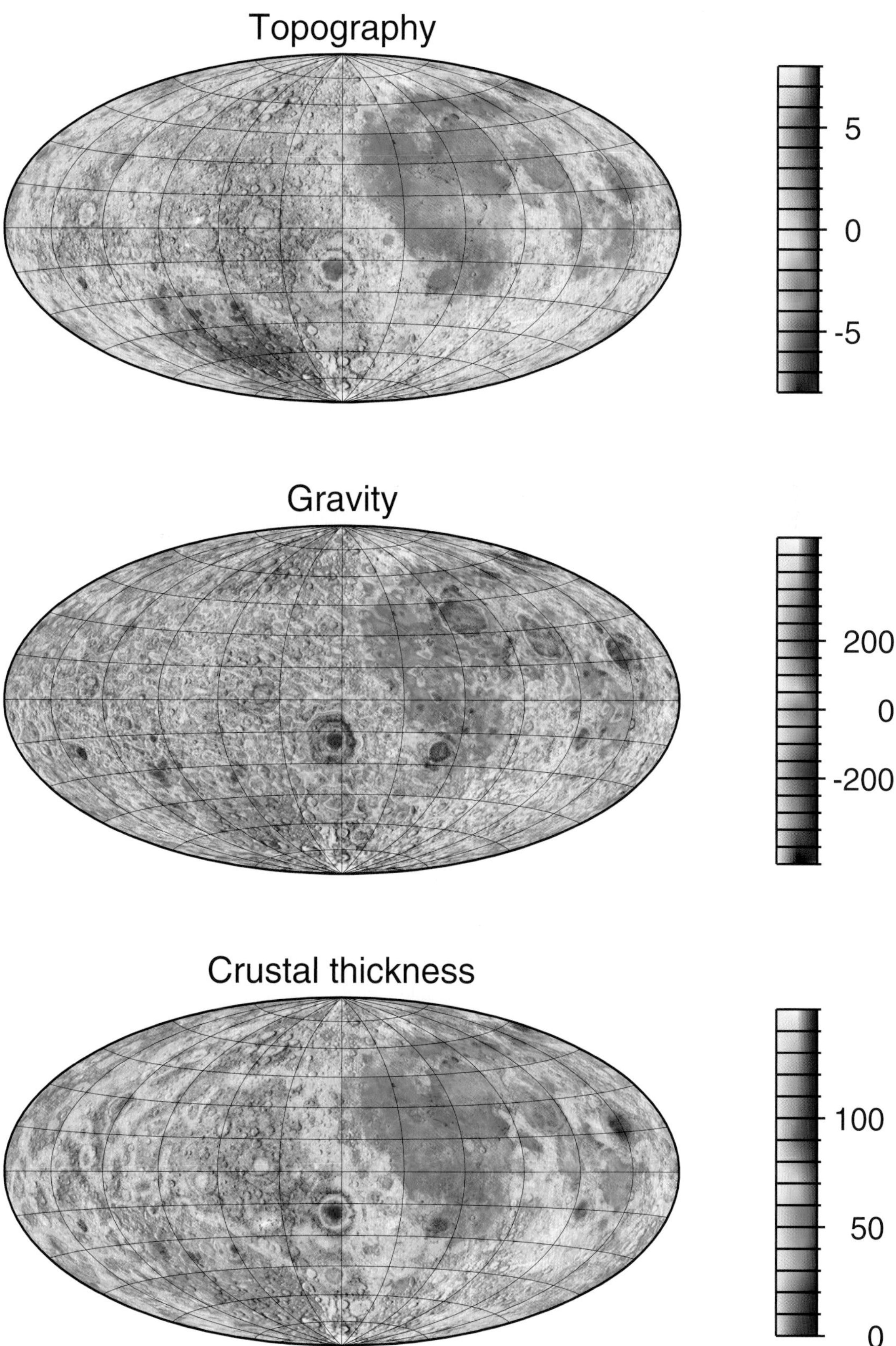

FIGURE 6.16

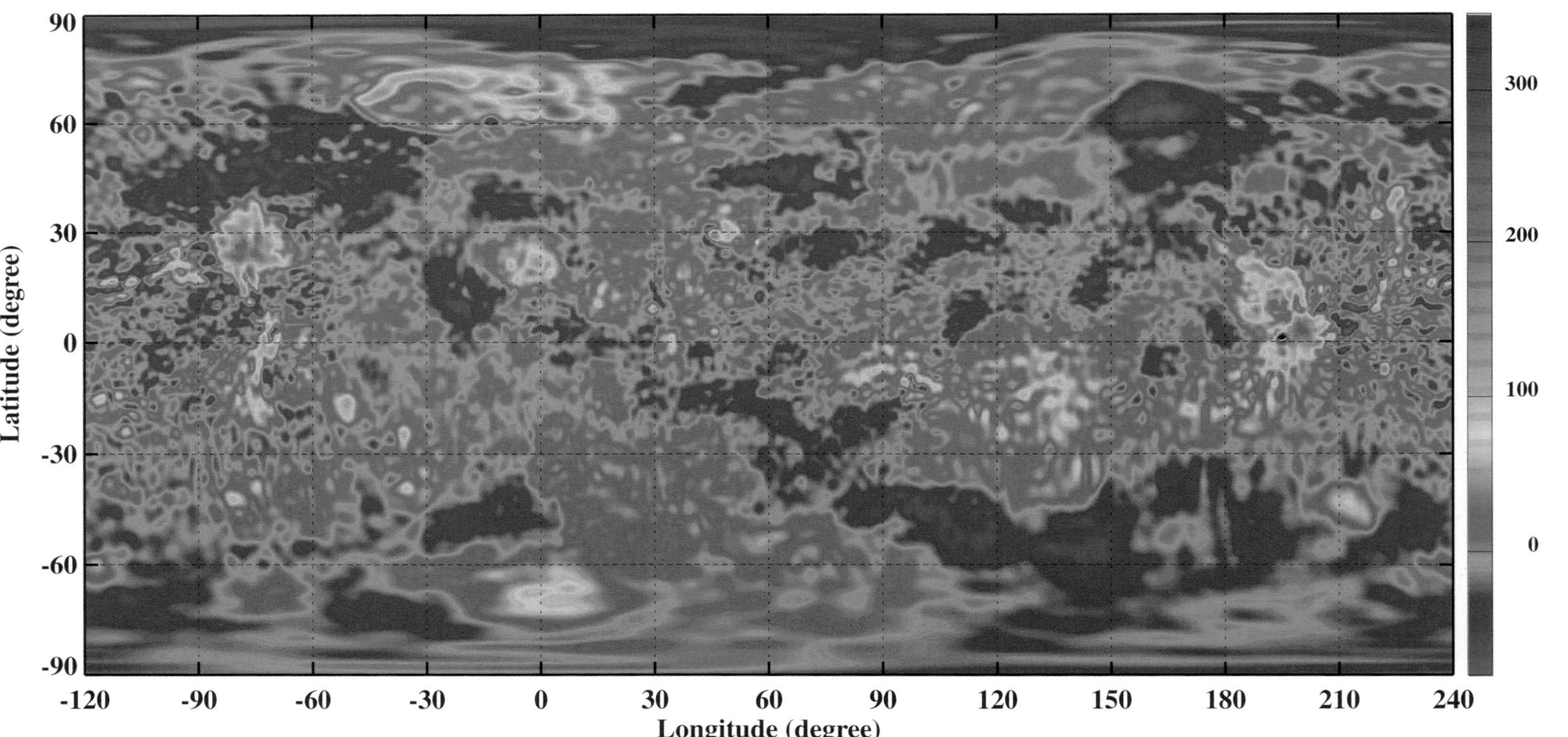
Latitude (degree)
90
60
30
0
-30
-60
-90
-120
-90
-60
-30
0
30
60
90
120
150
180
210
240
Longitude (degree)
300
200
100
0
Acceleration (mgal)

FIGURE 6.19

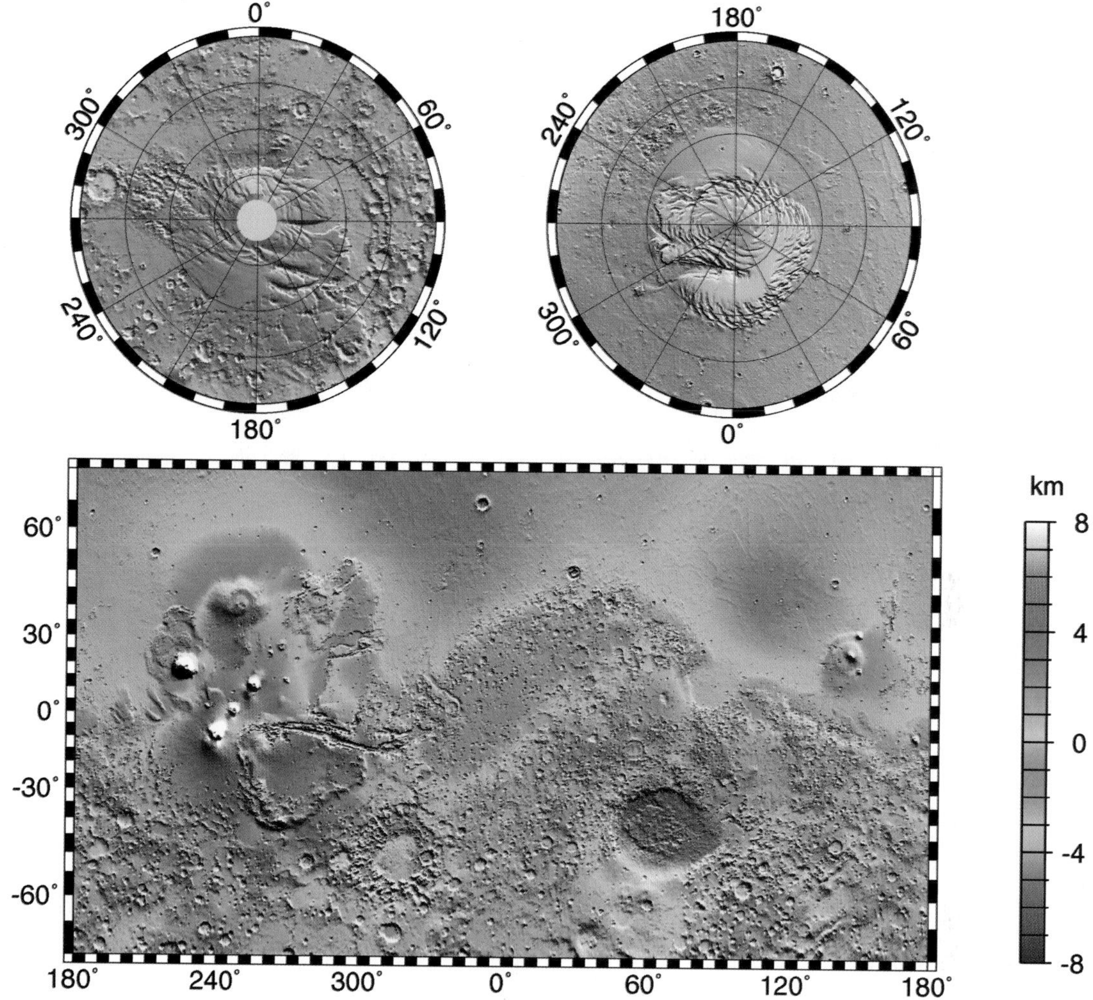
0°
300°
60°
240°
120°
180°
180°
240°
120°
300°
60°
0°
60°
30°
0°
-30°
-60°
180°
240°
300°
0°
60°
120°
180°
km
8
4
0
-4
-8
FIGURE 6.20a

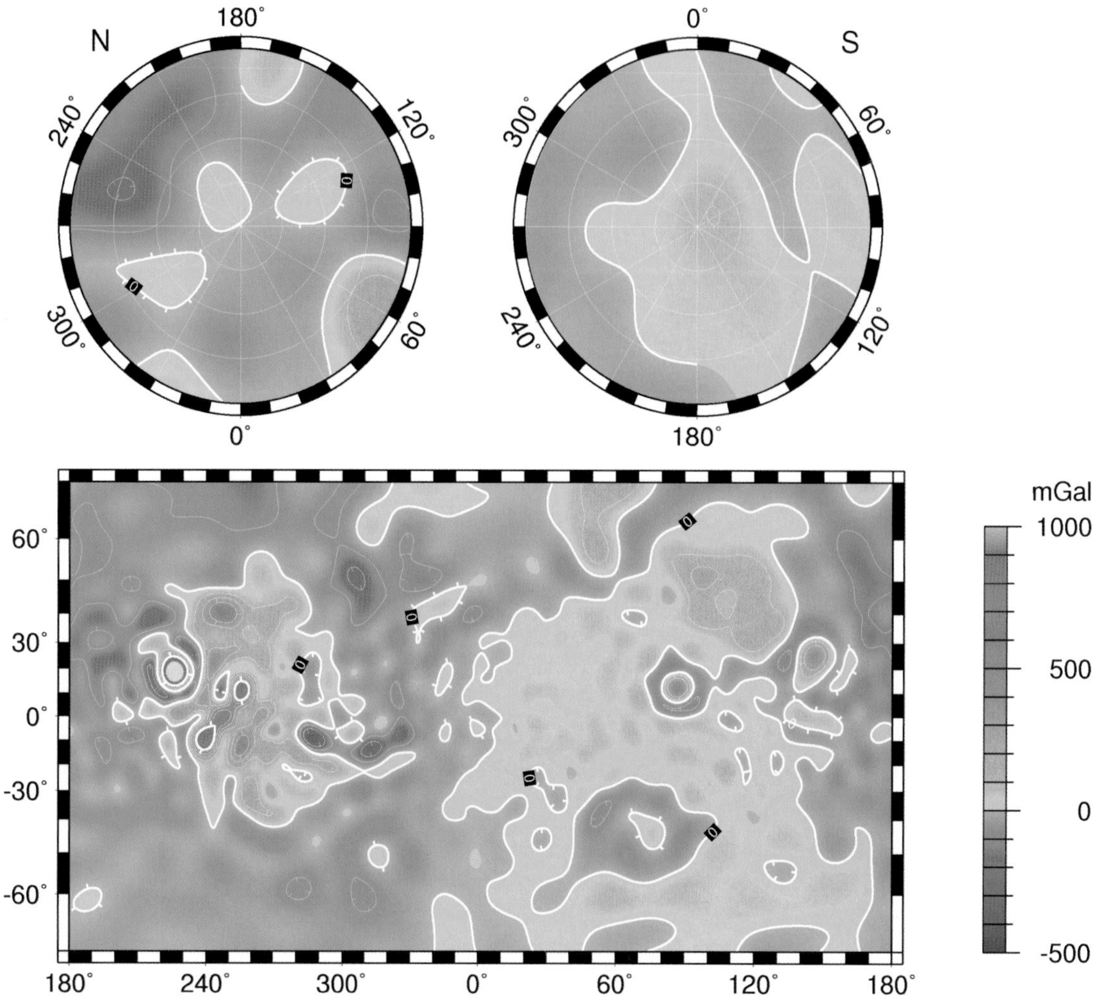

180°
N
240°
120°
300°
60°
0°
0°
S
300°
60°
240°
120°
180°
mGal
1000
500
0
-500
60°
30°
0°
-30°
-60°
180°
240°
300°
0°
60°
120°
180°

FIGURE 6.20b

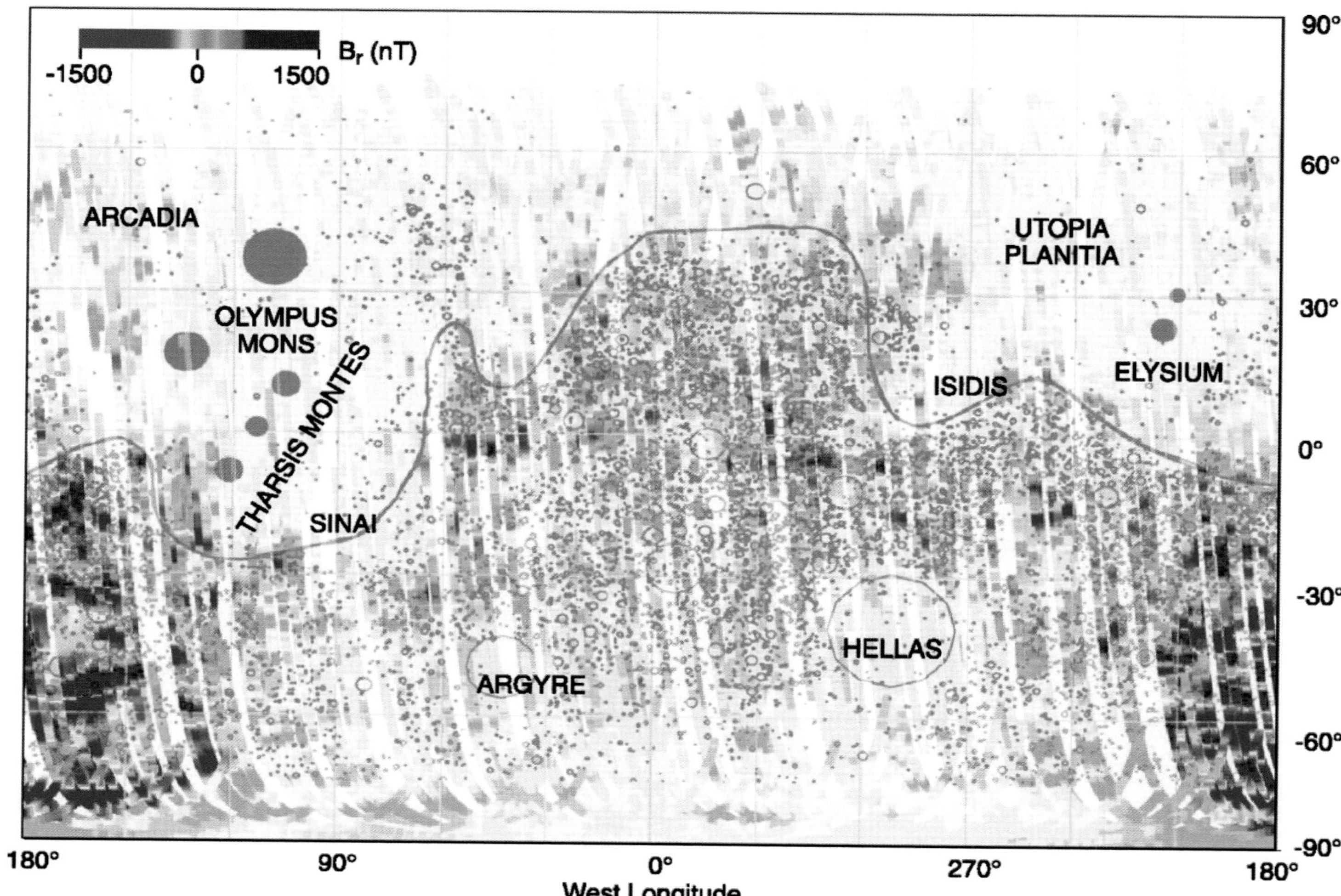
-1500
0
1500
B_r (nT)
ARCADIA
OLYMPUS MONS
THARSIS MONTES
SINAI
ARGYRE
HELLAS
ISIDIS
UTOPIA PLANITIA
ELYSIUM
90°
60°
30°
0°
-30°
-60°
-90°
180°
90°
0°
270°
180°
West Longitude

FIGURE 7.24

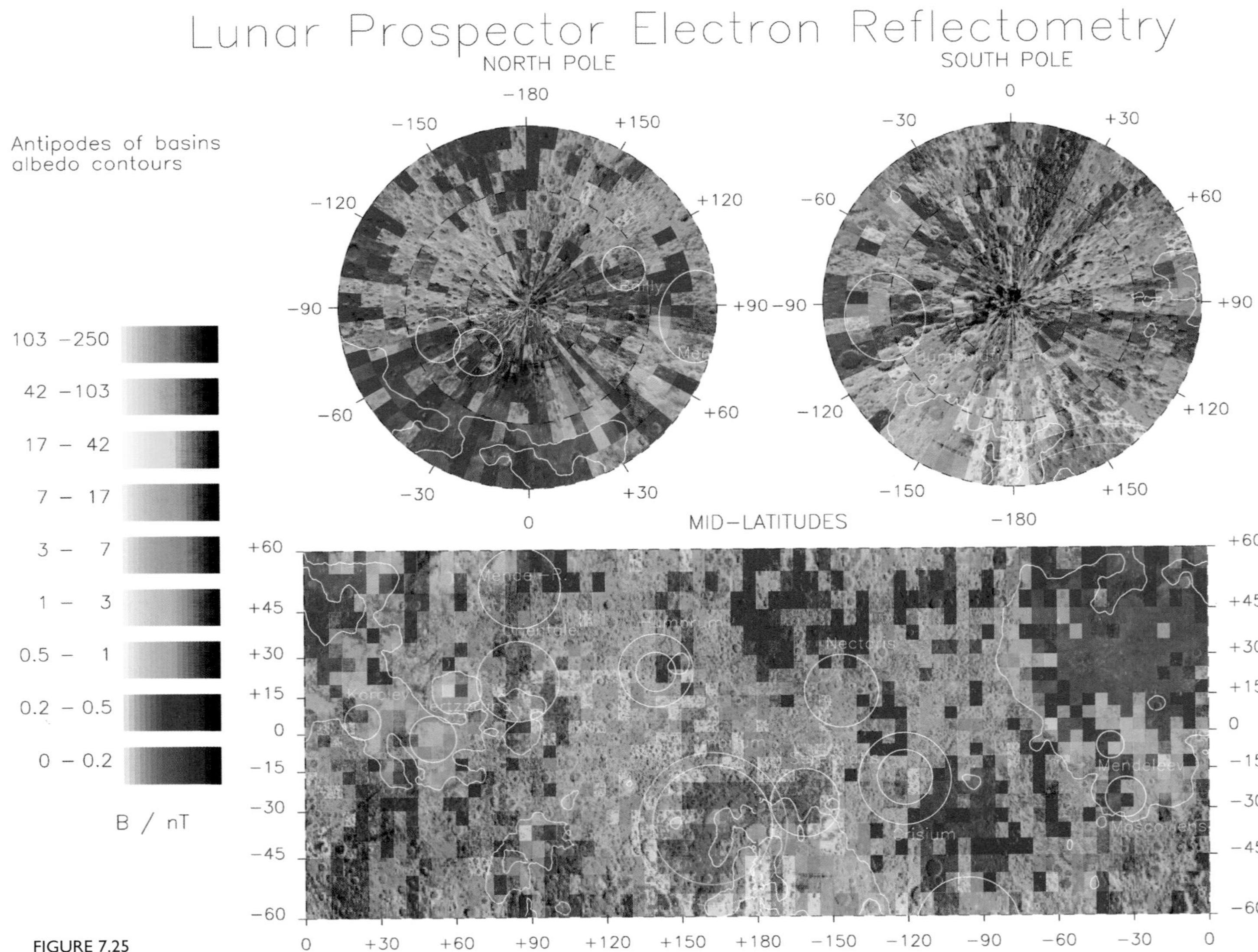
Lunar Prospector Electron Reflectometry
NORTH POLE
SOUTH POLE
Antipodes of basins
albedo contours
103 −250
42 −103
17 − 42
7 − 17
3 − 7
1 − 3
0.5 − 1
0.2 − 0.5
0 − 0.2
B / nT
MID−LATITUDES
−180
−150
+150
−120
+120
−90
+90
−60
+60
−30
+30
0
+45
+15
−15
−45

FIGURE 7.25

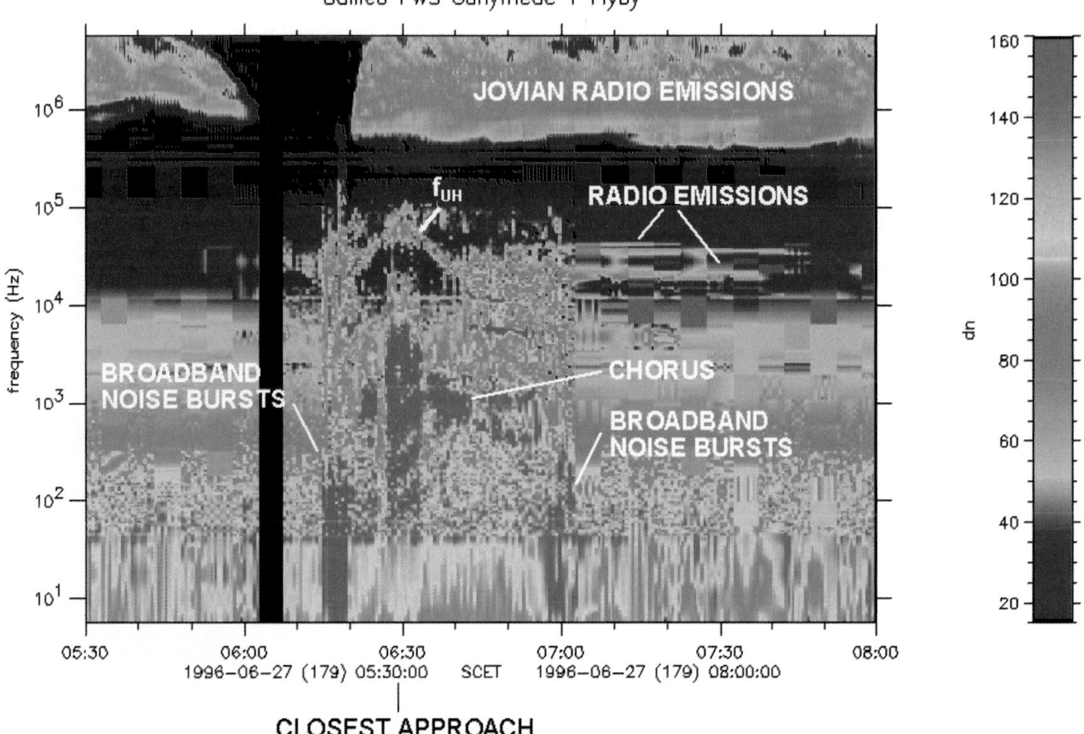

FIGURE 7.46b

FIGURE 8.7

FIGURE 10.1a

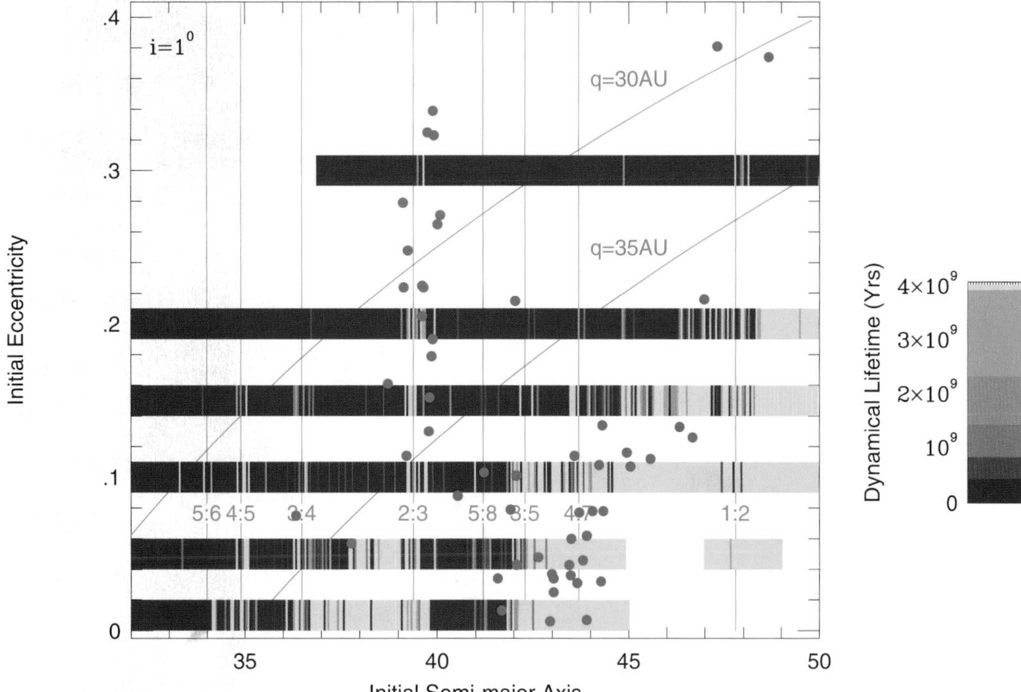

FIGURE 10.4a

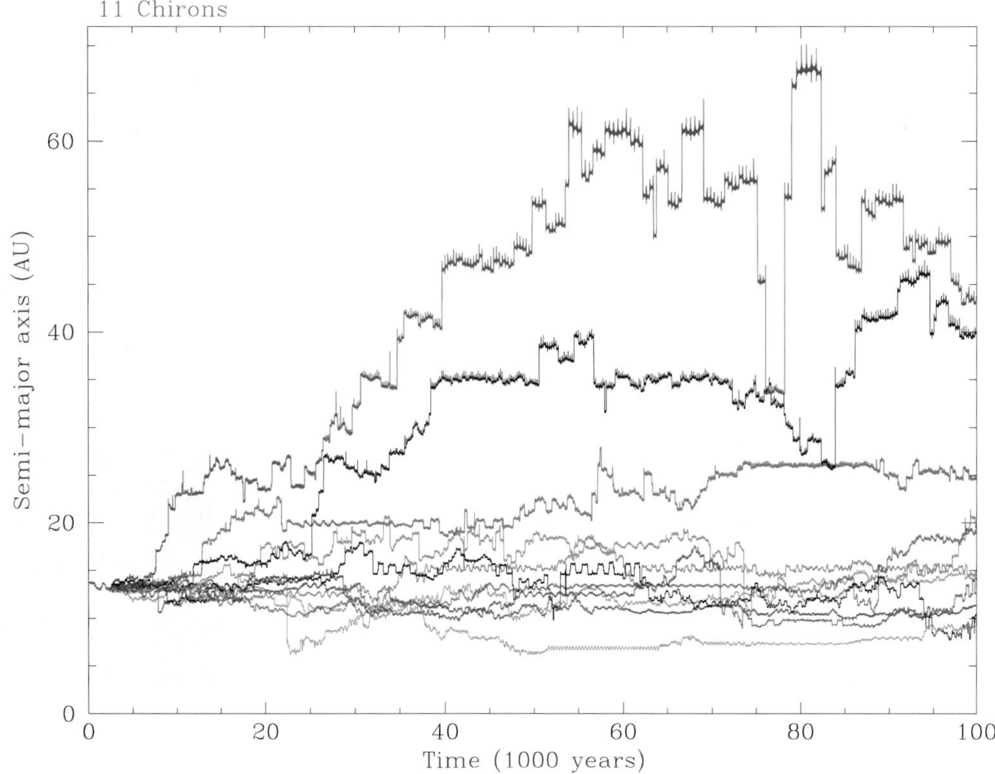

FIGURE 10.4b

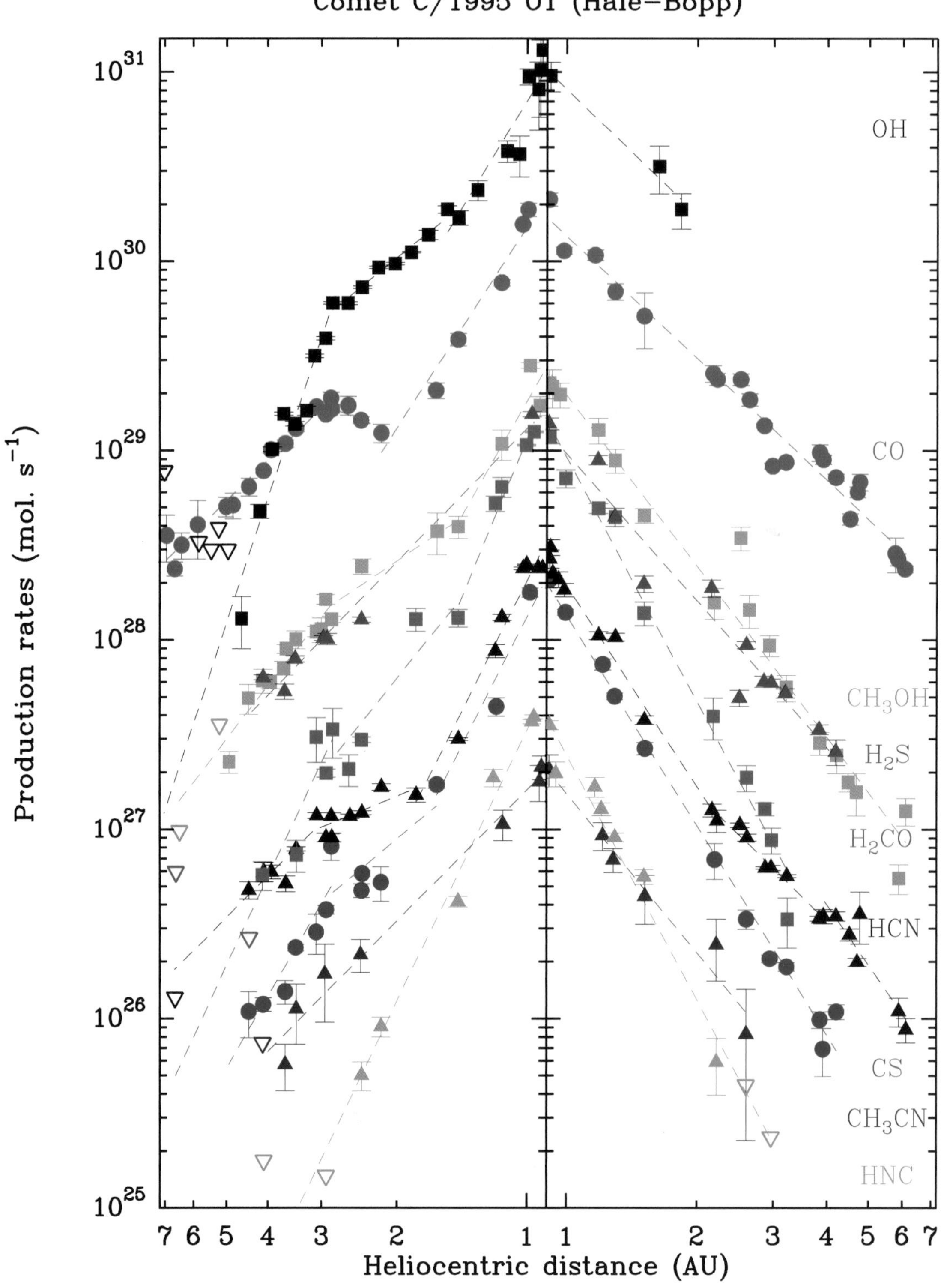
Comet C/1995 O1 (Hale–Bopp)
Production rates (mol. s^{-1})
Heliocentric distance (AU)
OH
CO
CH_3OH
H_2S
H_2CO
HCN
CS
CH_3CN
HNC
10^{31}
10^{30}
10^{29}
10^{28}
10^{27}
10^{26}
10^{25}
7 6 5 4 3 2 1 1 2 3 4 5 6 7

FIGURE 10.5

FIGURE 11.5

(a) t = 0

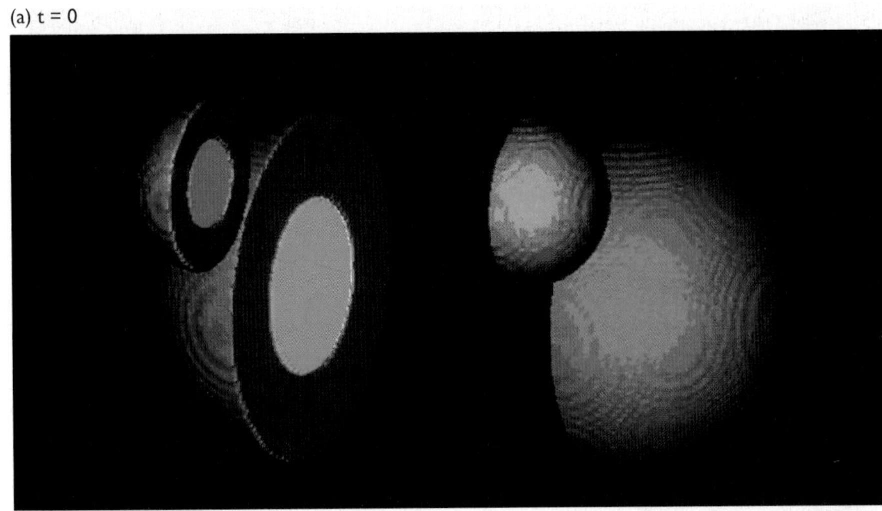

(b) t = 804 s

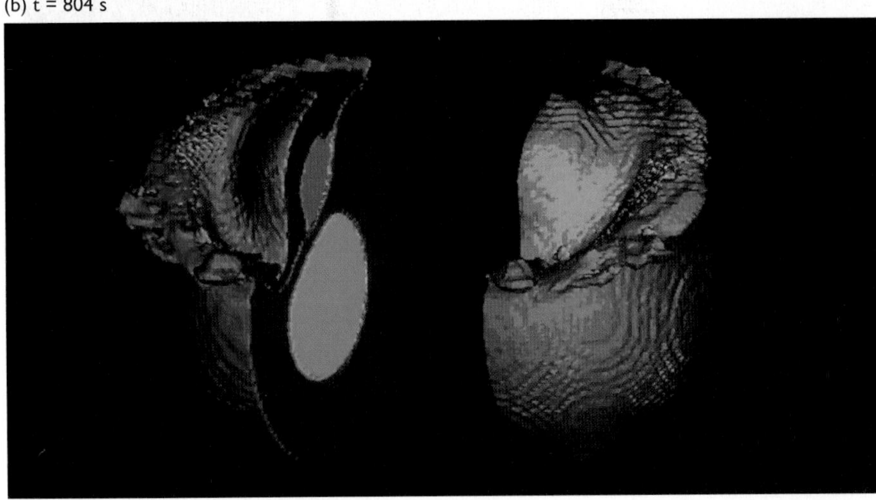

(c) t = 1600 s

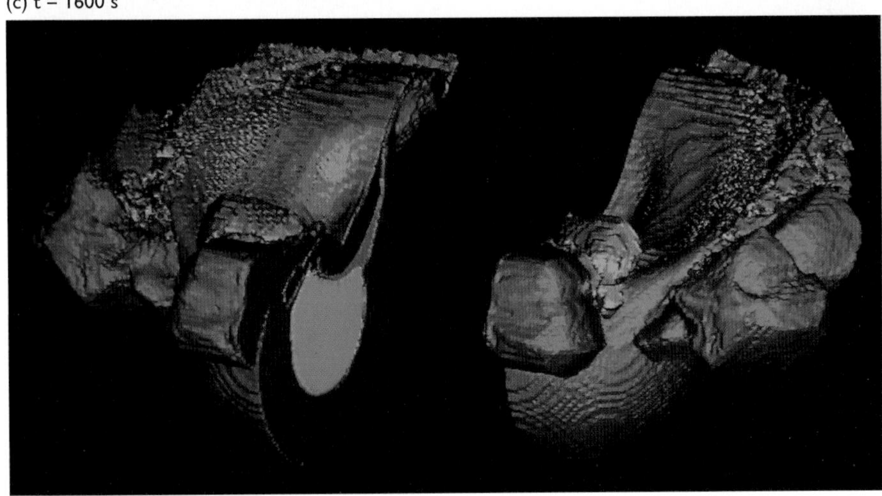

FIGURE 12.18

FIGURE 13.9

1 Introduction

SOCRATES: *Shall we set down astronomy among the subjects of study?*
GLAUCON: *I think so, to know something about the seasons, the months and the years is of use for military purposes, as well as for agriculture and for navigation.*
SOCRATES: *It amuses me to see how afraid you are, lest the common herd of people should accuse you of recommending useless studies.*

Plato, *The Republic VII*

Since ancient times, people have been intrigued by the wonders of the night sky, the Moon and the Sun. Many ancient peoples were in particular fascinated by a few brilliant 'stars', which move along particular trajectories among the far more numerous fixed stars. The Greeks called these objects *planets*, or wandering stars. Old drawings and manuscripts by people from all over the world, such as the Chinese, Greeks and Anasazis, attest to the interest in phenomena such as comets and solar eclipses. The copernican–keplerian–galilean–newtonian revolution in the sixteenth and seventeenth centuries completely changed our concepts of the dimensions and dynamics of the Solar System, including the relative sizes and masses of the bodies and the forces that make them orbit about one another.

Gradual progress was made over the next few centuries, but the next revolution had to await the space age. In October of 1959, the Soviet spacecraft Luna 3 returned the first pictures man had ever seen of the backside of Earth's Moon, and the age of planetary exploration began. Over the subsequent three decades, spacecraft visited all eight known terrestrial and giant planets in the Solar System, returning data concerning the planets, their rings and moons. Spacecraft images of many objects showed details which could never have been guessed from Earth-based pictures. Spectra from ultraviolet to infrared wavelengths revealed the presence of new gases and solid-state features on planets and moons, while radio detectors and magnetometers explored the giant magnetic fields surrounding many of the planets. Two comets and four asteroids have thus far been explored by spacecraft, and there have been several missions to study the Sun and the solar wind. The planets and their satellites have become familiar to us as individual bodies. The vast diversity of planetary and satellite surfaces, atmospheres and magnetic fields has surprised even the most imaginative researchers. Unexpected types of structure were observed in Saturn's rings, and whole new classes of rings and ring systems were seen. Some of the new discoveries have been explained, whereas others remain mysterious. The vast outer regions of the Solar System are so poorly explored that many bodies remain to be discovered, possibly including some of planetary size.

In this book, we shall discuss what has been learned and some of the unanswered questions that remain at the forefront of Solar System research today. A wide variety of topics in planetary sciences is covered, including the orbital, rotational and bulk properties of planets, moons and smaller bodies; gravitational interactions, tides and resonances between bodies; chemistry and dynamics of planetary atmospheres, including cloud physics; planetary geology and geophysics; planetary interiors; magnetospheric physics; asteroids, meteorites and comets; planetary ring dynamics. In the penultimate chapter of this book, we combine this knowledge of current Solar System properties and processes with astrophysical data and models of ongoing star and planet formation to develop a model for the origin of our Solar System. We conclude with an introduction to the new and rapidly blossoming field of extrasolar planets.

1.1 Inventory of the Solar System

What is the *Solar System*? Our naturally geocentric view gives a highly distorted picture, thus it is better to phrase the question as: What is seen by an objective observer from afar? The *Sun*, of course; the Sun has a luminosity 4×10^8 times as large as the total luminosity (reflected plus emitted) of Jupiter, the second brightest object in the Solar System. The Sun also contains over 99.8% of the mass of the known Solar System. The Solar System can thus be thought of as the Sun plus some debris. However, the planets are not insignificant in all aspects, for over 98% of the angular momentum in the Solar System lies in orbital motions of the planets. Moreover, the Sun is a fundamentally different type of body from the planets, a ball of plasma powered by nuclear fusion in its core, whereas the smaller bodies in the Solar System are composed of molecular matter, some of which is in the solid state. In this book, we shall focus on the debris in orbit about the Sun. This debris is comprised of the giant planets, the terrestrial planets and numerous and varied smaller objects (Figs. 1.1 and 1.2).

1.1.1 Giant Planets

Jupiter dominates the planetary system, with a mass exceeding twice that of all other known planets combined, over 300 Earth masses ($M_\oplus$). Thus as a second approximation, the Solar System can be viewed as the Sun, Jupiter and some debris. The largest of this debris is *Saturn*, with a mass of nearly 100 $M_\oplus$. Saturn, like Jupiter, is mostly composed of hydrogen (H) and helium (He); both planets possess a heavy element 'core' of $\gtrsim 10\ M_\oplus$. The third and fourth largest planets are *Neptune* and *Uranus*, each having a mass roughly one-sixth that of Saturn. These planets belong to a different class, with their compositions being dominated by water (H_2O), ammonia (NH_3), methane (CH_4) and 'rock', high temperature condensates consisting primarily of silicates and metals, and topped off with small (1–4 $M_\oplus$) H–He dominated atmospheres. The four largest planets are known collectively as the *giant planets*; Jupiter and Saturn are *gas giants*, with radii of ~ 70 000 km and 60 000 km respectively, whereas Uranus and Neptune are *ice giants* (although the 'ices' are present in fluid rather than solid form), with radii of ~ 25 000 km. These planets orbit the Sun at distances of approximately 5, 10, 20, and 30 AU, respectively. (One *astronomical unit*, 1 AU, is defined to be the semimajor axis of a massless (test) particle whose orbital period about the Sun is one year. The semimajor axis of Earth's orbit is slightly larger than 1 AU.) All four giant planets possess strong magnetic fields.

1.1.2 Terrestrial Planets and Other 'Debris'

The mass of the remaining known 'debris' totals less than one-fifth that of the smallest giant planet, and their orbital angular momentum is also much smaller. This debris consists of all of the solid bodies in the Solar System, and despite its small mass it contains a wide variety of objects which are interesting chemically, geologically and dynamically (and, in at least one case, biologically). The hierarchy continues within this group with two large terrestrial planets, *Earth* and *Venus*, each with a radius of about 6000 km, at approximately 1 and 0.7 AU from the Sun, respectively. Our Solar System also contains two small terrestrial planets: *Mars* with a radius of ~ 3500 km and orbiting at ~1.5 AU and *Mercury* with a radius of ~ 2500 km orbiting at ~ 0.4 AU. All four terrestrial planets have atmospheres; however, terrestrial planet atmospheres are minuscule by giant planet standards. Atmospheric composition and density varies widely among the terrestrial planets. Mercury's atmosphere is so thin that it has only recently been discovered. Earth and Mercury each have an internally generated magnetic field.

Following the terrestrial planets in mass are the seven major moons of the giant planets and Earth, and then the planet *Pluto*, whose *heliocentric distance*, the distance from the Sun, oscillates between 29 and 50 AU. Two planetary satellites, Jupiter's moon Ganymede and Saturn's moon Titan, are slightly larger than the planet Mercury, but because of their lower densities they are less than half as massive. Several satellites (Titan, Triton, Io, the Moon) and Pluto are known to possess atmospheres. *Asteroids*, which are minor planets having radii $\lesssim$ 500 km, are found primarily between the orbits of Mars and Jupiter. Smaller objects are also known to exist elsewhere in the Solar System, for example as small moons in orbit around planets, and comets. Most comets are thought to be 'stored' in the *Oort cloud*, a nearly spherical region at heliocentric distances of ~1–5 × 10^4 AU. Estimates of the total number of comets larger than one kilometer in radius in the entire Oort cloud range from ~10^{12} to ~10^{13}. Most short period comets may have spent the bulk of their lives in the *Kuiper belt*, a flattened disk beyond the orbit of Neptune (~ 35–500(?) AU). The total number of Kuiper belt objects larger than 1 km in radius is estimated to be ~10^8–10^{10}. The total mass and orbital angular momentum of bodies in the Kuiper belt and Oort cloud are uncertain by several orders of magnitude. The upper end of current estimates place as much mass in distant unseen icy bodies as is observed in the entire planetary system. Finally, there are the smallest bodies, which have been observed collectively, but not yet detected individually, such as ring particles and dust from comets and in the zodiacal cloud.

1.1.3 Satellite and Ring Systems

Natural satellites have been observed in orbit about most of the planets in the Solar System. The giant planets all have large satellite systems, consisting of large and/or medium-sized satellites and many smaller moons and rings (Fig. 1.1b). Most of the smaller moons orbiting close to the planet were discovered from spacecraft flybys. All major satellites, except Triton, orbit the respective planet in a *prograde* manner (i.e., in the direction that the planet rotates), close to the planet's equatorial plane. Small, close-in moons are also exclusively in low inclination, low eccentricity orbits, but small moons orbiting beyond the

(a)

(b)

FIGURE 1.1 (a) COLOR PLATE Images of the planets depicted to scale, in order of distance from the Sun. (Courtesy: Calvin J. Hamilton) (b) Images of the largest satellites of the four giant planets and Earth's Moon, which are depicted in order of distance from their planet. Note that these moons span a wide range of size, albedo and surface characteristics; most are spherical, but some of the smallest objects pictured are quite irregular in shape. (Courtesy: Paul Schenk)

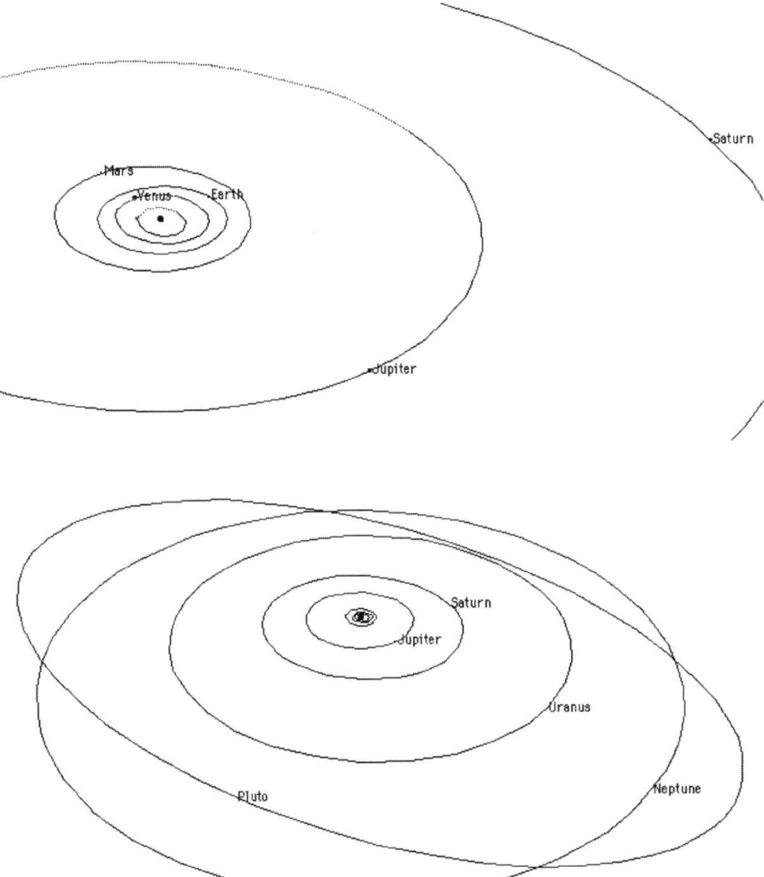

FIGURE 1.2 Orbits of the major planets in the Solar System. Note the relative closeness of the four terrestrial planets and the much larger spacings in the outer Solar System.

main satellite systems can travel around the planet in either direction, and their orbits are often highly inclined and eccentric. Earth and Pluto each have one large moon: our Moon has a little over 1% of Earth's mass, and Charon is believed to be over 10% the mass of Pluto. These moons may have been produced as the result of giant impacts on the Earth and Pluto, when the Solar System was a small fraction of its current age.

The four giant planets all have ring systems, which are primarily within about 2.5 planetary radii of the planet's center. However, in other respects, the characters of the four ring systems differ greatly. Saturn's rings are bright and broad, full of structure such as density waves, gaps, and 'spokes'. Jupiter's ring is very tenuous and composed mostly of small particles. Uranus has nine narrow opaque rings plus a broad region of tenuous dust. Neptune has four rings, two narrow ones and two faint broader rings; the most remarkable part of Neptune's ring system is the so-called ring arcs, which are bright segments within one of the narrow rings.

The orbital and bulk properties of the nine 'major' planets, the largest 'minor planets' or asteroids, and those of the major moons are listed in Tables 1.1–1.5.

1.1.4 Heliosphere

All planetary orbits lie within the *heliosphere*, the region of space containing magnetic fields and plasma of solar origin. The *solar wind* consists of plasma traveling outwards from the Sun, at supersonic speeds. The solar wind merges with the interstellar medium at the *heliopause*, the boundary of the heliosphere.

The composition of the heliosphere is dominated by solar wind protons and electrons, with a typical density of 5 protons cm^{-3} at 1 AU (decreasing with distance squared), and speed of $\sim$400 km s^{-1} near the solar equator and $\sim$400–800 km s^{-1} closer to the solar poles. In contrast, the local interstellar medium, at a density of less than 0.1 atoms cm^{-3}, contains mainly hydrogen and helium atoms. The Sun's motion relative to the mean motion of

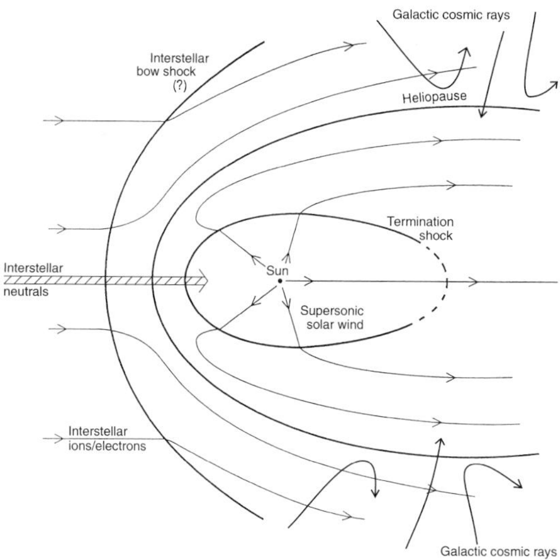

FIGURE 1.3 Sketch of the teardrop-shaped heliosphere. Within the heliosphere, the solar wind flows radially outward until it encounters the heliopause, the boundary between the solar wind dominated region and the interstellar medium. Weak cosmic rays are deflected away by the heliopause, but energetic particles penetrate the region down to the inner Solar System. (Adapted from Gosling 1999)

neighboring stars is roughly 16 km s^{-1}. Hence, the heliosphere moves through the interstellar medium at about this speed. The heliosphere is believed to be shaped like a teardrop, with a tail in the downwind direction (Fig. 1.3). Interstellar ions and electrons generally flow around the heliosphere, since they cannot cross the solar magnetic fieldlines. Neutrals, however, can enter the heliosphere, and as a result interstellar H and He atoms move through the Solar System, in the downstream direction, with a typical speed of ~15–20 km s^{-1}.

The heliopause has not (yet) been crossed by any of the outgoing spacecraft. In 1983/84 and 1992 the Voyager spacecraft detected strong signals at a frequency of 2–3 kHz, which are believed to have been generated by an interplanetary shock near the heliopause during a period of intense solar activity. Based upon these events, the heliopause is estimated to lie somewhere between 100 and 200 AU upstream in the interstellar wind. The location of the heliopause likely varies with the 11 year solar activity cycle.

1.2 Planetary Properties

All of our knowledge regarding specific characteristics of Solar System objects including planets, moons, comets, asteroids, rings and interplanetary dust, is ultimately derived from observations, either astronomical measurements from the ground or Earth-orbiting satellites, or from close-up (often *in situ*) measurements obtained by interplanetary spacecraft. One can determine the following quantities more or less directly from observations:

(1) Orbit
(2) Mass, distribution of mass
(3) Size
(4) Rotation rate and direction
(5) Shape
(6) Temperature
(7) Magnetic field
(8) Surface composition
(9) Surface structure
(10) Atmospheric structure and composition

With the help of various theories, these observations can be used to constrain planetary properties such as bulk composition and interior structure, two quantities which are crucial elements in modeling the formation of the Solar System.

1.2.1 Orbit

In the early part of the seventeenth century, Johannes Kepler deduced three 'laws' of planetary motion directly from observations:

(1) All planets move along elliptical paths with the Sun at one focus.
(2) A line segment connecting any given planet and the Sun sweeps out area at a constant rate.
(3) The square of a planet's orbital period about the Sun, P_{orb}, is proportional to the cube of its semimajor axis, a, i.e., $P_{orb}^2 \propto a^3$.

A keplerian orbit is uniquely specified by six orbital elements, a (semimajor axis), e (eccentricity), i (inclination), ω (argument of periapse; or ϖ for the longitude of periapse), Ω (longitude of ascending node), and f (true anomaly). These orbital elements are defined graphically in Figure 2.1 and discussed in more detail in Section 2.1. The first of these elements are more fundamental than the last: a and e fully define the size and shape of the orbit, i gives the tilt of the orbital plane to some reference plane, the longitudes ϖ and Ω determine the orientation of the orbit, and f (or t_ϖ, the time of periapse passage) tells where the planet is along its orbit at a given time. Alternative sets of orbital elements are also possible, for instance an orbit is fully specified by the planet's position and velocity relative to the Sun at a given time (again six independent scalar quantities), provided the masses of the Sun and planet are known.

TABLE 1.1 Planetary Mean Orbits.

Planet	Symbol	a (AU)	e	i (deg)	Ω (deg)	ϖ (deg)	L (deg)
Mercury	☿	0.387 098 80	0.205 631 75	7.004 99	48.3309	77.4561	252.2509
Venus	♀	0.723 332 01	0.006 771 77	3.394 47	76.6799	131.5637	181.9798
Earth	⊕	1.000 000 83	0.016 708 617	0.0	0.0	102.9374	100.4665
Mars	♂	1.523 689 46	0.093 400 62	1.849 73	49.5581	336.6023	355.4333
Jupiter	♃	5.202 758 4	0.048 495	1.303 3	100.464	14.331	34.351
Saturn	♄	9.542 824 4	0.055 509	2.488 9	113.666	93.057	50.077
Uranus	⛢	19.192 06	0.046 30	0.773	74.01	173.01	314.06
Neptune	♆	30.068 93	0.008 99	1.770	131.78	48.12	304.35
Pluto	♇	39.481 7	0.248 8	17.142	110.3	224.1	238.9

All data are from Yoder (1995).

TABLE 1.2 Terrestrial Planets: Geophysical Data.

	Mercury	Venus	Earth	Mars
Mean radius R (km)	2440 ± 1	$6051.8(4 \pm 1)$	$6371.0(1 \pm 2)$	$3389.9(2 \pm 4)$
Mass ($\times 10^{23}$ kg)	3.302	48.685	59.736	6.4185
Density (g cm^{-3})	5.427	5.204	5.515	$3.933(5 \pm 4)$
Flattening ϵ			1/298.257	1/154.409
Semimajor axis			6378.136	3397 ± 4
Sidereal rotation period	58.6462 d	−243.0185 d	23.934 19 h	24.622 962 h
Mean solar day (in days)	175.9421	116.7490	1	1.027 490 7
Polar gravity (m s^{-2})			9.832 186	3.758
Equatorial gravity (m s^{-2})	3.701	8.870	9.780 327	3.690
Core radius (km)	~1600	~3200	3485	~1700
Figure offset ($R_{CF} - R_{CM}$) (km)		0.19 ± 0.01	0.80	2.50 ± 0.07
Offset (lat./long.)		$11^\circ/102^\circ$	$46^\circ/35^\circ$	$62^\circ/88^\circ$
Obliquity to orbit (deg)	~0.1	177.3	23.45	25.19
Sidereal orbit period (yr)	0.240 844 5	0.615 182 6	0.999 978 6	1.880 711 05
Escape velocity v_e (km s^{-1})	4.435	10.361	11.186	5.027
Geometric albedo	0.106	0.65	0.367	0.150

All data are from Yoder (1995).

Kepler's laws (or more accurate versions thereof) can be derived from Newton's laws of motion and of gravity, which were formulated later in the seventeenth century (Section 2.1). Relativistic effects also affect planetary orbits, but they are small compared to the gravitational perturbations that the planets exert on one other (cf. Problem 2.15).

All planets and asteroids revolve around the Sun in the direction of solar rotation. Their orbital planes generally lie within a few degrees of each other and close to the solar equator. For observational convenience, inclinations are generally measured relative to the Earth's orbital plane, which is know as the *ecliptic plane*. Dynamically speaking, the best choice would be the *invariable plane*, which passes through the center of mass and is perpendicular to the angular momentum vector of the Solar System. The Solar System's invariable plane is nearly coincident with the plane of Jupiter's orbit, which is inclined by 1.3° relative to the ecliptic. In this book, we shall follow standard conventions and measure inclinations of heliocentric orbits with respect to the ecliptic plane and inclinations of planetocentric orbits relative to the planet's equator. The

TABLE 1.3 Giant Planets: Physical Data.

	Jupiter	Saturn	Uranus	Neptune
Mass (10^{24} kg)	1898.6	568.46	86.832	102.43
Density (g cm^{-3})	1.326	0.6873	1.318	1.638
Equatorial radius (1 bar) (km)	71 492 ± 4	60 268 ± 4	25 559 ± 4	24 766 ± 15
Polar radius (km)	66 854 ± 10	54 364 ± 10	24 973 ± 20	24 342 ± 30
Volumetric mean radius (km)	69 911 ± 6	58 232 ± 6	25 362 ± 12	24 624 ± 21
Flattening ϵ	0.064 87	0.097 96	0.022 93	0.0171
	±0.000 15	±0.000 18	±0.0008	±0.0014
Rotation period	$9^h55^m27^s.3$	$10^h39^m22^s.4$	17.24 ± 0.01 h	16.11 ± 0.01 h
Hydrostatic flattening[a]	0.065 09	0.098 29	0.019 87	0.018 04
Equatorial gravity (m s^{-2})	23.12 ± 0.01	8.96 ± 0.01	8.69 ± 0.01	11.00 ± 0.05
Polar gravity (m s^{-2})	27.01 ± 0.01	12.14 ± 0.01	9.19 ± 0.02	11.41 ± 0.03
Obliquity (deg)	3.12	26.73	97.86	29.56
Sidereal orbit period (yr)	11.856 523	29.423 519	83.747 407	163.723 21
Escape velocity v_e (km s^{-1})	59.5	35.5	21.3	23.5
Geometric albedo	0.52	0.47	0.51	0.41

All data are from Yoder (1995).

[a] Hydrostatic flattening as derived from the gravitational field and magnetic field rotation rate.

Sun's equatorial plane is inclined by 7° with respect to the ecliptic plane. Among the nine major planets, only Pluto's orbit is tilted significantly, with $i = 17°$. Similarly, most major satellites orbit their planet close to its rotational plane. Comets, many asteroids, and minor satellites may have much larger orbital inclinations. In addition, some comets, minor satellites and Neptune's large moon Triton orbit the Sun/planet in a *retrograde* sense (opposite to the Sun's/planet's rotation). The observed 'flatness' of most of the planetary system is explained by planetary formation models that hypothesize that the planets grew within a disk which was in orbit around the Sun (Chapter 12).

1.2.2 Mass

The mass of an object can be derived from the gravitational force that it exerts on other bodies.

- Orbits of moons: The orbital periods of natural satellites, together with Newton's generalization of Kepler's third law (cf. Chapter 2), can be used to solve for mass. The result is actually the sum of the mass of the planet and moon (plus, to a good approximation, the masses of moons on orbits interior to the one being considered), but except for the cases Pluto/Charon and Earth/Moon, the secondaries' masses are very small compared to that of the primary. The major source of uncertainty in this method results from measurement errors in the semimajor axis; timing errors are negligible.
- What about planets without moons? The gravity of each planet perturbs the orbits of all other planets. Because of the large distances involved, the forces are much smaller, so the accuracy of this method is not high. Note, however, that Neptune was discovered as a result of the perturbations that it forced on the orbit of Uranus. This technique is still used to provide the best (however crude) estimates of the masses of the largest asteroids. The perturbation method can actually be divided into two categories: short-term and long-term perturbations. The extreme example of short-term perturbations are single close encounters between asteroids. Trajectories can be computed for a variety of assumed masses of the body under consideration and fit to the observed path of the other body. Long-term perturbations are best exemplified by masses derived from periodic variations in the relative positions of moons locked in stable orbital resonances (Chapter 2).
- Spacecraft tracking data provide the best means of determining masses of planets and moons visited, as the Doppler shift of the transmitted radio signal can be measured very precisely. The long time baselines afforded by *orbiter* missions allow much higher accuracy than *flyby* missions. The best estimates for the masses of some of the outer planet moons are those obtained by combining accurate short-term perturbation measurements from Voyager images with Voyager tracking data and/or resonance constraints from long timeline ground-based observations.

TABLE 1.4 Planetary Satellites: Orbital Data.

Planet		Satellite	a (10^3 km)	Orbital period (days)	Rot. period (days)	e	i (deg)
Earth		Moon	384.40	27.321 661	S	0.054 900	5.15[a]
Mars	I	Phobos	9.3772	0.318 910	S	0.015 1	1.082
	II	Deimos	23.4632	1.262 441	S	0.000 33	1.791
Jupiter	XVI	Metis	127.98	0.294 78	S	<0.004	~0
	XV	Adrastea	128.98	0.298 26	S	~0	~0
	V	Almathea	181.37	0.498 18	S	0.003	0.40
	XIV	Thebe	221.90	0.674 5		0.015	0.8
	I	Io	421.77	1.769 138	S	0.041	0.040
	II	Europa	671.08	3.551 810	S	0.010 1	0.470
	III	Ganymede	1 070.4	7.154 553	S	0.001 5	0.195
	IV	Callisto	1 882.8	16.689 018	S	0.007	0.281
	XIII	Leda	11 160	241		0.148	27[a]
	VI	Himalia	11 460	251	0.4	0.163	175.3[a]
	X	Lysithea	11 720	259		0.107	29[a]
	VII	Elara	11 737	260	0.5	0.207	28[a]
	XII	Ananka	21 280	610		0.169	147[a]
	XI	Carme	23 400	702		0.207	163[a]
	VIII	Pasiphae	23 620	708		0.378	148[a]
	IX	Sinope	23 940	725		0.275	153[a]
	XVII	S/1999 J1	23 960	719		0.28	147[a]
		S/1975 J1 and S/2000J2–J11[b]					
Saturn	XVIII	Pan	133.583	0.575 0			
	XV	Atlas	137.64	0.601 9		~0	~0
	XVI	Prometheus	139.35	0.612 986		0.002 4	0.0
	XVII	Pandora	141.70	0.628 804		0.004 2	0.0
	XI	Epimetheus	151.422	0.694 590	S	0.009	0.34
	X	Janus	151.472	0.694 590	S	0.007	0.14
	I	Mimas	185.52	0.942 421 8	S	0.020 2	1.53
	II	Enceladus	238.02	1.370 218	S	0.004 5	0.02
	III	Tethys	294.66	1.887 802	S	0.000 0	1.09
	XIV	Calypso(T−)	294.66	1.887 802		~0	~0
	XIII	Telesto(T+)	294.66	1.887 802		~0	~0
	XII	Helene(T+)	377.42	2.736 915		0.005	0.2
	IV	Dione	377.71	2.736 915	S	0.002 2	0.02
	V	Rhea	527.04	4.517 500	S	0.001	0.35
	VI	Titan	1 221.85	15.945 421		0.029 2	0.33
	VII	Hyperion	1 481.1	21.276 609	C	0.104 2	0.43
	VIII	Iapetus	3 561.3	79.330 183	S	0.028 3	7.52
	IX	Phoebe	12 952	550.48	0.4	0.163	175.3[a]
		S/2000 S1–S12[b]					

TABLE 1.4 (*cont.*)

Planet		Satellite	a (10^3 km)	Orbital period (days)	Rot. period (days)	e	i (deg)
Uranus	VI	Cordelia	49.752	0.335 033		0.000	0.1
	VII	Ophelia	53.764	0.376 409		0.010	0.1
	VIII	Bianca	59.165	0.434 577		0.001	0.2
	IX	Cressida	61.777	0.463 570		0.000	0.0
	X	Desdemona	62.659	0.473 651		0.000	0.2
	XI	Juliet	64.358	0.493 066		0.001	0.1
	XII	Portia	66.097	0.513 196		0.000	0.1
	XIII	Rosalind	69.927	0.558 459		0.000	0.3
	XIV	Belinda	75.255	0.623 525		0.000	0.0
		S/1986 U10	76.417	0.638		~0	~0
	XV	Puck	86.004	0.761 832		0.000	0.3
	V	Miranda	129.8	1.413	S	0.0027	4.22
	I	Ariel	191.2	2.520	S	0.0034	0.31
	II	Umbriel	266.0	4.144	S	0.0050	0.36
	III	Titania	435.8	8.706	S	0.0022	0.10
	IV	Oberon	582.6	13.463	S	0.0008	0.10
	XVI	Caliban	7 169	580		0.08	140[a]
	XX	Stephano	7 948	674		0.24	143[a]
	XVII	Sycorax	12 213	1289		0.51	153[a]
	XVIII	Prospero	16 568	2019		0.44	152[a]
	XIX	Setebos	17 681	2239		0.57	158[a]
Neptune	III	Naiad	48.227	0.294 396		0.00	4.74
	IV	Thalassa	50.075	0.311 485		0.00	0.21
	V	Despina	52.526	0.334 655		0.00	0.07
	VI	Galatea	61.953	0.428 745		0.00	0.05
	VII	Larissa	73.548	0.554 654		0.00	0.20
	VIII	Proteus	117.647	1.122 315		0.00	0.55
	I	Triton	354.76	5.876 854	S	0.00	156.834
	II	Nereid	5 513.4	360.136 19		0.751	7.23[a]
Pluto	I	Charon	19.636	6.387 23		0.076	96.2[a]

Data are from Yoder (1995), with updates from http://ssd.jpl.nasa.gov.
i = orbit plane inclination (with respect to the parent planet's equator).
Abbreviations: T, Trojan-like satellite which leads (+) or trails (−) by ~60° in longitude the primary satellite with same semi-major axis; S, Synchronous rotation; C, Chaotic rotation.
[a] measured relative to the planet's heliocentric orbit, because the Sun (rather than the planetary oblateness) controls the local invariable plane of these distant satellites.
[b] see http://ssd.jpl.nasa.gov/sat_elem.html

- The best (rough) estimates of the masses of some of Saturn's small inner moons were derived from the amplitude of spiral density waves they resonantly excite in Saturn's rings or of density wakes that they produce in nearby ring material. These processes are discussed in Chapter 11.
- Crude estimates of the masses of some comets have been made by estimating nongravitational forces, which result from the asymmetric escape of released gases and dust (Chapter 10), and comparing them with observed orbital changes.

TABLE 1.5 Planetary Satellites: Physical Properties.

Satellite	Radius (km)	Mass (10^{23} g)	Density (g cm^{-3})	Geom. albedo	$V(1,0)$
Earth	6378	59 742	5.515	0.367	−3.86
Moon	1737.53 ± 0.03	734.9	3.34	0.12	+0.21
Mars	3394	6419	3.933	0.150	−1.52
MI Phobos	13.1 × 11.1 × 9.3(±0.1)	$1.08(\pm0.01) \times 10^{-4}$	1.90 ± 0.08	0.06	+11.8
MII Deimos	(7.8 × 6.0 × 5.1)(±0.2)	$1.80(\pm0.15) \times 10^{-5}$	1.76 ± 0.30	0.07	+12.89
Jupiter	71 492	1.8988×10^{7}	1.326	0.52	−9.40
JXVI Metis	(30 × 20 × 17)(±2)			0.06	+10.8
JXV Adrastea	(10 × 8 × 7)(±2)			0.1	+12.4
JV Amalthea	(125 × 73 × 64)(±2)			0.09	+7.4
JXIV Thebe	(58 × 49 × 42)(±2)			0.05	+9.0
JI Io	1821.3 ± 0.2	893.3 ± 1.5	3.53 ± 0.006	0.61	−1.68
JII Europa	1565 ± 8	479.7 ± 1.5	3.02 ± 0.04	0.64	−1.41
JIII Ganymede	2634 ± 10	1482 ± 1	1.94 ± 0.02	0.42	−2.09
JIV Callisto	2403 ± 5	1076 ± 1	1.85 ± 0.004	0.20	−1.05
JXIII Leda	~5				+13.5
JVI Himalia	85 ± 10				+8.14
JX Lysithea	12				+11.7
JXVI Elara	40 ± 10				+10.07
JXII Ananke	~10				+12.2
JXI Carme	~15				+11.3
JVIII Pasiphae	~18				+10.33
JIX Sinope	~14				+11.6
Saturn	60 268	5.6850×10^{6}	0.687	0.47	−8.88
SXVIII Pan	~10	$\sim4 \times 10^{-5}$		0.5	+9.5
SXV Atlas	(18.5 × 17.2 × 13.5)(±4)			0.9	+8.4
SXVI Prometheus	74 × 50 × 34(±3)	$0.001(4^{+8}_{-7})$	0.27 ± 0.16	0.6	+6.4
SXVII Pandora	(55 × 44 × 31)(±2)	$0.001(3^{+8}_{-7})$	0.42 ± 0.28	0.9	+6.4
SXI Epimetheus	(69 × 55 × 55)(±3)	0.0055 ± 0.0003	0.63 ± 0.11	0.8	+5.4
SX Janus	(99.3 × 95.6 × 75.6)(±3)	0.0198 ± 0.0012	0.65 ± 0.08	0.8	+4.4
SI Mimas	198.8 ± 0.6	0.375 ± 0.009	1.14 ± 0.02	0.5	+3.3
SII Enceladus	249.1 ± 0.3	0.73 ± 0.36	1.12 ± 0.55	1.0	+2.1
SIII Tethys	529.9 ± 1.5	6.22 ± 0.13	1.00 ± 0.02	0.9	+0.6
SXIV Calypso	15 × 8 × 8(±4)			0.6	+9.1
SXIII Telesto	15(2.5) × 12.5(5) × 7.5(2.5)			0.5	+8.9
SXII Helene	16 ± 5			0.7	+8.4
SIV Dione	560 ± 5	10.52 ± 0.33	1.44 ± 0.06	0.7	+0.8
SV Rhea	764 ± 4	23.1 ± 0.6	1.24 ± 0.04	0.7	+0.1
SVI Titan	2575 ± 2	1345.5 ± 0.2	1.881 ± 0.005	0.21	−1.28
SVII Hyperion	(185 × 140 × 113)(±10)			0.19–0.25	+4.6
SVIII Iapetus	718 ± 8	15.9 ± 1.5	1.02 ± 0.10	0.05–0.5	+1.5[a]
SIX Phoebe	(115 × 110 × 105)(±10)			0.06	+6.89

TABLE 1.5 (*cont.*)

Satellite	Radius (km)	Mass (10^{23} g)	Density (g cm^{-3})	Geom. albedo	$V(1,0)$
Uranus	25 559	8.6625×10^5	1.318	0.51	−7.19
UVI Cordelia	13 ± 2			0.07	+11.4
UVII Ophelia	16 ± 2			0.07	+11.1
UVIII Bianca	22 ± 3			0.07	+10.3
UIX Cressida	33 ± 4			0.07	+9.5
UX Desdemona	29 ± 3			0.07	+9.8
UXI Juliet	42 ± 5			0.07	+8.8
UXII Portia	55 ± 6			0.07	+8.3
UXIII Rosalind	29 ± 4			0.07	+9.8
UXIV Belinda	34 ± 4			0.07	+9.4
S/1986 U10	∼20				
UXV Puck	77 ± 3			0.07	+7.5
UV Miranda	$240(0.6) \times 234.2(0.9) \times 232.9(1.2)$	0.659 ± 0.075	1.20 ± 0.14	0.27	+3.6
UI Ariel	$581.1(0.9) \times 577.9(0.6) \times 577.7(1.0)$	13.53 ± 1.20	1.67 ± 0.15	0.34	+1.45
UII Umbriel	584.7 ± 2.8	11.72 ± 1.35	1.40 ± 0.16	0.18	+2.10
UIII Titania	788.9 ± 1.8	35.27 ± 0.90	1.71 ± 0.05	0.27	+1.02
UIV Oberon	761.4 ± 2.6	30.14 ± 0.75	1.63 ± 0.05	0.24	+1.23
UXVI Caliban					∼9.0
UXX Stephano					∼10.4
UXVII Sycorax					∼7.5
UXVIII Prospero					∼11.3
UXIX Setebos					∼10.3
Neptune	24 764	1.0278×10^6	1.638	0.41	−6.87
NIII Naiad	∼29			0.06	+10.0
NIV Thalassa	∼40			0.06	+9.1
NV Despina	74 ± 10			0.06	+7.9
NVI Galatea	79 ± 12			0.06	+7.6
NVII Larissa	$104 \times 89(\pm 7)$			0.06	+7.3
NVIII Proteus	$218 \times 208 \times 201$			0.06	+5.6
NI Triton	1352.6 ± 2.4	214.7 ± 0.7	2.054 ± 0.032	0.7	−1.24
NII Nereid	170 ± 2.5			0.2	+4.0
Pluto	1137 ± 8	132 ± 1	1.94 ± 0.12	0.44–0.61	-0.6^a
P1 Charon	586 ± 13	15 ± 1	1.5 ± 0.2	0.375 ± 0.08	1.3

Most data are from Yoder (1995), with updates from http://ssd.jpl.nasa.gov. Data on Pan: Showalter *et al.* (1986), Showalter (1991). Pluto: Tholen and Tedesco (1994), Buie, Tholen, and Wasserman (1997). Metis, Adrastea, Amalthea and Thebe: Thomas *et al.* (1998). $V(1,0)$ is the visual equivalent magnitude at 1 AU and zero phase angle. The apparent visual magnitude at phase angle ϕ, A_v, can be calculated from: $A_v = V(1,0) + C\phi + (5\log_{10})(r_\odot r_\oplus)$, with C the phase coefficient in magnitudes per degree.
[a] variable, depending on viewing aspect (number quoted is for $A_{0,v} = 0.2$); Pluto's brightness changes due to rotation (0.3 mag over 6.39 days) and changes in heliocentric distance.

The gravity field of a nonspherically symmetric mass distribution differs from that of a point source of identical mass. Such deviations, combined with the knowledge of the rotation period, can be used to estimate the degree of central concentration of mass in rotating bodies (Chapter 6). The deviation of the gravity field of an asymmetric body from that of a point mass is most pronounced, and thus most easily measured, closest to the body (Chapter

2). To determine the precise gravity field, one can make use of both spacecraft tracking data and the orbits of moons and/or eccentric rings.

1.2.3 Size

The sizes and shapes of objects in the Solar System differ substantially. The size of an object can be measured in various ways:

- The diameter of a body is the product of its angular size (measured in radians) and its distance from the observer. Solar System distances are simple to estimate from orbits; however, limited resolution from Earth results in large uncertainties in angular size. Thus, other techniques often give the best results for bodies not imaged at close distances by interplanetary spacecraft.
- The diameter of a Solar System body can be deduced by observing a star as it is occulted by the body. The angular velocity of the star relative to the occulting body can be calculated from orbital data, including the effects of the Earth's orbit and rotation. Thus, the duration of an occultation at a particular observing site gives the length of a chord of the body's projected silhouette. Three well-separated chords suffice for a spherical planet. Many chords are needed if the body is irregular in shape, and observations of the same event from many widely spaced telescopes are necessary. This technique is particularly useful for small bodies like asteroids and satellites, which cannot be resolved in Earth-based telescopic images and have not been visited by spacecraft. Occultations of sufficiently bright stars are rare and require appropriate predictions as well as significant observing campaigns (in order to obtain enough chords, even if some sites are clouded out); thus, occultation diameters exist only for a few small bodies.
- Radar echoes can be used to determine radii and shapes (Chapter 9). The radar signal strength drops as $1/r^4$ ($1/r^2$ going to the object and $1/r^2$ returning to the antenna), so only relatively nearby objects may be studied with radar. Radar is especially useful for solid planets, asteroids and cometary nuclei.
- An excellent way to measure the radius of an object is to send a lander, and triangulate using an orbiter. This method, as well as the radar technique, also works well for terrestrial planets and satellites with substantial atmospheres.
- The size and the albedo of bodies can be estimated by combining photometric observations at visible and infrared wavelengths. At visible wavelengths one measures the sunlight reflected off the object, while at infrared wavelengths one observes the thermal radiation from the body itself (see Chapters 3 and 9 for a detailed discussion).

We note that one can trivially determine the mean density of an object once its mass and size are known. The density of an object gives a rough idea of its composition, although compression at the high pressures which occur in planets and large moons must be taken into account, and the possibility of significant void space in small bodies should also be considered. The low density ($\sim$1 g cm^{-3}) of the four giant planets, for example, implies material with small mean molecular weight. Terrestrial planet densities of 3.5–5.5 g cm^{-3} imply rocky material, including some metal. Most of the satellites around the giant planets have densities between 1 and 2 g cm^{-3}, suggesting a combination of ices and rock. Comets have densities of the order of 1 g cm^{-3} or less, indicative of rather loosely packed dirty ices. In addition to the density, one can also calculate the escape velocity from the mass and size of the object; the escape velocity, together with temperature, can be used to deduce the possible presence of an atmosphere.

1.2.4 Rotation

The rotation of an object can be determined in various ways:

- The most straightforward way to determine a planetary body's rotation axis and period is to observe how markings on the surface move around with the disk. Unfortunately, not all planets have such features; moreover, if atmospheric features are used, winds may cause the period to vary with latitude.
- Planets with significant magnetic fields trap charged particles within their magnetospheres. These charged particles are accelerated by electromagnetic forces and emit radio waves. As magnetic fields are not uniform in longitude, and as they rotate with (presumably the bulk of) the planet, these radio signals have a periodicity equal to the planet's rotation period. For planets without detectable solid surfaces, the magnetic field period is viewed as more fundamental than the periods of cloud features, as the latter can change with latitude and time.
- The rotation period of a body can often be determined by periodicities observed in its *lightcurve*, which gives the total disk brightness as a function of time. Lightcurve variations can be the result of differences in albedo or, for irregularly shaped bodies, in projected area. Irregularly shaped bodies produce lightcurves with two very similar maxima and two very similar minima per revolution, whereas albedo variations have no such preferred symmetry. Thus, ambiguities of a factor of two sometimes exist in spin periods determined by lightcurve analysis. Most asteroids have double-peaked lightcurves, indicating that the major variations are due to shape, but the peaks are

distinguishable from each other because of minor variations in hemispheric albedo and local topography.

- The measured Doppler shift across the disk can give a rotation period and a crude estimate of the rotation axis, provided the body's radius is known. This can be done passively in visible light, or actively using radar.

The rotation periods of most objects orbiting the Sun are of the order of five hours to a week. Mercury and Venus, both of whose rotations have almost certainly been slowed by solar tides, form exceptions with periods of 59 and 243 days respectively. Six of the nine planets rotate in a prograde sense with *obliquities* (axial tilts) of 30° or less. Venus rotates in a retrograde direction with an obliquity of 177°, and the rotation axes of Uranus and Pluto are tilted nearly perpendicular to the ecliptic plane. Most planetary satellites rotate synchronously with their orbital periods as a result of planet-induced tides (Chapter 2).

1.2.5 Shape

Many different forces together determine the shape of a body. Self-gravity tends to produce bodies of spherical shape, a minimum for gravitational potential energy. Material strength maintains shape irregularities, which may be produced by accretion, impacts, or internal geological processes. As self-gravity increases with the size of an object, larger bodies tend to be rounder. Typically, bodies with mean radii larger than ~ 200 km are fairly round. Smaller objects may be quite oddly shaped (Fig. 1.4).

There is a relationship between the planet's rotation and its oblateness, since the rotation introduces a centrifugal pseudo-force, which causes the planet to bulge out at the equator and to flatten at the poles. A perfectly fluid planet would be shaped as an oblate spheroid. Polar flattening is greatest for planets which have a low density and rapid rotation. In the case of Saturn, the ratio of polar radius to equatorial radius, $R_{pol}/R_{eq} = 0.9$, and polar flattening is easily discernible on images of the planet.

The shape of an object can be determined from:

- Direct imaging, either from the ground or spacecraft.
- Length of chords observed by stellar occultation experiments at various sites (see Section 1.2.3).
- Analysis of radar echoes.
- Analysis of lightcurves. This only gives a rough measure of asymmetry, since there are substantial ambiguities between shape and albedo variations.
- The shape of the *central flash*, which is observed when the center of a body with an atmosphere passes in front of an occulted star. The central flash results from the focusing of light rays refracted by the atmosphere and can be seen only under fortuitous observing circumstances.

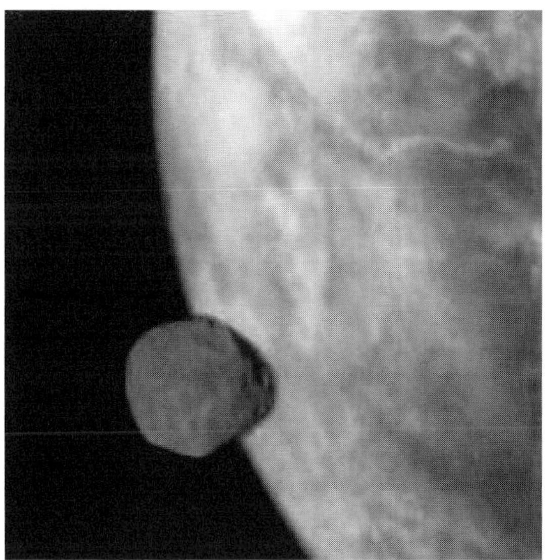

FIGURE 1.4 Image of the small irregularly-shaped moon Phobos against the background of the limb of the nearly spherical planet Mars. Phobos appears much larger relative to Mars than it actually is, because the Soviet spacecraft Phobos 2 was much closer to the moon than to the planet when it took this image.

1.2.6 Temperature

The equilibrium temperature of a planet can be calculated from the energy balance between solar insolation and reradiation outward (Chapter 3). However, for many planets internal heat sources provide a significant contribution. Moreover, there may be diurnal, latitudinal and seasonal variations in the temperature. The *greenhouse effect*, a thermal 'blanket' caused by an atmosphere which is more transparent to visible radiation (the Sun's primary output) than to infrared radiation from the planet, raises the surface temperature on some planets far above the equilibrium blackbody value. For example, because of the high albedo of its clouds, Venus actually absorbs less solar energy per unit area than does Earth; thus (as internal heat sources on these two planets are negligible compared to solar heating), the effective radiating temperature of Venus is lower than that of Earth. However, as a consequence of the greenhouse effect, Venus's surface temperature is raised up to ~ 730 K, well above the surface temperature on Earth.

Direct *in situ* measurements with a thermometer can provide an accurate estimate of the temperature of a body. The thermal infrared spectrum of a body's emitted radiation is also a good indicator of its temperature. Most solid and liquid planetary material can be characterized as a nearly perfect blackbody radiator with its emission peak at near- to mid-infrared wavelengths. Analysis of emitted radiation sometimes gives different temperatures at differing wavelengths. This could be due to a combination of temperatures from different locations on the surface, e.g., pole to equator differences, albedo variations, or volcanic hot spots such as those seen on Io. Also, the opacity of an atmosphere varies with wavelength, which allows us to probe different altitudes in a planetary atmosphere.

1.2.7 Magnetic Field

Magnetic fields are created by moving charges. Currents moving through a solid medium decay quickly (unless the medium is a superconductor, which is unreasonable to expect at the high temperatures found in planetary interiors). Thus, internally generated planetary magnetic fields must either be produced by a (poorly understood) *dynamo* process, which can only operate in a fluid region of a planet, or be due to *remnant ferromagnetism*, which is a result of charges which are bound to atoms of a solid locked in an aligned configuration. Remnant ferromagnetism is not believed to be a likely cause of large fields because, in addition to the fact that it is expected to decay away on timescales short compared to the age of the Solar System, it would require the planet to have been subjected to a nearly constant (in direction) magnetic field during the long period in which the bulk of its iron cooled through its Curie point. Very small planetary magnetic fields may also be induced by interaction between the solar wind (which is composed predominantly of charged particles) and conducting regions, either solid or fluid, within the planet.

A magnetic field may be detected directly using an *in situ* magnetometer or indirectly via effects of accelerating charges which consequently produce radiation (radio emissions). The presence of localized *aurorae*, luminous disturbances caused by charged particle precipitation in a planet's upper atmosphere, is also indicative of a magnetic field. The magnetic fields of the planets can be approximated by dipoles, with perturbations to account for their irregularities. All four giant planets, as well as Earth and Mercury, have magnetic fields generated in their interiors. Venus and comets have magnetic fields induced by the interaction between the solar wind and charged particles in their atmosphere/ionosphere, while Mars and the Moon have strong rather localized crustal magnetic fields.

1.2.8 Surface Composition

The composition of a body's surface can be derived from:

- Spectral reflectance data. Such spectra may be observed from Earth; however, spectra at ultraviolet wavelengths require being above the Earth's atmosphere.
- Thermal infrared spectra and thermal radio data. Though hard to interpret, these measurements contain information on a body's composition.
- Radar reflectivity. Such observations can be carried out on the ground or from orbiting spacecraft.
- X-ray and γ-ray fluorescence. These measurements may be conducted from a spacecraft in orbit around the planet (or, in theory, even a flyby spacecraft) if the body lacks a substantial atmosphere. Detailed measurements require landing a probe on the body's surface.
- Chemical analysis of surface samples. This can be performed on samples brought to Earth by natural processes (meteorites) or spacecraft, or (in less detail) by *in situ* analysis using spacecraft. Other forms of *in situ* analysis include mass spectroscopy and electrical and thermal conductivity measurements.

The compositions of the planets, asteroids and satellites show a dependence on heliocentric distance, with the objects closest to the Sun having the largest concentrations of dense materials (which tend to be refractory) and the smallest concentration of ices (which are much more volatile).

1.2.9 Surface Structure

The surface structure varies greatly from one planet or moon to another. There are various ways to determine the structure of a planet's surface:

- Structure on large scales can be detected by imaging, either passively in the visible/IR/radio or actively using radar imaging techniques. It is best to have imaging available at more than one illumination angle in order to separate tilt angle (slope) effects from albedo differences.
- Structure on small scales can be deduced from the radar echo brightness and the variation of reflectivity with *phase angle*, the angle between the illuminating Sun and the observer as seen from the body. The brightness of a body with a size much larger than the wavelength of light at which it is observed generally increases slowly with decreasing phase angle. For very small phase angles, this increase can be much more rapid, a phenomenon referred to as the *opposition effect* (Chapter 9).

1.2.10 Atmospheric Structure and Composition

Most of the planets and some satellites are surrounded by significant atmospheres. The giant planets Jupiter, Saturn, Uranus and Neptune are basically huge fluid balls, and their atmospheres are dominated by H_2 and He. Venus has a very dense CO_2 atmosphere, with clouds so thick that one cannot see its surface at visible wavelengths; Earth has an atmosphere consisting primarily of N_2 (78%) and O_2 (21%), and Mars has a more tenuous CO_2 atmosphere. Saturn's satellite Titan has a dense nitrogen-rich atmosphere, which is intriguing since it contains many kinds of organic molecules. Pluto and Neptune's moon Triton each have a tenuous atmosphere dominated by N_2 gas. Jupiter's volcanically active moon Io has a very tenuous atmosphere which consists primarily of SO_2 gas. Mercury and the Moon each have an extremely tenuous atmosphere ($\lesssim 10^{-12}$ bar); Mercury's atmosphere is dominated by atomic O, Na, and He, while the main constituents in the Moon's atmosphere are He and Ar.

The composition and structure (temperature–pressure profile) of an atmosphere can be determined from: spectral reflectance data at visible wavelengths, thermal spectra and photometry at infrared and radio wavelengths, stellar occultation profiles, *in situ* mass spectrometers and attenuation of radio signals sent back to Earth by atmospheric/surface probes.

1.2.11 Planetary Interiors

The interior of a planet is not directly accessible to observations. However, with the help of the observable parameters discussed above, one can derive information on a planet's bulk composition and its interior structure.

The *bulk composition* is not an observable quantity, except for extremely small bodies such as meteorites which we can actually take apart and analyze (Chapter 8). Thus, we must deduce bulk composition from a variety of direct and indirect clues and constraints. The most fundamental constraints are based on the mass and the size of the planet. Using only these constraints, together with material properties derived from laboratory data and quantum mechanical calculations, it can be shown that Jupiter and Saturn are composed mostly of hydrogen, simply because all other elements are too dense to fit the constraints (unless the internal temperature is much higher than is consistent with the observed effective temperature in a quasi-steady state). However, this method only gives definitive results for planets composed mostly of the lightest element. For all other bodies, bulk composition is best estimated from models which include mass and radius as well as the composition of the surface and atmosphere, the body's heliocentric distance (location is useful because it gives us an idea of the temperature of the region during the planet-formation epoch, and thus which elements were likely to condense) together with reasonable assumptions of cosmogonic abundances (Section 1.3 and Chapter 12).

The *internal structure* of a planet can be derived to some extent from its gravitational field and the rotation rate. From these two parameters, one can estimate the degree of concentration of the mass at the planet's center. The gravitational field can be determined from the orbits of satellites or rings (Chapter 2). Detailed information on the internal structure of a planet with a solid surface may be obtained if seismometers can be placed on its surface. The velocities and attenuations of seismic waves propagating through the planet's interior depend on density, rigidity and other physical properties (Chapter 6), which in turn depend on composition. The free oscillation periods of gaseous planets can, in theory, also provide clues to internal properties, just as helioseismology has greatly increased our data on the Sun's interior in recent years. Evidence of volcanism and plate tectonics constrain the thermal environment below the surface. Energy output provides information on the thermal structure of a planet's interior. Magnetic fields are produced by moving charges. While a small magnetic field such as the Moon's may be the result of remnant ferromagnetism, substantial planetary magnetic fields are believed to require a conducting fluid region within the planet's interior. Centered dipole fields are probably produced in or near the core of the planet, whereas highly irregular offset fields are likely to be produced closer to the planet's surface.

1.3 Formation of the Solar System

Questions concerning the formation of the Solar System are among the most intellectually challenging in planetary science. Observations provide direct information on the current state of the Solar System, but only indirect clues to its origin. Thus, even though placing a chapter on Solar System formation at the beginning of this book would make sense from a chronological perspective, we have chosen to defer such a discussion to near the end so that the reader can have more clues in hand when we attempt to piece together the puzzle. Nonetheless, it is useful for the student to begin with a brief overview of the currently accepted model of planetary formation because it provides a framework for interpreting unobservable planetary properties, such as the compositions of planetary interiors (Chapter 6), and it motivates the study of objects like meteorites (Chapter 8). While there is some component of circular reasoning to this arrangement, it emphasizes that scientific development is not linear, and that placing a new piece in the puzzle is aided by the perspectives we have from the pieces already in place, but also occasionally requires reorienting some previously accepted ideas.

The nearly planar and almost circular orbits of the planets in our Solar System argue strongly for planetary formation within a flattened circumsolar disk. Astrophysical models suggest that such disks are a natural byproduct of star formation from the collapse of rotating molecular cloud cores. Observational evidence for the presence of disks of Solar System dimensions around young stars has increased substantially in recent years, and infrared excesses in the spectra of young stars suggest that the lifetimes of protoplanetary disks range from 10^6 to 10^7 years.

Our galaxy contains many molecular clouds, most of which are several orders of magnitude larger than our Solar System. These are the sites in which star formation occurs at the current epoch. Even a very slowly rotating molecular cloud core has far too much spin angular momentum to collapse down to an object of stellar dimensions, so a significant fraction of the material in a collapsing core falls onto a rotationally supported disk orbiting the pressure-supported (proto)star. Such a disk has the same initial elemental composition as the growing star. At sufficient distances from the central star, it is cool enough for $\sim$1–2% of this material to be in solid form, either remnant interstellar grains or condensates formed within the disk. This dust is primarily composed of rock-forming compounds within a few AU of a 1 $M_\odot$ star, whereas in the more distant regions the amount of ices (H_2O, CH_4, CO, etc.) present in solid form is comparable to that of rocky solids.

During the infall stage, the disk is very active and probably highly turbulent, as a result of the mismatch of the specific angular momentum of the gas hitting the disk with that required to maintain keplerian rotation. Gravitational instabilities and viscous and magnetic forces may add to this activity. When the infall slows substantially or stops, the disk becomes more quiescent. Interactions with the gaseous component of the disk affect the dynamics of small solid bodies, and the growth from micrometer-sized dust to kilometer-sized planetesimals remains poorly understood. Meteorites (Chapter 8), asteroids (Chapter 9) and comets (Chapter 10), most of which were never incorporated into bodies of planetary dimensions, best preserve a record of this important period in Solar System development.

The dynamics of larger solid bodies within protoplanetary disks are better characterized. The primary perturbations on the keplerian orbits of kilometer-sized and larger planetesimals in protoplanetary disks are mutual gravitational interactions and physical collisions. These interactions lead to accretion (and in some cases erosion and fragmentation) of planetesimals. Eventually, solid bodies agglomerated into the terrestrial planets in the inner Solar System, and planetary cores several times the mass of the Earth in the outer Solar System. These massive cores were able to gravitationally attract and retain substantial amounts of gaseous material from the solar nebula. In contrast, terrestrial planets were not massive enough to attract and retain such gases, and the gases in their current thin atmospheres are derived from material that was incorporated in solid planetesimals.

The planets in our Solar System orbit close enough to each other that the final phases of planetary growth could have been the merger or ejection of planets on unstable orbits. However, the low eccentricities of the orbits of the outer planets imply that some damping process, such as accretion/ejection of numerous small planetesimals or interactions with residual gas within the protoplanetary disk, must also have been involved.

As we learn more about the individual bodies and classes of objects in our Solar System and simulations of planetary growth become more sophisticated, theories about the formation of our Solar System are being revised and (we hope) improved. The detection of planets around other stars has presented us with new challenges to develop a unified theory of planet formation which is applicable to all stellar systems. We will discuss these theories in more detail in Chapters 12 and 13.

Further Reading

A good nontechnical overview of the planetary system, complete with many beautiful color pictures, is given by:

Beatty, J.K., C.C. Peterson, and A. Chaikin, Eds., 1999. *The New Solar System*, 4th Edition. Sky Publishing Co., Cambridge, MA and Cambridge University Press, Cambridge, England.

A terse, but detailed overview including reproductions of paintings of various Solar System objects by the authors, is provided by:

Miller, R., and W.K. Hartmann, 1993. *The Grand Tour, A Traveler's Guide to the Solar System*. Workman Publishing, New York.

An overview of the planetary system emphasizing atmospheric and space physics is given by:

Encrenaz, T., J.-P. Bibring, and M. Blanc, 1995. *The Solar System*, 2nd Edition. Springer-Verlag, Berlin.

Two good overview texts aimed at undergraduate nonscience majors are:

Morrison, D., and T. Owen, 1996. *The Planetary System*. Addison-Wesley Publishing Company, New York.

Hartmann, W.K., 1999. *Moons and Planets*, 4th Edition. Wadsworth Publishing Company, Belmont, CA.

Chemical processes on the planets are covered in some detail by:

Lewis, J.S., 1997. *Physics and Chemistry of the Solar System*, 2nd Edition. Academic Press, San Diego.

The following encyclopedias on planetary sciences form a nice complement to this book:

Encyclopedia of Planetary Sciences (1999). Eds. J.H. Shirley and R.W. Fairbridge. Chapman and Hall, London.

Encyclopedia of the Solar System (1999). Eds. P. Weissman, L. McFadden, and T.V. Johnson. Academic Press, San Diego.

Planetary data tables can be found in:

Yoder, C.F., 1995. Astrometric and Geodetic Properties of Earth and the Solar System. In *Global Earth Physics, A Handbook of Physical Constants*. AGU Reference Shelf 1, American Geophysical Union.
Updated information: http://ssd.jpl.nasa.gov

Problems

1.1.**E** Because the distances between the planets are much larger than planetary sizes, very few diagrams or models of the Solar System are completely to scale. However, imagine that you are asked to give an astronomy lecture/demonstration to your niece's second grade class, and you decide to illustrate the vastness and near emptiness of space by constructing a scale model of the Solar System using ordinary objects. You begin by selecting a (1 cm diameter) marble to represent the Earth. What other objects can you use, and how far apart must you space them? Proxima Centauri, the nearest star to the Solar System, is 4.2 light years distant; where, in your model, would you place it?

1.2.**E** The satellite systems of the giant planets are often referred to as 'miniature solar systems'. In this problem you will make some calculations comparing the satellite systems of Jupiter, Saturn and Uranus to the planetary system.

(a) Calculate the ratio of the sum of the masses of the planets with that of the Sun, and similar ratios for the jovian, saturnian and uranian systems.

(b) Calculate the ratio of the sum of the orbital angular momenta of the planets to the rotational angular momentum of the Sun. You can assume circular orbits at zero inclination for all planets, and ignore the effects of planetary rotation and the presence of satellites.

(c) Repeat the calculation in (b) for the jovian, saturnian and uranian systems.

(d) Calculate the orbital semimajor axes of the planets in terms of solar radii, and the orbital semimajor axes of Jupiter's moons in jovian radii. How would a scale model of the jovian system compare to the model of the planetary system in Problem 1.1?

1.3.**I** A planet which keeps the same hemisphere pointed towards the Sun must rotate once per orbit in the *prograde* direction (i.e., in the same direction as the planet's orbit).

(a) Draw a diagram to demonstrate this fact. The rotation period (in an inertial frame) or *sidereal day* for such a planet is equal to its orbital period, whereas the length of a *solar day* on such a planet is infinite.

(b) Earth rotates in the prograde direction. How many times must Earth rotate per orbit in order for there to be 365.24 *solar days* per year? Verify your result by comparing the length of Earth's sidereal period (Table 1.2) to the length of a mean solar day.

(c) If a planet rotated once per orbit in the *retrograde* direction (opposite to the direction of its orbit), how many solar days would it have per orbit?

(d) Determine a general formula relating the lengths of solar and sidereal days on a planet. Use your formula to determine the lengths of solar days on Mercury, Venus, Mars and Jupiter.

(e) For a planet on an eccentric orbit, the length of either the solar day or the sidereal day varies on an annual cycle. Which one varies and why? Calculate the length of the longest such day on Earth. This longest day is how much longer than the mean day of its type?

1.4.**I** For the same reasons that the length of a mean solar day is not exactly equal to the sidereal rotation period of Earth (Problem 1.3), the length of the month is not equal to the sidereal orbital period of the Moon about the Earth. What, physically, does a month refer to? Calculate the length of an average (astronomical) month.

1.5.**I** As you may have already guessed by now if you have read the two previous problems, the length of the year is not exactly equal to the time it takes Earth to complete one orbit about the Sun. The common usage of year is the mean length of time over which seasons repeat, this is called the *tropical year*. The change in seasons is primarily a result of Earth's motion about the Sun, but seasons are also affected by a gradual change in the direction of Earth's spin axis. The principal cause of the precession of Earth's spin axis is torques on Earth's equatorial bulge exerted by the Moon and the Sun. The resulting *lunisolar precession* has a period of $\sim 26\,000$ years. (Torques exerted by the other planets also affect the direction of Earth's spin axis. These torques are important because they induce quasi-periodic variations in Earth's obliquity, but their influence on the precession rate of Earth's axis is small.)

(a) Draw a diagram of the system and use it to derive a formula relating the tropical year, the *sidereal year* and the precession period. The sidereal year is longer than the tropical year. Use this fact to deduce the direction of Earth's lunisolar precession.

(b) Compute the length of the sidereal year in terms of tropical years and in terms of days. Note that although the fractional difference between the tropical and sidereal years is much smaller than the difference between solar and sidereal days, it is still almost twice as large as the difference between the Julian and Gregorian calenders.

2 Dynamics

No human investigation can be called real science if it cannot be demonstrated mathematically.

Leonardo da Vinci

In 1687, Isaac Newton showed that the relative motion of two spherically symmetric bodies resulting from their mutual gravitational attraction is described by simple conic sections: ellipses for bound orbits and parabolas and hyperbolas for unbound trajectories. However, the introduction of additional gravitating bodies produces a rich variety of dynamical phenomena, even though the basic interactions between pairs of objects can be straightforwardly described. In this chapter, we describe the basic orbital properties of Solar System objects (planets, moons, minor bodies and dust) and their mutual interactions. We also provide several examples of important dynamical processes which occur in the Solar System and lay the groundwork for describing some of the phenomena which are discussed in more detail in other chapters of this book.

2.1 The Two-Body Problem

2.1.1 Kepler's Laws of Planetary Motion

By careful observation and analysis of the orbits of the planets, Johannes Kepler deduced his three 'laws' of planetary motion:

(1) All planets move along elliptical paths with the Sun at one focus. We can express the *heliocentric distance*, $r_\odot$ (i.e., the planet's distance from the Sun), as:

$$r_\odot = \frac{a(1-e^2)}{1+e\ \cos f}, \tag{2.1}$$

with a the *semimajor axis* (average of the minimum and maximum heliocentric distances). The *eccentricity* of the orbit, $e \equiv (1 - b_m^2/a^2)^{1/2}$, where $2b_m$ is the minor axis of the ellipse. The *true anomaly*, f, is the angle between the planet's *perihelion* (closest heliocentric distance) and its instantaneous position. These quantities are displayed graphically in Figure 2.1a.

(2) A line connecting a planet and the Sun sweeps out area, $\mathcal{A}$, at a constant rate:

$$\frac{d\mathcal{A}}{dt} = \text{constant}. \tag{2.2}$$

Note that the value of this constant rate differs from one planet to the next.

(3) The square of a planet's orbital period about the Sun (in years), P_{yr}, is equal to the cube of its semimajor axis (in AU), a_{AU}:

$$P_{yr}^2 = a_{AU}^3. \tag{2.3}$$

2.1.2 Newton's Laws of Motion and the Universal Law of Gravitation

Although Kepler's laws were originally deduced from careful observation of planetary motion, they were subsequently shown to be derivable from Newton's laws of motion together with his universal law of gravity. Consider a body of mass m_1 at instantaneous location $\mathbf{r}_1$ with instantaneous velocity $\mathbf{v}_1 \equiv d\mathbf{r}_1/dt$ and hence momentum $m_1\mathbf{v}_1$. The acceleration produced by a net force $\mathbf{F}_1$ is given by Newton's second law of motion:

$$\frac{d(m_1\mathbf{v}_1)}{dt} = \mathbf{F}_1. \tag{2.4}$$

Newton's third law states that for every action there is an equal and opposite reaction; thus, the force on each object of a pair due to the other object is equal in magnitude but opposite in direction:

$$\mathbf{F}_{12} = -\mathbf{F}_{21}, \tag{2.5}$$

where $\mathbf{F}_{ij}$ represents the force exerted by body j on body i. Newton's universal law of gravity states that a second body of mass m_2 at position $\mathbf{r}_2$ exerts an attractive force on the first body given by

$$\mathbf{F}_{g12} = -\frac{Gm_1m_2}{r^2}\hat{\mathbf{r}}, \tag{2.6}$$

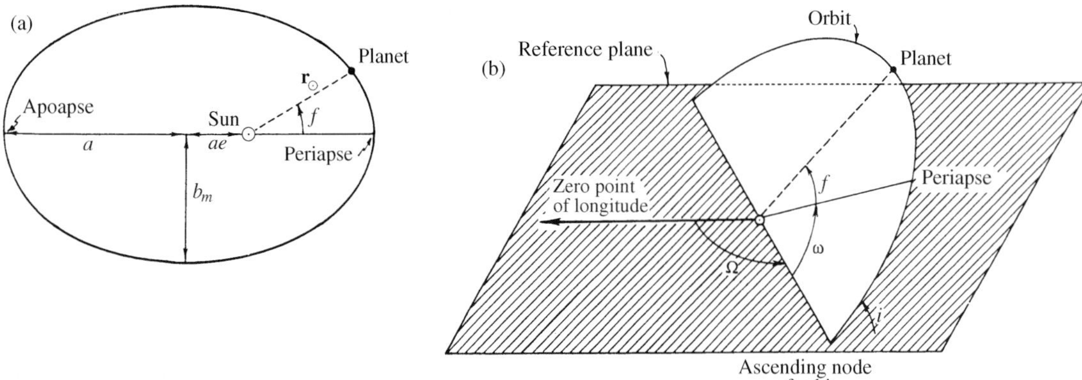

FIGURE 2.1 (a) Geometry of an elliptic orbit. The Sun is at one focus and the vector $\mathbf{r}_\odot$ denotes the instantaneous heliocentric location of the planet (i.e., $r_\odot$ is the planet's distance from the Sun). The semimajor axis of the ellipse is a, e denotes its eccentricity, and b_m is the ellipse's semiminor axis. The true anomaly, f, is the angle between the planet's perihelion and its instantaneous position. (b) Geometry of an orbit in three dimensions; i is the inclination of the orbit, Ω is the longitude of the ascending node, and ω is the argument of periapse. (Adapted from Hamilton 1993)

where $\mathbf{r} \equiv \mathbf{r}_1 - \mathbf{r}_2$ is the vector distance from particle 2 to particle 1, G is the gravitational constant, and $\hat{\mathbf{r}} \equiv \mathbf{r}/r$.

The equation for the relative motion of two mutually gravitating bodies can be derived from Newton's laws to be (Problem 2.1):

$$\mu_r \frac{d^2\mathbf{r}}{dt^2} = -\frac{G\mu_r M}{r^2}\hat{\mathbf{r}}, \tag{2.7}$$

where μ_r is the reduced mass and M is the total mass:

$$\mu_r \equiv \frac{m_1 m_2}{m_1 + m_2}, \tag{2.8a}$$

$$M \equiv m_1 + m_2. \tag{2.8b}$$

Thus, the relative motion is completely equivalent to that of a particle of reduced mass μ_r orbiting a *fixed* central mass M. Newton's generalization of Kepler's laws for the two-body problem is as follows:

(1) The two bodies move along elliptical paths, with one focus of each ellipse located at the center of mass, $\mathbf{r}_{cm} = (m_1\mathbf{r}_1 + m_2\mathbf{r}_2)/M$, of the system.

(2) A line connecting two bodies (as well as lines from each body to the center of mass) sweeps out area at a constant rate. This is a consequence of the conservation of angular momentum, $\mathbf{L}$:

$$\frac{d\mathbf{L}}{dt} = 0 \tag{2.9}$$

where

$$\mathbf{L} = \mathbf{r} \times m\mathbf{v}. \tag{2.10}$$

(3) The orbital period of a pair of bodies about their mutual center of mass is given by

$$P_{orb}^2 = \frac{4\pi^2 a^3}{G(m_1 + m_2)}. \tag{2.11}$$

Equation (2.11) reduces to Kepler's third law in the limit as $m_2/m_1 \to 0$. The derivations of (the newtonian generalization of) Kepler's laws is the topic of Problem 2.2.

2.1.3 Elliptical Motion, Orbital Elements, the Orbit in Space

The Sun contains more than 99.8% of the mass of the known Solar System. The gravitational force exerted by a body is proportional to its mass (eq. 2.6), so to an excellent first approximation we can regard the motion of the planets and many other bodies as being solely influenced by a fixed central point-like mass. For objects such as the planets, which are bound to the Sun and hence cannot go arbitrarily far from the central mass, the general solution for the orbit is the ellipse described by equation (2.1). The orbital plane, although fixed in space, can be arbitrarily oriented with respect to whatever reference plane we have chosen. This reference plane is usually taken to be either the Earth's orbital plane about the Sun, which is called the *ecliptic*, or the equatorial plane of the largest body in the system, or the plane perpendicular to the total angular momentum of the system. The *inclination*, i, of the orbit is the angle between the reference plane and the orbital plane; i can range from 0° to 180°. Conventionally, secondaries orbiting in the same direction as the primary rotates are defined to have inclinations from 0° to 90° and are said to be on *prograde* orbits. Secondaries orbiting in the opposite direction are defined to have $90° < i \leq 180°$ and said to be on *retrograde* orbits. For heliocentric orbits, the Earth's orbit rather than the Sun's equator is taken as reference. The intersection of the orbital and reference planes is called the *line of nodes*, and the orbit pierces the reference plane at two locations – one as the body passes upward through the plane (the *ascending node*) and one as it descends (the *descending node*). A fixed direction in the reference plane is chosen, and the angle to the direction of the orbit's ascending node is called the *longitude of the ascending node*, Ω. The angle between the line to the ascending node and the line to the direction of *periapse* (the point on the orbit when the two bodies are closest, which is referred to as perihelion for orbits about the Sun and *perigee* for orbits about the Earth) is called the *argument of periapse*, ω. For heliocentric orbits, Ω and ω are measured eastward from the *vernal equinox*[1]. Finally, the true anomaly, f, specifies the angle between the planet's periapse and its instantaneous position. Thus, the six *orbital elements*, a, e, i, Ω, ω and f, uniquely specify the location of the object in space, cf. Figure 2.1b. The first three quantities, a, e, and i, are often referred to as the *principal orbital elements*, as they describe the size, shape and tilt of the orbit.

For two bodies with known masses, specifying the elements of the relative orbit and the positions and velocities of the center of mass is equivalent to specifying the positions and velocities of both bodies. Alternative (sets of) orbital elements are often used for convenience. For example, the *longitude of periapse*,

$$\varpi \equiv \Omega + \omega, \tag{2.12a}$$

is often used in place of ω. The time of perihelion passage, t_ϖ, is commonly used in place of f. The mean angular frequency,

$$n \equiv \frac{2\pi}{P_{orb}} \tag{2.12b}$$

and the longitude,

$$\lambda = n(t - t_\varpi) + \varpi, \tag{2.12c}$$

are also used frequently to specify orbital properties. See Danby (1988) for a more detailed discussion of the various sets of commonly used orbital elements.

[1] The hour circle (i.e., great circle through the celestial poles) of the vernal equinox is the great circle that crosses the equator at the location of the Sun on the first day of spring. This is also the zero-point of the *right ascension*, which is a coordinate used by observers to describe the apparent location of a body in the sky.

2.1.4 Elliptical, Parabolic and Hyperbolic Orbits

The *centripetal force* necessary to keep an object of mass μ_r in a circular orbit of radius r with speed v_c is

$$\mathbf{F}_c = \frac{\mu_r {v_c}^2}{r}\hat{\mathbf{r}}. \tag{2.13}$$

Equating this to the gravitational force exerted by the central body of mass M, we find that the *circular velocity* for the orbit is:

$$v_c = \sqrt{\frac{GM}{r}}. \tag{2.14}$$

The total energy of the system, E, is a conserved quantity:

$$E = \frac{1}{2}\mu_r v^2 - \frac{GM\mu_r}{r} = -\frac{GM\mu_r}{2a}, \tag{2.15}$$

where the first term on the right is the kinetic energy of the system and the second term is the potential energy of the system. If $E < 0$, the absolute value of the potential energy of the system is larger than its kinetic energy, and the system is *bound*. The body will orbit the central mass on an elliptical path. If $E > 0$, the kinetic energy is larger than the absolute value of the potential energy, and the system is *unbound*. The relative orbit is then described mathematically as a hyperbola. If $E = 0$, the kinetic and potential energies are equal in magnitude, and the relative orbit is a parabola. By setting the total energy equal to zero, we can calculate the *escape velocity* at any separation:

$$v_e = \sqrt{\frac{2GM}{r}} = \sqrt{2}\, v_c. \tag{2.16}$$

For circular orbits, it is easy to show that both the kinetic energy and the total energy of the system are equal in magnitude to half the potential energy (Problem 2.5b).

As we have noted above, the relative orbit in the two-body problem is either an ellipse, parabola or hyperbola depending on whether the energy is negative, zero or positive, respectively. These curves are known collectively as *conic sections*, and the generalization of equation (2.1) is:

$$r = \frac{\zeta}{1 + e\cos f}, \tag{2.17}$$

where r and f have the same meaning as in equation (2.1), e is the *generalized eccentricity* and ζ is a constant. For a parabola, $e = 1$ and $\zeta = 2q$, where q is the pericentric separation, i.e., the distance of closest approach. For a hyperbola, $e > 1$ and $\zeta = q(1 + e)$. For all orbits, the three orientation angles i, Ω and ω are defined as in the elliptical case.

2.1.5 Gravitational Potential

For many applications, it is convenient to express the gravitational field in terms of a potential, $\Phi_g(\mathbf{r})$, defined as:

$$\Phi_g(\mathbf{r}) \equiv -\int_{\infty}^{\mathbf{r}} \frac{\mathbf{F}_g(\mathbf{r}')}{m} \cdot d\mathbf{r}'. \tag{2.18}$$

By inverting equation (2.18), one can see that the gravitational force is the gradient of the potential and

$$\frac{d^2\mathbf{r}}{dt^2} = -\nabla\Phi_g. \tag{2.19}$$

In general, $\Phi_g(\mathbf{r})$ satisfies Poisson's equation:

$$\nabla^2\Phi_g = 4\pi\rho G. \tag{2.20a}$$

In empty space, $\rho = 0$, so $\Phi_g(\mathbf{r})$ satisfies Laplace's equation:

$$\nabla^2\Phi_g = 0. \tag{2.20b}$$

2.2 The Three-Body Problem

Gravity is not restricted to interactions between the Sun and the planets or individual planets and their satellites, but rather all bodies feel the gravitational force of one another. The motion of two mutually gravitating bodies is *completely integrable* (i.e., there exists one independent integral or constraint per degree of freedom), and the relative trajectories of the two bodies are given by simple conic sections, as discussed above. However, when more bodies are added to the system, additional constraints are needed to specify the motion; not enough integrals of motion are available (Problem 2.6), so the trajectories of even three gravitationally interacting bodies cannot be deduced analytically except in certain limiting cases. The general three-body problem is quite complex, and little progress can be made without resorting to numerical integrations. Fortunately, various approximations based upon large differences between the masses of the bodies and nearly circular and planar orbits (which are quite accurate for most Solar System applications) simplify the problem sufficiently that some important analytic results may be obtained.

If one of the bodies is of negligible mass (e.g., a small asteroid, a ring particle or an artificial satellite), its effects on the other bodies may be ignored; the simpler system that results is called the *restricted three-body problem*, and the small body is referred to as a *test particle*. If the relative motion of the two massive particles is a circle, we refer to the situation as the *circular restricted three-body problem*. An alternative to the restricted three-body problem is *Hill's problem*, in which the mass of one of the bodies is much greater than the other two, but there is no restriction on the masses of the two small bodies relative to each other. An independent simplification is to assume that all three bodies travel within the same plane, the *planar three-body problem*. Various, but not all (cf. Problem 2.7), combinations of these assumptions are possible. Most of the results presented in this section are rigorously true only for the circular restricted three-body problem, but they are valid to a good approximation for many configurations which exist in the Solar System.

2.2.1 Jacobi's Constant and Lagrangian Points

Our study of the three-body problem begins by considering an idealized system in which two massive bodies move on circular orbits about their common center of mass. A third body is introduced which is much less massive than the smallest of the first two, so that to good approximation it has no effect on the orbits of the other bodies. Our analysis is performed in a noninertial frame which rotates about the z-axis at a rate equal to the orbital frequency of the two massive bodies. The origin is given by the center of mass of the pair, and the two bodies remain fixed at points on the x-axis, x_1 and x_2. We choose units such that the distance between the two bodies, sum of the masses and the gravitational constant are all equal to one; this implies that the angular frequency of the rotating frame also equals unity (Problem 2.8).

By analyzing a modified energy integral in the rotating frame, Jacobi deduced the following constant of motion for the (massless) test particle in the circular restricted three-body problem:

$$C_J = x^2 + y^2 + \frac{2m_1}{|\mathbf{r}-\mathbf{r}_1|} + \frac{2m_2}{|\mathbf{r}-\mathbf{r}_2|} - v^2. \tag{2.21}$$

In equation (2.21), $|\mathbf{r}-\mathbf{r}_i|$ is the distance from mass m_i to the test particle; the velocity of the test particle, v, is measured in the rotating frame and C_J is known as *Jacobi's constant*. By convention, $m_1 \geq m_2$; in most Solar System applications, $m_1 \gg m_2$.

For a given value of Jacobi's constant, equation (2.21) specifies the magnitude of the test particle's velocity (in the rotating frame) as a function of position. As v^2 cannot be negative, surfaces at which $v = 0$ bound the trajectory of a particle with fixed C_J. Such *zero-velocity surfaces*, or in the case of the planar problem *zero-velocity curves*, are quite useful in discussing the topology of the circular restricted three-body problem (Fig. 2.2).

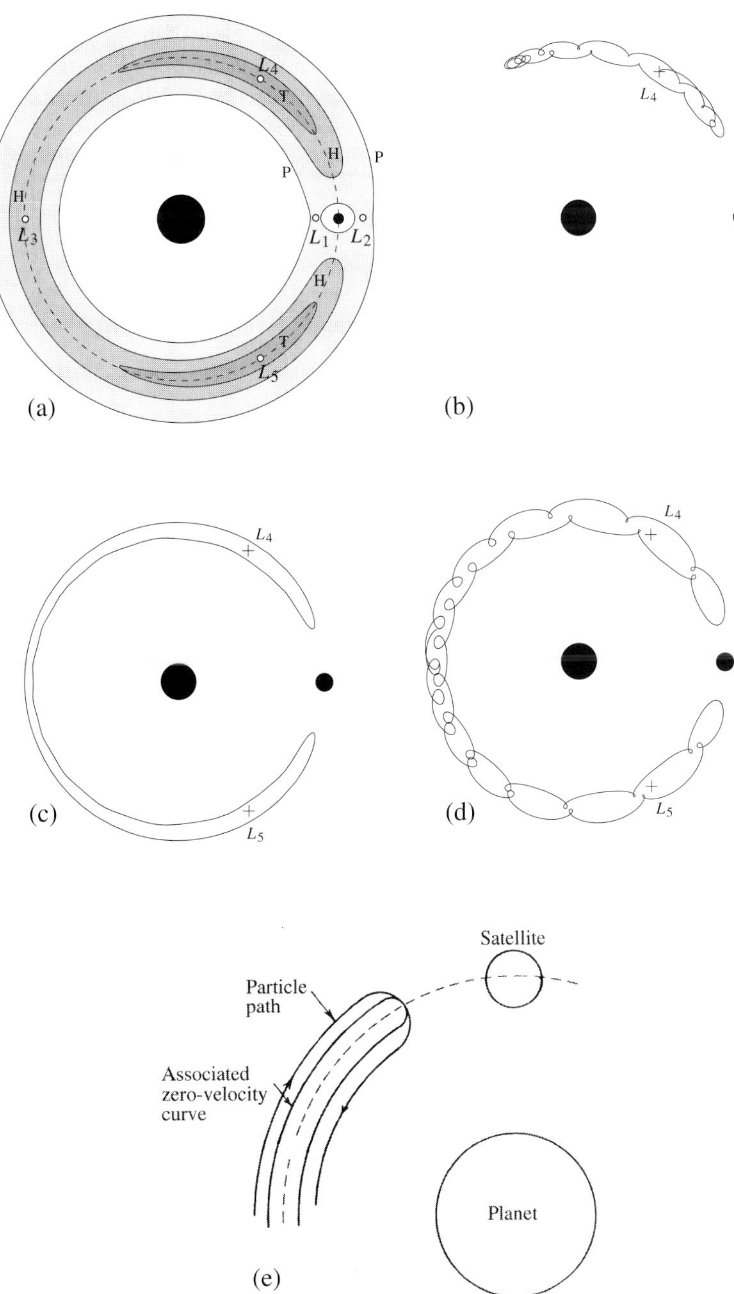

FIGURE 2.2 (a) Schematic diagram showing the Lagrangian equilibrium points and various zero-velocity curves for three values of the Jacobi constant, C_J, in the circular restricted three-body problem. The frame is centered on the primary and corotates with the secondary, and the mass ratio $m_1/m_2 = 100$. The locations of the Lagrangian equilibrium points L_1–L_5 are indicated by small open circles. The dashed line denotes a circle of radius equal to the planet's semimajor axis. The letters T (tadpole), H (horseshoe) and P (passing) denote the type of orbit associated with the curves. The regions enclosed by each curve (shaded) are excluded from the motion of a test particle that has the corresponding C_J. The critical horseshoe curve actually passes through L_1 and L_2 and the critical tadpole curve passes through L_3. Horseshoe orbits can exist between these two extremes. (Courtesy Carl Murray)
(b) Example of a tadpole orbit of a test particle viewed in the frame rotating at the orbital frequency of the two massive bodies. (Adapted from Murray and Dermott 1999)
(c) Similar to (b), but for a horseshoe orbit with small eccentricity. (Adapted from Murray and Dermott 1999)
(d) As in (c), but the particle has a larger eccentricity. (Adapted from Murray and Dermott 1999)
(e) Schematic diagram showing the relationship between a horseshoe orbit and its associated zero-velocity curve. The particle's velocity in the rotating frame drops as it approaches the zero-velocity curve, and it cannot cross the curve. (Adapted from Dermott and Murray 1981)

Lagrange found that in the circular restricted three-body problem there are five points where test particles placed at rest would feel no net force in the rotating frame. Three of these so-called *Lagrangian points* (L_1, L_2 and L_3) lie along a line joining the two masses m_1 and m_2. Zero-velocity curves intersect at each of the three collinear Lagrangian points, which are saddle points of C_J. The other two Lagrangian points (L_4 and L_5) form equilateral triangles with the two massive bodies (Fig. 2.2). The two triangular Lagrangian points together form the zero-velocity 'curve' with the smallest value of C_J. All five Lagrangian points are in the orbital plane of the two massive bodies.

Particles displaced slightly from the three collinear Lagrangian points will continue to move away; hence these locations are unstable. The triangular Lagrangian points are potential energy maxima, which are stable for sufficiently large primary to secondary mass ratio due to the

Coriolis force. Provided that the most massive body has at least $\sim$ 27 times the mass of the secondary (which is the case for all known examples in the Solar System except the Pluto–Charon system), the Lagrangian points L_4 and L_5 are linearly stable points. The precise ratio required for stability is $(29 + \sqrt{621})/2$. (See Danby (1988) for a derivation and further details.) If a particle at L_4 or L_5 is perturbed slightly, it will start to *librate* about these points (i.e., oscillate back and forth, without circulating past the moon). Note with irony that particles located at L_4 or L_5 have such a low value of C_J that Jacobi's integral does not exclude them from any location within the plane, nonetheless they remain stable indefinitely at these potential energy maxima! The L_4 and L_5 points are important in the Solar System. For example, the *Trojan asteroids* are located near Jupiter's triangular Lagrangian points, and the asteroid 5261 Eureka is a martian Trojan. There are also small moons in the saturnian system near the triangular Lagrangian points of Tethys and Dione. The L_4 or L_5 points in the Earth–Moon system have been suggested as possible locations for a future space station.

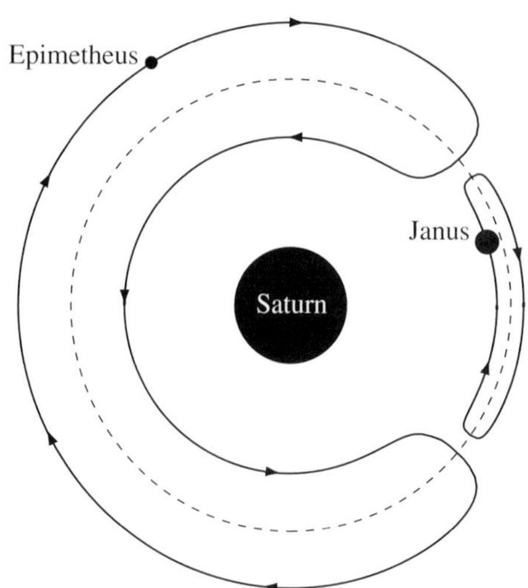

FIGURE 2.3 A schematic diagram of the librational behavior of the Janus and Epimetheus coorbital system in a frame rotating with the average mean motion of both satellites. The radial extent of the librational arcs is exaggerated; the ratio of the radial widths of the arcs is equal to the Janus/Epimetheus mass ratio ($\sim$ 0.25). (Murray and Dermott 1999)

2.2.2 Horseshoe and Tadpole Orbits

Consider a moon on a circular orbit around a planet. A particle just interior to the moon's orbit has a higher angular velocity, and moves with respect to the moon in the direction of corotation. A particle just outside the moon's orbit has a smaller angular velocity, and moves relative to the moon in the opposite direction. When the outer particle approaches the moon, the particle is pulled towards the moon and consequently loses angular momentum. Provided the initial difference in semimajor axis is not too large, the particle drops to an orbit lower than that of the moon. The particle then recedes in the forward direction. Similarly, the particle on the lower orbit is accelerated as it catches up with the moon, resulting in an outward motion towards the higher, and therefore slower, orbit. Orbits like these encircle the L_3, L_4 and L_5 points and appear shaped like horseshoes in the rotating frame (Fig. 2.2c), thus they are called *horseshoe orbits*. Saturn's small moons Janus and Epimetheus execute just such a dance, changing orbits every 4 years (Fig. 2.3). As Janus and Epimetheus are comparable in mass, Hill's approximation is more accurate than is the restricted three-body formalism used above, but the dynamical interactions are essentially the same.

Since the Lagrangian points L_4 and L_5 are stable, material can librate about these points individually; such orbits are called *tadpole orbits* after their asymmetric elongated shape in the rotating frame. The tadpole libration width at L_4 and L_5 is proportional to $(m_2/m_1)^{1/2}r$, and the horseshoe width varies as $(m_2/m_1)^{1/3}r$, where m_1 is the mass of the planet, m_2 the mass of the satellite, and r the distance between the two objects. For a planet of Saturn's mass, $M_♄ = 5.7 \times 10^{29}$ g, and a typical moon of mass $m_2 = 10^{20}$ g (a 30 km radius object, with density of $\sim$1 g cm^{-3}) at a distance of 2.5 saturnian radii, the tadpole libration half-width is about 3 km, and the horseshoe half-width about 60 km.

2.2.3 Hill Sphere

The approximate limit to a secondary's (e.g., planet's or moon's) gravitational dominance is given by the extent of its *Hill sphere*,

$$R_H = \left(\frac{m_2}{3(m_1 + m_2)}\right)^{1/3} a, \tag{2.22}$$

where m_2 is the mass of the secondary and m_1 the primary's (e.g., Sun or planet) mass. A test particle located at the boundary of a planet's Hill sphere is subject to a gravitational force from the planet comparable to the tidal difference between the force of the Sun on the planet and that on the test body. The Hill sphere stretches out to the L_1 point, and is essentially equivalent to the Roche lobe (Section 11.1) in the limit $m_2 \ll m_1$. Planetocentric orbits which are stable over long periods of time are those well within the boundary of a planet's Hill sphere; all known natural satellites lie in this region. Stable heliocentric orbits are always well outside the Hill sphere of any planet

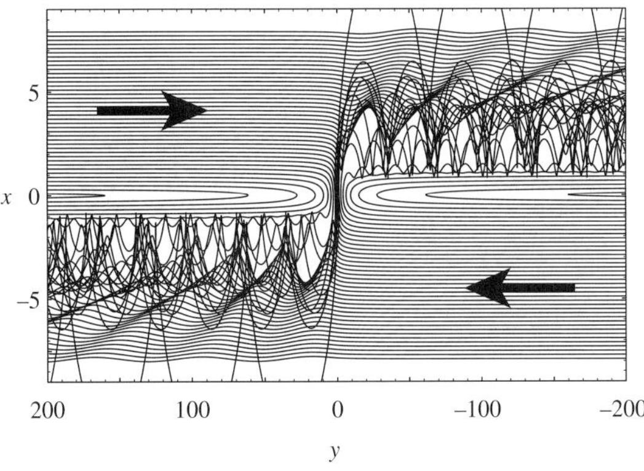

FIGURE 2.4 Particle trajectories obtained by solving the scaled form of Hill's equations are shown relative to the position of the planet. The perturbing mass is located at the origin and the L_1 and L_2 points are at $y = 0$, $x = \pm 1$. The particles were all started with $dx/dt = 0$ (i.e., circular orbits) at $y = \pm 200$. The arrows indicate their direction of motion before encountering the perturber. The primary is located at $y = 0, x = -\infty$. In an inertial frame, the planet and the test particles all move from right to left. (Adapted from Murray and Dermott 1999)

(Fig. 2.4; cf. eq. 2.28). Comets and other bodies which enter the Hill sphere of a planet at very low velocity can remain gravitationally bound to the planet for some time as *temporary satellites* (Fig. 2.5).

2.3 'Planetary' Perturbations and Resonances

Within the Solar System, one body typically produces the dominant gravitational force on any given object, and the resultant motion can be thought of as a keplerian orbit about a primary, subject to small perturbations by other bodies. In this section, we consider some important examples of the effects of these perturbations on the orbital motion.

2.3.1 Perturbed Keplerian Motion

Classically, much of the discussion of the evolution of orbits in the Solar System used perturbation theory as its foundation. Essentially, the method involves writing the potential as the sum of a part that describes the independent keplerian motion of the bodies about the Sun, plus a part (called the *disturbing function*) that contains the *direct terms* which account for the pairwise interactions among the planets and minor bodies and the *indirect terms* associated with the back-reaction of the planets on the Sun. For example, if m_2 and m_3 are two point masses in orbit about a common primary at instantaneous locations $\mathbf{r}_2$ and $\mathbf{r}_3$

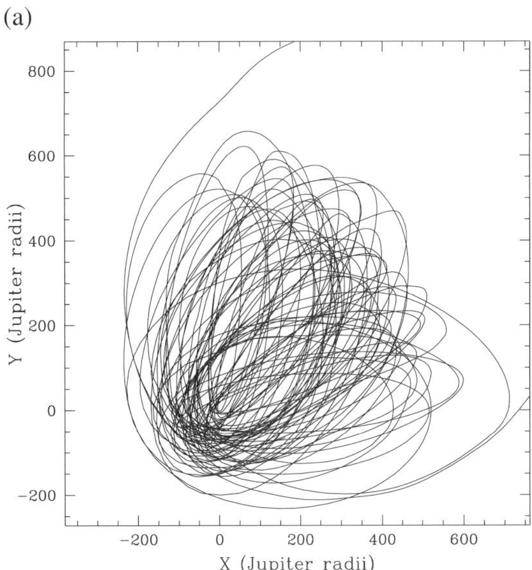

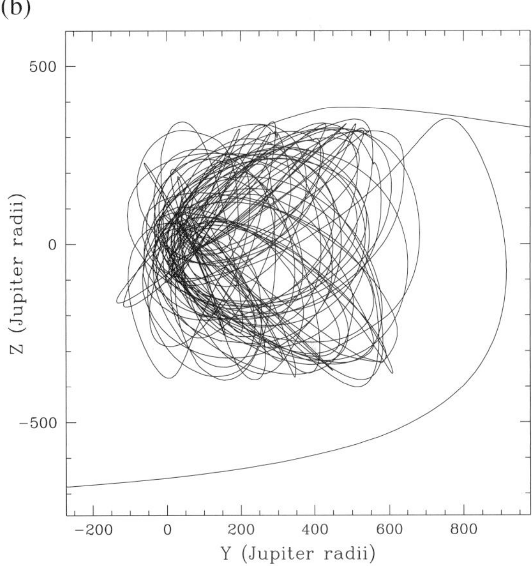

FIGURE 2.5 Trajectory relative to Jupiter of a test particle initially orbiting the Sun that was temporarily captured into an unusually long-duration unstable orbit about Jupiter. (a) Projected into the plane of Jupiter's orbit about the Sun. (b) Projected into a plane perpendicular to Jupiter's orbit. (Kary and Dones 1996)

relative to the primary, then the disturbing function, $\mathcal{R}$, for the action of m_2 on m_3 may be written as

$$\mathcal{R} = -Gm_2 \left(\frac{1}{|\mathbf{r}_2 - \mathbf{r}_3|} - \frac{\mathbf{r}_2 \cdot \mathbf{r}_3}{r_2^3} \right), \tag{2.23}$$

where the first quantity in the parentheses is the direct term and the second is the indirect term. The force on m_3 due to m_2 is given by the gradient of $\mathcal{R}$. In a many-planet system, the disturbing function can be expressed as the sum of terms each having the form of the expression on the right hand side of equation (2.23). A detailed discussion of the disturbing function, Fourier expansions thereof and applications to planetary dynamics is presented by Brouwer and Clemence (1961).

In general, one can expand the disturbing function in terms of the small parameters of the problem (such as the ratio of the planetary masses to the Sun's mass, the eccentricities and inclinations, etc.) as well as the other orbital elements of the bodies, including the mean longitudes (i.e., the locations of the bodies in their orbits) and attempt to solve the resulting equations for the time-dependence of the orbital elements. However, in the late nineteenth century Poincaré showed that these perturbation series are often divergent and have validity only over finite timespans. The full significance of Poincaré's work has only become apparent in the last few decades, in part because of long-term direct integrations on computers of the trajectories of bodies in the Solar System. What is often found in practice is that for some initial conditions the trajectories are *regular* with variations in their orbital elements that seem to be well-described by the perturbation series, while for other initial conditions the trajectories are found to be *chaotic* and are not as confined in their motions (Figure 2.6a). The evolution of a system which is chaotic depends so sensitively on the system's precise initial state that the behavior is in effect unpredictable, even though it is strictly determinate in a mathematical sense.

There is a key feature of the chaotic orbits that we will use here as a definition of *chaos*: Two trajectories which begin arbitrarily close in phase space (positions and velocities, or a more complicated set of orbital elements, can be considered to be the coordinates of such a phase space) within a chaotic region will typically diverge exponentially in time. Within a given chaotic region, the timescale for this divergence does not typically depend on the precise values of the initial conditions! Thus, if one computes the distance, $d(t)$, between two particles having an initially small separation, $d(0)$, it can be shown that for regular orbits, $d(t) - d(0)$ grows as a power of time t (typically linearly) whereas for chaotic orbits

$$d(t) \sim d(0)e^{\gamma_c t}, \tag{2.24}$$

where γ_c is the *Lyapunov characteristic exponent* and γ_c^{-1} is the *Lyapunov timescale* (Fig. 2.6b). From this definition of chaos, we see that chaotic orbits show such a sensitive dependence on initial conditions that the detailed long-term behavior of the orbits is lost within several Lyapunov timescales. Even a fractional perturbation as small as 10^{-10} in the initial conditions will result in a 100% discrepancy in about 20 Lyapunov times. However, one of the interesting features of much of the chaotic behavior seen in simulations of the orbital evolution of bodies in the Solar System is that the timescale for large changes in the prin-

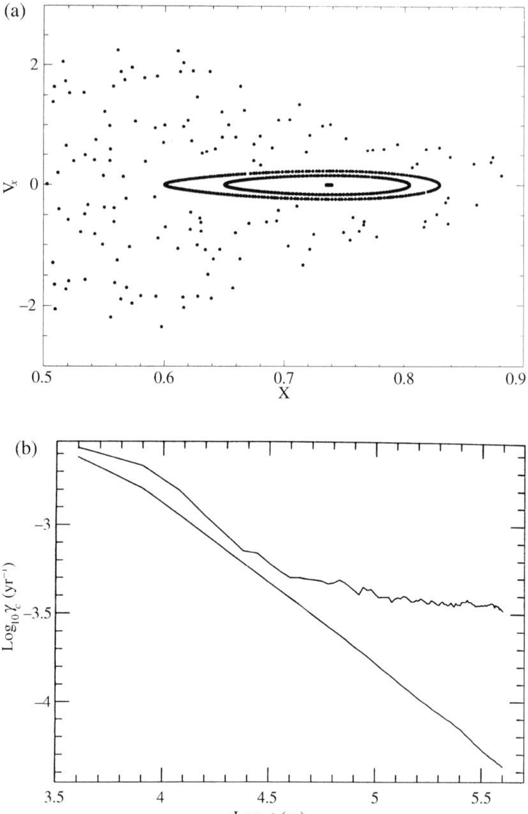

FIGURE 2.6 (a) *Surface of section* for the trajectories of four different test particles in the planar circular restricted three-body problem. The dots display the x coordinate and x velocity of the particles each time the particles pass through the $y = 0$ plane with positive y velocity. The four particles have the same value of C_J but different initial conditions. Three of the trajectories are regular and produce well-defined *quasiperiodic* patterns on the plot. The unconnected dots all represent the trajectory of the fourth particle, which is on a chaotic orbit and therefore is less confined in phase space. (Adapted from Duncan and Quinn 1993) (b) Distinction between regular (lower curve) and chaotic trajectories as characterized by the Lyapunov characteristic exponent, γ_c. Both trajectories are near the 3:1 resonance with Jupiter, and they have been integrated using the elliptic restricted three-body problem. For chaotic trajectories, a plot of $\log \gamma_c$ versus $\log t$ eventually levels off at a value of γ_c that is the inverse of the Lyapunov timescale for the divergence of initially adjacent trajectories, whereas for regular trajectories, $\log \gamma_c \to 0$ as $t \to \infty$. (Adapted from Duncan and Quinn 1993)

cipal orbital elements is often many orders of magnitude larger than the Lyapunov timescale.

In dynamical systems like the Solar System, chaotic regions do not appear randomly, but rather they are associated with trajectories in which the ratios of characteristic frequencies of the original problem are sufficiently well approximated by rational numbers, i.e., near resonances. The simplest of these resonances to visualize are so-called *mean motion resonances*, in which the orbital periods of two bodies are commensurate (e.g., have a ratio of the form $N/(N+1)$, where N is an integer). Some examples of the consequences of this type of resonance are given below. In Section 2.4, we define secular resonances and indicate their relationship to the stability of the Solar System.

2.3.2 Resonances

Although perturbations on a body's orbit are often small, they cannot always be ignored. They must be included in short-term calculations if high accuracy is required, e.g., for predicting stellar occultations or targeting spacecraft. Most long-term perturbations are periodic in nature, their directions oscillating with the relative longitudes of the bodies or with some more complicated function of the bodies' orbital elements. Small perturbations can produce large effects if the forcing frequency is commensurate or nearly commensurate with the natural frequency of oscillation of the responding elements. Under such circumstances, perturbations add coherently, and the effects of many small tugs can build up over time to create a large-amplitude, long-period response. This is an example of *resonance forcing*, which occurs in a wide range of physical systems.

An elementary example of resonance forcing is given by the one-dimensional simple harmonic oscillator, for which the equation of motion is:

$$m\frac{d^2x}{dt^2} + m\omega_o^2 x = F_d \cos\omega_d t, \tag{2.25}$$

where m is the mass of the oscillating particle, F_d is the amplitude of the driving force, ω_o is the natural frequency of the oscillator and ω_d is the forcing frequency. The solution to equation (2.25) is:

$$x = \frac{F_d}{m(\omega_o^2 - \omega_d^2)} \cos\omega_d t + C_1 \cos\omega_o t + C_2 \sin\omega_o t, \tag{2.26a}$$

where C_1 and C_2 are constants determined by the initial conditions. Note that if $\omega_d \approx \omega_o$, a large-amplitude, long-period response can occur even if F_d is small. Moreover, if $\omega_o = \omega_d$, equation (2.26*a*) is invalid. In this (resonant) case, the solution is given by:

$$x = \frac{F_d}{2m\omega_o} t \sin\omega_o t + C_1 \cos\omega_o t + C_2 \sin\omega_o t. \tag{2.26b}$$

The t in the middle of the first term at the right hand side of equation (2.26*b*) leads to secular (i.e., steady rather than periodic) growth. Often this linear growth is moderated by the effects of nonlinear terms which are not included in the simple example provided above. However, some perturbations have a secular component.

2.3.2.1 *Examples of Orbital Resonances*

Almost exact orbital commensurabilities exist at many places in the Solar System. Io orbits Jupiter twice as frequently as Europa does, and Europa in turn orbits Jupiter in half of the time that Ganymede takes. Conjunction (closest approach) always occurs at the same position of Io's orbit (its perijove). How can such commensurabilities exist? After all, the rational numbers form a set of measure zero on the real line, which means that the probability of randomly picking a rational from the real number line is 0! The answer lies in the fact that *orbital resonances* may be held in place by stable 'locks', which result from nonlinear effects not represented in the simple mathematical example of the harmonic oscillator. Differential tidal recession (Section 2.6) brings moons into resonance, and nonlinear interactions between the moons can keep them there. The stabilizing mechanisms are beyond the scope of this book; see Peale (1976) for an explanation.

Other examples of resonance locks include the Hilda and Trojan asteroids with Jupiter, Neptune–Pluto, and several pairs of moons orbiting Saturn, such as Janus–Epimetheus, Mimas–Tethys and Enceladus–Dione. Resonant perturbations can also force material into highly eccentric orbits, which may lead to collisions with other bodies; this is believed to be the dominant mechanism for clearing the Kirkwood gaps in the asteroid belt (see below).

Spiral density waves can result from resonant perturbations by a moon on a self-gravitating disk of particles. Density waves are observed at many resonances in Saturn's rings; they explain most of the structure seen in Saturn's A ring. The vertical analog of density waves, bending waves, are caused by resonant perturbations perpendicular to the ring plane from a satellite in an orbit which is inclined to the ring. Spiral bending waves excited by the moons Mimas and Titan have been observed in Saturn's rings. We discuss these manifestations of resonance effects in more detail in Chapter 11.

2.3.2.2 *Resonances in the Asteroid Belt*

There are obvious patterns in the distribution of asteroidal semimajor axes which appear to be associated with mean motion resonances with Jupiter (Fig. 9.1a). At these resonances, a particle's period of revolution about the Sun is a small integer ratio multiplied by Jupiter's orbital period. The Trojan asteroids travel in a 1:1 mean motion resonance with Jupiter, as described above. These asteroids execute small amplitude (tadpole) librations about the L_4 and L_5 points 60° behind or ahead of Jupiter and therefore never suffer a close approach to Jupiter. Another example of a protection mechanism provided by a resonance is the Hilda group of asteroids at Jupiter's 3:2 mean motion resonance and the asteroid 279 Thule at the 4:3 resonance. The Hilda asteroids have a libration about 0° of their critical argument, $3\lambda' - 2\lambda - \varpi$, where λ' is Jupiter's longitude, λ is the asteroid's longitude, and ϖ is the asteroid's longitude of perihelion. In this way, whenever the asteroid is in conjunction with Jupiter ($\lambda = \lambda'$), the asteroid is close to perihelion ($\lambda' \approx \varpi$) and well away from Jupiter.

Using resonances to explain the Kirkwood gaps in the main asteroid belt and the general depletion of the outer belt proves to be more difficult than understanding the protection mechanisms at other resonances. A feature subject to much investigation has been the gap at the 3:1 mean motion resonance. Early investigations found that most orbits starting at small eccentricity were regular and showed very little variation in eccentricity or semimajor axis over timescales of 5×10^4 yrs. In the 1980s, Jack Wisdom showed that an orbit near the resonance could maintain a low eccentricity ($e < 0.1$) for nearly a million years and then have a sudden increase in eccentricity to $e > 0.3$. This illustrates an important feature which often occurs in simulations to be discussed later: A particle can remain in a low-eccentricity state for hundreds of Lyapunov times before 'jumping' relatively quickly to high eccentricity.

The outer boundaries of the chaotic zone as determined by Wisdom's work coincide well with the boundaries of the 3:1 Kirkwood gap as seen in the numbered minor planets and the Palomar–Leiden survey (PLS) of asteroids (Fig. 2.7). Since asteroids which begin on near-circular orbits in the gap acquire sufficient eccentricities to cross the orbit of Mars and in many cases that of the Earth, the perturbative effects of the terrestrial planets are believed capable of clearing out the 3:1 gap in a time equivalent to the age of the Solar System. There is also a strong correlation between the libration widths of other resonances and regions of a–e space depleted of asteroids (Fig. 2.8).

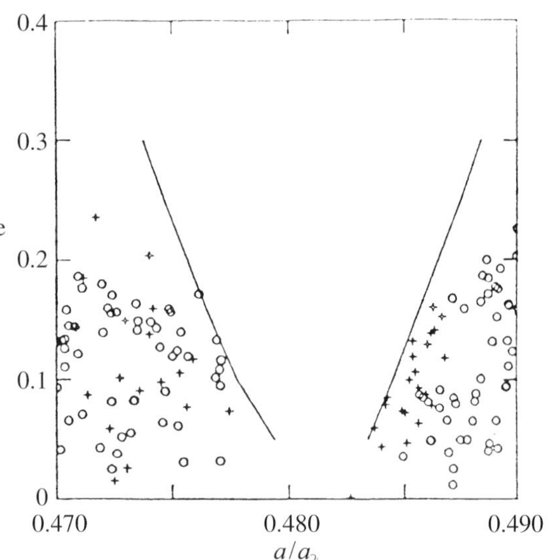

FIGURE 2.7 The outer boundaries of the chaotic zone surrounding Jupiter's 3:1 mean motion resonance in the a–e plane are shown as lines. Locations of numbered asteroids are shown as circles and Palomar–Leiden survey (PLS) asteroids (whose orbits are less well determined) are represented as plus signs. Note the excellent correspondence of the observed 3:1 Kirkwood gap with theoretical predictions. (Adapted from Wisdom 1983)

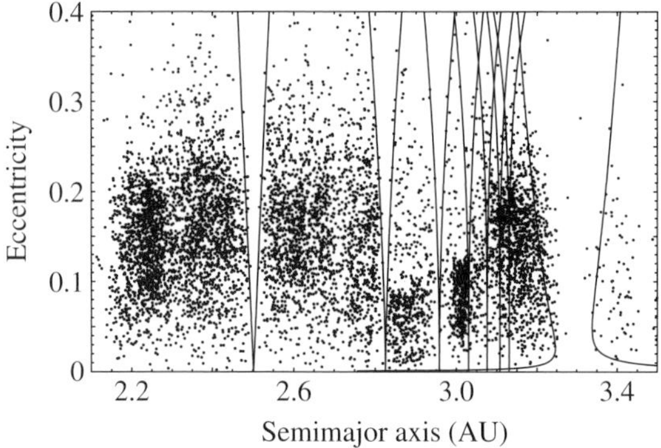

FIGURE 2.8 The maximum libration widths in a–e space of the strong jovian resonances superimposed on the distribution of asteroids in the main belt. Note the correspondence between the widths of the gaps and the widths of the resonances. (Murray and Dermott 1999)

2.3.3 The Resonance Overlap Criterion and Jacobi–Hill Stability

For nearly circular and coplanar orbits, the strongest mean motion resonances occur at locations where the ratio of test particle orbital periods to the massive body's period are of the form $N{:}(N \pm 1)$, where N is an integer. At these locations, conjunctions (closest approaches) always occur at the same phase in the orbit, and tugs add coherently. (The locations of these strong resonances are shifted slightly when the primary is oblate, cf. Section 2.5 and Chapter 11.) The strength of these first-order resonances increases as N grows, because the magnitudes of the perturbations are larger closer to the secondary. First-order resonances also become closer to one another near the orbit of the secondary (Problem 2.12). Sufficiently close to the secondary, the combined effects of greater strength and smaller spacing cause resonance regions to overlap; this overlapping can lead to the onset of chaos as particles shift between the nonlinear perturbations of various resonances. The region of overlapping resonances is approximately symmetric about the planet's orbit, and has a half-width, Δa_{ro}, given by:

$$\Delta a_{ro} \approx 1.5 \left(\frac{m_2}{m_1}\right)^{2/7} a, \tag{2.27}$$

where a is the semimajor axis of the planet's orbit. The functional form of equation (2.27) has been derived analytically, whereas the coefficient is a numerical result.

Zero-velocity surfaces can also be used to prove the stability of certain orbits in the circular restricted three-body problem. For example, in the planetary case, where $m_2 \ll m_1$, a test particle initially on a circular planar orbit separated in semimajor axis from the secondary by an amount larger than

$$\Delta a_J = 2\sqrt{3}\left(\frac{m_2}{3m_1}\right)^{1/3} a \tag{2.28}$$

can never approach to within the Hill sphere of the planet, and remains in an orbit inferior or superior to that of the planet forever. The relationship between the stability criteria given by equations (2.27) and (2.28) is analyzed in Problem 2.13.

2.4 Long-Term Stability of Planetary Orbits

We turn now to one of the oldest problems in dynamical astronomy: whether or not the planets will continue indefinitely in almost circular, almost coplanar orbits.

2.4.1 Secular Perturbation Theory

To study the very long-term behavior of planetary orbits, a fruitful approach involves averaging the disturbing function over the mean motions of the planets, resulting in what is known as the *secular part of the disturbing function*. If the disturbing function is further limited to terms of lowest order, the equations of motion of the orbital elements of the planets can be expressed as a coupled set of first-order linear differential equations. This system can then be diagonalized to find the proper modes, which are sinusoids, and the corresponding eigenfrequencies. The evolution of a given planet's orbital elements is, therefore, a sum of the proper modes. With the addition of higher order terms, the equations are no longer linear. It is, however, sometimes possible to find an approximate solution of a form similar to the linear solution, except with shifted proper mode frequencies and terms involving combinations of the proper mode frequencies. We discuss the long-term validity of this method in the next subsection.

2.4.2 Chaos and Planetary Motions

The above discussion implies that if the mutual planetary perturbations were calculated to first order in the masses, inclinations and eccentricities, the orbits could be described by a sum of periodic terms, indicating stability. This is still the case if the perturbations are expanded to somewhat higher orders. However, although perturbation expansions are done in powers of small parameters, the existence of resonances between the planets introduces small divisors into the expansion terms (cf. eq. 2.26). Such small divisors make some high order terms in the power series unexpectedly large and destroy the convergence of the series. There are two separate points in the construction of the secular system at which resonances can cause nonconvergence of the expansion. The first is in averaging over mean motions. Mean motion resonances between the planets can introduce small divisors, leading to divergences when forming the secular disturbing function. Second, there can be *secular resonances* between the proper mode frequencies, e.g. apse precession rates, leading to problems trying to solve the secular system using an expansion approach.

Mathematical stability (in the sense that planetary orbits would remain well separated and the system would remain bound for infinite time) can be proven for a system of extremely small but nonetheless finite mass planets with orbits similar to those in our Solar System. However, the set of initial conditions for which the proof *does not apply* is everywhere dense, i.e., there is always a point in phase space arbitrarily close to a given choice of initial conditions for which the proof does not guarantee stability. The system might not remain stable if it were subjected to perturbations, even if these perturbations were arbitrarily small. Mathematical theorems are thus of very limited use when discussing astronomical stability.

From an astronomical viewpoint, stability implies that the system will remain bound (no ejections) and that no mergers of planets will occur for the possibly long but finite period of interest, and that this result is robust against (most if not all) sufficiently small perturbations. In the remainder of this discussion, we shall only be concerned with stability in an astronomical sense.

The analytical complexity of the perturbation techniques and the development of ever faster computers has led to the investigation of Solar System stability by purely numerical models. Early integrations of the orbits of the giant planets on million-year timescales compared well with perturbation calculations, showing quasi-periodic behavior for the four major outer planets. Pluto's behavior, however, was sufficiently different to inspire further study. It was found that the angle $3\lambda - 2\lambda_{\Psi} - \varpi$ is in libration with a period of 20 000 yr, where λ and λ_{Ψ} are the mean longitudes of Pluto and Neptune, respectively, and ϖ is the longitude of perihelion of Pluto. This 2:3 mean motion resonance acts to prevent close encounters of Pluto with Neptune and hence protects the orbit of Pluto. However, numerical integrations show that Pluto's orbit is not quasi-periodic. There is evidence for the existence of very long period changes in Pluto's orbital elements, with a Lyapunov exponent of $\sim (20\ \mathrm{Myr})^{-1}$.

Numerical integrations that include the terrestrial planets show a surprisingly high Lyapunov exponent of $(5\ \mathrm{Myr})^{-1}$. Such large Lyapunov exponents certainly suggest chaotic behavior. However, the apparent regularity of the motion of the Earth and Pluto, and indeed the fact that the Solar System has survived for 4.5 billion years, implies that the chaotic regions must be narrow. What the chaotic motion does mean is that there is a horizon of predictability for the detailed motions of the planets. Thus, the exponential divergence of orbits with a 4–5 Myr timescale shown by the calculations implies that an error as small as 10^{-10} in the initial conditions will lead to a 100% discrepancy in longitude in 100 Myr. It is also worth bearing in mind the lessons learned from integration of test particle trajectories, namely that the timescale for macroscopic changes in the system can be many orders of magnitude longer than the Lyapunov timescales. Thus, the apparent stability of the current planetary system on billion-year timescales may simply be a manifestation of the fact that the Solar System is in the chaotic sense a dynamically young system.

2.5 Orbits About an Oblate Planet

Thus far we have approximated Solar System bodies as point masses for the purpose of calculating their mutual gravitational interactions. Self-gravity causes most sizable celestial bodies to be approximately spherically symmetric. Newton showed that the gravitational force exerted by a spherically symmetric body exterior to its surface is identical to the gravitational force of the same mass located at the body's center (cf. Problem 2.16); thus, the point-mass approximation is adequate for most purposes. There are, however, several forces which act to produce distributions of mass which deviate from spherical symmetry. In the Solar System, rotation, physical strength and tidal forces produce important departures from spherical symmetry in some bodies. The gravitational field of an aspherical body differs from that of a point-mass, with the largest deviation generally being found near the body's surface. In this section, we include the effects of a body's deviation from spherical symmetry in computing the gravitational force that it exerts. This is most conveniently done by using the newtonian gravitational potential, $\Phi_g(\mathbf{r})$, which is defined in equation (2.18).

2.5.1 Gravitational Potential of an Axisymmetric Planet

Most planets are very nearly axisymmetric, with the major departure from sphericity being due to rotationally induced polar flattening. Thus, as $\Phi_g(\mathbf{r})$ satisfies Laplace's equation (2.20*b*) in free space, the gravitational potential exterior to a planet can be expanded in terms of Legendre polynomials (instead of the complete spherical harmonic expansion, which would be required for the potential of a body of arbitrary shape):

$$\Phi_g(r,\phi,\theta) = -\frac{Gm}{r}\left[1 - \sum_{\mathrm{n}=2}^{\infty} J_{\mathrm{n}} P_{\mathrm{n}}(\cos\theta)\left(\frac{R}{r}\right)^{\mathrm{n}}\right]. \tag{2.29}$$

Equation (2.29) is written in standard spherical coordinates, with ϕ the longitude and θ representing the angle between the planet's symmetry axis and the vector to the particle (i.e., the colatitude). The terms $P_{\mathrm{n}}(\cos\theta)$ are the Legendre polynomials, given by the formula:

$$P_{\mathrm{n}}(x) = \frac{1}{2^{\mathrm{n}}\mathrm{n}!}\frac{d^{\mathrm{n}}}{dx^{\mathrm{n}}}\left(x^2-1\right)^{\mathrm{n}}. \tag{2.30}$$

The *gravitational moments*, J_{n}, are determined by the planet's mass distribution (Section 6.1.4). If the planet's mass is distributed symmetrically about the planet's equator, then the J_{n} are zero for odd n.

Let us consider a small body, e.g. a moon or ring particle, which travels around a planet on a circular orbit in the equatorial plane ($\theta = 90°$) at a distance r from the center of the planet. The centripetal force must be provided by the radial component of the planet's gravitational force (cf. eq. 2.13), so the particle's angular velocity n satisfies:

$$rn^2(r) = \left.\frac{\partial \Phi_g}{\partial r}\right|_{\theta=90^\circ}. \tag{2.31}$$

If the particle suffers an infinitesimal displacement from its circular orbit, it will oscillate freely in the horizontal and vertical directions about the reference circular orbit with radial (epicyclic) frequency $\kappa(r)$ and vertical frequency $\mu(r)$ respectively, given by:

$$\kappa^2(r) = r^{-3}\frac{\partial}{\partial r}[(r^2 n)^2], \tag{2.32}$$

$$\mu^2(r) = \left.\frac{\partial^2 \Phi_g}{\partial z^2}\right|_{z=0}. \tag{2.33}$$

2.5.2 Precession of Particle Orbits

Using the equations (2.29–2.33), one can show that the orbital, epicyclic and vertical frequencies can be written as:

$$n^2 = \frac{Gm}{r^3}\left[1 + \frac{3}{2}J_2\left(\frac{R}{r}\right)^2 - \frac{15}{8}J_4\left(\frac{R}{r}\right)^4 + \frac{35}{16}J_6\left(\frac{R}{r}\right)^6 + \cdots\right], \tag{2.34}$$

$$\kappa^2 = \frac{Gm}{r^3}\left[1 - \frac{3}{2}J_2\left(\frac{R}{r}\right)^2 + \frac{45}{8}J_4\left(\frac{R}{r}\right)^4 - \frac{175}{16}J_6\left(\frac{R}{r}\right)^6 + \cdots\right], \tag{2.35}$$

$$\mu^2 = 2n^2 - \kappa^2. \tag{2.36}$$

For a perfectly spherically symmetric planet, $\mu = \kappa = n$. Since planets are oblate, μ is slightly larger than the orbital frequency, n, and κ is slightly smaller. The oblateness of a planet therefore causes periapse longitudes of particle orbits in and near the equatorial plane to precess in the direction of the orbit and lines of nodes of nearly equatorial orbits to regress. Orbits about oblate planets are thus not keplerian ellipses. However, as the trajectories are nearly elliptical, they are often specified by instantaneous keplerian orbital elements. Note that

$$\frac{d\varpi}{dt} = n - \kappa, \tag{2.37}$$

$$\frac{d\Omega}{dt} = n - \mu. \tag{2.38}$$

2.6 Tides

The gravitational force arising from the pull of external objects varies from one part of a body to another. These differential tugs produce what is known as the *tidal force*. The net force on a body determines the acceleration of its center of mass, but tidal forces can deform a body, and can produce torques which alter its rotation state. Time-variable tidal forces such as those experienced by moons on eccentric orbits can result in flexing which leads to internal heating.

Tidal forces are important to many aspects of the structure and evolution of planetary bodies. For example, on short timescales, temporal variations in tides (as seen in the frame rotating with the body under consideration) cause stresses which can move fluids with respect to more rigid parts of the planet, for example the ocean tides with which we are familiar. Such stresses can even cause seismic disturbances. (While the evidence that the Moon causes some earthquakes is weak and disputable, it is clear that the tides raised by the Earth are a major cause of moonquakes.) On long timescales, tides change the orbital and spin properties of planets and moons. Tides also determine the equilibrium shape of a body located near any massive object; note that many materials which behave as solids on short timescales are effectively fluids on very long geological timescales, e.g., the Earth's mantle. In some cases, tidal forces are so strong that they exceed a body's cohesive force, and the body fragments.

2.6.1 The Tidal Force and Tidal Bulges

Consider a nearly spherical body of radius R, centered at the origin, which is subject to the gravitational force of a point mass, m, at $\mathbf{r}_o$, where $r_o \gg R$. The specific (per unit mass) tidal force is given by:

$$\mathbf{F}_T(\mathbf{r}) = -\frac{Gm}{|\mathbf{r}_o - \mathbf{r}|^3}(\mathbf{r}_o - \mathbf{r}) + \frac{Gm}{r_o^3}\mathbf{r}_o. \tag{2.39}$$

For points along the line joining the center of the body to the point mass (which we take to be the x-axis), equation (2.39) reduces to:

$$F_T(x) = -\frac{Gm}{(x_o - x)^2} + \frac{Gm}{x_o^2} \approx \frac{2xGm}{x_o^3}. \tag{2.40}$$

Equation (2.40) states that, to lowest order, the tidal force varies proportionally to the distance from the center of the body and inversely to the cube of the distance from the

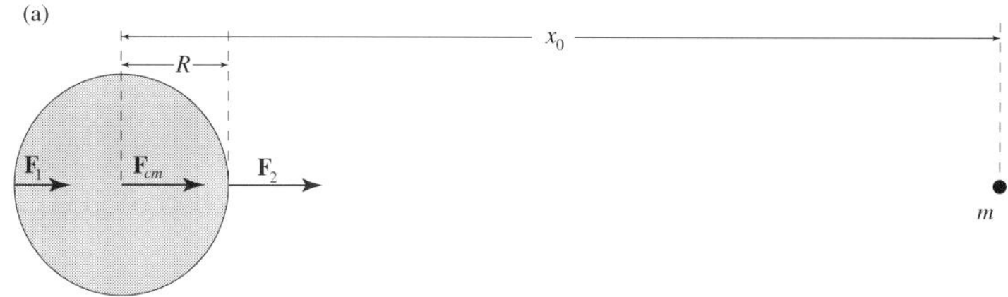

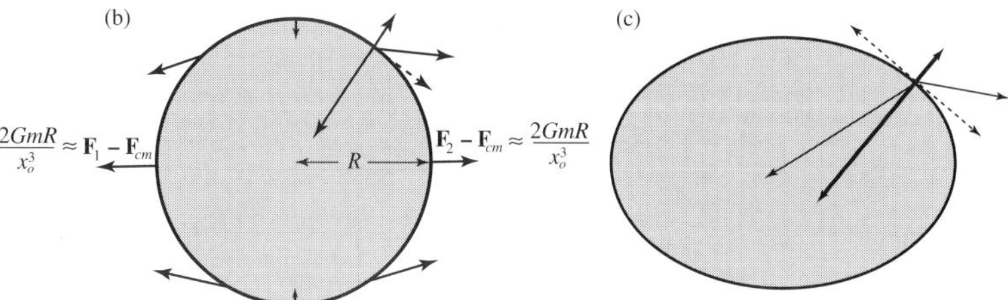

FIGURE 2.9 Schematic illustration of the tidal forces of a moon on a deformable planet. (a) Gravitational force of the moon on different parts of the planet. (b) Differential force of the moon's gravity relative to the force on the planet's center of mass. (c) Response of the planet's figure to the moon's tidal pull.

perturber. The portion of the body with positive x coordinate feels a force in the positive x-direction and the portion at negative x is tidally pulled in the opposite direction (Fig. 2.9). Note from equation (2.39) that material off the x-axis is tidally drawn towards the x-axis.

If the body is deformable, it responds by becoming elongated in the x-direction. For a perfectly fluid body, the degree of elongation is that necessary for the body's surface to become an equipotential, when self-gravity, centrifugal force due to rotation and tidal forces are all included in the calculation.

The gravitational attraction of, for example, the Moon and Earth on each other thus causes tidal bulges which rise along the line joining the centers of the two bodies. The near-side bulge is a direct consequence of the greater gravitational attraction closer to the other body, whereas the one at the opposite side results from the weaker attraction at the far side of the object than at its center. The differential centrifugal acceleration also contributes to the size of the tidal bulges.

The Moon spins once per orbit, so that the same face of the Moon always points towards the Earth and the Moon is always elongated in that direction. The Earth, however, rotates much faster than the Earth–Moon orbital period. Thus, different parts of the Earth point towards the Moon and are tidally stretched. Water responds much more readily to these varying forces than does the 'solid Earth', resulting in the tidal variations in the water level seen at ocean shorelines (Problem 2.20). As the combined effects of terrestrial rotation and the Moon's orbital motion imply that the Moon passes above a given place on Earth approximately once every 25 hours (Problem 2.21), there are almost two tidal cycles per day, and the principal tides that we see are known as the *semidiurnal tide*. The Sun also raises semidiurnal tides on Earth, with a period of 24 hours and an amplitude just under half those of lunar tides (Problem 2.22a). Tidal amplitudes reach a maximum twice each (astronomical) month, when the Moon, Earth and Sun are approximately aligned, i.e., when the Moon is 'new' or 'full'. Tides are also larger when the Moon is near perigee and when the Earth is near perihelion (which occurs in early January).

Strong tides can significantly affect the physical structure of bodies. Generally, the strongest tidal forces felt by Solar System bodies (other than Sun-grazing or planet-grazing comets) are those caused by planets on their closest satellites. Near a planet, tides are so strong that they can rip a fluid (or weakly aggregated solid) body apart. In such a region, large moons are unstable, and even small moons, which could be held together by internal strength, are unable to accrete due to tides. The boundary of this region is known as *Roche's limit*. Inside Roche's limit, solid material remains in the form of small bodies, and we see rings instead of large moons. The derivation of Roche's limit is outlined in Section 11.1.

2.6.2 Tidal Torque

Tidal dissipation causes secular variations in the rotation rates and orbits of moons and planets. Although the total angular momentum of an orbiting pair of bodies is conserved in the absence of an external torque, angular momentum can be transferred between rotation and orbital motions via tidal torques. Orbital angular momentum is given by equation (2.10). The rotational angular

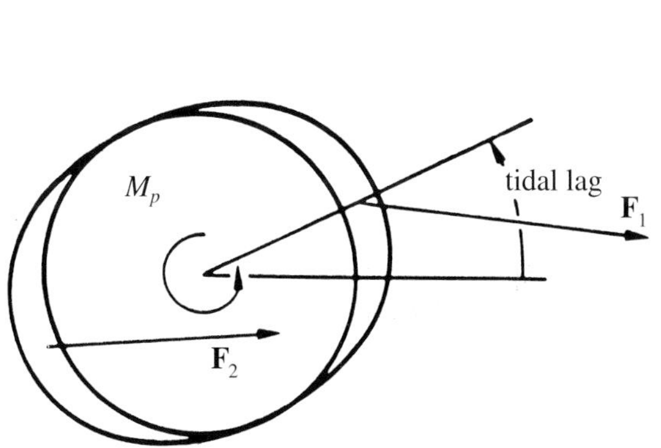

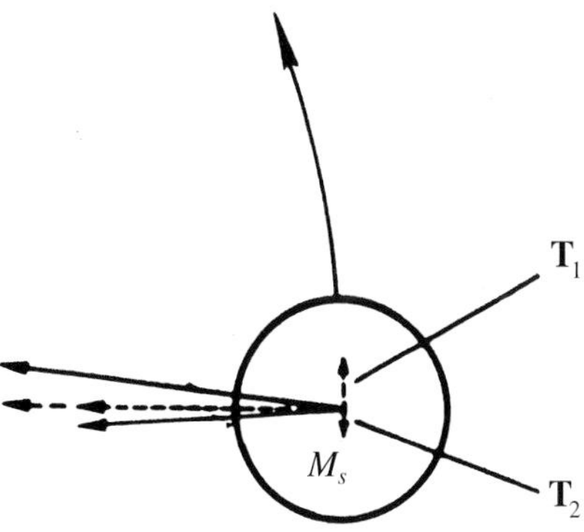

FIGURE 2.10 Schematic illustration of the tidal torque that a planet exerts on a moon orbiting in the prograde direction with a period longer than the planet's rotation period. Dissipation within the planet causes the tidal bulges that the moon raises on the planet to be located at places on the planet that were nearest to and farthest from the moon at a slightly earlier time. Although there is a temporal lag in the tidal bulges, for a moon on a slow prograde orbit the bulges lead the position of the moon. The asymmetries in the planet's figure imply that its gravity is not a central force, and thus it can exert a torque on the moon. The far-side bulge exerts a retarding torque, T_2, upon the moon, but the near-side bulge exerts a larger positive torque, T_1, on the moon, so the moon receives a net positive torque, and its orbit evolves outwards.

momentum of a rigidly rotating body is given by:

$$\mathbf{L} = \overleftrightarrow{\mathbf{I}} \cdot \boldsymbol{\omega}_{rot}, \tag{2.41a}$$

where $\overleftrightarrow{\mathbf{I}}$ is the *moment of inertia tensor* of the body and $\boldsymbol{\omega}_{rot}$ is its spin angular velocity. The kinetic energy of rotation is given by

$$E_{rot} = \frac{1}{2}\boldsymbol{\omega}_{rot} \cdot \overleftrightarrow{\mathbf{I}} \cdot \boldsymbol{\omega}_{rot}. \tag{2.41b}$$

The components of the moment of inertia tensor are

$$I_{jk} = \iiint \rho(\mathbf{r})(r^2\delta_{jk} - x_j x_k)d\mathbf{r}, \tag{2.42a}$$

where $\rho(\mathbf{r})$ represents density and δ_{jk} is the Kronecker δ, i.e., $\delta_{jj} = 1$, $\delta_{jk} = 0$ if $j \neq k$. The moment of inertia of a body about a particular axis is a scalar given by:

$$I = \iiint \rho(\mathbf{r})r_c^2 d\mathbf{r}, \tag{2.42b}$$

where r_c is the distance from the axis, and the integral is taken over the entire body. The moment of inertia of a uniform density sphere of radius R and mass m about its center of mass is given by:

$$I = \frac{2}{5}mR^2. \tag{2.43}$$

Centrally condensed bodies have smaller moment of inertia ratios, $I/(mR^2)$. The moment of inertia ratios for the planets are listed in Table 6.1.

If planets were perfectly elastic, they would respond immediately to varying forces, and tidal bulges raised by a satellite would point directly towards the moon responsible. However, the finite response time of a planet's figure causes the tidal bulges to lag behind, at locations on the planet which pointed towards the moon at a slightly earlier time (Figure 2.10). Provided the planet's rotation period is shorter than the moon's orbital period and the moon's orbit is prograde, this tidal lag causes the nearer bulge to lie in front of the moon, and the moon's greater gravitational force on the near-side bulge than on the far-side bulge acts to slow the rotation of the planet. The reaction force upon the moon causes its orbit to expand. Satellites in retrograde orbits (e.g., Triton) and satellites whose orbital periods are less than the planet's rotation period (e.g., Phobos) spiral inwards towards the planet as a result of tidal forces (cf. Problem 2.23). The torque on a satellite from the tidal bulge which it raises on its primary is given by:

$$\dot{L}_{s(p)} = \frac{3}{2}\frac{k_T}{Q_p}\frac{Gm_s^2 R_p^5}{r^6}\frac{(\omega_{rot_p} - n_s)}{|\omega_{rot_p} - n_s|}, \tag{2.44}$$

where k_T is the tidal Love number (which measures the elastic deformation of a body in response to a tidal perturbation) of the primary, Q is the dissipation factor, ω_{rot} is the rotational angular velocity, n is the orbital angular velocity, r is the distance between primary and secondary and the subscripts p and s refer to the primary and secondary respectively.

The above arguments remain valid if 'moon' is replaced by 'Sun', or if moon and planet are interchanged. Indeed, the stronger gravity of planets means that they have a much greater effect on the rotation of moons than vice versa. Most, if not all, major moons have been slowed to a syn-

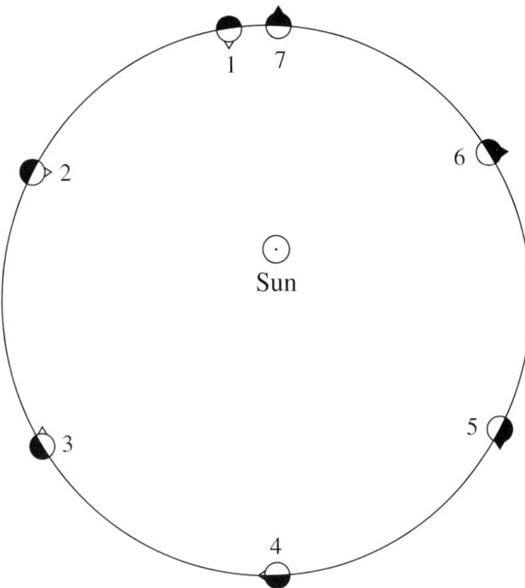

FIGURE 2.11 Mercury's rotation and solar day. Mercury's 3:2 spin-orbit resonance implies that the same axis is always aligned with the direction to the Sun at perihelion, and that the mercurian (solar) day lasts twice as long as the mercurian year.

chronous rotation state, in which the same hemisphere of the moon always faces the planet; thus, no tidal lag occurs.

Evidence exists for the tidal slowing of Earth's rotation on a variety of timescales. Growth bands observed in fossil bivalve shells and corals imply that there were 400 days per year approximately 350 million years ago. Eclipse timing records imply that the day has lengthened slightly over the past two millennia. Precise measurements using atomic clocks also show variations in the Earth's rotation rate; however, care must be taken to separate secular tidal effects from the short-term periodic influences. Most of the secular decrease in Earth's rotation rate is caused by tides raised by the Moon, but at the present epoch $\sim 20\%$ is due to solar tides (Problem 2.22b).

The Pluto–Charon system has evolved even further. Charon is roughly one-ninth as massive as Pluto, by far the largest satellite-to-planet mass ratio observed in the Solar System (Table 1.5). The Pluto–Charon system has reached a stable equilibrium configuration, in which each of the bodies spins upon its axis in the same length of time they orbit about their mutual center of mass (cf. Problem 2.28). Thus, the same hemisphere of Pluto always faces Charon, and the same hemisphere of Charon always faces Pluto.

Solar tides have resulted in a stable spin-orbit lock for nearby Mercury, but one that is more complicated than the synchronous state which exists for most planetary satellites. Mercury makes three revolutions upon its axis every two orbits about the Sun (Fig. 2.11). The reason that equilibrium exists at exactly one and one-half rotations per orbit is that Mercury has a small permanent (nontidal) deformation and a highly eccentric orbit. It is energetically most favorable for Mercury's long axis to point towards the Sun every time the planet passes perihelion, a configuration consistent with the observed 3:2 spin-orbit resonance. Were it not for Mercury's permanent deformation, solar tides would slow Mercury's rotation further. However, because of the substantial eccentricity of Mercury's orbit, synchronous rotation would probably never be achieved. According to Kepler's second law, a planet orbits the Sun much faster at perihelion than at aphelion. The variation is so large for Mercury that for a short time each orbit the planet's angular velocity about the Sun is even faster than its rate of spin. During this brief interval, the tidal bulge raised on Mercury by the Sun trails the Mercury–Sun line, so the Sun's gravity acts to speed up Mercury's rotation rate. As tidal effects on spin vary inversely with the sixth power of the distance between the two objects (eq. 2.44), the short interval during which Mercury's rotation is rapidly accelerated is almost able to balance the much greater fraction of the time during which the Sun's tides act to slow the planet's spin. If Mercury's spin period were increased to roughly 70 days (significantly above its current value of 59 days, yet still well below the planet's 88 day orbital period), a balance between addition of spin angular momentum by solar tides near perihelion and removal during the remainder of the orbit would be achieved even in the absence of a permanent deformation.

Solar gravitational tides are probably the principal reason that Venus rotates very slowly, but they do not explain why that planet spins in the retrograde direction. Solar heating produces asymmetries in Venus's massive atmosphere, and the Sun's gravitational pull on such an asymmetric mass distribution probably prevents Venus's solar day from becoming excessively long. Tidal forces slow the rotation rates of the other planets, but at rates too small to be significant, even over geologic time.

Temporal variations in tidal forces can lead to internal heating of planetary bodies. The location of the tidal bulges of a moon having nonsynchronous rotation vary as the planet moves in the sky as seen from the moon. A synchronously rotating moon on an eccentric orbit is subjected to two types of variations in tidal forces. The amplitude of the tidal bulge varies with the moon's distance from the planet, and the direction of the bulge varies because the moon spins at a constant rate (equal to its mean orbital angular velocity), whereas the instantaneous orbital angular velocity varies according to Kepler's second law.

Since planetary bodies are not perfectly rigid, these variations in tidal forces change the shape of the moon. Since bodies are not perfectly fluid either, moons dissipate energy as heat while they change in shape. Internal stresses caused by variations in tides on a body that is on an eccentric orbit or that is not rotating synchronously with its orbital period can therefore result in significant tidal heating of some bodies, most notably in Jupiter's moon Io. If no other forces were present, this process would lead to a decay of Io's orbital eccentricity. As Io orbits exterior to synchronous, the tides raised on Jupiter by Io cause Io to spiral outwards (eq. 2.44). However, there exists a 2:1 mean motion resonance lock between Io and Europa. Io passes on some of the orbital energy and angular momentum it receives from Jupiter to Europa, and Io's eccentricity is increased as a consequence of this transfer (Problem 2.29). This forced eccentricity maintains a high tidal dissipation rate and consequently there is a large internal heating of Io, which displays itself in the form of active volcanism (Section 5.5.5.1).

2.7 Dissipative Forces and the Orbits of Small Bodies

Sections 2.1–2.6 above describe the gravitational interactions between the Sun, planets and moons. Solar radiation, which provides an important force for small ($\lesssim$1 m) particles in the Solar System, has been ignored. Three effects can be distinguished:

(1) *Radiation pressure*, which pushes particles (primarily micrometer-sized dust) outwards from the Sun.

(2) *Poynting–Robertson drag*, which causes centimeter-sized particles to spiral inward towards the Sun.

(3) The *Yarkovski effect*, which changes the orbits of meter to kilometer-sized objects due to uneven temperature distributions at their surfaces.

The solar wind produces a *corpuscular drag* similar in form to the Poynting–Robertson drag; corpuscular drag is most important for submicrometer particles. We discuss each of these effects in the next four subsections and then examine the effect of gas drag. Perturbations on the motion of dust in planetocentric orbits caused by solar radiation are analyzed in Section 11.5.1. The motion of charged dust particles in planetary rings is discussed in Section 11.5.2. Nongravitational forces resulting from asymmetric mass loss by comets are considered in Section 10.2.3.

2.7.1 Radiation Force (micrometer-sized particles)

The Sun's radiation exerts a repulsive force, $\mathbf{F}_{rad}$, on all bodies in our Solar System. This force is given by:

$$\mathbf{F}_{rad} \approx \frac{L_\odot A}{4\pi c r_\odot^2} Q_{pr} \hat{\mathbf{r}}, \tag{2.45}$$

where A is the particle's geometric cross-section, $L_\odot$ the solar luminosity, $r_\odot$ the heliocentric distance, c is the speed of light, and Q_{pr} the *radiation pressure coefficient*. The radiation pressure coefficient accounts for both absorption and scattering and is equal to unity for a perfectly absorbing particle. Relativistic effects produced by the Doppler shift between the frame of the Sun and that of the particle are generally small, and have been omitted from equation (2.45), but they will be considered in Section 2.7.2. The parameter β is defined as the ratio between the forces due to the radiation pressure and the Sun's gravity:

$$\beta \equiv \frac{F_{rad}}{F_g} = 5.7 \times 10^{-5} \frac{Q_{pr}}{\rho R}, \tag{2.46}$$

with the particle's radius, R, in cm and its density, ρ, in g cm^{-3}. Note that β is independent of heliocentric distance and that the solar radiation force is only important for micrometer and submicrometer-sized particles. Extremely small particles are not strongly affected by radiation pressure, because Q_{pr} decreases as the particle radius drops below the (visible wavelength) peak in the solar spectrum (Fig. 2.12). The magnitude of the Sun's effective gravitational attraction is given by:

$$F_{g,eff} = \frac{-(1-\beta) G m M_\odot}{r_\odot^2}. \tag{2.47}$$

It is clear that small particles with $\beta > 1$ are repelled by the Sun's radiation, and thus quickly escape the Solar System, unless they are gravitationally bound to one of the planets. Dust released at a keplerian velocity from bodies on circular orbits is ejected from the Solar System if $\beta > 0.5$; critical values of β for dust released from bodies on eccentric orbits are calculated in Problem 2.31.

The importance of solar radiation pressure can, for example, be seen in comets (Section 10.4.1): Cometary tails always point in the anti-solar direction due to the Sun's radiation pressure. The dust tails are curved rather than straight as a result of the continuous ejection of dust grains from the comet, which itself is on an elliptical orbit around the Sun.

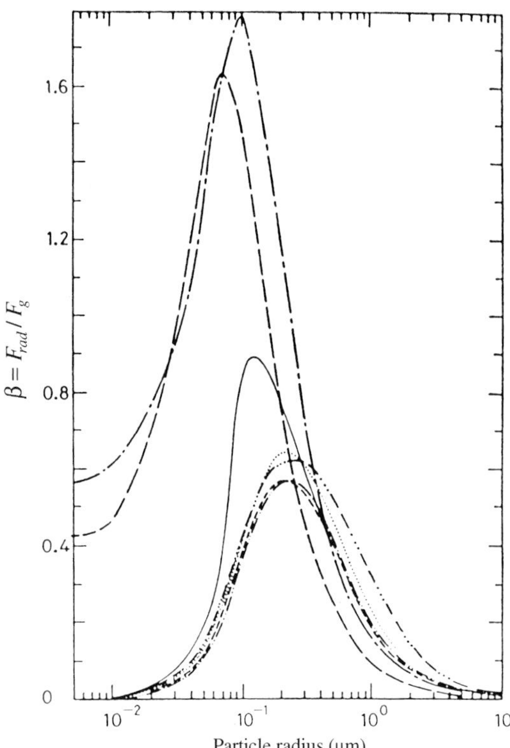

FIGURE 2.12 The relative radiation pressure force, $\beta = F_{rad}/F_g$, as a function of particle size for six cosmically significant substances and an ideal material: — — · · · — — ice at 100 K, – – – amorphous quartz, — — ·· — — obsidian, · · · basalt, — — — iron, — · — · — magnetite, ——— ideal material, $\rho = 3$ g cm^{-3}. Note that these values are for particles in orbit about the Sun. Grains orbiting stars of different mass, luminosity and/or spectral type would have different values of β. (Adapted from Burns *et al.* 1979)

2.7.2 Poynting–Robertson Drag (centimeter-sized grains)

A particle in orbit around the Sun absorbs solar radiation and reradiates the energy isotropically in its own frame. The particle thereby preferentially radiates (and loses momentum) in the forward direction in the inertial frame of the Sun (Fig. 2.13). This leads to a decrease in the particle's energy and angular momentum and causes dust in bound orbits to spiral sunward. This effect is called *Poynting–Robertson drag*.

Let us consider a perfectly absorbing, rapidly rotating, dust grain. The flux of solar radiation absorbed by the grain is equal to

$$\frac{L_\odot A}{4\pi r_\odot^2}\left(1 - \frac{v_r}{c}\right), \tag{2.48a}$$

where $v_r = \mathbf{v} \cdot \hat{\mathbf{r}}$ is the radial component of the particle's

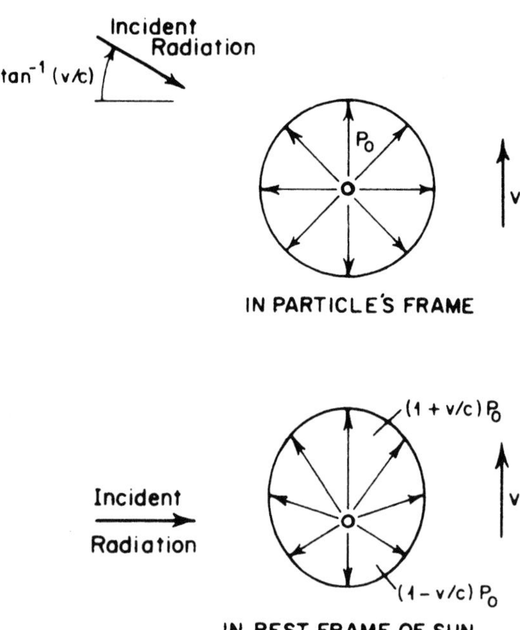

FIGURE 2.13 A particle in heliocentric orbit, that reradiates the solar energy flux isotropically in its own frame of reference, preferentially emits more momentum, p, in the forward direction as seen in the solar frame, because the frequencies and momenta of the photons emitted in the forward direction are increased by the particle's motion. (Adapted from Burns *et al.* 1979)

velocity (i.e., the component which is parallel to the incident beam of light). The second term in expression (2.48*a*) accounts for the Doppler shift between the Sun's rest frame and that of the particle; the transverse Doppler shift is of order $(v_\theta/c)^2 \ll 1$ and will be ignored here. The absorbed flux is reradiated isotropically and can be written as a mass loss rate in the particle's frame of motion (using $E = mc^2$):

$$\frac{L_\odot A}{4\pi c^2 r_\odot^2}\left(1 - \frac{v_r}{c}\right). \tag{2.48b}$$

As the particle moves relative to the Sun with velocity $\mathbf{v}$, there is a momentum flux from the particle as seen in the rest frame of the Sun, since the particle emits more momentum in the forward direction than in the backward direction (Fig. 2.13). The momentum flux is equal to

$$\frac{-L_\odot A}{4\pi r_\odot^2}\left(1 - \frac{v_r}{c}\right)\mathbf{v}, \tag{2.48c}$$

which can be generalized to the case in which the particle reflects and/or scatters some of the radiation impingent upon it via multiplication by Q_{pr}. The net force on the particle in this more general case is given by:

$$\mathbf{F}_{rad} = \frac{L_\odot Q_{pr} A}{4\pi c r_\odot^2}\left(1 - \frac{v_r}{c}\right)\hat{\mathbf{r}} - \frac{L_\odot Q_{pr} A v}{4\pi c^2 r_\odot^2}\left(1 - \frac{v_r}{c}\right)\hat{\mathbf{v}} \tag{2.49a}$$

$$\approx \frac{L_\odot Q_{pr} A}{4\pi c r_\odot^2}\left[\left(1 - \frac{2v_r}{c}\right)\hat{\mathbf{r}} - \frac{v_\theta}{c}\hat{\theta}\right]. \tag{2.49b}$$

The first term in equation (2.49*b*) is that due to radiation pressure and the second and third terms (those involving the velocity of the particle) represent the Poynting–Robertson drag.

From the above discussion, it is clear that small dust grains in the interplanetary medium disappear, with (sub-) micrometer-sized grains being blown out of the Solar System, while centimeter-sized particles spiral inward towards the Sun. Typical decay times (in years) for particles on circular orbits are given by:

$$\tau_{pr} \approx 400\frac{r_{AU}^2}{\beta}. \tag{2.50}$$

The *zodiacal light* is a band centered near the ecliptic plane that appears almost as bright as the Milky Way on a dark night. It is visible in the direction of the Sun, just after sunset or before sunrise. Particles that produce the bulk of the zodiacal light (at infrared and visible wavelengths) are between 20 and 200 μm, so their lifetimes at Earth's orbit are of the order of 10^5 years, which is much less than the age of the Solar System. A possible source for these dust grains is the asteroid belt, where numerous collisions occur between countless small asteroids; comets also contribute to the zodiacal light.

2.7.3 Yarkovski Effect (meter–kilometer-sized objects)

Consider a rotating body heated by the Sun. The evening hemisphere is typically warmer than the morning hemisphere, by an amount $\Delta T \ll T$. Let us assume that the temperature of the morning hemisphere is $T - \frac{\Delta T}{2}$, and that of the evening hemisphere is $T + \frac{\Delta T}{2}$. The radiation reaction upon a surface element dA, normal to its surface, is

$$dF = \frac{2\sigma T^4 dA}{3c}, \tag{2.51}$$

where σ is the Stefan–Boltzmann constant. For a spherical particle of radius R, the transverse reaction force in the orbit plane due to the excess emission on the evening side is:

$$F_Y = \frac{8}{3}\pi R^2 \frac{\sigma T^4}{c}\frac{\Delta T}{T}\cos\psi, \tag{2.52}$$

where ψ the particle's obliquity, i.e., the angle between its rotation axis and orbit pole. This process is referred to as the *Yarkovski effect*. The Yarkovski force is positive for an object which rotates in the prograde direction, $0° \leq \psi < 90°$, and negative for an object with retrograde rotation, $90° \leq \psi \leq 180°$. In the latter case, the Yarkovski force enhances the Poynting–Robertson drag. There is also an analogous seasonal Yarkovski effect which is produced by temperature differences between the spring/summer and the autumn/winter hemispheres.

The Yarkovski effect is important for bodies in the meter to kilometer size range. It may play an important role in transporting most meteorites to Earth. Asymmetric outgassing of comets produces a nongravitational force similar to the Yarkovski force. Nongravitational forces on comets are discussed in Section 10.2.3.

2.7.4 Corpuscular Drag (submicrometer-sized particles)

Particles with sizes much smaller than one micrometer are also subjected to a significant corpuscular 'drag' by solar wind particles. This effect can be calculated in a manner similar to the Poynting–Robertson drag above, except that the energy–momentum relation must be replaced by that of nonrelativistic particles:

$$p_{sw} = \frac{2E_{sw}}{v_{sw}}, \tag{2.53}$$

where v_{sw} is the solar wind velocity. The momentum and energy flux densities carried by the solar wind are roughly four and seven orders of magnitude less, respectively, than those carried by the electromagnetic radiation; hence, since pressure is proportional to the momentum flux density, the pressure of the solar wind is much smaller than the radiation pressure. However, the aberration angle (the change in apparent position of the source due to the motion of the receiving body) of the solar wind, $\tan^{-1}(v/v_{sw})$, is much larger than that for solar radiation, so the solar wind produces a significant drag force. The ratio of this *corpuscular drag* to the radiation drag, β_{cp}, can be expressed as:

$$\beta_{cp} = \frac{p_{sw}}{p_r}\frac{c}{v_{sw}}\frac{C_D}{Q_{pr}}, \tag{2.54}$$

where C_D is the corpuscular drag coefficient. Figure 2.14 shows a graph of β_{cp} as a function of particle radius. Corpuscular drag is more important than radiation drag for particles $\lesssim 0.1$ μm in size, and is the primary force behind these very small dust grains' inward spiral towards the Sun. See Burns *et al.* (1979) for a more detailed dis-

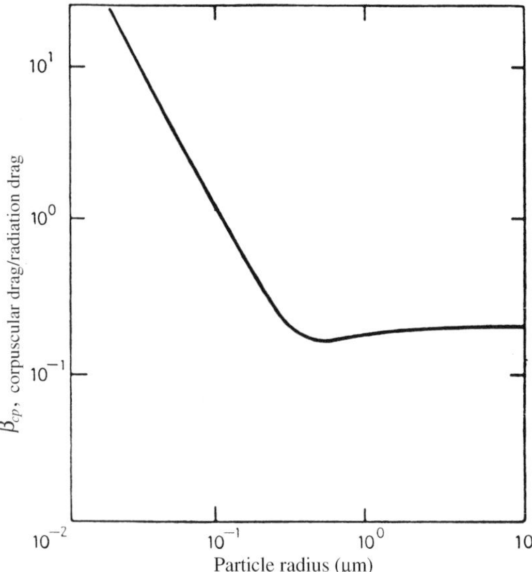

FIGURE 2.14 Ratio of corpuscular drag (caused by the solar wind) to that resulting from solar radiation, β_{cp}, is plotted against radius for grains composed of obsidian. (Adapted from Burns *et al.* 1979)

cussion of the effects of solar radiation and the solar wind on the motion of small particles.

2.7.5 Gas Drag

Although for most purposes interplanetary space can be considered to be a vacuum, there are certain situations in which interactions with gas can significantly alter the motion of solid particles. Two prominent examples of this process are planetesimal interactions with the gaseous component of the protoplanetary disk and orbital decay of ring particles as a result of drag caused by extended planetary atmospheres.

In the laboratory, gas drag slows solid objects down until their positions remain fixed relative to the gas. In the planetary dynamics case, the situation is more complicated. For example, a body on a circular orbit about a planet loses mechanical energy as a result of drag with a static atmosphere, but this energy loss leads to a decrease in semimajor axis of the orbit, which implies the body actually speeds up! Other, more intuitive effects of gas drag are the damping of eccentricities and, in the case where there is a preferred plane in which the gas density is the greatest, the damping of inclinations relative to this plane.

Objects whose dimensions are larger than the mean free path of the gas molecules experience *Stokes drag*:

$$F_D = -\frac{C_D A \rho v^2}{2}, \tag{2.55}$$

where v is the velocity of the body with respect to the gas, ρ is the gas density, A is the projected surface area of the body and C_D is a dimensionless drag coefficient, which is of order unity unless the Reynolds number is very small. Smaller bodies are subject to *Epstein drag*:

$$F_D = -A\rho v v_o, \tag{2.56}$$

where v_o is the mean thermal velocity of the gas. Note that as the drag force is proportional to surface area and the gravitational force is proportional to volume (for constant particle density), gas drag is usually most important for the dynamics of small bodies.

The gaseous component of the protoplanetary disk is believed to have been partially supported against the gravity of the Sun by a negative pressure gradient in the radial direction. Thus, less centrifugal force was required to complete the balance, and consequently the gas orbited less rapidly than the keplerian velocity. The 'effective gravity' felt by the gas was:

$$g_{eff} = -\frac{GM_\odot}{r_\odot^2} - \frac{1}{\rho_{gas}}\frac{dP}{dr_\odot}. \tag{2.57}$$

For circular orbits, the effective gravity must be balanced by centrifugal acceleration, $r_\odot n^2$. For estimated protoplanetary disk parameters, the gas rotated $\sim 0.5\%$ slower than the keplerian speed. The implications of gas drag for the accretion of planetesimals are discussed in Section 12.4.2.

Drag induced by a planetary atmosphere is substantially more effective for a given gas density than is drag in a primarily centrifugally supported disk. Because atmospheres are almost entirely pressure supported, the relative velocity between the gas and orbiting particles is large. As atmospheric densities drop rapidly with height, particle orbits decay slowly at first, but as they reach lower altitudes their decay can become very rapid (see Problem 2.34). Gas drag is the principal cause of orbital decay of artificial satellites in low Earth orbit.

FURTHER READING

A good introductory text:

Danby, J.M.A., 1988. *Fundamentals of Celestial Mechanics, 2nd Edition.* Willmann-Bell, Richmond, VA, 467 pp.

This book is not for the faint of heart, but it contains some important information that is difficult to locate elsewhere:

Brouwer, D., and G.M. Clemence, 1961. *Methods of Celestial Mechanics.* Academic Press, New York, 598 pp.

Useful individual articles include:

Burns, J.A., P.L. Lamy, and S. Soter, 1979. Radiation forces on small particles in the Solar System. *Icarus* **40**, 1–48.

Duncan, M.J., and T. Quinn, 1993. The long-term dynamical evolution of the Solar System. *Ann. Rev. Astron. Astrophys.* **31**, 265–295.

Peale, S.J., 1976. Orbital resonances in the Solar System. *Ann. Rev. Astron. Astrophys.*, **14**, 215–246.

An excellent overview of many important aspects of planetary dynamics is presented by:

Murray, C., and S. Dermott, 1999. *Solar System Dynamics*. Cambridge University Press. 592 pp.

Problems

2.1.**E** Consider two mutually gravitating bodies of masses m_1 and m_2 and positions $\mathbf{r}_1$ and $\mathbf{r}_2$.

(a) Write down the equations which govern the motion of these bodies.

(b) Using Newton's third law, show that the center of mass of the system moves at a constant velocity and that the relative position of the bodies, $\mathbf{r} \equiv \mathbf{r}_1 - \mathbf{r}_2$, changes according to:

$$\frac{d^2\mathbf{r}}{dt^2} = -\frac{GM}{r^2}\hat{\mathbf{r}}, \qquad (2.58)$$

where $M \equiv m_1 + m_2$. This reduces the two-body problem to an equivalent one-body problem.

2.2.**I** In this problem, you will finish the derivation of the newtonian generalization of Kepler's laws that you began in the previous problem.

(a) Derive the law of conservation of angular momentum for the system, $d(\mathbf{r} \times \mathbf{v})/dt = 0$, by taking the cross product of $\mathbf{r}$ with equation (2.58) and using various vector identities. By writing the expression for angular momentum in polar coordinates, deduce Kepler's second law and determine the constant rate of sweeping, $d\mathcal{A}/dt$.

(b) Take the dot product of $\mathbf{v}$ with equation (2.58) to deduce the conservation of energy per unit mass. Integrate your result to determine an expression for the specific energy of the system, E. Express your answer in polar coordinates and solve for dr/dt. Take the reciprocal, multiply both sides by $d\theta/dt$ and then use the magnitude of the specific angular momentum, L, to eliminate the angular velocity from your expression, yielding the following purely spatial relationship for the orbit:

$$\frac{d\theta}{dr} = \frac{1}{r}\left(\frac{2Er^2}{L^2} + \frac{2GMr}{L^2} - 1\right)^{-1/2}. \qquad (2.59)$$

Integrate equation (2.59) and solve for r. Set the constant of integration equal to $-\pi/2$, define $r_0 \equiv L^2/(GM)$, $e \equiv \left(1 + (2EL^2)/(G^2M^2)\right)^{1/2}$ and obtain:

$$r = \frac{r_0}{1 + e\cos\theta}. \qquad (2.60)$$

For $0 \le e < 1$, equation (2.60) represents an ellipse in polar coordinates. Thus, Kepler's first law is also precise in the two-body newtonian approximation, although the Sun itself is not fixed in space. Note that if $E = 0$ then $e = 1$ and equation (2.60) describes a parabola, and if $E > 0$ then $e > 1$ and the orbit is hyperbolic.

(c) Show that the semimajor and semiminor axes of the ellipse given by equation (2.60) are $a = r_0/(1 - e^2)$ and $b = r_0/(1 - e^2)^{1/2}$, respectively. Determine the orbital period, P, by setting the integral of $d\mathcal{A}/dt$ equal to the area of the ellipse, πab. Note that your result, $P = (4\pi^2 a^3/GM)^{1/2}$, differs from Kepler's third law by replacing the Sun's mass, m_1, by the sum of the masses of the Sun and the planet, M.

2.3.**E** A baseball pitcher can throw a fastball at a speed of ~150 km/hr. What is the largest size spherical asteroid of density $\rho = 3$ g cm^{-3} from which he can throw the ball fast enough that it:

(a) escapes from the asteroid into heliocentric orbit?

(b) rises to a height of 50 km?

(c) goes into a stable orbit about the asteroid?

2.4.**I** In this problem, you will make some calculations which would be useful for planning a spacecraft mission to Jupiter. To simplify matters, you may assume that both Earth and Jupiter travel on circular orbits.

(a) Calculate the minimum velocity (relative to Earth) at Earth's orbit of a small body whose orbit intersects both Earth's orbit and Jupiter's orbit.

(b) Calculate the minimum velocity necessary to launch a spacecraft from the surface of Earth to Jupiter, ignoring Earth's rotation.

(c) Calculate the minimum velocity necessary to launch a spacecraft from the Earth's equator to Jupiter, including Earth's rotation but ignoring its obliquity.

2.5.I The *virial theorem* states that the time-averaged potential energy of a bound self-gravitating system is equal to two times the negative of the time-average of the kinetic energy of the system,

$$\langle E_G \rangle = -2\langle E_K \rangle. \tag{2.61}$$

(a) State the virial theorem for the N-body problem in words and also in mathematical form.

(b) Verify the virial theorem for the case of two bodies on a circular orbit.

2.6.**I** The general solution to the N-body problem requires knowledge of $6N$ quantities at all times, representing the positions and velocities of every particle or some equivalent set of 'orbital elements'. In general, the system has 10 integrals of motion, six representing the location of the center of mass as a function of time, $\mathbf{x} + \mathbf{v}t$, three representing the angular momentum of the system, $\mathbf{L}$, and one representing the total energy of the system, E.

(a) The longitude at epoch provides an independent constraint for the two-body problem. One more (12–10–1) integral or independent constraint is required to completely specify the solution of the two-body problem. What is it?

(b) The restricted circular three-body problem includes an already-solved two-body problem (as the mass of the third body is ignored). The Jacobi constant is an integral of motion for the test particle in the restricted circular three-body problem. State two additional integrals of motion for the test particle in the planar restricted circular three-body problem.

2.7.**I** We listed several simplifications for the three-body problem in Section 2.2: planar, restricted, circular and Hill. If all of these simplifications were independent, there would be $2^4 = 16$ possibilities. However, there are actually only 12 distinct viable cases. Name these combinations, and state why they are possible, but the other 4 are not. (Note: Hill's original papers assumed coplanar orbits, but his calculations can be generalized to the noncoplanar case. We use this more general definition of Hill's problem here.)

2.8.**E** Show that if the two-body problem is analyzed using units in which the gravitational constant, the separation between the bodies and the sum of their masses are all set equal to one, then the orbital period of the two bodies about their mutual center of mass equals 2π.

2.9.**I** This problem concerns the stability of orbits in the planar circular restricted three-body problem. The ratio of the star's mass to that of the planet is 333. Choose a rotating coordinate system with the center of mass at the origin and the planet located at $x = 1$, $y = 0$.

(a) Calculate (approximately) the location of the equilibrium points.

(b) Which, if any, of these points are stable?

(c) In what region about the planet may moons have stable orbits? What regions must asteroids avoid in order to have stable orbits about the star? Be quantitative.

2.10.**I** One of the greatest triumphs of dynamical astronomy was the prediction of the existence and location of the planet Neptune on the basis of irregularities observed in the orbit of Uranus. The motion of Uranus could not be accurately accounted for using only (the newtonian modification of) Kepler's laws and perturbations of the then-known planets. Estimate the maximum displacement in the position of Uranus caused by the gravitational effects of Neptune as Uranus catches up and passes this slowly moving planet. Quote your results both in kilometers along Uranus's orbital path, and in seconds of arc against the sky as observed from Earth. For your calculations, you may neglect the effects of the other planets (which can be and were accurately estimated and factored into the solution) and assume that the unperturbed orbits of Uranus and Neptune are circular and coplanar, and neglect the affects of Uranus on Neptune. Note that although the displacements of Uranus in radius and longitude are comparable, the longitudinal displacement produces a much larger observable signature because of geometric factors.

(a) Obtain a very crude result by assuming that the potential energy released as Uranus gets closer to Neptune increases Uranus's semimajor axis and thus slows Uranus down. You may assume that Uranus's average semimajor axis during the interval under consideration is halfway between its semimajor axis at the beginning and the end of the interval. Remember to use the synodic (relative) period of the pair of planets rather than just Uranus's orbital period.

(b) Obtain an accurate result using numerical integration of Newton's equations on a computer.

2.11.**E** Explain how the Lyapunov characteristic exponent, γ_c, is used to distinguish between regular and chaotic trajectories. What is the value of γ_c for regular trajectories?

2.12.**E** A small planet travels about a star with an orbital semimajor axis equal to 1 in the units chosen. Calculate the locations of the 2:1, 3:2, 99:98 and 100:99 resonances of test particles with the planet.

2.13.**I** Two criteria for the stability of the orbits of test particles near the semimajor axis of a planet were introduced in Section 2.3.3: The resonance overlap criterion for the onset of chaos, $\Delta a_{ro} \approx 1.5(m_2/m_1)^{2/7}$; and the criterion for exclusion from close approaches to the planet resulting from the Jacobi integral, $\Delta a_J \geq 2\sqrt{3}(m_2/3m_1)^{1/3}$.

(a) Comment on the conceptual differences between these two stability criteria.
(b) For what value of m_2/m_1 are the two equal?
(c) Comment on the qualitative difference between orbits near the instability boundary for m_2/m_1 greater or less than this value.

2.14.**E** Einstein's *general theory of relativity* is conceptually quite different from Newton's theory of gravity, but the predictions of general relativity reduce to those of Newton's model in the low-velocity (relative to the speed of light, c), weak gravitational field (relative to that required for a body to collapse into a black hole) limit.

(a) Calculate v^2/c^2 for the following bodies:

(i) Mercury, using its circular orbit velocity
(ii) Mercury, using its velocity at perihelion
(iii) Earth, using its circular orbit velocity
(iv) Neptune
(v) Io, using its velocity relative to Jupiter
(vi) Metis, using its velocity relative to Jupiter.

(b) The *Schwarzchild radius* of a body,

$$R_{Sch} = \frac{2Gm}{c^2}, \tag{2.62}$$

is the radius inside of which light cannot escape from the body. Calculate the Schwarzchild radii of the following Solar System bodies, and the ratios of these radii to the sizes of the bodies in question and to the semimajor axes of their nearest (natural) satellites:

(i) Sun
(ii) Earth
(iii) Jupiter.

2.15.**E** As demonstrated in the previous problem, Newton's theory of gravity is quite accurate for most Solar System situations. The most easily observable effect of general relativity is the precession of orbits, because it is nil in the newtonian two-body approximation. The first-order (weak field) general relativistic corrections to newtonian gravity imply a precession of the periapse at the rate

$$\dot{\varpi} = \frac{3(Gm)^{3/2}}{a^{5/2}(1-e^2)c^2}. \tag{2.63}$$

Calculate the general relativistic precession of the periapse of the following objects:

(a) Mercury
(b) Earth
(c) Io.

Quote your answers in seconds of arc per year. Note: The average observed precession of Mercury's periapse is 56.00″ yr^{-1}, all but 5.74″ yr^{-1} of which is caused by the observations not being done in an inertial frame far from the Sun. Newtonian gravity of the other planets accounts for 5.315″ yr^{-1}. The small difference between the remainder and the value calculated above is within the uncertainties of the observations and calculations, and may also be affected by a very small contribution resulting from solar oblateness.

2.16.**I** Use multiple integration to show that the gravitational potential exterior to a spherically symmetric body is identical to that of a point-like particle of the same mass located at the center of the sphere. (Hint: Divide the sphere into concentric shells, and then subdivide the shells into rings that are oriented perpendicular to the direction from the center of the sphere to the point at which the potential is being evaluated. Determine the potential of each ring, then integrate over angle to deduce the potential of a shell and finally integrate over radius to determine the potential of the sphere.)

2.17.E Saturn is the most oblate major planet in the Solar System, with gravitational moments $J_2 = 1.63 \times 10^{-2}$, $J_4 = -9 \times 10^{-4}$ and $J_6 = 10^{-4}$. Calculate the orbital periods, apse precession rates and node regression rates for particles on nearly circular and equatorial orbits at 1.5 R♄ and 3 R♄:

(a) neglecting planetary oblateness entirely.
(b) including J_2, but neglecting higher-order moments.
(c) including J_2, J_4 and J_6.

2.18.E If the Earth–Moon distance was reduced to half its current value then:

(a) Neglecting solar tides, how many times as large as at present would the maximum tide heights on Earth be?
(b) Including solar tides, how many times as large as at present would the maximum tide heights on Earth be?

2.19.E Calculate the semimajor axis of a moon on a synchronous orbit around Mars. Express your answer in martian radii (R♂) from the center of the planet. Compare this distance with the orbits of Phobos (at 2.76 R♂) and Deimos (at 6.9 R♂). Describe the motion of each of these moons on the sky as seen from Mars.

2.20.I Estimate the amplitude of tides that the Moon raises on Earth. (Hint: Integrate the tidal force to compute a tidal potential. Next, calculate the tidal potential of a test particle at the sublunar point on the Earth's surface. Finally, determine the height by which the blob must be raised for the change in its gravitational potential relative to Earth to be equal in magnitude to its tidal potential.)

2.21.E Calculate the mean synodic period between Earth's rotation and the Moon's orbit.

2.22.E (a) Compute the ratio of the height of tides raised on Earth by the Moon to those raised by the Sun.
(b) Compute the ratio of the tidal torque on the Earth due to the Moon and to the Sun.

2.23.I Using equation (2.44), show that a moon starting inside synchronous orbit will impact the planet's surface after a time t_{impact}:

$$t_{impact} = \frac{2}{39} \frac{m_p}{m_s} \frac{Q_p}{k_T} \frac{a_s^{13/2}(0) - 1}{n_p^*}, \quad (2.64)$$

where $a_s \equiv r/R_p$ and $n_p^* \equiv \left(Gm_p/R_p^3\right)^{1/2}$. State all assumptions that you make. (Hint: Set up the orbit equations, use Kepler's laws to eliminate n_s and keep dimensional quantities until the final step.)

2.24.E If the Moon's mass was reduced to half its current value, then:

(a) The rate at which the Moon was receding from Earth would be how many times as large as at present?
(b) Neglecting solar tides, maximum tide heights on Earth would be how many times as large as at present?
(c) Including solar tides, maximum tide heights on Earth would be how many times as large as at present?

2.25.I If a moon and planet are both initially rotating faster than their mutual orbit period, then the tides raised by the planet on the moon slow the moon's rotation until it becomes synchronous. Tides raised by the moon on the planet cause a slowing of the planet's rotation rate. Both tides cause the moon and planet to move farther apart.

(a) Go over the details of this argument (using illustrations) and compare timescales for the various processes.
(b) Repeat for the case when a moon and planet are initially spinning more slowly than their mutual orbital period. What is the outcome in this case?
(c) Repeat for the case when the moon is on a retrograde orbit, i.e., its orbital angular momentum is antiparallel to the planet's spin angular momentum.

2.26.I Extrapolate the evolution of the Moon's orbit backwards in time using the data on fossil bivalve shells, conservation of angular momentum for the Earth–Moon system and a variant of equation (2.64), assuming that Q_p/k_T has remained constant. As tidal evolution was much more rapid when the bodies were closer, your result should imply that the Moon was quite close to the Earth substantially less than 4×10^9 years ago. This was considered a major problem until it was realized that a substantial fraction of the tidal dissipation in Earth today results from sloshing of waters in shallow seas, and that Q_p/k_T could have been much less in the past when Earth's continents were configured differently.

2.27.**I** The Moon is receding from Earth as a result of the lag of the tidal bulge it raises on Earth. This process will continue until the day is the same length as the month.

(a) Neglecting solar perturbations, what will this day/month period be? You may approximate the Earth by a homogeneous sphere in order to calculate its moment of inertia. You may use other approximations if they induce errors of $< 5\%$. Quote your answer in seconds.
(b) Compare the orbital radius of the Moon when the day and month are equal to the radius of the Earth's Hill sphere.
(c) Qualitatively, what will the effects of the Sun be on the tidal evolution of the Earth–Moon system; specifically, will the Moon stop its recession closer to or farther from Earth? Will this effect be large or small? Explain.

2.28.**D** The only possible true equilibrium state for a planet/moon system subject to tidal dissipation is one in which both bodies spin in a prograde sense with a period equal to their mutual orbital period.

(a) Draw a diagram of this system.
(b) Show that for a given total angular momentum of the system there are in general two possible equilibrium states, one in which the bodies are close to each other and most of the angular momentum resides in the planet's rotation, and the other where the bodies are far apart and the moon's orbital motion holds most of the angular momentum.
(c) Such a system may or may not be stable. Under what conditions is such a system stable?
(d) Can both equilibrium states exist for all planet/moon systems? Why or why not?

2.29.**D** Two moons are spiraling outwards, away from their planet, due to tidal forces. They become locked in a stable orbital resonance which requires them to maintain a constant ratio of orbital periods (see, e.g., Peale 1976). Calculate the energy available for tidal heating in equilibrium (i.e., assume that the moons' orbital eccentricities do not change). Your answer should depend on the masses of the two moons, m_I and m_{II}, and of the primary, m_p, on the angular velocities of the moons, n_I and n_{II}, and on (the z-components of) the tidal torques exerted by the planet on each of the moons, $\dot{L}_{I(p)}$ and $\dot{L}_{II(p)}$. (Hint: Use conservation of angular momentum and energy:

$$\frac{d}{dt}(L_I + L_{II}) = \dot{L}_{I(p)} + \dot{L}_{II(p)}, \qquad (2.65)$$

$$\frac{d}{dt}(E_I + E_{II}) = \dot{L}_{I(p)} n_I + \dot{L}_{II(p)} n_{II} - \mathcal{H}, \qquad (2.66)$$

where L_I and L_{II} are the orbital (technically plus rotational, however we may ignore these small factors) angular momenta of the moons, E_I and E_{II} are the orbital energies and $\mathcal{H}$ is the heating rate.)

2.30.**E** Calculate the orbital period around the Sun for a dust grain with $\beta = 0.3$ and semimajor axis $a = 1$ AU.

2.31.**E** A grain with $\beta = \beta_0$ travels on a circular orbit about the Sun. The grain splits apart into smaller grains with $\beta = \beta_n$. What are the eccentricities and semimajor axes of these new grains?

2.32.**I** (a) Show that dust released at perihelion from a body on an eccentric keplerian orbit will escape from the Solar System if the ratio of radiation pressure to the solar gravity it feels:

$$\beta \geq \frac{1-e}{2}. \qquad (2.67)$$

(b) Derive an analogous expression for the stability of a dust grain released at aphelion.

2.33.**I** Consider an icy ($\rho = 1$ g cm^{-3}) particle in Saturn's rings, located at 1.5 $R_♄$. The radius of the particle is 0.1 μm.

(a) Calculate the radiation pressure on the particle due to the Sun and due to Saturn's reflected light, at noon (Saturn's albedo is 0.46). (Hint: Use Figure 2.12 to estimate Q_{pr}.)
(b) Compare the radiation forces to the gravitational forces from the Sun and Saturn. Calculate the *appropriate* β. What will happen to the particle (e.g., blown in/out, stay in orbit)?

2.34.**I** This problem concerns the orbital decay of an artificial satellite in Earth orbit, but it is also applicable to, e.g., the decay of particles in the rings of Uranus as a result of that planet's extended atmosphere. The satellite has mass m and drag cross-section πR^2 and is initially on a circular orbit of semimajor axis a_o. The density of the atmosphere is given by the formula $\rho = \rho_o \exp[-(a-a_o)/H]$, where H is the scale height of the atmosphere. Assume the drag coefficient is 0.4 and that the Stokes drag formula is applicable, i.e., $F_D = 0.4\rho\pi R^2 v^2$.

(a) Assuming F_D is small, calculate the change during one orbit as a function of F_D.
(b) Calculate F_D as a function of a.
(c) Using the results from (a) and (b), calculate, approximately, the semimajor axis of the satellite's orbit as a function of time.

3 Solar Heating and Energy Transport

The three laws of thermodynamics:

(1) You can't win.
(2) You can't break even.
(3) You can't get out of the game.

Anonymous

Temperature is one of the most fundamental properties of planetary matter, as is evident from everyday experience such as the weather and cooking a meal, as well as from the most basic concepts of chemistry and thermodynamics. For example, H_2O is a liquid between 273 K and 373 K (at standard pressure), a gas at higher temperatures and a solid when it is colder; silicates undergo similar transitions at substantially higher temperatures and methane condenses and freezes at lower temperatures. Most substances expand when heated, with gases increasing in volume the most; the thermal expansion of liquid mercury allowed it to be the 'active ingredient' in most seventeenth to twentieth century thermometers. The equilibrium molecular composition of a given mixture of atoms often depends on temperature (as well as on pressure), and the time required for a mixture to reach chemical equilibrium generally decreases rapidly as temperature increases. Gradients in temperature and pressure are responsible for atmospheric winds (and, on Earth, ocean currents) as well as convective motions that can mix fluid material in planetary atmospheres and interiors. Earth's solid crust is dragged along by convective currents in the mantle, leading to continental drift. Temperature can even affect the orbital trajectory of a body, as we have seen in our discussion of the Yarkovski effect (Section 2.7.3).

Temperature, T, is a measure of the random kinetic energy of molecules, atoms, ions, etc. The energy, E, of a perfect gas is given by:

$$E = \frac{3}{2}NkT, \tag{3.1}$$

where N is the number of particles per unit volume, and k is *Boltzmann's constant*. The temperature of a (given region of a) body is determined by a combination of processes; solar radiation is the primary energy source for most planetary bodies, and reradiation to space is the primary loss mechanism. In this chapter, we summarize the mechanisms for solar heating and energy transport. We then use this background in our discussions of planetary atmospheres, surfaces and interiors in subsequent chapters.

3.1 Energy Balance and Temperature

Planetary bodies are heated primarily by absorbing radiation from the Sun, while they lose energy via radiation to space. Naturally, a point on the surface of a body is illuminated by the Sun during the day, while it radiates both day and night. The amount of energy incident per unit area depends both on its distance from the Sun and the local elevation angle of the Sun. As a consequence, most locales are coldest just before sunrise and hottest a little after local noon, while the polar regions are colder than the equator for bodies with obliquities $\lesssim 50$–$60°$.

Over the long term, most planetary bodies radiate almost the same amount of energy to space as they absorb from sunlight; were this not the case, planets would heat up or cool off. (The giant planets Jupiter, Saturn and Neptune are exceptions to this rule. These bodies radiate significantly more energy than they absorb, because their interiors are cooling or becoming more centrally condensed.) Although long-term global equilibrium is the norm, spatial and temporal fluctuations can be large. Energy can be stored from day to night, perihelion to aphelion or summer to winter, and can be transported from one location on a planet to another. We begin our discussion with the fundamental laws of radiation in Section 3.1.1 and factors affecting global energy balance in Section 3.1.2.

3.1.1 Thermal (Blackbody) Radiation

Electromagnetic radiation consists of photons at many wavelengths (Fig. 3.1). The frequency, ν, of an electro-

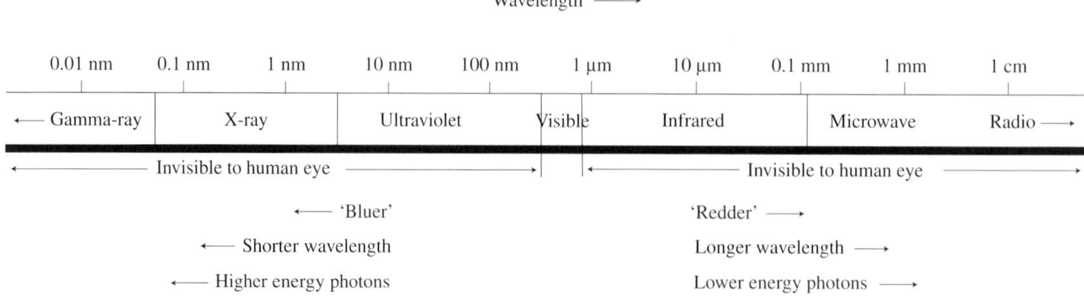

FIGURE 3.1 The electromagnetic spectrum. (Adapted from Hartmann 1989)

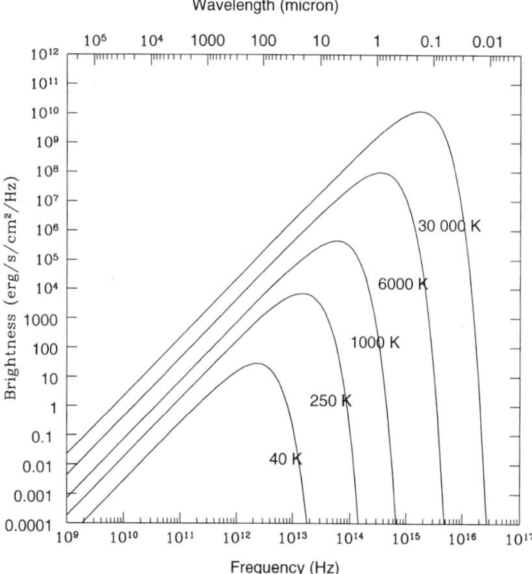

FIGURE 3.2 Blackbody radiation curves, $B_\nu(T)$, at various temperatures ranging from 40 K up to 30 000 K. The 6000 K curve is representative of the solar spectrum (cf. eq. 3.5).

magnetic wave propagating in a vacuum is related to its wavelength, λ, by:

$$\lambda\nu = c, \tag{3.2}$$

where c is the *speed of light in a vacuum*, 2.998×10^{10} cm s^{-1}.

Most objects emit a continuous spectrum of electromagnetic radiation. This *thermal emission* is well approximated by the theory of 'blackbody' radiation. A *blackbody* is defined as an object which absorbs all radiation that falls on it, at all frequencies and all angles of incidence; i.e., none of the radiation is reflected or scattered. A body's capacity to emit radiation is the same as its capability of absorbing radiation at the same frequency. The radiation emitted by a blackbody is described by *Planck's radiation law*:

$$B_\nu(T) = \frac{2h\nu^3}{c^2}\frac{1}{e^{h\nu/(kT)} - 1}, \tag{3.3}$$

where $B_\nu(T)$ is the *specific intensity* or *brightness* (erg cm^{-2} Hz^{-1} s^{-1} ster^{-1}), and h is Planck's constant. Figure 3.2 shows a graph of brightness as a function of frequency for bodies with temperatures between 40 and 30 000 K. Note that the brightness curve for a body like our Sun, with a surface temperature $\sim$5700 K, peaks at optical wavelengths, while those of the planets ($\sim$ 40–700 K) peak at infrared wavelengths. The brightness of most Solar System objects near their spectral peaks can be approximated quite well by blackbody curves.

Two limits of Planck's radiation law can be derived:

(1) *The Rayleigh–Jeans Law*: When $h\nu \ll kT$ (i.e., at radio wavelengths for temperatures typical of planetary bodies), the term $(e^{h\nu/(kT)} - 1) \approx h\nu/(kT)$, and equation (3.3) can be approximated by:

$$B_\nu(T) \approx \frac{2\nu^2}{c^2}kT. \tag{3.4a}$$

(2) *The Wien Law*: When $h\nu \gg kT$:

$$B_\nu(T) \approx \frac{2h\nu^3}{c^2}e^{-h\nu/(kT)}. \tag{3.4b}$$

Equations (3.4) are simpler than equation (3.3), and thus they can be quite useful in the regimes in which they are applicable.

The frequency, ν_{max}, at which the peak in the brightness $B_\nu(T)$ occurs, can be determined from the derivative of equation (3.3): $\partial B_\nu/\partial\nu = 0$. The result is known as the *Wien displacement law*:

$$\nu_{max} = 5.88 \times 10^{10} T, \tag{3.5a}$$

with ν_{max} in Hz. Similarly, the blackbody spectral peak in wavelength can be found by setting $\partial B_\lambda/\partial\lambda = 0$:

$$\lambda_{max} = \frac{0.29}{T}, \tag{3.5b}$$

with λ_{max} in cm (Problem 3.1). Note that $\lambda_{max} = 0.57\ c/\nu_{max}$, i.e., the brightness peak measured in terms of wavelength is blueward of the brightness peak measured in terms of frequency, due to the fact that $d\lambda \neq d\nu$.

The *flux density*, $\mathcal{F}_\nu$ (whose value is often quoted in erg s^{-1} cm^{-2} Hz^{-1} or in Jy[1]), of radiation from an object is given by (cf. Section 3.2.2.1):

$$\mathcal{F}_\nu = \Omega_s B_\nu(T), \tag{3.6}$$

where Ω_s is the solid angle into which the radiation is emitted. Just above the 'surface' of a source of brightness B_ν, the flux density is equal to (Problem 3.2):

$$\mathcal{F}_\nu = \pi B_\nu(T). \tag{3.7}$$

The *flux*, $\mathcal{F}$, is defined as the flux density integrated over all frequencies:

$$\mathcal{F} \equiv \int \mathcal{F}_\nu d\nu = \pi \int B_\nu(T) d\nu \ = \sigma T^4, \tag{3.8}$$

where σ is the *Stefan–Boltzmann constant*. This relationship is known as the *Stefan–Boltzmann law*.

3.1.2 Temperature

One can determine the temperature of a blackbody using Planck's radiation law by measuring a small part of the object's radiation (Planck) curve. This is usually not practical, since most bodies are not perfect blackbodies, but exhibit spectral features which complicate temperature measurements. It is common to relate the observed flux density, $\mathcal{F}_\nu$, to the *brightness temperature*, T_b, which is the temperature of a blackbody that has the same brightness at this particular frequency (i.e., replace T in eq. (3.3) by T_b). Conversely, if the total flux integrated over all frequencies of a body can be determined, the temperature that corresponds to a blackbody emitting the same amount of energy or flux $\mathcal{F}$ is referred to as the *effective temperature*, T_e:

$$\mathcal{F} = \sigma T_e^4. \tag{3.9}$$

The frequency range at which the object emits most of its radiation can be estimated via Wien's displacement law (eq. 3.5). This is typically at mid-infrared wavelengths (10–20 μm) for objects with temperatures of 150–300 K (inner Solar System), and far-infrared wavelengths (~ 60 μm) for bodies in the outer Solar System.

[1] 1 Jy ≡ 10^{-23} erg s^{-1} cm^{-2} Hz^{-1}

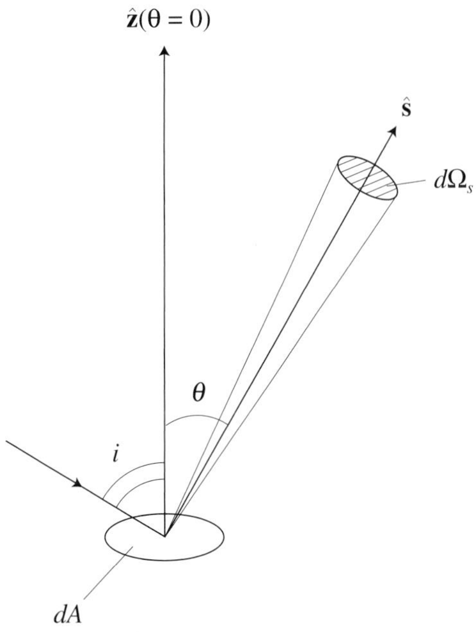

FIGURE 3.3 Sketch of the geometry of a surface element dA: $\hat{\mathbf{z}}$ is the normal to the surface, $\hat{\mathbf{s}}$ is a ray along the line of sight, and θ is the angle the ray makes with the normal to the surface.

3.1.2.1 *Bond Albedo, Geometric Albedo and Emissivity*

When an object is illuminated by the Sun, it reflects part of the energy back into space (which makes the object visible to us), while the remaining energy is absorbed. In principle, one can determine how much of the incident radiation is reflected into space at each frequency; the ratio between incident and reflected + scattered energy is called the *monochromatic albedo*, A_ν. Integrated over frequency, the ratio of the total radiation reflected or scattered by the object to the total incident light from the Sun is called the *Bond albedo*, A_b. The energy or flux absorbed by the object determines its temperature, as discussed further in Section 3.1.2.2. With regard to the albedo, it is important to consider how a unit surface element scatters light. The Sun's light is scattered off a planet and received by a telescope. The three angles of relevance are: the angle, i, that incident light makes with the normal to the planet's surface; the angle, θ, that the reflected ray received at the telescope (i.e., the ray along the line of sight) makes with the normal to the surface (Fig. 3.3); the *phase angle* or angle of reflectance, ϕ (Fig. 3.4).

The *phase integral*, q_{ph}, contains the phase dependence of the scattering:

$$q_{ph} \equiv 2 \int_0^\pi \frac{\mathcal{F}(\phi)}{\mathcal{F}(\phi = 0)} \sin\phi \, d\phi. \tag{3.10}$$

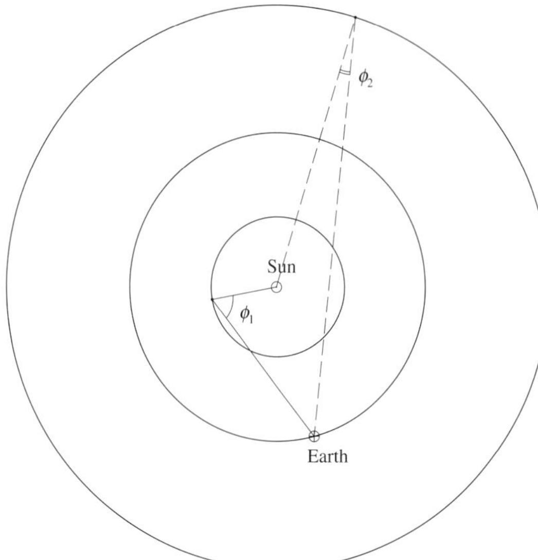

FIGURE 3.4 Scattering of light by a body that is illuminated by the Sun, with radiation received on Earth. For purely backscattered radiation, the scattering phase angle $\phi = 0$; in the case of forward scattered light, $\phi = 180°$. Two planets are indicated: one inside Earth's orbit, with phase angle ϕ_1, and one outside Earth's orbit with phase angle ϕ_2.

The phase integral can be measured from Earth for planets with heliocentric distances less than 1 AU (Mercury, Venus) as well as for the Moon, since the angle of reflectance, ϕ, varies between 0° and 180°. The outer planets are observed from Earth at phase angles close to 0°. Additional information on the phase integral can be recovered from Earth by using center-to-limb data, but only with help of spacecraft data can the full phase integral be determined.

We define the Bond albedo:

$$A_b \equiv A_0 q_{ph}, \tag{3.11}$$

with A_0 the *geometric albedo* or head-on reflectance:

$$A_0 = \frac{r_{AU}^2 \mathcal{F}(\phi = 0)}{\mathcal{F}_\odot}, \tag{3.12}$$

where $\mathcal{F}(\phi = 0)$ is the flux reflected from the body at phase angle $\phi = 0$. The heliocentric distance, r_{AU}, is expressed in AU, and $\mathcal{F}_\odot$, the solar constant, is defined as the solar flux at 1 AU:

$$\mathcal{F}_\odot \equiv \frac{L_\odot}{4\pi (r_{1AU})^2} = 1.37 \times 10^6 \text{ erg cm}^{-2} \text{ s}^{-1}, \tag{3.13}$$

with $L_\odot$ the solar luminosity. The combination $(\mathcal{F}_\odot / r_{AU}^2)$ is equal to the incident solar flux at heliocentric distance r_{AU}.

The geometric albedo can be thought of as the amount of radiation reflected from a body relative to that from a flat *Lambertian surface*, which is a diffuse perfect reflector at all wavelengths. Usually one determines a quantity referred to as $I/\mathcal{F}$ from planetary observations, where I is the reflected intensity at frequency ν, and $\pi\mathcal{F}$ is the incident solar flux density at the planet at the frequency of observation. By this definition, $I/\mathcal{F} = 1$ for a flat Lambertian surface when viewed at normal incidence, and $I/\mathcal{F}$ thus equals the geometric albedo at frequency ν when observed at a phase angle $\phi = 0$.

The reflectivity, A_ν, and emissivity, ϵ_ν, at frequency ν of a smooth, nonscattering sphere are complementary under the same viewing conditions:

$$(1 - A_\nu) = \epsilon_\nu. \tag{3.14}$$

If scattering is present, the sum of the reflectivity and emissivity remains unity when averaged over 4π steradians (conservation of energy), but not necessarily when viewed from a specific angle.

3.1.2.2 *Equilibrium Temperature*

Assuming the incoming solar radiation (insolation), $\mathcal{F}_{in}$, is balanced, on average, by reradiation outwards, $\mathcal{F}_{out}$, one can calculate the temperature of the object. This temperature is referred to as the *equilibrium temperature*. If indeed the temperature of the body is completely determined by the incident solar flux, the equilibrium temperature equals the effective temperature. Any discrepancies between the two numbers contains valuable information on the object. For example, the effective temperatures of Jupiter, Saturn and Neptune exceed the equilibrium temperature, which implies that these bodies possess internal heat sources (Sections 4.2 and 6.1.5). Venus's surface temperature is far higher than the equilibrium temperature of the planet indicates, a consequence of a strong greenhouse effect in this planet's atmosphere (Section 4.2). The effective temperature of Venus, which is dominated by radiation emitted from that planet's cool upper atmosphere, is equal to the equilibrium temperature, implying that Venus has a negligible internal heat source. We next show the average effect of insolation and reradiation for a rapidly rotating, spherical object of radius R, using approximate equations. At the end of this section, we provide the more precise equations that can be used for more detailed modeling.

The sunlit hemisphere of a (spherical) body receives radiation from the Sun:

$$\mathcal{F}_{in} = (1 - A_b)\frac{L_\odot}{4\pi r_\odot^2}\pi R^2, \tag{3.15}$$

with πR^2 the projected surface area for intercepting solar photons. A rapidly rotating planet reradiates energy from its entire surface (i.e., an area of $4\pi R^2$):

$$\mathcal{F}_{out} = 4\pi R^2 \epsilon \sigma T^4. \quad (3.16)$$

Note that the incoming solar radiation is primarily at optical wavelengths (Fig. 3.2), while thermal emission from planets is radiated primarily at infrared wavelengths. The emissivity, $\epsilon = \int \epsilon_\nu d\nu$, is usually close to 0.9 at infrared wavelengths, but can differ substantially from unity at radio wavelengths. From a balance between insolation and reradiation, $\mathcal{F}_{in} = \mathcal{F}_{out}$, one can calculate the equilibrium temperature, T_{eq}:

$$T_{eq} = \left(\frac{\mathcal{F}_\odot}{r_{AU}^2} \frac{(1 - A_b)}{4\epsilon\sigma} \right)^{1/4}. \quad (3.17)$$

Even though this simple derivation has many shortcomings, the disk-averaged equilibrium temperature from equation (3.17) gives useful information on the temperature well below a planetary surface. If ϵ is close to unity, it corresponds well with the actual (physical) temperature of subsurface layers that are below the depth where diurnal (day/night) and seasonal temperature variations are important, typically a meter or more below the surface. These layers can be probed at radio wavelengths, and the brightness temperature observed at these long wavelengths can be compared directly with the equilibrium temperature. In the derivation of equation (3.17), we omitted latitudinal and longitudinal effects of the insolation pattern. The magnitude of these effects depends on the planet's rotation rate, obliquity and orbit. Latitudinal and longitudinal effects are large, for example, on airless planets that rotate slowly, have small axial obliquities and/or travel on very eccentric orbits about the Sun.

In another limit for equilibrium temperatures, one can consider the subsolar point of a slowly rotating body. In this case, the surface areas πR^2 in equation (3.15) and $4\pi R^2$ in equation (3.16) should both be replaced by a unit area dA. It follows that the equilibrium temperature at the subsolar point is $\sqrt{2}$ times the disk-average equilibrium temperature for a rapidly rotating body. The subsolar temperature calculated in this way corresponds well with the measured subsolar surface temperature of airless bodies.

For more detailed modeling, consider the solar flux incident per unit surface area dA at location (α, δ) on a planet's surface:

$$\frac{\mathcal{F}_{in}}{dA} = \int_0^\infty (1 - A_\nu) \frac{L_\odot}{4\pi r^2} \cos[\alpha_\odot(t) - \alpha] \cos[\delta_\odot(t) - \delta] \, d\nu, \quad (3.18a)$$

where $(\alpha_\odot, \delta_\odot)$ are the coordinates of the Sun, and A_ν is the reflectivity at frequency ν. A surface element dA emits radiation according to the Stefan–Boltzmann law:

$$\mathcal{F}_{out} = \epsilon \sigma T^4 \, dA. \quad (3.18b)$$

The emissivity, ϵ_ν, depends upon wavelength; it usually varies from a few tenths up to unity for bodies that are large compared to the wavelength considered. Objects much smaller than the wavelength ($R \lesssim 0.1\lambda$) do not radiate efficiently.

3.2 Energy Transport

The temperature structure in a body is governed by the efficiency of energy transport. There are three principal mechanisms to transport energy: *conduction*, *radiation* and *convection*. Usually, one of these three mechanisms dominates and determines the thermal profile in any given region. Energy transport in a solid is usually dominated by conduction, while radiation typically dominates in space and tenuous gases. Convection is important in fluids and dense gases. All three mechanisms are experienced in everyday life, e.g., when boiling water on a stove: the entire pan, including the handle (in particular if metal), is heated by conduction, while the water in the pan is primarily heated through 'convection', up and down motions in the water. These motions are visible in the form of bubbles of vaporized water, which rise upwards because they are lighter than the surrounding water. Heat is transported from the Sun to planets, moons, etc. via radiation. Although it is obvious in these examples which transport mechanism dominates, this is not always easy to determine: in some parts of a planet's interior energy transport is dominated by convective motions, while in other parts conduction is by far the most efficient. In a planet's atmosphere we typically encounter all three mechanisms, though a particular mechanism is usually dominant in a certain altitude range. Usually all the energy transported to and from planetary bodies occurs via radiation. (Jupiter's moon Io is an exception to this rule; Io receives a substantial amount of energy from Jupiter via gravitational torques, cf. Sections 2.6 and 5.5.5.1.) In this section, we discuss all three principal mechanisms for energy transport and derive equations for the thermal profile in a surface or an atmosphere under the assumption that a particular mechanism to transport heat is dominant.

3.2.1 Conduction

Conduction is the transfer of energy by collisions between particles. Conduction provides the primary mechanism for heat transport near a planet's surface. It is also important in the very tenuous upper part of an atmosphere (the upper thermosphere), where the mean free path is very long, so that atoms exchange locations very rapidly and the conductivity is therefore large.

Sunlight heats a planet's surface during the day, and the heat is transported downwards from the surface mainly by conduction. The rate of flow of heat, the *heat flux*, $\mathbf{Q}$ (erg s^{-1} cm^{-2}), is determined by the *temperature gradient*, ∇T, and the *thermal conductivity*, K_T:

$$\mathbf{Q} = -K_T \nabla T. \tag{3.19}$$

The thermal conductivity is a measure of the material's physical ability to conduct heat. The *thermal heat capacity* or *molecular heat*, C_P, is defined as the amount of heat, Q, necessary to raise the temperature of one mole of matter by one degree Kelvin without changing the pressure (C_P: $dP = 0$) or volume (C_V: $dV = 0$). The *specific heat* is the amount of energy necessary to raise the temperature of one gram of material or gas by one degree Kelvin without changing the pressure or volume. This quantity is usually indicated by c_P and c_V, respectively.

$$m_{gm} c_P \equiv C_P \equiv \left(\frac{dQ}{dT}\right)_P, \tag{3.20a}$$

$$m_{gm} c_V \equiv C_V \equiv \left(\frac{dQ}{dT}\right)_V, \tag{3.20b}$$

where m_{gm} is a gram-mole[2]. The rate at which the subsurface layers gain heat is given by:

$$\rho c_P \frac{\partial T}{\partial t} = \frac{\partial Q}{\partial z}, \tag{3.21}$$

where ρ is the density of the material. Equations (3.19) and (3.21) together lead to the *thermal diffusion equation*:

$$k_d \frac{\partial^2 T}{\partial z^2} = \frac{\partial T}{\partial t}, \tag{3.22}$$

with the *thermal diffusivity*:

$$k_d \equiv \frac{K_T}{\rho c_P}. \tag{3.23}$$

[2] A gram-mole is the mass of a mole of molecules in units of grams. A mole contains $N_A \equiv 6.022 \times 10^{23}$ molecules (N_A is Avogadro's number), and its mass is numerically equal to its weight in amu. Thus a mole of carbon atoms has a mass of 12 g.

The amplitude and phase of the diurnal temperature variations, and the temperature gradient with depth in the crust, are largely determined by the *thermal inertia*,

$$\gamma_T \equiv \sqrt{K_T \rho c_P}, \tag{3.24}$$

which measures the ability of the surface to store energy, and the *thermal skin depth* of the material,

$$L_T \equiv \sqrt{\frac{2K_T}{\omega_{rot} \rho c_P}}. \tag{3.25}$$

The amplitude of diurnal temperature variations is largest at the surface, and decreases exponentially into the subsurface, with an *e*-folding scale length equal to the thermal skin depth. Moreover, since it takes time for the heat to be carried downwards, there is a phase lag in the diurnal heating pattern of the subsurface layers. The peak temperature at the surface is reached at noon, or soon thereafter, while the subsurface layers reach their peak temperature later in the afternoon. At night the surface cools off, and becomes cooler than the subsurface layers. Heat is then transported upwards from below. However, since the conductivity is a function of the temperature, at night the surface acts like an 'insulator', so that the subsurface heat is essentially trapped. Examples of this effect are illustrated in Figure 3.5, which shows the temperature as a function of depth and local time for Mercury. Figure 3.6 shows the surface temperature as a function of local time for a rocky body that orbits the Sun in a circular orbit at a heliocentric distance of 0.4 AU. Curves for various rotation periods from 2 hours up to 10^6 years are shown. This figure clearly demonstrates that the peak temperature is primarily determined by the heliocentric distance, while the night-side temperature also depends on the planet's rotation rate (and thermal inertia). Note the time delay in peak temperature from local noon when $P_{rot} = 2$ hours. This delay is caused by the rapid rotation rate combined with the relatively high thermal inertia.

Note that the thermal inertia (eq. 3.24) depends upon the product $K_T c_P$, and the thermal skin depth (eq. 3.25) depends upon the ratio K_T / c_P, as well as the angular rotation of the object, ω_{rot}. A typical value for the thermal skin depth on Mars and the Moon is ~4 cm, and on Mercury ~15 cm. When the thermal conductivity is low, the amplitude of the temperature wave is large, but it does not penetrate deeply into the crust. If the thermal conductivity is high, temperature variations are smaller near the surface, but penetrate to greater depths in the subsurface layers. Although the amplitude of diurnal variations in the subsurface layers may be small, seasonal effects can still be large on planets with significant axial obliquities,

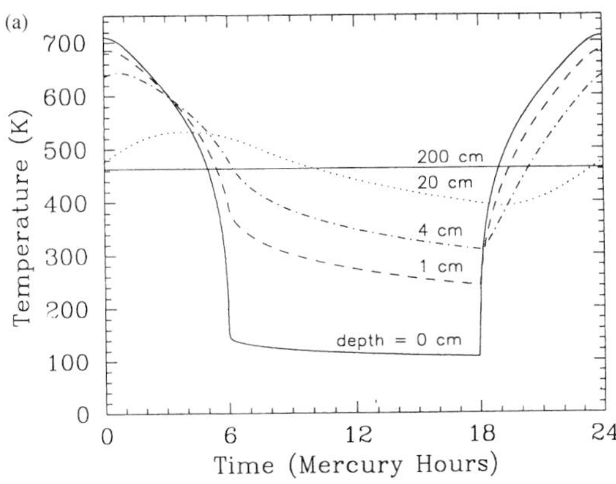

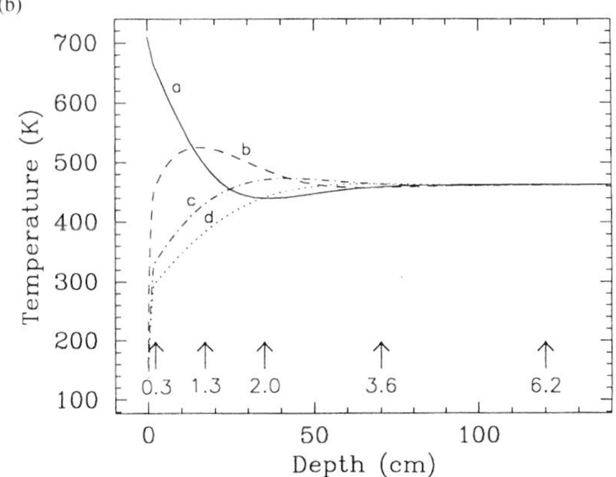

FIGURE 3.5 (a) The equatorial temperature structure of Mercury as a function of time after local noon, at various depths, for the subsolar longitude at perihelion (i.e., a 'hot' longitude). Different depths are probed at different radio wavelengths, as shown by the arrows at the bottom of graph (b) (wavelengths are in cm). (Mitchell 1993) (b) Vertical temperature profiles of the same region of Mercury at four different local times of day: a, noon; b, dusk; c, midnight and d, dawn. (Mitchell 1993)

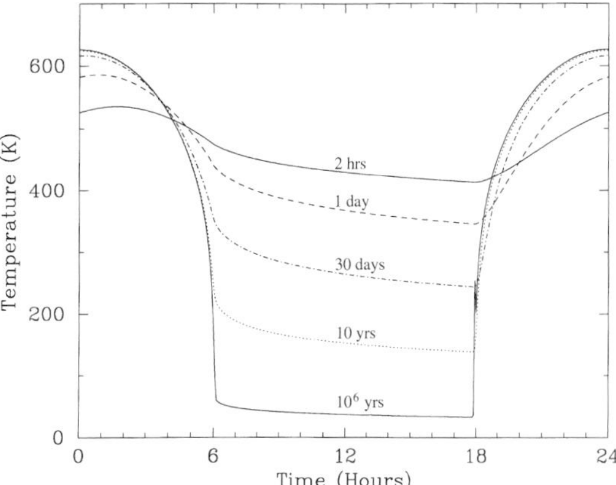

FIGURE 3.6 The surface temperature of a (typical) rocky body which orbits the Sun in a circular orbit at a heliocentric distance of 0.4 AU. Curves are shown for bodies with rotation periods (defined with respect to the Sun) of 2 hours, 1 day, 30 days, 10 years and 10^6 years. The calculations were performed assuming the following parameters: $A_b = 0.1$, $\epsilon_{ir} = 0.9$, $\rho = 2.8$ g cm^{-3}, $\gamma_T = 36.9$ mcal cm^{-2} K^{-1} s$^{-1/2}$, $\epsilon_r = 6.5$, and tan $\Delta = 0.02$. ϵ_r and tan Δ are defined in Section 3.2.2.5. (Courtesy: D.L. Mitchell)

such as Earth and Mars. Mercury has an interesting variation in temperature with longitude well below the depth where diurnal heating is important. Due to the 3/2 resonance between Mercury's rotation and orbital periods, in combination with Mercury's large orbital eccentricity, the average diurnal insolation varies significantly with longitude (as well as with latitude). Regions along Mercury's equator near longitudes $\lambda = 0°$ and $180°$ (the subsolar longitudes when the planet is at perihelion) receive on average approximately 2.5 times as much sunlight as at longitudes $90°$ and $270°$. The night time surface temperature is approximately 100 K, independent of longitude, but the peak (noon) surface temperature near Mercury's equator varies between 700 K at $\lambda = 0°$ and $180°$ to 570 K at $\lambda = 90°$ and $270°$. This nonuniform heating pattern produces longitudinal variations in the subsurface temperature, such that the temperature at depth, where the diurnal variations in temperature can be ignored, is higher at longitudes $0°$ and $180°$ ($T \sim 470$ K) than at $90°$ and $270°$ ($T \sim 350$ K). This effect can best be observed at radio wavelengths longwards of ~ 3 cm, where the influence of the diurnal heating cycle is minimal (Fig. 5.42a).

3.2.2 Radiation

The transport of heat in a planetary atmosphere is largely dominated by radiation in regions where the optical depth of the gas is not too large nor too small. This is usually the case in a planet's upper troposphere and stratosphere. The radiation efficiency depends critically upon the emission and absorption properties of the material involved. To develop equations of radiative transfer, one needs to be familiar with atomic structure and energy transitions in atoms and molecules, and with the radiation 'vocabulary' like specific intensity, flux density and mean intensity. In Section 3.2.2.1, we summarize the basic definitions needed to develop equations of radiative transfer. Sections 3.2.2.2 and 3.2.2.3 review the essential background of atomic and molecular structure, energy transitions and the Einstein A and B coefficients, which express the probability per unit time that emission or absorption of a photon takes place. The equations of radiative transfer in an atmosphere and surface are derived in Sections 3.2.2.4 and 3.2.2.5, respectively. The thermal structure in an atmosphere that is in radiative equilibrium is derived in Section 3.3.

3.2.2.1 *Definitions*

The energy and momentum of a photon are given by:

$$E = h\nu, \tag{3.26a}$$

$$\mathbf{p} = \frac{E}{c}\hat{\mathbf{s}}, \tag{3.26b}$$

where h is Planck's constant, c the speed of light, ν the frequency, and $\hat{\mathbf{s}}$ is a unit vector pointing in the direction of propagation.

The amount of energy crossing a differential element of area $\hat{\mathbf{s}} \cdot d\mathbf{A} = dA \cos\theta$, where θ is the angle between the surface normal and the direction of propagation, into a solid angle $d\Omega_s$ in time dt and frequency range $d\nu$, is given by (Fig. 3.3):

$$dE = I_\nu \cos\theta \; dt \; dA \; d\Omega_s \; d\nu, \tag{3.27}$$

where I_ν is the *specific intensity*, which has the dimensions of erg s^{-1} cm^{-2} str^{-1} Hz^{-1}. The specific intensity of radiation at frequency ν emitted by a blackbody is:

$$I_\nu = B_\nu(T). \tag{3.28}$$

The solid angle, $d\Omega_s$ (steradians = str, i.e., radians^2) is defined such that, integrated over a sphere, it is:

$$\oint d\Omega_s = \int_0^{2\pi} \int_0^{\pi} \sin\theta \; d\theta \; d\phi = 4\pi \text{ str}. \tag{3.29}$$

The *mean intensity*, J_ν, or the zeroth *moment* of the radiation field is equal to (using the conversion $\mu_\theta \equiv \cos\theta$):

$$J_\nu \equiv \frac{\oint I_\nu \, d\Omega_s}{\oint d\Omega_s} = \frac{1}{2}\int_{-1}^{1} I_\nu \; d\mu_\theta. \tag{3.30}$$

The *energy density*, u_ν, of the radiation is the amount of radiant energy per unit volume at frequency ν. Since, in vacuum, photons travel at the speed of light,

$$u_\nu = \frac{1}{c}\oint I_\nu \; d\Omega_s = \frac{4\pi}{c} J_\nu. \tag{3.31}$$

The net flux density, $\mathcal{F}_\nu$, in the direction z at frequency ν can be obtained by integrating over all solid angles (cf. 3.1.1):

$$\mathcal{F}_\nu = \oint I_\nu \cos\theta \; d\Omega_s. \tag{3.32a}$$

Note that for an isotropic radiation field the net flux density $\mathcal{F}_\nu = 0$, since $\oint \cos\theta \; d\Omega_s = 0$. Changing variables to $\mu_\theta \equiv \cos\theta$ and performing the integration over ϕ, equation (3.32a) becomes:

$$\mathcal{F}_\nu = 2\pi \int_{-1}^{+1} I_\nu \mu_\theta \, d\mu_\theta. \tag{3.32b}$$

Dividing equation (3.32b) by the solid angle results in the first moment of the intensity I_ν, or the *Eddington flux*:

$$\mathcal{H}_\nu \equiv \frac{1}{2}\int_{-1}^{+1} I_\nu \mu_\theta \, d\mu_\theta. \tag{3.32c}$$

The second moment of the radiation field is called the $\mathcal{K}$-integral:

$$\mathcal{K}_\nu \equiv \frac{1}{2}\int_{-1}^{+1} I_\nu \mu_\theta^2 \, d\mu_\theta. \tag{3.33a}$$

The $\mathcal{K}$-integral is related to the *radiation pressure*, p_r. The momentum flux, integrated over all frequencies along a ray's path in direction $\hat{\mathbf{s}}$, is equal to $d\mathcal{F}/c$. The radiation pressure is the component of momentum flux in the z direction:

$$p_r = \frac{\mathcal{F}\cos\theta}{c} = \frac{2\pi}{c}\int_{-1}^{+1} I\mu_\theta^2 \, d\mu_\theta = \frac{4\pi}{c}\mathcal{K}. \tag{3.33b}$$

In an isotropic radiation field, $I_\nu(\mu_\theta) \equiv I_\nu$, and the radiation pressure is equal to:

$$p_r = \frac{4\pi}{3c} I_\nu. \tag{3.33c}$$

3.2.2.2 *Atomic Structure and Line Transitions*

Emission and absorption of photons by atoms or molecules involve a change in the energy state of a particle. Each atom consists of a nucleus (protons plus neutrons) surrounded by a 'cloud' of electrons. In the semi-classical Bohr theory, the electrons orbit the nucleus such that the centripetal force is balanced by the Coulomb force:

$$\frac{m_e v^2}{r} = \frac{Zq^2}{r^2}, \tag{3.34}$$

where m_e and v are the mass and velocity of the electron, respectively, r the radius of the electron orbit (assumed circular), Z the atomic number and q the electric charge. Electrons are in orbits such that the angular momentum:

$$m_e v r = n\hbar, \tag{3.35a}$$

or the radius

$$r = \frac{n^2\hbar^2}{m_e Z q^2}, \tag{3.35b}$$

where n, an integer, is the *principal quantum number*, and $\hbar \equiv h/2\pi$. The radius of the lowest energy state ($n = 1$) for the hydrogen atom ($Z = 1$) is called the *Bohr radius*:

$r_{Bohr} = \hbar^2/(m_e q^2)$. The principal quantum number is the sum of the radial and azimuthal quantum numbers: $n = n_r + k$. These quantum numbers define the semimajor and semiminor axes, a and b, of an electron's orbit: $a/b = n/k$. The energy of orbit n is given by:

$$E_n = -\frac{Zq^2}{r} + \frac{Zq^2}{2r} \approx -\frac{\mathcal{R}Z^2}{n^2}, \tag{3.36}$$

where $\mathcal{R} \equiv \mu_r e^4/\hbar^2$, which is the *Rydberg constant* in the case of the hydrogen atom. The reduced mass, μ_r, was defined in equation (2.8a), where m_1 and m_2 in this case represent the mass of the electron and nucleus, respectively. The frequency of various transitions can be calculated using equations (3.26) and (3.36). An example of energy levels in the hydrogen atom is given in Figure 3.7. The transitions between the ground state and higher levels are called the Lyman series, where Ly α is the transition between level 1 and 2, Ly β between level 1 and 3, etc. The Balmer, Paschen, and Brackett series indicate transitions between levels 2, 3 and 4 with higher levels, respectively. If the electron is in energy state $n > \infty$, the atom is *ionized*. For hydrogen in the ground state, photons with energies $\geq$ 13.6 eV, or wavelengths shorter than 912 Å (the *Lyman limit*) may *photoionize* the atom (Problem 3.16).

The orbital angular momentum of an electron in an atom is $l\hbar$, with l an integer with a value $l = k - 1$, or $0, 1, \ldots, (n-1)$. The quantity l can only have $2l + 1$ quantized directions m_l. So, if $n = 1 \rightarrow l = 0$, $m_l = 0$. If $n = 2 \rightarrow l = 0$, $m_l = 0$, or $l = 1$ with $m_l = -1$, 0, or 1. The total orbital angular momentum of the electrons, i.e., the vectorial sum of all l, is denoted by $\mathbf{L}$; $|\mathbf{L}| = \sqrt{L(L+1)}\hbar$. Electrons also possess a spin angular momentum, $s\hbar$, with $s = 1/2$. The spinning electron acts like a bar magnet, where the magnetic polarity is opposite to its spin direction. The total spin angular momentum, $\mathbf{S}$, is the vectorial sum of the spin quantum numbers, s, of all electrons. Orbital and spin angular momentum interact to produce a total angular momentum, $\mathbf{J} = \mathbf{L} + \mathbf{S}$. Since $\mathbf{L}$ and $\mathbf{S}$ are oriented in particular directions with respect to each other, only discrete values of $\mathbf{J}$ are possible. The presence of magnetic and/or electric fields may cause a further splitting of $\mathbf{J}$ into $2J + 1$ values (*Zeeman* and *Stark effects*; see e.g., Herzberg, 1944). *Pauli's exclusion principle* states that each electron orbit is specified by a unique set of quantum numbers, so that the total number of sublevels for energy level n is given by:

$$2n^2 \equiv g_n, \tag{3.37}$$

where g_n is defined as the *statistical weight* or *degeneracy* of level n.

The multiplet lines of atomic spectra often exhibit a further splitting, a *hyperfine structure*. Such structure is induced by properties of the atomic nucleus. Hyperfine structure is induced by two properties: The *isotope effect*, caused by the fact that different isotopes of the atom have different masses, resulting from differences in the number of neutrons in the nucleus, and by the nuclear spin, similar to the electron spin. Nuclear spin produces splittings in a single atom, and the atomic lines are produced by an appropriate electron transition so that the orientation of $\mathbf{J}$ changes relative to the nuclear spin. The most famous example of such a transition is the 21 cm line in atomic hydrogen, where the nuclear and electron spins are either aligned (higher energy state; i.e., the magnetic polarity of the two is anti-parallel), or opposite to each other (lower energy state; magnetic polarity is parallel). Transitions are rare for a given atom, only once every 10^7 years, but the 21 cm line in interstellar space is easy to observe because atomic hydrogen is so abundant. For molecules the energy levels become even more numerous, since rotation and vibration of the nuclei with respect to each other lead to molecular lines. For a complete treatment of atomic and molecular structure the reader is referred to e.g., Herzberg (1944).

Transitions between energy levels result in the absorption or emission of a photon with an energy ΔE_{ul} equal to the difference in energy between the two levels u and l. However, transitions are only possible between certain levels: they follow specific selection rules (for more details on these rules see references given above). The energy difference between electron orbits and the frequency of the transition decrease with increasing n (Fig. 3.7). Whereas electronic transitions involving the ground state ($n = 1$) may be observed at ultraviolet or optical wavelengths, transitions at high n, (hyper)fine structure in atomic spectra and molecular rotation and rotation-vibration transitions occur at infrared or radio wavelengths, since spacing between energy levels is much smaller. Because each atom/molecule has its own unique set of energy transitions, one can use measurements of absorption/emission spectra to identify particular species in an atmosphere or surface. According to the *Heisenberg uncertainty principle*, a photon can be absorbed/emitted with an energy slightly different from ΔE_{ul}, which results in a finite width line profile, Φ_ν, discussed more fully in Section 4.3.2.

3.2.2.3 *Einstein A and B Coefficients*

The probability per unit time for emission or absorption of a photon can be expressed by the Einstein A and B coefficients. The coefficient A_{ul} is the probability per unit time for spontaneous emission, from the upper, u, to the

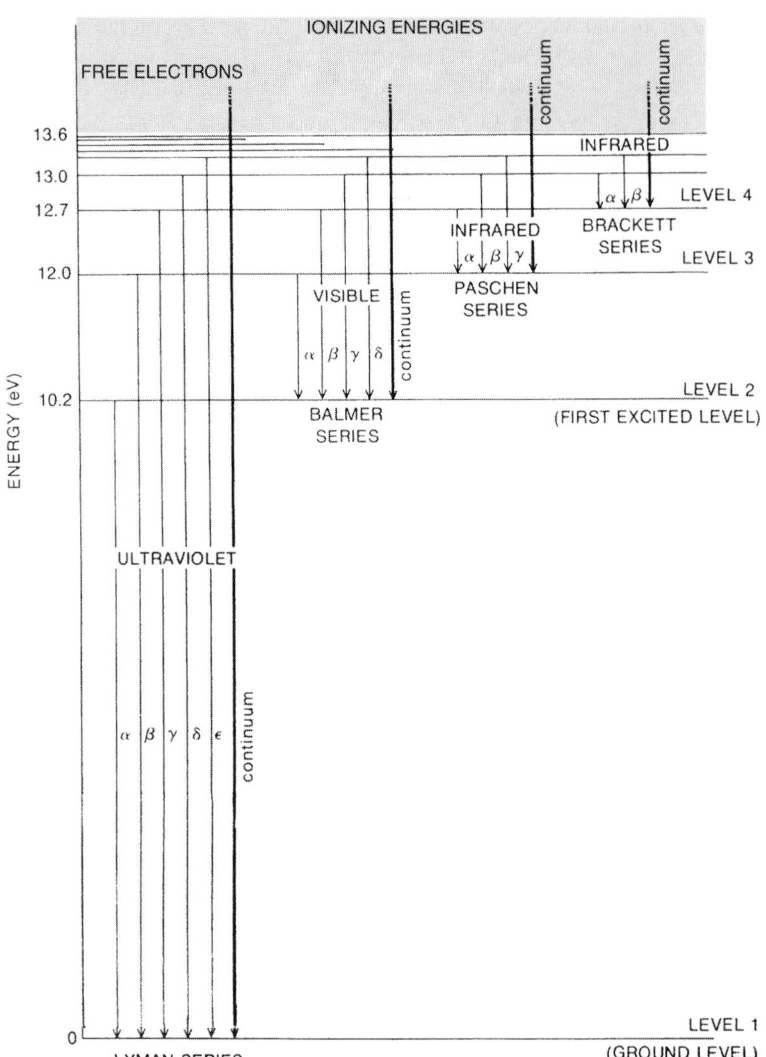

FIGURE 3.7 The energy levels of hydrogen and the series of transitions among the lowest of these energy levels. (Adapted from Pasachoff and Kutner 1978)

lower, l, energy level; $B_{lu}J_\nu$ is the probability per unit time for absorption, and $B_{ul}J_\nu$ the probability per unit time for stimulated emission. All probabilities are expressed as frequencies. Because of the finite width of the line profile, Φ_ν, the mean intensity, J_ν, for the probabilities given above should be integrated over the line profile: $\int J_\nu \Phi_\nu \, d\nu$.

In *thermodynamic equilibrium*, the following three equations are valid:

(1) The radiation field obeys:

$$I_\nu = J_\nu = B_\nu(T). \tag{3.38a}$$

(2) There is equilibrium between the rate of absorption and emission:

$$N_l B_{lu} J_\nu = N_u A_{ul} + N_u B_{ul} J_\nu. \tag{3.38b}$$

(3) The number density of atoms in energy state i, N_i, is determined by the temperature of the gas:

$$N_i \propto g_i e^{-E_i/kT}, \tag{3.38c}$$

with g_i the statistical weight of level i.

The ratio N_l/N_u is given by *Boltzmann's equation*:

$$\frac{N_l}{N_u} = \frac{g_l}{g_u} e^{\Delta E_{ul}/kT}, \tag{3.39a}$$

where the energy difference between the upper and lower energy levels is given by $\Delta E_{ul} = E_u - E_l$. Note that at low temperatures, most atoms are in the ground state, while higher energy levels are populated at higher temperatures. A more general form of the Boltzmann law is:

$$N_i = \frac{N g_i}{Z_p} e^{-E_i/kT}, \tag{3.39b}$$

with N the total number of atoms and Z_p the *partition function*:

$$Z_p = \sum_j g_j e^{-E_j/kT}. \tag{3.39c}$$

To validate Planck's radiation law (eq. 3.3) with the above relations, the *Einstein relations* become:

$$g_l B_{lu} = g_u B_{ul} \tag{3.40a}$$

$$A_{ul} = \frac{2h\nu^3}{c^2} B_{ul}. \tag{3.40b}$$

These relations do not depend on temperature, so are valid whether or not the medium is in thermodynamic equilibrium.

The mass absorption and emission coefficients, κ_ν and j_ν, (erg g^{-1} s^{-1} str^{-1} Hz^{-1}) become:

$$\kappa_\nu \rho = \frac{\Delta E_{ul}}{4\pi}(N_l B_{lu} - N_u B_{ul})\Phi_\nu \tag{3.41a}$$

$$j_\nu \rho = \frac{\Delta E_{ul}}{4\pi} N_u A_{ul} \Phi_\nu, \tag{3.41b}$$

where Φ_ν is the line profile. With help of quantum mechanics it can be shown that:

$$\int \kappa_\nu d\nu = \int j_\nu d\nu = \frac{\pi q^2}{m_e c} f_{osc}, \tag{3.42}$$

with f_{osc} the oscillator strength of the transition. Note that we have added stimulated emission as a negative absorption coefficient in equation (3.41a). For more details the reader is referred to e.g., Rybicki and Lightman (1979).

3.2.2.4 *Equation of Radiative Transfer in an Atmosphere*

When the primary mechanism of energy transport in an atmosphere is the absorption and re-emission of photons, the temperature–pressure profile in the atmosphere is governed by the equations of radiative energy transport. The change in intensity, dI_ν, due to absorption and emission within a cloud of gas is equal to the difference in intensity between emitted and absorbed radiation:

$$dI_\nu = j_\nu \rho\, ds - I_\nu \alpha_\nu \rho\, ds. \tag{3.43}$$

In equation (3.43), j_ν is the emission coefficient due to scattering and/or thermal excitation: $j_\nu = j_\nu$(scattering)+ j_ν(thermal excitation). The quantity α_ν is the mass extinction coefficient. Absorption (including stimulated emission) and scattering both contribute to the extinction: $\alpha_\nu = \kappa_\nu + \sigma_\nu$, where κ_ν and σ_ν are the mass absorption and mass scattering coefficients, respectively.

When $\hat{\mathbf{z}}$ is the coordinate in the direction of the normal to the planet's surface, and θ the angle between $\hat{\mathbf{s}}$ and $\hat{\mathbf{z}}$ (Fig. 3.3), $ds = \sec\theta\, dz$, with $\mu_\theta = \cos\theta$, and equation (3.43) becomes:

$$\mu_\theta \frac{dI_\nu}{d\tau_\nu} = -I_\nu + S_\nu, \tag{3.44}$$

where the *optical depth*, τ_ν, is defined as the integral along the vertical of the extinction coefficient:

$$\tau_\nu \equiv \int_{z_1}^{z_2} \alpha_\nu(z)\rho(z)\, dz. \tag{3.45}$$

The *source function*, S_ν, is defined as:

$$S_\nu \equiv \frac{j_\nu}{\alpha_\nu}. \tag{3.46}$$

The formal solution to equation (3.44) is (for $\mu_\theta = 1$):

$$I_\nu(\tau_\nu) = I_\nu(0)e^{-\tau_\nu} + \int_0^{\tau_\nu} S_\nu(\tau'_\nu)e^{-(\tau_\nu - \tau'_\nu)}\, d\tau'_\nu. \tag{3.47}$$

For the more general case, the optical depth should be replaced by the slant optical depth, i.e., τ/μ_θ. If S_ν is known, equation (3.47) can be solved for the radiation field. In practice the situation is often more complicated because S_ν usually depends upon the intensity I_ν (e.g., through scattering) and/or the temperature of the medium, which may, in part, be determined by I_ν. If S_ν does not vary with optical depth, equation (3.47) reduces to:

$$I_\nu(\tau_\nu) = S_\nu + e^{-\tau_\nu}(I_\nu(0) - S_\nu). \tag{3.48}$$

If $\tau_\nu \gg 1$, then $I_\nu = S_\nu$; i.e., the intensity of the emission is completely determined by the source function. If $\tau_\nu \ll 1$, then $I_\nu \rightarrow I_\nu(0)$; i.e., the intensity of the radiation is defined by the incident radiation.

In the remainder of this subsection we examine the equation of radiative transfer, or, more explicitly, the source function, S_ν, for four 'classic' cases. These examples help to better understand the theory of radiative transfer.

(1) Consider a nonemitting cloud of gas along the line of sight, thus $j_\nu = 0$. Suppose there is a source of radiation behind the cloud. The incident light $I_\nu(0)$ is reduced in intensity according to equation (3.48), and the resulting observed intensity becomes:

$$I_\nu(\tau_\nu) = I_\nu(0)e^{-\tau_\nu}. \tag{3.49}$$

This relation is called *Lambert's exponential absorption law*. If the gas cloud is optically thin ($\tau_\nu \ll 1$), equation (3.49) can be approximated by: $I_\nu(\tau_\nu) = I_\nu(0)(1 - \tau_\nu)$. If the cloud is optically thick ($\tau_\nu \gg 1$), the radiation is reduced to near zero.

(2) Assume the material to be in *local thermodynamic equilibrium* (LTE), so the scattering coefficient $\sigma_\nu = 0$ and $\kappa_\nu = \alpha_\nu$. If the material is in equilibrium with the radiation field, the amount of energy emitted must be equal to the amount of energy absorbed, as described by *Kirchoff's law*:

$$j_\nu = \kappa_\nu B_\nu(T), \tag{3.50}$$

where Planck's function, $B_\nu(T)$, describes the radiation field in thermodynamic equilibrium. In this case, the source function is given by:

$$S_\nu = B_\nu(T). \tag{3.51}$$

The energy levels of the atoms/molecules are populated according to Boltzmann's equation (3.39), and the absorption coefficient becomes (Problem 3.17):

$$\kappa_\nu \rho = \frac{\Delta E_{ul}}{4\pi} N_l B_{lu} \left(1 - e^{-\Delta E_{ul}/kT}\right) \Phi_\nu. \tag{3.52}$$

The factor in parentheses is due to stimulated emission (eq. 3.41).

(3) Assume that j_ν is due to scattering only, and that we receive sunlight reflected towards us: $I_\nu(\cos\phi)$, with ϕ the scattering or phase angle, i.e., the angle between the scattered radiation (along the line of sight) and the incident radiation (see Fig. 3.4). In Earth-based observations, the phase angle ϕ for the giant planets is always close to zero (backscattered light), while $\phi = 180^\circ$ for forward scattered light. Scattering removes radiation from a particular direction and redirects or introduces it into another direction. If photons undergo only one encounter with a particle, the process is referred to as *single scattering*; multiple scattering refers to multiple encounters. The angular distribution of the scattered radiation is given by the *scattering phase function*, $\mathcal{P}(\cos\phi)$, which itself is normalized such that integrated over a sphere:

$$\frac{1}{4\pi}\int \mathcal{P}(\cos\phi)\, d\Omega_s = \frac{\sigma_\nu}{\alpha_\nu} \equiv \varpi_\nu. \tag{3.53}$$

The term ϖ_ν is referred to as the *albedo for single scattering*, and thus represents the fraction of radiation lost due to scattering. The single scattering albedo is equal to unity if the mass absorption coefficient, κ_ν, is equal to zero. The source function can be written:

$$S_\nu = \frac{1}{4\pi}\int I_\nu \mathcal{P}(\cos\phi)\, d\Omega_s. \tag{3.54}$$

Common scattering phase functions are:

- Isotropic scattering: $\mathcal{P}(\cos\phi) = \varpi_\nu$, in which case the source function becomes: $S_\nu = \varpi_\nu J_\nu$.
- Rayleigh scattering: $\mathcal{P}(\cos\phi) = \frac{3}{4}(1 + \cos^2\phi)$, which is representative for scattering of sunlight by air molecules.
- First-order anisotropic scattering: $\mathcal{P}(\cos\phi) = \varpi_\nu(1 + q_{ph}\cos\phi)$, with $-1 \le q_{ph} \le 1$. The scattering is isotropic if $q_{ph} = 0$; the radiation is backscattered if $q_{ph} > 0$, and scattered in the forward direction if $q_{ph} < 0$. Radiation is scattered predominantly in the forward direction if the particles are similar in size or slightly larger than the wavelength of the scattered light.

(4) In the more general situation of LTE and isotropic scattering, the source function, S_ν, becomes (Problem 3.18):

$$S_\nu = \varpi_\nu J_\nu + (1 - \varpi_\nu) B_\nu(T), \tag{3.55}$$

where $1 - \varpi_\nu = \kappa_\nu/\alpha_\nu$, the fraction of radiation lost due to absorption.

Problems (3.18)–(3.21) contain exercises related to radiative transfer in an atmosphere. For example, in Problem (3.20) the student is asked to calculate the hypothetical brightness temperature of Mars at different frequencies. The temperature depends upon the optical depth in the planet's atmosphere; if the optical depth is near zero or infinity, the problem essentially simplifies to case (1) discussed above. If the optical depth is closer to unity, the radiation from the planetary disk as well as the atmosphere contributes to the observed intensity.

3.2.2.5 *Equation of Radiative Transfer in a Surface*

The equilibrium temperature of a planet can be determined from a balance between incoming solar radiation and reradiation outwards (eq. 3.15 + 3.16 or 3.18, with eq. 3.44). The heat is transported downwards primarily by conduction (Section 3.2.1), and the thermal structure in the crustal layers can be determined if the albedo, emissivity, thermal inertia and thermal skin depth of the material are known. The brightness temperature of a planet, i.e., the temperature of a blackbody that would emit the same amount of energy at a particular wavelength, can be calculated by integrating the equation of radiative transfer (eq. 3.44) through

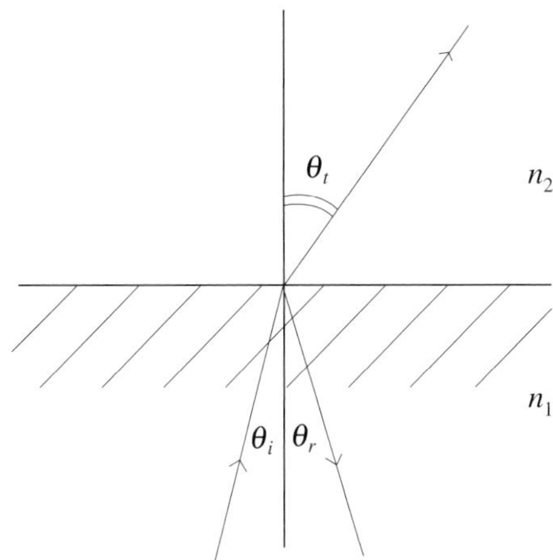

FIGURE 3.8 Geometry of refraction and reflection of radiation at the interface of two media with refractive indexes $n_2 < n_1$. Measured relative to the surface normal, θ_i and θ_r are the angles of incidence and reflection, respectively; θ_t is the angle for a ray transmitted (refracted) through the medium.

the crust layers. The mass absorption coefficient, κ_ν, is usually expressed as:

$$\kappa_\nu = \frac{2\pi\sqrt{\epsilon_r}\tan\Delta}{\rho\lambda}, \tag{3.56}$$

where ϵ_r is the real part of the complex dielectric constant. The *loss tangent* of the material, $\tan\Delta$, is the ratio of the imaginary part to the real part of the dielectric constant. Note that the dielectric constant is usually wavelength dependent. It has been determined empirically that the loss tangent increases approximately linearly with the density of the material. For the Moon $\tan\Delta/\rho \approx 0.007$–$0.01$ at wavelengths between a few mm and 20 cm.

The opacity is often expressed in terms of the *electrical skin depth* of the material, L_e, which is equivalent to the depth at which unit optical depth is reached:

$$L_e = \frac{\lambda}{2\pi\sqrt{\epsilon_r}\tan\Delta}. \tag{3.57}$$

The electrical skin depth is typically of the order of ten wavelengths. Thus, at infrared wavelengths one probes the surface layers, while at radio wavelengths one can probe up to a few meters down into the crust. Radio observations therefore sample the entire region of diurnal temperature variations, and by modeling the solar heating and outward radiation one can, through comparison with radio data, constrain the thermal and electrical properties in the upper few meters of the crust. For example, via such methods it was noticed that Mercury's surface is largely devoid of basalt (Section 5.5.2).

We discussed in Section 3.2.1 how the subsurface layers of a planet heat up during the day, when heat is transported downwards through the crust by conduction. The emission from the subsurface layers is transported upwards and transmitted through the surface into space. If θ_i is the angle with respect to the surface normal at which radiation from below the surface impinges upon the surface (Fig. 3.8), one can use *Snell's law of refraction* to relate θ_i to θ_t, the transmission or emission angle (toward the observer), and θ_r, the direction of propagation of the reflected component:

$$\sin\theta_i = \sin\theta_r, \tag{3.58a}$$

$$\frac{\sin\theta_t}{\sin\theta_i} \equiv n = \frac{n_1}{n_2} = \sqrt{\frac{\epsilon_{r1}}{\epsilon_{r2}}}, \tag{3.58b}$$

with n_1 and n_2 the indexes of refraction, and ϵ_{r1} and ϵ_{r2} the real parts of the dielectric constants for the two media. Since we are concerned with radiation from below a planet's surface into space ($n_2 = 1$), $n = n_1 = \sqrt{\epsilon_r}$, and $\sin\theta_t > \sin\theta_i$. The subsurface radiation transmitted through the surface into space is equal to $(1 - R_{0,p}(\theta_t))$, with $R_{0,p}(\theta_t)$ the ratio of the amount of energy in the reflected wave to that in the incident wave; $R_{0,p}(\theta_t)$ is the *Fresnel reflection coefficient* at the crust–vacuum interface for radiation at frequency ν.

The Fresnel reflection coefficients for each sense of polarization of the outgoing thermal emission, assuming a perfectly smooth surface, are given by:

$$R_\parallel = \frac{\tan^2(\theta_i - \theta_t)}{\tan^2(\theta_i + \theta_t)}, \tag{3.59a}$$

$$R_\perp = \frac{\sin^2(\theta_i - \theta_t)}{\sin^2(\theta_i + \theta_t)}, \tag{3.59b}$$

where $R_\parallel$ and $R_\perp$ are linearly polarized in and normal to the plane of incidence, respectively. Assuming that the subsurface emission is unpolarized, the emissivity at frequency ν, $\epsilon_\nu(\theta_t)$, is given by:

$$\epsilon_\nu(\theta_t) = 1 - \frac{1}{2}R_\perp(\theta_t) - \frac{1}{2}R_\parallel(\theta_t). \tag{3.60}$$

Using the equations of radiative transfer, assuming that the source function, S_ν, is given by the Planck function $B_\nu(T)$, the brightness temperature of polarization p at frequency ν can be expressed as:

$$T_{B_p}(\theta_t) = \left(1 - R_{0,p}(\theta_i)\right)\int_0^\infty \frac{\rho_\nu(z)\alpha_\nu(z)T_b(z)\, e^{(-\tau_\nu(z)/\sqrt{1-\epsilon_r(z)^{-1}\sin^2\theta_i})}}{\sqrt{1-\epsilon_r(z)^{-1}\sin^2\theta_i}}\,dz, \tag{3.61}$$

with $T_b(z)$ the subsurface brightness temperature at depth z:

$$T_b(z) = \frac{h\nu/k}{e^{(h\nu/kT(z))} - 1}, \tag{3.62}$$

and $T(z)$ the physical temperature at depth z. Although the thermal radiation from a planet's subsurface layers is usually unpolarized, the emergent radiation is polarized. The polarization increases strongly towards the limb of the planet, and the total emergent radiation decreases sharply at viewing angles larger than $\sim 70°$, an effect known as *Fresnel limb darkening*. Surface roughness decreases the polarization and limb darkening.

The Fresnel coefficient at normal incidence for a wave which hits the surface from free space becomes (Problem 3.22):

$$R_0 = \left(\frac{1 - \sqrt{\epsilon_r}}{1 + \sqrt{\epsilon_r}}\right)^2. \tag{3.63}$$

The reflected radiation for a wave from free space is linearly polarized in a plane normal to the plane of incidence at the *Brewster angle* of incidence ($R_{\parallel} = 0$):

$$\tan\theta_i = n = \sqrt{\epsilon_r}. \tag{3.64}$$

If the incoming plane wave is linearly polarized in the plane of incidence, there is no reflected wave at the Brewster angle.

3.2.3 Convection

Thus far, we have discussed the transport of heat by conduction and radiation. In dense atmospheres, molten interiors of planets, and the early solar nebula in which the planets formed, transport of heat is often most efficient by large scale fluid motions, in particular by convection. *Convection* is the motion in a fluid caused by density gradients which result from temperature differences. Consider a parcel of air in a planet's atmosphere that is slightly warmer than its surroundings. In order to re-establish pressure equilibrium, the parcel expands, and thus its density decreases below that of its surroundings. This causes the parcel to rise. Since the surrounding pressure decreases with height, the rising parcel expands and cools. If the temperature of the environment drops sufficiently rapidly with height, the parcel remains warmer than its surroundings, and thus continues to rise, transporting heat upwards. This process is an example of convection. For convection to occur, the temperature has to decrease with decreasing pressure (thus outwards in a planetary environment) at a sufficiently rapid rate that the parcel remains buoyant.

The temperature structure of an atmosphere in which energy transport is dominated by convection follows an adiabatic lapse rate (eq. 3.73). The temperature gradient in a planet's troposphere, that region in the atmosphere where we live and most clouds form, is usually close to an adiabat. Since the derivation of this lapse rate follows from the equation of hydrostatic equilibrium and the first law of thermodynamics, we discuss these two concepts in the next two subsections. The atmospheric structure is subsequently derived in Section 3.2.3.3. In Section 3.3 we derive the thermal structure for an atmosphere that is in radiative equilibrium.

3.2.3.1 *Hydrostatic Equilibrium*

The relationship between temperature, pressure and density in a planetary atmosphere and in a planet's interior is governed by a balance between gravity and pressure, called *hydrostatic equilibrium*. Consider a 'slab' of material of thickness dz and density ρ. The z-coordinate is taken to be positive going outward (decreasing pressure). This slab exerts a force due to its weight on the next slab down. Per unit area, this force becomes a pressure. So the increase in pressure across the slab, ΔP, is simply how much a column of height Δz and density ρ weighs:

$$\Delta P = -g_p \rho \Delta z. \tag{3.65a}$$

In general, both the density and gravitational acceleration, g_p, change with altitude, and the equation of hydrostatic equilibrium in differential form is:

$$\frac{dP}{dz} = -g_p(z)\rho(z). \tag{3.65b}$$

The relationship between temperature, pressure and density (*equation of state*) in a planetary atmosphere is usually well approximated by the *ideal gas law*:

$$P = NkT = \frac{\rho R_{gas} T}{\mu_a} = \frac{\rho k T}{\mu_a m_{amu}}, \tag{3.66}$$

where N is the particle number density, R_{gas} the universal gas constant ($R_{gas} = N_A k$, with N_A Avogadro's number), μ_a the mean molecular weight (in atomic mass units), and $m_{amu} \approx 1.67 \times 10^{-24}$ the mass of an atomic weight unit, which is slightly less than the mass of a hydrogen atom.

3.2.3.2 *Thermodynamics: First Law*

The first law of thermodynamics is an expression for the conservation of energy:

$$dQ = dU + P\,dV, \tag{3.67}$$

where dQ is the amount of heat absorbed by the system from its surroundings, dU the change in internal energy (sum of potential plus kinetic energy), and $P\,dV$ is the work done by the system on its environment, such as an expansion of the system; P is the pressure and dV the change in volume V. The thermal heat capacities, C_P and C_V, were defined above (eq. 3.20), and become:

$$C_V = \left(\frac{\partial U}{\partial T}\right)_V, \tag{3.68a}$$

$$C_P = \left(\frac{\partial U}{\partial T}\right)_P + P\left(\frac{\partial V}{\partial T}\right)_P. \tag{3.68b}$$

In the following we assume that V, the *specific volume*, contains one gram of molecules. Differentiating the ideal gas law gives:

$$dV = \frac{k}{\mu_a m_{amu} P} dT - \frac{kT}{\mu_a m_{amu} P^2} dP. \tag{3.69}$$

In an ideal gas, the difference between the two thermal heat capacities (erg mole^{-1} K^{-1}) or specific heats (erg g^{-1} K^{-1}) is given by:

$$C_P - C_V = R_{gas}, \tag{3.70a}$$

$$m_{gm}(c_P - c_V) = R_{gas}, \tag{3.70b}$$

with m_{gm} the mass of a gram-mole (see eq. 3.20) and R_{gas} the universal gas constant. If, in an ideal gas, a parcel of air moves *adiabatically*, i.e., no heat is exchanged between the parcel of air and its surroundings ($dQ = 0$), the first law of thermodynamics requires that (Problem 3.24):

$$c_V\,dT = -P\,dV, \tag{3.71}$$

$$c_P\,dT = \frac{1}{\rho}\,dP. \tag{3.72}$$

3.2.3.3 *Adiabatic Lapse Rate*

For an atmosphere that is marginally unstable to convection, we can use the thermodynamic relations above with the equation of hydrostatic equilibrium (eq. 3.65) and the definition for the ratio of the specific heats, $\gamma \equiv c_P/c_V = C_P/C_V$, to obtain its temperature structure, or the *dry adiabatic lapse rate* (Problem 3.25):

$$\frac{dT}{dz} = -g_p/c_P = -\frac{\gamma - 1}{\gamma}\frac{g_p \mu_a m_{amu}}{k}. \tag{3.73}$$

Typical values for γ are 5/3, 7/5, 4/3, for monatomic, diatomic, and polyatomic gases respectively. The dry adiabatic lapse rate on Earth is roughly 10 K/km. We show in Section 4.4.1 that the latent heat of condensation acts to decrease the adiabatic gradient in cloud-forming regions of moist atmospheres.

Convection is extremely efficient at transporting energy whenever the temperature gradient or lapse rate is *superadiabatic* (larger than the adiabatic lapse rate). Energy transport via convection thus effectively places an upper bound on the rate at which temperature can increase with depth in a planetary atmosphere or fluid interior. Substantial superadiabatic gradients are only possible when convection is suppressed by gradients in mean molecular weight or by the presence of a flow-inhibiting boundary, such as a solid surface.

3.3 Radiative Equilibrium in an Atmosphere

Energy transport in a planet's stratosphere, the region above the tropopause (Section 4.2), is usually dominated by radiation. If the total radiative flux is independent of height, the atmosphere is in *radiative equilibrium*. In this section, we derive the thermal profile for an atmosphere that is in radiative equilibrium.

3.3.1 Thermal Profile

An atmosphere is in radiative equilibrium if the total flux, $\mathcal{F} = \int \mathcal{F}_\nu\,d\nu$, is constant with depth:

$$\frac{d\mathcal{F}}{dz} = 0. \tag{3.74}$$

The temperature structure in such an atmosphere can be obtained from the *diffusion equation*, an expression for the radiative flux at altitude z. In the following we derive the diffusion equation in an optically thick atmosphere which is approximately in LTE: $I_\nu \approx S_\nu \approx B_\nu(T)$. We assume the atmosphere to be in monochromatic radiative equilibrium: $d\mathcal{F}_\nu/dz = 0$.

Integration of equation (3.44) over a sphere yields (Problem 3.26a):

$$\frac{d\mathcal{F}_\nu}{d\tau_\nu} = 4\pi(B_\nu - J_\nu), \tag{3.75}$$

where $\mathcal{F}_\nu$ is the flux density across a layer in a stratified atmosphere. Both $\mathcal{F}_\nu$ and the mean intensity J_ν were defined in Section 3.2.2.1. Multiplying equation (3.44) by μ_θ and integrating over a sphere yields the following relation between J_ν and $\mathcal{F}_\nu$ (Problem 3.26b):

$$\frac{4\pi}{3}\frac{dJ_\nu}{d\tau_\nu} = -\mathcal{F}_\nu. \tag{3.76}$$

Setting $d\mathcal{F}_\nu/d\tau_\nu = 0$ in equation (3.75), differentiating and using equation (3.76), we find (Problem 3.26c):

$$\frac{dB_\nu}{d\tau_\nu} = -\frac{3}{4\pi}\mathcal{F}_\nu. \tag{3.77}$$

Integrating over frequency yields the total radiative flux, or the *radiative diffusion equation*:

$$\mathcal{F}(z) = -\frac{4\pi}{3\rho}\frac{\partial T}{\partial z}\int_0^\infty \frac{1}{\alpha_\nu}\frac{\partial B_\nu(T)}{\partial T}d\nu. \tag{3.78}$$

Equation (3.78) can be simplified by the use of a mean absorption coefficient, such as the *Rosseland mean absorption coefficient*, α_R:

$$\frac{1}{\alpha_R} \equiv \frac{\int_0^\infty \frac{1}{\alpha_\nu} \frac{\partial B_\nu}{\partial T} d\nu}{\int_0^\infty \frac{\partial B_\nu}{\partial T} d\nu}. \tag{3.79}$$

With this simplification, we write the radiative diffusion equation as:

$$\mathcal{F}(z) = -\frac{16}{3} \frac{\sigma T^3}{\alpha_R \rho} \frac{\partial T}{\partial z}. \tag{3.80}$$

Note that flux travels upwards in an atmosphere if the temperature gradient dT/dz is negative (i.e., when the temperature decreases with altitude).

Using equations (3.9) and (3.80), the atmospheric temperature profile becomes:

$$\frac{dT}{dz} = -\frac{3}{16} \frac{\alpha_R \rho}{T^3} T_e^4. \tag{3.81}$$

If an atmosphere is in both hydrostatic and radiative equilibrium and its equation of state is given by the perfect gas law, then its temperature–pressure relation is:

$$\frac{d \ln T}{d \ln P} \approx -\frac{3}{16} \frac{P}{g_p} \left(\frac{T_e}{T}\right)^4 \alpha_R. \tag{3.82}$$

Equations (3.81) and (3.82) are approximate, since they use a mean absorption coefficient. Both T and $B_\nu(T)$, as well as the abundances of many of the absorbing gases, vary with depth in a planetary atmosphere. The best approach to solving for the temperature structure is to solve the transport equation (3.44) at all frequencies, together with the requirement that the flux, $\mathcal{F}$, is constant with depth (eq. 3.74).

3.3.2 Atmosphere Heated from Below: Greenhouse Effect

The surface temperature of a planet can be raised substantially above its equilibrium temperature if the planet is overlain by an atmosphere that is optically thick at infrared wavelengths, a situation referred to as the *greenhouse effect*. Sunlight, which has its peak intensity at optical wavelengths (Fig. 3.2 and eq. 3.5 for a blackbody of temperature ~5700 K), enters the atmosphere, which is relatively transparent at visible wavelengths, and heats the surface. The warm surface radiates its heat at infrared wavelengths. This radiation does not immediately escape into interplanetary space, but is absorbed by air molecules, especially CO_2, H_2O and CH_4. When these molecules de-excite, photons at infrared wavelengths are emitted in a random direction. The net effect of this process is that the atmospheric (and surface) temperature is increased until equilibrium is reached between solar energy input and the emergent planetary flux (Fig. 3.9). In this section we calculate the greenhouse effect in an atmosphere that is in radiative equilibrium. Since we are concerned with atmospheric heating at infrared wavelengths, we consider the planet itself as the energy source. Thus the atmosphere is heated from below, and direct solar illumination of atmospheric gases is ignored.

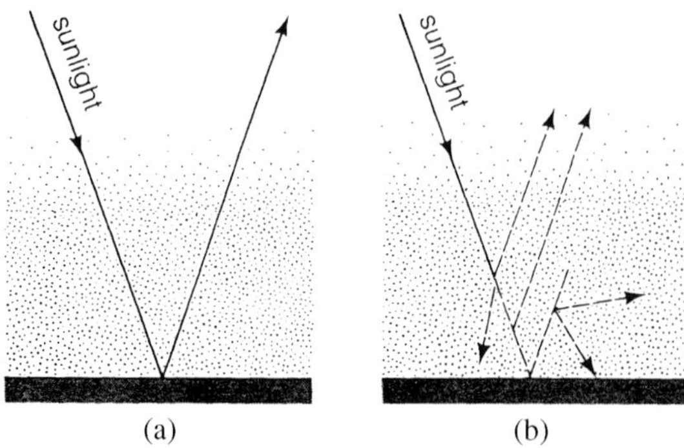

FIGURE 3.9 Transmission of sunlight through (a) a hypothetical transparent atmosphere and (b) an atmosphere in which absorption and re-emission of photons is important. (Hartmann 1993)

For this analysis, it is convenient to use the two-stream approximation:

$$I_\nu = (I_\nu^+ + I_\nu^-), \tag{3.83}$$

where I_ν^+ is the upward and I_ν^- the downward radiation at frequency ν. The net flux density across a layer becomes:

$$\mathcal{F}_\nu = \pi(I_\nu^+ - I_\nu^-). \tag{3.84}$$

We consider an atmosphere in monochromatic radiative equilibrium ($d\mathcal{F}_\nu/dz = 0$) and LTE, which is heated from below (i.e., $I_\nu^- \equiv 0$ at the top of the atmosphere). The upward intensity at the ground, $I_{\nu g}^+$, can be expressed with help of equations (3.83), (3.84) and (3.75) (Problem 3.27):

$$I_{\nu g}^+ \equiv B_\nu(T_g) = B_\nu(T_1) + \frac{1}{2\pi}\mathcal{F}_\nu, \tag{3.85}$$

where T_1 is the air temperature just above the ground. The downward and upward intensities at the top of the atmosphere are:

$$I_{\nu 0}^- \equiv 0 = B_\nu(T_0) - \frac{1}{2\pi}\mathcal{F}_\nu, \tag{3.86a}$$

$$I_{\nu 0}^+ = B_\nu(T_0) + \frac{1}{2\pi}\mathcal{F}_\nu = 2B_\nu(T_0), \tag{3.86b}$$

where T_0 is the temperature of the upper boundary, usually referred to as the *skin temperature*. Thus, the upward intensity at the top of the atmosphere is twice as large as that emitted by an opaque blackbody at temperature T_0. With help of equation (3.86*a*), the solution to equation (3.77) is given by:

$$B_\nu(\tau) = B_\nu(T_0)\left(1 + \frac{3}{2}\tau_\nu\right). \tag{3.87}$$

Integration over frequency and conversion to temperature via the Stefan–Boltzmann law (eq. 3.8) yields:

$$T^4(\tau) = T_0^4\left(1 + \frac{3}{2}\tau\right). \tag{3.88}$$

The total radiant flux from a body can be obtained by integrating equation (3.86*b*) over frequency. This flux translates into the effective temperature, T_e:

$$T_e^4 = 2T_0^4. \tag{3.89}$$

If the temperature of a body is determined exclusively by the incident solar flux, the effective and equilibrium temperatures are equal, i.e., $T_e = T_{eq}$. At the top of the atmosphere, where the optical depth is zero, the temperature $T_0 \approx 0.84T_e$. By combining equations (3.88) and (3.89), we see that the temperature, $T(\tau)$, is equal to the effective temperature, T_e, at an optical depth $\tau(z) = 2/3$. Thus, continuum radiation is received from an effective depth in the atmosphere where $\tau = 2/3$ (Problem 3.28).

The ground or surface temperature, T_g, can be obtained from equation (3.85):

$$T_g^4 = T_1^4 + \frac{1}{2}T_e^4. \tag{3.90a}$$

Note that there is a discontinuity: The surface temperature, T_g, is higher than the air temperature, T_1, just above it. In a real planetary atmosphere, conduction reduces this difference. Equation (3.90*a*) can be rewritten using equations (3.87) and (3.89):

$$T_g^4 = T_e^4\left(1 + \frac{3}{4}\tau_g\right), \tag{3.90b}$$

where τ_g is the optical depth to the ground. Equation (3.90*b*) shows that the surface temperature in a radiative atmosphere can be very high if the infrared opacity of the atmosphere is high. The greenhouse effect is particularly strong on Venus, where the surface temperature reaches a value of 733 K, well above the equilibrium temperature of $\sim$240 K. The greenhouse effect is also noticeable on Earth and, to a lesser extent, Titan and Mars.

Calculations performed for an atmosphere assumed to be in radiative equilibrium may produce superadiabatic lapse rates (e.g., in a planet's troposphere). In such cases, convection develops and drives the atmospheric structure to an adiabat, and the temperature structure can better be calculated assuming *radiative–convective equilibrium*, where the convective layer (troposphere) supplies the same amount of upward radiative flux as would have been produced under radiative equilibrium, while the temperature structure in the convective layer follows an adiabat. The temperature thus calculated is somewhat cooler near the surface compared to radiative equilibrium calculations and warmer at higher altitudes.

Icy material allows sunlight to penetrate several centimeters or more below the surface, but is mostly opaque to reradiated thermal infrared emission. Thus, the subsurface region can become significantly warmer than the equilibrium temperature would indicate. In analogy with atmospheric trapping of thermal infrared emission, this process is known as the *solid-state greenhouse effect*. This process may be important on icy bodies, such as the Galilean satellites and comets.

Further Reading

Bernath, P.F., 1995. *Spectra of Atoms and Molecules*. Oxford University Press, Oxford.

Chamberlain, J.W., and D.M. Hunten, 1987. *Theory of Planetary Atmospheres*. Academic Press, Inc., New York.

Herzberg, G., 1944. *Atomic Spectra and Atomic Structure*. Dover Publications, New York.

Rybicki, G.B., and A.P. Lightman, 1979. *Radiative Processes in Astrophysics*. John Wiley and Sons, New York.

Salby, M.L., 1996. *Fundamentals of Atmospheric Physics*. Academic Press, New York.

Shu, F.H., 1991. *The Physics of Astrophysics. Vol. I: Radiation*. University Science Books, Mill Valley, CA.

Townes, C.H., and A.L. Schawlow, 1955. *Microwave Spectroscopy*. McGraw-Hill, New York.

Problems

3.1.**E** (a) Write Planck's radiation law (eq. 3.3) in terms of λ rather than ν.

(b) Use your expression from part (a) to derive equation (3.5*b*).

3.2.**E** A sphere of radius R, at a distance r from the observer (along the z-axis), has a uniform brightness B. The specific intensity is equal to B if the ray intersects the sphere, and zero otherwise. Use equation (3.32) to express the flux $\mathcal{F}$ in terms of brightness B and angle θ_c, where θ_c is the angle at which a ray from the observer is tangent to the sphere (i.e., $2\theta_c$ is the angle subtended by the sphere as seen from the observer). Show that on the surface of the sphere, $\mathcal{F} = \pi B$ (eq. 3.7).

3.3.I (a) Find the unidirectional energy flux in space due to the 2.7 K background radiation. (Hint: A blackbody in thermal equilibrium with the radiation field will reach the same temperature as the radiation.) Express your answer in cgs units (erg cm^{-2} s^{-1}). Note that as the background radiation is (almost) isotropic, the net flux summed over all directions is zero.

(b) Galactic stars (other than the Sun) occupy a total solid angle of $\sim 10^{-14}$ str as seen from the Solar System. Assuming a typical stellar effective temperature of 10 000 K (this is weighted towards the blue stars, which radiate most of the energy), what is the unidirectional energy flux from galactic stars?

(c) Find the solar energy flux at Earth's orbit, at Neptune's orbit and at a typical Oort Cloud distance of 25 000 AU.

3.4.I (a) A dust grain is heated by the interstellar radiation field, which has a temperature T_{ISM}. Thus, if the grain were a perfect blackbody, its temperature would also be T_{ISM}. Do you expect the grain to be warmer or colder than this value, and why?

(b) Assume that a dust grain is heated entirely by the interstellar radiation field at UV wavelengths, and that the UV flux is 2×10^5 photons cm^{-2} s^{-1} Å^{-1}. The bandwidth of the radiation is 1000 Å, and the mean photon energy is 9 eV per photon. If the grain radiates with an efficiency of 0.1%, calculate its temperature. Compare your answer with part (a).

3.5.E Calculate the equilibrium temperature for the Moon:

(a) Averaged over the lunar surface. (Hint: Assume the Moon to be a rapid rotator.)

(b) As a function of solar elevation, assuming the Moon to be a slow rotator.

3.6.I Calculate the equilibrium temperature of the Moon as a function of latitude, assuming the Moon to be a rapid rotator with zero obliquity.

3.7.E Is the spectrum of emitted thermal radiation from a planet broader or narrower than the blackbody spectrum? Why?

3.8.E A body emits and absorbs radiation of any given frequency with the same efficiency. Given this fact, why does the temperature of a planet depend on its albedo?

3.9.E Thermal emissions by solid bodies can be analyzed to provide information on the temperature at a few wavelengths below the surface. Observations of the night hemisphere of Mercury longward of $\sim$10 cm yield temperatures close to the diurnal equilibrium temperature. As the radiation is obviously able to escape directly from the regions being observed, why does it not get much colder than this during the long mercurian night?

3.10.I (a) Assuming slow rotation and neglecting any internal heat sources, calculate the equilibrium temperature of each planet as a function of solar elevation angle. Assume the emissivity $\epsilon = 1$.

(b) Assuming rapid rotation and neglecting both internal heat sources and planetary obliquity, calculate the equilibrium temperature of each planet as a function of latitude.

3.11.E (a) Neglecting internal heat sources, calculate the average equilibrium temperature for all nine planets.

(b) At which wavelength would you expect the blackbody spectral peak of each planet?

3.12.E Jupiter's effective temperature is observed to be 125 K. Compare the observed temperature with the equilibrium temperature calculated in the previous problems. What could be responsible for the difference? (Hint: Consider the assumptions which are involved in the derivation of the equilibrium temperature.)

3.13.E Mercury's observed surface temperature is 100 K at night, and 700 K at the subsolar point (noon) at perihelion. Compare these temperatures with the values that you calculated in Problems 3.10 and 3.11, and comment on your result.

3.14.I Calculate the expected increase in the global average temperature of the Earth at a full Moon compared to a new Moon (neglecting eclipses). Which effect is larger, the change in position of the Earth or radiation reflected and emitted from the Moon?

3.15.I Show that in an isotropic radiation field $I_\nu = J_\nu$.

3.16.E (a) Calculate the wavelength and energy of photons corresponding to the Lyman α,

Balmer β and Brackett α emission from a hydrogen atom.

(b) Calculate the wavelength and energy of photons necessary to ionize a hydrogen atom from the electronic ground state.

3.17.**I** (a) Derive equation (3.52) from equation (3.41) for a cloud of atomic hydrogen gas in LTE.

(b) Calculate the relative importance of stimulated emission in the Ly α line ($n = 2 \rightarrow 1$) if the temperature of the gas is 100 K. (Hint: Which energy levels are populated?)

(c) Calculate the relative importance of stimulated emission in the Ly α line ($n = 2 \rightarrow 1$) if the temperature of the gas is 10^4 K.

(d) Calculate the relative importance of stimulated emission in the Ly α line ($n = 2 \rightarrow 1$) if the temperature of the gas is 10^6 K.

3.18.**I** Derive an expression for the source function, S_ν, and intensity, $I_\nu(\tau_\nu)$, for

(a) an atmosphere in LTE, where scattering can be ignored,

(b) an atmosphere in LTE and in which scattering is isotropic.

3.19.**I** (a) Consider an optically thick ($\tau \gg 1$) cloud at a geocentric distance d. The cloud emits thermal radiation. Express the observed intensity, I_ν, from the center of this cloud, as a brightness, $B_\nu(T)$.

(b) The brightness temperature, T_b, is the temperature of a blackbody which has the same brightness at this frequency. Derive the relationship between T_b, τ, and T using the Rayleigh–Jeans approximation. Under what conditions does this approximation hold?

(c) Suppose the cloud is spherical, and θ is the angle between the line of sight and the normal to the surface. Determine the center-to-limb variation in the observed intensity.

(d) Answer parts (a)–(c) for an optically thin cloud, $\tau \ll 1$.

3.20.**E** The solid surface of Mars is opaque at all wavelengths and is surrounded by an optically thin atmosphere. The atmosphere absorbs in a narrow spectral region: its absorption coefficient is large at a wavelength $\lambda_0 = 2.6$ mm and negligibly small at other radio wavelengths; thus for most radio wavelengths λ_1: $\alpha_{\lambda 0} \gg \alpha_{\lambda_1}$. Assume the temperature of Mars's surface $T_s = 230$ K, and of its atmosphere $T_a = 140$ K. (Note: For both parts of this problem you may use approximations appropriate for radio wavelengths, even though it is not fully correct at mm wavelengths.)

(a) What are the observed brightness temperatures at λ_1 and at λ_0?

(b) What is the observed brightness temperature at λ if the optical depth of the atmosphere at this wavelength is equal to 0.5?

3.21.**E** Consider a hypothetical planet of temperature T_p, surrounded by an extensive atmosphere of temperature T_a, where $T_a < T_p$. The atmosphere absorbs in a narrow spectral line; its absorption coefficient is large at frequency ν_o and is negligibly small at other frequencies, such as ν_1: $\alpha_{\nu_o} \gg \alpha_{\nu_1}$. The planet is observed at frequencies ν_o and ν_1. Assume that the Planck function does not change much between ν_o and ν_1.

(a) When observing the center of the planet's disk, at which frequency, ν_o or ν_1, will the brightness temperature be highest? Will the same be true when observing near the limb? Make a sketch of the 'observed' spectra. (Note: The limb is defined as the outer edge of the planet, including its atmosphere.)

(b) Answer the same question if $T_a > T_p$.

3.22.**I** Derive the Fresnel coefficient at normal incidence for a wave which hits the surface from free space (eq. 3.63) and determine the emissivity $\epsilon_\nu(\theta_t)$ at normal incidence. (Hint: Use equations (3.58)–(3.60).)

3.23.**I** Assume that you are observing a point on the surface of the asteroid Ceres, from viewing angles θ of 0° up to 90° (θ is the angle between the line of sight and the normal to the surface). The dielectric constant of the surface material is $\epsilon_r = 6$ (note that $\epsilon_r = 1$ for free space). Plot the Fresnel coefficient for each sense of polarization, as well as the total emissivity as a function of θ. Comment on your results.

3.24.**I** (a) Show that, in an ideal gas, equation (3.70) holds.

(b) Derive equations (3.71) and (3.72) from equations (3.67)–(3.70).

3.25.**I** Derive equation (3.73) for the dry adiabatic lapse rate, using the thermodynamic relations together with the equation for hydrostatic equilibrium.

3.26.I (a) Derive equation (3.75) for an optically thick atmosphere that is approximately in LTE ($S_\nu \approx B_\nu$). (Hint: Integrate equation (3.44), and use the definitions for $\mathcal{F}_\nu$ and J_ν as given in equations (3.30) and (3.32).)

(b) Derive equation (3.76) by multiplying equation (3.44) by μ_θ and integrating over a sphere. (Hint: Use the definitions for all three moments of the radiation field (eqs. 3.30–3.33). Assume the radiation is isotropic, so you can use equation (3.33*c*) and the relation $I_\nu = J_\nu$ (Problem 3.15).)

(c) Assuming monochromatic radiative equilibrium ($d\mathcal{F}_\nu/dz = 0$), and equations (3.75) and (3.76), derive equation (3.77).

3.27.I (a) Derive equation (3.85) for an atmosphere in LTE and in monochromatic radiative equilibrium. (Hint: Use equation (3.75) and the two-stream approximation.)

(b) Prove that the intensity at the top of a radiative atmosphere is twice that emitted by an opaque blackbody at the temperature of the top of the atmosphere (eq. 3.86*b*).

3.28.I Consider a rapidly rotating planet with an atmosphere in radiative equilibrium. The planet is located at a heliocentric distance $r_\odot = 2$ AU; its Bond albedo $A_b = 0.4$ and emissivity $\epsilon = 1$ at all wavelengths. Assume that the temperature of the planet is determined entirely by solar radiation.

(a) Calculate the effective and equilibrium temperatures.

(b) Calculate the temperature at the upper boundary of the atmosphere, where $\tau = 0$.

(c) Show that continuum radiation from the planet's atmosphere is received from a depth where $\tau = 2/3$.

(d) If the optical depth of the atmosphere $\int_0^\infty \tau(z)dz = 10$, determine the surface temperature of this planet.

3.29.E The radiation pressure coefficient, Q_{pr}, is given by:

$$Q_{pr} = Q_{abs} + Q_{sca}(1 - \langle \cos\varphi \rangle), \qquad (3.91)$$

where $\langle \cos\varphi \rangle = 1$ in the case of pure forward scattering, and $\langle \cos\varphi \rangle = -1$ for backscattered (reflected) radiation. Write down expressions for Q_{pr} and the net force on the particle due to radiation pressure and Poynting–Robertson drag for a particle in orbit around the Sun for the following (Hint: See Section 2.7):

(a) Translucent particle.

(b) Perfect absorber.

(c) Particle which does not absorb radiation and scatters in the forward direction only.

(d) Particle which is a perfect reflector (no absorption, backscattered light only).

4 Planetary Atmospheres

In the small hours of the third watch, when the stars that shown out in the first dusk of evening had gone down to their setting, a giant wind blew from heaven, and clouds driven by Zeus shrouded land and sea in a night of storm.

Homer, *The Odyssey*, ~ 800 BCE

The *atmosphere* is the gaseous outer portion of a planet. Atmospheres have been detected around all planets and several satellites, and each is unique. Some atmospheres are very dense, and gradually blend into fluid envelopes which contain most of the planet's mass. Others are extremely tenuous, so tenuous that even the best vacuum on Earth seems dense in comparison. The composition of a planetary atmosphere varies from solar-like hydrogen/helium atmospheres for the giant planets to atmospheres dominated by nitrogen, carbon dioxide or sulfur dioxide for terrestrial planets and satellites. However, even though all atmospheres are intrinsically different, they are governed by the same physical and chemical processes. For example, clouds form in most atmospheres, although the condensing gases may be very different. The upper layers of an atmosphere are modified by solar radiation, resulting in photochemistry, the details of which differ from planet to planet. Variations in temperature and pressure lead to winds, which can be steady, turbulent, strong or weak. In this chapter the various processes operating in planetary atmospheres are discussed, and the characteristics of the individual atmospheres are summarized.

4.1 Density and Scale Height

The relationship between temperature, pressure and density in a planetary atmosphere is governed by a balance between gravity and pressure: to a first approximation the atmosphere is in hydrostatic equilibrium (Section 3.2.3.1). Using the equations for hydrostatic equilibrium (eq. 3.65) and the ideal gas law (eq. 3.66), one can express the atmospheric pressure as a function of altitude:

$$P(z) = P(0)e^{-\int_0^z dr/H(r)}, \tag{4.1}$$

with the *pressure scale height*:

$$H(z) = \frac{kT(z)}{g_p(z)\mu_a(z)m_{amu}}, \tag{4.2}$$

where $g_p(z)$ is the acceleration due to gravity at altitude z, $\mu_a m_{amu}$ is the molecular mass and k is Boltzmann's constant. The pressure scale height, H, is equal to the distance over which the pressure decreases by a factor e. Small values imply a rapid decrease of atmospheric pressure with altitude. Similarly, we can express the density as a function of altitude:

$$\rho(z) = \rho(0)e^{-\int_0^z dr/H^*(r)}, \tag{4.3}$$

where $\rho(0)$ is the number density at altitude $z = 0$[1]. The *density scale height, H^**, is:

$$\frac{1}{H^*(z)} = \frac{1}{T(z)}\frac{dT(z)}{dz} + \frac{g_p(z)\mu_a(z)m_{amu}}{kT(z)}, \tag{4.4}$$

where we have neglected the (usually small) terms that result from gradients in μ_a and g_p. Note that for an isothermal (region of an) atmosphere, $H^*(z) = H(z)$. Approximate pressure scale heights for the planets and the Moon are shown in Tables 4.1 and 4.2. It is interesting to note that H is of the order of 10–20 km for most planets, since the ratio $T/(g_p\mu_a)$ for the giant and terrestrial planets is similar. Only in the tenuous atmospheres of Mercury, Pluto and various moons is the scale height larger (Problem 4.1).

4.2 Thermal Structure

The *thermal structure* of a planet's atmosphere, dT/dz, is primarily governed by the efficiency of energy transport, as discussed in Section 3.2. This process depends largely on the opacity or optical depth in the atmosphere, which is determined by a variety of physical and chemical processes. Astronomers may be used to stellar atmospheres, which are heated from below and which are so hot that most elements are in atomic form, and one certainly would not

[1] The location of the $z = 0$ 'plane' is selected for convenience. For Earth, mean sea level is often used. For other terrestrial planets and satellites, mean radius of the surface is typically selected, and the 1 bar pressure level is usually chosen for giant planets.

TABLE 4.1 Basic Atmospheric Parameters for the Terrestrial Planets, the Moon and Pluto.

Parameter	Mercury	Venus	Earth	Moon	Mars	Pluto	Reference
Mean heliocentric distance (AU)	0.387	0.723	1.000	1.000	1.524	39.48	1
Surface temperature (K)	100–725	733	288	277	215	~45	2, 3
Surface pressure (bar)	5×10^{-15}	92	1.013	3×10^{-15}	0.0056	1.5×10^{-5}	2, 3
Geometric albedo $A_{0,v}$	0.138	0.84	0.367	0.113	0.15	~0.44–0.61	1, 3, 4, 5, 6
Bond albedo	0.119	0.75	0.29	0.123	0.16	~0.4	1, 2, 4, 5
Equilibrium temperature (K)	446	238	263	277	222	40	Calc.[a] ($\epsilon = 0.9$)
Scale height (km)	13–95	16	8.5	65	18	33	Calc.[a]
Exobase temperature (K)	600	275, 1020	~1000		350	58	2

[a] Calculated (ϵ = emissivity)

1: Yoder (1995). 2: Chamberlain and Hunten (1987). 3: Clarke *et al.* (1992). 4: Veverka *et al.* (1988). 5: Moroz (1983). 6: Stern and Yelle (1999).

TABLE 4.2 Basic Atmospheric Parameters for the Giant Planets.

Parameter	Jupiter	Saturn	Uranus	Neptune	References
Mean heliocentric distance (AU)	5.203	9.543	19.19	30.07	1
Effective temperature (K)	124.4 ± 0.3	95.0 ± 0.4	59.1 ± 0.3	59.3 ± 0.8	2
Geometric albedo ($A_{0,v}$)	0.52	0.47	0.51	0.41	1
Geometric albedo ($A_{0,ir}$)	0.274 ± 0.013	0.242 ± 0.012	0.208 ± 0.048	0.25 ± 0.02	3
Phase integral	1.25 ± 0.10	1.42 ± 0.10	1.40 ± 0.14	1.25 ± 0.10	3
Bond albedo	0.343 ± 0.032	0.342 ± 0.030	0.290 ± 0.051	0.31 ± 0.04	3
Energy balance[a]	1.67 ± 0.09	1.78 ± 0.09	1.06 ± 0.08	2.61 ± 0.28	2
Equilibrium temperature (K)	113	83	60	48	Calc.[b] ($\epsilon = 0.9$)
Scale height (km)	18	35	20	19	Calc.[b]
Adiabatic lapse rate (K/km)	1.9	0.84	0.85	0.86	4
Temperature ($P = 1$ bar) (K)	165.0	134.8	76.4	71.5	5
Tropopause temperature (K)	111	82	53	52	5
Tropopause pressure (mbar)	140	65	110	140	5
Exobase temperature (K)	900–1300	800	750	750	6, 7
Stratosphere temperature (K)	140–180	140–150	90	170	6, 7
Mesosphere temperature (K)	160–170	150	140–150	140–150	4

[a] Ratio (energy radiated into space)/(solar energy absorbed).

[b] Calculated (ϵ = emissivity).

1: Yoder (1995). 2: Hubbard *et al.* (1995). 3: Conrath *et al.* (1989). 4: Chamberlain and Hunten (1987). 5: Lindal (1992). 6: Atreya (1986). 7: Bishop *et al.* (1995).

expect clouds to form. In contrast, planetary atmospheres consist of molecular gases and are primarily heated from the top. To determine the thermal structure in such an atmosphere one has to be aware of the following processes, most of which will be discussed in detail throughout this chapter:

(1) The top of the atmosphere is irradiated by the Sun. Some of this radiation is absorbed and scattered in the atmosphere. This process, together with radiative losses and conduction, basically defines the temperature profile in the upper part of the atmosphere.

(2) Energy from internal heat sources for the giant planets and reradiation of absorbed sunlight by a planet's surface modify the atmospheric temperature profile.

(3) Chemical reactions in an atmosphere change the atmospheric composition, which leads to changes in the opacity and thermal structure.

(4) Clouds and/or photochemically produced haze layers not only change the atmospheric opacity, but also change the temperature locally through release (cloud formation) or absorption (evaporation) of latent heat.

(5) Volcanoes and geyser activity on some planets and satellites may modify the atmosphere substantially.

(6) On the terrestrial planets and satellites, chemical interactions between the atmosphere and the crust or ocean influence a planet's atmosphere.

(7) On Earth the atmospheric composition, and hence opacity and thermal structure, is influenced by biochemical and anthropogenic processes.

Even though the composition of planetary atmospheres varies drastically from one planet/satellite to another, aside from the most tenuous atmospheres the temperature structure is qualitatively similar. Figure 4.1 shows the temperature–pressure profiles for Earth, Venus, Mars, Titan and the giant planets. Moving upwards from the surface or, for the giant planets, from the deep atmosphere, the temperature decreases with altitude: this part of the atmosphere is called the *troposphere*. It is in this part of the atmosphere that condensable gases, usually trace elements, form clouds. The atmospheric temperature typically reaches a minimum at the *tropopause*, near a pressure level of ~ 0.1 bar. Above the tropopause the temperature structure is inverted. This region in the atmosphere is called the *stratosphere*. At higher altitudes one finds the *mesosphere*, an almost isothermal region. The *stratopause* forms the boundary between the stratosphere and mesosphere. In contrast to the generic temperature profile, on Earth and Titan the *mesopause* forms a second temperature minimum. On Earth, the temperature above the tropopause, in the stratosphere, increases with altitude as a result of photochemistry (Section 4.2.2.1 and 4.6.1.1). The temperature reaches a maximum at the stratopause, above which it decreases with altitude in the mesosphere to reach a second temperature minimum at the mesopause. Above the mesopause absorption of solar radiation shortwards of ~ 1000 Å causes the temperature to increase with altitude. This region is referred to as the *thermosphere*. On Earth, the temperature rises to over 1000 K at altitudes close to ~ 500 km, and is isothermal above this level.

The outer part of the atmosphere is referred to as the *exosphere:* collisions between gas molecules in this part of the atmosphere are rare, and the rapidly moving molecules have a relatively large chance to escape into interplanetary space. The *exobase*, at the bottom of the exosphere (~ 500 km on Earth), is the altitude above which the product of the collisional cross-section and the integrated atmospheric number density is equal to one mean free path length for a fast atom (eq. 4.76).

4.2.1 Heat Sources and Transport of Energy

4.2.1.1 *Heat Sources*

All planetary atmospheres are subject to solar irradiation, which heats an atmosphere through absorption of solar photons. Since the solar 5700 K blackbody curve peaks at 0.5 μm, most of the Sun's energy output is in the visible wavelength range. These photons heat up a planet's surface (terrestrial planets) or layers in the atmosphere where the optical depth is moderately large (typically near the cloud layers). Reradiation of sunlight by a planet's surface or by dust particles in its atmosphere occurs primarily at infrared wavelengths, and forms a source of heat imbedded in or below the atmosphere. Internal heat sources may also heat the atmosphere from below, and this is important for the giant planets.

Solar heating of an upper atmosphere is usually most efficient at EUV (extreme ultraviolet) wavelengths, even though the number of photons in this wavelength range is very low. EUV photons have typical energies between 10 and 100 eV (1000 to 100 Å), which is usually more energy than needed to ionize constituents in an atmosphere (Section 4.6.2). The excess energy in an ionization process is carried off by electrons created in the process, referred to as *photoelectrons*. Photoelectrons collide with and excite/ionize other particles, either directly or via *bremsstrahlung*[2] induced by Coulomb collisions.

In addition to heating processes triggered by sunlight, an upper atmosphere can be heated substantially by *charged particle precipitation*: charged particles that enter the atmosphere from outside (solar wind). Charged particle precipitation is confined to high latitudes on planets with intrinsic magnetic fields, the *auroral zones* (Sections 4.6.4 and 7.4). Direct particle precipitation can heat an atmosphere substantially more than the photoelectrons mentioned above. Although most of the heating is localized, a planet-wide heating may be caused

[2] Bremsstrahlung or free–free emission is electromagnetic radiation resulting from the acceleration of a charged particle in the Coulomb field of another charged particle.

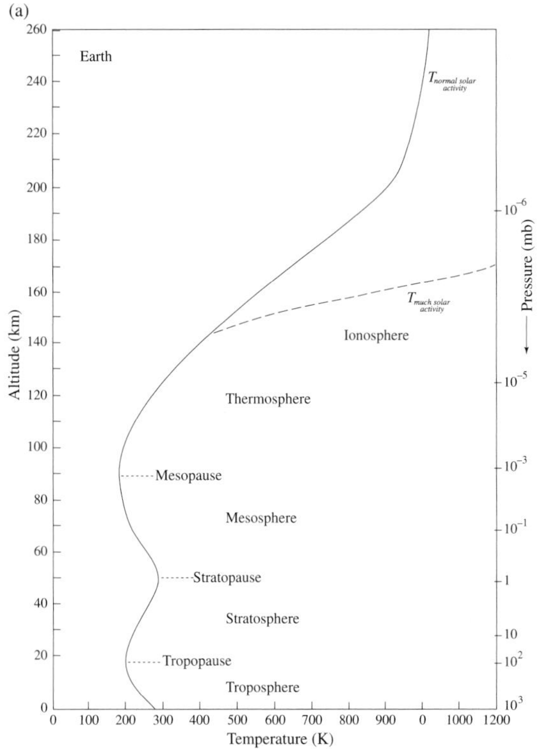

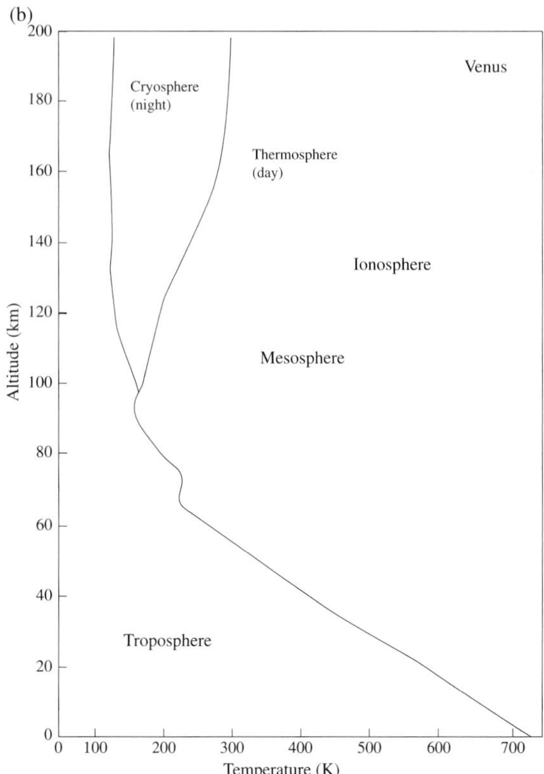

FIGURE 4.1 The approximate thermal structure of the atmospheres of (a) Earth, (b) Venus, (c) Mars, (d) Titan, and (e) Jupiter, Saturn, Uranus, and Neptune. The temperature–pressure profile in each planet's atmosphere is shown as a function of altitude (terrestrial planets) or pressure (gaseous planets). The profile for Venus is derived from Seiff (1983); for Mars from Barth *et al.* (1992); for Titan from Hunten *et al.* (1984) and Yelle *et al.* (1997). The profiles for the giant planets were taken from Lindal (1992).

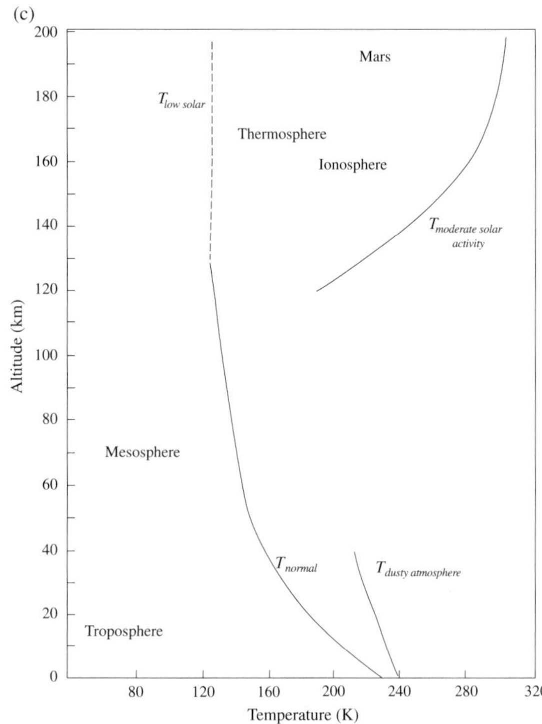

by thermospheric winds induced by auroral particle precipitation.

An additional heating mechanism of a planet's thermosphere is *Joule heating*, resulting from electric currents in a planet's ionosphere. The dissipation of electrical energy occurs through charged particle collisions (see Section 4.6.3).

4.2.1.2 *Energy Transport*

The temperature structure in an atmosphere is governed by the efficiency of energy transport. There are three distinct mechanisms to transport energy: conduction, convection and radiation. Each of these mechanisms is discussed in detail in Section 3.2.

Conduction:

Energy is transferred by collisions between particles. This type of transport is important in the very upper part of the

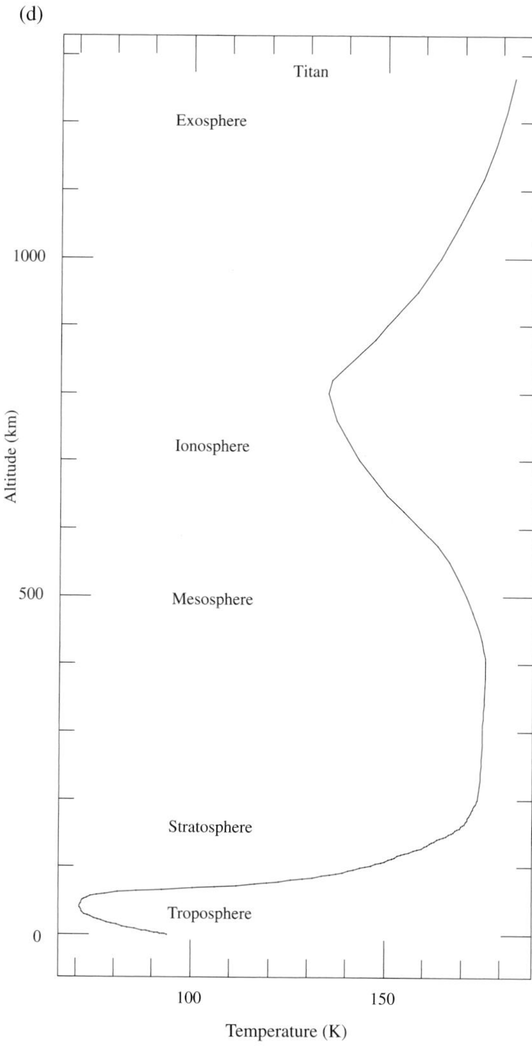

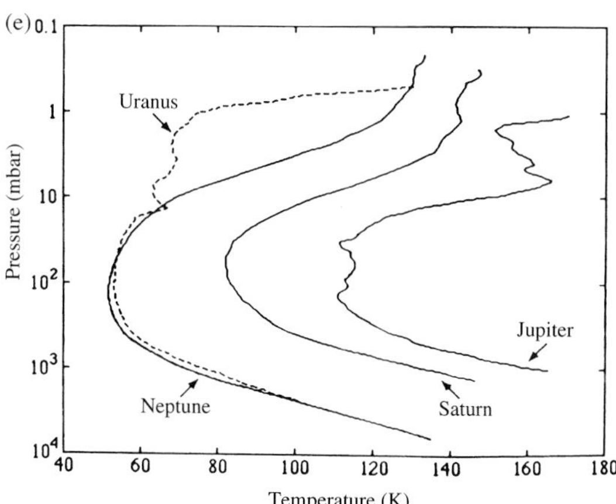

FIGURE 4.1 Continued.

thermosphere and in the exosphere, and very near the surface if one exists. Collisions tend to equalize the temperature distribution in an atmosphere, resulting in a nearly isothermal profile in the exosphere.

Convection:

Energy transport in a planet's troposphere is usually governed by convection, which leads to an adiabatic temperature profile. The *dry adiabatic lapse rate* was derived in Section 3.2.3.3 (eq. 3.73); the formation of clouds decreases the temperature gradient due to the latent heat of condensation (cf. Section 4.4.1). Convection is extremely efficient at transporting energy whenever the lapse rate becomes superadiabatic. Energy transport via convection thus effectively places an upper bound to the rate at which the temperature can decrease with height. Substantial superadiabatic gradients are only possible when convection is suppressed, for example by gradients in the mean molecular weight or by the presence of a flow-inhibiting boundary, such as a surface.

Radiation:

When the most efficient way for energy transport is by absorption and re-emission of photons, the thermal profile in an atmosphere is governed by the equations of radiative energy transport. The temperature structure for an atmosphere in radiative equilibrium was derived in Section 3.3.1 (eq. 3.81).

The thermal structure in any part of an atmosphere is governed by the most efficient mechanism to transport energy. In Chapter 3, the thermal structure was calculated for an atmosphere in which energy transport was either by convection (eq. 3.73) or radiation (eq. 3.81). Which process is most efficient depends upon the atmospheric temperature gradient, $-dT/dz$. Under equilibrium conditions, the temperature gradient can never exceed the dry adiabatic lapse rate. If the temperature gradient in a radiative atmosphere, $(-dT/dz)_{rad}$, is larger than that in a convective atmosphere, $(-dT/dz)_{conv}$, the atmosphere is superadiabatic, and convection will drive the temperature gradient into an adiabatic lapse rate. If $(-dT/dz)_{rad} < (-dT/dz)_{conv}$, energy is transported by radiation and the atmosphere is in radiative equilibrium. Superadiabatic temperature profiles can exist in an atmosphere if there is a stabilization mechanism, such as density gradients. In the tenuous upper parts of the thermosphere, energy transport is dominated by conduction. At deeper layers, down to a pressure of ~ 0.5 bar, an atmosphere is usually in radiative equilibrium, and below that convection dominates.

4.2.2 Observed Thermal Structure

The upper part of an atmosphere, at $P \lesssim 10$ μbar, is typically probed at UV wavelengths, or via stellar occultations at UV, visible and IR wavelengths. At optical and IR wavelengths one 'sees' the radiative part of an optically thick atmosphere, while convective regions at $P \gtrsim 0.5$–1 bar can be investigated at IR and radio wavelengths. These numbers differ slightly from planet to planet, since the opacity at different wavelengths depends upon the precise composition. Tables 4.1 and 4.2 show a comparison between the (observed) effective temperature and the (calculated) equilibrium temperature for all planets. The atmospheric structure is shown graphically in Figure 4.1 for many planets. As shown, the observed effective temperatures of Jupiter, Saturn and Neptune are substantially larger than the equilibrium values, which implies the presence of internal heat sources (Section 6.1.5). For the terrestrial planets, the observed surface temperature often exceeds the equilibrium value because of the greenhouse effect (Section 3.3.2).

Our knowledge of the thermal structure of the terrestrial planets is heavily based on *in situ* measurements by probes and/or landers, although temporal variations in the martian atmosphere have been derived from microwave measurements of various transitions and isotopes of CO. The thermal structure of the giant planets has been derived by inverting IR spectra, combined with UV and radio occultation profiles obtained from the Voyager spacecraft. At deeper levels in the atmosphere, typically at pressures $P > 1$–5 bar, where no direct information on the temperature structure can be obtained, one usually assumes the temperature to follow an adiabatic lapse rate. *In situ* observations by the Galileo probe in Jupiter's atmosphere showed the temperature lapse rate to be close to that of a dry adiabat.

4.2.2.1 *Terrestrial Planets and Titan*

Earth

The average temperature on Earth is 288 K, and the average surface pressure at sea level is 1.013 bar. This temperature is ~35 K warmer than the equilibrium value, a difference which can be attributed to the greenhouse effect (Section 3.3.2), due most notably to the presence of water vapor, carbon dioxide (CO_2) and a variety of trace gases including ozone (O_3), methane (CH_4) and nitrous oxide (N_2O). The troposphere extends up to an altitude of ~20 km at the equator, decreasing to ~10 km above the poles. Above the tropopause, the temperature in the stratosphere increases with altitude as a result of the formation and presence of ozone (O_3), which absorbs both at UV and IR wavelengths. Above the stratopause at ~50 km the temperature decreases with altitude, caused by a decrease in O_3 production and an increase in the CO_2 cooling rate to space. This region is referred to as the mesosphere. A second temperature minimum is found at around ~80–90 km, known as the mesopause. The temperature structure in Earth's stratosphere–mesosphere is unusual; massive atmospheres other than Earth's and Titan's show a single temperature minimum. In Earth's thermosphere, above ~80–90 km, the temperature increases with altitude, due to absorption of UV sunlight (O_2 photolysis and ionization) and the fact that there are too few atoms/molecules to cool the atmosphere efficiently by emitting IR radiation. Most of the IR emission originates from O and NO, molecules which radiate less efficiently than CO_2. At the base of the thermosphere, the CO_2 emission becomes strong enough to radiate the remaining energy. The temperatures in the upper thermosphere vary approximately from 800 K during the night up to 1000 K or more during the day.

Venus

Venus's atmosphere can be divided into three distinct regions: the lower, middle and upper atmosphere. The lower atmosphere, or troposphere, extends from the ground up to the top of the cloud layers, at ~70 km. The surface temperature and pressure are 733 K and 92 bar, respectively. The equilibrium temperature for a rapidly rotating Venus is only ~240 K; the observed excess in temperature is the result of a strong greenhouse effect due primarily to the large concentration of CO_2 gas. At infrared wavelengths, where the top of the cloud layers are probed, the observed temperature is ~240 K, in agreement with the temperature expected from solar irradiation alone. The mean lapse rate from the surface up to the base of the cloud layers at ~45 km is ~7.7 K km^{-1}, slightly less than the mean adiabatic lapse rate of 8.9 K km^{-1}. No temperature variations (diurnal, latitudinal or temporal) have been measured in excess of ~5 K.

The middle atmosphere, the mesosphere, extends from the top of the cloud layers up to ~100 km. The temperature lapse rate decreases sharply at ~60 km, and the atmosphere is nearly isothermal between 80 and 100 km altitude. At higher altitudes, in the thermosphere, there is a distinct difference in temperature between the day and night sides. Above about 100 km on the day side the temperature begins to rise, reaching about 300 K at 170 km. The night side is much colder, 100–130 K, and is often referred to as the *cryosphere*. The temperature rise in the thermosphere is much less than that on Earth, despite Venus's closer proximity to the Sun. The relatively small rise in temperature is attributed to the large concentration of CO_2 gas, a very efficient cooling agent.

Mars

The average surface pressure on Mars is 6 mbar, and the mean temperature ~215 K. However, due to the planet's low atmospheric pressure and hence low thermal inertia, its obliquity and eccentric orbit, the surface temperature displays large latitudinal, diurnal and seasonal variations. At the equator, the surface temperature drops to ~200 K at night and peaks at ~300 K during the day. The temperature at the winter pole is ~130 K, while the summer pole temperature is ~190 K. Mars Pathfinder typically measured temperatures from 195 K up to 265 K during the course of a day. Although the temperature in the lower 50 km of the atmosphere as measured by Mars Pathfinder in 1997 was very similar to that measured by the Viking spacecraft in 1976, ground-based microwave measurements which probe the 0–70 km altitude range have shown that the atmospheric temperature varies substantially on timescales of months to years, variations which appear strongly correlated with the amount of dust entrained in the atmosphere (Fig. 4.2). At altitudes above ~60 km the Viking profile is typically ~15 K warmer than the profile measured by Mars Pathfinder. Pronounced pressure variations are induced by the condensation of a significant fraction of Mars's CO_2-dominated atmosphere onto the planet's seasonal polar caps (see Section 4.5.1.3).

The adiabatic lapse rate for a clear CO_2 atmosphere in radiative–convective equilibrium is 5 K km^{-1}, but the observed lapse rate for Mars is seldom steeper than 3 K km^{-1}. Like Venus, Mars lacks a stratosphere. At altitudes above ~120 km, in Mars's thermosphere, the temperature is almost isothermal with a value of ~160 K. As on Venus, the low temperature can be explained by the efficiency of CO_2 as a cooling agent.

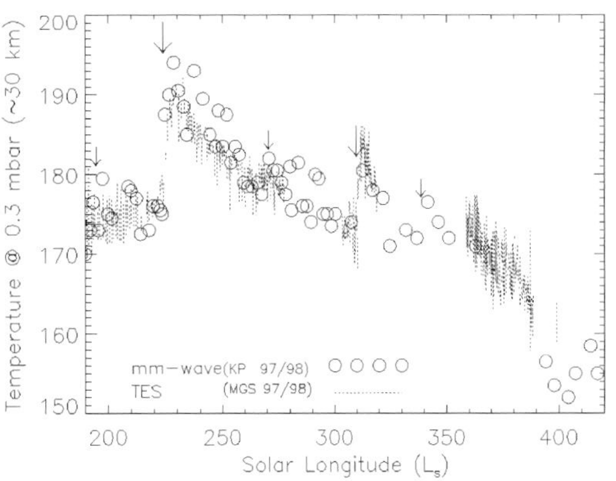

FIGURE 4.2 Ground-based millimeter and submillimeter spectra of Mars atmospheric CO yield vertical profiles of average temperatures for low-to-mid latitudes. This graph presents 1997–1998 Mars atmospheric temperatures at the 0.3 mbar atmospheric pressure level, retrieved from millimeter CO observations (open circles) and from thermal IR sounding observations with the Thermal Emission Spectrometer on the Mars Global Surveyor (MGS) orbiter (dashed vertical lines for 30° S to 30° N latitude range). The 20 K seasonal variation (with Mars solar longitude, L_s) is forced by Mars's eccentric orbit, through reduced solar flux and enhanced dust storm heating of the atmosphere near perihelion ($L_s = 251°$). The arrows indicate discrete global heating events due to abrupt increases in atmospheric dust loading. The large jump at $L_s = 226°$ corresponds to the Noachis regional dust storm in November of 1997. (Clancy *et al.* 2000)

Titan

The temperature structure in Titan's atmosphere resembles that of Earth: the surface pressure is 1.46 bars, with an average surface temperature of 93 K, and the troposphere is characterized by an adiabatic lapse rate. The tropopause is located near an altitude of 40 km, at 140 mbar, with a temperature of ~70 K. Above the tropopause, the temperature rises up to 177 K at an altitude of 320 km ($P \approx$ 50–60 μbar). In the mesosphere the temperature decreases with increasing altitude, due to radiative cooling into space in the rotation–vibration bands of several hydrocarbons. At lower altitudes the photons from these constituents cannot escape into space, and the same species act as heat sources. A second temperature minimum is therefore encountered at the mesopause, ~130–140 K at an altitude of ~500–600 km ($P \approx$ 0.1–0.3 μbar), above which the temperature rises to 175 K as a result of solar EUV heating. The temperature in the thermosphere stays relatively low due to radiative cooling by HCN, which is produced in the thermosphere as a byproduct of ionospheric chemistry. In the exosphere, at altitudes above ~1000 km ($P \lesssim 10^{-3}$–10^{-4}), the temperature profile is isothermal at 175 K.

4.2.2.2 *Giant Planets*

The various parameters which characterize the thermal structure of the giant planets' atmospheres are summarized in Table 4.2. The observed effective temperatures of Jupiter, Saturn and Neptune are significantly higher than expected from solar insolation alone. This excess emission, tabulated as an energy balance, implies that Jupiter, Saturn and Neptune emit roughly twice as much energy as they receive from the Sun. This has been attributed to internal heat sources, which are discussed in more detail in Sections 6.1.5 and 6.4. For Uranus, the upper limit to an internal heat source is 13% of the solar energy absorbed by the planet. It is not known why Uranus's internal heat source is so different than that of the other three giant planets.

The thermal profiles for all four giant planets likely follow adiabats in the troposphere; the tropopause occurs at a pressure level between ~100 and 200 mbar. Tropopause temperatures vary from ~ 50 K for Uranus and Neptune to 105 K for Jupiter. At higher altitudes, in the stratosphere, the temperature increases with height. The stratopause is reached at a pressure level of approximately 1 mbar and a temperature of ~150 K. Above the stratopause lies a near-isothermal region, the mesosphere. At pressures $\lesssim$ 1 μbar, the temperature increases dramatically with altitude (cf. the high exospheric temperatures listed in Table 4.2).

It is interesting to note that all four giant planets have almost the same mesospheric temperatures despite the fact that the solar and planetary heat sources differ by factors of ~ 30. Methane gas and dust or smog of photochemical origin in the stratosphere absorb at IR (3.3 μm) and UV wavelengths, while methane (CH_4), ethane (C_2H_6) and acetylene (C_2H_2) are efficient coolants in the stratosphere/mesosphere at wavelengths between 8 and 14 μm. At 150 K, the peak of the Planck function is near 19 μm, and barely overlaps with the 12.2 μm ethane band. If the atmosphere were colder, there would be no overlap, so cooling would be less efficient. As the 12.2 μm transition is closer to the peak of the Planck function for an atmosphere warmer than 150 K, cooling becomes more efficient as the temperature rises. Thus, there is an effective thermostat which keeps the mesosphere close to 150 K.

4.3 Atmospheric Composition

The composition of a planetary atmosphere can be measured either via remote sensing techniques, or *in situ* using mass spectrometers on a probe or lander. In a mass spectrometer the atomic weight and number density of the gas molecules is precisely measured. However, molecules are not uniquely specified by their mass (unless it is measured far more precisely than is currently possible using spacecraft) and isotopic variations further complicate the situation. Hence, the atmospheric composition has to be determined from a combination of *in situ* measurements, observations via remote sensing techniques and/or theories regarding the most probable atoms/molecules to fit the mass spectrometer data. *In situ* measurements have been made in the atmospheres of Venus, Mars, Jupiter and the Moon (and, of course, Earth). These data contain a wealth of information on atmospheric composition, since trace elements and atoms/molecules which do not exhibit observable spectral features, such as helium, nitrogen and the noble gases, can be measured with great accuracy. A drawback of such measurements, besides the cost, is that they are performed only along the path of the probe/lander at one specific moment in time; so the data, though extremely valuable, may not be representative of the atmosphere as a whole at all times.

Spectral line measurements are performed either in reflected sunlight or from a body's intrinsic thermal emission. The central frequency of a spectral line is indicative of the composition of the gas (atomic and/or molecular) producing the line, while the shape of the line contains information on the abundance of the gas, as well as the temperature and pressure of the environment. Ground-based instruments have improved considerably over the past decades, and, depending on spectral line strengths, small amounts of trace gases (volume mixing ratios of $\lesssim 10^{-9}$ in the giant planet atmospheres) or the composition of extremely tenuous atmospheres (Mercury: $P \lesssim 10^{-12}$ bar) can be measured. At the same time, high angular resolutions can be obtained ($\lesssim 0.5''$ from the ground via conventional observing techniques, and up to an order of magnitude better via adaptive optics, speckle and interferometric techniques, and from HST), so the spatial distribution of the gas over the disk can be measured. In addition, the line profile may contain information on the altitude distribution of the gas (through its shape) and the wind velocity field (through Doppler shifts).

In this section we discuss spectra and spectral line profiles, followed by a description of the atmospheric composition of the planets and various satellites. The basic principles of atomic and molecular line transitions were discussed in Section 3.2.2.

4.3.1 Spectra

Spectra include emission and absorption lines resulting from transitions between energy levels in atoms or molecules (Section 3.2.2). Generally, in astrophysics, one sees absorption lines when atoms/molecules absorb photons at a particular frequency from a beam of broad-band radiation, and emission lines when they emit photons. In the case of absorption, the intensity at the center of the line, $\mathcal{F}_{\nu_o}$ is less than the intensity from the background continuum level, $\mathcal{F}_c$: $\mathcal{F}_{\nu_o} < \mathcal{F}_c$ (Fig. 4.3); for emission lines $\mathcal{F}_{\nu_o} > \mathcal{F}_c$. For planets, we see the effect of atomic and molecular line absorption both in spectra of reflected sunlight (at UV, visible and near-infrared wavelengths) and in thermal emission (at infrared and radio wavelengths) spectra (Fig. 4.3). Planets, moons, asteroids and comets are visible because sunlight is reflected off their surface, cloud layers or atmospheric gases. Sunlight itself displays a large number of absorption lines, the *Fraunhofer ab-*

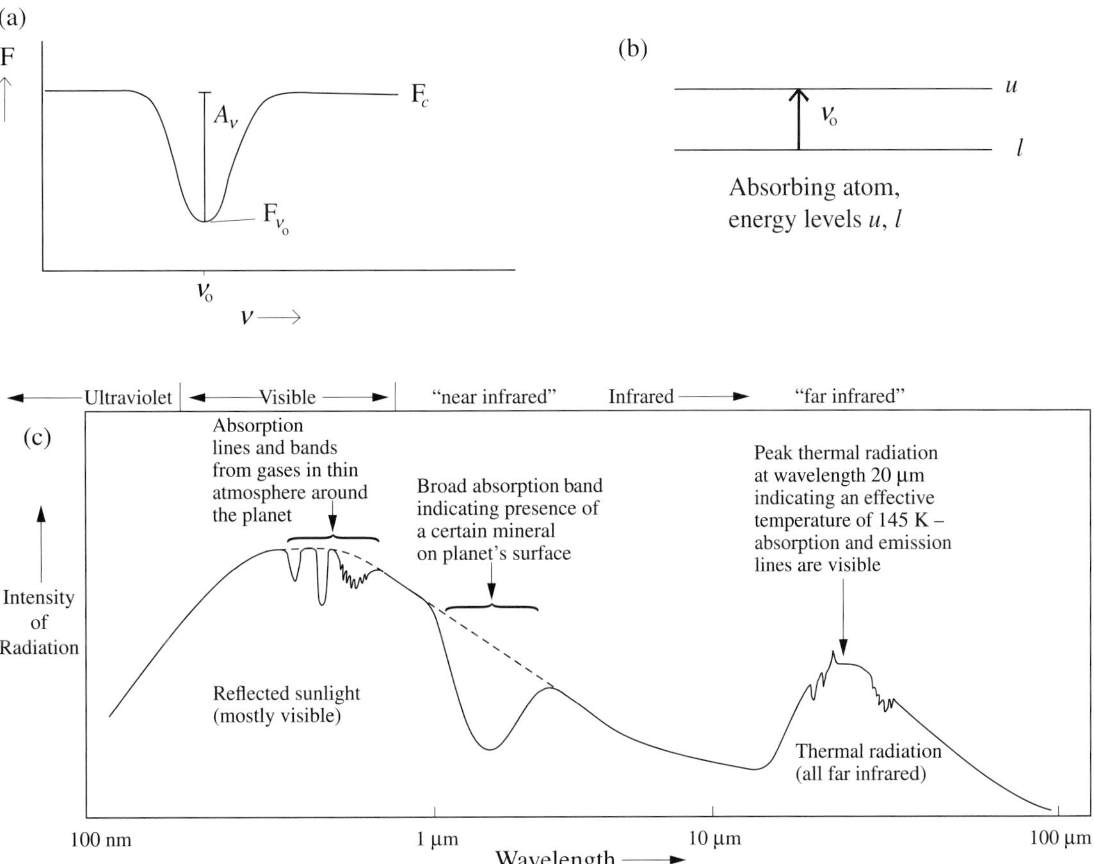

FIGURE 4.3 (a) Sketch of an absorption line profile. The flux density at the continuum level is $\mathcal{F}_c$, and at the center of the absorption line at frequency ν_o, the flux density is $\mathcal{F}_{\nu_o}$. The absorption depth is A_ν. (b) Sketch of the upper, u, and lower, ℓ, energy levels in an atom giving rise to the absorption line in (a). (c) Example of a spectrum from a hypothetical planet with an effective temperature of 145 K. The spectrum is shown from ultraviolet through far-infrared wavelengths. At the shorter wavelengths, the Sun's reflected spectrum is shown. The dashed line shows the spectrum if there were no absorption lines and bands. The spectrum is already corrected for the Sun's Fraunhofer line spectrum. At infrared wavelengths, the planet's thermal emission is detected, where both absorption and emission lines might be present. Note the hyperfine structure of the molecular bands. (Figure adapted from Hartmann 1989)

sorption spectrum, since atoms in the outer layers of the Sun's atmosphere (photosphere) absorb part of the sunlight coming from the deeper, warmer layers. If all of the sunlight hitting a planetary surface is reflected back into space, the planet's spectrum is shaped like the solar spectrum (Fig. 4.4), aside from an overall Doppler shift induced by the planet's motion (eq. 4.12); the spectrum thus exhibits the solar Fraunhofer line spectrum. Atoms and molecules in a planet's atmosphere or surface may absorb some of the Sun's light at specific frequencies, which becomes visible as additional absorption lines in the planet's spectrum (Fig. 4.4). For example, Uranus and Neptune are greenish-blue because methane gas, abundant in these planets' atmospheres, absorbs in the red part of the visible spectrum, so primarily bluish sunlight is reflected back into space.

As in the case of the Sun, most of the thermal emission from a planetary atmosphere comes from deep warmer layers and may be absorbed by gases in the outer layers. In the Sun's photosphere the temperature decreases with altitude, and the Fraunhofer absorption lines are visible as a decrease in the line intensity. Similarly, spectral lines

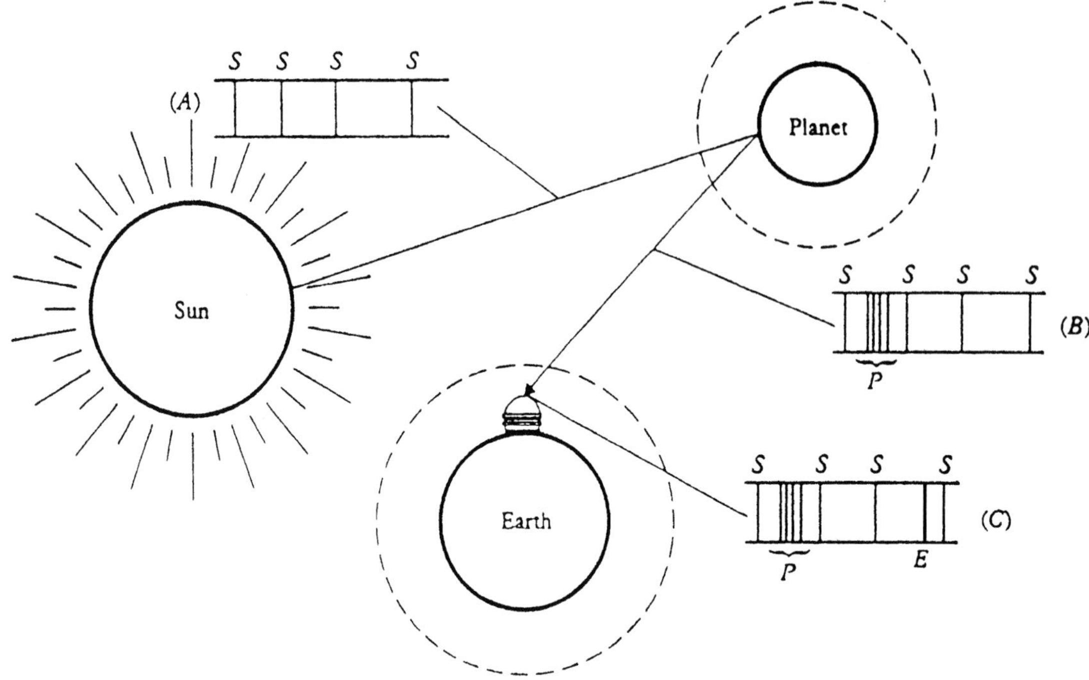

FIGURE 4.4 A sketch to help visualize the various contributions to an observed planetary spectrum. Sunlight, with its absorption spectrum (indicated by lines S) is reflected off a planet, where the planet's atmosphere may produce additional absorption/emission lines, P. Finally, additional absorption may occur in the Earth's atmosphere before the spectrum is recorded, indicated by lines E. (Adapted from Morrison and Owen 1996)

formed in a planet's troposphere are also visible as absorption profiles. When we take a spectrum of a planet's atmosphere, the optical depth at the center of the line is always largest, (much) larger than in the far wings or with respect to the continuum background. The measured line profile reflects the temperature at the altitude probed. In the troposphere, the temperature decreases with altitude, so that lines forming in the troposphere are seen in absorption against the warm continuum background. In contrast, if a line is formed above the tropopause, where the temperature is increasing with altitude, the line may be seen in emission against the cooler background. Thus, whether a spectral line is seen in emission or absorption depends upon the temperature–pressure profile in the region of line formation. Therefore, rather than speaking about emission or absorption lines, in atmospheric sciences one says that spectral lines are seen *in emission* when $\mathcal{F}_{\nu_o} > \mathcal{F}_c$ and *in absorption* when $\mathcal{F}_{\nu_o} < \mathcal{F}_c$.

Energy transitions triggered by electronic excitation/de-excitation of atoms and molecules are observed primarily at optical and UV wavelengths. Energy transitions induced by vibrations of atoms can be observed at infrared and submillimeter wavelengths, and those resulting from a rotation of molecules can be observed at radio wavelengths. The detailed shape of a line profile is determined by the abundance of the element or compound producing the line, as well as the pressure and temperature of the environment. The absorption depth is defined as:

$$A_\nu \equiv \frac{\mathcal{F}_c - \mathcal{F}_{\nu_o}}{\mathcal{F}_c}, \tag{4.5a}$$

with $\mathcal{F}_c$ the flux density of the continuum background and $\mathcal{F}_{\nu_o}$ the flux density at the center of the absorption line (Fig. 4.3). With help of equation (3.48) this can be written as:

$$A_\nu = 1 - e^{-\tau_\nu}. \tag{4.5b}$$

Unfortunately, it is not always possible to resolve the line profile. For unresolved lines one measures the *equivalent width, EW*:

$$EW = \int_0^\infty A_\nu d\nu = \int_0^\infty \frac{\mathcal{F}_c - \mathcal{F}_\nu}{\mathcal{F}_c} d\nu$$
$$= \int_0^\infty (1 - e^{-\tau_\nu}) d\nu. \tag{4.6}$$

The equivalent width is equal to the area between the line and the continuum. For an absorption line, EW is equal to the width of a totally black line ($\mathcal{F}_\nu = 0$), with the same total flux absorbed as by the line. The absorption depth, A_ν, of the line is determined by the optical depth τ_ν:

$$\tau_\nu = \tau_{\nu_o}\Phi_\nu, \tag{4.7}$$

with the optical depth at the center of the line

$$\tau_{\nu_o} = \int_0^L N\alpha_{\nu_o} dl, \tag{4.8}$$

and the *line shape*

$$\Phi_\nu \equiv \frac{\alpha_\nu}{\alpha_{\nu_o}}. \tag{4.9}$$

The symbol α_ν corresponds to the extinction coefficient at frequency ν (Section 3.2.2.4), while α_{ν_o} is the extinction coefficient at the line center. (Note that $\alpha_\nu = \kappa_\nu$ when $\sigma_\nu = 0$.)

If the thermal structure in an atmosphere is known, spectral line observations can be used to derive the integrated density of the absorbing material. If the line itself is not resolved but its shape is known, measurements of EW can be used. For an optically thin line ($\tau \ll 1$), EW can be written with help of equations (4.6), (4.8) and (4.9):

$$EW \approx \int_0^\infty \tau_\nu d\nu = \int_0^\infty \tau_{\nu_o}\Phi_\nu d\nu = N_c \int_0^\infty \alpha_{\nu_o}\Phi_\nu d\nu, \tag{4.10}$$

with $N_c \equiv \int N dl$ the column density of the absorbing material. The equivalent width increases linearly with N_c as long as $\tau_\nu \ll 1$. When the optical depth increases, the line profile becomes saturated, and the equivalent width cannot continue to increase linearly with the column density. When $\tau \gg 1$, it can be shown that the equivalent width is proportional to the square root of N_c (Problems 4.3–4.5). A graph of the equivalent width as a function of column density is called a *curve of growth*, and can be used to determine the abundance of an element from its observed line width. Depending on the line profile, there may be a region in between the linear and square root regime where the curve of growth is essentially flat, i.e., the equivalent width is nearly independent of column density.

4.3.2 Line Profiles

The shape of emission and absorption lines is determined by the abundance of the element or compound, and by the pressure and temperature of the environment. Spectral lines are therefore used to determine the abundance of an element or compound at a specific altitude or its distribution over altitude, as well as the thermal structure of an atmosphere over the altitude range probed. Since the observed spectral line center depends upon the observed radial (along the line of sight) velocity of the gas molecules (Doppler shift, eq. 4.12), spectral line observations can also be used to determine the wind velocity field in a planetary atmosphere. In this section, we discuss the most common line profiles encountered in planetary atmospheres.

4.3.2.1 *Natural Damping: Lorentz Profile*

Emission and absorption lines always display some width. The narrowest profile that a line can have is given by the natural damping profile, which results from the finite lifetime of excited states. A natural broadened line profile is given by a *Lorentz* line shape:

$$\alpha_\nu = \alpha \frac{4\Gamma}{(4\pi)^2(\nu - \nu_o)^2 + \Gamma^2}, \tag{4.11}$$

where Γ is the reciprocal of the lifetime of all states giving rise to emission or absorption at frequency ν ($\Gamma \propto 1/\Delta t \propto \Delta\nu$), ν_o is the central frequency, and α is the spectrally integrated extinction coefficient (eq. 3.42):

$$\alpha = \int \alpha_\nu d\nu = \frac{\pi q^2}{mc} f_{osc},$$

with f_{osc} the oscillator strength.

4.3.2.2 *Doppler Broadening: Voigt Profile*

When an atom has a velocity v_r along the line of sight, the frequency of its emission and absorption lines is Doppler shifted by the amount

$$\Delta\nu = \frac{\nu v_r}{c}. \tag{4.12}$$

The Doppler shift is positive (*blue shifted*) if the atom moves towards the observer, and negative (*red shifted*) if the atom moves in the opposite direction. In an atmosphere, atoms and molecules move in all directions: the radial velocity can generally be expressed by a Maxwellian velocity distribution. The net effect of such a velocity distribution on the line shape is a broadening of the line profile. The shape of such a profile can be obtained by a convolution of the Lorentzian line profile with the velocity (Maxwellian) distribution.

The probability $P(v_r)dv_r$ of finding an atom with a radial velocity between v_r and $v_r + dv_r$ is given by a Maxwellian distribution:

$$P(v_r)dv_r = \frac{1}{\sqrt{\pi}} e^{-(v_r/v_{r_o})^2} \frac{dv_r}{v_{r_o}}, \tag{4.13}$$

where $v_{r_o} = \sqrt{2kT/(\mu_a m_{amu})}$ and the product $\mu_a m_{amu}$ is the molecular mass. The absorption profile can be written:

$$\alpha_\nu = \int_{-\infty}^{\infty} \alpha_{\nu_o}\left(\nu - \frac{\nu v_r}{c}\right) P(v_r) dv_r, \tag{4.14}$$

which results in the *Voigt profile* for absorption:

$$\alpha_\nu = \alpha \frac{1}{\sqrt{\pi}\Delta\nu_D} H(a, x), \tag{4.15}$$

with the *Doppler width* $\Delta\nu_D$:

$$\Delta\nu_D \equiv \frac{v_{r_o}\nu_o}{c}. \tag{4.16}$$

The Doppler width is the full width of the line at half power divided by a factor of $2\sqrt{\ln 2}$ (Problem 4.8). The *Voigt function*, $H(a, x)$, is defined:

$$H(a, x) \equiv \frac{a}{\pi} \int_{-\infty}^{\infty} \frac{e^{-y^2} dy}{(x-y)^2 + a^2}, \tag{4.17}$$

with $x \equiv (\nu - \nu_o)/\Delta\nu_D \equiv (\lambda - \lambda_o)/\Delta\lambda_D$; $y \equiv v_r/v_{r_o}$; and $a \equiv \Gamma/(4\pi\Delta\nu_D)$. Note that $H(a, x = 0) = 1$ and $\int_{-\infty}^{\infty} H(a, x) dx = \sqrt{\pi}$.

When Doppler broadening dominates, the Voigt profile can be represented schematically by:

$$\alpha_\nu \approx e^{-x^2} + \frac{a}{\sqrt{\pi} x^2}. \tag{4.18}$$

The first term in equation (4.18) represents the core of a Doppler broadened line, which is Gaussian up to a width of $\sim 3\Delta\nu_D$. The second term is due to the wings from the natural broadening profile.

4.3.2.3 *Pressure or Collisional Broadening*

In a dense gas collisions between particles dominate, and perturb the energy levels of the electrons such that photons with a slightly lower or higher frequency can cause excitation/deexcitation. This leads to a broadening of the line profile. The line shape can be expressed by a Lorentzian profile (eq. 4.11), with Γ the reciprocal of the lifetime of all states giving rise to emission/absorption at frequency ν. In a collision-dominated environment $\Gamma = 2/t_c$, where t_c is the mean time between molecular collisions. In contrast to the narrow Lorentz profile, where Γ is the reciprocal of the lifetime in absence of collisions, the collision-broadened profile is sometimes referred to as the *Debye line shape*.

When pressure increases, the relative time molecules spend in collisions with other molecules increases, and the effects of these collisions become more important. When a molecule undergoes a collision, its geometry is temporarily altered, which changes the central frequency of the range of photons that the molecule can absorb. In an ensemble of molecules, the frequency shifts in each molecular absorption cause a broadening in the line width of each transition. As the line widths increase, individual lines may (partially) overlap each other; when the line widths become comparable to the average frequency spacing between individual lines, the individual line characters are lost. Instead, the molecular absorption can be represented by one broad absorption line. In 1945, Van Vleck and Weisskopf derived an expression for this absorption, based upon a quantum mechanical treatment of absorbing molecules in a collisional environment. The *Van Vleck–Weisskopf* line profile is given by:

$$\alpha_\nu = \alpha \left(\frac{\nu}{\nu_0}\right)^2 \left(\frac{4\Gamma}{(4\pi)^2(\nu - \nu_o)^2 + \Gamma^2} + \frac{4\Gamma}{(4\pi)^2(\nu + \nu_o)^2 + \Gamma^2}\right). \tag{4.19}$$

Note that the Van Vleck–Weisskopf profile at high frequencies is equal to the Debye (or Lorentzian) line shape. At low frequencies (radio wavelengths) the last term, caused by the so-called negative resonance terms, causes an asymmetry in the line shape (see Fig. 4.5), such that the absorptivity in the high frequency tail is larger than in the low frequency tail at equal distances from the center frequency. This asymmetry in the Van Vleck–Weisskopf line shape mimics observed line profiles. The Van Vleck–Weisskopf line profile is widely used to model line profiles in planetary atmospheres if the line width is determined by pressure broadening.

In the derivation of their line shape, Van Vleck and Weisskopf assumed that the relative time between collisions is much larger than the time spent in collisions. This assumption breaks down at pressures above $P \gtrsim 0.5$–1 bar, as shown by an apparent mismatch between the theory and observed line profiles. In the mid-sixties, Ben Reuven attempted to improve the line profile through a much more complex quantum mechanical treatment, in which he assumed the molecules to undergo collisions constantly, and by including the effects of coupling between adjacent transitions, which becomes important when the individual lines overlap. Since intermolecu-

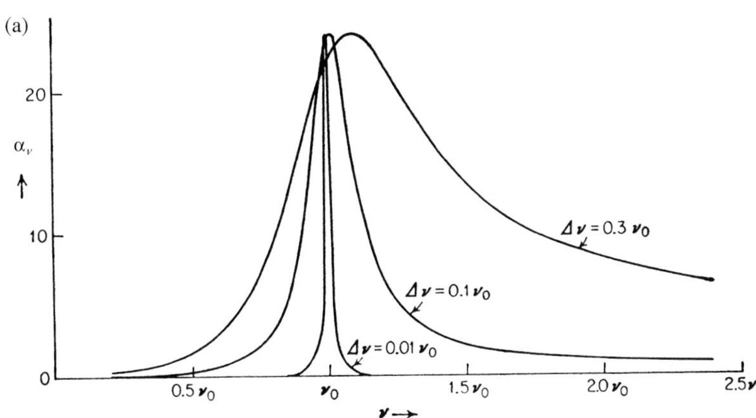

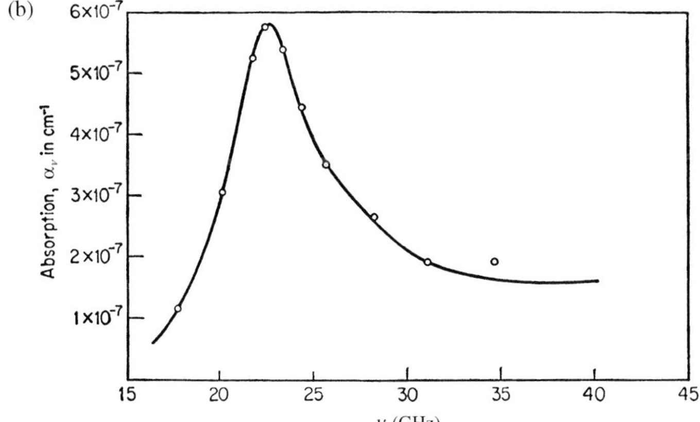

FIGURE 4.5 (a) Calculations of the Van Vleck–Weisskopf (VVW) line profile for various line widths $\Delta\nu$, where $\Delta\nu$ is the half-width at half-power of the line $(1/(2\pi t_c))$. (From Townes and Schawlov 1955). (b) Comparison of the observed line profile for water vapor in air (open circles: 10 g of H_2O per cubic meter) with a VVW line profile. The VVW profile fits the data quite well, although there is a small discrepancy at the higher frequencies. (Adapted from Becker and Autler 1946)

lar forces are still poorly known, the various coefficients in the *Ben Reuven line profile* are usually determined empirically.

We note that the center of a spectral line is generally formed higher up in an atmosphere than the wings of the line. The observed profile, therefore, may consist of a combination of profiles: a Doppler broadened profile at high altitudes (typically at pressures $P < 0.1$ mbar), a Van Vleck–Weisskopf profile down to about the 1 bar level, and a Ben Reuven shape at larger depths. The precise pressure level at which the Doppler or pressure broadened profile dominates is determined for each molecular species and planetary atmosphere separately.

4.3.3 Observations

The atmospheric composition of various bodies is given in Tables 4.3–4.5. In our discussion of planetary atmospheres, unless specified otherwise, we use *volume mixing ratios*, i.e., the fractional number density of particles in a given volume, often indicated as a percentage or a ratio of the particular species with respect to the total number of particles in a given volume. For the giant planets, the ratio of the particular species to the total number of hydrogen molecules in a given volume is sometimes used. Trace elements are usually given as parts per million (ppm) or billion (ppb) by volume.

4.3.3.1 *Earth, Venus and Mars*

Earth's atmosphere consists primarily of N_2 (78%) and O_2 (21%). The most abundant trace gases are H_2O, Ar and CO_2, but many more have been identified (Table 4.3). The composition of the atmospheres of Mars and Venus is dominated by CO_2, roughly 95–97% on each planet; nitrogen gas contributes approximately 3% by volume; the most abundant trace gases are Ar, CO, H_2O and O_2. On Venus we find small amounts of SO_2, H_2SO_4 (sulfuric acid) and HCl. Ozone is identified on both Earth and Mars. The differences in composition among the three planets must be tied in with the historical evolution of the bodies, the sur-

TABLE 4.3 Atmospheric Composition of Earth, Venus, Mars and Titan.

Constituent	Earth[a]	Venus	Mars	Titan	References
N_2	0.7808	0.035	0.027	0.90–0.97[b]	1, 2, 3, 4
O_2	0.2095	0–20 ppm	0.13 ppm		1, 2, 5
CO_2	345 ppm	0.965	0.953	10 ppb	1, 2, 4
CH_4	3 ppm			0.005–0.04	2, 3, 4, 6
H_2O	<0.03[c]	50 ppm	<100 ppm[c]	0.4 ppb	1, 7, 5
Ar	0.009	70 ppm	0.016	0.0–0.06[b]	2, 3, 4
CO	0.2 ppm	50 ppm	700 ppm	10 ppm	2
O_3	10 ppm		0.01 ppm		1, 2
HCN				0.1 ppm	4
HC_3N				10–100 ppb	8
C_2H_2	8.7 ppb			2 ppm	4, 9
C_2H_6	13.6 ppb			10 ppm	4, 9
C_3H_8	18.7 ppb			0.5 ppm	4, 9
C_2H_4	11.2 ppb			0.1 ppm	4, 9
C_4H_2				1 ppb	4
CH_3C_2H				30 ppb	8
C_2N_2				10–100 ppb	8
NO	<0.01 ppm		3 ppm		2
N_2O	0.35 ppm				1
SO_2	< 2[c] ppb	60 ppm			9, 10
H_2	0.5 ppm	?	10 ppm	0.002	2, 4
HCl		0.5 ppm			5
HF		5 ppb			5
COS		250 ppb			5
He	5 ppm	12 ppm			2
Ne	18 ppm	7 ppm	2.5 ppm	< 0.01	2, 8
Kr	1 ppm	0.2 ppm	0.3 ppm		2, 5
Xe	0.09 ppm	<0.1 ppm	0.08 ppm		2

All numbers are given in volume mixing ratios, as a fraction or ppm (part per million) or ppb (part per billion).

[a] There are numerous species known to be present in Earth's atmosphere at the $\lesssim$ ppb levels not mentioned in this table.

[b] Uncertain.

[c] Variable.

1: Salby (1996). 2: Chamberlain and Hunten (1997). 3: Yelle (1991). 4: Coustenis and Lorenz (1999). 5: Hunten (1999). 6: Dowling (1999). 7: Coustenis *et al.* (1998). 8: Atreya (1986). 9: Seinfield and Pandis (1998). 10: Fahd and Steffes (1992).

face temperature, volcanic and tectonic activity, as well as the biogenic evolution.

Figure 4.6 shows coarse infrared spectra of Earth, Venus and Mars between 5 and 100 μm; all three spectra were taken from space, although with different spacecraft. Each spectrum displays a pronounced CO_2 absorption profile at 15 μm (wavenumber 667 cm^{-1}), showing that CO_2 gas is present in the troposphere of all three planets. On Earth, there is a small emission spike at the center of the CO_2 absorption line, indicative of CO_2's presence in the Earth's stratosphere, where, in contrast to the troposphere, the temperature is increasing with altitude. Other prominent features in the terrestrial spectrum are ozone at 9.6 μm (1042 cm^{-1}) and methane at 7.66 μm (1306 cm^{-1}). Note the emission spike at the center of the ozone line, like in the CO_2 absorption band. Numerous water lines at

TABLE 4.4 Composition of Planets and Satellites with Tenuous Atmospheres[a].

Planet	Constituent	Abundance (cm^{-3})	References
Mercury	O	4×10^4	1
	Na	3×10^4	
	He	6×10^3	
	K	500	
	H	23 (suprathermal)	
		230 (thermal)	
	Ca	~ 30	2
Moon	He	2×10^3 (day)–4×10^4 (night)	1, 3, 4
	Ar	1.6×10^3 (day)–4×10^4 (night)	
	Na	70	
	K	16	
Pluto	N_2		5
	CO	trace	
	CH_4	trace	
Triton	N_2		6
	CH_4	trace	
Io[b]	SO_2	10^{11}–10^{12}	7
	SO	trace	
	Na		8
	K		
	O		

[a] See also Section 4.3.3.3. All number densities quoted are at the surface.
[b] We only list the neutral species detected in Io's atmosphere; see Table 7.3 for ionized species in Io's plasma torus.
1: Hunten *et al.* (1988). 2: Bida *et al.* (2000). 3: Sprague *et al.* (1992). 4: Strom (1999). 5: Stern and Yelle (1999). 6: Stone and Miner (1989). 7: Lellouch *et al.* (1990, 1995). 8: Bouchez *et al.* (1999).

wavelengths longwards of 20 μm, and shortwards of 7.7 μm, make the Earth's atmosphere almost opaque in these spectral regions. Water lines, though reduced in strength, are also seen in the spectra of Mars and Venus. The CO_2 absorption lines on the three planets can be used to constrain the thermal structure in the atmosphere; in clear conditions, if no other absorbers are present, the planetary surface is probed in the far wings (continuum) of the line (for Venus, the cloud deck rather than surface is probed), while higher altitudes are probed closer to the line center.

Since the emission/absorption lines in planetary atmospheres depend largely on the temperature structure probed, spectra taken at different locations may appear very different, even if similar concentrations of the absorbing gases are involved. An example of this effect is shown for Mars in Figure 4.7, where, as in Figure 4.6, a pronounced CO_2 absorption feature is seen at mid-latitudes. In the polar regions, however, CO_2 shows up in emission. Assuming that one probes the planetary surface in the wings of the lines, blackbody curves were fitted to the background level. The background level of the spectrum at mid-latitudes is considerably depressed from a blackbody curve at 280 K, expected for Mars's surface. This is caused by dust in the martian atmosphere, which absorbs sunlight and heats up the atmosphere (by conduction), while it partially shields the surface from direct sunlight. The CO_2 feature, therefore, is not as pronounced in this figure as it would be under dust-free conditions. The

TABLE 4.5 Atmospheric Composition of the Giant Planets.

Gas	Element	Sun	Jupiter	Saturn	Uranus	Neptune	References
Major gases							
H_2	H	0.835	0.864 ± 0.006	0.963 ± 0.03	0.85 ± 0.05	0.85 ± 0.05	1, 2, 3, 4
He	He	0.195	0.157 ± 0.004	0.034 ± 0.03	0.18 ± 0.05	0.18 ± 0.05	1, 2, 3, 4
Condensable gases							
H_2O	O	1.70×10^{-3}	2.6×10^{-3}	$>1.70 \times 10^{-3}$?	$>1.70 \times 10^{-3}$?	$> 1.70 \times 10^{-3}$?	5, 6
	in stratosphere		$(2–20) \times 10^{-9}$	detection	detection	detection	6, 7
	Galileo, 19 bar		$(6 \pm 3) \times 10^{-4}$				6
CH_4	C	7.94×10^{-4}	$(2.1 \pm 0.2) \times 10^{-3}$	$(4.5 \pm 2.2) \times 10^{-3}$	0.024 ± 0.01	0.035 ± 0.010	4, 8, 9, 10
NH_3	N	2.24×10^{-4}	$(2.60 \pm 0.3) \times 10^{-4}$	$(5 \pm 1) \times 10^{-4}$	$< 2.2 \times 10^{-4}$	$< 2.2 \times 10^{-4}$	6, 8
	Galileo, 8 bar		$(8 \pm 1) \times 10^{-4}$				6
H_2S	S	3.70×10^{-5}	$(2.22 \pm 0.4) \times 10^{-4}$?	$(4 \pm 1) \times 10^{-4}$?	3.7×10^{-4}?	1×10^{-3}?	6, 8
	Galileo, 16 bar		$(7.7 \pm 0.5) \times 10^{-5}$				6
Noble gases							
^{20}Ne	Ne	2.3×10^{-4}	$(2.30 \pm 0.2) \times 10^{-5}$				6
^{36}Ar	Ar	6.1×10^{-6}	$(1.5 \pm 0.3) \times 10^{-5}$				6
^{84}Kr	Kr	1.84×10^{-9}	$(5.0 \pm 1) \times 10^{-9}$				6
^{132}Xe	Xe	8.9×10^{-11}	$(2.3 \pm 0.5) \times 10^{-10}$				6
Disequilibrium species							
PH_3	P	7.50×10^{-7}	6×10^{-7}	$(7 \pm 3) \times 10^{-6}$			6
GeH_4			$(7 \pm 2) \times 10^{-10}$	$(4 \pm 4) \times 10^{-10}$			6
AsH_3			$(2.2 \pm 1.1) \times 10^{-10}$	$(3 \pm 1) \times 10^{-9}$			6
CO			$\sim 2 \times 10^{-9}$	$(1 \pm 0.3) \times 10^{-9}$	$< 1 \times 10^{-8}$	1×10^{-6}	6, 11
CO_2			detection	3×10^{-10}		detection	6, 7
HCN					$< 1 \times 10^{-10}$	1×10^{-9}	11
Photochemical species							
C_2H_2			$(3–10) \times 10^{-8}$	$(2.1 \pm 1.4) \times 10^{-7}$	1×10^{-8}	6×10^{-8}	6, 11
C_2H_4			$(7 \pm 3) \times 10^{-9}$			detection	6, 12
C_2H_6			$(1–5) \times 10^{-6}$	$(3 \pm 1) \times 10^{-6}$	$< 1 \times 10^{-8}$	2×10^{-6}	6, 11
C_3H_4			$(2.5 \pm 2) \times 10^{-9}$				6
C_3H_8			detection				6
C_4H_2			9×10^{-11}	9×10^{-11}			6, 13
C_6H_6			$(2 \pm 2) \times 10^{-9}$	2.5×10^{-10}			6
CH_3C_2H				6×10^{-10}			13

The elements O, C, N, S and P are in the form of H_2O, CH_4, NH_3, H_2S and PH_3 on the giant planets, respectively.
The abundance of H_2S is inferred rather than measured directly, except for the second value for Jupiter, which was measured directly by the Galileo probe.
All numbers are volume mixing ratios, relative to H_2; H_2 is measured relative to the entire atmosphere.
Mole fractions may be calculated by dividing the species number density by the atmospheric number density, thus the mole fraction of He in Jupiter's atmosphere for instance is 0.136.
The solar composition numbers are from Anders and Grevesse (1989) and Grevesse *et al.* (1991).
NH_3 on Jupiter: see also text, and de Pater *et al.* (2000).
CO abundance on Jupiter: $\sim 5 \times 10^{-9}$ at $P > 200$ mbar; $\sim 0.8 \times 10^{-9}$ at $P < 200$ mbar (Bézard *et al.*, 1999).
He is poorly constrained on Saturn. A reanalysis of the Voyager data suggests a He mixing ratio possibly as high as 0.11–0.16 (Conrath and Gautier, 2000).
1: Niemann *et al.* (1998). 2: Atreya (1986). 3: Conrath *et al.* (1989). 4: Gautier *et al.* (1995). 5: Carlson *et al.* (1992). 6: Atreya *et al.* (1999). 7: Bezard *et al.* (1999). 8: de Pater and Mitchell (1993). 9: Courtin *et al.* (1984). 10: Gautier and Owen (1989). 11: West (1999). 12: Encrenaz *et al.* (1999). 13: de Graauw *et al.* (1997).

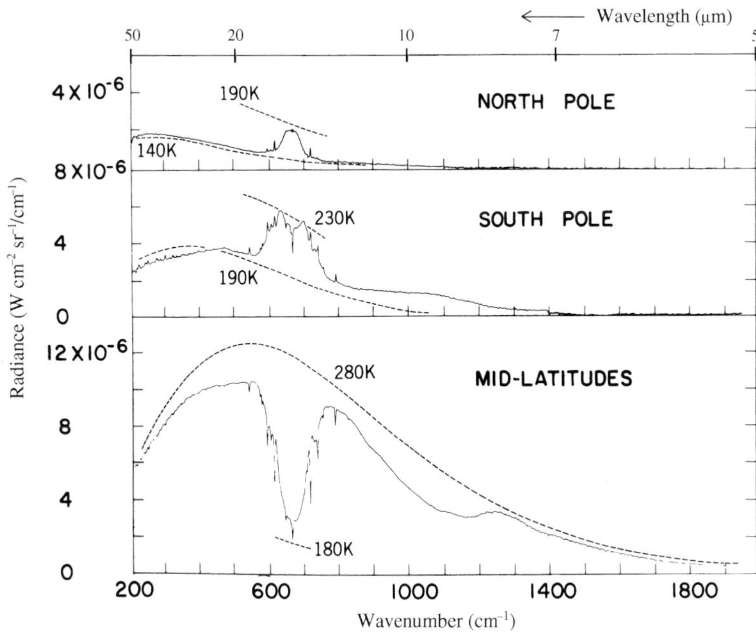

FIGURE 4.7 Thermal infrared spectra of the poles and mid-latitudes on Mars. The data were taken by the Mariner 9 spacecraft when it was late spring in the southern hemisphere. Blackbody curves at various temperatures are indicated for comparison. Note that the CO_2 feature is seen in emission in the polar spectra, while in absorption at mid-latitudes. (Adapted from Hanel *et al.* 1992)

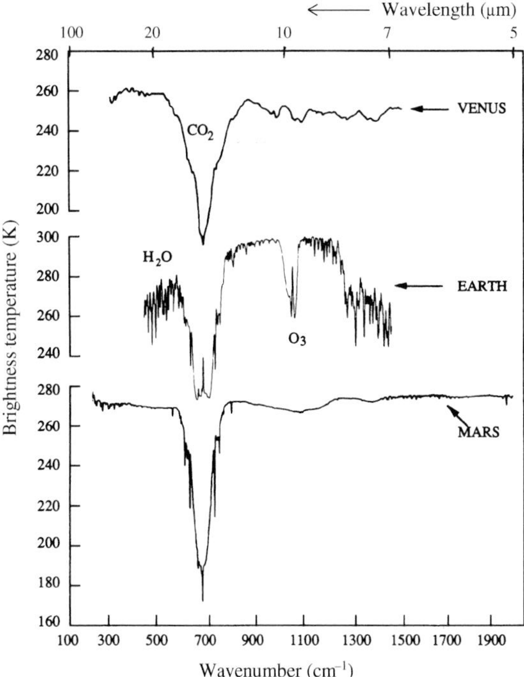

FIGURE 4.6 Thermal infrared emission spectra of Venus, Earth and Mars. The Venus spectrum was recorded by Venera 15, the spectrum of the Earth by Nimbus 4 and that of Mars by Mariner 9. (Adapted from Hanel *et al.* 1992)

broad absorption feature is caused by suspended dust particles. The background level of the spectrum taken from the north polar region can be fitted with a 145 K blackbody curve, the condensation temperature of CO_2 at an atmospheric pressure of 8 mbar. Since the atmosphere above the surface is warmer, CO_2 gas appears in emission. The continuum background temperature for the south polar spectrum cannot be fitted with a single blackbody curve; it can be matched quite well using the sum of two curves: one at 140 K which covers about 65% of the field of view, and one at 235 K covering the rest. This phenomenon has been attributed to sublimation of CO_2-ice above the summer pole: roughly 1/3 of the area was already free of ice, while 60–70% was still frozen. In addition to CO_2, one can also see several water lines in emission in the south polar spectrum.

Microwave observations of carbon monoxide (CO) on Mars and Venus form an important probe of the thermal structure of their atmospheres. The ^{12}CO line is optically thick, whereas the ^{13}CO line is optically thin. Disk-averaged line profiles for the two planets are shown in Figure 4.8. The ^{12}CO line on Mars is formed high up in the atmosphere, where it is cold; hence the core of the line is seen in absorption against the continuum background. Emission wings appear at either side of the line, where the atmosphere just above the surface is probed. As a consequence of the surface emissivity ($\epsilon < 1$), the brightness temperature of the surface is somewhat less than the kinetic temperature of the atmosphere just above it. The wings of the line are therefore seen in emission against the continuum background. On Venus, the continuum emission at millimeter wavelengths arises from within the planet's

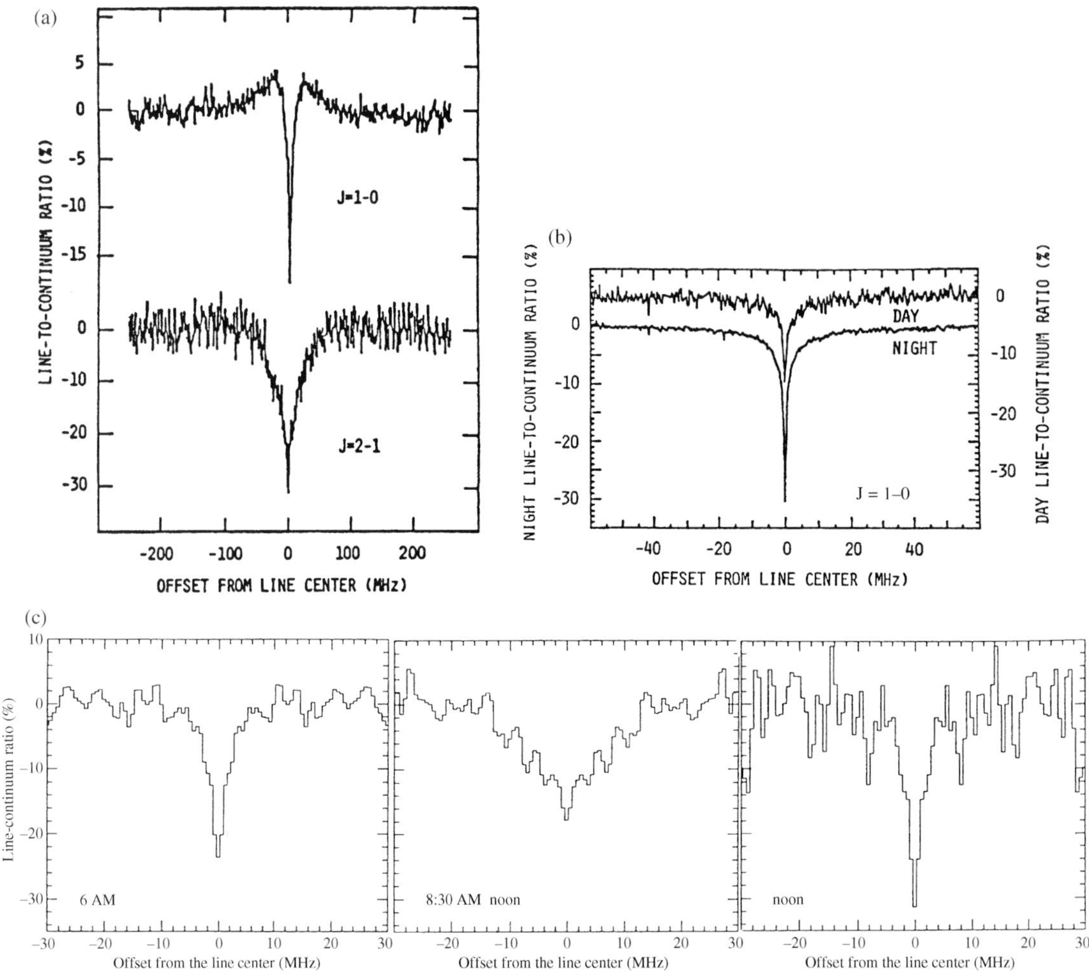

FIGURE 4.8 Carbon monoxide radio spectra of Mars and Venus. (a) Full disk radio spectra of Mars in the CO $J = 1$–0 and $J = 2$–1 transitions. (Schloerb 1985) (b) Full disk radio spectra of Venus in the CO $J = 1$–0 line. Spectra of Venus's day and night sides are shown. (Schloerb 1985) (c) Carbon monoxide spectra at different locations on Venus. The locations are indicated by local venusian times. (de Pater *et al.* 1991a)

main cloud deck. The CO lines on Venus are formed in the mesosphere, well above the cloud layers, where the temperature is decreasing with altitude. The CO lines on Venus are therefore also seen in absorption against the warm continuum background.

Observations of the different line transitions and isotopes of the CO lines allow retrieval of both the CO abundance and atmospheric temperature structure. The CO abundance in Mars's atmosphere appears to be quite stable over time, while the thermal structure varies drastically, depending on the amount of dust entrained in the atmosphere. Microwave observations of the CO line in Venus's atmosphere reveal little variability in the temperature structure, but there are substantial diurnal variations in the CO abundance, as displayed in Figure 4.8c: The line is deep and narrow on the night side, but broad and shallow on the day side, except at local noon, where the spectrum is similar to a night side spectrum. This suggests that the line is formed high up in the atmosphere at the night side and at local noon, where the pressure is low; it is formed at lower altitudes at the day side, except at local noon. Thus, the CO abundance must be largest at high altitudes at the night side and at local noon, despite the fact that CO is formed upon photodissociation of CO_2 on the day side (Section 4.6.1.2). This apparent discrepancy between expectations and observations may be explained by rapid day-to-night winds, transporting the CO from the day to the night side (Section 4.5.5.1).

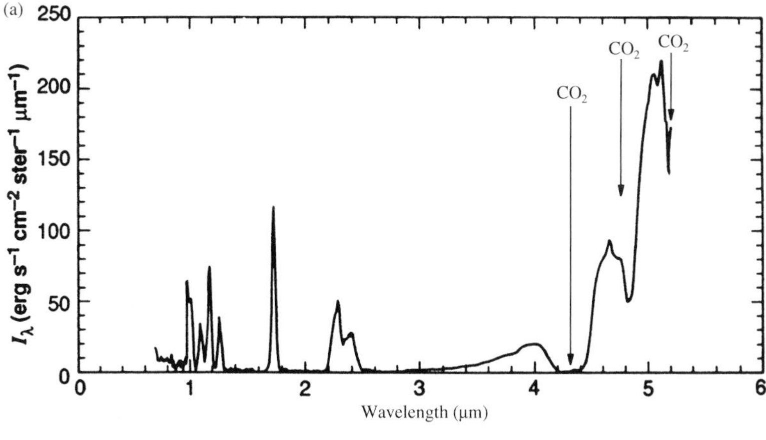

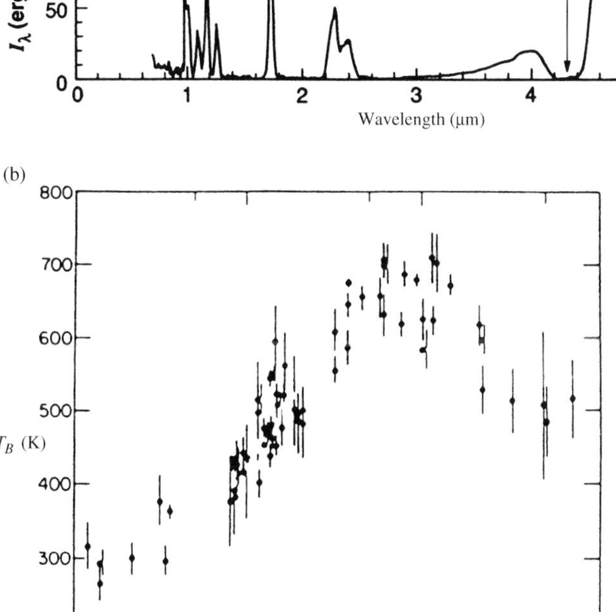

FIGURE 4.9 (a) A near-infrared spectrum of the dark side of Venus, taken by the Galileo spacecraft. At wavelengths $\lambda >$ 2.8 μm, Venus's sulfuric acid clouds are opaque and thermal emission from the clouds ($T_b \approx 235$ K) is received. Note the CO_2 absorption bands, which reduce the intensity of the blackbody curve at ~4.3, 4.8 and 5.2 μm. At several specific wavelengths shortwards of 2.8 μm, the clouds are rather transparent, allowing one to probe deeper warmer layers in the atmosphere, shown in the form of emission lines. (Adapted from Carlson *et al.* 1991) (b) A microwave spectrum of Venus. At millimeter wavelengths the planet's cloud layers are probed, while the surface is probed longwards of ~6 cm. (Adapted from Muhleman *et al.* 1979)

Although Venus's atmosphere and cloud deck are optically thick, one can probe through the clouds layers at radio wavelengths and, on Venus's night side, at several infrared wavelengths shortwards of ~2.5 μm. Figure 4.9a shows an average spectrum obtained near the center of the night side hemisphere with the Near Infrared Mapping Spectrometer (NIMS) on the Galileo spacecraft. Several emission features show up shortwards of 2.5 μm; at these wavelengths Venus's atmosphere and clouds are relatively transparent, so that deep warm atmospheric layers or the planet's surface are probed. At wavelengths longwards of 2.8 μm, thermal emission from Venus's clouds is observed, and shows strong CO_2 absorption bands (compare Fig. 4.6). In the absorption bands one probes higher, cooler altitudes (compare also Section 4.4.3.1).

A microwave spectrum of Venus is shown in Figure 4.9b. Roughly half of the microwave opacity in Venus's atmosphere is attributed to CO_2 gas, while prime suspects for the other half are H_2SO_4 and SO_2 (sulfur dioxide). At wavelengths longwards of ~7 cm, Venus's atmosphere is transparent and its surface is probed. Both the SO_2 and H_2SO_4 abundances are largely confined to the lower atmosphere, within and below the cloud layers. Sulfur dioxide gas may provide a source for Venus's sulfuric acid cloud layers (Section 4.6.1.2). Both ground-based and spacecraft measurements indicate that the SO_2 abundance near the cloud tops varies over time, by over an order of magnitude (Fig. 4.10). These variations may be correlated with volcanic eruptions (Section 5.5.3), or the variations may hint at changes in the eddy diffusion coefficient (Section 4.7).

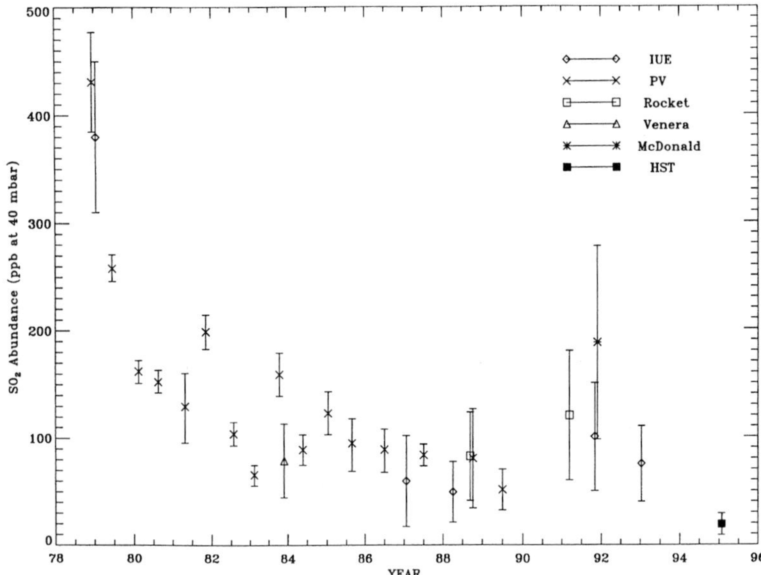

FIGURE 4.10 The SO_2 abundance in Venus's atmosphere as a function of time. The measurements were made by a variety of telescopes/spacecraft, as indicated on the figure. (PV stands for Pioneer-Venus; McDonald for McDonald Observatory.) (Esposito *et al.* 1997)

4.3.3.2 *Titan*

Titan's atmosphere is dominated by N_2 gas (Table 4.3), which shows up as strong UV line emissions. Infrared spectra of the satellite have revealed a large number of trace constituents including CH_4, CO_2, CO, and a variety of hydrocarbons and nitriles. Infrared spectra at different locations of Titan's disk are shown in Figure 4.11. The spectra display a smooth continuum background, with a large number of strong emission peaks. The identification of the various peaks is indicated on the figure. All lines must form in the stratosphere, where the temperature is rising with increasing altitude. Infrared spectra together with radio occultation profiles have been used to determine the mean molecular weight and temperature structure of the atmosphere.

4.3.3.3 *Planets and Satellites with Tenuous Atmospheres*

In recent years it has become clear that all planets and major satellites possess an atmosphere of some kind, though many of these atmospheres are so tenuous that to us these regions are a mere vacuum. Continuous 'bombardment' of energetic particles (solar wind, magnetospheric plasma) and micrometeorites kick up atoms and molecules from a planet's surface in a process called 'sputtering' (see Section 4.8.2). The particles kicked up from the surface usually have too low a velocity to escape the body's gravitational field, and form a 'corona' or atmosphere around the body. We find such atmospheres around Mercury and the Moon, many of the icy satellites and Saturn's rings. Other processes which can lead to the formation (or modification) of an atmosphere are volcanoes (e.g., Io), geysers (e.g., Triton), and sublimation of ices (e.g., Mars, Pluto, Triton). Below we briefly summarize observations of tenuous atmospheres detected around several of the smaller planets and larger satellites. Because of the low gravity of these bodies, the atmospheric scale height is usually large (Table 4.1, Problem 4.1), resulting in an extensive atmosphere.

Mercury and the Moon

Mercury has an extremely tenuous atmosphere with a surface pressure $\lesssim 10^{-12}$ bar. The atmosphere was first detected from space when oxygen, helium and hydrogen atoms were discovered using the airglow spectrometer on board the Mariner 10 spacecraft. Later, ground-based telescopes discovered the presence of sodium (Na), potassium (K) and calcium (Ca) atoms which have strong resonance lines at visible wavelengths. The major constituents discovered in Mercury's atmosphere are O, Na and He, with number densities near Mercury's surface of a few thousand atoms cm^{-3} for He up to a few tens of thousands of atoms cm^{-3} for O and Na (see Table 4.4). Both K and H have been detected at levels of a few hundred atoms cm^{-3} and calcium is another order of magnitude below this. Sodium, potassium, calcium and oxygen likely originate on the planet's surface, and have been kicked up into the atmosphere through 'sputtering' (Section 4.8.2). In contrast, H and He, major constituents of the solar wind, are probably captured from the solar wind (Section 4.8.2).

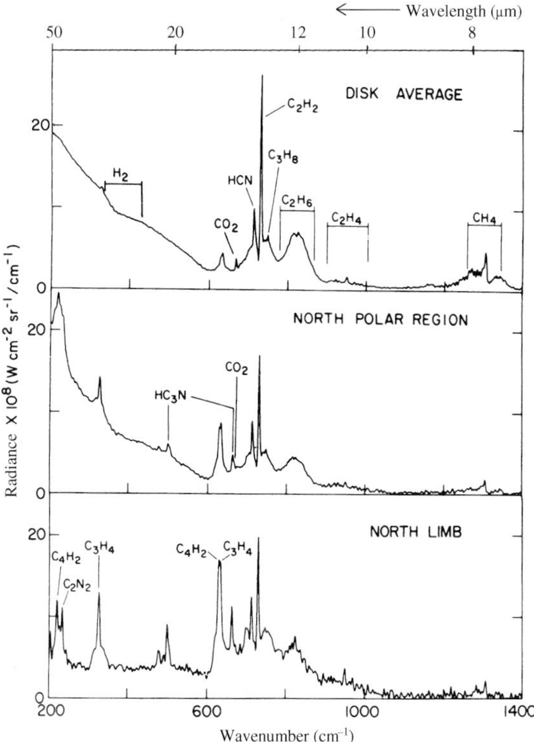

FIGURE 4.11 Thermal infrared spectrum at several locations on Titan. Note the numerous emission (i.e., stratospheric) lines from hydrocarbons and nitriles, which vary from one location to the next, superposed on a smooth continuum. (Samuelson *et al.* 1983)

Mass and UV spectrometers on the Apollo spacecraft have detected He and Ar on our Moon, with a surface density of a few thousand atoms cm^{-3} on the day side and an order of magnitude larger on the night side (Table 4.4). Ground-based spectroscopy revealed the presence of Na and K at levels of a few tens of atoms cm^{-3}. As on Mercury, the Moon's atmosphere is in part formed from sputtering by micrometeorites and energetic particles, and by capturing particles from the solar wind.

Pluto and Triton

Pluto and Triton are in many respects quite similar: they are alike in size, both extremely cold (Pluto: ~ 40 K for ice-covered regions, maybe up to 55–60 K for darker surface areas; Triton: 38 ± 4 K), although the surface temperature is high enough to partly sublime ices as N_2, CH_4 and CO_2, constituents which have been detected on the surface of both bodies via infrared spectroscopy. The amount of gas in an atmosphere surrounding these bodies can be calculated from the equations for vapor pressure equilibrium (Section 4.4.1). When Pluto occulted a 12th magnitude star in 1988, the gradual rather than abrupt disappearance and later re-appearance of the star revealed the presence of an atmosphere, with a surface pressure between 10 and 18 μbar. Because nitrogen ice is more abundant than methane and carbon dioxide ice by a factor of roughly 50, nitrogen gas is believed to be the dominant constituent of Pluto's atmosphere, while CO and CH_4 exist as trace gases. Airglow and occultation measurements of Triton by the UV spectrometer on board the Voyager spacecraft have revealed that Triton's atmosphere is also dominated by N_2, with a trace of CH_4 gas near the surface (mixing ratio $\sim 10^{-4}$). Triton's surface pressure is similar to that on Pluto, $\sim$13–19 μbar, and the atmosphere formed largely in the same way, from subliming gases and, as detected on Triton, from geyser activity (see Section 5.5.8).

Io

Io's atmosphere consists primarily of sulfur dioxide. Whereas this gas was first detected above a volcanic hot spot by the Voyager spacecraft, the presence of a global, relatively stable SO_2 atmosphere was established from the ground at radio wavelengths. The surface pressure is between 3 and 40 nanobars, covering 5–20% of the surface (note: all ground-based observations are made on the hemisphere of Io that is being illuminated by the Sun). The data suggest its origin to be primarily volcanic, probably supplemented by subliming SO_2 frost on Io's surface (global vapor pressure equilibrium). Sulfur monoxide has been detected at the $\sim$5–10% level, in agreement with photochemical models of SO_2. Other gases which have been detected are Na, K and O, while the presence of Cl is inferred from observations of Cl^+ in Io's plasma torus (Section 7.4.4.4).

Icy satellites

Europa, its surface covered by water-ice (Section 5.5.5.2), appears to have an oxygen atmosphere, as might be expected from sputtering processes. Such processes knock off H_2O molecules from Europa's surface, which, upon dissociation, break up into hydrogen and oxygen. Hydrogen escapes the small gravity field of Europa, leaving an oxygen-rich atmosphere behind. HST measurements of atomic oxygen suggest a surface pressure of molecular oxygen, assumed to be more abundant than atomic oxygen near the surface, of $\sim 10^{-11}$ bar. A tenuous oxygen atmosphere has been detected on Ganymede by HST, while Galileo observed Lyman α emissions. High signal-to-noise UV spectra obtained with HST of several satellites revealed the presence of oxidized gases, likely trapped as microscopic inclusions in the icy surfaces. In particular, SO_2 was found on Callisto and Europa, while absorption bands near 280 nm suggest SO_2 and/or OH on several uranian satellites and Triton. Ganymede reveals the presence of O_2, O_3 (strongest at the poles), SO_2 and CO_2. Ozone has further been detected on the saturnian satellites Rhea and Dione. Saturn's rings, which consist predominantly of icy particles (Section 11.3.2.4), are enveloped by a hydroxyl (OH) atmosphere. HST observations of the rings during ring plane crossing (when the rings were seen edge-on) suggest an OH density of ~ 500 molecules cm^{-3}. The infrared detector NIMS on board the Galileo spacecraft identified, tentatively, a carbon dioxide atmosphere on Callisto, with a surface pressure of $\sim 10^{-11}$ bar.

4.3.3.4 *Giant Planets*

All four giant planets have deep atmospheres, composed primarily of molecular hydrogen ($\sim$85–95% by volume) and helium. Observations at different wavelengths probe different altitudes. At UV wavelengths, where the opacity is mainly provided by Rayleigh scattering (Section 3.2.2.4), one typically probes the upper atmosphere at microbar pressure levels. Reflected sunlight from the cloud tops (and above) dominates at visual and near-IR wavelengths, while at longer wavelengths the thermal (blackbody) emission is observed. Most thermal IR radiation comes from altitudes near and above the cloud layers, although at around 5 μm one can probe to within or below some cloud layers on Jupiter (a few bars). The continuum opacity at far-IR and (sub)millimeter wavelengths is mainly provided by collision-induced absorption of molecular hydrogen gas, and by absorption/scattering of cloud particles. The latter depends on particle size: If the particles are much smaller than the wavelength, they are relatively transparent, and if they are comparable to or larger than the wavelength, the clouds become opaque. The deep layers of the atmospheres can be probed at radio wavelengths. The main source of opacity at radio wavelengths is ammonia gas, which has a broad absorption band at 1.3 cm. At wavelengths longwards of 1.3 cm, the opacity decreases roughly with λ^{-2}, and ever increasing depths are probed.

The composition of the giant planets' atmospheres is shown in Table 4.5. The mixing ratios of the elements are given with respect to H_2 (volume mixing ratios); the elemental mixing ratios for the Sun are given for comparison. Although He cannot be measured directly except by using a probe (such as the Galileo probe on Jupiter), good estimates of its abundance can be obtained by fitting simultaneously the thermal infrared spectra and radio occultation profiles. Occultation profiles yield information on the atmospheric scale height, H (eq. 4.2). The geometry of an occultation is sketched in Figure 4.12, where the Earth is moving along y, as indicated. The star is far away, in the direction x. The refractive bending angle, $\theta(r_1)$, of a ray passing at distance r_1 is smaller than that of a ray passing closer to the planet, at r. At Earth the two rays diverge by a distance dy, so the intensity received on Earth is decreased compared to pre-occultation levels. The bending angle is proportional to the refractivity in the atmosphere, which depends on the density. It can be shown that the intensity received on Earth has decreased by a factor of 2 when the refractive bending angle, $\theta(r)$ is equal to $H/r_\oplus$, which is proportional to $T/(\mu_a m_{amu})$ (see Hunten and Veverka, 1976 for details). While the thermal infrared spectra can be inverted to yield the thermal profile and abundance of the absorbing species, the occultations can be used to determine the mean molecular weight of the atmosphere. Since the mean molecular weight is mostly determined by H_2 and He, a fit to the infrared and radio occultation data yield an estimate for the He mixing ratio. Stellar occultations typically probe regions of the order of $\sim 10^{-2}$ μbar at UV, several tens μbar up to 1 mbar at visible wavelengths, and mbars up to 1 bar at radio wavelengths. The detailed mixing ratios for He vary from planet to planet. Helium is depleted compared to the solar value on Jupiter (by $\sim$20%) and Saturn (by $\sim$80%)[3]. This depletion is attributed to the immiscibility of He in metallic hydrogen at pressures of 1–3 Mbar, resulting in a 'raining out' of He towards the core (Section 6.4).

In an atmosphere of solar composition, one expects the dominant trace gases to be water vapor, methane, ammonia, hydrogen sulfide and neon. The gases H_2O, CH_4, NH_3 and H_2S have energy transitions at visible, ultraviolet and/or infrared wavelengths. Thermal infrared spectra and spectra taken in reflected sunlight have revealed the presence of H_2O, CH_4 and NH_3 on Jupiter. Moreover, the presence of these three gases as well as of H_2S and the noble gases He, Ne, Ar, Kr, and Xe has been confirmed via *in situ* measurements by the Galileo probe. The mixing ratios for some of these gases, however, were very different than was anticipated. The abundances of H_2O, H_2S and Ne were (much) smaller than was predicted (factors of $\sim$2–10), while Ar, Kr and Xe were detected at levels roughly 2.5 times solar. The latter enhancement is similar to that measured for CH_4, both from the ground and by the Galileo probe. At the deepest levels ($\sim$15–20 bar) H_2S (directly measured by the Galileo probe) and NH_3 (via measurements of the attenuation of the probe's radio signal) were measured at $\sim$2.5 and 3.5 times solar S and

[3] The He value on Saturn is, at present, rather poorly constrained. A reanalysis of the Voyager data suggests the mixing ratios might be as high as 0.11–0.16. We expect it will be measured more accurately by the infrared and radio instruments on board Cassini.

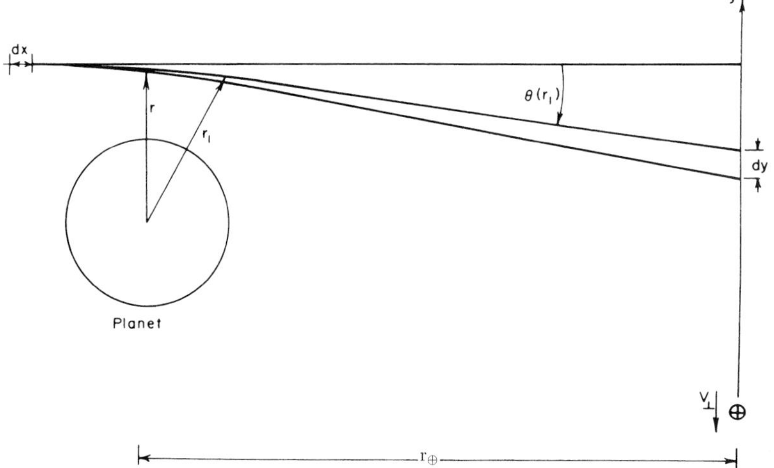

FIGURE 4.12 Geometry of a typical stellar occultation by a planet. The refractive bending angle $\theta(r_1)$ of a ray passing at distance r_1 is smaller than that of a star passing closer to the planet, at distance r. The Earth is moving along the y-axis, as indicated. (Adapted from Hunten and Veverka 1976)

N, respectively. The high NH_3 abundance is incompatible with ground-based radio data unless the abundance decreases to below solar in the upper troposphere, at $P \lesssim 2$ bar. Usually a decrease in NH_3 is explained via a reaction with H_2S (cf. Section 4.4.3.4); however, the amount of H_2S detected by the Galileo probe is roughly an order of magnitude too small to explain a decrease in NH_3 as required to explain the ground-based radio data. Although the mixing ratios for the condensable gases as measured by Galileo may be peculiar to the probe's path[4], based on dynamical considerations it is believed that the mixing ratios at the deepest levels probed are representative of the global values in Jupiter's deep atmosphere. At this time the physics/chemistry leading to the altitude profile of ammonia gas ($\sim$3.5 times solar at $P > 8$ bar; $\sim$ 0.5 times solar at $P < 2$ bar) is not understood. If indeed all species, including the heavier noble gases (Ne is expected to rain out with He because of its immiscibility in metallic hydrogen), are enriched by a factor of $\sim$2.5, this has interesting implications for giant planet formation theories, as further discussed in Chapter 12.

Both CH_4 and NH_3 have been detected on Saturn, but Uranus and Neptune have only revealed absorption by CH_4. To illustrate these points, Figure 4.13 shows spectra of the four giant planets at different wavelengths. Figures 4.13a and b show reflection spectra of the four giant planets, where numerous absorption lines of CH_4 gas are visible. Figure 4.13c shows thermal infrared spectra for Jupiter and Saturn, planets which exhibit numerous lines in their thermal spectra as well. The prominent broad absorption features in the latter spectra at 28.2 μm (354 cm^{-1}) and 16.6 μm (602 cm^{-1}) are caused by collision-induced absorption of molecular hydrogen. Ammonia absorption features are most prominent in the jovian spectrum, and completely absent in spectra taken of Uranus and Neptune. As discussed in the next section, even if NH_3 gas were present in large quantities in the deep atmospheres of Uranus and Neptune, it would condense out in the upper troposphere, preventing detection at infrared or visible wavelengths. Absorption features of CH_4 have been detected in infrared spectra of all four planets. A strong CH_4 absorption band at 7.7 μm (1304 cm^{-1}) shows up in the spectra of Jupiter and Saturn (Fig. 4.13c). The center of this band appears in emission, suggesting that CH_4 is present both below and above the tropopause. Saturn's spectrum between 11.8 μm (850 cm^{-1}) and 8.3 μm (1200 cm^{-1}) is dominated by absorption features of phosphine (PH_3). One of the lines (8.9 μm or 1118 cm^{-1}) is also clearly visible in Jupiter's spectrum, but the rest of the lines are masked on Jupiter by NH_3 lines. Emission lines of acetylene (C_2H_2) and ethane (C_2H_6) are identified on all four planets, and are particularly strong on Neptune. Since the lines appear in emission rather than absorption, they are indicative of stratospheric emissions. In certain regions on Jupiter, which are relatively clear of clouds and absorbing gases, absorption lines of water vapor, germane (GeH_4), and deuterated methane (CH_3D) are seen. Prominent emission lines of CO and HCN have been detected on Neptune in the 1 and 2 mm wavelength bands (Fig. 4.13d), revealing their presence in Neptune's stratosphere. Such emissions have not been seen from any of the other three giant planets.

Pressure levels greater than about one bar, within and well below the visible cloud layers, can be probed at radio wavelengths. At radio wavelengths, the opacity is controlled by NH_3, and, to a lesser extent, H_2O (plus H_2S

[4] The probe entered an infrared 'hot spot', a very 'dry' region in the atmosphere.

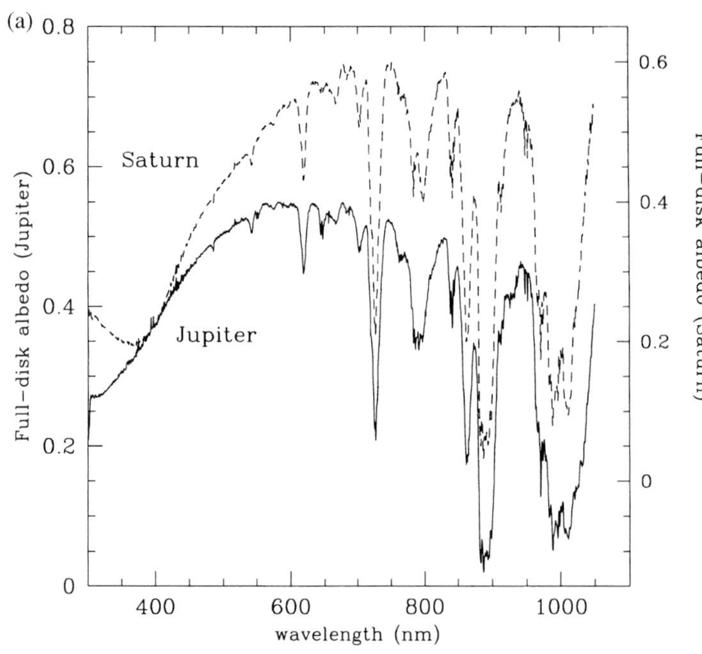

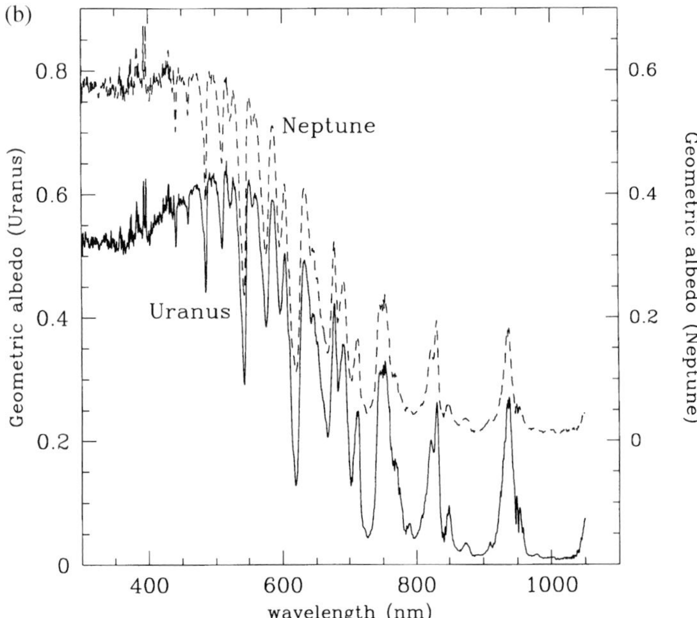

FIGURE 4.13 (a) Full-disk albedo spectra of Jupiter (solid curve, scale on left) and Saturn (dashed curve, scale on right) are dominated by CH_4 absorption lines. (b) Full-disk albedo spectra of Uranus (solid curve, scale on left) and Neptune (dashed curve, scale on right) are dominated by CH_4 absorption lines. (Adapted from Karkoschka 1994)

and possibly PH_3 on Uranus and Neptune). Microwave spectra for all four planets are shown in Figure 4.14, with various model atmosphere calculations superposed. The absorption lines are pressure-broadened to such an extent that the radio measurements are quasi-continuum data. Assuming the thermal profile in a giant planet's troposphere to be adiabatic, the microwave data can be inverted to yield an altitude profile of NH_3 gas, which indirectly provides an estimate for the mixing ratio of H_2S (Section 4.4.2.3).

Table 4.5 shows that the heavy elements ($Z > 3$) are usually enriched in the giant planets' atmospheres, by factors that increase with heliocentric distance. This observation yields information on the formation history of the giant planets (Chapter 12). Ammonia gas is likely enhanced compared to the solar N/H ratio on Jupiter and Saturn;

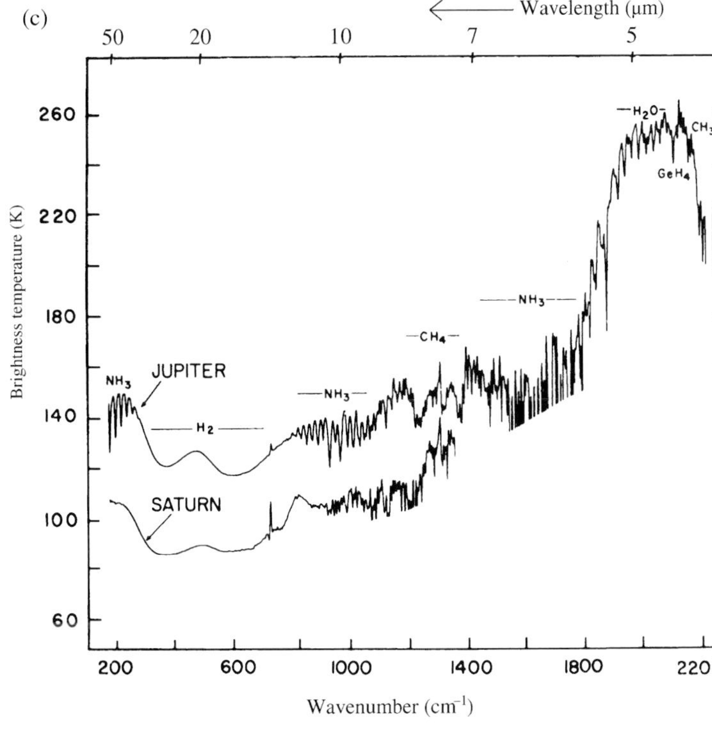

FIGURE 4.13 Continued. (c) Thermal infrared spectra of Jupiter and Saturn taken with IRIS on Voyager. (Hanel *et al.* 1992)

however, it may be depleted on Uranus and Neptune. A large global nitrogen depletion on any giant planet is inconsistent with current theories of planet formation. Even if there were a lack of nitrogen-bearing solids in the primitive solar nebula during the accretion phase of the outer planets, one would expect a solar N/H ratio, simply from the composition of solar nebula gas. A possible explanation for the apparent depletion in NH_3 gas is that the reservoir of nitrogen on Neptune and Uranus is N_2 rather than NH_3 gas, despite the fact that N_2 gas is expected to convert into NH_3: Carbon and nitrogen are present in the form of CO and N_2 at high temperatures and pressures, i.e., in the deep interiors of the giant planets. Rapid upward convection may inhibit the reactions CO $\rightarrow$ CH_4 and $N_2 \rightarrow NH_3$ (such reactions are strongly inhibited in the cold, low density outer regions of the protoplanetary disk, cf. Section 12.4.3), resulting in CO and N_2, rather than CH_4 and NH_3 in the upper layers of a planet's atmosphere. Molecular nitrogen is dissociated in the upper atmosphere (Section 4.6.1) and may form HCN through chemical reactions with hydrocarbon products. This scenario may explain the presence of CO and HCN on Neptune, a planet with obvious convective activity (Section 4.7.2). The nondetection of CO and HCN on Uranus has been attributed to the lack of an internal heat source, and therefore less vigorous convective motions.

Hitherto we discussed species in the planets' atmospheres whose origin is within the planet itself. Several species have been detected in the planets' stratospheres which clearly have been brought in from outside. In particular, water vapor has been detected in the stratospheres of all four planets (Table 4.5), which cannot have come up from these planets' deep atmospheres because the temperature is too low. This stratospheric water must have fallen in from outside. All four planets are surrounded by rings and moons; these, as well as meteoritic material, likely supplied the water. More recently the CO altitude profile on Jupiter was found to be 'inverted': i.e., the CO abundance appears to be higher in the stratosphere than in the troposphere. This observation is also attributed to infall from outside.

4.4 Clouds

When the temperature in the Earth's atmosphere drops below the condensation temperature of H_2O, water vapor condenses or freezes out; the numerous water droplets and/or ice crystals make up clouds. On other planets clouds are made of different condensable gases: we find, for example, clouds of NH_3, H_2S and CH_4 on the giant planets, CO_2 on Mars and H_2SO_4 on Venus. Clouds may modify the surface temperature and atmospheric structure considerably by changing the radiative energy balance.

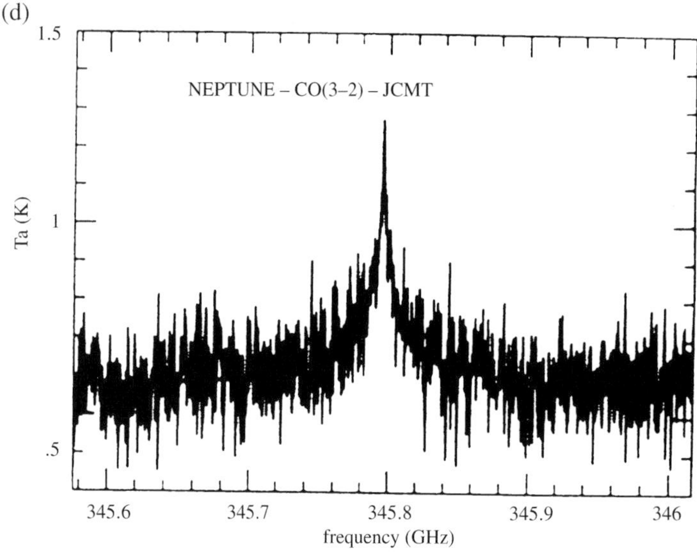

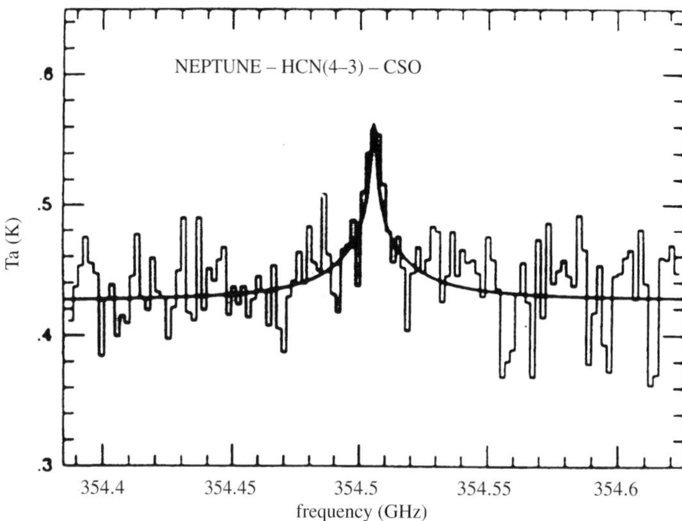

FIGURE 4.13 Continued. (d) CO and HCN emission spectra on Neptune. The *y*-axis is in units of antenna temperature Ta. (Marten *et al.* 1993)

Clouds are highly reflective, thus they decrease the amount of incoming sunlight, which leads to a cooling of the surface. The high opacity of clouds results in an absorption of incoming sunlight, and thus a heating of the immediate environment. Clouds can also block the outgoing infrared radiation, which increases the greenhouse effect. The thermal structure of an atmosphere is influenced by cloud formation, as the atmosphere is warmed by the latent heat of condensation. Clouds further play a major role in the meteorology of a planet, in particular in the formation of storm systems (Section 4.5). Cloud formation is discussed in the following subsections.

4.4.1 Wet Adiabatic Lapse Rate

Earth's atmosphere contains a small amount of water vapor (Table 4.3). The air is said to be *saturated* if the abundance of water vapor (or, in general, any condensable species under consideration) is at its maximum vapor partial pressure. In saturated air, evaporation (or *sublimation*, i.e., evaporation directly from the solid phase) is balanced by condensation. If water vapor is added, droplets condense out. Under equilibrium conditions, air at a certain temperature cannot contain more water vapor than indicated by its *saturated vapor pressure curve*, sketched in Figure 4.15a. Water to the right of the solid curves (e.g., point A) is completely vaporized, while liquid water exists between the

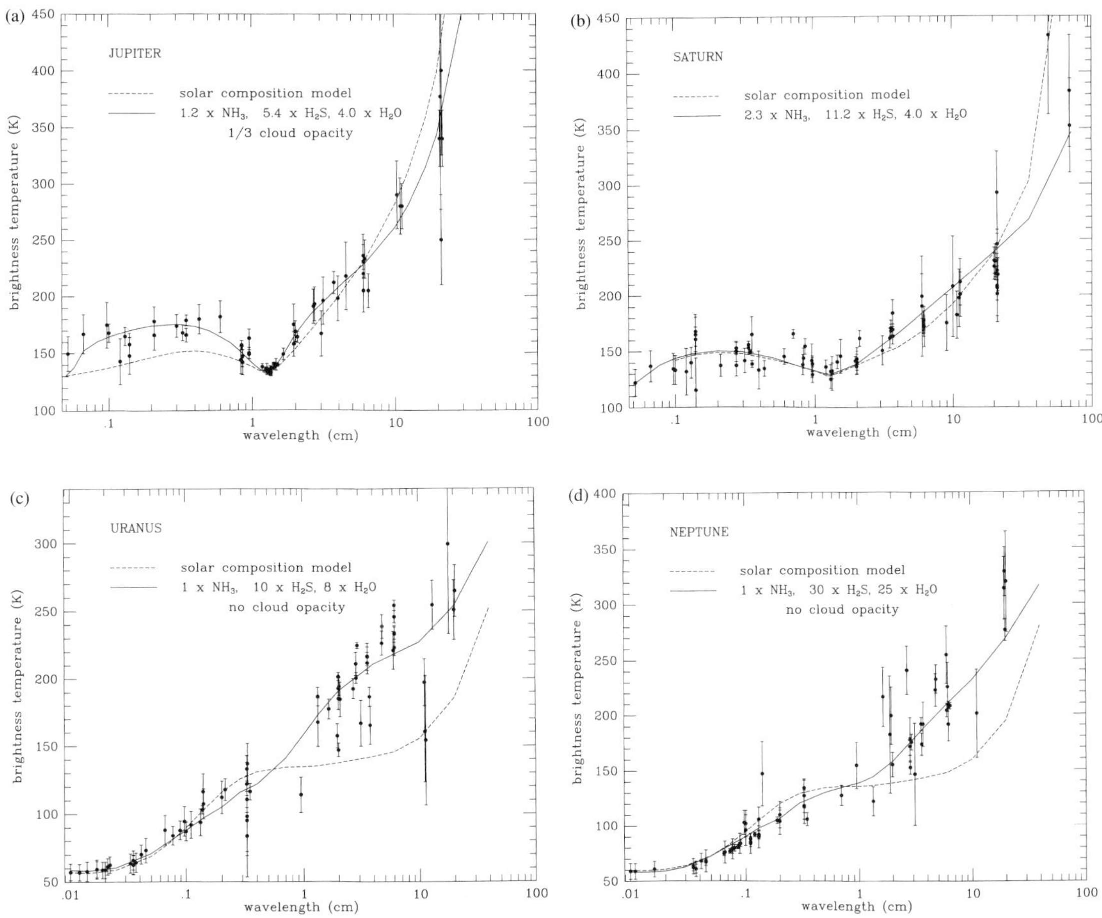

FIGURE 4.14 Microwave spectra of (a) Jupiter, (b) Saturn, (c) Uranus and (d) Neptune. Model calculations for an atmosphere assumed to be in thermochemical equilibrium are superposed. The dashed lines are calculations for a solar composition atmosphere. The solid lines are calculations for an atmosphere in which the condensable gases in the planets' deep atmospheres (i.e., below the levels of cloud formation) were enhanced; the enhancements are indicated in the figure. On all giant planets, the mixing ratio of NH_3, the main source of opacity, is reduced at higher altitudes through the formation of a cloud of NH_4SH and, at higher levels, NH_3 ice (Section 4.4.3.4). The formation of NH_4SH removes essentially all of the NH_3 in the upper troposphere of Uranus and Neptune. (Models after de Pater and Mitchell 1993)

two solid curves (e.g., at point B), and water-ice is present at the left side of the solid curves (e.g., at point C). The solid lines indicate the saturated vapor curves for liquid (to the right) and ice (to the left). Along these lines evaporation (sublimation) is balanced by condensation. The symbol T_{tr} indicates the triple point of water where ice, liquid and vapor coexist. Consider a parcel of air at point A, with a vapor pressure of 10 mb and a temperature of 15 °C. If the parcel is cooled, condensation starts when the solid line is first reached, at point D. Upon further cooling, the partial vapor pressure decreases along the line D–D′–T_{tr}. At 3 °C (point D′) the water vapor pressure is 7.6 mbar. Further chilling to −10 °C results in the formation of ice, where ice first forms upon crossing the second solid line. The vapor pressure above the ice is 2.6 mbar at −10°C.

Consider a moist parcel of air rising upward in the Earth's troposphere. The variation in temperature with altitude is sketched in Figure 4.15b, where the symbols A, D and T_{tr} correspond to the same points as on Figure 4.15a. The parcel of air cools adiabatically as it rises from A to D. In the Earth's troposphere, the temperature gradient is ~10 K km^{-1}, while the pressure drops according to equa-

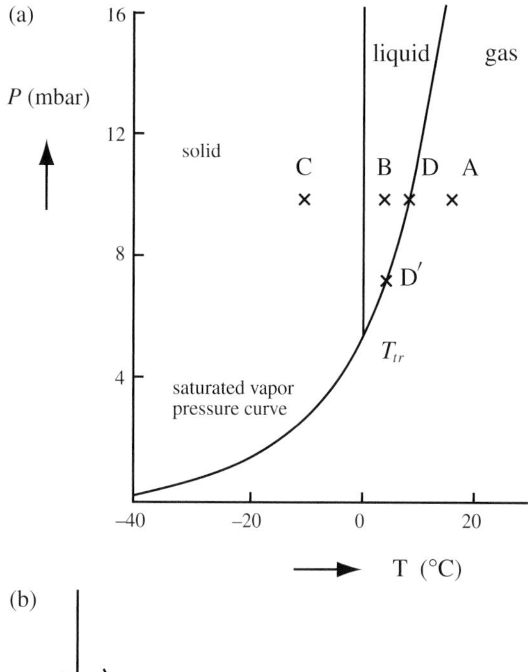

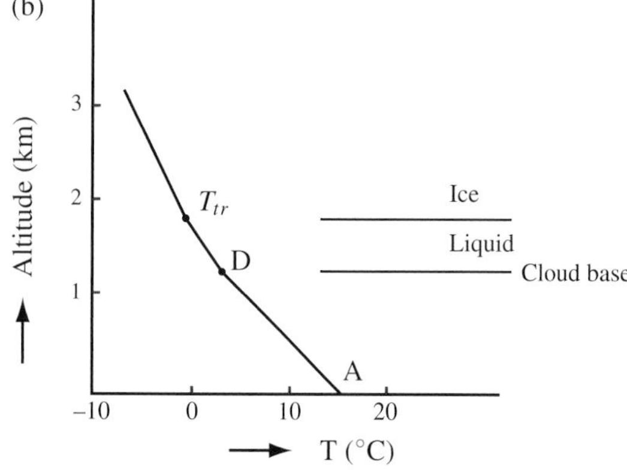

FIGURE 4.15 (a) Saturation vapor pressure curve for water. The y-axis indicates the partial pressure for H_2O vapor at the temperature (in °C) indicated along the x-axis. Water vapor is in the gas state when in point A; the saturated vapor pressure of water in point B is 7.6 mbar (indicated by point D′), and above C there is only 2.6 mbar of H_2O vapor. (b) Ideal sketch of the temperature structure in the Earth's atmosphere. In the lower troposphere, the air follows a dry adiabat. A wet air parcel rising up through the atmosphere starts to condense at point D′, when the water vapor inside the air parcel exceeds the saturated vapor curve (see panel a). The temperature profile in the atmosphere follows the wet adiabat in this region, and steepens even more above T_{tr}, when water freezes into ice crystals.

tion (4.1). At point D, the air parcel is saturated, and liquid water droplets condense out. The condensation process releases heat: the *latent heat of condensation*. This warms the air parcel, which therefore continues to rise. The release of latent heat also decreases the atmospheric lapse rate, shown by the change in slope from D to T_{tr}. At T_{tr} the atmospheric temperature is 0 °C (273.16 K), and water-ice forms, reducing the lapse rate even more.

The saturated vapor curve is calculated assuming a balance between evaporation and condensation between the vapor and the liquid or solid. The calculations are based upon a flat interface between the vapor and the liquid/ice. In the case of a small round droplet, the surface area is relatively large and evaporation/sublimation may exceed condensation at the condensation temperature. This excess evaporation/sublimation is larger for smaller droplets. If no solid or liquid material is present to act as condensation nuclei, air may become *supersaturated*. Air on Earth often needs to be supercooled by up to 20 K to start the condensation process if no condensation nuclei are present.

Relative humidity is the ratio of the measured partial pressure of the vapor relative to that in saturated air, multiplied by 100. The relative humidity in terrestrial clouds is usually $100 \pm 2\%$, although considerable departures from this value have been observed. The humidity can be as low as 70% at the edge of a cloud, caused by turbulent mixing or entrainment of drier air. In the interior layers, the humidity can be as high as 107%.

The saturation vapor pressure at temperature T is given by the *Clausius–Clapeyron equation of state*:

$$P = C_L e^{-L_s/R_{gas}T}, \tag{4.20}$$

where L_s is the latent heat, R_{gas} the gas constant, and C_L a constant. The thermodynamic equations discussed in Section 3.2.3.2 (eqs. 3.71, 3.72) are altered slightly by the inclusion of the release of latent heat:

$$c_V dT = -P dV - L_s dw_s, \tag{4.21a}$$

$$c_P dT = \frac{1}{\rho} dP - L_s dw_s, \tag{4.21b}$$

with w_s the mass of water vapor that condenses out per gram of air. The temperature gradient in a convective atmosphere becomes:

$$\frac{dT}{dz} = -\frac{g_p}{c_P + L_s dw_s/dT}. \tag{4.22}$$

The latent heat is effectively added to the specific heat c_P, resulting in a decrease from the dry adiabatic lapse rate. The lapse rate in the presence of clouds is commonly referred to as the *wet adiabatic lapse rate*. On Earth, the wet lapse rate is 5–6 K km^{-1} (Problem 4.14), slightly more than half the dry rate. Note that the wet adiabatic gradient can never exceed the dry lapse rate. Values for L_s and C_L for various gases can be found in e.g., Atreya (1986) and the *CRC Handbook of Physics and Chemistry*.

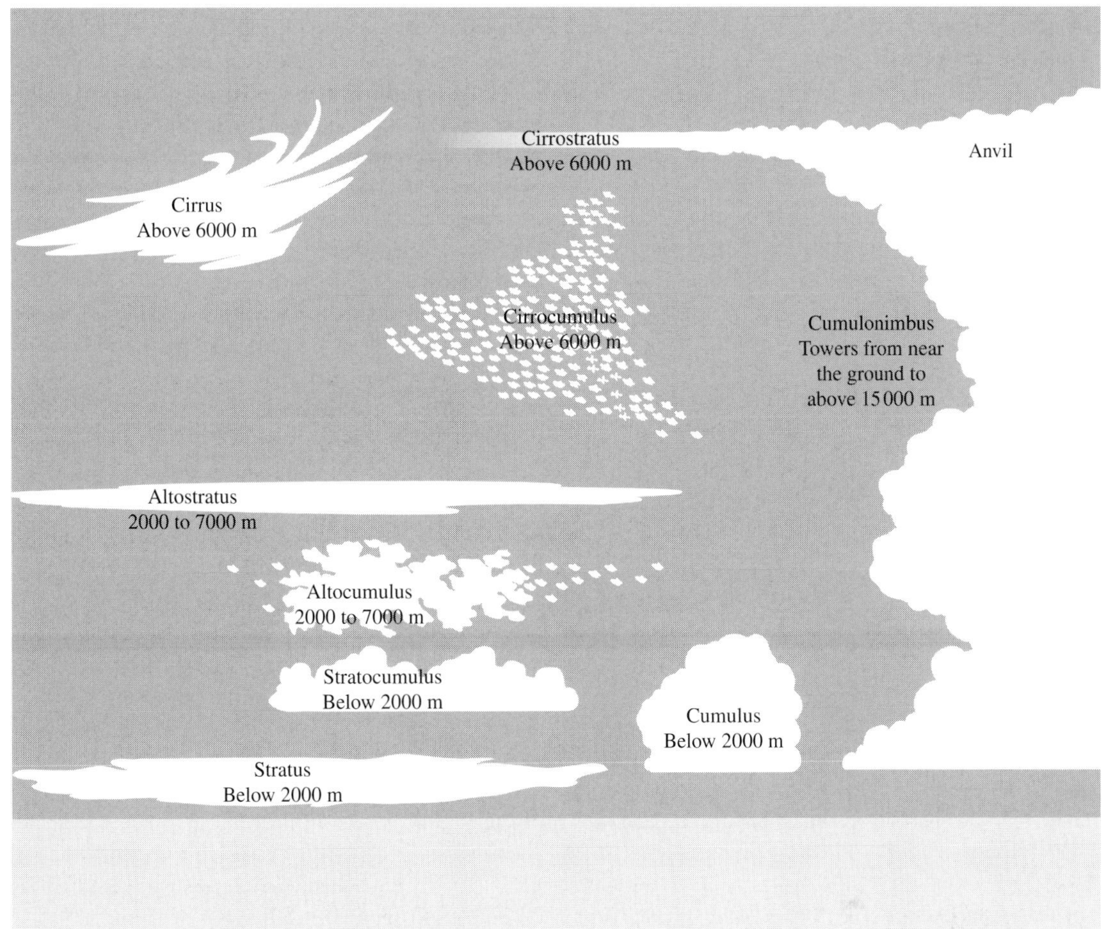

FIGURE 4.16 Schematic drawing of a variety of clouds familiar in Earth's atmosphere. (Adapted from Williams 1992)

4.4.2 Clouds on Earth

A parcel of air in the Earth's troposphere can be lighter than surrounding air because it is warmer or because it is humid compared to its surroundings (Problem 4.12). Such a parcel convects upwards (Section 3.2.3). When the temperature of this rising air drops below the condensation or freezing temperature of water vapor, water droplets or ice crystals form. The tiny (typically up to ten μm across) water droplets and/or ice crystals form a cloud. Clouds form primarily in the troposphere, that region in the atmosphere where the temperature decreases with altitude. Since condensation/freezing occurs at a particular temperature, the bottom of a cloud is usually flat.

4.4.2.1 *Shape*

Clouds come in a large variety of shapes (Fig. 4.16), determined primarily by the degree of stability of atmospheric air. When the air is stable, extensive flat (stratified) cloud layers are formed, referred to as *stratus* in the lower troposphere, or *altostratus* in the middle troposphere. *Cumulus* clouds form in unstable air. Since individual bubbles of air are rising, cumulus clouds look puffy. Small puffy clouds form in shallow layers of unstable air, whereas towering cumulus clouds, which can produce thunderstorms, form in deep layers of unstable air. In the lower troposphere one sometimes observes *stratocumulus* (stratified cumulus) clouds. *Cirrus* and/or *cirrocumulus* are found at high tropospheric altitudes, above ~ 6 km. These clouds consist of ice crystals, and sometimes show *mares' tails*, which are long, extended patterns formed by falling ice crystals. Clouds usually do not rise up through the tropopause into the inversion layer, since convection stops when the temperature is inverted. An exception is formed by *anvils* – giant thunderstorms which rush upwards through the tropopause into the stratosphere. These storm systems are discussed in more detail in Section 4.5.

4.4.2.2 *Formation and Precipitation*

Clouds are often related to precipitation: Droplets and ice crystals fall under the influence of gravity, while atmospheric viscosity resists free fall. When the gravitational acceleration is balanced by the atmospheric viscosity, the water droplets fall at the *terminal velocity*, v_∞, given by equating Stokes drag to gravity:

$$v_\infty = \frac{2 g_p \rho_d R^2}{9 \nu_v}, \tag{4.23}$$

where ρ_d is the particle's density, R its radius, and ν_v the atmospheric viscosity. The terminal velocity is proportional to the size (R^2) of the rain drops. Rain droplets are typically a few millimeters in size, several orders of magnitude larger than the size of a droplet just condensed out.

The simplest way for a cloud droplet to form is through direct condensation of vapor, where several vapor molecules collide by chance. Such droplets usually evaporate immediately, unless they are formed under supersaturated conditions or on condensation nuclei (Section 4.4.1). After a droplet has formed, it grows relatively rapidly (~ 0.1 μm s^{-1}) through condensation up to a radius of ~ 20 μm, which is still much smaller than observed droplets from precipitating clouds. Growth to larger sizes occurs through collisions with smaller particles, while the droplet falls downwards. The rate at which a droplet grows is proportional to its projected surface area and its velocity with respect to other droplets (e.g., Section 12.5). A small fraction of the droplets grows much faster than the rest of the droplets, and reaches the ground in the form of rain, hail or snow. Although collisions help droplets to grow, they also limit the size of precipitating droplets to a few millimeters, since larger droplets may break up during collisions.

4.4.3 Clouds on Other Planets

Clouds form on all planets with condensable gases where the temperature drops below the condensation or freezing temperature of such gases. Most terrestrial clouds consist of water droplets and ice crystals. Even though water vapor is a minor constituent of the Earth's atmosphere, it is the dominant, though not only, constituent of the clouds. Clouds on other planets also form from trace gases, each of which freeze or condense out when its saturated vapor pressure at the atmospheric temperature is exceeded. In this section, we review the clouds encountered on planets other than Earth.

4.4.3.1 *Venus*

At visible, UV and most infrared wavelengths, Venus's clouds are so thick ($\tau \gg 1$) that the planet's surface is completely hidden. The cloud particles consist of sulfuric acid, H_2SO_4, with some contaminants. At visible wavelengths, Venus looks like a bright yellow featureless disk. At UV wavelengths distinct markings are visible, with an overall V-shaped morphology (Fig. 4.17). Although the precise composition of UV absorbers in the cloud layers has yet to be identified, one expects sulfur- and chorine-bearing gases together with hazes to dominate the UV absorptions. Venus's main cloud layers span the altitude range between 45 and 70 km, with additional hazes up to 90 km, and down to 30 km. Based upon the microphysical properties of the cloud particles, the main cloud deck can be subdivided into a lower, middle and upper cloud layer. Particle sizes range from a few tenths of a micrometer up to ~ 35 μm. Most particles, however, are about 0.2 μm or 1 μm in radius; in addition the middle and lower clouds contain some particles with radii of ~ 4 μm. Thus, the particle size distribution is bi- or tri-modal. The larger particles are located in the lower and middle cloud layers between about 47 and 56 km altitude. The droplets are most likely formed at high altitudes (80–90 km), under the influence of solar UV light (Section 4.6.1.2). When the particles sediment out, they grow. However, since the temperature increases at lower altitudes, the droplets tend to evaporate below 45 km. Below 30 km altitude the temperature is too high for the droplets to exist.

At infrared (1–22 μm) and millimeter (~ 3 mm) wavelengths, the spatial distribution of Venus's brightness temperature is inhomogeneous (see Fig. 4.17). At millimeter wavelengths one probes approximately within Venus's cloud layers. The night side appears to be, on average, 10% brighter than the day side, which has been attributed to spatial variations in the H_2SO_4 vapor (i.e., cloud humidity). At mid-infrared wavelengths the temperature above Venus's cloud layers is measured. At these wavelengths the poles are generally bright, surrounded by a colder 'collar'. At near-infrared wavelengths, Venus's thermal emission can be measured on the night side hemisphere; reflected sunlight dominates the emission on the day side. Bright markings are visible, which are likely due to inhomogeneities in the venusian cloud deck. At places where the clouds are thinner, warm thermal emission from below the clouds can be observed (Section 4.3.3.1).

4.4.3.2 *Mars*

In the martian equatorial regions, clouds of water-ice are present near an altitude of 10 km. At night, the surface temperature is so low that water frost forms on the sur-

(a)

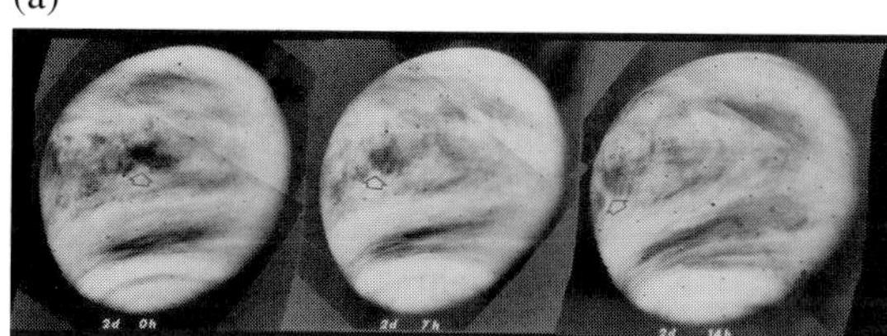

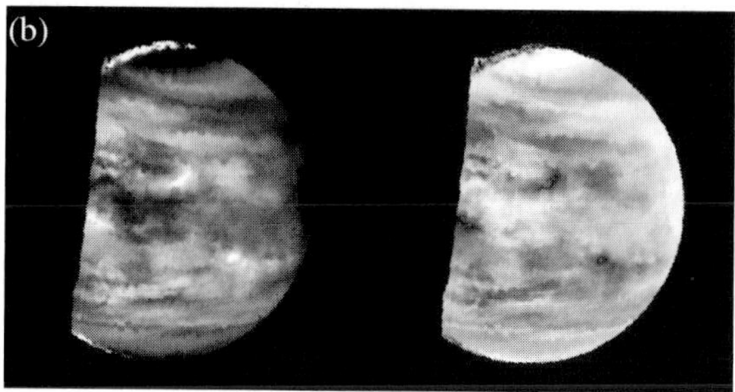

(c)

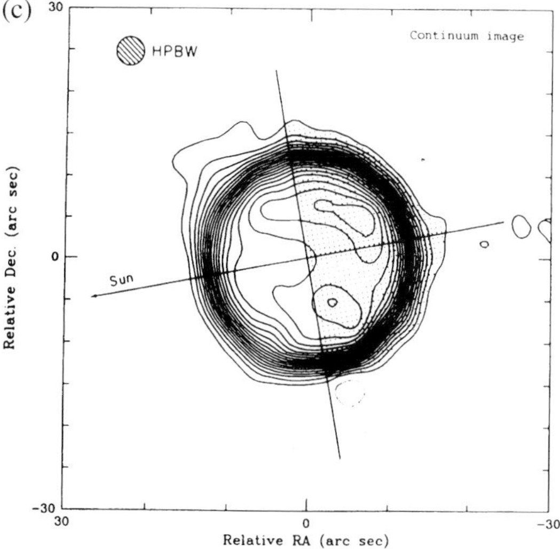

FIGURE 4.17 Images of Venus at different wavelengths. (a) Three images of Venus taken in ultraviolet light by the Mariner 10 spacecraft, 7 hours apart. A right-to-left motion of the cloud features can be seen (indicated by the tiny arrow). (Mariner 10/NASA: P14422) (b) Images at infrared wavelengths (night side only!), taken with NIMS on Galileo, 10 February 1990. On the left is an image at a wavelength of 2.3 μm, where the lower level clouds are probed (~ 50 km altitude). At this wavelength the clouds are partially transparent, and deeper hotter (brighter on image) regions are probed at places where the clouds are thin. On the right is an image at a wavelength of 4.56 μm, where the thermal emission from the clouds is observed. Note the bright pole, surrounded by a colder 'collar'. (NASA photo PIA00221) (Carlson *et al.* 1991) (c) An image at radio wavelengths. On the night side one probes deeper warmer layers than on the day side, presumably because the relative humidity is lower. This image represents an average over several days. (de Pater *et al.* 1991a)

face: the surface pressure on Mars is well below the saturated vapor pressure curve for liquid water, so water is present either as a vapor or a frost. Above the poles, the temperature never rises above the freezing point of water, so the poles are covered with a permanent water-ice cap. The sublimation temperature of CO_2 is well below that of water (~150 K). In addition, the surface pressure on both Mars and Earth is always below the saturated vapor pressure curve for liquid CO_2, and thus CO_2 is also present only as gas or a solid. Hence the popular name 'dry ice' for solid CO_2. Because of the low sublimation temperature, CO_2-ice clouds are present at higher altitudes than water clouds, typically near ~50 km, where the temperature is low enough for CO_2 to condense out. Since the sublimation temperature of CO_2 is in between the temperature measured at Mars's winter and summer poles, CO_2-ice sublimes from the summer pole, and it condenses out of the atmosphere above the winter pole.

4.4.3.3 *Titan and Triton*

Based upon the temperature structure and CH_4 mixing ratios of a few percent, a methane cloud layer is expected to form in *Titan's* middle troposphere. However, like Titan's surface, this cloud layer is hidden by a smog of hydrocarbons, which gives the satellite its featureless dark orange-brown appearance. At distinct infrared wavelengths, i.e., away from the methane absorption bands, it is possible to probe through the smog and 'see' the surface (Section 5.5.6.1). Infrared spectral measurements indicate, indirectly, the presence of methane clouds in the troposphere, covering $\lesssim 1\%$ of Titan's total surface area. Aerosol particles in the stratosphere are typically $\gtrsim 0.1$ μm in radius; they may grow to about a centimeter in size when they fall down (note that rain will fall slowly on Titan, because of the low gravity). The aerosol particles are thought to be the end products of methane photochemistry, and, through sedimentation, may slowly accumulate on the surface. The conditions at the surface are right for an ethane/methane solution to exist, possibly forming lakes or oceans on Titan.

Clouds and hazes are seen on *Triton*, but it is as yet unclear whether these result from simple condensation effects, surface eruptions or photochemistry.

4.4.3.4 *Giant Planets*

The compositions of the giant planets' atmospheres are dominated by H_2 and He. As on Earth, Venus and Mars, clouds on the giant planets also result from trace gases, in particular CH_4, NH_3, H_2S and H_2O. Unfortunately, the compositions of clouds are not easily determined directly by remote sensing techniques. Nor can we 'see' clouds directly below the upper cloud deck. Hence, the cloud layers discussed in this section are purely theoretical, based upon the (usually) known composition of the planets, together with laboratory measurements of the saturated vapor pressure curves of the condensable gases. Model calculations of the giant planet atmospheres suggest the existence of the following cloud layers (see Fig. 4.18, and Table 4.5):

(1) In each planet's deep atmosphere H_2O forms an *aqueous solution cloud*: liquid water with NH_3 and H_2S dissolved into it. This cloud is expected to form at temperatures above 273 K. The cloud forms when the water partial pressure exceeds the saturated vapor pressure. Hence, the precise altitude of the bottom or base level of the cloud depends upon the mixing ratio of water. If water is present at an abundance five times the solar O value, the base levels of the aqueous solution clouds on Jupiter and Saturn are at ~ 305 K and 325 K, respectively; on Uranus and Neptune the water abundance is more likely enhanced by a factor of 10–30 above solar O, resulting in a base level for the solution cloud near 400–450 K. Note that the aqueous cloud cannot exist at temperatures over 650 K, the *critical point* of water; above this temperature there is no first-order phase transition between gaseous and liquid H_2O. It becomes a supercritical fluid, a phase that is neither gas nor liquid.

(2) At $T \leq 273$ K, water-ice forms. Thus, the aqueous solution cloud is topped off by a water-ice layer.

(3) At $T \sim 230$ K, NH_3 and H_2S condense via a heterogeneous reaction:

$$NH_3 + H_2S \rightarrow NH_4SH. \tag{4.24}$$

The precise altitude or temperature/pressure level of the base of the NH_4SH cloud layer depends upon both the NH_3 and H_2S abundances. The less abundant of the two gases, NH_3 or H_2S, is effectively removed from the atmosphere through the reaction in equation (4.24); hence either NH_3 or H_2S is present above the NH_4SH cloud. This may explain why H_2S has never been detected on the planets Jupiter and Saturn via remote sensing techniques. H_2S gas has been detected at deeper levels *in situ* on Jupiter by the Galileo probe.

(4) At temperatures close to 140 K, NH_3 and H_2S each condense into their own ice clouds. Given our best estimates for the atmospheric composition of the four giant planets (Table 4.5), we expect NH_3-ice to form on Jupiter and Saturn and H_2S-ice on Uranus and Neptune. The composition of the NH_3-ice cloud on Jupiter has been confirmed by direct spectral measurements obtained with the ISO satellite.

(5) On Uranus and Neptune temperatures in the upper troposphere are low enough for CH_4, which has a condensation temperature ~ 80 K, to condense out. Thus, the upper visible cloud decks on Jupiter and Saturn are expected to consist of NH_3-ice, while the upper cloud deck on Uranus and Neptune consists of CH_4-ice.

Although the composition of the clouds cannot easily be measured directly via remote sensing techniques, altitude profiles of condensable gases give indirect information on the composition of cloud layers. For example, the atmospheric opacity at radio wavelengths is mainly provided by NH_3 gas. Thus, analysis of a microwave spectrum yields an altitude profile of NH_3 gas. As expected, the gaseous NH_3 abundance on Jupiter indeed decreases at the base of the postulated NH_4SH and NH_3-ice clouds, lending support to the existence of the proposed cloud layers. Note that the altitude profile of NH_3 gas also provides an independent measurement of the H_2S mixing ratio (assuming thermochemical equilibrium), since one molecule of NH_3 combines with one other molecule of H_2S gas to form NH_4SH (eq. 4.24). Similarly, radio occultation experiments on Voyager suggest a decrease in the atmospheric mean molecular weight (gaseous) on Uranus and Neptune at 1.2–1.3 bar, which is assumed to be the result of the formation of a CH_4-ice cloud. The altitude of the base of this cloud suggests that the CH_4 mixing ratio is ~ 30 times the solar C value (Problem 4.16). We note that observations at visible/infrared and radio wavelengths are complementary: At visible and near-infrared wavelengths the opacity near the cloud layers is largely due to scattering, preventing detection of gases at deeper levels, whereas clouds are largely transparent at radio wavelengths. Hence, at visible and near-infrared wavelengths one obtains information regarding the cloud tops, while radio data may provide information on the base of the cloud.

4.5 Meteorology

Everyone is familiar with 'weather', usually caused by a combination of Sun, winds and clouds. On Earth we have different seasons, and each season is associated with particular weather patterns, which vary with geographic location. One sometimes experiences long periods of dry sunny weather, while at other times we are threatened by long cold spells, periods of heavy rain, huge thunderstorms, blizzards, hurricanes or tornadoes. What is caus-

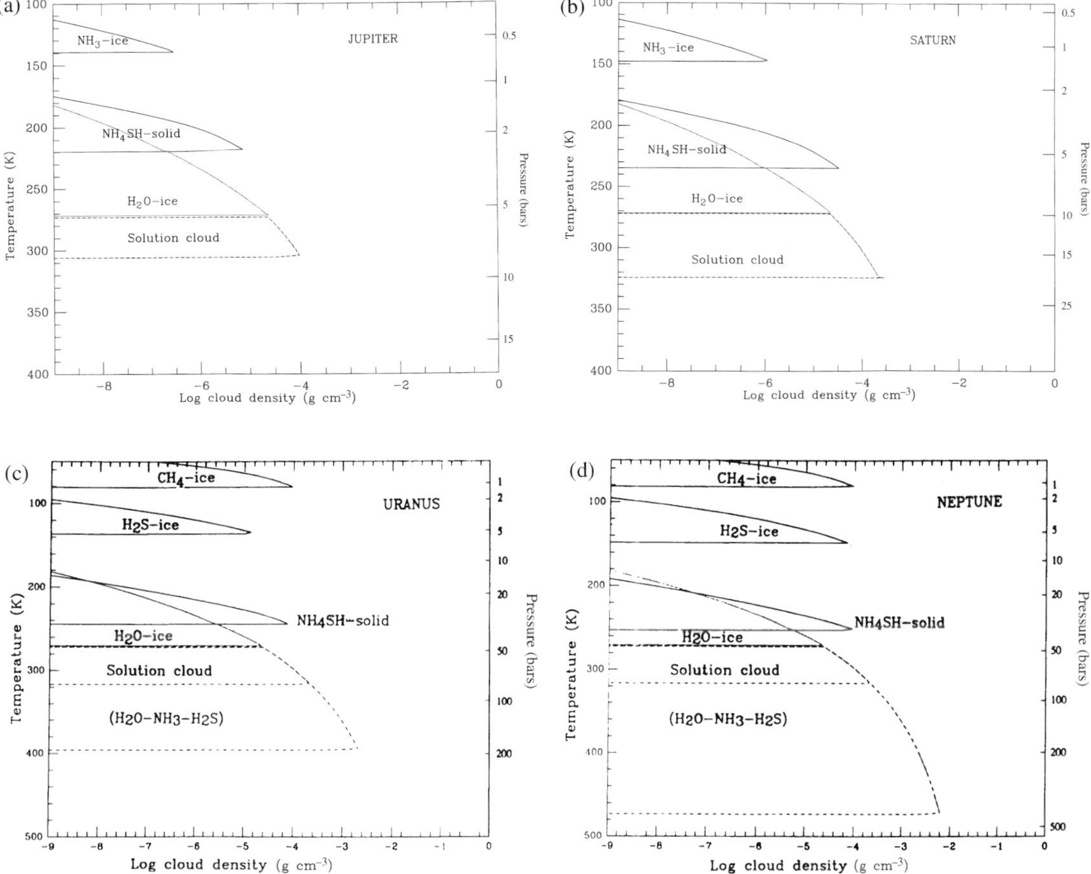

FIGURE 4.18 Sketch of the cloud layers in the giant planet atmospheres, as suggested by model atmosphere calculations (see Figure 4.14). Models are shown for (a) Jupiter, (b) Saturn, (c) Uranus, and (d) Neptune. The x-axis indicates the cloud density, the y-axis the temperature (left scale) and pressure (right scale). (de Pater *et al.* 1991b)

ing this weather, and what can we infer about weather on other planets? In this section, we summarize the basic motions of air as caused by pressure gradients (induced by e.g., solar heating) and the rotation of the body. We further discuss the vertical motion of air that causes it to heat or cool, which in the presence of condensable gases can lead to cloud formation and precipitation. For more detailed information on fluid dynamics, the reader is referred to, e.g., Salby (1996), which also includes an excellent review of meteorology.

4.5.1 Winds Forced by Solar Heating

Differential solar heating induces pressure gradients in an atmosphere, which trigger winds. Some examples of wind flows triggered directly by solar heating are the Hadley circulation, thermal tidal winds and condensation flows. Each of these topics is discussed below. The effects of planetary rotation on the winds are discussed in Section 4.5.3.

4.5.1.1 *Hadley Circulation*

If the planet's rotation axis is approximately perpendicular to the ecliptic plane, the planet's equator receives more solar energy than other latitudes. Hot air rises and flows towards regions with a lower pressure, thus towards the north and south. The air then cools, subsides and returns back to the equator at low altitudes. This atmospheric circulation is called the *Hadley cell circulation* (Fig. 4.19). For a slowly or nonrotating planet, such as Venus, there is one Hadley cell per hemisphere. If the planet rotates, the meridional winds are deflected, as discussed in detail in Section 4.5.3, and the Hadley cells break up. On Earth, there are three Hadley cells in each hemisphere. The middle cell in each hemisphere circulates in a thermodynamically indirect sense: the air rises at the cold end and sinks

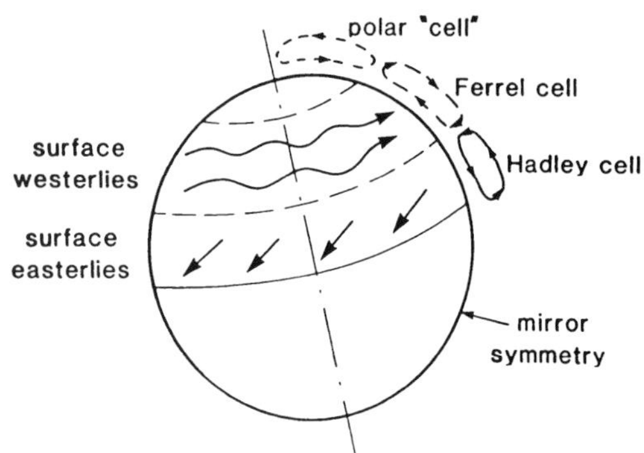

FIGURE 4.19 Sketch of the Hadley cell circulation on Earth. Three cells are indicated, with the surface winds caused by the Earth's rotation. The winds are indicated as easterly and westerly winds, that is winds blowing from the east and west, respectively. (Ghil and Childress 1987)

at the warm end of the pattern. The giant planets rotate extremely rapidly, and latitudinal temperature gradients lead to a large number of zonal winds. If the planet's rotation axis is not normal to the ecliptic plane, the Hadley cell circulation is displaced from the equator, and weather patterns can vary with season. Moreover, for a planet on an eccentric orbit (e.g., Mars) this may lead to large time-averaged differences between the polar regions.

4.5.1.2 *Thermal Tides*

If there is a large difference in temperature between the day and night side of a planet, air flows from the hot day side to the cool night side. Such winds are called *thermal tidal winds*. A return flow is necessary at lower altitudes. The effectiveness of these temperature differences can be judged by comparing the solar heat input, $\mathcal{F}_{in}$, with the heat capacity of the atmosphere. The solar heat input per day (see eq. 3.15):

$$\mathcal{F}_{in} = \pi R^2 (1 - A_0) \frac{F_\odot}{r_{AU}^2} t_d, \tag{4.25}$$

with t_d the length of the day. The heat capacity per unit area is $c_P P_o / g_p$, where P_o / g_p is the mass of the atmosphere per unit area of the surface. The heat, Q, necessary to raise the atmospheric temperature by ΔT is:

$$Q = c_P \frac{P_o}{g_p} 4\pi R^2 \Delta T. \tag{4.26}$$

If all of the solar heat is used to raise the atmospheric temperature by ΔT, $\mathcal{F}_{in} = Q$, and the fractional increase in the temperature becomes:

$$\frac{\Delta T}{T} = \frac{F_\odot (1 - A_0) g_p t_d}{4 P_o c_P T r_{AU}^2}. \tag{4.27}$$

The fractional change, $\Delta T / T$, is less than 1% for planets with substantial atmospheres, e.g., $\sim 0.4\%$ for Venus; $< 0.002\%$ for Jupiter. The atmospheric temperature for planets which have tenuous atmospheres changes drastically from day to night. The fractional change for Mars, for example, is 38% (Problem 4.18). We therefore expect thermal tides only on Mars and planets/satellites with tenuous atmospheres. On Venus and Earth, thermal tides are present in the thermosphere, well above the visible cloud layers, where the day–night temperature difference is large.

4.5.1.3 *Condensation Flows*

On several planets, such as Mars, Triton and Pluto, gas condenses out at the winter pole and sublimes in the summer. Such a process drives *condensation flows*. At the martian summer pole, CO_2 sublimes from the surface, thus enhancing the CO_2 content of the atmosphere. At the winter pole this gas condenses, either directly onto the surface, or onto dust grains which then fall down due to their increased weight. Mars's atmospheric pressure varies by $\sim 20\%$ from one season to the next (Mars's eccentric orbit contributes to the large annual variation). On Triton and Pluto a similar process is believed to occur with condensable gases nitrogen and methane. A fresh layer of ice overlays most of Triton's (cold) equatorial regions, while no ice cover is seen in the warmer areas. The observed decrease in Pluto's albedo as it approached perihelion may be evidence for evaporation of a substantial amount of ground frost.

4.5.2 Wind Equations

Winds are induced by gradients in atmospheric pressure, and they are deflected by a planet's rotation. This is the basic concept behind the winds and storm systems encountered on Earth and seen on other planets. In this section we summarize the equations which describe air motions. In Section 4.5.3 we discuss specific examples of steady flows, while turbulent motions are discussed in Section 4.5.4.

4.5.2.1 *Inertial Frame*

Euler's equation describes the motion of an incompressible, inviscid fluid that results from pressure gradients and the gravity field:

$$\rho \frac{D\mathbf{v}}{Dt} = -\nabla P + \rho \mathbf{g}_p. \tag{4.28}$$

In equation (4.28) we have used the *material* or *advective derivative, D/Dt*, which is the time derivative calculated by an observer in an inertial frame of reference:

$$\frac{D}{Dt} \equiv \frac{\partial}{\partial t} + \mathbf{v} \cdot \nabla. \tag{4.29}$$

The first term on the right hand side of equation (4.29) is the local derivative, caused by temporal changes in the fluid, and the second term is the advective contribution, caused by a motion of the material with respect to the observer (advection is the horizontal propagation of the mean wind). When viscosity becomes important, Euler's equation must be replaced by the *Navier–Stokes equation*:

$$\frac{D\mathbf{v}}{Dt} = -\frac{1}{\rho}\nabla P + \mathbf{g}_p + \nu_v \nabla^2 \mathbf{v}, \tag{4.30}$$

where ν_v is the kinematic viscosity (cm^2 s^{-1}).

In addition to the equation of momentum (Navier–Stokes equation), the time evolution of the density, pressure, temperature and velocity of atmospheric winds are related through the (hydrodynamic) equations of continuity and energy. Mass conservation is described by the equation of continuity:

$$\frac{\partial \rho}{\partial t} = -\nabla \cdot (\rho \mathbf{v}), \tag{4.31a}$$

which can be rewritten using the material derivative:

$$\frac{D\rho}{Dt} = -\nabla \cdot (\rho \mathbf{v}) + \mathbf{v} \cdot \nabla \rho = -\rho \nabla \cdot \mathbf{v}. \tag{4.31b}$$

In an incompressible fluid, $D\rho/Dt = 0$, so that the divergence of the velocity is equal to zero:

$$\nabla \cdot \mathbf{v} = 0. \tag{4.32}$$

If there is no interchange of heat by conduction or radiation, the flow of gas is isentropic and adiabatic, and the above equations can, in many cases, be complemented with the ideal gas law and the adiabatic temperature lapse rate (eq. 4.22).

4.5.2.2 *Rotating Frame*

Since all planets rotate, it is convenient to express the equation of motion in a rotating frame of reference. Assume a particle has a velocity $\mathbf{v}'$ in a rotating frame of reference, at distance $\mathbf{r}$ ($= R\cos\theta$, with R the radius of the planet and θ the latitude) from the rotation axis, while the planet (i.e., frame of reference) rotates with an angular velocity ω_{rot}. The absolute velocity, $\mathbf{v}$, and its derivative can be written:

$$\mathbf{v} = \mathbf{v}' + \omega_{rot} \times \mathbf{r}, \tag{4.33}$$

$$\frac{D\mathbf{v}}{Dt} = \left(\frac{D\mathbf{v}}{Dt}\right)' + \omega_{rot} \times \mathbf{v}, \tag{4.34}$$

with $(D\mathbf{v}/Dt)'$ the material derivative of $\mathbf{v}$ relative to the rotating observer, with $\mathbf{v}$ as defined in equation (4.33). With these equations the Navier–Stokes equation in a rotating frame of reference becomes (Problem 4.19):

$$\rho\left(\frac{D\mathbf{v}'}{Dt}\right)' = -2\rho\omega_{rot} \times \mathbf{v}' - \nabla P + \rho \mathbf{g}_{eff} + \nu_v \nabla^2 \mathbf{v}', \tag{4.35}$$

with the effective gravity:

$$\mathbf{g}_{eff} = \mathbf{g}_p + \omega_{rot}^2 \mathbf{r}. \tag{4.36}$$

For most planets $\omega_{rot}^2 r \ll g_p$, and the effective gravity is well approximated by $\mathbf{g}_p$.

The *vorticity* of a velocity field is defined by:

$$\varpi_v \equiv \nabla \times \mathbf{v}. \tag{4.37}$$

For a particle or fluid element fixed to a planet's surface, i.e., rotating with the planet so $\mathbf{v}' = 0$, the vorticity is equal to twice the angular velocity (Problem 4.20):

$$\varpi_v = 2\omega_{rot}. \tag{4.38}$$

4.5.3 Horizontal Winds

In geophysical fluid dynamics it is common to simplify the equations of motion by assuming an incompressible inviscid fluid, and adopt the *shallow water approximation*; shallowness means that the vertical scale length is much smaller than the horizontal scale. We make the same assumptions here. In addition, to investigate flows resulting from the Navier–Stokes equation (eq. 4.35) under various circumstances, we adopt a locally cartesian coordinate system, in which y is the coordinate on the surface to the north, x is to the east and z is upwards, perpendicular to the surface. The angle θ is the planetocentric latitude. All variables are defined in the rotating frame of reference (we drop the primes from the equations). The wind velocities are generally expressed as u, v, and w along the x, y, and z coordinates respectively:

$$\frac{Du}{Dt} = 2\omega_{rot}(v\sin\theta - w\cos\theta) - \frac{1}{\rho}\frac{\partial P}{\partial x}, \tag{4.39a}$$

$$\frac{Dv}{Dt} = -2\omega_{rot} u \sin\theta - \frac{1}{\rho}\frac{\partial P}{\partial y}, \tag{4.39b}$$

$$\frac{Dw}{Dt} = 2\omega_{rot} u \cos\theta - g_p - \frac{1}{\rho}\frac{\partial P}{\partial z}. \tag{4.39c}$$

Atmospheres are usually in hydrostatic equilibrium: $\partial P/\partial z = -\rho g_p$, so the acceleration of winds in the vertical direction equals $2\omega_{rot} u \cos\theta$. Thus there are no vertical winds if the planet does not rotate ($\omega_{rot} = 0$), and/or the zonal wind velocity $u = 0$. However, even for rotating planets, ω_{rot} is usually very small. Since we use the shallow water approximation, $\partial P/\partial z \gg \partial P/\partial x, \partial P/\partial y$. In

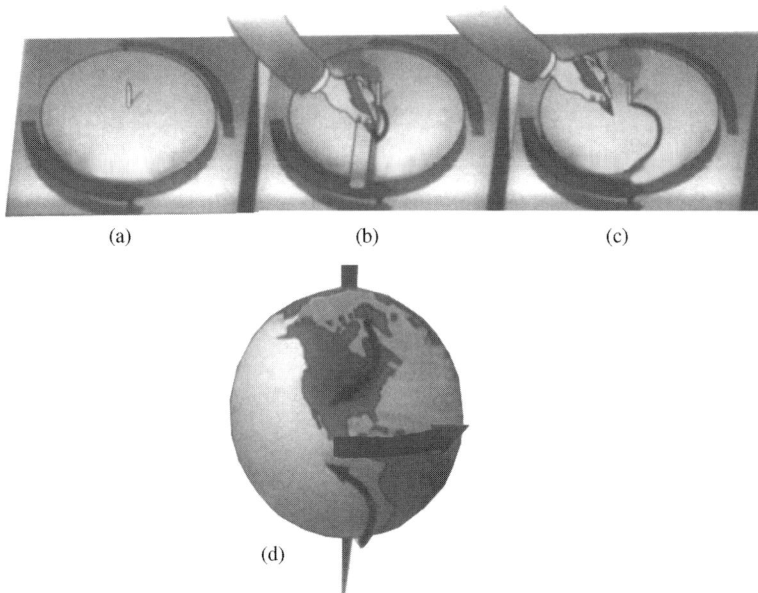

FIGURE 4.20 A schematic explanation of the Coriolis force: (a) A turntable which rotates counterclockwise. (b) Hold a ruler at a fixed position in inertial space and draw a 'straight' line on the turntable. (c) Even though you drew a straight line, the line on the turntable is curved. This is caused by the 'Coriolis' force. (d) The Coriolis force on the rotating Earth. The rotation of the Earth is indicated by the thick arrow. (Adapted from Williams 1992)

addition, as $w \ll u$ and v ($u/\ell \approx v/\ell \approx w/h$, where the vertical length scale, h, is much smaller than the horizontal length scale, ℓ), vertical winds can usually be ignored. Considering only winds in the two horizontal directions, equations (4.39) simplify to:

$$\frac{D\mathbf{v}}{Dt} = f_C \mathbf{v} \times \hat{\mathbf{z}} - \frac{1}{\rho}\nabla P, \tag{4.40}$$

where the symbol f_C represents the *Coriolis parameter*:

$$f_C \equiv 2\omega_{rot} \sin\theta = \varpi_v \sin\theta. \tag{4.41}$$

The Coriolis parameter is essentially the planet's vorticity normal to the surface at the latitude of interest.

4.5.3.1 *Coriolis Force*

Because planets rotate, winds cannot blow straight from a high to a low pressure area, but follow a curved path. This phenomenon can be visualized with help of a turntable (see Fig. 4.20). When one draws a straight line along a ruler, held fixed, to the edge of the rotating platform starting at the center, the line comes out curved, in the direction opposite to the platform's rotation. According to the same principle, winds on Earth (or any other prograde rotating planet) are deflected to the right on the northern hemisphere and to the left on the southern hemisphere, i.e., against the direction of the planet's rotation. On retrograde rotating planets the direction is reversed. This is called the *Coriolis effect*, and the 'fictitious force' causing the wind to curve is referred to as the *Coriolis force*.

The Coriolis force can also be explained in physical terms, using conservation of angular momentum. An air parcel at latitude θ has an angular momentum:

$$L = (\omega_{rot} R \cos\theta + v_{wind}) R \cos\theta, \tag{4.42}$$

where R is the planet's radius, and v_{wind} the wind velocity. If an air parcel initially at rest relative to the planet moves poleward while conserving angular momentum, then v_{wind} must grow in the direction of the planet's rotation to compensate for the decrease in $\cos\theta$. Thus, a planet's rotation deflects the wind perpendicular to its original direction of the motion, with an acceleration equal to $f_C\sqrt{u^2 + v^2}$. The direction of the wind is changed, but, since the acceleration is always perpendicular to the wind direction, no work is done and the speed of the wind is not altered. This acceleration is said to be due to the Coriolis force.

The Hadley cells on Earth cause the well known easterly (from the east) trade winds in the tropics ($\theta \sim$ 0–30°), where the return Hadley cell flow at low altitudes is deflected to the west. The mid-latitude jetstream in the Earth's stratosphere is the result of the Coriolis force with the Hadley cell circulation at high altitudes. On the giant planets, the Coriolis force is so strong that the Hadley cells are broken up in many pieces, and the circulation itself cannot be distinguished. We see the consequences of latitudinal temperature gradients in the form of zonal winds.

The importance of the Coriolis force is usually judged from the *Rossby number*, $\Re_o$, which is a measure of the ratio of the wind velocity v_{wind} and Coriolis term:

$$\Re_o \equiv \frac{v_{wind}}{f_C \ell}, \tag{4.43}$$

where ℓ is a length scale. Thus, when the Coriolis term is important, the Rossby number is small.

4.5.3.2 *Steady Horizontal Flow: Geostrophic and Cyclostrophic Balance*

It is convenient to write the wind equations in terms of forces tangential, $\hat{\mathbf{t}}$, and normal, $\hat{\mathbf{n}}$, (positive towards the left) to the flow, respectively, in a plane parallel to the surface (Fig. 4.21a). Consider a steady flow ($D\mathbf{v}/Dt = 0$), where the pressure gradient and the Coriolis force just balance each other, a situation referred to as *geostrophic balance*. The wind flows along *isobars*, lines of constant pressure, in a direction perpendicular to the pressure gradient (Fig. 4.21b). Under these circumstances, the velocity $\mathbf{v}$ follows directly from equation (4.40):

$$\mathbf{v} = \frac{1}{\rho f_C}(\hat{\mathbf{n}} \times \nabla \mathbf{P}). \tag{4.44}$$

The geostrophic approximation is usually valid if the Rossby number, $\Re_o \ll 1$. Prime examples of geostrophic winds are the trade and westerly jets in the Earth's troposphere, the Earth's stratospheric jet streams, and the zonal winds on the giant planets.

Equation (4.40) can be rewritten in terms of a tangential and a normal component, using (Fig. 4.21a):

$$\frac{d\mathbf{v}}{dt} = \frac{dv}{dt}\hat{\mathbf{t}} + v\frac{ds}{rdt}\hat{\mathbf{n}}, \tag{4.45a}$$

$$\frac{d\mathbf{v}}{dt} = \frac{dv}{dt}\hat{\mathbf{t}} + \frac{v^2}{r}\hat{\mathbf{n}}, \tag{4.45b}$$

and the Coriolis force:

$$f_C \mathbf{v} \times \hat{\mathbf{z}} = -f_C v \hat{\mathbf{n}}. \tag{4.46}$$

The equations for the forces tangential and normal to the flow become (where we use v as the flow velocity):

$$\frac{dv}{dt} = -\frac{1}{\rho}\frac{\partial \rho}{\partial s}, \tag{4.47}$$

$$\frac{v^2}{r} = -f_C v - \frac{1}{\rho}\frac{\partial P}{\partial n}. \tag{4.48}$$

Equation (4.48) describes the centrifugal acceleration. Any circular motion, including zonal winds moving around the surface of a planet (i.e., centered on the rotation axis), experiences a centrifugal force. At the equator, the centrifugal force for a zonal wind is perpendicular to the surface. At other latitudes, θ, the force can be written

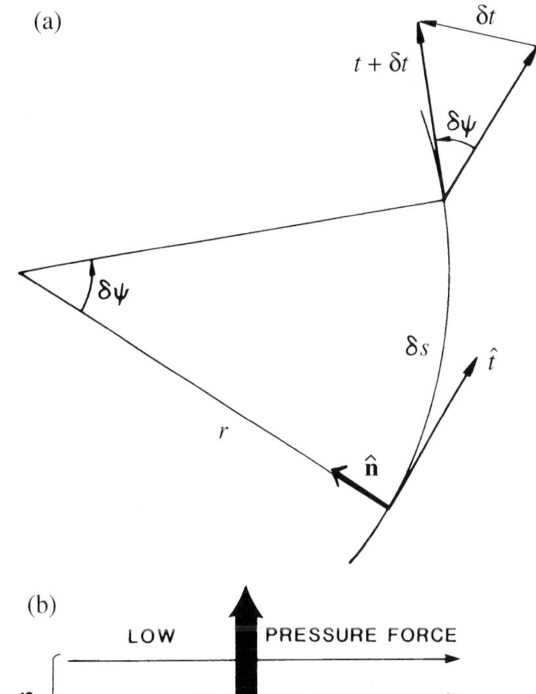

FIGURE 4.21 (a) Curvilinear coordinates, showing the differential change in the unit tangent vector $\hat{\mathbf{t}}$. (Chamberlain and Hunten 1987) (b) Geostrophic balance: the pressure and Coriolis forces balance each other and the wind flows along isobars. (Kivelson and Schubert 1987)

as a radial term plus a tangential term, where the latter is directed towards the equator: $v^2 \tan\theta / R$. Usually this term is very small compared to the geostrophic term $f_C v$, and can be neglected, in which case the flow can be described using the *geostrophic approximation*. If the centrifugal force $v^2/r \gg f_C v$, and if it balances the force induced by a meridional pressure gradient (Fig. 4.22), one speaks of *cyclostrophic balance*. The only terrestrial planet on which 'planet-wide' cyclostrophic balance is important is Venus, where the predominant horizontal pressure gradient is north–south, with pressure decreasing towards the poles. The winds near Venus's cloud tops are predominantly east-to-west, with a period of 4 days. Since Venus rotates only once in 240 days, these winds are in *superrotation*. In this case, the centrifugal force cannot be neglected, and in fact balances the pressure force, so the winds move along isobars. Cyclostrophic balance may be important on Saturn's satellite Titan, and near the equator of other bodies. Cyclostrophic balance is also satisfied on smaller scales, such as in storm systems and eddies (Section 4.5.4).

4.5.3.3 *Thermal Wind Equation*

In an atmosphere the temperature, pressure and density are related through the ideal gas law, the barometric law (eq. 4.1) and the temperature lapse rate (e.g., adiabatic lapse rate). The three parameters are related to altitude through the equation of hydrostatic equilibrium (eq. 3.65). For many purposes it is advantageous to use pressure, P, rather than altitude, z, as the vertical coordinate. We define the *geopotential* Φ_g:

$$\Phi_g = \int_o^z g_p dz = -\int_{P_o}^{P} \frac{dP}{\rho}. \tag{4.49}$$

Relative to surfaces of constant pressure, P, the geostrophic wind equation (4.44) can be written in isobaric coordinates:

$$\mathbf{v} = \frac{1}{f_C} \hat{\mathbf{z}} \times \left(\nabla \Phi_g\right)_P. \tag{4.50}$$

Differentiation with respect to P, and assuming hydrostatic equilibrium, yields for the vertical gradient of the geostrophic velocity:

$$\frac{\partial \mathbf{v}}{\partial \ln P} = \frac{R_{gas}}{\mu_a f_C} \hat{\mathbf{z}} \times (\nabla T)_P, \tag{4.51}$$

where R_{gas} is the universal gas constant and μ_a the mean molecular weight. Equation (4.51) is known as *thermal wind balance*, and thus follows from the combination of hydrostatic and geostrophic balance. The thermal wind equation for a horizontal flow couples the circulation to vertical stratification, since it gives an expression for the vertical *wind shear*, i.e., quick changes in the wind's velocity (or direction) with altitude. If there is a meridional temperature gradient $(\partial T/\partial y)_P$, the eastward velocity, u, along isobars increases with altitude. The thermal wind equations can be integrated in the vertical direction from P_0 to P to obtain the increase in zonal wind velocity over a range in altitude. Thus, the thermal wind equation relates the vertical wind shear to the horizontal temperature gradient along isobaric surfaces. Measurements of the wind velocity field can, therefore, be used to derive the vertical extent of the wind if the temperature gradient is known, or vice versa (Problem 4.21). Even if the horizontal temperature gradient is small, winds can be fierce if they extend to large depths; if the vertical extent of the winds is small, winds can only be strong if the meridional temperature gradient is large.

4.5.4 Storm Systems

Thus far we have discussed steady, large scale fluid motions. In this section we address turbulent motions. Laboratory experiments show that turbulent motions tend to arise when the dimensionless *Reynolds number*, $\Re_e$, is large:

$$\Re_e \equiv \frac{\ell v}{\nu_v} > 5000, \tag{4.52}$$

where ℓ and v are a characteristic length scale and velocity, respectively, and ν_v is the kinematic viscosity. A typical value for ν_v in the Earth's lower troposphere is ~ 0.1 cm^2 s^{-1}. Hence, for a length scale of one meter, the critical value for $\Re_e$ is already exceeded for velocities of the order of 5 cm s^{-1}. Thus atmospheres inevitably display some turbulence. In particular, turbulent motions are found in regions of high wind shear, regions which are strongly convective, and in regions over varying surface 'topography'. The flows can be modeled by allowing variations, perturbations and wave motions away from geostrophic balance.

4.5.4.1 *Convection*

The stability of air against convection can be judged by the vertical gradient of the *potential temperature*, $\partial \mathcal{T}/\partial z$, where the potential temperature is defined as the temperature a parcel of unsaturated air would have if compressed or expanded adiabatically to a pressure $P_0 = 1$ bar:

$$\mathcal{T} \equiv T \left(\frac{P_0}{P}\right)^{(\gamma-1)/\gamma}. \tag{4.53}$$

Potential temperature is conserved along an adiabatic path in (T, P) coordinates. The vertical gradient in the potential temperature is the difference between the actual and adiabatic lapse rate:

$$\frac{\partial \mathcal{T}}{\partial z} = \frac{\partial T}{\partial z} - \left(\frac{\partial T}{\partial z}\right)_{ad}. \tag{4.54}$$

If $\partial \mathcal{T}/\partial z < 0$, i.e., the lapse rate is superadiabatic, the atmosphere is unstable against convection and small-scale turbulence is triggered. This situation is referred to as *free convection*. If the gradient in potential temperature is close to zero, the large scale wind flows dominate local effects, and we speak about *forced convection*.

4.5.4.2 *Eddies*

In the previous sections, we showed that steady horizontal winds are common in planetary atmospheres. Examples include trade winds and jet streams on Earth and zonal winds on the giant planets. *Baroclinic eddies* may form in transition layers between two flows, as depicted in Figure

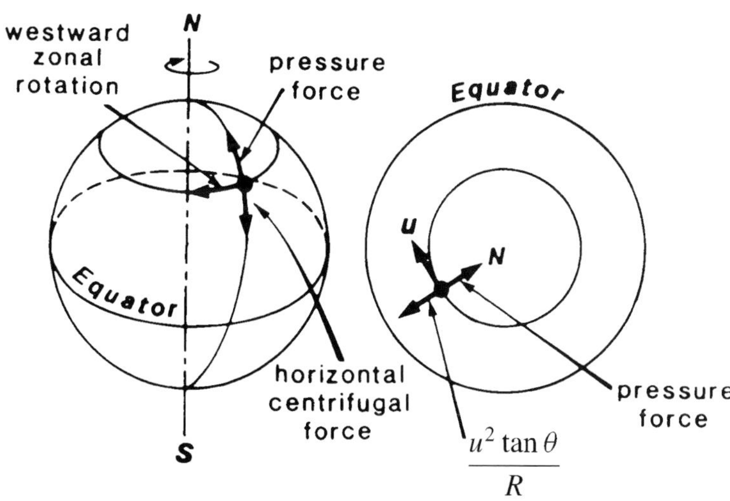

FIGURE 4.22 Cyclostrophic balance: the equatorward horizontal centrifugal force is balanced by a poleward pressure force. (Kivelson and Schubert 1987)

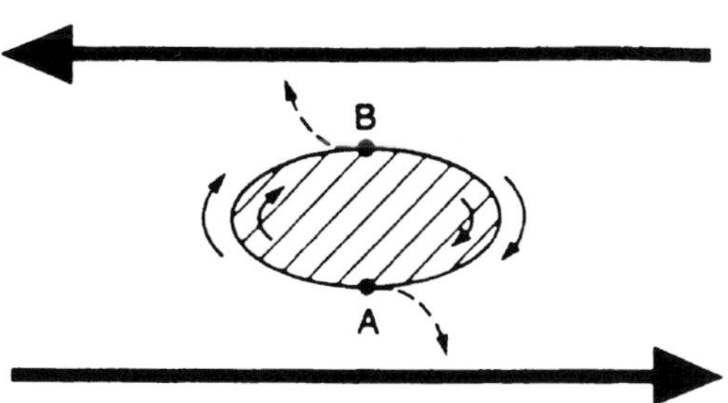

FIGURE 4.23 A sketch of an eddy in a zonal wind flow. (Adapted from Marcus 1993)

4.23. *Stationary eddies* may be induced by the local topography on a terrestrial planet; in contrast to baroclinic eddies, these eddies do not propagate. Stationary eddies are seen on Earth and Mars, both over mountains, and where a temperature difference occurs between oceans and continents. During the day, a cold sea breeze pushes itself under the warm air above the land, which leads to turbulence and the formation of stationary eddies. Mountains force air to flow up, resulting in turbulence and the formation of eddies. Baroclinic eddies form in the transition zone between Hadley cells on Earth and Mars, and in the transition region between zonal winds with different velocities on the giant planets.

As can be judged from Figure 4.23, only *prograde* (rotating in the direction of flow motion) baroclinic eddies survive. Observations show that long-lived baroclinic eddies are indeed always prograde. We define the *potential vorticity*, ϖ_{pv}:

$$\varpi_{pv} \equiv \frac{\varpi_v + f_C}{\ell}, \tag{4.55}$$

with ℓ the fluid depth. Potential vorticity is proportional to angular momentum around the vertical axis, and is conserved for a fluid element. The flow in eddies can be described by equation (4.48), where the winds flow along isobars and the pressure force is balanced by the combined Coriolis and centrifugal forces (Fig. 4.24). We speak about *cyclones* if the wind blows around a region of low pressure, while in an *anticyclone* the wind blows around a high pressure region. Cyclones and anticyclones are present on most planets.

If a fluid element shifts latitude, the Coriolis term, f_C, changes. If the fluid depth stays constant, the element must spin up or down to conserve ϖ_{pv}. Similarly, if the fluid element stays at the same latitude but its depth, ℓ, changes, the element also spins up or down. This is, for example, seen on Earth when a storm meets a mountain. The bottom of the storm is forced upwards, so the storm is compressed in altitude and expands horizontally, resulting in a spin down. When the system has crossed the mountain, the bottom of the storm descends and the storm forms a tall, rapidly spinning column of air. At the same time, the rising air cools adiabatically and water (or any other condensable gas) condenses out to form clouds. Rain often accompanies such storms, while the subsiding air at the lee side of the mountain only brings fierce dry winds, as clouds evaporate in subsiding air, which is dry because it lost its water when it first met the mountain.

4.5.4.3 *Hurricanes*

In the tropics on Earth we often encounter *hurricanes*, dangerous storms that lead to lots of destruction. What are hurricanes? Consider a parcel of humid air, which rises because it is lighter than the surrounding air (Problem 4.12). The rising air leaves behind an area of low pressure: Winds

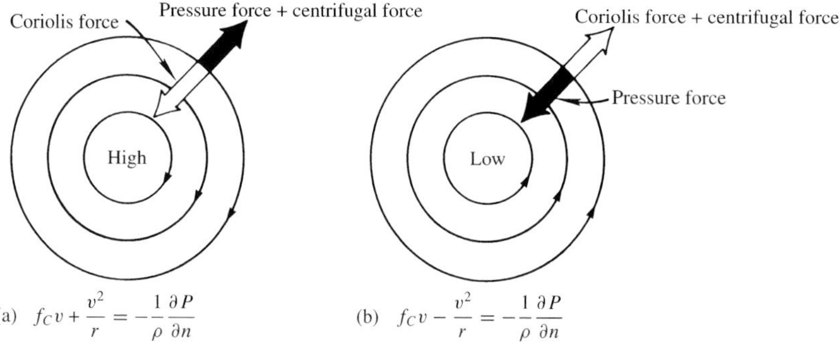

FIGURE 4.24 Isobars and wind flows around (a) a high pressure region (anticyclone) in the Earth's northern hemisphere and (b) a low pressure region (cyclone) in the Earth's northern hemisphere. (Kivelson and Schubert 1987)

rush in to equalize pressure, causing both air and water to flow towards the hurricane 'eye'. Water is carried off (return flow) at large depths in the ocean. While the humid air is rising, it cools and water droplets and/or ice crystals form when the temperature drops below the condensation temperature of water. The release of latent heat warms up the air parcel, thereby increasing the gradient in the potential temperature, $|\partial \mathcal{T}/\partial z|$, 'fueling' the upward motion of air. The storm system can reach large heights if there is no vertical wind shear in the atmosphere; otherwise the storm is ripped apart. At the tropopause the temperature gradient is inverted and convection is stopped. The high pressure build-up at the tropopause helps to pump the air away, aided by stratospheric winds. This situation is also seen in anvils (Fig. 4.16), the giant thunderstorms mentioned in Section 4.4.2.1. Hurricanes on Earth form in tropical regions above warm oceans, and break up when they move over land, where the winds and temperatures change. However, their fierce winds, rain and high waves can cause a considerable amount of damage in the coastal areas before they break up. The waves are induced by a combination of the high winds which cause water and air to flow towards the hurricane 'eye' (to equalize pressure), and the fact that the ocean is rather shallow near the coast, preventing water from flowing away from the 'eye', i.e., blocking the return flow of the water at large depths. As a result, water waves several meters high form; such waves can wipe out entire coastal areas.

4.5.4.4 *Lightning*

Lightning appears to be a common phenomenon in planetary atmospheres. It has been observed on Earth, Jupiter and possibly Venus. Electrostatic discharges on Saturn (SED) and Uranus (UED; Section 7.5.3) have been detected at radio wavelengths, and are probably caused by lightning. The basic mechanism for lightning generation in planetary atmospheres is believed to be collisional charging of cloud droplets, followed by gravitational separation of oppositely charged small and large particles, so that a vertical potential gradient develops. The amount of charges that can be separated this way is limited; once the resulting electric field becomes strong enough to ionize the intervening medium, a rapid 'lightning stroke' or *lightning* discharge occurs, releasing the energy stored in the electric field. For this process to work, the electric field must be large enough, roughly of the order of 30 volts per mean free path of an electron in the gas, so that an electron gains sufficient energy while traversing a mean free path to cause a collisional ionization. When that condition is met, the typical free electron causes an ionization at each collision with a gas molecule, producing an exponential cascade.

In Earth's atmosphere, lightning is almost always associated with precipitation, although significant, large scale electrical discharges also occur occasionally in connection with volcanic eruptions (and nuclear explosions). In a thunderstorm, hailstones that have grown to millimeter size and begun falling under the influence of gravity collide with 100 μm ice crystals, which are too small to fall efficiently. In each collision, $\sim 10^5$ electrons are preferentially transferred to the ice crystal. The continuing rainout of positively charged hailstones leads to a growing large-scale separation of charge and an increasing vertical electric field. When the field reaches its critical threshold value ($\sim 10^5$ V m^{-1}), the electrical resistance of the air breaks down and current, in the form of a lightning bolt, flows to cancel the charge separation. In analogy, lightning on other planets is only expected in atmospheres where both convection and condensation take place. Moreover, the condensed species, such as water droplets, must be able to undergo collisional charge exchange. It is possible that lightning on other planets (such as Venus) may also be triggered by active volcanism.

4.5.4.5 *Waves*

Disturbances in atmospheric pressure propagate in the form of waves, which mathematically can be derived from the hydrodynamic equations by adding small perturbation terms to the equations. Waves are important, since they 'communicate' changes in one region to another, and hence they may cause planet-wide weather patterns. There are several types of waves, most of which are discussed in detail in the books on fluid dynamics. The simplest waves are *sound* or *acoustic waves*, which are compressional waves, known as p modes in astronomy. *Surface gravity waves* are the familiar waves which propagate on the surface of water when a stone is dropped in. They are known as f modes in astronomy. While the compressibility of air provides the restoring force in case of a sound wave, the buoyancy provides the restoring force for gravity waves. If the wavelength of gravity waves is short compared to the radius of the planet, they behave like ordinary waves on the surface of a deep ocean. *Internal gravity waves* (g modes) are confined to a stably stratified portion of an atmosphere. All gravity waves propagate vertically as well as horizontally. Various *seismic waves* may travel through atmospheres as well (Section 6.2.1). *Atmospheric tides* are driven by solar heating (Section 4.5.3), as well as the gravitational forces of the Sun and the Moon. The *Rossby wave* or *planetary wave* results from changes in the balance between the relative vorticity of an air parcel and its planetary vorticity, through conservation of the potential vorticity (eq. 4.55). Such waves are triggered when an air parcel moves to higher or lower latitudes, so that its Coriolis term changes. For example, when an eastward moving parcel of air is deflected towards the equator, the Coriolis term (f_C) decreases, so it spins up cyclonically (eq. 4.55). Its trajectory is then deflected poleward, back to its original latitude. Upon overshooting, the parcel spins up anticyclonically, and is deflected equatorward. The variation in the Coriolis term with latitude exerts a torque on the displaced air, which provides the restoring force and enables air to move back and forth about its undisturbed latitude. The Rossby wave usually propagates slowly (few m s^{-1}) westward (on a prograde rotating planet), with respect to the mean zonal flow, although velocities of hundreds of meters per second can be reached.

4.5.5 Observations

4.5.5.1 *Terrestrial Planets*

Earth

Earth's global atmospheric wind system is characterized by a Hadley cell circulation with three cells per hemisphere, where the middle cell circulates in a thermodynamically indirect sense (see Fig. 4.19). The Coriolis force adds an east–west component to the meridional movement of the Hadley cell circulation, which effectively causes the Hadley cell to break up into its three separate cells. Air rises near the equator in the summer hemisphere and descends in the subtropics. The descending dry air dries the troposphere, inhibits convection and thus maintains the deserts common at subtropical latitudes. The easterly trade winds in the tropics are the low altitude return flows from the equatorial Hadley cell. At latitudes between about 10° and 60° on both hemispheres the global atmospheric circulation in the troposphere is characterized by westerly jet streams, which increase in strength with altitude up to the tropopause. These jets are zonal winds which flow from the west towards the east, in each hemisphere. Above the tropopause these jets weaken in strength, and then increase again with altitude above ~25 km (or $P \lesssim 30$ mbar), up to ~70 km altitude ($P \approx 0.05$ mbar). In the winter hemisphere these stratospheric/mesospheric jets are westerly winds, known as the *polar-night jet*, which reach speeds of 60 m s^{-1} in the lower mesosphere. In the summer hemisphere the stratospheric/mesospheric jets above the subtropics come from the east, reaching speeds of over 60 m s^{-1}. This wind pattern is globally and temporally averaged and rarely seen on individual days. The day-to-day wind patterns deviate significantly from this global circulation, as a consequence of local pressure highs and lows. Planetary waves cause large scale disturbances in the zonal flows away from zonal symmetry.

Along the equator the meridional temperature gradient is relatively small and the Coriolis force weak. The tradewinds drive the ocean currents, causing surface water to flow from east to west along the equatorial regions of the Pacific Ocean. The water is warmed up by the Sun, causing warm surface water to build up across the western Pacific. The water level here is typically 0.5 m higher than in the eastern Pacific. Evaporation of the warm moist air in the west leads to the formation of storm systems. The latent heat release inside organized convection cells is the primary source of energy for the circulation in the tropics. In addition to the zonal-mean Hadley cell, monsoon and Walker circulations occur in the tropics. *Monsoon* circulations are driven by horizontal gradients in surface temperature, where during the summer the subtropical land masses, such as India and northern Australia, are warmer than the surrounding oceans, a situation which is reversed during the winter. Thus air is rising above the landmasses during the summer (bringing rain), and subsiding during the winter (dry air). The *Walker circulation* in the tropics is driven by nonuniform heating, where air rises at longitudes where heating takes place (near Africa, Indonesia and South America) and sinks at longitudes where it is cooler (oceans).

The Walker circulation changes on interseasonal timescales. Every 3–5 years the trade winds weaken and the Walker cell circulation breaks down. The pressure difference between the eastern and western Pacific decreases, and less water is transported westward. Eventually, the warm water in the west, at a higher elevation, begins to move eastward, resulting in a warming of the sea surface temperature in the eastern and central Pacific, $\gtrsim 1.5\,^{\circ}$C above its normal temperature of $\sim 25\,^{\circ}$C. This phenomenon is known as *El Niño* (Spanish for little boy, named after the infant Jesus, since the warming of the ocean usually starts around Christmas time). Small changes in temperature produce large variations in the evaporation and latent heat release, due to the exponential dependence in the Clausius–Clapeyron equation (eq. 4.20). This brings about a large eastward shift of the convection cell patterns, causing air to rise above the central Pacific. This is accompanied by a change in surface pressure between the western (low $\rightarrow$ high pressure) and eastern Pacific (high $\rightarrow$ low pressure), a phenomenon known as the *El Niño Southern Oscillation*, which is propagated worldwide through planetary waves. El Niño has marked consequences for the global air circulation and weather patterns on Earth, as experienced through, e.g., abnormally wet or dry seasons.

In addition to the large scale circulations, weather on Earth is characterized by baroclinic eddies, which form both summer and winter in transition zones between easterly and westerly winds. These eddies may cause (large) storms on the planet, resulting sometimes in hurricanes and tornadoes. The development and movement of baroclinic eddies is also modified by El Niño. In contrast to moving baroclinic eddies, stationary eddies develop at places of varying local topography, such as mountain ranges and volcanoes, and from temperature differences between oceans and continents.

Venus

Venus's cloud deck is characterized by retrograde superrotating winds (in the same direction as the planet's rotation), which circle the planet with a 4-day period ($\sim$100 m s^{-1}) at an altitude of ~ 60 km. These winds decrease linearly with decreasing altitude, and reduce to 1 m s^{-1} at the surface. The winds are in cyclostrophic balance. In the thermosphere, strong day-to-night winds prevail, as a consequence of the large temperature gradient between Venus's thermosphere and cryosphere. Typical wind speeds are 100 m s^{-1}. The transition region between the retrograde winds in the troposphere and the day-to-night winds in Venus's thermosphere is not well studied. This transition takes place in the mesosphere, a region which can be probed at radio wavelengths in various line transitions of the CO molecule, and at infrared wavelengths using heterodyne spectrometry in the line core of the CO_2 molecule. Observations have confirmed the theoretically predicted thermal tides (day-to-night wind) in the upper mesosphere. It has been hypothesized that such winds may also explain the observed day-to-night variations in the CO abundance (Section 4.3.3.1).

Mars

The global atmospheric circulation on Mars resembles that on Earth. Like on Earth, air rises on Mars's summer hemisphere and subsides above the winter hemisphere. Since the warmest place does not coincide with the equator, the Hadley cells are not confined to the northern and southern hemispheres, but are displaced. Local topography with extreme altitude variations, from the deep Hellas basin up to the top of the Olympus and Tharsis ridge, leads to the formation of stationary eddies, while baroclinic eddies form over the winter hemisphere. Mars has substantial condensation flows, where CO_2 condenses out over the winter pole, and sublimes above the summer pole.

Since Mars's atmosphere is tenuous, it responds almost instantaneously to the solar heating, leading to strong winds across the terminator, the day–night line. These are the *thermal tide winds*, strong day-to-night winds, similar to the winds in Venus's thermosphere. If the winds near the surface have a preferred direction and speed, one might expect the formation of dunes. Dune fields have been observed on both Mars and Venus (Section 5.5). These fields yield clues to the wind direction and wind speed close to the surface. When these winds exceed $\sim$50–100 m s^{-1}, they may start local dust storms, either initiated by *saltation*, where grains start hopping over the surface, or when the dust is raised up in *dust devils*, due to convection in an atmosphere with a superadiabatic lapse rate. Such dust devils are frequently seen in desert areas on Earth. Over 100 dust devils have been identified on images taken with the Viking orbiter, with diameters up to 1 km and heights up to 6.8 km. Dust devils have also been observed by Mars Pathfinder. Once in the air, dust fuels the tidal winds, since the grains absorb sunlight and heat the atmosphere locally. Within just a few weeks, dust storms may grow so large that they envelop the entire planet. Such global storms may last for several months.

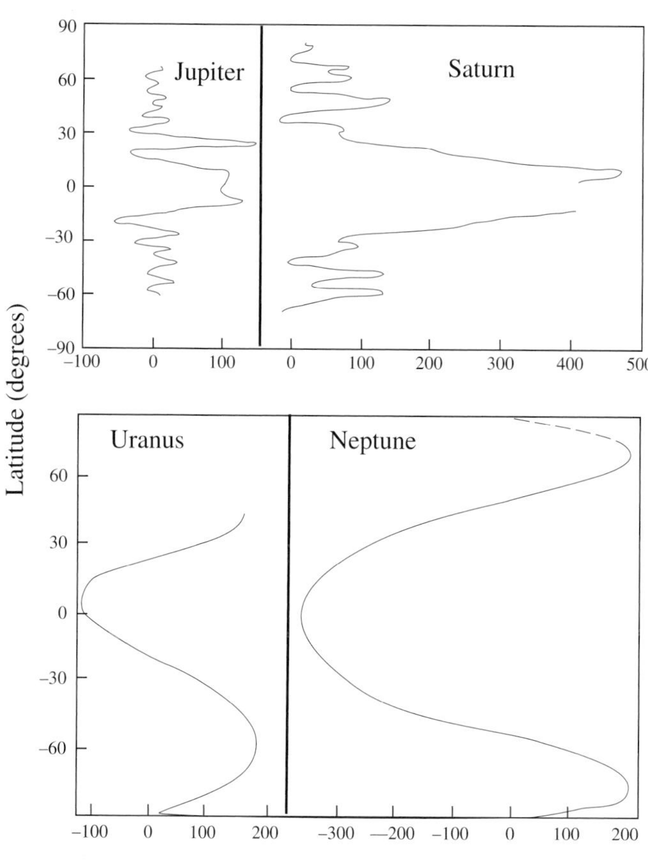

FIGURE 4.25 The zonal wind flow versus latitude on the giant planets Jupiter, Saturn, Uranus and Neptune. Velocities are measured relative to the planetary interiors. (Adapted from Ingersoll 1999)

4.5.5.2 *Giant Planets*

On terrestrial planets wind velocities are measured with respect to the planet's surface. Giant planets lack such a solid surface. The winds on these planets are measured with respect to the rotation rate of their magnetic fields. It is assumed that the fields are 'anchored' in the planet's interior, and that the rotation rates represent the 'true' rotation of the planet's interior.

High velocity zonal winds have been observed on all four giant planets (Fig. 4.25). Jupiter and Saturn have several jets in each hemisphere (5–6 on Jupiter, 3–4 on Saturn), of which the equatorial jet stream is by far the strongest: 100 m s^{-1} for Jupiter, and 500 m s^{-1} for Saturn, each in the eastward direction (faster than the planet rotates). Uranus has one jet stream near the equator in the westward direction, ~ -100 m s^{-1}, and one at higher latitudes towards the east, $\sim$100–200 m s^{-1}. Neptune shows a strong jet near the equator which flows in the westward direction, at ~ -500 m s^{-1}. It is interesting to note that the winds on Neptune usually lag behind the planet's rotation, in contrast to the wind systems on the other three giant planets.

The zonal wind patterns on the giant planets appear to be very stable in time. For example, on Jupiter, although the jets appear correlated with the white zones and brown belts (Fig. 4.28) on the planet, the banded structure sometimes changes morphology quite drastically (entire bands may disappear), while the zonal winds remain stable. The vertical extent of the winds is still a topic of debate. The winds may be confined to a relatively thin weather layer, as on Earth, or may extend very deep in the atmosphere. It has been suggested that the interiors of the planets may consist of large cylinders, each of which rotates at its own speed (Fig. 4.26). The zonal winds are the surface manifestation of these cylinders. Laboratory experiments in the 1920s showed the tendency of fluids in a rotating body to align with the rotation axis, and this model was applied to Jupiter and Saturn by F.J. Busse in the 1970s. This model gained much support, since the winds on both Jupiter and Saturn are quite symmetric around the equator.

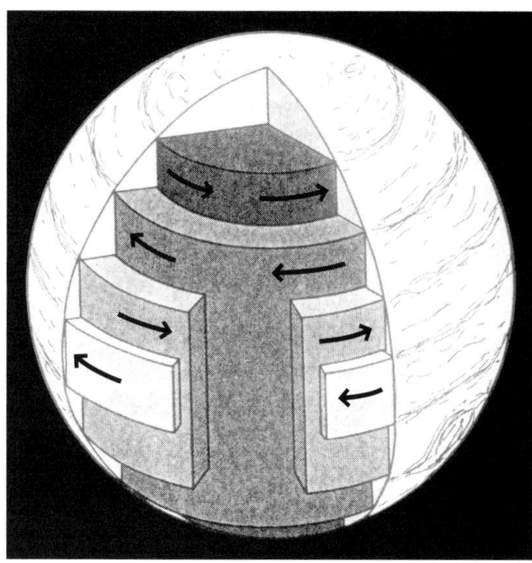

FIGURE 4.26 Possible large scale flow within the interiors of the giant planets. Each cylinder has a unique rotation rate, and the zonal winds on the giant planets may be the surface manifestation of these flows. (Adapted from Ingersoll 1990)

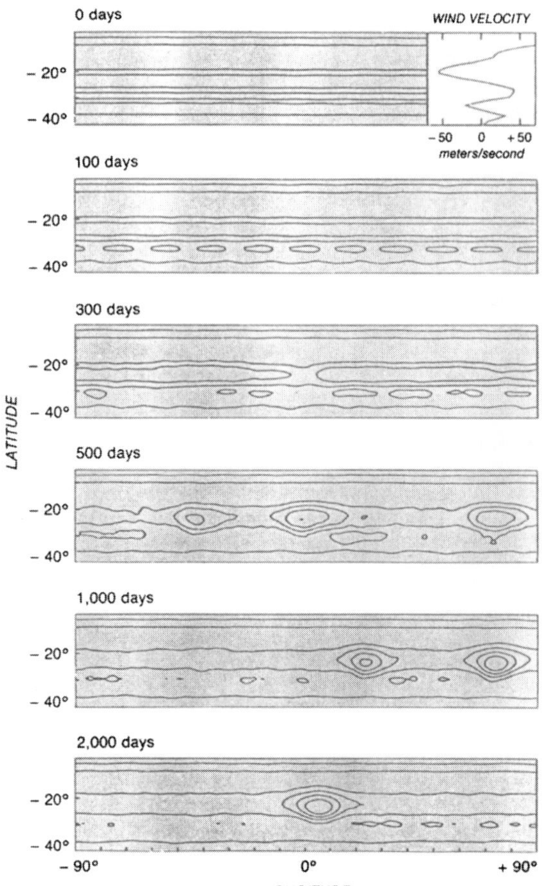

FIGURE 4.27 Computer simulation of the formation of eddies in Jupiter's atmosphere, modeled as a shallow weather layer. Note that the eddies combine until only one vortex is left. This could be the formation mechanism of the Great Red Spot. (Ingersoll 1990)

If the winds are entirely triggered by solar radiation combined with Jupiter's fast rotation, the winds pertain only to the upper atmospheric layers and decrease in strength with depth. The Galileo probe, however, measured an increase in wind speed with depth, indicative of internal heat being the driving force for the winds. Models are currently being developed to simulate the observed wind speeds. Computer simulations for a shallow-water flow for Jupiter are shown in Figure 4.27. Initially, the flow is unstable and breaks up into a series of small eddies. The eddies merge and only a few large eddies are left after a few hundred days. These finally merge into one eddy, after about 4–5 years. The final eddy appears to be stable for a many years. This may explain the longevity of the *Great Red Spot* (GRS) on Jupiter.

Figure 4.28 shows images of Jupiter at visible, infrared (5 μm) and radio wavelengths (2 cm). The white zone–brown belt structure is clearly visible on the optical image of the planet. Infrared wavelengths show the temperature at the top of the cloud layers, and there is a strong correlation between color on the optical image and temperature on the 5 μm image. The zones are generally slightly colder than the belt regions, indicative of more opaque clouds in the zones than the belts. At radio wavelengths the opacity is mainly provided by ammonia gas, whereas the clouds are transparent. As shown, the belt regions are brighter, which implies that ammonia gas is depleted in the belts compared to the zones. This is all indicative of gas rising in the zones. While the parcel rises ammonia gas freezes out (NH_4SH, NH_3-ice) when its partial pressure exceeds the saturated vapor curve. Dry air subsides in the belts, and this forces a simple convection pattern. Because the air above the belts is relatively dry, the ammonia-ice cloud layer is either absent or rather thin, and hence not seen at infrared wavelengths. The rising/subsiding air motions produce a latitudinal temperature gradient between belts and zones, which drives the zonal winds on the planet (through geostrophic balance with the Coriolis force).

Visual images of Jupiter (Fig. 4.28a) show the emergence of bright white plumes. These plumes indicate gas that is rising upwards and ammonia-ice, just condensed out, produces the bright white color. Jupiter also displays many short-lived eddies and a few long-lived storms, such

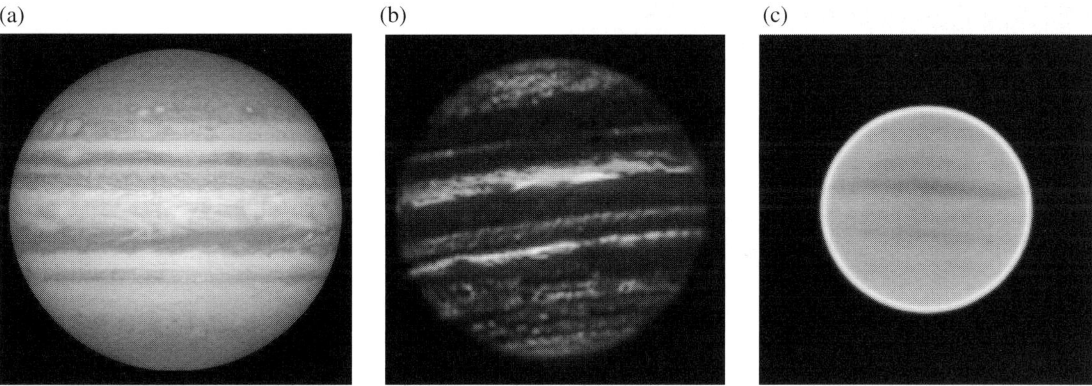

FIGURE 4.28 COLOR PLATE Images of Jupiter at (a) visible (HST), (b) thermal infrared (IRTF) and (c) radio wavelengths (VLA). At visible wavelengths, the planet's zones and belts show up as white and brown regions, respectively. At infrared and radio wavelengths we receive thermal radiation from the planet. The belts, as well as other brown regions, appear to be warmer than the zones. This suggests that the opacity in the belts is lower than in the zones, so deeper warmer layers in the planet are probed. The infrared (5 μm) and visible-light images were taken on 3 and 4 October 1995, respectively, just 2 months before the Galileo probe entry. The radio image (2 cm) was obtained 25 January 1996, about 1.5 months after probe entry. The position of the probe entry is indicated by an arrow on the visible-light and infrared images; the radio image is integrated over $\sim$ 6–7 hours, so that longitudinal structure is smeared out. The angular resolution on the latter image is $\sim 1.4'' \approx 0.044$ $R_{♃}$. (HST: Courtesy R. Beebe and HST/NASA; IRTF: Courtesy G. Orton; VLA: de Pater *et al.* 2001)

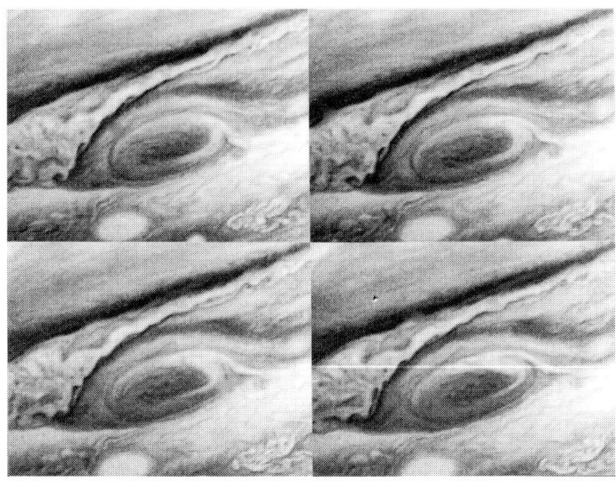

FIGURE 4.29 Time sequence of the Great Red Spot in blue light. The images are taken one Jovian rotation apart, on 2 and 3 February 1979 by the Voyager 1 spacecraft. Day-to-day changes in circulation are clearly seen. (Voyager 1/NASA P21148)

as the GRS and the big white ovals to the south. After having been visible for many decades, the white ovals merged in 1998. The circulation in the eddies is almost always clockwise in the northern hemisphere and anticlockwise in the southern hemisphere, indicative of high pressure regions. All of these features hint at a strongly convective planet.

Infrared images show that the GRS is much colder (Fig. 4.28b) than any other region on the planet. Because one is sensitive to temperature at these wavelengths, the data suggest that the opacity is high at high altitudes; i.e., the gases and cloud particles rise upwards to high altitudes, and the GRS is thought to be a giant hurricane. The system probably gets its energy from condensing gases below (latent heat release); also, small eddies are sometimes seen to be eaten by the GRS, which provides another source of energy. On Earth hurricanes do not live long, since they break apart once they hit land. On gas giant Jupiter hurricanes can exist indefinitely. A red spot was first seen by Cassini in the late 1600s. There is some debate whether Cassini saw the GRS, since during a period in the nineteenth century no red spots were seen. However, we know from observations throughout the 1980s and 1990s that the appearance of the GRS can change drastically in prominence, size and shape to the point of essential disappearance, so the system could be very long-lived.

The Galileo probe entered Jupiter's atmosphere at an infrared 'hot spot', i.e., at a region where the infrared opacity was low so deep warm layers were probed. While air is rising it cools and condensable gases can rain out if the temperature drops significantly below the condensation temperature of that gas. It is believed that 'dried out' air descends in the hot spots, like in the zone-belt convection pattern discussed above. However, in the hot spots the air descends in the form of strong downdrafts, so that the air is dry even at the 2–10 bar pressure levels probed by Galileo. The precise probe measurements regarding abundances of the condensable gases are, however, far from understood at the present time.

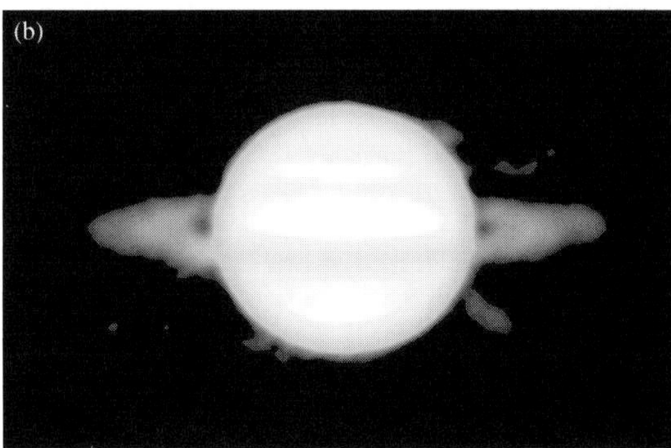

FIGURE 4.30 (a) Images of Saturn at different wavelengths. Upper panel: HST image at visible wavelengths taken in December 1994. Note the bright storm near the equator. (Courtesy: R. Beebe and NASA/HST) Lower panel: HST image at 8922 Å (methane absorption band). (Courtesy: A.S. Bosh and NASA/HST) At visible and near-infrared (CH_4 absorption band) wavelengths the planet is visible in reflected sunlight. In the lower image the equatorial bands are very bright, indicative of a relatively lower CH_4 gas abundance above the clouds. (b) VLA image at a wavelength of 2 cm (van der Tak *et al.* 1999). At these radio wavelengths the planet's thermal emission is observed. The bright bands are indicative of a lower atmospheric opacity in the region probed (NH_3 gas), so deeper warmer layers are probed (see also Figure 4.28).

Infrared and radio observations (2–20 cm) of Saturn show (time-variable) enhanced brightness temperatures at middle latitudes (Fig. 4.30), indicative of a decreased opacity. At IR wavelengths the opacity is caused by clouds, while the line-of-sight integrated NH_3 abundance provides most of the opacity at radio wavelengths. The Voyager infrared observations suggest that the ammonia ice cloud is thinner at mid-northern latitudes compared to other latitudes, while radio data from the 1980s suggest the base of the NH_4SH layer is deeper in the atmosphere in this band compared to other latitudes, so that the NH_3 mixing ratio here is low down to a considerable depth. Both observations indicate subsiding air motions, to depths of at least 5 bar.

Images of Saturn and Neptune show strong evidence of convection, in the form of the appearance of eddies and storms, as on Jupiter. No such storm systems have been seen on Uranus, indicative of a suppressed (compared to the other three giants) convection. Since Uranus is tipped on its side, one would expect (based upon a simple insolation model) that Uranus's poles would be warmer than its equator by ~ 6 K. However, Voyager infrared measurements indicate that the temperature is remarkably similar at all latitudes, which suggests a redistribution of heat in Uranus's atmosphere or interior. A similar redistribution of heat takes place in the interiors of the other three giant planets, where the equatorial region receives most of the heat, yet the infrared emission from the poles and equator are very similar. However, because Uranus's internal heat source is so small, it is hard to explain strong convection on Uranus, and this is amplified by the apparent lack of convective motions in the form of eddies and storm systems. On the other hand, large scale atmospheric motions do seem to be present, as infrared observations suggest rising air at latitudes between 20° and 40° on the southern hemisphere, and subsiding air at other latitudes. Radio observations are consistent with such a flow model, and imply the motions to extend down to depths of ~ 50 bar. Similar flows have been suggested for Neptune, since both the Voyager infrared measurements and the radio data for the two planets are very much alike. As shown in Figure 4.32, Neptune is very 'active': spots appear and disappear, as seen on Voyager images from 1989, and HST and ground-based infrared images throughout the 1990s.

4.6 Photochemistry

All planetary atmospheres are subjected to solar irradiation, which in addition to a general heating of the atmosphere through absorption of solar photons, can also change an atmosphere's composition considerably. Typically, absorption of photons at far-infrared and radio wavelengths ($\lambda \gtrsim 100$ μm) induces excitation of a molecule's lowest quantum states, i.e., of the rotational levels. Photons at infrared wavelengths ($\lambda \sim 2$–20 μm) can excite vibrational levels, while photons at visible and UV wavelengths may excite electrons to higher quantum states

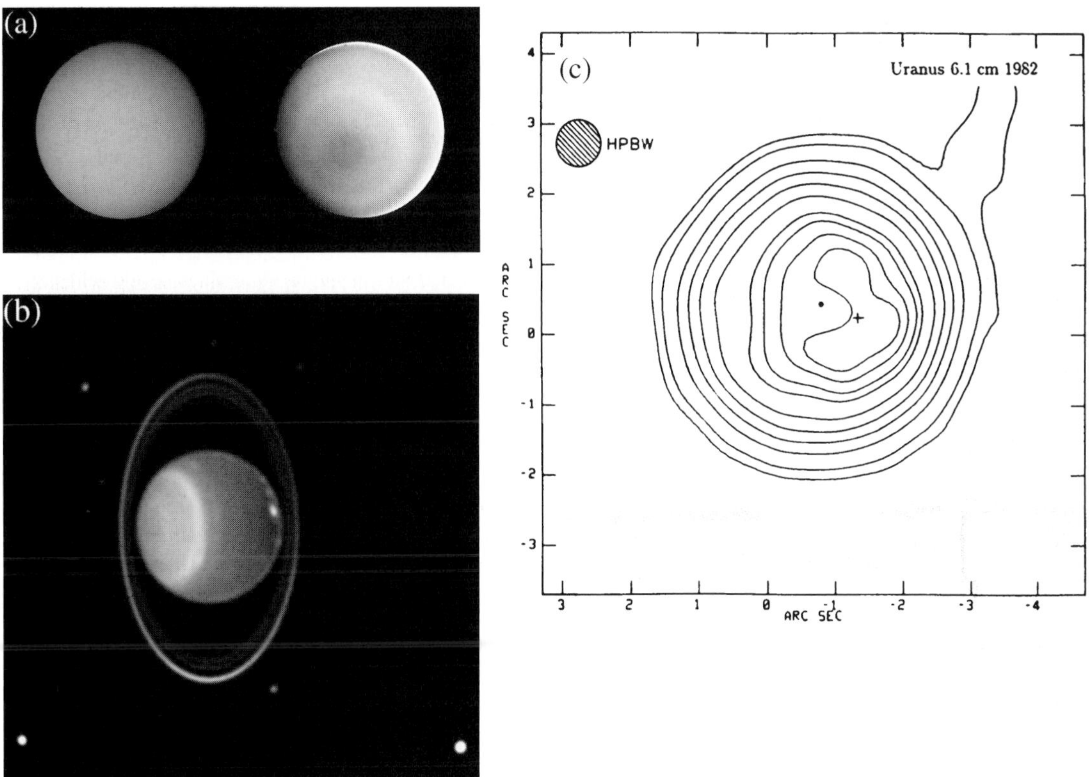

FIGURE 4.31 Images of Uranus at several different wavelengths. (a) Voyager 2 image at visible wavelengths, in natural grayscale (left) and with enhanced variations of albedo (right) (NASA/JPL). (b) HST image of Uranus at infrared wavelengths, taken in 1997. The image has been composed of a superposition of images in different filters. Many moons and the uranian ring system are visible. The bright features at high northern latitudes (near right limb of planet) are high-altitude clouds in Uranus's atmosphere. (Courtesy: E. Karkoschka and NASA/HST) (c) VLA image at a wavelength of 6.1 cm taken in 1982. (de Pater *et al.* 1989) The planet is visible in reflected sunlight at visible and infrared wavelengths, while the thermal emission is probed at radio wavelengths. Note that the brightest region on the latter image is centered around Uranus's pole (the cross on the figure) rather than the subsolar point (dot on figure). This is indicative of a relative lack of absorbing gases (H_2S, NH_3) above the pole.

within atoms and molecules. Photons with similar and higher energies may break up molecules (*photodissociation* or *photolysis* usually requiring $\lambda \lesssim 1$ μm) or *photoionize* atoms and molecules. Typically, the outer electron shell of an atom is photoionized by photons with $\lambda \lesssim 1000$ Å, while photons with $\lambda < 100$ Å can ionize the inner shell of an atom or molecule. The penetration depth of solar photons into an atmosphere depends upon the optical depth at the particular wavelength of radiation. Clouds, hazes, Rayleigh scattering and absorption by molecules/atoms affect the optical depth. As there are more solar photons at the top of an atmosphere than at deeper layers because of a strong absorption effect, most photochemical reactions occur at high altitudes. A solar spectrum is shown in Figure 4.33. Since the solar blackbody curve peaks at visible wavelengths, the number of UV photons decreases with decreasing wavelength. Note that the solar spectrum can be well represented by a blackbody spectrum with a temperature of 5777 K at wavelengths longwards from the peak in the spectrum. At shorter wavelengths absorption lines dominate the spectrum, while at UV wavelengths shortwards of 1500 Å emission lines dominate the spectrum (note the strong H Lyman α line).

When the photodissociation or ionization rate is balanced by the reverse process, we speak of *photochemical equilibrium*. Averaged over time, at least over a full planet day, planetary atmospheres are approximately in photochemical equilibrium. This assumption is used below to

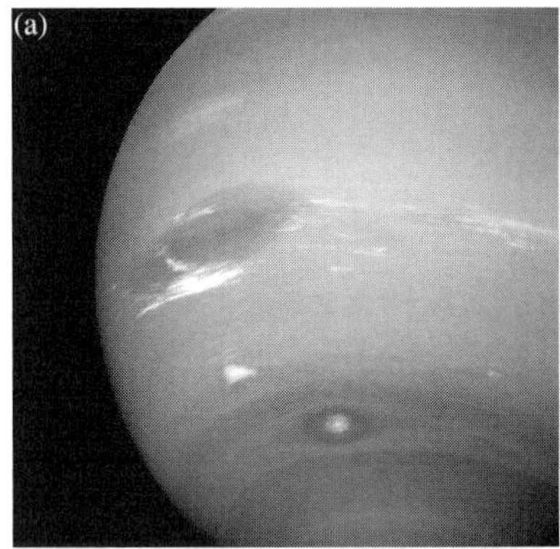

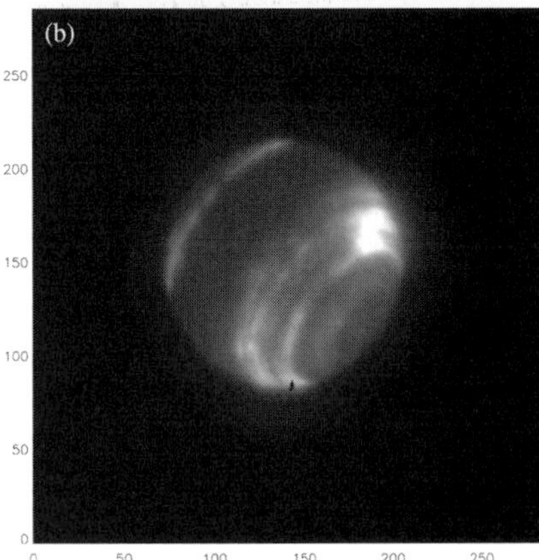

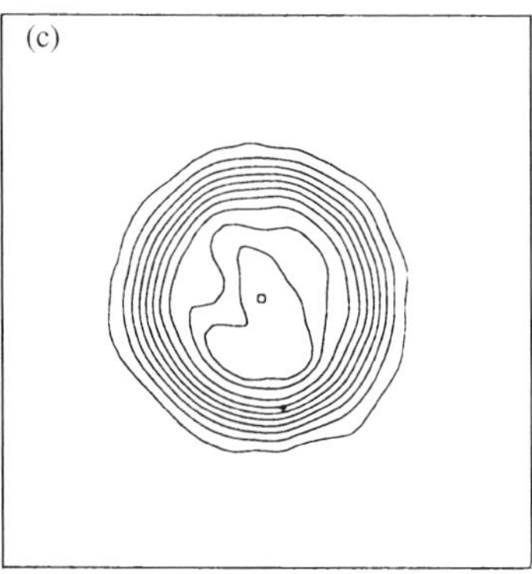

FIGURE 4.32 Images of Neptune at different wavelengths. (a) Voyager 2 image at visible wavelengths. Note the Great Dark Spot surrounded by white hazes. Multiwavelength observations suggest the spot to be located relatively deeply in the atmosphere, whereas the white hazes are at higher altitudes. This spot disappeared in the early 1990s. The Little Dark Spot and Scooter, the bright white cloud feature at latitudes in between the dark spots, are also clearly visible on this image. (b) Keck adaptive optics image at a wavelength of 1.6 μm. The resolution in this image is 0.04″, similar to that of HST at visible wavelengths. (Courtesy: W.M. Keck Observatory Adaptive Optics Team) (c) VLA image of Neptune at a wavelength of 3.6 cm. (Hofstadter 1993) The planet is visible in reflected sunlight at visible and infrared wavelengths, while the thermal emission is probed at radio wavelengths.

derive information on the altitude distribution of the atoms, molecules and ions in an atmosphere.

4.6.1 Photodissociation and Recombination

When absorption of photons leads to a break-up of molecules, we speak of *photodissociation*. As mentioned above, photodissociation typically takes place at high altitudes, whereas the reverse reaction, *recombination*, proceeds faster at lower altitudes. Nevertheless, through vertical transport, photodissociation and recombination are balanced over time. In this section we discuss the most important photochemical reactions on Earth, Venus, Mars, Titan and the giant planets.

4.6.1.1 *Oxygen Chemistry on Earth*

The reactions for photodissociation (reaction 1 below) and recombination (reactions 2–4) for oxygen in the Earth's atmosphere can be written:

Photodissociation: (1) $O_2 + h\nu \rightarrow O + O \quad \lambda < 1750$ Å.

The production rate at altitude z:

$$\frac{d[O]}{dt} = 2[O_2]J_1(z), \tag{4.56}$$

where a compound in square brackets, e.g., [O], refers to the number of O atoms per unit volume. The photolysis or photodissociation rate, J(z), is defined by:

$$J(z) = \int \sigma_{x_\nu} \mathcal{F}_\nu e^{-\tau_\nu(z)/\mu_\theta} d\nu, \tag{4.57}$$

where σ_{x_ν} is the photon absorption cross-section at frequency ν, μ_θ is the cosine of the angle between the solar direction and the local vertical, and $\mathcal{F}_\nu$ is the solar flux density expressed in photons cm^{-1} s^{-1} Hz^{-1} outside the atmosphere. The mean photolysis rate for oxygen at an altitude of 20 km is $J_1(20 \text{ km}) = 4.7 \times 10^{-14}$ s^{-1}; at an altitude of 60 km: $J_1(60 \text{ km}) = 5.7 \times 10^{-10}$ s^{-1}. Since the number density of solar photons decreases exponentially with optical depth (thus with decreasing altitude), and the number of oxygen molecules decreases with increasing altitude according to the barometric law (eq. 4.1), the concentration of oxygen atoms peaks somewhere in the middle of the atmosphere.

Recombination: the direct, two-body reaction

(2) $O + O \rightarrow O_2 + h\nu$,

is very slow, and oxygen recombination is therefore dominated by the three-body processes:

(3) $O + O + M \rightarrow O_2 + M$
(4) $O + O_2 + M \rightarrow O_3 + M$,

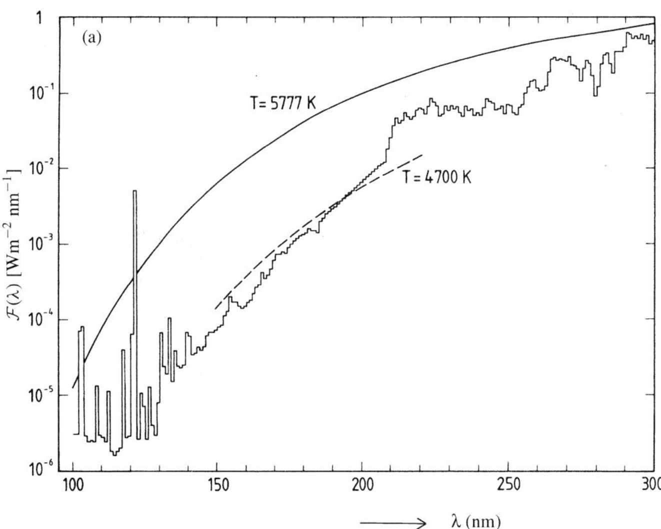

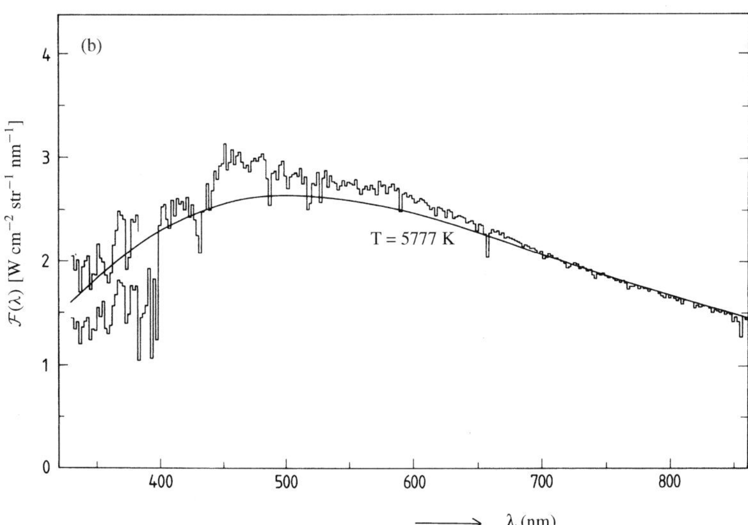

FIGURE 4.33 Solar spectrum at UV (panel a) and visible (panel b) wavelengths. Blackbody spectra at 5777 K and 4700 K are superposed. (Stix 1989)

where M is an atmospheric molecule which takes up the excess energy liberated in the reaction. Since the abundance of M and O_2 follows the barometric altitude distribution, reactions (3) and (4) are most effective at low altitudes, provided atomic oxygen is present. The production rate of $[O_2]$ in (3) is proportional to both [O] and [M]:

$$\frac{d[O_2]}{dt} = [O]^2[M]k_{r3}, \tag{4.58}$$

where the reaction rate, k_{r3}, (cm^6 s^{-1}) is the rate at which the three-body reaction proceeds (the subscript $r3$ stands for reaction 3). Reaction rates depend upon the collisional rates between molecules, which depend on the temperature of the medium. A two-body reaction rate (cm^3 s^{-1}), as in reaction (2), is usually expressed as:

$$k_{r2} = c_1 \left(\frac{T}{300}\right)^{c_2} e^{-E_o/kT}, \tag{4.59}$$

where E_o is the *activation energy* to overcome the potential barrier that reaction (2) might have. Note that the exponential in equation (4.59) is essentially a Boltzmann factor (Section 3.2.2), and recombination does not take place if $E < E_o$. The temperature, T, is expressed in kelvin; c_1 and c_2 are constants. An upper limit to k_{r2} is given by the gas-kinetic rate of collisions, k_{gk}:

$$k_{gk} = \sigma_x \bar{v}_o \approx 2 \times 10^{-10} \sqrt{\frac{T}{300}}, \tag{4.60}$$

where the collisional cross-section, σ_x, for atmospheric molecules is typically a few $\times$ 10^{-15} cm^2, and $\bar{v}_o$ is the mean thermal velocity of the particle ($\sqrt{2kT/m}$). For reaction (2) above the reaction rate is much lower, $k_{r2} < 10^{-20}$ cm^3 s^{-1}.

For a three-body interaction as in reactions (3) and (4), rate k_{r2} has to be multiplied by the chance a third molecule collides at the same time with the oxygen atoms/molecules. The duration of a collision is typically of the order of $2R/\bar{v}_o$, where R is the molecular radius, typically a few Å. The three-body reaction rate, k_{r3} for reaction (3) becomes:

$$k_{r3} = \frac{2R}{\bar{v}_o} k_{r2}^2 \approx 10^{-12} k_{r2}^2 \approx 4 \times 10^{-32} \frac{T}{300}. \tag{4.61}$$

For gas-kinetic collisions, reactions (2) and (3) are equal if the number of molecules, [M], per cubic centimeter, is given by:

$$[M] = \frac{k_{r2}}{k_{r3}} \approx 5 \times 10^{21} \text{ cm}^{-3}. \tag{4.62}$$

$[M] = 2.69 \times 10^{19}$ cm^{-3} at the Earth's surface; hence, unless the reaction rate for the two-body reaction is extremely slow, three-body interactions can be ignored. As mentioned above, k_{r2} for reaction (2) is very slow, and therefore the three-body reaction (3) is the dominant process.

Ozone, O_3, is produced in chemical reaction (4), where $k_{r4} = 6 \times 10^{-34} \left(\frac{T}{300}\right)^{-2.3}$ cm^6 s^{-1}. Ozone is very important in the Earth's atmosphere since it effectively blocks penetration of UV sunlight to the ground. To deduce the vertical distribution of ozone in our atmosphere, we must consider both the processes which lead to its formation and to its destruction. In a pure oxygen atmosphere the relevant reactions are the *Chapman reactions* (reactions 1, 4, 5 and 6). The formation of ozone was considered in reaction (4) above. In an oxygen atmosphere ozone is destroyed by:

Photodissociation: (5) $O_3 + h\nu \rightarrow O_2 + O \quad \lambda \lesssim 3100$ Å, with $J_5(60 \text{ km}) = 4.0 \times 10^{-3}$ s^{-1}, $J_5(20 \text{ km}) = 3.2 \times 10^{-5}$ s^{-1},

or the reaction:

(6) $O + O_3 \rightarrow O_2 + O_2$, with $k_{r6} = 8.0 \times 10^{-12} e^{-2060/T}$ cm^{3}s^{-1}.

The net change in the atomic oxygen and ozone number densities is:

$$\frac{d[O]}{dt} = 2J_1(z)[O_2] + J_5(z)[O_3] - k_{r6}[O][O_3] - k_{r4}[O][O_2][M] \tag{4.63}$$

$$\frac{d[O_3]}{dt} = k_{r4}[O][O_2][M] - k_{r6}[O][O_3] - J_5(z)[O_3], \tag{4.64}$$

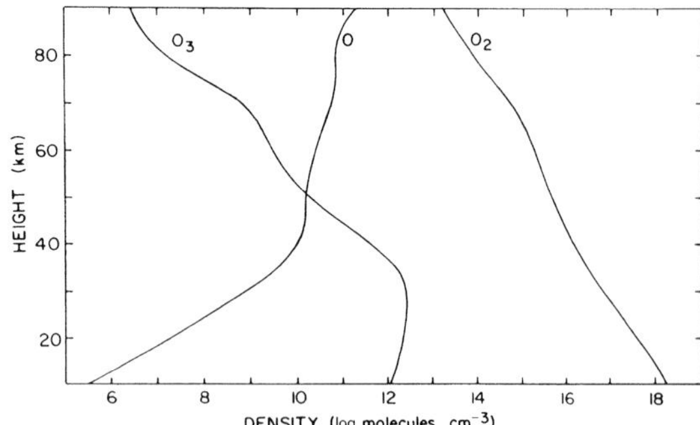

FIGURE 4.34 Graph of calculated densities of O, O_2 and O_3 in the Earth's atmosphere. (Chamberlain and Hunten 1987)

where the subscripts to J and k_r refer to reactions (1)–(6). In chemical equilibrium, the net change $\frac{d[O]}{dt}$ and $\frac{d[O_3]}{dt}$ is equal to zero. This leads to the altitude profiles of atomic oxygen and ozone:

$$[O] = \frac{J_1(z)[O_2]}{k_{r6}[O_3]} \tag{4.65}$$

$$[O_3] = \frac{k_{r4}[O][O_2][M]}{k_{r6}[O] + J_5(z)}. \tag{4.66}$$

The altitude distributions of atomic and molecular oxygen in our atmosphere together with ozone are shown in Figure 4.34. As expected from photolysis arguments, the number of oxygen atoms increases with increasing altitude up to a certain altitude, while the number of oxygen molecules is decreasing with altitude. Ozone peaks in number density at altitudes near 30 km.

In reality our atmosphere is not a pure oxygen atmosphere, and *catalytic* destruction of ozone takes place in addition to reactions (5) and (6) given above. Given that ozone protects life on Earth from the harmful solar UV photons, much research is currently devoted to the chemistry of production and destruction of ozone. Free hydrogen atoms are highly reactive with ozone, and readily destroy O_3:

(7) $H + O_3 \rightarrow OH + O_2$ with $k_{r7} = 1.4 \times 10^{-10} e^{-470/T}$ cm^3 s^{-1}.

Free atomic hydrogen on Earth is produced by chemical reactions starting from H_2O and CH_4, but it is not very abundant in the Earth's atmosphere.

Nitrogen oxides, chlorineand halomethanes form

'important' catalysts for the destruction of ozone. These molecules react with O_3, and are immediately regenerated through a reaction with atmospheric oxygen. The abundance of these molecules is thus constant:

(8a) $NO + O_3 \rightarrow NO_2 + O_2$ with $k_{r8a} = 1.8 \times 10^{-12} e^{-1370/T}$ cm^3 s^{-1}

(8b) $NO_2 + O \rightarrow NO + O_2$ with $k_{r8b} = 9.3 \times 10^{-12}$ cm^3 s^{-1}

(9a) $Cl + O_3 \rightarrow ClO + O_2$ with $k_{r9a} = 2.8 \times 10^{-11} e^{-257/T}$ cm^3 s^{-1}

(9b) $ClO + O \rightarrow Cl + O_2$ with $k_{r9b} = 7.7 \times 10^{-11} e^{-130/T}$ cm^3 s^{-1}.

Reactions with fluorine atoms are not important, since F is transformed rapidly into HF via a reaction with CH_4; it is not regenerated, in contrast to Cl. Bromineatoms may, like Cl, lead to the destruction of ozone. Nitrogen oxides in the Earth's stratosphere lead to a rapid decrease in the ozone abundance. Nitrogen oxides are generated by biological activities, engine exhausts, and industrial fertilizers close to the ground. Before these nitrogen oxides can destroy ozone, they must be brought up into the stratosphere. It is therefore not clear how harmful the production of NO_x in the troposphere is; it actually leads to a local production of ozone in the troposphere, through the formation of atomic oxygen. However, O_3 is not desirable in the lower troposphere because it is highly reactive/corrosive. While active research with respect to the formation and destruction of ozone is continuing, we feel it best to suppress the release of molecules which may lead to a destruction of ozone in the stratosphere.

The penetration of galactic and solar cosmic rays over the polar caps leads to free atomic nitrogen in an excited state, which leads to a rapid enhancement in the stratospheric NO abundance. Observations show that the O_3 abundance over the polar caps is indeed anticorrelated with NO. During high solar activity, the solar wind and heliomagnetic field are strong, and shield the Earth from galactic cosmic rays. This may explain a suspected 11-year cycle in the ozone abundance. Solar cosmic rays, however, occur in bursts during solar flares, and are out of phase with the galactic cosmic rays.

4.6.1.2 *Photochemistry on Venus and Mars*

The primary atmospheric constituent on Venus and Mars is CO_2, which is dissociated into CO and O under the influence of solar UV light:

Photodissociation: (10) $CO_2 + h\nu \rightarrow CO + O$
$\lambda < 1690$ Å
Recombination: (11) $CO + O + M \rightarrow CO_2 + M$.

Since the recombination rate k_{r11} is very slow ($k_{r11} \approx 2 \times 10^{-37}$ cm^6 s^{-1}), one expects large quantities of CO in the atmospheres of Venus and Mars. In addition, individual oxygen atoms may recombine into molecular oxygen. The reaction rate for the latter process is roughly 10^3–10^4 larger than k_{r11}. The creation of molecular oxygen is balanced by photodissociation (reaction 1). Despite the low recombination rate of CO (reaction 11), the observed CO, O and O_2 densities on both Venus and Mars are very low. An explanation for this observation on Mars may be rapid downward transport of CO, O and O_2, where recombination of CO and O proceeds faster in the presence of OH chemistry. Essentially, water vapor provides OH and H, while OH oxidizes CO:

Oxidation: (12) $CO + OH \rightarrow CO_2 + H$.

A series of chemical reactions results in the production of OH from H (reaction 12) and free oxygen (10), so that the net product is restoration of CO_2. On Venus, recombination of CO may be aided by catalytic reactions resulting from chlorine and sulfur chemistry on this planet. Since oxygen is formed on both Venus and Mars, one expects ozone to be present as well. Destruction of ozone is tied to the H_xO_y chemistry, resulting in a near total absence of ozone on the two planets, except near Mars's winter pole. Above Mars's winter pole, temperatures are so low that H_xO_y products freeze out, allowing accumulation of ozone.

Venus's clouds consist of sulfuric acid, formed by the reaction of sulfur dioxide with (photochemically produced) oxygen and water:

(13) $SO_2 + O \rightarrow SO_3$
(14) $SO_3 + H_2O \rightarrow H_2SO_4$.

Sulfuric acid readily condenses out in the upper troposphere of Venus, producing a dense smog of sulfuric acid droplets.

4.6.1.3 *Photochemistry on the Giant Planets*

In the deep layers of the H_2-rich atmospheres of the giant planets, N, C, S, O and P must be present in the form of NH_3, CH_4, H_2S, H_2O and PH_3. At higher altitudes, photolysis leads to a break-up of these molecules. Since Rayleigh scattering provides a large source of opacity at UV wavelengths, photodissociation is only important in the upper troposphere and stratosphere, at pressures $P \lesssim 0.3$ bar. Below we will summarize the effects of photodissociation and subsequent chemical reactions for NH_3, CH_4, H_2S and PH_3, gases which one might expect to be present in the upper troposphere and stratosphere, and which are thus susceptible to photolysis. For details on these processes the reader is referred to e.g., Atreya (1986).

(1) Ammonia, which is photolyzed by photons with $\lambda <$ 2300 Å, produces amidogen radicals, $NH_2(X)$. Roughly 30% of these radicals recycles back to NH_3, and the remainder forms hydrazine gas (N_2H_4) via a self-reaction (i.e., $NH_2(X)$ with $NH_2(X)$). Hydrazine gas has a low vapor pressure, and consequently we expect it to condense in the stratosphere, thus making up in part the dust and hazes in the upper atmospheres of the giant planets. Hydrazine is expected to photodissociate and/or react with H to form the condensable gas hydrazyl (N_2H_3). Reactions involving N_2H_3 ultimately lead to N_2, which is stable against photolysis. We note that molecular nitrogen may not only result from photolysis, but that in principle N_2 may be brought up from a planet's deep atmosphere by rapid vertical transport. The detection of CO and hydrogen cyanide (HCN) in Neptune's stratosphere has been explained by upward transport of CO and N_2 on this planet, where chemical reactions between hydrocarbons and nitrogen may lead to the formation of HCN. The observed NH_3 abundance in Jupiter and Saturn's atmospheres near and above the tropopause is well below saturation levels, as expected from photochemistry. On Uranus and Neptune, the tropopause temperatures are so low that NH_3 is completely frozen out well before an airparcel has risen up to the tropopause. We, therefore, do not expect nor see evidence of NH_3 photochemistry on the outer giants.
The detection of NH_3 gas in Jupiter's stratosphere after the impact of comet D/Shoemaker–Levy 9 presented a nice test-case of the NH_3 photodissociation rate. After the impact of one of the largest fragments (fragment K), NH_3 gas was detected above the impact site at close to saturated levels. Photolysis alone should reduce the NH_3 abundance by a factor of 6 in about one week. The observed decay of NH_3 emissions in the weeks following the impacts was consistent with the photochemical destruction rate.

(2) Methane gas is dissociated by UV photons with $\lambda <$ 1600 Å, and the photolysis products undergo a complicated series of chemical reactions. In the 10 mb–10 μb region, the hydrocarbons acetylene (C_2H_2), ethylene (C_2H_4), and ethane (C_2H_6) form. Ethylene is quickly lost by photolysis into C_2H_2 or recycled back to CH_4. Ethane is also converted to C_2H_2 or CH_4, but the reaction rate is about 10 times slower than that of C_2H_4. Hence, the mixing ratio of C_2H_4 is expected to be over an order of magnitude smaller than that of C_2H_6 and C_2H_2. The latter molecules are expected to be present on Jupiter at mixing ratios a few $\times\ 10^{-5}$ at pressure levels of 1 mb–10 μb. Hydrocarbons have been detected on all four giant planets. On Jupiter C_2H_6 has been measured at a mixing ratio 1–5 $\times\ 10^{-6}$, while C_2H_2 and C_2H_4 have mixing ratios 1–2 orders of magnitude less. Relatively copious amounts of hydrocarbons are present on Neptune (see Section 4.7.2.2), which has been attributed to rapid vertical transport on this planet (CH_4 is expected to condense near the tropopause; Section 4.4.3).
Catalytic reactions may convert acetylene into polyacetylenes ($C_{2n}H_2$). Models suggest mixing ratios of $\sim$0.1 ppm at $\sim$10 μb level on Jupiter. Higher order hydrocarbons are expected to form as well, of which propane (C_3H_8) is the most abundant, expected to be present at levels of 1 ppm at the 1 mb–10 μb region, decreasing to 10 ppb at 10 bar. Model calculations suggest methylacetylene (CH_3C_2H) to reach a peak of a few ppm at 1 μb, allene (CH_2CCH_2) and propylene (C_3H_6) are expected at $\sim$10 ppb level. The most abundant C_4 species is expected to be butane (C_4H_{10}), at $\sim$0.1 ppm at 100 μb, decreasing to 1 ppb at 10 bar. Butyne and butadiene (C_4H_6 isomers) as well as vinylacetylene (C_4H_4) are likely present at the 1 μb level at mixing ratios of $\sim$0.1 ppm. Most species are expected to decrease rapidly in abundance with decreasing altitude, so that typical mixing ratios are $\lesssim 10^{-14}$ at 10 bar.

(3) Photons at $\lambda <$ 3170 Å break up H_2S molecules, if there are any present in the upper troposphere and stratosphere. Photolysis and subsequent chemical reactions (see e.g., Atreya 1986) eventually lead to the formation of elemental sulfur (S_8), ammonium polysulfide ($(NH_4)_xS_y$) and hydrogen polysulfide (H_xS_y), which are, respectively, yellow, orange and brown in color. Suggestions have been made that these products may color the various bands on Jupiter and Saturn.

(4) Phosphine, PH_3, is photolyzed by photons with $1600 < \lambda < 2350$ Å. However, under equilibrium conditions one does not expect to see PH_3, since it is oxidized to P_4O_6 deep in the atmosphere where $300 < T < 800$ K, and the oxidized material is dissolved in water. Phosphine is, however, detected in the atmospheres of Jupiter and Saturn. At higher altitudes photolysis eventually leads to the formation of red phosphorus, P_4. This may color the Great Red Spot on Jupiter.

4.6.1.4 *Photochemistry on Titan and Icy Satellites*

Titan's atmosphere consists predominantly of nitrogen gas, with a small percentage of methane. As on Earth, N_2 is quite inert. The main source for dissociation of N_2 is impact by charged particles, as opposed to photolysis. Methane gas, at the other hand, is easily dissociated by solar photons. The methane photochemistry is in principle similar to that seen on the giant planets, although it is more efficient since there is far less free atomic hydrogen, as it readily escapes Titan's gravity field. As on the giant planets we expect the production of numerous hydrocarbons, as C_2H_2, C_2H_4 (which converts to C_2H_2), and C_2H_6. Because of the larger photolysis rate, the C_2H_6 production rate is much larger than that of C_2H_2. Photolysis of ethane, followed by a series of chemical reactions, is expected to produce higher order hydrocarbons, such as C_3H_8, C_4H_{10}, etc.

Reactions between N and CH_4 are expected to lead to the formation of e.g., HCN, CN, and more complex molecules such as cyanogen (C_2N_2), cyano acetylene (HC_3N), ethyl cyanide (C_2H_3CN) and possibly HCN polymers. Based upon the chemical reaction rates, the formation of the more complex molecules such as HC_3N and C_4H_2, as well as the formation of poly-acetylenes and HCN polymers are expected to thrive at low temperatures. Indeed, the more complex molecules have only been detected by the Voyager spacecraft near Titan's north pole, where it was winter at the time.

Because of the low stratospheric temperature, ethane and other photochemically produced complex molecules condense to form a heavy layer of smog in Titan's atmosphere. This layer of smog is densest above the winter pole, where complex molecules are more easily produced. Recent measurements at infrared wavelengths suggest the optical depth of the smog to be about three times larger above the winter than summer pole. Laboratory measurements by Sagan and coworkers in a simulated Titan atmosphere show the formation of a reddish-brown powder, called *tholins*, or mud.

The smog particles ultimately sediment out and fall to the ground. This has led to the speculation that Titan, if the same geological processes have operated throughout its history, might be covered by an ocean of liquid hydrocarbons, in particular liquid ethane, up to about one kilometer deep. A truly deep global ocean was ruled out by radar, and later by infrared measurements. More recently, disk-resolved images of Titan's surface at infrared wavelengths show very dark ($A_{ir} < 0.05$) and brighter ($A_{ir} \sim 0.10$–0.15) areas. The bright regions are probably composed of a rock/ice mixture, while the dark areas could be the elusive hydrocarbon oceans or lakes, though this has not been proven (Section 5.5.6.1).

Icy satellites must have atmospheres which contain a small amount of water vapor (Section 4.3.3.3), which readily dissociates to produce oxygen. Io's atmosphere must contain oxygen as well, from photodissociation of SO_2. In analogy with Earth, we therefore expect oxygen photochemistry on all these satellites. Indeed, HST has detected ozone on Ganymede and several saturnian satellites, although this ozone may be a direct product of the interaction of energetic magnetospheric particles with the water-ice on satellites' surfaces.

4.6.2 Photoionization: Ionospheres

UV photons at $\lambda \lesssim 1000$ Å can ionize atoms and molecules. Radiative recombination of atomic ions (e.g., $O^+ + e^- \rightarrow O + h\nu$) is very slow, in contrast to molecular recombination. Atomic ions are usually converted to molecular ions by ion–neutral reactions, and the molecular ions recombine. Photolysis in a tenuous atmosphere therefore leads to the formation of the *ionosphere*, a region characterized by the presence of free electrons. The electron density in the ionosphere is determined by both the ionization rate and how rapidly the ions recombine, whether directly by recombination, or indirectly via charge exchange and reactions followed by recombination. Each planet with a substantial atmosphere is expected to also have an ionosphere. In this section we first discuss the ionosphere on Earth, followed by a discussion and comparison of the ionospheres of Venus, Mars and the giant planets.

4.6.2.1 *Earth*

On Earth there are four distinct ionospheric layers: the D, E, F_1 and F_2 layers (Fig. 4.35). The nominal altitude of the D layer is at ~ 90 km, with a peak electron density of $\sim 10^4$ cm^{-3}. The E layer is concentrated around 110 km, and the peak electron density is an order of magnitude larger than in the D layer. The F_1 layer peaks at 200 km, with an electron density of $\sim 2.5 \times 10^5$ cm^{-3}, and the F_2 layer at 300 km, with a maximum electron density of $\sim 10^6$ cm^{-3}. The height and electron density of the various layers is highly variable in time, since they depend sensitively on the solar UV flux, which varies strongly during one day. The D and F_1 layers are generally absent during the night, while the electron densities in the E and F_2 layers are at least an order of magnitude smaller at night than during the day. Typical neutral densities at these altitudes can be calculated from the barometric law with allowance for diffusive separation, and are many orders of magnitude larger than the electron densities.

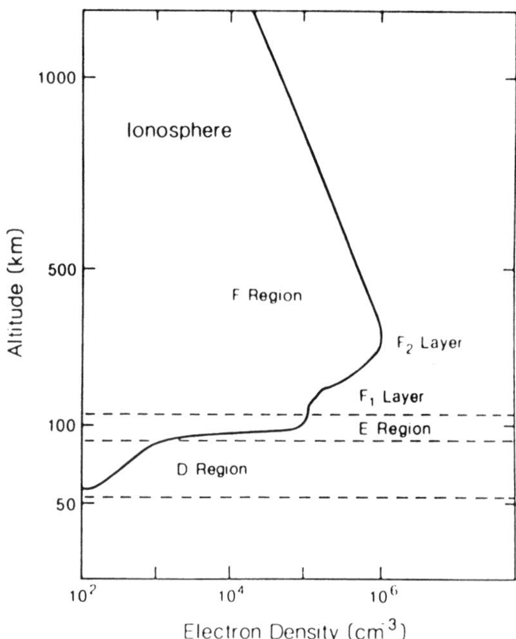

FIGURE 4.35 Sketch of the electron density in Earth's day side atmosphere, with the approximate locations of the ionospheric layers. (Russell 1995)

The ionospheric layers are distinct from each other, since the ionization and recombination processes are different for each layer. This is caused by changes in the composition and absorption characteristics of the atmosphere with altitude. In analogy with the physical processes that govern each ionospheric layer on Earth, ionospheric regions on other planets are sometimes denoted as a D, E, F_1 or F_2 region. Below we give a summary of the various processes which govern the different layers; for a detailed description the reader is referred to Chamberlain and Hunten (1987).

The *E layer* is characterized by direct photoionization of molecular oxygen:

$$(15)\ O_2 + h\nu \rightarrow O_2^+ + e^-, \qquad \lambda < 1027\ Å.$$

Solar coronal X-rays contribute as well, by ionizing O, O_2, and N_2, which leads to the production of O_2^+ and NO^+ ions via rapid charge exchange (16) or atom–ion interchange (17):

$$(16)\ N_2^+ + O_2 \rightarrow N_2 + O_2^+$$
$$(17)\ N_2^+ + O \rightarrow NO^+ + N.$$

Recombination occurs principally through dissociative recombination:

$$(18)\ O_2^+ + e^- \rightarrow O + O$$
$$(19)\ NO^+ + e^- \rightarrow N + O.$$

During the daytime the O_2^+ and NO^+ densities in the E layer are roughly equal.

The main ions formed in the *F_1 region* are from atomic oxygen and molecular nitrogen:

$$(20)\ O + h\nu \rightarrow O^+ + e^-, \qquad \lambda < 911\ Å$$
$$(21)\ N_2 + h\nu \rightarrow N_2^+ + e^-, \qquad \lambda < 796\ Å.$$

Dissociative recombination of N_2^+ ($N_2^+ + e^- \rightarrow N + N$) is very rare, since N_2^+ is rapidly converted into N_2 and NO^+ via reactions (16) and (17) above. Radiative recombination of atomic oxygen ions ($O^+ + e^- \rightarrow O + h\nu$) is very slow (reaction rate $k_r \approx 3 \times 10^{-12}$ cm^3 s^{-1}) compared to atom–ion interchange (roughly ten times faster):

$$(22)\ O^+ + O_2 \rightarrow O_2^+ + O$$
$$(23)\ O^+ + N_2 \rightarrow NO^+ + N,$$

followed by rapid dissociative recombination of O_2^+ and NO^+ (reactions (18) and (19) above, where $k_r \approx 3 \times 10^{-7}$ cm^3 s^{-1}).

The *F_2 region* is optically thin to most ionizing photons. Reaction (20) is the dominant ionization process. Since radiative recombination of O^+ is very slow, and the molecular density is low at these high altitudes (so reactions (22) and (23) are unimportant), the electron density is high in the F_2 region.

The dominant ionization process in the *D layer* is photoionization of N_2 and O_2 by X-rays. Photoionization of NO is important at these altitudes as well. The D layer is further characterized by the presence of negative ions, O_2^- and more complex ions, through electron attachment in a three-body reaction:

$$(24)\ O_2 + e^- + O_2 \rightarrow O_2^- + O_2,$$

at a rate of $\sim 5 \times 10^{-31}$ cm^6 s^{-1}. Although the third molecule could be N_2, the attachment rate is much lower with N_2 than O_2. These negative ions are destroyed, i.e., the electron is removed from the negative ion, by sunlight (photodetachment) or collisions (collisional detachment).

Although one would expect the ionosphere to be absent at night due to a lack of photoionizing photons, observations indicate the continued presence of an ionosphere, though with reduced electron and ion number densities. A night side ionosphere is, in part, explained by the fact that the recombination rate, especially for O^+, is slow relative to the planet's rotation. However, ionization may also be triggered by other processes, in particular precipitating electrons, micrometeorite bombardment, and UV photons from stars. These processes are an important source of ionization at night and in the polar regions.

4.6.2.2 *Venus and Mars*

The peak electron density in Mars's ionosphere is $\sim 10^5$ cm^{-3} at an altitude of $\sim$130 km. Venus's ionosphere

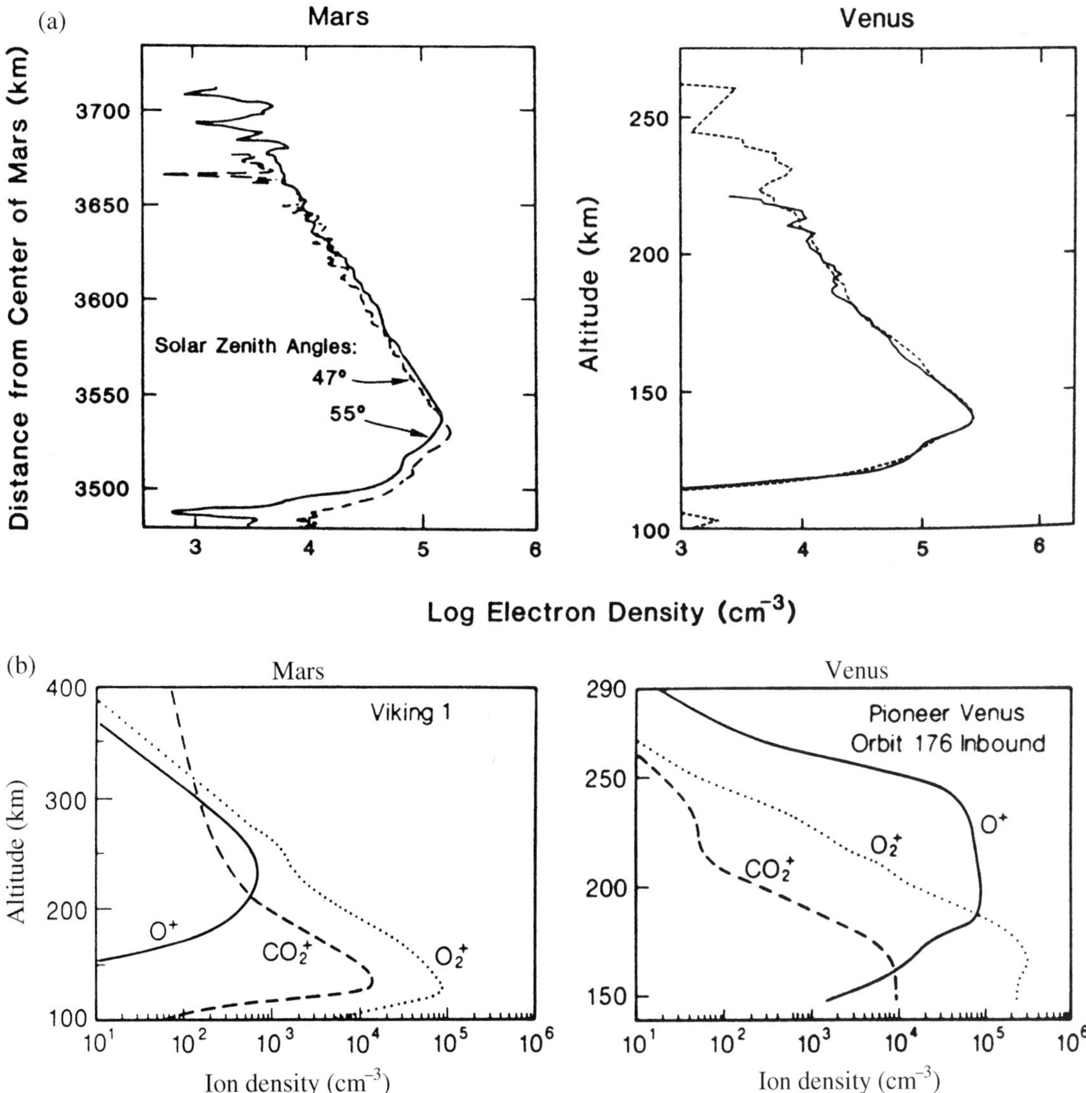

FIGURE 4.36 (a) The electron densities in the atmospheres of Mars and Venus. (Luhmann *et al.* 1992) (b) Observed ion densities on Mars and Venus. (Luhmann 1995)

reaches peak electron densities of roughly 3–5 × 10^5 cm^{-3} around 140 km. Graphs of ion and electron densities are shown in Figure 4.36. The physical processes in the ionospheres of both Venus and Mars best resemble those in the E layer of Earth's ionosphere, where O_2 is ionized directly by absorption of UV sunlight. The dominant constituent of the atmospheres of Venus and Mars up to ~150 km is CO_2, which is thus the main source for ionization:

$$(25)\ CO_2 + h\nu \rightarrow CO_2^+ + e^-, \qquad \text{at } \lambda < 900\ \text{Å}.$$

The dominant ambient ion, however, is O_2^+, which can be formed by various processes, all of which happen quickly:

atom–ion interchange: $(26)\ O + CO_2^+ \rightarrow O_2^+ + CO$

or charge transfer: $(27)\ O + CO_2^+ \rightarrow O^+ + CO_2$

quickly followed by: $(28)\ O^+ + CO_2 \rightarrow O_2^+ + CO.$

Both O_2^+ and CO_2^+ disappear via dissociative recombination:

$$(29)\ CO_2^+ + e^- \rightarrow CO + O$$
$$(30)\ O_2^+ + e^- \rightarrow O + O.$$

Peak electron densities on Mars's night side hemisphere are $\sim 5 \times 10^3$ cm^{-3} at 110–130 km altitude. In analogy with Earth, these electron densities might be explained by the relatively rapid rotation of the planet, although some contribution of direct ionization may be needed as well (precipitating electrons, meteor bombardment).

Despite Venus's slow rotation, it has a considerable night side ionosphere, with peak electron densities of $\sim 10^4$ cm^{-3} at an altitude of ~140 km. At night, the dominant ion at altitudes of ~150–170 km is O_2^+, while O^+ dominates at higher altitudes. Models suggest that fast horizontal transport (day → night winds, see Section 4.5.5.1) may carry O^+ from the day to the night side, where it descends to lower altitudes. Chemical reactions with CO_2 may produce the observed densities of O_2^+ (reaction (28)). When the solar wind pressure is high, the ionopause is low and the nightward ion flow is choked off. At these times the ionosphere is confined to altitudes less than 200 km, and precipitating low energy (~ 30 eV) electrons may be the dominant ionization agent.

4.6.2.3 *Giant Planets*

The main constituent of the giant planet atmospheres is molecular hydrogen. Direct photoionization yields:

(31) $H_2 + h\nu \rightarrow H_2^+ + e^-$, $\quad \lambda < 804$ Å
(32) $H_2 + h\nu \rightarrow H + H^+ + e^-$
(33) $H + h\nu \rightarrow H^+ + e^-$.

The H_2^+ concentration in the giant planet ionospheres, however, is very low since they undergo rapid charge transfer interactions with molecular hydrogen:

(34) $H_2^+ + H_2 \rightarrow H_3^+ + H$.

H^+ could form H_3^+ when interacting with two hydrogen molecules at once. However, above the ionospheric peak electron density the molecular densities are too low for this reaction to happen. Radiative recombination of H^+ is very slow, so that H^+ is expected to stay in the ionosphere as a terminal ion. This accounts for the observed high electron densities in the giant planets' ionospheres. Dissociative recombination of H_3^+ ions is very fast, so these ions are not expected to stay in the atmosphere for very long. Detections of H_3^+ emissions have been reported, however, in the auroral regions. The production of H_3^+ is enhanced in the auroral regions due to energetic particle precipitation. In addition to hydrogen atoms and molecules, the upper atmospheres of the giant planets contain many hydrocarbons and helium. Photoionization thresholds of these molecules, and subsequent chemical reactions are discussed in, e.g., Atreya (1986).

The ionospheric structure of the giant planets has been measured using radio occultation experiments with the Pioneer and Voyager spacecraft. Typical electron densities in Jupiter's ionosphere are of the order of 5–20 × 10^4 cm^{-3}. When the ionospheric density is plotted as a function of altitude (Fig. 4.37a), one can distinguish as many as eight ionospheric layers, under which there are a few narrow, dense layers at altitudes below ~1000 km ($z = 0$ at 1 bar). The data show considerable latitudinal, diurnal and temporal variations in the location and magnitude of the peak electron densities. On Saturn, typical peak electron densities of the order of a few × 10^4 cm^{-3} were recorded, roughly 1000–2000 km above the 1 bar level. Several sharp ionization layers around the 1000 km level are seen, with peak electron densities close to 10^5 cm^{-3}. Electron densities of a few thousand electrons cm^{-3} have been detected on Uranus, with peak electron densities up to a few × 10^5 cm^{-3} in a few sharply defined layers below 2000 km altitude. The ionosphere of Neptune is similar to that of Uranus, with local ionization layers enhanced in electron densities by up to an order of magnitude over the nominal values. The sharp ionization layers, detected in all giant planet ionospheres, suggest the presence of long-lived (atomic, perhaps metallic) ions.

4.6.3 Electric Currents

As discussed in the previous sections, there are numerous charged particles in an ionosphere, both positively charged ions and negatively charged electrons. These particles gyrate around magnetic field lines of a planet's internal or induced magnetic field (Chapter 7). Charge separation of electrons and ions (due to e.g., diffusion, Section 4.7) leads to electric fields. In the absence of collisions, ions and electrons move together, at velocity v, in the presence of an electric field, **E**, and magnetic field, **B**, (eq. 7.66):

$$v = \frac{c\mathbf{E} \times \mathbf{B}}{B^2}. \tag{4.67}$$

At high altitudes where the lifetimes of ions are large (e.g., in the F region of the Earth's ionosphere), the type of transport described by equation (4.67) becomes important. At lower altitudes the drift motion of charged particles is only momentary, until a collision with another particle deviates the electron/ion's path. The bulk motion (eq. 4.67) thus depends upon the ratio of the collision to cyclotron frequencies (Section 7.3.2). Typically, the drift or bulk motion for ions breaks down at altitudes below 200 km on Earth. As the ion–neutral collision cross-section is much larger than the electron–neutral collision cross-section, the bulk motion for electrons breaks down at lower altitudes, ~100 km on Earth.

The relative drift motion of ions and electrons under the influence of magnetic and electric fields induces a current, **j**, which can be expressed by the generalized Ohm's law (eq. 7.14). In an ionosphere where collisions between

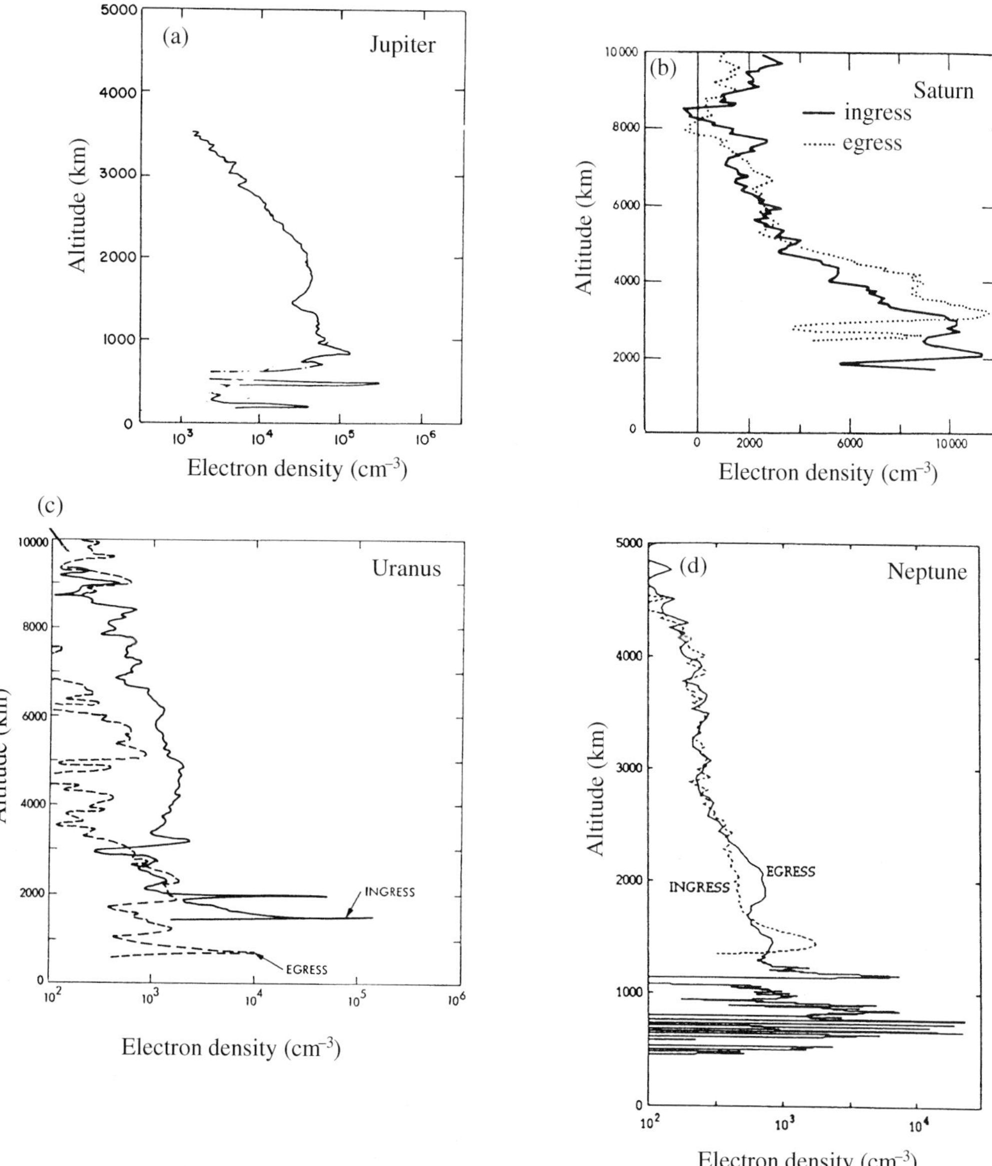

FIGURE 4.37 (a) Electron density in the ionosphere of Jupiter (Adapted from Atreya and Donahue 1976). (b) Electron density in the ionosphere of Saturn (Adapted from Kliore *et al.* 1980). (c) Electron density in the ionosphere of Uranus (Adapted from Lindal *et al.* 1987). (d) Electron density in the ionosphere of Neptune (Adapted from Lindal 1992). Altitudes are with respect to the 1 bar level.

particles are important, the conductivity is a tensor quantity rather than a scalar, because the magnetic field makes the medium anisotropic in its response to an applied electric field. In the frame moving with the neutral gas, and where the magnetic field is along the z-axis (normal to the surface), the conductivity consists of three components: the normal conductivity (σ_o) parallel to the magnetic field; the Pederson conductivity (σ_p) perpendicular to the magnetic field; and the Hall conductivity (σ_h) perpendicular to both the magnetic field and applied electric field. Each of the three conductivity components depends upon the number density, n_i, collisional frequency, ν_i, and cyclotron frequency, ω_i (the subscript i stands for ion, and e for electron):

(1) the normal, direct or longitudinal conductivity:

$$\sigma_o = \left(\frac{n_i}{m_i \nu_i} + \frac{n_e}{m_e \nu_e} \right) q^2, \tag{4.68}$$

(2) Pederson conductivity:

$$\sigma_P = \left[\frac{n_i \nu_i}{m_i(\nu_i^2 + \omega_i^2)} + \frac{n_e \nu_e}{m_e(\nu_e^2 + \omega_e^2)} \right] q^2, \quad (4.69)$$

(3) Hall conductivity:

$$\sigma_h = \left[\frac{n_e \omega_e}{m_e(\nu_e^2 + \omega_e^2)} - \frac{n_i \omega_i}{m_i(\nu_i^2 + \omega_i^2)} \right] q^2. \quad (4.70)$$

These conductivities induce the Birkeland, Pederson and Hall currents in an ionosphere, respectively. Generally, σ_o, which is parallel to the magnetic field, is much larger than the Pederson and Hall conductivities, and $\mathbf{E}_{\|} \ll \mathbf{E}_{\perp}$.

The presence of electric currents leads also to *Joule* or *frictional heating* of the ionosphere through the dissipation of electrical energy via charged particle collisions. The Joule heating rate, Q_J, due to currents transverse to the magnetic field is given by:

$$Q_J = \mathbf{j}_{\perp} \cdot \mathbf{E}_{\perp} = \frac{j_{\perp}^2}{\sigma_c}, \quad (4.71)$$

with the Cowling conductivity:

$$\sigma_c = \frac{\sigma_P^2 + \sigma_h^2}{\sigma_P}. \quad (4.72)$$

The height-integrated Pederson and Hall conductivities on Earth are about 20 mho[5] on the day side and 1 mho at night. Similar numbers may be expected on Jupiter and Saturn, although the uncertainties are very large. This implies that Joule heating may be comparable in magnitude to that caused by charged particle precipitation.

4.6.4 Airglow and Aurora

4.6.4.1 *Airglow*

Airglow is radiation that is continuously emitted by a planetary atmosphere. The day airglow results from direct scattering of sunlight and excitation/ionization by solar EUV photons and photoelectrons, followed by deexcitation and radiative recombination. Resonant scattering from atomic hydrogen has been detected from many planets, as well as from Saturn's rings and a torus enveloping Titan's orbit. Hydrogen is the most abundant species at very high altitudes because diffusive separation leaves the lightest element at the highest altitudes (see Section 4.7). This resonant scattering gives the planet an extensive *corona*, visible in Lyman α emission. The Earth's corona can be viewed from outside the atmosphere.

In addition to atomic hydrogen, airglow emissions from other atoms, e.g., O, He, H_2, O_2, N_2, and CO, have been observed on various planets. Jupiter is bright in H Ly α emission. It further displays an equatorial brightening above that expected from resonant scattering. This brightening peaks at a longitude of 100° (in System III, the magnetic longitude system), and is referred to as the *H Ly α bulge*. The bulge is seen at both the day and night sides, although at a reduced intensity at night. The H Ly α bulge implies a local enhancement in hydrogen atoms, which may be caused by magnetospheric effects (Section 7.4.4.2).

[5] 1 mho m^{-1} = 9 × 10^9 s^{-1} in gaussian units (1 mho = 1 siemens)

4.6.4.2 *Aurora*

Aurorae, commonly referred to as *northern lights*, can be observed on planets that have an intrinsic magnetic field (Chapter 7). The emissions result from the precipitation of charged particles originating outside the atmosphere. Aurorae occur in oval shaped regions roughly centered about the magnetic north and south poles of a planet. Depending upon the planet, auroral emissions can be observed at optical, infrared, UV and, for Jupiter and Earth, at X-ray wavelengths. These particles enter the atmosphere as field-aligned currents. The footpoints of field lines which thread through the 'storage place' of charged particles form an oval about the magnetic poles, which is referred to as the *auroral zone*. Atmospheric atoms, molecules and ions are excited upon interacting with the precipitating particles or by photoelectrons produced in the initial 'collision' with these particles. Upon deexcitation, the atmospheric species emit photons at different wavelengths.

Earth and Terrestrial Planets

On Earth, auroral phenomena are truly spectacular (Fig. 4.38). They are regularly seen in the night sky at high latitudes (auroral zone). Rapidly varying, colorful displays fill large fractions of the sky. The lights may appear diffuse, in arcs with or without ray patterns or may resemble draperies. The auroral emissions are studied both from the ground and from space, at wavelengths varying from X-rays up to radio wavelengths. Aurora are clearly related to disturbances of the geomagnetic field (Section 7.4.1.4), usually induced by variations in the solar wind (i.e., triggered by solar flares or coronal mass ejections). From the ground, the northern lights are generally only visible at high latitudes, but at times of strong disturbances (strong solar flares, solar mass ejections) the emissions are sometimes seen from more moderate latitudes. About once a century they can be seen from near-equatorial regions. The night side aurora displays discrete and diffuse components. The diffuse component is visible over a large region of the sky, while discrete 'arcs' are localized phenomena. The emissions generally vary rapidly, both in color, form and

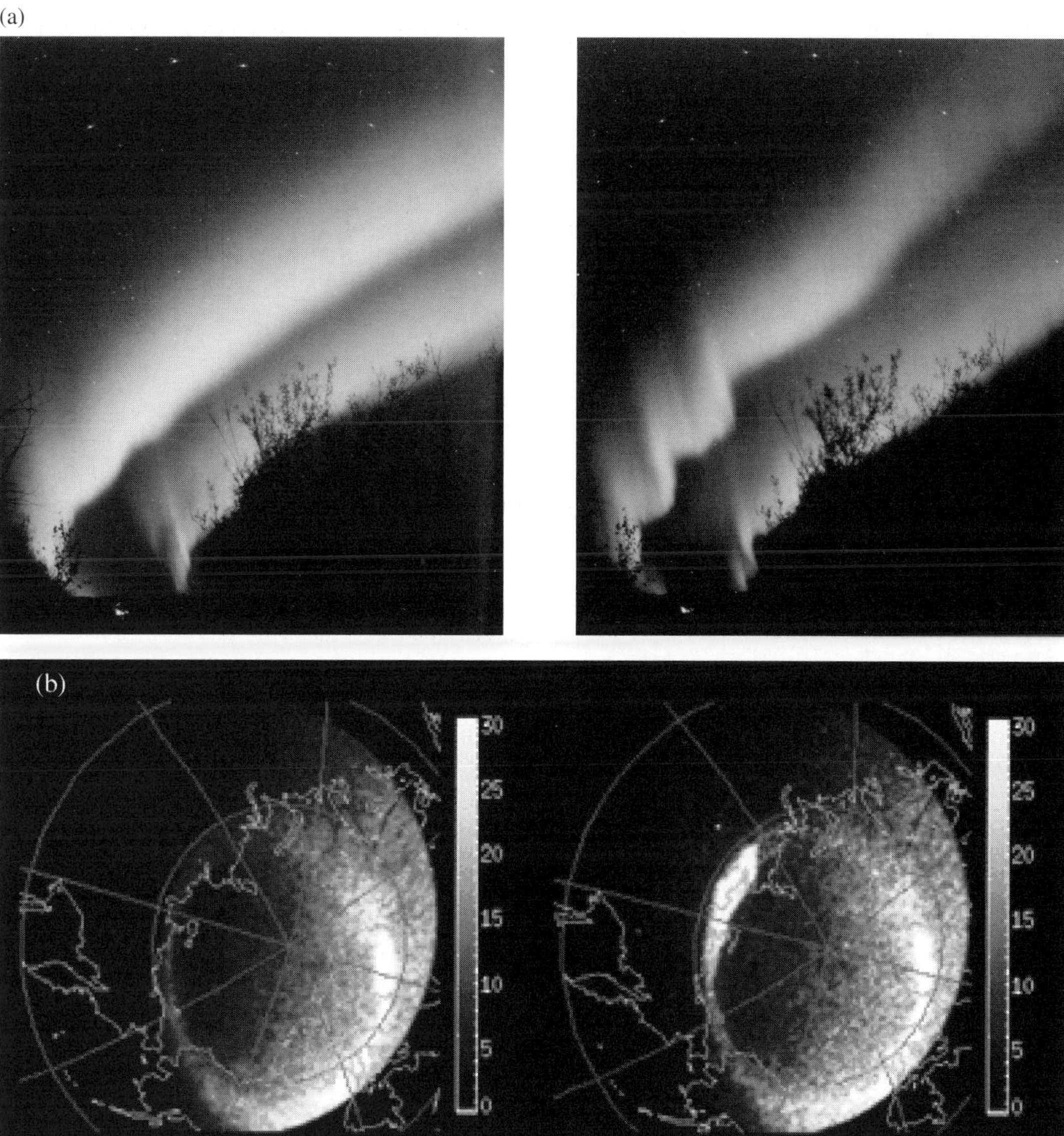

FIGURE 4.38 (a) COLOR PLATE Photographs of the northern (auroral) lights. Left photograph taken on 29 September 1996 at 11:25, from Fairbanks, Alaska. One can see a double arc below Big Dipper with a developing ray. A few minutes later (right image) the arcs have become unstable and develop curtains/drapes. Both exposures were for 10 seconds. (Courtesy: J. Curtis, Geophysical Institute, UAF) (b) Ultraviolet observations of auroral activations throughout the auroral zone on both dark and sunlit conditions on 9 April 1996 at 14:34 and 14:44 UT. The day side is to the right in both images. The day-side and night-side activities are essentially decoupled with continuous day-side activity coincident with distinct quiet and substorm periods on the night side. The scale is in units of photons cm^{-2} s^{-1}. (Courtesy: POLAR, Univ. of Iowa)

intensity. Although the emissions look quite disorderly, a specific pattern which consists of four 'scenes', referred to as *auroral substorms*, can be discerned in most auroral displays. An aurora usually starts out as a relatively dim arc oriented in the east–west direction. After a while, maybe a few hours, the arc moves towards the equator, becomes brighter and may develop rays. All of a sudden the emissions spread out over the entire sky, and move rapidly (up

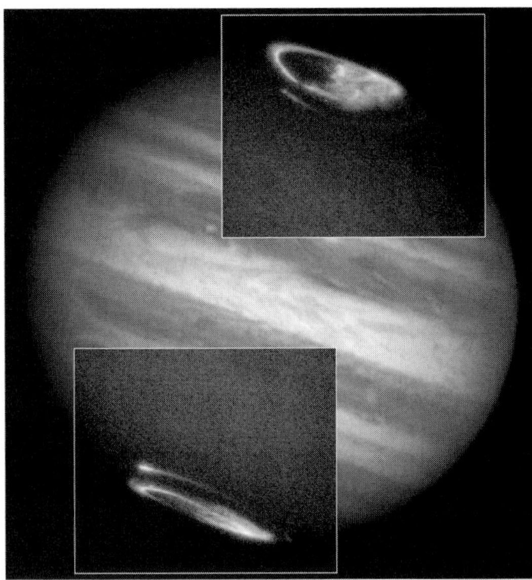

FIGURE 4.39 Composite HST image of Jupiter at visible wavelengths, with superposed northern and southern aurorae at UV wavelengths. Note the 'trail of light' left by Io, just outside the aurora. (Courtesy: J. Clarke and NASA/HST)

FIGURE 4.40 HST image of Saturn at UV wavelengths, which clearly shows Saturn's northern and southern aurorae. (Courtesy: J.T. Trauger and NASA/HST)

to tens of kilometers per second), while changing dramatically in form and intensity. This phase typically lasts a few minutes, after which the recovery phase begins, where the emissions weaken and the aurora becomes more diffuse.

The most prominent lines and bands in the terrestrial aurora are due to molecular and atomic nitrogen and oxygen. The visible display is dominated by the green (forbidden) and red oxygen lines at 5577 Å and 6300/6364 Å. Since the 5577 Å transitions are forbidden, they originate from high atmospheric altitudes, where collisions with other particles are relatively rare (above ~ 200 km). Blue emissions are produced by nitrogen (4278 Å and 3914 Å). Strong lines at UV wavelengths are also caused primarily by nitrogen and oxygen.

Weak auroral emissions have been observed on the night side of Venus with the Pioneer Venus Orbiter. The emissions are caused by excitation of atomic oxygen. HST and the 10-m Keck telescope detected aurorae of atomic oxygen on Ganymede at visible and UV wavelengths.

Giant Planets

Aurorae have been detected on all four giant planets. The aurora of Jupiter has been studied extensively at all wavelengths, from infrared to X-rays (Fig. 4.39). Emissions from ammonia, methane, ethane, acetylene and H_3^+ have been detected, as well as from the trace gases C_2H_4, C_3H_4 and C_6H_6. Within the auroral zone, there are regions where molecular emissions are enhanced: one such bright spot is centered near 180° in the northern hemisphere (fixed in the magnetic coordinate system; the magnetic pole is at 200°), and another near 0° in the southern hemisphere. The UV and H_3^+ aurora seem well correlated, as far as comparisons can be made. The H_3^+ aurora occurs at high magnetic latitudes, and is linked to distant regions in Jupiter's magnetic field ($\gtrsim$ 30 $R_♃$). The UV aurora occurs at all longitudes and extends equatorward of the H_3^+ aurora. It seems to be excited by charged particles which originate in the magnetosphere from field lines at distances $\gtrsim$ 15 $R_♃$. HST images (Fig. 4.39) show faint UV emissions extending roughly 60° in the wake or plasma flow direction beyond Io's magnetic footprint. Auroral emissions have also been found associated with the footpoints of Ganymede and Europa. These images provide direct evidence of charged particles entering Jupiter's ionosphere along the magnetic field lines that connect Io, Ganymede and Europa with Jupiter's ionosphere, since the emissions are intensified at and along the footprints of these magnetic field lines (Section 7.4.4). Such currents at Io's orbit have recently been detected directly by the Galileo spacecraft.

The aurora on Saturn are tightly confined to latitudes near 76° in the north and south (Fig. 4.40). The aurora are most promonent in the morning sector with patterns that appear fixed in local time. There are slow variations in locally bright emission features and in the mean latitudes of the auroral emissions.

4.7 Molecular and Eddy Diffusion

In the previous section we discussed how the atmospheric composition is changed by photochemical reactions. We showed that one can deduce the altitude distribution of photochemically derived constituents if the reaction and photolysis rates of the relevant reactions are known. Theoretically deduced altitude profiles for these constituents, however, seldom agree exactly with observations. This is, at least in part, caused by vertical movements of air parcels, known as *eddy diffusion*, and/or of individual molecules within the air, referred to as *molecular diffusion*.

4.7.1 Diffusion

The net vertical or upward flux, Φ_i, for minor constituent i in an atmosphere that is in hydrostatic equilibrium can be written as:

$$\Phi_i \equiv N_i v_u = -N_i D_i \left(\frac{1}{N_i}\frac{\partial N_i}{\partial z} + \frac{1}{H_i^*} + \frac{\alpha_i}{T(z)}\frac{\partial T(z)}{\partial z} \right) - N\mathcal{K}\frac{\partial (N_i/N)}{\partial z}, \tag{4.73}$$

where N is the atmospheric number density, N_i is the number density of constituent i, D_i is the molecular diffusion coefficient, $\mathcal{K}$ the eddy diffusion coefficient, H_i^* are the density scale heights for each atmospheric constituent i (cf. eq. 4.4), $T(z)$ is the atmospheric temperature at altitude z, and the coefficient α_i is the thermal diffusion parameter. The physical interpretation of the first three terms on the right hand side of this equation is as follows:

$(1/N_i)(dN_i/dz)$: Molecular diffusion caused by a gradient in the density N_i tends to smooth out density gradients, so that the mixing ratio of a constituent is constant with altitude. As an example, consider the O/O_2 mixing ratio in the Earth's atmosphere. The altitude profile of O/O_2 can be calculated according to the chemical reactions described in Section 4.6. Molecular diffusion is most effective at high altitudes ($z > 100$ km), where O atoms can be carried downwards to equalize the O/O_2 ratio with height. At lower altitudes, the oxygen atoms combine into molecules. This process causes the O/O_2 ratio at $z > 100$ km to be less than expected from the chemistry described in Section 4.6.

$(1/H_i^*)$: Molecular buoyancy diffusion, with H_i^* the density scale height (eq. 4.4) for individual constituents i, is a process which drives an atmosphere towards a barometric height distribution for each species i. Since the scale height varies with $1/\mu_a$, heavy molecules tend to stay closer to the surface. Collisions between particles slow the diffusion process down, so diffusion is only effective at high altitudes. As mentioned above, on Earth molecular diffusion is important at $z \gtrsim 100$ km. In contrast to molecular diffusion induced by dN_i/dz, molecular buoyancy diffusion would enhance the O/O_2 ratio at these high altitudes above that expected from local photochemical equilibrium considerations alone. The net effect of the two processes is that the O/O_2 mixing ratio is less than expected from photochemical considerations alone at altitudes around 100 km, and larger at altitudes above this level.

$(\alpha_i/T(z))(\partial T(z)/\partial z)$: Thermal molecular diffusion, with α_i the thermal diffusion parameter, is triggered by gradients in temperature (but note that there is a temperature dependence in $1/H_i^*$ as well).

The molecular diffusion coefficient D_i is inversely proportional to the atmospheric number density N: $D_i = b_i/N$, with b_i the binary collision parameter, which can best be determined empirically from data on diffusion, viscosity, and thermal conductivity. The maximum rate of diffusion occurs for complete mixing, $\partial(N_i/N)/\partial z = 0$, which leads to the concept of *limiting flux*, Φ_ℓ. For an atmosphere with small or no temperature gradients, where a light gas flows through the background atmosphere, the limiting flux can be written:

$$\Phi_\ell = \frac{N_i D_i}{H}\left(1 - \frac{\mu_{a_i}}{\mu_a}\right) \approx \frac{N_i D_i}{H}, \tag{4.74a}$$

with H the atmospheric (pressure) scale height. Using $D_i = b_i/N$, equation (4.74a) can be approximated by:

$$\Phi_\ell \approx \frac{b_i(N_i/N)}{H}. \tag{4.74b}$$

The binary collision parameter ($cm^{-1}\ s^{-1}$) can be approximated fairly well by:

$$b_i = C_b T^q, \tag{4.75}$$

where the parameters C_b and q for some gases are given in Table 4.6. The net outward flux is limited by the diffusion rate, and cannot exceed the limiting flux. The limiting flux depends only on the mixing ratio of constituent i and the pressure scale height. To calculate the limiting flux for hydrogen atoms in Earth's atmosphere (Problem 4.30), one needs to consider the upward flux of all hydrogen-bearing molecules (H_2O, CH_4, H_2) just below the homopause (i.e., ~100 km). The mixing ratio of all hydrogen-bearing molecules at the homopause is $\sim 10^{-5}$, which results in a limiting flux of $\sim 2 \times 10^8\ cm^{-2}\ s^{-1}$.

TABLE 4.6 Parameters C_b and q (cgs units) in Equation (4.75) for Various Gases.

Gas 1	Gas 2	C_b	q
H	H_2	145×10^{16}	1.61
	air	65×10^{16}	1.7
	CO_2	84×10^{16}	1.6
H_2	air	26.7×10^{16}	0.75
	CO_2	22.3×10^{16}	0.75
	N_2	18.8×10^{16}	0.82
H_2O	air	1.37×10^{16}	1.07
CH_4	air	7.34×10^{16}	0.75
Ne	N_2	11.7×10^{16}	0.743
Ar	air	6.73×10^{16}	0.749

All parameters from Chamberlain and Hunten (1987) and references therein.

4.7.2 Eddy Diffusion Coefficient

The last term on the right hand side of equation (4.73) is the *eddy* or *turbulent diffusion*, with the *eddy diffusion coefficient*, $\mathcal{K}$. Eddy diffusion is a macroscopic process, in contrast to molecular diffusion and mixing, discussed above. Eddy diffusion may occur if an atmosphere is unstable against turbulence, which happens when Reynolds number $\Re_e > 5000$ (eq. 4.52). The product $v\ell$ gives a crude estimate for the diffusion coefficient $\mathcal{K}$. Assuming $\ell \approx H$ and v is between 1 and 10^4 cm s^{-1}, $\mathcal{K}$ is of the order of 10^6 to 10^{10} cm^2 s^{-1}.

We find that eddy diffusion processes dominate the atmosphere below the *turbopause* or *homopause*, while molecular diffusion dominates above the homopause. On Earth, the homopause is located at an altitude of ~100 km. As mentioned above, at higher altitudes molecular diffusion become important. In (super)adiabatic atmospheres, free convection is usually the dominant motion involved in vertical mixing of the atmosphere. In subadiabatic regions, eddy diffusion might be driven by internal gravity waves or tides. The eddy diffusion coefficient is usually estimated from observed altitude distributions of trace gases in an atmosphere.

4.7.2.1 *Observational Constraints on $\mathcal{K}$ for Venus and Mars*

The observed abundances of CO, O and O_2 in the upper atmospheres of Venus and Mars are much smaller than estimated using local photochemistry (Section 4.6.1.2). On Mars, recombination of CO and O might proceed more rapidly via catalytic reactions with OH if CO and O are transported downwards via an efficient eddy diffusion mechanism. The observed abundances of CO and O can be matched if the eddy diffusion coefficient $\mathcal{K} \approx 10^8$ cm^2 s^{-1}, roughly two orders of magnitude larger than in the Earth's stratosphere. On Venus the observations can be matched if $\mathcal{K} \approx 10^5$ cm^2 s^{-1} between 70 and 95 km, and $\mathcal{K}$ is increasing at higher altitudes.

4.7.2.2 *Observational Constraints on $\mathcal{K}$ for the Giant Planets*

The eddy diffusion coefficient on the giant planets is often estimated from the height distribution of methane gas and/or the presence of compounds one would not expect in the upper atmosphere under equilibrium situations (i.e., GeH_4, PH_3, CO, HCN). Lyman α emission from the planets usually results from resonance scattering of solar Ly α photons by atmospheric hydrogen atoms. Since methane gas is a strong absorber of these photons, the Ly α emission arises from above the methane homopause, the altitude of which is determined by the eddy diffusion coefficient: a large value will raise the homopause, and hence diminish the Ly α emission because of the low number of hydrogen atoms. However, even though this method does provide an eddy diffusion coefficient at the methane homopause in principle, in practice photochemical reactions may deplete methane gas well below its homopause. Unless the proper photochemistry is taken into account, the eddy diffusion coefficient derived from Ly α scattering may be too low.

Compounds like GeH_4, PH_3 and CO are only expected deep ($T > 1000$ K) in the giant planet atmospheres. Since they have been detected on Jupiter and Saturn, there must be rapid vertical mixing, in order to bring the elements up against other transformation processes, such as oxidation. On Uranus and Neptune, methane gas freezes out in the troposphere, and not much of it is expected to be present above the tropopause. On Uranus, the methane mixing ratio at the tropopause is ~50% of the maximum value allowed by saturation. On Neptune, however, the abundance near the tropopause seems to be 10–100 times higher than the saturation value. Any methane gas brought up through the cold trap can undergo photochemical reactions, and end up as C_2H_2 and C_2H_6. A comparison between the observed and calculated mixing ratio of all these elements yields an estimate for the diffusion coefficient across the tropopause. The much larger concentration of hydrocarbons in Neptune's stratosphere compared to Uranus suggests much stronger convective motions on Neptune.

Despite many searches at radio wavelengths for emission/absorption lines in giant planet atmospheres, only CO and HCN have been detected in emission in Neptune's atmosphere (Fig. 4.13d); CO has also been seen in absorp-

TABLE 4.7 Eddy Diffusion Coefficient near the Homopause.

Planet	Eddy diffusion ($cm^2\ s^{-1}$)	Pressure (bar)
Earth	$(0.3–1) \times 10^6$	3×10^{-7}
Venus	10^7	2×10^{-8}
Mars	$(1–5) \times 10^8$	2×10^{-10}
Titan	$\sim 10^8$	6×10^{-10}
Jupiter	$\sim 10^6$	10^{-6}
Saturn	$\sim 10^8$	5×10^{-9}
Uranus	$\sim 10^4$	3×10^{-5}
Neptune	$\sim 10^7$	2×10^{-7}

All parameters from Atreya (1986) and Atreya *et al.* (1999).

tion. Since the lines are seen in emission, they must be formed above the tropopause. The observations suggest that CO is present both below and above the tropopause, thus it is likely brought up from deep depths by a fast upward convection. HCN is only seen in the stratosphere, above the tropopause. However, if HCN were brought up from below, it would condense out since the temperatures near the tropopause are well below its freezing point. HCN must thus be formed in the stratosphere from nitrogen and hydrocarbons. Nitrogen gas may either fall in from outside, e.g., the satellite Triton, or it may be brought up by fast convection as is CO. Neither HCN nor CO has been detected on Uranus. This, together with the large concentration of hydrocarbons on Neptune, hints at large eddy diffusion coefficients on Neptune.

Various estimates for the eddy diffusion coefficient on planets and satellites are summarized in Table 4.7.

4.8 Atmospheric Escape

A particle may escape from an atmosphere if its kinetic energy exceeds the gravitational binding energy, and if it moves along an upward trajectory without intersecting the path of another atom or molecule. The region from which escape can occur is referred to as the *exosphere*, and its lower boundary is called the *exobase*. The exobase is found at an altitude z_{ex} at which:

$$\int_{z_{ex}}^{\infty} \sigma_x N(z) dz \approx \sigma_x N(z_{ex}) H = 1, \tag{4.76}$$

where σ_x is the collisional cross-section. We assume the scale height, H, to be constant in the exosphere. Since the mean free path

$$\ell_{fp} = 1/(\sigma_x N), \tag{4.77}$$

the exobase is located at an altitude z_{ex}, where $\ell_{fp}(z_{ex}) = H$. Within the exosphere the mean free path for a molecule/atom is thus comparable to, or larger than, the atmospheric scale height, so an atom with sufficient upward velocity has a reasonable chance of escaping. In addition to the thermal or Jeans escape, discussed below, there are various nonthermal processes which can lead to atmospheric escape. These processes are discussed in the following subsections.

4.8.1 Thermal or Jeans Escape

The velocities of individual molecules of mass m in thermal equilibrium are given by a Maxwellian distribution function:

$$f(v)dv = N \left(\frac{2}{\pi}\right)^{1/2} \left(\frac{m}{kT}\right)^{3/2} v^2 e^{-mv^2/(2kT)} dv, \tag{4.78}$$

with v the particle's velocity, and N the local particle (number) density. This distribution formally extends up to infinite velocities, but due to the steep dropoff in the Gaussian distribution, there are practically no particles with velocities larger than about three times the mean thermal velocity $v_o = \sqrt{2kT/m}$. The ratio of the potential to the kinetic energy is referred to as the *escape parameter,* λ_{esc}:

$$\lambda_{esc} = \frac{GMm}{kT(R+z)} = \frac{(R+z)}{H(z)} = \left(\frac{v_e}{v_o}\right)^2. \tag{4.79}$$

At and below the exobase, collisions between particles drive the velocity distribution into a Maxwellian distribution, while above the exobase collisions are essentially absent and particles out in the tail of the Maxwellian velocity distribution, which have a velocity $v > v_e$, may escape into space. Integrating the upward flux in a Maxwellian velocity distribution above the exobase results in the *Jeans formula* for the rate of escape (atoms $cm^{-2}\ s^{-1}$) by thermal evaporation:

$$\Phi_J = \frac{N_{ex} v_o}{2\sqrt{\pi}} (1 + \lambda_{esc}) e^{-\lambda_{esc}}, \tag{4.80}$$

where the subscript ex refers to the exobase and λ_{esc} is the escape parameter at the exobase. Typical parameters for Earth are $N_{ex} = 10^5\ cm^{-3}$ and $T_{ex} = 900$ K. For atomic hydrogen $\lambda_{esc} \approx 8$, and $\Phi_J \approx 6 \times 10^7\ cm^{-2}\ s^{-1}$, which is a factor of 3–4 smaller than the limiting flux (eq. 4.74) for hydrogen atoms on Earth (Problem 4.30). Note that lighter elements/isotopes are lost at a much faster rate than heavier ones. Jeans escape can thus produce a substantial isotopic fractionation.

4.8.2 Nonthermal Escape

Jeans escape gives a lower limit to the escape flux from planetary atmospheres. *Nonthermal processes* often dominate the escape rate, and the major nonthermal processes are listed below under processes (1)–(6). All processes, except photodissociation, involve charged particles, i.e., ions and electrons. Neutral particles can gain sufficient energy in interactions (1)–(4) to escape into space. Charged particles are usually trapped in the planet's (intrinsic or induced) magnetic field, and cannot readily escape. Processes (4)–(6) discuss escape mechanisms involving charged particles. In this section, we adopt the following notation: i_2 = molecule, i, j = atoms, i^+, j^+ = ions, e^- = electron, and * indicates excess energy.

(1) *Dissociation and dissociative recombination*: when a molecule is dissociated by UV radiation or an impacting electron, or when an ion dissociates upon recombination, the end products may gain sufficient energy from the 'impact' to escape the planet's gravitational attraction:

$$i_2 + h\nu \rightarrow i^* + i^*$$

$$i_2 + e^{-*} \rightarrow i^* + i^* + e^-$$

$$i_2^+ + e^- \rightarrow i^* + i^*$$

(2) *Ion–neutral reaction*: a reaction between an atomic ion and a molecule, where a molecular ion and a fast atom are created:

$$j^+ + i_2 \rightarrow ij^+ + i^*$$

(3) *Charge exchange*: when a fast ion meets a neutral, charge exchange may take place, where the ion gets rid of its charge, but retains its kinetic energy. The new neutral (former ion) may have sufficient energy to escape the planet's gravitational attraction:

$$i + j^{+*} \rightarrow i^+ + j^*$$

This process plays an important role on Io, where fast sodium atoms are created by charge exchange with magnetospheric plasma (Section 7.4.4.3).

(4) *Sputtering*: when a fast atom or ion hits an atmospheric atom, the atom gains energy, and consequently may be able to escape the planet's gravitational attraction. Since it is much easier to accelerate an ion than an atom, sputtering is usually caused by fast ions. In a single collision, the atom is accelerated in the forward direction (conservation of momentum), a process generally referred to as 'knock-on'. Sputtering usually refers to a multiple sputtering process, involving a cascade of collisions. This process is important in thick and tenuous atmospheres, as well as on airless bodies. In the latter cases, the fast ions/atoms hit the surface directly, ejecting one or several atoms from the crust into space. Such atoms may not be fast enough to escape into interplanetary space, but rather get trapped in a 'corona', i.e., they stay gravitationally bound to the body. The atmospheres of the Moon and Mercury are partly formed by sputtering processes, and by meteoroid impacts (the latter probably dominate). The exobase for both bodies is at their surface. Sputtering is also important on outer planet satellites and rings. Subsequent collisions in a cascade can give the atoms sufficient energy to escape into interplanetary space:

$$i + j^{+*} \rightarrow i^* + j^{+*}$$

$$i + j^* \rightarrow i^* + j^*$$

(5) *Electric fields*: in addition to the particle interactions described above, charged particles can be accelerated by electric fields (ionospheric, magnetospheric) and move outwards along magnetic field lines. Molecular diffusion in an ionosphere causes a separation in altitude between the heavier ions and lighter electrons, which gives rise to a potential difference or electric field in the upper atmosphere. Other motions of ionized air, caused by e.g., solar heating and tidal interactions, may also induce electric fields and currents. Such fields accelerate charged particles, and these particles can then either collide with atmospheric particles and transfer momentum to the neutrals (process 3), or may themselves escape the planet's gravitational field. For the latter process to happen, the charged particle has to move outward along (open) magnetic field lines. The depletion of H^+ and He^+ above the Earth's magnetic poles is attributed to ion escape along open magnetic field lines.

(6) *Solar wind sweeping*: in the absence of an internal magnetic field, charged particles may interact directly with the solar wind, a process referred to as *solar wind sweeping*. The giant planets and Earth have strong intrinsic magnetic fields, and the trajectories of solar wind particles are deflected to flow around the field (cf., Section 7.1.4). Thus, there is no direct interaction between these planets and the solar wind. If the planet has an ionosphere, but no intrinsic magnetic field, such as Venus and comets, particle exchange between the solar wind and the planet's atmosphere occurs. Particles are captured

from the solar wind at the subsolar point, and lost to the wind near the limbs. Via this process, an airless planet may temporarily capture solar wind particles. For example, the hydrogen and helium atoms in Mercury's atmosphere, as well as the helium atoms on the Moon, originate in the solar wind. We also note that satellites embedded in a planetary magnetic field interact in a similar way with magnetospheric plasma.

4.8.3 Atmospheric Blowoff and Impact Erosion

4.8.3.1 *Hydrodynamic Escape*

An atmosphere experiences a 'blowoff' when escaping light gases drag heavier material along with them, even though the heavier constituents would not be able to escape according to the Jeans equation. The situation is in a way analogous to the solar wind, where the initially subsonic flow goes through a critical point where the velocity equals the speed of sound (Section 7.1.1). At larger distances from the Sun the velocity is supersonic. Chamberlain and Hunten (1987) derived an expression for *hydrodynamic escape* in an atmosphere. They assume that the light gas moves at speeds approaching the sonic value, in which case there are large drag forces with other constituents. By ignoring the terms in dT/dz and in $\mathcal{K}$ in equation (4.73), the outgoing flux of the heavier gas, Φ_2, becomes:

$$\Phi_2 = \frac{N_2}{N_1}\left(\frac{m_c - m_2}{m_c - m_1}\right)\Phi_1, \tag{4.81}$$

where the subscripts 1 and 2 refer to the light and heavy gas, respectively, and m_c is the crossover mass:

$$m_c = m_1 + \frac{NkT\Phi_1}{bg_p}, \tag{4.82}$$

with b the binary collision parameter (cf. Section 4.7.1). Φ_2 must be positive, so $m_2 < m_c$. If Φ_1 is equal to the limiting flux (eq. 4.74), $m_c = 2m_1$.

To maintain an atmosphere in a blowoff state requires a large source of energy at certain altitudes. Solar energy is usually not large enough to maintain any present day atmosphere in a blowoff state. An alternative heat source is that due to accretion of solid planetesimals during the planet formation epoch. Calculations show that the early atmospheres of Venus, Earth and Mars may have experienced periods of hydrodynamic escape.

4.8.3.2 *Impact Erosion*

Impact erosion can occur during or immediately following a large impact on a body that has a substantial atmosphere. For an impactor that is smaller than the atmospheric scale height, shock-heated air flows around the impactor and the energy is dispersed over a relatively large volume of the atmosphere. If the impactor is larger than an atmospheric scale height, however, a large fraction of the shock-heated gas can be blown off, since the impact velocity exceeds the escape velocity from the planet. The mass of the atmosphere blown into space, M_e, is given by:

$$M_e = \frac{\pi R^2 P_0 \mathcal{E}_e}{g_p}, \tag{4.83}$$

where R is the radius of the impactor, and P_0/g_p is the mass of the atmosphere per unit area. The atmospheric mass that can escape is thus the mass intercepted by the impactor multiplied by an enhancement factor $\mathcal{E}_e$:

$$\mathcal{E}_e = \frac{v_i^2}{v_e^2(1+\mathcal{E}_v)}, \tag{4.84}$$

where v_i and v_e are the impact and escape velocities, respectively, and $\mathcal{E}_v$ is the evaporative loading parameter, which is inversely proportional to the impactor's latent heat of evaporation. A typical value for $\mathcal{E}_v$ is ~ 20 (for meteors). Significant escape occurs when $\mathcal{E}_e > 1$. If $\mathcal{E}_e < 1$, evaporative loading is much larger than the energy gained by impact heating, and the gas does not have enough energy left to escape into space. In the case of a colossal cratering event, the ejecta from the crater may also be large enough and contain sufficient energy to accelerate atmospheric gas to escape velocities.

4.9 Evolution of Terrestrial Planet Atmospheres and Climate

The initial stages of planetary growth involve the accumulation of solid materials, but when a planet becomes massive enough, it may trap gases. The formation of planets and their atmospheres is discussed in detail in Chapter 12. In this section we argue that the atmospheres of the terrestrial planets cannot be primordial, but must have formed from outgassing of the hot forming planets. We further discuss the subsequent evolution of the 'climate' on Earth, Mars and Venus.

The atmospheres of the giant planets are composed primarily of hydrogen and helium, with traces of C, O, N and S in the form of CH_4, H_2O, NH_3 and H_2S, respectively. In contrast, the atmospheres of the terrestrial planets and satellites are dominated by CO_2, N_2, O_2, H_2O, and SO_2. The main difference between the giant and terrestrial planets is gravity, which allowed the giant planets to accrete large quantities of common species (e.g., H_2, He) which remain gaseous at Solar System temperatures. Moreover, the light elements H and He (if present originally) would have been able to escape from the shallow gravitational potential wells of the terrestrial planets.

The following chemical reactions can occur between H_2 and other volatiles within a planetary atmosphere:

$$
\begin{aligned}
CH_4 + H_2O &\longleftrightarrow CO + 3H_2 \\
2NH_3 &\longleftrightarrow N_2 + 3H_2 \\
H_2S + 2H_2O &\longleftrightarrow SO_2 + 3H_2 \\
8H_2S &\longleftrightarrow S_8 + 8H_2 \\
CO + H_2O &\longleftrightarrow CO_2 + H_2 \\
CH_4 &\longleftrightarrow C + 2H_2 \\
4PH_3 + 6H_2O &\longleftrightarrow P_4O_6 + 12H_2.
\end{aligned}
$$

A loss of hydrogen shifts the equilibrium towards the right, hence oxidizing material. We call an atmosphere *reducing* if a substantial amount of hydrogen is present, as on the giant planets, and *oxidizing* if little hydrogen is present, as on the terrestrial planets. If the atmospheres of the terrestrial planets were primordial in origin (accreted from a solar composition gas, similar to the formation of the giant planets; Section 12.6), and all of the hydrogen and helium subsequently escaped, the most abundant gases in these atmospheres would be CO_2 ($\sim$63%), Ne ($\sim$22%) and N_2 ($\sim$10%), with a small fraction of COS ($\sim$4%). In addition, one would expect solar concentrations for Ar, Kr and Xe. These abundances are quite different from what is observed. In particular, neon on Earth is present in minuscule amounts, about ten orders of magnitude less than predicted from this model. Similarly, Ar, Kr and Xe are present but at abundances over six orders of magnitude less than expected for a solar composition atmosphere. The noble gases are too heavy to escape via thermal processes if initially present, and could not be chemically confined to the condensed portion of the planet, as CO_2 is on Earth (see below). The observed small abundances of the noble gases form a major argument in the conclusion that the atmospheres of the terrestrial planets are secondary in origin, thus that they formed from outgassing (volcanoes) of the planets themselves and/or accreted volatile-rich asteroids and comets, and were modified by the interaction between the atmosphere and the crust, and (for Earth) biochemical reactions.

As discussed in Chapter 3, a planet's surface temperature is determined mainly by solar insolation, its Bond albedo and atmospheric opacity. The Sun's luminosity has slowly increased during the history of the Solar System. In the early history, 4.5 Gyr ago, the solar luminosity was probably 25–30% smaller than it is nowadays, implying lower surface temperatures for the terrestrial planets. In contrast, although the Sun's total energy output was less than it is at the present time, its X-ray and UV emission were much larger, and the solar wind was stronger and probably erratic during the Sun's T Tauri phase. In addition to changes in the solar flux, the planetary albedos may have fluctuated, as variations in a planet's cloud deck, ground-ice coverage and volcanic activity may significantly alter its albedo. Periodic variations (such as those caused by perturbations from other planets) or sudden modifications in the planet's orbital eccentricity and the obliquity of its rotation axis (e.g., due to a large impact) may have also played a role in climate evolution.

4.9.1 Earth

Even though the luminosity of the young Sun was smaller, the presence of sedimentary rocks and possible absence of glacial deposits on Earth about 4 Gyr ago suggest that the Earth may have been quite warm, maybe even warmer than today. If true, this could be attributed to an increased greenhouse effect, since the atmospheric H_2O, CO_2, CH_4 and possibly NH_3 content was likely larger during the early stages of outgassing. In fact, the long term cycling of CO_2 may be regulated in such a way that the Earth's surface temperature does not change too much over long time periods. Carbon dioxide is removed from the atmosphere–ocean system by silicate weathering, the *Urey weathering reaction*, a chemical reaction between CO_2, dissolved in water, with silicate minerals in the soil. The reaction releases Ca and Mg ions and converts CO_2 into bi-carbonate (HCO_3^-). The bi-carbonate reacts with the ions to form other carbonate minerals[6]. An example of such a chemical reaction for calcium is given by:

$$CaSiO_3 + 2CO_2 + H_2O \rightarrow Ca^{++} + SiO_2 + 2HCO_3^- \tag{4.85a}$$

$$Ca^{++} + 2HCO_3^- \rightarrow CaCO_3 + CO_2 + H_2O. \tag{4.85b}$$

Carbonate sediments on the ocean floor are carried downwards by plate tectonics (Section 5.3.2.2), and are transformed back into CO_2 in the high temperature/pressure environment of the Earth's mantle:

$$CaCO_3 + SiO_2 \rightarrow CaSiO_3 + CO_2. \tag{4.85c}$$

Volcanic outgassing brings the CO_2 back into the atmosphere. The weathering rate increases when the surface temperature is higher, while the surface temperature is related to the CO_2 content of the atmosphere through the

[6] On Earth, organisms in the oceans make shells of calcium carbonate.

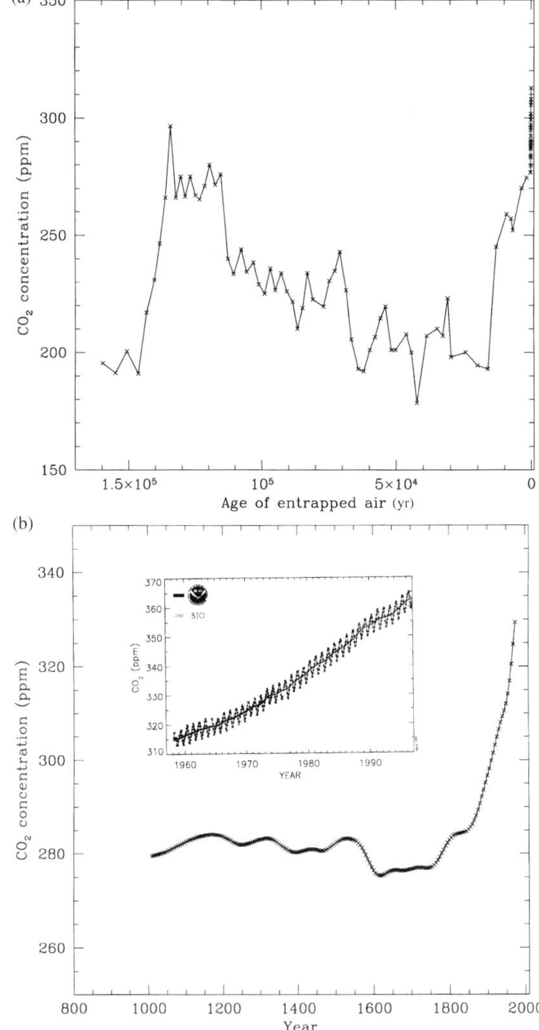

FIGURE 4.41 (a) Long-term time evolution of the CO_2 concentration in the Earth's atmosphere. This historical record is obtained through analysis of air bubbles trapped in the 2083 m long ice-core at the Russian Vostok station in East Antarctica. (Adapted from Barnola *et al.* 1987) (b) CO_2 concentration in the Earth's atmosphere over the past 1000 years. (Adapted from Etheridge *et al.* 1996) The insert is the CO_2 concentration from data obtained at Mauna Loa. The data before 1974 has been obtained by C.D. Keeling; the 1974–1997 data are from K.W. Thoning and P.P. Tans.

greenhouse effect. This effectively causes a self-regulation in the atmospheric CO_2 abundance on Earth.

The abundance of O_2 in the Earth's atmosphere is primarily due to photosynthesis of green plants, together possibly with past photodissociation of H_2O and subsequent escape of the hydrogen atoms. The slow disappearance of the rain forests on Earth may change the oxygen content of our atmosphere, as mankind transforms oxygen into carbon dioxide. The influence of mankind on the evolution of the Earth's atmosphere at the present time should not be underestimated. The CO_2 levels in our atmosphere are rising at an alarming rate (Fig. 4.41), which may lead to a global warming through the greenhouse effect. However, the increased temperature will also increase the weathering rate, and the rate of CO_2 accumulation is less than the rate at which it is released via the burning of fossil fuels. In addition, chemical reactions between atmospheric gases and pollutants may also influence the atmospheric composition, the consequences of which are hard to predict with any certainty. We have embarked on an inadvertent giant experiment with our home planet's atmosphere, which may have dire consequences for the future of life on Earth.

4.9.2 Mars

Mars's small size may be the primary cause for the difference in climate between Mars and Earth. Although liquid water cannot exist on Mars at the present time, the numerous channels on the planet are suggestive of running water in the past (cf. Section 5.5.4.1). This implies that Mars's atmosphere must have been denser and warmer in the past. Since the runoff channels are confined to the ancient, heavily cratered terrain, the warm martian climate did not extend beyond the end of the heavy bombardment era, about 3.8 Gyr ago. Estimates of Mars's early atmosphere suggest a mean surface pressure of the order of 1 bar, and temperature close to 300 K. Widespread volcanism, impacts by planetesimals and tectonic activity must have provided a large source of CO_2 and H_2O, whereas impacts by very large planetesimals may also have caused a loss of atmospheric gases through impact erosion. Mars has lost most of its CO_2 via carbonaceous (weathering) processes, adsorption onto the regolith, and/or condensation onto the surface. Since Mars does not show current tectonic activity, the CO_2 cannot be recycled back into the atmosphere. Without liquid water on the surface, weathering has ceased, and Mars has retained a small fraction of its CO_2 atmosphere. The present abundance of H_2O on Mars is largely unknown. Most of the H_2O might have escaped, but recent theories invoke large amounts of subsurface water-ice on the planet (Section 5.5.4).

Climatic changes may also be caused by changes in Mars's orbital eccentricity and obliquity. On Earth, these parameters vary periodically on timescales of $\sim 10^4$–10^5 yrs, and may be responsible for the succession of ice ages and ice-free epochs during the past million years. For Mars, these parameters have periods about ten times larger than for Earth, and the departures of the mean values are larger than for Earth. The polar regions receive more sunlight at a large obliquity, and large eccentricities increase the amount of sunlight falling on the summer hemisphere at perihelion. The layered deposits in Mars's polar region (Section 5.5.4) suggest that such periodic changes have taken place on Mars.

The Tharsis region of Mars consists of many volcanoes in a relatively small area, which appear to be roughly the same age. The eruptions of these volcanoes must have enhanced the atmospheric pressure and, via the greenhouse effect, its temperature. However, the sparsity of impact craters implies that the volcanic eruptions occurred well after the formation of the runoff channels on Mars's highlands.

4.9.3 Venus

Venus is very dry at the present time, with a volume mixing ratio of H_2O in the atmosphere of only 100 parts per million. This is about 10^5 times less H_2O than is present in the Earth's oceans. Various theories have been offered to explain the lack of water on Venus. It could have simply formed with very little water, because the minerals which condensed in this relatively warm region of the solar nebula lacked water. However, mixing of planetesimals between the accretion zones and cometary impacts may have provided similar amounts of volatiles to Venus and Earth, in which case it seems probable that there was an appreciable fraction of a terrestrial ocean on early Venus. The D/H ratio on Venus is $\sim$100 times larger than on Earth, which is persuasive evidence that Venus was once much wetter than it is now. But where did the water go? Water can be dissociated into molecular hydrogen and oxygen, either by photodissociation or other chemical reactions, and the hydrogen will escape into space. However, the current escape rate is only 10^7 H cm^{-2} s^{-1}, implying that only 9 m of water could have escaped during the planet's entire lifetime.

The classical explanation for Venus's loss of water is via the *runaway greenhouse effect*. Figure 4.42 shows the evolution of the surface temperatures for Earth, Mars and Venus, which each were assumed to have outgassed a pure water vapor atmosphere, starting from an initially airless planet. While the vapor pressure of water in the atmosphere is increasing, the surface temperature increases due to the enhanced greenhouse effect, thus there is a positive feedback between the increasing temperature and increasing opacity. On Venus, the temperature stayed well above the saturation pressure curve, which led to a runaway greenhouse effect, and all of Venus's water was accumulated in the atmosphere as steam. If one postulates an effective mixing process, the water is distributed throughout the atmosphere, and is photodissociated at high altitudes, with subsequent escape of the hydrogen atoms from the top of the atmosphere.

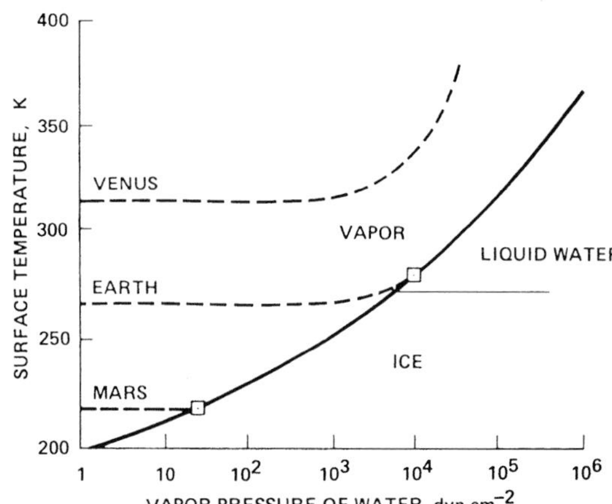

FIGURE 4.42 Evolution of the surface temperatures of Venus, Earth and Mars for a pure water vapor atmosphere. (Goody and Walker 1972)

An alternative model to explain Venus's loss of water is the *moist greenhouse effect*. This model, in contrast to the runaway greenhouse model, relies on moist convection. Venus may have had a surface temperature near 100 °C. Convection transported the saturated air upwards. Since the temperature initially decreases with altitude, water condensed out, and the latent heat of condensation led to a decrease in the atmospheric lapse rate, and increased the altitude of the tropopause. In this scenario, water vapor naturally reached high altitudes, where it was dissociated and escaped into space. With the accumulation of water vapor in the atmosphere, the atmospheric pressure increased, preventing the oceans from boiling. Because of the rapid loss rate of water from the top of the atmosphere, the oceans continued to evaporate. As long as there was liquid water on the surface, CO_2 and O_2 were removed from the atmosphere by weathering processes. Once the liquid water was gone, CO_2 could not form carbonate minerals, and accumulated in the atmosphere.

Further Reading

A superb book on weather at the nonscience major level is written by:

Williams, J., 1992. *The Weather Book*. USA Today, Vintage Books, New York.

A few relatively recent books on atmospheric physics (graduate student level) we can recommend are:

Atreya, S.K., 1986. *Atmospheres and Ionospheres of the Outer Planets and Their Satellites*. Springer-Verlag, Heidelberg.

Chamberlain, J.W., and D.M. Hunten, 1987. *Theory of Planetary Atmospheres*. Academic Press, Inc., New York.

Jacobson, M.Z., 1999. *Fundamentals of Atmospheric Modeling*. Cambridge University Press, New York, 656pp.

Salby, M.L., 1996. *Fundamentals of Atmospheric Physics*. Academic Press, New York.

Papers reviewing the origin and evolution of planetary atmospheres can be found in:

Atreya, S.K., J.B. Pollack, and M.S. Matthews, Eds., 1989. *Origin and Evolution of Planetary and Satellite Atmospheres*. University of Arizona Press, Tucson, Arizona.

A classical text on fluid dynamics is given by:

Pedlovsky, J., 1979. *Geophysical Fluid Dynamics*. Springer-Verlag, New York.

Individual planet atmospheres are published in the University of Arizona Press (Tucson, AZ) series:

Venus, by D.M. Hunten, L. Colin, T.M. Donahue, and V.I. Moroz, Eds., 1983, and *Venus II*, by S.W. Bougher, D.M. Hunten, and R.J. Phillips, Eds., 1997.

Mars, by H.H. Kieffer, B.M. Jakosky, C.W. Snyder, and M.S. Matthews, Eds., 1992.

Jupiter, by T. Gehrels, Ed., 1976.

Saturn, by T. Gehrels, and M.S. Matthews, Eds., 1984.

Uranus, by J.T. Bergstrahl, E.D. Miner, and M.S. Matthews, Eds., 1991.

Neptune and Triton, by P.D. Cruikshank, Ed., 1995.

Pluto and Charon, by S.A. Stern and D.J. Tholen, Eds., 1997.

A more up-to-date reference for Jupiter would be a review paper by:

Atreya, S.K., M.H. Wong, T.C. Owen, P.R. Mahaffy, H.B. Niemann, I. de Pater, P. Drossart, and Th. Encrenaz, 1999. A comparison of the atmospheres of Jupiter and Saturn: deep atmospheric composition, cloud structure, vertical mixing, and origin. *Planet. Space Sci.*, **47** 1243–1262.

Problems

4.1.E Calculate the approximate pressure scale height near the surfaces of Earth, Venus, Mars, Pluto, and Titan, and at the 1 bar levels of Jupiter and Neptune. Comment on similarities and differences.

4.2.E Although in some respects the Earth and Venus are 'twin planets', they have very different atmospheres. For example, the surface pressures on Earth and Venus are 1 bar and 95 bars, respectively. Calculate the mass of each atmosphere both in grams and as a fraction of each planet's total mass. Recalculate these values for Earth including Earth's oceans as part of its 'atmosphere'. (If all of the water above Earth's crust were spread evenly over the planet, this global ocean would be ~ 3 km deep.) Compare the calculations for the two planets and comment on the results.

4.3.E (a) Show that in an optically thin medium, the equivalent width is proportional to the column density of absorbing material, regardless of the shape of the line profile.

(b) Assume the line profile can be represented by a Voigt profile. Show that the equivalent width in an optically thin medium is:

$$EW \propto \Delta\nu_D N_c. \tag{4.86a}$$

4.4.I Derive an expression for the equivalent width in a saturated line. Assume a Voigt profile, with the difference in optical depth between the center of the line and the wings being $\sim 10^4$. The wings of the line can be ignored. Define a frequency $x_1 = (\nu_1 - \nu_o)/\Delta\nu_D$, where the optical depth $\tau_\nu = 1$. Inside of x_1 the line is fully saturated, and outside x_1 the line is optically thin. Show that the equivalent width is:

$$EW \propto \Delta\nu_D \sqrt{\log N_c}. \tag{4.86b}$$

Note that the equivalent width is practically insensitive to the number density of absorbing material.

4.5.I In an optically thick medium, the wings in the Voigt profile become important. Define the frequency $x_1 = (\nu_1 - \nu_o)/\Delta\nu_D$, such that the product of τ_{ν_o} with the absorbing wings (second term in equation 4.18) is equal to unity. Express the frequency x_1 in terms of a and τ_{ν_o}. At frequencies $|x| < |x_1|$ the medium is optically thick, while at $|x| > |x_1|$ the medium is optically thin. Show that the equivalent width is:

$$EW \propto \sqrt{N_c \Delta\nu_D}. \tag{4.86c}$$

4.6.E If you were to observe Jupiter's thermal emission at radio wavelengths, where one probes down to well below the planet's tropopause, would you expect limb brightening, darkening or no change in intensity when you scan the planet from the center to the limb? Explain your reasoning.

4.7.E Assume that you observe Saturn at infrared wavelengths in the line transitions of C_2H_2 and PH_3.

The C_2H_2 line is seen in emission, and the PH_3 line in absorption.

(a) Where in the atmosphere are these gases located?

(b) Do you expect the limb of the planet to be brighter or darker than the center of the planet in these lines?

4.8.I (a) The intensity of a Doppler broadened line profile is proportional to $e^{-(\Delta v/v_o)^2}$, where $v_o^2 = 2kT/m$, $\Delta v = |v - v_o|$, m is the mass of the molecules, T the temperature, k Boltzmann's constant, and v the velocity. Derive the equation for the full width at half power for the line profile:

$$\Delta \nu = \frac{v_o \nu_o}{c} 2\sqrt{\ln 2}. \qquad (4.87)$$

(b) Compare equation (4.87) to the Doppler width, $\Delta\nu_D$, in equation (4.16).

4.9.E (a) Calculate the full line width at half power (FWHP) in km s^{-1} for the hydrogen atoms in the upper atmosphere of Jupiter. You may assume a temperature $T = 10^3$ K, and that the upper atmosphere consists of H atoms only.

(b) Calculate the full line width at half power in km s^{-1} for Mars's atmosphere. Assume the atmosphere to consist entirely of CO_2 molecules, and the temperature to be 140 K.

(c) CO is a minor species in Mars's atmosphere; the mixing ratio CO/CO_2 is about 10^{-4}. What is the full width at half power in km s^{-1} for the CO molecules?

4.10.E Why do we believe that we can calculate fairly precisely the temperature vs. altitude profiles well below the observable clouds for Jupiter and Saturn? Sketch one of these profiles and describe how it is derived. Why would the assumptions made in deriving this profile be questionable if applied to Uranus?

4.11.I The SO_2 line on Io is observed in emission at 222 GHz. The contrast between the peak of the line and the continuum background is 18 K. The FWHP (full width at half power) is 600 kHz. Io's surface temperature is 130 K; the emissivity $\epsilon = 0.9$. The line strength is approximately $\alpha_{\nu_o} = 3.2 \times 10^{-22}(\frac{300}{T})^{5/2}$ cm per molecule. In the following you will derive approximate values of the optical depth, number density, temperature, and surface pressure, under the assumption that the atmosphere is optically thin. You can assume that the observations pertain to the center of the disk, and that the plane parallel atmosphere approximation applies.

(a) Assume that line broadening is due to Doppler broadening. Calculate the atmospheric temperature, T_A.

(b) Calculate the optical depth in Io's atmosphere, assuming the Rayleigh–Jeans approximation (eq. 3.4*a*) to be valid.

(c) Calculate the Doppler width $\Delta\nu_D$ from the FWHP.

(d) Using the answers above, determine the column density and surface pressure. (Hint: Convert the Doppler width $\Delta\nu_D \rightarrow \Delta\lambda_D$.)

You will find that your number is roughly an order of magnitude below published values. The reason for this discrepancy lies in the assumptions; better agreement can be reached for a more optically thick atmosphere and lower atmospheric temperatures.

4.12.E Consider a parcel of dry air in the Earth's atmosphere. Show that if you replace some portion of the air molecules (80% N_2, 20% O_2) by an equivalent number of water molecules, the parcel of air becomes lighter and rises.

4.13.E Calculate the dry adiabatic lapse rate (in K km^{-1}) in the atmospheres of the Earth, Jupiter, Venus and Mars. Assume the atmospheres of Venus and Mars to consist entirely of CO_2 gas; Earth is 20% O_2 and 80% N_2; Jupiter is 90% H_2 and 10% He. Make a reasonable guess for the value of γ in each atmosphere. (Hint: See Section 3.2.3.3.)

4.14.I Estimate (crudely) the wet adiabatic lapse rate (in K km^{-1}) in the Earth's lower troposphere, following steps (a)–(d) below.

(a) Determine c_P from the dry adiabatic lapse rate (see previous problem).

(b) Set $T = 280$ K and $P = 1$ bar. Near 280 K the saturation vapor pressure of water is roughly approximated by the Clausius–Clapeyron relation, with $C_L = 3 \times 10^7$ bar, and $L_S = 5.1 \times 10^{11}$ erg $mole^{-1}$. Calculate the partial pressure of H_2O in a saturated atmosphere at 280 K.

(c) As the concentration of water in a saturated atmosphere decreases with height much more rapidly than the total pressure, you may estimate the value of w_s (grams of water per gram of air) by multiplying the value of the partial pressure of water in bars by the ratio of the molecular mass of water to the mean molecular mass of air. Determine the value of w_s.

(d) Estimate the wet adiabatic lapse rate.

Note: Watch your units. The latent heat in equation (4.20) is given in erg mole^{-1}, whereas that in equation (4.22) is in ergs g^{-1}.

4.15.I The saturated vapor pressure curve for NH_3 gas is given by equation (4.20) with $C_L = 1.34 \times 10^7$ bar and $L_s = 3.12 \times 10^{11}$ erg mole^{-1}.

(a) Calculate the temperature at which ammonia gas condenses out if the NH_3 volume mixing ratio is 2.0×10^{-4}. Assume the atmosphere to consist of 90% H_2 and 10% He, and that the pressure is 1 bar. (Hint: Convert the volume mixing ratio to partial pressure.)

(b) Calculate the temperature at which ammonia gas condenses out if the NH_3 volume mixing ratio is 1.0×10^{-3}.

4.16.I The base of the methane cloud in Uranus's atmosphere is at a pressure level of 1.25 bars and temperature of 80 K. The saturation vapor pressure curve is given by equation (4.20), with $C_L = 4.658 \times 10^4$ bar and $L_s = 9.71 \times 10^{10}$ erg mole^{-1}. Derive the CH_4 volume mixing ratio in Uranus's atmosphere, assuming the composition of the atmosphere is 85% H_2 and 15% He. Compare your answer with the solar volume mixing ratio for carbon.

4.17.E Determine the terminal velocity of rain droplets in the Earth's atmosphere, assuming the viscosity $\nu_v = 0.134$ cm s^{-1}.

(a) Determine the velocity for rain drops 100 μm in radius.

(b) Determine the velocity for rain droplets 1 cm in radius.

(c) Compare your answer to the escape velocity from Earth.

4.18.E Calculate the fractional change in temperature, $\Delta T/T$, near the surface between local noon and midnight on Venus and Mars.

4.19.I Derive the Navier–Stokes equation in the rotating frame of reference using equations (4.29), (4.32) and (4.33).

4.20.E Show that a planet's vorticity is equal to twice its angular velocity (eq. 4.38).

4.21.I The zonal wind velocity on Jupiter can be measured at the cloud tops, from features in Jupiter's cloud deck. The cloud tops are at a pressure level of approximately 400 mbar. The wind speed is measured to be 100 m s^{-1} at a latitude of $\theta = 30°$. The meridional temperature gradient at this latitude $\frac{\partial T}{\partial \theta} \approx 3$ K deg^{-1}. Assume Jupiter's atmosphere to consist of 90% H_2 and 10% He. The average temperature in the range 0.4–4 bar can be taken as 150 K.

(a) Use the thermal wind equations to derive the depth (in bars) at which the zonal wind vanishes.

(b) How far (in km) below the cloud tops is this location?

4.22.I Consider a planet whose atmosphere can be approximated by an ideal gas. The planet's obliquity is very small, and the surface temperature varies smoothly from equator to pole. If the atmospheric density is a function of altitude only, then the pressure varies over the surface, and the gas is accelerated in the poleward direction.

(a) Show that the acceleration is $Dv/Dt = -R_{gas}\nabla T/\mu_a$, where R_{gas} is the gas constant and μ_a is the mean molecular weight.

(b) Using the parameters of Earth, with an equator to pole temperature difference of 60 K, calculate the time it would take a parcel to 'free fall' from the equator to the pole.

4.23.I Consider the planet described in the previous problem. Planetary rotation produces a Coriolis acceleration, which measured in the rotating frame of reference is given by: $(D\mathbf{v}/Dt) = -2\omega_{rot} \times \mathbf{v}$, where ω_{rot} is the planet's rotation rate. The Coriolis acceleration has a horizontal component everywhere except at the equator. Thus, moving air masses tend to go in circles of radius $\sim v/\omega_{rot}$. Using the free-fall velocity derived in the previous problem, estimate the characteristic radius of such motions on Earth, at a latitude $\theta = 30°$. Comment on your results; do you think the free-fall velocity is characteristic for wind velocities on Earth?

4.24.I Consider a storm on Earth at a geocentric latitude $\theta = 20°$. The height of the storm $\ell = 1$ km, and the radius is 100 km.

(a) Assume that the storm rotates with a velocity of 50 km per hour; i.e., the winds in the storm blow with this speed. Calculate the potential vorticity of the storm system.

(b) If the storm moves northwards to a latitude $\theta = 45^\circ$ and both the vertical and horizontal scale of the storm stays the same, calculate its new vorticity (wind speeds).

4.25.E (a) Calculate the minimum speed of air in the Earth's atmosphere to allow turbulence to develop at scales exceeding the scale height of the atmosphere (take $\nu_v = 0.134$ cm s^{-1}).

(b) If the flow velocity is 10 cm s^{-1}, what is the characteristic length scale on which we can expect turbulent motions?

4.26.E Explain briefly, qualitatively, why the density of ozone in Earth's atmosphere peaks near an altitude of 30 km. List the relevant reactions.

4.27.I Using the Chapman reactions (reactions 1, 4, 5 and 6), calculate the number density of O_3 molecules in Earth's atmosphere at altitudes of 20 and 60 km, using the number densities of [O] and [O_2] from Figure 4.34. The number density for [M] at $z = 0$ km is equal to 2.69×10^{19} cm^{-3}.

4.28.E Explain why the NH_3 mixing ratio in the stratospheres of Saturn and Jupiter is well below the mixing ratio based upon the saturated vapor curve.

4.29.E Cloud or haze layers of hydrocarbons are observed in the stratospheres of Uranus and Neptune. Explain why we see these hazes above rather than below the tropopause.

4.30.I The limiting flux is the maximum diffusion rate through a planetary atmosphere. The limiting flux can be calculated by assuming the same value for N_i/N for that part of the atmosphere which is well mixed (i.e., below the homopause, which for Earth is at $z < 100$ km). Consider hydrogen atoms in the Earth's atmosphere in all forms (H_2O, CH_4, H_2), at a fractional abundance $N_i/N \approx 10^{-5}$.

(a) Calculate the limiting flux of hydrogen-bearing molecules (and thus hydrogen) from the Earth's atmosphere. You can approximate the binary diffusion parameter using the quantities for H_2 in air (Table 4.6). (Hint: Calculate the limiting flux for an altitude $z = 100$ km; why?)

(b) Calculate the Jeans rate of escape for hydrogen atoms from Earth.

(c) Compare your answers from (a) and (b) and comment on the results.

4.31.E Suppose a body with a radius of 15 km hits a planet at a velocity of 30 km s^{-1}. Assume the evaporative loading parameter is ~ 20.

(a) Calculate the mass of the atmosphere blown into space if the impactor hits the Earth.

(b) Calculate the mass of the atmosphere blown into space if the impactor hits Venus.

(c) Calculate the mass of the atmosphere blown into space if the impactor hits Mars.

(d) Express the masses M_e from (a)–(c) as a fraction of each planetary atmosphere's mass. Comment on your results.

4.32.E Why is the $^{15}N/^{14}N$ ratio larger in Mars's atmosphere than it is in Earth's atmosphere?

4.33.I (a) Explain how the carbon cycle moderates climatic variations on Earth.

(b) Explain qualitatively why, if there were to be a major nuclear war, the temperature on Earth may drop to levels well below freezing, a scenario referred to as 'nuclear winter'.

5 Planetary Surfaces

I believe this nation should commit itself to the goal, before this decade is out, of landing a man on the Moon and returning him safely to the Earth.

USA President John F. Kennedy, in a speech before Congress, 25 May 1961

That's one small step for man, one giant leap for mankind.

Astronaut Neil Armstrong, 20 July 1969, as he became the first human to set foot on the Moon

The four largest planets in our Solar System are gas giants, with very deep atmospheres and no detectable solid 'surface'. All of the smaller bodies, the terrestrial planets, asteroids, moons and comets, have solid surfaces composed of rock and/or ice. Each of these bodies displays unique geological features, which yield clues about their formation as well as past and current geological activity. The surface reflectivity varies dramatically from one body to another; some surfaces have very low albedos (such as the maria on the Moon, carbonaceous asteroids, comet nuclei), while others are highly reflective (Europa, Enceladus). Large albedo variations may even be seen on a single object (Iapetus). Some bodies are almost completely covered by impact craters (Moon, Mercury, Mimas), while others show little or no evidence of impacts (Io, Europa, Earth). The terrestrial planets and many of the larger moons show clear evidence of past volcanic activity, and some (Earth and Io) are active even today. Past volcanic activity may be seen in the form of volcanoes of different shapes and size (Earth, Mars, Venus) or large solidified lava lakes (Moon). Most bodies, even small asteroids, display linear features like faults, ridges and scarps which are suggestive of past tectonic activity. None of the other objects, however, displays active tectonic activity through the motion of tectonic plates as the Earth does. Why are planetary surfaces so different superficially, and what similarities do they share? In this chapter, we review geological processes common to all planets. We start with a basic review of rocks and minerals, and discuss the crystallization of *magma* (molten rock at depth). Processes which 'shape' the surfaces (gravity, volcanism, tectonism, impacts) are discussed in Sections 5.3 and 5.4, and the surface characteristics of individual bodies are summarized in Section 5.5.

Although the interior structure of planets is discussed in Chapter 6, some basic terminology is required for this chapter. Figure 5.1 shows a sketch of the interior structure of the Earth, as it has been derived from seismological data. The Earth consists of a solid iron–nickel *inner core*, surrounded by a fluid metallic *outer core*. The core extends over roughly half the Earth's diameter. The outer ~ 3000 km is the rocky *mantle*, which itself is divided into a solid lower and partially molten upper mantle. The upper mantle consists of a cool elastic *lithosphere*, which 'floats' on a hot, highly viscous 'fluid', the *asthenosphere*. The lithosphere is an elastic layer which responds to a 'load'; the terminology is somewhat confusing, how-

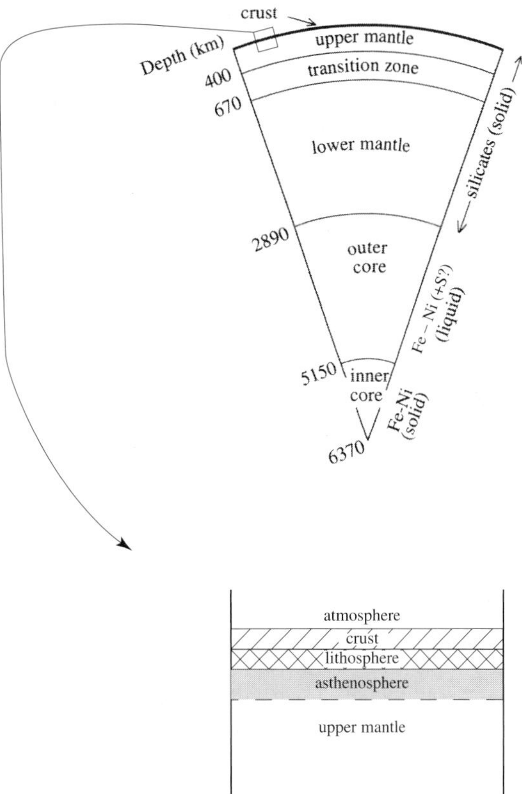

FIGURE 5.1 Sketch of the interior structure of the Earth. (Adapted from Putnis 1992)

ever, with the precise definition depending on the sub-field in geophysics. The asthenosphere is, although highly viscous, somewhat less viscous than the mantle below it. The outer 'skin' of the planet is the *crust*, a rather brittle layer, which is the topic of the present chapter. Earth's crust varies in thickness from ~6 km under the oceans up to ~50 km on the continents. Oceanic crust is denser than continental crust.

5.1 Mineralogy and Petrology

Petrology is the study of rocks, of their composition, structure and origin. Since solid planetary material consists of rock and ice, and rocks are made up of different minerals, it is essential for a planetary scientist to have some basic knowledge of rocks and minerals. In this section we start with a short summary of the basics of mineralogy, before reviewing the various types of rocks, how rocks are formed and where they are found. Additional information can be obtained from e.g., Press and Siever (1986, 1998), Putnis (1992), and Hartmann (1999).

5.1.1 Minerals

Minerals are solid chemical compounds that occur naturally, and that can be separated mechanically from other minerals that make up a rock. Each mineral is characterized by a specific chemical composition and a specific regular architecture of the atoms from which it is made. The forces that hold molecules together depend upon the electronic structure of the constituent atoms. Some atoms, for example Si, Mg, Fe, Ti, tend to give up electrons in their outer shells, creating a positively charged ion (a *cation*), while other atoms, in particular O, may adopt electrons, creating a negatively charged ion (an *anion*). Whether an atom gains or gives up electrons depends upon the electronic structure of the element. In particular, most atoms have one or more loosely bound *valence electrons* which can be 'shared' with other atoms to fill up electron shells, thereby lowering the energy state of the formed compound. Such interactions determine the chemical behavior of the elements. The valence of each atom is indicated in the periodic table of elements (Appendix C). The chemical bond between cations and anions is referred to as an *ionic bond*, caused by the electrostatic attraction between oppositely charged particles (Coulomb's law). Examples are the mineral halite (Na^+ with Cl^-) and magnesium oxide (Mg^{2+} with O^{2-}). Another type of bonding is *covalent bonding*, where the atoms share electrons in their outer shells. In a diamond each carbon atom is surrounded by four others in a regular tetrahedron (Fig. 5.2a), and these atoms are held together by a covalent bond. A third, but much weaker, bonding is due to the *Van der Waals force*, a weak electrical attractive force that exists between all ions and atoms in a solid. The strength of the bonds, often a combination of the three types mentioned above, determines the hardness of a mineral, indicated on the *Mohs scale of hardness*, from 0 to 10, where talc is 1 and diamond 10 (Table 5.1).

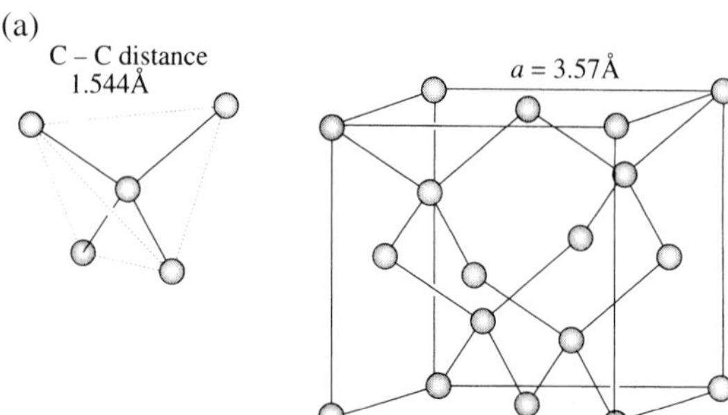

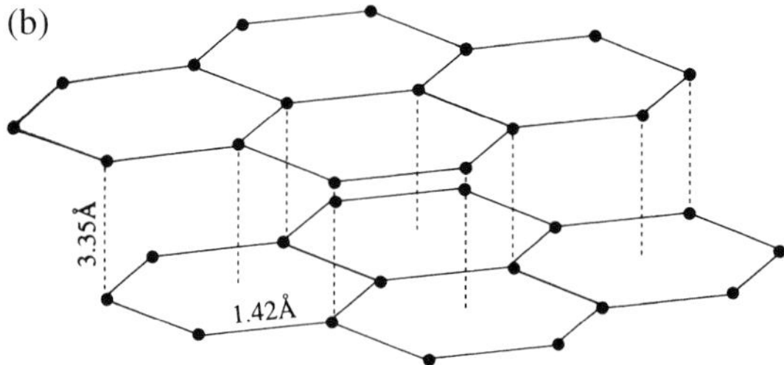

FIGURE 5.2 (a) Mineral structure of diamond: each carbon atom is surrounded by four others in a regular tetrahedron (shown by the dashed line, left figure). The diamond (right figure) is built up from these tetrahedra. (Putnis 1992) (b) The structure of graphite is made up of layers in which the carbon atoms lie at the corners of a hexagonal mesh. Within the layers the atoms are strongly bonded, while the layers are weakly bonded to each other. (Putnis 1992)

TABLE 5.1 Mohs Scale of Hardness[a].

Mineral	Scale number	Common objects
Talc	1	
Gypsum	2	Fingernail (2.5)
Calcite	3	Copper coin
Fluorite	4	
Apatite	5	Teeth
Orthoclase	6	Window glass (5.5)
Quartz	7	Steel file (6.5)
Topaz	8	
Corundum	9	
Diamond	10	

[a] After Press and Siever (1986).

A mineral is characterized by a combination of its chemical composition and crystalline structure. A different spatial arrangement of the atoms that make up a material can lead to a very different mineral, even if the chemical composition is the same. A classic example is graphite versus diamond, each of which consists exclusively of C atoms. Diamond, where the atoms are bonded in a covalent bond, is an extremely hard mineral in contrast to graphite, which is very soft. Graphite is made up of layers, where the C atoms form a hexagonal mesh in the layers, and Van der Waals forces bond the layers (Fig. 5.2b).

Minerals in the field can be identified through a combination of their *hardness*, *cleavage* or breakage along certain planes (e.g. mica), *fracture*, *density*, *color*, *luster* and *streak* (color of the powder that comes off when a mineral is scraped). Although there are several thousand known minerals, each with its own unique set of properties, most of them can be classified within a few major chemical classes. These classes are listed in Table 5.2. In addition to the native elements, such as Cu, Fe, Zi, etc., minerals can be made up of several different atoms organized in a regular crystalline structure, such as quartz, SiO_2, or olivine, $(Fe,Mg)_2SiO_4$. The notation (Fe,Mg) indicates that the elements Fe and Mg can be substituted for each other. Substitution of atoms in minerals takes place between elements of the same size and valence. The most abundant types of minerals on terrestrial planets are the *silicates*, minerals which contain silicon and oxygen, such as quartz, olivine, feldspar ($(K,Na)AlSi_3O_8$, $CaAl_2Si_2O_8$) and pyroxene (as e.g., $(Mg,Fe)_2Si_2O_6$). *Feldspars* make up about 60% of the surface rocks on Earth. They have a typical density of ~ 2.7 g cm^{-3}, and thus are relatively light. They, therefore, tend to float upwards in a magma, and end up relatively close to a planet's surface. Potassium-rich feldspars are referred to as *orthoclase* feldspars, while *plagioclase* feldspars are rich in sodium and/or calcium. *Quartz* is, like feldspars, very abundant on the Earth. Quartz has a density of ~ 2.7 g cm^{-3}, thus is also present on or near a planet's surface. *Pyroxenes* make up $\sim$10% of the Earth's crust. They contain a relatively large fraction of heavy elements, such as Mg and Fe, which makes them heavier than feldspars ($\rho \approx 2.8$ to 3.7 g cm^{-3}). Some common minerals in this class are augite ($Ca(Mg,Fe,Al)(Al,Si)_2O_6$), enstatite ($MgSiO_3$) and hypersthene ($(Mg,Fe)SiO_3$). *Olivine* ($(Fe,Mg)_2SiO_4$), an olive-colored mineral, is probably the heaviest of all silicates and, like the pyroxenes, tends to sink in magmas. It is therefore an important constituent of rocks formed at depth, and believed to be a major constituent of the Earth's mantle. *Amphibole* is a group of (Mg,Fe,Ca)-silicates, which are slightly less dense than pyroxenes, but have a more amorphous crystalline structure. They make up $\sim$7% of the Earth's crustal minerals. An example is *hornblende* ($(Ca, Na)_{2-3}(Mg, Fe, Al)_5(Si, Al)_8O_{22}(OH)_2$). *Micas* are sheet silicates of K, Al, and/or Mg. The most common examples of micas are biotite (black to dark brown; $K(Mg, Fe)_3AlSi_3O_{10}(OH, F)_2$) and muscovite (silvery, colorless or white, translucent; $KAl_2(AlSi_3O_{10})(OH, F)_2$).

TABLE 5.2 Chemical Classes of Minerals[a].

Class	Defining anions	Example
Native elements	none	Copper Cu, gold Au
Sulfides and similar compounds	S^{2-} similar anions	Pyrite FeS_2
Oxides and hydroxides	O^{2-} OH^-	Hematite Fe_2O_3 Brucite $Mg(OH)_2$
Halides	Cl^-, F^-, Br^-, I^-	Halite NaCl
Carbonates and similar compounds	CO_3^{2-}	Calcite $CaCO_3$
Sulfates and similar compounds	SO_4^{2-} similar anions	Barite $BaSO_4$
Phosphates and similar compounds	PO_4^{3-} similar anions	Apatite $Ca_5F(PO_4)_3$
Silicates and similar compounds	SiO_4^{4-}	Pyroxene $MgSiO_3$

[a] After Press and Siever (1986).

After silicates, the most abundant minerals are *oxides*, which are composed primarily of metals (in particular Fe) and oxygen. Common iron oxides include magnetite (Fe_3O_4), which has a black, metallic luster, hematite (Fe_2O_3), which is often reddish-brown or steely-gray and black, and limonite ($HFeO_2$). The color of limonite is yellowish-brown to dark brown, similar to rust. These minerals are believed to redden the surface of Mars. Ilmenite ($(Fe,Mg)TiO_3$), a black opaque mineral, and spinel ($MgAl_2O_4$) provide most of the opacity in the Moon's maria.

Other common minerals on Earth are pyrite (fool's gold, FeS_2) and troilite (FeS), both of which are probably abundant in planetary interiors (Section 6.1.2.3), since their high density (~ 5 g cm^{-3}) causes them to sink down in magma. Clay minerals are hydrous aluminum silicates, major erosion products on Earth and Mars, which also have been detected on carbonaceous asteroids (e.g., Ceres). Clay minerals may hold a lot of chemically bound water.

In the outer Solar System, beyond a heliocentric distance of ~ 4 AU, *ices* make up over half the mass of material which condensed in the solar nebula, and are thus important constituents of minerals in the outer Solar System. Important ices include water (H_2O), carbon dioxide (CO_2), ammonia (NH_3), and methane (CH_4). Much of the water is also found in the form of *hydrate* minerals (such as the hydrous aluminum silicates mentioned above), while water-ice can exist in the form of *clathrates*, where a guest molecule occupies a cage in the water-ice lattice. Other low-temperature condensates important in the outer Solar System are carbonaceous minerals, which color the surfaces of objects blackish and reddish-black (albedos ~ 2–8%). *In situ* measurements by the Giotto spacecraft at comet Halley revealed the presence of CHON particles, dust grains which are dominated by combinations of the elements H, C, N, and O (Section 10.3.5).

5.1.2 Rocks

Planetary surfaces are comprised of solid material, which is generally referred to as 'rocks', assemblages of different minerals. Rocks are classified on the basis of their formation history. We distinguish four major groups, each of which is discussed in more detail in the following subsections: primitive, igneous, metamorphic, and sedimentary rocks. Within these groups, the rocks can be further subdivided on the basis of the minerals which make them up, and/or on the basis of their texture, such as the size of the grains which make up the rock. Some rocks, such as breccias, can include material from various groups.

5.1.2.1 *Primitive Rocks*

Primitive rocks are formed directly from material which condensed out of the primitive solar nebula. These rocks have not undergone transformations due to high temperatures and pressures prevailing in the interiors of objects like the planets and larger moons and asteroids. Primitive rocks are common on the surfaces of many asteroids, and can be sampled from meteorites. The most primitive materials are the CHON particles in comets (Section 10.3.5) and chondritic meteoritic material. Chondrites have the same refractory element abundances as the Sun; their properties are discussed in detail in Chapters 8 and 9.

5.1.2.2 *Igneous Rocks*

Igneous rocks are the most common rocks on Earth and other bodies which have undergone melting. Igneous rocks are formed when a *magma*, i.e., a large amount of hot molten rock, cools. The physics and chemistry of a cooling magma, and the crystallization of minerals therein, is discussed in Section 5.2. Here we describe the end products of the melt, that is the various types of rocks that result from the cooling process. Rocks form from the magma either underground (*intrusive* or *plutonic* rocks) or above ground (*extrusive* or *volcanic* rocks). Magma deep underground cools slowly, and crystals have plenty of time to grow. The resulting intrusive rocks are therefore coarse-grained, and the minerals can easily be distinguished with the naked eye (e.g., common granite, Fig. 5.3). When magma erupts through the planetary crust, it cools rapidly through radiation into space. Volcanic rocks thus show a fine-grained structure, in which individual minerals can only be seen through a magnifying glass. In case of extremely rapid cooling, the rock may 'freeze' into a glassy material. Obsidian is a volcanic rock that cooled so fast that it shows no crystalline structure (Fig. 5.3). The minerals in these rocks are no longer in the form of crystals, but show an amorphous, glassy structure. The texture of the rock thus depends on how rapidly the magma cools, while the composition of the rock depends upon the minerals which crystallize from the melt.

Although the classification of the major rock groups is based upon their chemical and mineralogical composition, for practical purposes one can simply use the silica content of the rock. The two basic rock types are basalts and granites, where basalts contain a relatively low abundance of silica (40–50% by weight) and granites much more (~ 70% by weight). In addition to silica, basalts consist largely of heavy minerals, such as pyroxenes and olivines. Basalts are sometimes referred to as *basic* or *mafic* (from Mg, Fe) rocks. *Ultra-basic* (*ultra-mafic*) rocks have a very large percentage of heavy elements. In contrast, feldspars, in particular orthoclase (K-rich feldspars), and quartz are the dominant minerals in granite. Granites are therefore also referred to as *felsic* (from feldspar) or *silicic* rocks. Granites are usually light-colored, whereas basalts, and in particular the ultra-mafic basalts, are dark. Basaltic rocks are probably the most common rocks on planetary bodies, as they make up the lava (solidified magma) flows on bodies like the Earth and Moon. Although abundant on Earth, granite is less common on other planetary bodies.

Figure 5.4 shows a classification table of the various rock types in the form of a data cube. The x-axis shows the silica content of the rock, while the y-axis indicates the mineral content. The z-axis (towards the back of the cube)

FIGURE 5.3 Examples of different types of rocks (all rock photographs were taken by Floris van Breugel; most of the rock samples provided by M. Gennaro, Museum of Geology, UC Berkeley). The coin shown for scale is a USA dime (diameter 17.9 mm) (a) Four different types of granite: in each of them individual crystals (e.g., quartz, biotite, muscovite, plagioclase) are clearly visible. The rock in the middle back is granodiorite, and the rock in front is a red granite, which has been polished at the top. (b) Three different pieces of rhyolite: note that only small specks of grains are visible. (c) Pumice, a very light and froth-like rock. The piece on the right is 'glazed' over. (d) Four pieces of obsidian: the one on the right is pure black and the one on the left is black with brown stripes. The piece in the center back shows a frosted piece of rhyolite, while the piece in the front is like transparent, partly painted (brown and black) glass. (e) Mafic rocks: the two rocks in the back are basalt: a solid piece (upper left), and a more vesicular lava rock (upper right). The lower left rock is gabbro, and here individual crystals are much larger. This shows up in particular in the difference in texture, since all crystals are dark. In the lower right is andesite. (f) Sedimentary rocks: Two sandstones can be clearly recognized from the granular appearance (upper right: gray sandstone, lower left: Arizona pinkish sandstone), while the lower right rock is a piece of shale.

indicates grain size. The silica content determines whether a rock is felsic (granitic) or mafic (basaltic). The grain size is proportional to the time it took the rock to cool. Intrusive rocks consist of large grains, while volcanic rocks are fine-grained. Some pictures of rocks from the field are shown in Figure 5.3. The minerals which make up granite are typically several millimeters in size, and consist primarily of feldspars and quartz. Volcanic forms of granite are rhyolite, pumice and obsidian, in order of decreasing grain size (obsidian, a glass, does not contain crystals). Basalt is a fine-grained mafic volcanic rock, which consists primarily of the mineral pyroxene. The plutonic, coarse-grained mafic rock is called gabbro. Dunite is a very ultramafic rock, which consists almost entirely of the mineral olivine.

Extrusive rocks are directly correlated with volcanic activity. The type of eruption is determined by the viscosity (a measure of resistance to flow in a fluid) of the magma, and depends on the temperature, composition (in particular silica content), and the gas content of the melt. Typically, magma/lava with a high silica content, i.e., felsic lavas, has a high viscosity, which is increased even more if the gas content is high. Such eruptions are explosive, with typical temperatures of $\sim$1050–1250 K, and they form thick local deposits. In contrast, basaltic melts are very fluid, erupt with temperatures of 1250–1500 K, and flow fast and far; they form large lava beds and fill in lowlands, such as the maria on the Moon and the lava beds on Hawaii. These lava flows are usually dark in color, in contrast to the lighter felsic deposits.

Volcanic rocks can vary widely in appearance and density. These rocks are usually identified by a combination of their texture and composition. Violent explosions, where lava is ejected into the air and shattered in the process, usually due to a sudden release of gases, are more common in the silicic than basaltic magmas. Rocks which are blasted out during such a violent explosion are called *pyroclasts*, and vary in size from micrometer-sized *dust*, to

FIGURE 5.3 Continued. (g) Sedimentary rocks: A breccia is shown on the left, and a conglomerate on the right. The lower middle rock is chert, and the upper rock is dolomitized limestone. (h) Metamorphic rocks: Gneiss (left), marble (upper middle), schist (right), and slate (lower middle). (i) Evaporites: Gypsum is shown in the front, and the two other rocks are halite. The left one consists of nice halite crystals. (j) Iron oxides: Left: limonite (orange-blackish); middle: three pieces of hematite (dark-gray; one piece is polished); right: magnetite (black). (k) Quartz crystal.

millimeter-sized *ash*, and larger *bombs*. Many pyroclasts are glassy or fine-grained, caused by the fast cooling of the suddenly ejected magma. The silicic melts may have a high gas content, and because the gases usually cannot escape, pyroclastic fragments can be extremely light. Examples of such *vesicular rocks* are *pumice* or *volcanic foam*, a sponge-like glassy rock with a great number of bubbles or cavities (*vesicles*), formed by gas in the melt. Pumice is so light it floats in water. When pyroclasts fall down, the rocks may get cemented together under extreme heat into volcanic *tuff* and *breccias*.

5.1.2.3 *Metamorphic Rocks*

As there has been much reworking of the Earth's surface by great forces from within the interior, many rocks have been altered when subjected to high temperature and pressure, or when introduced to other chemically active ingredients. Rocks which have been altered are called *metamorphic* rocks. Such rocks are often named for a mineral constituent that is predominant in the rock, such as marble (from limestone or other carbonate rock), quartzite (from quartz), and amphibolite (amphiboles) (see Fig. 5.3 for some examples). Metamorphic rocks can be either *regional* or *contact* metamorphic rocks. The regional type consists of rocks (igneous, sedimentary, metamorphic) transformed many kilometers below the surface by extremely high temperatures and pressures. Large areas or regions of rock can be metamorphosed this way. Regional metamorphic rocks show *foliation*, a platy structure caused by a parallel alignment of the minerals, often perpendicular to the stress exerted on them. Examples are coarse-grained gneiss, transformed from granite, medium-grained schist from shale and/or granite, and fine-grained slate from shale. Contact metamorphic rocks are transformed near an igneous intrusion, largely by heat. Magma forces its way into layered rocks or penetrates cracks or cavities. If the stress on the rocks and the temperature are high enough, the rock is altered, metamorphosed. An example is hornfels, a very fine-grained silicate rock. Though not as common on Earth as regional or contact metamorphic rocks, rocks may also get altered as a result of impact-induced shocks at impact sites. Shocked quartz has been found at several impact sites.

5.1.2.4 *Sedimentary Rocks*

On planets that possess an atmosphere, material may be transported by winds, rain and water flows. Sedimentation is the final stage of this process, where material is deposited at some other location. These sediments may

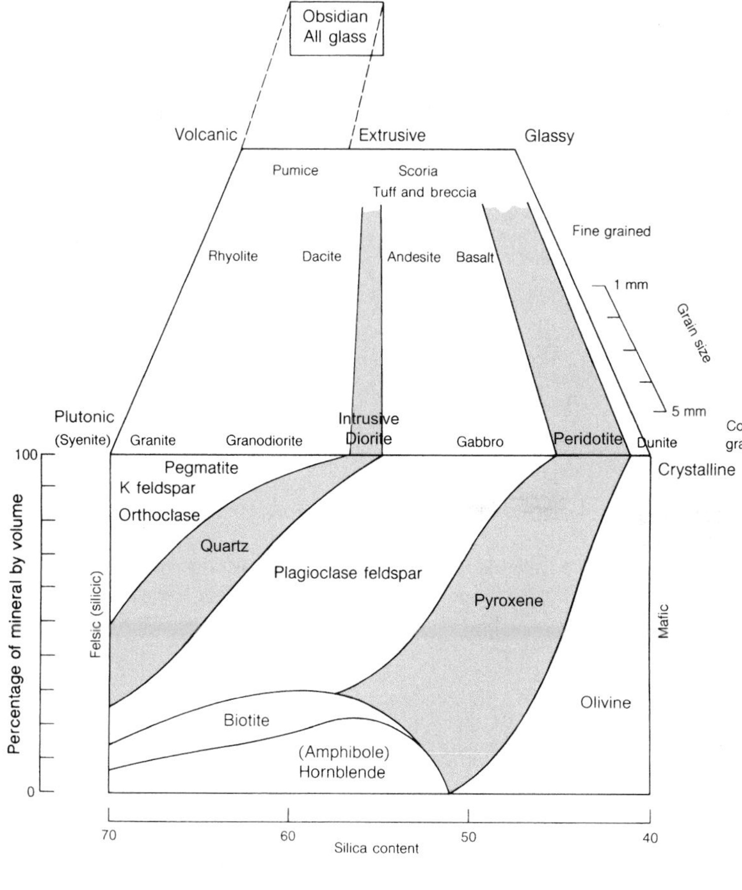

FIGURE 5.4 A classification cube for igneous rocks. The x-axis shows the silica content of the rocks (percent by volume), and the y-axis the percentage of a given mineral. The texture of the rocks is indicated as a function of grain size along the z-axis, at the top of the cube. A granite with a silica content of 70% contains about 25% quartz (SiO_2), less than 10% biotite, hornblende and plagioclase feldspar, and 50% K-feldspar or orthoclase. Fine-grained granites are called rhyolite. (Press and Siever 1986)

form new *sedimentary* rocks (Fig. 5.3). *Detrital* sediments have physically been transported from one place to another, for example by winds or water after erosion of rocks. The components of detrital sediments are fragments of pre-existing rocks and minerals, and are referred to as *clastic* (which, in Greek, means to break) fragments. While the rock fragments are transported, the minerals are sorted by size and weight with a variable efficiency. Due to the sorting process, rocks form with different textures, varying from coarse to very fine-grained rocks. Coarse-grained fragments, such as gravel, form *conglomerates* when cemented together. Medium-grained sands form *sandstone*, and fine-grained clays and silt may be cemented together into *mudstone* or *shale*. The individual grains in shale and sandstones are fairly round, as a result of the erosion. Shale and sandstone are the two most abundant sedimentary rocks on Earth. Shale makes up $\sim 70\%$ of the sediments on Earth, while sandstone makes up $\sim 20\%$. The remaining 10% consists predominantly of limestone, a chemical sediment discussed briefly below.

The composition of rocks may be altered through the interaction with other chemical constituents, such as those present in an atmosphere. Such sediments are referred to as *chemical* sediments. Prime examples are limestone ($CaCO_3$) and dolomite ($CaMg(CO_3)_2$), derived from the *Urey weathering reaction* (Section 4.9.1), a chemical reaction between CO_2, dissolved in water, with silicate minerals in the soil to form calcium carbonate or calcite, $CaCO_3$. We note that on Earth, however, most carbonates are from biological deposits, fossils of animal shells produced by organisms in the oceans. *Evaporite* is a rocky material from which liquid evaporated, leaving behind sediments such as halite or common salt (NaCl) and sulfate minerals, such as gypsum ($CaSO_4{\cdot}2H_2O$). Evaporites may bond other rocks together into a loose crumbly rock. Another type of sediments is formed by clay minerals, hydrous aluminum silicates such as hematite and limonite. They are abundant in erosion products on both Earth and Mars, as well as in carbonaceous material on asteroids and bodies in the outer Solar System.

5.1.2.5 *Breccias*

Breccias are 'broken rocks', that consist of sharp angular fragments which are cemented together. These rocks may originate from meteorite impacts, where the pieces are 'glued together' under the high temperature and pressure during and immediately following the impact. They therefore cover the bottom of many impact craters. Breccias may also form tectonically, e.g., along fault zones.

5.2 Crystallization of a Magma

The composition, pressure and temperature of a magma determine ultimately which minerals form when the magma cools. In this section we discuss phase diagrams of a melt and a general sequence of reactions that take place in a cooling magma.

5.2.1 Phases of the Magma

The states that the magma goes through as it cools and crystallizes can be shown on a *phase diagram*, similar to the phase diagrams discussed for water and other condensable gases in Section 4.4. For simplicity, let us assume the magma to crystallize under equilibrium conditions. The phase changes which occur in the magma are best predicted by making use of the *Gibbs free energy*, G, of the system:

$$G = H - TS, \tag{5.1}$$

where T is the temperature, and S the entropy of the system. Entropy is a quantity which measures the change in a mineral's state of order when it changes from one phase or structure to another. For a thermodynamically reversible process, the change in entropy is equal to the ratio of the amount of heat absorbed by the system, dQ, and T:

$$dS = \frac{dQ}{T}. \tag{5.2}$$

The *enthalpy*, H, is defined as the sum of the internal energy, U (potential energy stored in the interatomic bonding plus kinetic energy of the atomic vibrations), and the work done on the system, PV, with P the pressure and V the volume:

$$H = U + PV. \tag{5.3}$$

A common 'standard state' is to set the enthalpy of the regular elements to zero, by definition. The enthalpy of a mineral is then defined as the change it takes in enthalpy or 'heat' to form the mineral from the elements. If it takes energy to form the mineral, the reaction is *endothermic*, and the (formation) enthalpy (or change in enthalpy) is positive. If the reaction frees up energy, the reaction is *exothermic* and the (formation) enthalpy is negative. In Chapter 3 we discussed the first law of thermodynamics (eq. 3.67: $dQ = dU + PdV$), from which it follows that a change in the enthalpy:

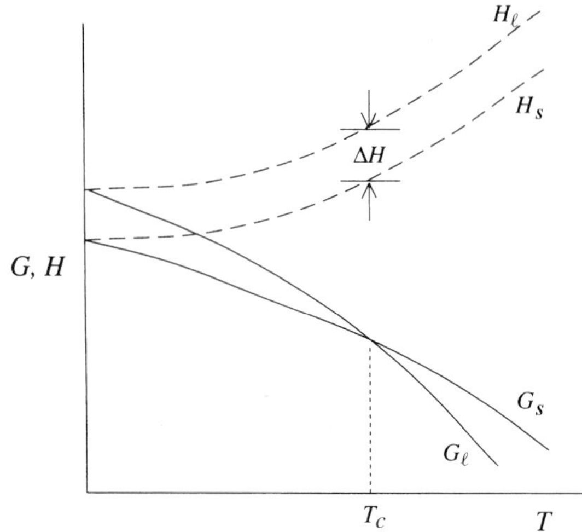

FIGURE 5.5 A graph of the variation in the Gibbs free energy, G, and the enthalpy, H, of the liquid (ℓ) and solid (s) phase of a solution as a function of temperature. The phase with the lowest free energy is the stable phase. At temperatures $T > T_c$, the mixture is in a liquid phase, while at $T < T_c$ the solution is solid. (Adapted from Putnis 1992)

$$dH = dQ + VdP. \tag{5.4}$$

The thermal heat capacity of the system, C_P, is defined as the amount of heat needed to raise the temperature of 1 mole of material by 1 K while keeping the pressure constant (equations 3.20, 3.68), which is also equal to the change in enthalpy:

$$\left(\frac{\partial H}{\partial T}\right)_P = \left(\frac{\partial Q}{\partial T}\right)_P \equiv C_P. \tag{5.5}$$

The enthalpy of the system at temperature T_1 is thus equal to:

$$H = H_0 + \int_0^{T_1} C_P dT, \tag{5.6}$$

with H_0 the enthalpy at $T = 0$ K.

For a thermodynamically reversible process, it follows from the equations above that the entropy of the system at temperature T_1 is given by:

$$S = S_0 + \int_0^{T_1} \frac{C_P}{T} dT, \tag{5.7}$$

with S_0 the entropy at $T = 0$ K. For a perfect crystal where $S_0 = 0$, all atoms would be in the ground state. For a reversible process the change in entropy after a sample is heated up and cooled down to the same temperature, would be zero. However, for any natural process, the change in entropy is larger than zero. This implies that if a mineral becomes more ordered in a transformation process, i.e., its entropy decreases, the heat liberated in the process must increase disorder in the environment. The change in entropy is defined by:

$$dS > \frac{dQ}{T}. \tag{5.8}$$

Since, at constant pressure, $dQ = dH$, a phase transition takes place if:

$$dH - TdS < 0, \tag{5.9}$$

which is the change in the Gibbs free energy. For a thermodynamically reversible process ($dQ = TdS$), changes in the Gibbs free energy as a function of pressure and/or temperature can be expressed as:

$$dG = VdP - SdT. \tag{5.10}$$

Figure 5.5 shows a graph of G as a function of temperature for the liquid, l, and solid, s, phase of a melt. The phase with the lowest free energy is the stable phase. At the critical temperature, T_c, the curves cross and upon further heating or cooling of the melt a phase transformation takes place. The phase transition brings about a change in the enthalpy, ΔH, which is the *latent heat of transformation* (see also Section 4.4.1):

$$\Delta H = T\Delta S. \tag{5.11}$$

We note that at a constant pressure, transformations or phase transitions only take place upon cooling or heating the sample, since the free energies of the two phases are equal at T_c. In practice transformations may be *reversible*, i.e., the phase changes upon cooling the sample and changes back when heating it, or *irreversible*, when the sample does not convert back to its original state when reversing the temperature gradient. Besides the temperature, the transformation also depends upon the kinetics or reaction rate of the processes involved. A prime example of an irreversible process is the case of diamond and graphite. Diamonds grow deep within the Earth's crust; when brought up to the surface, they are not transformed into graphite, even though the Gibbs free energy of graphite is lower than that of diamond. The reason for this is that the rate of transformation of diamond into graphite at room temperature and pressure is so small that little change occurs, even over billions of years.

Usually a melt consists of a variety of constituents, so that the total Gibbs free energy of the melt must be determined from the energies of the individual components. One needs to use here the minimal set of independent constituents that make up all the phases which describe and define all the reactions in the melt:

$$\begin{aligned} G &= \Delta G_{mix} + \sum(f_i G_i) \\ &= \Delta H_{mix} - T\Delta S_{mix} + \sum(f_i G_i), \end{aligned} \tag{5.12}$$

where f_i is the fractional concentration of the i-th component. The mix term, ΔG_{mix}, comes from a change in the entropy and enthalpy of the system when the various constituents are mixed. The entropy of the mixture, ΔS_{mix}, is related to the probability that a particular site in a crystalline lattice is taken up by atom or molecule i. From statistical arguments it can be shown that:

$$\Delta S_{mix} = -kN \sum(f_i \ln f_i), \tag{5.13a}$$

with k Boltzmann's constant, N the number of sites over which the atoms can be distributed, and $f_i N$ the number of components i. For one mole of sites, N is equal to Avogadro's number, N_A, and $kN = kN_A = R_{gas}$. If there are n structural sites over which substitution of atoms/molecules can take place, the entropy becomes:

$$\Delta S_{mix} = -nR_{gas} \sum(f_i \ln f_i). \tag{5.13b}$$

Since f_i is a fraction, $\Delta S_{mix} > 0$, and the Gibbs free energy is decreased, which thus favors the formation of a solid state. In the ideal solid solution, $\Delta H_{mix} = 0$, and ΔG_{mix} is entirely determined by the entropy of mixing. In a regular solid solution with two components, the enthalpy of mixing is given by:

$$\Delta H_{mix} = \alpha f_1 f_2, \tag{5.14}$$

with α the interaction parameter.

By defining the *chemical potential* or *partial mole free energy* of constituent i in the solid phase as:

$$\mu_i = G_i + R_{gas} T \ln f_i, \tag{5.15}$$

the Gibbs free energy in an ideal solid solution ($\Delta H_{mix} = 0$) becomes:

$$G = \sum(\mu_i f_i). \tag{5.16a}$$

In a nonideal solution, the fractions f_i in the chemical potential above are replaced by a quantity called the *activity*, a_i, which increases or decreases the effective concentration of component i to simulate the change in the strength of the bond between similar atoms. The Gibbs free energy can then be written:

$$G = \sum(\mu_i a_i). \tag{5.16b}$$

Figure 5.6 shows a free energy curve for a two-component solution, where the fractional composition of component

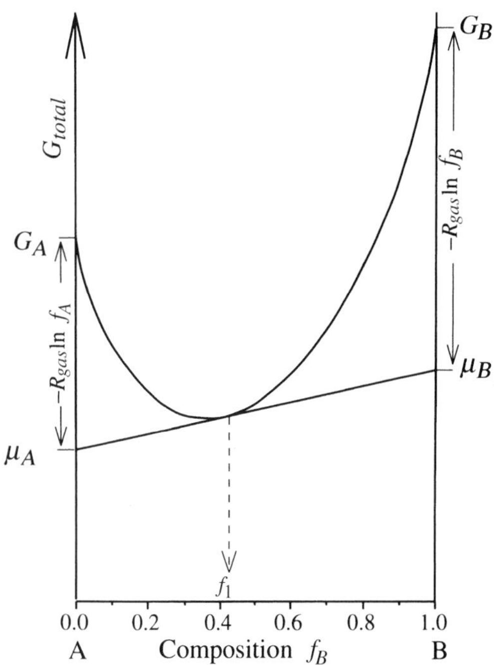

FIGURE 5.6 The Gibbs free energy curve for an ideal solid solution, which consists of components A and B. The x-axis shows the solution's fraction of component B. The free energy for the end members, G_A and G_B, are plotted along the y-ordinate. The tangent to the curve at any composition f_B (such as f_1 shown in the figure) intercepts the free energy axes at the chemical potential of the component, μ_A and μ_B, respectively. (Adapted from Putnis 1992)

B is plotted along the x-axis. The Gibbs free energy of the end-members, G_A and G_B, is indicated along the y-axis where the solution is 100% A (i.e., 0% B) and 100% B, respectively. The tangent to the curve at any composition f_B intercepts the free energy axes at the chemical potential of the component, μ_A and μ_B, respectively. The total free energy is then given by the sum $\mu_A f_A + \mu_B f_B$ (eq. 5.16). If we consider the free energy curves for a liquid and solid solution, phases can only coexist if the chemical potential of each component is the same in each phase.

An example of crystallization of a binary melt is given in Figure 5.7. The free energy, G, of the liquid and solid solution is shown at different temperatures, as a function of composition in panels (a)–(e). The resulting phase diagram is shown in panel (f). The composition of component B is plotted along the x-axis. At high temperatures, T_1 in panel (a), the free energy of the liquid is lower everywhere than for the solid, and the magma is completely molten. Upon cooling the melt, solid A starts to condense out when temperature T_2 is reached (panel b). At lower temperatures, T_3 (panel c), G_{solid} is below G_{melt} for a range of compositions. The coexisting compositions of the solid and liquid are defined by the common tangent, since here $\mu_{c1} = \mu_{c2}$ (see panel c). The equilibrium phase diagram in panel (f) consists of the locus of the common tangent points. It defines the composition of the coexisting phases at each temperature. Consider, for example, a melt of composition c_2 at point x in panel (f). Upon cooling of the melt, the vertical dashed line intersects the curve named *liquidus* at temperature T_3, and a solid of composition c_1 crystallizes out. Upon further cooling of the melt, under equilibrium conditions the composition of both the solid and liquid are adjusted or re-equilibrated continuously, and the compositions are given by the intersection of the horizontal line at temperature T with the curves liquidus and *solidus*. The composition of the liquid and solid follow essentially the curves liquidus and solidus to the right. At temperature T_x the entire melt has solidified into a solid with a composition equal to that of the original melt (c_2). If the original melt starts out with a higher concentration of B, the temperature of the melt has to be lowered more before the entire melt has solidified.

Phase diagrams can be very complicated; the solid and/or melt may become immiscible (insoluble in each other) at certain temperatures, solid phases may exist in different forms at different temperatures, the melting point of a substance may be lowered in the presence of a particular melt (*eutectic* behavior), intermediate products may be formed, etc. For example, the MgO–SiO_2 system sketched in Figure 5.8 contains several distinct solid phases. At very high temperatures, $T > 2270$ K, the melt exists as a single phase over the entire compositional range. At $T < 2270$ K, the melt becomes immiscible at the silica-rich end, so that silica-rich and magnesium-rich liquids coexist. At the magnesium-rich end in the diagram, periclase (MgO) crystallizes out at $T \lesssim 3070$ K. At $T \lesssim 2120$ K, in addition to periclase, forsterite (Mg_2SiO_4) crystallizes out in melts with $\lesssim 40\%$ SiO_2, and at 1830 K, in 40–60% SiO_2 melts, forsterite and enstatite ($MgSiO_3$) appear without periclase. Note the eutectic behavior of forsterite: pure forsterite melts at 2170 K, while in a mixture the melting point is lowered by 50 K. In more silica-rich melts silica (SiO_2) crystallizes as cristobalite and tridymite around 1770 K. Thus, depending on the composition and temperature, the melt may exist as one single liquid phase (high temperatures), or a number of different solid and/or liquid phases. Which solid phase is present depends upon the temperature and original composition of the cooling melt.

5.2.2 Magma

In the previous section we showed that the physics and chemistry of a cooling melt is very complex. As do binary

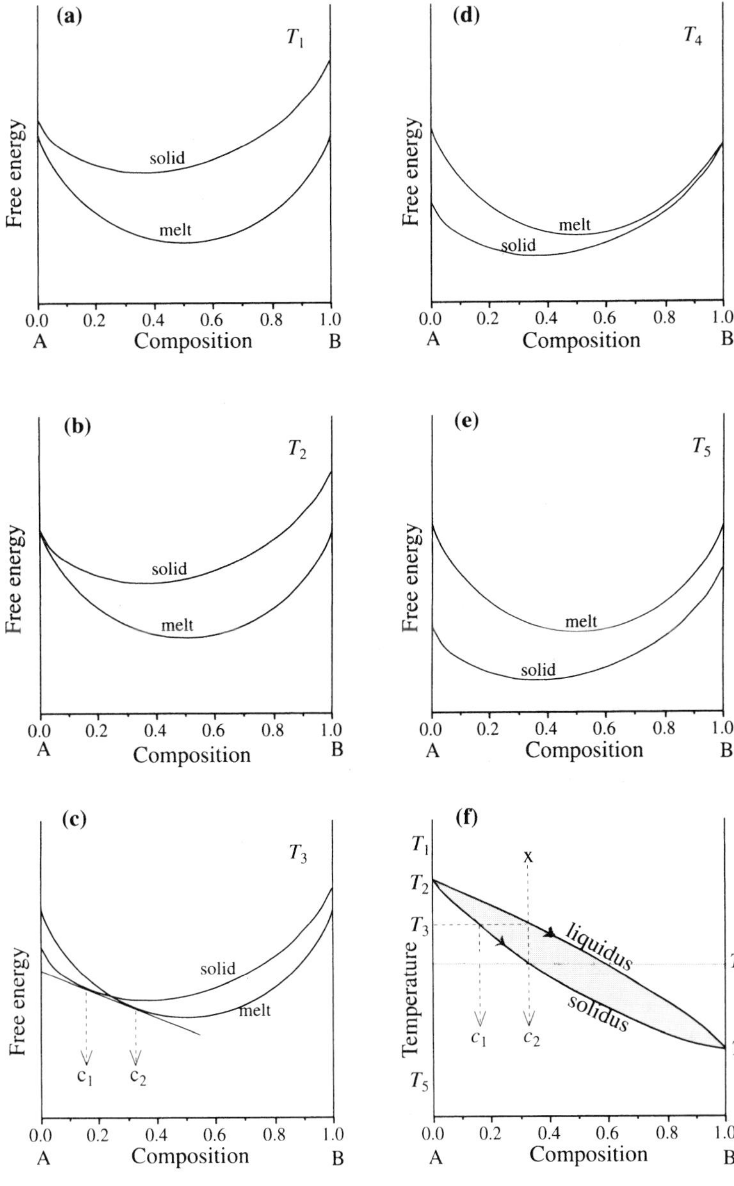

FIGURE 5.7 A sequence of free energy versus composition curves for a binary melt, at temperatures T_1 (highest) to T_5 (lowest). The resulting phase diagram is shown in panel f (see text for further explanation). (Adapted from Putnis 1992)

systems, magmas or molten 'rocks' crystallize over a large range of pressures and temperatures. Moreover, there are several solid solution mineral phases that crystallize over a range of conditions. Thus, while the melt is cooling, the crystallizing material as well as the magma is changing composition continuously. The minerals that crystallize from the magma, and the sequence in which they crystallize, depend on pressure, temperature and composition and how these vary as the system changes from a liquid to a solid phase. While the magma is cooling, existing crystals or nucleation seeds grow. The rate of growth is regulated, since the latent heat of crystallization warms the local environment. This heat has to be removed from the crystal to allow it to grow. On the other hand, if the temperature is reduced too quickly, the magma becomes very viscous and will not provide the crystal with enough material to continue to grow. Geologists study cooling magmas through analyses of igneous rocks and semi-molten lavas, and by conducting laboratory experiments on melting of rock and cooling of melts.

To complicate matters further, a magma usually does not cool under equilibrium conditions. The crystallized matter may not equilibrate with the magma, or heavy crystals may sink down in the magma, a process referred to

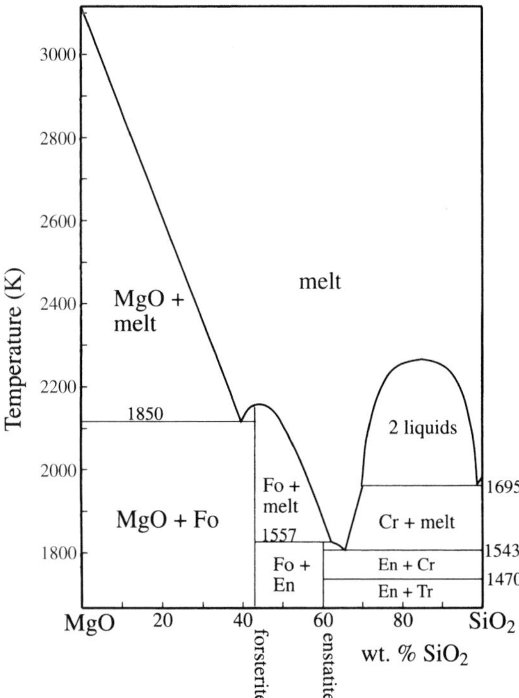

FIGURE 5.8 Phase diagram for the MgO–SiO_2 system, containing the phases periclase (MgO), forsterite (Mg_2SiO_4), enstatite ($MgSiO_3$) and silica (SiO_2) as cristobalite and tridymite. (Putnis 1992)

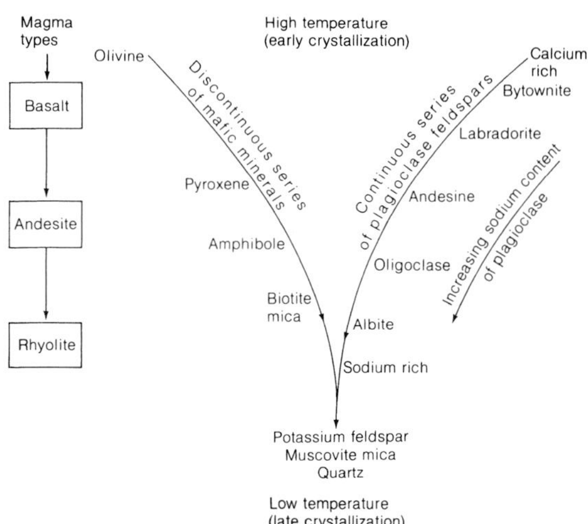

FIGURE 5.9 Bowen's reaction series of fractional crystallization and magmatic differentiation. The magma follows a double path of fractional crystallization: a discontinuous mafic series and a continuous series of plagioclase feldspars. (Press and Siever 1986)

as *differentiation*. In the latter case, the crystals are essentially removed from the melt. In the former case, if the crystals do not equilibrate, zoned crystals may form. Zoned crystals consist of a core rich in composition A, surrounded by a mantle (referred to as *rim*) which gradually grades to a composition richer and richer in B. An example is given by a plagioclase melt, where the mineral anorthite ($CaAl_2Si_2O_8$) has the higher melting temperature (like A in Fig. 5.7) than albite ($NaAlSi_3O_8$), resulting in plagioclase feldspars with calcium-rich cores and sodium-rich rims.

In 1928, Bowen derived from laboratory experiments a general scheme for the cooling of a magma with *fractional crystallization* and *magmatic differentiation*. His reaction series are summarized in Figure 5.9. Starting with a high temperature (ultra-)mafic magma, the first crystals which condense out are olivine crystals. Since they are heavy, they sink to the bottom of the magma chamber, and are thus removed from the melt through magmatic differentiation. We are left with a melt of basaltic composition. Upon further cooling, pyroxenes condense and differentiate out, leaving a melt of a more andesite composition. Then amphiboles and biotite micas appear, and the magma left behind is more and more silicic in composition. If the crystals had remained in the cooling melt, they may have reacted with the magma, and consequently been changed. For example, if the olivine crystals had not settled out, they would have been converted to pyroxenes through interaction with the cooling melt, in which case there would not be much olivine on Earth.

In the more silica-rich cooling melt, or when the above magma has become one of andesite composition, calcium-rich plagioclase (anorthite) condenses out. Upon further cooling, the crystals gradually become more and more sodium rich (albite). Since feldspars are relatively light, they tend to float upwards in the magma. The cooling magma becomes gradually more and more silicic, ending with granitic material (potassium feldspar, muscovite mica and quartz). At any time during the cooling process, if the melt reaches the surface, the resulting rocks have the composition of the melt 'frozen'. So, if the melt surfaces early on in the cooling process, basaltic rocks result. If the magma surfaces later during the cooling process, it contains relatively more silica.

Bowen's reaction series of fractional crystallization and magmatic differentiation is based upon laboratory experiments of melting and subsequent cooling of igneous rocks. The sequence of reactions taking place in a cooling magma in the Earth's mantle, however, is often far more complicated, and research is still going on. Usually there is partial rather than complete melting of rocks, and there is a wide range of temperatures even within one magma chamber. The differences in temperature may create chemical separation. While convective motions may mix magmas of different composition, some melts are immiscible, so that there may be melts with different compositions within one magma chamber. These melts each give rise to their own crystallization products.

In addition to the crystallization of a magma, it is at least as important for present day bodies to consider the melting process of materials. For example, on Earth in regions of subducting tectonic plates, the plates melt at a certain temperature and pressure. The presence of water lowers the melting temperature of some materials considerably. Water-rich silicic magmas may be produced by a remelting of the crust, which upon cooling may produce more granitic-type rocks.

5.3 Surface Morphology

The surfaces of planets, asteroids, moons and comets show distinct morphological features, such as mountain chains, volcanoes, craters, basins, (lava) lakes, canyons, faults, scarps, etc. Such features can result from *endogenic* (within the body itself) or *exogenic* (from outside) processes. In this section we summarize endogenic processes common on planetary bodies, with the resulting observable features. In Section 5.4 we discuss exogenic processes, in particular impact cratering. The last section of this chapter gives a brief summary of the features observed on a variety of Solar System bodies, and what these features tell us about the formation and evolution of these objects.

5.3.1 Gravity and Rotation

Gravity is ubiquitous. It wants to pull everything 'down', so that a nonrotating planet would be a perfect sphere, the equipotential of a stationary fluid body. On the largest scales the force which best competes with gravity is rotation, through the centripetal force. Polar flattening due to rotation is the cause of the largest deviation from sphericity for planet-sized bodies (cf. Section 6.1.4). The surface that is generated through the rotation of an ellipse about its minor axis is an equipotential surface called the *geoid* for Earth. The geoid can be approximated by an oblate spheroid, and is defined on Earth to be the mean sea level, which is the average of high and low tides. The precise figure of the geoid is discussed in more detail in Section 6.1.4.

The surface *topography* of a planet is measured with respect to the planet's geoid. Whether or not local structures on planetary surfaces survive the gravitational pull depends on the density and strength of the material. Small bodies, where the gravity field is weak, can maintain very nonspherical shapes (Figs. 9.6, 9.13, 9.15, and Figs. 5.60, 5.62h). Although downhill movements of material are induced by gravity, whether or not such movements occur depends on the steepness of the slope as compared to the *angle of repose*, which is the greatest slope that a particular material can support. The angle of repose depends mainly on friction. If one piles up sand in a sand box, the slope of the resulting hill is the same for small and large hills, but the slope is different if the mound is build out of fine sand, gravel, or pebbles. The angle of repose thus depends upon the type of material, the size and shapes of the 'granules', water and air content, and temperature. If the slope on a hill is steeper than the angle of repose, *mass movements* as landslides, mudflows or rockslides will occur. But even on slopes with angles less than the angle of repose, material can migrate downhill in *slumping* motions (e.g., landslides, avalanches), or as a slow continuous *creeping* motion (e.g., glaciers, lava flows). Such downhill migrations can be triggered by quakes (earthquakes on Earth, moonquakes on the Moon, etc.) caused by either internal or external processes. Precipitation and the presence of a fluid, such as water on Earth, also play a major role in downhill motions.

5.3.2 Tectonics

5.3.2.1 *Tectonic Features*

Any crustal deformation caused by motions of the surface, including those induced by stretching or compression of the crust, is referred to as *tectonic* activity. Many planetary bodies (terrestrial planets, most major satellites and asteroids) show evidence of crustal motions due to shrinking and/or expanding of the surface layers, commonly caused by heating or cooling of the crust aeons ago. Consider a forming planet as a hot, fluid ball of magma. The outer layers are in direct contact with cold outer space and thus cool off first by radiating the heat away, so that a thin crust forms over a hot magma. While the crust cools, it shrinks. Convection in the mantle may move 'hot plumes' around and heat the crust locally, leading to a local expansion of the crust. The interior cools off through convection and conduction, with volcanic outbursts at places where the crust is thin enough that the hot magma can burst through. The added weight of the magma on the crust may lead to local depressions, such as the lava-filled impact basins on the Moon, and 'corona' on Venus. Stretching and compressional forces on the crust result in folding and faulting of the rocks. Examples of common tectonic deformations are illustrated in Figure 5.10. *Folding* refers to an originally planar structure that has been bent. *Faulting* involves fractures. If the crust is cracked because it moved in response

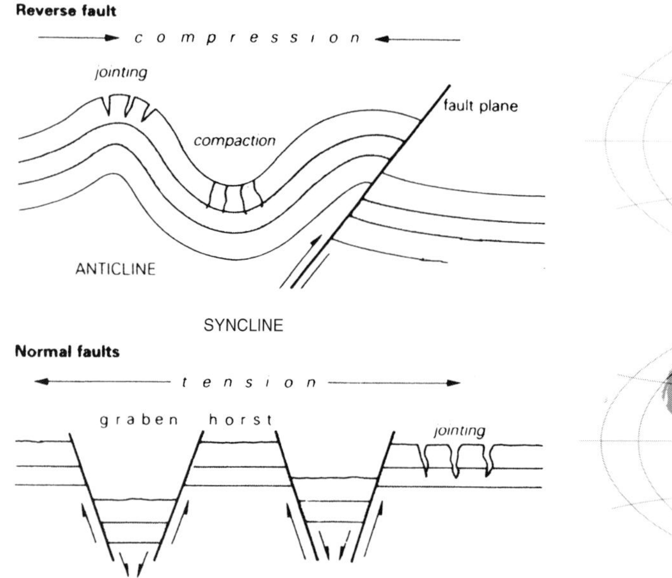

FIGURE 5.10 Cartoon showing common tectonic deformations such as faults, and the formation of grabens and horsts. (Greeley 1994)

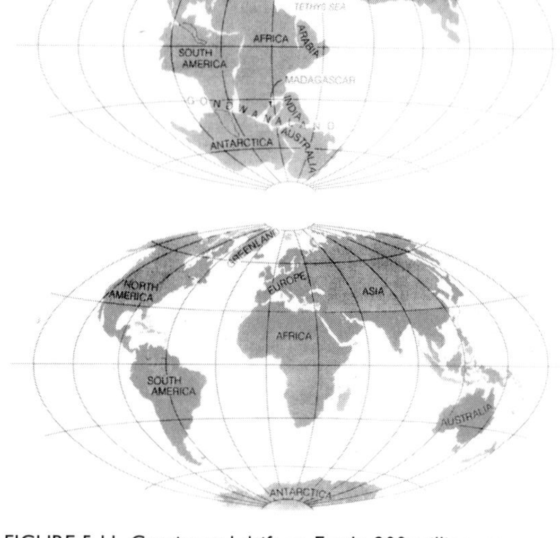

FIGURE 5.11 Continental drift on Earth: 200 million years ago the continents fit together as a 'jig-saw' puzzle, a supercontinent named Pangaea (top panel). (Press and Siever 1986)

to a compression or expansion of the crust, the cracks are called *faults*. The movement can be up and down as in *normal faults*, which result from tensional stresses, or in *reverse faults*, which result from compressional forces. In *strike-slip faults* the crustal motion is primarily in horizontal directions, such as seen when two tectonic plates slide alongside each other. If there is no crustal motion involved cracks are called *joints*.

Folding and faulting help shape a planet's surface, and the effects are visible as distinct geological features. Typical examples are *grabens* and *horsts*: A graben is an elongated fault block which has been lowered in elevation relative to surrounding blocks (Fig. 5.10), whereas a horst is a fault block which has been uplifted. *Scarps* are steep cliffs which can be produced by faulting or by erosion processes. On many bodies one has found *rilles*, which are elongated trenches either sinuous in shape or relatively linear, both of which are tectonic (from e.g., faulting) in origin. Folding and faulting processes can also lead to the formation of mountain ridges, such as found on many continents on Earth.

5.3.2.2 *Plate Tectonics*

A study of the shape and motions of the continents on Earth has lead to the concept of *plate tectonics*. The various continents seem to fit together as a 'jig-saw' puzzle (Fig. 5.11), and current theories suggest that roughly 200 Myr ago there was only one large landmass, *Pangaea*. Since that time the continents have moved away from each other, a process known as *continental drift*. This motion is induced by 'plate tectonics'. The lithosphere consists of about a dozen large plates, which 'float' on the asthenosphere. The plates move with respect to each other by a few, sometimes up to nearly 20, centimeters per year. The current motion of the plates can be measured either by using very-long baseline interferometry (VLBI) techniques, which use quasars as fixed radio sources, or by using the Global Positioning System (GPS), which is based upon a satellite laser-ranging technique. These techniques confirm the velocities as derived by geologic means from sedimentary layers on the ocean floor. The motion of the plates is caused by convection in the mantle, which gives rise to a large scale circulation pattern, where the plates 'ride' on top. Although the driving force for mantle convection and plate tectonics is still a hot topic of debate, it is clear that hot 'plumes' of magma rise upwards in the upper mantle and cold plates subduct at plate boundaries (Section 6.2.2.2). The end result is that plates recede from each other at the mid-oceanic rift (Fig. 5.12), where hot magma rises and fills the void. Through this process new ocean floor is created at the mid-oceanic rift, and the recession of plates is referred to as *sea floor spreading*. At other places plates bump into each other (compare a river full of logs), or slip past each other. These 'collisions' result in *earthquakes*. The San Andreas fault zone in California is a fault line or boundary where two plates slide past each

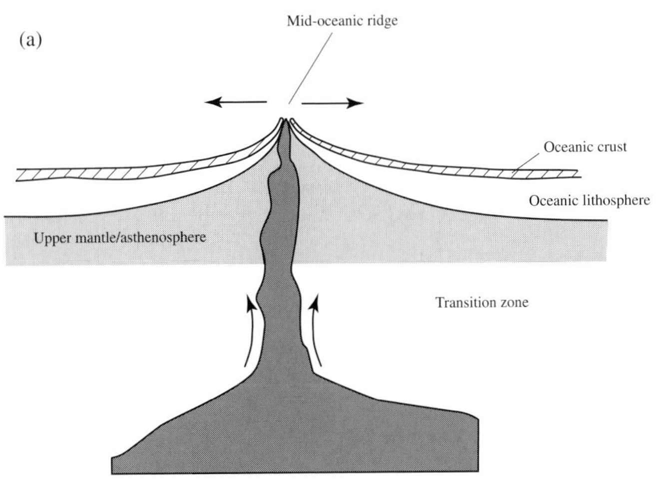

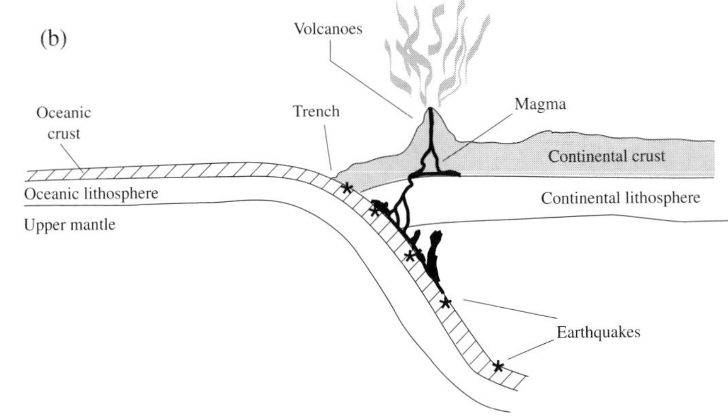

FIGURE 5.12 Schematic presentations of plate tectonics. (a) sea floor spreading: plates recede from each other at the mid-oceanic rift, where magma rises and fills the void. (b) Subduction zones: at convergent boundaries between two lithospheric plates, at least one of which is oceanic, the heavier oceanic plate is subducted. Volcanoes form near such subduction zones.

other, causing earthquakes. A similar earthquake prone fault line exists in Israel, where earthquakes have repeatedly destroyed old 'biblical' towns such as Bet She'an and Jericho.

At the mid-oceanic rift hot magma rises and fills the space between receding plates. As explained in Section 5.2, the resulting rocks are basaltic. All oceanic crust is made of dense basaltic material, in contrast to continental crust that is made of lighter (more granitic) material. Therefore, when an oceanic and continental plate collide or are pushed against each other, the oceanic plate may subduct, or dive under, the continental plate in *subduction zones*. Mountain ranges and volcanoes form at these intersections. Oceanic crust brought down is squeezed and heated, so that new metamorphic rocks may form, and at greater depths the crust melts. Because of the remelting of crust in a water-rich environment, solidification of the rising magma results in granitic rocks. Recycling of oceanic crust typically occurs on timescales of a few 10^8 years (Problem 5.5).

Plate tectonics is unique on Earth, and is not seen on any other body in our Solar System. The smaller planets, Mercury, Mars and the Moon, cooled off rapidly, and developed one thick lithospheric plate. Tectonic features on these planets involve primarily vertical movements, with features such as grabens and horsts. Venus shows evidence of local lateral tectonic movements, but not of plates. Based upon magnetic field measurements of Mars, it has been suggested that this planet may have had plate tectonics early in its history (Section 7.4.3).

5.3.3 Volcanism

Many planets and several satellites show signs of past volcanism, while a few bodies, in particular Earth and Io, are still active today (Figs. 5.13, 5.53). Volcanic outbursts can

(a)

(b)

(c)

FIGURE 5.13 Examples of volcanic activity on Earth: (a) Mount St. Helens in Washington State, on 10 April 1980. (Courtesy: USGS/Cascades Volcano Observatory) (b) Mount St. Helens after the big explosion, on 19 May 1982. (Courtesy: USGS/Cascades Volcano Observatory) (c) One of the numerous geysers in Yellowstone National Park, USA. (Courtesy: Wil van Breugel)

change a planet's surface drastically, both by covering up old features and by creating new ones. Volcanism can also affect the atmospheric composition, the weather and even create an atmosphere (Section 4.3.3.3). In this section we discuss what volcanic activity is, where we find it and how it changes the surface. We focus our discussion on Earth, one of the most volcanically active planets that has been studied in detail. In Section 5.5 the surface morphology of individual bodies in our Solar System is discussed and compared with Earth.

A prerequisite for volcanic activity is the presence of hot liquid material, magma, just below the crust. Possible sources of heat to melt rocks into magma include (Section 6.1.5): (*i*) Heat can be generated from accretion during the planet's formation and continuing differentiation of heavy and light material. (*ii*) Tidal interactions between various solid bodies can lead to substantial heating, such as is the case for Jupiter's moon Io. (*iii*) Radioactive nucleides form an important source of heat for all terrestrial planets.

Earth's upper mantle, or asthenosphere, consists of hot primarily unmolten rock under pressure, which behaves as a highly viscous fluid. The solid lithosphere and crust overlying the mantle can be compared to a lid on a pressure cooker or espresso machine. Magma formed within the hot rock, being less dense than the solid rocks surrounding it, rises up by buoyancy and is pushed out through any cracks or weakened structures in the surface. Volcanic activity on Earth is usually found at the boundary of two tectonic plates, and above hot thermal mantle 'plumes' at places where the crust is weak and the magma can break through. The chain of Hawaiian islands, for example, is formed where magma from a hot thermal plume breaks through the crustal plates.

As magma rises closer to the surface, volatile gases come out of solution when the pressure becomes low enough. These gases form bubbles, which grow as the pressure keeps dropping when the magma rises closer to the surface. The bubbles may create enough pressure near a vent that explosive eruptions result. Water vapor is the main constituent (70–95%) of volcanic gases on Earth. Other constituents are CO_2, SO_2, and traces of N, H, CO, S and Cl. Although sulfur is usually present in the form of SO_2, at lower temperatures one may also find H_2S. Sometimes volcanic gases yield metals, such as iron, copper, zinc, and mercury. Although water vapor is the main constituent of volcanic gases on Earth, its presence is not required for volcanic explosions. Vaporization of any subsurface liquid may lead to volcanism. The dominant volcanic gas on Io is sulfur dioxide, and many of the volcanic plumes are probably driven by the vaporization of liquid SO_2 (or S) in contact with molten sulfur (or silicates).

(a)

(b)

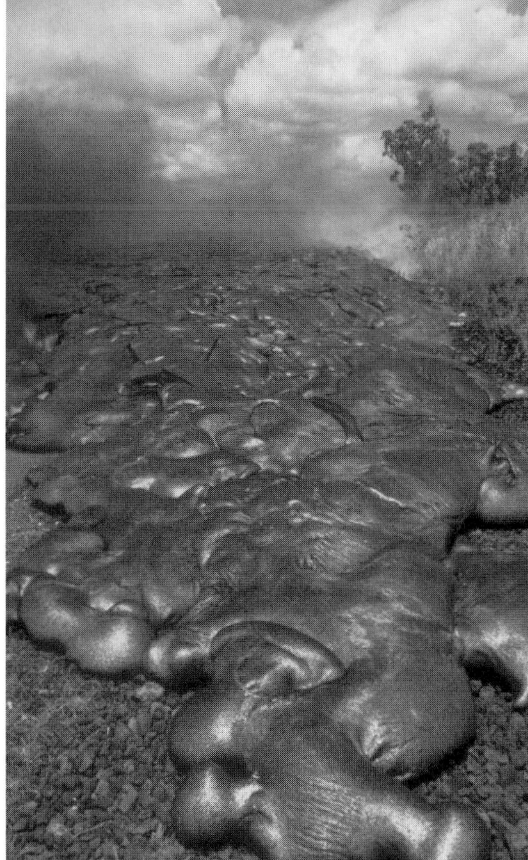

FIGURE 5.14 Lava flows on Hawaii: (a) Glowing 'a'a flow front advancing over pahoehoe on the coastal plain of Kilauea Volcano, Hawaii. (b) Toes of a pahoehoe flow advance across a road in Kalapana on the east rift zone of Kilauea Volcano, Hawaii. (Courtesy: USGS Volcano Hazards program)

Emissions of gas and vapor without the eruption of lava or pyroclastic matter often mark the last stages of volcanic activity. Vents that emit only gas and steam are referred to as *fumaroles*. Groundwater that is heated by the magma can

(a)

(b)

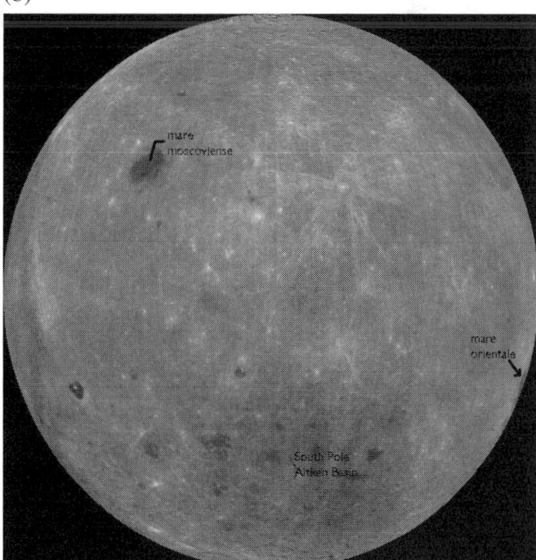

FIGURE 5.15 Images of the near (a) and far (b) side of the Moon taken by the Clementine spacecraft. (Courtesy: USGS)

produce *hot springs* and *geysers* (Fig. 5.13). Such springs and geysers are found in volcanic areas on Earth, such as Yellowstone National Park. They may occur on Europa, one of Jupiter's Galilean moons, and geysers of liquid nitrogen have been discovered on Triton, Neptune's largest moon (Fig. 5.64).

Lava erupting from volcanoes can be a basaltic melt of low viscosity, or a more felsic melt of high viscosity. Basaltic magma erupts at temperatures of 1250–1500 K, while the more silicic magma, such as rhyolite, erupts at lower temperatures, 1050–1250 K. Basaltic magma is very fluid, and can cover vast (many km) areas within hours, resulting in extensive lava flows, such as seen in Hawaii, on the Moon and the planets Venus and Mars. The basaltic flows in Hawaii are referred to as *'a'a* or *pahoehoe*. 'A'a is an Hawaiian word, which sounds like ah ah, outcries made when walking barefoot over this jagged form of lava. Pahoehoe is an Hawaiian word which means 'ropy'. A photograph for both lava flows is shown in Figure 5.14, where pahoehoe is the glistening smoother surface, in part covered by 'a'a, a very rough, jagged and broken lava flow. In contrast, the highly viscous rhyolitic lava flows very slowly, oozing out of a vent like toothpaste from a tube. Such volcanic events create *domes*, which may largely consist of obsidian. Such 'glass mountains' can, for example, be seen near Mono Lake and Mount Shasta in California.

Volcanic eruptions can emanate from long narrow fissures, or from a central vent or pipe. Fissure eruptions, such as occur along the oceanic ridge, cause large lava floods which can cover extensive areas. Such areas are known as *lava plains* or *plateaus*. A prime example of such plains are the lunar maria (Fig. 5.15). Localized eruptions give rise to the familiar volcanoes, often accompanied by large lava flows, which gush or ooze out and flow downhill. Volcanic activity can create many different features. *Cones*, in particular *cinder* and *spatter cones*, are cone-shaped hills up to a few hundred meters high, built up around vents that eject pyroclastic material as cinders, ash, and boulders. The profile of the cone is determined by the angle of repose. A *shield volcano* is a gently sloping volcanic mountain, built by low viscosity lava flowing out from a central vent (Fig. 5.16). These volcanoes may be very large; the largest known shield volcano is Olympus Mons on Mars, with a height of ~ 25 km and base diameter ~ 600 km. The largest shield volcano on Earth, Mauna Loa on Hawaii, measures about 9 km from the top to the bottom of the ocean floor and has a base diameter of ~ 100 km. *Composite cones* are built when a volcano emits (alternately) lava as well as pyroclasts. This is the most common type of large continental volcanoes, such as Vesuvius, Mount Etna and Mount St. Helens. Domes are produced by felsic lavas, lavas which are so viscous that they can barely flow.

Craters are found on the summits of most volcanoes. The craters are centered over the vent, and are produced when the central area collapses as the pressure that caused the eruption dissipates. Since the original crater walls are steep, they usually cave in after an eruption, enlarging the crater to several times the vent diameter. Craters can be hundreds of meters deep. A volcano in Italy, Mount Etna, has a central vent which is 300 m in diameter and over 850

(a)

(b)

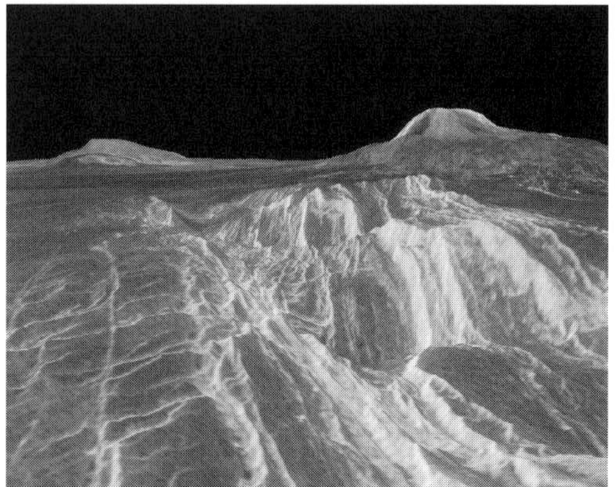

FIGURE 5.16 (a) The largest shield volcano in our Solar System: Olympus Mons on the planet Mars. (NASA/Mars Global Surveyor) (b) This three-dimensional, computer-generated view of the surface of Venus shows a portion of the western Eistla Regio. Gula Mons, on the right horizon, reaches 3 km, and Sif Mons, the volcano on the left horizon, has a diameter of 300 km and a height of 2 km. The image was produced from Magellan radar imaging and altimetry data. (NASA/Magellan PIA0200)

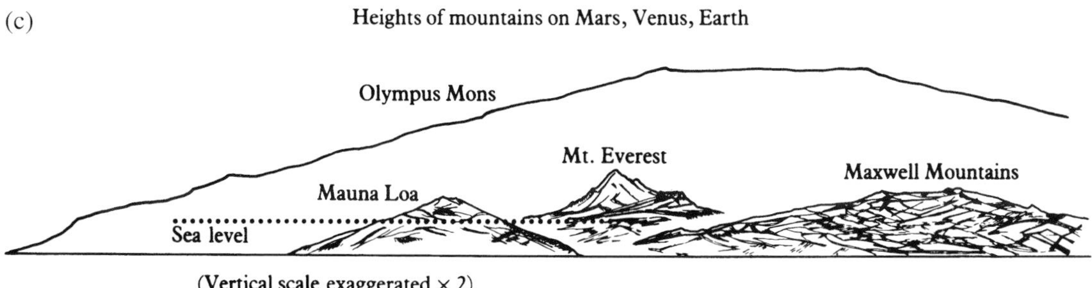

FIGURE 5.16 Continued. (c) A comparison of the the volcanoes/mountains on Mars (Olympus Mons), Earth (Mauna Loa and Mount Everest) and Venus (Maxwell Mountains). (Morrison and Owen 1996)

FIGURE 5.17 The Aniakchak Caldera in Alaska, USA, is an example of a resurgent volcanic caldera. The caldera formed about 3450 years ago; it is 10 km in diameter and 500–1000 m deep. Subsequent eruptions formed domes, cinder cones and explosion pits on the caldera floor. The picture is 125 × 160 km. (Courtesy: M. Williams, National Park Service 1977)

m deep. Small craters are termed *pit craters* ($\lesssim$ 1 km) or, if even smaller, *collapse depressions*. *Calderas* are large basin-shaped volcanic depressions, varying in size from a few kilometers up to 50 km in diameter. Such volcanic depressions are caused by collapse of the underlying magma chamber. After a volcanic eruption, the magma chamber is empty, causing the roof, i.e., crater floor, to collapse. Over time the crater walls erode, and lakes may form within the depression. Many years later (10^5–10^6 yrs) new magma may enter the chamber, push up the crater floor, and the entire process may start again. Many calderas show evidence of multiple eruptions. Examples of such *resurgent calderas* are the Yellowstone caldera in Wyoming, the Valles caldera in Mexico, and Crater Lake in Oregon (Fig. 5.17). After an explosion, when the lava flow cools and contracts, shrinkage cracks may form. Sometimes, when the source of a lava flow is cut off, and the outer layers have solidified, the lava is drained out, and *lava tubes* or *caves* result. Such caves are found, for example, in Hawaii and northern California. A sinuous rille, the Hadley Rille in Mare Imbrium on the Moon (Fig. 5.18), might be a lava tube where the roof has collapsed. Such volcanic structures, characterized as steep-sided troughs, are also named *lava channels*.

FIGURE 5.18 Hadley Rille, a typical sinuous rille on the Moon, close to the base of the Apennine mountains. The rille starts at a small volcanic crater and 'flows' downhill. The picture is 125 × 160 km. (NASA/Lunar Orbiter IV-102H3)

The characteristics and morphology of volcanic features are determined by the viscosity of the magma, its temperature, density and composition, the planet's gravity, lithospheric pressure and strength, and the presence and properties of the atmosphere. Just on Earth alone we recognize a wide variety of volcanic features, but our inventory may grow as we explore volcanism on other planets, moons and asteroids.

5.3.4 Atmospheric Effects on Landscape Morphology

The presence of a substantial atmosphere has profound effects on the landscape of a planetary body. If the atmospheric pressure and temperature at the surface are high enough for liquids, such as water, to exist, there may be oceans, rivers and precipitation, which will modify the landscape through both mechanical and chemical interactions. In 'dry' areas, e.g., currently on Mars and in deserts on Earth, winds displace dust grains and cause rocks to erode. Over time, these processes 'level' a planet's topography: high areas are gradually worn down, and low areas filled in. This process is called *gradation*, and material is displaced by *mass wasting*. The main driving force for gradation is gravity (Sections 5.3.1, 6.1.4). Although gradation and mass wasting occur on each body, the presence of an atmosphere and liquids on the surface enhance these processes and give rise to particular surface features, as discussed in more detail below. In addition to mass wasting, there are numerous chemical interactions between the crust and the atmosphere. These differ from planet to planet, depending on atmospheric and crustal composition, temperature and pressure, and the presence of life. Note that a massive atmosphere also protects a planet's surface from impacting debris, especially small and/or fragile projectiles, as well as from cosmic rays and ionizing photons.

In this section we summarize the most common morphological features on Earth as caused by water and wind, which serve as a basis for comparative studies with other planets. More details on the geological processes can be found in e.g., Press and Siever (1998).

5.3.4.1 *Water*

The atmospheric temperature and pressure on Earth is close to the triple point of water, so that water exists as a vapor, a liquid and as ice. Although at present Earth is unique in this respect, water may have flowed freely over Mars's surface during the first 0.5–2 Gyr after its formation. Europa is thought to have a large water ocean under its ice crust, and Titan may be covered by oceans and/or lakes of liquid hydrocarbons. All flows, whether water, hydrocarbons or lava, are downhill at a velocity determined by the flow's viscosity, the terrain and the planet's gravity. In addition to the liquid itself, the flow also transports solid materials, such as sediments eroded from rocks. The faster the river flows, the larger the particles or rocks it can transport. The largest particles usually stay close to the bottom, and may roll and slide over the stream bottom, while the finest particles (clay in the case of water flows) may be suspended throughout the flow. The finest particles are carried the farthest, a sifting process which produces the various sedimentary rocks described in Section 5.1.2.

Flow patterns contain information on the local topography and the surface characteristics of the underlying rocks. In analogy to Greeley (1994), we show a summary of Howard's (1967) classification scheme of drainage patterns on Earth in Figure 5.19. The various patterns are all associated with specific terrain. Dendritic patterns indicate gentle slopes, where water trickles down the small channels and accumulates in the larger rivers. Radial patterns are associated with dome-like features, often volcanic in origin. Annular features are associated with domes or basins. A comparison of flow features on other planets with Howard's classification scheme yields clues to the origin of the flow patterns and the local surface topography.

In addition to running water on a planet's surface, groundwater just below the surface may leave profound marks as well. Some rocks are dissolved in water (e.g., limestone, gypsum, salt), which leads to *karst topography*, which may show its appearance in the form of various sized sinkholes, solution valleys, or as haystacks or pinnacles. Seeping up of groundwater could also lead to certain types of drainage patterns as shown in Figure 5.19. Finally, dry lake beds or *playas* are associated with former lakes, swamps or oceans, while sea cliffs and beaches mark the shore lines of existing oceans.

The temperature on most bodies in our Solar System is well below freezing, which makes ice an important constituent of planetary surfaces. Although water-ice is the dominant form of ice throughout the Solar System, in the far outer reaches of the planetary system ices of methane, ammonia, or carbon dioxide may have similar roles as water-ice on Earth. For example, carbon dioxide on Mars freezes out above the winter pole, while it sublimes during the summer. Water-ice forms a permanent ice cap on Mars's north pole, while water vapor freezes near this planet's equator at night, subliming again during the day. The conditions are just right for methane-ice to exist on Titan's surface, while nitrogen-ice forms during the winter on Pluto and Triton. Ice may be considered a 'pseudo-plastic fluid' which moves downhill. On Earth we find *valley glaciers* and *ice sheets*, as well as the morphological features of past glaciation, such as U-shaped valleys, grooves and striations parallel to the flow, and amphitheater-shaped *cirques* at the head of the flow. Usually the ice contains dust and rocks, which are left behind when the ice melts or sublimes. The dust and rocks

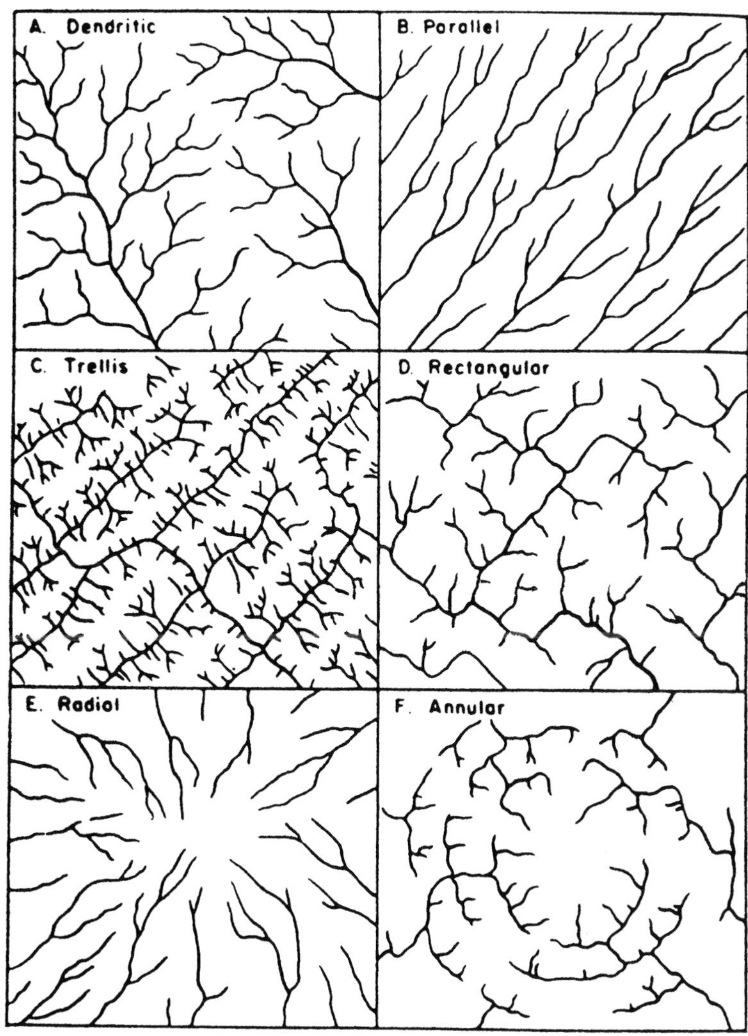

FIGURE 5.19 Short summary of Howard's (1967) classification scheme of drainage patterns on Earth. Dendritic: gentle regional slope at the time of drainage. Parallel: Moderate to steep slopes. Trellis: areas of parallel fractures. Rectangular: joints and/or faults at right angles. Radial: volcanoes, domes and residual erosion features. Annular: Structural domes and basins. (Adapted from Howard 1967)

may be deposited, carried away by melt water or blown away by winds. The morphology of the deposits contains information on the glaciers, and hence the surface topography and past climate.

5.3.4.2 *Winds*

Most planets with atmospheres show the effects of *aeolian* or wind processes. On Earth, such processes are most pronounced in desert and coastal areas. The winds transport material. The smallest particles, such as clay and silt ($\lesssim$ 60 μm) are *suspended* in the atmosphere. Larger dust and sand grains ($\sim$ 60–2000 μm) are transported via *saltation*, an intermittent 'jumping' and 'bounding' motion of the dust grains. Still larger grains are transported via *surface creep*, where particles are rolled or pushed over the ground. The amount of dust that the winds can displace depends upon the atmospheric density, viscosity, temperature and surface composition and roughness. When the wind blows over a large sandy area, it first ripples the surface and then builds dunes. Because the turbulence created by the wind increases with increasing surface roughness, the winds get stronger during this process (positive feedback). The amount of dust that winds can transport depends on the atmospheric density and wind strength. On Earth, half a ton of sand can be moved per day over a meter-wide strip of sand if the wind blows at 48 km per hour; the amount of sand transported by stronger winds increases more rapidly than the increase in wind speed. In large terrestrial dust storms, one cubic kilometer of air may carry up to 1000 tons of dust, and thus many millions of tons of dust can be suspended in the air if the dust storm covers hundreds or thousands of square kilometers. On planets with a low-density atmosphere, the winds need to be much stronger to transport material. For example the winds on Mars need to be about an order of magnitude stronger than on Earth to transport the same amount of material.

Winds both erode and shape the land. They erode the land through removal of loose particulates, thereby lowering or *deflating* the surface. When winds are loaded with sand, they can wear away and shape rocks through *sandblasting*. This causes erosion and rounding of rocks. The best known example of a wind-blown landform is the *dune* (Fig. 5.20a, b). Any obstacle to the wind, such as a large rock, can start the formation of a dune, where sand grains are deposited at the lee-side of the obstacle. The shape of dunes can be used to determine the local wind patterns. Dunes are found in desert and coastal areas, where winds are strong and there is an abundance of particulate material. Dunes have also been found on the planets Mars and Venus. On the latter planets, like on Earth, the presence of winds is also seen from *wind streaks*; an example of a wind streak on Venus is shown in Figure 5.20c.

5.3.4.3 *Chemical Reactions*

The interaction between a planet's atmosphere and its surface can lead to *weathering*, a process which depends upon the composition of both the atmosphere and surface rocks. Weathering on Earth is usually a two-fold process, consisting of *mechanical weathering* or fragmentation of rocks, together with *chemical weathering* or decay of the rock fragments. Many iron silicates, such as pyroxene, weather or oxidize slowly through the interaction with oxygen and water and get a rusty-iron color. Hydration is a more general process on planetary surfaces, since it only requires the presence of water. Hydrated minerals have been detected on several asteroids. Minerals, such as feldspar, partially dissolve when in contact with water and leave behind a layer of clay. Calcite and some mafic minerals may completely dissolve away. In the desert the products of chemical weathering, such as silica, calcium carbonate and iron oxides, may make up a hard surface crust, called *duricrust*. The Viking spacecraft noticed such a flaky crusty material on the surface of Mars, which probably formed from weathered iron silicates. Another type of chemical weathering is the slow evaporation of volatile material from rocks, leaving behind evaporates like salt, gypsum and borax (Section 5.1.2.4).

The presence of life may have profound effects on the surface morphology of a planet, as we know all too well from our own planet, Earth. Plants cover large fractions of continental crust, changing the albedo (even varying with season), atmospheric and soil composition as well as the local climate. Plants add matter to the soil and influence erosion. Mankind, of course, has large effects on the surface morphology, e.g., through building and mining projects, and through their influence on the atmospheric composition. Since these effects are (at present) only applicable to Earth, we will not discuss them in this book. Note, however, that even micro-organisms change the atmospheric and soil composition (locally), e.g., through metabolism.

5.4 Impact Cratering

Impact cratering involves the essentially instantaneous transfer of energy from the impactor to the target. If the target has a substantial atmosphere, such as on Earth, the object is seen as a fireball prior to impact and referred to as a *bolide*. The energy involved in the collision is the impactor's energy of relative motion at 'infinity' plus its potential energy, i.e., the target's gravitational escape energy, less energy lost as the impactor passes through the planet's atmosphere, if any. Typical impact velocities of large meteoroids (which are not significantly slowed by the atmosphere) on Earth are 10–40 km s^{-1}, although long-periodic comets may impact at speeds up to 73 km s^{-1} (Problem 8.1). Thus an average nickel–iron meteorite, 30 m in diameter, would impart an energy a few times 10^{23} ergs (Problem 5.8), or the equivalent of several million tons of TNT. We can compare this energy with the Richter magnitude scale for earthquakes, $\mathcal{M}_R$:

$$\log_{10} E = 12.24 + 1.44\mathcal{M}_R. \tag{5.17}$$

An impact of 10^{23} ergs would correspond to an earthquake of magnitude 7.7 on the Richter scale. This is a large earthquake. The collisions can thus be very energetic, such that the impacting body can create a hole (crater) much larger than its own size. Meteor Crater in Northern Arizona (Fig. 5.21), with a diameter of about 1 km and 200 m deep, probably formed within one minute by an impacting 30 m nickel–iron meteorite.

Impact craters are produced on all bodies in the Solar System that have a solid surface, thus the terrestrial planets, asteroids, satellites, and comets. They are the dominant landform on geologically inactive bodies without substantial atmospheres, which includes most small bodies. We only have to look at the Moon through a modest telescope or a good pair of binoculars to see that the object is covered by craters, which were formed by meteorites which hit the Moon over the past 4.4 Gyr. Earth has been subjected to a somewhat higher (due to gravitational focusing) flux of impacts, but most craters have disappeared on our planet as a result of plate tectonics and erosion.

(a)

(b)

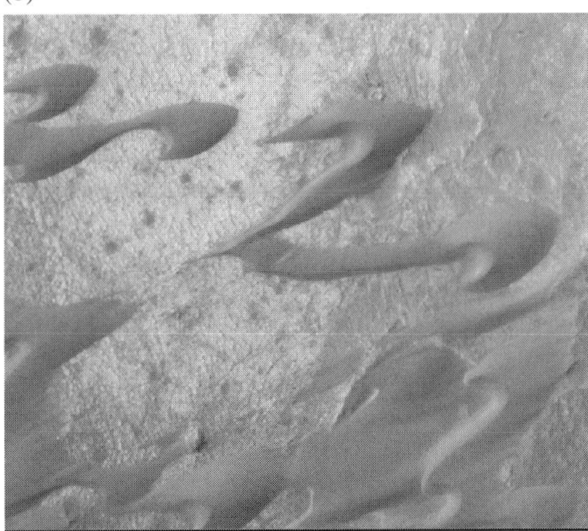

(c)

FIGURE 5.20 (a) Small barchan sand dune on Earth, in Peru. Prevailing wind is from the left to the right. The dune is moving over a surface of dark coarse granule ripples. The light sinuous streaks on the desert surface are the slip faces of small ripples, where light, finer grained sand is temporarily trapped. (USGS Interagency Report: Astrogeology 57, 'The Desert Landforms of Peru; A Preliminary Photographic Atlas'; February 1974) (b) Aerial photograph of the dark sand dunes of Nili Patera, Syrtis Major, on Mars. The shape of the dunes indicates that the wind has been steadily transporting dark sand from the right/upper right towards the lower left. The extent of the picture is 2.1 km. (NASA/Mars Global Surveyor MOC2-88) (c) Magellan radar image of the 30 km diameter Adivar crater and surrounding terrain on Venus. Crater ejecta appear bright due to the presence of rough fractured rock. A much broader area has been affected by the impact, particularly to the west of the crater. Radar-bright materials, including a jet-like streak just west of the crater, extend for over 500 km across the surrounding plains. A darker streak, in a horseshoe or paraboloidal shape, surrounds the bright area. These unusual streaks, seen only on Venus, are believed to result from the interaction of crater materials (the meteoroid, ejecta, or both) and high-speed winds in the upper atmosphere. The precise mechanism that produces the streaks is poorly understood. (NASA/Magellan PIA00083)

Clearly, impacts are an important aspect of planetary geology. Impact cratering has therefore been studied using astronomical and geological data on the crater morphology seen on our planet, on the Moon and on more remote Solar System objects. In addition, laboratory experiments of hypersonic (faster than the speed of sound) impacts, explosions (conventional and nuclear), as well as numerical computer simulations are of great value in studies concerning impacts and impact craters. Finally, impact theories could, in part, be tested and refined after astronomers witnessed a series of large impacts of comet D/Shoemaker–Levy 9 with Jupiter, an event discussed in Section 5.4.5.

5.4.1 Crater Morphology

Craters can be 'grouped' according to their morphology into four classes:

(1) *Microcraters* or *pits* (Fig. 5.22) are sub-centimeter scale craters caused by impacts of micrometeorites or high velocity cosmic dust grains on rocky surfaces.

(a)

(b)

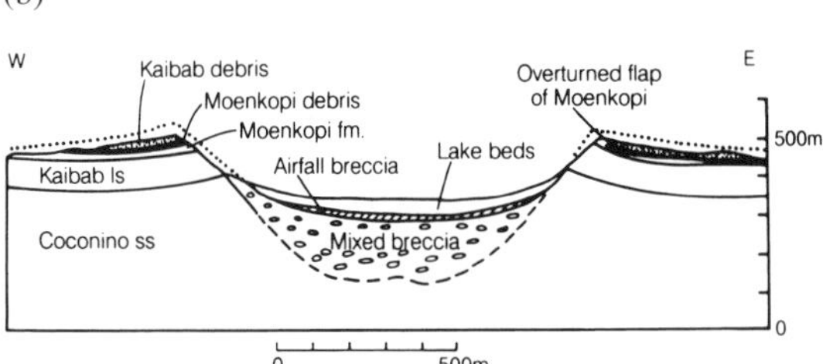

FIGURE 5.21 (a) Meteor crater in Arizona. The crater has a diameter of 1 km and is 200 m deep. (Courtesy: D. Roddy, USGS/NASA) (b) A cross-section through Meteor crater. (Melosh 1989, as derived from Shoemaker 1960)

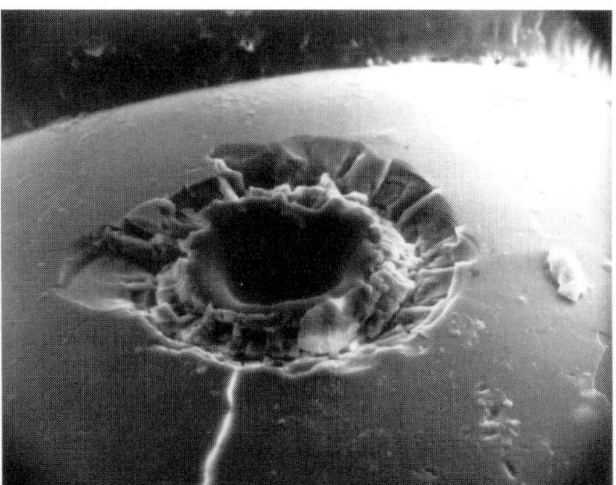

FIGURE 5.22 Microcrater or pit with a diameter of 30 μm. This is a scanning electron microprobe photograph of a glass sphere on the Moon, found by Apollo 11. (Courtesy: D. McKay, NASA S70-18264)

FIGURE 5.23 A small or simple crater: this is a photograph of the 2.5-km diameter crater Linné in western Mare Serenitatis on the Moon. (NASA/Apollo panoramic photo AS15-9353)

FIGURE 5.24 A close-up of the 75-km diameter complex lunar crater King, characterized by a relatively flat crater floor, a central peak and terraced walls. (NASA/Apollo Hasselblad photo AS16-122-19580)

Pits are only found on airless bodies. The central hole is often covered with glass.

(2) Small or *simple craters* (Fig. 5.23), roughly up to a few kilometers across, are bowl shaped. The depth (bottom to rim) of a simple crater is typically 1/5 its diameter, although variations do occur, depending on the strength of the surface material and the surface gravity.

(3) Large craters are more complex. They usually have a flat floor, and a central peak, while the inside of the rim is characterized by terraces (Fig. 5.24). *Complex craters* have diameters of a few tens up to a few hundred km. The transition size between small and large craters is ~18 km on the Moon and scales inversely with the gravitational acceleration, g_p, although it also depends on the strength of the target's surface material. Craters with dimensions between 100 and 300 km on the Moon, Mars and Mercury show a concentric ring of peaks, rather than a single central peak. The inner ring diameter is typically half the rim-to-rim diameter. The crater size at which the central peak is replaced by a peak-ring

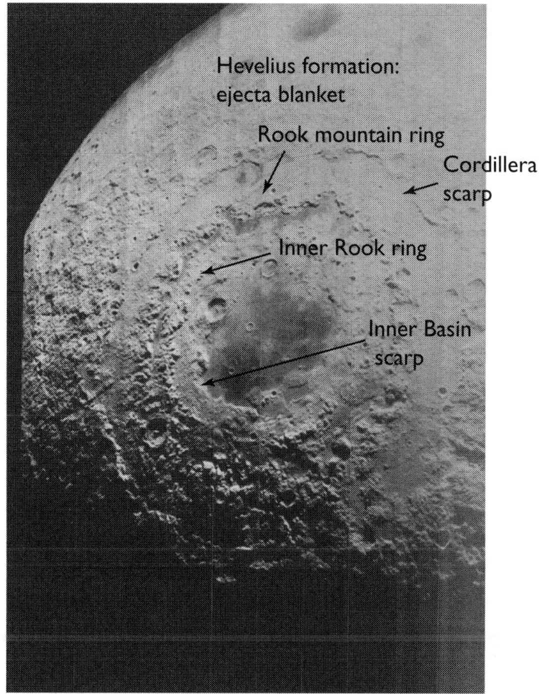

FIGURE 5.25 A photograph of the lunar multiring basin Mare Orientale. (NASA/Lunar Orbiter IV 194 M)

scales in the same way as the transition diameter between small and complex craters. There are no peak-ring craters on icy satellites.

(4) *Multiring basins* (Fig. 5.25) are systems of concentric rings, which cover a much larger area than the complex craters mentioned above. The inner rings often consist of hills in a rough circle, and the crater floor may be partly flooded by lava. The outer rings more clearly resemble crater rims.

5.4.2 Crater Formation

The formation of a crater is a rapid sequence of phenomena, which starts when the impactor first hits the target and ends when the last debris around the crater has fallen down. It helps to understand the process by identifying three stages: an impact event starts with the *contact and compression stage*, is followed by an *ejection* or *excavation stage* and ends with a *collapse and modification stage*. These three stages are sketched in Figure 5.26, and are discussed below for an airless planet. The influence of an atmosphere is summarized in Section 5.4.3, and the impact of Comet Shoemaker–Levy 9 with Jupiter is discussed in Section 5.4.5. More details on impact cratering can be found in the monograph by Melosh (1989).

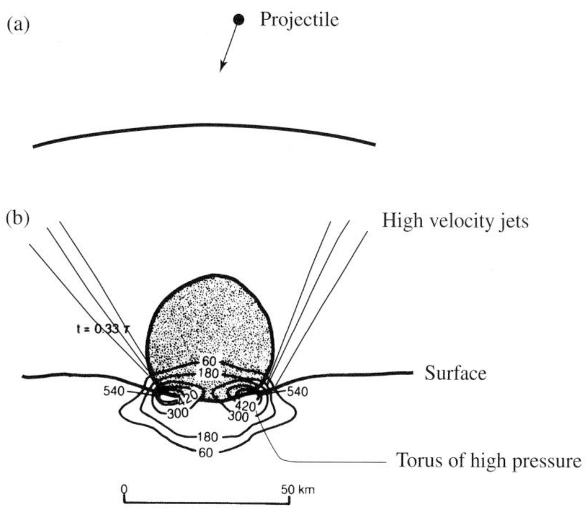

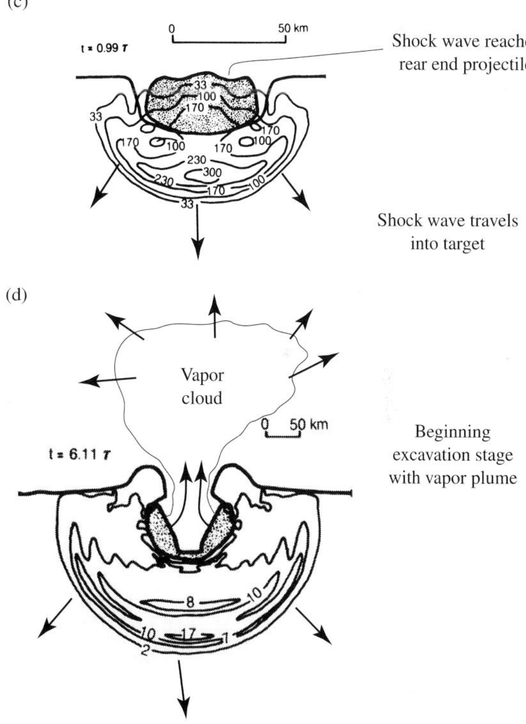

FIGURE 5.26 Schematic presentation of a hypervelocity impact, with times subsequent to impact given in units of τ, the ratio of the diameter of the projectile and its velocity. (a) Projectile on its way to the target. (b) A torus of extra-high pressure is centered on the circle of contact between the projectile and the target (perpendicular impact). Heavily shocked material squirts or jets outwards at velocities of many km s^{-1}. (c) Shock wave propagating into the target and projectile. The latter wave has reached the rear end of the projectile. The projectile will melt or vaporize (depending on the initial pressure) when decompressed by rarefaction waves. (d) Beginning of the excavation stage, preceded by the vapor plume leaving the impact site.

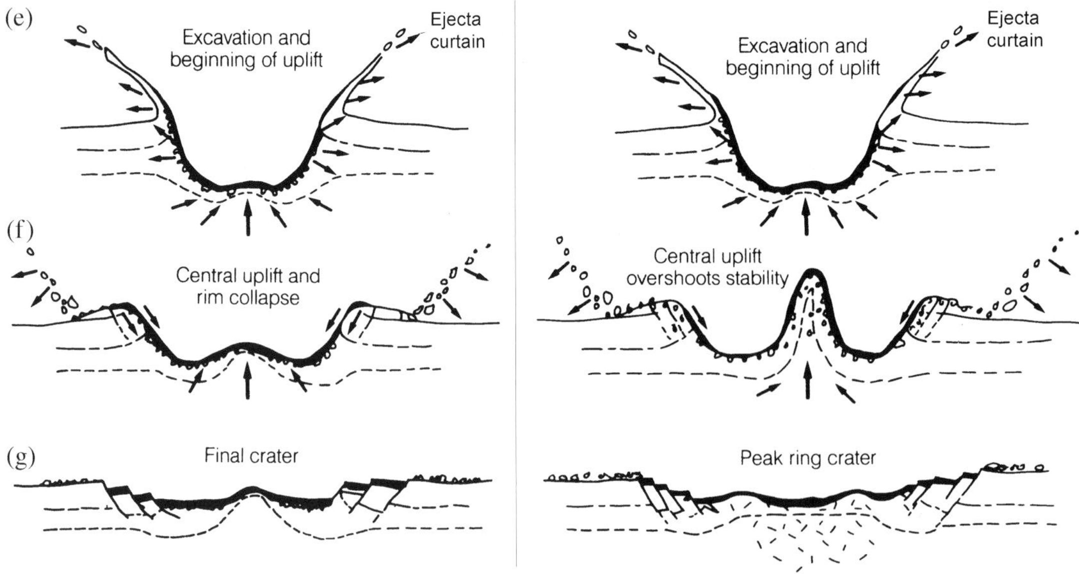

FIGURE 5.26 Continued. (e)–(g): Schematic illustration of the formation of central peaks (left side) and peak rings (right panels). (Adapted from Melosh 1989)

5.4.2.1 *Contact and Compression Stage*

Upon collision of a meteoroid with a planet, the relative kinetic energy is transferred to the body in the form of shock waves, one of which propagates into the planet, and another one into the projectile. The impact velocity of a typical rocky meteorite ($\rho = 3$ g cm^{-3}) with a planet like the Earth is of the order of 10 km s^{-1} (equal to or larger than the escape velocity from the planet). Since the velocity of seismic waves is only a few km s^{-1}, the impact velocity is hypersonic. The propagation of shock waves can be modeled numerically using the *Hugoniot equations*, relating the density, velocity, pressure and energy of the material across the shock front, augmented by an equation of state. The Hugoniot equations are derived from the conservation of mass, momentum and energy, and can be written:

$$\rho(v - v_p) = \rho_0 v, \tag{5.18a}$$

$$P - P_0 = \rho_0 v_p v, \tag{5.18b}$$

$$E - E_0 = (P + P_0)(V_0 - V)/2, \tag{5.18c}$$

where ρ and ρ_0 are the compressed and uncompressed densities, with V and V_0 the compressed and uncompressed specific (per unit mass) volumes, respectively; P_0 and P are the pressures in front of and behind the shock; v is the shock velocity, and v_p the particle velocity behind the shock, and E_0 and E the internal energies per unit mass in front of and behind the shock, respectively. This situation is sketched schematically in Figure 5.27.

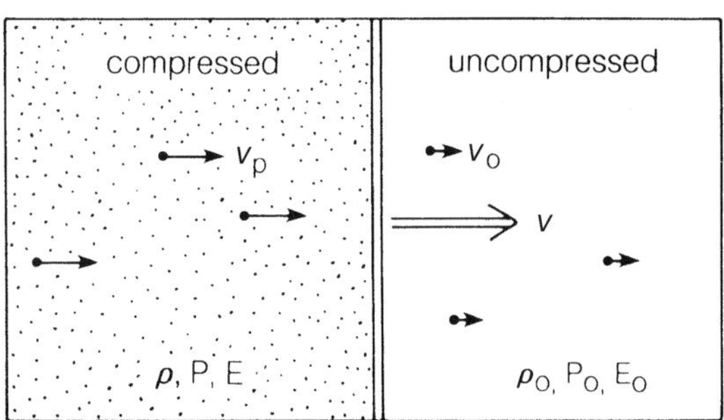

FIGURE 5.27 A sketch of a medium traversed by a shock front. The various quantities in the compressed and uncompressed medium are indicated (eq. 5.18). (Melosh 1989)

The pressure involved in a collision of a meteorite with Earth can be derived from the Hugoniot equations, and it follows that in low velocity impacts:

$$P \approx \frac{1}{2}\rho_0 c_s v, \tag{5.19a}$$

with c_s the sound velocity:

$$c_s = \sqrt{\frac{K_m}{\rho_0}}, \tag{5.20}$$

where K_m is the bulk modulus of material (see Section 6.2.1.1). In high velocity impacts:

$$P \approx \frac{1}{2}\rho_0 v^2. \tag{5.19b}$$

Rocks are thus easily compressed to pressures well over a few Mbar. Most rocks vaporize when suddenly decompressed from pressures exceeding ~ 600 kbar. The shock waves originate at the point of first contact, and compress the target and projectile material to extremely high pressures. A hemisphere of extra-high pressure (Fig. 5.26b) is centered on the point of contact between the projectile and target. As soon as this heavily shocked mixture of target and projectile material is decompressed by rarefaction waves, it squirts or *jets* outwards at velocities of many km s^{-1}. This jetting occurs nearly instantaneously with the projectile hitting the target, and is often finished by the time the projectile is fully compressed. The shock, propagating hemispherically into the target, can be detected as a seismic wave (Section 6.2).

The geometry of the shock wave system is modified by the presence of free surfaces on the target and the meteorite, i.e., the outer surface of the material which is in contact with air or the interplanetary medium. Free surfaces cannot sustain a state of stress, and therefore *rarefaction*, or release, waves develop behind the shock wave. When the shock wave in the projectile reaches its rear surface, a rarefaction wave is reflected from the surface. This rarefaction wave travels at the speed of sound through the shocked projectile, thereby decompressing the material to near-zero pressure. Since the initial pressure of the shocked material can be very high, the projectile may nearly completely melt or vaporize upon decompression, and shortly after the passage of the rarefaction wave leave the crater as a *vapor plume* or *fireball*. The duration of the contact and compression phase lasts as long as it takes the shock wave and subsequent rarefaction wave to traverse the projectile, which is typically 1–100 ms for meteoroids with sizes between 10 m and 1 km (Problem 5.9).

5.4.2.2 *Ejection or Excavation Stage*

As soon as the rarefaction wave from the rear of the projectile releases the high pressure, the projectile and target region vaporize if the initial pressure was high enough. The vapor plume or fireball expands adiabatically upward and outward, where a parcel of gas at distance r is accelerated as:

$$\frac{d^2r}{dt^2} = -\frac{1}{\rho_g}\frac{dP}{dr}, \tag{5.21}$$

with ρ_g the gas density. Meanwhile the shock wave propagates into the target, while it expands and therefore weakens. The shock wave gradually degrades to a stress wave, propagating at the local speed of sound. The rarefaction waves behind the shock decompress the material, and initiate a subsonic excavation flow, which opens up the crater.

The excavation of material may last for several minutes, depending on the energies involved. A rough estimate of the time it takes to excavate the crater can be obtained from the period of a gravity wave with a wavelength equal to the crater diameter, D (for craters whose excavation is dominated by gravity, i.e., craters larger than a few km):

$$t = \left(\frac{D}{g_p}\right)^{1/2}. \tag{5.22}$$

Material down to $\sim 1/3$ the depth of the transient cavity is excavated. Target material below the excavation depth is pushed downwards, whereas the strata above this depth are bent upwards, and are either excavated or lifted upwards to form the crater walls or *rim*. The rim height on relatively small craters (diameter $\lesssim 15$ km on the Moon) is typically $\sim 4\%$ of the crater diameter:

$$h_{rim} \approx 0.04D \quad \text{(small craters)}, \tag{5.23a}$$

and the depth (bottom to rim):

$$d_{br} \approx \frac{D}{5} \quad \text{(small craters)}. \tag{5.23b}$$

For larger craters these relationships break down, as gravity causes various morphological changes including rim and crater collapse.

Rocks and debris excavated from the crater are ejected at velocities which are much lower than those of the initial 'jets' of fluid-like material during the compression stage. The ejecta are thrown up and out along ballistic, nearly parabolic, trajectories (Fig. 5.28c; see also Section 2.1):

$$r = \frac{v_{ej}^2}{g_p}\sin 2\theta, \tag{5.24}$$

where $r \ll R$ is the distance (R denoting the planet's radius), v_{ej} is the ejection velocity, g_p the gravitational acceleration and θ the ejection angle with respect to the ground. The rock is in the air (or space) during a time t:

$$t = \frac{2v_{ej}}{g_p}\sin\theta. \tag{5.25}$$

The rarefaction waves cause material to move approximately upwards, while the original material velocities are directed radially away from the impact site. The excavation flow, therefore, forms an outward expanding *ejecta curtain*, which has the form of an inverted cone, whose sides make an angle of $\sim 45°$ with the target surface (Fig. 5.28). The ejecta velocities are highest (up to several km s^{-1}) early in the excavation process, thus near the impact site. More typical velocities during the process are in the 100 m s^{-1} range. The sides of the crater continue

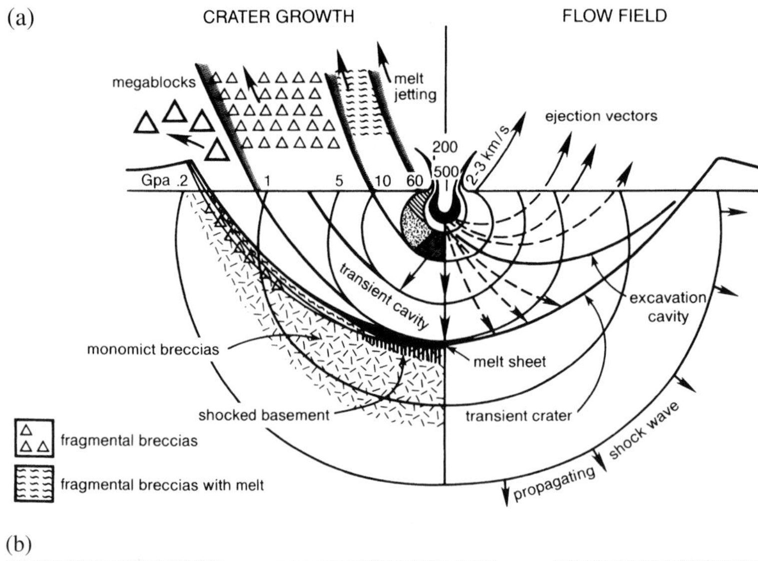

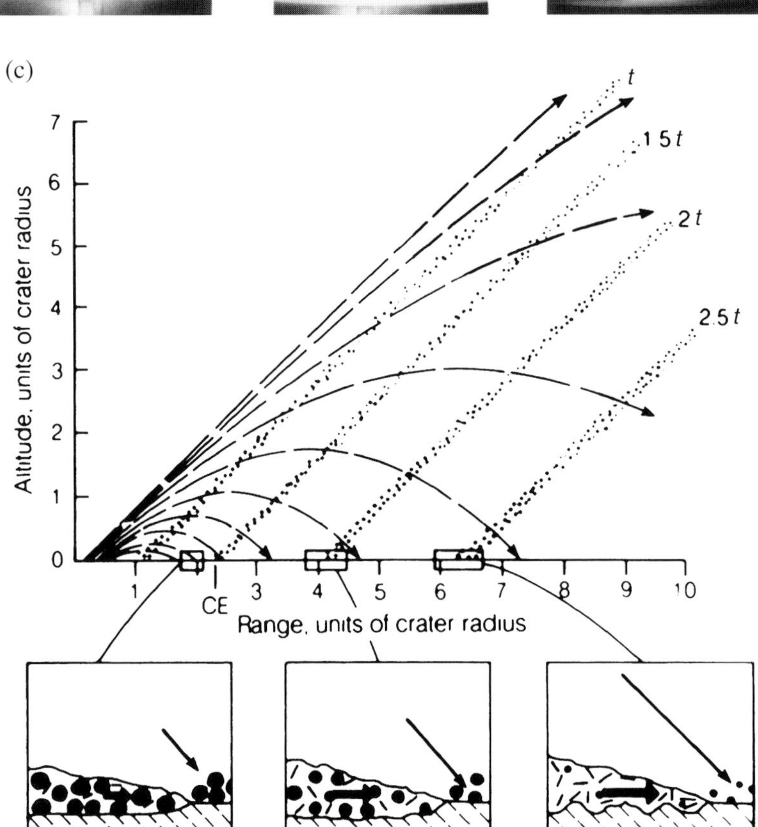

FIGURE 5.28 (a) Schematic cross-section through a growing impact crater before crater modification takes place. The left side shows various regimes of the melt zone and the formation of breccias. The right side shows the particles' flow field. (Taylor 1992) (b) The excavation flow forms an outward expanding ejecta curtain, which has the form of an inverted cone, whose sides make an angle of 45° with the target surface. The excavation flow in a large planetary impact is believed to be very similar to that from this small impact in a laboratory. (Courtesy: P.H. Schultz) (c) The trajectories (ballistic) of debris ejected from the crater at times 1, 1.5, 2 and 2.5 t, with t the crater formation time. (Melosh 1989)

to expand until all of the impact energy is dissipated by viscosity and/or carried off by the ejecta. The resulting crater is many times larger than the projectile which produced it. The crater is nearly hemispherical until the maximum depth is reached, after which it only grows horizontally. The ejecta form an *ejecta blanket* around the crater (Figs. 5.25, 5.26, 5.28), up to one or two crater radii from the rim, covering up the old surface. When the ejecta fall down, the outward momentum is retained, so that the material skids along horizontally before it stops, changing the morphology of the surface. The morphology of the ejecta blanket is defined by the subsurface material. On the Moon and most other bodies the ejecta blanket is formed by the material falling down. On Mars some ejecta blankets look morphologically more like mudflows, suggestive of a liquid substance, caused by, e.g., the presence of subsurface ice which liquefies when heated by an impact.

Some of the excavated rocks may, when they hit the surface, create *secondary craters*. Because the ejecta move along ballistic trajectories, the secondaries are closer to the primary crater on more massive planets (eq. 5.24; e.g., Problems 5.10, 5.11). The size of the secondary craters depends on the mass and impact velocity of the ejecta, as well as on the target material. Because the impact velocity of the ejecta is lower than that of the original impactors, the morphology of secondary craters is somewhat different, but there is an excavation of material from the newly produced craters. This newly excavated material might interact with neighboring cratering impacts, which leads to a 'herring-bone' pattern on the surface, or chevron-shaped dunes, between craters. The secondary craters are usually seen outside the ejecta blanket from the primary crater, and may be found many crater radii away from the primary crater. Relatively young craters on the Moon ($\lesssim 10^9$ yr) display bright *rays* emanating outwards from the primary. These rays may extend over a large surface area (Fig. 5.29), at least 10 times the crater radius from the primary. Many, but not all, secondary craters are associated with rays. The rays are bright because they are largely made up of material pulverized and/or melted by the impact. Over time the rays disappear, probably due to micrometeorite impacts which overturn the near-surface regolith.

At the end of the excavation stage, the crater is referred to as the *transient crater*. The shape of this crater depends on the meteroid's size, speed, composition, angle at which it struck, the planet's gravity and the material and structure of the surface in which the crater formed. The crater dimension scales approximately with the meteoroid's kinetic energy:

$$D \propto E^{1/3}. \tag{5.26a}$$

A more general scaling law, derived empirically, is given by (in mks units):

$$D = 1.8\rho_m^{0.11}\rho_p^{-1/3}g_p^{-0.22}(2R)^{0.13}E_k^{0.22}(\sin\theta)^{1/3}, \tag{5.26b}$$

where ρ_p and ρ_m are the densities of the planet (= target) and the meteoroid, respectively, R the projectile radius, E_k the impact (kinetic) energy, and θ the angle of impact from the local horizontal. It follows that a typical crater on Earth is roughly 10 times larger than the size of the impacting meteoroid.

5.4.2.3 *Crater Collapse and Modification*

After all of the material has been excavated, the crater is modified by geological processes induced by the planet's gravity, which tends to pull excess mass down, and by the relaxation of compressed material in the crater floor. The shape of the final crater depends on the original morphology of the crater, its size, the planet's gravity and the material involved. The four basic morphological classes of craters were summarized in Section 5.4.1. The transition between simple and complex craters is inversely proportional to the planet's gravity, and depends upon e.g., the material strength, melting point, and viscosity.

Complex craters are characterized by central peaks and terraced rims. For craters with sizes between about 15 and 80 km on the Moon, the height of the central peak, h_{cp}, typically increases with crater diameter, D, as:

$$h_{cp} \approx 0.0006D^2. \tag{5.27}$$

For larger craters, the central peak on the Moon tops out at about 3 km, so the central peak is usually lower than the rim (eq. 5.23*a*). The central peak width is approximately 20% of the crater diameter. In craters over 140 km in diameter, a ring of mountains develops in the crater, which replaces the central peak. This ring develops about halfway between the center and crater rim. In some craters both a central peak and peak ring are seen. The transition from craters with a central peak to peak rings varies from planet to planet in proportion to $1/g_p$. Sometimes complex craters are seen with a central *pit* rather than a peak. These are seen on Ganymede and Callisto for craters over 16 km in diameter. The formation of central pits might be caused by the unique properties of icy surfaces.

Shortly after the excavation process ends, the remaining debris in the crater moves downwards and back towards the center, while the crater floor undergoes a rebound of the compressed rocks, which probably leads to the formation of the central peak or mountain rings. The formation of the central peak might be analogous to the impact of a droplet into a fluid, shown in Fig. 5.30. After the inital crater has formed (panel d), the central peak

(a)

FIGURE 5.29 (a) Rays from the young lunar crater Tycho. The rays are visible over nearly an entire hemisphere. (Courtesy: UCO/Lick Observatory) (b) Basaltic rock from the lunar mare brought to Earth by Apollo 15. This rock, sample 15016, crystallized 3.3 billion years ago. The numerous vesicles (bubbles) were formed by gas that had been dissolved in the basaltic magma before it erupted. (NASA/Johnson) (c) A breccia from the lunar highlands, sample 67015, collected by Apollo 16 astronauts. This rock is termed *polymict* because it contains numerous fragments of pre-existing rocks, some of which were themselves breccias. It was compressed into a coherent rock about 4.0 billion years ago. (Left panel: NASA/Johnson; right panel (cut section): Courtesy: Paul Spudis)

(b)

(c)

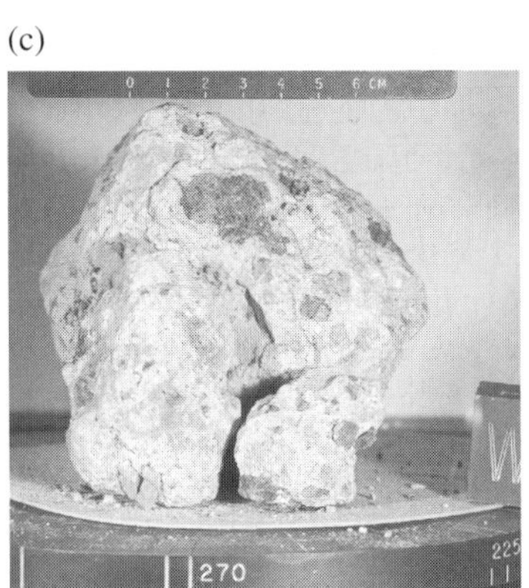

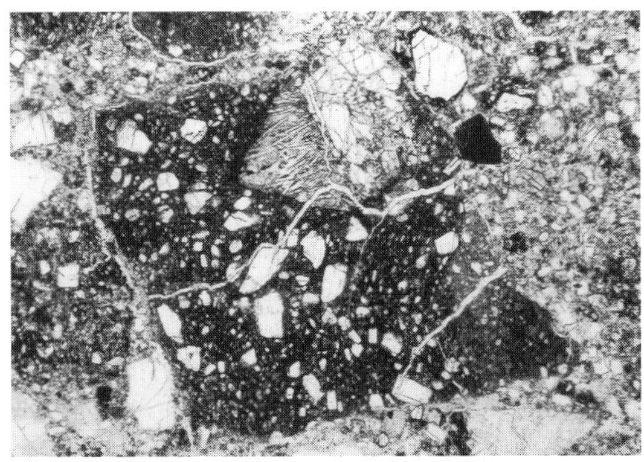

comes up (panel e) and grows (panels e–g). Collapse of the peak (panel h) triggers the formation of a second concentric ring (the first being the crater rim), which propagates outwards. The possibly analogous process on a solid surface is sketched in Fig. 5.26. The rebound starts before the crater has been completely excavated, and the central up-

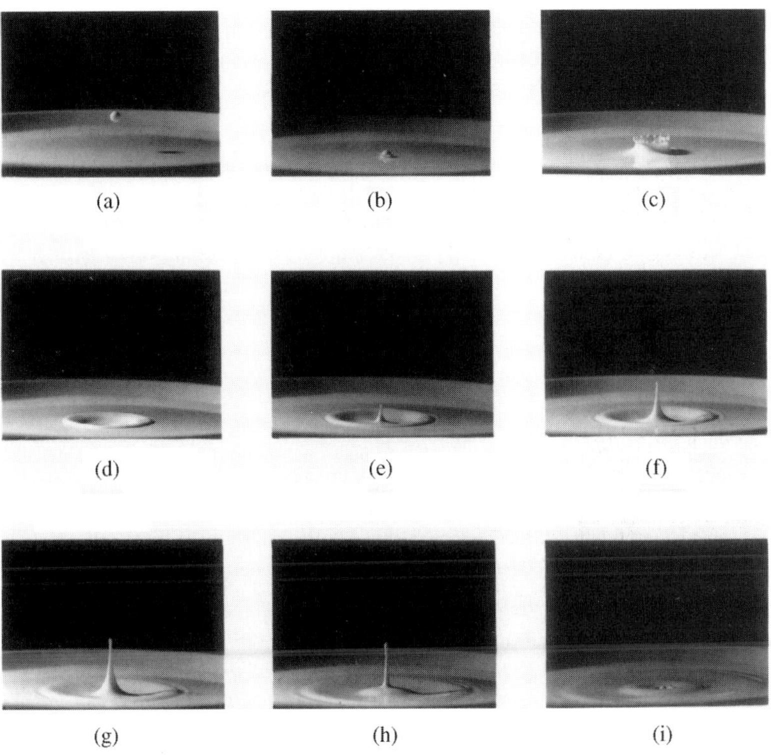

FIGURE 5.30 A series of photographs from the impact of a milk drop into a 50/50 mixture of milk and cream. An ejecta curtain forms immediately after impact (panel c), which creates the 'crater' wall. After the initial 'crater' forms (panel d), the central peak appears (panel e) and grows (panels e–g). Collapse of the peak (panel h) triggers the formation of a second, outward expanding ring (panel i). (Courtesy: R.B. Baldwin; photographs by Gene Wentworth of Honeywell Photograph Products)

lift 'freezes' to form the central peak. For larger craters the peak becomes too high and collapses, triggering an outward propagating ring, which 'freezes' into a ring of mountains. Although the details of this process are not completely understood, the peak ring in this process must form before the crater material soldifies.

As soon as all of the material has been excavated, the crater rim collapses or *slumps*, thereby increasing the crater diameter, filling in the floor of the crater, and shaping the wall in the form of terraces. Observations indicate that these terraces form before the rock, which acts as a liquid because of the 'shaking' from the impact, has 'solidified'. The entire collapse process typically takes several minutes.

The Orientale basin on the Moon (Fig. 5.25) is the youngest and best preserved multiring basin. The center of the basin is surrounded by four rings: the Inner Basin Ring, with a diameter $D \approx 320$ km; the Inner Rook Mountains ($D \approx 480$ km); the Rook Mountains ($D \approx 620$ km), and the Cordillera Mountains ($D \approx 920$ km). There may be an additional mountain range with $D \approx 1300$ km. The Rook and Cordillera Mountains are about 6 km high. The Rook Mountains may form the original crater rim, although the rim may also have been located interior to this ring. Rings appear to form outside the original crater rim in multiring basins, whereas peak rings in complex craters always form inside the crater rim. Moreover, peak rings are symmetric on the inside and outside slopes, whereas the outer rings on multiring basins are asymmetric, with steep inward-facing scarps and gentle backslopes. The ratio of the radii of successive rings is about $\sqrt{2}$, although no consensus has been reached about the validity or importance of this number. Multiring basins on other planets and satellites are similar, but different in details. The Valhalla structure on Callisto appears to have hundreds of rings, and the steep scarps face outwards rather than inwards. No universally accepted theory has been put forward to explain the formation of multiring basins. One possibility is that the rings form from a ripple effect, just like the peak rings in complex craters. Another theory which might account for the observational phenomena seen on different planets is the *ring tectonic theory*. This theory is based upon studies of crater collapse in a layered medium, where the material strength decreases with depth into the planet, such as for a planet where lithospheric plates overlie a fluid-like medium, like the Earth's asthenosphere. This theory is explained in detail by Melosh (1989).

Further modification of craters happens on long timescales, i.e., months–years–aeons. Erosion and micrometeorite impacts slowly erode the rim away, and smooth out or flatten the crater. On Earth, the lifetime against erosion of a 1 km impact crater is estimated to be only a million years or less. Isostatic adjustments (Section 6.1.4.2) may be important in large craters, where the crater floor may be uplifted to account for the deficiency in mass caused by the excavation of the crater. On icy satellites, craters are flattened or slowly disappear by plastic-like ice flows. Many large craters on Ganymede and Callisto may have disappeared and left vague discolored circular patches on the surface, called *palimpsests*. Volcanism and tectonic forces within the general area of the crater can modify a crater at a (much) later stage. Many impact basins on the Moon seem to have been flooded at some time. If material is excavated from the crater, for example by volcanism, the pressure on the underlying rocks decreases, which reduces the melting temperature of these rocks. The more buoyant 'magma' may then force its way through cracks or thin parts of the surface and flood the basin.

Analysis of craters on other bodies provides information regarding material on and below the surface of that body. The most direct examples are the central peak and ejecta blankets around the craters, which were formed from material originally below the surface. The number, shape and size of craters also yield information regarding the surface composition/materials and the impacting bodies, as discussed further below and in Section 5.5.

5.4.2.4 *Regolith*

Large impacts may fracture the crust down to many (20–30 km) kilometers below the surface, while small (down to submillimeter and micrometer-sized) impacts affect only the upper few millimeters to centimeters of the crust. The cumulative effect of meteoritic bombardment over millions or billions of years pulverizes the bedrock so that a thick layer of rubble and dust is created on airless bodies. This layer is called *regolith* (the Greek word for rocky layer), also termed 'soil' in a terrestrial analogue, though regolith does not necessarily contain organic matter. Most planetary bodies are covered by a thick layer of regolith. Since meteorites display a steep size distribution, the rate of 'gardening' or turnover decreases rapidly with depth on a planet. With present day meteoritic bombardment rates, within the past 10^6 years half of the regolith on the Moon has been overturned to a depth of 1 cm, while in most locations the uppermost millimeter has been turned over several tens of times. The thickness of the regolith depends upon the age of the underlying bedrock. In the ~ 3.5 Gyr old lunar maria the regolith is typically several meters thick, while it is well over ten meters deep in the 4.4 Gyr old lunar highlands. The ejecta from larger impacts have formed a 2–3 km thick *mega-regolith*, which consists of (many) meter-sized boulders. The regolith may be bonded at depth due to the higher pressure and temperature.

5.4.2.5 *Summary: Cratering on Airless Bodies*

The following features are recognized in connection with cratering events:

- Primary crater: the crater formed upon impact.
- Ejecta blanket: debris ejected from the crater up to roughly one crater diameter beyond the rim. The appearance of the ejecta blanket depends upon the subsurface properties of the target. For example, on the Moon we find boulders, but on Mars some ejecta blankets display evidence of fluid motions.
- Secondary impact craters and crater chains: these are caused by impacts of high velocity chunks of rock ejected from the primary crater.
- Rays: bright linear features 'radiating' outwards from the impact crater. Rays are formed by powdered and resolidified molten material, which was thrown out at high speed (several km s^{-1} on the Moon). They extend about 10 crater diameters out. Secondary craters may be clustered around the rays.
- Breccia and melt glasses: high temperature and pressure minerals, melt glasses and breccias (rocks of broken fragments cemented together) form and line the inside of a crater.
- Regolith: the rocks that were broken or ground down by (micro-)meteorites and secondary impacts.
- Focusing effects: large impacts can produce surface and body waves of sufficient amplitude that they may propagate through the entire planet. In planets with a seismic low velocity core the waves are 'focused' at the antipode (the point directly opposite the impact site). If the impact is very large, the waves still have enough energy to substantially modify terrain. This effect is, for example, clearly seen on Mercury. On icy bodies such focusing is not seen, since these bodies likely possess high rather than low seismic velocity cores.
- Erosion and disruption: sufficiently energetic impacts can substantially erode or even catastrophically disrupt the target. The energy required for such a disruptive impact is estimated to be comparable to the energy required to produce a crater whose diameter is about equal to the diameter of the target.

5.4.3 Impact Modification by Atmospheres

Hitherto we discussed impacts on airless bodies. If the target is enveloped by a dense atmosphere, such as Earth and Venus, impacts may be modified extensively. Projectiles may be completely vaporized while plunging through the atmosphere, and never hit the ground. Or the projectile might break up into many pieces. Atmospheric drag can slow down a small meteoroid, so that it merely hits the surface at the terminal velocity. Larger bodies can explode in the air, never creating an impact crater. These processes are described quantitatively in Section 8.3. In this section we consider the effects of atmospheric transit on the cratering process.

Not only does the atmosphere affect the projectile, the projectile can noticeably perturb the atmosphere during and after the explosion. For instance, consider the interaction (dynamically, chemically) with the upward expanding vapor cloud and the ejecta, where dust grains may be immersed in the atmosphere for a long time.

5.4.3.1 *Projectile's Flight Through the Atmosphere*

A meteorite is slowed down by atmospheric drag, where the aerodynamic force is proportional to the atmospheric ram pressure ($0.5\rho v^2$), and this drag may be substantial compared to a meteorite's internal strength. Meteors, therefore, often break up and fall down in clusters. A body that is broken apart by ram pressure is stopped very rapidly in the atmosphere, although such a cluster of bodies can also hit the ground and produce either a crater or a *strewn field*. The effect of such a break-up is usually reflected in the crater morphology.

The atmospheric mass displaced by the meteoroid traveling through the atmosphere is proportional to the body's cross-sectional area. The meteor is slowed substantially when it encounters an atmospheric mass about equal to its own mass. A meteor will only hit the ground in a hypervelocity impact if its radius, R, is larger than:

$$R \gtrsim 8 \times 10^6 \frac{P}{\rho g_p \sin\theta}, \tag{5.28}$$

where P is the atmospheric surface pressure (in bars) and ρ the density of the meteor. $P/(g_p \sin\theta)$ is the atmospheric mass per unit area displaced during the meteor's fall through the atmosphere, where the trajectory makes an angle θ with the surface. Smaller bodies will essentially be stopped above the ground. It follows that the radius of a typical stony meteorite must be over 30 m to hit the ground in a hypersonic impact (Problem 5.16). Meteors with diameters less than a few tens of meters can reach the ground, but at subsonic speeds. Micrometeors, with diameters ~10–100 μm, are slowed substantially in the upper atmosphere, where the mean free path length between molecules exceeds the diameter of the micrometeorite. On Earth, meteoroids which hit the Earth at speeds up to ~100 m s^{-1} produce a small hole or pit equal in size to the diameter of the meteoroid itself. At speeds of ~ 0.5 km s^{-1}, the hole produced is larger than that of the impacting meteoroid, while impacts at supersonic speeds produce craters as discussed in Section 5.4.2.

When a meteor plunges through the atmosphere at supersonic speeds, a bow shock forms in front of it and gases are considerably compressed. The shock waves of such meteors, even for projectiles which do not hit the surface, can be devastating and leave obvious marks. In 1908 Tunguska, Siberia, is thought to have been hit by a ~100 m sized meteor, which did not reach the ground. The shock wave flattened about 2000 km^2 of forest. On Venus, the Magellan spacecraft detected several radar-dark features, which were sometimes connected to impact events. Such features probably result from the blast waves from meteors which never hit the ground.

5.4.3.2 *Fireball*

As mentioned in the previous sections, immediately after a high velocity impact a hot plume of gases, the *fireball*, leaves the impact site, preceded by a shock wave in the lower atmosphere. The fireball expands adiabatically, and its radius can be calculated from its initial pressure, P_i, and volume, V_i, assuming PV^γ = constant, where γ is the ratio of specific heats. Since the vapor is hotter, and therefore less dense than its surroundings, the fireball rises, driven by buoyancy forces. It dissipates in the stratosphere. Fine dust brought up in the plume or by ejecta may stay suspended in the Earth's atmosphere for many months to a year, which may have profound climatic consequences through blocking of sunlight and trapping of infrared heat (Section 5.4.6). If the impactor exceeds one scale height in size (i.e., ~10 km on Earth; Section 4.8.3), a large fraction of the shock-heated air may be blown off into space.

5.4.4 Spatial Density of Craters

Although most airless bodies show evidence of impact cratering, the crater density varies substantially from object to object. Some surfaces, like portions of the Moon, Mercury and Callisto, appear *saturated* with craters, that is the craters are so closely packed that, on average, each additional impact obliterates an existing crater. The surface has reached a 'steady state'. Other bodies, like Io and Europa, have very few, if any, impact craters. Why is there such a large range in crater densities? The variations must be caused by the combined effect of impact frequency and crater removal, two topics which are discussed below.

5.4.4.1 *Cratering Rate*

The surface of the Moon shows the cumulative effect of impact cratering since the time that the Moon's crust solidified. The size–frequency distribution of lunar craters is shown in Figure 5.31a. For the Moon, we can date some surface areas precisely using radioisotope dating (Section

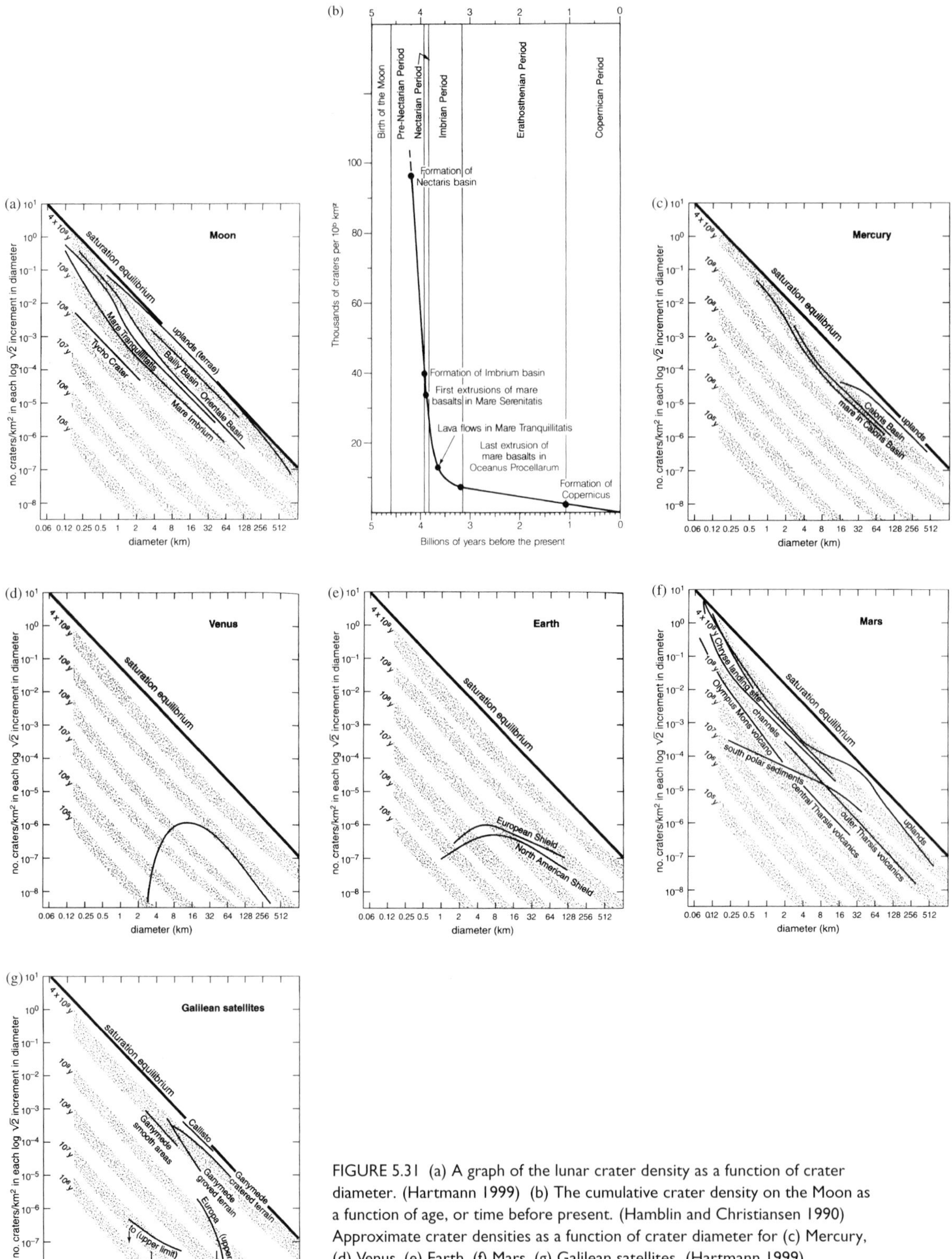

FIGURE 5.31 (a) A graph of the lunar crater density as a function of crater diameter. (Hartmann 1999) (b) The cumulative crater density on the Moon as a function of age, or time before present. (Hamblin and Christiansen 1990) Approximate crater densities as a function of crater diameter for (c) Mercury, (d) Venus, (e) Earth, (f) Mars, (g) Galilean satellites. (Hartmann 1999)

8.5) on rock samples returned to Earth by the various Apollo and Luna landers. Nine different missions returned a total of 382 kg of rocks and soil to Earth. A comparison of these absolute ages with graphs such as shown in Figure 5.31a have lead to the result in Figure 5.31b, where the cumulative crater density is shown as a function of age, or time before present. The very oldest regions, $\sim$4.45 Gyr, are saturated with craters (the lunar highlands), while younger regions (lunar maria, $\sim$3–3.5 Gyr) show a much lower crater density. This difference can be understood when considering formation processes (Chapter 12). Since there was a lot more debris around during the early history of the Solar System, the crater frequency was highest during the planet-forming era. This period is referred to as the *early bombardment era*. The crater frequency dropped off rapidly during the first billion years, to a roughly constant cratering flux during the past 3–3.5 Gyr. The present cratering rate on the Moon for craters with sizes $D > 4$ km is $\sim 2.7 \times 10^{-14}$ craters km^{-2} yr^{-1}, a rate which can be accounted for within a factor of $\sim$3 by the observed and extrapolated distributions of Earth-crossing asteroids and comets. One model suggests that the cumulative crater density for the Moon for craters with $D > 4$ km can be crudely approximated by:

$$N_{cum} \approx 2.68 \times 10^{-5} \left(t + 4.57 \times 10^{-7}(e^{qt} - 1)\right), \tag{5.29}$$

where t is the age of the surface in Gyr, and $q = 4.53$ Gyr^{-1}. Most of the impact melts in rocks returned from the Moon by the Apollo and Luna missions date from 3.8–3.9 Gyr before the present. These dates suggest that the Moon may have been subjected to a terminal cataclysm during this epoch, and that the cratering rate was less than given by equation (5.29) at earlier times.

The size distribution of craters (Figure 5.31) also provides a historical record of the size–frequency distribution of impactors. Assuming a model for the density and size distribution of impactors over time and throughout the Solar System, one can determine a body's age from graphs such as that shown in Figure 5.31, and using the Moon as the 'ground-truth'. This *crater-dating* technique is widely used, since it often provides the only means to assign an age to a body's surface. The simplest assumption regarding impact frequency and size distributions contains no variations with location or time. This assumption is clearly oversimplified. The population and velocity distribution of impactors varies within the Solar System. Less than one-fifth of the known Earth-crossing asteroids pass also inside Mercury's orbit. However, gravitational focusing by the Sun, the increased velocity of the impactor and Mercury's weaker gravitational field cause the cratering rate per unit area on Mercury's surface to be roughly half that on Earth. Since more asteroids cross Mars's orbit, the cratering rate on Mars is somewhat higher than that on Earth. While roughly two-thirds of the impact craters on Earth and Mars can be attributed to asteroidal impacts, most impact craters on the satellite systems around the giant planets are caused by comets. The gravitational focusing effect of the planets causes the impact rate to be largest for the satellites closest to the planet. The present cratering rate on Io is estimated to be roughly four times as large as that on Callisto, while that on Callisto is presently about twice that on the Moon. The cratering rate on the outer satellites of Saturn and Uranus is roughly half that on the Moon. In addition to variations in impact frequency with location, the cratering frequency and size distributions of the impactors have changed drastically over time for the Moon (Fig. 5.31b); it may have changed differently for other bodies.

5.4.4.2 *Crater Removal*

The gravity and material strength of the target body influence the size and shape of craters produced by impacts through what is called the viscous relaxation of the craters. Although the gravity and material strength of the body determine how long a crater can be recognized as an impact crater, there are numerous other processes which shape, modify and obliterate/remove craters. A major process which removes craters on Earth is plate tectonics, which 'recycles' the Earth's crust on timescales of several $\times 10^8$ years (Section 5.3.2.1). In addition, weathering and erosion will modify a crater. Volcanic activity is another important endogenic process which effectively removes craters from sight by blowing up part of the crust, and/or depositing material (lava, ejecta) on top of it. No impact craters have been seen on Jupiter's satellite Io, which has been attributed to Io's extreme active volcanism. Atmospheric (and oceanic) weathering slowly erodes craters, both mechanically (e.g., water and wind flows) and through chemical interactions. Melting of (sub-)surface material, which may be triggered by an impact, a volcanic outburst, or by a change in atmospheric temperature (e.g., greenhouse effect), can shorten the lifetime of an impact crater considerably. Any other endogenic changes in the crust, for example through (local) shrinkage or expansion of the crust, and/or the formation of mountains, remove or modify existing craters.

In addition to endogenic removal processes, craters can be destroyed and eroded by exogenic processes. Impacts may hit a pre-existing crater and destroy it, or later impacts can cover up previously formed impacts with ejecta blankets. Small craters are easier to obliterate than larger ones. The more densely cratered a surface is, the more craters are likely to be destroyed by a given impact. Eventually, the surface reaches a statistical equilibrium, where one new crater destroys, on average, one old crater. Further bombardment by the same population of projectiles will not cause any further secular changes in the size–frequency distribution of craters. Such a surface is *saturated* with craters. Because small craters are easier to destroy than large ones, saturation is first reached for the smallest craters.

Micrometeoritic impacts have an eroding effect on airless bodies, which is best described as *sandblasting* and *gardening*. A similar effect is caused by 'sputtering' of low energy ions (keV range) (solar wind, magnetosphere) (Section 4.8.2). As discussed in Section 5.4.2.4, micrometeoroid impacts and sputtering pulverize the bedrock over time and form regolith, which cover most, if not all, planetary bodies. Impacts by charged particles and UV-photons also contribute to this process, though these are more important in changing the local composition of the surface. If the surface consists, for example, of water-ice contaminated with carbon products, then charged particles and/or energetic photons may dissociate the molecules and/or knock off atoms/molecules from the ice into an atmosphere or corona (Section 4.8.2). The lowest weight atoms, such as H, may gain sufficient energy in this process to escape the gravitational attraction from the body, leaving the darker material behind. Icy surfaces, therefore, tend to darken over time, an effect clearly seen on the outer planets' satellites, centaur asteroids and comets.

5.4.4.3 *Stratigraphy*

Stratigraphy is the study of the time sequence of geological events and of dating the events with respect to each other. The simple counting of craters on a given surface yields a crude estimate of its age. The surfaces of bodies which formed 4.5 Gyr ago are saturated with craters. Since all bodies essentially formed 4.5 Gyr ago, we know that something happened to a body if its surface is less than 4.5 Gyr old. When, for example, the plains on the Moon (maria) were flooded by lava any pre-existing craters were covered and disappeared. When the lava solidified, new craters started to accumulate, and crater counts yield the age of the surface after it last solidified. We can thus date less heavily cratered surfaces relative to more densely cratered areas. On several bodies one can discern linear features which are clearly caused by an endogenic process. Sometimes these features cross or disect a preexisting crater, while others appear partially obliterated by an impact crater (Fig. 5.32). The stratigraphy of these features yields information on the chronology of the various events. Also, the study of the appearance of the craters themselves yields much information on the sequence of events involved in the impact. For example, the crater walls and/or central peak may have collapsed over time, and a good photograph of the crater reveals the sequence of these events.

The word stratigraphy originates from a study of the sequence and correlation of stratified rock layers. On other bodies there are uplifts of strata (e.g., the walls of an impact

(a)

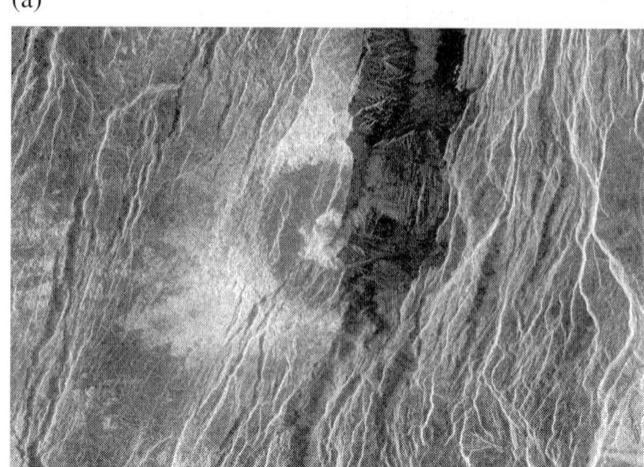

(b)

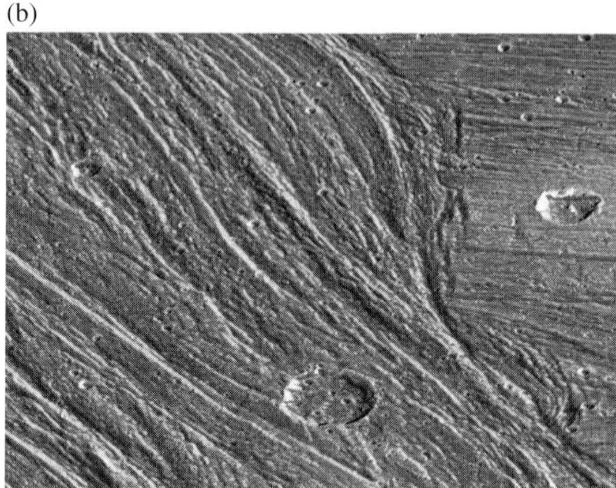

FIGURE 5.32 (a) A Magellan radar image of a 'half crater' on Venus, located in the rift between Rhea and Theia Montes in Beta Regio. The crater is 37 km in diameter, and has been cut by many fractures or faults since it was formed. The eastern half of the crater was destroyed during the formation of a fault valley that is up to 20 km wide and apparently quite deep. (NASA/Magellan PIA00100) (b) The Nippur Sulcus region, an example of bright terrain on Ganymede, shows a complex pattern of multiple sets of ridges and grooves. The intersections of these sets reveal complex age relationships. The Sun illuminates the surface from the southeast (lower right). In this image a younger sinuous northwest–southeast trending groove set cuts through and apparently destroys the older east–west trending features on the right of the image. The area contains many impact craters; the large crater at the bottom of the image is about 12 km in diameter. (NASA/Galileo Orbiter PIA01086)

FIGURE 5.33 An HST image of Comet D/Shoemaker–Levy 9 (SL9) about 2 months before the comet crashed into Jupiter. Note that each fragment A–W is a small comet, with its own tail. (Courtesy: Hal Weaver and T. Ed Smith, HST/NASA)

crater), and such morphologies yield information not only on the dating of the event, but also on the crustal layers and the forces involved to create the end result, as we observe it.

5.4.5 The Impact of Comet D/Shoemaker–Levy 9 with Jupiter

In 1993, Carolyn and Gene Shoemaker, in collaboration with David Levy, discovered their ninth comet, Comet Shoemaker–Levy 9 (or SL9). This comet was very unusual: it consisted of over 20 individual cometary fragments (Fig. 5.33) which orbited Jupiter. Orbital calculations showed that the comet was probably captured by Jupiter around 1930 in a loosely bound orbit, that was chaotically perturbed by the Sun. In July of 1992 the comet's perijove came within 1.3 $R_♃$ from Jupiter's center, well within Roche's limit (Section 11.1). As a consequence, the body was torn apart by the planet's strong tidal forces, after which the individual fragments developed cometary characteristics (coma and tail) and continued their journey around the planet. The comet's next perijove was below Jupiter's surface, and each fragment crashed into the planet. This happened over a 6-day period in July of 1994. Since the collision of SL9 with Jupiter was predicted about a year before it happened, astronomers all over the world had ample time to prepare for the event. Telescopes at wavelengths varying from X-rays to meter radio wavelengths were used to extract information on Jupiter and its surroundings, the comet and, in particular, on impacts. Simulations of the explosions after the events and of the tidal disruption of the parent body suggested that the individual pieces were less than a kilometer across, with densities of ~ 0.5 g cm^{-3}. The kinetic energy involved in an impact by such a body is a few times 10^{27} ergs (Problem 5.8), equivalent to the explosive energy of nearly a hundred million megatons of TNT. These were powerful impacts!

All SL9 fragments hit the hemisphere of Jupiter facing away from Earth, varying from $\sim 3^\circ$ up to 9° behind the east limb. The Galileo spacecraft, although still 1.6 AU away from Jupiter, had a direct view of the impact sites. It is the only large impact mankind has witnessed directly. Even though the impacts occurred at the backside of the planet, and Jupiter has no solid surface, much knowledge has been gained from the impacts that can be extrapolated and used to explain and predict effects from cratering events on terrestrial planets. In this section, we summarize some of the observations of the SL9 impacts. For further information the reader is referred to Noll *et al.* (1996), and Spencer and Mitton (1995).

The impact of fragment R was moderately sized and well observed. Superb data have been obtained by both the Keck (at a wavelength of 2.3 μm) and Palomar (at 3.2 and 4.5 μm) telescopes, as well as with the Galileo spacecraft (4.5 μm). This impact has been used as a generic impact, and indeed most features are visible in the other impacts. A series of Keck images is shown in Figure 5.34. These data were taken at a wavelength of 2.3 μm, near the center of a methane absorption band. Since methane gas in Jupiter's atmosphere absorbs sunlight at this wavelength, none of the solar photons gets reflected from Jupiter's cloud deck, and the planet appears dark. Any high altitude haze layers, which are located above most of Jupiter's methane gas, do reflect sunlight and appear bright against dark Jupiter. The images in Figure 5.34 show the southern hemisphere, with the south polar haze. The bright spots are former impact sites G, L and K (as marked on frame 1). The fact that they appear bright at these wavelengths suggests, correctly, that the material is located at high altitudes in Jupiter's atmosphere. The images shown in Figure 5.34 were taken from a movie, which shows two faint flashes on the east limb, at a latitude of –44° (panels 2–5). The flashes were followed by a dramatic brightening, displayed in panels 7 and 8. Panel 9 shows the new R-impact site rotating into view.

Lightcurves at 2.3, 3.2 and 4.5 μm are shown in Figure 5.35a and b, and the viewing geometry is sketched in Figure 5.36. The first flash occurred simultaneously at the three wavelengths, and lasted for about 40 seconds. This flash is likely the meteor trail of the fragment/bolide falling through the atmosphere, while it is still well above Jupiter's cloud layers. The second flash was first seen at 4.5 μm, and the last at 2.3 μm. The 2.3 μm flash lasted about 3 minutes, while at 4.5 μm it lasted only 30 seconds. This flash is attributed to the thermal emission from the rising fireball, which can be seen as soon as it peaks out above the 'limb', i.e., the local line of sight to the Earth; at this point the fireball is still in the pre-dawn shadow. The onset of this flash was abrupt, ~ 7 s, as expected from a fireball with a diameter of 100–200 km rising at a velocity of 13

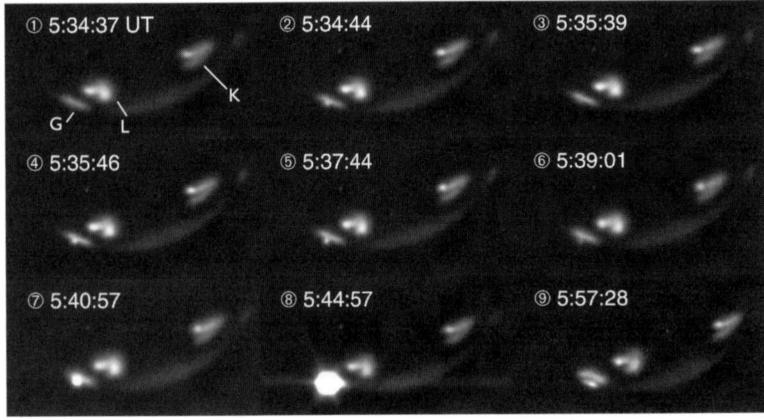

FIGURE 5.34 The impact of SL9 fragment R with Jupiter as observed at 2.3 μm by the Keck telescope. Each of the nine panels is a frame selected from a movie of the impact. The UT time is indicated on each panel. Former impact sites L and K are the bright spots east and west of the meridian, respectively, and the impact site of fragment G is just coming into view on the dawn limb. The planet is dark at this wavelength since methane gas in the atmosphere absorbs (incoming and reflected) sunlight. The impact sites are bright, since some of the impact material is located at high altitudes, above most of the absorbing methane gas. (Graham *et al.* 1995)

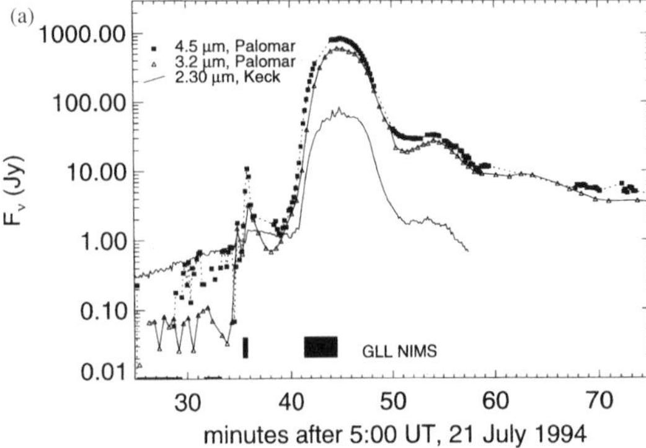

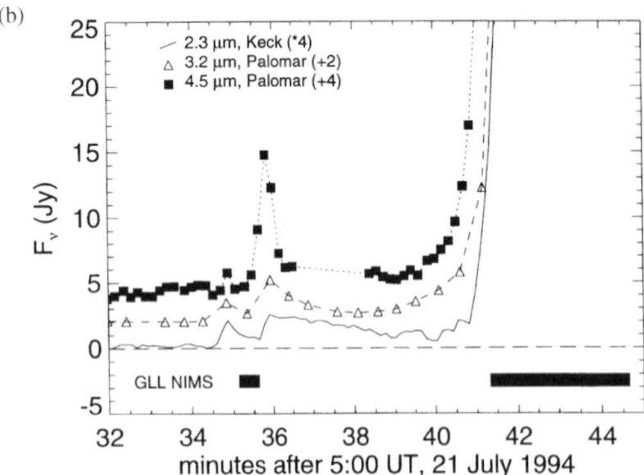

FIGURE 5.35 (a) Lightcurves of the SL9 R impact at 2.3, 3.2 and 4.5 μm, taken with the Keck and Palomar telescopes, respectively. (b) An expanded view of the two precursor flashes. (Nicholson 1996)

km s^{-1}. Since the atmospheric optical depth is largest at 2.3 μm, the 2.3 μm horizon is higher in the atmosphere than the 3.2 and 4.5 μm horizons, and hence the second flash was first seen at 4.5 μm and last at 2.3 μm. The decay in intensity of the flash represents the rapid cooling of the rising, expanding fireball. The Galileo spacecraft had a direct view of the impact site, and observed it at a wavelength of 4.5 μm. It saw its first flash after ground-based observers had seen their first flash, but before the second ground-based flash. Galileo was not sensitive enough to pick up the meteor trail, but because it had a direct view of the impact site it saw the fireball nearly immediately after the explosion. Other impacts were observed by different instruments on Galileo, at shorter wavelengths, where the initial bolide was also seen, but not until the bolide had dropped below the horizon for Earth-based astronomers.

The Hubble Space Telescope (HST) observed the 'plume' when the fireball peaked out above the shadow zone and was visible in reflected sunlight (Fig. 5.37). The plume reached a maximum height of 3200 km above Jupiter's cloud tops, before falling back onto the atmosphere. The infrared main event, characterized by the dramatic brightening six minutes after the first flash, lasted for about ten minutes, and was seen simultaneously by Galileo and ground-based observers. The thermal emission from the atmosphere, shock-heated by the plume debris that falls back down, caused the fantastic bright displays at the infrared wavelengths. The timescales involved agree with those expected from a ballistic plume. About half of the impact energy was invested in the plume.

Following the main event in Figure 5.35 are several shoulders in the emission. These might be caused by material 'bouncing' off the top of the atmosphere and reentering and shock-heating the atmosphere for a second or third time. The R-impact point rotated into view about half an hour after the impact, at which point the impact site becomes visible in reflected sunlight.

The explosions were modeled numerically by several

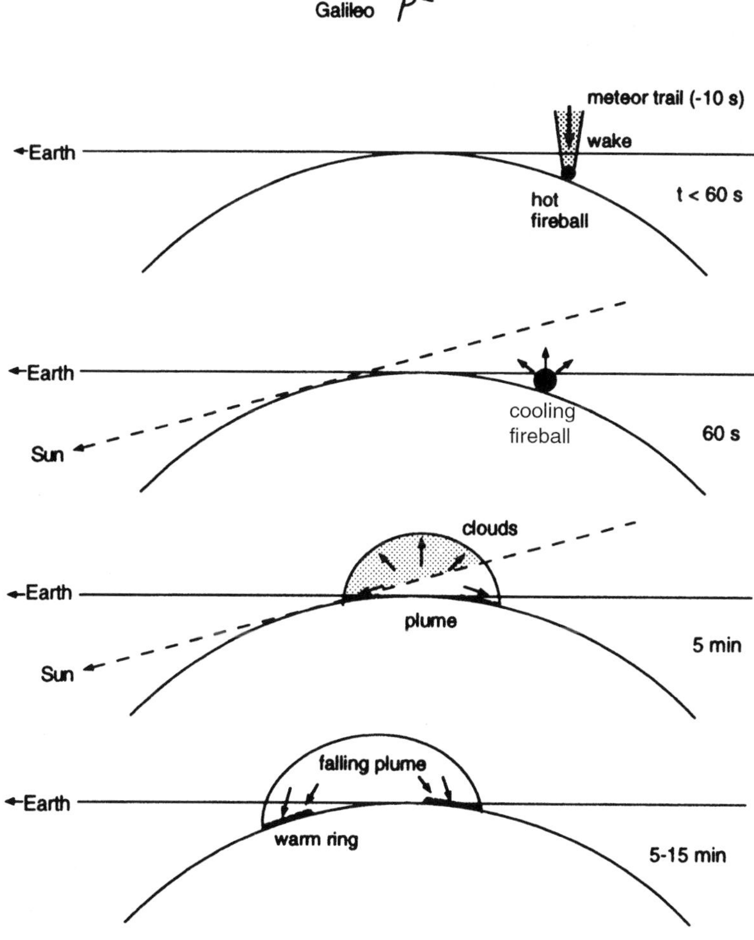

FIGURE 5.36 A cartoon illustrating the geometry during the SL9 impacts. (Adapted from Zahnle 1996)

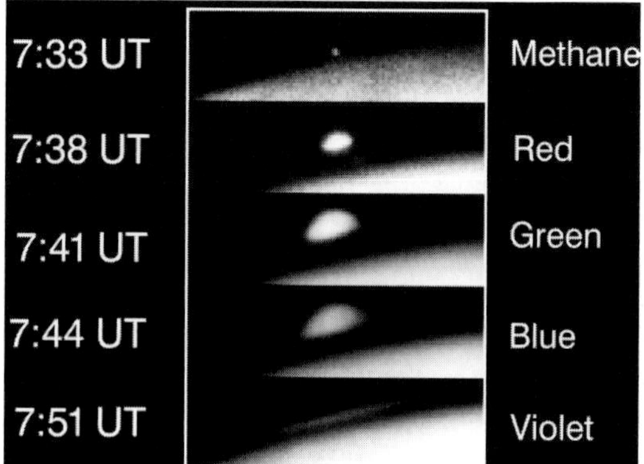

FIGURE 5.37 A sequence of images obtained with NASA's Hubble Space Telescope showing the plume rising, spreading and collapsing after the impact of SL9 fragment G. (Courtesy: H.B. Hammel, HST/NASA)

groups of scientists, and compared to the data. The bolide probably exploded at an altitude around the few bar level, though scientists have not yet reached consensus on this issue. Upon the explosion, the hot fireball rose up in the atmosphere, through the 'wake' of the bolide's entry path. The hot gas contained a fair fraction of the original cometary gas, mixed with a similar mass of shocked jovian air. It was followed by a trail of nearly pure jovian air from relatively deep atmospheric layers (at least several bars, maybe a few tens of bars deep). The inital fireball must have been very hot, probably over 10 000 K. Temperatures of $\sim$ 8000 K were measured by Galileo. While the fireball rose, it cooled. For an adiabatically cooling plume of hot gas, one would expect the decay time to be shortest at the shortest wavelength (Planck's law); however, this is opposite to what was oberved. The discrepancy can probably be solved if the proper sources of opacity are included in the calculations.

When the plume material fell back onto the atmosphere, its gravitational potential energy was converted to kinetic energy, the vertical component of which, upon reentry, was used to shock-heat the atmosphere. The horizontal velocity was preserved across the shock, so that the

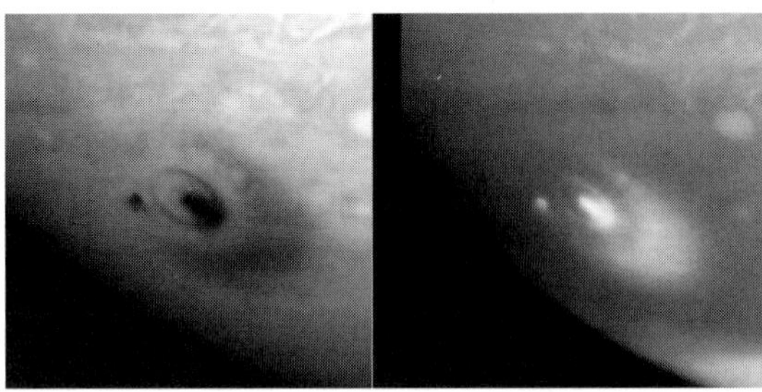

FIGURE 5.38 An HST image of Jupiter, on 18 July 1994, after the impact of SL9 fragment G, one of the largest fragments. The left image was taken through a green filter (5550 Å) and the right image through a near-infrared methane filter (8900 Å), approximately 1.75 hrs after fragment G impacted the planet. Note that the impact sites appear bright at near-infrared wavelengths in methane absorption bands and dark at visible wavelengths. The G impact site has concentric rings around it, with a central spot 2500 km in diameter. The thick outermost ring's inner edge has a diameter of 12 000 km. The small spot to the left of the G impact site was created by the impact of the smaller sized fragment D, a day earlier. (Courtesy: H.B. Hammel, HST/NASA)

'impact region' continued to expand radially, and impact sites cover areas with radii of several thousands of kilometers. The infrared data reveal temperatures of 500–1000 K during the main event, which are much lower than expected. They are also well below the temperatures measured in the 2.3 μm CO band (2000–5000 K), which is indicative of fast and effective radiative cooling, possibly by dust and hot molecules.

In HST images (Fig. 5.38) the impact site appears dark, with a special morphology: a brown dot at the point of entry, surrounded by a dark ring expanding at a velocity of $\sim$450–500 m s^{-1}. A larger crescent is visible to the southwest, and is interpreted as caused by 'impact ejecta'. The dark color has been attributed to carbonaceous material or 'soot'. The morphology can be accounted for by a combination of the ballistic trajectories of the plume material, the Coriolis force (Section 4.5.3.1) and the fact that the 'impact region' continued to expand radially for about 20–30 minutes after impact.

Many atomic and molecular emission lines have been observed during and after the various impacts, some of which originated from cometary gases and others from jovian air. For the interpretation of these data, during the impacts and the long-term evolution of the various species, one has to use detailed photochemical models in combination with models on the chemical composition of comets and Jupiter's atmosphere. Such calculations provide information on the photochemical processes during and after the impacts. Since such information is unique to this particular comet collision with Jupiter, we do not attempt to summarize the results here, but refer the reader to Noll *et al.* (1996).

5.4.6 Dangers of Impacts on Earth and the Extinction of the Dinosaurs

The well-preserved history of impacts on the Moon, together with the widely observed impact of Comet Shoemaker–Levy 9 with Jupiter, makes us uncomfortably aware of the chances and dangers of being hit by meteorites. Because the Earth's crust is continuously renewed, impact craters are relatively rare and hard to find or recognize. Yet studies of asteroids, comets and interplanetary dust suggest that Earth 'sweeps up' 10 000 tons of micrometeoritic material each year during its orbit around the Sun. Meteors, cm-sized material falling through Earth's atmosphere, are a familiar sight, in particular around August 10 (Perseids), December 11 (Geminids) and January 1 (Quadrantids). One calculation suggests that $\sim$7240 meteorites over 100 g in weight ($\gtrsim$2 cm in radius) make it to the ground each year, which translates into one fall per square kilometer every 100 000 years (Section 9.2.4). The Moon likely formed about 4.45 Gyr ago in a disk that was produced by an impact of a Mars-sized or larger body and the proto-Earth. At that time, just after the formation of the planets, there still were many planetesimals around which battered the surfaces of planets, and moderately large impacts were probably not uncommon. To assess the impact probabilities at the current epoch, we first investigate the evidence of moderately large (several kilometers in size) impactors that hit our planet over the past billion years.

It is well known from fossil records in rock strata that animal and plant life have been wiped out occasionally over the past half billion years, in processes we refer to as sudden 'mass extinctions'. One of the most dramatic mass extinctions occurred $\sim$65 million years ago, where large numbers of animal (in particular the dinosaurs) and plant species were suddenly removed from Earth. This event marks the end of the Cretaceous time period and the beginning of the Tertiary period, and is known as the KT boundary. The discovery by geologist Walter Alvarez of much higher than (10–100 times) normal levels of iridium in sediments deposited at the KT boundary, gave these researchers the first hint that the mass extinctions might have an extra-terrestrial origin. After a 10-year long search for a large impact crater formed at the same time as the KT boundary, the discovery of the Chicxulub crater on the Yucatán Peninsula in Mexico (Fig. 5.39) rewarded the Alvarez group with the last piece of evidence that the KT boundary and the extinction of the dinosaurs was indeed most likely triggered by an impact of a $\sim$10 km sized ob-

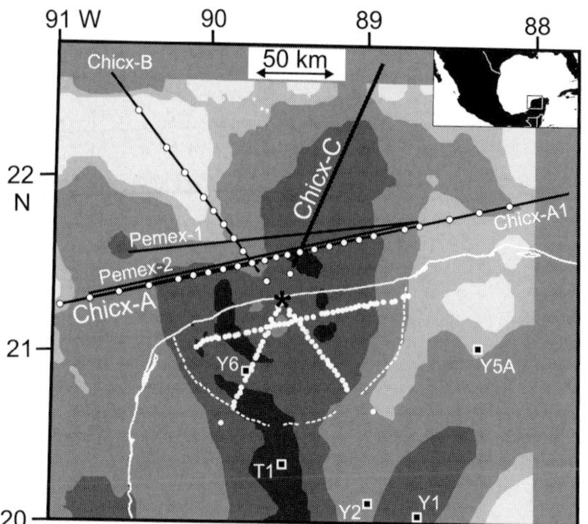

FIGURE 5.39 The Chicxulub seismic experiment. Solid lines show off-shore reflection lines, white dots show wide-angle receivers. Shading shows the Bouger gravity anomaly (Section 6.1.4.2); the crater is marked by a ~ 30 mGal circular gravity low. The dashed white lines mark the positions of the sink-holes in the carbonate 'platform'. Squares show well locations; Y6 is $\sim$1.6 km deep, and T1, Y1, Y2, and Y5a are 3–4 km deep. All radii are calculated using the asterisk as the nominal center. (J. Morgan *et al.* 1997)

ject. The Chicxulub crater is no longer visible at the surface, but can be studied by using gravity anomaly measurements. The crater appears to have a multiring basin morphology: a peak ring with a diameter $D \approx 80$ km, an inner ring with $D \approx 130$ km, and an outer ring with $D \approx 195$ km.

The energy involved in an impact of a 10 km sized body with an impact velocity of ~ 20 km s^{-1} is about 100 times as much as the energy involved in the impact of individual fragments of Comet Shoemaker–Levy 9, roughly 10 billion megatons of TNT, about 10^9 times more than the energy of one large hydrogen bomb. At the instant the bolide hit the surface, two shock waves propagated away from the impact site, as discussed earlier in this section. One shock wave propagated into the bedrock, while the other went backward, into the impactor. Immediately following this, a colossal plume of vaporized rock, the fireball, rose upward into space, launching dust and rocks on ballistic trajectories which carried them far around the Earth. In this particular case, this fireball is thought to have been followed by a second plume, driven by the sudden release of CO_2 gas from a layer of shocked limestone, located about 3 km below the surface. The cavity itself may have reached a depth of about 40 km before the center rose into a central peak. The peak must have grown so large and high that it collapsed, thereby triggering several outward expanding rings and ridges. Meanwhile the crater walls continued to expand outwards. The transient cavity is thought to have had a diameter of $\sim$100 km.

One expects that within a radius of a few thousand kilometers the atmosphere turned so hot from falling debris launched by the fireball and excavated from the crater, that plants and animals ignited and burned. Also large amounts of nitric acid (HNO_3) and sulfuric acid (H_2SO_4) were likely formed, raining down (acid rain), killing plants and animals and dissolving rocks over a large area around the impact site. Because the impact happened on a peninsula, a large tsunami wave, triggered by the seismic waves, spread outward and, upon hitting Florida and the Gulf coast, must have destroyed vast areas in Mexico and the United States. Fine dust, which had been brought up by the fireballs, settled down and stayed suspended in the atmosphere for many months before reaching the surface. This could have turned the sky dark over the entire Earth and inhibited sunlight from reaching the surface, so that the surface temperature dropped to well below freezing for many months on end. Once the sky cleared, temperatures may have risen to uncomfortably high levels due to enhanced levels of greenhouse gases such as H_2O and CO_2. This cycle of extreme hot–cold–hot temperatures over the entire Earth would have destroyed many animal and plant species in areas quite remote from the impact site. An alternative model on the consequences of the KT impact suggests that the greenhouse warming effect is small, but that the slow (over many years) production of sulfuric acid keeps the temperature low (by tens of degrees K) for many decades, which has a similar destructive effect on plant and animal life.

Impacts of increasingly larger sizes have greater potential for killing individual organisms and wiping out entire species; such impacts also become increasingly rare. Based upon the size distribution of asteroids and comets, one expects the Earth to be hit by a body 5 km in radius once every 10^8 years. Smaller bodies will hit more frequently: a body with a 2.5 km radius (density $\rho = 3$ g cm^{-3}, and velocity 20 km s^{-1}) will hit Earth on average once every 6×10^6 yr, and a body with a radius of 0.3 km ($\rho = 3$ g cm^{-3}, $v = 20$ km s^{-1}) once every 70 000 yr. Table 5.3 summarizes when an impactor of a given size last hit the Earth – some numbers are based upon data, others on statistical arguments (indicated by $\sim$ preceding time). The effect of the impact on the planet and on life are indicated in the last two columns.

The smallest debris to hit the Earth's atmosphere is slowed to benign speeds by gas drag or vaporized before it hits the ground. The largest impactors can melt a planet's entire crust and eliminate life entirely. An impact typically

TABLE 5.3 Impacts and Life[a].

Size[b]	Example(s)	Most recent	Planetary effects	Effects on life
Super colossal $R > 2000$ km	Moon-forming event	4.45×10^9 yr ago	Melts planet	Drives off volatiles Wipes out life on planet
Colossal $R > 700$ km	Pluto 1 Ceres (borderline)	$\gtrsim 4.3 \times 10^9$ yr ago	Melts crust	Wipes out life on planet
Jumbo $R > 200$ km	4 Vesta (large asteroid)	$\sim 4.0 \times 10^9$ yr ago	Vaporizes oceans	Life may survive below surface
Extra large $R > 70$ km	Chiron (largest active comet)	3.8×10^9 yr ago	Vaporizes upper 100 m of oceans	Pressure-cooks photic zone May wipe out photosynthesis
Large $R > 30$ km	Comet Hale–Bopp	$\sim 2 \times 10^9$ yr ago	Heats atmosphere and surface to $\sim$1000 K	Continents cauterized
Medium $R > 10$ km	KT impactor 433 Eros (largest NEA)	65×10^6 yr ago	Fires, dust, darkness; atmosphere/ocean chemical changes, large temperature swings	Half of species extinct
Small $R > 1$ km	$\sim$500 NEAs	$\sim$300 000 yr ago	Global dusty atmosphere for months	Photosynthesis interrupted Individuals die but few species extinct Civilization threatened
Peewee $R > 100$ m	Tunguska event	93 yr ago	Major local effects Minor hemispheric dusty atmosphere	Newspaper headlines Romantic sunsets increase birth rate

[a] Adapted from Lissauer (1999b) and Zahnle and Sleep (1997).

[b] Based on USDA classifications for olives and eggs.

destroys an area in proportion to the energy involved:

$$\mathcal{A} = 100E^{2/3}, \tag{5.30}$$

where E is the kinetic energy of the impact in MT, and $\mathcal{A}$ the area in km^2. The Tunguska event in 1908, an airblast caused by a 100 m sized stony body, flattened about 2000 km^2 of forests. In addition to the direct collisional destruction, the effects an impact can have on climate worldwide can be dramatic. If the amount of dust injected into the stratosphere is sufficient to produce worldwide an optical depth > 2 for several months, the surface temperature can be suppressed by $\sim$10 K globally. The amount of dust necessary to produce an optical depth of 2 is about 10^{16} g of dust, about a hundred times more than has been alofted by any of the large volcanic eruptions from the past century (e.g., Mt. Pinatubo in 1991). This much dust is injected into the stratosphere by an impact with an energy of 10^5–10^6 MT, i.e., an impact by a stony asteroid 1–2 km in diameter at a velocity of 20 km s^{-1}. There is a 1-in-10 000 chance that such a body will collide with Earth over the next hundred years.

5.5 Surface Geology of Individual Bodies

The surface characteristics of many bodies have been determined via remote sensing techniques such as imaging, photometry, polarimetry, thermal and reflectance spectra, radio and radar observations. These techniques are discussed in more detail in Section 9.3. Although the spatial resolution from the ground has been relatively poor ($\gtrsim 0.3''$) compared to the size of the object ($< 1''$ up to $1'$ for the largest body, Venus), one can now achieve spatial resolutions of order $\sim 0.05''$ with the Hubble Space Telescope (HST), or using speckle imaging and adaptive optics techniques on large ground-based telescopes. Our highest quality images, however, usually come from *in situ* spacecraft flybys or landers, such as the Mariner 10 at Mercury, Magellan at Venus, various spacecraft including the Apollo missions at the Moon, Viking, Mars Pathfinder and Mars Global Surveyor at the red planet, the NEAR (Near-Earth Asteroid Rendezvous) spacecraft at asteroids Mathilde and Eros, Galileo at Jupiter and two asteroids, the Voyager spacecraft at the giant planets, and Giotto at Comet Halley. A combination of remote sensing techniques from the ground and space have lead to quite detailed geological pictures of many bodies in our Solar System. It is most striking that each body has its own peculiarities, its own geology which often looks quite different from any other known body. Yet there are similarities too, and we recognize many of the geological features known to us from terrestrial processes such as volcanic and tectonic structures, atmospheric effects as winds and condensation flows, impact craters, etc. In this section, we summarize our present understanding of the terrestrial planets and many of the satellites. Geological properties of the few asteroids that have beeen imaged at high resolution are reviewed in Section 9.5.2, and the surface characteristics of Comet Halley are summarized in Section 10.6.

When viewed from space Earth looks much like other planets, though distinctly different in color because of the overwhelming presence of the oceans (Fig. 5.40). The continents are visible as brown land masses, and white ice sheets dominate the polar regions. The planet has been mapped from space at different wavelengths and also with radar techniques. These data are useful in comparison with images obtained via very similar techniques from other planets. Since we have used planet Earth as a prototype throughout this chapter, and have already shown many photographs of terrestrial features, we only discuss other Solar System bodies here and compare them to Earth, where applicable.

FIGURE 5.40 Image of the Earth taken by the Galileo spacecraft on 11 December 1990, at a distance of $\sim 2.5 \times 10^6$ km. India is near the top of the picture, and Australia is to the right of center. The white, sunlit continent of Antarctica is below. Picturesque weather fronts are visible in the South Pacific, lower right. (NASA/Galileo image PIA00122)

5.5.1 Moon

With the naked eye one can discern two major types of geological units on the Moon: the bright *highlands* or *terrae* which account for over 80% of the Moon's surface area and have an albedo of 11–18%, and the darker plains or *maria* with an albedo of 7–10% which cover 16% of the lunar surface (Fig. 5.15). The maria are concentrated on the hemisphere facing Earth. The dominant landforms on the Moon are impact craters. The highlands appear saturated with craters, varying in diameter from micrometers (Fig. 5.22) up to hundreds of kilometers in size (e.g., Orientale basin, Fig. 5.25). Some of the large younger craters show the bright rays and patterns of secondary craters. The highlands clearly date back to the heavy bombardment era 4.4 Gyr ago (Fig. 5.31). In contrast, the maria are less heavily cratered and must therefore be younger.

Some maria cover parts of seemingly older impact basins, e.g., Mare Imbrium within the Imbrium basin. Radioisotope dating (cf. Section 8.5) of rocks brought back by the various Apollo and Luna missions indicates that the maria are typically between 3.1 and 3.9 Gyr old. The lunar samples further indicate that the maria consist of fine-grained, sometimes glass-like basalt, rich in iron, magnesium and titanium. The rocks lack volatile material such as water and hydrated minerals. All of this together suggests that the mare basalts originated hundreds of kilometers below the surface, and must have been brought up by volcanic activity, 3.1–3.9 Gyr ago. The lava lakes cooled and solidified rapidly, as shown by the glassiness and small grain size of the minerals in the rocks. Data from the Clementine spacecraft show, as expected, that the lava-flooded maria are extremely flat, with slopes of less than 1 part in 10^3. Since the large impact basins were created by energetic impacts, the crust was probably fractured down to many kilometers below the surface. The hot magma may have oozed its way up through the cracks and flooded the low lying regions of the impact basins.

The most pronounced topographic feature on the Moon is the South Pole Aitken Basin, the oldest discernible impact feature. It is 2500 km in diameter, with a maximum depth of 8.2 km below the reference ellipsoid, the Moon's geoid, or 13 km from the rim crest to the crater floor. This is the largest and deepest impact basin known. The maria are typically topographic lows. The highest point on the Moon, 8 km above the reference ellipsoid, is in the highlands on the far side of the Moon, adjacent to the South Pole Aitken Basin. The lunar poles are of interest since some regions here are in permanent shadow, and may remain as cold as 40 K. In 1998, measurements made with a neutron spectrometer on board the 'Lunar Prospector' spacecraft demonstrated the presence of hydrogen, presumably contained in water-ice, in cold craters at the lunar poles. Neutrons from the solar wind continually bombard the Moon. If such neutrons interact with hydrogen, they lose speed. Lunar Prospector's neutron spectrometer detected increased numbers of thermal neutrons compared to the higher energy counterparts above the lunar poles, a signature expected if water-ice is present. The data suggest the ice to be present in the form of tiny crystals mixed in with the soil, at concentrations ranging from 0.3 to 1%. The spectrometer is sensitive down to half a meter depth below the surface, but the ice crystals are expected to extend deeper, since the surface is 'gardened' to a depth of a few meters over the past few billion years (Section 5.4.2.4). The data further suggest that the crystals must be dispersed over an area of 5000–20 000 square kilometers around the south pole, and 10 000–50 000 square kilometers around the north pole. The water was presumably brought in by comets and asteroids well after the Moon itself had formed (see also Mercury, Section 5.5.2).

The Apollo missions showed that the lunar crust in the highlands has been pulverized by the numerous impacts. Many of the rocks are breccias, different types of rock cemented together by impact processes. The breccias as well as the lunar soil contain numerous glasses, smooth round spheres or dumbbells, which are clearly products from impact processes. The rocks are usually pulverized further by micrometeorite impacts, which created a fine-grained layer of regolith, more than 15 m deep, which overlies the lunar highlands. The maria are also covered by regolith, but because the maria are younger the layer is only 2–8 m deep. The highlands generally lack rocks abundant in heavy minerals such as iron and titanium. The rocks are mainly *anorthosites*, composed of calcium-rich plagioclase feldspars, or anorthites. The contrast between highlands and maria suggests that the Moon, like Earth, must have undergone a 'global' differentiation process, where the heavier elements settled down, and the low-density elements accumulated near the top of the magma ocean to form the lunar crust. The Apollo missions revealed a very unusual chemical component in the lunar rocks, which is called *KREEP*, based upon the elements present in this component: potassium (K), rare-earth elements (REE) and phosphorus (P). Its age is estimated to be 4.36 Gyr, which has lead to the concept that KREEP may be the final crystallization product of a global magma system.

Although impact cratering is by far the most important geological process on the Moon, there is also clear evidence of volcanism and tectonics. Volcanism is most evident from the maria, but there are a few other features which show evidence of volcanic flows. The sinuous rille shown in Figure 5.18 has been interpreted as a collapsed lava channel. Tectonic features such as linear rilles, similar to graben faults, likely formed by expansion or contraction. There is no evidence for (past or present) plate tectonics. The Moon is believed to have been geologically inactive for the past 1.5–2 Gyr.

5.5.2 Mercury

The hemisphere of Mercury which has been imaged by Mariner 10 is very similar in appearance to the surface of the Moon. Like on the Moon, craters are the dominant landform. By far the largest feature observed on Mercury is the 1300 km diameter Caloris basin (Fig. 5.41b), a large ring basin analogous to the large impact basins on the Moon. The basin is surrounded by a 2 km high ring of mountains. Shock waves from the impact that produced the Caloris basin are believed to be responsible for the irregular or 'weird' terrain antipodal to Caloris, which consists of a chaotic formation of rock-blocks and hills. It is interesting to note that, as a consequence of Mercury's 2:3 orbital resonance around the Sun, every other orbit the Caloris basin is directly facing the Sun at perihelion (Section 2.6.2).

Although superficially Mercury and the Moon are very much alike, the two bodies differ dramatically in detail. Mercury is brighter (albedo 0.14, Tables 4.1 and 9.2) than the Moon, and does not show the contrast of maria versus highlands. Craters on Mercury are shallower than like-sized craters on the Moon, and secondary craters and ejecta blankets are closer to the primary craters of a given size because of the greater surface gravity on Mercury. Mercury's heavily cratered terrain is interspersed with smooth 'intercrater plains', resembling in some ways the maria on the

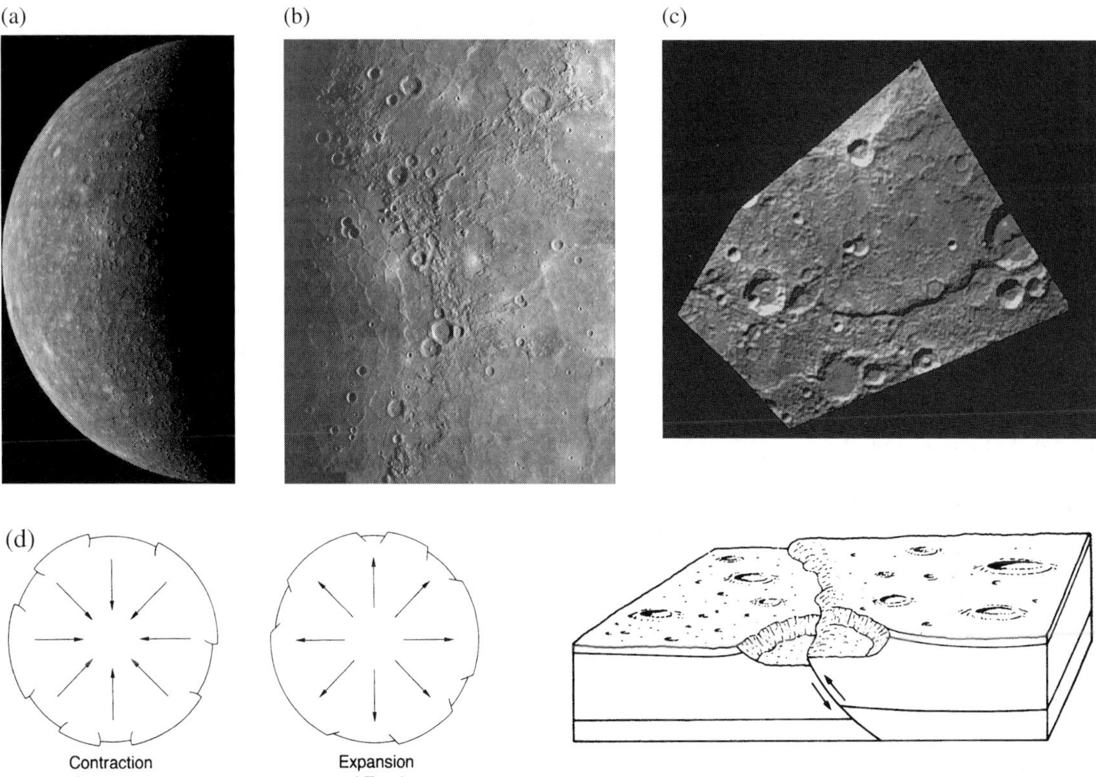

FIGURE 5.41 (a) Mariner 10's first image of Mercury acquired on 24 March 1974, from a distance of $\sim 5.4 \times 10^6$ km. (NASA/Mariner 10 PIA00437) (b) This mosaic shows the Caloris basin (located half-way in shadow on the morning terminator). Caloris basin is 1300 km in diameter and is the largest known structure on Mercury. (Courtesy: Calvin J. Hamilton, NASA/Mariner 10) (c) A mosaic of Mariner 10 images, featuring Discovery Rupes, the sinuous dark feature running through the craters at the center of the image. Many such features were discovered in the Mariner images of Mercury and are interpreted to be enormous (1–2 km high) thrust faults, or *scarps*, where part of the mercurian crust was pushed over an adjacent part by compressional forces. (Courtesy: P. Schenk, Lunar and Planetary Institute/NASA) (d) Schematic of the formation of scarps. (Hamblin and Christiansen 1990)

Moon. However, no lava flow features have been seen, and data at infrared and radio wavelengths suggest that Mercury lacks basaltic material that is rich in heavy elements like iron and titanium (Fig. 5.42a). The mineral ilmenite ((Fe,Mg)TiO_3) that makes up most of the opaque material in the Moon's maria appears to be largely absent on Mercury.

Mercury shows unique scarps on its surface: linear features hundreds of kilometers long, which range in height from a few hundred meters up to a few kilometers (Fig. 5.41c). These scarps probably were produced by a planet-wide contraction, like the wrinkles on the skin of a dried-out apple. The scarps suggest a global decrease of Mercury's radius by 1–2 km.

To explain Mercury's high uncompressed density (Section 6.3.2), it is generally believed that the planet's mantle has been stripped away by a giant impact early in its formation history (Section 12.6.1). Such an impact would 'boil away' the more volatile elements, leaving the planet as a hot ball of magma dominated by extreme refractory material. The planet cooled by radiating into space, while heavy material migrated/fell to the center, leaving Mercury's crust relatively devoid of heavy elements. The lava which covered or created the smooth plains must have originated close to the crust, in contrast to the Moon's maria, which were flooded by basalt from deep below the Moon's surface.

Since Mercury's obliquity is 0°, the poles do not receive much sunlight. In fact, the crater floors in some regions are in permanent shadow, and the temperature stays well below 100 K. Radar echoes indicate an unusual large reflection from the poles, which has been attributed to the presence of water-ice. Detailed radar images (Fig. 5.42b) show a high correlation of the icy regions with visual craters near the pole. It may seem very counter-intuitive

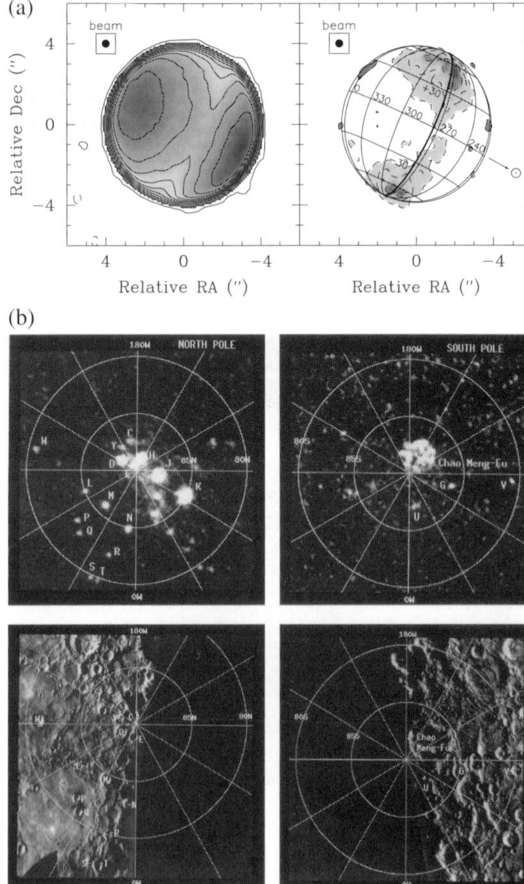

FIGURE 5.42 (a) Radio image of Mercury (left panel) at a wavelength of 3.6 cm, as observed with the VLA. Contours are at 42 K intervals (10% of maximum), except for the lowest contour, which is at 8 K (dashed contours are negative). The beamsize is 0.4″ or 1/10 of Mercury's radius. The geometry of Mercury during the observation, the direction to the Sun, and the morning terminator (dashed line) are superimposed on the image on the right. This image shows the residuals after subtracting a model image from the observed map. Note that the darker grays show peaks in emission in the image on the left, but peaks in depression on the right. One effectively probes a depth of ~70 cm at this wavelength (Figure 3.5). The two 'hot' longitudes (those facing the Sun at perihelion) are clearly seen in the left image (Section 3.2.1). The image on the right shows the thermal depressions at both poles and along the sunlit side of the morning terminator. Contour intervals (right image) are in steps of 10 K, which is roughly three times the rms noise in the image. The loss tangent of the material determined from these images appears to be much less than that on the Moon (Section 3.2.2.5). This may be evidence of a lack of basaltic material on Mercury's surface. (Mitchell and de Pater 1994) (b) Radar images of the polar regions of Mercury (top panels), compared to Mariner 10 photographs (bottom panels). The radar-bright features are interpreted as layers of ice. Their locations correspond to crater floors on the poles. The regions are in permanent shadow, and hence cold enough that ice is stable over billions of years. (Harmon *et al.* 1994)

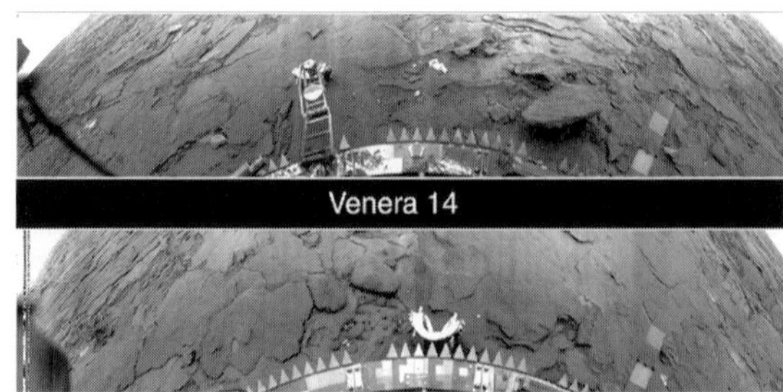

FIGURE 5.43 Venera 14 Lander images of the surface of Venus. The lander touched down at 13° S, 310° E on 5 March 1982. It survived on the surface for 60 minutes before succumbing to the planet's heat. Chemical analyses performed by the Venera landers indicate that most venusian rocks, including those shown here, are basaltic and therefore black or gray. They appear slabby or platy, and are separated by minor amounts of soil. Parts of the lander can be seen at the bottom of each picture (a mechanical arm in the upper picture, a lens cover on the lower one). The landscape appears distorted because Venera 14's wide angle camera scanned in a tilted sweeping arc. The horizon is seen in the upper left and right corners. (Courtesy: Carlé Pieters and the Russian Academy of Sciences)

to have water-ice on a planet so close to the Sun, which otherwise is very dry like the Moon. However, comets and volatile-rich asteroids continue to impact Mercury. Although the impactors' volatile material rapidly evaporates, not all may have escaped Mercury's gravitational attraction. Water molecules may 'hop' over the surface, until they hit the polar regions where they freeze and remain stable for long periods of time. At temperatures $T < 112$ K, water-ice is stable to evaporation over billions of years. Calculations do show, however, that a very large source of icy material is needed to build up a many meters ($\gtrsim 50$ m) thick layer of ice, as required to explain the radar data. Assuming an impacting distribution of meteoritic material as expected at Mercury during the heavy bombardment era, one might only build up a layer ~20 cm thick! However, a few giant comets or volatile-rich asteroids might significantly alter these results.

5.5.3 Venus

As discussed in Section 4.4, Venus is covered by a thick cloud deck, impenetrable to visible light. One can, however, probe through the cloud deck at a few particular infrared wavelengths and at radio wavelengths longwards of a few centimeters. We have learned a great deal about Venus's surface using radar observations carried out both from the ground and from space. The planet's surface has been mapped at high resolution (100 m) using radar experiments from the Magellan spacecraft, which orbited Venus for about 4 years. Several Venera spacecraft equipped with

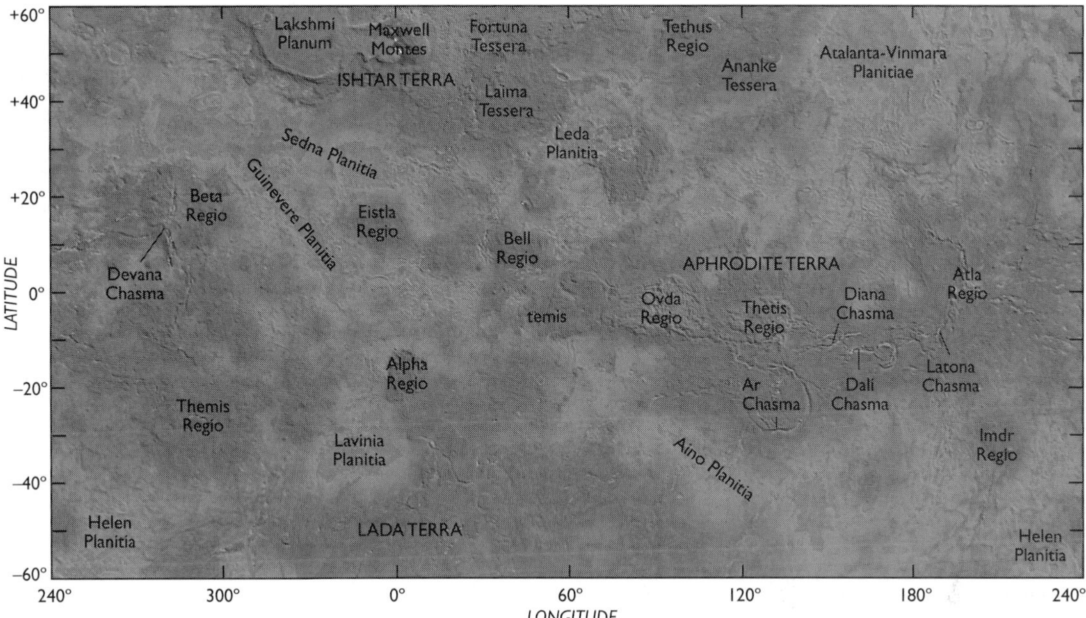

FIGURE 5.44 COLOR PLATE Global Mercator-projected view of Venus's surface as derived from Magellan's radar altimeter data. Maxwell Montes, the planet's highest mountain region, rises 12 km above the mean elevation. (Courtesy: Peter Ford, NASA/Magellan)

photographic cameras have landed on the surface and sent back *in situ* photographs (Fig. 5.43). In color the landscape on these images looks orange, because the dense, cloudy atmosphere scatters and absorbs the blue component of sunlight. The photographs reveal a dark surface, and slightly eroded rocks; the rocks are not as smooth, however, as typical terrestrial rocks. The compositions of rocks analyzed at the landing sites are similar to various types of terrestrial basalts.

Most of our information regarding Venus's surface can be attributed to radar measurements. A small fraction of the surface is covered by highlands, large continent-sized areas which are well (3–5 km) above the average surface level (Fig. 5.44). We recognize four major highlands (Ishtar and Aphrodite Terra, Alpha and Beta Regio), which together cover about 8% of the surface. Roughly 20% of the surface consists of lowland plains, and $\sim 70\%$ of rolling uplands. Overall, most of the surface lies within a kilometer of the mean planetary radius. The difference in elevation between the highest and lowest features is $\sim$13 km, which is very similar to the elevation contrast on Earth (Himalayas ~ 8 km above sea level, oceans ~ 5 km below sea level; volcano Mauna Loa ~ 9 km above the sea floor). This is not too surprising, since the gravity and lithospheric strength of the two planets are very similar, and these quantities determine how high a volcano can be without collapsing.

The radar reflection coefficient varies from roughly 0.14 in the lowlands to roughly 0.4 in the highlands. Regions with a high radar reflectivity usually correlate with low radio brightness temperatures, or low radio emissivities. The emissivities and radar reflectivity can be related to the dielectric constant of the material (Section 3.2.2.5). The disk-averaged value for the dielectric constant is ~ 5, a typical value for solid rocks (granite, basalt). There is no evidence of a dielectric constant as low as ~ 2, a value expected for porous surface material. This suggests that Venus's surface consists mostly of dry solid rock, and (parts of) the surface might be overlain at most by a few centimeters of soil or dust. The highlands show dielectric constants well over 20–30, indicative of inclusions of metallic and/or sulfide material, such as iron pyrites. Alternatively, the high dielectric constants may also be caused by the scattering effects of rocks embedded in dry soils.

Volcanism. The most detailed (spatial resolutions of ~ 0.2–1 km) almost global data set of Venus's surface features has been obtained by Magellan (e.g., Fig. 5.16b). In addition to the four major highlands mentioned above, the surface is dominated by complex plains, which like the highlands are probably of volcanic origin. Numerous small dome-like hills exist, which are suggestive of small shield volcanoes. In addition, there are a number of circular flattened domes, ~ 30 km across and $\sim$1 km high, which have a small pit at the summit. Magellan discovered many peculiar volcanic structures, including pancake-like domes, coronae, and arachnoids. Approximately 1100 volcanic constructs have been identified on Venus's surface.

Pancake-like domes (Fig. 5.45a) have diameters of $\sim$20–50 km, and their heights range from $\sim$100 to $\sim$1000 m. They may resemble terrestrial rhyolite–dacite domes ('glass' mountains, made of obsidian), i.e., may be composed of very thick (silica-rich?) lava flows that came from an opening on relatively level ground, which allowed the lava to flow in an even pattern outward from the opening. The complex fractures on top of the domes suggest that if the domes were created by lava flows, a cooled outer layer formed and then further lava flowing in the interior stretched the surface. Another interpretation is that the domes are the result of molten rock or magma in the interior that pushed the surface layer upward. The near-surface magma then withdrew to deeper levels, causing the collapse and fracturing of the dome surface. The bright margins possibly indicate the presence of rock debris on the slopes of the domes. Some of the fractures on the plains cut through the domes, while others appear to be covered by the domes. This indicates that active processes both pre-date and post-date the dome-like hills.

Coronae (Fig. 5.45c) are large circular or oval structures, with concentric multiple ridges, and diameters ranging from less than 100 m to over 1000 km. They are located primarily within the volcanic plains, and are believed to form over hot upwellings of magma within the venusian mantle. *Arachnoids* (Fig. 5.45d) look like multi-legged spiders sitting on webs of interconnecting fractures. The central region, 50–150 km in diameter, is usually depressed in altitude, as if collapsed, merging with radial lineaments on the outer flanks. Arachnoids are similar in form but generally smaller than coronae. They may be a precursor to coronae formation. The radar-bright lines extending for many kilometers may have resulted from an upwelling of magma from Venus's mantle which pushed up the surface to form 'cracks'. *Novae* have radial fracture patterns, without the central corona. Magellan images further reveal hundreds of lava channels on volcanic plains. Some of these are hundreds of kilometers long, and others are long narrow ($\lesssim$ 2 km wide) features, similar to the sinuous rilles on the Moon. These sometimes end in delta-like distributaries in the plains. One particular channel, Baltis Vallis, is 6800 km in length, the longest sinuous rille yet identified in the Solar System (Fig. 5.45e).

It is not known if Venus is still active today. Large temporal variations in the abundance of atmospheric sulfur dioxide have been attributed to possible volcanic outbursts, but no definitive proof (or disproof) of such activity exists (Section 4.3.3.1).

Tectonics. Magellan radar images show that Venus's surface has undergone numerous episodes of volcanism and tectonic deformations. The distribution of impact craters suggests that volcanic resurfacing is locally very efficient, but also very episodic. Prominent tectonic features include long linear mountain ridges and strain patterns, which can be parallel to one another, or they may cross-cut each other. They can extend over hundreds of kilometers. These tectonic deformations reflect the crustal response to dynamical processes in the mantle. The tectonic deformations clearly both pre- and post-date episodes of volcanic activity. Although there is no evidence that a planet-wide system of tectonic plates exist on Venus, there may have been some local 'tectonic plate' activity.

Erosion. The formation of sediments and erosion is less important on Venus than on Earth and Mars, which can largely be attributed to the fact that Venus's atmosphere is extremely dense and hot. The atmosphere prevents small meteorites from hitting the ground, a process that on airless bodies is the main source of erosion and regolith formation. The lack of water and thermal cycling on Venus limits weathering processes, and because there is little wind near the surface (wind velocities are $\lesssim$ 1 m/s), there is no erosion by wind. Weathering, winds and water erode rocks on Earth, while on Mars the main erosion process at present is due to high speed winds. Even though the wind velocities near the surface of Venus are extremely low, Magellan revealed several morphological features which must be caused by winds, such as wind streaks at the lee side of obstacles (Fig. 5.20c) and a dune field in the neighborhood of a large crater. The wind streaks and dune field yield information on the prevailing venusian winds. We note that, because Venus's atmosphere is 90 times as dense as the Earth's atmosphere, even slow winds may be able to displace a considerable amount of sand.

Impact craters. The number and size distribution of impact craters (Fig. 5.31) implies that Venus's surface is younger than that of Mars, but older than Earth's. Typical ages range from a few hundred million up to one billion years. Craters seem to be randomly distributed over the planet's surface, suggesting that most areas have quite similar ages. This suggests that Venus may have undergone major global resurfacing. Venus's lithosphere may be very thick (Section 6.3.3), perhaps $\sim$200 km, so that its internal heat produced by radiogenic processes cannot escape as fast as it is generated. At some point the thick lithosphere may rupture and sink through the overheated buoyant mantle. Once subduction starts, slabs of crust may sink at rates of 20–50 cm per year, so that the crust is completely renewed within 10^8 yr. Alternatively, the crust can

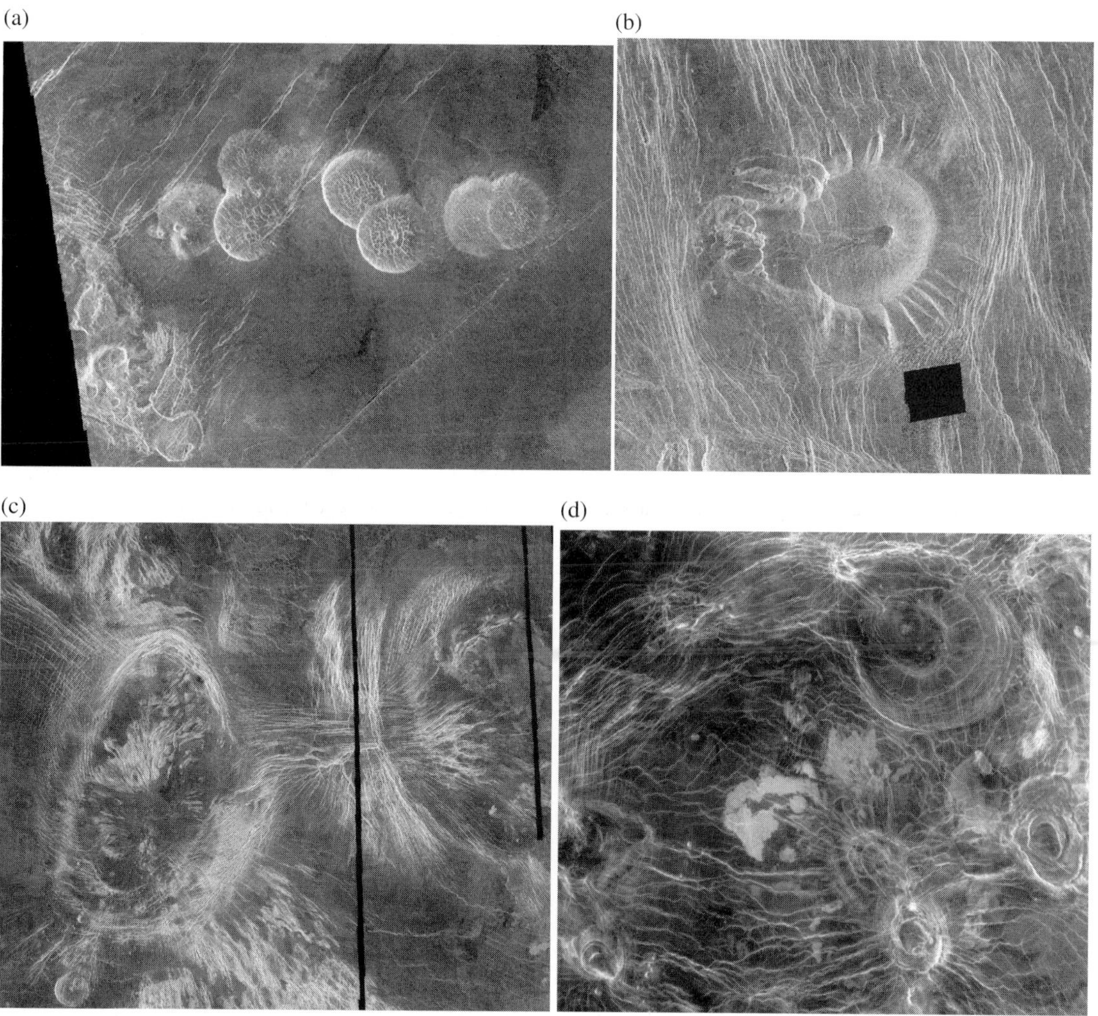

FIGURE 5.45 (a) This Magellan radar image shows seven pancakes on Venus's surface: circular, dome-like hills, averaging 25 km in diameter with maximum heights of 750 m. (NASA/Magellan PIA00215) (b) This peculiar volcanic construct on Venus, nicknamed 'the tick', is approximately 66 km across at the base and has a relatively flat, slightly concave summit 35 km in diameter. The sides of the edifice are characterized by radiating ridges and valleys that impart a fluted appearance. To the west, the rim of the structure appears to have been breached by dark lava flows that emanated from a shallow summit pit. A series of coalescing, collapsed pits are located 10 km west of the summit. The edifice and western pits are circumscribed by faint, concentric lineaments up to 70 km in diameter. A series of north/northwest trending graben are deflected eastward around the edifice; the interplay of these graben and the fluted rim of the edifice produce a distinctive scalloped pattern in the image. (NASA/Magellan PIA00089) (c) This mosaic of Magellan data on Venus shows two coronae. The structure on the left, Bahet Corona, is about 230 km long and 150 km across. A portion of Onatah Corona, over 350 km in diameter, can be seen on the right of the mosaic. Both features are surrounded by a ring of ridges and troughs, which in places cut more radially oriented fractures. The centers of the features also contain radial fractures as well as volcanic domes and flows. The two coronae may have formed at the same time over a single upwelling, or may indicate movement of the upwelling or the upper layers of the planet to the west over time. A 'pancake' dome is located just to the southwest of Bahet. (NASA/Magellan PIA00461) (d) This Magellan image features 'arachnoids' on radar-dark plains on Venus. Arachnoids are circular to ovoid features with concentric rings and a complex network of fractures extending outward. The arachnoids range in size from approximately 50 km to 230 km in diameter. (NASA/Magellan P37501)

be thin, where slices of crust occasionally peel off and sink down, while the crust is slowly renewed by global volcanic resurfacing events.

Impact craters larger than ~15 km are generally circular complex craters with (multiple) central peaks and peak rings (Fig. 5.45f, g). Smaller craters show multiple floors,

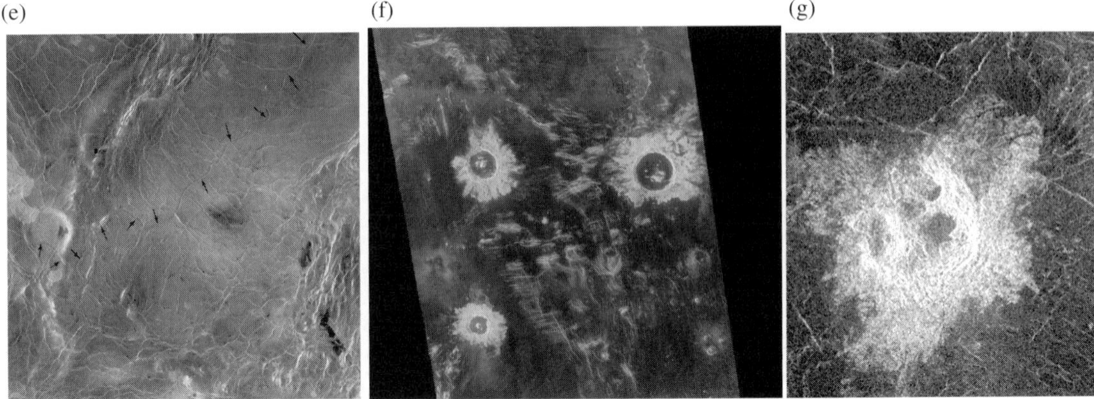

FIGURE 5.45 Continued. (e) A 600-km segment of sinuous channel Baltis Vallis (indicated by arrows) on Venus, the longest channel discovered in our Solar System to date. (NASA/Magellan PIA00245) (f) Three impact craters, with diameters that range from 37 to 50 km, located in a region of fractured plains, show many features typical of meteorite impact craters: rough (bright) material around the rim resembling asymmetric flower petals, and terraced inner walls and central peaks. Numerous domes, 1 to 12 km in extent, which are probably caused by volcanic activity, are seen in the southeastern corner of the mosaic. (NASA/Magellan PIA00214) (g) This Magellan image shows an irregular crater on Venus, approximately 14 km in diameter. The crater is actually a cluster of four separate craters that are in rim contact. The noncircular rims and multiple, hummocky floors are probably the result of the breakup and dispersion of an incoming meteoroid during passage through the dense venusian atmosphere. After breaking up, the meteoroid fragments impacted nearly simultaneously, creating the crater cluster. (NASA/Magellan PIA00476)

and often appear in clusters, suggestive of a break-up of meteorites in Venus's dense atmosphere. No craters with diameters less than 3 km have been seen. The projectiles which would have created such small craters must have been broken up or substantially slowed down in the atmosphere. The ejecta blankets around impact craters typically extend out to ~ 2.5 crater radii, like a bright pattern of flower petals. Such patterns are seen only on Venus, whose atmosphere is so thick that ejecta cannot travel very far. Still, the ejecta extend much farther than predicted from simple ballistic emplacements. The ejecta patterns are often asymmetric as a result of oblique impacts, where the missing sector is in the uprange direction. In several cases lava flows are connected with the ejecta. In a number of places radar-dark streaks are seen, sometimes surrounding an impact crater. These streaks are rather smooth areas, and may be caused by the deposition of fine material or pulverization of the surface by atmospheric shocks or pressure waves produced by an incoming meteorite, which itself may have broken up completely in the atmosphere, as was the case in the Tunguska event on Earth (Sections 5.4.3.1, 5.4.6).

5.5.4 Mars

Detailed telescopic drawings from the late nineteenth and early twentieth century showed 'canals' on the surface of Mars (Fig. 5.46b). At the same time, the martian polar caps were discovered. These white regions at the martian poles, which were largest in winter and smallest in summer time, are reminiscent of the polar ice caps on our own planet. The ice caps, canals, and seasonal changes observed in Mars's appearance were factors which suggested to scientists in the late nineteenth to early twentieth century that life might exist on Mars. Although we now know that the surface of the red planet is not inhabited, the questions of whether there is life underground or has been life in the past are central to Mars exploration programs today.

Mars's two moons, Phobos and Deimos, were discovered in 1877. These bodies are likely captured carbonaceous asteroids; they are discussed in detail in Section 9.5.2.

5.5.4.1 *Surface of Mars*

Impact craters. Images of the planet Mars are displayed in Figure 5.46a. The planet's surface shows a global asymmetry. One half of the planet, mostly in the southern hemisphere, is heavily cratered and elevated 1–4 km above the nominal surface level. The other hemisphere is relatively smooth, and lies at or below the nominal surface level. The geologic division between these two hemispheres is referred to as the *crustal dichotomy*. Superficially the highlands, being saturated with craters and interspersed with younger intercrater plains, resemble the Moon and smaller bodies in appearance. In detail, however, the martian terrain and its craters are quite different. Ejecta blankets of

(a)

(b)

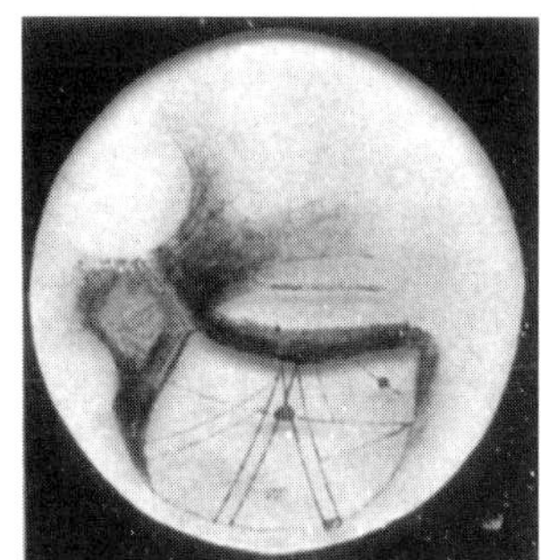

FIGURE 5.46 (a) Hubble Space Telescope images of Mars, taken on 26 February 1995, when Mars was near opposition. *Tharsis region*: A crescent-shaped cloud just right of the center identifies the immense shield volcano Olympus Mons. Warm afternoon air pushed up over the summit forms ice-crystal clouds downwind from the volcano. Farther to the east (right) a line of clouds forms over a row of three extinct volcanoes. *Valles Marineris region*: The 10 km high volcano Ascraeus Mons pokes through the cloud deck along the western (left) limb of the planet. Other interesting geologic features include (lower left) Valles Marineris, (near center) the Chryse basin made up of cratered and chaotic terrain, and (bottom) the oval-looking Argyre impact basin, which appears white due to clouds or frost. *Syrtis Major region*: The dark 'shark fin' feature left of center is Syrtis Major. Below it is the giant impact basin Hellas. (Courtesy: Philip James, Steven Lee, HST/NASA) (b) One of Percival Lowell's sketches of Mars, showing details of his 'canals'. He believed that most canals existed in pairs of two, as shown here.

many martian craters appear to have 'flowed' to their current positions (Fig. 5.47a), rather than traveled through space along ballistic trajectories. This strongly suggests that the surface was fluidized by the impacts. Hence, in contrast to the Moon and Mercury, Mars must have a significant fraction of water-ice in its crust, or at least have had subsurface ice during the (heavy) bombardment era. In addition, martian craters are usually shallower than those seen on the Moon and Mercury, and the craters (rocks and rim) show signs of atmospheric erosion, though not as much as on Earth.

Outflow channels. The oldest martian terrain contains numerous channels, similar in appearance to dendritic river systems on Earth (Fig. 5.47b), where water acts at slow rates over long periods of time. Individual segments are usually less than 50 km long and up to 1 km wide, while the entire dendritic system may be up to 1000 km long. Crater rims and volcanoes are sometimes eroded by these channels. In addition to these dendritic systems, there are other fluvial features on Mars, such as the large channel systems, or *outflow channels*, starting in the highlands and draining into the low northern plains (Fig. 5.47c). Some of these channels are tens of kilometers wide, several kilometers deep and hundreds of kilometers long. The presence of teardrop-shaped 'islands' in the outflow channels suggests vast flows of water, flooding the plains. While some of the martian channels may have been formed by lava flows, the morphology of most (dendritic and outflow) channels strongly suggests that they must have been carved out by water, i.e., by a sustained fluid flow. Since the martian drainage systems lack small-scale streams feeding into the larger valleys, the valleys may be carved primarily by groundwater flow rather than by runoff of rain. Mars Global Surveyor obtained several high resolution images which clearly show evidence of ground seepage and the possible accumulation of water in ponds of some crater floors (Fig. 5.47d).

The most tantalizing images of groundwater flow are shown in Figure 5.47f. These images show the result of fluid seepage and surface runoff. The 'head alcove', located just below the brink of a slope, seems to be the 'source' of a depositional apron just below it. The apron usually shows a main and some secondary channels emanating from the downslope apex of the alcove. These channels start broad and deep at their highest point and taper downslope. The features are clearly distinct from landforms involving 'dry' mass movements, such as granular flows or avalanches. In analogy to terrestrial landforms,

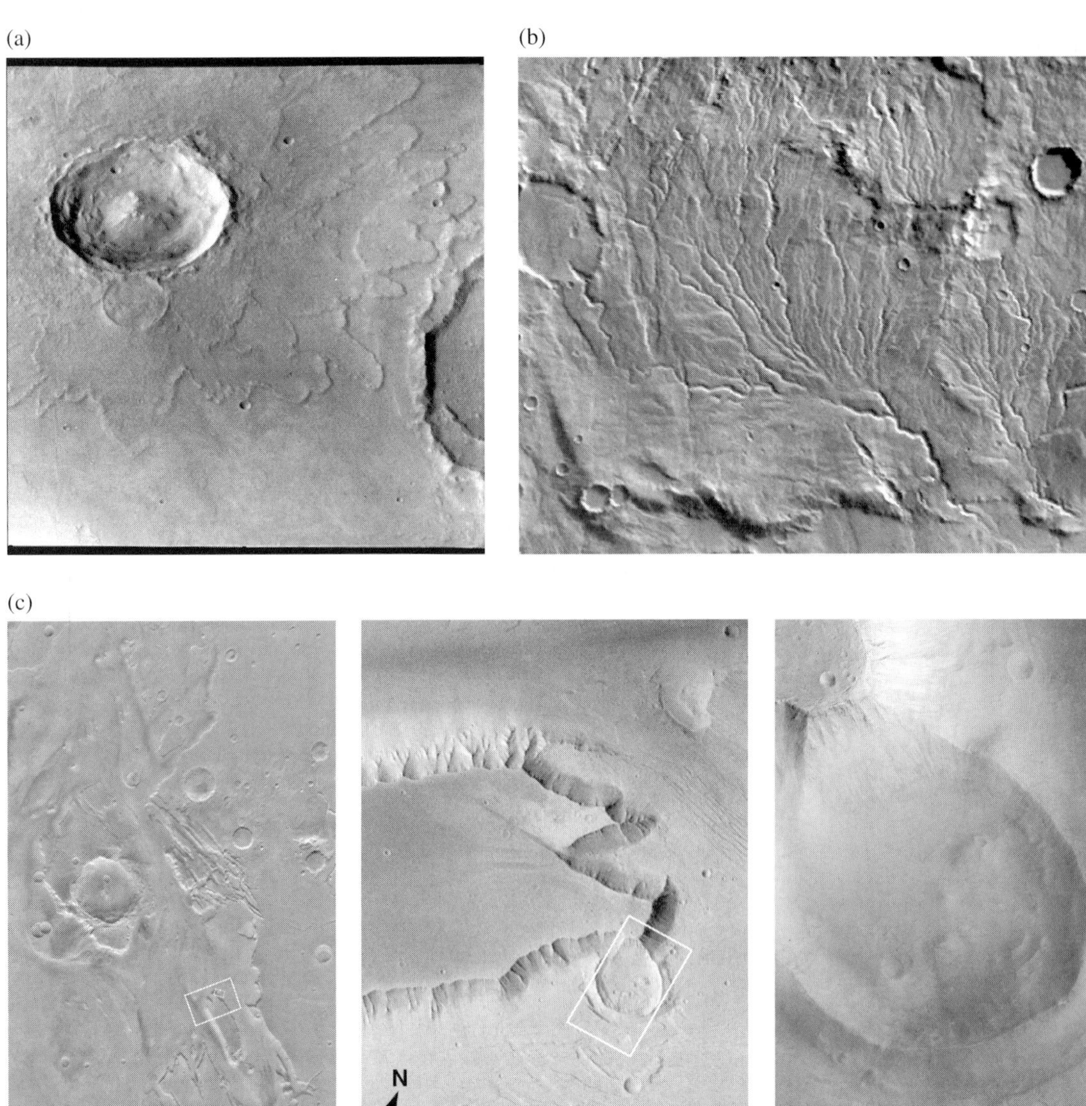

FIGURE 5.47 (a) The ejecta deposits around this martian impact crater Yuty (18 km in diameter) consist of many overlapping lobes. This type of ejecta morphology is characteristic of many craters at equatorial and mid-latitudes on Mars but is unlike that seen around small craters on the Moon. (NASA/Viking Orbiter image 3A07) (b) These channels on Mars resemble dendritic drainage patterns on Earth, where water acts at slow rates over long periods of time. The channels merge together to form larger channels. Because the valley networks are confined to relatively old regions on Mars, their presence may indicate that Mars once possessed a warmer and wetter climate in its early history. The area shown is about 200 km across. (Courtesy: Brian Fessler, image from the Mars Digital Image Map, NASA/Viking Orbiter) (c) Image of a portion of the Kasei Vallis outflow channel system. One can distinguish flow patterns creating 'islands' on the first image. The large white box shows the outline of the Viking 1 image shown in the middle panel, while the small white box outlines the area imaged by Mars Global Surveyor, shown at right. The large crater in the upper center of this overview scene is 95 km in diameter. The highest resolution image shows the 6 km diameter crater that was once buried by about 3 km of martian 'bedrock'. The crater was partly excavated by the Kasei Valles floods over a billion years ago. The crater is poking out from beneath an 'island' in the Kasei Valles. The mesa was created by a combination of the flood and subsequent retreat via small landslides of the scarp that encircles it. (USGS Viking 1 mosaic; Viking 226a08; MOC34504)

(d)

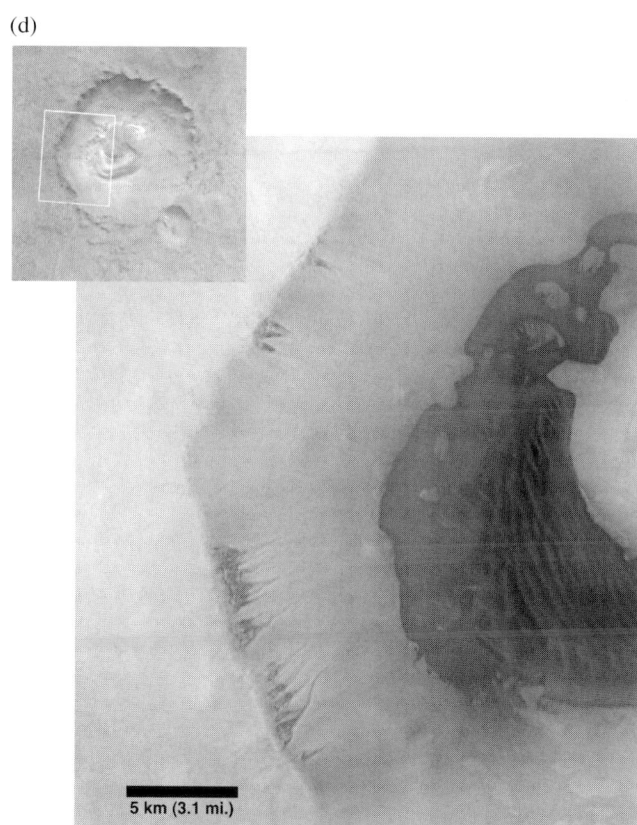

FIGURE 5.47 Continued. (d) The insert shows a global Viking Orbiter 2 image of a 50 km diameter crater. The box shows the location of the large image, a Mars Global Surveyor picture, showing an area approximately 25.1×31.3 km in size. Solar illumination is from the lower left. This image appears to show evidence of seepage (darker features at the edge of the crater) and ponding on the crater floor. (NASA/Mars Global Surveyor 46703) (e) Between the heavily cratered southern hemisphere on Mars and the smooth plains in the northern hemisphere is a third zone, characterized by a complicated mix of cliffs, mesas, buttes, straight-walled and sinuous canyons. This zone is known as the 'fretted terrain', displayed in the low-resolution image on the far left. Steep-walled valleys with lineations, ridges and grooves on their floors are common in this area. The general impression is that the ridges and grooves on the valley floors represent material that has been shed from the smooth canyon walls and was subsequently modified by wind. It is not clear whether any of this material is moving or flowing as it would in an ice-rich deposit (e.g., a glacier). The white boxes on the left images indicate the size of the middle and right panels. (NASA/Mars Global Surveyor MOC 46703)

(e)

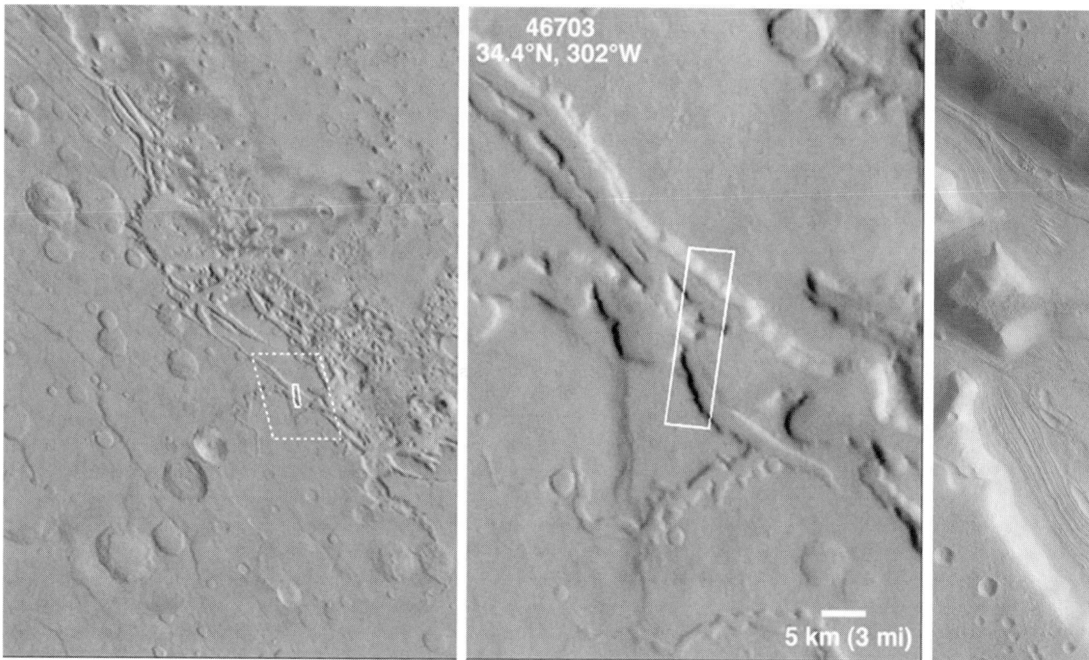

(f)

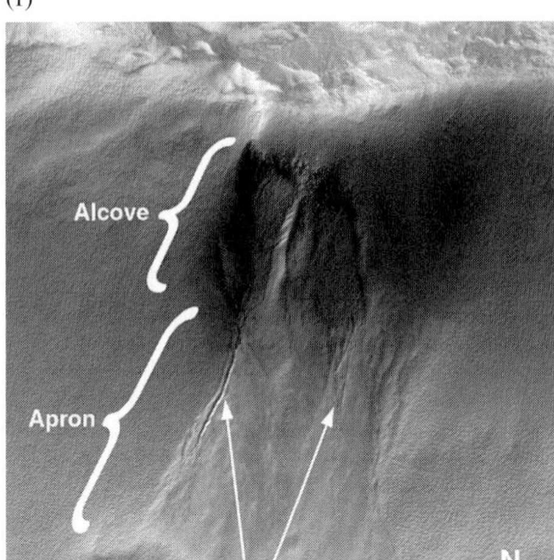

FIGURE 5.47 Continued. (f) Example of landforms that contain the martian gullies. These features are characterized by a theater-shaped alcove that tapers downslope, below which is an apron that broadens downslope. The apron appears to be made of material that has been transported downslope through the channels or gullies on the apron. On the right is a larger-scale view of some such channels. (M03_00537, M07_01873; Malin and Edgett 2000)

these martian gullies must involve a low-viscosity fluid like water. Though these gullies are not the only landforms requiring liquid water to form, intriguing is the observation that all these gullies must be very young: there are no impact craters superposed on these features, and in fact some of the features partly obscure eolian landforms – indicative of the very recent past. It might just be possible that even today subsurface liquid water reservoirs sometimes break through the crust in a short-lived outburst, transporting debris downslope before the water is evaporated. In support of this idea we may note that most of the features were seen at latitudes above 30°.

Flood plains. In 1997, Mars Pathfinder landed in Ares Vallis, a 'flood plain', where rover Sojourner moved around to analyze rocks (Fig. 5.48). The area appeared indeed consistent with a depositional plain as seen on Earth after catastrophic floods, with (semi-)rounded pebbles and here and there a possible conglomerate. Many of the rocks show albedo differences, where the bottom 5–7 cm is lighter than the top, such as could result from a higher soil level in the past, swept away by catastrophic floods.

Volcanism. Crater density studies of the smoother northern hemisphere imply that the plains are much younger than the highlands. Like the lunar maria, the plains appear to have formed by extensive flooding of basaltic lava, about 3–4 Gyr ago. Evidence of volcanism and tectonics is apparent from the giant shield volcanoes in the Tharsis region, including Olympus Mons, and from the incredible canyon system, Valles Marineris (Fig. 5.49). Despite the planet's small size the scale of these martian features dwarfs similar structures on Earth. Figure 5.16 shows a photograph of Olympus Mons, and a comparison in size with Mauna Loa, the largest volcano on Earth. The lower surface gravity and a colder, thicker lithosphere allows the existence of high mountains, which on Earth and Venus would have collapsed due to the larger surface gravity (and tectonic plate movements on Earth). The Tharsis region is about 4000 km wide and rises 10 km above Mars's mean surface level. Three giant shield volcanoes rise another 15 km higher, while Olympus Mons, the largest volcano in our Solar System, located to the west of the Tharsis ridge, rises 18 km above the surrounding high plains to a total height of 27 km. Its base is nearly 600 km wide. The Tharsis region is surrounded by linear rilles and fractures.

Tectonics. Valles Marineris is a tectonicly formed canyon system extending eastwards of Tharsis for 4000 km. The canyons of the Valles Marineris system are 2–7 km deep, and over 600 km wide at its broadest section. The martian canyons probably formed by faulting, processes

(a)

(b)

(c)

FIGURE 5.48 (a) The view from Mars Pathfinder out on 'Twin Peaks'. The Twin Peaks are modest-size hills to the southwest of the Mars Pathfinder landing site. The peaks are approximately 30–35 m tall. North Twin is approximately 860 m from the lander, and South Twin is about a kilometer away. (NASA/Mars Pathfinder PIA01001) (b) This high resolution photo of the surface of Mars was taken by Viking Lander 2 at its Utopia Planitia landing site on 18 May 1979. It shows a thin coating of water-ice on the rocks and soil. The ice seen in this picture is extremely thin, perhaps only tens of micrometers thick. (NASA/Viking lander 2 PIA00533) (c) This image shows the Mars Pathfinder's rover, Sojourner, deploy the Alpha Proton X-ray Spectrometer (APXS) against the rock 'Stimpy'. (NASA/Mars Pathfinder PIA00968)

FIGURE 5.49 COLOR PLATE This image is a mosaic of the Valles Marineris hemisphere of Mars. The center of the scene shows the entire Valles Marineris canyon system, more than 3000 km long and up to 8 km deep, extending from the arcuate system of graben to the west, to the chaotic terrain to the east. Many huge ancient river channels begin in the chaotic terrain and run north into a basin, the dark area in the extreme north of this picture. The three Tharsis volcanoes (dark spots), each about 25 km high, are visible to the west. (Courtesy: USGS, NASA/Viking)

which on Earth result in long straight walls or fault scarps, similar to the canyon walls of Valles Marineris. Some deep branching side valleys suggest that erosion by water, likely by seepage of groundwater, also played a role in the formation of this canyon system, while at other places the canyons have been widened by landslides. The presence of layered sediments in the canyons also suggests that the canyons may have been partly filled by water in the past. Mars Global Surveyor noticed that the walls of the canyons are layered (Fig. 5.50), and sometimes consist of alternate dark ledges and brighter slopes. These materials seem to come out of the walls, and thus must have been deposited and buried long before there was a Valles Marineris. They are seen now because of the faulting and erosion that opened up and widened the Valles Marineris troughs.

Winds. In situ photographs of the martian surface taken by the Viking landers and Mars Pathfinder (Fig. 5.48) show a reddish landscape where big boulders are scattered on a surface of finer-grained soil. The sandy areas sometimes show wind-swept dunes (Fig. 5.20b). The regolith material is thought to be an iron-rich clay, limonite (like rust; Section 5.3.4.3), which gives Mars its reddish color. Near the surface this clay is cemented together by evaporite materials, such as salt, to form a hard crust, called *duricrust*. As discussed in Section 4.5.5.1, the micrometer-sized regolith particles are carried around the planet by fierce winds during giant dust storms. Mars Pathfinder noticed *flutes*, scallop-shaped depressions, and narrow longitudinal grooves in rocks, likely caused by dust-loaded wind. Giant planet-wide dust storms lead to large scale variations in the appearance of the planet through variations in its albedo, which led to theories of life and vegetation changing with seasons in the late 1800s.

Frost, polar caps. The atmospheric pressure at Mars's surface is about 6 mbar, and the temperature varies

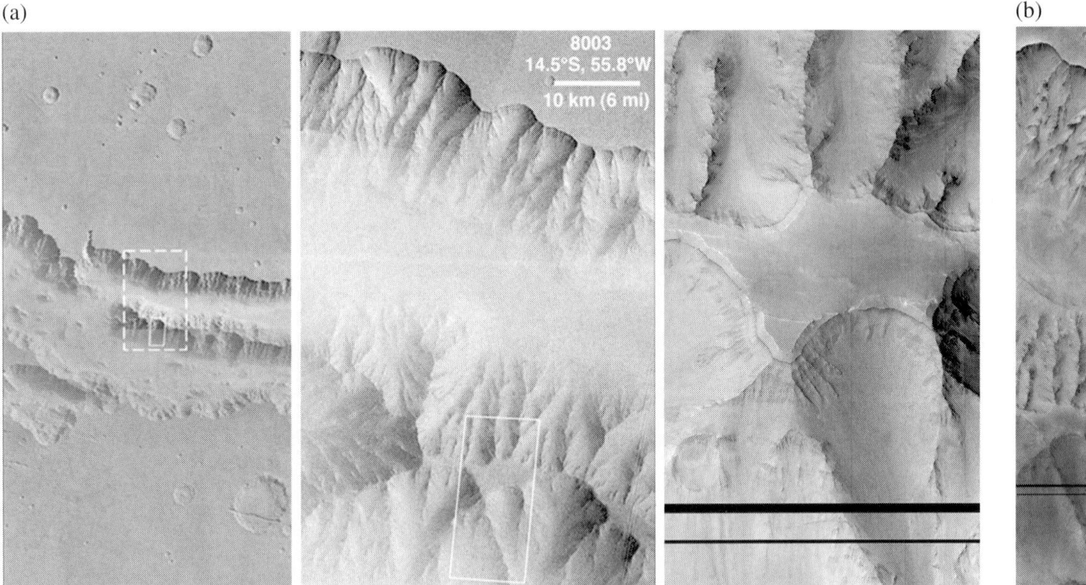

FIGURE 5.50 (a) Close-up photographs of Coprates Chasma in the eastern Valles Marineris system which show many layered outcrops in various locations across the surface of the red planet. The images on the left give context images of the high resolution Mars Global Surveyor image on the right. The white boxes give the approximate size and location; the color plate section (b) shows an enlarged view of the image on the right, extending about three times farther upwards than the image displayed here. The highest terrain in the image is the relatively smooth plateau near the center of the right frame (bottom of the image in (b)). Slopes descend to the north and south from this plateau in broad, debris-filled gullies with intervening, rocky spurs. Multiple rock layers, varying from a few meters to a few tens of meters thick, are visible in the steep slopes on the spurs and gullies. (NASA/Mars Global Surveyor, MOC 8003) (b) COLOR PLATE Enlarged color view of the right panel in Figure 5.50a.

between ~130 K and 300 K (Section 4.2.2.1). Due to the low surface pressure, water exists only in vapor or ice form (Section 4.4.3.2). At night water vapor freezes out and forms a thin layer of frost on Mars's surface (Fig. 5.48b), which sublimes immediately after sunrise. At the poles water is permanently frozen. In the winter the temperature drops below the freezing point of carbon dioxide, which condenses out to form a (seasonal) polar cap of dry ice (Section 4.5.1.3). The CO_2 gas either condenses out directly onto the surface or in the air on condensation nuclei, such as dust grains, which then fall down to the surface. In winter time the ice sheet extends out to a latitude of $\sim 60^\circ$. During the summer the ice sublimes, leaving the dust behind. Over time this has produced a layered structure of dust and ice, as clearly seen in the Viking and Mars Global Surveyor photographs (Fig. 5.51). The dry ice in the northern ice cap sublimes completely away during the summer, leaving behind a permanent ice cap of water-ice, ~1000 km in diameter. In the south the CO_2 never completely sublimates away, leaving a permanent southern cap ~ 350 km in diameter. Images obtained with Mars Global Surveyor show that this residual cap is not just a simple residue of CO_2 frost, but a geological feature indicative of depositional and ablational events unique to Mars's south pole. The northern cap is surrounded by large dune fields, indicative of differences in dust storms between the north and the south poles. The difference between the ice caps on the two poles has been attributed to periodic variations in the orbital eccentricity, obliquity and season of perihelion of Mars. At the moment the northern summer is hotter but shorter than the southern summer.

5.5.4.2 *Signs of (Past) Life?*

The surface morphology shows that Mars must have had much liquid water in the past, and therefore it must have had a very different climate, as discussed in Section 4.9.2. When Mars had running water on its surface, the climate may have been suitable for life to develop. The Viking landers, therefore, searched for life via a number of different experiments. In addition to simple cameras, experiments looked for organic chemicals and metabolic activity in the atmosphere and soil, for example through the addition of nutrients to the soil and looking for chemical byproducts resulting from living organisms (life as we know it). No signs of life were detected. The soil was completely devoid of organic molecules. This could have been

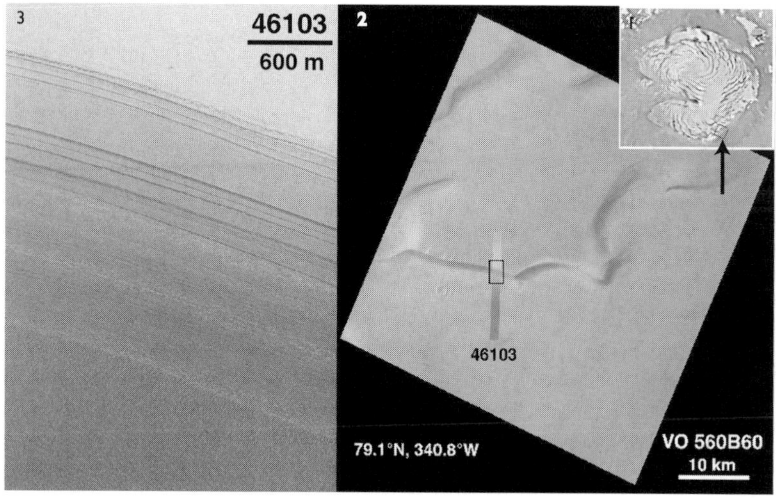

FIGURE 5.51 The martian north and south polar regions are covered by large areas of layered deposits that consist of a mixture of ice and dust. This picture shows Mars's north polar cap (Viking image), with higher resolution Mars Global Surveyor images of a slope along the edge of the permanent north polar cap (arrow) that has dozens of layers exposed in it. The image shows many more layers than were visible to the Viking Orbiters in the 1970s. The layers appear to have different thicknesses (some thinner than 10 m) and different physical expressions. Some of the layers form steeper slopes than others, suggesting that they are more resistant to erosion. All of the layers appear to have a rough texture that might be the result of erosion and/or redistribution of sediment and polar ice on the slope surface.The highest resolution scene covers an area approximately 2.9 km by 1.5 km. (NASA/Mars Global Surveyor, MOC 46103)

expected in retrospect, since the martian soil is directly exposed to solar UV radiation, which will break up any organic molecules. The experiments designed to search for metabolic activity gave some positive results, which are now attributed to unfamiliar reactive chemical states in martian minerals, produced by solar UV radiation.

The meteorite ALH 84001, an igneous rock found in Antarctica which clearly formed on Mars 4.5 Gyr ago (Section 8.2) has some chemical and morphological features which some scientists believe to be traces of microbial life on ancient Mars. The rock was probably ejected into space by an impact about 16 Myr ago, and fell in Antarctica 13 000 yrs ago, as judged from the time that cosmic ray exposure stopped. This rock contains tiny globules of carbonate minerals, which may have been deposited in the cracks by martian groundwater laden with CO_2. In and near these globules scientists have found PAHs (polycyclic aromatic hydrocarbons; an example of these are mothballs), constituents which can be formed through the transformation of dead organisms when exposed to mild heat. However, PAHs are abundant in the interstellar medium, and they can quite easily be formed through a reaction of CO/CO_2 with hydrogen in the presence of minerals, such as magnetite. Tiny (< 0.1 μm long) perfect crystals of pure magnetite have been found in the globules, such as are made by bacteria on Earth. When viewed through an electron microscope, the globules show elongated bacteria-like shapes, as if fossilized microbes; however, they are much smaller (~ 0.1 μm) than terrestrial analogs. The existence of such seemingly incompatible minerals as iron oxide and iron sulfide close together suggests microbial action, unless the globules are formed under extreme high temperatures. The temperature under which the globules formed, however, is very controversial. Some scientists think they formed close to zero °C, others think they must have formed at well over several hundred °C.

5.5.5 Satellites of Jupiter

The Galilean satellites, discovered by Galileo in 1610, are Jupiter's four large moons (Fig. 5.52a). They range in size from Europa, which is slightly smaller than Earth's Moon, to Ganymede, the largest moon in our Solar System. Ganymede is slightly larger than the planet Mercury, but less than half as massive. Compositionally the Galilean satellites represent a diverse grouping, ranging from rocky Io, which has a bulk composition similar to that of the terrestrial planets, to Ganymede and Callisto, which are ~50% rock and 50% water-ice by mass. Geologically, the Galilean satellites are even more diverse: Io is the most volcanically active body in the Solar System, Europa may be covered by a vast ocean topped off by a layer of water-ice, Ganymede has a very diverse and complex geological history and generates its own internal magnetic field, while Callisto's surface is saturated with craters over 10 km in size, but seems to lack smaller craters.

(a)

(b)

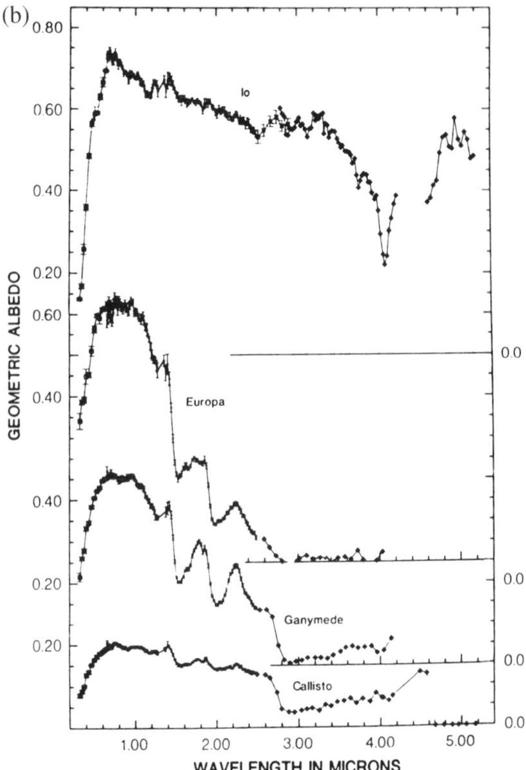

FIGURE 5.52 (a) COLOR PLATE The Galilean satellites: Io, Europa, Ganymede and Callisto, shown from left to right in order of increasing distance from Jupiter. All satellites have been scaled to a resolution of 10 km per picture element. The Io and Ganymede images were acquired in June 1996, the Europa images in September 1996, and the Callisto images in November 1997. (NASA/Galileo orbiter PIA01299) (b) Spectra of the Galilean satellites. (Clark *et al.* 1986)

5.5.5.1 *Io*

The mass and density of Io are very similar to those of Earth's Moon, but the surfaces of the two bodies look vastly different. This difference is due primarily to the dynamical environments that the Moon and Io occupy. Another important difference is that Io is far richer in moderately volatile elements, such as sodium and sulfur, than is the Moon. (Despite containing fewer volatiles, the Moon is less dense than Io because it is also depleted in iron, see Sections 6.3.1, 12.11.) Reflectance spectra (Fig. 5.52b) show that Io's crust is dominated by sulfur-bearing species, such as SO_2-frost and metastable polymorphs of elemental sulfur mixed in with other species. These can also explain many of the colors which give Io its spectacular visual appearance (Fig. 5.52a). The detection of Cl^+ in Io's plasma torus (Section 7.4.4.4) suggests that ordinary table salt (NaCl) may also be present on Io's surface.

The Moon occupies a relatively benign environment, with negligible external stresses or tidal heating. The only major resurfacing process currently operating on the Moon is impact cratering. The Moon's surface is dominated by impact craters together with endogenic features (volcanic basins and mountains) created billions of years ago when the Moon was still warm. In contrast, Io is constantly flexed and heated by tidal forces from Jupiter (Section 2.6.2), producing a global heat flux 40 times larger than the terrestrial value. The vast amount of tidal heat deposited within Io's interior is too large to be removed by conduction or solid-state convection. Melting therefore occurs, and lavas erupt through the surface via giant volcanoes. Over 200 volcanic calderas, most of them over 20 km in size, are distributed over Io's surface. The lava flows from the calderas are hundreds of kilometers long, which implies that the lavas have low viscosities, like basaltic lavas on Earth. Some of the calderas are several kilometers deep, and mountains can rise up to 10 km in altitude.

Nine volcanic plumes were observed by Voyager 1, and most of these were still active eight months later, during the Voyager 2 encounter (Fig. 5.53a). Galileo has seen ten volcanic plumes, six of which are at locations where Voyager did not see plumes. All of the plumes are concentrated in the equatorial region. The plumes can be subdivided into two classes: *Prometheus-type plumes* are typically $\sim$100 km high and a few hundred kilometers across at their base. They are active for several years at least. Prometheus-type plumes are surrounded by bright whitish deposits, attributed to SO_2-frost. These plumes are so-called low-temperature eruptions, probably driven by the vaporization of SO_2 in contact with molten sulfur. The second class of plumes is named after the volcano Pele, one of the few volcanoes that turned off in the four months between the flybys of Voyager 1 and Voyager 2. *Pele-type plumes* are $\sim$300 km high, and over 1000 km wide at their base. These plumes are short-lived (days to weeks) and surrounded by orange to red-black deposits. These high-temperature events may be driven by liquid sulfur, which could be heated to over 1000 K by hot, possibly molten silicates at depths of a few kilometers. The high temperature brings about a phase change in the sulfur (liquid $\rightarrow$ gas), driving the volcano.

(a)

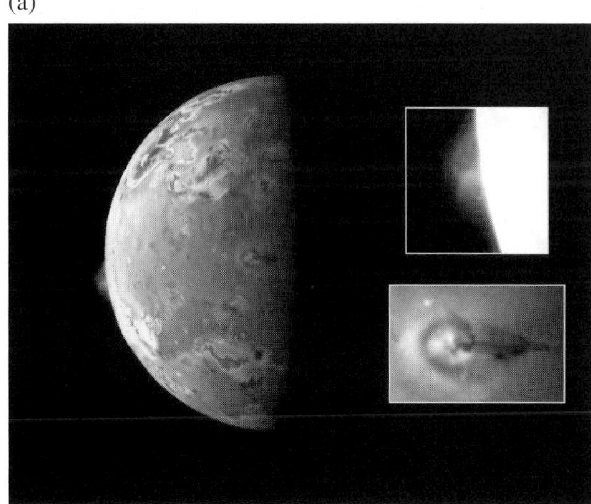

(b)

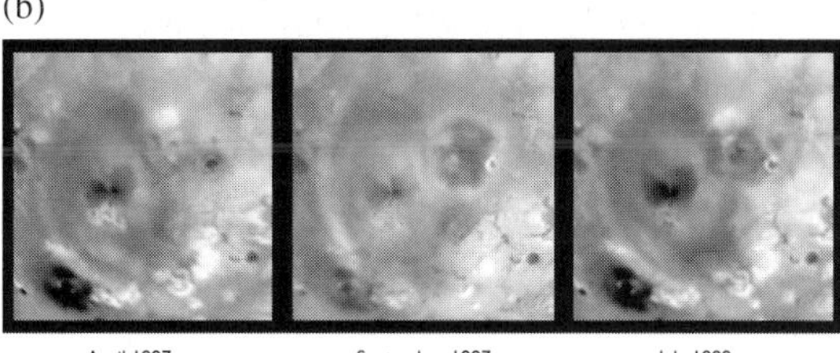

FIGURE 5.53 COLOR PLATE (a) This Galileo image from 28 June 1997 shows two volcanic plumes on Io. A 140 km high plume is seen on the bright limb of Io (see inset at upper right), erupting over caldera Pillan Patera. The second plume, Prometheus, is seen near the terminator (see inset at lower right). The shadow of the 75 km high plume extends to the right of the eruption vent, near the center of the bright and dark rings. The plumes have a blue color, so the plume shadow is reddish. (NASA/Galileo Orbiter PIA00703) (b) The two images of Jupiter's volcanic moon, Io, on the left and in the middle, taken on 4 April 1997 and 19 September 1997, respectively, show the results of a dramatic event that occurred during this five-month interval. A new dark spot, 400 km in diameter, surrounds Pillan Patera (Figure 5.53a). Pele, which produced the larger plume deposit southwest of Pillan, has also changed, perhaps due to interaction between the two large plumes. The image on the right was acquired on 2 July 1999. It shows changes that have taken place on the surface since the 1997 eruption. The red material from Pele, which probably contains some form of sulfur, has started to cover, but has not yet entirely obscured, the dark material around Pillan. This image also shows that a small, unnamed volcano to the right of Pillan has erupted, depositing dark material surrounded by a yellow ring. Some of the color differences between the three images are caused by differences in lighting conditions when the images were taken. The apparent change in brightness of the dark feature in the lower left corner (Babbar Patera) and of parts of Pele's red plume deposit, are thought to be due to changes in illumination. However, such illumination changes cannot explain the dramatic changes seen at Pillan. (NASA/Galileo PIA02501)

The typical plume morphology suggests that material is ejected on ballistic trajectories, at velocities of ~ 0.5 km s^{-1} for Prometheus-type plumes and up to 1 km s^{-1} for Pele-type volcanoes, a factor of 5–10 higher than typical vent velocities on Earth. The volcanoes resurface Io at an incredible rate of $\sim$1–10 cm yr^{-1}. Many regions around volcanic vents had undergone drastic changes in the time between the Voyager and Galileo flybys (1979 to 1996), but even over timescales of months changes can be seen. A blackish area about 40 000 km^2 around the volcano Ra Patera formed within a 5 month period between two Galileo encounters (Fig. 5.53b).

All geological features on Io (Fig. 5.54) are connected to the satellite's strong volcanic activity. There are rugged kilometers-high mountains, layered materials forming plateaus, and many irregular depressions or volcanic calderas. One can also clearly discern dark lava flows and bright deposits of SO_2 frost and/or other sulfurous materials, which have no discernible topographic relief. No impact craters have been identified on Io, which is not surprising when considering these rapid resurfacing processes.

Although much of our knowledge regarding Io's volcanoes is based upon the Voyager and Galileo data, ground-based data are crucial to provide information on the time variability of hot spots, which are usually associated with volcanoes. Low-temperature volcanism appears to be correlated with thermal hot spots having temperatures of a few hundred up to ~ 650 K, while high-temperature events are characterized by hot spots of 900–1200 K. Some observations of hot spots suggest the temperature at their centers to be over 1700 K, indicative of volcanism driven by ultramafic magmas, a style of volcanism that on Earth only occurred early in its history. Speckle or adaptive optics (AO) images at infrared wavelengths can isolate hot spots (Fig. 5.55a), while fast photometry during eclipse observations (when Io disappears or re-appears from Jupiter's shadow) or during occultations of Io by other satellites, also yields highly accurate information on the location, size and temperature (if at more than one wavelength) of hot spots (Fig. 5.55b). The Near Infrared Mapping Spectrometer (NIMS) on Galileo has 'seen' over 50 hot spots in the vicinity of volcanoes, while speckle images from the ground (usually at an even higher spatial resolution, except for the close Galileo encounters) show $\sim$16 hot spots at once (Fig. 5.55a). Hot spots appear and disappear, as clearly shown by the sequence of speckle images in Figure 5.55a. All data together suggest that the hottest regions are usually the smallest in size (a few kilometers), while cooler regions can be a few hundred kilometers in extent. The hot spot information is important for further understanding the

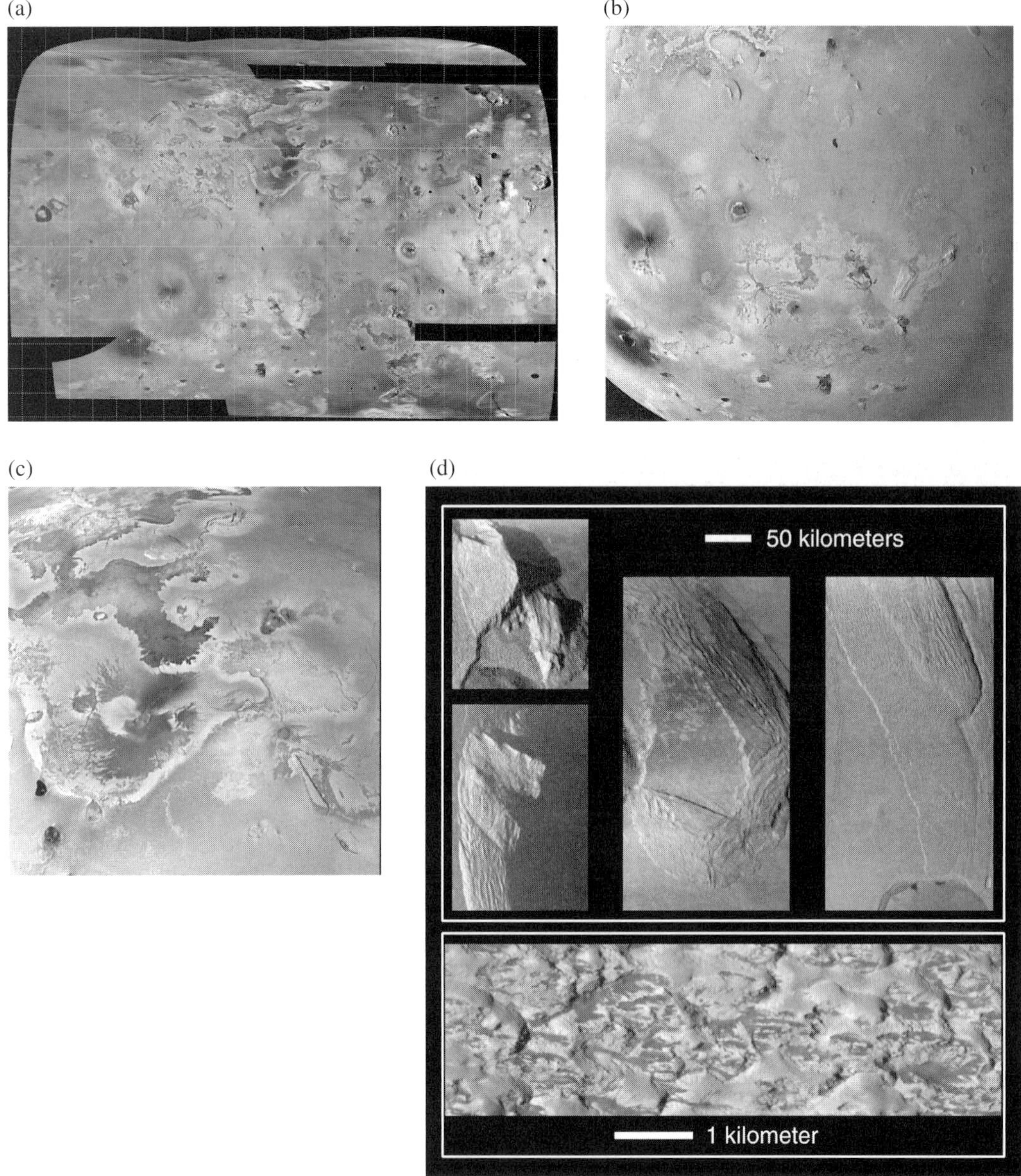

FIGURE 5.54 (a) Mosaic of images taken 6 November 1996 showing more than half of Io's surface. Many different types of topographic features can be discerned. The grid lines mark 10° intervals of latitude and longitude. (NASA/Galileo Orbiter PIA01108) (b) This Galileo view of Io reveals a great variety of landforms. There are rugged mountains several kilometers high, layered materials forming plateaus, and many irregular volcanic calderas. There are also dark lava flows and bright deposits of SO_2 frost or other sulfurous materials, which have no discernable topographic relief at this scale. Several of the dark, flow-like features correspond to hot spots, and may be active lava flows. The large oval on the left hand side is the fallout deposit from Pele, the largest volcanic eruption plume on Io. The image covers an area about 2390 km wide, and the smallest features that can be discerned are 3 km in size. The image was taken 6 November 1996. (NASA/Galileo Orbiter PIA01106) (c) A high resolution image of Io reveals immense lava flows and other volcanic landforms. Several high-temperature volcanic hot spots have been detected in this region, suggesting active silicate volcanism in lava flows or lava lakes. The large dark lava flow in the upper left region of the image is more than 400 km long, similar to ancient flood basalts on Earth and mare lavas on the Moon. The image was taken 6 November 1996, covers an area 1230 km wide and the smallest features that can be discerned are 2.5 km in size. (NASA/Galileo Orbiter PIA00537) (d) Unusual mountains on Jupiter's moon Io are shown in these images that were captured by NASA's Galileo spacecraft during its close Io flyby on 10 October 1999. The images in the top panels show four different mountains at resolutions of $\sim$ 500 m per pixel. The bottom picture is a close-up of another mountain. The Sun illuminates the surface from the left in all five images. (NASA/Galileo PIA02513)

(a)

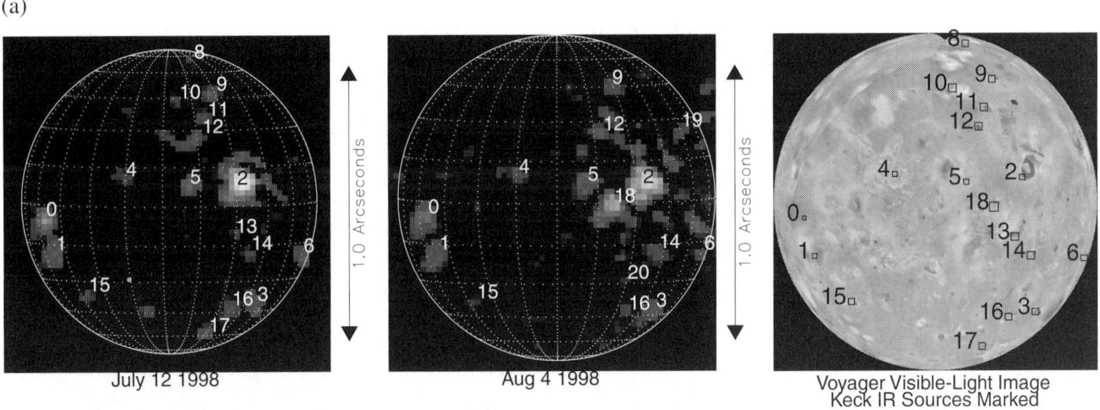

(b)

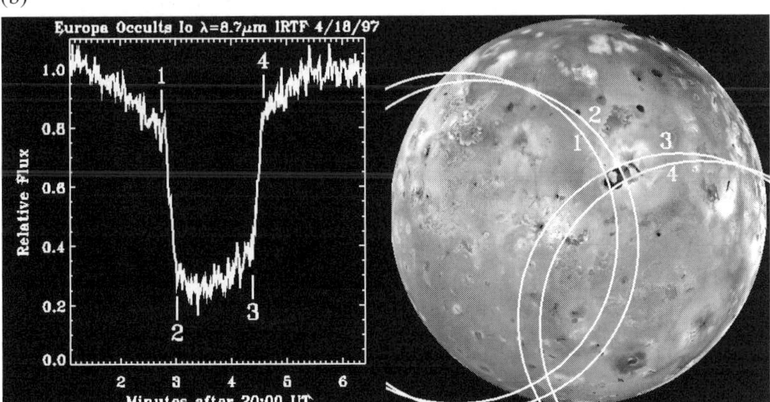

FIGURE 5.55 (a) COLOR PLATE Two near-infrared (2.3 μm) speckle images of Io, taken with the 10-m Keck telescope in 1998 on 12 July and 4 August, respectively. The resolution in this image is 0.04″, or 120 km on Io's surface. Numerous hot spots have been identified, which generally coincide with the dark areas seen at visible wavelengths; the positions of the infrared bright sources are indicated with squares on the Voyager-derived photograph on the right. From the full sequence of pictures a bright transient, spot 18, appeared between 12 and 28 July; after its discovery on 28 July, when it was nearly as bright as Loki (spot 2) it started fading to the intensity seen on 4 August, and disappeared within a few weeks. (Adapted from Macintosh *et al.* 2001) (b) A lightcurve of Io taken when the satellite got occulted by Europa on 18 April 1997. The observations were made at a wavelength of 8.7 μm with NASA's IRTF and the Palomar telescopes. This wavelength is close to the peak of the Planck function for Io's hot spots. As Io's disk is being covered by Europa, there is a steady decrease in the combined brightness from both satellites plotted in the figure (normalized to unity before the occultation). Between times 1 and 2 a bright hot spot, containing about 50% of the total flux, disappears behind Europa's limb and the intensity drops steeply. This source reappears between the times labeled 3 and 4. The image shows a Voyager-derived map of Io, with the outline of Europa's limb at these four times. The hot spot is clearly Loki Patera. (Goguen *et al.* 1997)

heating and cooling mechanism, the driving force behind the plumes, and the composition of Io's interior, its surface and subsurface layers.

5.5.5.2 *Europa*

Europa is slightly smaller and less dense than the Moon. Its surface is very bright and has the spectral properties of nearly pure water-ice. The combination of spectral and density information suggests that Europa is a mostly rocky body with a water-ice crust ~100 km thick. Part of the lower crust is probably liquid, as has been suggested from tidal heating models and details in the surface characteristics as imaged by Galileo. Europa's surface is extremely flat and topographic features are all less than 300 m high,

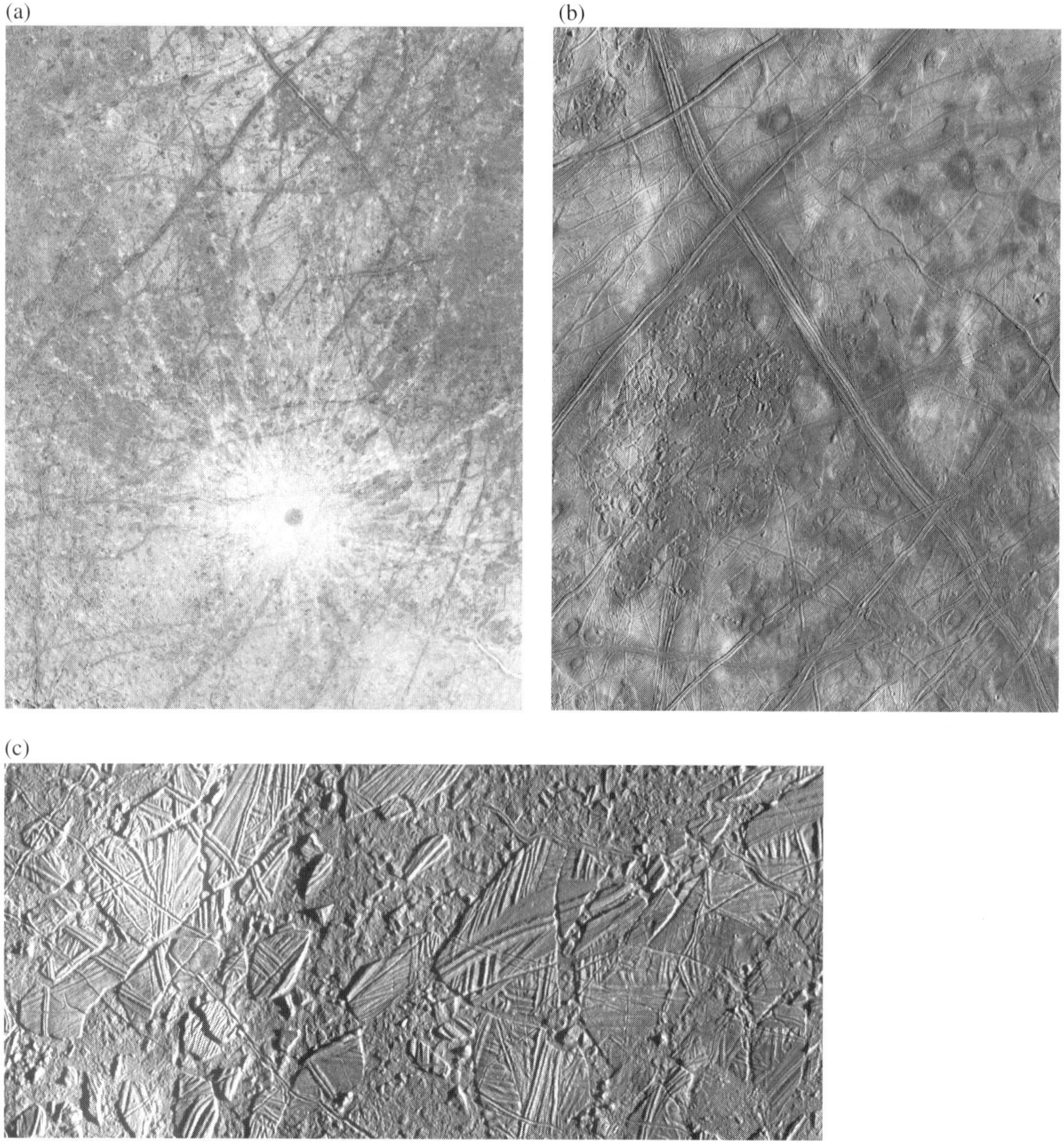

(a)

(b)

(c)

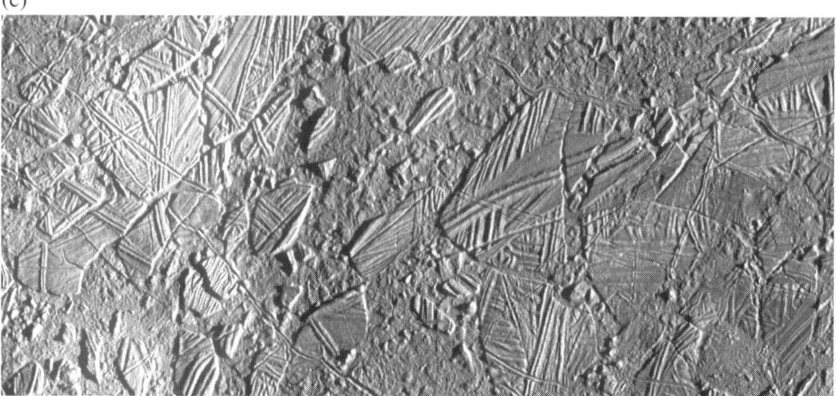

FIGURE 5.56 (a) The 26 km diameter impact crater Pwyll, just below the center of the image, is thought to be one of the youngest features on the surface of Europa. The diameter of the central dark spot is $\sim$40 km, and bright white rays extend for over a thousand kilometers in all directions from the impact site. Also visible in this image are a number of the dark lineaments which are called 'triple bands' because they have a bright central stripe surrounded by darker material. The order in which these bands cross each other can be used to determine their relative ages. The image is 1240 km across (NASA/Galileo Orbiter PIA01211) (b) This image of Europa shows surface features such as domes and ridges, as well as a region of disrupted terrain including crustal plates which are thought to have broken apart and 'rafted' into new positions. The image covers an area of Europa's surface about 250 by 200 km (the X-shaped ridges are seen north of Pwyll crater, Figure 5.56a). (NASA/Galileo Orbiter PIA01296) (c) COLOR PLATE View of a small region of the thin, disrupted, ice crust in the Conamara region of Europa (the disrupted terrain seen in Figure 5.56b) showing the interplay of surface color with ice structures. The white and blue colors outline areas that have been blanketed by a fine dust of ice particles ejected at the time of formation of the large crater Pwyll. A few small craters $\lesssim$500 m in diameter can be seen associated with these regions. These craters were probably formed by large, intact, blocks of ice thrown up in the impact explosion that formed Pwyll. The unblanketed surface has a reddish brown color that has been painted by mineral contaminants carried and spread by water vapor released from below the crust when it was disrupted. The original color of the icy surface was probably the deep blue color seen in large areas elsewhere on Europa. The colors in this picture have been enhanced for visibility. The image covers an area of 70 $\times$ 30 km. (NASA/Galileo Orbiter PIA01127)

(a)

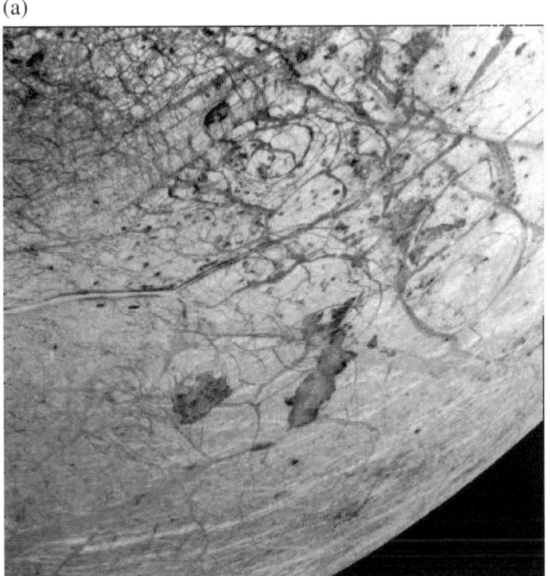

(b)

FIGURE 5.57 (a) Europa's southern hemisphere: The upper left portion of the image shows the southern extent of the 'wedges' region, an area that has undergone extensive disruption. Thera and Thrace Macula are the dark irregular features southeast of the ~1000 km long Agenor Linea. The image covers an area approximately 675 by 675 km, and the finest details that can discerned are about 3.3 km across. (NASA/Galileo Orbiter PIA00875) (b) This mosaic of Europa images shows a region that is characterized by mottled (dark and splotchy) terrain. The mottled appearance results from areas of the bright, icy crust that have been broken apart (known as 'chaos' terrain), exposing a darker underlying material. This terrain is typified by the area in the upper right hand part of the image. The smooth, gray band (lower part of image) represents a zone where the crust has been fractured, separated, and filled in with material derived from the interior. Some of the ridges have themselves been disrupted by the localized formation of domes and other features that may be indicative of thermal upwelling of water from beneath the crust. The mosaic covers an area of 365 km by 335 km. (NASA/Galileo Orbiter PIA01125)

with the youngest features being the highest. Galileo found only a few impact craters, suggestive of a very young surface (tens to at most hundreds of million years), which may still be undergoing active resurfacing.

Europa's surface is characterized by a vast network of ridges (Figs. 5.52a, 5.56, 5.57), many over 1500 km long, that appear brown against Europa's bright disk. These fractures have 'broken' Europa's icy crust into plates ~ 30 km across. The ridges may have formed by an expansion of the crust, or when two ice plates pulled slightly apart. Warmer, slushy or liquid material may have been pushed up through the crack, forming a ridge. The brownish color suggests the slush to consist in part of rocky material, hydrated minerals or clays, or salts. Alternatively, a compression or two plates pushing against each other may have formed a ridge in a similar way, by pushing up the edges of the plates. Many ridges appear to be double or triple. There is some evidence of (past) geyser-like or volcanic activity along these ridges, causing one or more lines of fresh ice on the ridges. The Galileo spacecraft also found regions resembling ice flows, similar in appearance to ice flows in the polar seas on Earth (Figs. 5.56, 5.57). Although these features are suggestive of a liquid ocean underneath Europa's ice crust, the presence of such an ocean has not been proven; the morphological features can also be explained by ice sheets 'floating' on softer/slushy ice below. Measurements of Jupiter's magnetic field around Europa (Section 7.4.4.7) provide additional evidence in favor of the presence of an ocean. Such an ocean, possibly warmed by volcanic activity, opens up a whole host of interesting speculations on the possible existence of life forms on Europa, in analogy with early microbial life forms on Earth which may have thrived on hot vents in the deep ocean.

5.5.5.3 *Ganymede and Callisto*

The outer two Galilean satellites represent a fundamentally different type of body from the inner two. Their low densities imply they contain substantial amounts of water-ice. We note, however, that the ice near the surfaces of these bodies is at such low temperatures that it behaves more like rock than the ice we are familiar with on Earth. The large size of Callisto and Ganymede imply high enough pressures that the ice must be significantly compressed, thus a larger ice fraction is suspected than would be for

(c)

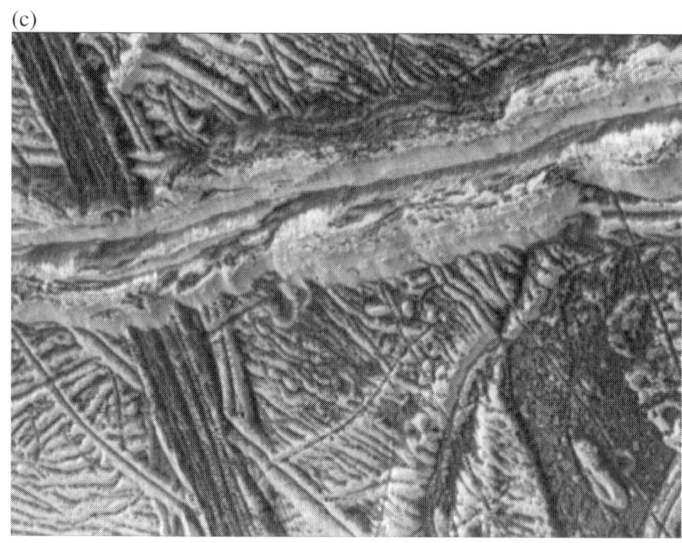

FIGURE 5.57 Continued. (c) This high resolution image (26 m/pixel) of Europa shows a dark, relatively smooth region surrounding a bright hill at the lower right hand corner of the image which may be a place where warm ice has welled up from below. Two prominent ridges are shown: the youngest ridge runs from left to top right and is ~ 5 km wide. This ridge has two bright, raised rims and a central valley. The inner and outer walls show bright and dark debris streaming downslope. An older ridge runs from top to bottom on the left side of the image. This dark ~ 2 km wide ridge is relatively flat, and has smaller-scale ridges and troughs along its length. This image covers an area approximately 15 km by 20 km. (NASA/Galileo Orbiter PIA01179)

smaller bodies of the same density. Ice is also observed in the spectra of both bodies (Fig. 5.52b), but it is far more contaminated than the ice on Europa's surface. The difference in density between Ganymede (1.94 g cm^{-3}) and Callisto (1.85 g cm^{-3}) is greater than can be accounted for by compression, so Callisto probably contains a slightly higher fraction of water-ice. The surfaces of the two bodies are vastly different (Fig. 5.52a). This difference may be the result of processes in the deep interiors of these bodies. Ganymede is clearly differentiated, possibly as the result of an ancient episode of tidal heating, whereas Callisto may remain in a more pristine state (Section 6.3.5.1).

Callisto's surface is dominated by craters and related impact features at large sizes (Fig. 5.59). The craters on both Callisto and Ganymede are usually flatter than on the Moon, and the largest impact basins have no central depressions and surrounding ring mountains. This has been attributed to the relatively low (compared to rock) viscosity of the icy crust. Both satellites have a relatively large number of bright circular features, some with and others without concentric rings with diameters of a few hundred kilometers. These features, called *palimpsests*, are similar to large impact basins, but lack the topographic relief, likely due to viscous surface flows. Both satellites also show many bright impact craters. These may be relatively young, with their high albedo resulting from fresh ice that was ejected from the impact site. In contrast, some craters on Ganymede show an unusually dark floor, and others show dark ejecta. The dark component on the floor of a crater may be residual material from the impactor that formed the crater, or the impactor may have punched through the bright surface to reveal a dark layer beneath. The dark material seen in some ejecta may be part of the impactor (dark asteroid or comet) strewn across the surface upon impact. Callisto's surface shows signs of weakness/crumbling at small scales, which may be produced by sublimation of a volatile component of the crust. This degradation appears to bury and/or destroy craters, and may explain the apparent dearth of (sub-)kilometer sized craters on Callisto's surface (Fig. 5.59).

The geology of Ganymede is very complex. At low resolution Ganymede resembles the Moon, in that both dark and light areas are visible (Fig. 5.52a). However, in contrast to the Moon, the dark areas on Ganymede's surface are the oldest regions, being heavily cratered, nearly to saturation. The lighter terrain is much less cratered, though more than the lunar maria; so it must be younger than the dark terrain, though probably still close to 4 Gyr in age. The light-colored terrain is characterized by a complex system of parallel ridges and grooves (Fig. 5.58a), up to tens of kilometers wide and maybe a few hundred meters high. These features are clearly of endogenic origin and may be of a similar origin (tensional) as the grabens on Earth. It is believed that this form of tectonism is typical for icy satellites, since similar patterns have been seen on a few other icy bodies (e.g., Enceladus and Miranda). One can generally see younger grooves or faults overlaying older sets of grooves, while smooth bands on the surface may be evidence of water-ice volcanic flooding of a fault valley.

5.5.5.4 *Jupiter's Small Moons*

Jupiter's other moons are all very small compared to the Galilean satellites. Their combined mass is about 1/1000 that of Europa, the smallest of the Galilean satellites.

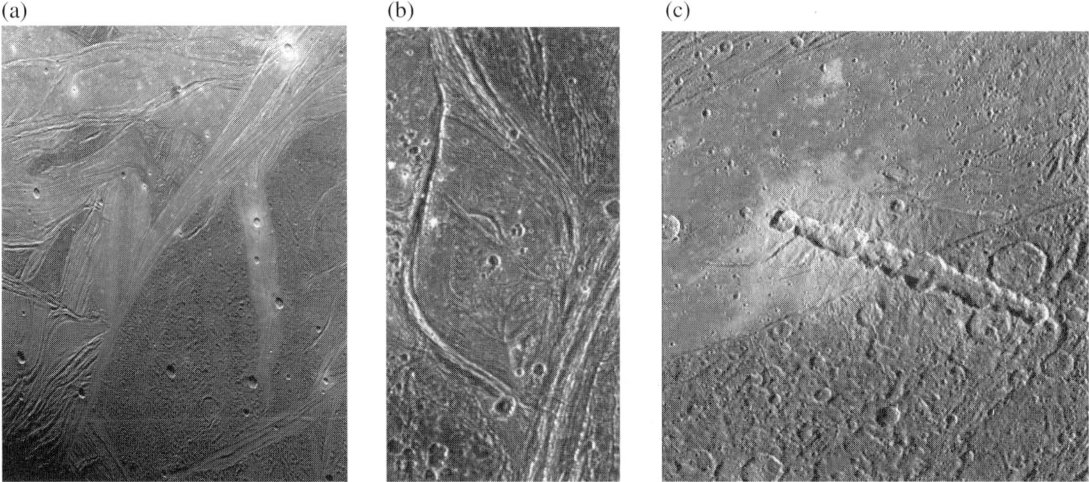

FIGURE 5.58 (a) View of the Marius Regio and Nippur Sulcus area of Ganymede showing the dark and bright grooved terrain which is typical of this satellite. The older, more heavily cratered dark terrain is rutted with furrows, shallow troughs perhaps formed as a result of ancient giant impacts. Bright grooved terrain is younger and was formed through tectonism, probably combined with icy volcanism. The image covers an area approximately 664 by 518 km at a resolution of 940 m/pixel. (NASA/Galileo Orbiter PIA01618) (b) Complex tectonism on Ganymede: The 80 km wide lens-shaped feature in the center of the image is located along the border of Marius Regio, a region of ancient dark terrain near Nippur Sulcus, an area of younger bright terrain (panel a). The tectonism that created the structures in the bright terrain nearby has strongly affected the local dark terrain to form unusual structures such as the one shown here. The lens-like appearance of this feature is probably due to shearing of the surface, where areas have slid past each other and also rotated slightly. The image covers an area about 63 km by 120 km, at 188 m/pixel. (NASA/Galileo Orbiter PIA01091) (c) This chain of 13 craters on Ganymede was probably formed by a comet which was pulled into pieces by Jupiter's tidal force as it passed too close to the planet. Soon after this break-up, the 13 fragments crashed onto Ganymede in rapid succession. The craters formed across the sharp boundary between areas of bright terrain and dark terrain. It is difficult to discern any ejecta deposit on the dark terrain. This may be because the impacts excavated and mixed dark material into the ejecta and the resulting mix is not apparent against the dark background. The image covers an area of 214 $\times$ 217 km, at 545 m/pixel. (NASA/Galileo orbiter PIA01610)

Four moons have been detected inside the orbit of Io: Amalthea is the largest of these moons (Fig. 5.60), discovered in 1892 by Edward Barnard. It is distinctly nonspherical in shape, with dimensions of $270 \times 170 \times 150$ km. Due to its location near Jupiter (at $\sim 2.5\ R_{♃}$), Amalthea probably formed in a hot environment and may contain very few volatiles. It is likely to be a fragment of a larger body disrupted by an impact. The moon is dark and red, and heavily cratered. It possesses two large craters, 90 and 75 km in diameter, and probably 8–15 km deep, in addition to two mountains. The local relief on small Amalthea is about 20 km. The red color might be caused by a coating of sulfur (compounds) spewed out by or eroded off Io. There is some evidence that Amalthea, like Io (Chapter 7), interacts with the local plasma in Jupiter's magnetic field.

The other inner satellites of Jupiter, Thebe, Metis and Adrastea, are also dark and red, in particular Metis and Adrastea are very red, just like Jupiter's ring. Thebe, located at $\sim 3\ R_{♃}$, is $\sim$100 km across. Metis and Adrastea are substantially smaller, and they are located near the outer edge of Jupiter's dusty main ring. All four of Jupiter's inner moons likely form the source for the ring's primarily dust-sized material (Sections 11.3.1, 11.6.3).

Jupiter's outer moons are much further away from Jupiter than the Galilean satellites. They are in highly eccentric and inclined orbits. From their orbits, the moons can be divided into two groups: an outer retrograde set and an inner prograde one. The spectra of these outer moons are similar to outer belt (and Trojan) asteroids, so they may be the remnants of two or more asteroids captured by Jupiter.

5.5.6 Satellites of Saturn

5.5.6.1 *Titan*

Titan, Saturn's largest satellite, was discovered in 1655 by Christiaan Huygens. The satellite, being larger than our Moon, is similar in size to Ganymede, Callisto and Mercury. Titan is surrounded by a dense atmosphere (1.44 bar surface pressure) of nitrogen gas and a small percentage of methane gas. The atmosphere is characterized by a dense smog layer of photochemical origin (Section

(a)

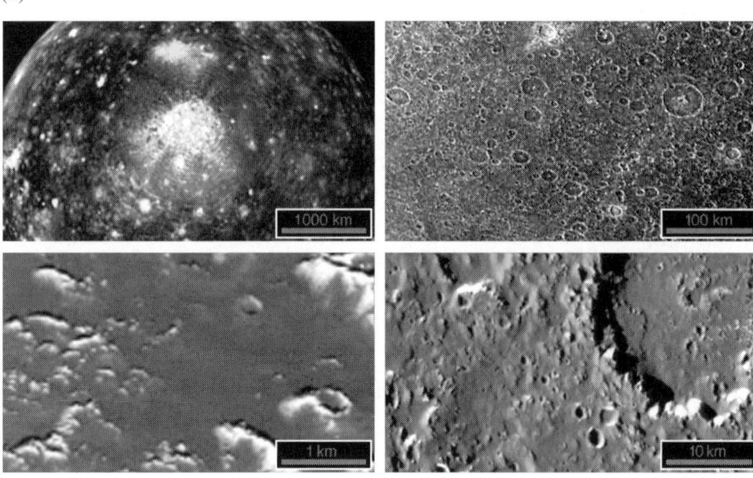

FIGURE 5.59 (a) Four views of Callisto, at increasing resolution. In the global view (top left; 4400 × 2500 km) the surface shows many small bright spots (Valhalla is seen at the center), while the regional view (top right; 10 × higher resolution) reveals the spots to be the larger craters. The local view (bottom right; again 10 × higher resolution) not only brings out smaller craters and detailed structure of larger craters, but also shows a smooth dark layer of material that appears to cover much of the surface. The close-up frame (bottom left) presents a surprising smoothness in this highest resolution (30 m per pixel; area covered 4.4 × 2.5 km) view of Callisto's surface. (NASA/Galileo Orbiter PIA01297) (b) The global image on the right shows the 1700 km wide Asgard multiring structure on Callisto. The young, bright-rayed crater Burr on the northern part of Asgard is about 75 km across. A third type of impact crater, a 55 km wide dome crater (Doh, left image) is located in the bright central plains of Asgard. Dome craters contain a central mound instead of a bowl shaped depression. This type of crater could represent penetration into a slushy zone beneath the surface of the Asgard impact. (NASA/Galileo Orbiter PIA01648)

(b)

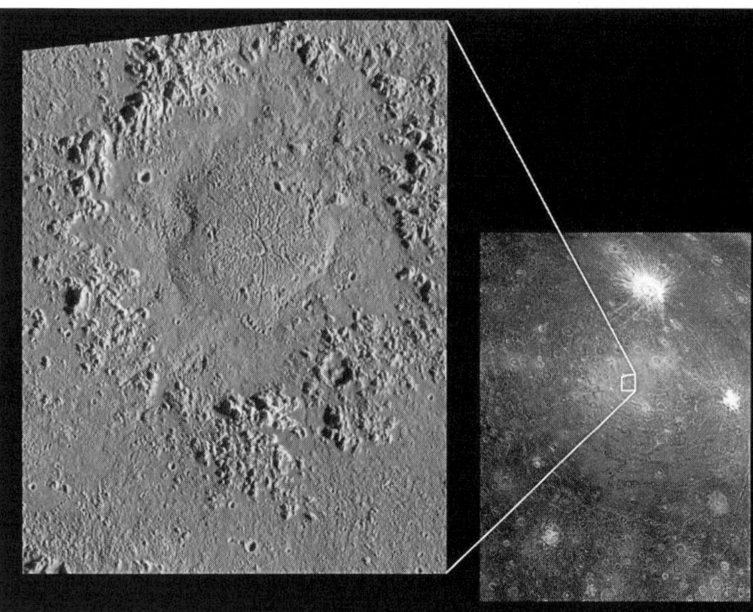

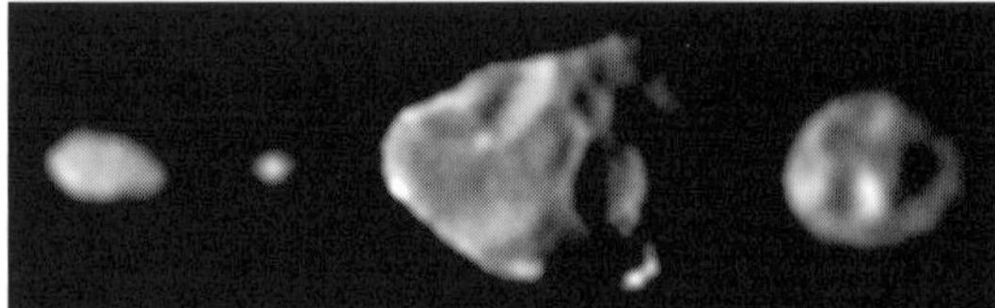

FIGURE 5.60 Four small, irregularly shaped moons that orbit Jupiter in the zone that extends between the planet's ring and the larger Galilean satellites. The moons are shown in their correct relative sizes, with north approximately up in all cases. From left to right, arranged in order of increasing distance from Jupiter, are Metis (longest dimension is approximately 60 km), Adrastea (20 km), Amalthea (247 km across), and Thebe (116 km). (NASA/Galileo Orbiter PIA01076)

4.4.3.3), which makes it impossible to probe the satellite's surface at visible wavelengths. Voyager images show the satellite as an orange ball. Spectra of the atmosphere show the smog layer to consist of various hydrocarbons, nitriles and some carbon oxide compounds. When the aerosol particles, typically 0.2–1 μm across initially, precipitate out of the atmosphere, they grow in size to about a centimeter, because the particles fall down slowly compared to raindrops on Earth due to the lower gravity, and hence have ample time to grow. The temperature and pressure at Titan's surface are such that ethane (C_2H_6), the most abundant photolysis product of methane gas, remains liquid. This has

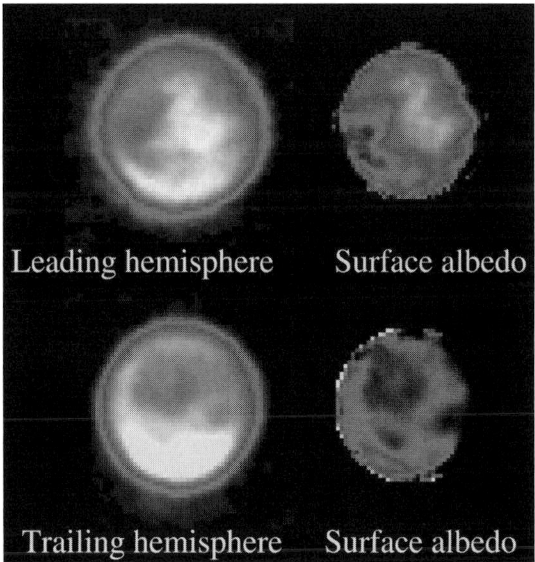

FIGURE 5.61 COLOR PLATE Speckle images of Titan's leading (top) and trailing (bottom) hemispheres, obtained at a wavelength of 2.3 μm with the 10-m Keck telescope. The resolution is 0.04″, or 260 km on Titan's surface. The left side shows the original images, where one can see both the atmospheric haze (in particular near the southern pole) and surface features. The right side shows the recovered surface reflectance, after the atmospheric haze was modeled and subtracted. (Adapted from Gibbard *et al.* 2001)

lead to widespread speculations of Titan possibly being covered by a ∼1 km deep, global ocean of liquid ethane, with some methane, nitrogen and other smog products dissolved into it. Radar observations, however, show weak echoes which would not have been possible if the surface were completely covered by a deep ocean, although the echoes can be produced even if large areas on the surface are flooded.

Although Titan's surface cannot be probed at visible wavelengths, the smog layer is transparent at longer wavelengths, so that the surface can be probed at infrared wavelengths outside the methane absorption bands. Several groups of scientists have made use of these 'windows' to image Titan's surface using speckle and adaptive optics techniques. Speckle images at 2.3 μm of both the leading and trailing hemispheres are shown in Figure 5.61. The images show clear variations in surface albedo, from $A_{ir} < 0.05$ to $A_{ir} > 0.15$, at scales varying from a few hundred up to several thousands of kilometers across. The dark regions are consistent with (though not proven to be) liquid hydrocarbons, while the bright areas are suggestive of a mixture of rock and water-ice. The Huygens probe on the Cassini spacecraft is expected to land near one of the putative oceans on the leading hemisphere.

5.5.6.2 *Medium-sized Saturnian Satellites*

After the discovery of Titan, eight other satellites (Fig. 5.62) were found prior to 1900. These moons range in radius from a litle over 100 km (Phoebe, Hyperion) up to 750 km (Rhea). Most of Saturn's satellites are rather bright, with albedos, A_V, ranging from about 0.3 up to 1.0. *Phoebe*, Saturn's outermost moon whose retrograde orbit suggests capture origin, forms an exception with $A_V \approx 0.06$. *Iapetus* shows one bright side, its trailing hemisphere with $A_V \approx 0.5$ and a dark leading hemisphere with $A_V \approx 0.05$. All of the regular satellites show the presence of water-ice in their surface spectra.

Detailed images taken by various spacecraft show that each satellite has its own unique characteristics. The surfaces of Mimas, Tethys and Rhea are heavily cratered. *Mimas* is characterized by one gigantic crater near the center of its leading hemisphere, about 135 km in diameter, one-third the moon's own size. The crater is about 10 km deep, and the central peak ∼6 km high. The impacting body must have been ∼10 km across, and must have almost broken the satellite apart. *Tethys* displays a ∼2000 km long complex of valleys or troughs, Ithaca Chasma, which stretches three-quarters of the way around the satellite. This system must be caused by tectonic activity, which might have been triggered by the large impact that produced the 400 km diameter crater, Odysseus, on Tethys's leading hemisphere. Both *Dione* and *Rhea* exhibit large variations in surface albedo, although the variations are much less extreme than on Iapetus. The trailing hemispheres of both satellites are relatively dark and covered by whispy, white streaks, perhaps snow or ice. Their leading hemispheres are bright, bland and heavily cratered, although on Dione the crater density varies quite substantially from one region to another. This implies that extensive resurfacing must have taken place. Mimas, Dione and Rhea all have fractures on their surface which manifest themselves as narrow shallow troughs. The whispy streaks may have formed by frost and ice extruded along such fractures. *Hyperion* is oddly shaped, ∼400 × 250 × 200 km, saturated with craters, and it displays a chaotic rotation. Its irregular shape implies that it is a collisional remnant of a larger body.

Enceladus, in orbit around Saturn between Mimas and Tethys, is a remarkable and enigmatic satellite. Parts of this moon are heavily cratered, but large regions on the surface show virtually no impact craters at all. Thus portions of Enceladus's surface must geologically be very young. The youngest parts are probably no more than 100 Myr old. Enceladus displays regions of grooved terrain, just like Ganymede. In addition, its surface reflectivity is very high, about 100%, implying fresh, uncontaminated ice.

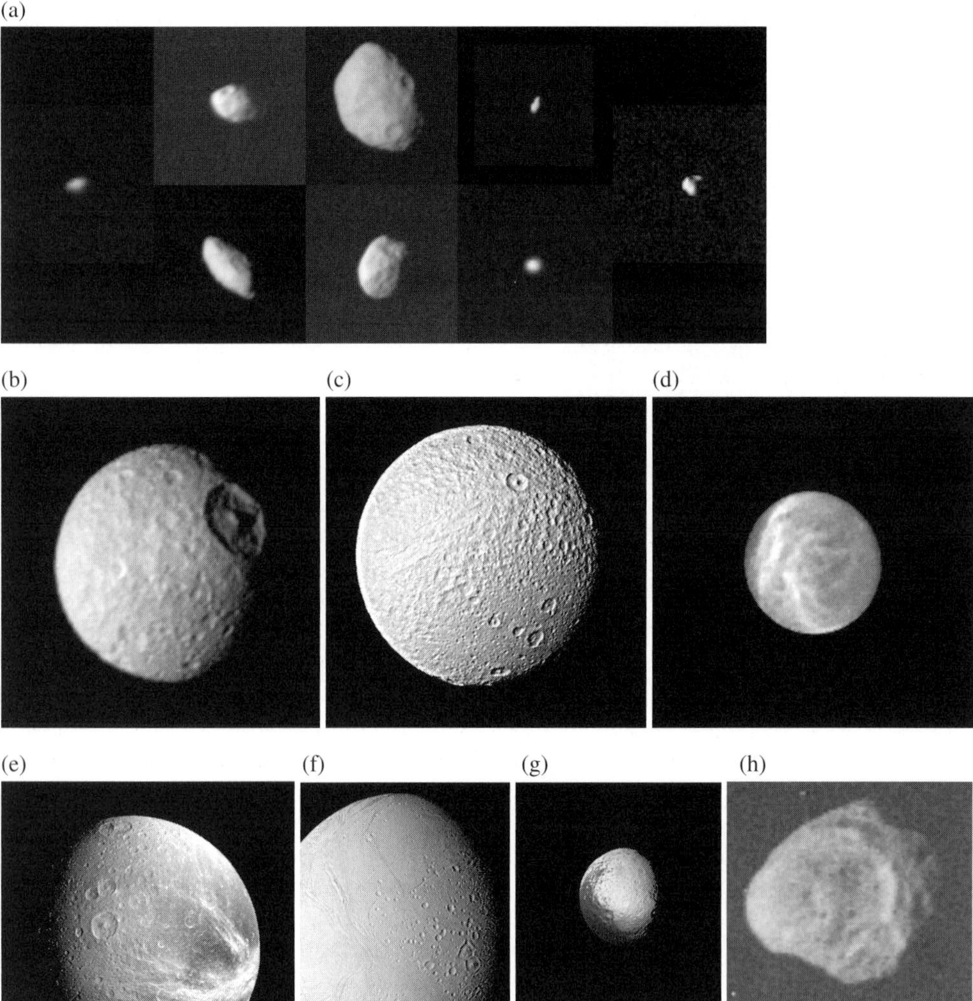

FIGURE 5.62 (a) This family portrait shows the smaller satellites of Saturn as viewed by Voyager 2 during its swing through the saturnian system. The images have been scaled to show the satellites in true relative sizes. This set of small objects ranges in size from ~10 km mean radius to nearly 100 km mean radius. They are probably fragments of somewhat larger bodies broken up during the bombardment period that followed accretion of the saturnian system. (NASA/Voyager 2 PIA01954) (b) The cratered surface of Saturn's moon Mimas. The largest crater, named Herschel, is more than 100 km in diameter and displays a prominent central peak. (NASA/Voyager 1 PIA01968) (c) A Voyager 1 image of Saturn's moon Tethys. The large crater in the upper right lies almost on the huge trench system that girdles nearly three-fourths of the circumference of the satellite. The trench itself is seen in this image as a linear set of markings to the lower left of the crater. (NASA Voyager 2 PIA01397) (d) Bright streaks and blotches are visible against a darker background on the surface of Saturn's satellite Rhea. Even the dark areas, thought to be water frost and ice, are fairly bright with about 50% reflectance. The bright streaks may be related to impacts that threw out pulverized ice grains from beneath the ice-covered surface. (NASA/Voyager 1 PIA01372) (e) Many large impact craters are seen in this view of the saturnian moon Dione. Bright radiating patterns probably represent debris rays thrown out of impact craters; other bright areas may be topographic ridges and valleys. Also visible are irregular valleys that represent old fault troughs degraded by impacts. (NASA/Voyager 1 PIA01366) (f) The surface of Enceladus is moderately cratered, and areas have been transected by strips of younger grooved terrain. Resurfacing has consumed portions of craters such as those near the bottom center of this picture. This suggests that Enceladus has experienced internal melting even though it is only ~490 km in diameter. The grooves and linear features indicate that the satellite has been subjected to considerable crustal deformation as a result of this internal melting. The largest crater visible here is about 35 km across. (NASA/Voyager 2 PIA01950) (g) Very dark material covers the side of Iapetus that leads in the direction of orbital motion around Saturn, whereas the bright material occurs on the trailing hemisphere and at the poles. The bright terrain is made of dirty ice, and the dark terrain is surfaced by carbonaceous molecules. (NASA/Voyager 2 PIA00348) (h) Hyperion is a small oddly shaped satellite of Saturn, presumably a fragment of a larger body that was disrupted by an energetic impact. (Courtesy: Calvin J. Hamilton, NASA/Voyager 2)

Although no volcanic landforms have been detected on the satellite, many features seem to have formed from water flows. The satellite may be resurfaced by a continuous internal process, in which case Enceladus may still contain liquid. This requires a substantial heat source, the cause of which is still a puzzle. Primordial heat or radioactive decay of elements cannot explain the origin of such a heat source. Tidal heating resulting from orbital eccentricities excited by its resonance with Dione appears to be marginally adequate. Enceladus appears to be associated with the tenuous E ring, since the maximum E ring brightness coincides with the satellite's orbit (Section 11.3).

Iapetus is a bizarre body, with its trailing hemisphere $\sim$10 times brighter than the leading hemisphere. Its icy trailing hemisphere and polar regions are very similar to the cratered surface of Rhea. The black material on its leading hemisphere still puzzles scientists. The material is reddish, and might consist of organic, carbon-bearing compounds. It looks like a coat of dark material without any brighter markings on top of it. Thus it must either be very thick, such that impacting bodies cannot penetrate it, or it must continuously be replenished. The low density of the satellite hints at a thin black coating. The question then arises whether it is of internal or external origin. Being on the leading side, Iapetus may just sweep up 'dirt' from Saturn's magnetosphere, such as dust from the dark satellite Phoebe. A few researchers argue for an internal origin, based upon the presence of a few black crater floors on the bright side of the moon.

5.5.6.3 *Small Satellites of Saturn*

In addition to the nine satellites described above, Saturn has a large number of smaller moons, many of which were discovered by the Voyager spacecraft. The smallest of these moons are $\sim$ 20–30 km across. Small moonlets have further been discovered during ring plane crossings from ground-based observations. When the Earth travels through Saturn's ring plane, the rings are seen 'edge-on' and are therefore practically invisible. During this time it is possible to detect tiny moons in Saturn's main ring system. During the 1995 ring plane crossing, both HST and ground-based images revealed the presence of 'clumps' or transient phenomena in the rings. All of Saturn's small moons are oddly shaped, and as reflective as Saturn's larger satellites. Two of Saturn's small moons, *Janus* and *Epimetheus* share the same orbits, and change places every 4 years. *Calypso* and *Telesto* are located at the L_4 and L_5 Lagrangian points of Tethys's orbit. *Helene* resides close to a similar spot, the leading L_4 point in Dione's orbit. *Atlas* is a small moon orbiting just outside the A ring. *Prometheus* and *Pandora* are the inner and outer shepherds of the F ring. Not much is known about the detailed surface morphology of Saturn's small moons. For more information on the dynamical consequences of the orbits of many of Saturn's small satellites, we refer the reader to Chapters 2 and 11.

5.5.7 Satellites of Uranus

Uranus has 21 known moons. Four small uranian moons, which were discovered from ground-based images in the late 1990s, have eccentric and highly inclined orbits located far from the planet. All of the other moons orbit in or near the plane of the planet's equator, which is tilted by 98° with respect to Uranus's orbit about the Sun. Most of our knowledge concerns the five largest moons (Fig. 5.63), which were discovered prior to the space age. Their radii vary from 235 km for Miranda, the innermost large moon, to almost 800 km for Titania (slightly larger than Rhea).

Miranda's surface is bizarre. The smallest and innermost of the five classical moons, Miranda is almost as large as Enceladus. Some areas of Miranda's surface are extremely heavily cratered, as expected for a small cold moon. Other regions have only a few craters and a surprising endogenic terrain including subparallel sets of bright and dark bands, scarps and ridges, with very sharp boundaries between differing types of terrain. There is no good explanation for this great diversity in terrain types. One theory invokes tidal heating, possibly due to chaotic excitation of orbital eccentricities upon passage through resonances. The surface features may result from an incomplete differentiation and convection pattern. It has also been suggested that Miranda was disrupted by a catastrophic impact early in its history and then reaccreted with parts falling into various positions. The cratered terrain, however, could probably not be preserved through such an event. The differences in terrain could also have resulted from a sinking of material in regions where the heavy core material reaccreted on the outside and subsequently sank to the center.

Next in distance from Uranus is *Ariel*. Ariel shows clear signs of local resurfacing, but nothing as striking as Miranda. The age of terrain varies significantly, but the entire surface appears to be younger than the oldest terrains on each of the three outer classical moons. There exists a global system of faults and evidence of flows attributed to ice volcanism. Although *Umbriel* is similar in size to Ariel, this moon is heavily cratered and appears to have the oldest surface in the uranian system. There is little or no evidence of tectonic activity. Most of *Titania* is heavily cratered; however, some patches of smoother material with fewer craters imply local resurfacing. An extensive net-

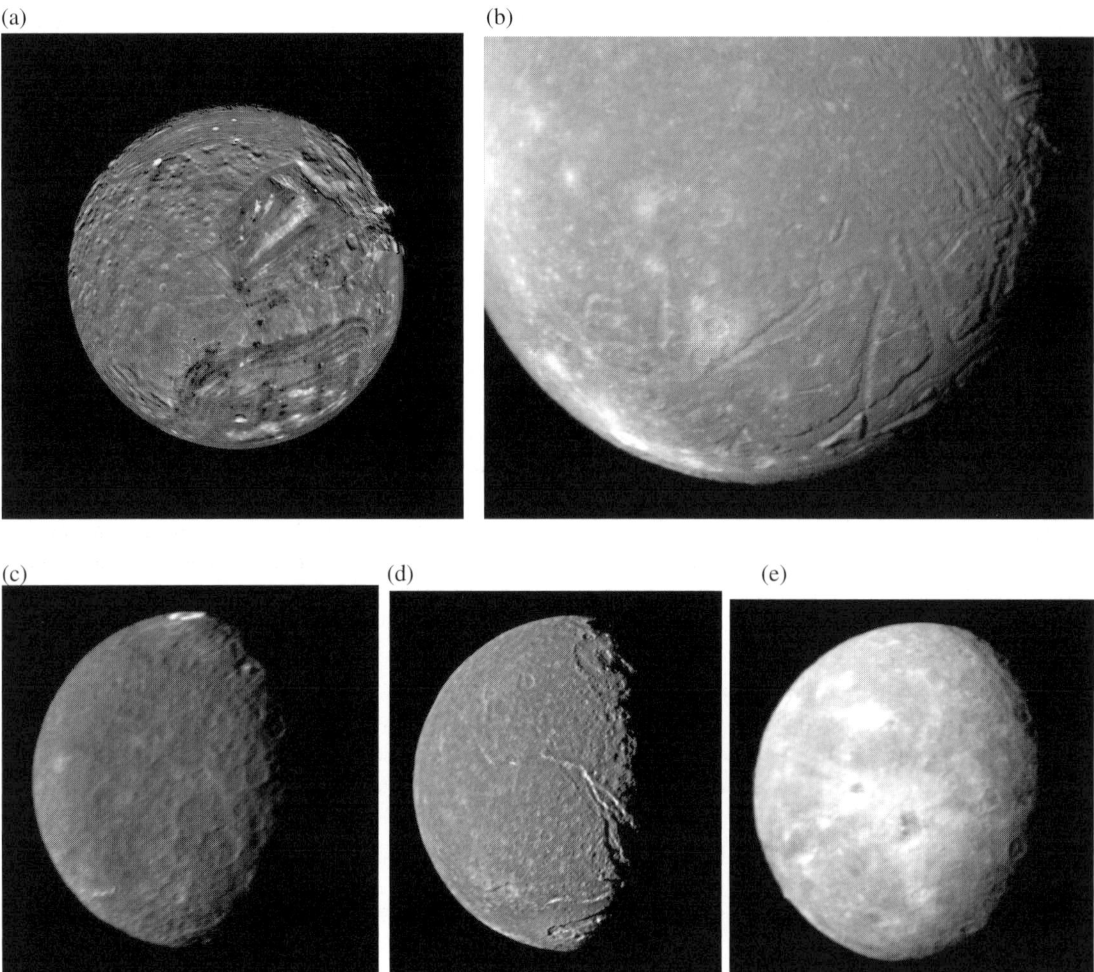

FIGURE 5.63 (a) Uranus's satellite Miranda displays two strikingly different types of terrain. An old, heavily cratered rolling terrain with relatively uniform albedo, contrasted by a young, complex terrain which is characterized by sets of bright and dark bands, scarps and ridges, as seen most distinctly in the 'chevron' feature. (NASA/Voyager 2 PIA01490) (b) Most of Ariel's visible surface consists of relatively heavily cratered terrain transected by fault scarps and fault-bounded valleys (graben). Some of the largest valleys, which can be seen near the terminator (at right), are partly filled with younger deposits that are less heavily cratered. Bright spots near the limb and toward the left are chiefly the rims of small craters. Although Ariel has a radius of only ~ 600 km, it has clearly experienced a great deal of geological activity in the past. (NASA/Voyager 2 PIA00041) (c) This is the most detailed Voyager image of Umbriel, the darkest of Uranus's larger moons and the one that appears to have experienced the least geological activity. Umbriel is heavily cratered but lacks the numerous bright-ray craters seen on the other large uranian satellites. The prominent crater on the terminator (upper right) is about 110 km across and has a bright central peak. The strangest feature in this image (at top) is a curious bright ring, the most reflective area seen on Umbriel. The ring is about 140 km in diameter and lies near the satellite's equator. The nature of the ring is not known, although it might be a frost deposit, perhaps associated with an impact crater. (NASA/Voyager 2 PIA00040) (d) Titania, Uranus's largest satellite, is heavily cratered and displays prominent fault valleys that are up to 1500 km long and as much as 75 km wide. In valleys seen at right-center, the sunward-facing walls are very bright, perhaps indicative of younger frost deposits. An impact crater more than 200 km in diameter is seen at the very bottom of the disk, while an even larger impact crater is visible at top. (NASA/Voyager 2 PIA00039) (e) Several large impact craters in Oberon's icy surface surrounded by bright rays similar to those seen on Jupiter's moon Callisto. Quite prominent near the center of Oberon's disk is a large crater with a bright central peak and a floor partially covered with very dark material. This may be icy, carbon-rich material erupted onto the crater floor sometime after the crater formed. Another striking topographic feature is a large mountain, about 6 km high, peeking out on the lower left limb. (NASA/Voyager 2 PIA00034)

work of faults cuts the surface of Titania. *Oberon*'s surface is dominated by craters, but there are several high-contrast albedo features and signs of faulting.

All of Uranus's small moons discovered by Voyager orbit the planet inside of Miranda's orbit. The smallest satellites, with radii of ~15 km, are *Cordelia* and *Ophelia*, two ring shepherds which control the inner and outer edges of the outermost ring. The largest of the small satellites is *Puck*, with a radius of ~ 80 km. It is darker than any of the five large satellites, slightly irregular in shape, and heavily cratered. The other small satellites are also quite dark. The dark surfaces suggest that the satellites consist of an undifferentiated mixture of icy, silicate and carbonaceous material, or that the bodies' surfaces darkened over time as the result of micrometeorite impacts and sputtering, subliming ices away but leaving the darker material behind (Section 4.8.2). If the former speculation is right, these satellites consist of a chemically very primitive material.

The densities of the four largest satellites of Uranus are significantly higher than those of Saturn's satellites of comparable size. Uncompressed densities of these moons are estimated to be ~1.45–1.5 g cm^{-3}. These high densities have led to the suggestion that the circum-uranian disk out of which these moons presumably formed had most of its oxygen in the form of CO rather than H_2O, and thus was depleted in water relative to the saturnian nebula. While plausible, this suggestion must be regarded as quite tentative owing to all the bizarre ways in which other bodies are believed to have gotten their anomalous densities.

5.5.8 Satellites of Neptune

Before the Voyager flyby Neptune was known to possess two moons, *Triton* and *Nereid*. Both of these bodies occupy 'unusual' orbits. *Triton* orbits Neptune at 14.0 R_ψ, has a very small eccentricity ($e < 0.0005$), but its orbit is inclined 159° with respect to Neptune's equator, while Neptune's rotation axis is inclined by 28.8°. Because of these tilts, Triton's poles point occasionally to the Sun during a complicated cycle of seasons within seasons, which lasts about 600 years. *Nereid* orbits Neptune on a prograde orbit, inclined by ~ 27°, with a semimajor axis of 219 R_ψ. The orbit has the largest eccentricity of any known moon, with $e = 0.76$. Nereid is a small moon, with a radius of ~170 km, and is reasonably round. Six more small satellites have been found in the Voyager images. These six satellites orbit Neptune in a prograde manner, on low-inclination circular orbits.

Triton is by far the largest moon in the neptunian satellite system (Fig. 5.64), with a size somewhat smaller than Europa. It has a tenuous atmosphere of nitrogen, with a trace of methane gas (mixing ratio ~ 10^{-4}; Section 4.3.3.3). Triton has the lowest observed surface temperature of any body in the Solar System, 38 ± 4 K. The polar cap on the southern hemisphere is bright, with an albedo of ~ 0.9, while the equatorial region is somewhat darker and redder. Most of the surface is covered with a thin layer of nitrogen and methane ice, although it does not completely hide the underlying terrain. The western (trailing) hemisphere of Triton looks like a 'cantaloupe'; a dense concentration of pits or dimples, criss-crossed by ridges or fracture systems. This terrain may have a long history of repeated fracturing and some form of viscous icy volcanism. It is the oldest terrain on Triton, but, since it is much less cratered than the satellites of Uranus and Saturn, it must be geologically very young. The leading hemisphere consists of a smoother surface, with large calderas and/or lava lakes. This terrain is probably a few billion years old. The icy substance which produced the smooth plains was less viscous than that seen on the trailing hemisphere, which suggests a chemically different composition (there may be less ammonia involved). Multiple levels of cooling and stagnation are apparent near the ice lava lakes. At several places it seems like a volcanic fluid cooled to form a solid lid, after which the fluid underneath drained away causing parts of the lid to founder and melt. The newer lava lakes at lower altitudes undergo a similar process, forming layers of 'lids' which look like collapsed calderas. It is not clear if the features seen on Triton are formed by drainage of lava lakes or from collapsed calderas. The polar regions are covered with ice, presumably N_2-ice, which evaporates in the spring (see Section 4.5.1.3). The ice has a slightly reddish tint, indicative of organic compounds. In these regions a large number of relatively dark (10–20% lower albedos than the surroundings) streaks have been found. At least two of these streaks appear to be active geyser-like plumes of liquid nitrogen. The plumes rise ~8 km before the winds sweep the materials westwards across the planet. Horizontal plumes may reach lengths of over a hundred kilometers. Although the heating mechanism for the geysers has not yet been understood, it is suspected that sunlight plays a major role, since all four geysers detected are in the south polar region where the surface was continuously illuminated by the Sun. The geysers may result from a solid-state, subsurface ice greenhouse effect, where the subsurface ice is heated by a few degrees, leading to an enhanced subsurface nitrogen vapor pressure, resulting in geyser-like eruptions of gas and dust.

Voyager discovered six satellites within and near Neptune's ring system. The largest of these, *Proteus*, with a radius of 200 km, is slightly larger than Nereid. The radii

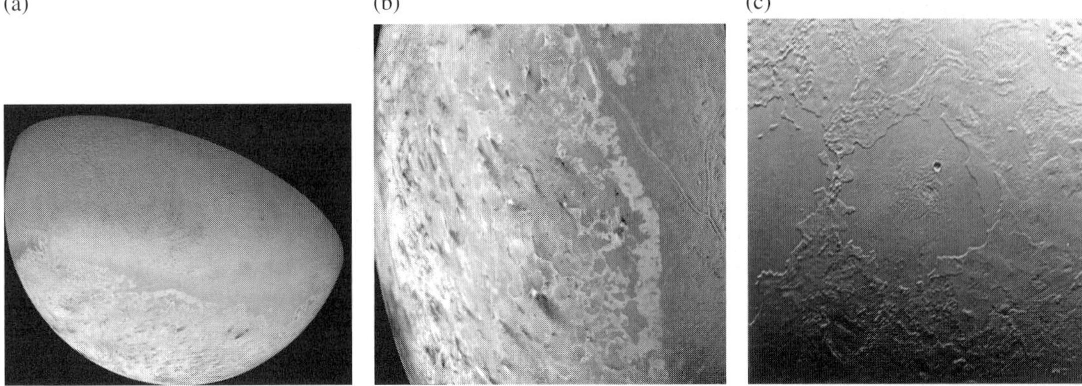

FIGURE 5.64 (a) COLOR PLATE A global color mosaic of Neptune's largest moon, Triton. Color was synthesized by combining images taken through orange, violet, and ultraviolet filters, which were displayed as red, green, and blue images and combined to create this color version. Triton's surface is covered by nitrogen-ice, while the pinkish deposits on the south polar cap are believed to contain methane-ice, which would have reacted under sunlight to form pink or red compounds. The dark streaks overlying these pink ices are believed to be an icy and perhaps carbonaceous dust deposited from huge geyser-like plumes, some of which were found to be active during the Voyager 2 flyby. The bluish-green band visible in this image extends all the way around Triton near the equator; it may consist of relatively fresh nitrogen frost deposits. The greenish areas include what is called the cantaloupe terrain, whose origin is unknown, and a set of 'cryovolcanic' landscapes apparently produced by icy-cold liquids (now frozen) erupted from Triton's interior. (NASA/Voyager 2 PIA00317) (b) This image of the south polar terrain of Triton reveals about 50 dark plumes or 'wind streaks' on the icy surface. The plumes originate at very dark spots, generally several km in diameter and some are more than 150 km long. The spots, which clearly mark the source of the dark material, may be vents or geysers where gas has erupted from beneath the surface and carried dark particles into Triton's nitrogen atmosphere. Southwesterly winds then transported the erupted particles, which formed gradually thinning deposits to the northeast of most vents. (NASA/Voyager 2 PIA00059) (c) This view of Triton is about 500 km across. It encompasses two depressions, possibly old impact basins, that have been extensively modified by flooding, melting, faulting and collapse. Several episodes of filling and partial removal of material appear to have occurred. The rough area in the middle of the bottom depression probably marks the most recent eruption of material. Only a few impact craters dot the area, which shows the dominance of internally driven geologic processes on Triton. (NASA/Voyager 2 PIA01538)

of the other satellites vary between 27 and 100 km. All of these satellites are dark and irregular in shape.

5.5.9 Pluto and Charon

Pluto is by far the smallest of the nine major planets, with a diameter about 2/3 the size of our Moon. As discussed in more detail in Sections 10.2.2 and 10.7.2, Pluto might be better classified as a large Kuiper belt object than a planet. Its moon, *Charon*, is about half Pluto's size. The mean density of Pluto and Charon combined is $\sim$2 g cm^{-3}, which is significantly higher than that of a rock/water-ice mixture in cosmic proportion. Since Pluto and Charon are so close together on an angular scale, it is difficult to extract information on each of the bodies separately. In the mid-1980's Charon's apparent orbit was such that the satellite passed in front of and behind Pluto, enabling scientists to obtain data on Pluto and Charon separately, and to model the spatial distribution of Pluto's surface albedo. It appeared that Pluto's poles were much brighter than its equator. In more recent years, high resolution images of the planet have been obtained directly with the Hubble Space Telescope (Fig. 5.65), which confirmed the brightness difference between the poles and the equator.

Infrared spectra of Pluto, obtained in the early 1990's, suggest its surface to be covered by nitrogen-ice with traces of methane and carbon monoxide. During the decades that the planet approached perihelion, which occurred in 1989, the temperature slowly increased and it is believed that more and more ice sublimed from the surface and formed the tenuous atmosphere discovered roughly a decade earlier. The sublimation of ice explains why Pluto's brightness decreased as the body approached perihelion. The fact that Pluto's equatorial regions are darker than its poles also agrees with this general view of ices subliming from the warmest areas on the planet.

In contrast to Pluto, Charon's spectrum is characterized by absorption lines of water-ice, and is devoid of methane features. Though the methane lines were the first features discovered on Pluto, we now know that methane-ice is only a trace constituent in the otherwise nitrogen dominated ice cover. Because of Charon's lower surface gravity, the satellite probably lost its nitrogen–methane–carbon monoxide ice through sublimation and subsequent escape.

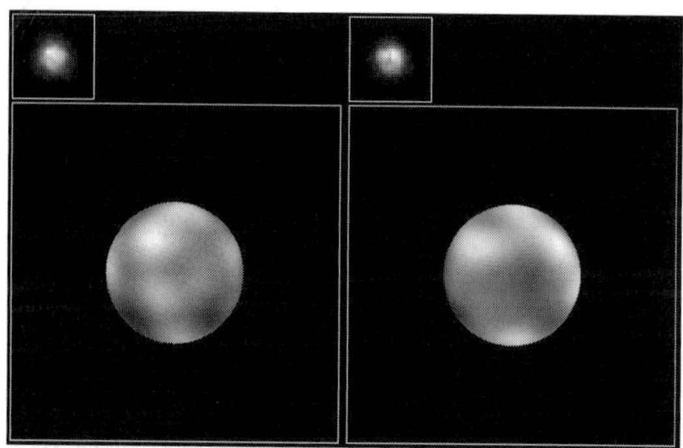

FIGURE 5.65 Hubble Space Telescope images of the surface of Pluto, taken over a 6.4 day period in late June and early July 1994. The two smaller inset pictures at the top are the actual images from HST. North is up. Each pixel is more than 150 km across. At this resolution, HST discerns roughly 12 major 'regions' where the surface is either bright or dark. The larger images (bottom) are from a global map constructed through computer image processing performed on the Hubble data. The tile pattern is an artifact of the image enhancement technique. The two views show opposite hemispheres of Pluto. Most of the surface features, including the prominent northern polar cap, are likely produced by the complex distribution of frosts that migrate across Pluto's surface with its orbital and seasonal cycles and chemical byproducts deposited out of Pluto's nitrogen–methane atmosphere. (Courtesy: Alan Stern, Marc Buie, NASA/HST and ESA, PIA00825)

It is feasible that a fraction of Charon's sublimated ice molecules, after escaping the moon's gravity, were trapped by Pluto and enhanced Pluto's ice coverage.

Further Reading

An easy to read, yet accurate description of the geology of many planetary bodies can be found in:

Beatty J.K., C.C. Peterson, and A. Chaikin, Eds., 1999, *The New Solar System*, 4th Edition. Sky Publishing Co., Cambridge, MA and Cambridge University Press, Cambridge, England.

Background material on mineral phases and cooling of a magma is summarized by:

Putnis, A., 1992. *Mineral Science*. Cambridge University Press, Cambridge.

Examples of books on general geology/geophysics of the Earth and other planetary bodies are:

Catermole, P., 1994. *Venus, The Geological Story*. Johns Hopkins University Press, Baltimore.

Frankel, C., 1996. *Volcanoes of the Solar System*. Cambridge University Press, Cambridge.

Greeley, R., 1994. *Planetary Landscapes*. Chapman and Hall, New York, London.

Hamblin, W.K., and E.H. Christiansen, 1990. *Exploring the Planets*. Macmillan Publishing Company, New York.

Hartmann, W.K., 1999. *Moons and Planets*, 4th Edition. Wadsworth Publishing Company, Belmont, CA.

Press, F., and R. Siever, 1986. *Earth*, 4th Edition. W.H. Freeman and Company, New York.

Press, F., and R. Siever, 1998. *Understanding Earth*, 2nd Edition. W.H. Freeman and Company, New York.

Stacey, F. 1992. *Physics of the Earth*. Brookfield Press.

An excellent monograph of impact cratering is written by:

Melosh, H.J., 1989. *Impact Cratering, A Geologic Process*. Oxford Monographs on Geology and Geophysics, no. 11. Oxford University Press, New York.

A pre-Galileo view of Io is given by:

Spencer, J.R., and N.M. Schneider, 1996. Io on the eve of the Galileo mission. *Annu. Rev. Earth Planet. Sci.* **24**, 125–190.

The impact of Comet Shoemaker–Levy 9 with Jupiter has been described in two books, which came out within two years after the impact:

A lay level book is written by:

Spencer, J.R., and J. Mitton, 1995. *The Great Comet Crash*. Cambridge University Press, Cambridge.

A collection of review papers can be found in:

Noll, K.S., H.A. Weaver, and P.D. Feldman, Eds., 1996. *The Collision of Comet Shoemaker–Levy 9 and Jupiter*. Space Telescope Science Intitute Symposium Series 9, IAU Colloquium 156, Cambridge University Press, Cambridge.

A pleasant 'novel' about the history of the development of the impact theory on the extinction of the dinosaurs is written by:

Alvarez, W., 1997. *T.Rex and the Crater of Doom*. Princeton University Press, Princeton, NJ.

Problems

5.1.E (a) Determine the approximate mineral composition and texture of dacite with a silica content of 60%. (Hint: use Figure 5.4.)

(b) Determine the approximate silica content, mineral composition and texture of obsidian, pumice, rhyolite and granite. Explain the similarities and differences.

(c) Determine the approximate silica content, mineral composition and texture of gabbro and basalt. Compare with your answer in (b) and comment on your result.

5.2.I Derive the relation in equation (5.10) by differentiating equation (5.1), using the first law of thermodynamics (eq. 3.67) and assuming that the process is reversible (i.e., use eq. 5.7).

5.3.E Calcium carbonate, $CaCO_3$, exists in two polymorphic forms: calcite and aragonite. In the following you will calculate which form (or phase) of calcium carbonate exists at a temperature of 298 K and pressure of 1 atm. The enthalpy at a temperature of 298 K and pressure of 1 atm for calcite is -12.0737×10^{12} erg mol^{-1} and for aragonite it is -12.0774×10^{12} erg mol^{-1}. The entropy for calcite is 91.7×10^7 erg mol^{-1} K^{-1}; the entropy for aragonite is 88×10^7 erg mol^{-1} K^{-1}.

(a) Calculate the change in enthalpy for the transformation aragonite → calcite at a temperature of 298 K and 1 atm pressure. Is the reaction endothermic or exothermic?
(b) Calculate the change in entropy for this same transformation (aragonite → calcite).
(c) Determine which form of $CaCO_3$ is the stable form at a temperature of 298 K and pressure of 1 atm.

5.4.I Consider the system jadeite + quartz → albite ($NaAlSi_2O_6 + SiO_2 \rightarrow NaAlSi_3O_8$). The temperature dependence of the thermal heat capacity is usually expressed by:

$$C_p = c_1 + c_2 T + c_3 T^{-2} + c_4 T^{-1/2}, \qquad (5.31)$$

with c_1, c_2, c_3, and c_4 constants. For the above system these constants, as well as the entropy at room temperature and 1 atm pressure are given in Table 5.4. The change in enthalpy for the transformation jadeite + quartz → albite at 298 K is $+12.525 \times 10^{10}$.

(a) Calculate the stable phase(s) for the system at a temperature of 298 K.
(b) Calculate the stable phase(s) for the system at a temperature of 1000 K.

5.5.E If the lithospheric plates move, on average, at a speed of 6 cm yr^{-1}, what would be a typical recycling time of terrestrial crust? (Hint: Calculate the recycling time based upon the motion of one plate over the Earth's surface; how would your answer change if you have several plates moving over the surface?)

5.6.E The Voyager 1 spacecraft detected nine active volcanoes on Io. If we assume that, on average, there are nine volcanoes active on Io, and that the average eruptive rate per volcano is 50 km^3 yr^{-1}, calculate:

(a) the average resurfacing rate on Io in cm yr^{-1},
(b) the time it takes to completely renew the upper kilometer of Io's surface.

5.7.E Suppose a child leaves a toy truck on a deserted sandy beach facing the ocean, and returns many years later to retrieve it. The beach is characterized by strong winds, which usually blow inland. Sketch the dune which formed around the toy truck.

5.8.E (a) Calculate the kinetic energy and pressure involved when the Earth gets hit by a stony meteoroid ($\rho = 3.4$ g cm^{-3}) which has a diameter of 10 km, and which has a zero velocity at a very large distance from Earth.
(b) Calculate the kinetic energy involved were the same meteoroid to hit Jupiter instead of Earth, assuming the body has zero velocity at a large distance from Jupiter.
(c) Calculate the kinetic energy involved when a fragment of Comet D/Shoemaker–Levy 9 ($\rho = 0.5$ g cm^{-3}, $R = 0.5$ km) hits Jupiter at the planet's escape velocity.
(d) Express the energies from (a)–(c) in magnitudes on the Richter scale, and compare these with common earthquakes.

5.9.E (a) The compression stage typically lasts a few times longer than the time required for the impacting body to fall down a distance equal to its own diameter. Calculate the duration of the compression stage for a 10 m sized and a 1 km sized meteoroid, which impact Earth at a velocity of 15 km s^{-1}.
(b) Estimate the pressure involved in these collisions, assuming the meteoroids are stony bodies with a density $\rho = 3$ g cm^{-3}.

5.10.E Consider the impact between an iron meteoroid ($\rho = 7$ g cm^{-3}) with a diameter of 300 m and the Moon.

(a) Calculate the kinetic energy involved if the meteoroid hits the Moon at a velocity of 12 km s^{-1}.
(b) Estimate the size of the crater formed by a head-on collision, and one where the angle of impact with respect to the local horizontal is 30°.
(c) If the rocks are excavated from the crater with typical ejection velocities of 500 m s^{-1}, calculate how far from the main crater one may find secondary craters.

TABLE 5.4 Constants for the System Jadeite + Quartz → Albite[a].

Mineral	$S_0 \times 10^9$	$c_1 \times 10^{10}$	$c_2 \times 10^5$	$c_3 \times 10^{10}$	$c_4 \times 10^{10}$
Albite	2.074	0.4521	−1.336	−1276	−3.954
Jadeite	1.335	0.3011	1.014	−2239	−2.055
Quartz	0.415	0.1044	0.607	34	−1.070

[a] From Putnis (1992).

5.11.E Repeat the same questions as in Problem 5.10 for Mercury. Comment on the similarities and differences.

5.12.E After the Moon has been hit by the meteorite from Problem 5.10, many rocks are excavated from the crater during the excavation stage.

(a) If the ejection velocity is 500 m s^{-1}, calculate how long the rock will be in flight, if its ejection angle with respect to the ground is 25°, 45°, and 60°.

(b) Calculate the maximum height above the ground reached by the three rocks from (a).

5.13.E The secondary craters related to a crater of a given size on Mercury typically lie closer to the primary crater than do the secondary craters of a similarly sized primary on the Moon. Presumably, this is the result Mercury's greater gravity reducing the distance which ejecta travel.

(a) Verify this difference quantitatively by calculating the 'throw distance' of ejecta launched at a 45° angle with a velocity of 1 km s^{-1} from the surfaces of Mercury and the Moon.

(b) Typical projectile impact velocities are greater on Mercury than they are on the Moon. Why doesn't this difference counteract the surface gravity effect discussed above?

5.14.E (a) Determine the diameter of the crater produced when a stony ($\rho = 3$ g cm^{-3}) meteoroid, 1 km across, hits the Earth at a velocity of 15 km s^{-1}, at an angle of 45° (ignore the Earth's atmosphere). Use a density of 3.5 g cm^{-3} for the Earth's surface layer.

(b) Determine the crater diameter from (a) if the meteoroid is 10 km across.

(c) Determine the diameter of craters produced by both meteoroids if they had hit the Moon ($v = 15$ km s^{-1}) instead of Earth.

5.15.I (a) Calculate the average crater density (km^{-2}) for craters over 4 km in size for a portion of the lunar highlands that is 4.44 Gyr old.

(b) Calculate the average crater density (km^{-2}) for craters over 4 km in size for a region on the lunar surface that is 1 Gyr old.

(c) Calculate the approximate age of the lunar maria if the average crater density for craters over 4 km in diameter is 9×10^{-5} craters km^{-2}.

5.16.E (a) Determine the minimum radius of an iron meteoroid ($\rho = 8$ g cm^{-3}) to impact the Earth at hypersonic speed.

(b) Determine the minimum radius for a similar meteoroid of similar composition to make it through Venus's atmosphere.

(c) Calculate the approximate crater size the meteoroids produce on both planets, if the impacting velocity is equal to the escape velocity from the planet. Have craters smaller than this size been observed on Venus and Earth?

5.17.E Calculate the surface area in square kilometers which would be devastated by the blast wave created when a 500 m diameter stony body ($\rho = 3$ g cm^{-3}) hits the Earth. Compare your answer to the size of the city, county or country you live in.

5.18.E (a) Calculate the surface temperature of Pluto and Charon at perihelion, assuming both bodies are rapid rotators and in equilibrium with the solar radiation field. (Hint: See Section 3.1.2.2.)

(b) Calculate the escape velocity from Pluto and Charon, and compare these numbers with the velocity of N_2, CH_4 and H_2O molecules.

(c) Given your answers in (a) and (b), explain qualitatively the differences in surface ice coverage for Pluto and Charon.

6 Planetary Interiors

– at somewhere between 0.6 and 0.5 of the radius, measured from the surface, a very marked and remarkable change in the nature of the material, of which the Earth is composed, takes place.

R.D. Oldham, 1913

In the previous two chapters, we discussed the atmospheres and surface geology of planets. Both of these regions of a planet can be observed directly from Earth and/or space. But what can we say about the deep interior of a planet? We are unable to observe the inside of a planet directly. For the Earth and the Moon we have seismic data, where the reflection and refraction of waves that penetrate deeply below the surface can be used to retrieve information on the interior structure (Section 6.2). For all other bodies we must derive the interior structure through a comparison of observable characteristics predicted from models with remote observations. The relevant observations are the body's mass, size (and thus density), its rotational period and geometric oblateness, gravity field, characteristics of its magnetic field (or absence thereof), the total energy output, and the composition of its atmosphere and/or surface. In addition, laboratory data on the behavior of materials under high temperature and pressure are invaluable for models of the interiors of planets, including our Earth. Quantum mechanical calculations are used to deduce the behavior of elements (especially hydrogen) at pressures inaccessible in the laboratory. In this chapter we discuss the basics of how one can infer the interior structure of a body from the observed quantities. As expected, there are large differences between the interior structure of the giant and the terrestrial planets. However, even among the gas giants and among the solid bodies there are noticeable differences in interior structure.

6.1 Modeling the Interior Structure of a Planet

For all planetary bodies, except the Earth and the Moon, we depend completely on remote observations, in conjunction with theoretical models, to extract information on the interior structure of the body. Obvious observations are the mass, size and shape of a body. The mass and size yield an estimate for the average density, which can be used directly to derive some first-order estimates on the body's composition. For small bodies, a density $\rho \lesssim 1$ g cm^{-3} implies an icy and/or porous object, while large planets of this density consist primarily of hydrogen and helium. A density $\rho \approx 3$ g cm^{-3} suggests a rocky object, while higher (uncompressed[1]) densities indicate the presence of heavier elements, in particular iron, one of the most abundant heavy elements in space (Table 8.1). The shape of a body depends upon its size, density, material strength, rotation rate and history. An object is approximately spherical if the weight of the mantle and crust exerted on its inner parts is high enough to deform the body. Any nonrotating 'fluid-like' body will take on the shape of a sphere, which corresponds to the lowest energy state. Note that the term 'fluid-like' in this context means deformable over geologic time (i.e., $\gtrsim$ millions of years), also referred to as *plasticity* or *rheidity*. The shape or *figure* of a planet depends upon the plasticity and the rotation rate of the body. Rotation flattens a deformable object somewhat, changing its figure to an *oblate spheroid*, the equilibrium shape under the combined influence of gravity and centrifugal forces. A rocky body with a typical density $\rho = 3.5$ g cm^{-3} and material strength $\mathcal{S}_m = 2 \times 10^9$ dyne cm^{-2} is approximately spherical if $R > 350$ km; iron bodies with $\rho = 7.9$ g cm^{-3} and $\mathcal{S}_m = 4 \times 10^9$ dyne cm^{-2} are spherical if $R > 220$ km (Problem 6.2).

In this chapter, we discuss the interior structure of bodies large enough to be in *hydrostatic equilibrium* (cf. Section 3.2.3.1). To calculate the balance between gravity and pressure, one must know the gravity field as well as an equation of state which relates the temperature, pressure and density in a planet's interior. The equation of state depends upon the constituent relations of the various materials the planet is made of. In addition, the sources, losses and transport mechanism(s) of the heat inside a planet

[1] Uncompressed density: the density a planet would have if material was not compressed by the weight of overlying layers.

TABLE 6.1 Densities and Central Properties of the Planets and the Moon.

Planet	Radius (equatorial) (km)	Density ($g\ cm^{-3}$)	Uncompressed density ($g\ cm^{-3}$)	Central pressure (Mbar)	Central temperature (K)
Mercury	2 440	5.43	5.3	~0.4	~2 000
Venus	6 042	5.20	4.3	~3	~5 000
Earth	6 378	5.515	4.4	3.6	6 000
Moon	1 738	3.34	3.3	0.045	~1 800
Mars	3 390	3.93	3.74	~0.4	~2 000
Jupiter	71 492	1.33		~80	~20 000
Saturn	60 268	0.69		~50	~10 000
Uranus	25 559	1.32		~20	~7 000
Neptune	24 766	1.64		~20	~7 000

Data from Hubbard (1984), Lewis (1995), Hood (1987), Guillot (1999), and Yoder (1995).

are crucial to determining the object's thermal structure, which in turn is an important parameter in the derivation of a body's interior structure.

6.1.1 Hydrostatic Equilibrium

To first order, the internal structure of a spherical body is determined by a balance between gravity and pressure, and planetary interiors are said to be in *hydrostatic equilibrium* (see eq. (3.65)):

$$P(r) = -\int_r^R g_p(r')\rho(r')dr'. \tag{6.1}$$

With equation (6.1), the pressure can be calculated throughout the planet provided $\rho(r)$ is known. If the density is constant throughout the planet's interior, the pressure at the center of a planetary body, P_c, is given by (Problem 6.3a):

$$P_c = \frac{3GM^2}{8\pi R^4}. \tag{6.2}$$

This sets a lower limit to the central pressure, since the density usually decreases with distance r. This method provides good estimates for relatively small bodies, with a uniform density, such as the Moon, which has a central pressure of (only) 45 kbar. An alternative quick estimate can be obtained by assuming the planet consists of one slab of material, in which case the central pressure is a factor of two larger than the value obtained in the previous estimate (Problem 6.3b). Since this technique overestimates the gravity over most of the region of integration, the actual pressure at the center of the planet probably lies between these two values. On the other hand, if the planet is extremely centrally condensed, the density increases sharply towards the center of the planet, and the pressure calculated using the single slab model may still be too low compared to the actual value. We find that the central pressure of Earth calculated according to the single slab model agrees quite well with the actual value of 3.6 Mbar (Problem 6.3). The Earth is differentiated, and the increase in density towards the center just about compensates for our overestimate in gravity. Jupiter's central pressure is still underestimated by a factor of ~4, since this planet is very dense near its center (Jupiter's central pressure is roughly 80 Mbar).

An accurate estimate of a planet's internal structure requires assumptions regarding the planet's composition as well as knowledge of the equation of state and constituent relations of the material. It is also crucial to know the temperature structure throughout the interior, which is determined by internal heat sources, heat transport and heat loss mechanisms. The sources of heat are strongly tied to the planet's formation history. All of this information must be used to compute interior models, which can then be checked against observations and refined in an iterative manner.

6.1.2 Constituent Relations

In order to develop realistic models of a planet's interior structure, one needs to know the phases of the materials which make up a planet as functions of temperature and pressure. In Section 5.2 we showed that the state of the material, i.e., whether the material is in a solid, liquid, or vapor phase, depends upon the temperature and pressure of the environment. In principle, one can determine the melting temperature, T_m, as a function of pressure for any component by solving the equation (Section 5.2.1):

$$G_\ell(T_m, P) = G_s(T_m, P), \tag{6.3}$$

where G_ℓ and G_s are the Gibbs free energies for the liquid and solid phase of the material, respectively. The *Lindemann criterion* states that melting occurs when the thermal oscillations of the ions in the material lattice become a significant fraction of the equilibrium spacing of the ions in the lattice, i.e., when neighboring ions start to overlap:

$$\sqrt{\frac{\bar{\zeta^2}}{r_s^2}} \gtrsim 0.1, \tag{6.4}$$

where $\sqrt{\bar{\zeta^2}}$ is the average displacement of ions from their equilibrium positions and r_s is a measure of the equilibrium spacing of ions in the lattice (both in atomic units). The thermal oscillations of the ions are proportional to the temperature, and inversely proportional to the atomic number, Z:

$$\bar{\zeta^2} \propto \frac{T}{Z^2}. \tag{6.5}$$

Lindemann's criterion is sometimes used together with an equation of state to extrapolate melting points to high pressure; however, for some materials agreement with the experimental data is poor.

Based upon statistical arguments and computer simulations it has been shown that in a fully pressure-ionized medium the melting temperature can be approximated by:

$$T_m \approx \frac{Z^2}{150 r_s}. \tag{6.6}$$

One can use equations (6.4) and (6.6) to derive an approximate relationship between the melting temperature, T_m, and pressure, P, for various elements, as illustrated in Figure 6.1.

The derivation of empirical constituent relations is relatively easy at low pressure, where the chemical reactions and phase transitions are well known for many materials. However, the pressures and temperatures in planetary interiors can be very high, and in such an environment it is difficult to predict whether a material will be in a solid or liquid phase. Moreover, as discussed in Section 5.2, mixtures of elements undergo different chemical reactions, depending on the temperature and pressure of the environment, eutectic behaviors play a role and phase diagrams may become quite complicated. A stable system is one where the Gibbs free energy is at a minimum. Typically,

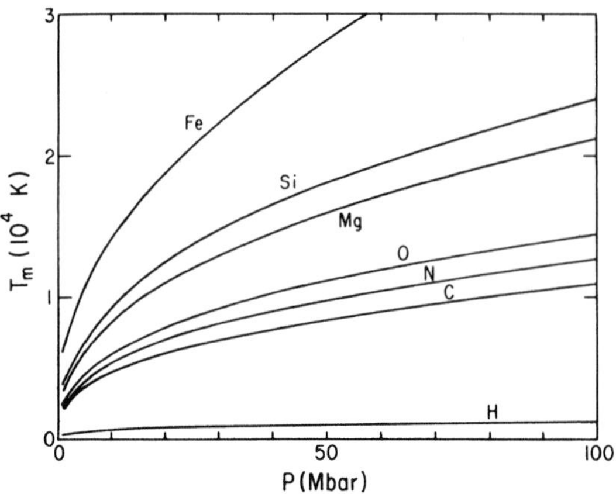

FIGURE 6.1 The calculated melting temperature as a function of pressure for various common elements. (Hubbard 1984)

above some critical temperature, a solution likes to be in a single liquid phase, while below this temperature there are several liquid and solid phases of different composition. It appears that under the temperature and pressure conditions encountered in the terrestrial planet interiors, one usually expects a chemical separation of the compounds (such as metals/siderophiles from the silicates/lithophiles).

Experiments under high pressure can be conducted using various techniques. By using a hydraulic press, rocks in the laboratory can be squeezed to pressures of ~100 kbar, and heated to roughly 1200–1400 K. Many measurements at pressures $\gtrsim$ 1 Mbar are conducted by means of shock wave experiments (Fig. 6.2a). Unfortunately, the duration of the high pressure state in such an experiment lasts from only a fraction of a microsecond up to at most a few milliseconds, and the rock sample is destroyed in the process. In these experiments, however, temperatures of many thousand K can be reached, close to the temperatures prevailing in planetary interiors. Static pressures of a little over one Mbar can be reached using a diamond anvil press (Fig. 6.2b). In these devices, material is squeezed between two diamonds, each ~ 350 μm across. The high pressure can be kept constant for weeks, months, or longer. Since the diamonds are transparent, the sample can be seen while being compressed and heated, and the temperature can be regulated, in contrast to that in shock experiments. It is not yet possible to conduct experiments at pressures much above 5–10 Mbar. In this pressure regime one still has to rely completely on theoretical arguments.

In the following subsections we discuss the various phases of materials which make up the planets and relate our findings to what is known about planetary interiors. We give a detailed description of individual planetary interiors in Sections 6.2, 6.3 and 6.4.

(a)

(b)

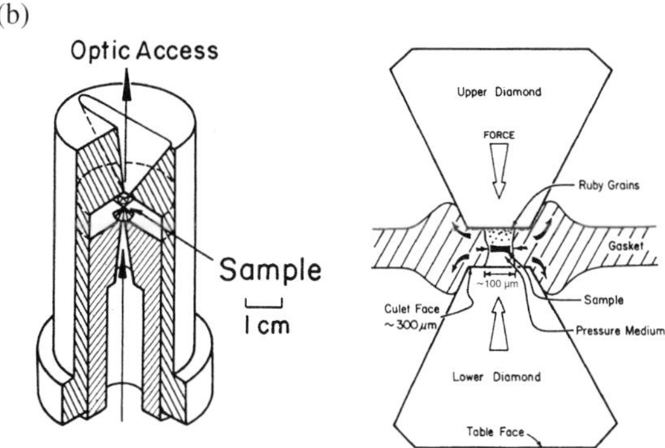

FIGURE 6.2 (a) Photograph of the 60-foot long, two-stage light-gas (usually H_2) gun at Lawrence Livermore National Laboratory. This gun is used to obtain the equation of state of various materials through shock wave experiments and to investigate impact events. (Courtesy: Lawrence Livermore National Laboratory) (b) Left: Sketch of a diamond anvil press. Two single-crystal gem quality diamonds are compressed in a piston-cylinder assembly. The sample is placed between the opposed points of the diamonds, and because of the small surface area, extremely high pressures ($\sim$ 1 Mbar) can be reached and maintained over long periods of time. Because the diamonds are transparent, the sample can be 'seen' at wavelengths extending from far-infrared to hard X-ray and γ-ray energies. Right: Details of the sample and pressure medium. Spectra of embedded fine-grained ruby-powder ($\lesssim 5$ μm grain size) allow precise determination of the temperature and pressure. (Jeanloz 1989)

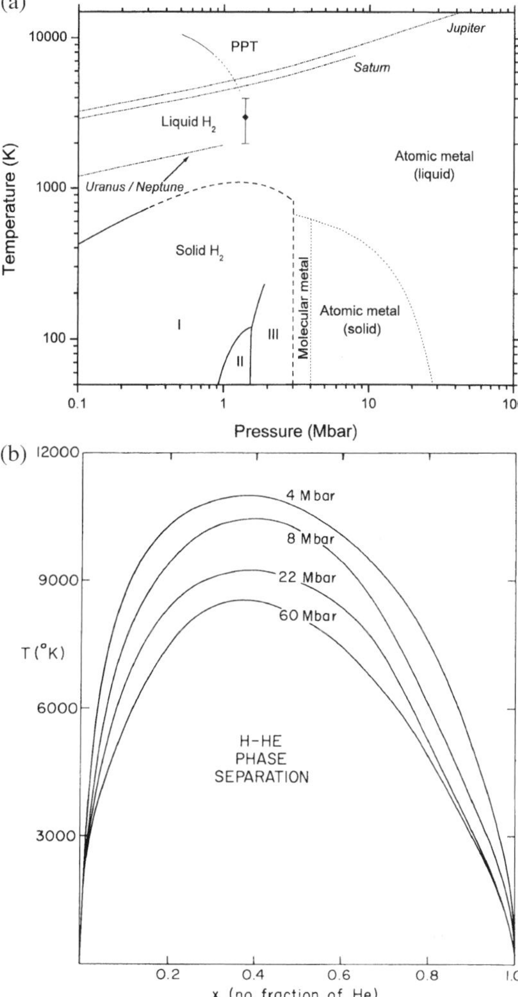

FIGURE 6.3 (a) Phase diagram of hydrogen at high pressures, showing the transition from molecular hydrogen into metallic hydrogen. At low temperature (unrealistic for planets within our Solar System) hydrogen is present as an electrically insulating solid (I, II or III) below $\sim$ 3 Mbar, changing to (solid) molecular metallic hydrogen at $\sim$ 3 Mbar, and liquid atomic metallic hydrogen at higher pressures and temperatures. At much higher temperatures hydrogen breaks up into a plasma or becomes highly degenerate (PPT). The adiabats of the giant planets are superposed as dot-dashed curves. (Diagram courtesy of Stephen A. Gramsch, Carnegie Institution of Washington. Adiabats for the planets were provided by William B. Hubbard.) (b) Phase separation between hydrogen and helium as a function of temperature at different pressures. The miscibility gap is below the curve in each case. (Stevensen and Salpeter 1976)

6.1.2.1 *Hydrogen and Helium*

Typical temperatures and pressures in the giant planets range from about 50–150 K at a planet's tropopause up to 10 000–20 000 K at their centers, while the pressure varies from near zero in the outer atmosphere up to 20–80 Mbar at the planet's center. A phase diagram for hydrogen over a wide range of pressures and temperatures is shown in Figure 6.3a. This diagram is based upon experiments and theoretical calculations by many different groups. At pressures less than $\sim$1 Mbar, hydrogen is mostly present in molecular form. At pressures over 1 Mbar, there is continuous transition from fluid molecular to atomic hydrogen. In Jupiter, dissociation ($H_2 \rightarrow 2H$) begins near

0.95 $R_♃$ and goes to completion at an estimated radius of $\sim 0.8\ R_♃$. Because of the energy required to dissociate H_2, the temperature is nearly constant over a relatively wide range of pressure and radius. At these high (Mbar) pressures the fluid is so densely packed that the separation between the molecules becomes comparable to the size of the molecules, so that their electron clouds start to overlap and an electron can hop or percolate from one molecule to the next. Shock wave experiments at pressures between 0.1 and 1.8 Mbar and temperatures up to ~ 5000 K show that the conductivity of fluid molecular hydrogen is increasing continuously and reaches the minimum conductivity of a metal at a pressure of 1.4 Mbar. Thus at $P \gtrsim 1.4$ Mbar hydrogen behaves like a metal, a phase referred to as *fluid metallic hydrogen* (Fig. 6.3a). In Jupiter, this transition occurs at a radius of $\sim 0.90\ R_♃$. Convection in this fluid is thought to create the magnetic fields observed to exist around Jupiter and Saturn. At $P \gtrsim 3$ Mbar, it is expected that the molecular hydrogen dissociates into an atomic metallic state. No measurements exist on the dissociation of metallic molecular hydrogen into metallic atomic hydrogen, however. At much higher temperatures, theoretical calculations predict that hydrogen may either become highly degenerate, or a plasma. The plasma phase transition (PPT on Fig. 6.3a) is highly controversial at the present time, although the various groups do agree that the phase of hydrogen changes at these high temperatures and pressures.

The giant planets consist primarily of a mixture of hydrogen and helium. In the interiors of both Jupiter and Saturn, hydrogen is expected to be in liquid metallic form. Also helium transforms into a liquid metallic state but at pressures much higher than encountered in the giant planets. Hydrogen and helium are only fully mixed if the temperature and pressure are high enough. At lower temperatures and pressures, the liquid phases of hydrogen and helium do not mix. Calculations on the miscibility of hydrogen and helium, as a function of temperature and helium abundance, are shown in Figure 6.3b. Curves for four different pressures are shown. Above the line the two phases are completely mixed, but below the line the phases separate out. We expect that helium and hydrogen are not fully mixed in the interiors of Jupiter and, in particular, Saturn. The observed helium abundance is slightly less than the solar He/H abundance in Jupiter's atmosphere, and it may be depleted by a factor of a few in Saturn's atmosphere. These depletions relative to the solar value have been attributed to separation within the metallic hydrogen region. Because the temperature at a given pressure in Saturn's interior is always less than in Jupiter, the effect of immiscibility must be stronger within Saturn than within Jupiter (cf. Section 6.4.2).

6.1.2.2 *Ices*

Water-ice is a major constituent of bodies in the outer parts of our Solar System. Depending on the temperature and pressure of the environment, water-ice can take on at least 15 different crystalline forms, many of which are shown in the water phase diagram displayed in Figure 6.4. Although the water molecule does not change its identity, the molecules are more densely packed at higher pressures, so that the density of the various crystalline forms varies from 0.92 g cm^{-3} for common ice (form I) up to 1.66 g cm^{-3} for ice VII, near the triple point with ices VI and VIII. Temperatures and pressures expected in the interiors of the icy satellites range from $\sim$ 50–100 K at the surface up to several hundred K at pressures of up to a few tens kbar in their deep interiors. So one might expect a wide range of ice-forms in these satellites. The adiabats of the giant planets are superposed on Figure 6.4. At the higher temperatures, above 273 K at moderate pressures, water is a liquid. The *critical point* of water (650 K) is indicated by a C; above this temperature there is no first-order phase transition between gaseous and liquid H_2O. Water becomes a supercritical fluid, a phase that is neither gas nor liquid.

Pure water is slightly ionized, with H_3O^+ and OH^- ions. At higher temperatures and pressures, the ionization is enhanced. Shock wave data on a mixture of water, isopropanol and ammonia ('synthetic Uranus') show that at pressures over 200 kbar the mixture ionizes to form an electrically conductive fluid. The conductivity of this mixture is essentially the same as that for pure water. The conductivity is high enough to explain the existence of the observed magnetic fields of Uranus and Neptune. At pressures over 1 Mbar the ice constituents dissociate, and the fluid becomes rather 'stiff', i.e., the density is not very sensitive to pressure.

In addition to water-ice, one would expect the outer planets to contain substantial amounts of other 'ices', such as ammonia, methane and hydrogen sulfide. Phase diagrams of these ices might be as complex as those for water. Although experiments up to 0.5–0.9 Mbar have been carried out for some of these constituents, they are not as well studied under high pressures as is water-ice, and mixtures of these various ices are even less well characterized.

6.1.2.3 *Rocks and Metals*

The phase diagrams of magmas, i.e., molten rocks, were discussed in Section 5.2, and were shown to be very complicated. The various elements and compounds interact in different ways, depending on the temperature, pressure

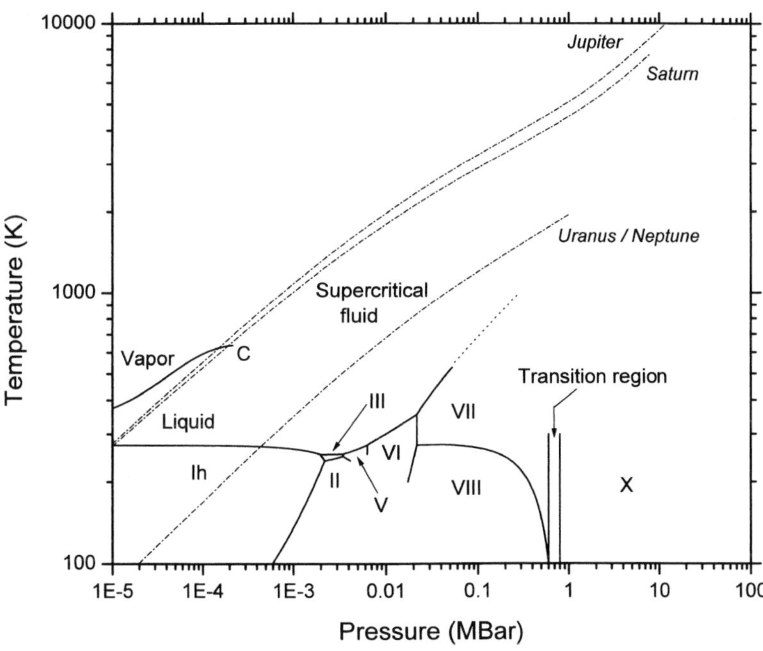

FIGURE 6.4 Phase diagram for water at temperatures and pressures relevant for the icy satellites as well as the giant planets (adiabats for the giant planets are superposed). The various crystal forms of ice are indicated by roman numerals I–X. The critical point of water is indicated by a C. (Diagram courtesy of Stephen A. Gramsch, Carnegie Institution of Washington. Adiabats for the planets were provided by William B. Hubbard)

and composition of the magma. From terrestrial rocks, models of phase diagrams, laboratory experiments and discontinuities in the density profile of the Earth's interior (Section 6.2), we have obtained a reasonable idea of the composition of the Earth's mantle and core. The primary minerals of the Earth's upper mantle are olivine, $(Mg,Fe)_2SiO_4$, and pyroxene, $(Mg,Fe)\ SiO_3$. Experiments have shown that at higher pressure the molecules get more closely packed, which leads to a reorganization of the atoms in the lattice structure. This rearrangement of atoms/molecules into a more compact crystalline structure involves an increase in density, and characterizes a phase change. A common example is carbon, which at low pressure is present in the form of graphite, and at high pressure as diamond. At a depth of $\sim$50 km ($\sim$15 kbar) basalt changes into *eclogite*, i.e., the structure of some minerals in the basalt changes, such as the conversion of pyroxene to garnet. Since this transformation takes place only at several 'spots' near the Earth's surface, where cold oceanic lithospheric plates descend down (Section 6.2.2.2), there is no global seismic discontinuity measured at this depth. Global seismic discontinuities are measured at a depth of $\sim$400 km, where the mineral olivine changes into a spinel structure, and at a depth of $\sim$660 km, where spinel is decomposed into magnesium oxide or periclase, MgO, and perovskite, $MgSiO_3$. Perovskite is found to be stable even at very high pressures, and may be the dominant 'rock' in the Earth's interior. Whereas the first phase change (olivine $\rightarrow$ spinel-structure) is an exothermic reaction, the second one (spinel-structure $\rightarrow$ perovskite) is endothermic.

The primary constituent in the Earth's core is almost certainly iron (Section 6.2.2.3), though the density of the Earth's core is $\sim$5–10% lower than the density of pure iron. This suggests that the core is not pure iron, but Fe is mixed with a lower density material, possibly sulfur, oxygen or hydrogen (Section 6.2.2.3). The phase diagram of iron is well known for pressures up to $\sim$200 kbar (Fig. 6.5), where four different solid phases can be distinguished: α-Fe is the phase that is stable at room temperature and ambient pressure. At high temperature it changes to γ-Fe, and to δ-Fe just below the melting point. At higher pressure α-Fe changes into the hexagonal-close-packed ϵ-Fe. The melting curve for iron has been determined up to 2 Mbar using laser heating in a diamond anvil cell. Shock compression experiments have been used at higher pressures. Results of these experiments are indicated on Figure 6.5, together with two theoretical melting curves. Although it is clear that much progress has been made in understanding the behavior of iron under high pressure, the details of the various experiments are not (yet) in perfect agreement.

In the primordial nebula at temperatures below $\sim$700 K, Fe reacts with H_2O and H_2S to form FeO and FeS. Within a larger body, iron oxide is largely incorporated with magnesium silicates in rocks as olivine and pyroxene. Iron sulfide, however, is expected to settle in the planet's core with iron. This could explain the lower

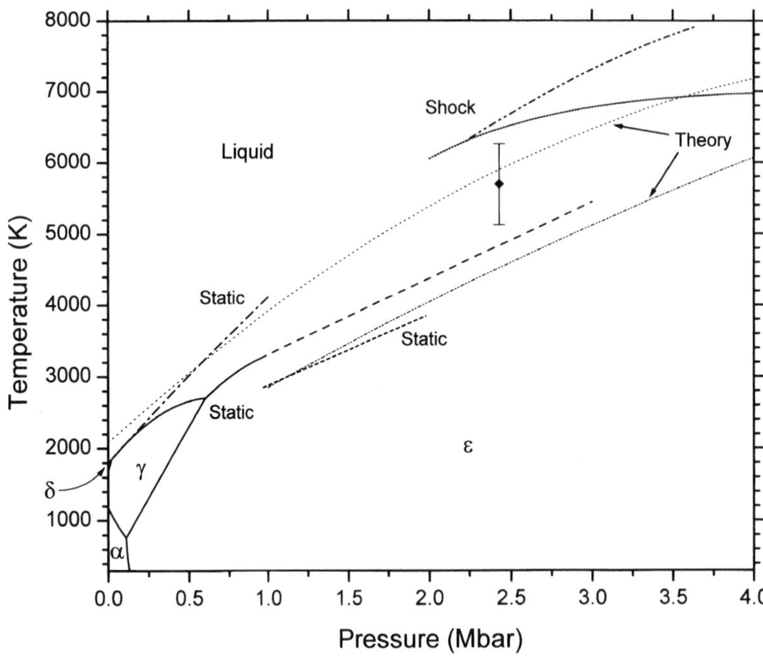

FIGURE 6.5 Phase diagram of iron under high temperature and pressure, based upon: (*i*) Experiments using a diamond anvil cell ('static' measurements). The solid (measured) line is extrapolated to higher temperatures and pressures via the dashed line. (*ii*) Shock wave experiments at higher temperature and pressure, as indicated by the two lines and the diamond. (*iii*) The thin dotted lines are theoretical calculations of the melting curve of iron for conditions appropriate to Earth's core. The data were assembled from many different sources. (Diagram courtesy of Stephen A. Gramsch, Carnegie Institution of Washington)

density of the Earth's core compared to that of pure iron. The alloy of iron and sulfur has a *eutectic* behavior, which means that the mixture melts completely into Fe + FeS at a temperature well below the melting temperature of each mineral separately. At the 1 bar pressure level, Fe melts at a temperature of 1808 K, FeS at 1469 K, while an iron–sulfur alloy (mixture) of 27% sulfur and 73% iron (solar mixture, which is close to a eutectic mixture) melts at 1262 K. At pressures as high as 100 kbar, the melting temperature of this eutectic mixture is depressed by nearly 1000 K compared to the melting temperature of pure iron.

6.1.3 Equation of State

The equation of state is an expression which relates the pressure, density, temperature and composition, $P = P(\rho, T, f_i)$. In planetary atmospheres at pressures below ~ 50 bar one can use the perfect gas law:

$$P = NkT. \tag{6.7}$$

At higher temperatures and pressures this simple equation is not adequate, because the molecules can no longer be treated as infinitesimally small spheres. Intermolecular spacings decrease to ∼1–2 Å, where the van der Waals forces become important. At higher pressures liquids and solids may be formed. The equation of state is usually derived from measurements at room temperature, sometimes augmented by measurements at higher temperature and pressure such as obtained by, e.g., shock wave experiments. To match all the different aspects of the measurements, the data are usually fit with rather complex forms; for example, for condensed matter one uses:

$$P = c_1 f\,(1 + 2f)^{2.5}\left(1 + c_2 f + c_3 f^2 + \cdots\right), \tag{6.8a}$$

with

$$f = \frac{1}{2}\left(c_0 \rho^{2/3} - 1\right) \tag{6.8b}$$

where c_n are constants to be fitted to the data.

In a hydrogen-rich environment, like the giant planets, the electron clouds of the hydrogen molecules start to overlap at high pressures, which increases the conductivity. At pressures of ∼1.4 Mbar hydrogen enters a molecular metallic state, which is referred to as *metallic hydrogen* (Section 6.1.2.1). In the extreme limit of a fully pressure ionized gas ($P \gtrsim 300$ Mbar), hydrogen is presumably present in atomic form, the electrons become degenerate and the pressure is independent of temperature. Although the pressures where electrons become degenerate are much higher than encountered in planetary interiors, a short discussion on the mass–density relation for this situation is enlightening for our discussion of the giant planets in Section 6.4 (see also Section 13.1):

Let us express the pressure–density relation in the form of a power law, as is often done to simplify the equation of state:

$$P = K\rho^{1+1/n}, \tag{6.9}$$

where K and n are the *polytropic constant* and *polytropic index*, respectively. At very low pressures, $P \to 0$, $n \approx \infty$, while in the limit of high pressures $n = 3/2$ and $P \propto \rho^{5/3}$.

For an incompressible planet, the relation between planetary mass and radius is given by:

$$M \propto R^3. \tag{6.10}$$

When a sufficient amount of matter is added to the planet, the material is compressed and the radius increases more slowly than indicated by equation (6.10). When the internal pressure becomes so large that electrons become degenerate, it can be shown that the size of the body decreases when more mass is added (Problem 6.4):

$$M \propto \frac{1}{R^3}. \tag{6.11}$$

This situation is encountered in *white dwarf* stars. There is a maximum size that a cold sphere of matter can reach, and adding more mass to the body makes it shrink (cf. Fig. 6.22). Jupiter's size is near-maximal for a solar composition planet, i.e., it lies at the boundary between equations (6.10) and (6.11). Cool massive brown dwarfs, discussed in Section 13.1, are slightly smaller than Jupiter.

6.1.4 Gravity Field

The gravity field of a planet contains useful information on the internal density structure. A planet's gravity field can be determined to quite high accuracy by tracking the orbits of spacecraft close to a planet or from the rate of precession of the periapses of moons and rings orbiting the planet. For a nonrotating fluid body in hydrostatic equilibrium, the gravitational potential is:

$$\Phi_g = -\frac{GM}{r}. \tag{6.12}$$

In reality, planets are not perfect spheres, and the equipotential surface at a planet's surface is called the *geoid* (on Earth the geoid is measured at the mean sea level, which is the average of high and low tides). The gravitational potential of an axisymmetric body can be represented by a *Fourier-like series* of the form:

$$\Phi_g(r, \theta, \phi) = -\frac{GM}{r}\left[1 - \sum_{\mathrm{n}=2}^{\infty}\left(\frac{R_e}{r}\right)^{\mathrm{n}} J_{\mathrm{n}} P_{\mathrm{n}}(\cos\theta)\right], \tag{6.13}$$

where θ is the colatitude, R_e the equatorial radius of the planet, J_{n} the *gravitational moments*, and $P_{\mathrm{n}}(\cos\theta)$ are *Legendre polynomials*, as specified in Section 2.5.1. The origin is chosen to be the center of mass, thus the gravitational moment $J_1 = 0$. For a nonrotating fluid body in hydrostatic equilibrium, the potential is spherically symmetric and the moments $J_{\mathrm{n}} = 0$, so that equation (6.12) is valid. Rotating fluid bodies in hydrostatic equilibrium have $J_{\mathrm{n}} = 0$ for all odd n. This has been found to be a very good approximation for the giant planets. However, nonzero odd moments, as well as nonaxisymmetric components of the gravity field that cannot be represented in the form of equation (6.13), have been measured for the terrestrial planets.

For rotating bodies, the effective gravity is less than the gravitational attraction calculated for a nonrotating planet, since the centrifugal force induced by rotation is directed outwards from the planet (Section 2.1.4). The equipotential surface must thus be derived from the sum of the gravitational potential, Φ_g, and the rotational potential Φ_q:

$$\Phi_g(r, \theta, \phi) + \Phi_q(r, \theta, \phi) = \text{constant}, \tag{6.14}$$

where Φ_g exterior to the planet's surface is given by equation (6.13) and Φ_q is defined:

$$\Phi_q = \frac{1}{2} r^2 \omega_{rot}^2 \sin^2\theta = \frac{1}{3} r^2 \omega_{rot}^2 \left(1 - P_2(\cos\theta)\right). \tag{6.15}$$

The planetocentric distance of the equipotential surface, $r(\theta, \phi)$, and the surface gravity, $g_p(\theta)$, as a function of colatitude θ can then be obtained by solving equation (6.14). We find for the surface gravity as a function of latitude, θ':

$$g_p(\theta') = g_p(0)(1 + C_1 \sin^2\theta' + C_2 \sin^4\theta'). \tag{6.16}$$

For Earth, the constants $C_1 = 5.278\,895 \times 10^{-3}$ and $C_2 = 2.3462 \times 10^{-5}$.

The rotation period discussed above refers to the rotation of the bulk of the planet. Note that for planetary bodies with optically thick atmospheres, such as the giant planets and Venus, measurements from the rotation of features on their 'disks' yield the rotation period of atmospheric winds rather than the interior. For the giant planets, observations of the nonthermal radio emission are useful diagnostics of the rotation period of the interior of a planet (Chapter 7), while radar techniques can be used to determine the rotation rate for terrestrial-like bodies surrounded by a dense atmosphere.

6.1.4.1 *Gravitational Moments*

A rotating planet which consists of an incompressible fluid of uniform density takes on the form of a *Maclaurin spheroid*. From the solution for the equipotential surface (eq. 6.14) it follows that the second harmonic, J_2, is related to the centrifugal and gravitational potentials:

$$J_2 = \frac{1}{2} q_r, \tag{6.17}$$

with

$$q_r \equiv \frac{\omega_{rot}^2 R^3}{GM}. \tag{6.18}$$

Usually the density in a planet increases towards the center, in which case it can be shown that:

$$J_2 < \frac{1}{2} q_r. \tag{6.19}$$

The ratio between J_2 and q_r is expressed by the *response coefficient* Λ_2:

$$\Lambda_2 \equiv \frac{J_2}{q}. \tag{6.20}$$

The coefficient Λ_2 contains information on the spatial distribution of the mass in a planet's interior: Rotating planets with high density cores have small values of Λ_2, whereas bodies with a more homogenous density distribution have larger values of Λ_2. For an incompressible fluid of uniform density, $\Lambda_2 = 0.5$ (eq. 6.17).

From a solution of the equipotential surface for a planet in hydrostatic equilibrium, it can be shown that the geometric oblateness, ϵ, is related to the rotation period and the second harmonic:

$$\epsilon \equiv \frac{R_e - R_p}{R_e} \approx \frac{3}{2} J_2 + \frac{q_r}{2}, \tag{6.21}$$

where R_e and R_p represent the planet's equatorial and polar radius, respectively. From equation (6.17) it follows that ϵ and J_2 are of order q_r. It can be shown that for rapidly rotating planets in hydrostatic equilibrium higher order zonal harmonics are proportional to:

$$J_{2\mathrm{n}} \propto q^{\mathrm{n}}; \tag{6.22}$$

thus the higher order zonal harmonics for such planets are small compared to J_2. Yet these higher zonal harmonics make up the differences between the real geoid and a spinning fluid planet, and hence contain important clues to the interior structure of a planet.

For a sufficiently rapidly rotating planet in hydrostatic equilibrium, the moment of inertia (eq. 2.42) is related to J_2 and q. Although the relationship between the gravitational moments and the internal density distribution is in fact determined by a complicated integro-differential equation, one can derive an approximate algebraic formula to illustrate the relations between the moment of inertia and the rotation rate and gravitational moment, J_2:

$$\frac{I}{MR^2} \approx \frac{\frac{3}{2} J_2}{J_2 + \frac{1}{3} q_r}. \tag{6.23}$$

If the density ρ is uniform throughout the planet, $I = 0.4MR^2$ (Problem 6.5). $I/MR^2 = 0.667$ for a hollow sphere. The ratio $I/MR^2 < 0.4$ if ρ increases with depth in the planet, with $I/MR^2 = 0.0$ if all the mass is concentrated at the center. Usually the density increases towards a planet's center, both because dense compounds tend to sink and because material gets compressed at higher pressure.

Table 6.2 shows the J_n, q_r, Λ_2 and I/MR^2 values for all of the planets. Since the relation between the gravitational moments and the internal density distribution was derived for a rotating planet in hydrostatic equilibrium, one has to be careful with the interpretation of these values. For example, the q_r-values are very small for Mercury, Venus and the Moon, since the rotation periods of these three bodies are very long. The observed zonal harmonics can therefore not be used in an evaluation of the internal density structure of the planet, since nonhydrostatic effects (e.g., mantle convection) have a much larger contribution to the J_2 than does the rotational effect. The zonal harmonics are largest for Saturn and Jupiter, both rapidly rotating giants. The Λ_2 and I/MR^2 are smallest for the four giant planets, indicative of a pronounced density increase towards the center of these planets.

6.1.4.2 *Isostatic Equilibrium*

The surface gravity can be measured from small changes in the orbital parameters of artificial or natural satellites and radar altimetry measurements. Deviations in the measured gravity with respect to the geoid provide information on the structure of the crust and mantle. In the eighteenth century, it was already recognized that the measured surface gravity field of Earth does not deviate substantially from an oblate spheroid, even in the proximity of high mountains, despite the large land masses which make up the mountains. This observation lead to the concept of *isostatic equilibrium*, which is based upon the theory of hydrostatic equilibrium (Section 3.2.3.1), and says that a floating object displaces its own weight of the substance on which it floats. Mountains are compared to icebergs floating in water, since the rigid surface layers, the *lithosphere*, 'float' on a hot, highly viscous 'fluid' layer, the *asthenosphere*. Figure 6.6a shows a schematic presentation of the outermost layers or shell of the Earth. The lithosphere is topped off with a lighter *crust*, which is relatively light and thick for continents, and denser and thinner under the oceans. An iceberg floats because the volume submerged is lighter than the volume of water displaced. Similarly, a mountain in isostatic equilibrium is compensated by a deficiency of mass underneath, since the volume of the mountain submerged in the upper mantle is lighter than the mantle ma-

TABLE 6.2 Gravitational Moments and the Moment of Inertia Ratio.

Body	J_2 ($\times 10^{-6}$)	J_3 ($\times 10^{-6}$)	J_4 ($\times 10^{-6}$)	J_5 ($\times 10^{-6}$)	J_6 ($\times 10^{-6}$)	q_r	Λ_2	$I/(MR^2)$	Refs
Mercury	60 ± 20					1.0×10^{-6}	60	0.33	1
Venus	4.46 ± 0.03	-1.93 ± 0.02	-2.38 ± 0.02			6.1×10^{-8}	73	0.33	1
Earth	1 082.627	-2.532 ± 0.002	-1.620 ± 0.003	-0.21	0.65	3.45×10^{-3}	0.314	0.33	1
Moon	203.43 ± 0.09					7.6×10^{-6}	26.8	0.393	1, 2
Mars	$1\,960.5 \pm 0.2$	31.5 ± 0.5	-15.5 ± 0.7			4.57×10^{-3}	0.429	0.366	1
Jupiter	$14\,736 \pm 1$	0	-587 ± 5	0	31 ± 20	0.089	0.165	0.254	1
Saturn	$16\,298 \pm 10$	0	-915 ± 40	0	103 ± 50	0.155	0.105	0.210	1
Uranus	$3\,343.4 \pm 0.3$	0	-28.9 ± 0.5			0.029	0.113	0.23	1
Neptune	$3\,411 \pm 10$	0	-35 ± 10			0.026	0.131	0.23	1
Io[a]	$1\,863 \pm 90$							0.375 ± 0.005	3
Europa	438 ± 9							0.348 ± 0.002	4
Ganymede	127 ± 3							0.311 ± 0.003	5
Callisto	34 ± 5							0.358 ± 0.004	6

1: Yoder (1995). 2: Konopliv *et al.* (1998). 3: Anderson *et al.* (1996a). 4: Anderson *et al.* (1998a). 5: Anderson *et al.* (1996b). 6: Anderson *et al.* (1998b).

[a] J_2 was determined from C_{22} assuming hydrostatic equilibrium, in which case $J_2 = (10/3)C_{22}$.

terial displaced. The total mass of the material displaced is equal to the total mass of the mountain. Similarly, ocean and impact basins in isostatic equilibrium have extra mass at deeper layers. The deficiency or addition of mass at deeper layers can be calculated by assuming isostatic equilibrium (see Fig. 6.6b). *Airy's hypothesis* assumes one and the same density for all the crustal layers, ρ_c, and a larger value for the fluid mantle material, ρ_m ($\rho_c < \rho_m$). Isostatic equilibrium is reached by varying the height of the crust under the various regions A, B and C (Problem 6.7). In *Pratt's hypothesis* it is assumed that the depth of the base level of the crust is the same for all regions (dashed line in Fig. 6.6b), and that isostacy is reached because the densities of the columns A, B and C are different (Problem 6.8).

Before one can use gravity measurements to determine if a region is in isostatic equilibrium, a number of corrections need to be applied to the data. In addition to the variations in surface gravity as a function of latitude (eq. 6.16), one must take the height, h, above the geoid into account, i.e., the altitude at which the measurements were made ($h \ll R$, with R the radius of the planet). If one assumes there is only air present above sea level up to altitude h, this correction is know as the *free-air correction*, and is equal to $2h/R$ (Problem 6.9). This term has to be subtracted from the theoretical reference geoid. If there is a large slab of rock at the latitude of the measurement around the entire planet, the free-air correction factor should be decreased by the graviational attraction of the rocks, $2\pi G\rho h$, where ρ is the density of the rock of height h above the geoid. This correction is known as the *Bouger correction*, which itself can then be corrected using a terrain correction factor, δg_T, based upon a topographic map. The total correction factor to be applied to the measurements is known as the *Bouger anomaly*, g_B, and is thus the difference between the observed gravity, $g_p(obs)$, and the value expected at the point of measurement:

$$g_B = g_p(obs) - g_p(\theta)\left(1 - \frac{2h}{R}\right) - 2\pi G\rho h + \delta g_T. \tag{6.24}$$

Lateral variations in the density distribution thus result in gravity anomalies, such as the free-air and Bouger anomalies. In addition, such variations cause deviations in the measured geoid with respect to a spheroid calculated for a rotating, spherically symmetric, fluid planet. Such deviations are expressed in terms of the *geoid height anomaly* (geoid radius minus spheroid radius), Δh_g, which is related to the measured anomaly in the gravitational potential, $\Delta\Phi_g$:

$$g_p(\theta)\Delta h_g = -\Delta\Phi_g. \tag{6.25}$$

For an isostatic density distribution, equation (6.25) can be approximated by:

$$\Delta h_g = -\frac{2\pi G}{g_p(\theta)}\int_0^D \Delta\rho(z) z\, dz, \tag{6.26}$$

with $\Delta\rho(z)$ the anomalous density at depth z and D the compensation depth, below which there are no horizontal gradients in density. Depth is measured positive going down, and $z = 0$ corresponds to the geoid surface. The effective gravitational attraction is always normal to the geoid, so that there is a trough in the geoid where there is a negative gravity anomaly or positive potential anomaly

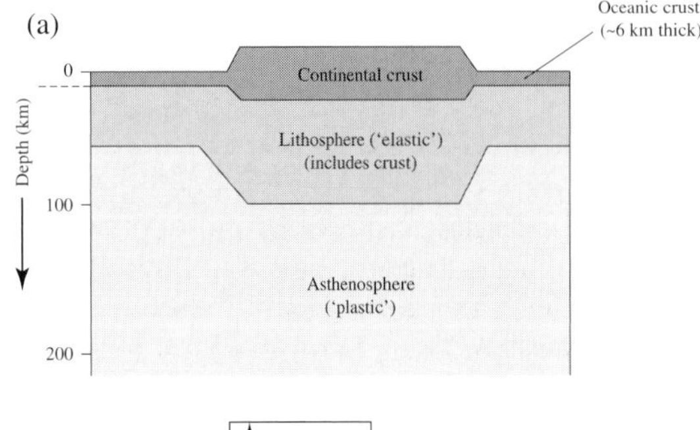

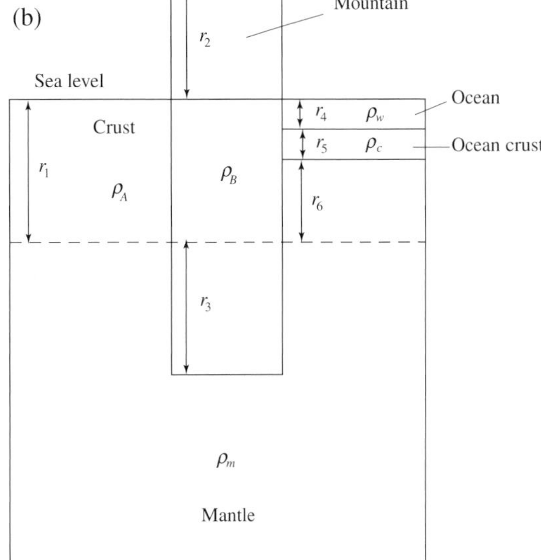

FIGURE 6.6 (a) Schematic representation of the outermost layers on Earth: the lithosphere is topped off with a lighter crust, which is relatively light and thick for continents, and denser and thinner under the oceans. (b) Isostatic equilibrium: The mountain is compensated by a mass deficiency of height r_3, and the ocean crust, r_5, by a layer of extra mass, r_6 (see Problems 6.7, 6.8).

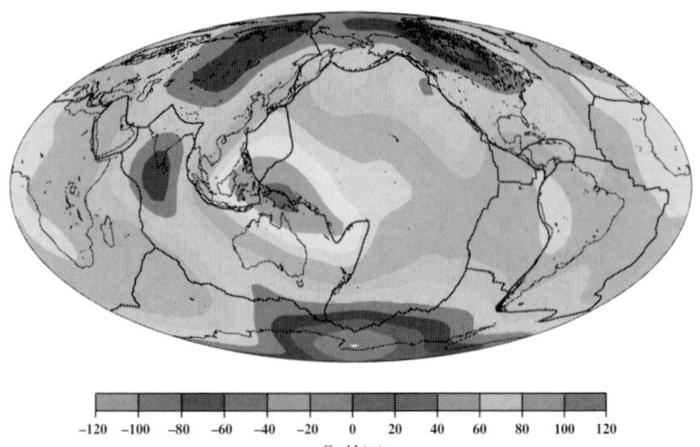

FIGURE 6.7 COLOR PLATE The observed geoid (degrees 2–15) superposed on a map of the planet Earth. Contour levels are from −120 m (near the South Pole) to +120 meters (just north of Australia). (Lithgow-Bertelloni and Richards 1998)

(mass deficit), and there is a bulge in the geoid if there is a positive gravity anomaly or negative potential anomaly (mass excess).

A gravity map of a planet is usually represented as a contour plot of the difference between the measured field (the real geoid) and that expected from equation (6.14) for a fluid rotating body in hydrostatic equilibrium. Such a map shows the elevations and depressions of the surface equipotential with respect to the mean planetary surface. The degree to which surface topography and gravity are correlated can be interpreted in terms of how much or little isostatic compensation is present, information which can be used to increase our knowledge of a planet's lithosphere and mantle. However, to derive this information a number of assumptions need to be made. One needs to know the local topography to high accuracy in order to calculate the Bouger anomaly, even if the small terrain factor δ_T is ignored. Also, the density of the underlying structures must be known (Airy and Pratt hypotheses). Finally, hitherto we have tacitly assumed static structures; however, we know that there is convection in the mantle, with regions of upwelling and subsiding motions. The columns of rising material are hotter, and therefore less dense than neighboring regions, while the opposite is true in columns of sinking material. Additionally, the surface is influenced: there are ridges or mountains above columns of rising material, whereas we find depressions above columns of sinking material. The combined effect usually shows up as small positive gravity and geoid height anomalies above regions of rising material, because the surface deflections introduce a larger positive anomaly than the negative one introduced by the low density. At locations where material is sinking, however, the effect on gravity and geoid height is strongly influenced by the viscosity in the mantle. So whether the net geoid anomaly is positive or negative depends on the subtle cancellation of the 'dynamic' surface topography and the density effect in the mantle, where the surface topography is influenced by the viscosity structure in the mantle.

An example of the geoid shape for Earth is shown in Figure 6.7, superposed on a map of the Earth. This is a relatively low order ($J_2 - J_{15}$) gravity map, comparable to that obtained for other planets. There is no correlation between topography and the geoid, because of the high degree of isostatic compensation in the Earth's mantle. As we will see in our discussion of the individual planets (Section 6.3), Earth is quite unique in this respect. Most other bodies reveal a correlation between gravity field and topography. This must be related to the presence of a high plasticity upper mantle on Earth, allowing material to flow and reach isostatic equilibrium in a geologically short time. Bodies with thicker 'solid' mantles show a much lower degree of isostatic compensation.

6.1.5 Internal Heat: Sources and Losses

In Section 4.2.2 we compared the equilibrium temperatures of the giant planets, i.e., the temperature the planet would have if heated by solar radiation only, with their observed effective temperatures (Table 4.2). This comparison showed that Jupiter, Saturn and Neptune are warmer than can be explained from solar heating alone, which led to the suggestion that these planets possess internal heat sources. The Earth must also have an internal source of heat, as deduced from its measured heat flux of 75 erg cm^{-2} s^{-1}. In this section we discuss possible sources of internal heat, as well as mechanisms to transport the energy and ultimately lose it to space. Heat flow parameters for all planets are summarized in Table 6.3.

6.1.5.1 *Heat Sources*

Gravitational

Accretion of material during the formation of planets is likely one of the largest sources of heat (Section 12.6; Problems 12.27, 12.28). Bodies hit the forming planet with roughly the escape velocity, yielding an energy for heating of GM/R per unit mass. The increase in energy per unit volume is equal to $\rho c_P \Delta T$, with c_P the specific heat (per gram material), ρ the density and ΔT the increase in temperature. The gain in energy at distance r must be equal to the difference between the gravitational energy acquired at r, $GM(r)/r$, and the energy which is radiated away, σT^4, (Section 3.1) over a time dt during which the body accreted a layer of thickness dr:

$$\rho c_P(T(r) - T_0)dr = \left(\frac{GM(r)}{r} - \sigma(T^4(r) - T_0^4)\right)dt, \tag{6.27}$$

with T_0 the initial temperature of the accreting material. If accretion is rapid, much of the heat is 'stored' inside the planet before it has time to radiate away into space, since subsequent impacts 'bury' it. The ultimate temperature structure inside a planet further depends on the size of accreting bodies and the internal heat transfer (Problems 12.19–12.22).

Giant Planets. The internal heat sources of the giant planets Jupiter, Saturn and Neptune are attributed to gravitational energy, either from gradual escape of primordial heat generated during the planet's formation, and/or from previous or ongoing differentiation. If we consider an initially hot adiabatic planet, which gradually cools down, we can estimate the time it takes to cool down to the present temperature. The luminosity of the planetary body, L, consists of three components: L_v is reflected sunlight (mainly at optical wavelengths), L_{ir} is incident sunlight absorbed by the planet and re-emitted at infrared wavelengths, and L_i is the planet's intrinsic luminosity. For the terrestrial planets L_i is very small, but for the gas giants, Jupiter, Saturn and Neptune, the internal luminosity is comparable to L_{ir}. The effective temperature T_e is obtained by integrating the emitted energy over all infrared wavelengths, and it thus consists of both L_{ir} and L_i. The equilibrium temperature T_{eq} is the temperature the planet would have in the absence of internal heat sources (cf. Section 3.1.2.2). The intrinsic luminosity of the planet is thus equal to:

$$L_i = 4\pi R^2 \sigma (T_e^4 - T_{eq}^4). \tag{6.28}$$

If we assume that the internal heat flux merely represents the leakage of the primordial heat stored during the planet's formation period, we can express the rate of change in the mean internal temperature dT_i/dt:

$$\frac{dT_i}{dt} = \frac{L_i}{c_V M}, \tag{6.29a}$$

where M is the planet's mass and c_V the specific heat at constant volume. Assuming the luminosity to have stayed constant over time, the cooling time, Δt, can be approximated:

$$\Delta t \approx \frac{\Delta T_i M c_V}{L_i}. \tag{6.29b}$$

For metallic hydrogen, $c_V \approx 2.5k/m_{amu}$ erg g^{-1} K^{-1}, with k Boltzmann's constant and m_{amu} the atomic mass unit.

Jupiter's excess luminosity is consistent with the energy released from gravitational contraction/accretion in the past. In contrast, detailed models of Saturn's interior structure, including evolutionary tracks (such as those discussed below for Uranus and Neptune), show that primordial heat alone is not sufficient to explain Saturn's excess heat. The additional heat loss can be explained from a differentiation process involving helium, a process which at the same time explains the observed depletion in the helium abundance in Saturn's atmosphere compared to solar values. Since Saturn is less massive and therefore colder in its deep interior than is Jupiter, helium has likely been immiscible in Saturn's metallic hydrogen region for a few billion years, while it became immiscible in Jupiter's interior only 'recently', i.e., when the temperature dropped to levels where the hydrogen and helium phases separate out.

TABLE 6.3 Heat Flow Parameters.

Body	T_e (K)	T_{eq} (K)	H_i (erg cm^{-2} s^{-1})	L/M (erg g^{-1} s^{-1})	References
Sun	5770		6.2×10^{10}	1.9	1
Carbonaceous chondrites				4×10^{-8}	1
Mercury		446			3
Venus		238			3
Earth		263	75	6.4×10^{-8}	1, 3, 4
Moon		277	~26	$\sim 1.3 \times 10^{-7}$	3, 5
Mars		222	40	9×10^{-8}	1, 3, 4
Io			~2500	$\sim 10^{-5}$	5
Jupiter	124.4	113	5440	1.8×10^{-6}	1, 2, 3
Saturn	95.0	83	2010	1.5×10^{-6}	1, 2, 3
Uranus	59.1	60	<42	$<4 \times 10^{-8}$	1, 2, 3
Neptune	59.3	48	433	3.2×10^{-6}	1, 2, 3

1: Hubbard (1984). 2: Hubbard *et al.* (1995). 3: Tables 4.1, 4.2. 4: Carr (1999). 5: Lodders and Fegley (1998).

Helium, therefore, has steadily 'rained out' of the metallic hydrogen region towards Saturn's core. The energy release from this process per unit drop in internal temperature can explain Saturn's observed L_i.

For Uranus and Neptune the specific heat can be approximated by $c_V \approx 3k/(\mu_a m_{amu})$ erg g^{-1} K^{-1}, with μ_a the mean atomic weight of the interior material, which is ~ 5 (amu) for icy material. The drop in temperature for Uranus and Neptune over the age of the Solar System is thus ~ 200 K (Problem 6.14), which is small compared to the internal temperature expected for an adiabatic planet (a few thousand K). However, although this implies that Neptune's heat flow could in principle arise from the planet's formation or accretional energy during the time of its formation, the thermal evolution model for Neptune appears inconsistent with this picture. Thermal evolution is characterized by equation (6.29a), where $L_i = dE_i/dt$, with E_i the internal heat reservoir. This equation can be written as:

$$Cf\frac{dT_e}{dt} = -(T_e^4 - T_{eq}^4), \tag{6.30}$$

with C a constant that characterizes the thermal inertia of the interior, and f is that fraction of the internal heat reservoir that gives rise to the observed luminosity. This fraction f is less than unity if convection is (partly) inhibited, such as may happen if there is a stable stratification in the interior. Equation (6.30) can be solved to yield the observed luminosity, depending on f and the initial ratio $(T_e/T_{eq})_i$, just after the planet formed. The relationship for f and $(T_e/T_{eq})_i$ is shown in Figure 6.8 for Uranus and Neptune, where the upper and lower curves correspond to the upper and lower limits to the observed present-day luminosities. Thus the lower curve for Uranus corresponds to a planet which reached a zero heat flow 4.5 Gyr after it formed. Any parameter choice to the left of this curve is acceptable, judged from the zero heat flow argument, but unacceptable because Uranus's magnetic field is thought to be sustained by a convective dynamo in its mantle (Section 6.4.3). It seems impossible to find a parameter choice which satisfies the thermal evolution curves for both Uranus and Neptune.

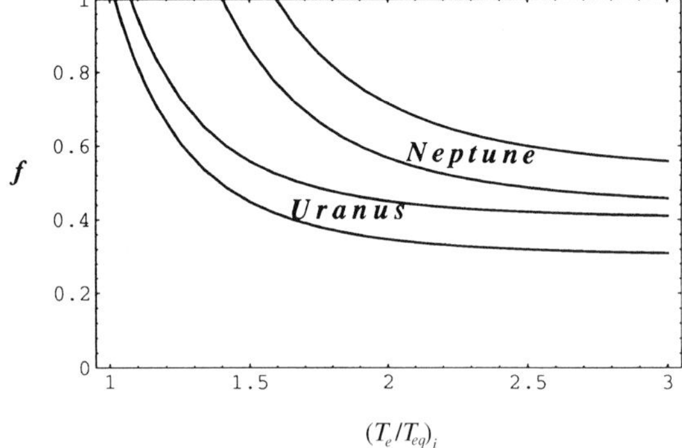

FIGURE 6.8 The relationship between the fraction, f, of the initial heat reservoir that may drive the planet's observed luminosity and the initial internal temperature. The latter quantity is given by the initial ratio between the effective and equilibrium temperatures, $(T_e/T_{eq})_i$. The upper and lower curves correspond to upper and lower bounds on the observed present-day luminosity. (After Hubbard *et al.* 1995)

It is thought that the internal temperature for both planets was very high initially, due to planetary accretion, which suggests f to be substantially less than unity, of the order of 0.4 for Uranus and 0.6 for Neptune, which corresponds to assuming that convection is inhibited inside a radius of 0.6 $R_♅$ for Uranus and inside 0.5 $R_♆$ for Neptune. This difference seems small, but nevertheless would lead to the observed differences in the intrinsic luminosities for the two planets.

Terrestrial Planets. As shown by equation (6.27), the interior of a planet can become very hot if accretion is rapid. For the 10^8 year accretion times estimated for the terrestrial planets, there is, however, little heating (cf. Problem 12.19), and it alone cannot account for the heat budget of the terrestrial planets and smaller bodies (though accretion with burying of heat would result in more sensible numbers; see Section 12.6.2). The smaller asteroids and satellites cannot have gained enough energy through accretion alone to melt their interiors allowing heavy material to sink down; yet the interiors of some of these bodies are differentiated (Section 5.2). As discussed below, the decay of radioactive elements and, for some solid bodies, tidal and ohmic heat provide additional important energy sources.

Radioactive Decay

Radioactive decay has been proposed as an important source of heat in the interiors of the terrestrial planets, satellites, asteroids and the icy bodies in the outer Solar System. If radioactive decay of elements is a major contributor to the heat flow measured today, the elements must have long half-lives, roughly of the order of a billion years. ^{235}U, ^{238}U, ^{232}Th, and ^{40}K have lifetimes approximately 0.71, 4.5, 13.9, and 1.4 Gyr respectively. These isotopes are present in the Earth's crust at levels of a few parts per million, and produce on average an energy of ~10 erg cm^{-2} s^{-1}. These elements are about two orders of magnitude less abundant in the mantle, but because the volume of the mantle is so much larger than that of the crust, the heat produced in the mantle significantly influences the total heat output. Only ~ 20% of the Earth's radioactive heating occurs in the crust. The heat generation from radioactive elements was larger in the past, and it appears that the total energy liberated by these elements over the first ~1–2 Gyr would have been enough to melt Earth and Venus (note that we do not say that Earth differentiated as a result of radioactive heating; we merely show how important radioactive heating must have been early on). The heat generated by short-lived radionuclides, in particular ^{26}Al with a half-life of 0.74 Myr, was high at an early stage, but disappeared quickly. At present the total heat released by radioactive decay in carbonaceous chondrites is about 4×10^{-8} erg g^{-1} s^{-1}, which is only slightly less than the total heat flow from Earth (6.4×10^{-8} erg g^{-1} s^{-1}). It may have been an order of magnitude larger at the time of planet formation (Problems 12.24, 12.25). At present roughly half of the Earth's heat output is thought to be due to radioactive heat production, while the remainder may be due to secular cooling.

Tidal and Ohmic Heat

Temporal variations in tidal forces can lead to internal heating of planetary bodies, as discussed in detail in Section 2.6.2. For most objects this potential source of energy is much smaller than that due to gravitational contraction or radioactive heating. For the moons Io and Europa, however, tidal heating is a major source of energy (see Sections 2.6.2, and 5.5). Tidal heating is also likely to have been important in the past for the moons Triton, Enceladus and possibly Ganymede.

Ohmic heating results from the dissipation of an induced electric current, such as the processes discussed in Section 7.4.4.6. This possible source of heat may be important for Io, depending on its conductivity. Ohmic heating may also have been sufficient to melt asteroid-sized planetesimals in some regions of the protoplanetary disk during the young Sun's active T-Tauri phase (Section 12.3.4).

6.1.5.2 *Energy Transport and Loss of Heat*

Energy transport determines the temperature gradient in a planet's interior, just like it does in a planetary atmosphere. The temperature gradient is determined by the process which is most effective in transporting heat. The three mechanisms by which heat is transported are conduction, radiation and convection (cf. Section 3.2). Conduction and convection are important in planetary interiors, while radiation is important in transporting energy from a planet's surface into space, and in a planet's atmosphere where the opacity is small but finite.

Conduction and Radiation

The heat flux, $\mathbf{Q}$ (erg s^{-1} cm^{-2}), is given by the Fourier heat law (eq. 3.19; Section 3.2.1):

$$\mathbf{Q} = -K_T \nabla T, \tag{6.31}$$

where K_T is the *thermal conductivity*. Conduction is the most effective way to transport energy in solid materials, such as the crustal layers of the terrestrial planets and throughout the interiors of smaller bodies like asteroids and satellites. The energy may be transported by free electrons or heavier particles, or by photons or phonons. The latter correspond to waves excited by vibrations in the crystal lattice. Good heat conductors are materials with a large number of free electrons, such as metals. In metal-poor materials, such as silicates, heat transport is dominated by phonons, where the energy is carried by propagating elastic lattice waves. The propagation of these waves decreases with increasing temperature, because the anharmonicity in the crystal increases at higher temperatures, which causes the waves to scatter. At high temperatures radiative transport of heat by photons also becomes significant. At lower temperatures the photon mean free path is small and the total energy in the photon field is small compared to the vibrational thermal energy; but the energy in the photon field is proportional to T^3, so that it becomes appreciable at high temperatures. The thermal conductivity in uncompressed silicate can therefore be best written as the sum of K_L, the lattice thermal conductivity, and K_R, the radiative thermal conductivity:

$$K_T = K_L + K_R, \tag{6.32a}$$

where

$$K_L = \frac{4.184 \times 10^7}{30.6 + 0.21T}, \tag{6.32b}$$

$$K_R = 0 \qquad (\text{for } T < 500), \tag{6.32c}$$

$$K_R = 230(T - 500) \qquad (\text{for } T > 500). \tag{6.32d}$$

A quick calculation of the importance of temperature changes due to conduction can be obtained via the thermal diffusivity, k_d (Section 3.2.1):

$$\ell \approx \sqrt{k_d t}, \tag{6.33}$$

where ℓ is a length scale over which changes in the temperature gradient become significant over a timescale t. The thermal diffusivity is given by equation (3.22):

$$k_d \equiv \frac{K_T}{\rho c_P}. \tag{6.34}$$

In Problem (6.16) the thermal diffusivity and scale length ℓ are calculated for a typical rocky body. One finds that, over the age of the Solar System, the temperature gradient has been affected by conduction over only a few hundred kilometers. Hence, the temperature structure in small bodies, such as asteroids and satellites, can be modified by thermal conduction over the age of the Solar System, but planet-sized bodies cannot have lost much of their primordial heat via conduction.

Convection

As discussed in Section 3.2, convection is caused by rising and sinking motions of the material: hot material rises towards cooler regions at higher altitudes, and cold material sinks down. Convection can only proceed if the rate at which energy is liberated by buoyant forces exceeds that at which energy is dissipated by viscous forces. This criterion is expressed in terms of the *Rayleigh number*, $\Re_a$:

$$\Re_a = \frac{\alpha_T (T_1 - T_2) g_p \rho_0 \ell^3}{k_d \nu_v} > \Re_a^{crit}, \tag{6.35}$$

where $\Re_a^{crit}$ is a critical value, usually of the order 500–1000. The variation in density over this layer of thickness ℓ is approximated by:

$$\rho = \rho_0(1 - \alpha_T T), \tag{6.36}$$

with α_T the thermal expansion coefficient, and T_1 (T_2) is the temperature at the bottom (top) of the layer, with $T_1 > T_2$. The Rayleigh number can only exceed the critical value if the viscosity, ν_v, is finite. Since the viscosity is small in fluids and gases, convection is likely the primary mode of energy transport in regions where the planet is gaseous or fluid, as in giant planets and the liquid outer core of Earth. Many materials, including rocks, can deform under an applied rate of strain. This means that, although on 'short' timescales materials such as mantle rocks behave like solids, over geological time periods (several million to hundreds of millions of years) these 'rocks' flow in response to the applied rate of strain, and the viscosity of this material is finite. Energy may therefore be transported by solid-state convection in planetary mantles, if the characteristic timescale for convection is small compared to geological timescales. It has been shown empirically that, if $\Re_a \gg \Re_a^{crit}$, the ratio of the heat flux carried by convection plus conduction to that carried by conduction alone, the *Nusselt number*, is approximately equal to:

$$\mathcal{N}_u \equiv \frac{Q(\text{convection} + \text{conduction})}{Q(\text{conduction})} \approx \left(\frac{\Re_a}{\Re_a^{crit}}\right)^{0.3}. \tag{6.37}$$

The viscosity of rocks depends strongly on temperature. At low temperatures, the viscosity is essentially infinite, and the material behaves as a solid. At temperatures above roughly 1100–1300 K, the viscosity of rock is low enough that the material 'flows' over geological timescales. This explains why the cold outer layers of a planet, the *lithosphere*, are made of solid rock where energy is transported by conduction, while deeper in the planet (mantle), energy is mainly transported by convection.

Heat Loss

On solid bodies heat is generally lost by conduction upwards through the crust and radiation from the surface into space. At deeper levels, for planets that are large enough that solids 'flow' over geological timescales, the energy may be transported by convection, a process which, when it occurs, typically transports more heat than conduction. In the upper 'boundary' layers heat transport is again by conduction upwards through the rigid lithosphere and crust. However, conduction alone may not be sufficient to 'drain' the incoming energy from below. On Earth additional heat is lost through tectonic activity along plate boundaries, and via hydrothermal circulation along the mid-ocean ridge. Heat may also be lost during episodes of high volcanic activity, either in volcanic eruptions, or through vents or hot spots. The latter source of heat loss is dominant on Io; in fact, the total heat outflow through Io's hot spots may, at the present time, even exceed the generation of heat due to tidal dissipation.

In the giant planets heat is transported by convection throughout the mantle and most of the troposphere, while radiation to space plays an important role at higher altitudes, in the stratosphere and lower thermosphere. It is possible that radiative zones exist in the mantles of Jupiter and Saturn at temperatures roughly between 1200–1400 K and 3000 K, i.e., at pressure levels of several (tens of) kbar. Similarly, a radiative window may exist in Uranus around 10 kbar ($>$ 1200 K) (Section 6.4.2).

Measurements of the heat flow on solid bodies are very difficult, since the outer parts of a planet from which we can measure the temperature or luminosity, are also heated up by the Sun during the day, so that the crustal layers display the effects of the diurnal solar insolation, as discussed in Section 3.2.1. The temperature gradient in planetary atmospheres, in the stratosphere and upper troposphere, is determined primarily by sunlight, while greenhouse phenomena complicate matters in the lower troposphere. For the giant planets Jupiter, Saturn and Neptune the internal source of energy is comparable in magnitude to the energy received from the Sun, and it is possible to determine the internal luminosity from measurements of the effective temperature (cf. Section 6.1.5.1). For terrestrial planets, if the thermal conductivity in the crustal layers is known, the internal heat flow can in principle be determined via measurements of the temperature gradient in the upper layers of the planet's crust. The temperature gradient can, in principle, be obtained by drilling holes (Earth, Moon), or by using remote sensing techniques. A typical value for the heat flux from Earth is 75 erg cm^{-2} s^{-1}, which corresponds to an intrinsic luminosity $L_i = 3.84 \times 10^{20}$ erg s^{-1}, roughly 5000 times smaller than L_{ir} and L_v combined (Problem 6.13). The heat flux from the Moon is roughly half that of Earth, as determined from boreholes in the lunar surface. In contrast to boreholes or mines, which can go down 100–1000 m, remote sensing techniques only sample the upper few meters of the crust at best, and it is impossible to determine the temperature gradient below the region sensitive to diurnal heating. Since the internal heat flow from bodies in the inner Solar System is several orders of magnitude smaller than the energy the body receives from the Sun, it is impossible at the present time to measure the temperature gradient via remote sensing techniques with such accuracy that the influence on the temperature gradient due to an internal heat source can be distinguished from that caused by solar insolation effects.

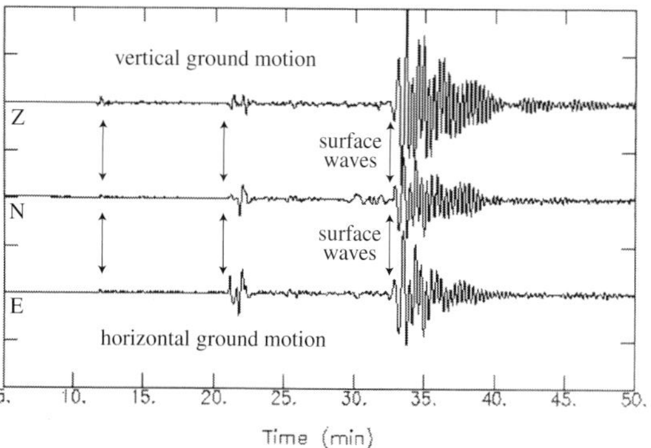

FIGURE 6.9 A recording from a seismograph showing P, S and surface waves from a distant earthquake. Vertical (Z) and horizontal (N = to the north, E = to the east) motions are shown. (UC Berkeley Seismological Laboratory)

6.2 Seismic Tomography and the Earth's Interior

Whereas the structure of the Earth's crust can be determined by *in situ* experiments, such as drilling, details on the Earth's deep interior have to be obtained via indirect means. The most powerful method to obtain information on the Earth's interior comes from *seismology*, the study of the passage of elastic waves through the planet. Such waves are induced either by *earthquakes*, meteoritic impacts, volcanic or man-made explosions. The waves are detected by *seismometers*, sensitive instruments that measure the motion of the ground on which they are located. A seismograph recording of vertical and horizontal ground motions is shown in Figure 6.9. With a number of seismometers spread around Earth, the waves are studied from many different 'viewing' points, and the data can be used to derive the structure of the Earth's interior via *seismic tomography*, like CAT-scanning the interior.

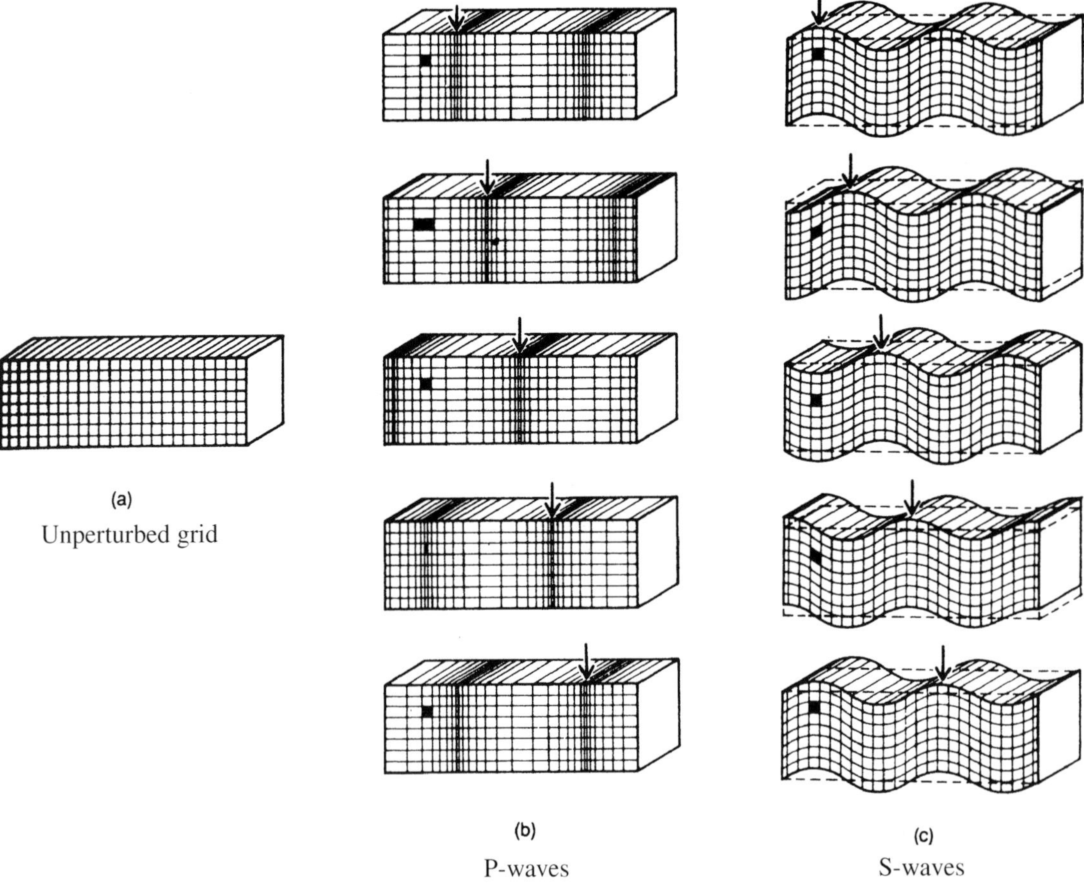

FIGURE 6.10 Seismic P and S waves: In P waves the particles oscillate back and forth along the direction of motion, while the particle motion in S waves is transverse to the direction of motion. (Fowler 1990)

6.2.1 Seismology

6.2.1.1 *Seismic Waves*

Body waves are seismic disturbances that travel through a planet's interior, whereas *surface waves* propagate along the surface. Body waves obey Snell's law, and are reflected and transmitted at interfaces where the material density changes. We distinguish between P and S waves (see Fig. 6.10). *P waves* are Primary, Push or Pressure waves, where the individual particles of the material are oscillating back and forth in the direction of wave propagation. They are longitudinal waves, similar to ordinary sound waves, and involve compression and rarefaction of the material as the wave passes through it. The first P waves travel rapidly and arrive at a seismic station well before the first S waves. *S waves* are the Secondary, Shake or Shear waves, which have their oscillations transverse to the direction of propagation. They are analogous to the waves you can make on a rope, and to electromagnetic waves. They involve shearing and rotation of the material as the wave passes through it.

Surface waves are confined to the near-surface layers on Earth (Fig. 6.11). These waves have a larger amplitude and longer duration than body waves, and their velocity is smaller than that of body waves. In *Rayleigh waves* the particle motion is a vertical ellipse. These waves show many similarities to waves on water. The amplitude of the wave decreases with depth, just like water waves. In *Love waves* the motion is entirely horizontal, but transverse to the propagation of the waves.

Any motion of the medium is due to the P and/or S waves, where the wave equations for the two types of waves can be deduced from the theory of elasticity (see e.g., Fowler 1990). The compressional wave equation for P waves is:

$$\frac{\partial^2 \Phi}{\partial t^2} = v_P^2 \nabla^2 \Phi, \tag{6.38a}$$

(a)

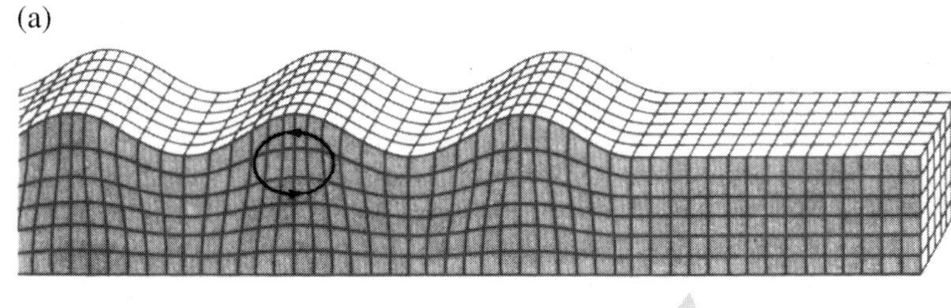

(b)

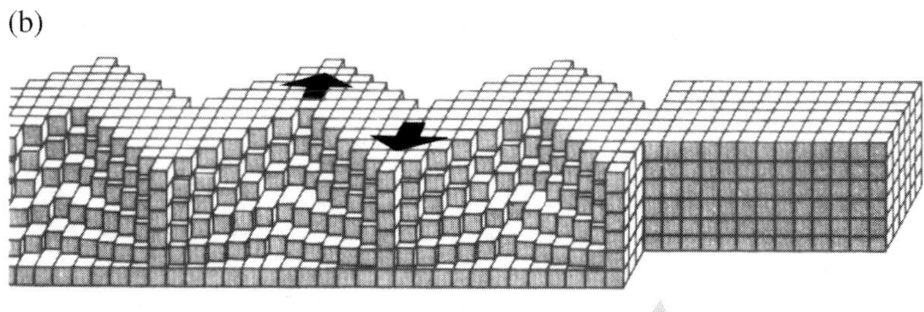

FIGURE 6.11 Seismic surface waves: (a) Rayleigh waves and (b) Love waves. (Fowler 1990)

and the rotational wave equation for S waves:

$$\frac{\partial^2 \bar{\psi}}{\partial t^2} = v_S^2 \nabla^2 \bar{\psi}, \tag{6.38b}$$

with the P wave and S wave velocities v_P and v_S. The displacement of the medium, $\mathbf{x}$, can be expressed as the sum of the gradient of the scalar potential, Φ, and the curl of the vector potential, $\bar{\psi}$:

$$\mathbf{x} = \nabla\Phi + \nabla \times \bar{\psi}. \tag{6.39}$$

From the theory of elasticity it can further be shown that the P wave and S wave velocities are related to thermodynamic properties of the medium:

$$v_P = \sqrt{\frac{K_m + \frac{4}{3}\mu_{rg}}{\rho}}, \tag{6.40a}$$

and

$$v_S = \sqrt{\frac{\mu_{rg}}{\rho}}, \tag{6.40b}$$

where ρ is the density, μ_{rg} the rigidity or *shear modulus*, and K_m the bulk or (adiabatic) *incompressibility modulus* of the material at constant entropy S:

$$K_m \equiv \rho \left(\frac{\partial P}{\partial \rho} \right)_S. \tag{6.41a}$$

If a planet's interior is adiabatic and chemically homogeneous, the bulk modulus of the material becomes:

$$K_m \approx \rho \frac{dP}{d\rho}. \tag{6.41b}$$

From the equations above it follows that K_m, v_P and v_S are related as follows:

$$\frac{K_m}{\rho} = v_P^2 - \frac{4}{3} v_S^2. \tag{6.42}$$

The bulk modulus is a measure of the stress or pressure needed to compress a material, thus it involves a change in volume of the material. The shear modulus is a measure of the stress needed to change the shape of the material, without changing its volume. Note that v_P depends both on K_m and μ_{rg}, since P waves involve a change of both volume and shape of the material, while v_S depends only on μ_{rg}, since S waves do not involve a change in volume (see Fig. 6.10). Since $K_m > 0$, P waves travel faster than S waves: $v_P > v_S$. The wave velocities are measured by timing how long it takes the wave to travel from the source (*hypocenter*) of the earthquake to the seismograph. (The *epicenter* is the point on the surface vertically above the hypocenter.) In a liquid, $\mu_{rg} = 0$, and therefore, in contrast to P waves, the S waves cannot propagate through liquids. Because there are no detectable seismic phases corresponding to S wave propagation in the Earth's outer core, we know that the Earth's outer core is liquid (Fig. 6.12). In fact, when an S wave is incident on the outer core, part of it is reflected and part is transmitted as a P wave (S $\rightarrow$ P wave conversion). Both K_m and μ_{rg} depend on the density, but increase faster than ρ. Since the density increases towards the center of a planet, the velocities of both P and S waves increase with depth. In addition,

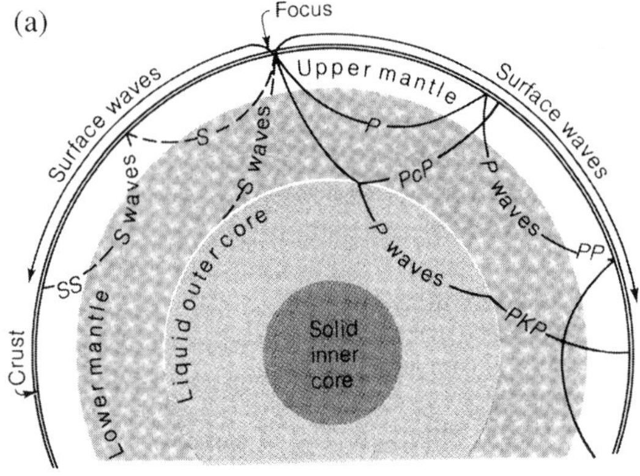

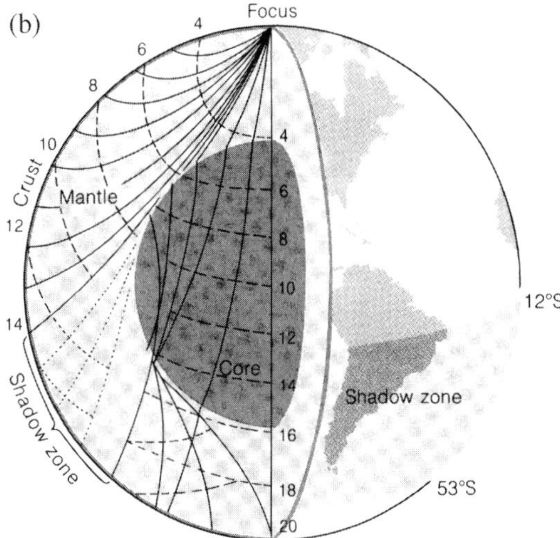

FIGURE 6.12 (a) A schematic representation of the propagation of seismic waves on Earth superimposed on a sketch of the interior structure as emerged from seismological experiments. Surface waves propagate along the near-surface layers, while the S and P waves propagate through the Earth's interior. Waves reflected from the Earth's crust are called SS or PP waves. When an S wave is incident on the outer core, part of it is reflected and part is transmitted as a P wave; S waves themselves cannot propagate through liquid material. P waves are refracted (PKP waves) as well as reflected (PCP waves) at the core–mantle interface. Since changes in the wave velocity take place continuously throughout the Earth's interior (in addition to abrupt changes at interfaces), the wave paths are curved rather than straight. (Press and Siever 1986) (b) A different representation of seismic waves, which shows the presence of a *shadow zone*, a region which cannot be reached by P waves because they are deflected by the Earth's core. (Press and Siever 1986)

because the density changes along a wave's path, the path is curved (upwards, since ρ increases with depth) according to Snell's law. Since v_P and v_S can be determined as a function of depth from seismic data, K_m/ρ is known as a function of depth, which allows for a direct determination of the density as a function of depth.

6.2.1.2 *Free Oscillations*

When a rope is fastened at one end and moved up and down with an appropriate frequency, *standing waves* can be created. These waves do not seem to propagate along the rope, but are stationary. Such waves can be induced in any elastic body. In contrast to other waves, standing waves may persist for a long time. In daily life one makes use of this physical phenomenon when playing a musical instrument, like a violin, an organ, or just ringing a bell. The waves with the longest periods of oscillation generally persist the greatest length of time; these wavelengths correspond to low 'tones' in music. In a very similar way, standing waves can be induced in any planet by a large quake, and the planet can 'ring as a bell' for many days. The Earth can vibrate in an infinite number of ways, or *modes*. Some patterns of induced waves are shown in Figure 6.13. The particle motion is either radial (up and down) or tangential to the Earth's surface, causing *spheroidal* (S) and *toroidal* (T) modes, respectively. A few examples of such modes are shown graphically in Figure 6.13a. The modes are indicated by ${}_nS_m$, where there are n nodes in the interior of the planet and m on the surface. In the ${}_0T_2$ mode the two hemispheres twist in opposite directions, while in the ${}_1T_2$ mode the outside sphere twists like in the ${}_0T_2$ mode and the inner sphere twists in the opposite direction. The surface toroidal oscillations are equivalent to Love waves, while the surface spheroidal oscillations are equivalent to Rayleigh waves. The simplest spheroidal oscillation is a purely radial expansion and contraction of the Earth as a whole, as indicated by the ${}_0S_0$ mode. The ${}_0S_1$ mode is not really a free oscillation mode: the Earth as a whole is shifted in one direction. Such a shift is not possible for an earthquake, but could be triggered by a meteorite impact. Higher order modes become more complicated, as already shown by the ${}_0S_2$ mode. Typical periods for free oscillations on Earth are of the order of 3.5 minutes for ${}_0S_{40}$ and ${}_0T_{40}$ modes, increasing to 44 minutes for ${}_0T_2$ and 54 minutes for the ${}_0S_2$ mode. Adjacent areas on the Earth's surface move thus in opposite directions (except for ${}_0S_0$, which has no nodes), as shown in Figure 6.13b, and are separated by lines or *nodes*, which do not move at all.

Free oscillations have been observed for the Earth, Sun and Moon. Several research groups are actively working towards detecting such oscillations on Jupiter, but thus far without much success, although there has been one report on a possible detection of normal modes in Jupiter. Several searches for seismic waves on Jupiter were carried out during and following the impacts of Comet D/Shoemaker–

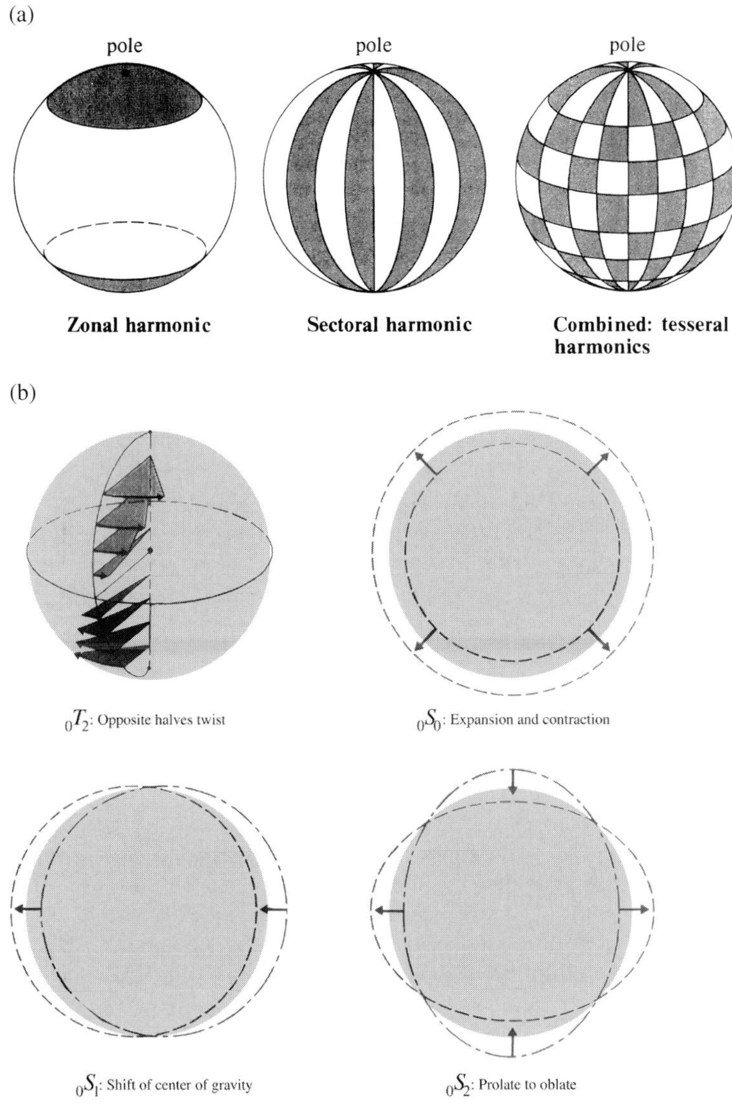

FIGURE 6.13 (a) Surface movements of some free oscillations on Earth. The light and dark areas have displacements which are in opposite senses at any instant. They are separated by *nodes*, lines where there is never any movement. (b) Examples of modes of oscillation. In toroidal modes, T, the movements are tangential to the Earth's surface, and in spheroidal modes, S, the movement is predominantly radial. (Brown and Mussett 1981)

Levy 9 with Jupiter (Section 5.4.5), but no seismic waves have been detected.

6.2.2 Interior Structure of the Earth

6.2.2.1 *Density Profile*

From the equation of hydrostatic equilibrium (eq. 6.1) and the definition of the bulk modulus K_m in an adiabatic and chemically homogeneous planet (eq. 6.41a), one derives the density distribution as a function of distance, r (Problem 6.17):

$$\frac{d\rho}{dr} = -\frac{GM\rho^2}{K_m r^2}. \tag{6.43}$$

Since K_m/ρ can be determined from the seismic wave velocities v_P and v_S (eq. 6.40), one can integrate the *Adams–Williams equation* (eq. 6.43) from the surface inwards to determine the Earth's density structure, assuming a *self-compression model*, where the density at each point is assumed to be due to compression by the layers above it. The mass, $M(r)$, is the mass within radius r, which is equal to:

$$M(r) = M_\oplus - 4\pi \int_r^{R_\oplus} \rho(r) r^2 dr, \tag{6.44}$$

with $M_\oplus$ and $R_\oplus$ the mass and radius of the Earth, respectively. The result of this calculation depends on which density is chosen in the top layer. It may not be too surprising that it appears impossible to find a density structure in this self-compression model which satisfies the seismic wave velocities. These velocities show clear 'jumps' at particular depths, such as at the core–mantle boundary, indicative of abrupt changes in density (see Fig. 6.14 discussed below). But even when the Adams–Williams equation is integrated separately over the mantle and the core, with different boundary conditions so that the seismic data can largely be reproduced, the resulting density distribution predicts a moment of inertia that is smaller than the measured value. This suggests that there must be more mass in the Earth's mantle than predicted by our simple model, and thus that some of our assumptions in the model are invalid. The most significant simplification we make is to ignore any chemical or phase changes in the Earth's interior, other than the differences in density between the crust, mantle and core. The seismic velocity profiles show that there are numerous density 'jumps' in the mantle, which are presumably caused by chemical and/or phase changes. Although our model is also based upon an adiabatic temperature profile, deviations in the temperature structure from an adiabatic model are found to be very small. When the chemical and phase changes are included in the model, the density distribution inside the Earth can be determined to quite high accuracy from seismic wave velocities, in particular when including additional constraints from free oscillations, the moment of inertia ratio I/MR^2, and the total mass of the Earth.

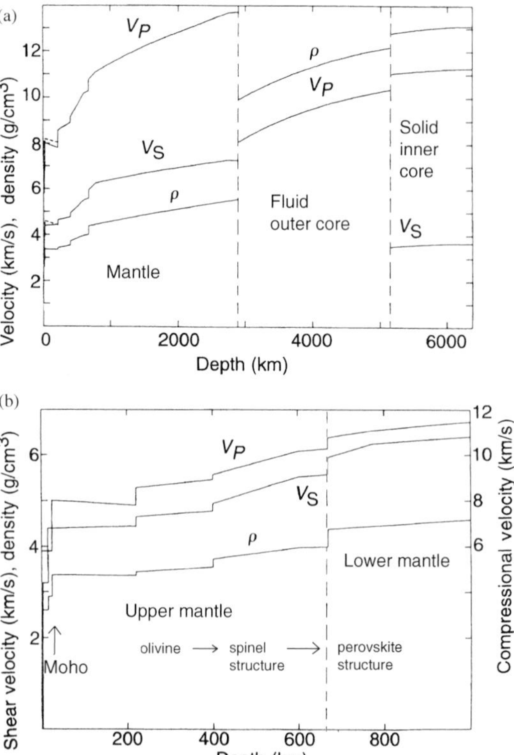

FIGURE 6.14 (a) The seismic P and S wave velocities and the density in the Earth's mantle and core. (b) An expansion of the uppermost 1000 km from (a). (After Pieri and Dziewonski 1999)

A diagram of wave velocities and densities as a function of depth in the Earth is shown in Figure 6.14. In the outer 3000 km, the density as well as v_P and v_S increase with depth. An abrupt boundary appears at ~ 3000 km, where the S waves disappear altogether, and P waves slow down considerably. This boundary is interpreted as the *interface between the solid mantle and liquid outer core*, since S waves cannot propagate through liquids. In addition to the disappearance of S waves, analysis of Earth's free oscillations excited by large earthquakes also suggests the presence of a liquid outer core. Model fits to geodetic observations of the nutation (wobble) of the Earth's rotation axis form a third piece of evidence that the outer core is a liquid, since the viscosity of the outer core must be relatively low to explain the data. At ~ 5200 km depth, there is another discontinuity, interpreted as the *boundary between the liquid outer core and a solid inner core*. At this boundary the P wave velocity increases. Analysis of seismic free oscillations shows conclusively that the inner core must have some rigidity, i.e., be solid, with a density about 0.5 g cm^{-3} higher than that of the outer core. Although S waves disappear at the outer core boundary, they can travel through the inner core.

The boundary between the crust and the mantle, within the lithosphere, is called the *Mohorovičić* or *Moho discontinuity*. The continental crust consists mostly of granitic rocks, with gabbro at the bottom, while the oceanic crust consists entirely of basaltic rocks. The thickness of the crust varies from roughly 35 km for typical continents to ~ 65 km under a mountainous area and ~ 6 km for the oceans. The crust is the upper part of the *lithosphere*, which 'floats' on a plastic-like layer called the *asthenosphere*. The lithosphere is cool and rigid, while the asthenosphere consists of material which, over geologic time, has a high plasticity. The lithosphere is often modeled as an elastic layer; it bends under loads, such as glacier caps, whereas the asthenosphere flows. The shear wave velocity decreases somewhat at the transition between the lithosphere and asthenosphere. As further shown in Figure 6.14b, the wave velocities in the outer 670 km of the Earth increase (stepwise) with depth. The velocities change in regions where the density changes, which are attributed, for example, to phase changes of the material: olivine → spinel-structure → perovskite (Section 6.1.2.3).

6.2.2.2 *Mantle Dynamics*

The lithosphere is broken into quasi-rigid plates, and the movement of these tectonic plates with sea floor spreading and continental drift (Section 5.3.2.2) suggests these plates to 'ride on top' of a convection pattern in the Earth's mantle. These plates may be regarded as the upper thermal boundary layer for mantle convection. Although seismic tomographic maps support such a circulation pattern, it is not known whether this pattern is driven by rising material below the mid-oceanic ridge, or by the sinking slabs of cold lithospheric plates at the outer edge of the oceanic plates, where they bump into continental plates. It is clear, though, that both mechanisms influence the convection pattern; and above all, that this type of convection is only possible due to a low viscosity zone in the upper mantle over which the plates glide easily. This low viscosity zone, or 'asthenosphere', concept dates from nineteenth century investigations of isostatic support of mountains and is supported by modern evidence for low seismic velocities at depths of 100–300 km, as well as by studies of postglacial rebound and dynamic compensation of the Earth's gravity field. In recent years researchers have been able to numerically model this plate tectonic style of convection by combining a pronounced low viscosity zone in the upper mantle with a particular plastic yield stress of the lithosphere, so that weak plate boundaries form above places of subsidence and upwelling. Despite these advances in our understanding of plate tectonics and mantle convection, there still is an ongoing debate to what depth this circulation pattern extends: does it go down to the core–mantle boundary, as suggested by some seismic velocity models, or are there separate convection patterns in the upper and lower mantle, as suggested by geochemical data?

It is generally assumed that when Earth formed, its composition was largely homogeneous and similar to that found in volatile-poor chondritic meteorites. The upper mantle, however, seems depleted in a number of elements (e.g., rare earth elements such as neodymium, and the noble gases) compared to the crust and the lower mantle. Estimates of the concentration of these elements in the upper mantle come from the mid-oceanic ridge, whereas some researchers believe that the lower mantle may be sampled from certain ocean island basalts. These oceanic islands (like the Hawaiian islands and Iceland) are known as volcanic 'hot spots'. They are interpreted as being formed by hot crystalline mantle material, which wells up like a plume inside a narrow ($\lesssim$ 100 km) column. According to the model, the location of these plumes is approximately fixed relative to each other, while the lithospheric plates move over the hot plumes. Because of the plate movement, chains of volcanic islands form, like the Hawaiian chain of islands, where the far end of the big island is still volcanically active (Kilauea). In contrast, the magma which fills the oceanic ridge is less hot and slowly oozes up to fill the void left by the receding oceanic plates. It is therefore thought that the mid-ocean ridge is made up of upper mantle material, while certain oceanic islands have been inferred to form from mantle plumes originating in the lower mantle, possibly near the core–mantle boundary. If global mixing between the lower and upper mantle were to take place, one would expect the same concentration of rare earth elements and noble gases throughout the mantle, and thus also in the mid-oceanic ridge and ocean island basalts. Since this has not been observed, the geochemical data appear to support theories on convection cells which are separate between the lower and upper mantle. Moreover, variations in the isotope abundances as measured from island to island suggest the lower mantle to be quite heterogenous, which suggests that convection in the lower mantle is much less efficient than in the upper mantle; this difference in convection efficiency can be explained if the viscosity in the mantle increases with depth.

Detailed three-dimensional seismic velocity models have been derived for the Earth's mantle using seismic tomography. Examples are shown in Figure 6.15. Figure 6.15a shows anomalies in the S wave velocities at different depths. The blue color indicates areas with above-average seismic velocities, and the red areas indicate below-average velocities. Since the seismic speed decreases with increasing temperature, the blue areas are interpreted as cold regions and the red areas as warmer (an interpretation that completely ignores variations in rock types, as observed at the surface). At a depth of 175 km, an extended cold region is seen over North America into South America, and between Europe and Indonesia, across southern Asia. These cold regions correspond to the stable part of the continents ('cratons') which have been tectonically inert for long periods of time ($\gtrsim 10^9$ years) and are therefore cold, and as such 'fast' in the tomography. Clear hot/cold structures are seen at many depths. At some places the structures can be correlated over large depths, as shown in Figure 6.15b. On geological timescales (millions of years) the hot material rises, while cold material sinks down. The cold areas correspond indeed to sub-

(a)

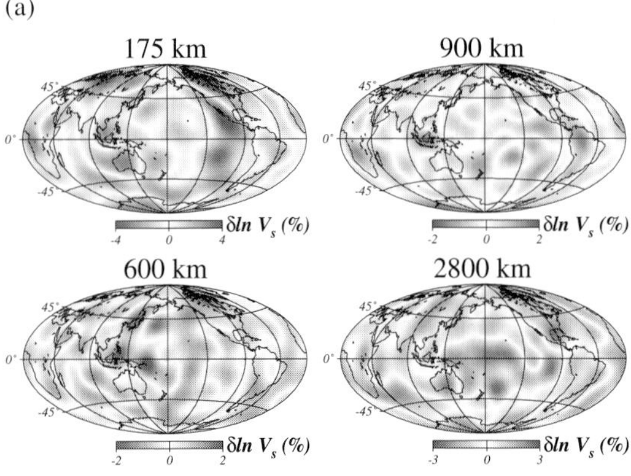

(b)

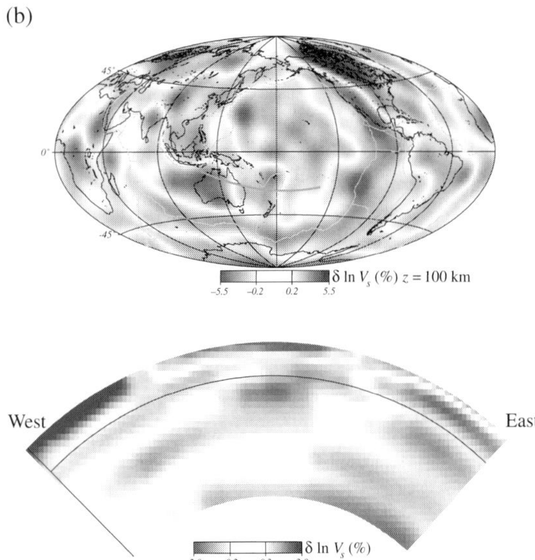

FIGURE 6.15 COLOR PLATE Seismic tomography maps of the Earth. (a) Four depth sections through the seismic model SAW12D derived by Li and Romanowicz (1996). (b) Plots of the seismic model SAW18B16 (Megnin and Romanowicz 1999). The top map shows the model at 100 km depth, and the bottom plot shows a cross-section through the green line in the South Pacific at the latitude of the Tonga Kermadec subduction zone. The line at 670 km in the cross-section corresponds to the separation between upper and lower mantle, while the vertical slice extends down to the core–mantle boundary.

ducting tectonic plates, typically at the outer eges of continental plates. The plot in Figure 6.15b shows a seismic model at 100 km depth (upper), and a vertical slice through the Earth's crust and mantle down to the core–mantle boundary, in the South Pacific at the latitude of the 'Tonga Kermadec' subduction zone (along the green line). In the cross-section, the cold regions are associated with the westward dipping slab. At the eastern end of the cross-section, the warm area corresponds to the Pacific 'superswell'. The line in the cross-section corresponds to the separation between the upper and lower mantle at a depth of 670 km. The maps in Figure 6.15b suggest that some of the hot–cold patterns extend all the way down to the core–mantle boundary, where the cold (blue) streaks suggest that lithospheric plates have descended continuously through the mantle over perhaps 40–50 Myr in geologically young areas, such as the Mariana subduction zone, and up to 180–190 Myr in older areas such as the Tonga subduction zone (Fig. 6.15b). There are also examples of seismic models, however, where descending slabs are stopped ∼1300 km above the core–mantle boundary.

Seismic velocity gradients in the lowermost ∼ 200 km of the mantle, i.e., near the core–mantle boundary, are anomalously low. This region is termed the seismic D'' region. It is very heterogeneous, which might, in part, result from the interaction between the rocky mantle and the liquid metal outer core, or from being a 'graveyard' for subducting slabs. Laboratory experiments suggest that the oxides from the rocky mantle undergo vigorous reactions with the liquid metal, suggesting the mantle to slowly dissolve into the outer core. Outer core material may condense onto the inner core, a process which could liberate energy (from latent heat) (Problem 6.21). In addition to the heterogeneity, narrow ($\lesssim$ 40 km wide) hot (velocities of $\gtrsim$ 10% below average) regions are seen at or near the core–mantle boundary, indicative of 'plumes' of magma. Since metals conduct heat much better than rock, spatial variations of the metal alloys at the core–mantle boundary might trigger the formation of these hot plumes. The plumes cool off the Earth's core by transferring energy outwards. These plumes probably surface in the form of oceanic islands, as mentioned above. This view thus suggests that some processes on the Earth's surface are connected to those happening at the core–mantle boundary.

The models based upon seismic tomography and those based upon geochemical data are thus seemingly incompatible. However, there is a theory which may satisfy both sets of data: Computer models suggest that convection is modified by boundaries where a phase change takes place, very similar to inhibition of convection through a layered atmosphere (Section 4.2.1.2). The subducting slabs of material at plate boundaries sink down and mass accumulates at the boundary. When a critical mass is reached, the material suddenly plunges through the boundary, like an 'avalanche'. Computer models, supported by seismic tomography maps, suggest that such avalanches might have occurred and may fall all the way through the mantle to the core–mantle boundary. There may thus be a partially layered convection, but with some intermittent mixing of materials throughout the entire mantle so that the distinction between layered and whole mantle convection becomes vague. Research in this area is very lively, there are many competing theories, and our view is continuously changing.

6.2.2.3 *Earth's Inner Core and its Rotation*

The Earth's inner core is solid, as discussed above, with an atomic weight close to that of iron. Since iron is one of the most abundant heavy elements in space (as a result of stellar nucleosynthesis; Section 12.2), and since the metal in metallic meteorites is usually iron-dominated, it seems most reasonable that the inner core is an iron-rich alloy. Since the density of the outer core is $\sim$5–10% less than that of pure iron (or Ni–Fe), there must be some lighter contaminant present. Possible contaminants are sulfur (Section 6.1.2.3), oxygen and/or hydrogen. The latter two only form an alloy with iron under high pressure; so, if these are the contaminants in the core, they must have gotten there after most of the planet had formed. It is possible that oxygen from the Earth's mantle has slowly infiltrated the core via chemical reactions at the core–mantle boundary. It is generally assumed, however, that the core is made up of an Fe–Ni alloy with sulfur.

Seismic waves pass through the inner core slightly faster when they follow a north–south track than in the east–west direction. The source of the anisotropy is not known, but it is probably related to the crystalline structure of the inner core. At the high pressures prevailing in the Earth's core, iron forms hexagonal close packed crystals, which may line up in a certain orientation as a result of solid-state convection. The seismic 'fast-track' axis is tilted by $\sim 10^\circ$ compared to the rotation axis, and recent studies have shown that this axis traces out a circle around the geographic north pole, with the inner core appearing to rotate slightly (up to 1° per year) faster than the rest of the planet. Why the inner core should spin at a different rate than the rest of the solid Earth is a complex problem, possibly related to connections between the fluid outer core and the Earth's magnetic field. When fluid from the outer core sinks down, it is spun up (conservation of angular momentum), and magnetic field lines in the fluid are dragged forward (compare with the magnetic braking effect, discussed in Section 12.4.2). Since the field lines thread through the inner core, the inner core is slightly accelerated in its rotation. An alternative model is that the rotation rate of Earth's mantle is gradually slowed by tidal torques from the Moon and the Sun, and that these torques are only weakly coupled to the inner core because the viscosity of the outer core is low.

6.2.2.4 *Earth's Magnetic Field*

The existence of a magnetic field around any planet poses constraints on interior models of that planet. As discussed in Chapter 7, it is generally believed that magnetic fields originate from a magnetohydrodynamic dynamo, where in essence the magnetic field is produced by electric currents, similar to magnetic fields created by electric currents running through a copper wire. The details of this process, however, are still beyond our grasp, although computer models have made significant progress in modeling the Earth's magnetic field, including its field reversals.

The geological history of the Earth's magnetic field is most apparent from the orientation of iron-bearing crystals along the mid-ocean ridge, where for millions of years magma has risen through the crust causing sea floor spreading. Thus the youngest material is found directly on top of the ridge, and the crustal age increases with increasing distance from the ridge. So in essence, the ocean floor is lined with 'strata', and the geological record can be established by investigating the material at different distances from the mid-ocean ridge. Basaltic magma is rich in iron, and any iron-bearing minerals in a cooling magma or lava can become magnetically aligned with the Earth's field. Upon cooling below a magnetic 'freezing-in' (Curie) temperature, this orientation gets locked in. Thus, by analyzing the iron-bearing minerals as a function of distance from the mid-ocean ridge, one can determine the geological history of the Earth's magnetic field. The crystals make up a striped pattern, where the amplitude variations, seen as 'stripes', parallel the ridge. The amplitude variations leading to the striped pattern have been interpreted as caused by changes in the magnetization direction, where the magnetic orientation changes direction from one stripe to the next. These changes in magnetic orientation are indicative of sudden reversals in the orientation of the Earth's magnetic field. The primary evidence for such 'field reversals', however, comes from paleomagnetic studies, where the magnetism and age of old rocks in many parts of the world have been determined. From these records scientists have been able to reconstruct the history of the Earth's magnetic field, and it has become apparent that magnetic field reversals were quite common on geological timescales; they appear to have occurred at irregular intervals of between 10^5 and a few million years.

If a planet has an internal magnetic field, it must contain an internal region that is fluid, convective and electrically conductive. Since the Earth's outer core is fluid and iron is its major constituent, our planet's magnetic field most likely originates in the outer core. With the discussion of other planet interiors, we will see that the presence or absence of a magnetic field may be used to constrain our knowledge of the interior stucture of a planet.

6.3 Interior Structure of Other Terrestrial Bodies and Moons

Seismic data are only available for the Earth and the Moon. One can use these bodies as prototypes, and develop models for the internal structure of other solid bodies, which mimic the observable parameters as summarized in the introduction of this chapter. In this section we summarize our current understanding of the Moon, the terrestrial planets and the largest satellites in the outer Solar System. Asteroids and comets are discussed in Chapters 9 and 10, respectively. The interior structure of the giant planets is described in Section 6.4.

6.3.1 Moon

Measurements of the Moon's moment of inertia show a value $I/MR^2 = 0.3932 \pm 0.0002$, only slightly less than the value of 0.4 expected for a homogeneous sphere. A model which fits the measured I/MR^2 together with the lunar seismic data suggests the average density of the Moon to be 3.344 ± 0.003 g cm^{-3}, with a ~ 50 km thick lower density crust ($\rho = 2.85$ g cm^{-3}), and an iron core ($\rho = 8$ g cm^{-3}) with a radius $R \lesssim 220$–450 km. The latter estimates on the Moon's core are also consistent with measurements of the large localized magnetic field in the Moon's crust (Section 7.4.3.3).

A laser ranging device and a microwave transmitter on board the Clementine spacecraft, launched in 1994, provided detailed measurements on the Moon's gravity field and surface topography. Gravity and geoid height anomalies are determined from microwave Doppler tracking, and topography is determined from laser ranging. Topography and gravity anomaly maps are shown in Figure 6.16. The lunar gravity models typically show an elevated equipotential surface at low latitudes and gravitational lows closer to the poles. Assuming a crustal density of 2.8 g cm^{-3}, the Bouger gravity correction factor was calculated and subtracted from the free-air anomalies to allow determination of the subsurface density distribution. If all deviations in the Bouger anomaly are attributed to variations in the thickness of the crust, these data yield a map of the crustal thickness. Such a map for the lunar crustal thickness is shown in Figure 6.16c, where a density of 3.3 g cm^{-3} was assumed for the lunar mantle, and a reference value of 64 km for the lunar crust.

The maps displayed in Figure 6.16 show that the highlands are gravitationally smooth, indicative of isostatic compensation such as seen over most regions on Earth. The lunar basins, however, show a broad range of isostatic compensations. Some basins on the near side show clear gravity highs (e.g., the Imbrium basin), circular features called *mascons* (from mass concentrations). A photograph of the Moon, annotated with many of the geologically interesting features, is shown in Figure 5.15.

(a)

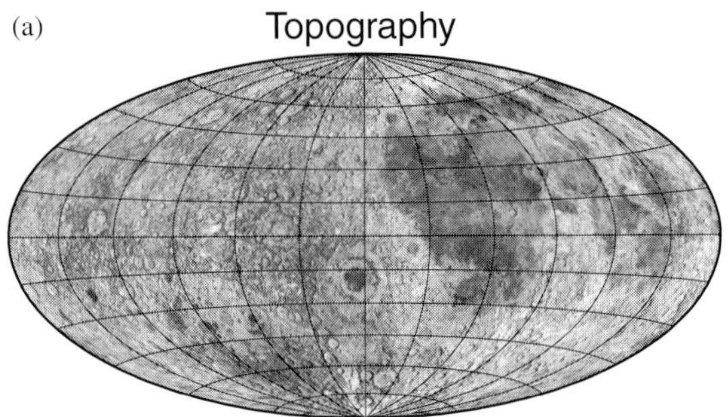

(b)

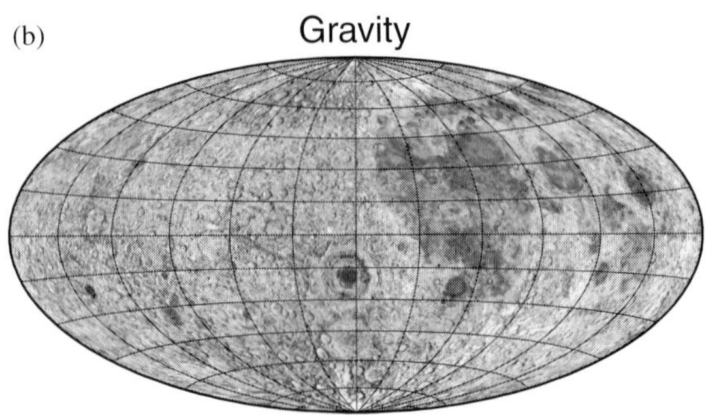

(c)

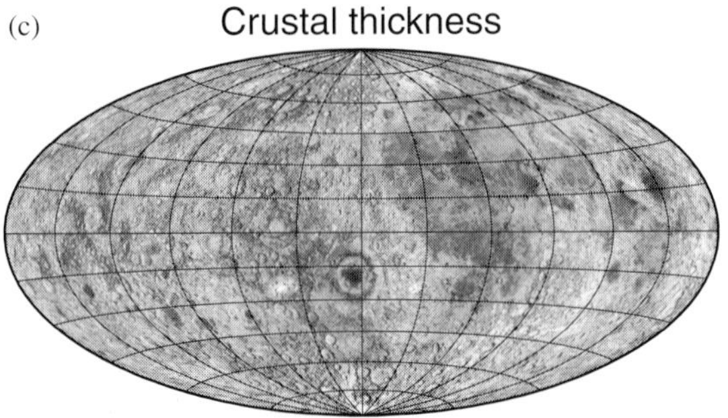

FIGURE 6.16 COLOR PLATE A map of the lunar topography, free-air gravity anomaly and the crustal thickness. (a) Topography model GLGM-2 based upon Clementine data (Smith *et al.* 1997). (b). Gravity model LP75G (Konopliv *et al.* 1998). This model also includes data from Lunar Prospector. (c) Single layer Airy compensation crustal thickness model using topography model GLTM-2 and gravity model LP75D. This latter model corrects for the presence of mare basalt in the major impact basins in the determination of crustal thickness. (Hood and Zuber 1999)

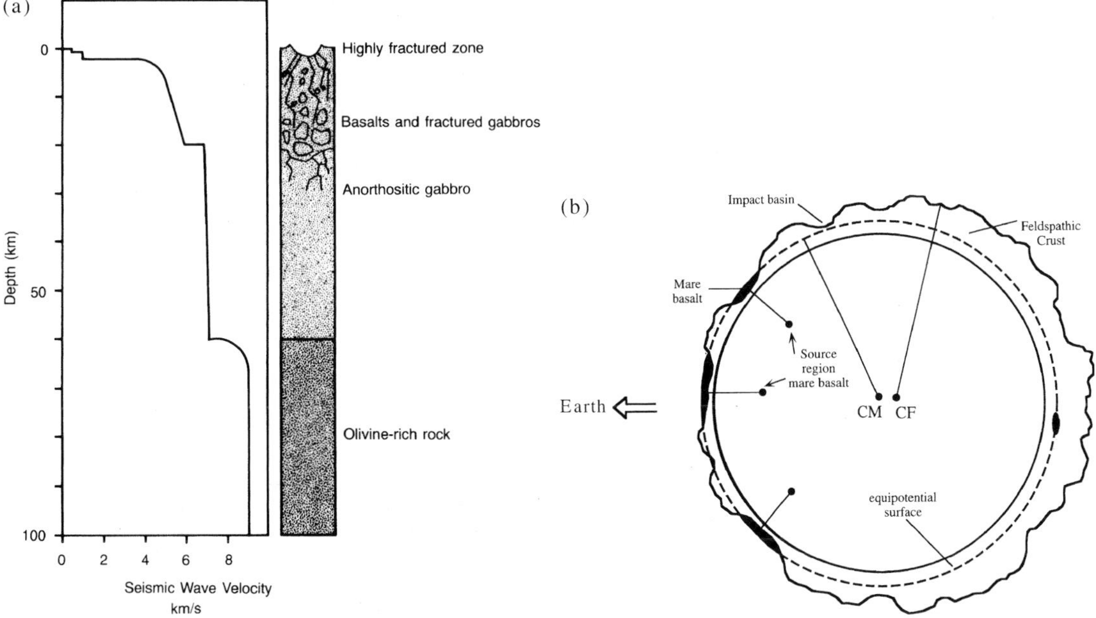

FIGURE 6.17 (a) Lunar seismic velocity profile, measured by Apollo equipment. The interpretation of the measurements is shown on the right. (After Hartmann 1993, with data from Taylor 1975) (b) Sketch of the interior structure of the Moon, in the equatorial plane, that shows the displacement toward the Earth of the center-of-mass (CM) (greatly exaggerated) relative to the center of the Moon's figure (CF). (Taylor 1999)

The gravity highs seen in the lunar basins suggest that the Moon's lithosphere here was very strong when the area was flooded by lava, as well as subsequent to that time. Both the lava and local uplift of mantle material following the impact must have caused the gravitational highs. The high gravity areas are surrounded by a ring of low gravity, suggesting flexure of the lithosphere in response to the flooding and mantle uplift following the impact. Mare Orientale shows a modest gravity high at the center, surrounded by a broad asymmetric ring of a gravitational low. The Aitken basin adjacent to it, on the other hand, is nearly fully compensated, or relaxed. There is no correlation between isostatic compensation of the basins with their size or age, so the lithosphere must have displayed large spatial variations in strength at the time the basins were formed and flooded. The gravitational differences can, to first approximation, be interpreted as caused by variations in crustal thickness. Assuming a constant crustal density, the far-side crust is, on average, thicker (68 km) than the near-side crust (60 km). The crust is usually thinner under the maria, with a minimum near 0 km at Mare Crisium, 4 km under Mare Orientale and 20 km under the Aitken basin. The thickest crust, 107 km, is on the far side, beneath a topographic high.

The interior structure of the Moon has largely been determined from seismic measurements of *moonquakes* at various Apollo landing sites. Many moonquakes are caused by tides raised by Earth, while others are triggered by meteoritic impacts, which are more numerous than on Earth since the Moon does not have a substantial atmosphere (Section 5.4.3). In contrast to the Earth where quakes originate close to the surface, moonquakes originate both deep within the Moon (down to ~1000 km) and close to its surface. The free oscillations require a long time to damp, implying that the Moon has little water and other volatiles. Figure 6.17 shows a lunar seismic velocity profile (panel a) and a sketch of the interior structure of the Moon (panel b), as deduced from the various seismic experiments. Whether the velocity profile is stepwise or gradual cannot be determined from the data. The velocity of both S and P waves increases in the first few tens of kilometers, to $v_S \approx 4.5$ km s^{-1} and $v_P \approx 8$ km s^{-1}. This region is the lunar crust, which varies in thickness from less than a few kilometers over maria up to over 100 km over the highlands, as mentioned above. The crust is on average thicker on the far side, which results in an offset of the Moon's center-of-mass from its geometric center of 1.68 km ± 50 m in the Earth–Moon direction, with the center of mass closer to Earth (Fig. 6.17b).

There appears to be a slight seismic velocity inversion at a depth of 300–500 km, which might suggest a chemical gradient with depth, such as a relative increase in the ratio Fe/(Fe + rock). The region below the crust down to ~ 500 km is called the upper mantle. Its composition is dominated by olivine. The middle mantle, down to ~ 1000 km, is dominated in composition by olivine and pyroxene, like the Earth's mantle. Attenuation of S waves at great depth, ~ 1000 km, may imply the Moon's lower mantle to be partially molten. No moonquakes have been detected below this depth. At deeper layers, below ~ 1400 km, the seismic velocity of P waves decreases by a factor of ~ 2, indicative of the presence of an iron-rich core. Although this last measurement was derived from only one, relatively weak meteoritic impact, the observation that many of the lunar samples seemed to have solidified in a strong magnetic environment ~ 3–4 Gyr ago, while the Moon itself does not now have a magnetic field, is also suggestive that the Moon has a small metallic core. A core with a radius $R \lesssim 300$–400 km is consistent with all available data.

6.3.2 Mercury

Unfortunately, there is not much data on Mercury that can be used to infer its interior structure. Photographs and other measurements by the Mariner 10 spacecraft, together with radar experiments and Earth-based observations at far-infrared and radio wavelengths, provide all the available data. Radar data revealed the unique commensurability of Mercury's rotation and orbital period (Section 2.6.2), with the upshot that Mercury spins around its axis in 59 days. The high bulk density, $\rho = 5.44$ g cm^{-3} (5.3 g cm^{-3} uncompressed density), of a planet about 20 times less massive than the Earth, implies that $\sim 60\%$ of the planet's mass consists of iron (Problem 6.23), which is twice the chondritic percentage. Static models suggest the iron core extends out to 75% of the planet's radius. The outer 600 km is the mantle, composed primarily of rocky material. The outermost 200 km of Mercury, at temperatures below 1100–1300 K, is the lithosphere (Fig. 6.18a). The moment of inertia of such a planet would be $I/MR^2 = 0.325$. Unfortunately, this value cannot be compared with observations of the gravitational moment J_2, because, even if the planet is in hydrostatic equilibrium, its figure is not determined by its present rotation since Mercury is tidally despun (Section 6.1.4.1).

The absence of a large rocky mantle has led to the theory that the planet was hit by a large object towards the end of its formation. The impact ejected and possibly vaporized much of Mercury's mantle, leaving an iron-rich planet. When the large iron core started to cool, it contracted and the rigid outer crust collapsed and formed the unique scarps seen all over Mercury (Section 5.5.2). Far-infrared and radio observations indicate a general lack of basaltic iron- and titanium-rich material; the abundance of these elements is even lower on Mercury's surface than in the lunar highlands. This suggests that deep-seated widespread volcanism shut off early in Mercury's formation history. If true, it suggests a very slow cooling of the iron core, so that the core may still be partially molten. If the core consists of a mixture of Fe and FeS, the melting temperature of the core is much lower than for pure iron, so that it is even more likely that the core is partially molten. A molten convective outer core could sustain a magnetohydrodynamic dynamo system that could produce the magnetic field detected around Mercury (Section 7.4.2).

6.3.3 Venus

Venus is very similar to Earth in size and mean density, suggesting similar interior structures for the two planets. This argument is strengthened by the observations of the Soviet landers that show that the mean density of Venus's crust is 2.7–2.9 g cm^{-3}, as expected for matter condensed from differentiated material. The mean uncompressed planetary density is slightly ($\sim 3\%$) less than that of the Earth. This slight difference may, in fact, be related to a major difference between the two planets: Venus, in contrast to the Earth, does not possess an intrinsic magnetic field, which implies the absence of a convective metallic region in its mantle and/or core. Could the core be completely frozen? The slightly lower mean density of Venus compared to Earth hints at the absence of some abundant heavy element. Based upon theories on the origin of the Solar System, Venus may contain less sulfur than Earth, since it formed in a warmer region of the solar nebula. If true, Venus's core contains less iron sulfide than the terrestrial core. Since iron sulfide lowers the melting temperature of iron, the absence of this alloy may lead to a completely frozen core. Alternatively, the difference between planetary densities may be the result of stochastic disruptions of differentiated planetesimals. In contrast to a frozen core, it might be possible that the core is liquid, but that there is no convection. On Earth a major source of energy driving convection in the liquid outer core comes from the phase transition between the solid inner core and liquid outer core. Solidification of core material liberates energy, driving convection in the Earth's core. If this phase boundary is absent on Venus, there may be no convection, even if the core is completely liquid. Measurements of the venusian gravitational field as estimated from Doppler tracking of Magellan and Pioneer Venus Orbiter spacecraft data seem to support the hypothesis that Venus's core is liquid. An alternative explanation for the absence of con-

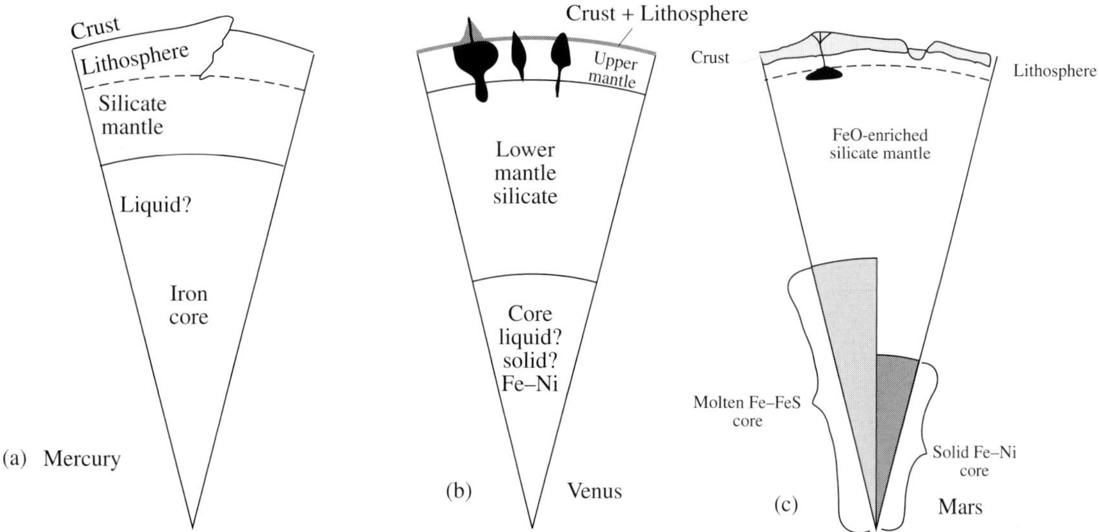

FIGURE 6.18 Sketch of the interior structure of (a) Mercury, (b) Venus, and (c) Mars. We show two equally plausible model stuctures for Mars's core.

vection in a liquid core would be provided if the mantle were hotter than the core, a situation which would inhibit rather than trigger convection in the core. Why might the mantle be hotter than the core?

The lowlands and highlands on Venus are sometimes compared to the ocean floors and continents on Earth, even though the composition and detailed topography of the venusian regions are quite different from those on Earth. The largest difference between the two planets is the absence of planet-wide tectonic plate activity on Venus. Although Venus's mantle is probably very similar to that of the Earth (Fig. 6.18b), Venus's higher surface temperature suggests the lithosphere to be about half as thick as on Earth (Problem 6.25). The lighter lithosphere may be too buoyant for tectonic plate activity, i.e., if plates exist, they may not be able to subduct. On Earth plate tectonic activity is a major avenue of heat loss. In the absence of plate movements, hot spot and volcanic activity may be more important. It is generally believed, however, that the current surface heat flow on Venus is less than the current internal heat production, so that the interior is essentially heating up. Assuming that most of the radioactive heating occurs in the mantle rather than the core, this may lead to a hot mantle overlying a relatively cooler core.

With a relatively thin lithosphere, one might expect the local topography on the planet to be isostatically compensated. Measurements of the gravity field, however, do show a high correlation between local topography and gravity highs and lows (Fig. 6.19), which suggests that Venus's lithosphere might be thicker than theoretical expectations, and/or that Venus lacks the low viscosity zone, asthenosphere, in its upper mantle. What produces this low viscosity zone in Earth's mantle? Recycling of water into the upper mantle at subduction zones is a plausible cause. Since Venus lacks water, it does not have a low viscosity zone and hence no plate tectonics. Numerical models such as those leading to tectonic plate convection on Earth show that under venusian conditions the lithospheric plate may completely subduct, i.e., show a 'catastrophic resurfacing' event every $\sim 10^8$ years or so (Section 5.5.3).

6.3.4 Mars

Mars is intermediate in size between the Moon and Earth. Its mean density is 3.93 g cm^{-3}, slightly larger than expected for a planet composed entirely of chondritic material. Mars's low surface temperature suggests a relatively thick lithosphere (Problem 6.25), which is consistent with the observed absence of tectonic plate activity. A significant fraction of the early heat loss was probably via volcanic activity, as suggested by the presence of several large shield volcanoes. These large volcanoes suggest further that the erupting magma had a low viscosity, such as measured for lavas rich in FeO (25% by mass). It can be shown that the magma rose up from a depth of ~ 200 km (Problem 6.26), which is within the planet's mantle, a region which might correspond to the Earth's asthenosphere. Mars's mantle may thus be enriched in FeO compared to its

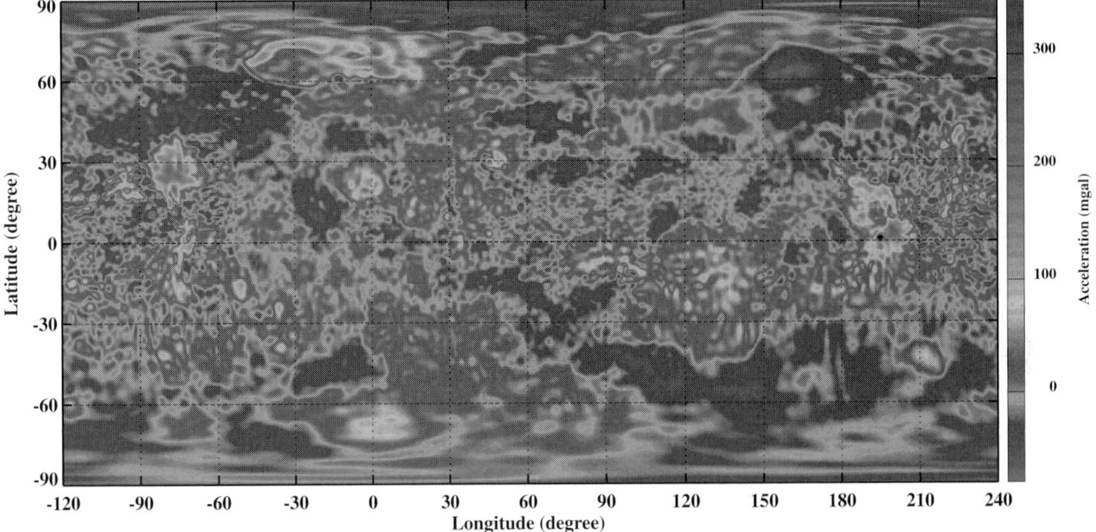

FIGURE 6.19 COLOR PLATE When compared with a topography map of Venus (Figure 5.44), this free-air gravity map appears to be highly correlated with Venus's topography. Note the gravity and topography highs at e.g., the Beta Regio and Atla Regio, while the low lying planitiae show gravity lows. (Konopliv *et al.* 1999)

surface, but even its surface is rich in iron oxides. Mars's red color has been attributed to relatively large quantities of rust, Fe_2O_3, which as measured by the Viking landers and Mars Pathfinder is $\sim$18% by mass of the surface minerals. Such large quantities of iron oxides have not been seen on any other planet.

A map of Mars's global topography reveals a striking $\sim$5 km difference in elevation between the northern and southern hemispheres (Fig. 6.20a). After removing a best fit reference ellipsoid, which is displaced $\sim$3 km to the south from the center of mass, most hemispheric asymmetries disappear. A global gravity map (Fig. 6.20b) shows that departures from the martian geoid are highly correlated with topography, indicating that topographic features are not fully isostatically compensated. This may be attributed to the thickness of the rigid lithosphere. The Tharsis region, for example, shows a clear gravity high of $\gtrsim$ 1000 mGal, whereas the Hellas basin shows a gravity low.

In contrast to Mercury, Venus and the Moon, the gravitational moments of Mars are largely representative of the hydrodynamic response of a planet to rotation, except for a slight effect due to the presence of the Tharsis region. The measured moment of inertia ratio is $I/MR^2 = 0.377 \pm 0.001$, which after correction for the Tharsis region leaves $I/MR^2 = 0.365$ as completely determined by the response of a planet in hydrodynamic equilibrium to rotation. Mars's interior structure has been derived from its average density and moment of inertia (Fig. 6.18c). The planet's core may be like the Earth's core, solid Ni–Fe, with a density of 8 g cm^{-3}. Alternatively, the core may consist of a liquid Fe–FeS mixture, in which case the uncompressed density is 6 g cm^{-3}, and the radius of the core is larger than in the model with the denser core (see Fig. 6.18c). As yet, there are no models that explain the formation of the core with the observed absence of an internal magnetic field. The uncompressed mantle density is 3.55 g cm^{-3}, which is clearly larger than that of Earth (3.34 g cm^{-3}), indicative of a larger iron abundance. Estimates based upon models to fit the observed parameters, such as I/MR^2, yield mass abundances in the range 16–21% FeO, or more than twice the value in the Earth's upper mantle (7.8%). Analysis of Mars meteorites (Chapter 8) yields an absolute FeO concentration of 18% by weight for the martian mantle, a value which agrees with the range of values derived above. Although the iron content is much larger than in Earth's mantle, it is still depleted compared to chondritic material (the chondrite-normalized value is 0.39).

In addition to a higher concentration of FeO compared to Earth, Mars also has higher abundances of relatively volatile material, such as Na, P, K, and Rb. The planet probably accreted from material with a higher volatile concentration than the planetesimals which formed Earth, which led to more extensive oxidation of metal, thus

(a)

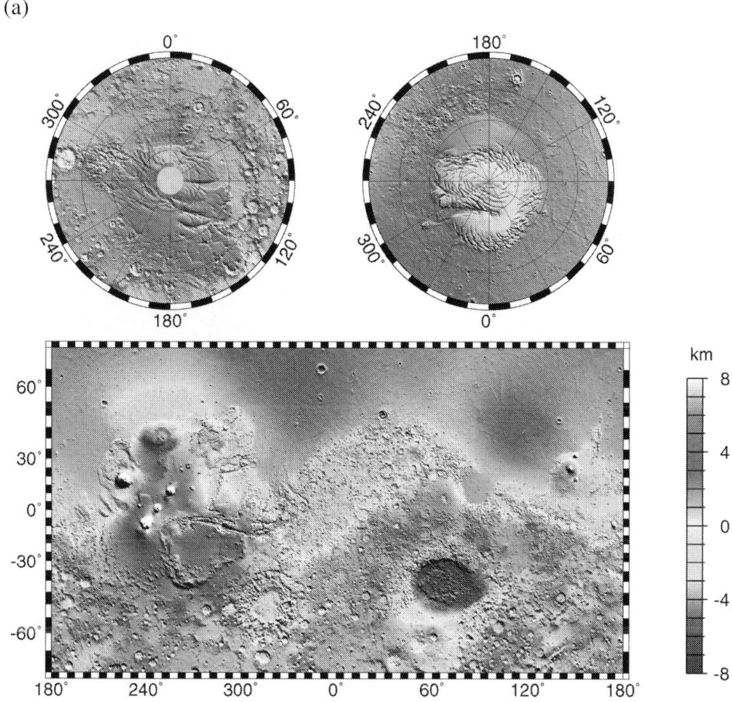

(b)

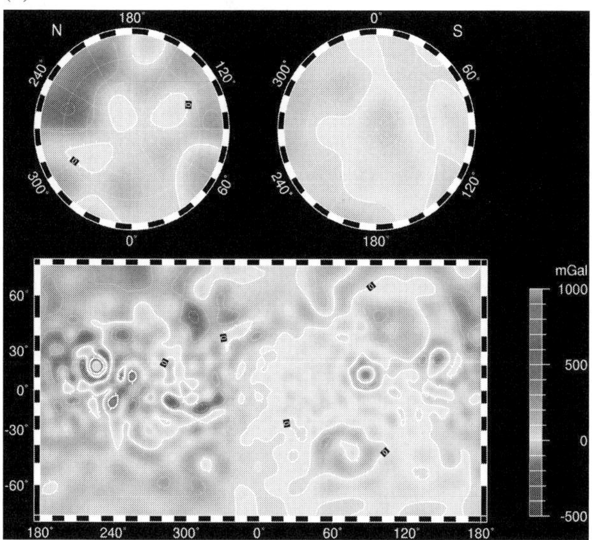

FIGURE 6.20 COLOR PLATE Relationship between gravity and local topography on Mars. The projections are Mercator to 70° latitude and stereographic at the poles with the south pole at left and north pole at right. (a) Map of Mars's global topography. Note the elevation difference between the northern and southern hemispheres. The Tharsis volcano–tectonic province is centered near the equator in the longitude range 220° E to 300° E and contains the vast east–west trending Valles Marineris canyon system and several major volcanic shields including Olympus Mons (18° N, 225° E). The Hellas impact basin (45° S, 70° E) has the deepest topography on Mars, a total relief of over 9 km. (NASA/Mars Global Surveyor PIA02031) (b) A vertical gravity map of Mars in mgals based on radio tracking of Mars Global Surveyor. Note the correlations (volcanoes, Hellas basin) and lack of correlations (overall north–south symmetry) with the global topography map in (a). (NASA/Mars Global Surveyor PIA02054)

higher concentrations of FeO and other siderophile elements compared to Earth. Separation of a FeS phase depleted Fe relative to chondritic material (or the solar Fe value), and also depleted some siderophile elements which have strong chalcophile affinities (such as Co, Ni, Cu).

6.3.5 Solid Bodies in the Outer Solar System

Bodies in the outer Solar System are composed of ice as well as rock. Solid bodies that accreted in regions where H_2O existed in the form of ice are expected to be composed of 35–50% rock + iron and 50–65% water-ice. In areas where ammonia also condensed out, the ice content should be 2% higher, while in the far outer regions where methane condensed out, 76% of the body should be made up of ices and only 24% of rock and iron. Typical densities of such rock/ice bodies are 1–2 g cm^{-3}. This is indeed the density as measured for many of the outer Solar System bodies.

6.3.5.1 *Galilean Satellites*

Even though the Galilean satellites formed in the outer Solar System, not all Galilean satellites are composed of a mixture of rock and ice. Io's mean density is 3.53 g cm^{-3}, indicative of a rock + iron composition. Spectra reveal the presence of SO_2 frost, rather than water-ice, on Io's surface; SO_2 gas is one of the major gases spewn out by Io's volcanoes (Section 5.5.5). With so much active volcanism, the satellite's interior must be very hot. This has been attributed to a tidal interaction with Jupiter (Section 2.6). Current volcanic models predict Io's crust to consist of a layer of solid silicates of varying thickness, in total $\sim$ 30 km or more. This layer lies on top of an $\gtrsim$ 8 km thick asthenosphere of molten silicates. The surface is overlain with sulfur and sulfur compounds, or a mixture of silicate and sulfur lavas. Below the asthenosphere may be a liquid layer of unknown depth (probably thin) and composition, overlying a hot solid mantle. Measurements of Io's moment of inertia by the Galileo spacecraft indicate a ratio $I/MR^2 = 0.37 \pm 0.007$, under the assumption that Io is in rotational and tidal equilibrium. This number implies that Io's mass is concentrated towards its center. Io's core is assumed to consist either of pure iron with a core radius of 36%, or of a eutectic mixture of Fe–FeS with a core radius of 52%. The core is surrounded by a hot silicate mantle (Fig. 6.21a).

Europa has a mean density of 3.02 g cm^{-3}, indicative of a rock/ice composition where the rocky mantle/core takes up 92% of the mass (Fig. 6.21b). Europa's moment of inertia ratio $I/MR^2 = 0.347 \pm 0.014$ suggests a differentiated, centrally condensed body. The rocky dehydrated mantle is overlain by a $\sim$140 km thick H_2O crust, while the satellite probably has an inner metallic core. Galileo images (Fig. 5.52) strongly suggest that the crust consists of thin ice sheets or plates floating on an ocean of soft ice; the images remind one of the ice in the Arctic regions on Earth (Section 5.5.5.2). The Galileo magnetometer data revealed disturbances in the magnetic field around Europa which are consistent with that expected from a body covered by a salty ocean (Section 7.4.4.7).

Ganymede has an average density $\rho = 1.94$ g cm^{-3}, suggestive of a rock/ice mixture. Its moment of inertia, $I/MR^2 = 0.311 \pm 0.003$, implies that its mass is heavily concentrated towards the center. Galileo discovered an intrinsic magnetic field for this satellite, suggestive of a liquid metallic core (Section 7.4.4.7). Best fits to the gravity, magnetic field and density data are obtained with three-layer internal models, where each layer is $\sim$ 900 km thick. These models consist of a liquid metallic core, surrounded by a silicate mantle, which is topped off by a thick ice shell (Fig. 6.21c). It is possible that the ice is liquid at a depth of $\sim$150 km (2 kbar), where the temperature (253 K) corresponds to the minimal melting point of water.

Callisto's average density, $\rho = 1.85$ g cm^{-3}, suggests this moon must contain a larger percentage of ice than Ganymede, consistent with the trend of increasing ice abundance from the inner to the outermost Galilean satellite. Its moment of inertia was measured by Galileo and found to be slightly less than expected for a homogeneous body ($I/MR^2 = 0.358$). Callisto is likely partially, but incompletely, differentiated, with an icy crustal layer (few hundred km) and an ice/rock mantle which is slightly denser towards the center of the satellite. The magnetometer on board the Galileo spacecraft discovered magnetic field disturbances which, like for Europa, suggest the presence of a salty ocean on Callisto (Section 7.4.4.7). As on Ganymede, such an ocean may exist at a depth of $\sim$150 km.

6.3.5.2 *Satellites of the Outer Planets*

The other satellites of Saturn, Uranus, Neptune and Pluto have low mean densities, between $\sim$1 and $\sim$2 g cm^{-3}, as expected for bodies that consist of ice/rock mixtures. Since only a small amount of heat is necessary to melt the ices, the satellites are probably differentiated into rocky cores with an ice/silicate mantle and icy crust, though we do not know to what extent the satellites of the outer planets are differentiated. Depending on the size of the object and its environment (tidal heating), the mantle could be solid or fluid-like. Saturn's satellite Titan has a large fraction of rock, 65%, whereas the fraction of rock in the other saturnian satellites is typically between 20 and 50%. Interestingly enough, the satellites of Uranus and Neptune may

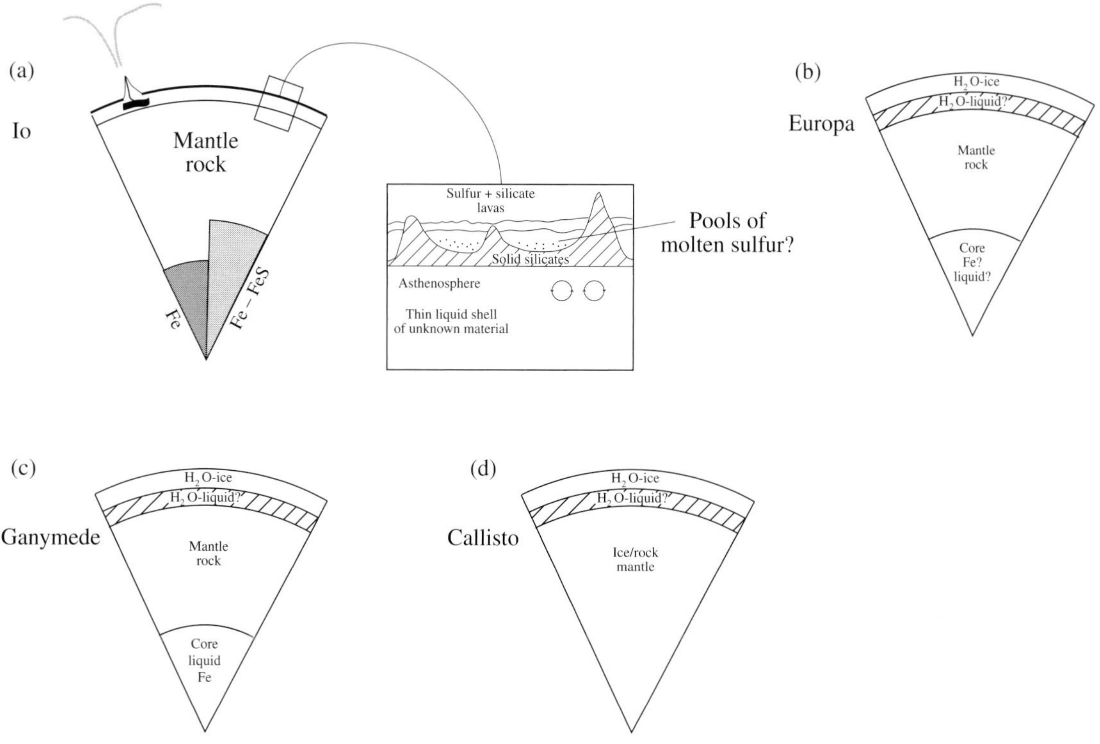

FIGURE 6.21 Interior structure of the four Galilean satellites.

have slightly higher densities than comparably sized jovian and saturnian satellites, which imply a larger rock fraction.

6.4 Interior Structure of the Giant Planets

6.4.1 Modeling the Giant Planets

The bulk density of the giant planets implies that they consist primarily of light elements. In Section 6.1.3 we showed that a sphere of material will reach a maximum size when matter is added, and that the radius of the body decreases when still more material is added. Below we derive this maximum radius for a sphere of pure hydrogen gas. Consider the equation of state as given by equation (6.9) with a polytropic index $n = 1$, so that $P = K\rho^2$. In this particular case, the radius is independent of mass. Thus, when more mass is added to the body, the material gets compressed such that its radius does not change. The advantage of a polytropic index $n = 1$ is that the equations can be solved analytically, and the results appear to be in good agreement with calculations based upon more detailed pressure–density relations. Integration of the equation of hydrostatic equilibrium (eq. 6.1) leads to the following density profile:

$$\rho = \rho_c \left(\frac{\sin C_K r}{C_K r} \right), \tag{6.45}$$

with ρ_c the density at the center of the body, and

$$C_K = \sqrt{\frac{2\pi G}{K}}. \tag{6.46}$$

The radius of the body, R, is defined by $\rho = 0$, thus $\sin(C_K R) = 0$, and $R = \pi / C_K$. With a value for the polytropic constant $K = 2.7 \times 10^{12}$ cm^5 g^{-1} s^{-2}, as obtained from a fit to a more precise equation of state, we find that the planet's radius $R = 7.97 \times 10^4$ km. This number is thus independent of the planet's mass. The radius of this hydrogen sphere is slightly larger than Jupiter's radius, $R_{♃} = 7.19 \times 10^4$ km, and significantly larger than Saturn's radius, $R_{♄} = 6.03 \times 10^4$ km. This suggests that the two planets are composed primarily, but not entirely, of pure hydrogen.

Using experimental data at low pressure and theoretical models at very high pressure, it can be shown that the maximum radius for cold self-gravitating spheres of heavier elements can be approximated by:

$$R_{max} = \frac{Z \times 10^5}{\mu_a m_{amu} \sqrt{Z^{2/3} + 0.51}} \text{ km}, \tag{6.47}$$

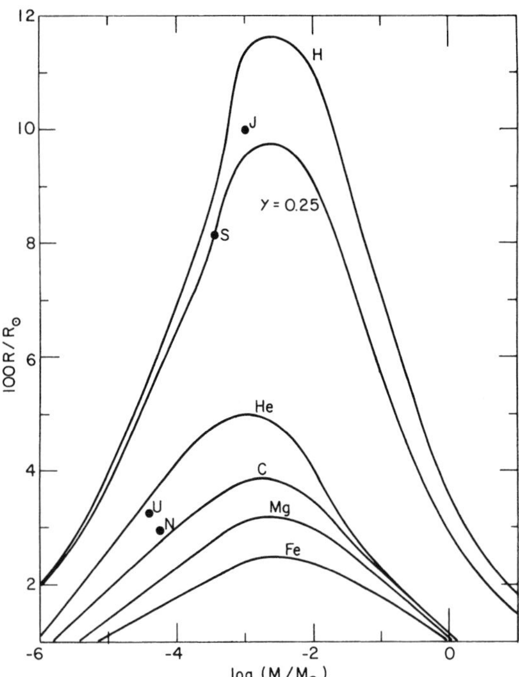

FIGURE 6.22 Graphs of the mass–density relation for spheres of different materials at zero temperature, as calculated numerically using precise (empirical) equations of state. The second curve from the top is for a mixture of 75% H, 25% He by mass; all other curves are for planets composed entirely of one single element. The locations of the giant planets are indicated on the graph. (Stevenson and Salpeter 1976)

where Z is the atomic number and $\mu_a m_{amu}$ the atomic mass. With this equation one finds that R_{max} for a pure hydrogen sphere is 82 600 km, for pure helium $R_{max} =$ 35 000 km, and for heavier material R_{max} is smaller. Figure 6.22 shows graphs of the mass–density relation for spheres of zero-temperature matter for different materials, as calculated numerically using precise (empirical) equations of state. The locations of the giant planets are indicated on the graph. We note that the radius of a giant planet is usually indicated by the distance from a planet's center out to the average 1-bar level along the planet's equator. The observed atmospheric species form a boundary condition on the choice of elements to include in models of the interior structure of the giant planets. The atmospheric composition of Jupiter and Saturn is close to a solar composition, and the location of these planets on the radius–mass graphs in Figure 6.22 shows that the composition of Jupiter and Saturn's interiors must indeed be close to solar.

In precise models of a giant planet's interior, the observed rotation rate and inferred density (from measurements of the mass and radius) are used to fit the gravity field. The models are further constrained by the characteristics of the planet's magnetic field, interior heat source, and the nearby presence of massive satellites. The mass of the planets can be accurately determined from spacecraft tracking data, and the radius can be measured from star occultations and/or spacecraft images. The rotation rate of a planet is more complicated to determine, since timing of albedo features gives information on the rotation rate of atmospheric phenomena. The rotation period of the gas giants is assumed to be equal to the rotation period of their magnetic field, which can be determined via ground-based (Jupiter) or spacecraft radio observations (Chapter 7). To first order, these values are consistent with measurements of the planets' oblateness and gravitational moments. The distribution of material inside the planet, $\rho(r)$, can then be derived from the response coefficient, Λ_2, or the moment of inertia ratio, I/MR^2.

As expected, spacecraft measurements show that all gravitational moments, J_n, for odd n are very small for the giant planets. The second moment, J_2, is related to the planet's rotation, while higher order moments indicate further deviations in a planet's shape from a spinning oblate spheroid. Since the Voyager encounters, the moment J_4 has been measured for all four gas giants, while for Jupiter and Saturn J_6 has also been estimated. These moments provide valuable constraints for detailed models of these planets' interiors. The observed strong internal magnetic moments require the presence of an electrically conductive and convective fluid medium in the planets' interiors, while any internal heat source for these planets is most likely gravitational in origin (Section 6.1.5.1).

Since the solution to the modeling technique outlined above is nonunique, there are substantial variations from one model to the next. Overall, the models show that the total amount of heavy ($Z > 2$) material in the giant planets is roughly between 10 and 30 $M_\oplus$, some of which has to reside near the core to satisfy the gravitational moments. Below we discuss the most likely models for the four planets, but keep in mind that research is still ongoing and that details may change in the future.

6.4.2 Interior Structure of Jupiter and Saturn

The interior structures of Jupiter and Saturn are constrained by their mass, radius, rotation period, oblateness, internal heat source and gravitational moments J_2, J_4 and J_6. Most models for Jupiter predict a relatively small dense core containing 5–10 $M_\oplus$. It is typically assumed to consist of an inner rocky/iron core, surrounded by an ice-rich outer core, though it could also be homogeneous. The masses of the cores are difficult to estimate, but it is clear that both Jupiter and Saturn have a total of ~15–30

$M_\oplus$ high-Z ($Z > 2$) material distributed among the core and surrounding envelope, so that the envelope is enriched in high-Z material by approximately a factor of 3–5 compared to a solar composition planet. Such enhancements in the high-Z component are consistent with observations of Jupiter's atmosphere (Section 4.3.3.4), which suggest that carbon, nitrogen and sulfur are enhanced by factors of the order of 2–3 compared to the solar elemental values. The core probably consists of relatively high quantities of iron and rock, some of which was there initially when the planet formed from solid body accretion, while more may have been added later via gravitational settling (iron and rock may be immiscible in metallic hydrogen, and therefore drained out to the core). The mantle material probably contains a relatively large amount of 'ices' of H_2O, NH_3, CH_4 and S-bearing materials.

Saturn's interior structure is quite similar to that of Jupiter, although there are marked differences. As these planets contain a similar amount of heavy materials, roughly 15–30 $M_\oplus$, and Saturn is smaller and less massive than Jupiter, the relative abundance of high-Z ($Z > 2$) material is larger on Saturn. These findings are consistent with observations of the chemical composition of Saturn's atmosphere which suggest C, N, and S to be approximately 2–3 times more enhanced than on Jupiter. Saturn's core is smaller than Jupiter's core, and may consist of $\sim$1 $M_\oplus$ of a mixture of rock/iron and ices. Models of planetary formation explain these different abundances in a natural way (Section 12.7). A schematic of the interior structure of both planets is shown in Figure 6.23.

Jupiter possesses a strong magnetic field, and Saturn is surrounded by a somewhat weaker one (Section 7.4). As discussed in Section 6.1.2.1, theories and laboratory experiments suggest hydrogen in the interiors of Jupiter and Saturn to be present in the form of liquid metallic hydrogen, a highly conductive fluid. In analogy to models of the magnetohydrodynamic dynamo theory in the Earth's outer liquid core (Section 7.7), one expects electromagnetic currents and therefore magnetic fields to be generated in the metallic hydrogen region. Since Saturn is less massive than Jupiter, the extent of the metallic hydrogen region is smaller, which probably explains its weaker magnetic field.

Both planets emit much more energy than they receive from the Sun. The excess heat has been attributed to the presence of an internal source of heat. As discussed in Section 6.1.5.1, the source of this internal heat is most likely gravitational. For Jupiter, models suggest that most of the heat originates from gravitational contraction/accretion in the past, while for Saturn a substantial part (roughly half) of the excess heat is attributed to the release of gravitational energy while helium is raining out onto the core. After the planet had fully formed, Saturn started to cool off. Ever since the time when its interior reached the point (near 2 Mbar and 8000 K) that helium became immiscible in metallic hydrogen, helium has slowly been drained out of the planet's outer envelope 'raining' down onto the core. This theory explains both the excess energy from this planet, and the low observed helium abundance in Saturn's atmosphere. A similar process may have 'recently' started on Jupiter, given the somewhat lower than solar elemental abundance of helium in Jupiter's atmosphere.

The envelopes of both planets are almost fully convective, and the thermal structure is thus most likely very close to adiabatic below the tropopause. However, these planets may have a small 'radiative window' in their atmospheres where energy transport is dominated by radiation rather than convection. Such a 'window' could exist because collision-induced absorption by hydrogen, the main source of opacity in these atmospheres, displays a minimum at a wavelength near 1 μm. The Planck radiation curve in this wavelength region for temperatures at a few thousand K (eq. 3.5), and detailed calculations suggest that energy transport in the atmospheres of Jupiter and Saturn might be dominated by radiation at temperatures between 1200 and 3000 K.

6.4.3 Interior Structure of Uranus and Neptune

Although Uranus and Neptune are usually referred to as giant planets together with Jupiter and Saturn, the composition of these planets differs substantially from that of Jupiter and Saturn. Models of the structure of Uranus and Neptune are far less well constrained than are models of Jupiter and Saturn. This is in part because we have better data for the largest planets. More importantly, the sizes and masses of Jupiter and Saturn imply that these planets must consist predominantly of the lightest elements, hydrogen and helium. In contrast, planets with radii and masses equal to those of Uranus and Neptune could either be composed mostly of 'ices' (condensable volatiles such as H_2O, CH_4, NH_3, H_2S) or of a mixture of hydrogen, helium and 'rock' (refractory material, including Si, Mg, Fe, etc.), or some intermediate combination. In contrast to Jupiter and Saturn, hydrogen and helium make up a relatively small fraction of the mass of Uranus and Neptune, a few $M_\oplus$ in total, while models of their interior structure show that the total mass of the $Z > 2$ elements is very similar, in absolute numbers, to that of Jupiter and Saturn. This increase in high-Z abundance relative to hydrogen and helium with increasing heliocentric distance agrees also with

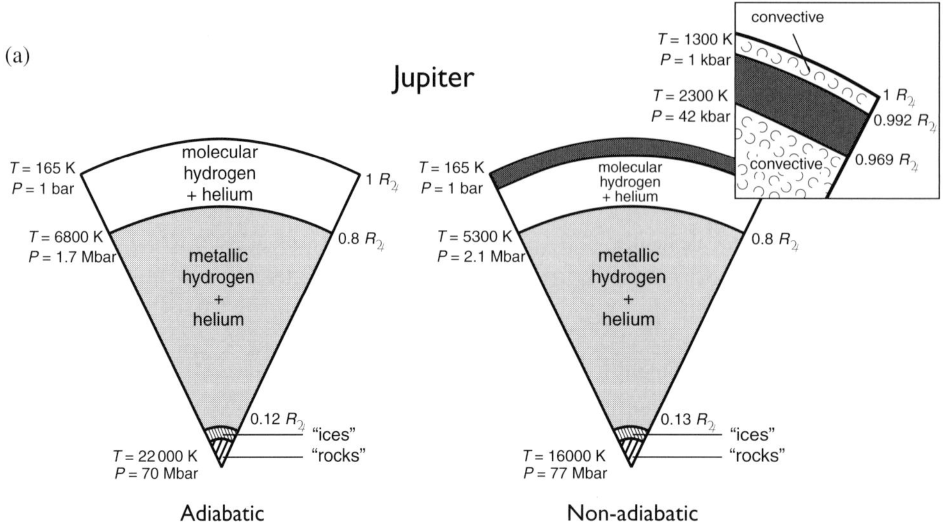

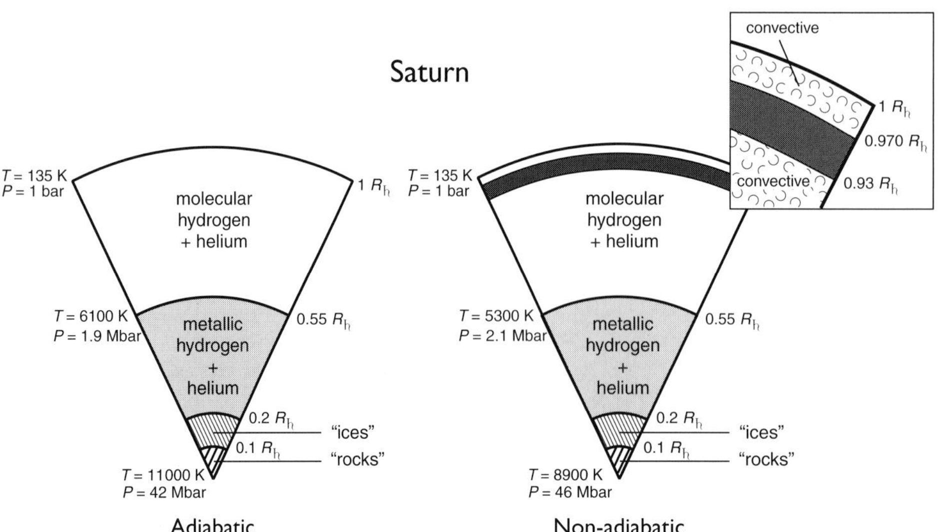

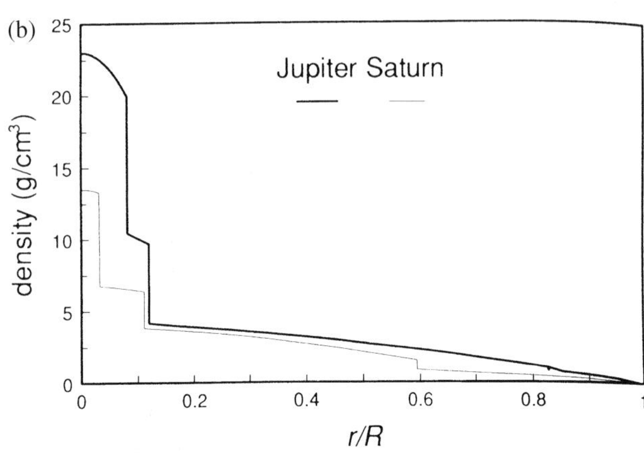

FIGURE 6.23 Models of the interior structure of the four giant planets. (a) Models of the interior stucture of Jupiter (top) and Saturn (bottom), assuming a fully convective hydrogen–helium envelope (adiabatic models), and assuming the presence of a radiative window in the molecular envelope (nonadiabatic models). (Guillot *et al.* 1995) (b) Models of the density as a function of normalized radius for Jupiter and Saturn. The transition from molecular to metallic hydrogen is responsible for the small density change in Saturn at 0.6 $R_♄$, and in Jupiter at 0.8 $R_♃$. The relative size of the presumed two-layer rock–ice core is about the same for both planets, although Jupiter's core is more dense owing to the much greater overlying pressure. (Marley 1999)

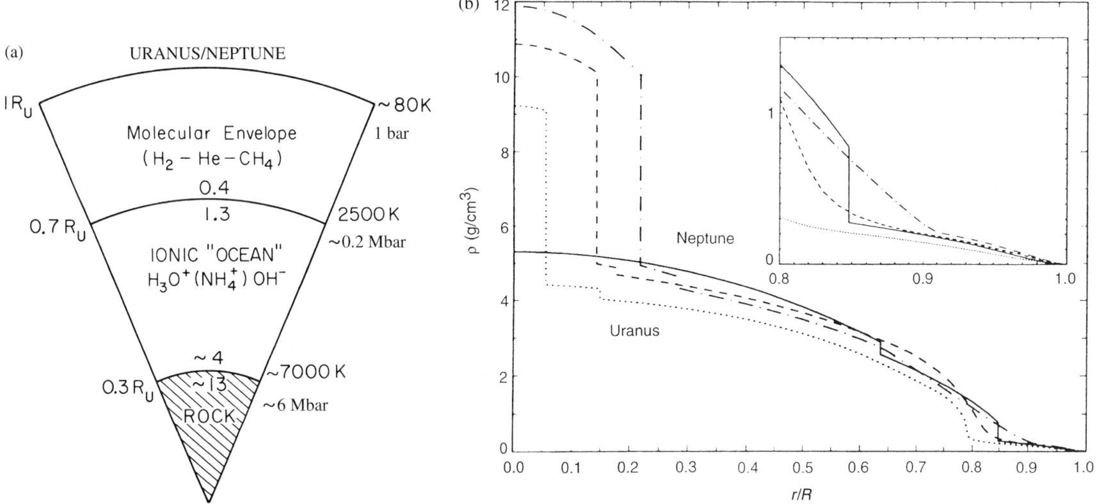

FIGURE 6.24 (a) Schematic representation of the interiors of Uranus and Neptune. (After Stevenson 1982) (b) Models of the density as a function of normalized radius for three Neptune and one Uranus interior models. The solid, dashed and dot-dashed curves represent the range in possible Neptune models, where the core could be absent or extend out to a fractional radius of 20%. The dotted curve represents a single Uranus model. Because of Neptune's larger mass, it is denser than Uranus at each fractional radius. The inset shows the region of transition from a hydrogen-rich atmosphere to the icy mantle in more detail. (Marley 1999)

observations of their atmospheric abundances, which suggest that both carbon and sulfur in Uranus's and Neptune's atmospheres are enhanced by factors of several tens compared to the solar elemental values (Section 4.3.3.4).

Neptune is 3% smaller than Uranus, while its mass is 15% larger, resulting in a density that is 24% larger than Uranus's bulk density. Interior models for the two planets are not well constrained at present. The best fitting models include a small dense core, $\sim$ 1 $M_\oplus$, which most likely is of terrestrial composition; models without a core, however, satisfy the available data as well. Based upon the constituent relations, the iron may be in the solid phase, whereas the rocky component may be liquid, in which case the core would be differentiated. The mantle consists of 'ices', and comprises $\sim$ 80% of the mass of these planets. The outer 5–15% of the planets' radii make up a hydrogen- and helium-rich atmosphere (Fig. 6.24). Since both planets have an internal magnetic field, their interiors must be electrically conductive and convective. Although the pressure and temperature are high in the planets' interiors, the pressure is too low throughout most of the planet for metallic hydrogen to form. The conductivity in Uranus's and Neptune's interiors must thus be attributed to other materials. Because of the high temperature and pressure, the icy mantles are probably hot, dense, liquid, ionic 'oceans' of water with some methane, ammonia, nitrogen and hydrogen sulfide, where the conductivity is high enough to set up an electromagnetic current system that can generate the observed magnetic fields (see Section 6.1.2.2).

Neptune must contain an internal heat source, whereas Uranus has only a very small source, if any at all (as deduced from a comparison between the equilibrium and observed effective temperatures). The observed upper limit for Uranus is consistent with the heat flow expected from radioactive decay alone. Why does Neptune possess an internal heat source, and Uranus not? With current evolutionary models (Chapter 12) it seems unreasonable to assume a 'cold' start for Uranus and a 'hot' start for Neptune. Both planets were presumably hot initially from the accretion process. It is possible that, in contrast to Neptune, convection is inhibited in Uranus's interior due to (subtle) differences in density gradients. The absence of obvious large scale atmospheric turbulence in Uranus's atmosphere compared to the violent activity observed in Neptune's atmosphere (appearance and disappearance of dark and bright spots) does hint at a difference in convection patterns. The observed differences in the hydrocarbon abundances, HCN and CO in the planets' stratospheres also suggest vigorous vertical transport in Neptune's atmosphere, in contrast to Uranus (Section 4.7.2.2). Hence, we are inclined to believe that the difference in internal heat loss is due to an inhibition of convection in Uranus's interior, which might be caused by compositional gradients. Models show that all observed characteristics can be reconciled if convection is inhibited within 0.6 $R_♅$ of Uranus's outer radius, and within 0.5 R_Ψ for Neptune. The compositional gradients leading to inhibition of convection might have arisen from late accretion of large (e.g., 0.1–1 $M_\oplus$) sized planetesimals, which break up upon impact and only partially mix with material already present. Such an impact, if oblique, has also been postulated to explain Uranus's large obliquity.

Despite the fact that the ionic oceans on Uranus and Neptune are thought to extend deep within the planets' interiors, the absence of convection not only inhibits mass and energy transport from the core to outer layers on the planet, but it also prevents the generation of magnetic fields within 0.5–0.6 planet radii. This theory, therefore, predicts the magnetic fields of Uranus and Neptune to originate within a thin shell of the highly conductive ionic ocean, between 0.5–0.6 and 0.8 planetary radii, via a process similar to that which generates the fields around Earth, Jupiter and Saturn. This would explain the presence of large quadrupole and higher order moments (same magnitude as the dipole moment) in the representation of Uranus's and Neptune's fields. Magnetic fields induced near a planet's core are primarily dipolar in character (see Chapter 7).

The helium abundance in Uranus and Neptune's atmospheres is similar to that expected for a solar composition atmosphere, in contrast to the helium abundance on Jupiter and Saturn, which is found to be smaller than on the Sun. If indeed hydrogen is confined to the outer layers of Uranus and Neptune, metallic hydrogen will not form, and hence helium will not be separated from hydrogen. However, some models indicate substantial amounts of hydrogen and helium mixed in with the icy material at deep levels in the planets' interiors. At pressure levels $\gtrsim$ few Mbar, hydrogen will become metallic, and helium should separate out. However, these high pressure levels are only reached in the deep interior of the planets, a region we just argued where convection may be inhibited. Hence, there is no mass (or energy) transfer between these deep layers and the atmosphere, and the effect of helium differentiation, if present, may not be noticed in the atmosphere.

In contrast to inhibition of convection and mass transfer, suggestions of equilibration between core, mantle and atmospheric material have been made based upon measurements of the D/H ratio. The observed D/H ratio is $\sim 1.2 \times 10^{-4}$, about an order of magnitude higher than primordial values and D/H ratios on Jupiter and Saturn, and similar to terrestrial numbers and comets. These high values presumably originate from the icy compounds in the planets, either from the ice in the deep interiors of the planets, which requires mass transfer and thus convection (which is inhibited nowadays, but could have been different in the past), or in the original icy planetesimals which accreted to form Uranus and Neptune. In the latter case one does not need to invoke efficient mixing processes between the deep interior and the outer layers to explain the high observed D/H numbers, since the planetesimals deposited ice with a high D/H throughout the forming planet. The upshot of both scenarios is that the D/H ratio in Uranus and Neptune's atmospheres may not be a good indicator of the degree of mixing in these planets' deep interiors.

FURTHER READING

Although this book contains mostly material pertaining to the surface structure of the Earth, the chapters on seismology and the Earth's interior form good introductory chapters to the more advanced books listed below:

Press, F., and R. Siever, 1998. *Understanding Earth*, 2nd Edition. W.H. Freeman and Company, New York.

Standard textbooks on the physics of the interior of the Earth:

Brown, G.C., and A.E. Mussett, 1993. *The Inaccesible Earth*, 2nd Edition. George Allen and Unwin, London.

Fowler, C.M.R., 1990. *The Solid Earth: An Introduction to Global Geophysics*. Cambridge University Press, New York.

A book about the Earth's core, magnetic field and magnetic field reversals:

Jacobs, J.A., 1987. *The Earth's Core*, 2nd Edition. Academic Press, New York.

Basic introduction to the physics of the Earth's interior at the graduate student level:

Poirier, J.P., 1991. *Introduction to the Physics of the Earth's Interior*. Cambridge University Press, New York.

A good book (although it is outdated at places) on the interior structure of all planets is:

Hubbard, W.B., 1984. *Planetary Interiors*. Van Nostrand Reinhold Company Inc., New York.

Interiors of the giant planets are discussed by:

Stevenson, D.J., 1982. Interiors of the giant planets. *Annu. Rev. Earth Planet. Sci.* **10**, 257–295.

A more up-to-date review of Neptune's interior is written by:

Hubbard, W.B., M. Podolak, and D.J. Stevenson, 1995. The interior of Neptune. In *Neptune and Triton*. Ed. D.P. Cruikshank. University of Arizona Press, Tucson, AZ. pp. 109–138.

Problems

6.1.E Compute the gravitational potential energy of:

(a) A uniform sphere of radius R and density ρ.

(b) A sphere of identical mass and radius to that above, but whose mass is distributed so that its core, whose radius is $R/2$, has twice the density of its mantle.

6.2.E Assume that material can be compressed significantly if the pressure exceeds the material strength. If material can be compressed considerably over a large fraction of a body's radius, the body will take on a spherical shape (the lowest energy state of a nonrotating fluid body).

(a) Calculate the minimum radius of a rocky body to be significantly compressed at its center. Assume an (uncompressed) density $\rho = 3.5$ g cm^{-3}, and material strength $\mathcal{S}_m = 2 \times 10^9$ dyne cm^{-2}.

(b) Calculate the minimum radius of the rocky body in (a) to be significantly compressed over a region containing about half the body's mass.

(c) Repeat for an iron body, which has an (uncompressed) density of $\rho = 8$ g cm^{-3}, and material strength $\mathcal{S}_m = 4 \times 10^9$ dyne cm^{-2}.

6.3.E Use the equation of hydrostatic equilibrium to estimate the pressure at the center of the Moon, Earth and Jupiter.

(a) Take the simplest approach, and approximate the planet to consist of one slab of material with thickness R, the planetary radius. Assume the gravity $g_p(r) = g_p(R)$, and use the mean density $\rho(r) = \rho$.

(b) Assume the density of each planet to be constant throughout its interior, and derive an expression for the pressure in a planet's interior, as a function of distance r from the center (Hint: You should get eq. 6.2).

(c) Although the pressure obtained in (a) and (b) is not quite right, it will give you a fair estimate of its magnitude. Compare your answer with the more sophisticated estimates given in Section 6.1.1, and comment on your results.

6.4.I For small bodies the relation between mass and size is given by equation (6.10). When more mass is added to a planet, the material will be compressed. When the internal pressure becomes very large, the matter becomes degenerate, as is the case for white dwarf stars. Consider the central pressure, P_c, of a white dwarf star, which can be calculated from equation (6.1), as you did in the previous problem. The polytropic constant n in the equation of state (eq. 6.9) is 2/3 in the limit of high pressure. Show that for a white dwarf star $M \propto R^{-3}$.

6.5.E (a) The moment of inertia is related to the density via equation (2.42). If ρ is a function of distance r, show that for a spherical planet:

$$I = \frac{8}{3}\pi \int \rho(r) r^4 \, dr. \tag{6.48}$$

(b) Calculate the moment of inertia expressed in units of MR^2 for planets for which the density distributions are given in Problems 6.1 (a) and (b).

6.6.E Show that the net gravitational plus centrifugal acceleration, $g_{eff}(\theta)$, on a rotating sphere is:

$$g_{eff}(\theta) = g_p(\theta) - \omega_{rot}^2 \cos^2\theta, \tag{6.49}$$

where ω_{rot} represents the spin angular velocity, g_p is the gravitational acceleration, and θ is the planetocentric latitude.

6.7.I Consider Figure 6.6b: the Earth's continental crust with a thickness r_1, a mountain on top with a height r_2, an ocean with a depth r_4, and the oceanic crust with a thickness r_5. Assume that the density of the crust is equal for continental crust, mountains and oceanic crust. The density of the crust is ρ_c, that of the mantle ρ_m, and water has a density ρ_w. You can further assume that $\rho_w < \rho_c < \rho_m$.

(a) Derive the relations between the quantities r_3 and r_2, and between r_6 and r_4, assuming isostatic equilibrium. The mountain is compensated by a mass deficiency of height r_3, and the ocean crust by a layer of extra mass of height r_6. These assumptions are referred to as *Airy's hypothesis*.

(b) If the mountain is 6 km high, calculate the height of the mass deficiency r_3. Assume $\rho_w = 1$ g cm^{-3}, $\rho_c = 2.8$ g cm^{-3} and $\rho_m = 3.3$ g cm^{-3}.

(c) If both the ocean and oceanic crust are 5 km deep, calculate r_6.

6.8.I An alternative hypothesis for isostatic equilibrium is *Pratt's hypothesis*. Pratt assumed that the depth of the base of the continental and oceanic crust, including mountains and oceans, is the same, i.e., equal to r_1 (see the dashed line on Figure 6.6b), and that isostatic equilibrium is reached by variations in the density between the three columns A, B and C, i.e., ρ_A, ρ_B, and ρ_C. (The density of water is $\rho_w = 1$ g cm^{-3}.)

(a) Derive a relation between the densities ρ_A, ρ_B and ρ_C with the heights r_1, r_2, and r_4.
(b) If the crust is 30 km thick and the mountain 6 km high, calculate the density of the mountain + crust, ρ_B, if $\rho_A = 2.8$ g cm^{-3}.
(c) If the ocean is 5 km deep, calculate ρ_C.

6.9.I (a) Derive the expression for the free-air correction, if the gravity measurements are made at an altitude h above sea level. You may assume $h \ll R$, where R is the radius of the planet.
(b) Show how the free-air correction factor at latitude θ changes if the gravitational attraction of a horizontally infinite slab of rock at latitude θ, of density ρ and height h, is included.

6.10.I To calculate the geoid anomaly above mountains and oceans, the density anomalies are measured with respect to a reference structure. Using the Airy hypothesis, calculate the geoid height anomaly at the equator above the mountain and the ocean from Problem 6.7, using the crust with density ρ_c as the reference structure.

6.11.I Approximate the interior structure of the Earth by two layers: a uniform density core, out to a radius of 0.57 $R_\oplus$, surrounded by a uniform density mantle. The Earth's average density is 5.52 g cm^{-3}, and its moment of inertia ratio $I/MR^2 = 0.331$. The planet is in hydrostatic equilibrium.

(a) Using the equation for the moment of inertia, determine the density of the Earth's core and of the Earth's mantle.
(b) Using the equation of hydrostatic equilibrium for a planet which consists of two layers (core plus mantle), determine the pressure at the center of the Earth, and compare with the values obtained in Problem 6.3.

6.12.E (a) Calculate the total amount of internal energy lost per year from Earth, assuming a heat flux of 75 erg cm^{-2} s^{-1}.
(b) Calculate the temperature of Earth if the 'heat flux' had been constant over the age of the Solar System, but none of the heat had escaped the planet. A typical value for the specific heat of rock $c_P = 1.2 \times 10^7$ erg g^{-1} K^{-1}.

6.13.E Compare the Earth's intrinsic luminosity (L_i) or heat flux (75 erg cm^{-2} s^{-1}) with the luminosity from reflection of sunlight (L_v), and from emitted infrared radiation (L_{ir}). Assume the Earth's albedo is 0.36. Note that you can ignore the greenhouse effect, and that the value of the infrared emissivity is irrelevant (Hint: Use Section 3.1.2). Comment on your results, and the likelihood of detecting Earth's intrinsic luminosity via remote sensing techniques from space.

6.14.I (a) Using the effective and equilibrium temperatures, calculate the decrease in the internal temperature of each of the giant planets over the age of the Solar System, assuming the planets' luminosities have not changed.
(b) Can you use this technique to estimate the temperature these planets had initially? If so, provide such an estimate, if not, explain why not.

6.15.I Calculate the ratio of internal radioactive heat to absorbed solar radiation for the planet Pluto. You may assume that Pluto consists of a 60:40 mixture of chondritic rock to ice, and that its albedo is 0.4.

6.16.E If energy transport is dominated by conduction, calculate the depth over which the temperature gradient is influenced significantly in a rocky object for the timescales specified below. Assume the surface layers consist of uncompressed silicates, and estimate the thermal conductivity using equation (6.32). Assume the heat capacity is given by $c_P = 1.2 \times 10^7$ erg g^{-1} K^{-1}, and the density $\rho = 3.3$ g cm^{-3}.

(a) Assume a temperature of 300 K and a timescale of 1 day (i.e., this gives you an approximate depth of diurnal temperature variations in Earth's crust).
(b) Assume a temperature of 300 K and a timescale equal to the age of the Solar System.
(c) Assume a temperature of 10 000 K and a timescale equal to the age of the Solar System.
(d) Comment on your results above. Do you think energy transport is by conduction only throughout the entire Earth?

6.17.**E** Derive the Adams–Williams equation (6.43) from the equation of hydrostatic equilibrium and the definition of the bulk modulus K.

6.18.**E** (a) Use Figure 6.14 to determine the time that it takes a P wave to travel from the epicenter of an earthquake to the farthest point on Earth.
(b) Use Figure 6.14 to estimate the time that P waves and S waves take to travel from the epicenter of an earthquake to a point on the globe that is 60° away. Ignore refraction, i.e., assume the waves propagate along straight paths.
(c) How, qualitatively, would refraction change the value that you computed in part (b). Explain using a diagram.

6.19.**I** Suppose a powerful jolt from below woke you up at 2 am sharp. Exactly 4 seconds later a strong horizontal shaking begins and you are knocked off the bed.

(a) Explain both types of motions.
(b) If the average velocity of P and S waves is 5.8 km s^{-1} and 3.4 km s^{-1}, respectively, calculate your distance from the epicenter. You can assume that the focus was at the surface.
(c) How many measurements like your own do you need to have to exactly pinpoint the epicenter – explain in words and pictures.

6.20.**E** The continents on Earth drift relative to each other at rates of up to a few centimeters per year; this drift is due to convective motions in the mantle.

(a) Assuming typical bulk motions in the mantle of Earth are 1 cm/year, calculate the total energy associated with mantle convection.
(b) Calculate the kinetic energy associated with the Earth's rotation.
(c) Calculate the kinetic energy associated with the Earth's orbital motion about the Sun.

6.21.**I** Fluid motions in the Earth's outer liquid core are believed to be responsible for the currents which produce Earth's magnetic field. By far the most likely source of energy for these motions is convection within the core. The core contains very little radioactive material, so some other process must be providing the buoyancy necessary for such convection. Three mechanisms have been suggested: (1) Energy loss to the mantle at the top of the core. (2) Latent heat released by the freezing out of nickel and iron at the boundary between the inner solid core and the outer fluid core. (3) Decrease in the density of material at the bottom of the outer core due to freezing out of denser portions of the liquid.
Process (2) is the easiest to study quantitatively, and even here several simplifying assumptions are needed. From modeling which includes the size of Earth's solid inner core and estimates of its age (1–3.6 × 10^9 years), a growth rate of 0.04 cm/year is estimated at the current epoch. Making the assumption that the latent heat of fusion is not affected by either megabar pressures or the fact that nickel/iron is freezing out from a solution which also contains more volatile elements, estimate the energy released by this process over the course of a year.

6.22.**I** (a) What is the current rate at which energy is being generated within Earth via radioactive decay? (You may assume that the Earth is made of chondritic material. State your answer in ergs/year.)
(b) How much solar energy is absorbed by Earth? Again, state your answer in ergs/year.
(c) Compare the energies calculated above with the average energy released in earthquakes per year, which is $\sim 10^{28}$ ergs, and comment.

6.23.**E** Mercury's mean density $\rho = 5.44$ g cm^{-3}. This value is very close to the planet's uncompressed density. If Mercury consists entirely of rock ($\rho =$ 3.3 g cm^{-3}) and iron ($\rho = 7.95$ g cm^{-3}), calculate the planet's relative abundance of iron by mass.

6.24.**E** Assume the density of 'rock' in the jovian satellites is equal to the mean density of Io and that the density of water-ice is 1 g cm^{-3}.

(a) What proportion of each of the other Galilean satellites consists of rock, and what proportion ice?
(b) In reality, this calculation is good for Europa, but underestimates the ice fraction for Ganymede and Callisto. Why?

6.25.**I** The lithosphere is defined as the solid outer layer of a planet's upper mantle. The upper mantle is solid if the temperature $T \lesssim 1200$ K. For Earth, the thermal conductivity is $K_T = 3 \times 10^5$ erg cm^{-1} s^{-1} K^{-1}, and the average heat flow is 75 erg cm^{-2} s^{-1}. Assume that these same parameters hold for Venus and Mars. Determine the thickness of the lithosphere for Earth, Venus and

Mars (Hint: What is the surface temperature for the three planets?). Comment on the differences, and discuss the potential for tectonic plate activity.

6.26.**E** Suppose that the shield volcanoes on Mars are produced from a partial melting of martian rocks at a temperature of ~1100 K. The density is then ~10% less than that of the surrounding rocks, and the magma will rise up through the surface and build up the 20 km high Tharsis region. Assuming the pressure on the magma chamber below the magma column is equal to the ambient pressure at that depth, calculate the depth of the magma chamber below the martian surface.

6.27.**E** Calculate the rotation period for Neptune from the observed oblateness 0.019, equatorial radius of 24 764 km and $J_2 = 3.5 \times 10^{-3}$. Compare your answer with a typical rotation period of 18 hours for atmospheric phenomena, and of 16.11 hours for Neptune's magnetic field. Comment on your results.

6.28.**I** Approximate Jupiter and Saturn by pure hydrogen spheres, with an equation of state $P = K\rho^2$, and $K = 2.7 \times 10^{12}$ cm^5 g^{-1} s^{-2}.

(a) Determine the moment of inertia for the planets.

(b) Assume the planets each have a core of density 10 g cm^{-3}, and the moment of inertia ratio $I/MR^2 = 0.253$ for Jupiter and 0.227 for Saturn. Determine the mass of these cores.

7 Planetary Magnetospheres and the Interplanetary Medium

The secret of magnetism, now explain that to me! There is no greater secret, except love and hate.
Johann Wolfgang von Goethe, in *Gott, Gemüt und Welt*

Most planets are surrounded by huge magnetic structures, known as *magnetospheres*, that are produced by the planets' internal magnetic fields. These magnetospheres are often more than 10–100 times larger than the planet itself, and therefore form the largest structures in our Solar System, other than the heliosphere. The solar wind flows around and interacts with these magnetic 'bubbles'. A planet's magnetosphere can either be generated in the interior of the planet via a dynamo process (Earth, giant planets, Mercury), or induced by the interaction of the solar wind with the body's ionosphere (Venus, comets). Large scale remnant magnetism is important on Mars, the Moon and some asteroids.

The shape of a magnetosphere is determined by the strength of its magnetic field, the solar wind flow past the field and the motion of charged particles within the magnetosphere. Charged particles are present in all magnetospheres, though the density and composition of the particles varies from planet to planet. The magnetospheric particles may originate in the solar wind, the planet's ionosphere or on satellites or ring particles whose orbits are partly or entirely within the planet's magnetic field. The motion of these charged particles gives rise to currents and large scale electric fields, which in turn influence the magnetic field and the particles' motion through the field. It is probably not surprising that the interaction of charged particles, magnetic and electric fields triggers very complex physical processes which are often not well understood.

Although most of our information is derived from *in situ* spacecraft measurements, atoms and ions in some magnetospheres have been observed from Earth through the emission of photons at ultraviolet and visible wavelengths. Accelerated electrons emit photons at radio wavelengths, observable at frequencies ranging from a couple of hertz to several gigahertz. Such radio emissions were detected from Jupiter in the early 1950s, and formed the first evidence that planets other than Earth might have strong magnetic fields.

7.1 The Interplanetary Medium

7.1.1 Solar Wind

7.1.1.1 *Quiescent Solar Wind*

The first suggestion of corpuscular radiation from the Sun, the *solar wind*, was made by L. Biermann in 1951 based on the observation that cometary ion tails point away from the Sun. The angle between the ion tail and the Sun–comet line is usually $\lesssim 5°$ (Problem 7.1). This angle, ϕ, is determined by the ratio of the comet's (transverse) orbital velocity, v_θ, and the speed of the solar wind particles, v_{sw}:

$$\tan\phi = \frac{v_\theta}{v_{sw} - v_r}, \tag{7.1}$$

where v_r is the radial component of the comet's velocity.

Observation of the Sun during a solar eclipse reveals the solar *corona* (Fig. 7.1), which consists of highly variable *streamers* and *filaments*. Closer to the Sun's surface, in the *chromosphere*, are bright *prominences*, small loop-like structures which are connected to *sunspots*, pairs of dark spots on the Sun's surface. These spots appear dark because the temperature here is lower ($T \approx 4000$ K) than the average photospheric temperature ($T \approx 5750$ K). The magnetic field strength in the spots is higher than in the surrounding regions, so that there is approximate pressure equilibrium between the spots and the surrounding regions. The spots in each pair show opposite polarity, as expected for the loop-like magnetic structures anchored to the sunspots. The number of sunspots varies on an 11-year cycle (Fig. 7.2), and the leading sunspot of a pair tends to show the same polarity throughout the cycle. This polar-

TABLE 7.1 Solar Wind Parameters at Earth's Orbit.

	Quiet solar wind	Fast solar wind	Magnetosheath
Density (protons cm^{-3})	5–8	8–12	8–18
Velocity (km s^{-1})	300–500	500–900	100
Electron temp. (K)	10^5	10^5	10^6
Magnetic field (γ)	3–10	8–16	8–20

FIGURE 7.1 The total solar eclipse of 3 November 1994 as seen from Chile. The dark center is the Moon as it passes between between us and the Sun. The darker coronal regions at the top and bottom of the lunar disk mark the Sun's north and south polar caps, respectively, and these polar regions contain faint plumes of coronal plasma outlining the poloidal magnetic field of the Sun. The bright white streamers extending outwards from the Sun are visible because sunlight is scattered to us by electrons in the streamers. The small bright features just outside the Moon's edge are solar prominences glowing red by fluorescence from hydrogen atoms. (Courtesy: High Altitude Observatory, NCAR, Boulder, Colorado USA)

ity as well as the overall magnetic field of the Sun reverses every 11 years.

In several regions, called *coronal holes* (described below), the solar magnetic field lines open up into interplanetary space, and the solar particles can flow freely away from the Sun. This expansion of the corona forms the solar wind. At Earth's orbit and beyond, during *solar maxima* (maximum number of sunspots) the solar wind has an average speed of ~400 km s^{-1}. During solar minima the average solar wind speed is ~400 km s^{-1} near the ecliptic region, but at higher solar latitudes the speed is typically 750–800 km s^{-1}. During this time there are large coronal holes near the Sun's poles from which the wind emanates.

The solar wind consists of a roughly equal mixture of protons and electrons, with a minor proportion of heavier ions. The density decreases roughly as the inverse square of the heliocentric distance. A typical ion density at Earth's orbit is 6–7 protons cm^{-3}; the temperature is $\sim 10^5$ K and the magnetic field strength a few $\times 10^{-5}$ G (Table 7.1).

In 1958, Eugene Parker calculated the velocity of the solar wind flow, assuming the particles flowed radially outward from the Sun, carrying the solar magnetic field as if it were 'frozen in'. The outward acceleration of the solar wind is primarily caused by the pressure difference between the corona and the interplanetary medium. Parker solved the equations of continuity and momentum, using the ideal gas law as the equation of state. His family of solutions is shown in Figure 7.3. Two of the solutions start at a supersonic velocity in the corona, which disagrees with observations, and therefore can be rejected. A third solution stays subsonic regardless of distance to the Sun. In this solution, the velocity reaches a maximum at a critical radius, R_c, and decreases at larger distances (see e.g., Hundhausen's chapter in Kivelson and Russell, 1995 for a detailed discussion):

$$R_c = \frac{GM_\odot}{2c_s^2} \tag{7.2}$$

where c_s is the *speed of sound*:

$$c_s = \sqrt{\frac{\gamma P}{\rho}} = \sqrt{\frac{\gamma kT}{\mu_a m_{amu}}}, \tag{7.3}$$

γ is the ratio of specific heats ($\gamma \approx 5/3$ for the solar wind), and T the temperature ($T \approx 2 \times 10^6$ K in the solar corona, and $T \approx 10^5$ K at $r = 1$ AU). The sound speed at Earth's orbit is thus approximately 60 km s^{-1}. The critical radius is approximately equal to a solar radius: $R_c \approx R_\odot$. Since the decrease in the solar wind velocity beyond R_c has not been observed, this is not the right solution for the solar wind either. The fourth curve starts at zero velocity and turns supersonic at the critical radius. This solution satisfies all observed quantities.

The expansion of the solar wind is radially outwards, but each fluid-like element of the wind effectively carries

(a)

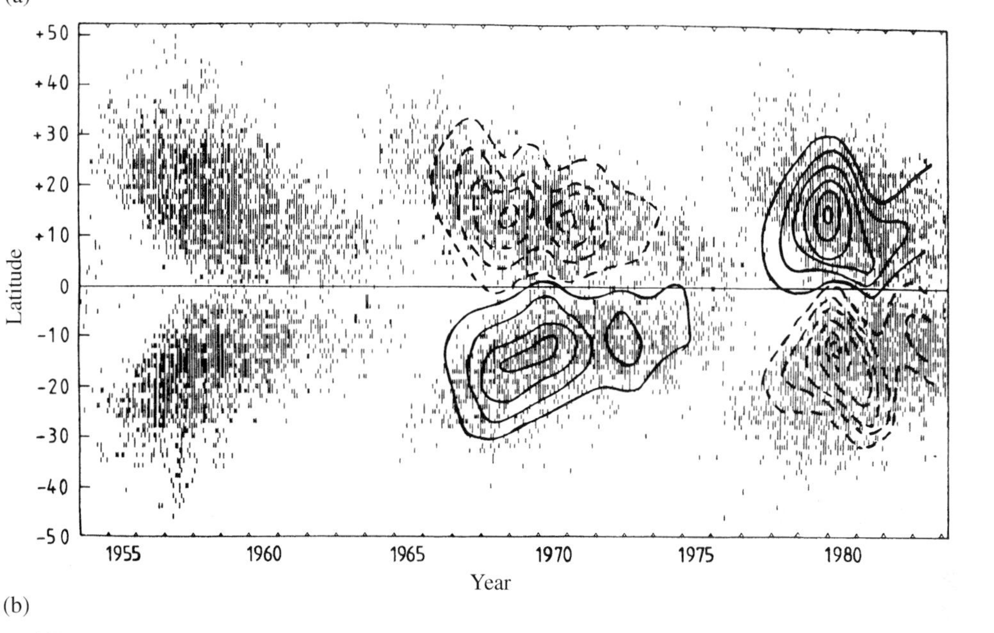

(b)

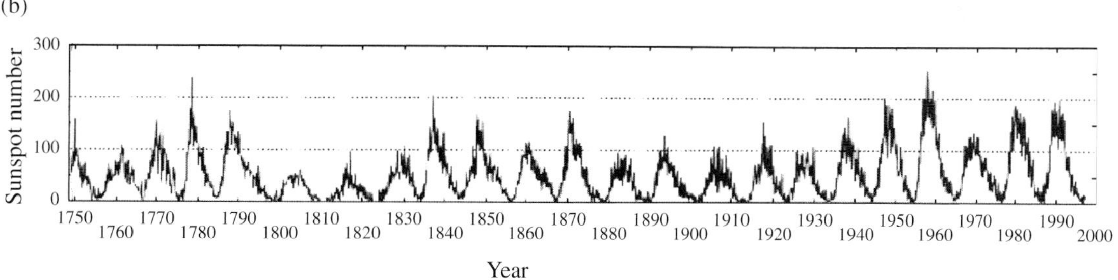

FIGURE 7.2 (a) The spatial distribution of sunspots as a function of time, which shows the familiar butterfly diagram. The 11-year solar sunspot cycle is clearly visible. The contours outline the radial magnetic field, at levels of $\pm$ 0.27 G, $\pm$ 0.54 G. Solid lines indicate positive magnetic field values, dashed lines negative values. (Stix 1987) (b) The number of sunspots as a function of time. As in panel (a), the 11-year solar sunspot cycle is clearly visible, while it is clear from this panel that the number of sunspots during a minimum and maximum also varies over long time scales.

a specific magnetic field line, which is rooted at the Sun. Consequently, the solar wind magnetic field takes on the approximate form of an Archimedean spiral, as displayed in Figure 7.4. The radial and azimuthal components of the field are roughly equal at Earth's orbit, with a strength of a few $\times$ 10^{-5} G. Since the total magnetic flux through any closed surface around the Sun must be zero, inward and outward magnetic fluxes must balance each other. Spacecraft measurements have shown that the inward and outward fluxes are distributed in a systematic way such that there are interplanetary sectors with predominantly outward fluxes and others with predominantly inward fluxes. The different sectors are magnetically connected to different regions on the solar surface–generally different coronal holes. The flows from different coronal regions often have different speeds, and so 'collide' and produce spiral-shaped compressions in the solar wind. The entire magnetic field and stream structure rotates with the Sun. The structure seen in 1963/1964 by the spacecraft IMP-1 is shown in Figure 7.5. This sector structure changes over time as conditions on the Sun change. Note that, while the Sun rotates, the different sectors sweep by the planets and other bodies in our Solar System. The sudden reversals of magnetic field direction and stream structure are responsible for the 'disconnection' events seen in cometary ion tails (Section 10.5.2), as well as certain magnetospheric disturbances.

7.1.1.2 *Space Weather, Solar Flares and Coronal Mass Ejections*

X-ray images of the Sun reveal a tremendous amount of structure within the solar corona (Fig. 7.6a). The lumi-

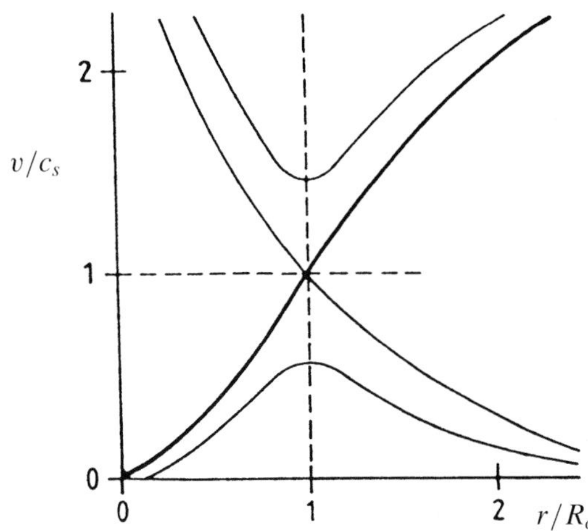

FIGURE 7.3 Parker's four solutions to the equations which describe the solar wind outflow. The heavy curve is the actual solar wind solution. (Adapted from Parker 1963)

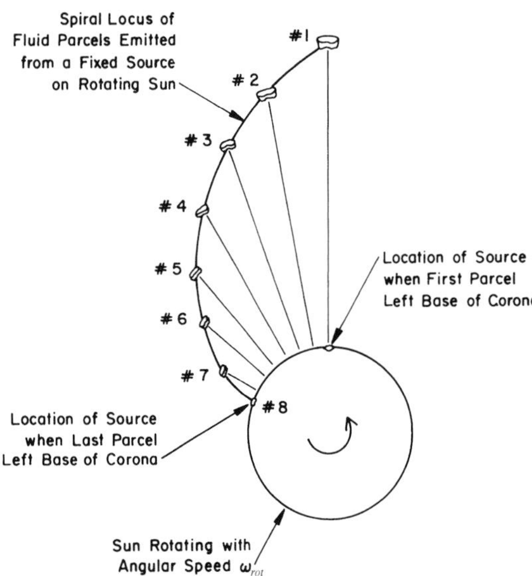

FIGURE 7.4 Archimedean spiral of solar wind particles streaming away from the Sun. (Hundhausen 1995)

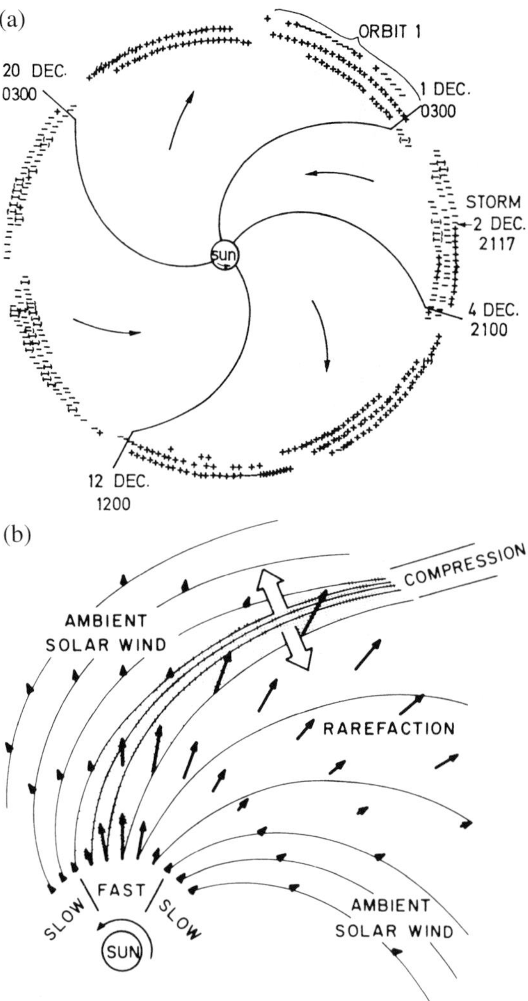

FIGURE 7.5 (a) Sector structure in the interplanetary field as observed by IMP-1 in 1963/1964. Plus signs indicate magnetic field directed outward from the Sun. (Wilcox and Ness 1965) (b) The interaction between a region of high speed solar wind with the ambient solar wind. The high speed solar wind is less tightly bound. (Hundhausen 1995)

nosity from the X-ray bright areas is produced by a dense hot thermal plasma, which is heated by energetic electrons trapped on closed magnetic field loops in the low corona. The dark areas indicate regions devoid of hot X-ray emitting plasma, and are referred to as *coronal holes*. Magnetic field lines in these regions have opened up into interplanetary space, and particles can freely escape into space, forming and/or 'feeding' the solar wind. Solar rotation causes these high speed winds to run into slower flows, causing a compression of material at the interface, which often coincides with a magnetic sector boundary (Fig. 7.5). These high speed winds are preceded by shock fronts, which accelerate particles locally, sometimes up to cosmic-ray-like energies.

Sometimes large prominences erupt in connection with *coronal mass ejections*, such as shown in Figure 7.7. This sequence of images shows that solar material, in part from the lower chromosphere, is injected into the solar wind at speeds much larger than that of the quiescent winds. Speeds may be up to 2000 km s^{-1}, though values of ~ 800 km s^{-1} are seen more frequently. The ejecta blasted outward at high speeds can produce leading interplanetary

(a)

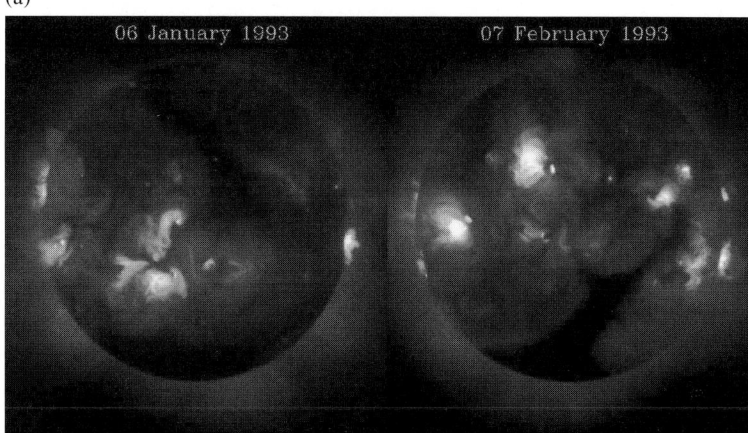

(b)

Energy Release Site
Soft X-ray Loop
Solar Surface
Hard X-ray Sources
Figure courtesy of Solar Data Analysis Center

FIGURE 7.6 (a) The Sun as observed by the Yohkoh spacecraft in soft X-rays. X-ray bright regions indicate heating to temperatures in excess of 2×10^6 K. These regions usually overlie sunspots or active regions. The very dark regions are the coronal holes, which are usually located above the solar poles. These coronal holes sometimes extend down to lower latitudes, as illustrated by these two images. (Courtesy: Yohkoh Science Team) (b) Expanded view of the extraordinary heating associated with a solar flare. The Sun's magnetic field traps the plasma in loops, and the X-ray bright regions represent plasma which has been heated and accelerated in the solar flare. (Courtesy: Solar Data Analysis Center, Goddard Space Flight Center) (c) This image of a prominence on the Sun was taken by the Transition Region and Coronal Explorer (TRACE) in FeIX (a line emitted by iron that is missing 8 electrons). It clearly shows the magnetic loop structures such as visualized in (b); the details on this picture are startling. (Courtesy: NASA)

(c)

shocks and compressions in the solar wind. The shocks are the major source of solar-activity related energetic particles in the Solar System, while the solar wind compressions disturb planetary magnetospheres. Coronal mass ejections sometimes, but not always, precede *solar flares*, events where the coronal X-ray and UV emissions suddenly brighten up, sometimes so dramatically that the intensity increases by several orders of magnitude. At a flare onset, the field lines reconnect. This gives rise to sudden impulses of energy release, including accelerated particles. During the main phase of the flare, reconnection of field lines continues, energizing the event, which becomes visible through hot UV and X-ray loops (Fig. 7.6b). Major solar flares may occur approximately once a week during years of maximum sunspot activity. Weak flares and coronal mass ejections typically happen a few times daily during times of maximum solar activity, and about once a week during years of minimum activity.

Interplanetary space is a region full of complicated and sometimes violent processes. It takes an ordinary solar wind particle typically a few days to reach the Earth, whereas the most energetic cosmic-ray-like particles in-

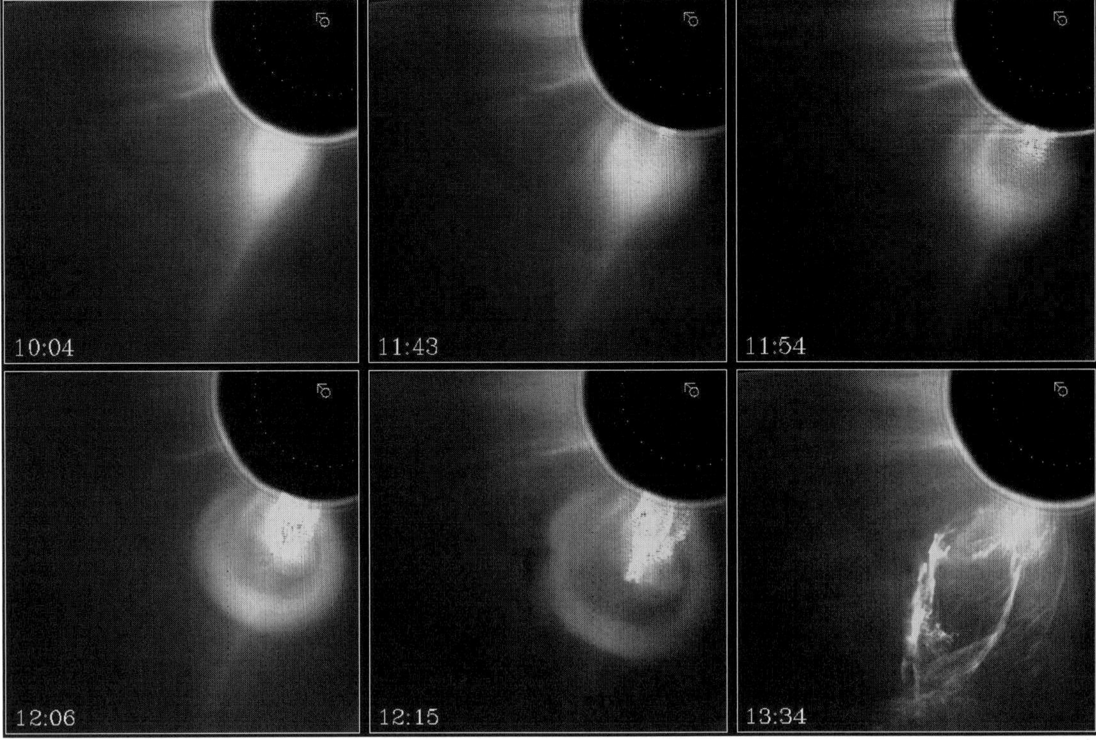

FIGURE 7.7 Sequence of images obtained on 18 August 1980, showing the formation of a coronal mass ejection through the distortion of a coronal streamer, which is triggered by the disruption of a solar prominence beneath it. The prominence material is blown outward along with the original streamer material; the bright, filamentary structures on the 13:34 frame are in fact the remnants of the prominence. A comparison of images taken before and after a coronal mass ejection often reveal the disappearance of a filament originally located along the magnetic neutral line straddled by the disrupted helmet streamer. Note, however, that not all filaments disappear this way, and not all mass ejections are accompanied by erupting prominences. The dark circle in the upper right corner of each frame is the occulting disk of the Solar Maximum Mission (SMM) coronagraph, used to take these images (on this instrument the occulting disk has a radius 60% larger than the solar disk). (Courtesy: J. Burkepile, High Altitude Observatory, NCAR)

jected through a solar flare may reach Earth several tens of minutes after the event. The response of Earth's environment to the continuously varying interplanetary medium is known as *space weather*, and the effects coronal mass ejections and solar flares have on the interplanetary medium and Earth's environment are referred to as *space weather storms*. All objects in the interplanetary medium are subjected to space weather. For example, the effect of space weather over millions or billions of years may have 'masked' the surface characteristics of asteroids, which makes it difficult to properly classify them (Section 9.4.1).

Space weather is of particular importance because of its effects on man-made satellite operations and communication systems and on some ground-based electronics. Fast coronal mass ejections may damage exposed electronic equipment directly, and may disrupt communications indirectly. Such solar wind disturbances trigger magnetic storms (Section 7.4.1.4) which energize the entire magnetosphere and drive large scale currents that can bring down power systems. The Earth's atmosphere gets heated by enhanced auroral precipitation and electric field-driven 'Joule heating' and expands, which increases the drag on near-Earth satellites and 'space junk', leading to changes in their orbits. Such changes may result in a (temporary) 'loss' of spacecraft. Radio communication systems on the ground rely on the reflection of radio waves by the Earth's ionosphere. Such communications are temporarily disrupted when the ionosphere is 'changed' by a solar flare or coronal mass ejection.

7.1.2 Maxwell's Equations

Charged particles and magnetic fields are intimately connected in a magnetized plasma, such as the solar wind and planetary magnetic fields. If the magnetic field points towards the observer, protons appear to spiral around the field in a clockwise motion; electrons gyrate anticlockwise. Currents[1], electric and magnetic fields are related through *Maxwell's equations*:

(i) Poisson's equation relates the electric field, **E**, to the charge density, ρ_c:

$$\nabla \cdot \mathbf{E} = 4\pi\rho_c. \tag{7.4}$$

(ii) The divergence of the magnetic field, **B**, is zero:

$$\nabla \cdot \mathbf{B} = 0, \tag{7.5}$$

or in other words, magnetic field lines do not begin or end.

(iii) Faraday's law describes the relationship between time varying magnetic and electric fields:

$$\frac{1}{c}\frac{\partial B}{\partial t} = -\nabla \times \mathbf{E}. \tag{7.6}$$

(iv) Ampere's law relates spatial variations in a magnetic field to currents, **J**, and time variable electric fields:

$$\nabla \times \mathbf{B} = \frac{4\pi}{c}\mathbf{J} + \frac{1}{c}\frac{\partial E}{\partial t}. \tag{7.7}$$

Usually, in planetary magnetic fields the term $(1/c)(\partial E/\partial t)$ is very small and Ampere's law can be approximated by:

$$\frac{4\pi}{c}\mathbf{J} \approx \nabla \times \mathbf{B}. \tag{7.8}$$

Equation (7.8) shows that a current induces a magnetic field that is in a direction *opposite* to the original field, thus weakening the magnetic field strength.

7.1.3 Magnetohydrodynamics (MHD)

Although the solar wind is a collisionless flow (Problem 7.3), the magnetic field 'binds' the particles together and, for most purposes, the solar wind can be considered a fluid whose flow can be described by the equations of hydrodynamics. Because electric and magnetic fields are important in a magnetized fluid, the hydrodynamic equations become slightly more complicated, and are replaced by the equations of magnetohydrodynamics (MHD). We assume that the plasma is an electrically neutral fluid, i.e., assuming that all the ions carry just single charges, the number density of ions and electrons is equal. The MHD equations thus describe the macroscopic behavior of a plasma, i.e., the electric, magnetic and gravitational fields, together with the plasma density and bulk-flow velocity. The equation of continuity (conservation of mass) reads:

$$\frac{\partial \rho}{\partial t} + \nabla \cdot (\rho \mathbf{v}) = 0, \tag{7.9}$$

where ρ is the total mass density. The equation of motion (conservation of momentum) is:

$$\rho\left(\frac{\partial \mathbf{v}}{\partial t} + \mathbf{v} \cdot \nabla \mathbf{v}\right) = -\nabla P + \mathbf{J} \times \mathbf{B} + \rho \mathbf{g}_p, \tag{7.10}$$

with P the thermal pressure of the plasma and $\mathbf{g}_p$ the gravitational acceleration. If the plasma is not neutral, an electrical force term should be added to the right hand side of equation (7.10).

Conservation of energy in the plasma implies:

$$\frac{\partial U}{\partial t} + \nabla \cdot \mathbf{J}_u = 0, \tag{7.11}$$

with U the total energy, which consists of the kinetic, magnetic and thermal energies, respectively:

$$U = \frac{1}{2}\rho v^2 + \frac{B^2}{8\pi} + \frac{P}{\gamma - 1}, \tag{7.12}$$

with γ the ratio of specific heats. The energy flux vector $\mathbf{J}_u$ is given by:

$$\mathbf{J}_u = \mathbf{v}\left(\frac{1}{2}\rho v^2 + \frac{\gamma P}{\gamma - 1}\right) + \frac{c}{4\pi}\mathbf{E} \times \mathbf{B}. \tag{7.13}$$

The various contributions to this flux vector are the transport of kinetic and thermal energy, and the work done by hydrostatic and radiation pressure (Poynting vector), respectively.

7.1.4 Interaction of the Solar Wind with Planets

All planetary bodies interact to some extent with the solar wind, as schematically indicated in Figure 7.8. For bodies without intrinsic magnetic fields, the interaction depends on the conductivity of the body. For rocky objects such as the Moon and most asteroids which are poor conductors, the solar wind particles hit the body directly and are absorbed. In contrast to bodies discussed later, no shock is formed upstream from the object. The interplanetary magnetic field lines diffuse through the body. The wake immediately behind the object is practically devoid of particles (Fig. 7.8a).

[1] Note that a current in a wire is caused by the flow of electrons from a negative to a positive voltage; however, the current through the wire is defined as a flow in the opposite direction, i.e., as given by a flow consisting of positively charged particles moving to the negative voltage. Currents induce a magnetic field (via the right-hand or screwdriver motion), which underlies the principle of the electromagnet. Equation (7.8) shows that such a magnetic field is actually oriented in a direction opposite to that of the originally present magnetic field.

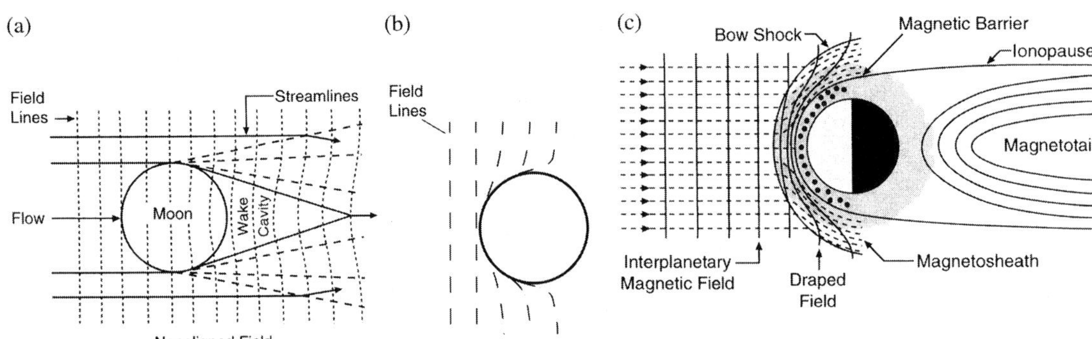

FIGURE 7.8 Interaction of the solar wind with various types of planetary bodies that do not possess internal magnetic fields. (a) A nonconducting body, (b) a conducting body, and (c) a body with an ionosphere. (Adapted from Luhmann 1995)

In a conductive medium, the motion of a body or plasma through the interplanetary magnetic field induces a current, **J**, as given by Ohm's law:

$$\mathbf{J} = \sigma_o \left(\mathbf{E} + \frac{\mathbf{v} \times \mathbf{B}}{c} \right), \tag{7.14}$$

where σ_o is the conductivity of the plasma, **v** the plasma velocity and **B** the interplanetary magnetic field strength. In a highly conducting plasma like the solar wind, $\mathbf{J}/\sigma_o \approx 0$, and thus:

$$\mathbf{E} + \frac{\mathbf{v} \times \mathbf{B}}{c} = 0. \tag{7.15}$$

It can be shown that this condition is equivalent to saying that there is no relative motion between the plasma and the interplanetary magnetic field, or in other words that the interplanetary magnetic field lines are *frozen into* the plasma, i.e., carried with the wind. If a planetary body is highly conductive, the interplanetary magnetic field lines drape around the body because the plasma flows around the conductor (Fig. 7.8b). The electric field (eq. 7.15) induces a current ($\mathbf{J} = \sigma \mathbf{E}$) in the body, which in turn disturbs the interplanetary magnetic field. The perturbations in the magnetic field and plasma flow propagate away as Alfvén waves (Section 7.6).

If a poorly conducting body has an atmosphere but no internal magnetic field, the solar wind interacts with the atmosphere. Atmospheric ions gyrate around the ambient magnetic field lines. Since the field is moving with the solar wind plasma, the atmospheric ion is accelerated and picked up by the solar wind. If the body has an extensive ionosphere (e.g., is a good conductor, $\sigma_o \neq 0$), currents are set up, which prevent the magnetic field from diffusing through the body (Fig. 7.8c). This situation gives rise to a magnetic configuration very similar to the magnetic 'cavity' created by the interaction of the solar wind with a magnetized planet, discussed below. The boundary between the ionosphere and the solar wind plasma is called the *ionopause*, and can in a sense be compared to the magnetopause discussed below. The ionopause is located where the ionospheric pressure is balanced by the outside plasma pressure (magnetic plus ram pressure).

For bodies which have an internal magnetic field, the solar wind interacts with the field around the object. It confines the magnetic field to a 'cavity' in the solar wind, referred to as a *magnetosphere*. The magnetic fields around the planets Mercury, Earth, Jupiter, Saturn, Uranus and Neptune are rooted in the planets' interiors, while fields around the 'nonmagnetic' planets Venus and Mars and comets are largely induced by the interaction of the solar wind with the planets' ionospheres. A planetary magnetic field resembles to first approximation a dipole field, similar to that generated by a bar magnet. A sketch of the Earth's magnetic field is given in Figure 7.9. The shape of the magnetosphere depends on the strength of the magnetic field and the solar wind flow past the field. The magnetospheric boundary is called the *magnetopause*. The solar wind pressure shapes the 'nose' of the field, while the solar wind flow stretches the field out into a tail, the *magnetotail*. The magnetotail consists of two lobes of opposite polarity, separated by a neutral sheet. A plasma or current sheet is located at the interface of the two lobes. Since the solar wind is faster than waves in the medium can propagate, a *bow shock* is formed upstream from the magnetosphere.

The approximate position of the magnetopause can be calculated from a balance between the ram pressure of the solar wind, P_{sw}, and the pressure inside the magnetosphere, P_m:

$$P_{sw} = P_m, \tag{7.16}$$

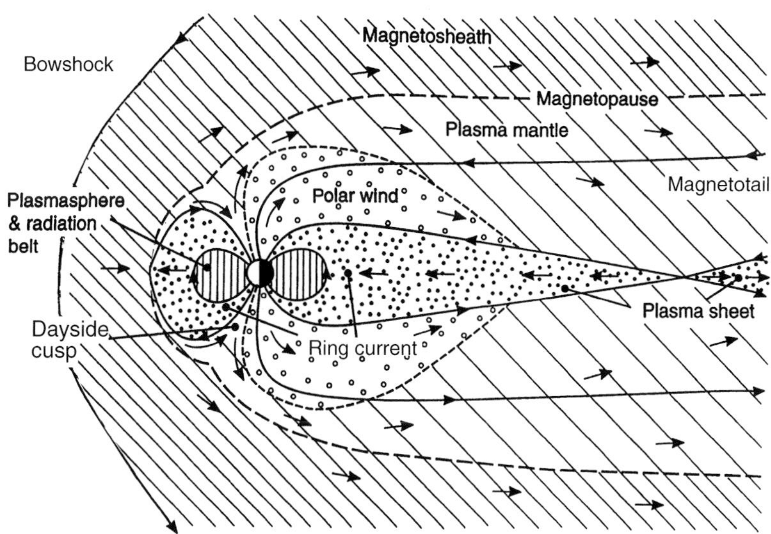

FIGURE 7.9 A sketch of the Earth's magnetic field, on which the various plasma regions are indicated. The solid arrowed lines indicate magnetic field lines, the heavy long-dashed line the magnetopause, and the arrows represent the direction of the plasma flow. Diagonal hatching indicates plasma in, or directly derived from, the solar wind/magnetosheath. Outflowing ionospheric plasma is indicated by open circles; the solid dots indicate hot plasma accelerated in the tail, and the vertical hatching shows the corotating plasmasphere. (Adapted from S. Cowley 1995)

where

$$P_{sw} \approx \rho v^2, \tag{7.17}$$

and

$$P_m = \frac{B^2}{8\pi} + P, \tag{7.18}$$

with P the thermal gas pressure. Since the pressure in the magnetosphere is dominated by the magnetic field strength, a balance in pressure leads to the approximate equality:

$$\left(\rho v^2\right)_{sw} \approx \left(\frac{B^2}{8\pi}\right)_m, \tag{7.19}$$

where the parameters on the left hand side pertain to the solar wind, those on the right hand side to the magnetosphere. A typical *standoff distance* of the magnetopause for Earth is 6–15 $R_\oplus$ (Problem 7.4).

At the sunside of the magnetosphere, the interaction of the supersonic solar wind with the magnetic field induces a bow shock. The creation of this shock is analogous to the formation of a bow shock in front of a speedboat on a lake. In both cases the quiescent flow (solar wind or water) does not 'know' about the obstacle, since the relative velocity between the obstacle and medium is larger than the speed of the relevant waves (surface waves in the case of water, and magnetosonic waves in the case of the solar wind). The presence of the shock causes the solar wind to decelerate enough that it can smoothly flow around the magnetospheric obstacle.

The standoff distance of the Earth's bow shock, $\mathcal{R}_{bs}$ is determined empirically:

$$\mathcal{R}_{bs} = \mathcal{R}_{mp}\left(1 + 1.1\frac{(\gamma - 1)\mathcal{M}_o^2 + 2}{(\gamma + 1)\mathcal{M}_o^2}\right), \tag{7.20}$$

with $\mathcal{R}_{mp}$ the standoff distance of the magnetopause, γ the ratio of specific heats and $\mathcal{M}_o$ the magnetosonic Mach number, defined as:

$$\mathcal{M}_o \equiv \frac{v_{sw}}{\sqrt{c_A^2 + c_s^2}}, \tag{7.21}$$

with c_A the *Alfvén speed*:

$$c_A = \frac{B}{\sqrt{4\pi\rho}}. \tag{7.22}$$

Typical values for the standoff distance of the Earth's bow shock are 10–20 $R_\oplus$ (Problem 7.4). The minimum distance, Δ, between the bow shock and the magnetopause is given by:

$$\Delta = 1.1\frac{\rho_{sw}}{\rho_{bs}}\mathcal{R}_{mp}, \tag{7.23}$$

with ρ_{bs} the density just behind the shock. The solar wind is slowed down in the shock and flows around the magnetosphere. The region immediately behind the bow shock is called the *magnetosheath*. Plasma processes at the bow shock convert some of the 'bulk' energy in the flow into thermal energy of the plasma. The solar wind is subsonic in this region, and the particles flow around the planet's magnetosphere. The bow shock is usually quite thin.

7.1.5 Shocks

In the supersonic solar wind countless discontinuities and shocks occur, induced by, e.g., the interaction of the wind with a magnetized planet, with an ionosphere (planet, comet) and through coronal mass ejections and solar flares. A shock wave changes the state of the medium through which it travels. In the frame of reference of the shock, the upstream velocity is supersonic, while downstream from the shock the velocity is subsonic and the density of the medium is higher. In an unmagnetized plasma, the density downstream of the shock can be up to four times higher than in the upstream medium (Problem 7.5). Since the solar wind is a collisionless plasma, the shocks are also collisionless. Mass, momentum and energy flux are conserved across a shock:

$$[\rho v_\perp] = 0, \tag{7.24}$$

$$\left[\rho v_\perp \mathbf{v} + P + \frac{B_\parallel^2}{8\pi} - \frac{B_\perp B_\parallel}{4\pi}\right] = 0, \tag{7.25}$$

$$\left[\rho v_\perp \left(\frac{\gamma}{\gamma-1}\frac{P}{\rho} + \frac{1}{2}v^2\right) - \frac{B_\parallel}{4\pi}(B_\perp v_\parallel - B_\parallel v_\perp)\right] = 0, \tag{7.26}$$

$$[B_\perp v_\parallel - B_\parallel v_\perp] = 0, \tag{7.27}$$

$$[B_\perp] = 0, \tag{7.28}$$

The brackets in equations (7.24)–(7.28) indicate the difference between the enclosed quantity from one side of the shock to the other. The flow across the shock is indicated by a $\perp$ sign, and parallel to the shock by a $\parallel$ sign. These shock or jump equations are referred to as the *Rankine–Hugoniot relations*. The Rankine–Hugoniot equations relate the density, pressure, temperature and magnetic field strength before and after the shock, like in the solar wind and the magnetosheath. Both *discontinuities* and *shocks* are described by the jump equations.

If there is no mass flow across the shocked region, $v_\perp = 0$. The discontinuity is called *tangential* when the magnetic field is aligned along the discontinuity and $B_\perp = 0$; there is a *contact discontinuity* when $B_\perp \neq 0$. When there is a mass flow across the discontinuity, $v_\perp \neq 0$, and we distinguish three different situations: (i) If $v_\perp = \pm\, c_A$, and $[v_\perp] = 0$, it can be shown that the flow normal to the discontinuity is the same at both sides; the discontinuity is only visible in the velocity and magnetic field components tangential to the discontinuity. This discontinuity is referred to as the *Alfvén shock*, although strictly speaking it is a discontinuity rather than a shock. (ii) When $v_\perp > c_A$ and $[\rho] > 0$, the shock is a *fast shock*. The critical velocity in this case is the magnetosonic fast-mode velocity, $\sqrt{c_A^2 + c_s^2}$. From the jump equations it follows that the magnetic field strength $B_\parallel$ and B^2 increase from the unshocked into the shocked medium. (iii) When $v_\perp < c_A$, and $[\rho] > 0$, the shock is a *slow shock*.

A coronal mass ejection, which travels faster than the local magnetosonic velocity, can drive an interplanetary shock ahead of it. The density and velocity in the shocked or disturbed solar wind are higher than in the quiescent wind. When a solar wind shock hits a planet's bow shock and magnetopause, multiple shocks and waves are induced within the magnetosphere. In this way energy is transferred from the solar wind to the planet's magnetic field. Due to the countless fluctuations in the solar wind, the interaction between the solar wind and a magnetosphere is thus very dynamic. Table 7.1 lists typical parameters measured in the quiet and disturbed solar wind and in the Earth's magnetosheath.

7.2 Magnetic Field Configuration: Mathematical Description

7.2.1 Overview

The solar wind pressure causes a compression of Earth's magnetosphere at the sunside, while the solar wind flow past the Earth's magnetic field stretches the field lines radially behind the planet, and forms the *magnetotail* (Fig. 7.9). Close to the planet where the magnetic field is hardly deformed by the solar wind, we find the *radiation belts* or *Van Allen belts*, regions in the magnetosphere where charged particles are trapped. These particles gyrate around and bounce up and down magnetic field lines, while they drift around the Earth (Section 7.3.1). Under normal equilibrium circumstances these particles cannot escape from the magnetosphere. The thin region in the magnetotail which separates field lines of opposite polarity is referred to as the *neutral sheet* or *current sheet*: due to the field line reversal the magnetic field strength is minimal in this region, and to maintain pressure equilibrium across the magnetosphere, the plasma density in the neutral sheet is maximal. Moreover, *field line reconnection* or *annihilation* may take place in the neutral sheet. A sketch of this process is shown in Figure 7.10, where plasma flows into the neutral X point from above and below, and out towards the sides. Energy released from magnetic reconnection accelerates/energizes the plasma. Magnetic field lines in the *polar caps* are connected to the interplanetary magnetic field (IMF), where the amount of interconnection depends on the IMF orientation with respect to the planetary field.

The magnetic field strength of planets is usually given in terms of its *magnetic dipole moment*, $\mathcal{M}_B$, expressed

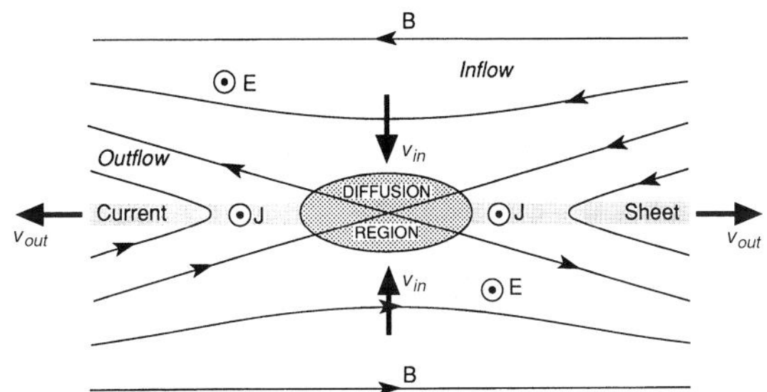

FIGURE 7.10 Magnetic reconnection according to Sweet's mechanism, occurring on an X-type magnetic neutral line. Plasma and magnetic field flow in from the top and bottom of the figure (v_{in}), and out towards the sides (v_{out}). Plasma is not tied to the magnetic field in the diffusion region. (Adapted from Hughes 1995)

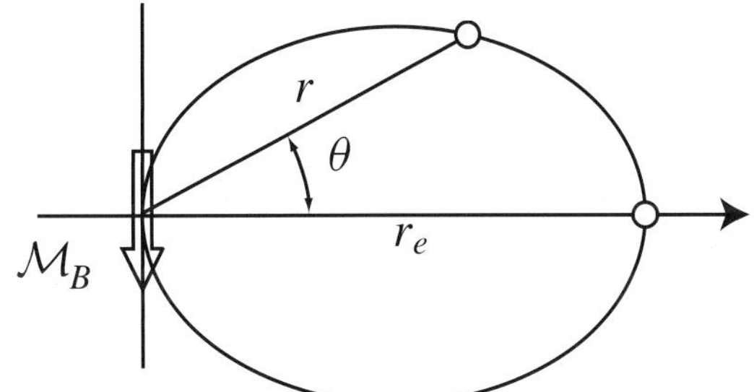

FIGURE 7.11 Schematic sketch of a field line in a dipole magnetic field. (Adapted from Roederer 1970)

in gauss cm^{-3}. Typical parameters for the Earth and other magnetospheres are summarized in Table 7.2. Note that the 'surface' magnetic field strengths (the 1-bar level for the giant planets) for Earth, Saturn, Uranus and Neptune are all $\sim$0.3 G; however, because the radii of the giant planets are much larger than that of Earth, their magnetic dipole moments are 25–500 times larger than the terrestrial moment. Jupiter has by far the strongest magnetic dipole moment, nearly 20 000 times stronger than that of the Earth. The north magnetic poles on Earth and Mercury are near the planets' south poles; thus field lines exit the planets in the south, and enter in the northern hemisphere. The north magnetic poles of Jupiter and Saturn are in the northern hemispheres.

The standoff distance of the magnetopause for Earth and the giant planets is typically larger than 6–10 planetary radii. For Mercury, however, the magnetopause is thought to, at times, be pushed down to the surface, in which case the solar wind interacts directly with the planet's surface.

7.2.2 Dipole Magnetic Field

The magnetic field of a dipole in polar coordinates can be described by:

$$B_r = -\frac{2\mathcal{M}_B}{r^3}\sin\theta, \qquad (7.29a)$$

$$B_\theta = \frac{\mathcal{M}_B}{r^3}\cos\theta, \qquad (7.29b)$$

$$B_\phi = 0, \qquad (7.29c)$$

with r, θ and ϕ the coordinates in the radial, latitudinal (to the north) and azimuthal (to the east) directions, respectively. A field line lies in the meridional plane and is completely specified by the distance to its equatorial point, r_e, and its longitude or azimuth, ϕ (Fig. 7.11):

$$r = r_e\cos^2\theta, \qquad (7.30a)$$

$$\phi = \phi_o = \text{constant}. \qquad (7.30b)$$

A small arc length, ds, along the field line is given by:

$$ds = \sqrt{dr^2 + r^2 d\theta} = r_e\cos\theta\sqrt{4 - 3\cos^2\theta}\,d\theta. \qquad (7.31)$$

The magnetic field strength along a field line is expressed by:

$$B(\theta) = B_e\frac{\sqrt{4 - 3\cos^2\theta}}{\cos^6\theta}, \qquad (7.32)$$

where the magnetic field strength in the magnetic equator, B_e, is determined by the magnetic dipole moment, $\mathcal{M}_B$:

$$B_e = \mathcal{M}_B/r_e^3. \qquad (7.33)$$

7.2.3 Multipole Expansion of the Magnetic Field

Although a planet's magnetic field can to first order be approximated by a dipole, deviations from a simple dipole field are apparent for all magnetospheres. In analogy with the mathematical description of a planet's gravitational field, the planet's internal magnetic field can be described as the gradient of a scalar potential $\Phi_V(r, \theta', \phi)$:

$$\mathbf{B} = -\nabla\Phi_V, \qquad (7.34)$$

where r, θ' and ϕ are the planetocentric coordinates of the point in question: radial distance, planetocentric colatitude and east longitude, respectively. The potential is expressed by:

$$\Phi_V = R\sum_{n=1}^{\infty}\left(\frac{R}{r}\right)^{n+1} T_n, \qquad (7.35)$$

where R is the planet's radius, and the function T_n is given by:

$$T_n = \sum_{m=0}^{n}[g_n^m\cos(m\phi_i) + h_n^m\sin(m\phi_i)]P_n^m(\cos\theta'), \qquad (7.36)$$

TABLE 7.2 Characteristics of Planetary Magnetic Fields.

	Mercury	Earth	Jupiter	Saturn	Uranus	Neptune
Magnetic moment ($\mathcal{M}_\oplus$)	4×10^{-4}	1^a	20 000	600	50	25
Surface B at dipole equator (gauss)	0.0033	0.31	4.28	0.22	0.23	0.14
Maximum/minimum[b]	2	2.8	4.5	4.6	12	9
Dipole tilt and sense[c]	+14°	+10.8°	−9.6°	0.0°	−59°	−47°
Dipole offset (R_p)		0.08	0.12	∼0.04	0.3	0.55
Obliquity	0°	23.5°	3.1°	26.7°	97.9°	29.6°
Solar wind angle[d]	90°	67–114°	87–93°	64–114°	8–172°	60–120°
Magnetopause distance[e] (R_p)	1.5	10	42	19	25	24
Observed size of magnetosphere (R_p)	1.4	8–12	50–100	16–22	18	23–26

After Kivelson and Bagenal (1999).
[a] $\mathcal{M}_\oplus = 7.906 \times 10^{25}$ gauss cm^{-3}.
[b] Ratio of maximum to minimum surface magnetic field strength (equal to 2 for a centered dipole field).
[c] Angle between the magnetic and rotation axis.
[d] Range of angles between the radial direction from the Sun and the planet's rotation axis over an orbital period.
[e] Typical standoff distance of the magnetopause at the nose of the magnetosphere, in planetary radii.

with g_n^m and h_n^m the *Gauss coefficients*, which define the field configuration, and $P_n^m(\cos\theta')$ are the Schmidt-normalized associated Legendre polynomials:

$$P_n^m(x) \equiv N_{nm}(1-x^2)^{m/2}\frac{d^m P_n(x)}{dx^m}, \tag{7.37}$$

where $N_{nm} \equiv 1$ if $m = 0$ and $N_{nm} \equiv \left(\frac{2(n-m)!}{(n+m)!}\right)^{1/2}$ if $m \neq 0$.

The terms with $n = 1$ are called dipole terms, those with $n = 2$ quadrupole terms, $n = 3$ octupole terms, etc. Note that the terms decrease with distance from the planet as r^{-n}, so that the higher order terms are most important near a planet's 'surface'. For a magnetic field which consists of only the three dipole terms g_1^0, g_1^1, and h_1^1, its magnetic moment, $\mathcal{M}_B$, tilt angle with respect to the rotational axis, θ_B, and the longitude of the magnetic north pole, λ_{np}, are:

$$\mathcal{M}_B = \sqrt{(g_1^0)^2 + (g_1^1)^2 + (h_1^1)^2}, \tag{7.38}$$

$$\tan\theta_B = \sqrt{\left(\frac{g_1^1}{g_1^0}\right)^2 + \left(\frac{h_1^1}{g_1^0}\right)^2}, \tag{7.39}$$

$$\lambda_{np} = 360° - \tan^{-1}\left(\frac{h_1^1}{g_1^1}\right). \tag{7.40}$$

If the frame of reference is chosen such that the $\theta' = 0$ axis is aligned with the magnetic dipole axis and the $\phi = 0$ meridian is at the longitude of the magnetic north pole, the (1, 1) terms in equation (7.36) are zero, and deviations from a pure dipole field are more easily identifiable. The quadrupole terms (2, 0) and (2, 1) in this new frame of reference can be interpreted as a displacement of the main dipole. The quadrupole term (2, 2) causes an apparent longitude dependent tilt of the field, i.e., a warping of the magnetic equatorial plane, with a periodicity of 180°. The higher order terms have similar effects on the magnetic field configuration, although with different strengths and periodicities.

The above equations show how to mathematically approximate a magnetic field caused by current systems within the planet. In reality the field deviates from this ideal configuration, since currents in and near the magnetosphere itself induce electric and magnetic fields which distort the field lines. These effects increase with increasing r. So the total magnetic field is better represented by:

$$\mathbf{B}_{total} = -\nabla\Phi_V + \mathbf{B}_{external}. \tag{7.41}$$

Characteristic parameters for the magnetospheres of the planets are shown in Table 7.2. The magnetic field geometries of Jupiter and Saturn are similar to that of the Earth, where the rotation and magnetic axes are within 10° of each other, and the magnetic field is to first approximation dipolar. Uranus and Neptune have quite irregular magnetospheres, and their magnetic axes make large angles with the rotation axes of the planets. The large scale structure of individual planetary magnetospheres is discussed in detail in Section 7.4.

TABLE 7.3 Plasma Characteristics of Planetary Magnetospheres.

	Mercury	Earth	Jupiter	Saturn	Uranus	Neptune
Maximum density (cm^{-3})	1	1000–4000	>3000	~100	3	2
Composition	H^+	O^+, H^+, N^+, He^+	$O^{n+}, S^{n+}, SO_2^+, Cl^+$	O^+, H_2O^+, H^+	H^+	N^+, H^+
Dominant source	solar wind	ionosphere[a]	Io	rings, satellites	atmosphere	Triton
Production rate (ions s^{-1})	?	2×10^{26}	$>10^{28}$	10^{26}	10^{25}	10^{25}
Ion lifetime	minutes	days[a], hours[b]	10–100 days	1 month	1–30 days	1 day
Plasma motion controlled by:	solar wind	rotation[a] solar wind[b]	rotation	rotation	solar wind + rotation	rotation (+ solar wind?)

After Kivelson and Bagenal (1999).
[a] Inside plasmasphere.
[b] Outside plasmasphere.

7.3 Magnetospheric Plasma and Particle Motions

Magnetospheres are populated with charged particles: protons, electrons and ions. The dominant species detected in each magnetosphere are summarized in Table 7.3. We find primarily oxygen and hydrogen ions in the Earth's magnetosphere; in Jupiter's magnetosphere these are augmented with sulfur ions, in Saturn's magnetosphere with H_2O^+, and in Neptune's magnetosphere both H^+ and N^+ have been detected. Note the large variations in ion densities: in Uranus and Neptune's magnetospheres the maximum ion densities are only 2–3 protons cm^{-3}, while in the magnetospheres of Earth and Jupiter the densities measure over a few thousand cm^{-3}. All magnetospheres also contain large numbers of electrons; on average magnetospheric plasma is approximately neutral in charge. The particle distribution functions are almost Maxwellian, or thermal, though there is a pronounced high energy tail to this distribution, with particle energies ranging up to several hundreds of MeV in some magnetospheres.

The spatial distribution of plasma is determined by the sources and losses of the plasma, as well as the motions of the particles in the planet's magnetic field. In equilibrium situations, a particle's motion is completely defined by the magnetic field configuration, the planet's gravity field, centrifugal forces, and large scale electric fields. In this section we summarize the sources and sinks of plasma, and give a detailed treatment of the motions of charged particles in a stable magnetosphere.

7.3.1 Particle Motions: Adiabatic Invariants

The general motion of nonrelativistic charged particles in a magnetic field is governed by:

$$m\frac{d^2\mathbf{r}}{dt^2} = \mathbf{F} + \frac{q\mathbf{v}\times\mathbf{B}}{c}, \tag{7.42}$$

with q the elemental charge and c the speed of light. In the absence of external forces ($\mathbf{F} = 0$), this equation simplifies to the *Lorentz force*:

$$\mathbf{F}_L = \frac{q\mathbf{v}\times\mathbf{B}}{c}. \tag{7.43}$$

The Lorentz force leads to rapid circulation of the particle centered on a location known as the *gyrocenter*. External forces change the simple circular motion of the particle into three components, as depicted in Figure 7.12: (a) a gyration around field lines, (b) a bounce motion along field lines, and (c) a drift motion perpendicular to the field lines. The trajectory of the particle can be approximated by a circular motion of radius R_L around the instantaneous gyrocenter, plus a displacement of the gyrocenter, referred to as the *guiding center* $\mathbf{r}_g$:

$$\mathbf{r} = \mathbf{r}_g + \mathbf{R}_L. \tag{7.44}$$

7.3.1.1 *First Adiabatic Invariant*

The cyclotron or gyro motion around a field line is a circular motion around the guiding center. The Lorentz force balances the centripetal force of the particle's motion around the field line. The *gyro* or *Larmor* radius is (Problem 7.8):

$$|\mathbf{R}_L| = \frac{c\mathbf{p}_\perp}{qB} = \left|\frac{mc\mathbf{v}\times\mathbf{B}}{qB^2}\right|, \tag{7.45}$$

where $\mathbf{p}_\perp$ is the momentum perpendicular to the field line, $p_\perp = mv\sin\alpha$, with α the instantaneous *pitch angle* of

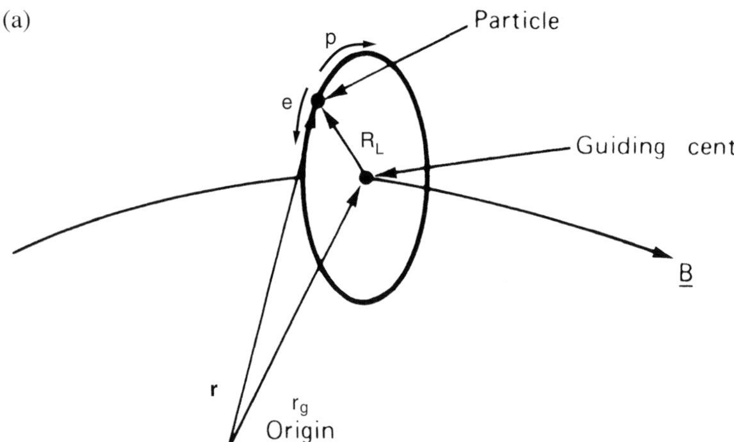

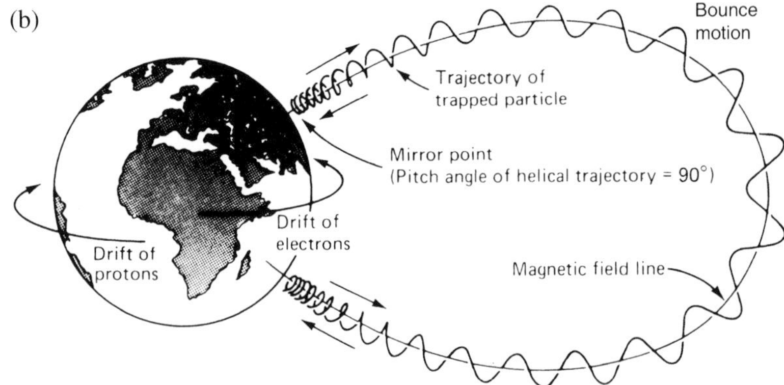

FIGURE 7.12 Basic motion of charged particles: (a) cyclotron (gyro) motion, (b) bounce motion along a field line and drift motion around the Earth. (Lyons and Williams 1984)

the particle, i.e., the angle between the direction of motion and the local magnetic field line. The particle 'orbits' the field line at an angular frequency n:

$$n = \frac{v_\perp}{2\pi R_L}. \tag{7.46a}$$

The *cyclotron, gyro* or *Larmor frequency* is defined by (Problem 7.8):

$$\omega_B \equiv 2\pi n = \frac{v_\perp}{R_L} = \frac{qB}{mc}. \tag{7.46b}$$

Typical values for the Larmor radius and period for a ~100 keV electron in the Earth's inner radiation belt are ~100 m and a few μs, respectively; for a proton these numbers are larger by a factor of 1836, the ratio of their masses.

If changes in the magnetic field are small over one gyro radius and during one gyro period, the particle gyrates in a nearly static magnetic field, and the magnetic flux, Φ_B, through a particle's orbit is constant: $d\Phi_B/dt = 0$. From this one can derive a constant of motion, known as the *first adiabatic invariant*, μ_B (Problem 7.9):

$$\mu_B = \frac{p_\perp^2}{2mB}, \tag{7.47}$$

which is valid for nonrelativistic as well as relativistic particles. For nonrelativistic particles the first adiabatic invariant is equal to the particle's magnetic moment, μ_b, induced by its circular motion around the magnetic field lines. The adiabatic invariant can be written in terms of the particle's energy, E:

$$\mu_B = \frac{E \sin^2 \alpha}{B}. \tag{7.48}$$

For relativistic particles the momentum:

$$p = \gamma_r m_o v, \tag{7.49a}$$

with m_o the particle's rest mass and γ_r the relativistic correction factor:

$$\gamma_r \equiv \frac{1}{\sqrt{(1 - v^2/c^2)}}. \tag{7.49b}$$

For relativistic particles the first adiabatic invariant in terms of a particle's energy becomes (Problem 7.9):

$$\mu_B = \frac{(E^2 + 2m_o c^2 E)}{2m_o c^2 B} \sin^2 \alpha. \tag{7.50}$$

Note that in the nonrelativistic limit equation (7.50) reduces to equation (7.48).

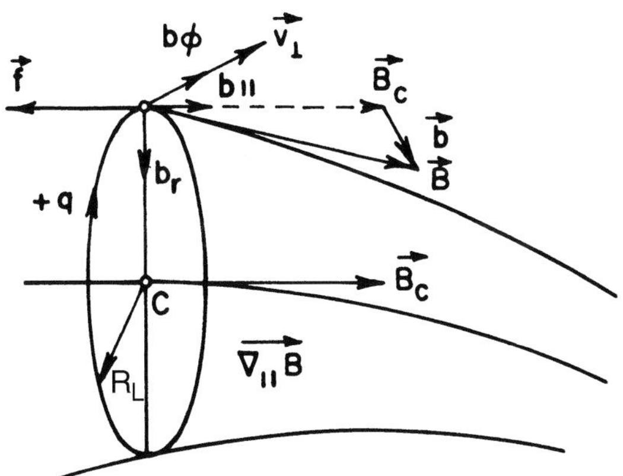

FIGURE 7.13 Effect of a field-aligned Lorentz force on a particle's helical motion along a field line. (Roederer 1970)

7.3.1.2 *Second Adiabatic Invariant*

Consider a particle's motion along the magnetic field line in the absence of electric fields, so that the particle's kinetic energy, E, is conserved. Conservation of the first adiabatic invariant shows that the pitch angle, α, increases when the particle moves to larger field strengths, until $\alpha = 90°$ at the *mirror point*. At this point the particle turns around, i.e., is 'reflected' back along the field line, as explained in physical terms below. The pitch angle is minimal at the magnetic equator. For particles confined to the planet's magnetic equator, the pitch angle $\alpha_e = 90°$. Particles which move along field lines with $v_\perp \neq 0$ are subject to a gradient in the field strength directed along the field lines, $\nabla_\parallel B$. This situation is sketched in Figure 7.13. Since the (dipole-like) field lines bunch up at higher latitudes, the particle experiences a field-aligned Lorentz force:

$$\mathbf{F}_{L_\parallel} = \frac{q\mathbf{v}_\perp \times \mathbf{b}_r}{c}, \tag{7.51}$$

where $\mathbf{b}_r$ is directed along the Larmor radius of the particle (Figure 7.13). $\mathbf{F}_{L_\parallel}$ is directed along $\mathbf{B}_c$, always in a direction opposite to $\nabla_\parallel B$. The Lorentz force decreases the particle's velocity along the field line, until the velocity reaches zero at the particle's mirror point, at which point the particle turns around and moves in the opposite direction. This instantaneous field-aligned Lorentz force causes the particles to bounce up and down the field lines, between the two mirror points.

If changes in the magnetic field are small over the time of one bounce period, one can derive a second constant of motion, known as the *second adiabatic invariant*, J_B, which is the integral of the particle's momentum along the field line, $p_\parallel = mv\cos\alpha$, between the mirror points s_m:

$$J_B = 2\int_{s_m}^{s'_m} p_\parallel \, ds. \tag{7.52}$$

A typical bounce period for a ~100 keV electron in the Earth's inner radiation belt is ~0.1 s. We define the integral $I_B \equiv J_B/(2mv)$, which is an invariant if the velocity v is constant:

$$I_B \equiv \frac{J_B}{2mv} = \int_{s_m}^{s'_m} \left(1 - \frac{B_s}{B_m}\right)^{1/2} ds, \tag{7.53}$$

with B_s and B_m the local field strength and the field strength at the particle's mirror point, respectively.

When a particle diffuses radially inwards, its energy increases since the first adiabatic invariant is constant. In equilibrium situations, the second invariant, $J_B = 2mvI_B$ is also constant. Thus, since the particle's kinetic energy, and hence v, changes, the integral I_B is changing as well. Since both the first and second adiabatic invariants are constant, their ratio should also be invariant:

$$K_B = \frac{J_B}{2\sqrt{m_o \mu_B}} = I_B\sqrt{B_m} = \int_{s_m}^{s'_m} [B_m - B_s]^{1/2} ds \\ = \text{constant}. \tag{7.54}$$

To calculate the equatorial pitch angle α_{e2} at position r_2, knowing α_{e1} at $r_1 > r_2$, one needs to evaluate the integral in equation (7.54). This can best be done by field line tracing or a numerical calculation. However, it is fairly easy to show that the pitch angle must increase while the particle is diffusing radially inwards (Problem 7.12). Hence, closer to the planet the equatorial pitch angles of the particles are generally closer to 90°.

7.3.1.3 *Third Adiabatic Invariant*

Particles drift around a planet along contours of constant magnetic field strength as discussed in more detail in Section 7.3.2. If changes in the magnetic field are small over timescales of one drift period, the magnetic flux, Φ_B, enclosed by the particle's orbit is a constant of motion, known as the *third adiabatic invariant*:

$$\Phi_B = \oint B\, ds. \tag{7.55}$$

In planetary magnetospheres, the first and second adiabatic invariants are usually conserved, but the third invariant is often violated.

7.3.2 Drift Motions in a Magnetosphere

Using equations (7.44) and (7.45), the position of a particle's guiding center, $\mathbf{r}_g$, can be written as:

$$\mathbf{r}_g = \mathbf{r} - \frac{mc}{qB^2}\mathbf{v} \times \mathbf{B}. \tag{7.56}$$

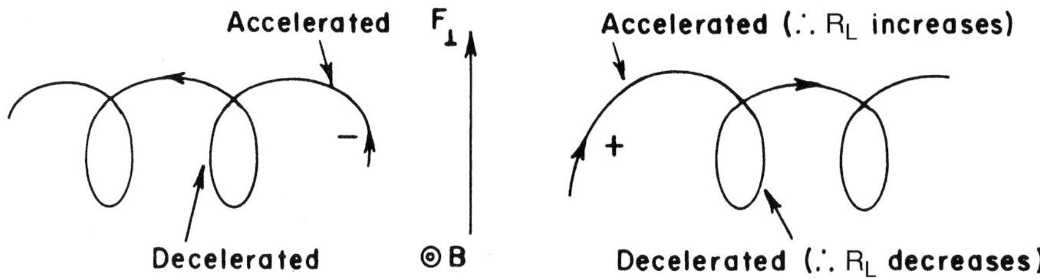

FIGURE 7.14 General motion of a particle in a magnetic field caused by a force perpendicular to the field line. (Roederer 1970)

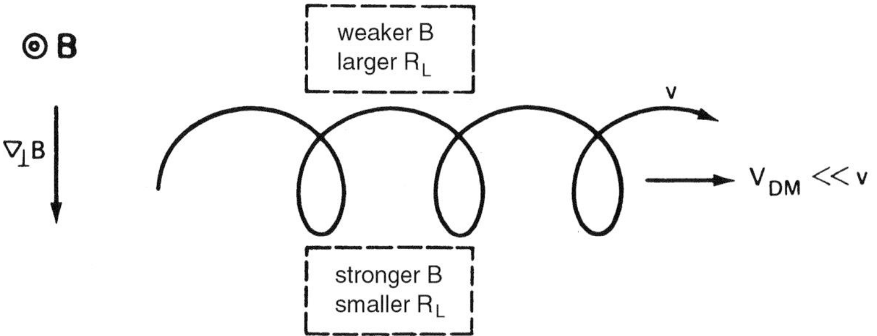

FIGURE 7.15 Drift motion of an equatorial ($\alpha_o = 90°$) particle due to a gradient in the magnetic field. (Lyons and Williams 1984)

Any instantaneous force $\mathbf{F}_i$ changes a particle's velocity:

$$d\mathbf{v} = \frac{1}{m}\int \mathbf{F}_i \, dt, \tag{7.57}$$

which results in a change in the position of the gyrocenter:

$$d\mathbf{r}_g = \frac{mc}{qB^2} d\mathbf{v} \times \mathbf{B} = -\frac{c}{qB^2}\mathbf{B} \times \int \mathbf{F}_i \, dt. \tag{7.58}$$

If $\mathbf{F}$ is the time average of $\mathbf{F}_i$, a change in the position of the guiding center leads to a drift in the particle's motion, with a drift velocity $\mathbf{v}_F$:

$$\mathbf{v}_F = \frac{c\mathbf{F} \times \mathbf{B}}{qB^2}. \tag{7.59}$$

Note that the drift velocity is perpendicular to both the force exerted on the particle and the magnetic field direction. In addition, protons and electrons drift in opposite directions for a given direction of $\mathbf{F}$. The general motion of a proton in a magnetic field under the influence of a force perpendicular to the field line is sketched in Figure 7.14. The drift motion of the particle is caused by the subsequent increases and decreases of the particle's Larmor radius.

7.3.2.1 *Gradient* **B** *Drift*

The field strength in a planetary magnetosphere decreases with increasing distance from the planet. The gradient in magnetic field strength induces a force:

$$\mathbf{F} = -\mu_b \nabla B, \tag{7.60}$$

with μ_b the magnetic moment, $\mu_b \equiv \mu_B/\gamma_r$. The force caused by the gradient in the magnetic field causes charged particles to drift around the planet. If changes in the magnetic field are small during a drift period, the magnetic flux enclosed by the drift orbit is conserved (the third adiabatic invariant, see Section 7.3.1.3). The drift velocity, $\mathbf{v}_B$, is equal to:

$$\mathbf{v}_B = \frac{\mu_b c \mathbf{B} \times \nabla B}{qB^2}. \tag{7.61}$$

Particles move in a direction perpendicular to the lines of force and perpendicular to the magnetic gradient. Note that the gradient in field strength consists of components parallel and perpendicular to the field lines: $\nabla B = \nabla_\parallel B + \nabla_\perp B$.

Consider a dipole field with particles confined to the magnetic equatorial plane, i.e., the equatorial pitch angle $\alpha_e = 90°$. Since the field strength is proportional to r^{-3}, the particles are subject to a force induced by the gradient in the field strength, perpendicular to $\nabla_\perp B$. This situation is sketched in Figure 7.15; the drift motion is caused by the subsequent increases and decreases in the Larmor radius of the particle due to the changes in field strength (note the difference with Figure 7.14). Protons and electrons drift in opposite directions, each along contours of constant magnetic field strength. In the Earth's magnetic field, electrons drift around the Earth in the eastward direction, protons towards the west. In the inner radiation belt it takes ~100 keV protons and electrons about 5–6 hours to

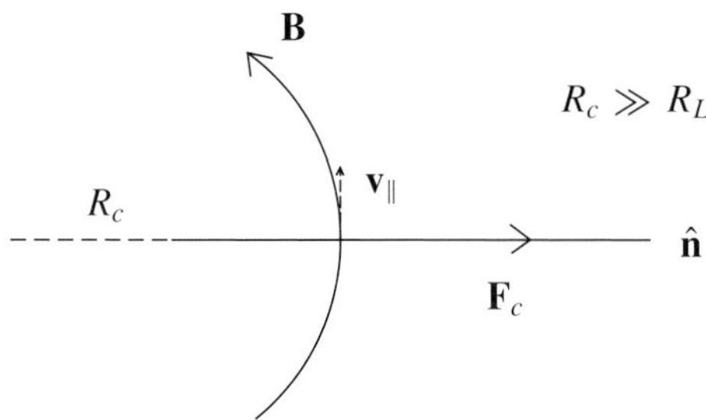

FIGURE 7.16 Effect of field line curvature on a particle's motion.

complete one drift orbit. The drift motion of these particles produces an electric current system known as the *ring current*. The magnetic field induced by this current (eq. 7.8) is oriented to the south, and thus reduces the magnetic field strength on the surface of the Earth.

The distance of a magnetic field line to the center of a planet is indicated by *McIlwain's parameter*:

$$\mathcal{L} \equiv \left(\frac{\mathcal{M}_B}{B_e}\right)^{1/3}, \tag{7.62}$$

with $\mathcal{M}_B$ the magnetic dipole moment and B_e the field intensity of the field line in the magnetic equator, or the minimum magnetic field strength along a given field line. For a centered dipole field, $\mathcal{L}$ represents the actual distance in planetary radii from the planet's center to the equatorial point of a field line. If radial diffusion is absent or very small, particles drift around the planet thereby tracing out *drift shells*. In the case of a pure dipole field, these drift shells are alike for particles at the same planetocentric distance, even if they have different equatorial pitch angles α_e. For a multipole magnetic field this is not the case and drift shells depend both on α_e and on the particle's starting point. This effect is called *shell degeneracy* or *shell splitting* (cf. Section 7.4.1.2).

7.3.2.2 *Field Line Curvature Drift*

Particles which move along field lines are subject to a field line curvature drift, in addition to the field aligned Lorentz force which causes the particle to bounce up and down field lines. If the guiding center follows the curved field line (Fig. 7.16), the centripetal force is equal to:

$$\mathbf{F}_c = \frac{m\mathbf{v}_\parallel^2}{R_c}\hat{\mathbf{n}}, \tag{7.63}$$

where $\hat{\mathbf{n}}$ is a unit vector outwards, along the direction of the field line's radius of curvature, R_c. The field line's radius of curvature is usually much larger than the particle's Larmor radius, $R_c \gg R_L$. This results in the curvature drift:

$$\mathbf{v}_C = \frac{mcv_\parallel^2}{qR_cB^2}\hat{\mathbf{n}} \times \mathbf{B}. \tag{7.64}$$

This drift motion is perpendicular to both the field line's radius of curvature and the field line itself.

7.3.2.3 *Drift Induced by Electric Fields*

When there is a drop in voltage or an electric field in a magnetosphere, particles are influenced by a force:

$$\mathbf{F} = q\mathbf{E}, \tag{7.65}$$

which results in a drift velocity:

$$\mathbf{v}_E = \frac{c\mathbf{E} \times \mathbf{B}}{B^2}. \tag{7.66}$$

Charged particles move in a direction perpendicular to the field lines, $\mathbf{B}$, and the electric field, $\mathbf{E}$; protons and electrons move in the same direction. Large scale electric fields play a significant role in planetary magnetospheres (Section 7.3.3). The $\mathbf{E} \times \mathbf{B}$ force, together with the ∇B and curvature drift forces, dominate the drift motion of particles in a magnetosphere.

7.3.2.4 *Drift Induced by the Gravitational Field*

Particles in a planetary magnetic field are also subject to forces from the planet's gravitational field, $\mathbf{F} = m\mathbf{g}_p$:

$$\mathbf{v_g} = \frac{mc}{qB^2}\mathbf{g}_p \times \mathbf{B}. \tag{7.67}$$

This gravitational 'perturbation' causes particles to move perpendicular to both the gravitational force and the magnetic lines of force. This drift motion is usually small compared to that due to electric fields and gradients in the magnetic field strength.

7.3.2.5 *Total Drift*

For a magnetic field with negligible currents, $\nabla \times \mathbf{B} = 0$, and the magnetic field gradient is related to the field line curvature by the equations:

$$\frac{1}{R_c} = \frac{\nabla_\perp B}{B}, \tag{7.68}$$

and

$$\hat{\mathbf{n}} = -\frac{R_c \nabla_\perp B}{B}. \tag{7.69}$$

In this case the total zeroth-order drift velocity, $\mathbf{v}_D$, can be written:

$$\mathbf{v}_D = \frac{c\mathbf{B}}{qB^2} \times \left(-\mathbf{F} + \frac{m}{2B}(v_\perp^2 + 2v_\parallel^2)\nabla_\perp B\right), \tag{7.70}$$

where $\mathbf{F}$ represents all external forces (e.g., electric fields, gravity, etc).

7.3.3 Electric Fields

The overall drift motion of charged particles in a magnetosphere is governed by the gradient in the magnetic field strength, field line curvature, and by the presence of electric fields. The first two forces cause particles to drift around the planet on $B =$ constant contours. Electric fields aligned along magnetic field lines accelerate electrons and protons/ions in opposite directions, resulting in a field aligned *current*. Parallel electric fields can only exist along field lines if there is a continuous supply of charged particles. Such conditions can occur in auroral regions (e.g., 'Birkeland' currents; Section 4.6.3) and in Io's plasma torus (Section 7.4.4). Electric fields perpendicular to magnetic field lines cause both ions and electrons to drift in the same direction, and thus the potential difference is not decreased by a current and such large scale electric fields can be stable. Two large scale electric fields which are present in each planetary magnetosphere are the corotational and convection electric fields.

7.3.3.1 *Corotational Electric Field*

The rotation of a planet's magnetic field induces an electric field in the radial direction: the *corotational electric field*. For a neutral particle or an observer in the magnetosphere, the magnetic field moves with a velocity $\mathbf{v} = \omega_{rot} \times \mathbf{r}$, with ω_{rot} the spin angular velocity of the planet and $\mathbf{r}$ the distance. This induces an electric field (Ohm's law, assuming the conductivity is large):

$$\mathbf{E}_{cor} = -\frac{\mathbf{v} \times \mathbf{B}}{c} = -\frac{(\omega_{rot} \times \mathbf{r}) \times \mathbf{B}_o}{cr^3}, \tag{7.71}$$

with B_o the surface dipole magnetic field strength. The direction of the electric field depends upon the direction of the magnetic field and the sense of rotation of the planet. For the Earth, the corotational electric field is directed inwards, for the giant planets outwards.

7.3.3.2 *Convection Electric Field*

The solar wind flowing past the Earth's field lines pulls them back, forming the magnetotail. This situation is sketched in Figure 7.17, where the interplanetary field is directed southwards, thus approximately antiparallel to the Earth's magnetic field. Reconnection (Fig. 7.10) must take place between the interplanetary and geomagnetic field lines. This gives rise to the creation of open field lines, which have one end attached to one of the polar regions on Earth, while the other end stretches out into interplanetary space. The interplanetary part of the field line is swept back around the Earth's magnetic field by the solar wind. The plasma on this flux tube senses an electric field $\mathbf{E} \propto \mathbf{v}_{sw} \times \mathbf{B}_{sw}$. For Earth, this field is directed from dawn to dusk. The field lines move in the antisolar direction through the locations numbered in Figure 7.17, and form the magnetotail. The return flow of the magnetic flux is achieved through reconnection in the tail. In this noon–midnight meridian view, two originally open field lines (number 6 on Fig. 7.17) reconnect in the tail and form a new closed field line. The Earth's rotation carries the field lines back to the sunside, while the open field lines (number 7′ on Fig. 7.17) continue to flow down the tail. The path of the numbered flux tube feet is shown on the inset in Figure 7.17. Because this circulation reminds one of thermal convection cells, the dawn-to-dusk electric field generated by this process is referred to as the *convection electric field*. On Earth, this field is directed from the dawn to the dusk side; on planets such as Jupiter, where the magnetic field is directed northwards, the convection electric field is directed from dusk to dawn.

The solar wind thus pulls the magnetic field lines back, in the antisolar direction, to form the magnetotail. The fast flowing solar wind is present on the interplanetary side of the magnetotail, while inside the tail there is magnetospheric plasma. When two fluids in contact try to slip past each other via a tangential discontinuity, the *Kelvin–Helmholtz instability* can be induced, which manifests itself as ripples on a surface. Other examples of the Kelvin–Helmholtz instability include a flag waving in the wind, and the ripples on a lake induced by the wind blowing over its surface. The fast solar wind tries to flow past the magnetosphere, thereby inducing ripples in the magnetospheric boundary or magnetopause. These ripples induce a component in $\mathbf{B}$ perpendicular to the solar wind flow; hence, locally an electric field is set up, which enhances the convection electric field. This mechanism also allows solar wind particles to diffuse into the planetary magnetosphere (Section 7.3.4).

As a result of the convection electric field, protons pile up at the dawn and electrons at the dusk side of the Earth's magnetosphere. The accumulation of charges at these positions is consistent with that induced by the magnetopause current, $\mathbf{J} \propto \nabla \times \mathbf{B}$. The total voltage drop, Φ_{conv}, between the dusk and dawn sides on Earth can be estimated:

$$E_{conv} = \frac{\Phi_{conv}}{2R_{pc}} = v_{pc}B_{pc}, \tag{7.72}$$

where the subscript pc stands for the open field lines across the polar cap, the area around the magnetic pole within the auroral oval. This results in a potential drop of 20–200 kV, with an average value of $\sim$50 kV (Problem 7.13). The drift motion of charged particles in the magnetosphere due to the convection electric field is in the sunward direction at low magnetic latitudes and towards the center plane (cur-

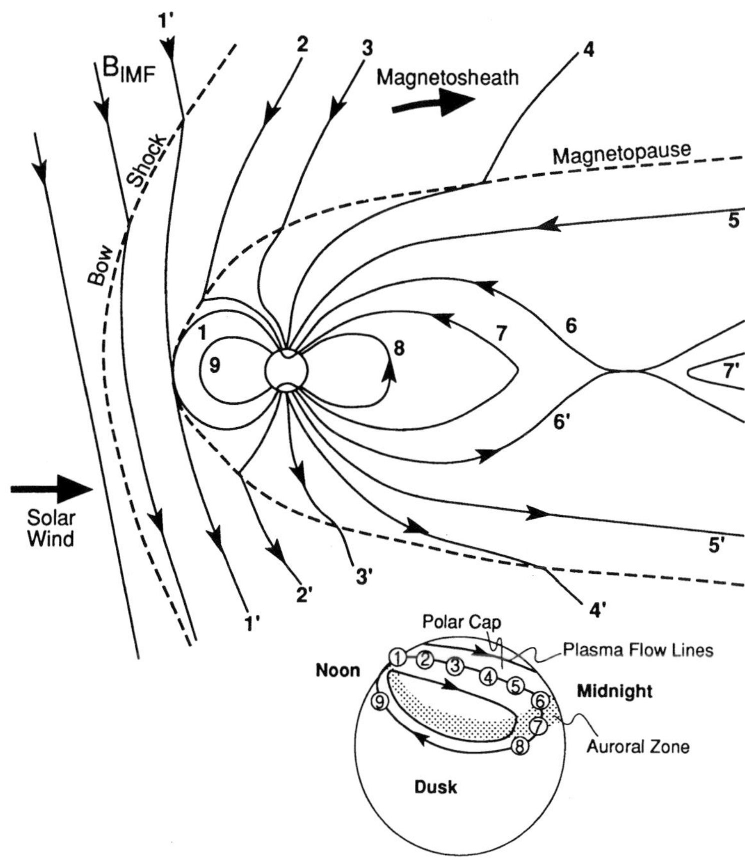

FIGURE 7.17 Interaction of the solar wind (southward directed magnetic field) with the Earth's magnetosphere. Reconnection takes place at the 'nose' of the magnetosphere, between field lines I and I′, and in the geomagnetic tail between 6 and 6′. The inset shows the path of the feet of the numbered field lines in the northern hemisphere. (Hughes 1995)

rent sheet) at high latitudes (Fig. 7.9). Hence, particles entering the magnetotail drift towards the Earth and sunside magnetopause. This dawn-to-dusk electric field thus induces a large scale global circulation of magnetospheric plasma. While particles drift towards a planet, their first and second adiabatic invariants are conserved, while the third invariant is usually violated. Hence the particle's energy increases inversely with the increase in magnetic field strength. The increase in energy is tapped from the electric field, and thus from the solar wind. Particles gain a considerable amount of energy this way. A typical solar wind particle has an energy of $\sim$10 eV ($T \sim 10^5$ K), while the interplanetary magnetic field strength at Earth is $\sim 5 \times 10^{-5}$ G. This translates into a first adiabatic invariant $\mu_B \approx 0.2$ MeV G^{-1}, which implies that solar wind electrons entering the Earth's magnetosphere gain energy by adiabatic diffusion alone up to nearly 0.2 MeV at the Earth's surface (Problem 7.14).

7.3.3.3 *Particle Drift*

A comparison of the strength of the corotational and convection electric fields shows whether magnetospheric circulation is driven primarily by the solar wind or the planet's rotation. In addition, the gradient in the magnetic field and field line curvature gives rise to another global particle drift. We can express the total drift for magnetospheric particles in the form:

$$\mathbf{v}_D = \frac{\mathbf{B} \times \nabla\Phi_{eff}}{B^2}, \tag{7.73}$$

where Φ_{eff} is the effective potential due to the convection and corotational electric fields and ∇B, respectively:

$$\Phi_{eff} = \Phi_{conv} + \Phi_{cor} + \Phi_{\nabla B}, \tag{7.74a}$$

where

$$\Phi_{conv} = -E_0 r \sin\phi, \tag{7.74b}$$

$$\Phi_{cor} = \frac{-\omega_{rot} B_0 R^3}{r}, \tag{7.74c}$$

$$\Phi_{\nabla B} = \frac{\mu_B B_0 R^3}{q r^3}. \tag{7.74d}$$

In the above equations, B_0 is the surface magnetic field strength at the magnetic equator, R is the planet's radius, r the planetocentric distance of the particle, and E_0 is the dawn-to-dusk electric field. The coordinate ϕ is measured in the magnetic equator from the planet–Sun direction to the dusk side. Since the convection potential, Φ_{conv}, is proportional to r and the potential due to the planet's rotation is proportional to r^{-1}, corotation dominates near the planet and solar wind induced convection is more important at larger distances. In magnetospheres of rapidly rotating planets with strong magnetic fields, such as Jupiter and Saturn, plasma circulation is likely dominated by the planet's rotation, while the solar wind controls the plasma flow in the smaller fields around more slowly rotating planets, such as Mercury (Problem 7.15). At Earth, the inner magnetosphere is controlled by rotation, while the outer magnetosphere is driven by the solar wind.

Given the various drift motions of charged particles in the Earth's magnetosphere, one can predict a particle's trajectory once its energy and initial location are known. Drift paths for protons and electrons that are injected into the Earth's magnetosphere at the dusk meridian with an energy of 1 keV, are displayed in Figure 7.18. Electric fields cause protons and electrons to drift in the same direction. The dawn-to-dusk electric field causes a general motion towards the Sun, while the corotational electric field causes the particles to drift around the Earth on closed equipotential contours. The gradient B drift causes protons to move westward and electrons eastward around the Earth. The gradient B drift motion of the electrons is in the same direction as the drift due to the corotational electric field. For low energy protons (energy < 1 keV), the electric field drift dominates, while for high energy protons (energy $>$ 100 keV), the gradient B drift is more important. Protons with intermediate energies may not completely orbit the Earth. The orbits in Figure 7.18b are for protons, injected with an energy of 1 keV at several locations along the dusk meridian. Starting at 3, 4 or 5 $R_\oplus$, the corotational field takes them eastwards around the Earth in orbits similar to those of electrons. The energy-dependent gradient B drift is negligible at all times. Protons which start between 5 and 7 $R_\oplus$ get sufficiently accelerated in their eastward drift that eventually the gradient B drift takes over and turns them around westwards, on the same evening side of the Earth. The proton is then decelerated, and at some point the electric field drift takes over again. These particles thus follow closed drift paths which do not encircle the Earth. At larger distances the corotational field always dominates (Problem 7.16).

7.3.4 Particle Sources and Sinks

7.3.4.1 *Sources of Plasma*

There are several sources for magnetospheric plasma, and the relative contributions of each source vary from planet to planet. Charged particles can in principle originate in cosmic rays, the solar wind, the planet's ionosphere or on satellites/rings which are partially or entirely embedded in the magnetosphere. Although ionospheric particles are usually gravitationally bound to the planet, some charged particles escape along magnetic field lines into the magnetosphere (Section 4.8.2). Sputtering by micrometeorites and charged particles may cause ejection of atoms and molecules from moons/rings into space (Section 4.8.2); if such particles become ionized, they enrich the magnetospheric plasma. A planetary magnetosphere is embedded in the solar wind; simple entry of solar wind particles into the magnetosphere would populate a planet's magnetic field with solar wind plasma. The detection of protons and electrons, and sometimes helium nuclei (Earth, Jupiter) in each magnetosphere strongly suggests the solar wind to be a rich source of plasma. Both solar and galactic cosmic rays can also enter the magnetosphere. Their access is mainly at high latitudes, and is energy dependent. Interplanetary particles can enter a magnetosphere via the following processes:

(1) Interplanetary magnetic field lines are directly connected to the planet via 'open' magnetospheric field lines, and thus interplanetary particles can spiral down the field lines into a planet's polar cusp. During periods of enhanced solar activity, numerous particles come down along the field lines into the atmosphere, where they enhance atmospheric ionization and cause auroral displays.

(2) Whenever the interplanetary field has a component antiparallel to the planetary field, magnetic reconnection may occur (Fig. 7.10): field lines merge together, and solar wind particles can enter the magnetosphere through the magnetic neutral points. Field line reconnection may occur at the day-side magnetopause (Section 7.4.1) and in the magnetotail neutral sheet.

(3) Charged particles can diffuse or gradient/curvature drift from the solar wind into the magnetosphere. The particles may also enter the magnetosphere as a result of the Kelvin–Helmholtz instability. These processes are discussed in more detail in Section 7.4.1.

7.3.4.2 *Particle Losses*

Moons, rings and atmospheres are both sources and sinks of magnetospheric plasma. Particles which hit the surface of a solid body are generally absorbed and lost from the magnetosphere. Similarly, if a particle enters the collisionally thick part of an atmosphere, it gets 'captured' and won't return to the magnetosphere. Charged particles carry

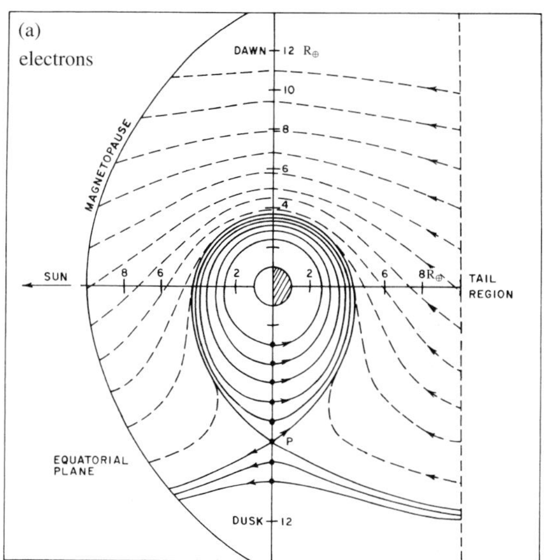

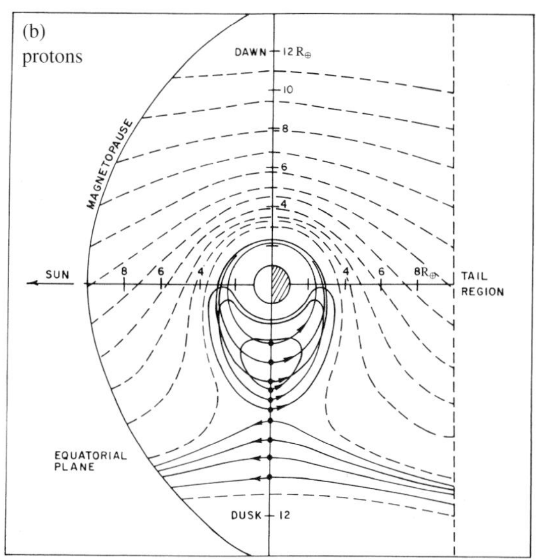

FIGURE 7.18 Motion of charged particles with $\alpha_e = 90°$ in the Earth's magnetic field. Broken lines: equipotentials for the dawn-to-dusk electric field. These curves represent the drift paths (in the direction of the arrows) for 'zero' energy particles in the convection region. Solid lines: drift paths of electrons (panel a) and protons (panel b) injected with an energy of 1 keV along the dusk meridian (at the dots). (Roederer 1970)

out a helical bounce motion along magnetic field lines, as sketched in Figure 7.12 and discussed in detail below. The particle is reflected at the mirror point, and if this point lies in the ionosphere/atmosphere, the numerous collisions with atmospheric particles 'trap' the magnetospheric particle in the atmosphere. The location of this mirror point depends upon the particle's initial pitch angle, α_e, which is the angle between the direction of motion of a particle in the magnetic equator and the local magnetic field line there. We define a particle's *loss cone*, α_l, as the smallest pitch angle an equatorial particle can have without being absorbed. We will show below that particles with equatorial pitch angles $|\alpha_e| \leq |\alpha_l|$, or $\sin^2 \alpha_e < \sin^2 \alpha_l$, have their projected mirror points within the atmosphere. These particles are thus lost from the magnetosphere.

Another loss process is induced by charge exchange of magnetospheric ions. The ions are forced to corotate with the planet's magnetic field. Their velocity is therefore very high, and increases to values much higher than the keplerian velocity of a neutral particle at large planetocentric distances (Problem 7.11). If this ion undergoes a charge exchange with a neutral, the newly formed ion picks up the corotation speed and stays trapped in the magnetosphere. The former ion, however, becomes a fast neutral and can be flung out of the system by its large centrifugal force if the corotation speed exceeds the escape velocity. This is a primary loss mechanism for magnetospheric ions.

7.3.5 Particle Diffusion

Diffusion of particles in and through a magnetosphere is an important process: without particle diffusion the radiation belts would probably be empty, unless there is an *in situ* source. *Radial diffusion* displaces particles across field lines, while *pitch angle diffusion* moves a particle's mirror point along field lines. Whereas the first mechanism transports particles from their place of origin to other regions in a magnetosphere, pitch angle scattering can be regarded as a principal means to lose particles. It causes the particles' pitch angle distribution to spread, forcing a number of particles into the loss cones.

It is convenient to express the particle density in terms of its *phase space density* $f_p(x, y, z, p_x, p_y, p_z)$. *Liouville's theorem* states that f_p is constant along the dynamical path of a particle:

$$f_p = \frac{dN}{dx\,dy\,dz\,dp_x\,dp_y\,dp_z} = \text{constant}. \tag{7.75}$$

By reorienting the coordinate system such that the z-coordinate is in the direction of the particle's motion, $dz = v\,dt$, $dx\,dy = dA$, and $dp_x\,dp_y\,dp_z = p^2\,dp\,d\Omega_s$, with p the particle's motion and Ω_s the solid angle. Since the differential energy flux, j, (which is measured by a detector in space), is:

$$j = \frac{dN}{dA\,dt\,d\Omega_s\,v\,dp}, \tag{7.76}$$

we can relate the phase space density to the differential energy flux:

$$f_p = \frac{j}{p^2} = \text{constant}. \tag{7.77}$$

Particle diffusion can be represented by:

$$\frac{\partial f_p}{\partial t} = \sum_{ij} \frac{\partial}{\partial J_i}\left(D_{ij}\frac{\partial f_p}{\partial J_i}\right), \tag{7.78}$$

with J_i the action variable (closely associated to the adiabatic invariants μ_B, J_B and/or Φ_B) and D_{ij} the tensoral diffusion coefficient. In radial diffusion the third adiabatic invariant, Φ_B, is violated; if particles diffuse in pitch angle, μ_B and/or J_B are violated.

In the case of pure radial diffusion, the time derivative in the phase space density becomes:

$$\frac{\partial f_p}{\partial t} = \mathcal{L}^2\frac{\partial}{\partial \mathcal{L}}\left(\frac{1}{L^2}D_{\mathcal{LL}}\frac{\partial f_p}{\partial \mathcal{L}}\right) + Q - S, \tag{7.79}$$

with Q and S the source and loss terms respectively, and $D_{\mathcal{LL}}$ the radial diffusion coefficient. The diffusion coefficient is often approximated by $D_{\mathcal{LL}} = D_o\mathcal{L}^n$, where the value of n is indicative of a particular type of diffusion process.

Radial diffusion is driven by large scale electric fields in a magnetosphere. One example is the solar wind induced convection field, which plays a dominant role in particle diffusion in the quiescent terrestrial magnetic field. We also discussed the corotational electric field. Any plasma instabilities in the neutral sheet and temporal variations in the magnetic field induce stochastically varying electric fields (eqs. 7.6, 7.7). Sudden impulses in the magnetic field may dominate particle diffusion in the Earth's magnetosphere. We note that fluctuating electric and magnetic fields cause the particles to 'random walk' through the magnetosphere, so that they may diffuse inwards and outwards at times. One of the primary loss mechanisms is diffusion or scattering into a particle's loss cone. With a particle source somewhere in the outer magnetosphere, and a sink in the planet's atmosphere, the overall diffusion of particles is inwards. This has typically been observed *in situ* by spacecraft.

It has been shown that if diffusion is driven by sudden magnetic impulses, the diffusion coefficient, $D_{\mathcal{LL}} = D_o\mathcal{L}^n$, shows an $\mathcal{L}$-dependence with $n = 10$. Random variations of potential electric fields lead to radial diffusion with $n = 6$. Winds in a planet's upper atmosphere/ionosphere induce fluctuating electric fields which probably form the driving mechanism for particle diffusion in Jupiter's inner radiation belts; the diffusion coefficient in this case has a dependence with $n \approx 3$.

When particles diffuse radially inwards, conservation of the first and second adiabatic invariants shows that the particle's equatorial pitch angle increases; thus we expect particles to be more confined to the magnetic equator closer to the planet. In contrast to this slow change in a particle's pitch angle, stochastic variations have also been observed. Such a stochastic pitch angle diffusion can be caused by collisions with other particles (Coulomb scattering with atmospheric particles, charge exchange) or via wave–particle interactions.

7.4 Magnetospheres of Individual Bodies

7.4.1 Earth

A sketch of the Earth's magnetosphere was shown in Figure 7.9. At the sunside, at ~15 $R_\oplus$, we find the bow shock, which results from the interaction of the supermagnetosonic solar wind with the Earth's magnetosphere. The turbulent subsonic region behind the bow shock is the magnetosheath, which is shielded from the Earth's magnetic field by the magnetopause, a boundary at ~10 $R_\oplus$ which separates the solar wind plasma from the terrestrial magnetic field. The solar wind drags the field lines back and forms the magnetotail. The midplane, where magnetic fields of opposite polarity meet, is a region of field reversal, where reconnection takes place. The near-zero magnetic pressure here is balanced by a higher plasma pressure. This region is referred to as the current sheet, embedded in the plasma sheet. Closer to Earth we find the plasmasphere and the radiation or Van Allen belts, regions of stable particle trapping.

As described in detail in Section 7.3, charged particles in a magnetosphere describe helical paths around field lines, being reflected at their mirror points. The particles' drift around the Earth is caused by gradients in the magnetic field strength and curvature of the field lines. Their drift orbits are modified by the presence of electric fields, in particular the corotational and convection fields. For a pure dipole field, the particle trajectories can be expressed analytically; as planetary magnetic fields are observed to be complex, the particle trajectories need to be described numerically. In addition, the magnetosphere responds continuously to changes in the solar wind, complicating detailed modeling of the field and plasma therein.

7.4.1.1 *Nondipolar Magnetic Field and the South Atlantic Anomaly*

In the inner radiation belts, the most important departures from a dipole field result from the higher order moments in the Earth's internal magnetic field. On the Earth's surface,

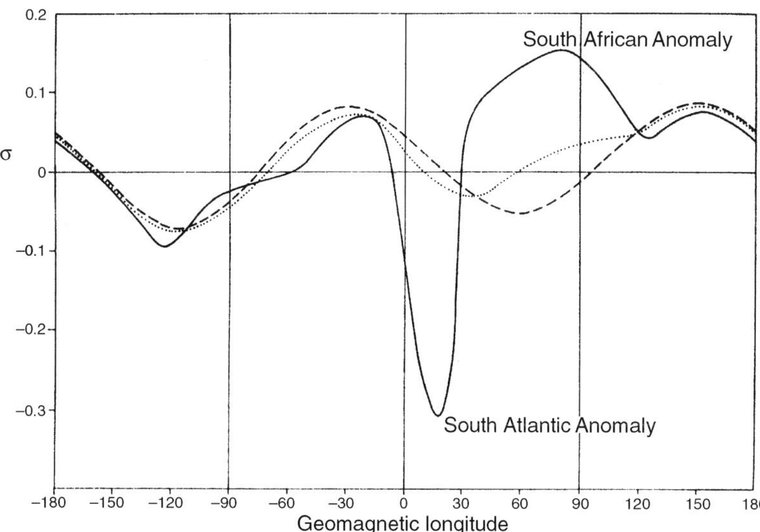

FIGURE 7.19 Deviation of a particle's drift path in the Earth's magnetic field, compared to the drift orbit in a pure dipole field, at distances $\mathcal{L} = 1$ (——),2 (. . . . . .) and 7 (- - - - -). $\sigma = \mathcal{L}^2 \left(\frac{\partial^2 B/\partial s^2}{\partial^2 B/\partial s^2|_{dipole}} - 1 \right)$. (Adapted from Roederer 1972)

the magnetic field is weakest near the east coast of South America, a feature of the Earth's field which is known as the *South Atlantic Anomaly*. Because particles drift around the Earth on paths along which the magnetic field is constant, they venture much closer to Earth's surface in the South Atlantic Anomaly than at other longitudes. Figure 7.19 shows the deviation of drift contours of equatorially confined particles in the Earth's magnetosphere (including only Earth's internal magnetic field) compared to that in a pure centered dipole field, at distances $\mathcal{L} = 1$, 2 and 7. It is clear that the largest deviations occur close to the planet, and the region of the South Atlantic Anomaly stands out as one where the particle drift orbits are at much lower altitudes. The fact that energetic Van Allen belt particles pass Earth at such low altitudes in the South Atlantic Anomaly can substantially affect the performance of spacecraft in low Earth orbit as they pass through this region (Problem 7.18). At $L = 7$, the largest variations are caused by the offset of the dipole field from the center of the Earth (note: this ignores the field modifications by magnetospheric currents, which are substantial at this distance).

As discussed in Section 7.3, particles may get lost in the planet's atmosphere when they venture too close to Earth. Along a particle's drift orbit, its loss cone varies with geocentric longitude. A particle's loss cone is largest near the South Atlantic Anomaly, and therefore all particles which have an equatorial pitch angle which is less than the loss cone at the South Atlantic Anomaly are removed from the magnetosphere during their drift around Earth. When a particle drifts around a planet in an offset dipole field, the particle's mirror point may lie well above the planet's atmosphere at certain longitudes, but within the atmosphere at others. If the field is displaced towards the north or south, the atmospheric loss cone is asymmetric; obviously the largest loss cone a particle encounters during its drift orbit regulates the trapped particle distribution. This loss cone is referred to as the *drift loss cone*.

7.4.1.2 *Magnetospheric Currents and Their Effects on the Magnetosphere*

The magnetic gradient and curvature drift cause protons in the Earth's magnetic field to drift westwards and electrons eastwards around Earth, which induces a large scale current: the *ring current*. *Partial ring currents* flow partway around the Earth in the middle magnetosphere. The ends of the partial rings are connected to the ionosphere via field-aligned or *Birkeland currents*, where currents in the ionosphere complete the circuit. In the magnetopause we find the *Chapman–Ferraro currents* in the eastward direction, currents which are named after Chapman and Ferraro who first proposed their existence and who first modeled the geomagnetic cavity within the high speed solar wind. The convection electric field induces the dawn-to-dusk *tail current* across the magnetotail, near the equatorial plane. The various current systems are sketched schematically in Figure 7.20. Since electric currents induce magnetic fields, the various current systems can cause significant perturbations in the geomagnetic field compared to the internal field structure. For example, the ring current induces a southward field inside the particle orbits, which weakens the magnetic field here, as verified from measurements on Earth's surface. The Chapman–Ferraro current induces a northward field, which strengthens the geomagnetic field on the day side, whereas the westward tail current tends to weaken the magnetospheric field in the tail. These latter

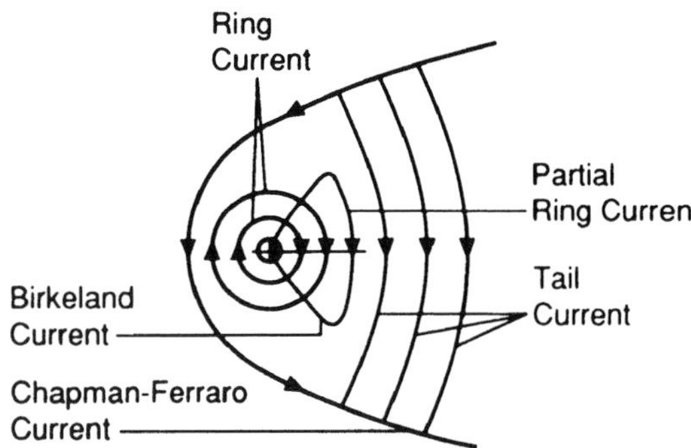

FIGURE 7.20 Major current systems in the Earth's magnetic field. In this figure one looks down from the north pole onto the magnetic equatorial plane. The Sun is to the left. (Wolf 1995)

two currents thus cause a systematic difference between the day and night side magnetosphere, which affects particle drift orbits. Moreover, the effect is different for particles with different pitch angles. Particles with small equatorial pitch angles drift farther from the Earth on the night side than on the day side, while the opposite is true for particles which are confined to the magnetic equatorial plane. This phenomenon of *drift-shell splitting* was discussed in Section 7.3.2.1.

7.4.1.3 *Magnetospheric Plasma*

The Earth's magnetic field is populated mainly by protons, electrons and ions of oxygen, helium and nitrogen. This composition indicates that the particles originate from both the solar wind and the ionosphere. The various plasma regions are indicated in Figure 7.9. Magnetosheath plasma penetrates to low altitudes in the 'polar cusps', which are effectively dents or gaps in the magnetopause. The open field lines in the polar caps provide a route for magnetosheath plasma to enter the magnetosphere, and thus populate the high latitude regions. Upon reflection at their mirror points, cusp particles can enter the geomagnetic tail and form the plasma mantle (together with magnetosheath plasma which 'leaks' through the tail magnetopause). As a result of the solar wind induced convection electric field, there is a large scale magnetospheric 'circulation', such that particles at low latitudes are driven in the direction of the Sun, with an anti-solar flow at higher latitudes. The particles in the plasma mantle thus drift towards the midplane, while the particles in this current sheet either drift towards the Earth (when Earthwards from the neutral X point, where tail reconnection is occurring – Fig. 7.17), or away from it. In the latter case, the particles ultimately join the solar wind. Plasma in the current sheet is accelerated and fills the plasma sheet with hot plasma, as discussed below. A *polar wind*, consisting of low energy protons and singly ionized oxygen atoms, flows out from the high latitude ionosphere. Like the polar wind, the *plasmasphere* inside $L \lesssim 4$ is filled with cold ionospheric plasma.

Plasma Sheet and Current Sheet

Typical plasma parameters in the current sheet, the central part of the plasma sheet, are ~ 0.3 particles cm^{-3}, with ion energies of a few up to a few tens keV, and electron energies a factor of 2–3 lower. This is much higher than the typical energy of magnetosheath (few hundred eV) or suprathermally escaped ionospheric (few eV) plasma. Hence, the plasma in the plasma sheet and current sheet must be energized somehow. Since the convection electric field is parallel to the tail current, this field energizes the particles carrying the tail current. Thus, the particles in the current sheet derive their energy from a slowing down of the solar wind particles. The energy input is $\sim$2–20% of the total energy flux carried by the solar wind through an area equal to the cross-section of the day-side magnetopause (Section 7.3.3.2; Problem 7.13). To understand how the energy is transferred to the magnetospheric particles in the plasma sheet, one has to consider the particle trajectories in the magnetotail. Since the magnetic field changes dramatically across the current sheet (B changes sign and is ≈ 0 at the center), conservation of the adiabatic invariants is no longer valid. Above and below the current sheet the particles undergo their normal helical motion around the field lines. However, when they approach the region of field reversal, their circular motion changes depending on the direction of the field. This causes an oscillatory motion of the particle within the current sheet, whereby the particle gets essentially trapped in this region. Electrons move towards the dawn and protons towards the dusk side. The dawn-to-dusk convection electric field accelerates the particles in this direction, thus energizing them. If the field B is indeed zero in the midplane, the particles are trapped and continually energized, until they reach the edge of the current sheet and are lost from the magnetotail. Energetic particles are then only found in the current sheet, not in the broader plasma sheet. However, the magnetic field has a small northward component in the midplane, which deflects the particles in the current sheet towards the Earth and out of the midplane. The energy input of particles in the current sheet thus forms an important energization mechanism for all particles in the plasma sheet. We note that only particles entering the current sheet at the Earth side of reconnection points are energized and trapped in the magnetosphere, while particles entering the current sheet in the tail side of the reconnection point, are carried down the tail and lost from the magnetotail.

Plasmasphere

The *plasmasphere* is located inside $\mathcal{L} \lesssim 4$, the same region in space as occupied by the radiation or Van Allen belts. The plasmasphere is separated from the plasma sheet by the *plasmapause*. The plasmasphere is filled with cold dense plasma from the ionosphere. The particle motion is dominated by the corotational and convection electric fields. As pointed out in Section 7.3.3, the drift trajectories of protons with ~keV energies, starting at 3–5 $R_{\oplus}$ on the dusk side, are closed paths which do not encircle the Earth. This leads to a bulge in the plasmasphere on the dusk side (Fig. 7.18), which is in qualitative agreement with observations. Since the strength of the convection electric field is determined by the solar wind, its strength fluctuates continuously in response to changes in the interplanetary magnetic field and solar wind. Therefore, the size and shape of the plasmasphere are not constant over time. If the convection electric field is suddenly increased, the plasmapause moves inwards, so that particles which were originally inside the plasmasphere are now on trajectories which cause them to drift towards the day-side magnetopause. In addition, the bulge of the plasmasphere rotates somewhat towards local noon.

The more energetic particle drifts are dominated by the gradient in the magnetic field (Section 7.3.2.1), which drives positively charged particles towards the west, and electrons towards the east. This gives rise to the ring current, as discussed above (Sections 7.3.2.1, 7.4.1.2). One can show that the more energetic particles therefore cannot penetrate as close to the Earth as the lower energy (colder) particles, so that the hotter plasma from the plasma sheet does not penetrate the plasmasphere. However, radial diffusion of plasma, caused by fluctuations in electric fields, such as the convection electric field, tends to decrease any radial gradients in the particle populations.

In addition to a reduction of the plasmasphere in size in response to an increase in the convection electric field, the plasmapause is deformed, moving in closer to Earth at the night side, but farther away from Earth at the dawn and dusk sides. This deformation induces field-aligned or Birkeland currents to flow up from the ionosphere at the dawn side, and down into the ionosphere at dusk (Fig. 7.20). These form the partial ring current in the Earth's middle magnetoshere, and induce a dusk-to-dawn directed electric field across the inner magnetosphere, which effectively shields the inner magnetosphere from the convection (dawn-to-dusk) electric field. The intensity of ionospheric electric fields is indeed much smaller equatorwards of the auroral zone compared to that at higher latitudes.

Van Allen Belts

The Earth's radiation belts, discovered by James A. Van Allen during one of the first spacecraft flights, are a region in space filled with energetic particles which can penetrate deep into dense materials and thus cause damage to spacecraft instruments and humans. All the particles in the Van Allen belts contribute to the ring current, discussed above. The Van Allen belts consist of two main belts, as indicated on Fig. 7.21a. The inner belt, centered around $\mathcal{L} \approx 1.5$, is characterized by highly energetic protons and electrons (Fig. 7.21b), whereas the outer belt is devoid of energetic protons. There is a region at $\mathcal{L} \approx 2.2$ where the number density of energetic electrons is a minimum, caused by the interaction of the electrons with whistler mode waves (Fig. 7.21b; Section 7.6.3). A third, less intense, belt exists at $\mathcal{L} \approx 2$; here we find multiply charged ions of oxygen, with smaller amounts of nitrogen and neon, and very little carbon. All of these ions have energies (well) over 10 MeV.

There are a variety of sources which may contribute to the particle population in the Earth's radiation belts. The Earth's ionosphere and the solar wind form two important sources. However, as pointed out earlier, typical solar wind particles have too low an energy to contribute to the most energetic particle (hundreds of MeV) distribution in the inner magnetosphere. Protons with such high energies may originate from energetic neutrons, produced in the atmosphere through collisions with cosmic rays. When these neutrons escape from the atmosphere, they may decay while en route through the magnetosphere. This decay process leads to the creation of very energetic protons (few 100 MeV) and electrons of lower energies (few 100 keV). Alternatively, these highly energetic protons might also be injected and/or accelerated by interplanetary shock waves, such as those induced by coronal mass ejections or solar flares, which propagate through the geomagnetic field (Section 7.1.5). The Solar Anomalous Magnetospheric Particle EXplorer (SAMPEX) discovered a belt of ultrarelativistic (peaking in energy at 8–15 MeV) electrons ($2.5 < \mathcal{L} < 5$), the origin of which has been attributed to an interplanetary shock event on 21 February 1994. The ultrarelativistic electrons were seen to form a belt right after the interplanetary shock event, that persisted and migrated inward for years afterward.

The relative abundances of the multiply charged ions of C:N:O which make up the radiation belt at $\mathcal{L} = 2$ strongly suggest that these ions originate from anomalous cosmic rays (ACR). Interstellar atoms near the Sun may get ionized by solar UV radiation or by collisions with energetic interplanetary particles. Once ionized, the particles are swept away from the Sun with the solar wind. At the heliospheric boundary these particles are accelerated again,

(a)

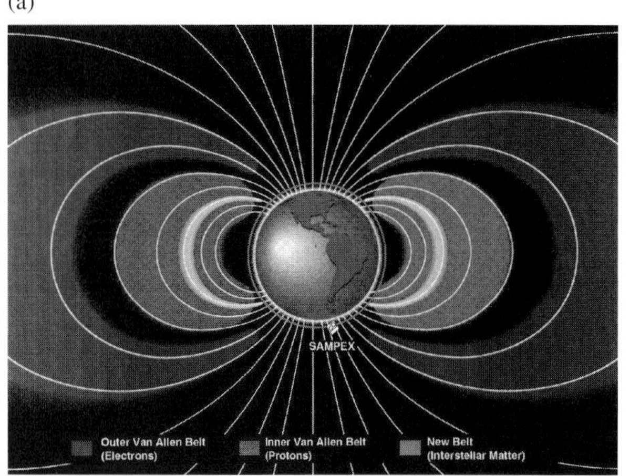

FIGURE 7.21 (a) Schematic representation of the Earth's Van Allen belts, where the inner and outer belts are characterized by a bimodal distribution in electron energies. A narrow belt within the inner belt (at $\mathcal{L} = 2$) is comprised of anomalous cosmic rays (ACR) trapped from the interstellar medium. (Mewaldt *et al.* 1997) (b) Spatial distribution of energetic electrons and protons in the Van Allen belts during a quiescent phase in solar wind activity. (Wolf 1995) (c) The variation in trapped magnetospheric 'ACR' (upper line) and interplanetary ACR fluxes, as observed by various instruments on SAMPEX (Solar Anomalous Magnetospheric Particle EXplorer) and ACE (Advanced Composition Explorer). (Courtesy: R.A. Mewaldt and R.S. Selesnick; NASA)

(b)

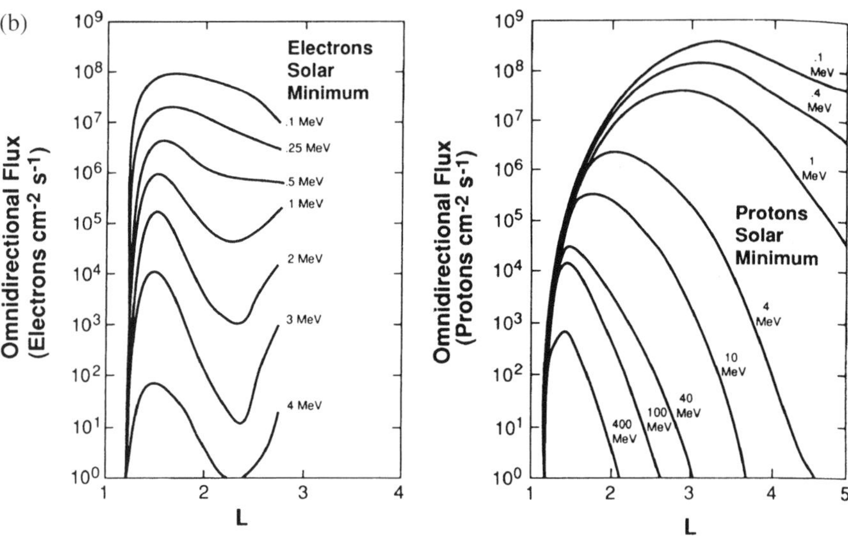

(c)

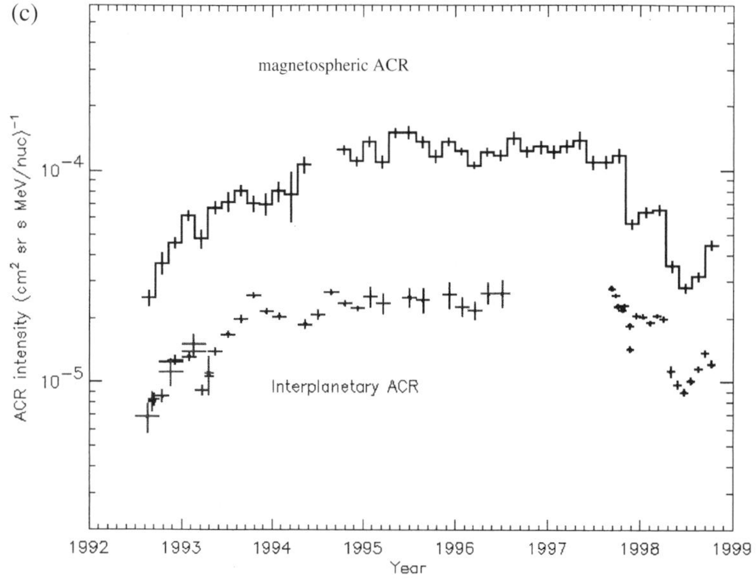

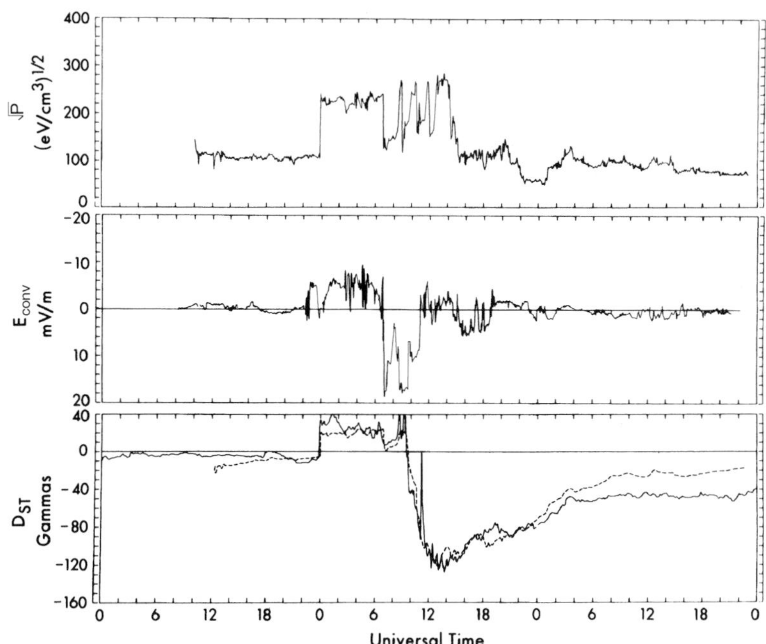

FIGURE 7.22 Effects of a magnetic storm on Earth (bottom panel), as recorded in the D_{st} index. The top panel shows the solar wind dynamic pressure, and the middle panel the solar wind dusk-to-dawn electric field. (Adapted from McPherron 1995)

after which they can re-enter the heliosphere as low energy cosmic rays. At Earth such anomalous cosmic rays, heavy element ions with energies below a few 100 MeV (which is low compared to galactic cosmic rays) have been detected by spacecraft (e.g., SAMPEX). The singly ionized particles penetrate the magnetosphere, and if they lose more electrons by ionizing collisions in the upper atmosphere, they can become trapped in the radiation belts. Spacecraft observations of the ACR abundance show a strong correlation in abundance with the solar cycle (i.e., anti-correlated with the solar sunspot number). SAMPEX observations showed an immediate response in the heavy ion density with the interplanetary ACR flux, resulting in long term (years) variations by factors of up to 3–4 in the ACR and heavy ion radiation belt population (Fig. 7.21c).

7.4.1.4 *Magnetospheric Storms and Substorms*

It is probably not surprising that with increases in solar wind intensity, in particular when the interplanetary magnetic field (IMF) turns southwards (i.e., opposite to the Earth's field, Fig. 7.17), large changes are induced in the magnetosphere. The ordered sequence of changes which are induced when the IMF turns southward and when the energy flow from the solar wind to the geomagnetic field increases are referred to as *magnetospheric substorms*. One of the first visible phenomena during a substorm is enhanced auroral activity, with an equatorward movement of the auroral zone (Section 4.6.4). A substorm typically lasts for about an hour, and may repeat on a $\sim$few hour timescale. When a long duration (hours to days) period of unusually large southward IMF, sometimes accompanied by periods of high dynamic pressure in the solar wind, occurs (typically during passage of a coronal mass ejection (CME) generated disturbance), a large reduction in Earth's surface field is measured. Such geomagnetic or magnetospheric disturbances are known as *magnetic storms*. A magnetic storm is characterized by the D_{st} index, the instantaneous worldwide average of the disturbance in the magnetic field strength. A typical example of a magnetic storm signature is shown in Figure 7.22. Since a strong CME disturbance in the solar wind is usually preceded by an interplanetary shock followed by enhanced density and velocity, the field strength first increases when the disturbance hits the magnetosphere, inducing an increase in the magnetopause current. Several hours (up to over 25 hrs) later the field strength, D_{st}, decreases dramatically during the storm's *main phase*, which typically lasts for about a day. The main phase is caused by an increase in the ring current, resulting from an enhanced particle flow towards the Earth, that weakens the Earth's surface field. The subsequent recovery phase can last for many days, and is caused by a gradual loss of particles from the radiation belts, due e.g., to pitch angle scattering into the loss cone by enhanced wave activity and radial diffusion. Although these large scale phenomena can be explained in a qualitative way, no theories have yet been developed which explain the detailed sequence of all events triggered during a magnetospheric substorm.

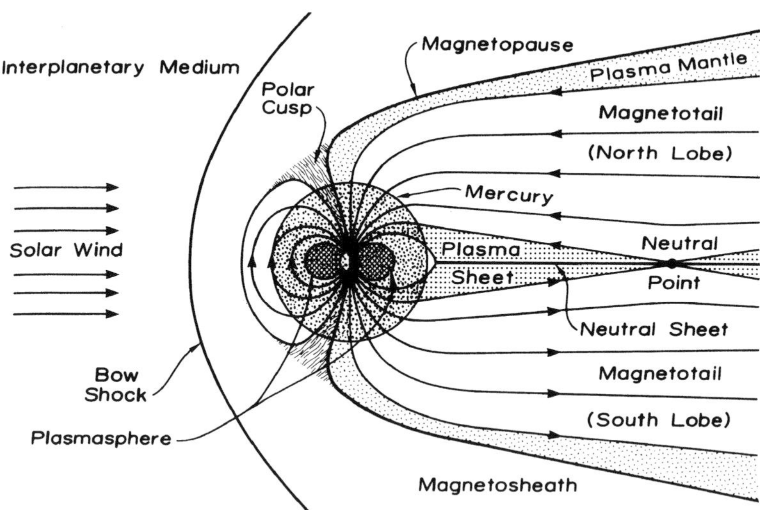

FIGURE 7.23 Mercury scaled such that its magnetosphere occupies the same volume as the Earth's magnetosphere. A large fraction of the planet's inner magnetosphere and radiation belts would reside inside the planet. (Russell *et al.* 1988)

As mentioned here and in Section 7.4.1.3, at times of high magnetic activity, the plasmapause is closer to the Earth than at quiescent times. After a magnetic storm, during the recovery phase, the outer plasmasphere refills gradually, over a period of a couple of days. The Earth's ionosphere is the primary source of oxygen and nitrogen, while both the ionosphere and solar wind contribute to the proton density. During quiescent times, the O^+/H^+ ratio in the plasma sheet is low, while during periods of enhanced magnetic activity, the concentration of oxygen ions is nearly as high as that of H^+. The ring current O^+ contribution also increases. This suggests that the particle population in the magnetosphere is dominated by solar wind particles during quiescent times, while during active times a large fraction of the particles comes from the Earth's ionosphere.

7.4.2 Mercury

Based on two close encounters of the planet by the Mariner 10 spacecraft, we know that Mercury possesses a small Earth-like magnetosphere. The magnetic axis is within 10° of Mercury's rotational axis. Under usual solar wind conditions, Mercury's intrinsic field is strong enough to stand off the solar wind well above its surface (1.3–2.1 $R_{☿}$). However, at times of increased solar wind pressure, the interplanetary particles may impinge directly onto Mercury's surface.

Mercury occupies a much larger fractional volume of its magnetosphere than Earth and the giant planets do. This implies that the stable trapping regions we see in other planetary magnetospheres, the radiation belts, cannot form. A sketch of Mercury's magnetosphere is shown in Figure 7.23, where, in dotted lines, the radiation belts and plasmasphere in the Earth's magnetic field are indicated. As shown, these would lie below Mercury's surface.

Another important difference between Mercury and Earth is the near-absence of an atmosphere and ionosphere. An ionosphere usually affects a planet's electric and magnetic fields, thus indirectly the transport of charged particles. It further serves as a source of plasma in a magnetosphere (Section 7.3.4). The solar wind is expected to be the primary source of Mercury's magnetospheric plasma, although planetary ions released or sputtered from the surface make some contribution according to ground-based observations of emissions from sodium and other elements (Section 4.8.2). The plasma densities in the plasma sheet are higher than in the Earth's magnetosphere by roughly the difference in solar wind density at Mercury and Earth's orbits, i.e., by a factor of $\sim$10. The plasma sheet almost touches Mercury's surface near midnight. The convection electric field may cause particles to diffuse from the magnetotail towards the planet, where it is swept out into the solar wind through the magnetopause at the day side. However, these theories are just extensions from our knowledge of the terrestrial magnetosphere. We will not be able to accurately describe Mercury's magnetic field until an appropriately instrumented orbiter samples its space environment for a long period.

7.4.3 Venus, Mars and the Moon

7.4.3.1 *Venus*

Venus does not possess an internal magnetic field, but the interaction of the solar wind with Venus's ionosphere induces a magnetic field that produces an obstacle to the so-

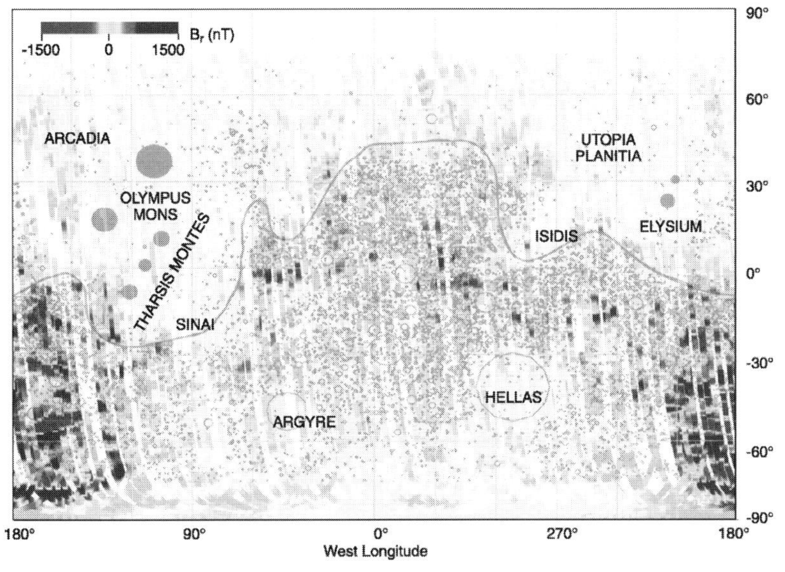

FIGURE 7.24 COLOR PLATE A map of crustal magnetic field sources on Mars superimposed on a map showing the distribution of craters over 15 km in diameter and the dichotomy boundary (solid line). The measured radial (vertical) component of the magnetic field associated with the crustal sources is illustrated using a color scale that reveals the location of significant magnetic sources detected regardless of intensity. Note that the region where magnetic crustal sources appear have high crater densities, and that there is an absence of magnetic imprints within the Hellas, Argyre and Isidis impact basins. No magnetic signatures have been found over Elysium, Olympus Mons or Tharsis Montes. (Acuna *et al.* 1999)

lar wind flow. A bow shock forms upstream of Venus's ionosphere, decelerating the solar wind to sub-Alfvénic and sub-sonic speeds. The magnetic field lines and solar wind plasma accumulate behind the shock, in a magnetosheath. A distinct boundary, the *ionopause*, separates the plasma and field lines in the magnetosheath from the planet's ionosphere. There is pressure equilibrium across the ionopause, with predominantly thermal pressure in Venus's ionosphere and the solar wind pressure outside. The ionized flow in the magnetosheath can interact directly with Venus's dense atmosphere through charge exchange (Section 4.8.2), where a fast solar wind ion takes over an electron from a slow atmospheric neutral; thus a fast neutral and slow ion are created. Energy is taken away from the solar wind and deposited in the atmosphere, while mass (a heavy ion) is added to the solar wind. Photoionization by UV sunlight ionizes atmospheric atoms, thus also 'mass-loads' the solar wind, and slows it down. The interplanetary field lines in the magnetosheath are draped around Venus and form an induced magnetotail behind the planet. The tail consists of two lobes of opposite polarity, and is very similar in appearance to the Earth's magnetotail, except that its 'polarity' is controlled by the draped IMF orientation, like a comet's tail. Without a dipole field, there are no durably trapped particles around Venus.

7.4.3.2 *Mars*

The interaction of the solar wind with Mars is quite similar to that with Venus. As on Venus, an ionopause separates the planet's ionosphere from the interplanetary plasma. Interplanetary magnetic field lines are draped around Mars to form a magnetosheath. However, the ionospheric pressure of Mars is much smaller than that of Venus, and the ionopause may be a more permeable boundary to the solar wind plasma, as indeed measured by the electron reflectometer on Mars Global Surveyor (MGS). The most surprising result from the magnetic field experiment on MGS was the detection of localized very intense magnetic fields. The strongest field measured $\sim$16 mG at orbital altitude (100 km), which, in combination with the ambient ionospheric pressure, is strong enough to stand off and deflect the solar wind at Mars. Thus, the bow shock is asymmetric when this region of strong magnetic field rotates through the sunlit side of the planet.

The localized magnetic fields on Mars are clearly of a crustal origin, i.e., caused by remnant crustal magnetism . Most sources are located in the heavily cratered highlands (Fig. 7.24), south of the crustal dichotomy. There further is an apparent lack of magnetic sources over impact basins. This all suggests that during the first few hundred million years Mars had a magnetic dynamo moment comparable to, or larger than, Earth's dynamo at present. When the ancient crust formed and cooled below the Curie point, the iron-rich crust of Mars got magnetized. Of course, reheating of the crust by volcanism or large impacts after the dynamo ceased to exist destroyed evidence that there once was a dynamo. One can therefore argue that the martian dynamo ceased to exist a few hundred million years after the planet formed.

Figure 7.24 further shows an interesting 'striped' magnetic pattern at west longitudes between about 120° and 210°, where bands of alternating magnetic polarity up to 2000 km long are seen. On Earth such magnetic lineations

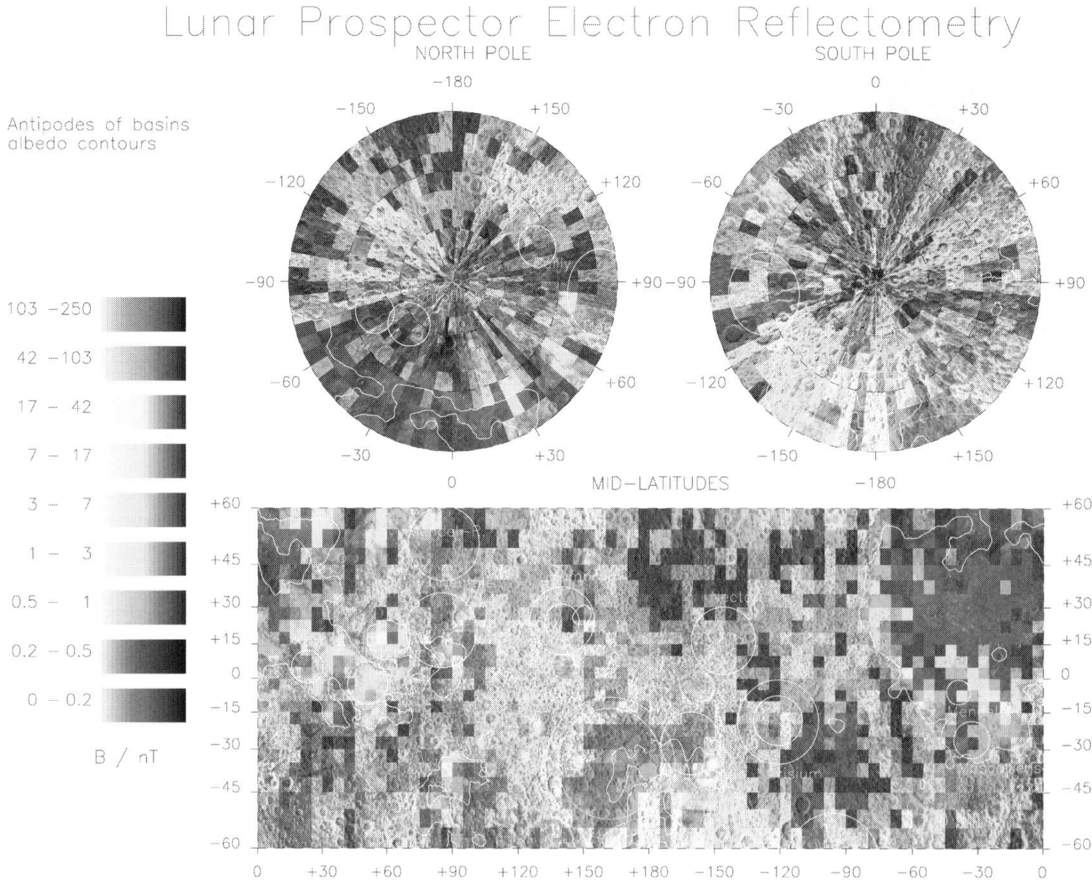

FIGURE 7.25 COLOR PLATE The Moon shows localized patches of remnant crustal magnetic fields, with surface strengths ranging from less than 2 μG to more than 2500 μG. The largest concentrations of strong crustal fields are located diametrically opposite to the four largest post-Nectarian impact basins: Imbrium, Orientale, Serenitatis, and Crisium. (Mitchell *et al.* 2001)

are found along the mid-ocean rift, and are associated with sea floor spreading and repeated reversals of the Earth's dipole field. Hence, this observation suggests that Mars may have had plate tectonics during the first few hundred million years, where like on Earth magma from below filled the void created by two plates moving apart. Upon cooling through the Curie point, the new crust would have Mars's magnetic polarity imprinted. The horizontal scale length of the magnetic lineations on Mars is about 100 km. If the plates moved apart at a rate of about 8 cm year^{-1}, the frequency of magnetic field reversals on Mars would be comparable to that seen on Earth. But of course, both the rate of plate movement and frequency of field reversals may have been very different 4.5 Gyr ago, both for Mars and the Earth. A drawback of this interpretation is that no spreading center has been found, such as that along the mid-ocean ridge where the magnetic polarity is symmetric at either side.

An alternative explanation for the stripy magnetic field pattern is that the once intact magnetized crust 'broke up' into a series of long narrow 'plates', similar perhaps in appearance to the linear fractures or rilles near the Tharsis region (Section 5.5.4.1). Magma from below would fill the cracks and dipole magnetic field patterns would bridge the gaps between broken plates (like when breaking a bar magnet in pieces). One thus would end up with hundreds to thousands of kilometer long 'tracks' of dipolar-like fields. The regions in between these tracks are similar to magnetospheric 'cusp' regions (as in the Earth's polar regions). Solar wind plasma can enter these cusp regions freely, while it is excluded from the crustal dipole-like magnetic field regions. Initial results from MGS indeed suggest that solar wind plasma penetrates down to low altitudes (below 200 km) in magnetic cusp regions while no such plasma is present at these altitudes above areas of strong crustal fields.

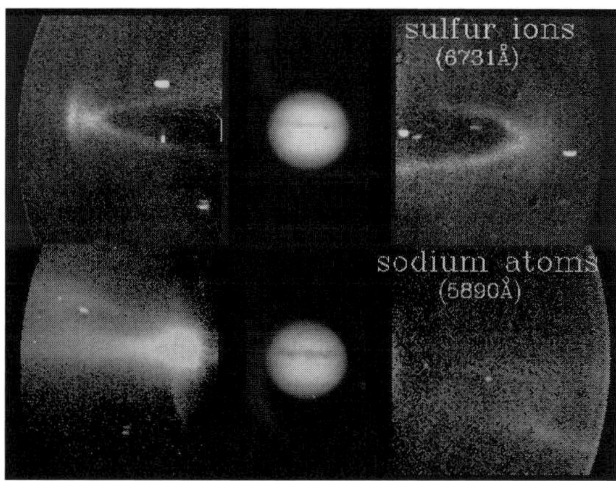

FIGURE 7.26 An image of Jupiter (through a neutral density filter) and Io's neutral sodium cloud (bottom) and plasma torus imaged in S^+ (top). (Courtesy: N.M. Schneider and J.T. Trauger)

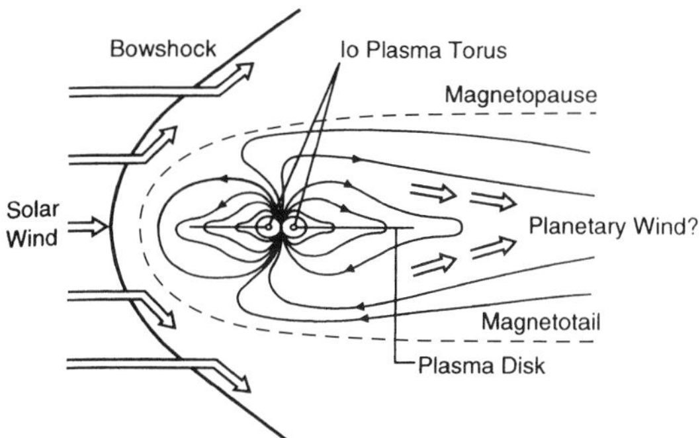

FIGURE 7.27 A sketch of Jupiter's magnetosphere. (Kivelson and Bagenal 1999)

7.4.3.3 *Moon*

The Moon shows strong localized patches of surface magnetic fields of a few tens up to 2500 μG. There is a strong correlation of these patches with the antipodal regions of large young impact basins, such as the Crisium, Serenitatis and Imbrium basins (Fig. 7.25; Section 5.5.1). This correlation suggests that the crustal magnetization is associated with the formation of large impact basins. The hypervelocity impacts that formed these basins likely produced a plasma cloud that surrounded the Moon within about 5 minutes of impact. This compressed and amplified any pre-existing magnetic field at the antipode. Seismic waves and impact ejecta arrive at the antipode within tens of minutes of time, so that the crust here is shocked and magnetized in the amplified field. Long (3.9–3.6 Gyr) ago, the Moon may initially have had a magnetic dynamo field with a surface field strength ~ 0.1–1 G, as suggested from paleomagnetic data obtained from returned Apollo samples. Such a field would be strong enough to explain the high crustal magnetic fields in many of the young large impact basins.

7.4.4 Jupiter

The presence of a magnetic field surrounding Jupiter was first postulated in the late 1950s after the detection of nonthermal radio signals from the planet (Section 7.5). Later, but well before spacecraft traversed Jupiter's magnetosphere, neutral sodium atoms were detected in the vicinity of Io through optical emissions, soon followed by ground-based detections of potassium and ionized sulfur (Fig. 7.26). With the passage of the Pioneer, Voyager, Ulysses and Galileo spacecraft through Jupiter's magnetosphere, the planet's magnetic field, the plasma environments of the Galilean satellites, and in particular Io's neutral and plasma clouds, were studied *in situ*. Intensive monitoring programs from the ground enriched the available data base, and help in understanding the complex relationships between the volcanically active moon Io, its neutral cloud and plasma torus. Since we have a wealth of data for Jupiter, both *in situ* spacecraft data and remote from the ground, we discuss this planet's magnetosphere in much more detail than that of any other planet (except Earth).

7.4.4.1 *Magnetic Field Configuration*

The general form of Jupiter's magnetosphere resembles that of Earth, but its dimensions are over three orders of magnitude larger. If visible to the eye, Jupiter's magnetosphere would appear several times larger in the sky than the Moon. The magnetotail extends to beyond Saturn's orbit; at times, Saturn is engulfed in Jupiter's magnetosphere. A graphical representation of the magnetosphere is shown in Figure 7.27. The outer magnetosphere is a large disk-shaped region. Plasma in this region, 'forced' to rotate with the planet by the corotational electric field, is pushed outwards by centrifugal forces, forming the plasma- or magnetodisk. Analogous to the Earth's Van Allen belts, there are *radiation belts* close to the planet, at $\mathcal{L} \lesssim 2.5$. These are filled with electrons, protons and helium ions.

When subtracting a best-fit displaced dipole from Jupiter's magnetic field, deviations from a dipole field can readily be identified. Regions of weak and strong magnetic field strength occur on the surface. In Figure 7.28, several field lines emanating from the surface are sketched. The *flux tubes* A and B have the same cross-sectional areas in the magnetic equatorial plane. However, flux tube B is anchored in a region of weak magnetic field, while flux tube A is connected to a strong field. The foot of flux tube B is therefore much larger than for flux tube A, so

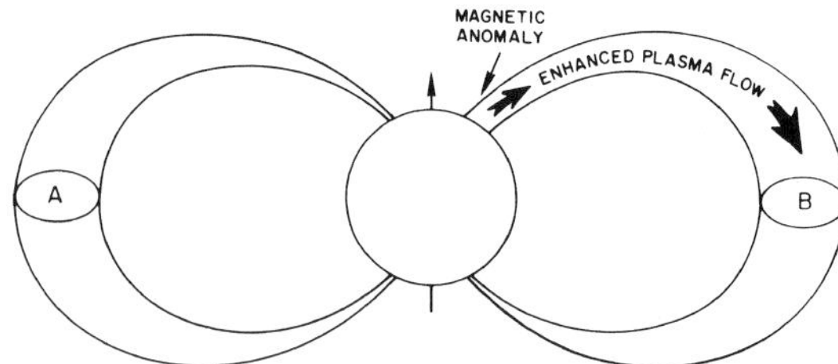

FIGURE 7.28 Flux tubes emanating from Jupiter's 'surface': the flux tubes A and B have the same cross-sectional areas at the equator. Flux tube B is anchored in a region of weak magnetic field, A to strong magnetic field. (Hill *et al.* 1983)

that the flow of ionospheric plasma through flux tube B is enhanced compared to that through flux tube A. This is the essence of the *magnetic anomaly model* developed to explain many observed phenomena in Jupiter's magnetosphere. In this model, the magnetic anomaly is the depression in magnetic field strength in Jupiter's northern hemisphere, centered near a jovian longitude of 260°. This region is referred to as the *active sector*.

Jupiter's magnetic anomaly affects the distribution of plasma in the jovian magnetosphere. As mentioned above, the plasma flow from the ionosphere is enhanced in the active sector compared to that at other longitudes. In addition, the height-integrated conductivity of the jovian ionosphere is enhanced in the active sector, because the Pederson conductivity (cf. Section 4.6.3), which is proportional to B^{-1}, is enhanced. The ionization rate due to particle bombardment is largest in the active sector because of the reduced mirror altitude, which leads to a larger particle loss cone. This explains why jovian aurora are usually brightest at longitudes in the active sector. In addition to the longitudinal asymmetry of aurora, the magnetic anomaly model explains many phenomena with observed longitudinal asymmetries. For example, there is a well known 'clock' modulation of relativistic electrons in the interplanetary medium. These electrons seem to originate at Jupiter. The 'clock' modulation can be explained since release of electrons into interplanetary space occurs primarily through the tail. There is a maximum in the energetic electron population escaping into interplanetary space when the active sector faces the tail.

7.4.4.2 *Hydrogen Bulge*

An interesting observational feature which might be explained by the magnetic anomaly model is the *hydrogen bulge*. The resonantly scattered hydrogen Ly α line is enhanced at longitudes about 180° away from the active sector, near the magnetic equator (or more specifically the particle's drift equator). This suggests the presence of a 'mountain of atomic hydrogen', generally referred to as the hydrogen bulge. This feature is unique for Jupiter. As discussed in more detail below, ionization of neutrals and thus mass loading into the Io plasma torus is largest in the active sector. Centripetal forces push the plasma outwards, inducing an electric field pattern that corotates with the planet. This field induces a corotating convection pattern, such that plasma moves outwards in the active sector and inwards at longitudes ~180° away from it. The inward convection causes hot magnetospheric plasma to impinge on Jupiter's atmosphere and dissociate CH_4 and H_2 into atomic hydrogen, leading to the formation of the hydrogen bulge.

7.4.4.3 *Io's Neutral Clouds*

Although sodium and potassium are only trace elements of Io's neutral cloud, the atoms are easily excited by resonant solar scattering and are readily observed from the ground. The main constituents of the neutral clouds are oxygen and sulfur atoms, but observations of these are more difficult. The sodium cloud is shaped like a banana and pointed in the forward direction (Fig. 7.29). Since the neutral atoms follow keplerian orbits, those closest to Jupiter travel fastest, which results in the forward pointed banana-shaped cloud. The ultimate source of material in Jupiter's magnetospheric neutral clouds is the satellite Io. Sublimation of SO_2-frost from the surface, volcanism and sputtering create an atmosphere around the satellite (Fig. 7.29). The surface pressure of Io's SO_2 atmosphere is a few nanobars, enough to call the atmosphere collisionally thick. Ions corotating with Jupiter's magnetosphere have typical velocities of 75 km s^{-1}. They readily overtake Io, which orbits Jupiter at a speed of 17 km s^{-1}. The ions interact with Io's atmosphere through a collisional cascade process. Some of the atmospheric molecules escape directly into the neutral cloud, others first form a 'sputter corona' around the satellite (see also Section 4.8.2).

The extent of the neutral cloud is determined by the lifetime of the individual atoms (Problem 7.20). A neutral can get ionized through photoionization, electron impact ionization, elastic collisions or charge exchange. Photoionization of neutrals near Io is very slow: typical lifetimes for O and S are a few years, for SO_2 it is about a year, and for Na it is about one month. Hence photoionization is not important compared to the other three processes listed, as shown below. It is relatively straightforward to calculate the rates for these 'collisional' processes in the low density regions far from Io, but it is more complicated in the immediate vicinity of the satellite. Lifetimes of neutrals against electron impact ionization depend upon the density and temperature of the electrons. The particle lifetimes are

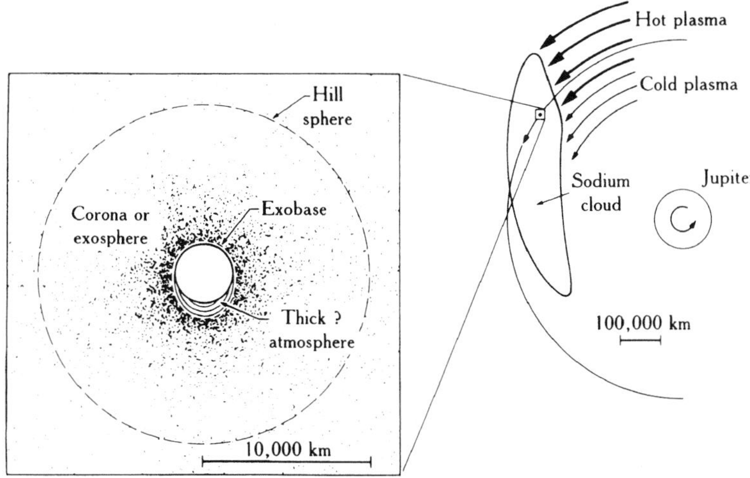

FIGURE 7.29 A schematic representation of Io's sodium cloud and atmosphere/corona. The right side shows the sodium cloud and the corotating plasma, indicated by arrows. The left side (enlarged by a factor of 40 compared to the right side) illustrates Io's immediate vicinity: the thick atmosphere is shown by contours (on the sunlit hemisphere). The thin corona or exosphere is indicated by dots. The dashed line shows the location of the Hill sphere. (Schneider *et al.* 1987)

generally quite short; sodium and potassium typically last for about an hour, oxygen for 55 hrs and sulfur for about 10 hrs. The newly created ion is accelerated to corotational velocities and, although lost from the neutral cloud, will add to the mass and density of the Io plasma torus, discussed in detail in the next section. Since the plasma is cooler inside of Io's orbit, electron impact ionization is less effective here, causing Io's neutral cloud to lie preferentially inside Io's orbit.

Depending on the plasma density, elastic collisions and charge exchange can also terminate the life of a neutral particle. An oxygen atom is more likely to be ionized by charge exchange (typical lifetime is 18 hrs) or an elastic collision (13 hrs) than by electron impact ionization (55 hrs). In an elastic collision, a fast ion impinges on a slow neutral, sending it off at a high velocity nearly perpendicular to the plasma flow (the exact outcome depends upon the impacting angle). In a charge-exchange reaction the corotating ion strips off an electron from a neutral. The neutral is ionized and accelerated to corotational speed, while the former ion becomes a very fast neutral, flying out through the magnetosphere.

Detailed images in the neutral sodium line show jets, fans and loops; these are created by fast sodium atoms, particles traveling at tens of kilometers per second. On a much larger scale, Jupiter appears to be enveloped by a giant disk-shaped sodium cloud, extending out to a few hundred $R_♃$. This cloud is likely formed by fast neutrals flung out from the magnetosphere via charge exchange or elastic collisions between particles in the Io plasma torus and neutral clouds, as discussed in more detail below. Thus, although electron impact ionization, elastic collisions and charge exchange form sinks for slow neutrals, the latter two processes also provide a source of fast neutrals. Another major source of fast neutrals may be dissociative recombination and dissociation of sodium bearing molecules (such as, perhaps, NaCl).

7.4.4.4 *Io Plasma Torus*

Both *in situ* and ground-based observations at visible and UV wavelengths have provided a wealth of information on the heavy ions in Jupiter's magnetosphere. The ions are concentrated in a torus around the planet, centered near Io's orbit at 5.9 $R_♃$ (Fig. 7.26). The plane of symmetry is the centrifugal equator, which goes through the field lines where they are farthest away from Jupiter. This plane makes roughly an angle of 3° with the magnetic equator and 7° with the rotational equator. The exact equilibrium point depends upon the mass and energy of the ion. The vertical extent of the torus depends upon the temperature and mass of the ions. To first approximation the plasma density $N(z)$ decreases exponentially with distance, z, away from the equator:

$$N(z) = N_o e^{-z/H}, \tag{7.80}$$

with the scale height $H = \sqrt{2kT_i/(3m_i n^2)}$, n is the orbital rotation rate and m_i and T_i the mass and temperature of ion species i.

The main species detected in Io's plasma torus are ions of oxygen (up to O^{3+}), sulfur (up to S^{4+}), chlorine (Cl^+, Cl^{2+}), and sulfur dioxide (SO_2^+). The excitation process, through which the ions can be detected, is through collisions with electrons. Both forbidden (at optical wavelengths) and allowed transitions have been observed. The various line intensities have been used to determine the electron density and temperature in the torus. Typical maximum electron densities are a few thousand cm^{-3}. Estimates to maintain the plasma torus as observed suggest a production rate of $\sim 10^{28}$–10^{29} ions per second.

Measurements of the particle density and temperature in the plasma torus show that the torus is divided into two regions: the cold inner torus (a few eV) inside of 5.7 $R_{♃}$, which drops off sharply inside 5.3 $R_{♃}$, and the hot outer torus ($\sim$ 80 eV) up to 7–8 $R_{♃}$. The hot outer torus is much more spread out in latitude than the cold inner torus (Problem 7.21). While emissions from the cold inner torus are confined to the optical wavelength range, both optical and UV emissions have been observed from the hot outer part of the torus. Since Io is the source of this plasma, material must be transported both inward and outward from the satellite.

Radial transport outward from Io is rapid (10–100 days), aided by the centrifugal forces on the plasma. Details of this process are not understood. Some researchers favor centrifugally driven interchange, where denser flux tubes slip outwards, changing places with less-dense flux tubes. A relatively thin sheet of warm (10–100 eV) plasma, dominated by sulfur and oxygen ions, fills the plasma sheet. Densities decrease from several thousand particles cm^{-3} in the torus to a few particles cm^{-3} near 20 $R_{♃}$. The outward flowing plasma in the plasma sheet generates a dawn-to-dusk directed electric field over Jupiter's magnetosphere (note: opposite to solar wind induced dusk-to-dawn field), which has a number of observable effects, as discussed below. Radial transport inward from Io is slow, allowing ample time for the ions to cool. This diffusion is probably driven by currents in Jupiter's ionosphere. The peak in torus density near 5.7 $R_{♃}$, forming the 'ribbon' (left side in Fig. 7.26), has not been explained.

Ground-based observations of the torus via emissions from S^+ ions (Fig. 7.26) show variations in intensity and phenomenology over time. Part of the variability can be attributed to the changing viewing geometry of the planet's magnetic field, but the time-averaged brightness distribution of the torus shows a clear east–west asymmetry in intensity and location. The torus exhibits a clear maximum in intensity at the dusk (west, receding) side, and a minimum at the dawn (east, approaching) ansa. The peak intensity at the dawn ansa is always located on Io's average $\mathcal{L}$-shell, while the maximum intensity at the dusk side is shifted inwards towards Jupiter by $\sim$ 0.4 $R_{♃}$. These features are interpreted as being due to the dawn-to-dusk electric field, generated by the anti-sunward flow of plasma that originates in the Io torus and flows down the tail. The electric field accelerates particles during one half of their drift orbit, causing an inward motion of the dusk end of the torus. The torus brightens here due to adiabatic compression of the electrons that excite the torus emissions. The particles lose their energy during the second half of their drift orbits, so the dawn side stays unaffected.

In addition to the east–west asymmetries seen in the torus, both the intensity and phenomenology of the torus change over time. The ribbon-like feature in Figure 7.26 shows up very clearly in some years, while it is nearly absent in other years. The torus and ribbon are always brightest in the 'active sector'. The intensity of the neutral torus and its jet-like features of fast sodium are also highly time variable.

7.4.4.5 *Positive Feedback Mechanism?*

Increased volcanic activity on Io would enhance the neutral cloud density, and hence the Io plasma torus. The charged particles in the torus form one component of particles which collide with the satellite and replenish the neutral clouds with particles through sputtering. An increased plasma density thus increases the sputtering rate, leading to a positive feedback mechanism, or runaway model to supply the plasma torus with material. However, even though the ion density in the plasma torus has been observed to increase following an outburst in the sodium emission, the magnetosphere somehow imposes a stabilization mechanism, preventing the plasma torus from runaway growth.

7.4.4.6 *Magnetosphere–Ionosphere Coupling*

Coupling between the magnetosphere, plasma torus and Jupiter's ionosphere occurs via field-aligned currents, which connect the plasma torus to Jupiter's ionosphere, as shown graphically in Figure 7.30. The existence of such currents has been 'shown' directly via HST images of auroral emissions along the wake of the footpoint of Io's flux tube, and they have been detected *in situ* by the Galileo spacecraft (Section 4.6.4.2). Through these currents, the ionosphere tries to enforce corotation throughout the magnetosphere. Indeed, inside of $\sim$ 5 $R_{♃}$ the magnetosphere is observed to be in rigid corotation with the planet, but at larger distances there is a significant departure from corotation. At jovian distances of 6–10 $R_{♃}$ the plasma lags behind corotation by $\sim$1–10%, while beyond 20 $R_{♃}$ the azimuthal flow is at a constant speed of about 200 km s^{-1}.

There is a complicated three-way coupling between the ionosphere, magnetosphere and plasma torus. Heating by sunlight causes the weakly ionized air in the ionosphere to flow, resulting in currents and electric fields which affect particle motions in the magnetosphere. Conversely, mass loading and radial motions of particles in the magnetosphere drive currents in the ionosphere, and particle precipitation modifies the ionospheric conductivity. Any

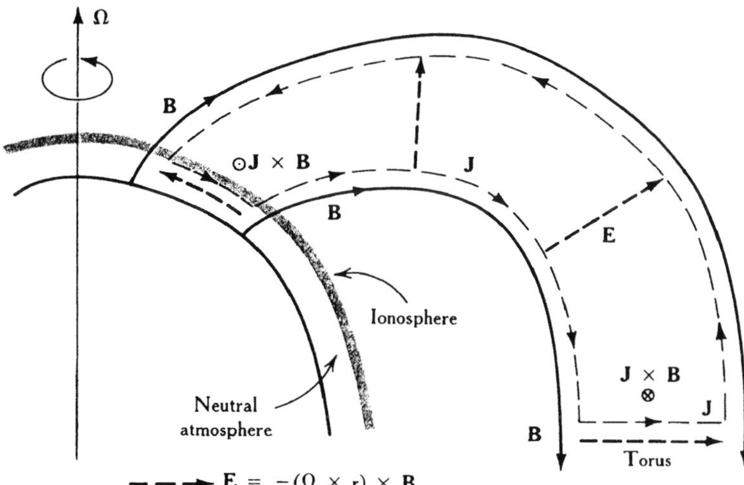

FIGURE 7.30 Sketch of the coupling between Jupiter's magnetosphere and ionosphere along Io's flux tube: the Birkeland current system between Io and Jupiter's ionosphere. (Bagenal 1989; Adapted from Belcher 1987)

currents in the magnetosphere and ionosphere can be expressed by Ohm's law (eq. 7.14). In the ionosphere the electrical conductivity, σ_o, perpendicular to the magnetic field lines is large, but just above the ionosphere in the magnetosphere $\sigma_o \approx 0$. The conductivity along the magnetic field lines, however, is large, and the field lines can be considered as equipotentials ($\mathbf{E} \cdot \mathbf{B} = 0$). Any plasma injected into the magnetosphere is accelerated up to corotation with the planet, since the newly charged particles feel the corotational electric field (directed outwards for Jupiter), which induces an $\mathbf{E} \times \mathbf{B}$ drift around the planet. Thus the particles are accelerated to corotational speeds. In the rest frame of the plasma, $\mathbf{E} = -(\mathbf{v} \times \mathbf{B})/c = 0$. Since the gyro-radii of ions and electrons are different, there is a small radial displacement between the charges, inducing an electric field in a direction opposite to the corotational field (thus inwards for Jupiter), and a radial current outwards. This radial current forms the Birkeland current, and closes in Jupiter's ionosphere (Fig. 7.30). The $\mathbf{J} \times \mathbf{B}$ force enforces corotation. The inward directed electric field weakens the corotational field, resulting in a small net electric field seen by the corotating plasma. In response to this electric field, the corotating plasma slows down (Ohm's law, $\sigma_o = 0$), while the electric field, mapped into the ionosphere since the field lines can be considered equipotentials, induces strong currents in the ionosphere (Ohm's law, $\sigma_o \neq 0$). Hence, the plasma cannot be slowed down in the ionosphere, and corotation breaks down. In general, whenever $\mathbf{J}/\sigma_o$ becomes significant in an ionosphere, corotation may break down. Such situations may be induced, for example, if the plasma density in a magnetosphere increases significantly locally, e.g., by a mass loading process, or if the conductivity in the ionosphere changes significantly, e.g., by particle precipitation or a meteorite impact. Interestingly, the impacts of Comet D/Shoemaker–Levy 9 with Jupiter in 1994 (Section 5.4.5) did not cause any observable changes in the torus emissions.

7.4.4.7 *Galilean Satellites*

The Galilean satellites, ranging in size from ~10% smaller than our Moon (Europa) to ~50% times larger than our Moon (Ganymede), orbit Jupiter while embedded deep in the planet's magnetosphere (distances vary from 6 $R_♃$ for Io up to ~27 $R_♃$ for Callisto, whereas our Moon orbits Earth at a distance of ~60 $R_\oplus$). The immediate environments of the Galilean satellites have been observed *in situ* by the Galileo spacecraft. One of the large surprises was the discovery of pronounced changes in the jovian magnetic field in the neighborhood of all four Galilean satellites. The spacecraft observed a large depression in the ambient magnetic field strength when crossing Io's plasma wake, which could be reconciled if Io possesses a magnetic field of its own, with a surface magnetic field strength of ~17 mG, anti-aligned with Jupiter's magnetic field. This interpretation is not unambiguous, however, since the magnetometer data would show similar magnetic disturbances induced by current systems in Io's vicinity.

Ganymede, in contrast, is a moon that has been unambiguously shown to possess an intrinsic magnetic field, as attested to by both the Galileo magnetometer and plasma wave results. The magnetometer data reveal a magnetic field with an equatorial surface strength of 7.6 mG, tilted 10° with respect to its spin axis. The dipole moment is anti-aligned with Jupiter's magnetic moment. The plasma wave experiment detected strong whistler-mode wave activity (Fig. 7.46), phenomena which have only been seen at bodies surrounded by magnetospheres with closed field lines.

The ambient magnetic field near Europa changed in strength and direction when Galileo passed the satellite. The signature can be modeled as the electromagnetic response to Jupiter's time varying magnetic field provided there is a layer of electrically conducting material below Europa's surface. Although Jupiter's magnetic field itself is roughly constant over time, the projection of the field into Europa's equatorial plane as seen from the satellite varies with Jupiter's rotation. Such a time varying **B** field can induce currents in an electrically conductive medium (eq. 7.8). Such currents, in turn, generate a secondary magnetic field, which is observed as a distortion in Jupiter's **B** field. The magnetic field distortions as measured during eight close passes of Galileo to Europa, together with images of Europa's cracked icy surface, suggest that the electrically conducting shell is likely a ~100 km deep salty ocean immediately below Europa's crust (Sections 5.5.5.2, 6.3.5.1).

In contrast to the pronounced signatures detected by the magnetometer when Galileo passed Io, Europa and Ganymede, only a small change in the field strength was detected near Callisto. Like in the case of Europa, the latter signature can be simulated well if Callisto has a deep salty ocean, a plausible model given constraints on its interior structure (Section 6.3.5.1).

7.4.5 Saturn

Saturn's magnetosphere is intermediate in extent between those of Earth and Jupiter. The strength of the magnetic field at the equator is a little less than that found on the Earth's surface; remember, however, that Saturn is roughly 10 times larger than Earth and it is roughly 10 times further away from the Sun; both factors lead to a substantially larger magnetosphere around Saturn than around Earth. Most remarkable is the nearly perfect alignment between Saturn's magnetic and rotational axes. The center of its dipole field is slightly shifted towards the north, by 0.04 $R_♄$. The outer magnetosphere resembles a magnetodisk in the tail region, but is more like Earth's field at the sun side.

Five icy satellites (Mimas, Enceladus, Tethys, Dione and Rhea) are located in Saturn's inner magnetosphere, between 3 and 9 $R_♄$. Inside 2.27 $R_♄$ we find Saturn's ring system. The ring particles are predominantly made of water-ice (Chapter 11). Sputtering by charged particles and meteorite bombardment eject water products from the satellites and rings into the magnetosphere. Dissociation and ionization result in an oxygen-ion rich plasma torus, at a density of a few cm^{-3}, that was observed by both the Pioneer and Voyager spacecraft. Outwards from about 6 $R_♄$ the plasma lags behind corotation by ~ 30%, which may be caused by a local production of plasma in the vicinity of the icy satellites.

Although we mentioned the satellites and rings as sources of plasma, they also act as sinks (Section 7.3.4). The satellites and ring particles can absorb particles which impinge on them. Inspection of charged particle data from spacecraft flybys usually shows clear absorption signatures in the vicinity of a satellite, or when crossing $\mathcal{L}$-shells traversed by a satellite. In fact, the presence of Jupiter's ring was first noticed in charged particle data from Pioneer 11, which showed clear absorption dips. Similarly, one of several interpretations of an ambiguous proton absorption signature in Pioneer 11 data when the spacecraft traversed Saturn's magnetosphere was the presence of a faint ring at ~ 2.8 $R_♄$, where much later the Voyager spacecraft located the G ring. Since Saturn's rings lie in the planet's rotational equator, which coincides with its magnetic equator, they are efficient absorbers of magnetospheric plasma. The intensity of all trapped particles drops dramatically at the outer edge of the A ring. Interior to this location there are practically no charged particles.

Titan orbits Saturn at a distance of ~ 20 $R_♄$. Because Saturn's magnetospheric boundary is moving in and out with respect to the planet, in response to variations in the solar wind ram pressure, the satellite is sometimes embedded in Saturn's magnetosphere, and at other times it is in the solar wind. Under quiet solar wind conditions, Titan is in Saturn's magnetosphere where the flow is sub-magnetosonic, and we do not expect the formation of a bow shock. The expected interaction of the plasma flow with Titan's dense atmosphere is controlled by atmospheric-ion mass loading of the plasma flow, with a subsequent slowing down of the flow and draping of field lines around Titan. Hence a 'magnetosphere' is induced, just like in the case of comets (Section 10.5), Venus and Mars, although its relative dimensions are much smaller. Because of the preferential pickup of ions on one hemisphere (due to electric fields and curvature drift), the magnetic field must be highly asymmetric. Hydrogen and nitrogen escape from Titan into Saturn's magnetosphere to form the Titan torus, a large plasma cloud around Saturn. Titan's atomic hydrogen cloud has been detected by the Voyager UV spectrometer; it extends radially from 8 to 25 $R_♄$, and has an average density of 10–20 cm^{-3}. The density of nitrogen atoms is only ~ 0.4 cm^{-3}.

7.4.6 Uranus

The axis of Uranus's magnetic field makes an angle of ~ 60° with the planet's rotation axis. This is much larger than that of Mercury, Earth, Jupiter and Saturn, which are

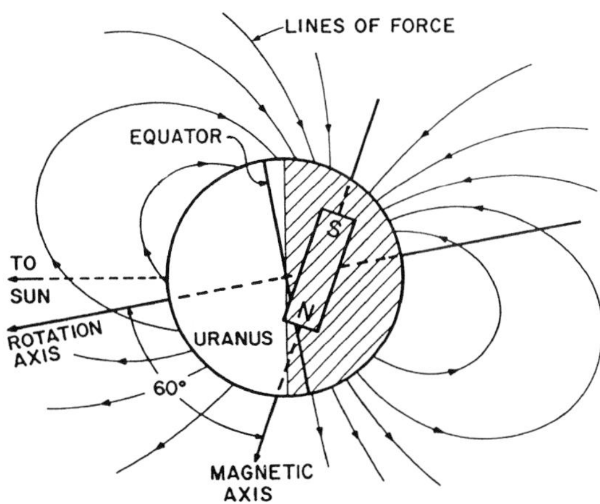

FIGURE 7.31 A sketch of Uranus's offset dipole magnetic field. (Ness *et al.* 1991)

all aligned to within $\sim 10°$. The magnetic center is displaced by $\sim 0.3\ R_{♅}$ from the planet's center, resulting in a configuration as sketched in Figure 7.31. Note that particles mirroring at the sun side have a much larger chance of getting lost in the atmosphere than particles mirroring in the opposite hemisphere. Since Uranus's rotation axis is nearly perpendicular to the ecliptic plane, the magnetic field geometry as viewed from the solar wind is quite Earth-like. The magnetotail wobbles about the planet–Sun line as the planet rotates, and with it the planet's magnetic polarity, as depicted in Figure 7.32. The stagnation point of the magnetopause is at $\sim 18\ R_{♅}$, so all five major satellites of Uranus lie in Uranus's magnetosphere, although far from its magnetic equator. It is therefore assumed that the satellites do not contribute much material to Uranus's magnetosphere.

Indeed, Uranus's magnetosphere does not contain much plasma at all. Protons and electrons have been observed at densities ~ 0.1–$1\ \mathrm{cm}^{-3}$. The primary source of plasma is ionization of Uranus's extended neutral hydrogen corona, whereas the solar wind is likely a second source of plasma. Due to the geometry of the planet's rotation and magnetic field direction, plasma motions in the magnetosphere are quite complicated. The plasma corotates with Uranus (~ 17 hr) and is circulated throughout the magnetosphere within a couple of days by solar wind driven convection.

7.4.7 Neptune

Neptune's magnetic axis makes an angle of 47° with its rotation axis, which has an obliquity of $\sim 30°$. The center of the magnetic dipole is displaced by $0.55\ R_{\Psi}$ with respect to the planet's center, even more than in Uranus's case. The misalignment between Neptune's rotational and magnetic axes brings about a magnetic field configuration which is unique. While the field is rotating with the planet, two extreme situations are encountered, as sketched in Figure 7.33. At times the field is similar to that of the Earth, Jupiter and Saturn, where the magnetic field in the tail is separated into two lobes of opposite polarity, separated by the plasma sheet. Half a rotation later the field topology is that of a 'pole-on' configuration, with the magnetic pole directed towards the Sun. The magnetic field topology in this case is very different, with a cylindrical plasma sheet, separating planet-ward field lines on the outside and field lines pointing away from the planet on the inside. The magnetic pole is facing the Sun, and the solar wind flows directly into the planet's polar cusp. The large offset in the planet's dipole field (which can be translated into quadrupole moments), together with the presence of high order moments result in large variations in the 'surface' magnetic field strength for both Uranus and Neptune (Table 7.2).

Although one would expect Neptune's satellite Triton to be a source of hydrogen and nitrogen ions in Neptune's magnetosphere, observed densities were very low, $\lesssim 0.1\ \mathrm{cm}^{-3}$ up to a few tens cm^{-3} close ($\lesssim 2\ R_{\Psi}$) to the planet. The H^+ and N^+ escape Triton via sputtering processes as neutrals and ions, respectively. The N^+ moves inward from Triton and was detected even at $\mathcal{L} \approx 1.2$. A hydrogen cloud surrounds Triton with a density of $\sim 500\ \mathrm{cm}^{-3}$, extending inwards to $L \approx 8$. This cloud serves as a source for H^+, which moves inwards from its place of formation.

Theories predict the cumulative effect of solar wind driven convection to result in a net sunward transport of plasma in the magnetic equator. However, since convection is strongest when the field configuration is like that of Earth, the convection is strongly longitude dependent, such that plasma may move either toward or away from the planet, depending on its longitude. Convection also produces longitude dependent variations in plasma density. The spacecraft trajectory was not well suited for detecting any longitudinal asymmetries in the plasma density. However, the observations indicate only inward transport of plasma, inconsistent with any of the convection models proposed to date.

7.5 Radio Emissions

Planetary magnetospheres can be observed remotely via emissions from their neutral or plasma tori (Section 7.4.4), as well as via radio emissions produced by electrons gyrating around magnetic field lines. Strong radio signals were first detected from Jupiter, in the early 1950s, at a frequency of 22.2 MHz (wavelength $\lambda = 1.35$ m). The emis-

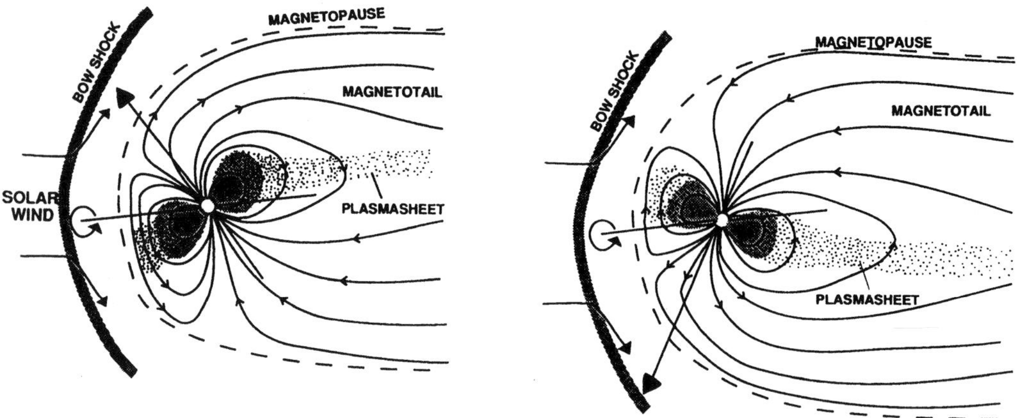

FIGURE 7.32 A sketch of Uranus's magnetosphere at the epoch of Voyager 2's encounter with the planet in 1986. The left and right panels are separated by half a planetary rotation. (Bagenal 1992)

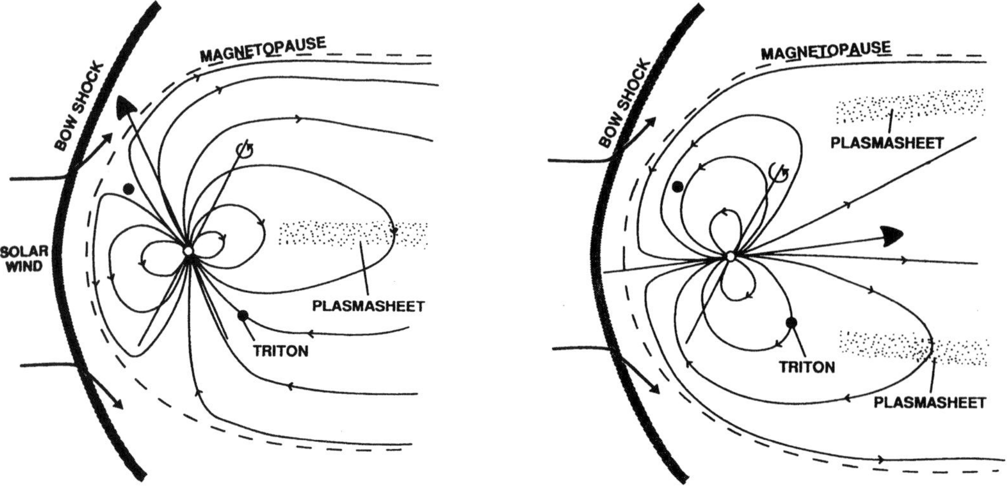

FIGURE 7.33 A sketch of two extreme situations of Neptune's magnetic field configuration at the epoch of Voyager 2's encounter with the planet in 1989. The left and right panels are separated by half a planetary rotation. (Bagenal 1992)

sion was sporadic in character and confined to frequencies less than 40 MHz. Jupiter's microwave emission was subsequently recorded at 3 cm wavelength. The measured flux density corresponded to a blackbody temperature of 140 K. Observations at longer wavelengths revealed Jupiter's synchrotron radiation. Until the Voyager spacecraft missions, Jupiter was the only planet, besides Earth, from which we had received nonthermal radio emissions. Now we know that all four giant planets, as well as Earth, are strong radio sources at low frequencies (kilometric wavelengths).

7.5.1 Low Frequency Radio Emissions

Low frequency radio emission (kilometric wavelengths) is usually attributed to electron cyclotron maser radiation, emitted by keV (nonrelativistic) electrons in the auroral regions of a planet's magnetic field at the frequency of gyration around the magnetic field lines. This radiation is also referred to as *auroral radio emissions*. The radiation can escape its region of origin only if the local cyclotron frequency is larger than the electron plasma frequency (Section 7.6). If this condition is not met, the waves are locally trapped and amplified, until the radiation reaches a region from where it can escape.

Electron cyclotron radiation is emitted at the frequency of gyration around the magnetic field line (eq. 7.46). The radiation is emitted in the shape of a hollow cone dipole pattern, where the lobes are bent in the forward direction (Fig. 7.34). The radiation intensity is zero along the axis

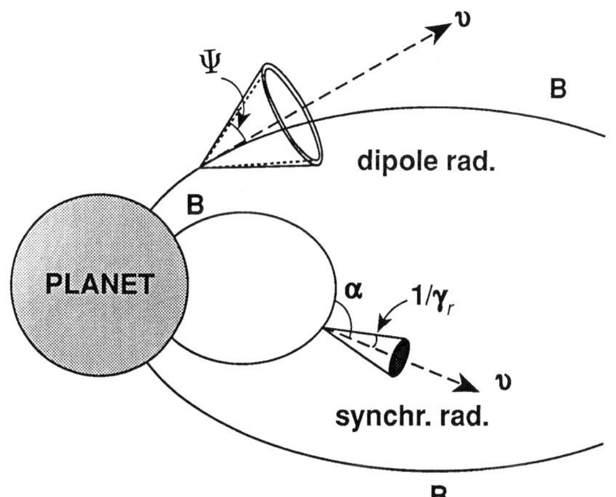

FIGURE 7.34 Radiation patterns in a magnetic field. The hollow cone pattern caused by cyclotron (dipole) radiation from nonrelativistic electrons near the auroral zone is indicated. The electrons spiral outwards along the planet's magnetic field lines. The hollow cone opening half-angle is given by Ψ. At low magnetic latitudes, in the radiation belts, the filled radiation cone of a relativistic electron is indicated. The angle between the particle's instantaneous direction of motion and the magnetic field, commonly referred to as the particle's pitch angle, α, is indicated on the sketch. The emission is radiated into a narrow cone with a half-width of $1/\gamma_r$.

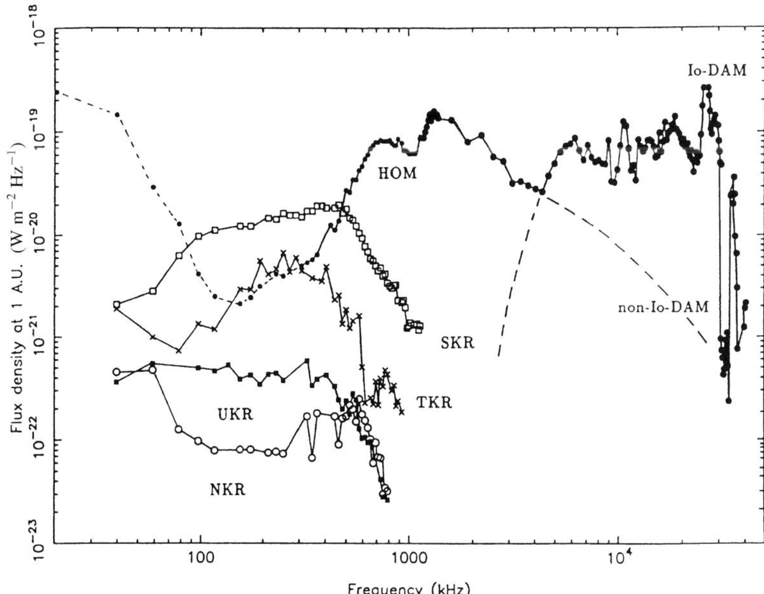

FIGURE 7.35 A comparison of the peak flux density spectrum of the auroral radio emissions of the four giant planets and Earth. TKR refers to terrestrial kilometric radiation, HOM and DAM are emitted at Jupiter, SKR at Saturn, UKR at Uranus and NKR at Neptune. All emissions are scaled such that the planets appear to be at a geocentric distance of 1 AU. (Zarka 1992)

of the cone, in the direction of the particle's motion, and reaches a maximum at an angle Ψ. Theoretical calculations show that Ψ is very close to 90°. Observed opening angles, however, can be much smaller, down to $\sim 50°$, which has been attributed to refraction of the electromagnetic waves as they depart from the source region.

Averaged normalized spectra of the auroral radio emissions from the four giant planets and Earth are displayed in Figure 7.35. All data are adjusted to an observer located at a distance of 1 AU from the source. Jupiter is the strongest low frequency radio source, followed by Saturn, Earth, Uranus and Neptune. The low frequency radio emissions are usually displayed in the form of a dynamic spectrum: a graph of the emission intensity as a function of frequency and time (Fig. 7.41).

7.5.2 Synchrotron Radiation

Jupiter is the only planet from which we receive synchrotron radiation in addition to low frequency radio emissions. In contrast to the low frequency emissions which are produced by keV electrons, synchrotron radiation is produced by relativistic (MeV energies) electrons. At these high energies, the emissions are radiated at higher (GHz) frequencies (wavelengths from a few cm up to ~ 1 m). The emission characteristics are very different from those of the cyclotron radiation. For relativistic electrons, $v \approx c$, and the radiation is strongly beamed in the forward direction (Fig. 7.34) within a cone of opening angle $1/\gamma_r$, where γ_r is the relativistic correction factor: $\gamma_r = 2E$ for relativistic electrons, with E the energy in MeV. The radiation

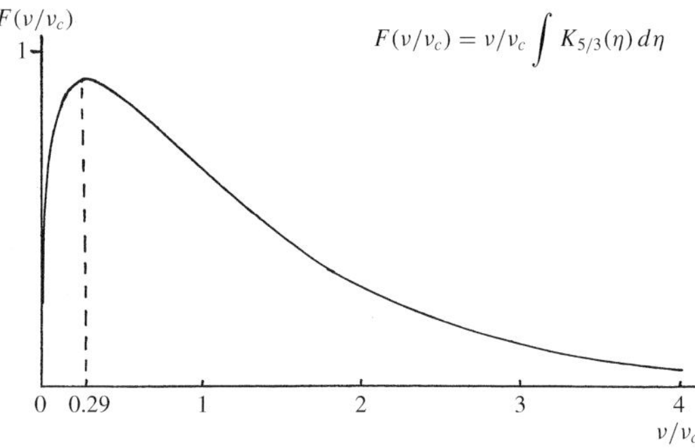

FIGURE 7.36 Power spectrum of the synchrotron radiation emitted by a single electron trapped in a magnetic field. The $K_{5/3}$ is a modified Bessel function. (Ginzburg and Syrovatskii 1965)

is emitted over a wide range of frequencies (Fig. 7.36), and shows a maximum at 0.29 ν_c, with the critical frequency, ν_c:

$$\nu_c = \frac{3}{4\pi}\frac{q\gamma_r^2 B_\perp}{m_e c} = 16.08 E^2 B_\perp, \tag{7.81}$$

where the energy E is in MeV and the field strength $B_\perp$ (component perpendicular to the line of sight) in G. For emission received at 20 cm, the typical energy of the radiating electrons is ~18 MeV if the field strength is 1 G (Problem 7.22). At lower field strengths and/or higher observing frequencies, the typical energy increases. Hence one probes a different electron population when observing at different frequencies. Furthermore, since the magnetic field strength decreases approximately with $\mathcal{L}^{-3}$, we observe different electron distributions at different distances from the planet.

The energy distribution of the electrons is often expressed by a power law:

$$N(E)dE \propto E^{-\zeta} dE. \tag{7.82}$$

The intensity of the radiation at frequency ν can be obtained by integrating the radiation along the line of sight:

$$\mathcal{F}_\nu = \Omega_s \int\int N(E, L) B_\perp F\left(\frac{\nu}{\nu_c}\right) dE\, d\ell, \tag{7.83}$$

where $F(\nu/\nu_c)$ represents the power spectrum of the electrons (Fig. 7.36) (see e.g., Ginzburg and Syrovatskii, 1965, for a full treatment of synchrotron radiation). If the energy spectrum of the electrons follows a power law as in equation (7.82), the flux density of the observed radiation depends on frequency:

$$\mathcal{F}_\nu \propto \nu^{-(\zeta-1)/2}. \tag{7.84}$$

7.5.3 Radio Observations

7.5.3.1 *Earth*

The terrestrial kilometric radiation (TKR) has been studied both at close range and at larger distances by many Earth-orbiting satellites. The radiation is very intense; the total power is on average 10^8 W. It originates in the Earth's auroral regions at low altitudes and high frequencies, and spreads to higher altitudes and lower frequencies. The emission is confined to frequencies less than 800 kHz, which corresponds to a magnetic field strength of 0.28 G (eq. 7.46), close to the field strength measured at Earth's surface. The lower frequency cutoff is around 100 kHz. TKR events are highly circularly polarized, where the sense of polarization depends upon the orientation of the magnetic field. The emission is fixed in local time, centered around 10 pm; thus it originates in the night-side auroral regions. Some events appear to come from the day-side polar cusps. The intensity is highly correlated with the presence of geomagnetic substorms, thus it is indirectly modulated by the solar wind. The dynamic spectra consist of narrowband rapidly drifting features.

7.5.3.2 *Jupiter*

Synchrotron Radiation

Synchrotron radiation from Jupiter is received at wavelengths between a few cm and a few m (frequencies $\gtrsim$ 40 MHz). The variation in total intensity during one jovian rotation is indicated in Figure 7.37. The orientation of Jupiter's magnetosphere is indicated at the top. The maxima in intensity occur approximately at a magnetic latitude of the Earth $\phi_m = 0$, and the minima occur where $|\phi_m|$ is largest. Such data imply that most of the radiating electrons are confined to the magnetic equatorial plane, thus forming a ring of radiating particles around Jupiter, such as sketched in Figure 7.37b. Figure 7.38a displays a two-dimensional image of Jupiter's synchrotron radiation at 20 cm, with a resolution of 0.3 $R_♃$. The two main radiation peaks, L and R, result from the line-of-sight integration through the ring of radiating electrons sketched in Figure 7.37b. The high latitude regions, Ln, Ls, Rn, and Rs suggest the additional existence of less intense rings of emission at latitudes of ~35°. Since Jupiter's synchrotron radiation is optically thin, one can use tomography to extract the three-dimensional distribution of the radio emissivity from the data; this is shown in Figure 7.38b. In particular the latter image clearly shows that most of the radiation is concentrated to the magnetic equator, which, due to the higher order moments in Jupiter's field, is warped like a potato-chip.

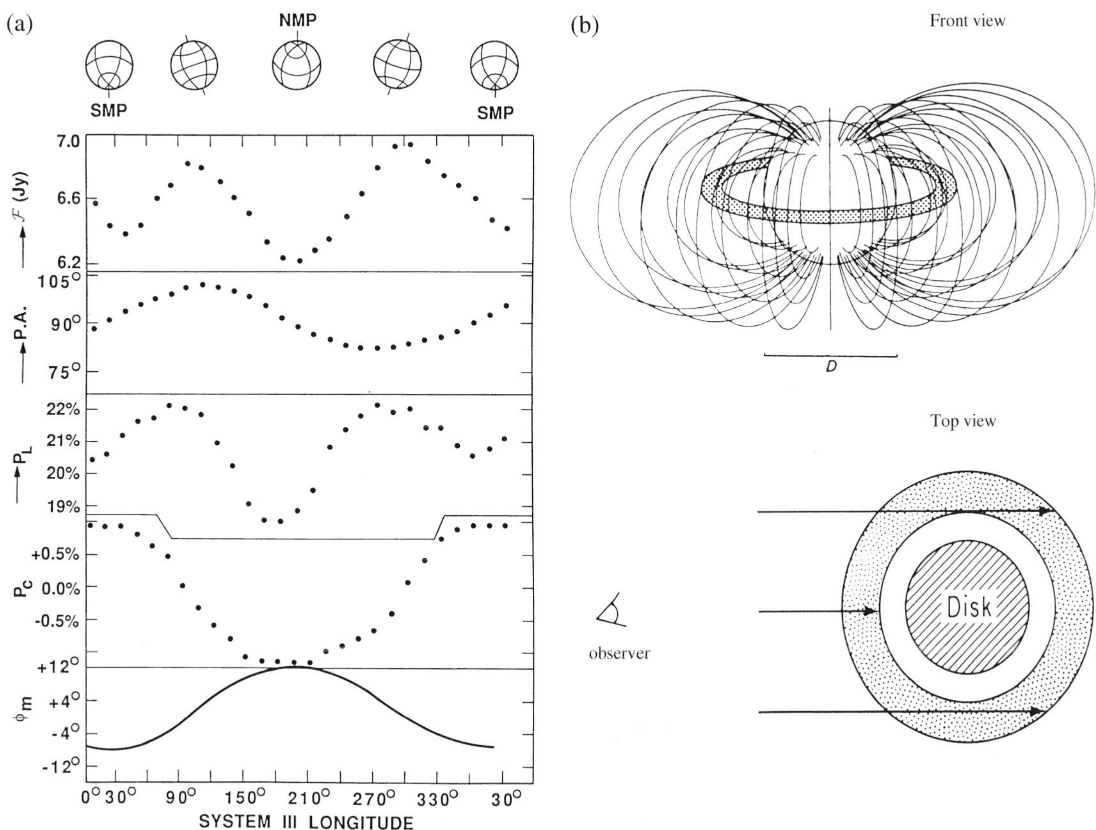

FIGURE 7.37 (a) An example of the modulation of Jupiter's synchrotron radiation due to the planet's rotation. The orientation of the planet is indicated at the top; the different panels show subsequently the total flux density $\mathcal{F}$, the position angle P.A. of the electric vector, the degree of linear and circular polarization P_L and P_C, and the magnetic latitude of the Earth, ϕ_m. This latitude can be calculated with: $\phi_m = D_E + \theta_B \cos(\lambda - \lambda_{np})$, with D_E the declination of the Earth, θ_B the angle between Jupiter's magnetic and rotational axes, λ the central meridian longitude, and λ_{np} the central meridian longitude of the magnetic north pole. (de Pater and Klein 1989; Adapted from de Pater 1980) (b) A schematic representation of the energetic electrons in Jupiter's magnetic field, as seen from the front and top, as indicated. (de Pater 1981)

Superposed on Figure 7.38a are magnetic field lines at $\mathcal{L} = 1.5$ and $\mathcal{L} = 2.5$. These show that the radiation from the main radiation regions L and R comes from electrons which are confined to the magnetic equatorial plane, and which are at $\mathcal{L} \approx 1.5$. In contrast, the high latitude emissions are produced by small pitch angle electrons at their mirror points, and these electrons reside on $\mathcal{L}$ shells between 2 and 2.5. It has been suggested that the moon Amalthea (or its dusty ring) might interact with the electron population such as to scatter electrons into field-aligned beams, and that these electrons produce the emissions.

Figure 7.39 shows a radio spectrum of Jupiter's synchrotron radiation from 74 MHz to 5 GHz. Model calculations, such as those superposed on the data, suggest that Jupiter's electron spectrum is not a simple $N(E) \propto E^{-\zeta}$ power law, but that it consists of two power laws ($N(E) \propto E^{-\zeta_1}(1 + E/z)^{-\zeta_2}$, with $\zeta_1 \approx 0.5$, $\zeta_2 \approx 3$, and $z \approx 100$), consistent with *in situ* measurements by the Pioneer spacecraft. Radial diffusion of the radiating electrons, pitch angle scattering, absorption by moons and ring, etc. were all included in the model.

The total flux density of the planet varies significantly over time, as shown in Figure 7.40. Variations on timescales of years (panel a), between 1965 and 1987, seem to be correlated with solar wind parameters, in particular the solar wind ram pressure, suggesting that the solar wind is influencing the supply and/or loss of electrons into Jupiter's inner magnetosphere. This correlation breaks down, however, after 1987. When Comet D/Shoemaker–

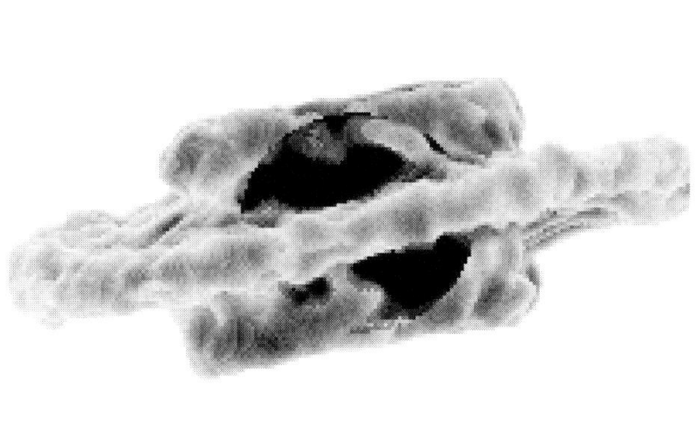

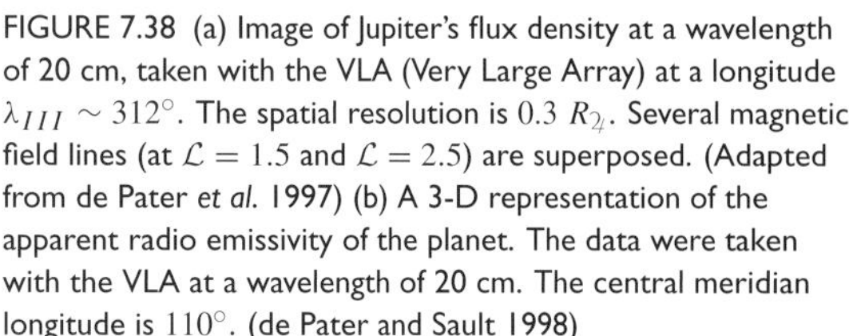

FIGURE 7.38 (a) Image of Jupiter's flux density at a wavelength of 20 cm, taken with the VLA (Very Large Array) at a longitude $\lambda_{III} \sim 312^\circ$. The spatial resolution is $0.3\ R_{♃}$. Several magnetic field lines (at $\mathcal{L} = 1.5$ and $\mathcal{L} = 2.5$) are superposed. (Adapted from de Pater *et al.* 1997) (b) A 3-D representation of the apparent radio emissivity of the planet. The data were taken with the VLA at a wavelength of 20 cm. The central meridian longitude is 110°. (de Pater and Sault 1998)

Levy 9 collided with Jupiter (Section 5.4.5), large temporary changes were detected in the synchrotron radiation (Fig. 7.40b, c). The total flux density increased by $\sim 20\%$, the radio spectrum hardened, and the spatial brightness distribution changed considerably during the week of impacts. These changes were brought about by a complex interaction of the radiating particles with shocks and electromagnetic waves induced in the magnetosphere by the atmospheric explosions of the series of cometary impacts (see e.g., de Pater *et al.* 1997, and Brecht *et al.* 2001).

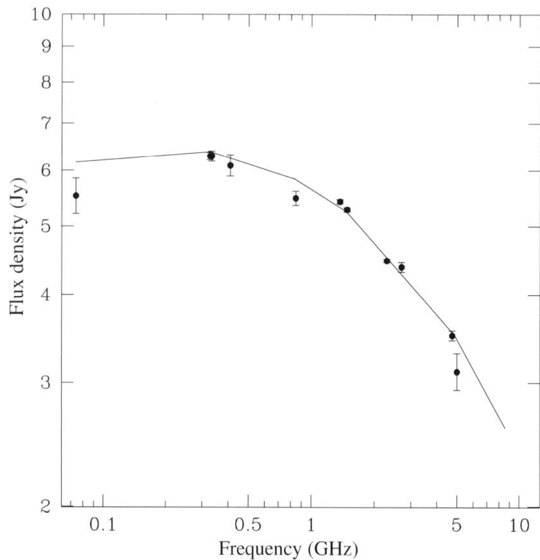

FIGURE 7.39 Jupiter's synchrotron radiation radio spectrum from 74 MHz up to $\gtrsim 5$ GHz. The solid line is a model. (Adapted from de Pater 1999)

Decametric (DAM) and Hectometric (HOM) Radio Emissions

At frequencies below 40 MHz ($\lambda > 7$ m), Jupiter is a strong emitter of sporadic low frequency emissions. The radiation is characterized by a complex, highly organized structure in the frequency–time domain. One of the most striking features is that, on timescales of minutes, DAM displays a series of arcs, like 'open' or 'closed parentheses', shown in Figure 7.41. One can distinguish greater arcs between 1 and 40 MHz, and lesser arcs at frequencies below 20 MHz. Within one storm, the arcs are all oriented the same way. The intensity of DAM as observed from Earth shows a modulation with a period of 11.9 years, Jupiter's orbital period, likely caused by the changing declination of Earth with respect to Jupiter. Striking is the modulation in DAM intensity and its activity level with jovian longitude and the position of Io with respect to the observer (Io phase). Figure 7.42 shows a graph of the dependence of DAM activity on jovian longitude and Io phase. Although not all storms are related to Io's position relative to the observer, the most intense ones are.

DAM emission is modulated on timescales of seconds. The emission drifts in frequency by up to ± 150 kHz s^{-1}. In addition to these *L-bursts* (Long-lived bursts), there are also *S-bursts*, emission features of a short duration, $\lesssim 0.01$ s, which drift rapidly in frequency at a rate between -5 and -45 MHz s^{-1}. The observed drift rate in these S-bursts can be explained if mono-energetic electrons with ener-

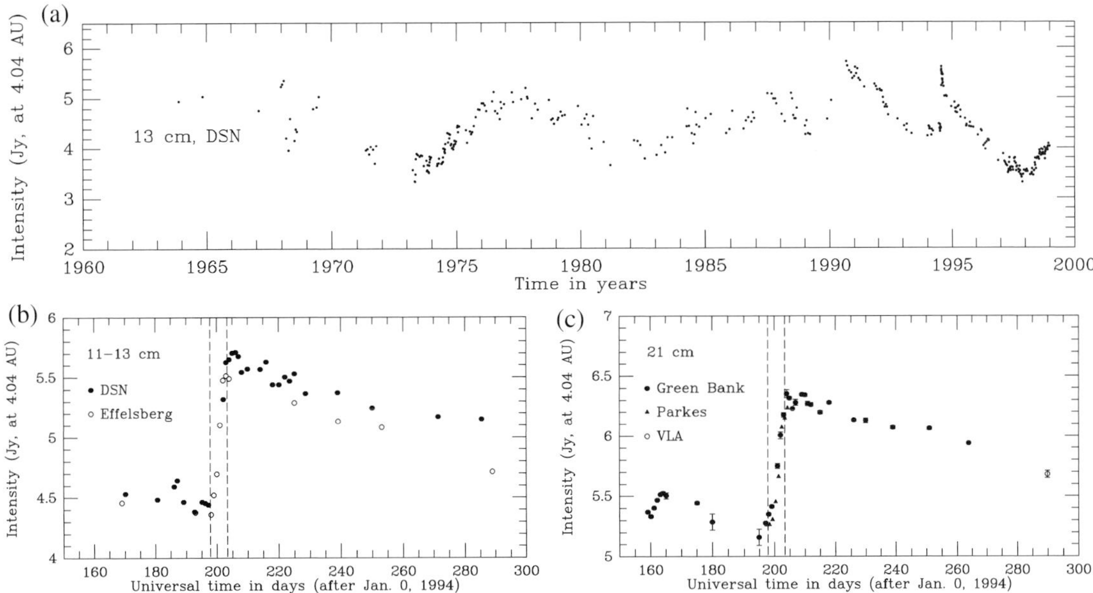

FIGURE 7.40 Time variability in Jupiter's radio emission. (a) Graph of the radio intensity at a wavelength of 13 cm from 1963 through 1998. The data after 1969.2 are NASA/JPL Jupiter Patrol observations made by M.J. Klein using the antennas of NASA's Deep Space Network. The data up to 1968.7 were taken with the Parkes and Nancey radio telescopes. The sudden increase in intensity in 1994 was caused by the impact of Comet D/Shoemaker–Levy 9 with Jupiter. (b, c) An expanded view of the SL impact period is shown in the two lower panels, where data taken at 11–13 cm (panel b) and 21 cm (panel c) are shown. Note that the intensity of the 11 cm data is slightly less than that at 13 cm (see Figure 7.39). The impact of Comet D/Shoemaker–Levy 9 with Jupiter occurred 16–22 July 1994, and is indicated by vertical dashed lines on panels (b) and (c). (The data in the latter panels were taken by Klein *et al.* (1995), Bird *et al.* (1995), and Wong *et al.* (1996))

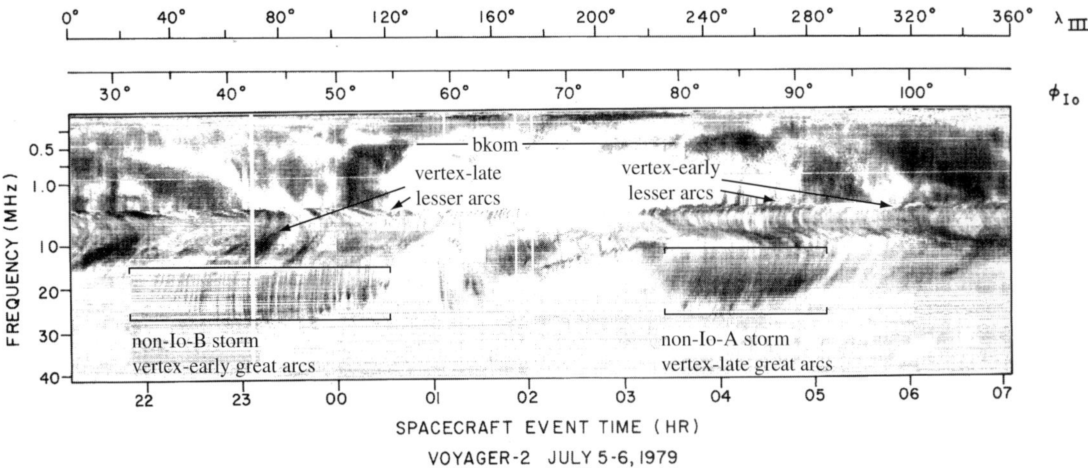

FIGURE 7.41 A representative dynamic spectrum of Jupiter's low frequency radio emissions during one complete rotation of the planet. The darkness of the gray shading is proportional to the intensity of the emission. The emission is plotted as a function of frequency (along y-axis) and time (along x-axis). The differently 'arced' patterns at longitudes less than 180° and at longitudes above 180° is typical of the Voyager data before the encounter with Jupiter. Jovicentric longitude (λ_{III}) and Io's orbital phase (ϕ_{Io}) are indicated at the top. (Carr *et al.* 1983)

gies of a few keV move outwards from Jupiter, along the magnetic field lines (Birkeland currents). Since the field strength decreases with increasing jovicentric distance, the gyrofrequency decreases rapidly in time. DAM emis-

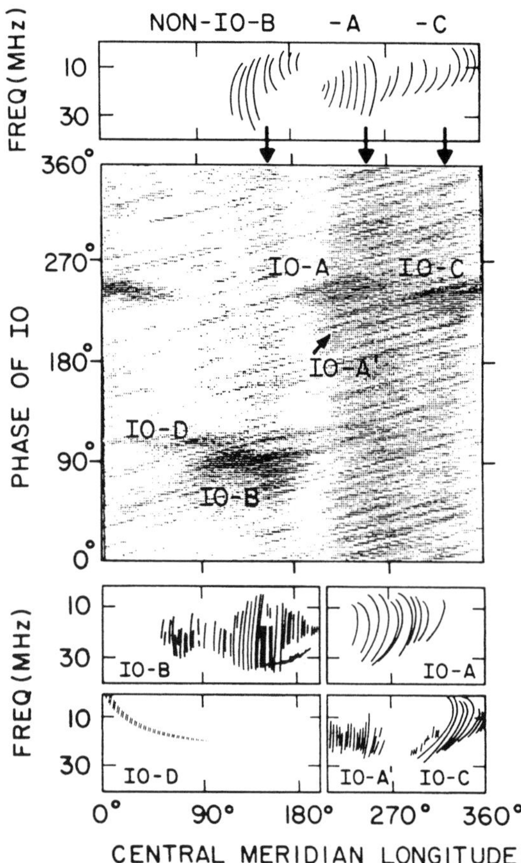

FIGURE 7.42 Dependence of DAM activity at a frequency of about 20 MHz on Jupiter's central meridian longitude (CML) and Io's phase. The Io-related sources are labeled in the intensity plot. Schematic illustrations of the various plots are shown in the lower (Io-related events) and upper (non-Io-related events) panels. (Carr *et al.* 1983)

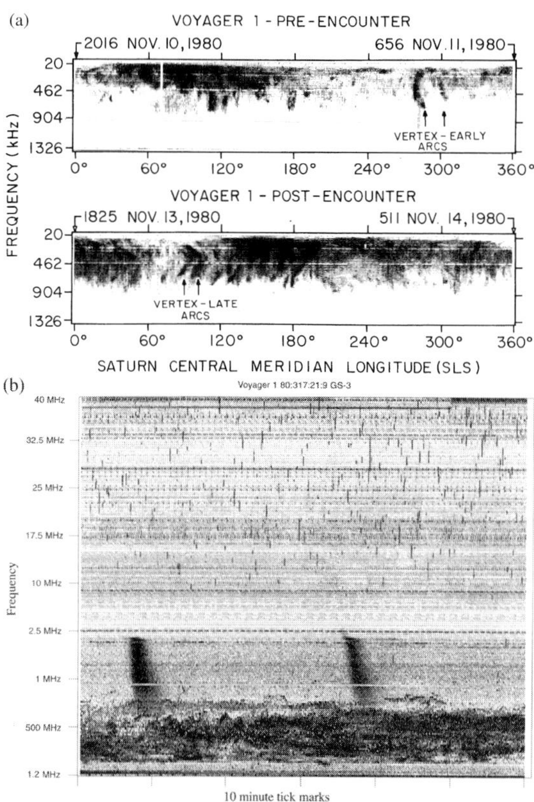

FIGURE 7.43 (a) Dynamic spectra of Saturn's SKR emission. Data are shown from one rotation of Saturn before (top) and after (bottom) Voyager 1's closest approach. The beginning and ending spacecraft acquisition times are marked on the figure, as well as two vertex-early and two vertex-late arcs. (Thieman and Goldstein 1981) (b) A dynamic spectrum of Saturn during one hour near Voyager 1's closest approach. The SED's are clearly visible, in particular between 10 and 40 MHz on this figure. They appear as short streaks parallel to the frequency scale. At lower frequencies the emission is dominated by SKR emission; the two prominent bursts between about 0.7 and 2 MHz are solar type III bursts, radio emissions generated by streams of energetic electrons from the Sun. (Courtesy: D. Evans)

sion is anti-correlated with infrared H_3^+ emission (Section 4.6.4.2). The latter emission is triggered by particle precipitation into Jupiter's ionosphere. The anti-correlation suggests that DAM is caused by the return flow of upward propagating electrons, reflected back to Io at regions of high magnetic field strength.

Kilometric Radio Emissions (KOM)

From a few kHz up to 1 MHz ($\lambda \sim 0.1$–100 km) the Voyager and Ulysses spacecraft detected both (broadband) bursty and (narrow-band) smooth kilometric radiation from Jupiter. The bursty component originates at high magnetic latitudes, while the smooth component has been detected from the outer plasma torus. The recurrence period for the latter events is a little longer than expected from strict corotation with the planet; the source seems to lag behind by 3–5%, as indeed expected at jovicentric distances of 8–9 $R_♃$. Continuum emission has been detected below ~ 25 kHz.

7.5.3.3 *Saturn*

Saturn's nonthermal radio spectrum, displayed in Figure 7.43, consists of three components: (i) Saturn *kilometric radiation* (SKR) at frequencies 3 kHz–1.2 MHz; (ii) low frequency *continuum emission* at 0.3–100 kHz; and (iii) Saturn *electrostatic discharges* (SED).

Saturn kilometric radiation was detected by Voyager from a distance of 3 AU from the planet. A broad band of emission extends up to ~ 1 MHz. The upper cutoff of 1.2 MHz suggests the radiation to be confined to magnetic field lines where the field strength is less than 0.43 G. SKR emission is sometimes organized in arc-like structures, reminiscent of the decametric arcs seen in Jupiter's radio emission. The radio source is located in the polar regions, where the field lines tend to open up into the interplanetary medium and where intense particle precipitation and auroral activity occur. The SKR intensity is correlated with the solar wind ram pressure, indicative of a continuous mass transfer of the solar wind into Saturn's low-altitude polar cusps. Most remarkable was the detection by both Voyager spacecraft of Saturn electrostatic discharges: strong, unpolarized, impulsive events, which last for a few tens of milliseconds and occurred at any frequency of the Voyager radio experiment (20 kHz to 40 MHz). Structure in individual bursts can be seen down to the time resolution limit of 140 microseconds, which suggests a source size less than 40 km (Problem 7.24). SED episodes occur approximately every $10^h 10^m$, distinctly different from the $10^h 39^m.4$ periodicity in SKR, but close to the rotation period of storms near Saturn's equator. In contrast to SKR, the SED source is fixed relative to the planet–observer line. The emissions are likely electrostatic discharge events, and may form the radio counterpart of lightning flashes in Saturn's atmosphere, and are observable because the shadow of the rings substantially reduce the ion density or the regions that it shields from sunlight.

7.5.3.4 *Uranus*

In contrast to Saturn, Uranus's radio emission is dominant on the night-side hemisphere. Both smooth and bursty components can be distinguished in the emissions; however, the detailed morphology of the emissions is much more complicated than on Saturn. Many of the components are visible in the spectrogram shown in Figure 7.44a. Impulsive bursts of radio emission, similar to the SED events of Saturn, have also been detected and are referred to as UED or Uranus electrostatic discharge events. They were fewer in number and less intensive than the SEDs. If these emissions are caused by lightning, the lower frequency cutoff (7 MHz at the day side, 0.9 MHz at night) suggests peak ionospheric electron densities of $\sim 6 \times 10^5$ cm^{-3} over the day-side hemisphere, and $\sim 10^4$ cm^{-3} above the night side. These numbers agree with ionospheric densities determined by other experiments on board the Voyager spacecraft.

7.5.3.5 *Neptune*

Neptune's radio emissions are quite similar to those received from Uranus (Fig. 7.44b). Both smoothly varying and bursty emission components have been detected at frequencies up to 1.3 MHz. The cutoff frequency of 1.3 MHz corresponds to a magnetic field strength of 0.46 G, indicative of the presence of high order magnetic moments in the magnetic field configuration (Problem 7.25). The source of the impulsive emissions is likely located just above the south magnetic pole, within the auroral region. The radiation is emitted into a thin hollow cone with a wide opening angle ($\sim 80°$). Some researchers believe the source to be a flickering search light rotating with the planet; others advocate the source to be fixed in local time, like Saturn's SKR. In the latter scenario the source will 'turn on' or intensify when a particular (active) longitude sector passes by the planet's dawn terminator.

7.6 Waves in Magnetospheres

Perturbations in a medium induce waves which propagate at specific velocities. Magnetospheres consist of magnetic fields and contain copious amounts of plasma, so numerous wave phenomena are induced inside a magnetosphere. In this section we briefly discuss the low frequency magnetohydrodynamic (MHD) waves, followed by the higher frequency plasma waves. We will not derive equations nor dive into the plasma physics, but merely give the reader a flavor of the rich variety of waves encountered in magnetospheres and the solar wind. For a more complete treatment of MHD and plasma waves the reader is referred to e.g., Chapters 2, 11 and 12 in Kivelson and Russell (1995).

7.6.1 General Wave Theory (Summary)

The most common wave is the sound wave in a gas. In this case, the perturbation is a change in pressure (compressional wave), which propagates through the gas. In a magnetized plasma the dynamics is controlled both by the electromagnetic field and the thermal plasma. Perturbations in the electromagnetic field are equally important to those induced in the gas. Usually perturbations are assumed to be small, so the governing equations can be linearized (see e.g., Kivelson and Russell, 1995) and solved for the wave properties. The perturbations propagate as plane waves of the form:

$$e^{i(\mathbf{k}\cdot\mathbf{x}-\omega_o t)} = \cos(\mathbf{k}\cdot\mathbf{x}-\omega_o t) + i\sin(\mathbf{k}\cdot\mathbf{x}-\omega_o t), \tag{7.85}$$

where $\mathbf{k}$ is the wave vector ($k = 2\pi/\lambda$, with λ the wavelength) and ω_o the angular wave frequency ($\omega_o = 2\pi\nu$, with ν the frequency of the wave). The argument of the exponential is constant if:

$$x = x_o + \frac{\omega_o t}{k}. \tag{7.86}$$

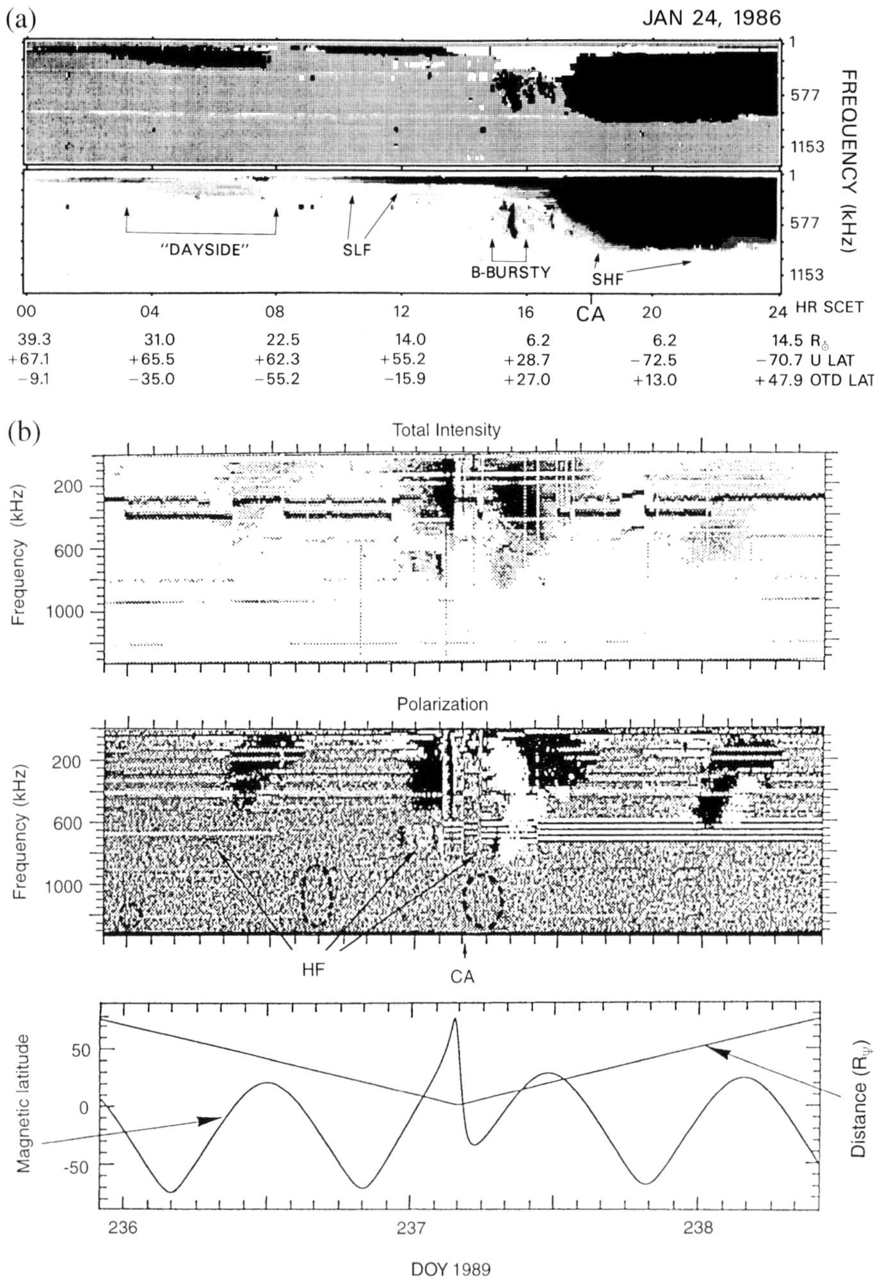

FIGURE 7.44 (a) Dynamic spectra of Uranus's low frequency radio emissions. The various components in the emission are labeled. The top panel shows the emission polarization, with white (black) corresponding to right (left) hand circularly polarized signals. The bottom panel shows the intensities of the emissions. The distance to Uranus ($R_♅$), planetocentric latitude (U LAT) and magnetic latitude in an offset-tilted dipole model (OTD LAT) is indicated at the bottom. (Desch *et al.* 1991) (b) Dynamic spectrum from Neptune during 60 hours around Voyager 2's closest approach. The upper panel shows the intensity, where increasing darkness represents increasing intensity. The middle panel shows the polarization: white indicates LH and black RH polarization. The bottom panel shows the magnetic latitude and distance of the spacecraft at the time of the observations. The abbreviation CA stands for closest approach, and HF for higher frequency emissions. (Zarka *et al.* 1995)

Thus the solutions are constant at a position which moves with the *phase velocity* of the wave:

$$v_{ph} = \frac{dx}{dt} = \frac{\omega_o}{k}. \tag{7.87a}$$

The phase velocity of waves depends upon the dielectric properties of the medium: v_{ph} is inversely proportional to the index of refraction of the medium, n:

$$n = \frac{ck}{\omega_o}, \quad \text{and} \quad \nu\lambda = \frac{c}{n}. \tag{7.87b}$$

Thus if the frequency and wavelength of a wave are known, the product $\nu\lambda$ provides information on the dielectric properties of the medium.

The linearized MHD equations can be solved under various circumstances to yield the *dispersion relation*, an equation which relates ω_o to $\mathbf{k}$. An example of a dispersion relation for electromagnetic waves in free space is that the waves propagate at the speed of light,

$$\frac{\omega_o}{k} = c. \tag{7.88}$$

Sound waves in a gas propagate at the speed of sound, c_s. The dielectric constant of a medium is often a function of ω_o, and the wave shows a small spread in frequency about the central frequency ω_o. A pulse travels through the medium without distortion at the *group velocity*:

$$v_g = \nabla_k\, \omega_o = v_{ph} + \nabla_k v_{ph}. \tag{7.89}$$

In a nondispersive medium the phase and group velocity of a wave are identical.

7.6.2 MHD Waves

Dispersion relations in a magnetized plasma depend on the magnetic field strength, the plasma density and the direction of wave propagation. Magnetohydrodynamic waves are the lowest frequency waves that occur in a plasma: the wave frequency, ω_o, is well below the natural frequencies encountered in a plasma: $\omega_o \ll \omega_{Bi},\ \omega_{pi}$, with ω_{Bi} the ion cyclotron frequency (eq. 7.46) and ω_{pi} the plasma frequency for ions (protons):

$$\omega_{pi} = \left(\frac{4\pi N_i q^2}{m_i}\right)^{1/2}, \tag{7.90}$$

with N_i the ion density, m_i the ion mass, and q the electric charge.

To simplify the equations, one can classify a plasma as 'cold' or 'warm'. In a cold plasma the thermal pressure can be ignored, whereas this is not the case in a warm plasma. In a cold plasma we find two different MHD waves: The *shear Alfvén wave*, which is the regular Alfvén wave, causes field lines to bend. The perturbations are all perpendicular to $\mathbf{B}$, like the propagation of a wave along a guitar string. These Alfvén waves propagate at a velocity:

$$v_{ph} = c_A \cos\theta, \tag{7.91a}$$

with c_A the Alfvén velocity defined in equation (7.22), and θ the angle between the wave vector $\mathbf{k}$ and the magnetic field $\mathbf{B}$. A second wave, a *compressional wave*, propagates at the Alfvén velocity:

$$v_{ph} = c_A. \tag{7.91b}$$

In contrast to the Alfvén wave, the latter wave changes the density of the fluid and the magnetic pressure, and is therefore referred to as a compressional wave, like a sound wave. In a compressional wave the perturbations in the field are under an oblique angle, so that the separation between field lines varies. Moreover, the phase velocity in equation (7.91b) is larger than for the shear Alfvén wave; the compressional wave is therefore also referred to as the *fast mode* wave.

In a warm magnetized plasma, the plasma pressure is comparable in strength to the magnetic pressure, in which case three wave solutions are obtained: The shear Alfvén wave, with the same properties as in a cold plasma (eq. 7.91a), and two compressional waves which carry changes in plasma and magnetic pressure, characterized by the dispersion relation:

$$\left(\frac{\omega_o}{k}\right)^2 = \frac{1}{2}\left(c_s^2 + c_A^2 \pm \sqrt{(c_s^2 + c_A^2)^2 - 4c_s^2 c_A^2 \cos^2\theta}\right). \tag{7.92}$$

We distinguish a *fast mode* (+ sign in eq. 7.92) and a *slow mode* (− sign in eq. 7.92) compressional or *magnetoacoustic* wave. The fast mode wave can propagate in any direction, although its speed depends on θ; it is maximal for propagation perpendicular to $\mathbf{B}$ ($\theta = 90°$). The slow mode compressional wave cannot propagate perpendicular to $\mathbf{B}$. Compressional waves act to reduce field-aligned gradients in plasma pressure, while Alfvén waves act to reduce the bending in magnetic field lines. Plasma flow across the field can increase the bending of field lines. The associated field perturbations create field-aligned currents that act to reduce the additional curvature of the field line. In a magnetosphere, Alfvén waves which propagate along field lines rapidly damp out, unless the waves reflect back against the ionosphere and the frequency is such that standing waves are induced (Problem 7.26). When a perturbation in the solar wind hits the magnetopause, shear Alfvén

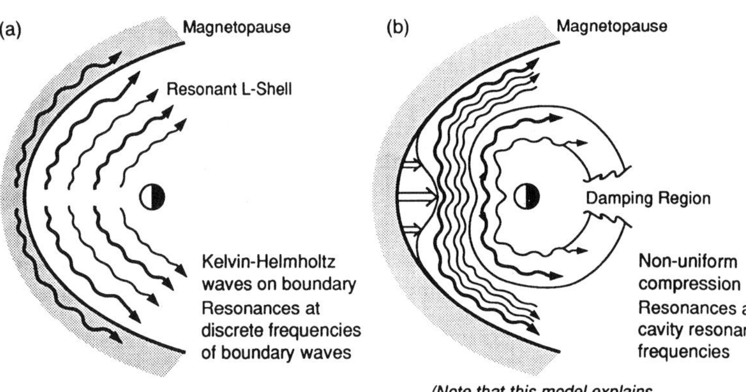

FIGURE 7.45 Wave perturbations throughout the day-side magnetosphere produced by (a) Kelvin–Helmholtz instablity or waves on the surface, and (b) a compression of the nose of the magnetic field. (Kivelson 1995)

and compressional waves propagate inwards into the magnetosphere, generally with decreasing wave amplitude; a large wave amplitude, however, is found at resonant $\mathcal{L}$-shells (Fig. 7.45). Through the propagation of waves, solar wind energy is transferred to a magnetosphere.

7.6.3 Plasma Waves

The dispersion relations for waves at higher frequencies are more complicated than for MHD waves. In addition to the general Maxwell equations, the fluid equations (conservation of mass, momentum and energy) for each species have to be considered. It follows from the dispersion relations that a plasma can support a large variety of electromagnetic (waves involving perturbations in both **E** and **B**), electrostatic (waves involving perturbations in **E** only) and magnetoacoustic waves. Waves can propagate parallel or perpendicular to **B**; longitudinal waves propagate along **E** ($\mathbf{k} \times \mathbf{E} = 0$), while transverse waves have $\mathbf{k} \cdot \mathbf{E} = 0$. Many waves are excited at or in resonance with the natural plasma frequencies: the electron and ion cyclotron frequencies ω_{Be} and ω_{Bi} (eq. 7.47), the electron and ion plasma frequencies ω_{pe} and ω_{pi} (eq. 7.90), and the upper and lower hybrid resonance frequencies ω_{UHR} and ω_{LHR}:

$$\omega_{UHR} = \sqrt{\omega_{pe}^2 + \omega_{Be}^2}, \tag{7.93a}$$

$$\omega_{LHR} = \sqrt{|\omega_{Be}\omega_{Bi}|}. \tag{7.93b}$$

Waves which propagate along **B** have their energy primarily in the wave magnetic field and are circularly polarized, while waves which propagate perpendicular to the field have their energy primarily in the electric field component and are linearly polarized. In the case of perpendicular propagation, the dispersion relation for *ordinary* (O) waves does not depend on the magnetic field **B**; waves with dispersion relations which do depend on **B** are referred to as *extra-ordinary* (X) waves. O-mode waves are linearly polarized, whereas X-mode waves are elliptically polarized. Waves are in resonance with the plasma when the index of refracion $n^2 \to \infty$, while waves do not propagate above or below the *cutoff frequency*, which is at the frequency where $n^2 \to 0$.

The maximum frequency at which plasma can respond to the presence of wave fields is the electron plasma frequency (Problem 7.27). Waves at higher frequencies escape from the magnetosphere, and are observed remotely at radio wavelengths (Section 7.5). Plasma waves at lower frequencies are generated and trapped locally; these can be and have been observed *in situ* by spacecraft. Waves are often named after the region they have been observed in (e.g., auroral hiss), the frequency range at which they are observed (radio waves, kilometric radiation), or the natural frequency of the plasma if this is close to the observed wave frequency (upper and lower hybrid waves, electron and ion cyclotron waves). Sometimes they are named after the 'sound' they make as received through a radio receiver (whistlers, chorus, hiss, lion roar). Observed waves are usually presented in the form of dynamic spectrograms, like the low frequency radio waves. Examples of auroral hiss, chorus and whistlers are shown in Figure 7.46.

Plasma waves have been observed in the magnetospheres of Earth and the four giant planets. Despite the large differences in size, magnetic orientation, energy sources and plasma content, many of the same types of wave modes are present in all five magnetospheres, although the relative and absolute intensities of the emissions differ from planet to planet. Whistler mode waves have been observed in all five magnetospheres, as well as in Ganymede's magnetic field. These waves may be triggered by a variety of processes, including lightning discharges in an atmosphere. The propagation time of the waves depends upon the group velocity. Since the higher frequency waves propagate fastest, the frequency

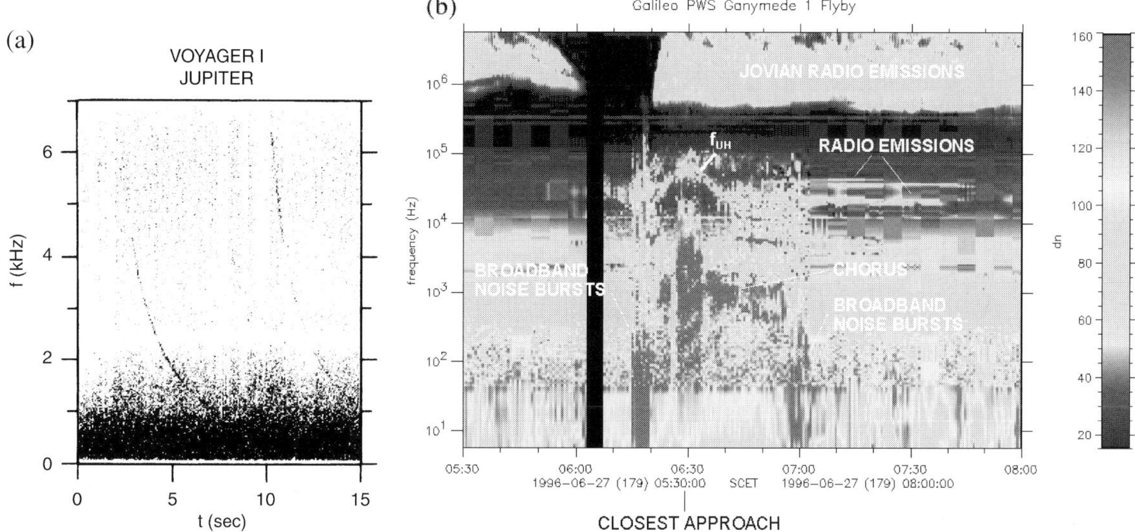

FIGURE 7.46 (a) Spectrogram of whistler mode waves in Jupiter's magnetosphere, as observed by Voyager 1. Increasingly intense waves are represented by increasingly darker shading in the spectrogram. (Kurth 1997) (b) COLOR PLATE An electric field spectrogram taken during one of Galileo's close approaches to Ganymede. This spectrogram shows a very strong interaction between Ganymede and the jovian magnetosphere, and provides strong evidence of a small magnetosphere surrounding Ganymede. The band of noise labeled f_{UH} is at the upper hybrid resonance frequency and has been used to determine a plasma density of approximately 100 particles per cubic centimeter. The broadband bursts at the beginning and end of the interaction period are typical of the plasma wave signature for a magnetopause, or boundary of a magnetosphere. The banded emissions after closest approach are electron cyclotron harmonic emissions which are known at Earth to contribute to the generation of the aurora. The bright, broadband emission centered on closest approach and the emissions identified as 'chorus' in the spectrogram are whistler-mode emissions. The maximum frequency of these emissions translates into a maximum magnetic field strength traversed by Galileo of about 4 mG. The narrowband radio emissions extending primary to the right of the Ganymede interaction in the spectrogram are the first known radio emissions from a planetary satellite; these are similar to radio emissions studied at Earth and the outer planets, including Jupiter. (Gurnett *et al.* 1996)

of the waves reaching an observer decreases over time (Fig. 7.46b), reminiscent of a whistling sound with a decreasing pitch. The rate at which the frequency decreases contains information on the plasma density. Broadband whistler noise is often called hiss or, if bursty, chorus. Earth, Jupiter and Neptune show distinct evidence of lightning induced whistlers. Electron (ECH) and ion (ICH) cyclotron emissions are also present in all magnetospheres, Voyager could only detect the ion cyclotron waves at places where the magnetic field strength was although strong enough to shift the wave mode into the observable frequency range. Broadband electrostatic noise (BEN) is often found at the magnetopause and plasma sheet boundaries, including the outer edge of the Io plasma torus. Jupiter's inner magnetosphere is characterized by a rich spectrum of plasma waves. Whistler mode hiss and chorus dominates the warm outer Io plasma torus. ICH has been detected from the inner cold torus, while VLF (very low frequency radio) hiss originates at the boundary between the warm and cold torus. This may be related to the Birkeland currents.

Plasma and waves can interact and transfer energy and momentum, i.e., a wave can either increase the energy of a plasma or take energy out. Pitch angle scattering and diffusion of particles by plasma waves is dominated by resonant interactions at harmonics of the particle's cyclotron frequency (for near equatorial particles the bounce period may be important as well). Whistler mode waves are probably the most important waves which trigger pitch angle diffusion and scattering, primarily because these waves are present throughout the magnetosphere. In the Earth's magnetosphere, the interactions of whistlers with energetic particles has lead to a separated inner and outer radiation belt by removing the energetic electrons from the region in between these two belts (Fig. 7.21). Under certain conditions the transfer of energy and momentum can lead to plasma instabilities.

7.7 Generation of Magnetic Fields

7.7.1 Magnetic Dynamo Theory

Magnetic fields around planets cannot be caused by

permanent magnetism in a planet's interior. The *Curie point* for iron is near 800 K; at higher temperatures iron loses its magnetism. Since planetary interiors are much warmer, all ferromagnetic materials deep inside a planet have lost their permanent magnetism. In addition, a permanent magnetism would gradually decay away, since the Ohmic dissipation time is given by:

$$t_D = \frac{4\pi\sigma_o \ell^2}{c^2}, \tag{7.94}$$

with σ_o the electrical conductivity, c the speed of light and ℓ a characteristic length scale. For Earth, the dissipation time turns out to be of the order of 10^4–10^5 years. The pattern of remnant magnetism observed near mid-ocean rifts implies that Earth's magnetic field has reversed its direction frequently during geologically recent times (Section 6.2.2.4); a permanent magnet would not behave in this manner. Paleomagnetic evidence from the magnetization of old rocks implies that the Earth's magnetic field has existed for at least $\sim 3.5 \times 10^9$ yr, and that its strength has usually been within a factor of 2 from its present value (except during field reversals), so there must be a mechanism which continuously produces magnetic field. Electric currents in the interior of a planet form the only plausible source of planetary magnetism. The outer core of the Earth is liquid nickel–iron, thus highly conductive; similarly, the interiors of Jupiter and Saturn are fluid and conductive (metallic hydrogen), and Uranus and Neptune have large ionic mantles. Electric currents are likely to exist in all these planets, and it is generally assumed that planetary magnetic fields are generated by a *magnetohydrodynamic dynamo*, a process which reinforces an already present magnetic field.

Electric currents are set up in a fluid which moves with respect to a magnetic field, provided the electrical conductivity is nonzero (eq. 7.14). Magnetic field lines tend to convect and distort their shape with the local motion of the fluid, while they diffuse through it (Fig. 7.47a). Slowly evolving magnetic fields are described by the magnetic induction equation, derived from Maxwell's equations:

$$\frac{\partial \mathbf{B}}{\partial t} = \nabla \times (\mathbf{v} \times \mathbf{B}) + \frac{c^2}{4\pi\sigma_o}\nabla^2 \mathbf{B}. \tag{7.95}$$

The first term on the right represents convection of field lines with the moving fluid, and the second term is diffusion of the field through the fluid. Obviously, fluid motions cannot lead to the spontaneous appearance of magnetic field if the field strength $B = 0$ to begin with. Equation (7.95), supplemented with the full equations of motion, governs the behavior of the dynamo magnetic field.

Consider a planet as an idealized rotating sphere of conducting fluid with an internal heat source. The heat source sets up a thermal gradient, which causes convection of the fluid. The inner parts of the planet rotate faster than the outer parts, due to the conservation of angular momentum. Hence the fluid sphere has a differential rotation. In a convection cell (Fig. 7.47a) there is a radially outward motion of the fluid along the axis of the convective eddy, with a general return flow in the surrounding region. Thus we find that the fluid is converging to the axis of the cell at the bottom, and diverging at the top. The Coriolis force associated with the converging part of the flow causes the fluid to turn locally around the axis of the convecting eddy, in the same direction as the overall rotation, similar to the motion in a cyclone. The divergence at the top of the eddy reduces the cyclonic motion, and may even reverse its direction.

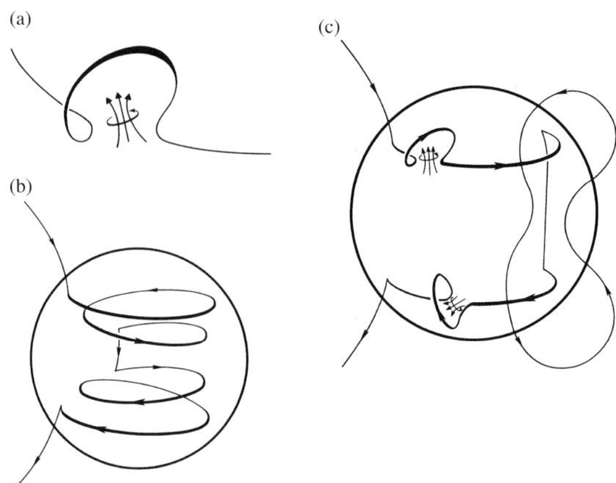

FIGURE 7.47 (a) Magnetic field lines tend to convect and distort their shape with the local motion of the fluid. Within the convection cell, a radially outward motion of the fluid occurs along the axis of the eddy, with a general return flow in the surrounding region. Fluid, therefore, is converging at the bottom and diverging at the top. (b) Differential rotation of a planet causes magnetic field lines to coil up. (c) Convective eddies can maintain the overall dipole field by regenerating the poloidal field: the convective fluid motion causes a loop in the toroidal field, in the same direction as in the poloidal field. (Levy 1986)

The magnetic field lines of a planet are rooted in the core. Differential rotation of the conducting part of the planet causes the magnetic field lines to coil up (Fig. 7.47b): this toroidal field can be a hundred times stronger than the original poloidal field (the polar field lines). Both the poloidal and toroidal fields, however, decay due to resistive dissipation of associated electric currents. The convective eddies can maintain the overall dipole field by regenerating the poloidal field, as sketched in Figure 7.47c: the convective fluid motion causes a loop in the toroidal field, in the same direction as the poloidal field, and therefore reinforces the planet's magnetic field.

This is the essence of the magnetic dynamo theory.

Although this theory was proposed near the middle of the twentieth century, it was impossible to solve the complicated nonlinear magnetohydrodynamic equations that govern the geodynamo until the mid-1990s, when two researchers developed a self-consistent three-dimensional computer code that describes the dynamic evolution of Earth's magnetic field, using realistic properties of Earth's core. This simulation shows that indeed the geodynamo model works, including the spontaneous reversals in field polarity (Section 7.7.2). The geodynamo theory, involving a conductive, convective fluid, tells us that a rotating planet must have an internal magnetic field if its interior contains a heat source to drive convection. We have seen that this statement is true for the giant planets, Earth and Mercury.

In analogy to the planet's gravitational field, one can further surmise that the dipole terms of a planet's magnetic field are induced deep in the body's interior, whereas the higher order terms must have an origin closer to the surface. This suggests that Uranus and Neptune's magnetic fields are probably induced in the ionic mantle regions, in contrast to the fields of Jupiter, Saturn, Earth and Mercury which originate deep in the planets' interiors. The interior structures of the Galilean satellites, discussed in Section 6.3.5.1, also seem consistent with the observed characteristics of internal magnetic fields, as discussed in this chapter. The interior structure of Io and Ganymede, being conductive, could support the generation of an internal magnetic field, while the highly nondipolar magnetic fields of Europa and Callisto are most likely induced in salty oceans below their crusts.

7.7.2 Variability in Magnetic Fields

If we subtract a dipole field from the Earth's magnetic field, we are left with only higher order moments of the field. It looks like about a dozen, continental sized patches of field are scattered around the globe. Both the dipole field and these higher order patches change over time: the magnetic dipole moves westwards and completes a complete circle in about 2000 years. The magnetic patches seem to develop and disappear on timescales of about 1000 years. At irregular intervals, between 10^5 and a few million years, the field reverses its polarity, as has been deduced from the magnetic polarity in rock samples of the oceanic ridge and volcanic lava flows (Section 6.2.2.4). It is clear that the Earth's magnetic field is not a static phenomenon; it is continuously changing, although the timescales involved are long compared to human lifetimes. With the field generated by electric currents in the Earth's interior, variability in the strength and pattern can be understood, in principle. Neither the field reversals, nor other excursions of the magnetic pole have been explained adequately, although three-dimensional computer simulations do predict such field reversals to take place. The reversals take place so quickly, on a geological timescale, that it is hard to find rocks preserving a record of such changes. From the (sparse) records, it seems that the intensity of the field decreases by a factor of 3–4 during the first few thousand years of a field reversal, while the field maintains its direction. The field then swings around a few times by $\sim 30°$, and finally moves to the opposite polarity. Afterwards, the intensity increases again to its normal value. The details of the changes in magnetic field configuration during these episodes, however, are not clear.

Whether similar changes occur in magnetic fields around other planets is not known. We do know that the Sun's magnetic field reverses every 11 years, as discussed in Section 7.1. One theory of the magnetic field data of Mars's crustal field suggests that Mars may have had tectonic plate movements and a magnetic dynamo including field reversals during the first few hundred million years after the planet formed (Section 7.4.3). Among the planets, only Jupiter's magnetic field can be monitored from Earth, via its radio emissions. However, only half a century has elapsed since the first detection of Jupiter's radio emission; we need a much longer time baseline to infer the presence of changes in its magnetic field. Moreover, the radio emissions depend on both the magnetic field configuration and particle distribution; it is not a simple task to disentangle the two.

Further Reading

Good review papers on planetary magnetospheres are can be found in the two encyclopedias mentioned in Chapter 1, and in *The New Solar System*, also referenced in Chapter 1. In addition we recommend the paper by:

Bagenal, F., 1992. Giant planet magnetospheres. *Annu. Rev. Earth Planet. Sci.* **22**, 289–328.

Two papers which discuss the magnetospheric effects resulting from the impact of Comet Shoemaker–Levy 9 with Jupiter are:

de Pater, I., F. van der Tak, R.G. Strom, and S.H. Brecht, 1997. The evolution of Jupiter's radiation belts after the impact of comet D/Shoemaker–Levy 9, *Icarus* **129**, 21–47.

Brecht, S.H., I. de Pater, D.J. Larson, and M.E. Pesses, 2001. Modification of the jovian radiation belts by Shoemaker–Levy 9: An explanation of the data. *Icarus* **151**, 25–38.

Basic books on magnetospheric physics and plasma physics:

Chen, F.F., 1974. *Introduction to Plasma Physics*, Plenum Press, New York.

Keen, B.E., Ed., 1974. *Plasma Physics*. The Institute of Physics, London.

Kivelson, M.G., and C.T. Russell, Eds., 1995. *Introduction to Space Physics*. Cambridge University Press, Cambridge.

Kurth, W.S., and D.A. Gurnett, 1991. Plasma waves in planetary magnetosphere. *J. Geophys. Res.*, **96**, 18,977–18,991.

Lyons, L.R., and D.J. Williams, 1984. *Quantitative Aspects of Magnetospheric Physics*. Reidel Publishing Company, Dordrecht.

Roederer, J.G., 1970. *Physics and Chemistry in Space 2: Dynamics of Geomagnetically Trapped Radiation*. Springer-Verlag, Berlin.

Schulz, M., and L.J. Lanzerotti, 1974. *Physics and Chemistry in Space 7: Particle Diffusion in the Radiation Belts*. Springer-Verlag, Berlin.

Shu, F.H., 1992. *The Physics of Astrophysics, Vol. II: Gas Dynamics*. University Science Books, Mill Valley, CA.

Problems

7.1.E Consider a comet on a keplerian orbit around the Sun. The perihelion distance is 0.3 AU, and aphelion distance is 15 AU. Assume a solar wind velocity of 400 km s^{-1}. Calculate the angle between the ion tail and the Sun–comet line both at perihelion and aphelion.

7.2.E Calculate the speed of sound and the Alfvén velocity in the solar wind at Earth's orbit. Compare your answers with the quiet solar wind velocity.

7.3.I (a) Calculate the mean free path between collisions in the solar wind, assuming quiescent solar wind properties (Table 7.1). Compare this number with the size of the Earth's magnetosphere and the thickness of the bow shock.

(b) If the mean free path between collisions is much larger than the typical size of the system, the plasma is collisionless. Is the solar wind a collisionless plasma by this definition? Explain why the solar wind is often treated as a fluid nevertheless.

7.4.I Calculate the approximate standoff distance of the Earth's magnetopause and bow shock, assuming quiescent solar wind properties (Table 7.1). Approximate the Earth's magnetic field as a dipole field, with a surface magnetic field strength of 0.3 G. You may ignore the plasma density in the outer magnetosphere.

7.5.I Show that in an ideal unmagnetized gas passing through a shock the following equation holds:

$$\frac{\rho_1}{\rho_2} = \frac{(\gamma+1)P_1 + (\gamma-1)P_2}{(\gamma-1)P_1 + (\gamma+1)P_2}, \tag{7.96}$$

where the subscript 1 refers to conditions before the shock (i.e., solar wind) and subscript 2 after the shock (i.e., magnetosheath). Calculate the ratio ρ_2/ρ_1 for a strong shock ($P_2 \gg P_1$) in a monatomic gas.

7.6.I Assume that the magnetic field and velocity in the solar wind parallel to the bow shock are both zero. The solar wind density is 5 protons cm^{-3}, the magnetic field 5×10^{-5} G and the temperature is 2×10^5 K. The solar wind velocity is 400 km s^{-1}. Calculate the density, velocity, temperature and magnetic field strength in the magnetosheath.

7.7.E The Gauss coefficients for Jupiter's magnetic field are: $g_1^0 = 4.218$, $g_1^1 = -0.664$ and $h_1^1 = 0.264$. Calculate the magnetic dipole moment, the angle between the magnetic and rotational angle and the longitude of the magnetic north pole.

7.8.I Derive the equations for the Larmor radius and frequency by balancing the centripetal force with the Lorentz force.

7.9.I (a) If changes in the magnetic field are small over one gyro radius and period, the magnetic field flux through the particle's orbit is constant. Derive the equation for the first adiabatic invariant (eq. 7.47).

(b) Express the first adiabatic invariant in terms of energy rather than momentum, both for relativistic and nonrelativistic particles (eqs. 7.48 and 7.50).

(c) Show that the two expressions for relativistic and nonrelativistic particles are equal at low energies.

7.10.I Consider a proton in the Earth's magnetic field at a distance $\mathcal{L} = 3$. Assume the Earth's field can be approximated by a dipole field with a magnetic moment $\mathcal{M}_B = 7.9 \times 10^{25}$ G cm^3, which is centered at the center of the planet and aligned with the rotation axis.

(a) Calculate the magnetic field strength at the particle's mirror point if the equatorial pitch angle of the particle $\alpha_e = 60°$, $\alpha_e = 30°$ and $\alpha_e = 10°$.

(b) Calculate the latitude of the mirror points for the three cases in (a).

(c) Compare your answers in (a) with the surface magnetic field strength, and comment on the results.

7.11.**I** Consider a proton and an electron in the Earth's magnetic field at a distance $\mathcal{L} = 2$. Both particles are confined to the magnetic equator, and have an energy of 1 keV. Assume the properties of Earth's magnetic field described in Problem 7.10.

(a) Calculate the gyro period for both particles.

(b) Calculate the drift orbital period for both particles.

(c) Calculate the orbital velocity around the Earth, and compare this with the keplerian velocity of a neutral particle at the same geocentric distance ($\mathcal{L} = 2$; assume a circular orbit). Describe what would happen if the ion undergoes charge exchange.

7.12.**I** Consider a particle in a dipole field at a distance $\mathcal{L} = 6$, which diffuses radially inwards until $\mathcal{L} = 2$. The latitude of the mirror points is indicated by θ_m, and the field strength by B_{m_1} at $\mathcal{L} = 6$ and B_{m_2} at $\mathcal{L} = 2$, respectively.

(a) Assume the magnetic latitude θ_m is the same for the mirror points B_{m_1} and B_{m_2}. Calculate the relative strength of the magnetic field at the mirror points: B_{m_1}/B_{m_2} and at the equator: B_{e_1}/B_{e_2}.

(b) Calculate the increase in the particle's kinetic energy when diffusing inwards from $\mathcal{L} = 6$ to $\mathcal{L} = 2$.

(c) Write an expression for the invariant $K_B = I_B\sqrt{B_m}$, in terms of B_{m_1} and B_{e_1} at both positions $r_1 = \mathcal{L} = 6$ and $r_2 = \mathcal{L} = 2$. Based upon this expression do you think B_{m_1} and B_{m_2} will be at the same latitude θ_m? Explain your result.

7.13.**E** (a) Calculate the total potential drop across the Earth's magnetosphere induced by the convection electric field.

(b) Calculate the potential drop over a distance equal to the diameter of the tail (50 $R_\oplus$) in the undisturbed solar wind ($v_{sw} = 400$ km s^{-1}, $B_{sw} = 5\ \gamma$).

(c) Compare your answers in (a) and (b) and determine the fraction of the IMF magnetic flux that reconnects with the geomagnetic field. (Hint: The potential drop would be equal to your answer in (b), if all IMF flux reconnects with the geomagnetic tail.)

7.14.**E** Assume Jupiter's magnetic field can be approximated by a dipole field with a magnetic moment $\mathcal{M}_B = 1.5 \times 10^{30}$ G cm^{-3}. The solar wind magnetic field strength near Jupiter is $\sim 2 \times 10^{-6}$ G. Calculate the gain in energy for a solar wind proton which enters Jupiter's magnetosphere and diffuses inwards until $\mathcal{L} = 1.5$. (Hint: Calculate the proton's energy assuming a solar wind velocity of 400 km s^{-1}.)

7.15.**E** (a) Estimate the critical radius, R_c, expressed in Earth radii, inside of which corotation dominates over convection, for particles near the dawn and dusk sides in the Earth's magnetosphere (see Problem 7.10 for typical geomagnetic parameters).

(b) Calculate the critical radius, R_c, expressed in jovian radii, inside of which corotation dominates over convection, for particles near the dawn and dusk sides in Jupiter's magnetosphere, using the magnetic field parameters as specified in Problem 7.14. Comment on the differences between the critical radii for Earth and Jupiter.

7.16.**I** (a) Calculate the effective potential for protons with an energy of 50 keV at the dawn and dusk sides in the Earth's magnetosphere, at $\mathcal{L} = 2$ and $\mathcal{L} = 4$.

(b) Calculate the effective potential for protons with an energy of 200 keV at the dawn and dusk sides in the Earth's magnetosphere, at $\mathcal{L} = 2$ and $\mathcal{L} = 4$.

(c) Calculate the effective potential for electrons with an energy of 200 keV at the dawn and dusk sides in the Earth's magnetosphere, at $\mathcal{L} = 2$ and $\mathcal{L} = 4$.

(d) Calculate the drift velocity for the above mentioned particles, and comment on your results (describe in words what happens to the particles drifting in the geomagnetic field).

7.17.**I** Show that the ring current tends to weaken the Earth's surface magnetic field.

7.18.**E** Describe how the orbit of a spacecraft which orbits the Earth 2000 km above its surface is influenced by the South Atlantic Anomaly. (Hint: Compare the keplerian orbital velocity with the orbital velocity of the particles.)

7.19.**I** (a) Calculate the location of the ionopause above Venus's surface. Assume the atmospheric pressure to drop with altitude according to the barometric law (Chapter 4), with a surface pressure of 90 bar. Make a reasonable estimate for the temperature based upon the various atmospheric profiles in Chapter 4. Assume a solar wind velocity of 400 km s^{-1}, solar wind density of 10 protons cm^{-3} and an interplanetary magnetic field strength of 10^{-4} G.

(b) Calculate the location of the ionopause above Mars's surface, assuming a surface pressure of 6 mbar, and estimate the atmospheric temperature from profiles given in Chapter 4. Assume that the solar wind density and interplanetary magnetic field strength scale with heliocentric distance as $r_{\odot}^{-2}$. (What is the value for the solar wind velocity?)

(c) Compare your results from (a) and (b), and comment on the comparison.

7.20.E (a) Calculate the energy involved when an ion in the Io plasma torus impacts Io. (Hint: Calculate the orbital velocity of Io and the ion, assuming both orbit Jupiter at a planetocentric distance of 6 $R_♃$.)

(b) Calculate the size of Io's neutral sodium cloud, if the lifetime for sodium atoms is a few hours.

(c) Explain in words why the banana-shaped neutral sodium cloud is pointed in the forward direction.

7.21.I (a) Calculate the scale height for sulfur and oxygen ions in Io's cold inner and hot outer torus.

(b) Estimate the ion production rate from Io if the ion density is 3000 particles cm^{-3}, and the torus stretches radially from 5.3 to 7.5 $R_♃$; assume the latitudinal extent to be equal to twice the scale height.

7.22.E (a) Calculate the typical energy of electrons which radiate at a wavelength of 20 cm, if the magnetic field strength is 0.8 G.

(b) Calculate the typical energy of electrons which radiate at a wavelength of 6 cm if the magnetic field strength is 0.8 G.

(c) If the energy distribution of the radiating electrons is flat, show the wavelength dependence of the observed radiation.

(d) The above numbers are valid for Jupiter's magnetosphere. Assuming the planet's magnetic field to resemble a dipole field with the parameters described in Problem 7.14, calculate the location in the magnetosphere from which the synchrotron radiation in (a) and (b) is emitted.

7.23.E Jupiter emits strong polarized bursts of radiation at frequencies $\nu \leq 40$ MHz. This radiation is thought to originate close to Jupiter's cloudtops, and is attributed to cyclotron radiation.

(a) Why is the radiation cyclotron and not synchrotron radiation?

(b) Calculate the magnetic field strength at Jupiter's cloudtops.

7.24.E Structure in SED bursts can be distinguished on timescales of 140 μs. Derive a size scale for the source of this emission.

7.25.E Neptune's magnetic dipole moment is 2.14×10^{27} G cm^3. The cutoff frequency of Neptune's decametric radiation is at 1.3 MHz.

(a) Calculate the magnetic field strength which corresponds to the cutoff frequency.

(b) Compare the magnetic field strength calculated in (a) with Neptune's surface dipole magnetic field strength. Explain the difference.

7.26.I Derive an expression for the frequency of a standing Alfvén wave, as a function of magnetic field strength and plasma density. Assume ℓ is the length of the field line, and n is the number of harmonics.

7.27.E Calculate the six natural plasma frequencies for a magnetosphere in which the magnetic field strength is 0.3 G and plasma density is 20 protons cm^{-3}. Assume the energy of the particles to be 20 kHz, and that there are only protons and electrons ($N_p = N_e$).

7.28.I The dispersion relation for electromagnetic waves is given by:

$$\omega_o^2 = \omega_{pe}^2 + k^2c^2. \tag{7.97}$$

(a) Express the index of refraction, n, in terms of ω_o and ω_{pe}.

(b) Assume there are two antennas in space. One of the antennas acts as a transmitter, and the other receives the signals. Show that by sweeping the signals in frequency, one can derive the plasma density between the two antennas.

8 Meteorites

I could more easily believe two Yankee professors would lie than that stones would fall from from heaven.
Attributed (probably incorrectly) to USA President Thomas Jefferson, 1807

A *meteorite* is a rock which has fallen from the sky. It was a *meteoroid* (or, if it was large enough, an asteroid) before it hit the atmosphere and a *meteor* while heated to incandescence by atmospheric friction. A meteor that explodes while passing through the atmosphere is termed a *bolide*. Meteorites that are recovered following observations of falling meteors are called *falls*, whereas those recognized in the field which cannot be associated with observed falls are referred to as *finds*.

The study of meteorites has a long and colorful history. Meteorite falls have been observed and recorded since ancient times (Fig. 8.1). Iron meteorites were an important raw material for some primitive societies. However, during the Renaissance it was difficult for many people (including scientists and other natural philosophers) to believe that stones could possibly fall from the sky, and reports of meteorite falls were sometimes treated with as much skepticism as UFO reports are given today. The extraterrestrial origin of meteorites became commonly accepted following the study of some well observed and documented falls in Europe around the year 1800. The discovery of the first four asteroids, celestial bodies of sub-planetary size, during the same period added to the conceptual framework which enabled scientists to accept extraterrestrial origins for some rocks.

Meteorites provide us with samples of other worlds that can be analyzed in terrestrial laboratories. The overwhelming majority of meteorites are pieces of small asteroids, which never grew to anywhere near planetary dimensions. Some meteorites, like the iron-rich bodies, presumably come from an asteroid's deep interior, while others are samples from asteroid crusts. As small objects cool more rapidly than do large ones, these bodies either never got very hot, or cooled and solidified early in the history of the Solar System. Many meteorites thus preserve a record of early Solar System history that has been wiped out on geologically active planets like the Earth. The clues that meteorites provide to the formation of our planetary system are the focus of most meteorite studies and also of this chapter.

8.1 Basic Classification and Fall Statistics

The traditional classification of meteorites is based upon their gross appearance. Many people think of meteorites as chunks of metal, because metallic meteorites appear quite different from ordinary terrestrial rocks. Museums also tend to specialize in metal meteorites, because most people find these odd-shaped pieces of nickel–iron interesting to look at. Metallic meteorites are made primarily of iron, with a significant component of nickel and smaller amounts of several other *siderophile* elements (elements which readily combine with molten iron, such as gold and silver); thus metal meteorites are referred to as *irons*. Meteorites that do not contain large chunks of metal, many of which are difficult for the untrained eye to distinguish from terrestrial rocks, are known as *stones*. Meteorites which contain comparable amounts of macroscopic metallic and rocky components are called *stony-irons*.

A more fundamental classification scheme is based on the thermal history of meteorites. All irons and stony-irons, as well as some stones known as *achondrites* (in contrast to chondrites, which are described below), come from *differentiated* parent bodies (i.e., bodies that have undergone density-dependent phase separation). Such bodies experienced an epoch in which they were mostly molten, and much of their iron sank to the center, taking with it siderophile elements. The bulk compositions of achondrites are enriched in *lithophile* and/or *chalcophile* elements; such elements tend to concentrate in the silicate and oxide phases of a melt, and their abundances are also enhanced in the Earth's crust. Relative to a solar mixture of refractory elements, achondrites are significantly depleted in iron and siderophile elements. *Primitive meteorites* have not melted; they are composed of material condensed

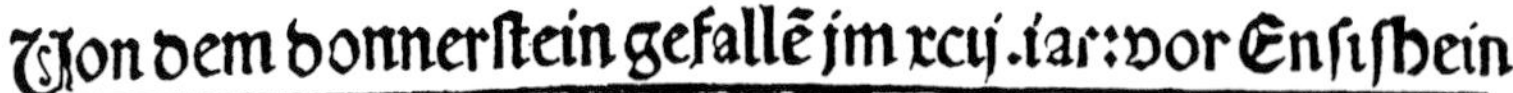

FIGURE 8.1 Woodcut depicting the fall of a meteorite near the town of Ensisheim, Alsace on 7 November 1492. A literal translation of the German caption reads 'of the thunder-stone (that) fell in 92 year outside of Ensisheim'. This meteorite is the oldest recorded fall from which material is still available.

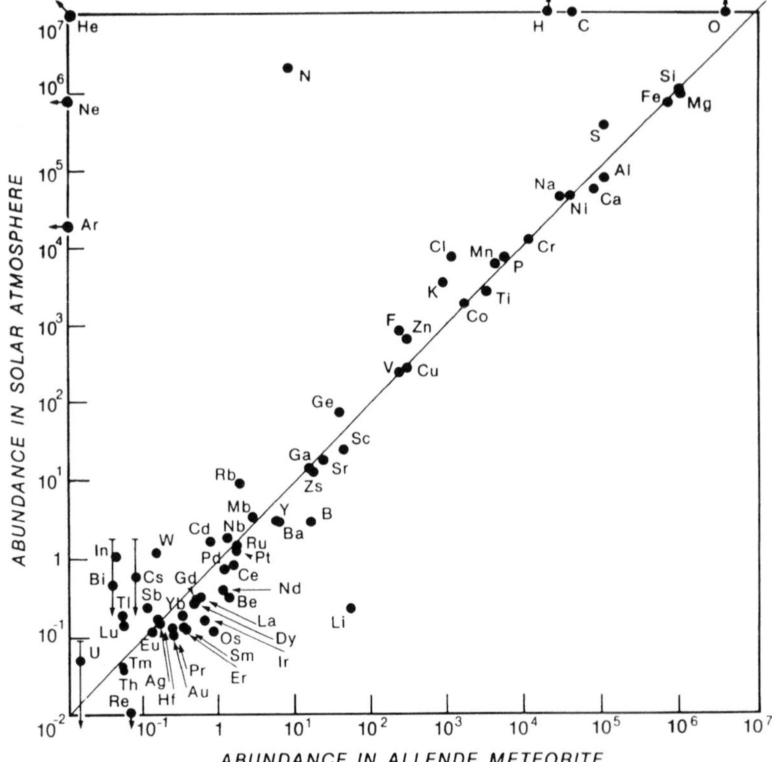

FIGURE 8.2 The abundance of elements in the Sun's photosphere plotted against their abundance in the Allende CV3 chondrites. Most elements lie very close to the curve of equal abundance (normalized to silicon). Several volatile elements lie above this curve, presumably because they are depleted in meteorites (rather than being enriched in the Sun), whereas only lithium, which is depleted in the solar photosphere because it is destroyed by nuclear reactions near the base of the Sun's convective zone, lies substantially below the curve.

directly from the solar nebula and surviving interstellar grains, modified in some cases by aqueous processing (implying that liquid water was once present) and/or thermal processing (implying that the body was quite warm at some time). Silicates, metals and other minerals are found in close proximity of one another within primitive meteorites. Primitive meteorites are called *chondrites* because most of them contain small, nearly spherical, igneous inclusions known as *chondrules*, which solidified from melt droplets. Some chondrules are glassy, implying that they cooled extremely rapidly. Apart from the most volatile elements, the composition of all chondrites is remarkably similar to that of the solar photosphere (Fig. 8.2). As meteorite compositions are easier to measure than are abundances in the Sun, analysis of chondrites provides the best estimates of the average Solar System composition of most elements (Table 8.1). Densities of meteorites vary from about 2.3 g cm^{-3} for volatile-rich porous achondrites to between 7 and 8 g cm^{-3} for irons.

Nine different classes of chondrites have been cataloged on the basis of composition and mineralogy; these classes are grouped into three sets. The most volatile-rich

TABLE 8.1 The Anders and Ebihara Table of Cosmic Abundances of the Elements Based on Analyses for CI Chondrites[a].

	Element	Atomic abundance (atoms/10^6 Si)	Orgueil concentration
1	H	2.72×10^{10}	20.2 mg/g
2	He	2.18×10^9	56 nL/g
3	Li	59.7	1.59 μg/g
4	Be	0.78	26.7 ng/g
5	B	24	1.25 μg/g
6	C	1.21×10^7	34.5 mg/g
7	N	2.48×10^6	3180 μg/g
8	O	2.01×10^7	464 mg/g
9	F	843	58.2 μg/g
10	Ne	3.76×10^6	203 pL/g
11	Na	5.70×10^4	4830 μg/g
12	Mg	1.075×10^6	95.5 mg/g
13	Al	8.49×10^4	8620 μg/g
14	Si	1.00×10^6	106.7 mg/g
15	P	1.04×10^4	1180 μg/g
16	S	5.15×10^5	52.5 mg/g
17	Cl	5240	698 μg/g
18	Ar	1.04×10^5	751 pL/g
19	K	3770	569 μg/g
20	Ca	6.11×10^4	9020 μg/g
21	Sc	33.8	5.76 μg/g
22	Ti	2400	436 μg/g
23	V	295	56.7 μg/g
24	Cr	1.34×10^4	2650 μg/g
25	Mn	9510	1960 μg/g
26	Fe	9.00×10^5	185.1 mg/g
27	Co	2250	509 μg/g
28	Ni	4.93×10^4	11.0 mg/g
29	Cu	514	112 μg/g
30	Zn	1260	308 μg/g
31	Ga	37.8	10.1 μg/g
32	Ge	118	32.2 μg/g
33	As	6.79	1.91 μg/g
34	Se	62.1	18.2 μg/g
35	Br	11.8	3.56 μg/g
36	Kr	45.3	8.7 pL/g
37	Rb	7.09	2.30 μg/g
38	Sr	23.8	7.91 μg/g
39	Y	4.64	1.50 μg/g
40	Zr	10.7	3.69 μg/g
41	Nb	0.71	250 ng/g
42	Mo	2.52	920 ng/g
44	Ru	1.86	714 ng/g
45	Rh	0.344	134 ng/g
46	Pd	1.39	557 ng/g

TABLE 8.1 Continued.

	Element	Atomic abundance (atoms/10^6 Si)	Orgueil concentration
47	Ag	0.529	220 ng/g
48	Cd	1.69	673 ng/g
49	In	0.184	77.8 ng/g
50	Sn	3.82	1680 ng/g
51	Sb	0.352	155 ng/g
52	Te	4.91	2280 ng/g
53	I	0.90	430 ng/g
54	Xe	4.35	8.6 pL/g
55	Cs	0.372	186 ng/g
56	Ba	4.36	2270 ng/g
57	La	0.448	236 ng/g
58	Ce	1.16	619 ng/g
59	Pr	0.174	90 ng/g
60	Nd	0.836	462 ng/g
62	Sm	0.261	142 ng/g
63	Eu	0.0972	54.3 ng/g
64	Gd	0.331	196 ng/g
65	Tb	0.0589	35.3 ng/g
66	Dy	0.398	242 ng/g
67	Ho	0.0875	54 ng/g
68	Er	0.253	160 ng/g
69	Tm	0.0386	22 ng/g
70	Yb	0.243	166 ng/g
71	Lu	0.0369	24.3 ng/g
72	Hf	0.176	119 ng/g
73	Ta	0.0226	17 ng/g
74	W	0.137	89 ng/g
75	Re	0.0507	36.9 ng/g
76	Os	0.717	590 ng/g
77	Ir	0.660	473 ng/g
78	Pt	1.37	953 ng/g
79	Au	0.186	145 ng/g
80	Hg	0.52	390 ng/g
81	Tl	0.184	143 ng/g
82	Pb	3.15	2430 ng/g
83	Bi	0.144	111 ng/g
90	Th	0.0335	28.6 ng/g
92	U	0.0090	8.1 ng/g

[a] See Anders and Ebihara (1982) for details and sources.

chondrites contain up to several percent carbon by mass, and are known as *carbonaceous chondrites*. Carbonaceous chondrites are divided into four major subgroups which differ slightly in composition and are denoted CI, CM, CO and CV. The most common primitive meteorites are the

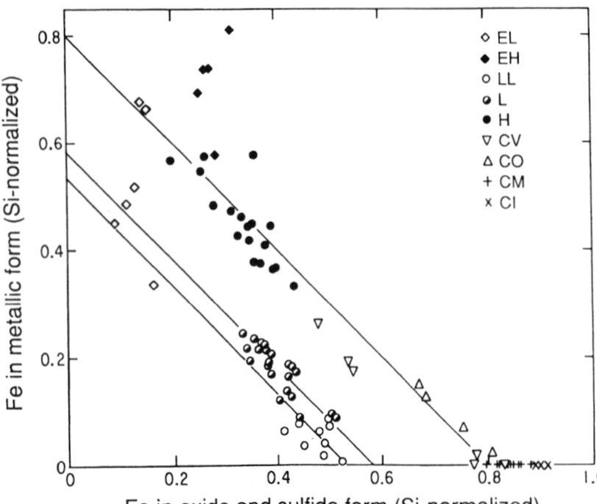

FIGURE 8.3 Plot showing the ratio of metallic iron to silicon as a function of Fe/Si in oxidized phases (mainly silicates) for various chondritic meteorites. The bulk Fe/Si for the meteorites determines the position of the diagonal, while the oxidation state of the Fe determines the location of a given point on the diagonal. (Kerridge and Matthews 1988)

ordinary chondrites, which are subclassified primarily on the basis of their Fe/Si ratio: H (high Fe), L (low Fe) and LL (low Fe, low metal; i.e., most of the iron that is present is oxidized). The third group of primitive meteorites, *enstatite chondrites*, is named after the dominant mineral in this group ($MgSiO_3$); enstatite chondrites are also divided on the basis of their iron abundance, and subgroups are denoted EH and EL. The iron abundance and its mineralogical location in various meteorites is diagrammed in Figure 8.3.

Although chondritic meteorites have never been melted, they have been processed to some extent in 'planetary' environments (i.e., in asteroid-like parent bodies) via thermal metamorphism, shock, brecciation (breaking up and reassembly) and chemical reactions often involving liquid water. Chondrites are assigned a *petrographic type* ranging from 1 in the most volatile-rich primitive meteorites to 6 in the most thermally equilibrated chondrites (Fig. 8.4). Type 3 chondrites appear to be the least altered in planetary environments, and provide the best data on the conditions within the protoplanetary disk. Types 2 and 1 show progressively more aqueous alteration. Type 1 are devoid of chondrules, which either never were present or have been completely destroyed by aqueous processing. In contrast, the degree of metamorphic alteration increases in higher numbered types above type 3, with type 7 sometimes being included to denote meteorites that have undergone partial melting.

Iron meteorites are classified primarily on the basis of their abundance of nickel and of the moderately volatile trace elements germanium and gallium. Compositional differences are correlated to observed differences in structure. Crystallization patterns seen in irons also depend on the rate at which the meteorite cooled. These intriguing structures, such as the famous *Widmanstätten pattern* (see Fig. 8.5), thus provide information about meteorite parent bodies. There are two classes of stony-irons: *pallasites*, which are closely related to irons, and *mesosiderites*, which are related to achondrites (Fig. 8.6). Pallasites consist of networks of iron-nickel alloy containing nodules of olivine typically ~ 5 mm in size. Pallasites are of igneous origin, and they probably formed at the interface between a region of molten metal and a magma chamber in which olivine could form and sink to the bottom, e.g., at a core–mantle interface. Mesosiderites contain a mixture of metal and magmatic rocks similar to *eucrite* achondrites. Several different types of achondrites exist, presumably from different regions of differentiated (or at least locally melted) asteroids of a variety of compositions and sizes. A few achondrites are from larger bodies: the Moon and Mars (Section 8.2).

The relative abundances of meteorites that have been collected and cataloged are not representative of the meteoroids whose orbits cross that of Earth. The overwhelming majority of meteorites collected after being observed to fall are stones, most of which are chondrites (Table 8.2). An even larger majority of stones has been found on Antarctic ice sheets, possibly because stones are more fragile than irons and fragments of a disrupted meteor are counted as one body when observed to fall, but presumably they are often multiply counted if found in Antarctica, where ice flows destroy evidence of strewn fields. Denser iron meteorites may also be more likely to sink through the ice, enhancing the relative abundance of stones. Irons are much more prevalent among non-Antarctic finds, as they are more resistant to weathering and they are also more easily identifiable as meteorites, or at least as unusual objects, to nonexperts. The total mass of cosmic debris impacting the Earth's atmosphere in a typical year is 10^{10}–10^{11} g. Micrometeorites $\sim$100 μm in radius account for most of the mass during most years, although the infrequent impacts of kilometer-sized and larger bodies dominate the flux averaged over very long timescales.

8.2 Source Regions

Meteorites are identified by their extraterrestrial origin. In theory, a rock could be knocked off the Earth, escape the Earth's gravity and orbit the Sun for a period of time before re-impacting Earth. Such a body would be a meteorite, despite its terrestrial origin. However, it is extremely difficult

FIGURE 8.4 Photographs of various chondritic meteorites. The scale bars are labeled in both centimeters and inches, unless otherwise noted. (a) Brownfield H3.7 ordinary chondrite, which fell in Texas in 1937. Very small chondrules, plus highly reflective metal and sulphide grains, can be picked out. (b) Parnallee LL3 ordinary chondrite, which fell in India in 1857. the cut surface clearly shows well-deliniated chondrules and slightly larger clasts. Parnallee is of very similar metamorphic grade to Brownfield, but it has a much coarser texture. (c) A cut surface of the ALH (Allan Hills, Antarctica) 77278 ordinary chondrite, which measures 8 cm across. Numerous round chondrules can be seen in this photograph. The sides of the small white square are 1 cm long. (NASA/Johnson) (d) Beardsley H5 ordinary chondite and matrix. (e) Dhurmsala LL6 chondrite resolution, indicate that it experienced thermal metamorphism. (f) Abee EH4 enstatite chondrite, which fell in Canada in 1953. The cut surface shows clearly the metal-rich and brecciated texture of Abee. The scale bar is labeled in centimeters.

FIGURE 8.4 Continued. (g) Vigarano CV3 carbaceous chondrite, which fell in Italy in 1910. Vigarano has a beautifully deliniated chondrites, as well as large CAIs. (h) Murchison CM2 carbonaceous chondrite, which fell in September 1969, in Victoria, Australia. Murchison is rich in indigenous (i.e. not terrestrial contaminent) amino acids and other organic molecules.

TABLE 8.2 Meteorite Classes and Numbers (as of 1997).

	Falls	Fall frequency %	Finds non-Antarctic	Finds Antarctic
Chondrites	803	86.1	1700	8497
Carbonaceous chondrites	33	3.5	28	160
Achondrites	73	7.8	49	391
Martian meteorites	4	0.4	2	6
Lunar meteorites	0		2	11
Stony-irons	12	1.3	57	29
Irons	45	4.8	681	65

Data from McSween (1999).

FIGURE 8.5 A cut acid-etched surface of the Maltahöhe iron meteorite. This surface shows a Widmanstätten pattern, an intergrowth of several alloys of iron and nickel which formed by diffusion of nickel atoms into solid iron during slow cooling within an asteroid's core. (Courtesy: Jeff Smith)

to accelerate rocks to escape speed from Earth without vaporizing them. Also, it would be very difficult to identify such rocks as meteorites, the only evidence being cosmic ray tracks and rare isotopes produced by cosmic rays; no such meteorite has ever been identified. However, *tektites*, which are rocks formed from molten material knocked off the Earth's surface by impacts, have been found in several locations. The geographic distribution of tektite finds and the lack of cosmic ray tracks in tektites imply that these rocks did not escape and quickly fell back to Earth's surface.

A few Antarctic meteorites are *anorthositic breccias* which, based upon a comparison with Apollo samples, are clearly of lunar origin. More than a dozen achondrite meteorites, including four falls (representing a total of $\sim 0.5\%$ of all known meteorite falls) are in the *SNC* class

FIGURE 8.6 Photographs of representative meteorites in the two major classes of stony-iron meteorites. (a) Estherville mesoderite, which fell in Iowa, USA, in 1879. Note Esterville's brecciated structure, with large pods of iron–nickel metal mixed with a seemingly random jumble of stony clasts. (b) Eagle Station pallasite, which was found in Kentucky, USA, in 1880. A continuous network of iron–nickel metal acts as a frame holding grains of forsteritic olivine. The scale bars are labeled both in centimeters and inches.

FIGURE 8.7 COLOR PLATE A cut surface of the martian meteorite ALH 77005. This rock contains dark olivine crystals and light-colored pyroxene crystals, plus some patches of impact melt, which were probably produced when it was ejected from the surface of Mars. The cube is 1 cm on a side. (NASA/Johnson)

(shergotites, nakhlites and chassigny, named after the places they were found), which are generally believed to be of martian origin. Figure 8.7 shows an SNC found in Antarctica. These rocks are young, with many having crystallization ages of $\sim 1.3 \times 10^9$ years. (Meteorite ages are measured using radiometric dating techniques, which are described in detail in Section 8.5.) There is evidence of a shock event (which reset some but not all radiometric clocks in these meteorites) $\sim 2 \times 10^8$ years ago. One SNC meteorite has a cosmic ray age of 2×10^6 years, indicating recent breakup of a parent $\gtrsim 1$ meter in diameter. The most convincing evidence for martian origin of the SNC meteorites is the similarity between their noble gas abundances and those measured in the martian atmosphere by the Viking landers. A 4.5×10^9 year old martian meteorite has also been recovered from Antarctica. This rock, ALH84001, is being extensively studied because it possesses several intriguing characteristics (including magnetite) which have been interpreted by some researchers as evidence for ancient life on Mars. Theoretical studies indicate that it is much easier for impacts to eject unvaporized rocks from Mars than from Earth because the velocity required to escape from Mars is less than half of Earth's escape velocity.

The overwhelming majority ($> 99\%$) of meteorites are from bodies of sub-planetary size. Two classes of such bodies are known in heliocentric orbits: comets and asteroids. Although the dividing line between these two sets of objects is not as sharp as once believed, either using orbital or physical characteristics, most objects are clearly in one class or the other. Are meteorites extinct comets depleted of volatiles? A few of the primitive chondrites may be, but 'evolved' objects (achondrites, irons) must come from asteroids. Comparison of the spectra of reflected light from several types of meteorites with asteroid spectra

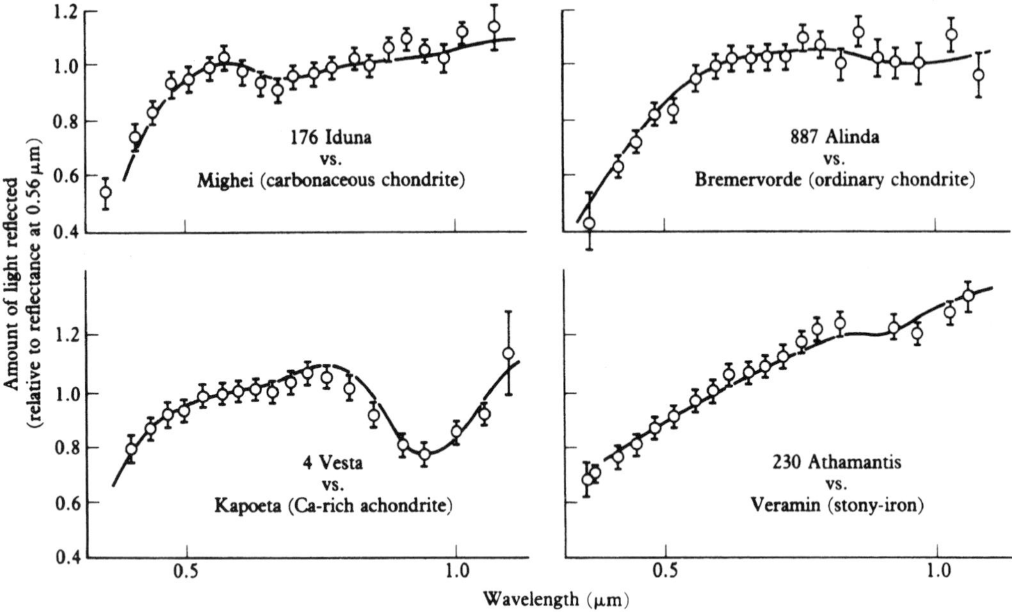

FIGURE 8.8 A comparison of the reflection spectra of four asteroids (points with error bars) with meteorite spectra as determined in the laboratory. (Morrison and Owen 1996)

yields many close correspondences (Fig. 8.8). This enables us to determine many compositional facts about asteroids otherwise very difficult or impossible to obtain via remote sensing from Earth.

The pre-impact orbits have been determined for only four meteorite falls, all of which are ordinary chondrites. All four orbits had perihelia near Earth's orbit, penetrated the asteroid belt, but remained well interior to Jupiter's orbit (Fig. 8.9).

The typical time for a large object to reach Earth from source bodies in the main asteroid belt is much longer than the age of the Solar System, except near certain strong resonances. Poynting–Robertson drag (Section 2.7.2) can move small meteorites to the vicinity of resonances which can transport them into Earth-crossing orbits with characteristic lifetimes of 10^7 years. Earth-crossing asteroids provide another source of meteoroids with travel times of $\sim 10^7$ years. These short intervals are reasonable, but note that a source is required for Earth-crossing asteroids, whose dynamical lifetime is far less than the age of the Solar System (Section 9.1).

8.3 Fall Phenomena: Atmospheric Entry to Impact

Meteoroids encounter the Earth's atmosphere at speeds ranging from 11 km s^{-1} to 73 km s^{-1}, with typical velocities of $\sim$15 km s^{-1} for bodies of asteroidal origin and $\sim$ 30 km s^{-1} for cometary objects (Problem 8.1). At such velocities, meteoroids have substantial kinetic energy per unit mass, enough energy to completely vaporize the bodies if it were converted to heat.

In the rarefied upper portion of a planet's atmosphere, gas molecules independently collide with the rapidly moving meteoroid. Interactions with this tenuous gas are dynamically insignificant for large meteoroids, but are able to retard the motion of tiny bodies substantially (cf. eq. 8.3*a*). *Micrometeoroids* smaller than $\sim$10–100 μm are able to radiate the heat they acquire from this drag rapidly enough that they can reach the ground. These very small particles are known as *micrometeorites*. Micrometeorites, such as the one pictured in Figure 8.10, have been recovered using specially equipped jets flying in the stratosphere (these grains are sometimes referred to as *Brownlee particles*) and also from sediments found in certain locations in Greenland, Antarctica and the ocean bottom. Laboratory analysis has confirmed the extraterrestrial origin of this dust, and suggests that much of it is likely to have been incorporated in cold, volatile-rich bodies such as comets.

The surface of a meteor is heated by radiation from the atmospheric shock front that it produces. Meteors rarely get significantly hotter than 2000 K, because this temperature is sufficient to cause iron and silicates to melt (Fig. 8.11). The liquid evaporates and/or just falls off the

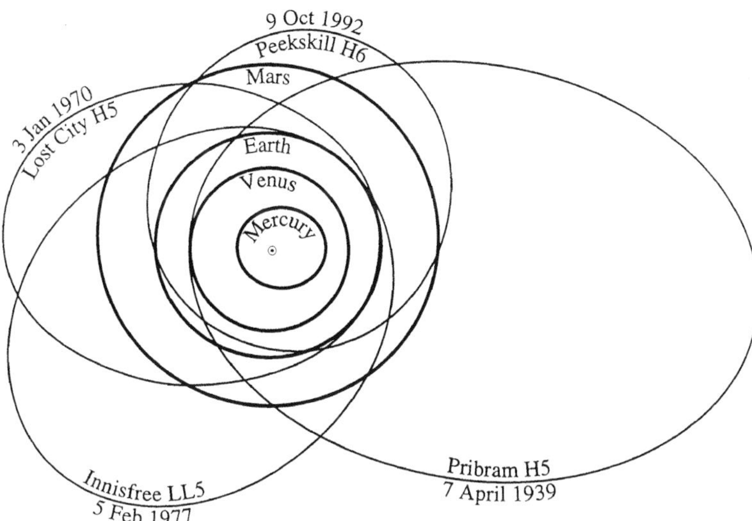

FIGURE 8.9 Pre-impact orbits of the four recovered meteorites that were photographed well enough to allow for the calculation of accurate trajectories. (Adapted from Lipschutz and Schultz 1999)

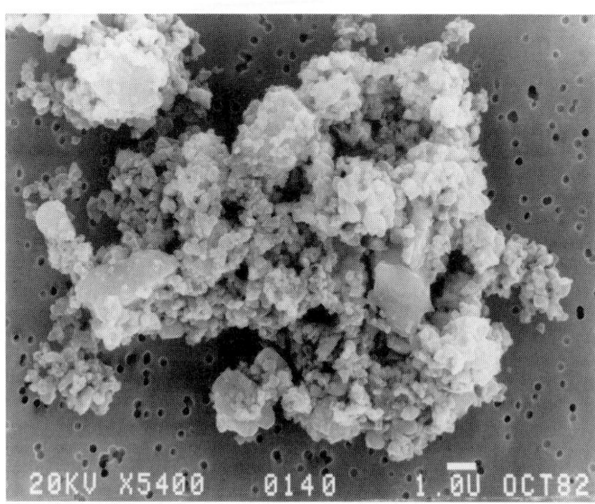

FIGURE 8.10 Scanning electron microscope image of a micrometeorite that was once an interplanetary dust particle. Note the fluffy fractal-like structure (compare with Figure 12.9). (Courtesy: Donald Brownlee)

meteor, so ablation provides an effective thermostat. The decrease in the meteor's mass is given by:

$$Q\frac{dm}{dt} = -\frac{1}{2}C_H\rho_g A v^3\left(\frac{v^2 - v_c^2}{v^2}\right), \tag{8.1}$$

where Q is the heat of ablation, and C_H is the heat transfer coefficient. The value of C_H is approximately 0.1 above an altitude of 30 km in Earth's atmosphere and varies inversely with atmospheric density at lower altitudes. The heat of ablation is $\sim 5 \times 10^{10}$ ergs g^{-1} for stony and iron meteorites. Ablation is only important for velocities $v > v_c$, where the critical velocity, v_c, above which radiative cooling cannot release energy rapidly enough to prevent material at the surface of the meteor from melting, is approximately 3 km s^{-1} for iron meteorites. Most visible meteors completely vaporize in the atmosphere. The light emitted by the hot atmospheric gas and ablated meteoritic material forms the familiar meteors in our atmosphere. Most visible meteors are from centimeter-sized bodies. The initial mass necessary for part of a meteoroid to make it to the ground depends on its initial velocity and composition. The surface of a meteor can become hot enough to melt rock to a depth of ~1 mm (~1 cm for irons, which have a higher thermal conductivity). However, interior to this hot outer skin the temperature stays just under 0 °C (Problem 8.6), so a meteorite is cold a few minutes after hitting the surface of Earth. The crust is melted and greatly altered (becoming magnetized for iron-rich meteorites), but the interior of the meteorite is basically undisturbed.

Larger cosmic intruders continue to travel at *hypersonic* (substantially supersonic) speeds as they penetrate into denser regions of the planet's atmosphere, and therefore they induce a shock in the gas in front of them. The leading face of a meteor is subject to an average pressure, P, given approximately by the formula:

$$P \approx \frac{C_D\rho_g v^2}{2}, \tag{8.2}$$

where v is the velocity of the meteor, C_D is the drag coefficient (which is approximately unity for a sphere) and ρ_g is the local density of the atmosphere. When this pressure exceeds the tensile strength of the meteor, the meteor is

(a)

(b)

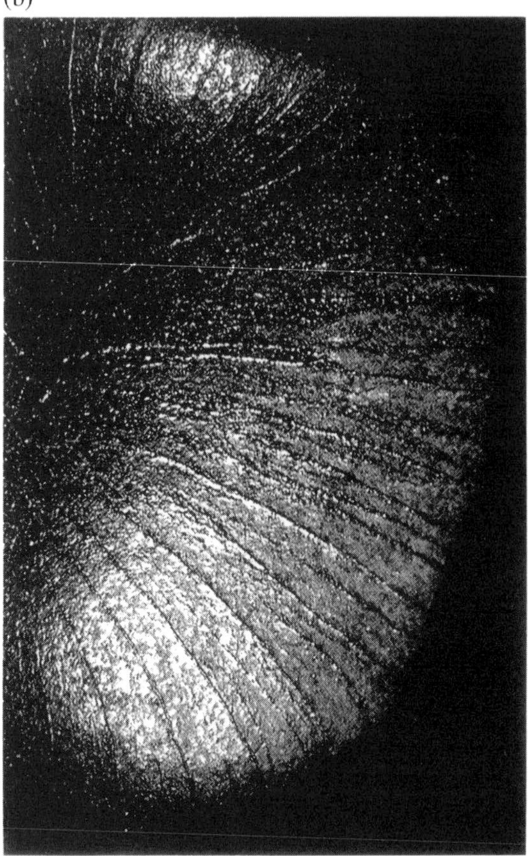

FIGURE 8.11 (a) Carbonaceous chondrite meteorite ALH 77307, which was found in Antarctica. The meteorite's rounded surface was sculpted during its passage through Earth's atmosphere. The cracked surface was produced by subsequent weathering. The cube is 1 cm on a side. (NASA/Johnson) (b) The Lafayette (Indiana) nakhlite (martian meteorite) shows an exquisitely preserved fusion crust. During the stone's rapid transit through the atmosphere, air friction melted its exterior. The lines trace beads of melted rock streaming away from the leading edge. (Courtesy: Smithsonian Institution)

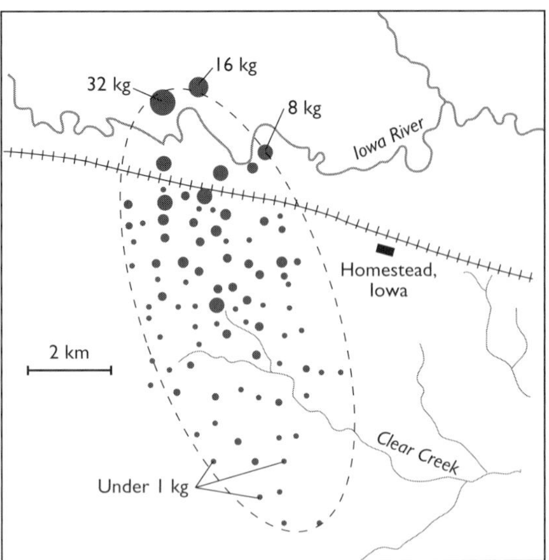

FIGURE 8.12 Illustration of the strewn field in the Homestead, Iowa, USA, meteorite shower, which occurred on 12 February 1875. Note that larger objects landed at the far end of the ellipse. (Adapted from O. Farrington)

likely to fracture and disperse (cf. Problem 8.3), with the resulting debris scattered over a *strewn field* (Fig. 8.12). The meteor is slowed by this pressure, but it continues to be accelerated by the planet's gravity, so its velocity varies as:

$$\frac{d\mathbf{v}}{dt} = -\frac{C_D \rho_g A v}{2m}\mathbf{v} - g_p \hat{\mathbf{z}}, \tag{8.3a}$$

where m is the mass of the meteor, A is its projected surface area in the direction of motion, g_p is the gravitational acceleration, and $\hat{\mathbf{z}}$ is the unit vector pointing in the upwards direction. A meteor loses a substantial fraction of its initial kinetic energy if it passes through a column of atmosphere with a mass equal to its own.

For shallow angles of incidence, the meteoroid may skip off the atmosphere back into space, just like skipping stones on water. In 1996 a meteor was observed to skip off our atmosphere, and slightly less than one orbit later it re-entered the atmosphere and landed in southern California.

The fate of centimeter- to meter-sized meteoroids depends primarily on their vertical velocity at atmospheric entry, v_{z0}. Rapidly moving ($v_{z0} > 15$ km s^{-1}) meteoroids of this size tend to be ablated away, whereas slowly moving ($v_{z0} < 10$ km s^{-1}) similar objects tend to be aerobraked to the *terminal velocity*, v_∞, at which gravitational acceleration balances atmospheric drag:

$$v_\infty = \frac{2 g_p m}{C_D \rho_g A}. \tag{8.3b}$$

Stony meteors in the ~1–10 m size range tend to break

up in the atmosphere; the fragments are aerobraked to terminal velocity (eq. 8.3b) and are often recovered in many pieces spread out over strewn fields. Stony bodies between ~10 and 100 m in size continue at high speed to deeper and denser levels in the atmosphere, where they can be disrupted by ram pressure of the dense atmosphere. The giant (~ 5×10^{23} erg) explosion that occurred over Tunguska, Siberia in 1908 was probably produced by the disruption of a stony boloid roughly 60 m in diameter about 10 km above the Earth's surface (see also Section 5.4.6). Meteoroids larger than ~100 m generally reach the surface with high velocity, because even if they are flattened out by ram pressure from their interactions with the atmosphere, they still collide with an amount of gas totalling much less than their own mass. In contrast, iron meteorites are very cohesive, and iron boloids over a broad size range can reach Earth's surface moving with sufficient velocity to produce impact craters, such as the famous Meteor Crater in Arizona (Fig. 5.21). Cratering of planetary surfaces by hypervelocity impacts is described in detail in Section 5.4. Impact erosion of planetary atmospheres is discussed in Section 4.8.3.

8.4 Chemical and Isotopic Fractionation

Meteorites provide the oldest and most primitive rocks available for study in terrestrial laboratories. Analysis of meteorites yields important clues as to how the planets themselves formed. The primary information on early Solar System conditions obtained from meteorites is given by their chemical and isotopic composition, and variations thereof between different parts of individual meteorites and among meteorites as a group. In this section, we provide some of the basic chemistry background necessary to understand these results. In Section 8.5, we discuss how some observed variations can be used to determine the ages of meteorites. The mineralogical structure of meteorites also yields information on conditions in the early Solar System, as does remnant ferromagnetism detected in some meteorites; these subjects are reviewed in Sections 8.6 and 8.7.

8.4.1 Chemical Separation

In a sufficiently hot gas or plasma, atoms become well-mixed, with locations independent of species. Provided diffusion has had sufficient time to erase initial gradients and turbulence is large enough to prevent gravitational settling of massive species, a gas is generally well-mixed at the molecular level. Solid bodies, however, tend to mix very little, retaining the molecular composition that they aquired when they solidified. When solids form from a gas or from a melt, molecules tend to group with mineralogically compatible counterparts, and distinct minerals are formed, which have elemental compositions that can be considerably different from the bulk composition of the mixture (Section 5.2). Condensation from a gas can produce small grains, which form a heterogeneous mixture. In contrast, crystallization from a melt allows greater separation of materials, producing samples with large scale heterogeneities. On an even larger scale, the combination of chemical separation in a melt and density-dependent settling results in planetary differentiation (Section 5.2). However, under most circumstances, the isotopic composition of each element usually remains uniform across mineral phases.

Analysis of the most primitive meteorites known implies that to a good approximation the material from which the planets formed was well-mixed over large distance scales on both the isotopic and elemental level. Gross chemical differences result primarily from temperature variations within the protoplanetary disk. Exceptions to isotopic homogeneity have been used to determine the age of the Solar System, and to show that at least a small amount of pre-solar grains survived intact and never melted or vaporized before being incorporated into planetesimals (Sections 8.6 and 8.7).

8.4.2 Isotopic Fractionation

Although different isotopes of the same element are chemically identical, there are several physical and nuclear processes that can produce isotopic inhomogeneities. Sorting out these different processes is essential to use the isotope data obtained from meteorites. One explanation for isotopic differences between grains is that they came from different reservoirs which were never mixed, e.g., interstellar grains which formed in distinct parts of the galaxy from material with different nucleosynthesis histories. This explanation can have profound consequences for our understanding of planetary formation, so other processes must also be considered.

Isotopes can be separated from one another by mass-dependent processes. These processes can rely on gravitational forces, such as the preferential escape of lighter isotopes from a planet's atmosphere, or as the result of molecular forces, such as the preference for deuterium (as opposed to ordinary hydrogen) to bond with heavy elements, which is a consequence of a slightly lower energy resulting from deuterium's greater mass. Mass-dependent fractionation is easy to identify for elements such as oxygen which have three or more stable isotopes, because the degree of

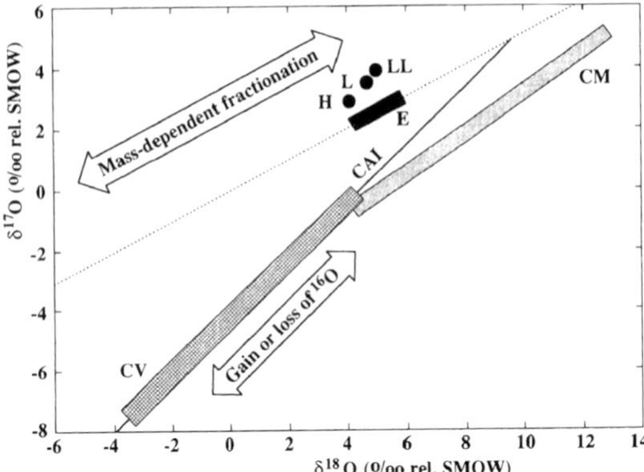

FIGURE 8.13 Plot showing the distribution of the three stable oxygen isotopes in various Solar System bodies relative to the standard (terrestrial) mean ocean water (SMOW). The dotted line represents the mass-dependent fractionation pattern observed in terrestrial samples. (Kerridge 1993)

fractionation is proportional to the difference in mass. One can plot differences in the $^{17}O/^{16}O$ ratio against the corresponding difference in $^{18}O/^{16}O$. Mass-dependent fractionation leads to points along a line of slope 0.52 (Fig. 8.13; the fractionation slope is slightly larger than 1/2 because $17/16 > 18/17$). The deviations of meteorite data from this line are believed to result from incomplete mixing of material from differing nucleosynthesis sources (cf. Section 8.7.2).

Nuclear processes can also lead to isotopic variations. Paramount among these is radioactive decay, which can transform chemical variations into isotopic differences. Cosmic rays produce a variety of nuclear reactions. Energetic particles from local radioactivity may also induce nuclear transformations.

8.5 Radiometric Dating

There are several different ages that may be assigned to a given meteorite. All of these meteorite ages are determined by *radionuclide dating*.

The most fundamental age of a meteorite is its *formation age*, which is often referred to simply as the age. The formation age is the length of time since the meteorite was last molten or gaseous. Most meteorites have ages of 4.53–4.57×10^9 years, but a few are much younger. Techniques for determining meteorite formation ages are based upon radioactive decay of long-lived radioactive isotopes. These techniques are presented in Section 8.5.2. The relative ages of individual chondrites can be measured more precisely than their absolute ages using short-lived extinct radioactive isotopes. This technique, which can be used to determine an upper bound of 2×10^7 to the *formation interval* of chondritic meteorites, is described in detail in Section 8.5.3.

Some isotopes of noble gases such as helium, argon and xenon are produced by radioactive decay and build up in rocks as time progresses, but can be lost if the rock is fractured or heated. The abundance of such radiogenic gases can be used to determine the *gas retention age* of the rock. Usually this gas retention age is less than the formation age, but for some rocks these two ages are equal. As lighter noble gases diffuse more readily than the heavy ones, some events can lead to the loss of most of a rock's helium, but little of its argon, resulting in a meteorite with a younger helium retention age than its argon retention age.

The time between the final epoch of nucleosynthesis that the material underwent prior to incorporation in the solar nebula and condensation can also be estimated from isotopic measurements. The techniques used for obtaining such estimates are described in Section 8.5.4. Finally, nuclear reactions produced by cosmic rays can be used to determine how long the meteorite existed as a small body in space, and the time at which it reached Earth (cf. Section 8.5.5). Additional information on radiometric dating can be found in Tilton's article on p. 249 of Kerridge and Matthews (1988).

8.5.1 Radioactive Decay

Many naturally occurring isotopes are *radioactive*, that is they spontaneously decay into isotopes of other elements that are usually of lesser mass. Radioactive decay rates can be accurately measured; thus the abundances of decay products provide precise clocks that can be used to reconstruct the history of many rocks. The most common types of radioactivity are *β decay*, whereby a nucleus emits an electron, and *α decay*, in which a helium nucleus (composed of two protons and two neutrons) is emitted. When an atomic nucleus undergoes β decay, a neutron within the nucleus is transformed into a proton, so the atomic number increases by one; the total number of nucleons (protons plus neutrons) remains fixed, so the atomic mass number of the nucleus does not change and the actual mass of the nucleus decreases very slightly. An example of β decay is the following transformation of an isotope of rubidium into one of strontium: $^{87}_{37}\mathrm{Rb} \rightarrow {}^{87}_{38}\mathrm{Sr}$; the number shown to the upper left of the element symbol is the atomic weight of the isotope (number of nucleons), and the number to the lower left (which is usually omitted because it is redundant with the name of the element) is the atomic number (number of protons). When a nucleus undergoes α decay, its atomic number decreases by two and its atomic mass decreases by four; an example of α decay from uranium to thorium is ${}^{238}_{92}\mathrm{U} \rightarrow {}^{234}_{90}\mathrm{Th}$. Some heavy nuclei decay via *spontaneous fission*, which produces at least two nuclei more massive than helium, as well as smaller debris. Spontaneous fission is an alternative decay mode for ${}^{238}_{92}\mathrm{U}$ (and also of the now almost extinct ${}^{244}_{94}\mathrm{Pu}$), leading typically to xenon and lighter byproducts. Proton-rich nuclei can undergo *inverse β decay* (positron emission), decreasing their atomic number by one; a closely related process is *electron capture*, whereby an atom's inner electron is captured by the nucleus; both of these processes convert a proton into a neutron. An example of a decay that can occur via positron emission or electron capture is ${}^{40}_{19}\mathrm{K} \rightarrow {}^{40}_{18}\mathrm{Ar}$; however potassium 40 ($\beta$) decays into calcium (${}^{40}_{20}\mathrm{Ca}$) eight times more frequently than it decays into argon.

The time required for an individual radioactive nucleus to decay is not fixed; however, there is a characteristic lifetime for each radioactive isotope. The probability that a nucleus will decay in a specified interval of time does not depend on the age of the nucleus, so the number of atoms of a given radioactive isotope remaining in a sample drops exponentially if no new atoms of this isotope are produced. The timescale over which this process occurs can be characterized by the mean lifetime of a species, τ_m, or the *decay constant*, τ_m^{-1}. The abundance of a 'parent' species at time t is related to its abundance at t_0 as:

$$N_p(t) = N_p(t_0)e^{-(t-t_0)/\tau_m}. \tag{8.4}$$

Alternatively, the *half-life* of the isotope, $t_{1/2}$, which represents the time required for half of a given sample to decay, can be used to quantify the decay rate. The relationship between these quantities is:

$$t_{1/2} = \ln 2\, \tau_m. \tag{8.5}$$

The half-lives of isotopes commonly used to date events in the early Solar System are given in Table 8.3. Many isotopes produced by radioactive decay are themselves unstable. In many cases, these 'daughter' isotopes have shorter half-lives than their 'parents'. A sequence of successive radioactive decays leading to a stable or nearly stable isotope is referred to as a *decay chain*. Two decay chains which are important in meteorite evolution are:

TABLE 8.3 Half-lives of Selected Isotopes.

Parent	Measurable stable daughter(s)	Half-life $t_{1/2}$
Long-lived radionuclides		
^{40}K	^{40}Ar, ^{40}Ca	1.25 Gyr
^{87}Rb	^{87}Sr	48.8 Gyr
^{147}Sm	^{143}Nd, ^{4}He	106 Gyr
^{187}Re	^{187}Os	46 Gyr
^{232}Th	^{208}Pb, ^{4}He	14 Gyr
^{235}U	^{207}Pb, ^{4}He	0.704 Gyr
^{238}U	^{206}Pb, ^{4}He	4.47 Gyr
Extinct radionuclides		
^{22}Na	^{22}Ne	2.6 yr
^{26}Al	^{26}Mg	0.72 Myr
^{41}Ca	^{41}K	0.1 Myr
^{53}Mn	^{53}Cr	3.6 Myr
^{60}Fe	^{60}Ni	1.5 Myr
^{107}Pd	^{107}Ag	6.5 Myr
^{129}I	^{129}Xe	17 Myr
^{182}Hf	^{182}W	9 Myr
^{244}Pu	${}^{131-136}$Xe	82 Myr

$$\begin{aligned}
&{}^{238}_{92}\mathrm{U} \xrightarrow[t_{1/2}=4.51\times10^{9}\mathrm{y}]{\alpha} {}^{234}_{90}\mathrm{Th} \xrightarrow[21.4\mathrm{d}]{\beta} {}^{234}_{91}\mathrm{Pa} \xrightarrow[6.75\mathrm{h}]{\beta} {}^{234}_{92}\mathrm{U}\\
&\xrightarrow[2.47\times10^{5}\mathrm{y}]{\alpha} {}^{230}_{90}\mathrm{Th} \xrightarrow[8\times10^{4}\mathrm{y}]{\alpha} {}^{226}_{88}\mathrm{Ra} \xrightarrow[1600\mathrm{y}]{\alpha} {}^{222}_{86}\mathrm{Rn}\\
&\xrightarrow[3.8\mathrm{d}]{\alpha} {}^{218}_{84}\mathrm{Po} \xrightarrow[3\mathrm{m}]{\alpha} {}^{214}_{82}\mathrm{Pb} \xrightarrow[27\mathrm{m}]{\beta} {}^{214}_{83}\mathrm{Bi} \xrightarrow[19.7\mathrm{m}]{\beta} {}^{214}_{84}\mathrm{Po}\\
&\xrightarrow[1.64\times10^{-4}\mathrm{s}]{\alpha} {}^{210}_{82}\mathrm{Pb} \xrightarrow[21\mathrm{y}]{\beta} {}^{210}_{83}\mathrm{Bi} \xrightarrow[5\mathrm{d}]{\beta} {}^{210}_{84}\mathrm{Po}\\
&\xrightarrow[183\mathrm{d}]{\alpha} {}^{206}_{82}\mathrm{Pb}\ (\mathrm{stable}),
\end{aligned} \tag{8.6a}$$

$$\begin{aligned}
&{}^{235}_{92}\mathrm{U} \xrightarrow[t_{1/2}=7.1\times10^{8}\mathrm{y}]{\alpha} {}^{231}_{90}\mathrm{Th} \xrightarrow[25.5\mathrm{h}]{\beta} {}^{231}_{91}\mathrm{Pa} \xrightarrow[3.25\times10^{4}\mathrm{y}]{\alpha} {}^{227}_{89}\mathrm{Ac}\\
&\xrightarrow[21.6\mathrm{y}]{\beta} {}^{227}_{90}\mathrm{Th} \xrightarrow[18.5\mathrm{d}]{\alpha} {}^{223}_{88}\mathrm{Ra} \xrightarrow[11.43\mathrm{d}]{\alpha} {}^{219}_{86}\mathrm{Rn} \xrightarrow[4\mathrm{s}]{\alpha} {}^{215}_{84}\mathrm{Po}\\
&\xrightarrow[1.8\times10^{-3}\mathrm{s}]{\alpha} {}^{211}_{82}\mathrm{Pb} \xrightarrow[36\mathrm{m}]{\beta} {}^{211}_{83}\mathrm{Bi} \xrightarrow[2.15\mathrm{m}]{\alpha} {}^{207}_{81}\mathrm{Tl}\\
&\xrightarrow[4.8\mathrm{m}]{\beta} {}^{207}_{82}\mathrm{Pb}\ (\mathrm{stable}).
\end{aligned} \tag{8.6b}$$

Note that the first decay in each of these chains takes of order 10^9 years, but subsequent decays are much more rapid.

8.5.2 Dating Rocks Containing Radioactive Isotopes

With the passage of time, the abundance of radioactive 'parent' isotopes in a rock, $N_p(t)$ decreases, as these atoms decay into 'daughter' species (or 'granddaughter' etc. isotopes if the initial decay products are unstable with short half-lives, e.g., expressions (8.6)). The abundance of the

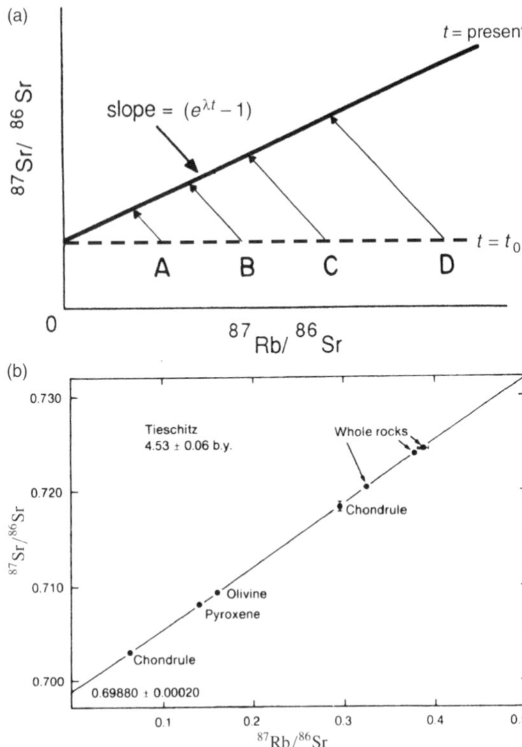

FIGURE 8.14 (a) Schematic isochron diagram of the ^{87}Rb–^{87}Sr system. Phases A, B, C and D have identical initial ^{87}Sr/^{86}Sr ratios at t = 0, but differing ^{87}Rb/^{86}Sr ratios. Assuming that the system remains closed, these ratios evolve as shown by the arrows to define an isochron for which t is the age of the rock. (b) ^{87}Rb–^{87}Sr isochron for the unequilibrated H3 chondrite Tieschitz meteorite. (Taylor 1992)

daughter species can be expressed as

$$N_d(t) = N_d(t_0) + \xi(1 - e^{-(t-t_0)/\tau_m})N_p(t_0), \qquad (8.7)$$

where the *branching ratio*, $0 < \xi \leq 1$, represents the fraction of the parent isotope that decays into the daughter species under consideration. In most cases, $\xi = 1$.

The current abundances, $N_d(t)$ and $N_p(t)$, are measurable quantities. The initial abundance of the parent, $N_p(t_0)$, can be expressed in terms of the measured abundance and the age of the rock, $t - t_0$, using equation (8.4). However, this combination of equations (8.4) and (8.7) yields a single equation for two unknowns (($t - t_0$) and $N_d(t_0)$). If we could determine independently the 'initial' (nonradiogenic) abundance of the daughter isotope (its abundance when the rock solidified), $N_d(t_0)$, then we could determine both the initial abundance of the parent and the age of the rock, t–t_0.

Chemical separation during a rock's solidification epoch can create an inhomogeneous sample that can be analyzed to determine both initial abundances and the age of the rock. Consider two samples within the rock containing different ratios of the parent and daughter elements. Assume that the system can be considered closed (i.e., no migration since the rock solidified). Analysis of the samples provides relationships between $N_d(t_0)$ and $t - t_0$ for each sample. Presumably $t - t_0$ is the same for both, and at the time the rock solidified each element was isotopically well-mixed. Thus, by measuring the abundances of a nonradiogenic isotope of the same element as the daughter isotope (e.g., ^{86}Sr for Rb-Sr dating), we can compute the ratio of $N_d(t_0)$ in the same samples, which provides us with the third equation needed to solve for the three unknowns. In practice, several samples are usually analyzed and solutions are obtained graphically using an *isochron diagram* (Fig. 8.14). Since the abundance ratio of strontium isotopes was constant, whereas the abundance ratio ^{87}Rb/^{86}Sr varied between minerals as a result of chemical fractionation, a plot showing the relationship between these abundance ratios would have yielded a constant at $t = t_0$ (dashed line in Figure 8.14a). As the rock aged, however, the abundance ratio ^{87}Sr/^{86}Sr has increased in proportion to the ^{87}Rb/^{86}Sr ratio. This has produced a current (measurable) slope on the isochron diagram of

$$\frac{d\left(^{87}\mathrm{Sr}/^{86}\mathrm{Sr}\right)}{d\left(^{87}\mathrm{Rb}/^{86}\mathrm{Sr}\right)} = e^{(t-t_0)/\tau_m} - 1 \qquad (8.8)$$

(Problem 8.14a). The slope on an isochron diagram thus gives the age of the rock, and the intercept represents the initial abundance of the daughter species. Isochron dating is preferred because if the system was disturbed, then the data do not fall along a straight line, and the age is known to be unreliable.

Operationally, the most difficult part of applying equation (8.8), and the part that leads to the largest uncertainty, is the measurement of the abundances of daughter and parent isotopes at the present epoch. Additional uncertainties come from imperfect knowledge of $t_{1/2}$ (especially for isotopes that β decay and have long half-lives) and ξ.

The long-lived isotopes of uranium, ^{235}U and ^{238}U, initiate decay chains that culminate in the lead isotopes ^{207}Pb and ^{206}Pb, respectively (expressions 8.6). These decay chains can be analyzed individually to estimate ages via the techniques described above for rubidium/strontium dating, using the nonradiogenic lead isotope ^{204}Pb as a stable comparison. However, more accurate dates can be obtained using the two chains in tandem. Two techniques are commonly used for this analysis, one of which only requires measurement of the lead isotopes and is described in Problem 8.14b. Radiometric dates for chondritic meteorites cluster tightly around 4.56×10^9 years. Differentiated meteorites are usually slightly younger. These results and their implications are discussed in more detail in Section 8.7.

8.5.3 Extinct-Nuclide Dating

Absolute radiometric dating requires that a measurable fraction of the parent nucleus remains in the rock. For rocks that date back to the formation of the Solar System, this implies long-lived parents, which due to their slow decay rates cannot give highly accurate ages. The *relative ages* of rocks which formed from a single well-mixed reservoir of material can be determined more precisely using the daughter products of short-lived radioisotopes that are no longer present in the rocks. Additionally, extinct nuclei can provide estimates of the time between nucleosynthesis and rock formation.

A widely used extinct chronometer is ^{129}I, which β decays into ^{129}Xe with a half-life of 17 million years. Because xenon is a noble gas, it is much rarer than iodine in meteorites, so even though only a small fraction of the iodine originally in the meteorite was radioactive, the amount of ^{129}Xe in the rock may have undergone a large percentage increase. By taking the ratio of the excess (that beyond the typical Solar System abundance for xenon) ^{129}Xe to the abundance of ^{127}I, one can estimate the fraction of iodine that was radioactive at the time that the meteorite solidified. Assuming that iodine was isotopically well-mixed in the protoplanetary disk, this fraction declined with time in a predictable manner. As the ratio ^{129}Xe/^{127}I is the same to within a factor of two for all primitive meteorites surveyed, they formed within an interval of $\sim 2 \times 10^7$ years. As some ^{129}Xe may escape, the actual *formation interval* may have been even shorter. Other extinct nuclei imply rapid formation and differentiation of some planetesimals (Section 8.7).

8.5.4 Interval Between Final Nucleosynthesis and Condensation

8.5.4.1 *Heavy Elements Produced by r-Process Nucleosynthesis*

The *r-process nucleosynthesis* of heavy elements occurs when there is a sufficiently high flux of neutrons present that slightly unstable nuclei do not have time to β decay before further neutron capture. Neutron-rich nuclei are thus produced by this rapid process, which is believed to occur in some types of supernovae. The ratio of certain similar isotopes produced in r-process nucleosynthesis can be calculated fairly accurately.

In order to determine the time that elapsed between the last r-process nucleosynthesis of Solar System material and the condensation of meteorites, consider the initial Solar System abundances of two chemically similar r-process pairs of short-lived and long-lived isotopes which are believed to be produced at the same rate (^{244}Pu, ^{238}U) and (^{129}I, ^{127}I). The abundance ratios at the time of Solar System formation inferred from meteorites are: ^{244}Pu/^{238}U $= 0.007$ ($t_{1/2}(^{244}$Pu$) = 8.2 \times 10^7$ yr); ^{129}I/^{127}I $= 0.0001$ ($t_{1/2}(^{129}$I$) = 1.7 \times 10^7$ yr). The ratio of half-lives is 5, thus the ratio of relative abundances should be 5 if nucleosynthesis occurred at a uniform rate prior to formation of the Solar System. The observed ratio of 70 implies that nucleosynthesis stopped $\sim 8 \times 10^7$ years before condensation (Problem 8.17). This is approximately equal to the time it takes for a molecular cloud at the Sun's distance from the galactic center to pass from one spiral arm to the next. Most stars are formed when molecular clouds are compressed as they pass through spiral arms. The lifetimes of massive stars capable of r-process nucleosynthesis are only a few million years, so presumably the last injection of fresh r-process isotopes occurred during or soon after the Solar System-forming cloud's penultimate passage through a spiral arm.

Isotopic abundances in pre-solar grains, which are found in chondritic meteorites and which were formed in stellar atmospheres and outflows, can sometimes be used to indicate the age and mass of the star in which they originated. Primitive meteorites thus contain information on the nucleosynthesis history of our galaxy. As some of the grains must have formed in stars that were already several billion years of age when our Solar System formed, studies of such grains also yield a lower bound to the age of the galaxy of nine billion years.

8.5.4.2 *Light Elements with Short Half-lives*

Correlations between aluminum abundance and ^{26}Mg/^{24}Mg excess have been detected within chondritic meteorites. The excess cannot be the result of mass-dependent fractionation because the relative abundance of the nonradiogenic isotopes of magnesium, ^{25}Mg/^{24}Mg, is normal. This is strong evidence that ^{26}Al ($t_{1/2} = 720\,000$ years for β decay into ^{26}Mg) was present in (at least some regions of) the early Solar System with an abundance of $\sim 5 \times 10^{-5}$ that of the stable isotope ^{27}Al. Note that, because aluminum is fairly abundant (Table 8.2), this fraction of radioactive aluminum could have provided a substantial heat source for melting planetesimals in the early Solar System (Problem 8.18).

The ^{26}Mg from ^{26}Al decay has been detected in many large inclusions that are otherwise isotopically normal. The correlation between ^{26}Mg excess and aluminum abundance implies that 'live' ^{26}Al was present in the early Solar System, which requires the interval of time between nucleosynthesis of the ^{26}Al and the formation of solid bodies within the Solar System to have been at most a few million years. Note that this is far shorter than the interval since last r-process nucleosynthesis discussed in Section 8.5.4.1. This is consistent with theoretical models that ^{26}Al is produced in a much wider array of astrophysical environments than are r-process isotopes. Gamma ray observations show that ^{26}Al is 'plentiful' in the galaxy today, implying that it can be formed in abundance. The ^{26}Al that we see evidence for in chondrites also may have been produced by high-energy particles released from the active young Sun which bombarded grains that approached to within a few solar radii. Excess ^{41}K, from decay of ^{41}Ca ($t_{1/2} = 10^5$ years), has also been detected in meteorites. The correlation between ^{41}K and ^{26}Mg observed in some meteorite samples indicate a common source for their parent nuclides ^{41}Ca and ^{26}Al.

Note that some isotopic anomalies indicate survival of interstellar grains rather than radioactive decay within the Solar System. Almost pure concentrations of the rare heavy isotope of the noble gas neon, ^{22}Ne, possibly from sodium decay (^{22}Na, $t_{1/2} = 2.6$ years!) have been detected within small, carbon-rich phases within some primitive meteorites. It is believed that grains containing this neon condensed within outflows from carbon-rich stars and possibly novae outbursts.

8.5.5 Cosmic Ray Exposure Ages

Galactic cosmic rays are extremely energetic particles (mostly protons + some heavier nuclei) that possess enough energy to produce nuclear reactions in particles they collide with. Cosmic rays and the energetic secondary particles that they produce have a mean interaction depth of $\sim$1 meter in rock, thus they do not affect the bulk of material in any sizable asteroid. The amount of cosmic rays that a meteorite has been exposed to indicates how long it has 'been on its own', or at least near the surface of an asteroid.

To determine *cosmic ray ages*, we examine certain rare isotopes of noble gases, e.g., ^{21}Ne and ^{38}Ar, that in meteorites are almost exclusively produced by cosmic rays. As production rates of these gases vary with depth, we need to determine depth independently. Short-lived isotopes such as ^{10}Be are produced by cosmic rays. Their abundance is determined by the equilibrium between production and decay, and thus this abundance tells us how far the meteorite was below the surface. Using this estimate of depth, we can determine age from stable (or long-lived) isotopes produced by cosmic rays.

Typical cosmic ray exposure ages are 10^5–10^6 years for carbonaceous chondrites, 10^7 years for other stones, 10^8 years for stony-irons and 10^8–10^9 years for irons. The differences in age are due to material strength, which governs how quickly a body breaks up or a surface erodes. An additional factor which may contribute to the larger ages of irons and stony-irons is that the Yarkovski force, which can help move small objects into Earth-crossing orbits, is weaker on these meteoroids than it is on stony meteoroids. The relative numbers of meter-sized meteoroids of different classes injected by collisions in space can be estimated by dividing the number of falls by the typical cosmic ray ages of meteorites of that class. Note that the strength of iron meteorites helps them survive better than chondrites at many stages: in space, during atmospheric entry and on the ground. Irons are also easier to identify as meteorites (except in Antarctica, where any rock on an ice sheet is distinctive and of suspect origin), making them more overrepresented in meteorite collections. However, cosmic ray ages imply that even fall statistics vastly overestimate the percentage of small fragments produced in the inner Solar System that are composed primarily of iron. The nonrandom cosmic ray ages deduced imply that certain meteorite groups experienced major break-ups that generated a large fraction of the members of each of these groups in a single event.

The *terrestrial age* of a meteorite is the time since it fell, i.e., how long the meteorite has been on Earth. This age is important in determining weathering and contamination. Weathered appearance gives a clue to the approximate age. The best method of determining the terrestrial age of a meteorite is to measure the relative abundances of two short-lived cosmic-ray produced radioisotopes (two are required in order to eliminate the effects of variations of the cosmic ray flux with depth).

8.6 Physical Characteristics of Chondrites

Chondrites contain very nearly (within a factor of two) solar abundance ratios of all but the most volatile elements. Isotopic ratios are even more strikingly regular; almost all differences can be accounted for by radioactive decay (excesses of daughter isotopes), cosmic ray-induced *in situ* nucleosynthesis or mass fractionation (cf. Section 8.4.2). However, slight deviations from this rule show that the material within the solar nebula was not completely mixed at the atomic level.

Chondrites have not been melted since their original accretion $\sim 4.56 \times 10^9$ years ago. Although chondrites represent well-mixed isotopic and elemental (except for volatiles) samples of the material in the protoplanetary disk, they are far from uniform on small scales. In addition to chondrules, many primitive meteorites contain *CAIs*, which are refractory inclusions that are rich in calcium and aluminum. Chondrules and CAIs are embedded within a dark fine-grained *matrix* that is present in all chondrites. Chondrites formed with different percentages of inclusions (CAIs, chondrules) and differing amounts of moderately volatile elements.

Chondrules are small ($\sim$ 0.1–2 mm), rounded igneous rocks (i.e., they solidified from a melt) composed primarily of refractory elements. They range from 0 to 80% of the mass of a chondrite, with abundances depending on compositional class (CI chondrites do not contain any chondrules) and petrographic type. Chondrules are totally absent in petrographic type 1 (they may have been destroyed by aqueous processes) and are substantially degraded by recrystallization resulting from thermal metamorphism in types 5 and 6; the most pristine chondrules are found in type 3 chondrites. Mineralogical properties imply that chondrules cooled very quickly, dropping from a peak temperature of $\sim$1900 K to $\sim$1500 K over a period ranging from 10 minutes to a few hours. The observed strong correlation of chondrule properties (size and compositions) within individual meteorites implies that chondrules were not well-mixed before incorporation into larger bodies.

Compound chondrules appear to be two or more chondrules joined together. Most compound chondrules may be produced by collisions of partially molten objects. Other compound chondrules consist of a primary chondrule that is entirely entrained in a larger secondary chondrule. The secondary chondrules are believed to have formed from the heating and melting of fine-grained dust that had accreted onto the surface of the primary. Many chondrules have melted rims, providing additional evidence for multiple heating events.

CAIs are light-colored round inclusions, typically a few millimetres in size. They are composed of very refractory minerals, including substantial amounts of Ca and Al and abundances of high-Z elements that are greatly enhanced relative to bulk chondrites. CAIs are most abundant in CV chondrites, and are also seen in most other classes of primitive meteorites. They are among the oldest objects formed in the Solar System. Many CAIs have melted rims which imply secondary processing. These rims resolidified in a more oxygen-rich environment than the original condensation. The interiors of the CAIs were not heated to anywhere near melting during this secondary processing, implying very short duration heating. Clearly, the protoplanetary disk was an active and sometimes violent place. Dust (often similar to the matrix) is seen in CAI rims. Some chondrules are surrounded by analogous rims. Parent body processing could well have affected some of these rims.

The characteristics of the fine-grained matrix material that makes up the bulk of most chondrites vary with compositional class and petrographic type. Mean grain sizes in type 3 chondrites vary from 0.1 to 10 μm, with larger grains being more common in carbonaceous chondrites than in ordinary chondrites. Olivine and pyroxine are the most common minerals in the majority of ordinary chondrites and in carbonaceous chondrites of petrographic type 3 and higher; serpentine is most common in carbonaceous chondrite types 1 and 2; magnetite is found in a variety of classes and types. Chondrite matrices appear to contain material from a wide variety of sources including pre-solar grains, direct condensates from the protoplanetary disk, and dust from fragmented chondrules and CAIs; some of these grains have been altered by post-accretional aqueous or thermal processing.

8.7 Meteorite Clues to the Formation of the Solar System

The smaller bodies in the Solar System have not been subjected to as much heat or pressure as the planet-sized bodies, and thus they remain in a more pristine state. Meteorites provide detailed information about environmental conditions and physical and chemical processes in the early Solar System. This information pertains to timescales, thermal and chemical evolution, mixing, magnetic fields and grain growth within the protoplanetary disk. Processes identified include evaporation, condensation, localized melting and fractionation both of solids from gas and among different solids.

Meteorites definitively date the origin of the Solar System to about 1 part in 1000. Chondritic solids formed within a period of $\lesssim$ 20 million years at the beginning of Solar System history. The age of the Solar System, based on ^{207}Pb/^{206}Pb dating of CAIs in the Allende CV3 meteorite, is $4.563 \pm 0.004 \times 10^9$ years; other isotope systems and other chondrites yield similar ages. Meteorites from differentiated parent bodies are often a bit younger, but usually not very much. Almost all meteorites are thus older than known Moon rocks ($\sim$ 3–4.45×10^9 yr), and terrestrial rocks ($\lesssim 4 \times 10^9$ yr, although they contain grains of the durable mineral zircon up to 4.4×10^9 yr old).

Most elements in most meteorite groups are identical in isotopic composition, aside from variations which may plausibly be attributed to mass fractionation, nuclear decay or cosmic rays; thus, matter within the solar nebula must have been relatively well-mixed. Differences in isotopic composition between individual meteorites and between meteorites and the Earth yield information on the place of formation of the individual molecules and grains out of which the meteorites have been made. Evidence is growing, based upon the D/H and $^{15}N/^{14}N$ ratios, that some grains must come from the cold interstellar molecular clouds, while isotopic ratios in other elements strongly suggest production by r-process nucleosynthesis, such as occurs in supernova explosions (cf. Section 12.2). Hence, meteorites seem to contain interstellar materials in addition to that processed/formed in our own Solar System.

8.7.1 Meteorites from Differentiated Bodies

Isotopic anomalies found in some achondrites imply rapid differentiation and recrystallization of planetesimals. Excess ^{60}Ni, which is the stable decay product of ^{60}Fe ($t_{1/2} = 1.5 \times 10^6$ yr), is correlated with iron abundances in eucrite achondrite meteorites. Eucrites originate from the asteroid 4 Vesta and/or other differentiated planetesimals, which were once molten. Live ^{60}Fe must thus have been present when the planetesimal resolidified. This implies differentiation within a few to several ^{60}Fe half-lives after nucleosynthesis. Signatures left by ^{26}Al and ^{53}Mn imply that other planetesimals formed and differentiated within 3–5 Myr of the solidification of the oldest known Solar System materials.

The cooling rate of a rock, as well as the pressure and the gravity field that it was subjected to while cooling, can be deduced from the structure and composition of its minerals. Thus, we can estimate the size of a meteorite's original parent body. By knowing what size bodies melted in the early Solar System (and possibly where they accreted relative to other bodies by the presence or lack of volatiles in the meteorite), we can get a better idea of the type(s) of heat sources responsible for differentiation.

Radiation damage produced in meteoritic minerals by ^{244}Pu fission tends to be annealed out at high temperatures. The temperature beneath which damage is not annealed out varies among minerals. The difference in fission track density among a set of minerals with different annealing temperatures can be interpreted in terms of a cooling history. The retention of radiogenic noble gases in meteoritic minerals (Section 8.5) is analogous to fission track retention and can also be used to estimate a cooling history.

The composition and texture of differentiated meteorites such as achondrites, irons and pallasites reflect igneous differentiation processes (i.e., large scale melting) within asteroid-size parent bodies. It appears as if some small bodies $\lesssim 100$ km in radius differentiated. Neither accretion energy nor long-lived radioactive isotopes are adequate heat sources. Possible heat sources include the extinct radionucleides ^{26}Al and ^{60}Fe, as well as *electromagnetic induction heating*. Electromagnetic induction heating could occur when the meteorite parent bodies passed through a current created by the massive T-Tauri phase solar wind (Section 12.3). Although there are many uncertainties in this mechanism, maximum heating would occur in bodies nearest the Sun, and probably for bodies between 50 and 100 km in radius. The largest $^{26}Al/^{27}Al$ concentrations observed in primitive meteorites would be sufficient to melt chondritic composition planetesimals as small as 5 km in diameter. The largest abundance of ^{60}Fe deduced in achondrites is a substantially weaker heat source, but it would have been adequate to melt planetary bodies larger than a few hundred kilometers in diameter.

8.7.2 Primitive Meteorites

Most meteorites that are observed to fall are chondrites, which have never been differentiated and thus preserve a better record of conditions within the protoplanetary disk than the differentiated meteorites. Indeed, some of the grains in chondrites predate the Solar System, and thus also preserve a record of processing in stellar atmospheres, winds, explosions and the interstellar medium. These grains may have been affected by passage through a hot shocked layer of gas during their entry into the protoplanetary disk. The precursors to chondrules and CAIs formed by agglomeration of pre-solar grains and solar nebula condensates. The agglomerates were subsequently heated to the point of melting. Many had their rims melted at a later time, and/or were fragmented as the result of high speed collisions. Ultimately, they were incorporated into planetesimals in which they were subjected to nonhydrous processing at 700–1700 K (especially petrographic types 4–6) and/or hydrothermal processing at lower temperatures (primarily types 1 and 2). Some chondrites also show evidence for shock processing in the geologically recent past.

The differences in bulk composition among chondrites are closely related to the volatility of the elements. In almost all cases, the more volatile elements are depleted; however, very refractory materials may be depleted relative to silicon in some enstatite and ordinary chondrites. Small inclusions of very refractory material (CAIs) are seen in chondrites, but bulk meteorites of such refractory

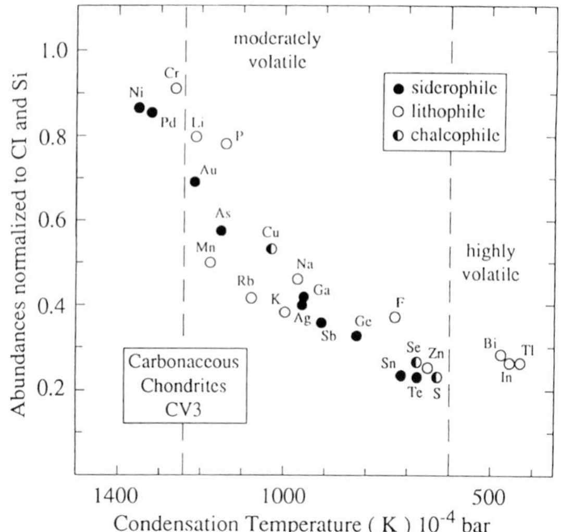

FIGURE 8.15 The abundances of moderately volatile elements in bulk CV chondrites compared to their abundances relative to silicon in CI chondrites are plotted against the condensation temperatures of the elements in a solar composition gas. Note the gradual decrease in abundance with decreasing condensation temperature, and the lack of dependence of abundances on the geochemical character of the elements. (Palme and Boynton 1993)

composition have not been found. Magnesium silicates and iron–nickel metal, which are not significantly depleted in primitive meteorites, condense at ~1300 K. A gradual depletion pattern with increasing volatility is seen in most chondrites (e.g., Fig. 8.15). If each meteorite was formed in equilibrium at a unique temperature, then relative abundances of elements with condensation temperatures above this value would be solar, and more volatile elements would be almost absent. The only elements that would be depleted by tens of percent would be those that either condense at temperatures very near the local equilibration temperature (the saturation vapor pressures of most compounds vary rapidly with temperature near their 50% condensation temperature, cf. Section 4.4), or those that form refractory compounds only with rarer elements. The gradual depletion patterns which are observed (Fig. 8.15) imply that the constituents of individual meteorites condensed in a broad range of environments. Grains could be brought together from a variety of locations to produce such admixtures, or most of the material that formed the terrestrial planets and asteroids cooled to around 1300 K while the gaseous and solid components remained well-mixed, and then gas was subsequently lost as material cooled further. Significant condensation in the asteroid region continued down to $\lesssim$ 500 K before the gas was completely removed. Elemental depletions in bulk terrestrial planets and differentiated asteroids are consistent with this conclusion, although they are less well constrained.

In contrast to meteorite bulk abundances, the compositions of CAIs imply equilibrium temperatures as high as ~1600 K. The structure of chondrules and the melted rims of CAIs evidence even higher temperatures during transient heating events. The peak temperatures reached by chondrules were 1800–2200 K. CAIs reached temperatures slightly lower than those of chondrules, but remained hot much longer. At nebular densities, molecular hydrogen dissociates at temperatures comparable to the peak values experienced by chondrules and CAIs. The energy required to dissociate H_2 could have provided an effective thermostat, preventing the molten chondrules and CAIs from being heated to even higher temperatures.

The presence of significant amounts of moderately volatile elements show that chondrules were not molten under equilibrium conditions, implying local, brief heating. The volatile elements would have been lost had the chondrules remained molten for more than a few minutes, and if chondrule precursors were fluffy dustballs (as most models assume), then the heating interval must have been similarly short. The textures and mineral chemistries imply cooling rates of 50–1000 K per hour for chondrules and 2–50 K per hour for CAIs. The rapidity of chondrule heating and cooling implies local processes were responsible, as large regions of the nebula cannot cool quickly.

A major conceptional gap in meteoritics is how the epoch of global nebular cooling and settling of solids related to the very rapid heating episodes that produced chondrules and CAIs. The mineral chemistry of chondrites implies substantial variations in the *fugacity* (fractional abundance) of oxygen, fO_2, within the solar nebula. Enstatite chondrites formed in a highly reducing environment, whereas CAI rims solidified in environments with fO_2 up to 10^4 times that corresponding to solar composition. More reducing conditions could have been created locally by evaporation of C-rich dust (which would have sequestered oxygen in CO) and/or removal of H_2O into icy planetesimals. Oxidizing conditions probably resulted from enhancements in the local ratio of O-rich dust and ice to H-rich gas within the protoplanetary disk and/or photochemical destruction of H_2O.

The origin of chondrules and CAIs remains enigmatic. Possible mechanisms for chondrule heating include drag during passage through an accretion shock either upon entry into the protoplanetary disk or during subsequent disk processing, flares or lightning within the disk, and heating by intense sunlight coincident with their removal from the vicinity of the Sun by the powerful T-Tauri phase solar wind. Melting during entry into the protoplanetary disk

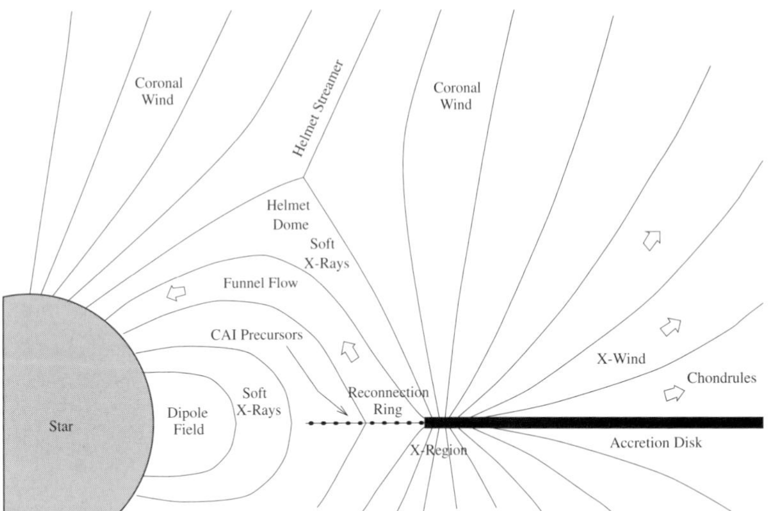

FIGURE 8.16 Schematic drawing of the magnetic field and gas flow in the stellar wind model for the production of CAIs and chondrules. (Shu *et al.* 1997)

requires pre-existing clumps much larger than observed interstellar grain sizes, and would probably be evidenced by substantially greater isotopic variations than have been detected. The disk mechanisms all suffer from the energetics problem of concentrating a substantial amount of energy more than 1 AU from the Sun.

Lightning within the protoplanetary disk might have provided the energy to rapidly melt chondrules and CAIs. Models suggest that lightning analogous to that produced in terrestrial clouds (Section 4.5.4.4) could have occurred within the asteroidal region of the disk. Size-segregated motions of chondritic composition particles could be produced by gas convection or by the component of solar gravity perpendicular to the midplane of the protoplanetary disk. These moving particles presumably would transfer charge as do ice particles in thunderstorms within Earth's atmosphere. Large scale charge separation would build up until the nebular gas breaks down. Observations of lightning in terrestrial volcanic plumes and in outer planet atmospheres show that lightning can be produced in a wide variety of environments.

The nebular shock wave model of chondrule formation envisions gas that is overrun by a shock and becomes abruptly heated, compressed and accelerated. A *shock front* is a sharp discontinuity between hot, compressed, supersonic gas and cooler, less dense, slower-moving gas (Section 7.1). Shocks may have been produced within the protoplanetary disk by, e.g., energetic outbursts from the protosun or by dense clumps of gas falling onto the disk. The timescales for shock heating are seconds, in accord with observations of chondrules and CAIs. However, radiative cooling in an optically thin region of the protoplanetary disk would also have occurred in seconds, producing different mineral characteristics than those observed. Shocks which produced large numbers of chondrules within dense, dust-rich regions of the disk may have allowed cooling on a slower timescale consistent with observations.

Sufficient energy for the thermal processing required for the formation of chondrules and CAIs is readily available close to a protostar. Chondrules, CAIs and their rims could form when solid bodies are lifted out of the relatively cool shaded region of the protoplanetary disk into direct sunlight by the aerodynamic drag of a magnetocentrifugally driven T-Tauri phase stellar wind. Such winds have been theorized to explain the substantial bipolar outflows observed to accompany star formation and the loss of angular momentum by the star and inner disk required for accretion to proceed (cf. Section 12.3.3). In this model, two-thirds of the matter reaching the inner edge of a protoplanetary disk is accreted by the protostar, whereas one-third is ejected into interstellar space in a powerful stellar wind (Fig. 8.16). For reasonable parameters of the bipolar flow, the peak temperatures reached by solid bodies resemble those needed to melt CAIs and chondrules. The high velocity expulsion of these bodies from the vicinity of the star yields rapid cooling, although it is not clear whether or not it would be sufficiently quick to be in agreement with the observed chondrule properties. Rims could have formed when very rapidly moving dust grains (which are better coupled to the wind than are larger particles) impacted chondrules and CAIs. Small solid grains would be carried by the wind off into interstellar space. Large particles would fall back to the inner portion of the protoplanetary disk and would return to the vicinity of the protostar. Some of these large particles would be accreted by the

FIGURE 8.17 Image of a tiny pre-solar silicon carbide (SiC) grain (1 μm across) extracted from the Murchison meteorite. This extremely high resolution image was obtained using a secondary electron microscope. (Courtesy: Scott Messenger)

protostar, while others would be cycled again. Chondrule-sized objects would be given just enough velocity for them to fall back to the disk in the planetary region. Chondrule and CAI sizes would vary with position within the disk (the smaller objects being thrown the farthest) and with their time of formation as the protostellar wind evolved, thus the correlation of chondrule properties within individual meteorites could be understood.

The general high degree of uniformity of isotope ratios implies the solar nebula was for the most part well-mixed, but the small violations of this rule tell us that some things didn't mix or never vaporized. Oxygen isotopic ratios show relatively large variations which cannot be explained by mass-dependent fractionation within primitive chondrites and between groups of meteorites (Fig. 8.13); these data are usually taken to imply distinct reservoirs that were incompletely mixed during nebular processes, although non-mass-dependent fractionation, via e.g., photochemical processes, can occur in certain circumstances. Why is oxygen special? Oxygen is the only element (apart from sulphur, which only has two stable isotopes and thus doesn't provide useful evidence because mass-dependent fractionation cannot be distinguished from other effects) common in both high temperature and low temperature phases. The oxygen isotopic variations could have been produced if the gas was well-mixed and the grains were well-mixed, but no isotopic equilibrium existed between the two in portions of the solar nebula where the grains never completely evaporated.

There is no strong evidence that any macroscopic ($\gtrsim 1$ mm) grains are of pre-solar origin. However, tiny carbon-rich grains (most of which are $\ll 10$ μm) that clearly represent surviving interstellar material have been found in some chondrites (Fig. 8.17). Unambiguous proof of the pre-solar origins of these grains comes from their isotopic compositions, which are anomalous both in trace elements such as Ne and Xe and in the more common elements C, N and Si. The survival of these grains, and their ability to retain noble gases, constrains the thermal and chemical environments that they experienced on their journey from interstellar cloud to meteorite parent bodies. Some of the grains were clearly never heated above 1000 K, and must have been much cooler during any episode in which they were exposed to an oxygen-rich environment.

Some refractory inclusions contain isotopic anomalies in several elements that are strongly correlated according to theoretical models of particular nucleosynthesis processes. These anomalies imply that unvaporized interstellar grains formed in various differing environments survived entry into the protoplanetary disk.

The D/H ratios observed in organic material within carbonaceous chondrites are far higher than solar values. (D/H is more than 1000 times the solar value in some grains.) Fractionation of this magnitude is difficult if not impossible to achieve within the warm solar nebula. Rather, these complex hydrocarbons or their precursor molecules are believed to be formed by ion–molecule reactions in very cold interstellar clouds (which are capable of producing high D/H ratios). This constraint requires either the survival of substantial amounts of interstellar grains or that the gas containing the hydrocarbon molecules did not get sufficiently hot for isotopic equilibration to occur. This implies that some of the material in meteorites was never heated to the $\gtrsim 1500$ K temperatures experienced by chondrules and CAIs.

Remnant magnetism in carbonaceous chondrites suggests that a magnetic field of strength 1–10 G existed at some locations within the protoplanetary disk. The magnetic field is anisotropic from grain to grain, so magnetization presumably occurred prior to the incorporation of the grains into the meteorite parent body. Possible sources for such magnetism include a dynamo within the protoplanetary disk, the solar field carried outward by solar wind, or the solar field itself if the chondrules cooled through the Curie point while in the vicinity of the protosun.

In addition to being fascinating in their own right, meteorites provide us with a wealth of extremely detailed data on conditions during the planet forming epoch. Some of these data are easy to interpret. For instance, radiometric dates provide an estimate of the age of the Solar System accurate to one part in a thousand. Other data give potentially very valuable clues but are subject to more diverse interpretations. The near but not total homogeneity of isotopic composition among meteorites tells us that substantial mixing of pre-solar material occurred, but that some interstellar grains survived, and also that certain short-lived radionuclides were present in the material out of which the planets formed. The local mineralogical and compositional heterogeneity of primitive meteorites implies an active dynamic environment within the protoplanetary disk, but specific models to explain these data remain quite controversial research topics. Future refinements in laboratory analysis techniques and in theoretical modeling should advance our understanding of meteorite clues to the origin of the Solar System considerably over the coming years and decades.

FURTHER READING

A compendium of articles written from various different viewpoints:

Hewins, R.H., R.H. Jones, and E.R.D. Scott, Eds., 1996. *Chondrules and the Protoplanetary Disk*. Cambridge University Press, Cambridge. 346 pp.

Kerridge, J.F., 1993. What can meteorites tell us about nebular conditions and processes during planetesimal accretion? *Icarus* **106**, 135–150.

This collection of review chapters is quite comprehensive, but some articles are getting out of date:

Kerridge, J.F., and M.S. Matthews, Eds., 1988. *Meteorites and the Early Solar System*. University of Arizona Press, Tucson, 1269 pp.

Palme, H., and W.V. Boynton, 1993. Meteoritic constraints on conditions in the solar nebula. In *Protostars and Planets III*. E.H. Levy and J.I. Lunine, Eds., University of Arizona Press, Tucson, pp. 979–1004.

Podosek, F.A., and P. Cassen, 1994. Theoretical, observational, and isotopic estimates of the lifetime of the solar nebula. *Meteoritics* **29**, 6–25.

Shu, F.H., H. Shang, and T. Lee, 1996. Toward an astrophysical theory of chondrites. *Science* **271**, 1545–1552.

A nice nontechnical summary, including many good photographs can be found in:

Wasson, J.T., 1985. *Meteorites: Their Record of Early Solar-System History*. W.H. Freeman, New York.

Wood, J.A., 1988. Chondritic meteorites and the solar nebula. *Annu. Rev. Earth Planet. Sci.* **16**, 53–72.

Zinner, E. 1998. Stellar nucleosynthesis and the isotopic composition of presolar grains from primitive meteorites. *Annu. Rev. Earth Planet. Sci.* **26**, 147–188.

Problems

8.1.E Calculate the speed at which meteoroids with the following heliocentric orbits encounter the Earth's atmosphere:

(a) An orbit very similar to that of Earth.
(b) A parabolic orbit, with perihelion of 1 AU and $i = 180°$.
(c) A parabolic orbit, with perihelion of 1 AU and $i = 0°$.
(d) A parabolic orbit, with perihelion of 1 AU and $i = 90°$.
(e) An orbit with $a = 2.5$ AU, $e = 0.6$ and $i = 0°$.
(f) An orbit with $a = 2.5$ AU, $e = 0.6$ and $i = 30°$.

8.2.I (a) Calculate the kinetic energy of a meteoroid of radius 1 cm and density 1 g cm^{-3} moving at 20 km s^{-1}.
(b) Assume that this meteoroid enters the Earth's atmosphere and radiates away 0.01% of its kinetic energy as visible light over a period of 5 seconds. What is the rate at which it radiates during this period? State your answer both in erg s^{-1} and in watts.
(c) At what visual magnitude would this meteor appear to an observer at a distance of 100 km?

8.3.I (a) Calculate the pressure on a meteor traveling at 10 km s^{-1} at an altitude of 100 km above the Earth's surface.
(b) Repeat for a meteor at the same speed 10 km above Earth's surface.
(c) Repeat your calculations in parts (a) and (b) for a meteor traveling at 30 km s^{-1}.
(d) The tensile strengths of comets are of order 10^4 dyne cm^{-2}, the strengths of chondrites are roughly 3×10^7 dyne cm^{-2}, stronger stony objects have strengths approximately 10^8 dyne cm^{-2}, whereas iron impactors have effective strengths of about 10^9 dyne cm^{-2}. Compare these tensile strengths to the pressures calculated in parts (a)–(c) and comment.

8.4.E (a) Calculate the size of an iron meteor (density $\rho = 8$ g cm^{-3}) that passes through an amount

of atmospheric gas equal to its own mass en route to the surface of the Earth. You may assume a spherical meteorite, vertical entry into the atmosphere, and neglect ablation.
(b) Repeat your calculation for a chondritic meteorite of density $\rho = 4$ g cm^{-3}.
(c) Repeat your calculation in part (a) for an entry angle of 45°.

8.5.E Calculate the terminal velocity near the Earth's surface for falling rocks of the following sizes and densities:

(a) $R = 10$ cm, $\rho = 8$ g cm^{-3}.
(b) $R = 10$ cm, $\rho = 2$ g cm^{-3}.
(c) $R = 100$ cm, $\rho = 2$ g cm^{-3}.
(d) $R = 100$ μm, $\rho = 2$ g cm^{-3}.

8.6.E (a) Calculate the equilibrium temperature of a meteoroid of mass M, density ρ and albedo A in the vicinity of the Earth.
(b) Evaluate your result for a chondrite with $M = 10^9$ g, $\rho = 2.5$ g cm^{-3} and albedo $A = 0.05$ and for an achondrite with $M = 10^6$ g, $\rho = 3$ g cm^{-3} and albedo $A = 0.3$.

8.7.E Use equation (8.4) to verify that τ_m is indeed the average lifetime of the isotope.

8.8.E Calculate the fractional abundance of ^{234}U in naturally occurring uranium ore. (Hint: Use expression (8.6*a*).)

8.9.I (a) Use tabulated data on nuclear decay (from e.g., the *CRC Handbook*) in order to write the decay chain from ^{244}Pu to ^{232}Th.
(b) Continue this decay chain until a stable isotope is reached.

8.10.E (a) Use the decay chains given in expressions (8.6) to estimate lower bounds on the abundance of elements 84–91 in terrestrial uranium ore.
(b) Why are your values only lower bounds?
(c) Why are your estimates less trustworthy for elements with atomic number ≤ 86 than they are for elements with higher atomic numbers?

8.11.I Estimate the amount of (naturally occurring) ^{244}Pu present on Earth at the present epoch. (Hint: You may assume that the abundance ratio of plutonium to uranium 4.55×10^9 years ago was $^{244}\text{Pu}/^{238}\text{U} = 0.007$ and that the Earth has a chondritic abundance of uranium.) Comment on the fact that Pu is often considered not to occur in nature.

8.12.I All elements with atomic numbers 1–112, 114, 116 and 118 have either been discovered in nature or produced in the laboratory. In this exercise, you will determine a complicated set of answers to the question of how many of these elements are naturally occurring.

(a) How many elements have at least one stable isotope? What are the atomic numbers of these elements? Note: Ignore the possibility of proton decay, which may make all isotopes unstable on *very* long timescales ($t_{1/2} > 10^{30}$ yr).
(b) Which other elements have isotopes that are so long-lived that they have survived in measurable quantities on Earth?
(c) Which elements not represented above are produced on Earth as the result of radioactive decay of naturally occurring elements? Justify your answer by writing down the appropriate decay chains. Note: Some isotopes have more than one possible decay path.
(d) A very small amount of the plutonium isotope ^{244}Pu has almost certainly survived since the Earth formed (Problem 8.11). Does ^{244}Pu decay add any more elements to your list of naturally occurring elements?
(e) Spontaneous fission of uranium produces a variety of daughter nuclei. Although the dominant massive element produced is xenon, the distribution is broad and contains small amounts of two elements without stable isotopes. Which elements?
(f) Spontaneous fission releases neutrons which can initiate nuclear reactions in uranium ore. A few billion years ago, the concentration of ^{235}U was sufficient to initiate chain reactions, producing *natural nuclear reactors*, which have left isotopic signatures that have been detected in some rich uranium ore. Although the concentration of ^{235}U is no longer sufficient to produce chain reactions, some neutrons released by spontaneous fission are absorbed by ^{238}U and others initiate fission in ^{235}U. Can these reactions add any elements to your list? What about neutron capture by ^{244}Pu?

8.13.I Determine the least abundant element represented in each of parts (a)–(d) of the previous problem and the total quantity of this element on Earth.

You may assume chondritic abundances for elements in parts (a) and (b).

8.14.I (a) Derive equation (8.8). (Hint: Differentiate equations (8.4) and (8.7) with respect to $N_p(t_0)$.)

(b) The primary decay modes of uranium isotopes ^{235}U and ^{238}U initiate decay chains which ultimately produce the lead isotopes ^{207}Pb and ^{206}Pb respectively; intermediate decay products are relatively short-lived and thus contain insignificant amounts of material (cf. expressions 8.6). At the time that a rock solidifies, an isochron plot of ^{207}Pb/^{204}Pb vs. ^{206}Pb/^{204}Pb is represented by a point. How does an isochron plot of these ratios appear at subsequent times? Be quantitative, and derive an equation analogous to equation (8.8). You may neglect the small fraction of uranium ($< 10^{-4}$) which decays via spontaneous fission. (Hint: The ratio between the abundances of the uranium isotopes is the same in all minerals, but the ratio of uranium to lead varies.) This technique for determining the age of a rock is known as *lead-lead dating*.

8.15.I The following isotopic abundances (atoms/10^6 Si atoms) are measured in CI carbonaceous chondrites and meteorite X respectively:

Isotope	CI Chondrites	Meteorite X
^{204}Pb	0.0612	0.1224
^{206}Pb	0.603	?
^{207}Pb	0.650	2.63
^{235}U	6.49×10^{-5}	1.59×10^{-2}
^{238}U	8.49×10^{-3}	?

Assume that uranium and its daughters decay exclusively by α and β decay, and that no mass-dependent fractionation has occurred. Determine the values of each of the quantities represented by question marks, and calculate the age of meteorite X.

8.16.E In this problem you will calculate the age of a rock using actual data on the abundances of rhenium and osmium, which are related via the decay $^{187}\text{Re} \rightarrow {}^{187}\text{Os}$ ($t_{1/2}(^{187}\text{Re}) = 4.16 \times 10^{10}$ yr).

(a) The following list summarizes some measurements of Re and Os isotope ratios for different minerals within a particular rock:

^{187}Re/^{188}Os	^{187}Os/^{188}Os
0.664	0.148
0.669	0.148
0.604	0.143
0.484	0.133
0.512	0.136
0.537	0.138
0.414	0.128
0.369	0.124

Plot the results on a piece of graph paper with ^{187}Re/^{188}Os along the horizontal axis and ^{187}Os/^{188}Os along the vertical axis.

(b) Draw a straight line that goes as closely as possible through all the points and extend your line to the vertical axis to determine the initial ratio of ^{187}Os/^{188}Os.

(c) Draw and label several lines representing theoretical isochrones for a rock with the same initial ratio of ^{187}Os/^{188}Os as the rock being studied. Use these lines to estimate the age of the rock.

8.17.I Estimate the time interval between the last r-process nucleosynthesis of Solar System material and the condensation of chondrites. (Hint: Write down a formula for the abundance pair ratio of (^{244}Pu/^{238}U)/(^{129}I/^{127}I) as a function of time assuming no ongoing nucleosynthesis. Use your formula to determine the time required for this ratio to grow from its 'steady-state' value of 5 to the observed value of 70.)

8.18.I Calculate the abundance of ^{26}Al (in grams per gram of chondritic material and as a ratio to the abundance of ^{27}Al in chondrites) required to generate sufficient heat to melt a chondritic mixture of magnesium silicates and iron initially at 500 K. You may assume that the asteroid is sufficiently large that negligible heat is lost during the period in which most of the ^{26}Al decays.

9 Asteroids

I have announced this star as a comet, but since it is not accompanied by any nebulosity and, further, since its movement is so slow and rather uniform, it has occurred to me several times that it might be something better than a comet.

Giuseppe Piazzi, 24 January 1801, commenting on the object that he had discovered 23 days earlier, which was later determined to be the first known minor planet, 1 Ceres

Asteroids are minor planets that orbit the Sun at distances ranging from inside Mercury's orbit to outside the orbit of Neptune. Most known asteroids, however, are concentrated in the *asteroid belt*, between the orbits of Mars and Jupiter. Asteroids with well determined orbits are designated by a number, in chronological order, followed by a name, e.g., 1 Ceres, 7 Iris, 324 Bamberga. Well over 10 000 asteroids have been permanently catalogued, and thousands more are added each year. Asteroids exhibit a large range of sizes, with the largest asteroid, 1 Ceres, being $\sim$470 km in radius. Ceres was discovered in 1801. The next largest asteroids are 2 Pallas, 4 Vesta, and 10 Hygiea, ranging in radius from about 250 to 203 km. (The 20 largest asteroids are listed in Table 9.1.) Smaller asteroids are more numerous; the number of asteroids with radii between R and $R + dR$ scales roughly as $R^{-3.5}$, implying that most of the mass in the asteroid belt is contained in a few large bodies. The total mass in the asteroid belt is $\sim 5 \times 10^{-4}\ M_{\oplus}$.

A study of asteroids, meteorites and comets provides unique information regarding the formation of our Solar System. Asteroids and comets can be viewed as remnant planetesimals which have undergone relatively little endogenic geological evolution, although some asteroids melted early in the history of the Solar System and there has been a substantial collisional evolution among bodies in the asteroid belt. This collisional evolution complicates interpretation of the data, but it can also work to our advantage. For example, during a collision an entire object can be broken up, which provides us with samples of the cores of bodies (the iron meteorites).

Over the past two decades our knowledge of asteroids has increased dramatically, as a result of the increased sensitivity of optical and infrared detectors, the exploitation of radar techniques, the availability of data from the Earth-orbiting observatories IRAS (Infrared Astronomical Satellite) and HST (Hubble Space Telescope), and four asteroid flybys by spacecraft. The NEAR (Near-Earth Asteroid Rendezvous) Shoemaker spacecraft went into orbit around 433 Eros in February 2000. The Spacewatch program has already vastly increased our database on small Earth-approaching asteroids, which pose a potential hazard to life on Earth; this program is currently being expanded. With the advent of large optical telescopes, improved infrared imaging devices, speckle imaging and adaptive optics techniques, as well as a tremendous (at least an order of magnitude) improvement in radar sensitivity (the Arecibo radio telescope upgrade), the observational data base of asteroids is expected to continue to grow substantially.

9.1 Orbits

Figure 9.1a shows the distribution of the semimajor axes for the orbits of nearly 4000 asteroids. Most asteroids are located in the *main asteroid belt*, at heliocentric distances between 2.1 and 3.3 AU. The spread in the eccentricities of the main belt asteroids appears to be described well by a Rayleigh distribution (like a Maxwellian distribution in one dimension, eq. 4.13), suggesting some kind of quasi-equilibrium situation:

$$N(e) \propto \frac{e}{e_*} \exp\left(\frac{-e^2}{e_*^2}\right), \tag{9.1}$$

where the mean eccentricity $e_* \approx 0.14$. The large eccentricities of many asteroids imply that the perihelia and aphelia of the asteroids occupy a significantly wider zone than do their semimajor axes (Fig. 9.1b). The mean inclination of asteroid orbits to the ecliptic plane is 15°; the standard deviation of asteroidal inclinations is larger than that of a Rayleigh distribution with the same mean value.

Several gaps and concentrations of asteroid semimajor axes can be distinguished in Figure 9.1a. The gaps were first noted in 1867 by Daniel Kirkwood and are known as the *Kirkwood gaps*. The Kirkwood gaps coincide with resonance locations with the planet Jupiter. As discussed in Section 2.3.2.2, if an asteroid orbits the Sun with a

TABLE 9.1 Twenty Largest Asteroids ($a < 6$ AU).

#	Name	Diam. (km)	Tax. Class	a (AU)	e	i (deg)	Ω (deg)	ω (deg)	M (deg)	Period (yr)	Rotation (hr)
1	Ceres	933	G?	2.769	0.0780	10.61	80.0	71.2	287.3	4.607	9.075
2	Pallas	525		2.770	0.2347	34.81	172.6	309.8	273.8	4.611	7.811
4	Vesta	510	V	2.361	0.0906	7.14	103.4	150.1	43.3	3.629	5.342
10	Hygiea	429	C	3.138	0.1201	3.84	283.0	316.1	33.0	5.656	27.659
511	Davida	337	C	3.174	0.1784	15.94	107.3	339.0	244.5	5.656	5.130
704	Interamnia	333	F	3.064	0.1475	17.30	280.4	92.2	276.8	5.364	8.727
52	Europa	312	C	3.101	0.1002	7.44	128.6	337.0	92.6	5.460	5.631
15	Eunomia	272	S	2.644	0.1849	11.76	292.9	97.5	327.9	4.299	6.083
87	Sylvia	271	PC	3.490	0.0820	10.87	73.1	273.3	248.8	6.519	5.183
3	Juno	267	S	2.668	0.0258	13.00	169.9	246.7	115.4	4.359	7.210
16	Psyche	264	M	2.923	0.1335	3.09	149.9	227.5	318.7	4.999	4.196
31	Euphrosyne	248	C	3.146	0.2290	26.34	30.7	63.1	341.0	5.581	5.531
65	Cybele	240	C	3.437	0.1044	3.55	155.4	109.8	20.1	6.372	4.041
107	Camilla	237	C	3.484	0.0842	9.93	173.5	296.0	139.7	6.503	4.840
624	Hektor	233	D	5.181	0.0246	18.23	342.1	178.0	2.9	11.794	6.921
88	Thisbe	232	C	2.767	0.1638	5.22	276.3	35.3	259.0	4.603	6.042
451	Patientia	230	C	3.062	0.0709	15.24	89.0	343.2	269.4	5.358	9.727
324	Bamberga	228	C	2.681	0.3409	11.14	327.8	43.4	189.6	4.390	29.43
48	Doris	225	C	3.110	0.0693	6.54	183.4	262.8	278.8	5.485	11.89
532	Herculina	225	S	2.771	0.1764	16.36	107.4	75.1	199.4	4.613	9.405

All data are from Yoder (1995).

period commensurate to that of Jupiter, the asteroid's orbit is strongly affected by the cumulative gravitational influence of Jupiter. Perturbations by the giant planet produce chaotic zones around the resonance locations, where asteroid eccentricities can be forced to values high enough to cross the orbits of Mars and Earth. These asteroids may then be removed by gravitational interactions and/or collisions with the latter planets. Secular resonances, such as the ν_6 resonance with Saturn located near the inner edge of the asteroid belt, where the apse precession rate of asteroids is equal to Saturn's apse precession rate, can also excite asteroids onto high eccentricity orbits. In some cases, eccentricities can be excited to such high values that the asteroids ultimately collide with the Sun, unless they are tidally shattered or vaporized during close approaches to the Sun as their periapses approach the solar photosphere. In contrast, the asteroid population is actually enhanced at the 3/2 and 4/3 resonances with Jupiter, which are located exterior to the main asteroid belt. The *Hilda asteroids*, which complete three orbits during two jovian years, and 279 Thule, which orbits the Sun four times every three jovian years, are locked in resonances with Jupiter that prevent close approaches to the giant planet. Close approaches to Jupiter lead to collisions or removal of the small bodies from the Solar System.

As of September 2000, a total of ~700 asteroids have been discovered near Jupiter's L_4 and L_5 triangular Lagrangian points. These bodies are known as the *Trojan asteroids*. The dynamics of the Trojans are discussed in Section 2.2.1. Trojan asteroids are more distant from both the Sun and the Earth than are main belt asteroids, and they have low albedos. Thus, Trojans are much more difficult to detect than are main belt asteroids of comparable size. Extrapolation of the observed distribution suggests that the total population of Trojan asteroids over 15 km in size is roughly half that estimated for main belt asteroids. However, as the Trojans lack bodies comparable in size to the larger main belt objects (624 Hector, by far the largest Trojan, has a mean radius of ~100 km), their total mass is much lower than that of main belt asteroids. The Trojans, the 4/3 and 3/2 librators, along with the main belt asteroids which are not near resonances, occupy the only known stable (for the age of the Solar System) orbits between the major planets. Orbits well inside the orbit of Mercury or well outside that of Neptune would also be stable, and small zones of stability probably exist at or near

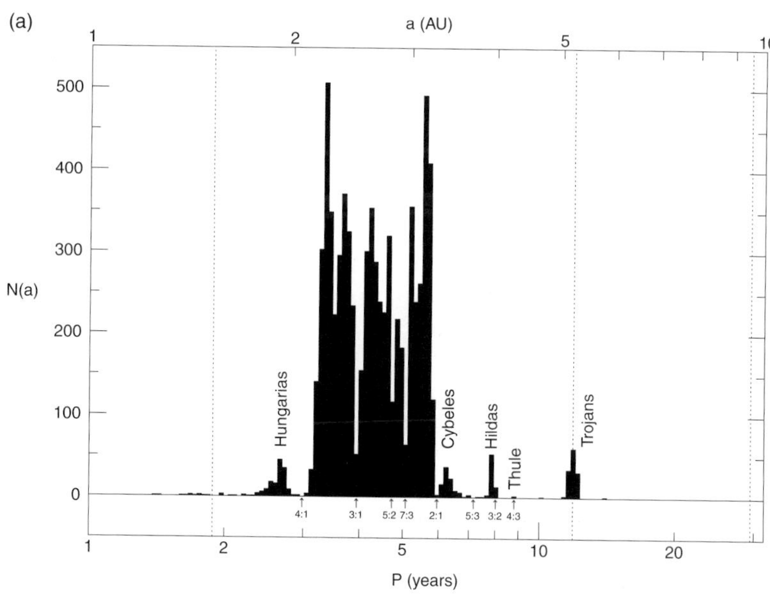

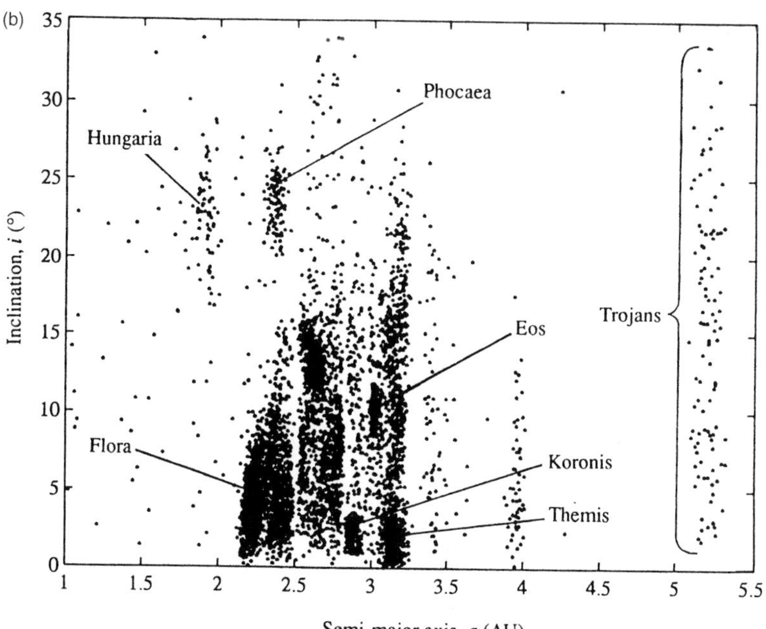

FIGURE 9.1 (a) Histogram of all numbered asteroids brighter than 15th magnitude versus orbital period (with corresponding semimajor axes shown on the upper scale); the scale of the abscissa is logarithmic. The planets Mars, Jupiter and Saturn are represented by dashed vertical lines. As asteroids with shorter orbital periods are better lit and pass closer to Earth, there is a strong observational bias favoring objects plotted towards the left of the plot. Note the prominent gaps in the distribution for orbital periods 1/4, 1/2, 2/5, 3/7 and 1/3 that of Jupiter. (Courtesy: A. Dobrovolskis) (b) Coordinates of asteroids in a–i space. (Kowal 1996)

the triangular Lagrangian points of giant planets other than Jupiter, as well as between the orbits of Uranus and Neptune. No asteroids with orbits completely interior to that of Mercury have been observed. Dozens of objects have been discovered outside Neptune's orbit, and are referred to as members of the Kuiper belt (Sections 10.2.2, 10.7.2). Pluto occupies a 2/3 mean motion resonance with Neptune, which appears to be chaotic (Section 2.3), but the chaos is so mild that Pluto's orbit is stable for billions of years, and maybe much longer. Two Trojan asteroids have been found in the L_5 Lagrangian point with Mars (5261 Eureka and 1998 VF$_{31}$), and near-Earth asteroid 3753 Cruithe is a coorbital horseshoe librator with Earth, but these orbital locks are most likely of geologically recent origin.

Many asteroids are currently on orbits that are unstable over time intervals much shorter than the age of the Solar System; such asteroids need an ongoing replenishment source from more stable orbits. Hundreds of known asteroids cross inwards of the orbit of Mars. Asteroids with perihelia between 1.017 and 1.3 AU are referred to

(c)

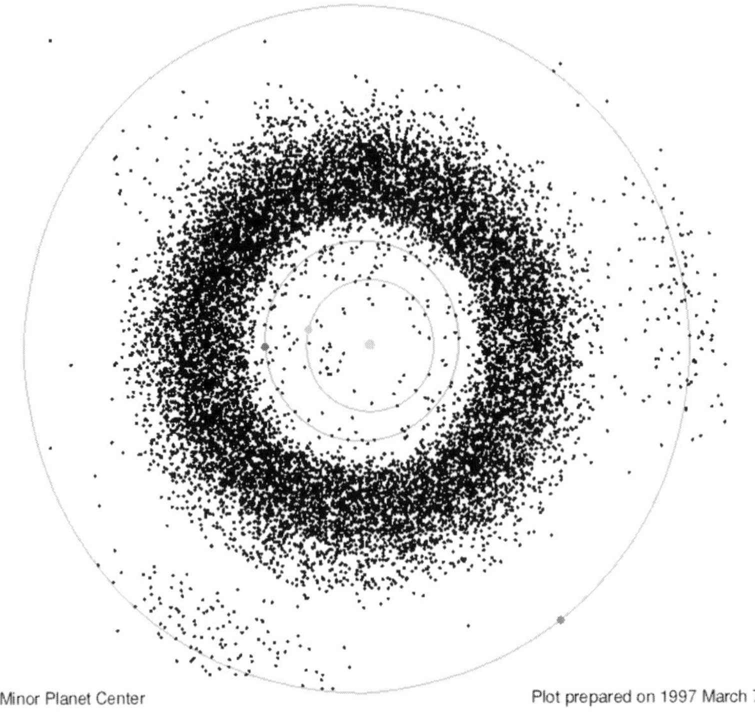

FIGURE 9.1 Continued. (c) Locations projected onto the ecliptic plane of approximately 7000 asteroids on 7 March 1997. The orbits and locations of the Earth, Mars and Jupiter are indicated, and the Sun is represented by the dot in the center. The vast majority of the asteroids depicted are in the main asteroid belt, but Trojans are shown leading and trailing the position of Jupiter; Aten, Apollo and Amor asteroids are seen in the inner Solar System, crossing the orbits of Mars and Earth. (Courtesy: Minor Planet Center)

as *Amor* asteroids, named after one of the prominent members of this group, 1221 Amor; their radii range up to $\sim$15 km. Many asteroids are also Earth-crossers. Asteroids with perihelia $q < 1.017$ AU and semimajor axes $a > 1$ AU are called *Apollo* asteroids, after a classic example in this group, 1862 Apollo. The largest detected Apollo asteroids have radii of 4–5 km. The innermost asteroids, the *Atens*, have semimajor axes $a < 1$ AU. The three groups of asteroids, Amor, Apollo, and Aten, plus comets which cross the Earth's orbit, are often referred to collectively as near-Earth Objects (NEOs), or, together with their kin in the outer Solar System, as planet-crossing objects. The dynamical lifetime of the near-Earth asteroids is relatively short, $\sim 10^7$ years. The primary loss mechanisms for NEOs are injection to interstellar space, and collision with or tidal or thermal destruction by the Sun; nonetheless, a small but significant fraction collides with planets and moons (Section 5.4.6).

Since the boundaries of the chaotic zones near resonance locations are gradual, bodies placed near the boundaries can linger for billions of years before they become Earth-crossing asteroids. However, this potential source for near-Earth asteroids is very small. Asteroids can also be perturbed into resonance zones as a result of collisions with other asteroids. Collisions within the asteroid belt are often disruptive (see below), and the orbits of the smaller fragments are typically altered the most. Thus, the population of near-Earth objects is expected to be skewed towards smaller objects relative to the main belt. Another potential source of Earth-crossing asteroids is extinct comets, which have developed nonvolatile crusts and ceased activity. A few bodies in Earth-crossing orbits that have been classified as asteroids are associated with meteor streams, suggesting a cometary origin. Also, Comet 2P/Encke is in an orbit typical of Earth-crossing asteroids. One NEO, 1979 VA, has been found which appears to be the same as the active comet 107P/1949 Wilson–Harrington. In 1949 this object was identified as having a tail, but in all apparitions since 1979 there has been no evidence of any cometary activity. This object is the first to have been 'seen' to transform from an active cometary state to a dormant asteroidal state.

Several objects are known to orbit the Sun between the orbits of Jupiter and Neptune (Section 10.2.2); examples include 944 Hidalgo ($a = 5.8$ AU), 2060 Chiron ($a = 13.7$ AU), 1994 TA ($a = 17.5$ AU), 5145 Pholus ($a = 20.2$ AU), and 7066 Nessus ($a = 24.9$ AU). Collectively, these bodies are referred to as *centaurs*. All of these objects are in chaotic planet-crossing orbits, which have dynamical lifetimes of 10^6–10^8 yrs. The centaurs may represent a transition between Kuiper belt objects and short-period comets (Section 10.2.2). The Kuiper belt objects, centaurs and other planet-crossers blur the distinction between asteroids and comets. Volatile-rich asteroids can become comets if they are brought close enough to the Sun, and comets look like asteroids if they outgas all of their near-surface volatiles and become inert. When an object is not outgassing over at least part of its orbit, it is usually considered to be an asteroid. However, many 'dormant' comets may be present among asteroids. In this text, we adopt the traditional observational definition that an object is a comet if, and only if, a coma has been observed. Asteroids on highly eccentric and/or inclined orbits are prime suspects to be inert comets. In contrast to the above definition for comets, however, we follow the tradition of discussing Kuiper belt objects together with comets because they orbit the Sun in the region believed to be the place of formation and 'storage' for most short-period comets (Section 10.2.2). Hidalgo has long been suspected of being a dormant comet. Both its orbit and spectral class (reddish-black, D type; see Section 9.4) are typical of comets. Chiron has a neutral black color, like a C type asteroid, and is less reddish than most comets. It has an orbit which crosses those of both Saturn and Uranus. Calculations show that this object must pass close to Saturn every 10^4–10^5 years. When this happens, the orbit will be perturbed significantly. Thus, Chiron orbits the Sun on a highly chaotic orbit. Chiron was therefore thought to be of cometary origin, a suspicion which was confirmed in 1987–1988, when the object brightened during its approach to perihelion (Fig. 9.2). It developed a true coma, and is now classified as a comet. Interestingly enough, after a maximum in the late 1980s, Chiron's brightness reached a low around 1996, when the object was at perihelion. However, its activity had not ceased, and it is now believed that Chiron may always have had some degree of activity. Both 5145 Pholus and 7066 Nessus are very red (Fig. 9.17), redder than most known asteroids and comets. It has been suggested that the red color may be caused by the presence of exposed organic materials (Section 10.6.5).

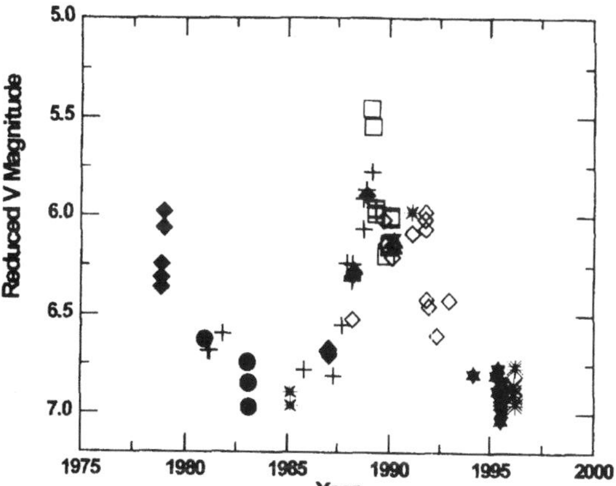

FIGURE 9.2 The 'outburst' of 2060 Chiron: a graph of the magnitude of Chiron as a function of time, which clearly shows Chiron's dramatic increase in brightness starting in 1987. The data were taken by different groups, as symbolized by the different symbols used. (Lazzaro *et al.* 1997)

9.2 Size Distribution and Collisional Evolution

Most of the main belt asteroids that have been discovered and numbered to date are ~10–30 km in radius. This is near the lower size limit of broad asteroid surveys. Most of the mass, however, is concentrated in the largest bodies. The orbits of asteroids cross one another, and main belt asteroids must undergo collisions on a timescale short compared to the age of the Solar System (Problem 9.5). Indeed, many phenomena are suggestive of a collisionally evolved population of objects.

9.2.1 Size Distribution

The size distribution of asteroids can be approximated by a power law valid over a finite range in radius. Size distributions can be given in differential form:

$$N(R) = N_o \left(\frac{R}{R_o}\right)^{-\zeta} \qquad (R_{min} < R < R_{max}), \quad (9.2a)$$

where R is the asteroid radius, and $N(R)dR$ the number of asteroids with radii between R and $R + dR$. The size distribution can also be presented in cumulative form:

$$N_>(R) \equiv \int_R^\infty N(R')dR' = \frac{N_0}{\zeta - 1}\left(\frac{R}{R_o}\right)^{1-\zeta}, \quad (9.2b)$$

where $N_>(R)$ is the number of asteroids with radii larger than R.

Collisions play a major role in the asteroids' evolution, except possibly for the largest bodies, which are believed to date back to the epoch of planet formation. Theoretical

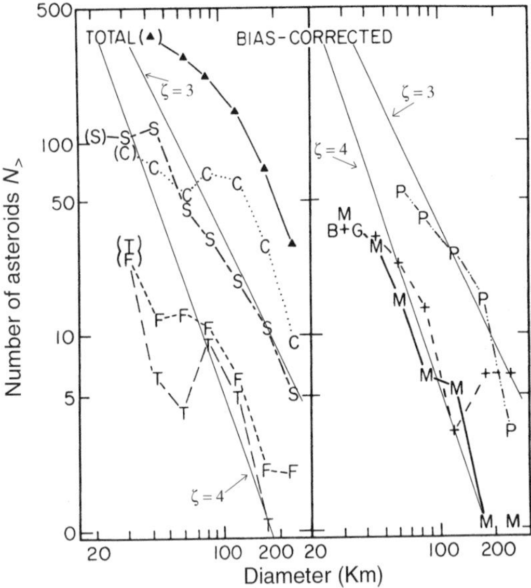

FIGURE 9.3 The relative size distribution as observed for asteroids, with superposed lines which correspond to power law size distributions with $\zeta = 3$, and $\zeta = 4$ (eq. 9.2). (Adapted from Gradie *et al.* 1989)

calculations imply that a population of collisionally interacting bodies evolves towards a power law size distribution with $\zeta = 3.5$, provided the disruption process is self-similar (not size dependent). A slope of $\zeta = 3.5$ implies that most of the mass is in the largest bodies and most of the surface area is in the smallest bodies (Problem 9.3). Figure 9.3 shows a comparison between the relative size distributions observed for asteroids and power law distributions for various slopes ζ. The overall asteroidal size distribution is consistent with that expected from a collisional evolution. The flatter slope ($\zeta \approx 3$) observed for mid-sized main belt asteroids has been attributed to complex variations in asteroidal properties with size. The dropoff in $N(R)$ at small sizes is caused by an observational bias against the detection of small asteroids. The size distributions of individual asteroid classes sometimes show prominent bumps, which are normally not expected for an evolved population of objects. These bumps could be remnants of an original population, or they may be caused by a difference in internal strength of bodies between different sizes. Near-Earth asteroids, most of which are probably fragments of asteroids in the main belt, appear to have a steeper size distribution ($\zeta \gtrsim 4$), i.e., most of the mass, as well as the surface area, is in the smallest bodies. Smaller fragments from collisions within the main belt receive a greater velocity kick and thus are preferentially transferred to chaotic regions surrounding resonances, from which their eccentricities can be excited sufficiently for them to become planet crossers.

9.2.2 Collisions

Random asteroid velocities are typically a few kilometers per second, which is much larger than the escape velocities of asteroids. (The escape velocity from Ceres $v_e \approx 0.6$ km s^{-1}.) Thus, most collisions are expected to be erosive or disruptive. The final outcome of a collision depends on the relative velocity and strength/size of the object (Section 12.5.3.2). Large asteroids and iron–nickel bodies have the greatest resistance to disruption. In super-catastrophic collisions, the colliding bodies are completely shattered, and the fragments are dispersed into independent, yet similar, orbits. Groups of asteroids with nearly the same orbital parameters (a, e, and i) are called *Hirayama families*, in honor of the Japanese astronomer who discovered the first asteroid families. Individual families are named after their largest member. At least eight large families have been discovered and more than 100 smaller suspected families have been tabulated. These families often share similar spectral properties, further suggesting a common origin in a single body that has undergone a catastrophic collision. However, the genetic relationship among bodies within the smaller families is often controversial. Only five families are well established based upon both orbital and cosmochemical perspectives: the Eos, Koronis, Themis, Flora, and Nysa/Hertha families.

Collisions between asteroids and (micro)meteorite impacts on asteroids lead to erosion of the bodies. Dust thrown off asteroids in collision and/or sputtering processes, as well as dust produced in catastrophic disruptions of asteroids, 'pollute' the environment. As discussed in more detail in Section 9.2.4, these processes are important sources of interplanetary dust. Dust originating from a break-up of asteroids has been discovered by the IRAS satellite in the form of pairs of *dust bands*. If a body is completely shattered, but the velocity of the individual fragments is not large enough to disperse them, some or most of the fragments may coalesce back into a single body and become a rubble-pile. They may also form a binary or multiple asteroid system, where fragments are of similar size and mass, or an object surrounded by one or more small satellites. Such systems, however, are usually not long-lived. Tidal interactions between a pair of asteroids lead to orbital evolution on a timescale of $\sim 10^5$ years. Satellites inside the synchronous orbit evolve inwards, so that the system ultimately becomes a rubble-pile compound asteroid. Satellites outside the synchronous orbit evolve outwards, and ultimately escape the gravitational attraction from the primary.

There is evidence of compound or binary asteroids, and of satellites in orbit around asteroids. In 1993, the Galileo spacecraft image of 243 Ida and its satellite Dactyl proved unambiguously that asteroids can have satellites – but how common is this phenomenon? About 200 asteroids were imaged with high spatial resolution (0.05–0.15″) using an adaptive optics system on the CFHT and Keck in Hawaii. This search identified moons in orbit about three different asteroids, as well as one double asteroid. A 13 km sized (diameter) satellite was found in a low eccentricity orbit around 45 Eugenia, at a distance of 1190 km and period of 47 days. Similarly sized satellites were found to orbit 762 Pulcova in 4 days, and 87 Sylvia in 3.6 days. 90 Antiope appears to be a double asteroid, where the two 80–85 km sized components orbit each other in 16.5 days at a distance of 170 km. Radar observations discovered companions around two NEOs, each at a distance of a few kilometers (Table 9.3). The Kuiper belt object 1998 WW_{31} is also a binary pair. We expect that more sensitive searches at higher spatial resolution will reveal more companions around asteroids and KBOs. Such studies are important since the orbits of asteroidal satellites can be used to estimate the masses of the bodies (Section 9.3.7).

The idea of multiplet-asteroids is not new, nor limited to these few direct detections. As discussed in Section 6.1, only the largest asteroids are expected to be spherical in shape, an expectation which has been verified by observations. But even a few large objects are quite irregular, such as the Trojan asteroid 624 Hector, with short and long diameters $\sim$150 × 300 km. It has been suggested that this asteroid is a compound body, formed when two round Trojans collided at low speed. As typical collisions between Trojan asteroids occur at $\sim$5 km s^{-1}, this type of gentle encounter is only likely to occur between two fragments of a super-catastrophic collision. Alternatively, a large collision could produce a single coherent large elongated object if most of the ejecta escape.

Roughly 10% of the largest ($\gtrsim$20 km in diameter) known impact craters on Earth (e.g., the Clearwater Lakes crater pair, which are 32 and 22 km across) and Venus and $\sim$2% on Mars are doublets, and must have been formed by the nearly simultaneous impact of objects of comparable size. The separation of the craters, as well as the crater size, is too large to be caused by an asteroid tidally disrupted or fragmented just before impact. It has been proposed that these doublet craters are caused by the impact of binary asteroids. In contrast, the crater chains on the surfaces of Ganymede (Fig. 5.58c), Callisto and the Moon are the result of an impact by a body which was tidally disrupted (by Jupiter or Earth) just before impact, indicative of a rubble-pile nature of the original body. Detailed radar observations show 216 Kleopatra to be shaped like a dumbbell and the NEOs 4769 Castalia and 4179 Toutatis to be bifurcated (Section 9.5); the latter bodies look like contact binaries. The lightcurves (Section 9.3.2) of the NEOs 1994 AW_1, 1991 VH, 1996 FG_3 and 3071 Dionysus appear to consist of two components with different periods and amplitudes. The components are consistent with models of eclipsing/occulting binary asteroids, separated by up to a few asteroid diameters. There is growing evidence that many of the NEOs are compound bodies, and that (contact) binary systems among these bodies are common. It is possible that binary systems are more common among the NEOs than main belt asteroids because close encounters with Earth could tidally disrupt a compound body, similar to the tidal disruption as has been witnessed for Comet D/Shoemaker–Levy 9 by Jupiter (Section 5.4.5). Such a system could evolve into a binary system, and impact the planet on a later return.

We suspect that the present asteroid population is just a small fraction of what it once was. Collisions may gradually grind the bodies to smaller and smaller fragments; the smallest being removed by Poynting–Robertson drag and radiation forces. Other fragments might have been placed into chaotic orbits. The few large asteroids that exist today may, by chance, have escaped catastrophic collisions. If such large bodies fragmented at all during their lifespan, they may have coalesced back into a single body. Re-accumulation is more difficult for asteroids less than $\sim$50 km in size. With the diversity of asteroids observed today it is almost impossible to develop a consistent theory that can explain all aspects. We note here the existence of the large metallic asteroid 16 Psyche, which could be the remnant of a very large differentiated object, stripped down to its core by numerous impacts. At the opposite extreme, 4 Vesta seems to have preserved its thin basaltic crust.

9.2.3 Rotation

Over 80% of all planetary bodies rotate with a period between 4 and 16 hours (Fig. 9.4). Asteroids spinning faster than about 2 hours would throw loosely attached material off their equator, and hence are unlikely to survive at all (Problem 9.11). There appears to be a clear correlation between rotation period and asteroid size: asteroids with radii less than about 5 km spin much faster than larger bodies (Fig. 9.4b). The tiny ($\sim$15 m radius) asteroid 1998 KY_{26} has a rotation period of only 10.7 minutes, implying that it is a single coherent body without a regolith. The shortest rotation period observed for asteroids larger than 1 km is 2.35 hours, slightly longer than the theoretical lower limit of 2 hours. The spin rates of asteroids with radii over 60 km are most likely governed by their collisional history, since their distribution can be fit by a Maxwellian distribution (Fig. 9.4a). Smaller asteroids show a relative excess of slowly rotating bodies, and their distribution cannot be matched with a Maxwellian. This may be a result of their formation in catastrophic disruption events. A few asteroids with exceptionally long rotation periods may have been tidally despun by (as yet undetected) moons (cf. Section 2.6). The asteroidal spin vectors appear to be distributed isotropically.

The moment of inertia tensor (eq. 2.42) of any body can be diagonalized (made into the form $I_{jk} = 0$ for $j \neq k$) if the proper coordinate axes are chosen; these axes are referred to as the *principal axes* of the body. The principal axes of a homogeneous triaxial ellipsoid are simply the axes of the ellipsoid itself, and the principal axes of a spherically symmetric body can be chosen to lie along any three mutually orthogonal lines which pass through the center of the body. The lowest energy state for a given rotational angular momentum is simple rotation about a body's axis of greatest moment of inertia (the short axis), cf. equation (2.41). Simple rotation about a body's axis of least moment of inertia (long axis) is also possible, but requires more energy and thus is secularly unstable (often on very long timescales) to energy dissipation resulting from rotationally induced stresses within the body. Rotation about the axis of intermediate moment of inertia is unstable on a dynamical (very short) timescale. If a body does not spin about one of its principal axes, then the rotational angular momentum is not parallel to the instantaneous axis of rotation (eq. 2.41*a*), and as a consequence the axis of rotation varies, i.e., the body undergoes torque-free precession; in layman's terms, it wobbles. This wobble is described quantitatively by Euler's equations; see Goldstein (1980) or another text on classical mechanics for details.

Internal stresses caused by precessional wobble damp the rotation of planetary bodies down towards the lowest energy state. A good way to visualize this process is to imagine that the asteroid is a collection of rigid balls connected together by springs. The springs oscillate as variations in the axis of rotation alter internal stresses, and mechanical energy is lost to heat because the springs are damped via friction. The damping timescale, t_{damp}, depends on the density, radius and rigidity of the body, ρ, R and μ_{rg}, a shape-dependent factor K_3^2 which varies from ~ 0.01 for a nearly spherical body to ~ 0.1 for a highly elongated one, the ratio of energy contained in the internal oscillations of the body to that lost per cycle, f_Q, and the rotational frequency ω_{rot} as:

$$t_{damp} \approx \frac{\mu_{rg} f_Q}{\rho K_3^2 R^2 \omega_{rot}^3}. \tag{9.3a}$$

For nominal asteroidal parameters, the damping timescale in billions of years is given by:

$$t_{damp} \sim \frac{0.7}{R^2} \frac{2\pi}{\omega_{rot}^3}, \tag{9.3b}$$

where the radius of the asteroid is in kilometers and its orbital period, $2\pi\omega_{rot}^{-1}$, is in days. The uncertainty/spread in damping times given by equation (9.3*b*) is estimated to be approximately a factor of 10. Precessional motions are thus damped very rapidly for large bodies which rotate rapidly, but small slow rotators can remain in complex rotational states for long periods of time (Problem 9.13). Nutation of these bodies is also easiest to excite via collisions, or in the case of comets as a result of outgassing. Several small asteroids and comets are believed to wobble substantially.

9.2.4 Interplanetary Dust

The interplanetary medium contains countless microscopic dust grains. Faint reflections of sunlight off dust grains in the interplanetary medium produce two observable phenomena: the *zodiacal light* and the *gegenschein*. The zodiacal light is visible just after sunset and before sunrise, in the direction of the Sun. Dust particles, concentrated in the invariable plane, scatter sunlight in the forward direction, and the zodiacal light is about as bright as the Milky Way. The gegenschein is visible in the anti-solar direction as a faint glow, caused by backscattered light from interplanetary dust particles.

The presence of interplanetary dust is further 'seen' in the form of *meteors*, streaks of light in the night sky, caused by centimeter-sized dust grains which, when falling down, are heated to incandescence by atmospheric friction (Section 5.4.3, Section 8.3). Under excellent conditions, one may see 5–7 meteors per hour. On some nights there are many more which appear to come from a single point in space: such events are called *meteor showers*, and are generally named after the stellar constellation that contains the radiant point, e.g., the Perseids on 11 August and the Leonids on 17 November. The latter sometimes displays spectacular *meteor storms*, with up to 150 000 meteors per hour! Many of the meteor showers are associated with cometary orbits: The debris left behind by the outgassing comet is intercepted by the Earth when it intersects the comet's path.

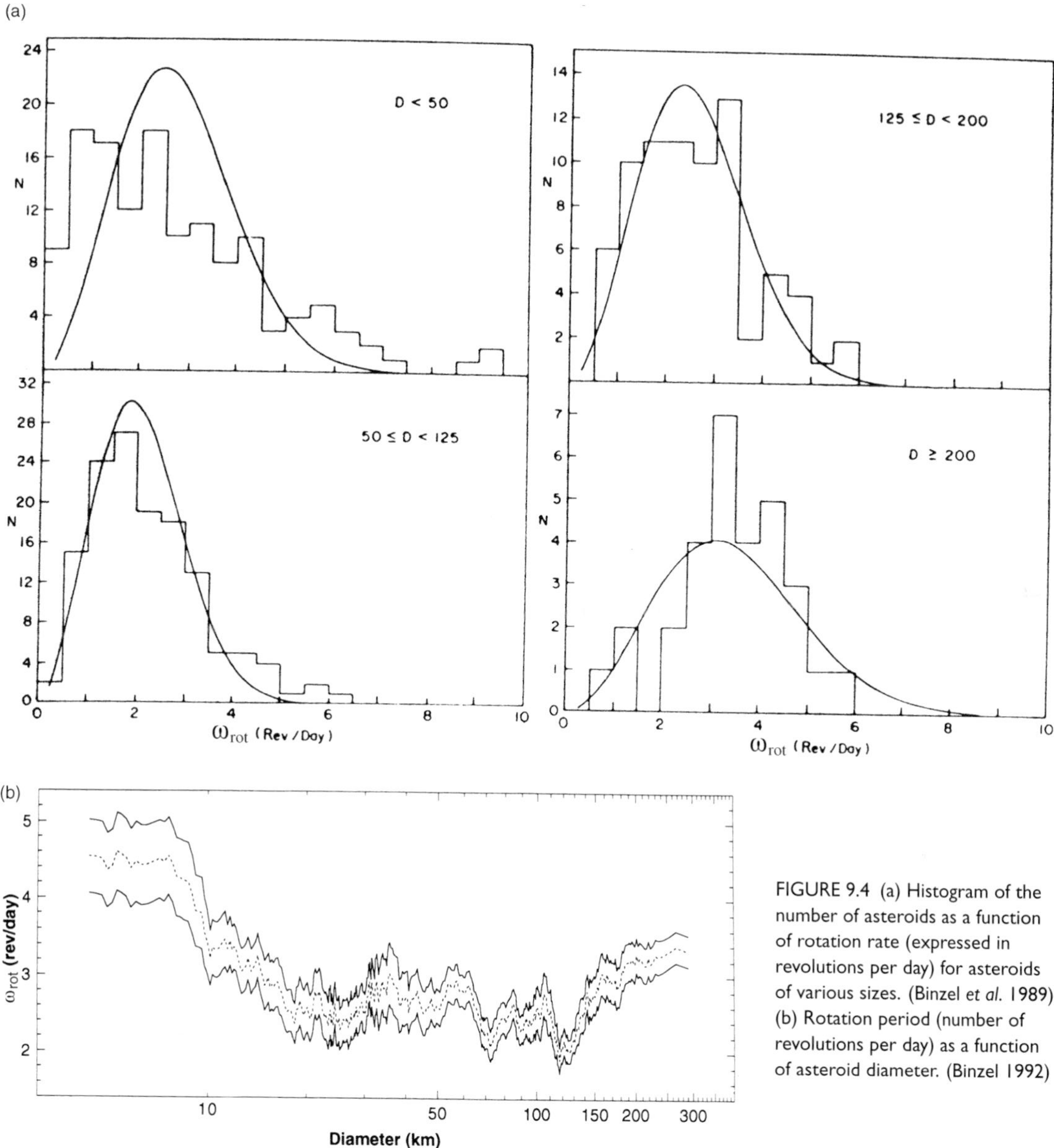

FIGURE 9.4 (a) Histogram of the number of asteroids as a function of rotation rate (expressed in revolutions per day) for asteroids of various sizes. (Binzel *et al.* 1989) (b) Rotation period (number of revolutions per day) as a function of asteroid diameter. (Binzel 1992)

Studies show that the Earth 'accretes' about 4×10^7 kg of interplanetary dust per year! Interplanetary dust particles have been collected by high-altitude aircraft in the Earth's stratosphere, the so-called *Brownlee particles* (Section 8.3). Analysis in the laboratory showed an overall similarity in composition of these particles with chondritic meteorites. Some particles show a much larger than terrestrial value for the D/H ratio, suggesting an interstellar origin (Section 10.7.1).

The IRAS satellite, sensitive to the thermal emission from 10–100 μm sized interplanetary dust particles, revealed considerable structure in the zodiacal dust cloud. In particular, the satellite detected a large number of bright emission streaks, attributed to bands of dust encircling the inner Solar System. Many of the *dust bands* are attributed to outgassing by comets, producing the meteor streams associated with the meteor showers mentioned above. A total of eight of these trails have been positively identified with short-period comets. Several dust bands appear as 'double' structures. Roughly seven pairs of such dust bands appear to straddle the plane of the ecliptic (Fig. 9.5). The IRAS color temperature (~ 200 K) and parallactic studies place these bands within the asteroid belt. These dust bands are seen as pairs, since the particles which make up the band

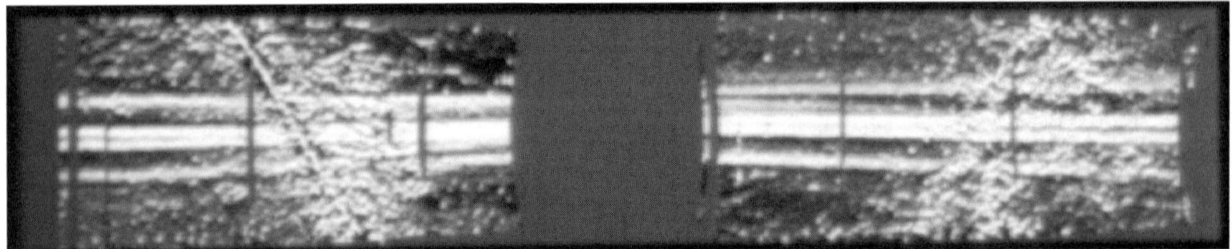

FIGURE 9.5 IRAS image at 25 μm of several zodiacal dust bands in the asteroid belt. Parallel bands are seen above and below the ecliptic, encircling the inner Solar System. The two central bands are connected to the Koronis–Themis families, and the 10° bands may be from the Eos family. The thin band between the center bands and the outer 10° bands is a type 2 dust trail. Type 1 dust trails originate from short-period comets; type 2 dust trails have not yet been understood. It could signify a relatively recent break-up of an asteroid. The diagonally shaped bands are the galaxy. (Courtesy: M.V. Sykes)

have the same orbital inclination (to the local Laplacian plane), but their nodes are uniformly distributed over all longitudes. Because each particle spends most of its time near the extreme of its oscillation (as does a swing or pendulum), a collection of dust particles with uniform inclination orbits is seen as a pair of bands straddling the plane of the ecliptic, with a separation depending on the inclination of the particles' orbits. These asteroidal dust bands likely result from erosive collisions in the asteroid belt, or from a catastrophic disruption of an asteroid. Asteroid families represent concentrations of asteroids in a–e–i space, and some of the dust bands have inclinations corresponding to major asteroid families.

As discussed in Section 2.7, (sub-)micrometer–centimeter sized material is removed from the Solar System by radiation pressure, and/or Poynting–Robertson or solar wind drag. Thus the orbits of particles in these dust bands change over time, and the bands are expected to fade away. Since these small dust grains are lost from the interplanetary medium, they must be resupplied on a frequent basis. Possible sources include collisions in the asteroid belt and refractory particles released as a result of outgassing by comets in the inner Solar System. The asteroid-related and cometary dust bands, as well as the meteor showers confirm the collisional hypothesis in the asteroid belt and cometary outgassing as sources for interplanetary dust. The distribution of velocities of meteoroids entering the Earth's atmosphere also implies a mixture of particles with cometary and asteroidal origin. It is not clear, however, which source dominates. In addition, calculations have shown that dust created in the Kuiper belt contributes to the interplanetary dust in the inner Solar System. Although $\sim 80\%$ of these grains should be ejected from the Solar System by the giant planets, a fair number of them evolve all the way in towards the Sun. Some of these particles collide with Earth, and represent (some of) the extraterrestrial grains scooped up by aircraft in the stratosphere.

Numerical integrations of dust particles originating in the asteroid belt show that 20–25% of these grains are temporarily trapped in corotational resonances just outside the orbits of the terrestrial planets. The Earth, therefore, is embedded in an extremely tenuous ring of asteroidal dust particles with a width of 0.4 AU. The ring is longitudinally nearly uniform, except for a cavity which contains the Earth. Poynting–Robertson drag introduces a phase lag in the equation of motion of the resonant particles, so that the Earth is closer to the edge of the cavity in the trailing orbital direction than in the leading direction. This is consistent with an observed 3% enhancement in the zodiacal brightness in the trailing direction compared to that in the leading direction.

9.3 Observing Techniques

High resolution images of two asteroids, 951 Gaspra and 243 Ida, have been obtained by the Galileo spacecraft, and the NEAR spacecraft imaged 253 Mathilde and 433 Eros in detail (Fig. 9.6). Some of the larger asteroids can be resolved by HST, or from the ground using speckle interferometry or adaptive optics (Fig. 9.7; cf. Section 9.3.4). The resolution in these images is high enough to allow detection of large scale shape irregularities and albedo features. The majority of the asteroids, however, cannot be optically resolved by Earth-based telescopes. For these asteroids, sizes are best estimated using stellar occultations, where an asteroid passes in front of a star and star light is temporarily blocked out during an interval of time, the duration of which depends on the speed and size of the object. A drawback of this technique is that many chords are required, as asteroids are usually irregular in shape. If such information is not available, there are a number of observing techniques that yield information on the size

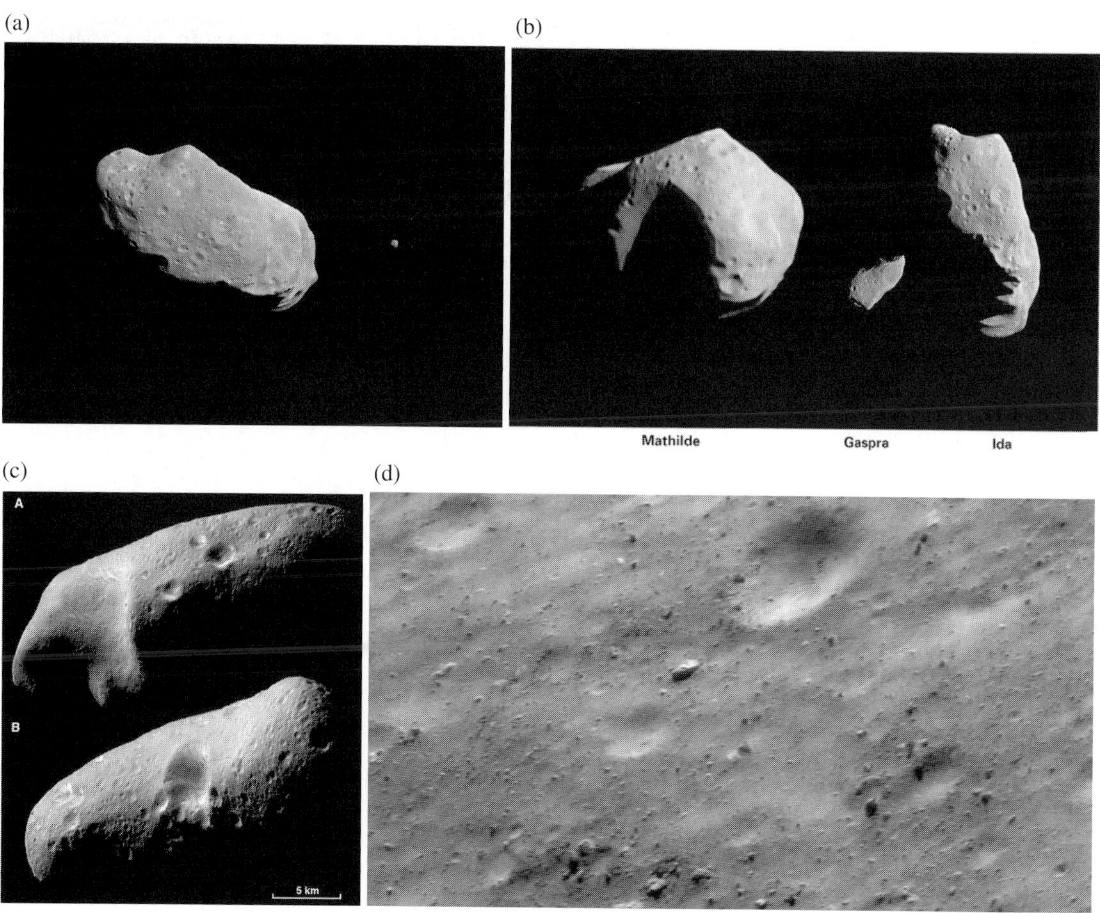

FIGURE 9.6 (a) Galileo image of S type asteroid 243 Ida and its moon Dactyl. Ida is about 56 km long. Dactyl, the small object to Ida's right, is about 1.5 km across in this view, and probably ~100 km away from Ida. (NASA/Galileo PIA00136) (b) Comparison of S type asteroids 951 Gaspra, 243 Ida and C type asteroid 253 Mathilde. The first two satellites were imaged by the Galileo spacecraft, Mathilde was photographed by the NEAR spacecraft. (NASA) (c) Mosaiced images of the near-Earth asteroid 433 Eros obtained by the NEAR Shoemaker spacecraft on 25 February 2000. These are two views of the asteroid as the body is rotating underneath the spacecraft, at a distance of 350 km. The resolution is about 25 m/pixel. The top mosaic shows wavy brightness banding exposed on the interior walls of the ~10 km wide saddle. The bottom image shows a similar banding in one of the craters near the limb at the left. Towards the right, the angle of illumination accentuates the quasi-linear troughs near the terminator. (Veverka *et al.* 2000) (d) This view of Eros is part of an image mosaic taken in the early hours of 26 October 2000, during NEAR Shoemaker's low-altitude flyover of Eros. Taken while the spacecraft's digital camera was looking at a spot 8 km away, the image covers a region about 800 m across. Rocks of all sizes and shapes are set on a gently rolling, cratered surface. Locally, fine debris or regolith buries the rocks. The large boulder at the center of the scene is about 25 m across. (NASA/NEAR Image 0147952603)

and/or shape of an object: radiometry, photometric light curves, polarimetry, and radar techniques. Each of these techniques is summarized below.

9.3.1 Radiometry

The size and albedo of an asteroid can be estimated by means of *radiometry*, a technique that makes use of a combination of the thermal infrared (~ 20 μm) emission from an object and a photometric measurement of sunlight reflected off the object at visible wavelengths. The total thermal emission is proportional to the product of absorbed insolation, $1 - A_b$, (A_b is the asteroid's Bond albedo) and the object's projected surface area, while the amount of reflected sunlight is equal to the body's visual albedo times its projected surface area (eqs. 3.15, 3.16). Asteroids are most likely in equilibrium with insolation, so the sum of reflected and emitted radiation must be equal to the solar radiation intercepted. Therefore, measurements of the thermal and reflected emissions are sufficient to determine

TABLE 9.2 Albedos and Phase Functions for Various Airless Bodies.

Body	$A_{0,v}$	$q_{ph,v}$	A_v	A_b	Ref.
Moon	0.113	0.611	0.069	0.123	1
Mercury	0.138	0.486	0.067	0.119	1
243 Ida	0.21	0.34	0.071	0.081	2
Dactyl	0.20	0.32	0.064	0.073	2
253 Mathilde	0.047	0.280	0.013		3
433 Eros	0.29	0.39	0.11	0.12	4
951 Gaspra	0.23	0.47	0.11		5
Phobos	0.071	0.300	0.021		6
Deimos	0.068	0.390	0.027		7

1: Veverka *et al.* (1988). 2: Veverka *et al.* (1996). 3: Clark *et al.* (1999). 4: Domingue *et al.* (2001). 5: Helfenstein *et al.* (1994). 6: Simonelli *et al.* (1998). 7: Thomas *et al.* (1996).

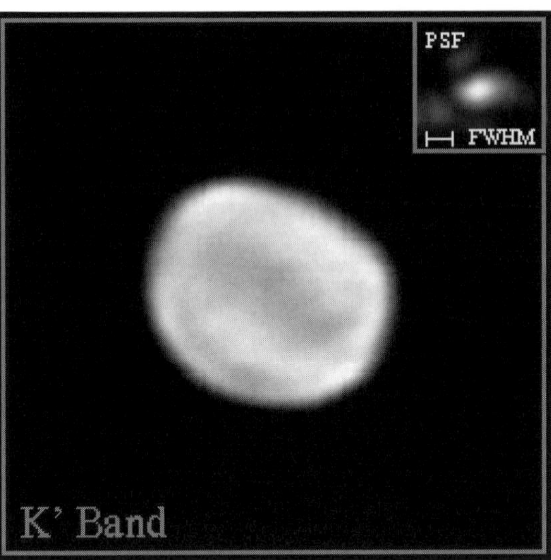

FIGURE 9.7 Keck adaptive optics image of 4 Vesta . Bright and darker bands are clearly visible. Image is taken at 2.1 μm. The resolution is 50 mas. (Courtesy: Keck Observatory Adaptive Optics Team)

the size and albedo of an object if the relation between visual geometric and Bond albedo is known.

For main belt and outer Solar System asteroids, measurements are usually made at small phase angles. To estimate the phase dependence of the albedo, the photometric and thermal properties of asteroid surfaces must be known. Most researchers have assumed that asteroids resemble the Moon, a spherical object covered with a thick loose particulate regolith, which has a low thermal inertia. A body with such properties radiates only a small percentage of the absorbed insolation from its dark hemisphere, so that the infrared radiation from large phase angles is small. Assuming similar characteristics for a couple of large asteroids and the Galilean satellites of Jupiter, however, showed that the radii and albedos for these objects disagreed with radii obtained from stellar occultations, suggesting differences in the angular distribution of the thermal emission between the Moon and asteroids. The Galileo flyby of 243 Ida provided direct measurements of the phase integral and Bond albedo for one asteroid (Table 9.2), and the values are quite different from those of the Moon. To simulate the differences, a 'fudge' or 'beaming' factor, η_ν, is added to equation (3.16) for a slowly rotating asteroid at phase angle $\phi \approx 0°$:

$$\mathcal{F}_{out} = \pi R^2 \eta_\nu \epsilon_\nu \sigma T^4. \tag{9.4}$$

If $\eta_\nu \approx 0.7$–0.8, the radii and albedos for asteroids with diameters over 50 km agree quite well for values derived independently from radiometry, polarimetry and stellar occultations.

For smaller objects, the diameters and albedos derived from different techniques do not always agree. There are a number of obvious ways in which the standard lunar model may break down, even after inclusion of the beaming factor η_ν: (i) the object may be very aspherical, (ii) the rate and sense of rotation and obliquity modify the 'observed' temperature, (iii) the asteroid may not have a regolith, in which case the thermal inertia is much higher and as much as 50% of the absorbed insolation may be radiated from the night-side hemisphere, or (iv) the asteroid's surface may be highly metallic. In the latter case, the infrared emissivity may be as low as 0.1, while the thermal conductivity is very high. The low emissivity would raise the surface temperature, and shift the bulk of thermal radiation to shorter wavelengths. Unless the emission is observed at the proper wavelengths, this shift may go unnoticed, and the derived asteroid properties may be grossly in error.

Asteroids are now being modeled by full thermophysical models, where the diurnal temperature variations in the subsurface layers are determined from a balance between insolation (eq. 3.18a), conduction of heat through the subsurface (eq. 3.22), and reradiation outward (eqs. 3.44, 3.61). This technique requires knowledge of the thermal inertia (eq. 3.24) and thermal skin depth of the material (eq. 3.25), parameters which are usually not known for asteroids. As mentioned above, surface roughness tends to increase the sunward thermal emission from asteroids, an effect modeled in the thermophysical model by inclusion of the parameter η_ν from equation (9.4). Hence, although full thermophysical models are clearly preferred over the standard lunar model, given the large uncertainties in the basic quantities the reliability of the thermophysical model

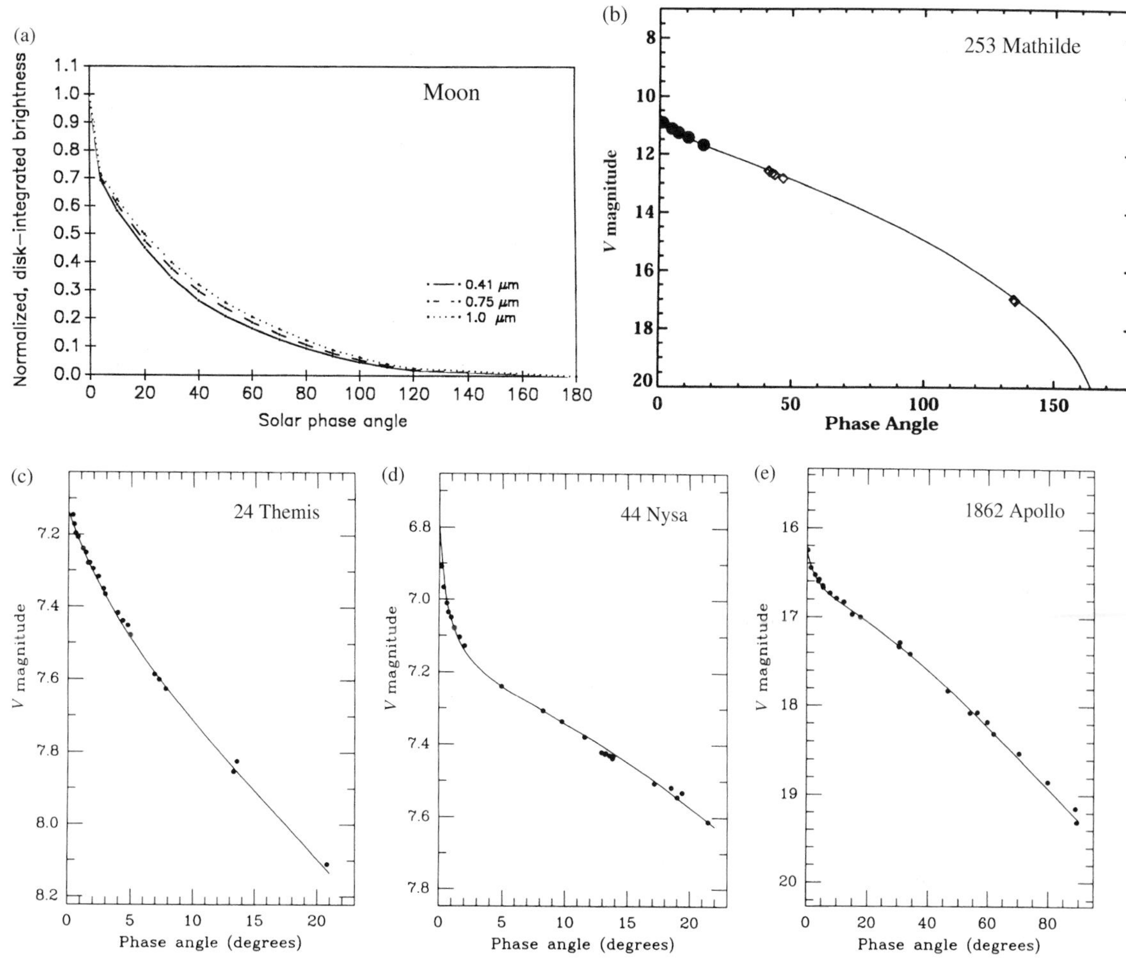

FIGURE 9.8 (a) Brightness of the Moon as a function of solar phase angle, at three different wavelengths. The data at small phase angles ($< 5^\circ$) were derived from Clementine data. (Buratti *et al.* 1996) (b) Phase function of 253 Mathilde, as determined from ground-based (filled circles) and NEAR whole-disk 0.70 μm data (diamonds). The data are compared with the best-fit model Hapke parameters. (Clark *et al.* 1998) Asteroid magnitude as a function of phase angle for 24 Themis (c), 44 Nysa (d) and 1862 Apollo (e). Note the opposition effect, the anomalous increase in intensity near $\phi = 0^\circ$. The data are compared to best-fit Hapke models. (Bowell *et al.* 1989)

may not be much higher. However, it does allow one to calculate an asteroid's temperature as a function of depth in the crust, and an asteroid's brightness both at infrared and radio wavelengths.

Graphs of an asteroid's brightness as a function of phase angle usually show an anomalous increase in intensity at phase angles $\phi \lesssim 2^\circ$, referred to as the *opposition effect* (see Fig. 9.8). This same effect is seen on the Moon (Fig. 9.8a), where data at $\phi < 0.5^\circ$ have been obtained by the Clementine spacecraft (because of the Moon's finite angular extent as seen from the Earth, such observations could not be made from Earth). The opposition effect for the Moon is very large; the intensity increases by $\sim 20\%$ from $\phi = 0.15^\circ$ down to $\phi = 0^\circ$. Laboratory experiments have shown that the opposition effect is seen from any particulate material. In the past this effect has been attributed to the disappearance and hiding (extinction) of shadows when the Sun and observer are located in the same direction as seen from the asteroid. It can be shown from the equations of radiative transfer in a semi-infinite stratified medium of particulate material, where scattering is dominated by single scattering and where the grains are large compared to the wavelength, that the intensity of the backscattered radiation indeed increases substantially

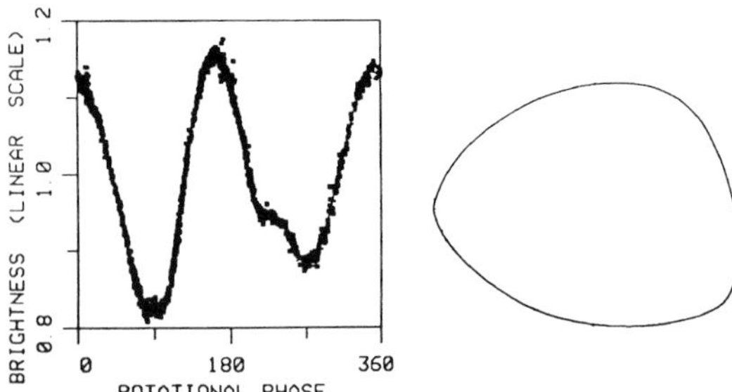

FIGURE 9.9 Photometric lightcurve of the asteroid 164 Eva. The lightcurve is shown on the left. The curve can be matched by the solid profile on the right: imagine this as a two-dimensional, geometrically scattering asteroid. As this asteroid rotates, it generates a lightcurve similar to that shown on the left. (Magnusson *et al.* 1989)

at zero phase angle. In some cases, however, the effect is much larger than can be explained from the disappearance of shadows alone. Another phenomenon that causes an opposition effect in particulate material is the *coherent-backscatter effect*, which is caused by multiple scattering in particulate material, so that the incoming waves are diffusely scattered in all directions of the medium. At zero phase angle the waves interfere constructively, and the reflected intensity can be amplified considerably, up to a factor of 2. Both the shadow-hiding and coherent-backscatter contribute to the opposition effect. For the Moon the coherent-backscatter effect results in the narrow peak near opposition (Fig. 9.8a) at phase angles $\phi < 2^\circ$, whereas the broader component at $\phi < 20^\circ$ can be explained by the shadow-hiding theory.

9.3.2 Photometric Lightcurves

Photometry is a photon-counting technique, where the brightness of the entire object is measured. Measurements taken over the course of one or more nights can be combined into a *photometric lightcurve*, which shows the variations in the body's brightness as the object rotates around its axis (Fig. 9.9). Such lightcurves have shown that the rotation periods of the majority of asteroids are between 4 and 16 hours. Irregularities in the asteroid's shape and variations in surface albedo both influence the shape of lightcurves. The visible brightness of an asteroid is proportional to the product of the asteroid's projected area and the average albedo of this region, whereas the infrared brightness varies as the product of the projected area and $(1 - A_b)$. Thus, by combining lightcurves at visible and at infrared wavelengths, one can distinguish amplitude variations caused by the object's shape from those resulting from albedo variations on its surface. Note that shape variations produce lightcurves with two maxima and two minima per rotation, whereas albedo variations alone usually produce lightcurves which are single-peaked. Lightcurves obtained at different viewing angles can be used to determine an asteroid's pole position and its sense of rotation. It takes, however, many years to gather data over enough different viewing aspects to determine the spin orientation of an asteroid.

A comparison of lightcurves for small main belt and near-Earth asteroids shows that the distribution of lightcurve amplitudes is similar, suggesting that the gross distributions of shapes of these bodies are similar. Comets, although providing a limited sample, typically show longer rotation periods and larger lightcurve amplitudes. This may be caused by differences in intrinsic strength and densities. Trojan asteroids typically show larger mean lightcurve amplitudes than the main belt and near-Earth asteroids, indicative of more elongated shapes.

9.3.3 Polarimetry

The linear polarization of reflected light depends upon the scattering geometry, surface refractive index and surface texture of the asteroid. When unpolarized light, e.g., sunlight, is reflected or scattered off a rough surface, it becomes (partially) linearly polarized. The degree of polarization is given by:

$$P_L = \frac{I_\perp - I_\parallel}{I_\perp + I_\parallel}, \tag{9.5}$$

where $I_\perp$ and $I_\parallel$ are the components of the intensity measured perpendicular and parallel to the plane of scattering. The polarization for the Moon and other bodies covered by relatively dark particulate material (such as asteroids) is almost constant over the disk, and depends only on phase angle ϕ. Figure 9.10a shows plots of the polarization as a function of phase angle for 1 Ceres. The polarization is usually negative at small phase angles, ϕ, and becomes positive at $\phi \gtrsim 15$–20°. This is indicative of rough, porous or particulate surfaces. The minimum degree of polarization, P_{Lmin}, reached at phase angle ϕ_{min}, the phase angle ϕ_0 where $P_L = 0$, and the slope of P_L versus ϕ (near ϕ_0), are diagnostic of the surface texture and optical properties of the object. The peak polarization at large phase angles would be very diagnostic for the object's surface texture, but is practically impossible to observe from Earth, except for Earth-crossing objects. The slope h of the polarization curve appears to be directly related to the geometric albedo A_0, irrespective of the nature of the surface. This is depicted in Figure 9.10b. There is a saturation effect at albedos $A_0 < 0.06$, because the polarization of these

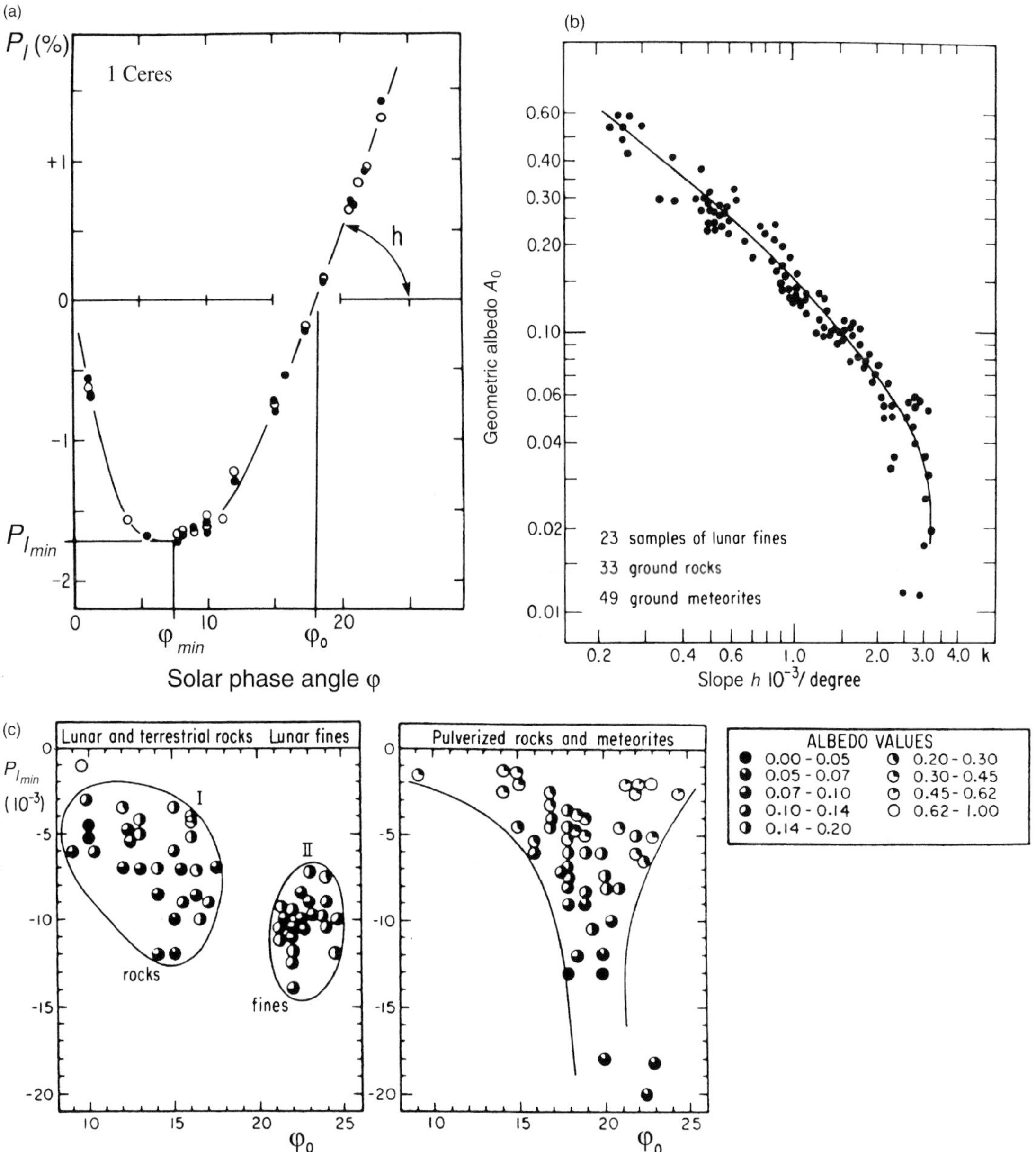

FIGURE 9.10 (a) Degree of linear polarization as a function of solar phase angle ϕ for asteroid 1 Ceres. Definitions for the various polarization parameters are shown. (Dollfus et al. 1989) (b) The relationship between the slope, h (see panel (a)), and the albedo $A_{0,\phi=5^\circ}$, normalized to the case of a white magnesium oxide surface. (Dollfus et al. 1989) (c) A graph of P_{Lmin} versus ϕ_0 for: left, lunar and terrestrial rocks (region I) and rock powders and lunar fines (region II); right, meteorites and rocks with grain sizes between 30 and 300 μm lie in between these two regions. (Dollfus et al. 1989)

dark objects reaches very large values, close to unity. The empirical relation at $A_0 > 0.06$ can be represented by:

$$\log A_0 = -C_1 \log h + C_2, \tag{9.6}$$

where the polarimetric slope, h, is measured in percent polarization per degree of phase angle. The constants $C_1 = 0.93$ and $C_2 = -1.78$. It is thus possible to determine the geometric albedo, and hence diameter of an object, from its polarization–phase curve.

Based upon measurements of rocks and pulverized rocks from meteorites, lunar and terrestrial samples, empirical relationships between the various polarization parameters have been derived. Figure 9.10c shows a plot of P_{Lmin} versus ϕ_0 for lunar and terrestrial rocks (region I) and rock powders and lunar fines (region II); meteorites and rocks with grain sizes between 30 and 300 μm lie between these two regions. Thus, a measurement of the polarization characteristics of an object can be used to determine, in addition to the geometric albedo and radius, also the surface texture of the body. Large bodies, such as Mercury, the Moon and Mars, are within region II, suggesting that these bodies are covered by a fine-grained regolith. Asteroids are located in between regions I and II, suggesting a mixture of pulverized rocks with coarser-grained material.

9.3.4 Adaptive Optics

Two factors limit the spatial resolution of telescopes. The first is optical diffraction, which limits the resolution (FWHM: full width at half maximum) at a wavelength λ to $\sim\lambda/D$ radians, with D the telescope diameter. For a 10-m telescope at a wavelength of 1 micrometer, this diffraction limit is $0.02''$ (arcseconds). The second limit is 'seeing', the blurring of images caused by turbulence in the Earth's atmosphere. For even the best telescopes at the best sites, this limits the resolution to 0.4–$1''$. The Hubble Space Telescope (HST, 2.4 m in diameter) was launched largely to overcome this limitation, and in addition to observe ultraviolet light to which the atmosphere is opaque.

Before entering the atmosphere, light from a far-away source forms a plane wave. The speed of light varies as the inverse of the refractive index (Section 7.6.1), and fluctuations in this quantity are essentially proportional to fluctuations in the atmospheric temperature. Such fluctuations are common at the interface between different layers, where winds produce turbulence. Hence light traveling through different parts of the atmosphere travels at different speeds, which produces a deformation in the originally plane wavefront (Fig. 9.11). The wavefront phase fluctuations also depend on wavelength, since the wave vector is inversely proportional to wavelength ($|\mathbf{k}| = 2\pi/\lambda$). The wavefront perturbations are therefore smaller at longer wavelengths, and less detrimental to image quality. The development of speckle imaging and adaptive optics techniques has made it possible to overcome this atmospheric 'seeing'. *Speckle imaging* corrects for the effects of atmospheric turbulence in software, via post-processing. At the telescope one takes many short time exposures, short enough to freeze the turbulent atmosphere during each exposure ($\lesssim$ 200–300 ms). The result is a set of specklegrams which contain diffraction-limited information about the observed object. These can be combined either in the image plane by aligning the brightest point on each specklegram, or by averaging the specklegram intensity autocorrelations. An example of an image obtained using speckle techniques was shown in Figure 5.61.

With Adaptive Optics (AO) one monitors atmospheric distortions in real time and compensates for the wavefront errors in the incident beam by means of a deformable mirror (Fig. 9.11). The AO correction requires a bright guidestar near the object of interest, which could either be a natural star or a laser beacon (this latter method is still under development)[1]. The corrugated wavefront from the guidestar can be compensated using a deformable mirror which, at each time, has the same deformation as the incoming wavefront but with only half the amplitude. Such a mirror essentially introduces a difference in the phase between different rays, since the rays travel different distances from the deformable mirror to the science instrument, i.e., rays are delayed by an amount equal to the time it takes the light to travel twice (back and forth) across the depth of the mirror deformation. This technique makes the wavefront planar again, and thus overcomes the effect of 'seeing' in the atmosphere. Examples of AO images obtained with the 10-m Keck telescope are shown in Figures 4.32b, 9.7 and 11.14c.

9.3.5 Interferometry

The resolution of a telescope can be improved by connecting the outputs of two antennas which are separated by a distance r_0, at the input of a receiver. Such a system is called an *interferometer*. Interferometers are common in radio astronomy (e.g., the VLA), and become more common at infrared wavelengths. The response of an interferometer to an unresolved source is an interference pattern, as sketched in Figure 9.12, where the maxima are separated by an angle λ/r radians, where r is the baseline length as projected on the sky. As shown in Figure 9.12 the single antenna output is essentially modulated on a scale λ/r. This angle is the *resolving power* of the interferometer in the direction of the projected baseline r. Assume that the interferometer is built along the east–west direction, and that you observe a source X. The Earth's rotation causes the baseline projected onto the sky at X to trace out an ellipse during the course of the day. The coordinates of

[1] The laser is capable of producing a guidestar over most locations in the sky by inducing a fluorescent glow in a layer of mesospheric sodium at an altitude of 90 km.

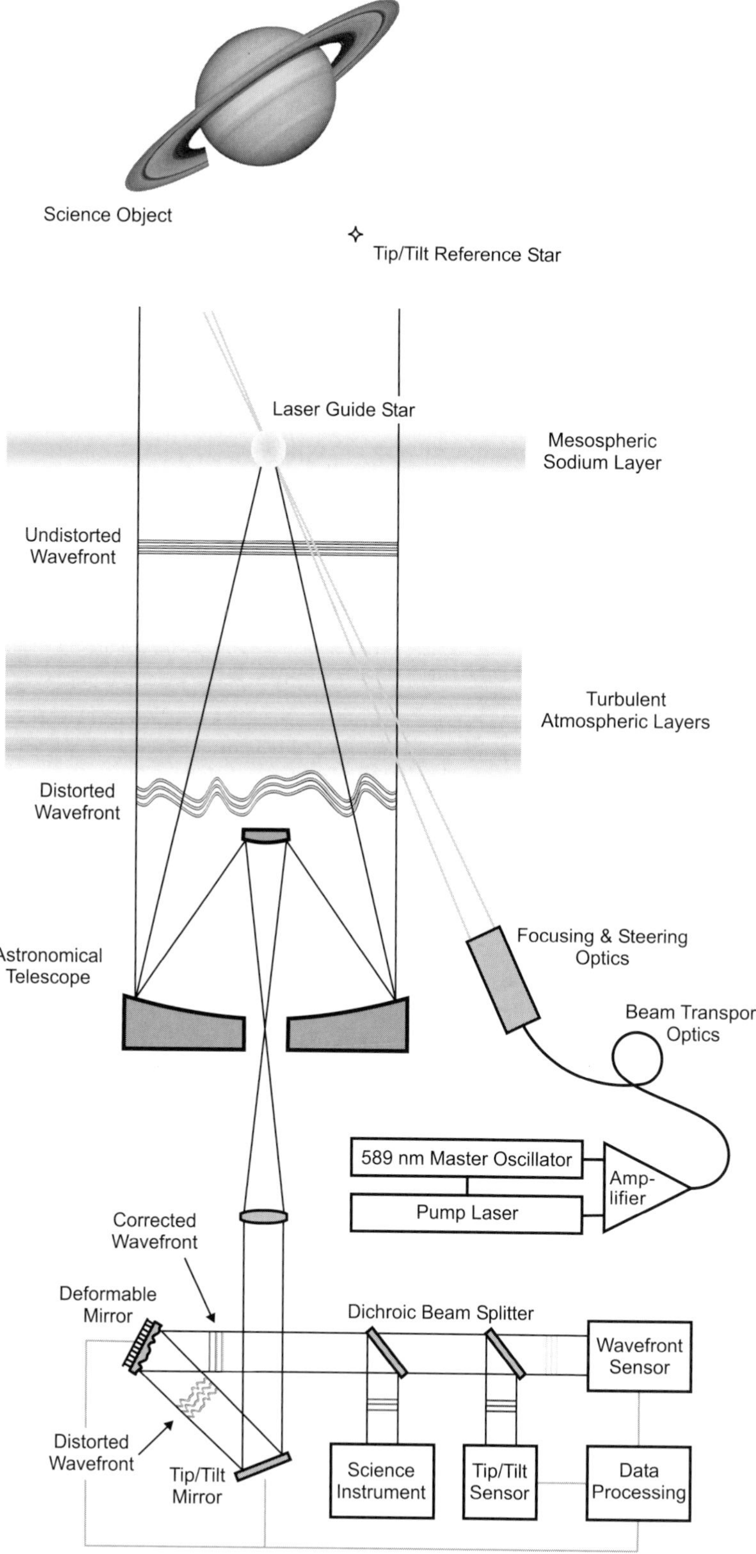

FIGURE 9.11 A schematic representation of the adaptive optics (AO) technique. The incoming undisturbed wavefront is distorted by atmospheric turbulence. By using a deformable mirror the wavefront can be compensated, so that one can image at the diffraction limit of the telescope. (Courtesy: Wolfgang Hackenberg and Andreas Quirrenbach)

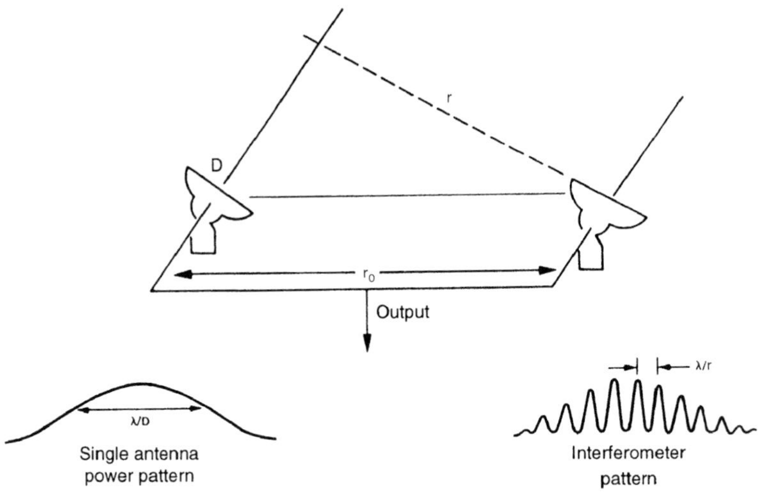

FIGURE 9.12 Top: Geometry of a two-element interferometer. Bottom: Antenna response for a single element of the interferometer (left) and the response of the interferometer (right) to an unresolved radio source. (Adapted from Gulkis and de Pater 1992)

this ellipse are generally referred to as the u (east–west on the sky) and v (north–south on the sky) coordinates in the (u, v)-plane. The parameters of the ellipse depend upon the declination of the radio source, the length and orientation of the baseline, and the latitude of the center of the baseline. The ellipse determines the angular resolution on the source.

The VLA consists of a Y-shaped track, with nine antennas along each of the arms. The antennas operate at centimeter wavelengths, and each antenna is connected with each of the others into an interferometer. With this instrument one can thus gather data from 351 individual interferometer pairs, which each trace out their own unique ellipse in the (u, v)-plane; or in other words each interferometer pair has its own instantaneous resolution along its projected baseline r. With such an array of antennas one can build up an image which shows both the large and small scale structure of a radio source. At short spacings the entire object can be 'seen', but details on the planet are washed out due to the low resolution of such baselines. At longer baselines details on the planet can be distinguished, but the large scale structure of the object gets resolved out, and hence is invisible on the image unless short spacing data are included as well. To recover the spatial brightness distribution of an object, one needs to measure both the amplitude and phase of fringes received (called the complex visibilities) by many interferometer pairs. This measurement forms the basis of mapping by Fourier synthesis in (radio) astronomy. The response of all the individual interferometer pairs, the visibility data, are then gridded into cells having uniform intervals in the (u, v)-plane. This grid of data is then Fourier transformed to give a map of the brightness distribution on the sky. Examples of radio interferometric images are shown in Figures 4.28c, 4.30b, 4.31c and 7.38.

9.3.6 Radar Observations

In a radar experiment, a signal with known properties (intensity, polarization, time/frequency spectrum) is transmitted for a short duration, usually equal to the roundtrip propagation time, and received by the same antenna for a comparable length of time. In contrast to these *monostatic radar experiments*, there are *bistatic experiments* where the signal is transmitted continuously by one telescope and received by a different telescope. Direct radar images can be obtained by receiving the return echo with an array of telescopes, such as the VLA. We will discuss below how monostatic data can also be used to derive the three-dimensional shape of small bodies.

The transmission frequency is usually continuously adjusted so that if the ephemeris of the object is well known and the target is a point source, then the echo received will be a spike at a particular frequency. Any frequency spreading of the signal can then be attributed to the object's rotation and geometry.

The radar flux received by an object per unit area (erg cm^{-2} s^{-1}) is equal to:

$$\mathcal{F} = \frac{P_t G_t}{4\pi r_\oplus^2}, \tag{9.7}$$

with P_t the transmission power, G_t the *gain* ('efficiency') of the transmitting telescope, and $r_\oplus$ the distance to the object (for which we adopted here the geocentric distance).

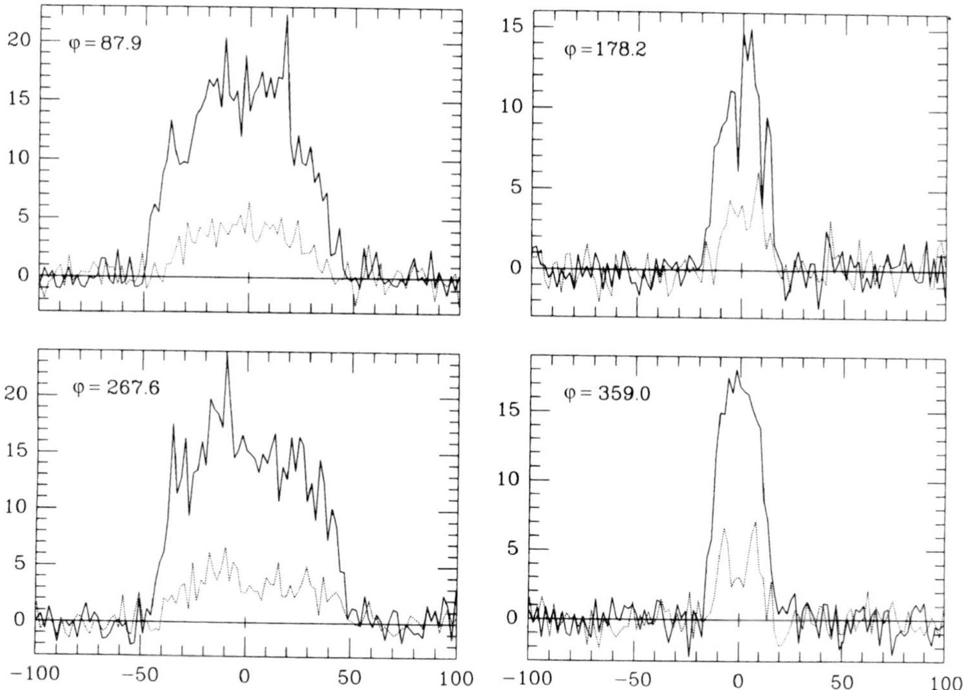

FIGURE 9.13 Radar echo spectrum of near-Earth asteroid 1620 Geographos obtained in 1994 at Goldstone at a transmitter frequency of 8510 MHz. The echo power is plotted in standard deviations versus Doppler frequency (in Hz) relative to the estimated frequency of echoes from the asteroid's center-of-mass. Solid and dotted lines are echoes in the OC and SC polarizations, respectively. These spectra are at rotational phases, φ, which correspond to bandwidth extrema. (Ostro *et al.* 1996)

The total flux received from the object (i.e., integrated over frequency), $\mathcal{F}_r$, can be written:

$$\mathcal{F}_r = \frac{\mathcal{F}\sigma_{xr} A_r}{4\pi r_\oplus^2} = \frac{\sigma_{xr} A_r P_t G_t}{16\pi^2 r_\oplus^4}, \tag{9.8a}$$

with A_r the receiving antenna's effective aperture, and σ_{xr} the radar cross-section of the object:

$$\sigma_{xr} = A_{rdr}\pi R^2, \tag{9.8b}$$

where A_{rdr} is the radar albedo and πR^2 the target's projected surface area. The radar cross-section is equal to the area of an equivalently backscattering metallic sphere, and $\sigma_{xr}/(4\pi)$ is the backscattered power per steradian per unit of flux incident at the target. Note that the power signal travels the distance between Earth and the object twice, so that the received power is diminished by the fourth power of the distance. It is therefore important to make radar measurements when the object is at opposition or inferior conjunction.

The geometric albedo $A_{0,\lambda}$ (at radio wavelengths) is related to the radar albedo:

$$A_{0,\lambda} = \frac{A_{r\,dr}}{4}. \tag{9.9}$$

The radar cross-section is thus a measure of the backscatter efficiency of the target, and its size. Typical radar cross-sections are of the order of 10% of the body's projected surface area. The radar albedo contains information on the composition and porosity of the surface layers. The radar albedo is related to the effective Fresnel coefficient at normal incidence, R_0 (eq. 3.63), and the radar backscatter gain g_r:

$$A_{r\,dr} = g_r R_0. \tag{9.10}$$

The radar backscatter gain is usually close to unity, in which case the radar albedo can be related directly to the dielectric constant, ϵ_r, of the material probed. The dielectric constant contains information on the composition and compactness of the surface layers (cf. Section 9.5).

If the transmitted wave is circularly polarized and the object is a smooth dielectric sphere, the sense of polarization is reversed upon reflection, and the received signal is called an OC (opposite circularly polarized) signal. Power in the same sense of polarization (SC) can arise from single backscattering from rough surfaces, multiple scattering or from subsurface refraction effects. The ratio SC/OC is a measure of the near-surface roughness at scales comparable with the observing wavelength. Main belt asteroids usually show low values for the polarization ratio ($SC/OC < 0.2$), whereas near-Earth objects often show higher numbers ($0.2 \lesssim SC/OC \lesssim 0.5$), indicative of a rougher surface.

The radar echo is Doppler broadened as a result of the rotation of the body. The full or maximum bandwidth, B_{max} (in Hz) depends upon the size of the object, its rotation period, P_{rot}, and the viewing geometry from the radar:

$$B_{max} = \frac{8\pi R}{\lambda P_{rot}} \sin\theta, \qquad (9.11)$$

with λ the radar wavelength and θ the aspect angle, which is the angle between the spin vector and the line of sight. The observed bandwidth is usually less than B_{max}, depending on the radar scattering properties of the target. OC echoes are dominated by specular reflections from surface slopes that are tilted in the direction of the radar system. For near-spherical objects, such as the Moon and Mercury, these slopes are concentrated near the center of the projected disk, where Doppler shifts caused by the object's rotation are relatively small. Diffuse echo power (OC and SC) is much less centrally concentrated but is relatively weak. Consequently, OC echoes from near-spherical objects consist of a 'specular spike' at small Doppler shifts and relatively weak wings that extend out to B_{max}. Equation (9.11) shows that the shape of a radar echo spectrum can be used to derive information on the object's size, shape, spin-state and surface characteristics. In contrast to radar observations of the Moon, Mercury or Mars, the observed spectral bandwidth of asteroids is not much smaller than the expected edge-to-edge bandwidth (Fig. 9.13), despite the fact that the ratio SC/OC is generally low. This suggests the objects to be very irregular on scales exceeding meters, yet smooth on centimeter scales.

Radar echos from NEOs which come very close to the Earth can be very strong. For such objects, high resolution both in time and frequency (Doppler delay) are possible, allowing the returned signal to be inverted to yield a three-dimensional image of the asteroid. In Figure 9.14 the relation between frequency and spatial information on a body is explained. The full radar signal inversion requires 15 parameter fits to the data; however, despite the large number of free parameters the results are quite robust. Several NEOs and main belt asteroids have been observed using radar over the past 10–15 years. Striking images have been obtained for the near-Earth asteroids 4769 Castalia and 4179 Toutatis (Fig. 9.15), which came to within a geocentric distance of 0.03 AU (August 1989) and 0.06 AU (December 1992), respectively. Both asteroids are shaped like a dumbbell or a contact-binary, and numerous craters can be distinguished on their surfaces. Radar images of 1620 Geographos and 433 Eros also clearly show an irregular elongated shape, and the presence of a few large craters.

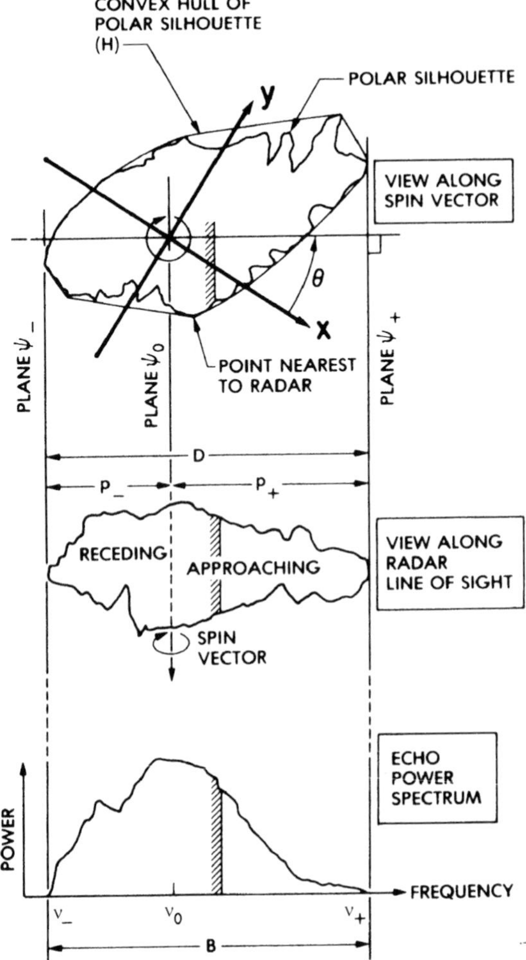

FIGURE 9.14 A cartoon on the geometric relationship between echo power and an asteroid's shape. The upper picture shows the convex hull H of the polar silhouette, or the asteroid shape as viewed from the pole. The middle panel shows a view along the radar line of sight, and the radar echo is shown at the bottom. The plane Ψ_o contains the line of sight and the asteroid's spin vector. The radar echo from any part of the asteroid which intersects Ψ_o has a Doppler frequency ν_o. The cross-hatched strip of power in the spectrum corresponds to echoes from the cross-hatched strip on the asteroid. The asteroid's polar silhouette can be estimated from echo spectra which are adequately distributed in rotational phase. (Ostro 1989)

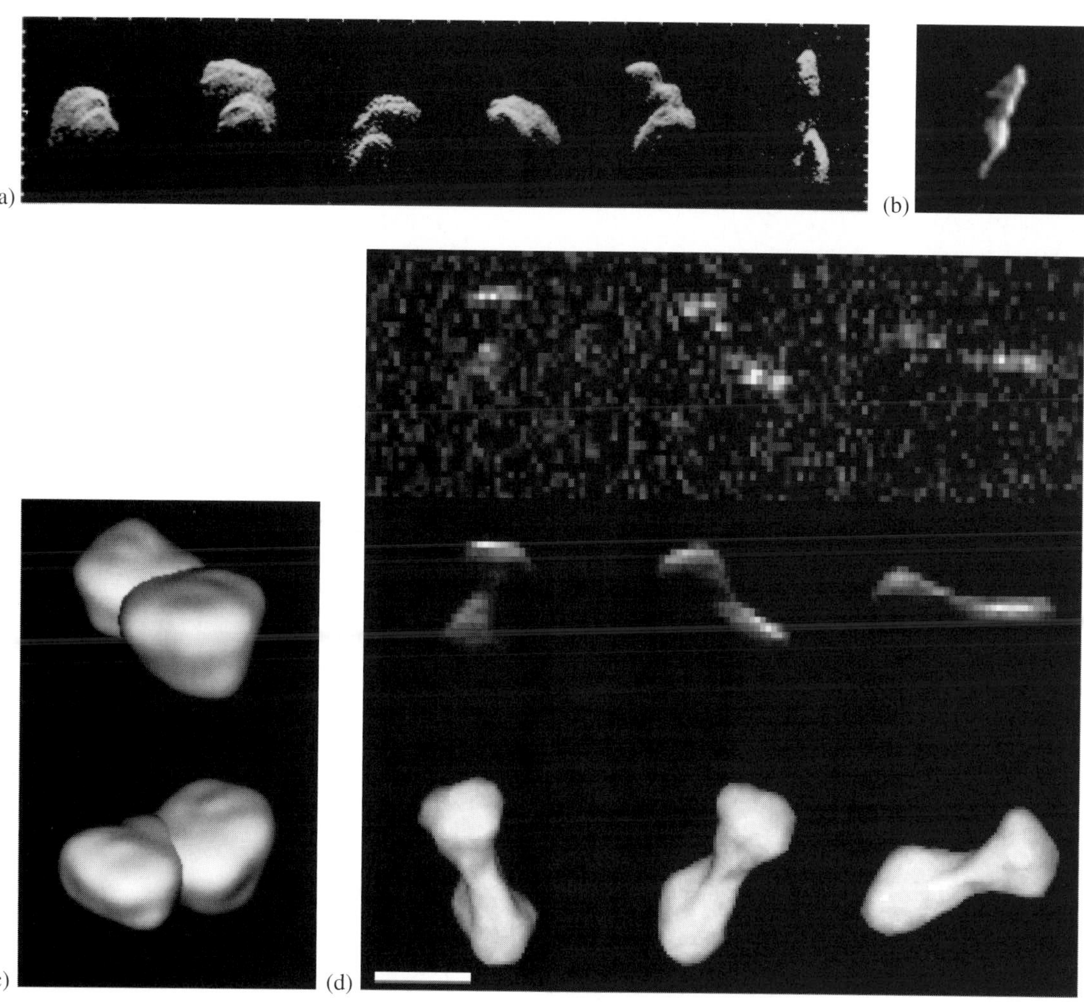

FIGURE 9.15 (a) Radar images of the near-Earth asteroid 4179 Toutatis, obtained with Arecibo from 14 to 19 December 1992. These images are different from regular visual views: They consist of echoes in time delay (range) increasing towards the bottom, and Doppler-frequency (line-of-sight velocity) towards the left. On the vertical axis, ticks are 2 μs apart, and on the horizontal axes they are 0.28 Hz apart (a radial velocity difference of 18 mm s^{-1}). The highest resolution views show craters with sizes ranging from 100 up to 600 m. Images like these are used in shape reconstruction algorithms. (Ostro *et al.* 1995) (b) Goldstone radar image of 1620 Geographos, taken in August–September 1994. The approximate spatial resolution is 100 m. (Ostro *et al.* 1996) (c) Two views of the shape of asteroid 4769 Castalia as reconstructed from radar images. (Hudson and Ostro 1994) (d) Shape of the main belt asteroid 216 Kleopatra as reconstructed from Arecibo radar images. The top row shows the actual images from 16 November 1999. The middle row shows the shape model, and the bottom the corresponding plane-of-sky views of the model. The scale bar represents 100 km. (Ostro *et al.*, 2000)

In a bistatic radar experiment, when the radar echo is received by an interferometer such as the VLA, the signal can be imaged in each frequency channel. Although these observations do not produce images at a higher spatial resolution (compare frequency bandwidth and baseline length of the interferometer), the measurements yield an unambiguous result regarding the sense of rotation of the object, since the Doppler red- and blue-shifted signals are displaced from the center of the object in opposite directions. An example of such observations for 324 Bamberga is shown in Figure 9.16.

9.3.7 Determination of an Asteroid's Mass and Density

To determine the mass of an object, one needs to observe a gravitational interaction, such as a natural or artificial satellite orbiting the primary, a binary system, or a gravi-

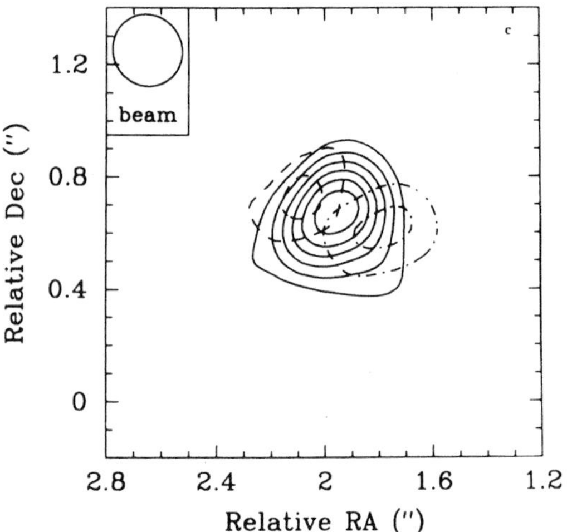

FIGURE 9.16 Radio images of 324 Bamberga at 8510 MHz, obtained with a bistatic radar system where the Goldstone antenna was used to transmit the signal and the VLA to receive and image the radar echo. Radar echoes are shown for 13 September 1991, in the center channel (solid contours), and channels at a Doppler frequency of −381 Hz (dashed contours; redshifted) and +381 Hz (dot-dashed contours; blueshifted). Contour levels are from 3 to 19 standard deviations, where one standard deviation is 5.5 mJy/beam. (de Pater *et al.* 1994)

tational encounter between two bodies. The most efficient gravitational interactions for asteroids are long-lasting encounters at small distances, or repeated encounters with similar geometries, so that the gravitational effects are accumulated. The gravitational perturbations can best be measured along the line of sight by radar ranging and Doppler shifts in the frequency of the radar echo (or radar transmitter if measured from a spacecraft which encounters or orbits the asteroid). The gravitational perturbation can also be measured from accurate astrometric measurements of the perturbed object in the plane of the sky.

Because it is very hard to measure the mass of asteroids, it has only been possible to determine the density of a few asteroids (Table 9.3). The mass of 4 Vesta was determined from its gravitational perturbation on the orbit of 197 Arete; the masses of 1 Ceres and 2 Pallas were derived from their mutual perturbations, and from the influence of Vesta on Ceres. The asteroid 10 Hygiea perturbed the orbit of 829 Academia, but the mass determination is uncertain by 50%. The masses of Phobos and Deimos, the two martian satellites which are thought to be captured C type asteroids, have been determined using the Viking spacecraft, while the masses of 243 Ida, 253 Mathilde and 433 Eros have been derived from data obtained with the Galileo and NEAR spacecraft. The masses of binary systems, such as Ida/Dactyl, 1996 FG_3 and 45 Eugenia have been determined from their orbits (Kepler's laws).

Asteroid densities appear to vary substantially from object to object, with values ranging approximately from 1.2 up to 4 g cm^{-3} (upper limit error bars). Although the densities have been measured for very few asteroids, it appears as if the carbonaceous bodies have densities in the 1.2–2 g cm^{-3} range, while the less primitive bodies have slightly higher densities, as might be expected based upon their composition and evolution history. The densities for several of the carbonaceous asteroids, and in particular Mathilde, are less than those of carbonaceous meteorites (2.2–2.4 g cm^{-3}), suggesting quite porous, possibly rubble-pile bodies.

9.4 Surface Composition

Asteroids, like meteorites, appear to be a compositionally diverse group of large rocks. Some asteroids contain volatile material as carbon compounds and hydrated minerals, while others appear to be almost exclusively composed of refractory silicates and/or metals. Many asteroids look like igneous rocks, but most appear to be primitive rocks, i.e., bodies which have undergone little processing after they formed/accreted in the primitive solar nebula. The composition of an asteroid, therefore, provides constraints on the environment in which it formed and on its thermal evolution.

9.4.1 Asteroid Taxonomy

Reflectance spectra at visible and near-infrared wavelengths yield a wealth of information on a body's surface composition. Some asteroid spectra are shown in Figure 9.17. The distribution in asteroidal albedo is bimodal, with pronounced peaks at 0.05 and 0.18. Based upon their spectra and albedos, asteroids are classified into *taxonomic* groups, as summarized in Table 9.4. The largest class of asteroids are the *carbonaceous* or *C type* asteroids; roughly 40% of the known asteroids are in this class. These bodies are very dark, with typical geometric albedos $A_0 \sim 0.04$–0.06. The spectra of C asteroids are flat (neutral in color) longwards of 0.4 μm. The C type asteroids sometimes exhibit absorption bands at UV wavelengths and near 3 μm. The latter feature is indicative of water, which is probably present in the form of hydrated silicates. The C type spectra are suggestive of carbon-rich material, because carbonaceous chondritic meteorites (CI, CM) have very similar reflectance spectra (Section 8.2, and Fig. 8.8). They are low temperature condensates, primitive objects which have undergone little or no heating. Asteroids with similar,

TABLE 9.3 Sizes and Densities of Asteroids and Martian Moons.

Body	Class	R (km)	ρ (g cm^{-3})	Ref.
1 Ceres	G	457	2.7 ± 0.14	1
2 Pallas	M	262	2.6 ± 0.5	1
4 Vesta	V	251	3.62 ± 0.35	2
10 Hygiea	C	215	2.05 ± 1	3
45 Eugenia	C	215	1.2	4
87 Sylvia	PC	135	1.6 ± 0.1	5
216 Kleopatra	M	$217 \times 94 \times 81$	>3.5	6
243 Ida	S	$28 \times 12 \times 7.4$	2.6 ± 0.5	7
253 Mathilde	C	$33 \times 24 \times 23$	1.3 ± 0.2	8
433 Eros	S	$31 \times 13 \times 13$	2.67 ± 0.03	9
762 Pulcova	C	140	1.8	10
1996 FG_3	C	1.4	1.4 ± 0.3	11
2000 DP_{107}	C	0.40	$1.6^{+0.7}_{-0.2}$	5
2000 UG_{11}	R	0.115		5
Phobos	C	$13.3 \times 11.1 \times 9.3$	1.9 ± 0.1	12
Deimos	C	$7.5 \times 6.1 \times 5.2$	1.8 ± 0.2	12

1: Millis *et al.* (1987). 2: Millis and Elliot (1979). 3: Scholl *et al.* (1987). 4: Merline *et al.* (1999). 5: Margot *et al.* (2000). 6: Ostro *et al.* (2000). 7: Thomas *et al.* (1996). 8: Veverka *et al.* (1997). 9: Yeomans *et al.* (2000). 10: Merline *et al.* (2000). 11: Mottola and Mahulla (2000), and Pravec *et al.* (2000). 12: Thomas (1999).

yet quantifiably different, spectra are grouped in classes B, F and G. In contrast to primitive C type asteroids, B, F and G asteroids have probably been heated sufficiently to cause some mineralogical changes.

The next largest class of catalogued asteroids (30–35%) are the *S type* or *stony* asteroids. These objects are fairly bright, with geometric albedos ranging from 0.14 to 0.17, and reddish. S type asteroids have a strong absorption feature shortwards of 0.7 μm, indicative of the presence of iron oxides. S class spectra with their weak to moderate absorption bands near 1 and 2 μm are suggestive of assemblages of iron and magnesium bearing silicates, like pyroxene ($(Fe,Mg)SiO_3$) and olivine ($FeMgSiO_4$), mixed with pure metallic nickel–iron. These bodies likely crystallized from a melt, and consequently S type asteroids are usually classified as igneous bodies. However, as discussed below, it is not yet clear whether S type asteroids are indeed igneous, or if they could be primitive bodies whose surface has been altered by *space weathering*; evidence in favor of the latter is growing (see below; and Section 7.1.1.2).

In addition to these two main classes, 5–10% of the asteroids have been classified as *D* and *P types*. The D and P type asteroids are quite dark, with $A_0 \approx 0.02$–0.07, and on average somewhat redder than S types. D asteroids are typically somewhat darker and redder than P types. Neither D nor P types exhibit spectral features. The red color might be caused by organic compounds. The D and P asteroids may represent even more primitive bodies than the carbonaceous C type asteroids. There is no meteoritic analogue to their spectra.

M type asteroids have a spectrum similar to P type asteroids, but with a higher albedo, $A_0 \approx 0.1$–0.2. The M type visible wavelength spectra lack silicate absorption features and are reminiscent of metallic nickel–iron. Their spectra are analogous to iron meteorites and enstatite chondrites, meteorites which consist of grains of nickel–iron, embedded in enstatite ($MgSiO_3$), a magnesium-rich silicate. Infrared spectra near 3 μm reveal absorption features diagnostic of *water of hydration* (i.e., water in all the forms of hydrated minerals) in most large ($R > 30$ km) and some small M class asteroids. These features are inconsistent with bodies being primarily iron–nickel metal, and suggest that many asteroids placed in the M class on the basis of their visible spectra are primitive objects, possibly akin to enstatite chondrite meteorites. A new *W class* has been proposed for these H_2O-rich asteroids.

A few very bright asteroids have been discovered with albedos in the range 0.25–0.6. These are *E type* asteroids, which display linear, flat or slightly reddish spectra. They may consist of enstatite or some iron-poor silicate. The visible wavelength spectra of M and E type asteroids suggest that all bodies in these classes have undergone sub-

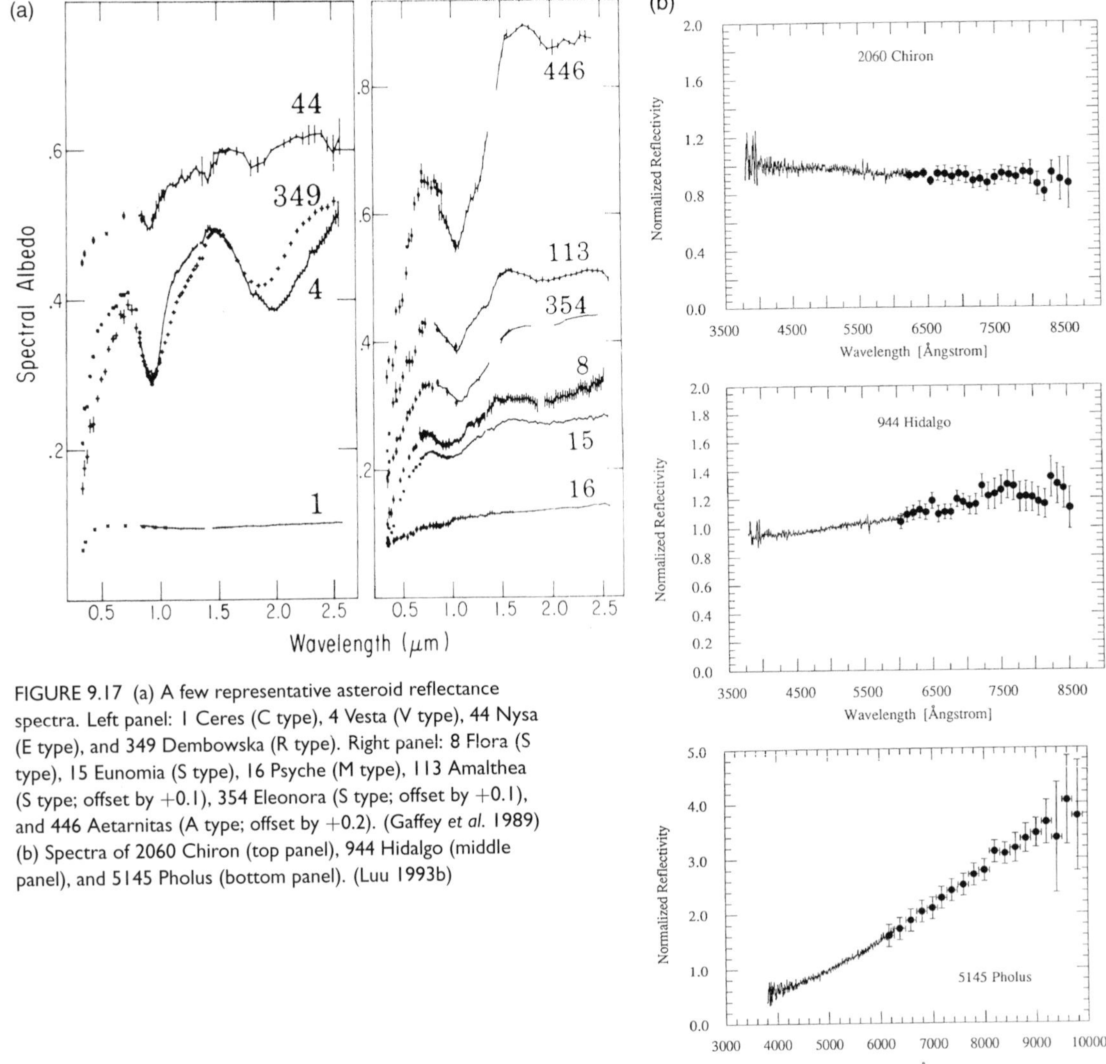

FIGURE 9.17 (a) A few representative asteroid reflectance spectra. Left panel: 1 Ceres (C type), 4 Vesta (V type), 44 Nysa (E type), and 349 Dembowska (R type). Right panel: 8 Flora (S type), 15 Eunomia (S type), 16 Psyche (M type), 113 Amalthea (S type; offset by +0.1), 354 Eleonora (S type; offset by +0.1), and 446 Aetarnitas (A type; offset by +0.2). (Gaffey *et al.* 1989) (b) Spectra of 2060 Chiron (top panel), 944 Hidalgo (middle panel), and 5145 Pholus (bottom panel). (Luu 1993b)

stantial thermal processing via a melt phase. If true, this picture is hard to reconcile with the discovery that a few members of each of these classes exhibit an absorption feature near 3 μm, suggestive of hydration, and thus indicative of primitive objects. One interpretation of the data is that both E and M classes actually contain two or more varieties of distinct objects, and thus that new types of asteroids will have to be introduced into the classification scheme. The reader is cautioned that some of the finer points about asteroid classification are changing quite rapidly, and articles and books on this subject may become outdated by the time that they are published!

In addition to the large main groups, there are a number of small asteroid classes, some of which contain only a few asteroids. Asteroids whose spectra do not fit into any established class are referred to as *U type* or unclassified. The spectrum of *R type* 349 Dembowska suggests the presence of large amounts of olivine with little or no metals. It resembles a silicate residue after metal has been extracted from the upper mantle of an incompletely differentiated body. Asteroid 4 Vesta (*V type*) is unique among large objects in that it seems to be covered by basaltic material. Its spectrum resembles that of eucrite meteorites (cf. Chapter 8): It displays pyroxene absorption bands near 1 and 2 μm, as well as a weaker absorption at 1.25 μm attributed to plagioclase feldspar ($(CaAl,NaSi)AlSi_2O_8$). In addition to Vesta, several small V type asteroids have been detected in orbits with similar orbital parameters as those of Vesta. The Deep Space 1 spacecraft acquired detailed near-infrared spectra of NEO 9969 Braille. These spec-

TABLE 9.4 Asteroid Taxonomic Types.

Low-albedo classes:	
C	Carbonaceous asteroids; similar in surface composition to CI and CM meteorites. Dominant in outer belt beyond 2.7 AU. UV absorption feature shortwards of 0.4 μm. Spectrum flat, slightly reddish longwards of 0.4 μm. Subclasses: B, F, G.
D	Red featureless spectrum, possibly due to organic material. Extreme outer belt and Trojans.
P	Spectrum is flat to slightly reddish; shape resembles that of M type asteroids. Outer and extreme outer belt.
K	Resembles CV and CO meteorites.
T	Moderate absorption feature shortwards of 0.85 μm; flat spectrum at longer wavelengths. Rare, unknown composition. Possibly highly altered C types.
Moderate-albedo classes:	
S	Stony asteroids. Major class in inner to central belt. Absorption feature shortward of 0.7 μm. Weak absorption bands near 1 and 2 μm.
M	Stony-iron or iron asteroids; featureless flat to reddish spectrum.
Q	Resembles ordinary (H, L, LL) chondrite meteorites. Absorption features shortwards and longwards of 0.7 μm. 1862 Apollo is type example.
A	Very reddish spectrum shortwards of 0.7 μm. Strong absorption feature near 1 μm.
V	Strong absorption feature shortwards of 0.7 μm, and near 1 μm. Similar to basaltic achondrites. Type example: 4 Vesta.
R	Spectrum intermediate between A and V classes. Similar to that of olivine-rich achondrites. Type example: 349 Dembowska.
High-albedo (>0.3) class:	
E	Enstatite asteroids. Concentrated near inner edge of belt. Featureless, flat to slightly reddish spectrum.

tra resemble that of 4 Vesta and eucrite meteorites. It is conceivable that Braille, other V type asteroids and eucrite meteorites were blasted off Vesta in the impact that created the 460 km wide crater discernible on high resolution images of this asteroid.

It is curious that no large asteroid class seems to match the meteorite spectra of ordinary chondrites, the most common meteoritic samples. The *Q type* asteroid 1862 Apollo is one of very few asteroids that matches an ordinary chondrite spectrum. Where are the parent bodies of these primitive meteorites? Are they hidden in the asteroid belt (i.e., too small to be detected)? Recent theories suggest that space weathering may alter the spectra of asteroids: the interaction of solar wind particles, solar radiation and cosmic rays with planetary bodies may induce chemical alterations in the surface material and therefore 'hide' the real composition of an asteroid. It has been suggested that space weathering has altered the original spectra of S type asteroids, and that S type asteroids are the parent bodies of ordinary chondritic meteorites. This idea seems supported by the observation that the smallest S type asteroids display a larger likeness to the ordinary chondrites than larger bodies. Since the smaller ones are presumably relatively young fragments, they have been exposed less long to space weathering than the larger parent bodies, and hence should better resemble spectra of ordinary chondrites, if the above theory is right. Observations of the larger S type asteroid, 433 Eros, by the NEAR spacecraft show that its bulk elemental composition and spectrum at visible and near-infrared wavelengths is very similar to that of ordinary chondrites, an observation which lends strong support to the theory of space weathering.

The 'alphabet soup' of asteroid classes (still growing and sometimes inconsistent) can be quite confusing to the nonexpert. The most important classes to remember are the very common C type, which are dark, gray and resemble carbonaceous chondrite meteorites, the brighter reddish S type which are common in the inner asteroid belt, the dark red D type asteroids common at large heliocentric distances and the M types (at least some of) which appear to be very rich in metals.

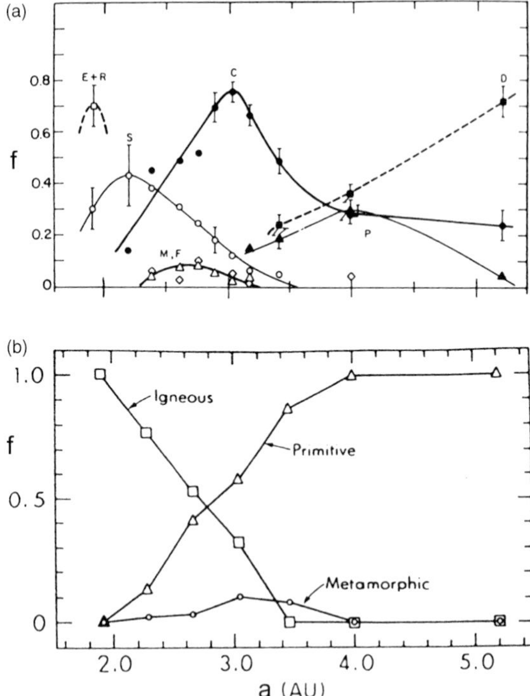

FIGURE 9.18 (a) Graph showing the relative distribution of the asteroid taxonomic classes as a function of heliocentric distance. The classes E, S, C, M, P and D are shown. Smooth curves are drawn through the data points for clarity. (Gradie *et al.* 1989) (b) Distribution of igneous, primitive and metamorphic classes as a function of heliocentric distance. (This figure assumes that S type asteroids are igneous bodies.) (Bell *et al.* 1989)

9.4.2 Spatial Distribution of Taxonomic Classes

There appears to be a clear trend among the taxonomic classes with heliocentric distance, as shown graphically in Figure 9.18. The high albedo E type asteroids are only found near the inner edge of the asteroid belt. S type asteroids prevail in the inner parts of the main belt, M types are seen in the central regions of the main belt, while the dark C type objects are primarily found near the outer regions of the belt. D and P asteroids are found only in the extreme outer parts of the asteroid belt and among the Trojan asteroids. U types are found throughout the asteroid belt. Figure 9.18b shows the distribution of igneous (assuming S type asteroids to be igneous bodies, see previous section) and primitive asteroids as a function of heliocentric distance. This figure shows a correlation with heliocentric distance: Igneous asteroids dominate at heliocentric distances $r_{AU} < 2.7$, and primitive asteroids at $r_{AU} > 3.4$. Metamorphic asteroids, which must have undergone some changes due to heating as characterized by their spectra, have been detected throughout the main belt. The strong correlation of asteroid classes with heliocentric distance cannot be a chance occurrence. It must be a primordial effect, although it is likely modified by subsequent evolutionary or dynamical processes. If asteroids are indeed remnant planetesimals, their distribution in space might provide insight into the temperatures, pressures and chemistry of the solar nebula. The difference in spatial distribution between the C type (primitive) and S type (igneous?) asteroids is qualitatively consistent with the temperature structure expected in the primitive solar nebula. P and D type asteroids, probably even more primitive than the C types, may have formed at even lower temperatures; unfortunately, not much is known about the composition of P and D type asteroids, as there are no spectral analogs in terrestrial meteorite collections. The highly reflective objects in the inner belt would be consistent with an even higher temperature than must have prevailed in the inner-central belt.

It is not known to what degree post-formation evolution, including space weathering, of asteroids masks the original distribution of asteroid compositions. Dynamical processes may have considerably changed the original distribution of bodies in the asteroid belt. However, the orderly arrangement of taxonomic classes with heliocentric distance clearly contains important clues to the formation history of our Solar System.

We caution the reader here about statistics of asteroid samples. The samples are (strongly) biased to nearby bright, high albedo objects. Thus, the S class objects are almost certainly over-represented relative to the C class asteroids, because they are, on average, brighter and closer to the Sun; an even stronger bias against detection of D class asteroids exists for the same reasons.

9.5 Structure

Most airless bodies are covered by craters and a regolith, which could be many meters thick, depending on the age of the body's surface and the size of the asteroid. The regolith consists of loose rocks and dust created by numerous meteorite impacts (Section 5.4.2.4). As the asteroids, like the terrestrial planets, the Moon and other satellites, have accumulated craters since their surface last solidified, the number and size distribution of craters contains information on the age of the asteroid's surface (Section 5.4.4). Unfortunately, very few asteroids have been imaged with sufficient detail that craters can be identified and counted. But despite the fact that we have detailed images of only a handful of asteroids, the combined data base of photometric and spectral observations, polarimetry and radar data contain a wealth of information on the near-surface characteristics of these bodies, such as composition, porosity and surface roughness. In the next section we discuss what we have learned from the ground-based data set, followed in Section 9.5.2 by brief descriptions of the images obtained from spacecraft and radar data.

9.5.1 Asteroid Surface Characteristics from Disk-Averaged Ground-Based Observations

The lunar regolith consists of fine-grained sub-millimeter material. In the lunar maria the regolith is typically several meters deep, and over 10 m in the lunar highlands. In addition, a many kilometers thick *mega-regolith* has been hypothesized for the deeper layers under the lunar highlands (Section 5.4.2.4). The mega-regolith is likely bonded at deeper layers, rather than in the form of loose regolith. We expect the larger asteroids also to be covered by a thick layer of regolith. There probably are systematic differences in the quantity of regolith produced on asteroids as a function of asteroid size; larger asteroids retain a larger fraction of impact ejecta and they have longer lifetimes against super-catastrophic collisions. Calculations by various groups predict asteroids over 100 km in size to be covered by a layer of regolith many meters thick, while

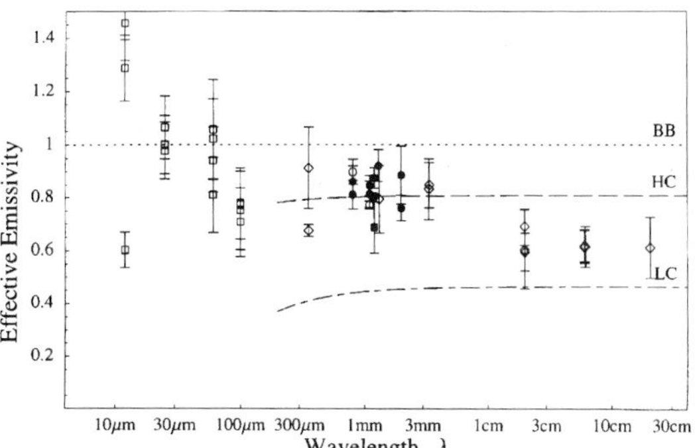

FIGURE 9.19 Microwave spectrum of 1 Ceres. The curves represent different models. Let $\epsilon_{eff} \equiv \mathcal{F}_\nu/\mathcal{F}_{nr}$, where $\mathcal{F}_{nr}$ is the flux density for a nonrotating sphere at phase angle $\phi = 0°$. In model BB, $\epsilon_{eff} = 1$. Model HC is for a rapidly rotating sphere, overlain by material with a high thermal conductivity, with $\epsilon_{eff} \approx 0.8$. In model LC the surface has a low thermal conductivity, with $\epsilon_{eff} \approx 0.46$. (Redman *et al.* 1998)

bodies less than 10 km in size may be covered by at most a few centimeters of regolith.

Evidence for regolith-covered asteroids comes from polarimetry, radiometry and from radio and radar observations (Section 9.3). The presence of the 'opposition effect' (brightening) near zero phase angle indicates that asteroids are covered by particulate material – both the shadow-hiding and coherent-backscatter effects only arise in particulate materials. The linear polarization characteristics of reflected sunlight suggest asteroids to be overlain by a regolith which contains a mixture of pulverized rocks and finer-grained material, though this latter material is much coarser than that found on the Moon. A comparison of asteroid radii derived from radiometric observations via thermophysical models with radii determined from star occultation or polarimetric techniques, shows that an asteroid is generally overlain by a layer of regolith. The detailed angular distribution of the scattering characteristics of the asteroid's particulate material, however, is often slightly different than that of the Moon, due in particular to differences in regolith size distributions. This became quite evident, for example, after measurements of the asteroids Ida and Mathilde (Table 9.2).

Electromagnetic radiation typically probes ~10 wavelengths deep in a body's surface (Section 3.2.2.5). Thus, observations at radio wavelengths probe much deeper into the asteroid's crust than measurements at visible wavelengths. Passive microwave observations yield the total thermal flux density of the object at wavelength λ. If the size of the object is known, this can be converted to a brightness temperature via the Rayleigh–Jeans law or Planck function. A microwave spectrum of asteroid 1

Ceres is shown in Figure 9.19. A comparison of multi-wavelength data with thermo-physical models provides information on the depth dependence of the density and temperature in the surface layers. Radio spectra of several main belt asteroids suggest that these bodies are typically overlain by a layer of fluffy (highly porous) dust that is a few centimeters thick; similar dust layers have been observed on the Moon and Mercury. This general structure of a fluffy layer of dust overlying a highly compacted regolith is likely caused by bombardment of small meteoroids, which maintains the top layer at a low density while compacting deeper layers.

Radar data contain information on the dielectric constant of the material probed. A typical value for the dielectric constant of terrestrial rocks is between 6 and 8. The dielectric constant for rock powders is lower. Laboratory measurements show that most lunar and terrestrial rock powders follow the *Rayleigh mixing formula*:

$$\frac{1}{\rho}\left(\frac{\epsilon_r - 1}{\epsilon_r + 2}\right) = \frac{1}{\rho_o}\left(\frac{\epsilon_{ro} - 1}{\epsilon_{ro} + 2}\right), \tag{9.12}$$

where ρ and ϵ_r are the density and dielectric constant of the rock powder, and ρ_o and ϵ_{ro} those of the parent rock. A typical density for terrestrial and lunar rocks is $\rho_o = 2.8$ g cm^{-3}. The radar albedo, dielectric constant, and rock porosity are thus related through equations (3.63), (9.10) and (9.12). Assuming asteroid surfaces to be similar to meteorites, the radar albedo provides a measure of regolith porosity. Figure 9.20c shows this relation graphically. Thus if one assumes, for example, that the meteoritic analogue of a particular asteroid is a stony-iron meteorite, a radar reflectivity of 0.2 suggests a 50% porosity of the regolith material. A radar albedo of 0.2 for a C type asteroid, however, would suggest that solid material is being probed rather than a layer of regolith.

Radar observations of near-Earth asteroids usually show a high polarization ratio, $SC/OC \approx 0.2$–0.5, indicative of a surface that is rough compared to the observing wavelength (centimeter scales), as expected for objects overlain by at most a thin layer of regolith. One near-Earth object, 1986 DA (object 6178) has an extremely high radar albedo, $A_{rdr} = 0.58$, which is indicative of a very metal-rich composition and at most a very thin layer of regolith.

9.5.2 Detailed Images of Individual Asteroids and Mars's Moons

Phobos and Deimos

The visual albedos of Mars's small moons Phobos and Deimos are $\sim$0.07, and the spectral properties of these moons are similar to those of carbonaceous meteorites. Phobos and Deimos, being so close to the asteroid belt, are therefore thought to be captured C type asteroids. Phobos orbits Mars at a distance of 2.76 $R_♂$ which is inside the synchronous orbit, while Deimos orbits Mars well outside synchronous orbit (6.92 $R_♂$). Both satellites were imaged at a spatial resolution $\gtrsim$3–100 m by the Viking spacecraft in 1977, and 20 years later by Mars Global Surveyor (Fig. 9.21). It is not surprising that both objects, being so small (Table 9.3), are very irregular in shape. Phobos is heavily cratered, close to saturation. The shape of the craters is very similar to that of lunar craters. The largest crater is $\sim$10 km in diameter; if the impactor had been slightly larger, Phobos would have been shattered to pieces. Intriguing are the linear depressions or grooves, typically 10–20 m deep, 100–200 m across and a few kilometers long, which could be traces of former fractures. The thermal inertia derived from thermal infrared measurements by Viking suggests that Phobos is covered by very loose fine-grained regolith, similar to lunar soil. The regolith may be over 100 m deep in places, as suggested by the depth of some grooves. Although initially some or all the ejecta may have been lost from this tiny moon ($v_e \approx 8$ m s^{-1}), it should have been re-accumulated quickly because it is trapped on a similar orbit around Mars as Phobos itself. Note that asteroids are not able to re-accrete crater ejecta in a similar manner.

Deimos's surface is rather smooth, and shows prominent albedo markings, varying from 6 to 8%. The images also show a concavity 11 km across, twice as large as the mean radius of the object. It is not clear whether this feature is a crater, or evidence that Deimos may consist of two or more objects, i.e., a rubble-pile. The craters on Deimos are partly or totally filled by sediments, material which moved downhill into the lower lying craters. Many craters may have been buried this way, which may explain why Deimos's surface is less heavily cratered than Phobos's surface.

951 Gaspra

951 Gaspra is a small irregularly shaped S type asteroid (Fig. 9.6b), which is unusually red and olivine-rich. Its geometric albedo is typical for S type asteroids, $A_0 = 0.22$. It probably belongs to the Flora family. Over 600 craters have been identified on Gaspra's surface, many relatively young and fresh. There is a relative lack of craters over 1.5 km in diameter, however, and the surface is also far from saturation. The density and crater size distribution suggest the surface of Gaspra to be a few $\times$ 10^8 years old, although the reader is cautioned that craters can only give good estimates of relative surface ages, and no 'ground-truth' radiometric ages are known anywhere in the asteroid

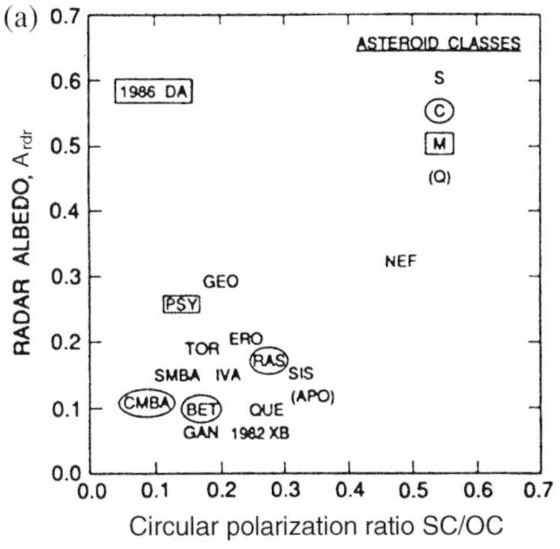

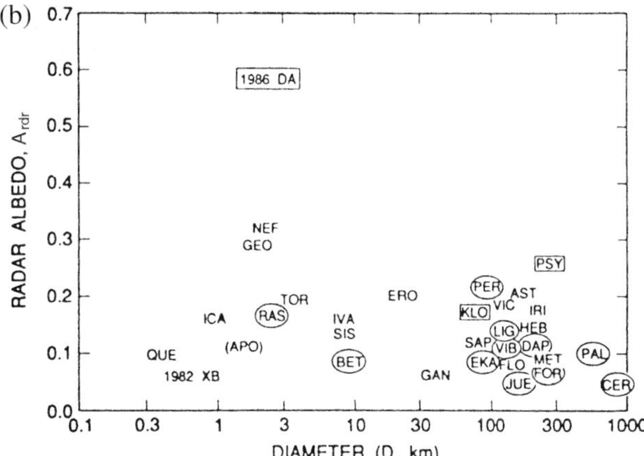

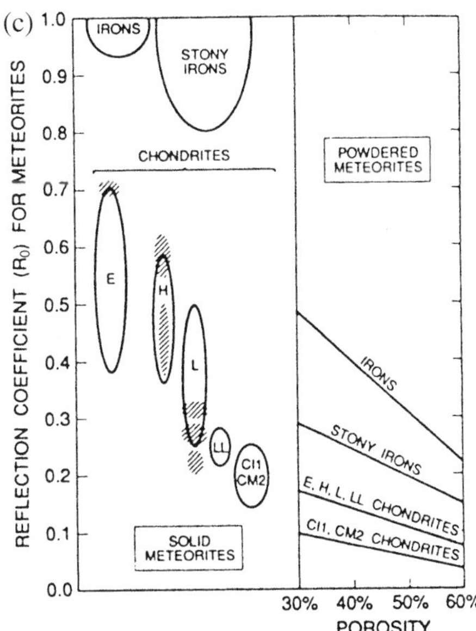

FIGURE 9.20 Graph of the asteroid radar albedo versus circular polarization ratio (a) and diameter (b). In (a) SMBA and CMBA indicate mean values for seven S type main belt asteroids and nine C type asteroids. (c) Reflection coefficient for meteorite types (left) and powdered meteorites (right) as a function of porosity. (Ostro *et al.* 1991)

belt. In addition to the numerous small craters, Gaspra displays grooves similar to the fractures seen on Mars's moon Phobos. The grooves and subtle color/albedo variations that cover the surface suggest that Gaspra is covered by a layer of regolith, many tens of meters thick. This is much thicker than expected for such a small, relatively young asteroid. Maybe the asteroid experienced several episodes of cratering, and the regolith results (in part) from previous cratering events, the craters of which have been erased by burial under ejecta from a particularly large impact.

243 Ida and Dactyl

243 Ida, an S type asteroid, is a member of the Koronis family. The largest surprise from the Galileo data was the small (mean radius of 0.7 km) almost round satellite Dactyl orbiting Ida at a distance of 85 km (Fig. 9.6a). Ida and Dactyl have very similar photometric properties, but distinct differences in spectral properties. Thus Ida and Dactyl's textures must be similar, but the satellite appears to have a slightly different composition, in particular it may contain more pyroxene than Ida. The differences are within the range of differences reported for other members of the Koronis family, and argue for compositional inhomogeneities of the Koronis parent body, possibly resulting from differentiation processes.

The crater density on Ida and its satellite is roughly five times larger than that on Gaspra for craters over 1 km in size. The craters are severely degraded, which hints at an old age. The density and crater size distribution suggest the surface of Ida to be $\sim$1–2 $\times$ 10^9 years old. Dactyl's life expectancy against impact disruption is only a few $\times 10^8$ yrs, which is thus much less than the age of Ida. It is very unlikely that Dactyl would have been captured in the past 10^8 years; so it has to be a 'primordial' companion, created at the same time as Ida from the Koronis parent body. It could also have formed as the result of a large impact on Ida, in analogy to the formation of our Moon (Section 12.11). It is possible that Dactyl has been disrupted several times during its history, but that the fragments, bound in orbit around Ida, re-accreted to form a 'new' satellite. It is also possible that the pair is less than 10^8 yrs, but that

(a) (b)

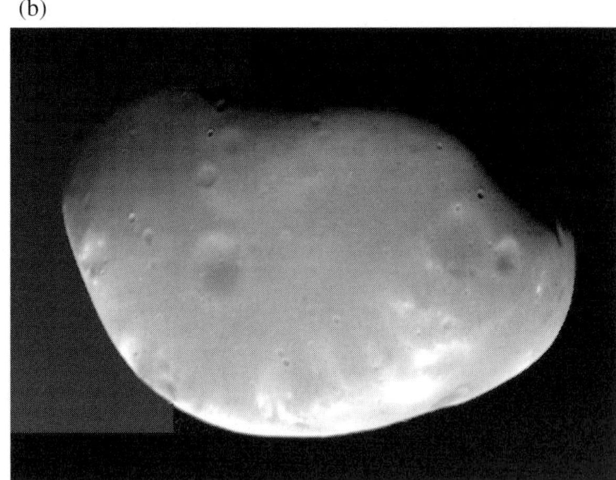

FIGURE 9.21 (a) Image of Phobos, the inner and larger of the two moons of Mars, taken by the Mars Global Surveyor in 1998. The Thermal Emission Spectrometer (TES) measured the brightness of thermal radiation at the same time the camera acquired this image. By analyzing the brightness, TES scientists could deduce the various fractions of the surface exposed to the Sun and their temperatures. This preliminary analysis shows that the surface temperature, dependent on slope and particle size, varies from a high of +25 °F (–4 °C) on the most illuminated slopes to –170 °F (–112 °C) in shadows. This large difference, and the fact that such differences can be found in close proximity, adds support to the notion that the surface of Phobos is covered by very small particles. (NASA/Mars Global Surveyor PIA01332) (b) Viking image of the martian satellite Deimos.

it underwent some heavy bombardment after the Koronis parent body broke up. This, however, does not explain the degraded condition of craters on Ida's surface, unless the sputtering by micrometeorites is much more severe in Ida's environment.

Based upon estimates of the thermal inertia, Ida is likely covered by a thicker layer of regolith than Gaspra; it is expected to be ∼ 50–100 m thick. Galileo detected about 17 large blocks of rock on its surface, 45–150 m across. In analogy to Earth, the Moon, Phobos and Deimos, these blocks could be impact ejecta. On Ida, however, it would also be possible that these blocks are fragments of the original Koronis parent body, which survived after the parent body broke up, and re-accreted onto Ida.

253 Mathilde

253 Mathilde is a C type asteroid, with a geometric albedo of 0.047 ± 0.005 (Fig. 9.6d). It is very homogeneous in albedo across its surface, suggesting that the entire body is probably homogeneous and undifferentiated, as expected for primitive bodies. (If the asteroid had a thin dark coating, impact ejecta and craters would result in albedo variations.) Mathilde's color is similar to that of CM carbonaceous chondrites, but Mathilde's density (1.3 ± 0.2 g cm^{-3}) is much lower than that of the meteorites, suggesting that the asteroid is very porous. This could indicate that Mathilde is a rubble-pile compound body. The fact that the asteroid rotates very slowly (P_{rot} = 17.4 days) raised suspicion that Mathilde might be a binary system, tidally despun. No satellite over 200–300 m in size has been found by NEAR, however.

Mathilde's phase integral (Table 9.3) is slightly lower than that for Phobos, indicating that Mathilde's surface displays more backscattering than Phobos's surface. Mathilde's phase curve (Fig. 9.8b) further indicates that its surface is rougher than that of Phobos. About 50% of Mathilde's surface was imaged by the NEAR spacecraft. Five large craters, with diameters between 20 and 33 km, were found. The largest crater may be 5–6 km deep, and was probably produced by an impactor 1–3 km across. Crater morphologies range from relatively deep fresh craters to shallow degraded ones. Such a range of morphologies is consistent with a surface in equilibrium with the cratering process. The number density and crater size distribution is similar to that seen on Ida, and suggests that Mathilde, like Ida, is 1–2 × 10^9 yrs.

No water or hydrated features have been seen on Mathilde's surface. The spectral properties of the asteroid match those of the Murchison meteorite (CM) after the meteorite had been heated substantially, so that it had lost all its water and hydrated minerals. This suggests that Mathilde either formed in a region where it was too cold to allow aqueous alterations on its surface, or that the asteroid was heated up so any hydrated material would have been driven off.

433 Eros

The S type near-Earth asteroid 433 Eros has been studied extensively by the orbiting NEAR Shoemaker spacecraft (Fig. 9.6c, d). Eros's visual geometric albedo $A_0 = 0.25 \pm 0.05$, similar to that of 951 Gaspra and 243 Ida. Eros has a smaller opposition surge, however, and its phase integral and Bond albedo are lower than for Gaspra and Ida. Eros's surface is heavily cratered, like Ida's, but craters with radii < 50 m are progressively depleted compared to a lunar-like surface. The asteroid's surface is further covered with roughly a million ejecta blocks having diameters between 8 and 100 m.

Eros's elemental composition and spectral properties were found to be similar to ordinary chondrites. Thus, space weathering may indeed 'mask' the surfaces of many asteroids, and the common S type asteroids are likely to be the parent bodies of ordinary chondrites. This would solve the long-standing question of the apparent absence of parent bodies for these common meteorites. Eros appears to be a primitive undifferentiated body with a density of 2.67 ± 0.03 g cm^{-3}; there is no indication that the asteroid is a rubble-pile composite object.

Radar Images of Asteroids

Images of several NEOs with resolutions as high as $\sim$100 m have been reconstructed from radar observations (Fig. 9.15). Several objects appear to be very elongated, with the major axis 2–3 times the short axis. The first of these which was imaged in detail, 4769 Castalia, appears to be bifurcated, which suggests Castalia is a contact-binary. The NEO 4179 Toutatis is also bifurcated, and the images show shallow craters and linear ridges. Its rotation spin is very interesting, in that it 'tumbles': it is composed of two types of motion with periods of 5.4 and 7.3 days, which combine in such a way that Toutatis's orientation with respect to the Sun never repeats. The NEO 1620 Geographos shows peculiar protuberances at the ends of its elongated body. These may result from the excavation process and subsequent deposition of impact ejecta on an elongated body, where the ratio of gravitational attraction and centripetal force varies drastically over the surface of the body.

The radar systems have also observed several rounder NEOs, such as 1998 KY_{26} and 1999 JM_8. The M class main belt asteroid 216 Kleopatra is the most reflective of the several dozen asteroids thus far observed with radar. Its reflectivity is similar to that of near-Earth asteroid 6178 (1986 DA). The radar measurements of 216 Kleopatra reveal a dumbbell shaped object with dimensions $217 \times 94 \times 81$ km. The surface properties of this asteroid are consistent with a regolith of metallic composition and a porosity of $< 60\%$, similar to that of lunar soil. Its shape, like that of other bifurcated bodies, suggests that the object once consisted of two separate bodies which may have 'fused' together via a gentle collision, perhaps from a low velocity infall of fragments after a disruption event, or from tidal decay of a binary system. Kleopatra's interior probably consists of a rubble-pile structure.

9.5.3 Magnetic Fields?

The interplanetary magnetic field changed abruptly towards a radial orientation in the direction of 951 Gaspra when the Galileo spacecraft was closest to the asteroid. These data hint at an intrinsic magnetic field, which is quite a puzzling result since an object as small as Gaspra is presumably solid throughout. However, certain meteorites, in particular iron and stony-iron objects, are distinctly magnetized. Gaspra may be a scaled-up version of these meteorites, and show remnant magnetism from the epoch of its formation. Magnetic disturbances were also recorded when Galileo flew past Ida. The measurements leave more ambiguity in the interpretation; it is not clear whether Ida is magnetized, like Gaspra, or whether the body is merely conductive.

9.6 Origin and Evolution of the Asteroid Belt

The bodies in the asteroid belt presumably grew from planetesimals, as did the terrestrial planets. Just like the other planets, when the bodies grow large enough, the accretional and radioactive heating lead to a (partial) melting of the material, resulting in *differentiation* (Sections 5.2.2, 12.6.2), i.e., the heavy materials sink down and form the core of the object, while the mantle is made up of lighter elements. Analysis of meteorites implies that some asteroids are differentiated (Section 8.7). The asteroids can thus be viewed as remnant planetesimals, all of which are distinct individual bodies, which failed to accrete into a single body. At present, the mass in the asteroid belt is too small by 3–4 orders of magnitude to allow the formation of a full-sized planet. It is unlikely that the mass has always been this small; thus many bodies have probably been removed from the asteroid belt by the gravitational influence of Jupiter (Section 12.9).

The heliocentric distribution of the various asteroid classes follows the general condensation sequence: High temperature condensates are found in the inner regions of the asteroid belt, while lower temperature condensates typically orbit the Sun at larger distances. Bound water, e.g., in the form of hydrated silicates, has only been found in a few C, M and E type asteroids; in contrast to expectations, it has not been seen in D or P types. It is currently believed that hydration of silicates did not occur in the nebula, but rather in the parent asteroidal bodies. Aqueous alteration can occur at temperatures as low as 300 K. The lack of hydrated silicates in D and P asteroids may be caused by insufficient heating to melt and mobilize the ice. The dependence on heliocentric distance of igneous, metamorphic and primitive asteroids (Fig. 9.18) suggests a heating mechanism for these bodies which declined rapidly in efficiency with heliocentric distance. Although both accretional and radioactive heating are more effective for larger objects, neither accretion nor decay of long-lived radioactive isotopes could have provided enough energy to have melted typical asteroidal-sized bodies. Short-lived radioactive nuclei such as ^{26}Al may have provided sufficient energy in the earliest days of Solar System formation. An alternative model involves heating by electromagnetic induction during the Sun's active T Tauri phase (Section 12.3.3). Although many aspects of this process remain highly uncertain, calculations suggest that this mechanism may melt bodies with sizes $\gtrsim$100 km in the inner asteroid belt.

Regardless of the heating mechanism, it is clear that asteroids preserve some kind of transition between primitive chondritic material in the outer parts of the asteroid belt and more evolved igneous material in the inner regions of the asteroid belt. This overall pattern was 'shaped' by compositional gradients in the original solar nebula.

The asteroid belt probably contained at least several times as much mass during the planet building epoch as it does nowadays. Ceres and Vesta appear to have not been totally disrupted, but most bodies with radii between 50 and 125 km have probably re-accumulated into rubble-pile bodies. Collisions between planetesimals may lead to fragmentation of the parent bodies, and may strip differentiated objects from their mantles, leaving just the metal-rich core of these bodies. Large M type asteroids may be such exposed asteroidal cores. Since the material strength of iron–nickel cores is much larger than that of the silicate mantles, the metal-rich asteroids survived many collisions, while the mantle fragments were likely efficiently destroyed. This would explain the apparent lack of olivine-rich (mantle material) asteroids in comparison with metal-rich bodies.

Further Reading

Binzel, R.P., 1992. 1991 Urey Prize lecture: Physical evolution in the Solar System – present observations as a key to the past. *Icarus* **100**, 274–287.

The following book contains many review papers written prior to the spacecraft encounters of asteroids (Galileo and NEAR):

Binzel, R.P., T. Gehrels, and M.S. Matthews, Eds., 1989. *Asteroids II*. University of Arizona Press, Tucson, 1258pp.

Background regarding asteroid dynamics can be obtained from:

Goldstein, H., 1980. *Classical Mechanics*, 2nd Edition. Addison Wesley, MA.

A thorough overview of the shadow-hiding and coherent-backscatter effects near opposition are presented in a series of papers by Hapke *et al.*, and Helfenstein *et al.*, which include:

Hapke, B., R. Nelson, and W. Smythe, 1998. The opposition effect of the Moon: Coherent backscatter and shadow hiding. *Icarus* **133**, 89–97.

Helfenstein, P., J. Veverka, and J. Hiller, 1997. The Lunar opposition effect: A test of alternative models. *Icarus* **128**, 2–14.

Two special issues of *Icarus* have been devoted to the Galileo encounters with Gaspra and Ida/Dactyl: *Icarus* 1994 **107** number 1; and *Icarus* 1996 **120**, number 1, respectively.

A special issue of Icarus devoted to the NEAR encounter with Mathilde appeared in: *Icarus* 1999 **140**, number 1.

Research papers regarding radar observations include:

Ostro, S.J., *et al.*, 1995. Radar images of asteroid 4179 Toutatis. *Science* **270**, 80–83.

Ostro, S.J., R.S. Hudson, M.C. Nolan, J-L. Margot, D.J. Scheeres, D.B. Campbell, C. Magri, J.D. Giorgini, and D.K. Yeomans, 2000. Radar observations of asteroid 216 Kleopatra. *Science* **288**, 836–839.

Comet/asteroid transition objects are discussed in:

Yeomans, D., 2000. Small bodies of the Solar System. *Nature* **404**, 829–832.

Problems

9.1.I Figure 9.1b shows the instantaneous positions of over 7000 asteroids.

(a) Explain why the Kirkwood gaps cannot easily be detected in this figure, in contrast to their clear visibility in Figure 9.1a.

(b) Discuss the dynamical causes of the radial and longitudinal structure which is apparent in Figure 9.1b.

9.2.E Calculate the locations of the 3:1, 5:2, 7:3, 2:1 and 3:2 resonances with Jupiter. (Hint: See Chapter 2.) Use Figure 9.1a to determine which of these resonances produce gaps and which have led to a concentration in the population of asteroids.

9.3.E (a) Show that if the exponent $\zeta = 4$ in equation (9.2), then the mass is divided equally among equal logarithmic intervals in radius.

(b) Show that if $\zeta < 3$, then most of the mass in the asteroid belt is contained in a few large bodies, in the sense that the largest factor of 2 in radius contains more mass than all smaller bodies combined.

(c) For which value of ζ does one find equal areas in equal logarithmic size intervals?

9.4.E If a differential size–frequency distribution of a group of objects that have equal densities can be adequately described by a power law in radius of the form $N(R) \propto R^{-\zeta}$, then it can also be described as a power law in mass of the form $N(m) \propto m^{-x}$. Derive the relationship between ζ and x.

9.5.I The number of asteroids within the main belt (2.1–3.3 AU) of radius larger than R (km), $N_>(R)$, is given approximately by equation (9.2*b*) with $\zeta - 1 = 2.5$, $R_o = 1$ km and $N_o/(\zeta - 1) = 10^5$. Typical asteroid orbits have inclinations of $\sim 15°$. For this problem, you may assume that the asteroids are uniformly distributed in semimajor axis within the main belt.

(a) How often does a particular 100 km radius asteroid get hit by any asteroid 1 km in radius or larger?

(b) How often does a particular X km radius asteroid get hit by any asteroid Y km in radius or larger?

(c) How often does any 100 km radius asteroid somewhere in the main belt get hit by any asteroid 1 km in radius or larger?

9.6.I (a) Calculate the mean optical depth of the main asteroid belt. (Hint: Divide the projected surface area of the asteroids by the area of the annulus between 2.1 and 3.3 AU. Use the size–frequency distribution given by equation (9.2) and assume (and justify) reasonable values for R_{min} and R_{max} if necessary.)

(b) Calculate the fraction of space (i.e., volume) in the main asteroid belt near the ecliptic which is occupied by asteroids.

9.7.E A 'typical' asteroid orbits the Sun at 2.8 AU, has an inclination of 15° and an eccentricity of 0.14. Calculate the typical collision velocity of two asteroids. (Hint: Compute the speed of a single asteroid relative to a circular orbit in the Laplacian plane of the Solar System and multiply by $\sqrt{2}$ to obtain the mean encounter velocity.)

9.8.E (a) Calculate the gravitational binding energy (in ergs) of a spherical asteroid of radius R (km) and density ρ (g cm^{-3}).

(b) For what size asteroid is the gravitational binding energy equal to the physical cohesion (the fracture stress of rock is $\approx 10^9$ dyne cm^{-2})?

(c) Are nonspherical asteroids more or less tightly bound gravitationally than spherical ones of the same mass?

(d) What is the radius of the largest coherent spherical asteroid that can be disrupted by 1000 MT of TNT equivalent explosives (1 MT TNT $= 4.18 \times 10^{22}$ ergs)?

(e) What if the asteroid is a rubble-pile (no strength)?

(f) Repeat parts (d) and (e) for 1 000 000 MT TNT.

9.9.I Asteroid A has radius R, density $\rho = 3$ g cm^{-3}. What is the size of the smallest asteroid B, also of density $\rho = 3$ g cm^{-3}, needed to catastrophically disrupt asteroid A? For your calculation, you may assume a collision velocity of 7.5 km s^{-1}, and that disruption requires a kinetic energy equal to gravitational plus physical binding energy. Comment on the sign and magnitude of errors induced by these assumptions.

9.10.I (a) Calculate the synodic period (time between conjunctions) of a pair of asteroids with orbital semimajor axes of 2.8 and 2.8028 AU.

(b) Assume that the asteroid with $a = 2.8028$ AU has periapse at 2.8 AU. Calculate its speed relative to local circular orbit when it is at periapse. If the other asteroid is on a circular orbit, this is the relative speed of the two bodies where their paths cross.

(c) Compare the speed that you calculated in part (b) with typical relative velocities of asteroids and also with the escape velocity from a 100 km radius rocky asteroid ($\rho = 2.5$ g cm^{-3}).

(d) Use your results from the above calculations to explain why members of asteroid families (which were formed via catastrophic

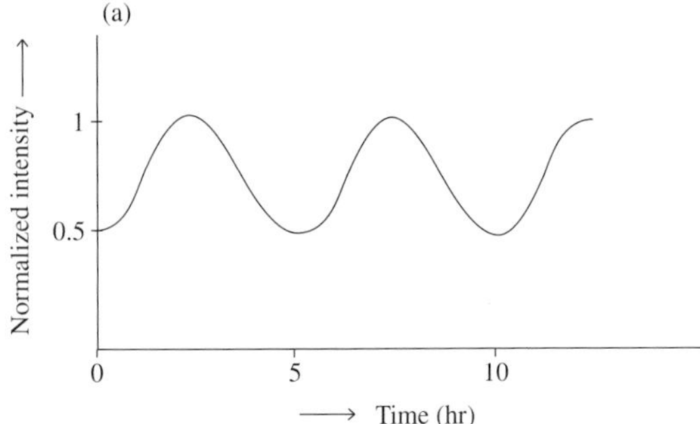

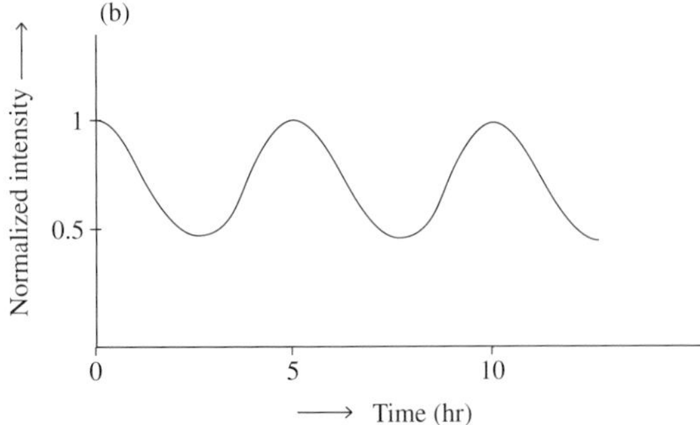

FIGURE 9.22 Fictitious lightcurves for Problem 9.15.

collisions) are not necessarily in close proximity to one another. What parameters are most useful in determining membership in asteroidal families?

9.11.E Consider a spherical asteroid with a density $\rho = 3$ g cm^{-3} and radius $R = 100$ km. The asteroid is covered by a layer of loosely bound regolith. What is the shortest rotation period this asteroid can have without losing the regolith from its equator?

9.12.I (a) Estimate the size of the largest asteroid from which you could propel yourself into orbit under your own power.

(b) What variables other than size must be considered?

(c) How would you be able to launch yourself into a stable orbit?

9.13.E Estimate the damping times for wobbles of asteroids with the following radii and rotation periods:

(a) $R = 200$ km, $2\pi/\omega_{rot} = 16$ hours.

(b) $R = 20$ km, $2\pi/\omega_{rot} = 4$ hours.

(c) $R = 2$ km, $2\pi/\omega_{rot} = 4$ hours.

(d) $R = 2$ km, $2\pi/\omega_{rot} = 4$ days.

(e) $R = 15$ m, $2\pi/\omega_{rot} = 10$ minutes.

9.14.E What techniques are used to estimate the diameters of asteroids? Briefly describe each technique, and comment on its advantages and disadvantages.

9.15.E (a) Consider the visual lightcurve in Figure 9.22a. Assume that the shape of the lightcurve at thermal infrared wavelengths is similar. What can you say about the asteroid's shape, size, albedo and rotation period?

(b) Consider the visual lightcurve of an asteroid as sketched in Figure 9.22a and the infrared curve in Figure 9.22b. What can you say about the asteroid's shape, size, albedo and rotation period?

9.16.I Consider an asteroid shaped like a triaxial ellipsoid with axes ratios $a = 2b = 3c$ which rotates about its short axis, and has one bright (high albedo) spot. Sketch the visual and infrared lightcurves for this asteroid.

(a) Assuming that the spot is at a pole and the asteroid's rotation axis is perpendicular to the line of sight.

(b) Assuming that the spot is at a pole and the asteroid's rotation axis is parallel to the line of sight.

(c) Assuming that the spot is along the equator at one tip of the long axis and the asteroid's rotation axis is perpendicular to the line of sight.

9.17.I (a) In order to calculate both the radius and the albedo of an unresolved asteroid, one needs observations at infrared as well as visible wavelengths. Explain in one sentence why this is the case.

(b) Assume the Bond and geometric albedos to be equal, the infrared emissivity to be 0.9, the asteroid to be spherical and rapidly rotating. The asteroid is at a heliocentric distance of 2.5 AU, and a geocentric distance of 1.5 AU. The infrared flux (note: integrated over frequency!) is 9.2×10^{-9} erg cm^{-2} s^{-1}, and the flux at visible wavelengths is 1.6×10^{-9} erg cm^{-2} s^{-1}. Determine the temperature of the asteroid, its Bond albedo and radius. (Hint: See Section 3.1.)

9.18.E Derive equation (9.11). (Hint: Compare the Doppler shift of radar waves reflected from the location on the body's surface which is moving away from the observer most rapidly with those reflected from the point moving towards the observer most rapidly.)

9.19.E (a) What features distinguish the spectra of S, C and D type asteroids from one another?

(b) Where do most members of each of these taxonomic classes orbit?

9.20.I Observations of asteroids A1 and A2 show that they are in circular orbits with periods of 4.4 and 6.0 yrs, respectively. Assume that each asteroid is spherical, with a radius of 50 km. The geometric albedos are measured to be: $A_{0_1} = 0.245$, and $A_{0_2} = 0.049$, respectively.

(a) Calculate the semimajor axes of the asteroids' orbits.

(b) Assume the asteroids are overlain by a layer of regolith so that the phase integral, $q_{ph,v}$, of the solar reflectivity is equal to that of the Moon. Estimate the visual albedo for the two asteroids from the measured geometric albedos.

(c) Calculate both the average subsurface temperature and the subsolar surface temperature for the two asteroids, assuming they are rapid rotators. (Hint: See Section 3.1.)

(d) What major taxonomic class would you assign to A1 and A2? Explain your answer in detail. (Hint: See Problem 9.19.)

9.21.I In the previous problem it was assumed that the phase integral for the asteroids A1 and A2 was similar to that of the Moon. If the phase integral has half the value of the Moon's phase integral, calculate the 'beaming factor' $\eta_{v=ir}$, assuming that all other quantities are equal.

9.22.E The observed radar cross-section for asteroid 324 Bamberga is 4500 km^2. Bamberga's radius is 120 km. Assume the radar backscatter gain $g_r = 1$.

(a) Determine Bamberga's radar albedo.

(b) What is the dielectric constant of Bamberga's regolith?

(c) Determine the regolith density and porosity of Bamberga's regolith. Assume the density and dielectric constant of the solid rock in Bamberga's crust are $\rho_o = 2.6$ g cm^{-3} and $\epsilon_o = 6.5$.

9.23.I Bamberga's geocentric distance is 0.83 AU and its heliocentric distance 1.78 AU at the time of the observations. Bamberga's Bond albedo is 0.10. The asteroid's total observed emission at 3.6 cm is 1.72 mJy.

(a) Determine Bamberga's brightness temperature T_b at 3.6 cm.

(b) Compare T_b with the expected equilibrium temperature of Bamberga a meter below its surface, and derive the radio emissivity.

(c) Compare the radio emissivity with the radar albedo from the previous problem, and comment on your results.

10 Comets

You see therefore an agreement of all the Elements in these three, which would be next to a miracle if they were three different Comets. ... Wherefore, if according to what we have already said it should return again about the year 1758, candid posterity will not refuse to acknowledge that this was first discovered by an Englishman.

Edmond Halley, 1752, *Astronomical Tables*, London

The generally unexpected and sometimes spectacular appearances of comets have triggered the interest of many people throughout history. A bright comet can easily be seen with the naked eye, and its tail can extend more than 45° on the sky (Fig. 10.1). Comets were once believed to be bad omens. An apparition of Comet Halley is depicted on the Bayeux Tapestry (Fig. 10.2), which commemorates the Norman conquest in 1066; the comet's appearance was considered a bad omen for King Harold of England, when William the Conqueror invaded England from Normandy.

The first detailed scientific observations of comets were made by Tycho Brahe in 1577. Brahe observed a very bright comet from two different locations; since he could not find any parallax shift, he concluded that the comet must be farther away than the Moon. Edmond Halley used Newton's gravitational theory to compute parabolic orbits of 24 comets observed up to 1698. He noted that the comet apparitions in 1531, 1607 and 1682 were separated by 75–76 years, and that the orbits were described by roughly the same parameters. He predicted the next apparition in 1758. It was noticed much later that this Comet Halley, as it was named afterwards, has returned 30 times from 240 BCE to 1986; records of all of these apparitions have been found with the exception of 164 BCE. Up through the early twentieth century, 3–4 comets were discovered each year, a rate which went up to 20–25 comets per year with the development of more powerful cameras (CCDs) in the 1980s.

The name comet is derived from the Greek word $\kappa\omega\mu\eta\tau\eta\zeta$ which means 'the hairy one', describing a comet's most prominent feature: its long *tail*. A schematic picture of a comet is shown in Figure 10.1c. The small *nucleus*, often only a few kilometers in diameter, is usually hidden from view by the large *coma*: a cloud of gas and dust roughly 10^4–10^5 km in diameter. Not seen with the naked eye is the large *hydrogen coma*, between one and ten million kilometers in extent, which surrounds the nucleus and visible gas/dust coma. Radiation pressure drives the tiny dust particles in a comet's coma outwards from the Sun (Section 2.7); these dust particles form the yellowish *dust tail*, which, like the coma, is seen in reflected sunlight. As the heliocentric distance of the dust particles increases, they slow down in their orbit (conservation of angular momentum), which results in a curvature of the dust tail in a direction opposite to the comet's motion. In contrast to the curved yellowish tail, some comets display a straight, usually bluish, tail in the anti-solar direction. This tail consists of ions which are bound to the interplanetary magnetic field lines, and which are dragged along with the solar wind. *Ion tails* can reach lengths up to 10^8 km. The bluish color is produced primarily by emission from CO^+ ions.

10.1 Nomenclature

Comets are named after their discoverer(s). Numbers follow names when one person or group discovers multiple comets. Comets are also given a designation, which includes the year of their (re)discovery or perihelion passage. The form of these designations changed in 1995, so the literature contains both formats. According to the old format, when a comet was discovered or recovered (seen for the first time during an apparition) it was given a provisional designation based on the year of discovery followed by a letter sequentially assigned as comets are discovered, e.g., 1994c represents the third comet discovered in 1994. Several years later, a final designation was given, in which the name of the comet was followed by the year of its perihelion passage, together with a Roman numeral based sequentially on perihelion date to distinguish it from other comets passing perihelion that same year, e.g., Comet Kohoutek C/1973 XII. Short-period comets are preceded by a P/, e.g., P/Halley and P/Encke. Deceased comets, e.g., comets which have collided with the Sun or one of the planets, are preceded by D/; the most famous such object is (was) D/Shoemaker–Levy 9 (Sections 5.4.5, 10.6.3). In contrast to the old system,

(a)

(c)

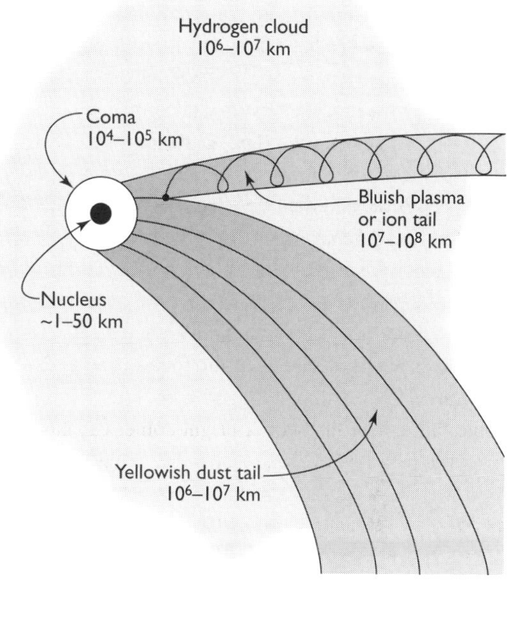

(b)

FIGURE 10.1 (a) COLOR PLATE C/West (1976 VI) as photographed 9 March 1976, by amateur astronomer John Laborde. Both the yellow or 'whitish' dust tail and the blue plasma tail (lower right) are apparent. (Courtesy: John Laborde) (b) C/Hale–Bopp (C/1995 O2) above Camp Mercier, Reserve Faunique des Laurentides, Canada on 1 April 1997. (Courtesy: Laurent Drissen) (c) Schematic picture of a comet, showing its nucleus, coma, tails and hydrogen cloud.

FIGURE 10.2 Section of the Bayeux Tapestry, commemorating the Norman conquest of England in 1066. P/Halley is shown; this comet was very bright at the time William the Conqueror invaded England from Normandy. The comet was considered a bad omen for King Harold of England. The Latin sentence 'Isti Mirant Stella' means 'they marvel at the star'. (Courtesy: Beatty *et al.* 1999)

the new system avoids duplication. The year of discovery (or recovery) is followed by a letter indicating the half-month in which the comet was first observed, followed by a number to distinguish it from other comets seen during the same period. The names of long-period comets are now preceded by C/, and short-period comets are given a number according to their initial discovery. Thus, 1P/1682 Q1 represents Halley's comet during its 1682 apparition, and indicates that it was initially spotted in the first half of October of 1682. Note that short-period comets receive a different designation for each apparition in both the old and new systems, but the name and number of a short-period comet remains unchanged. Details on cometary nomenclature can be found in Marsden and Williams (1997).

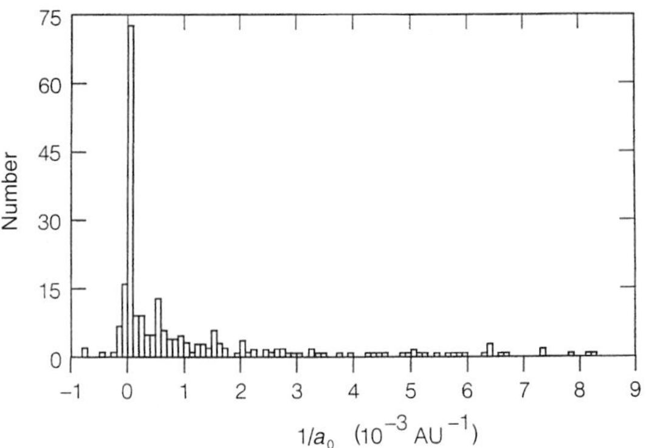

FIGURE 10.3 Distribution of 264 long-period comets as a function of the original inverse semimajor axis, $1/a_0$. (Mumma *et al.* 1993)

10.2 Cometary Orbits and Comet Reservoirs

Cometary orbits are eccentric ellipses, so that only a small fraction of their orbital period is spent in the inner planetary region; most of the time comets reside in the cold outer parts of our Solar System (Section 2.1). Comets with orbital periods > 200 years are classified as *long-period (LP) comets*, while those with periods < 200 years are called *short-period (SP) comets*. Comets whose orbits suggest that they have entered the inner Solar System, the planetary region, for the first time are referred to as *dynamically 'new' comets*. The most recent *Catalogue of Cometary Orbits* (Marsden and Williams 1997) contains 1582 orbits for 1548 cometary apparitions (this difference being because some comets have multiple components) of 937 specific comets (this difference being because orbits of periodic comets are given for each observed passage through perihelion), of which 191 are SP comets, 132 of these having been observed at more than one passage.

10.2.1 Oort Cloud

To deduce the source region of dynamically 'new' comets, Jan Oort plotted (in 1950) the distribution of the inverse semimajor axes, $1/a_0$, for all long-period comets observed up to that time. Figure 10.3 shows the distribution of the original $1/a_0$ for a much larger distribution of LP comets. The inverse semimajor axis is a measure of the orbital energy per unit mass, $M_\odot G/(2a_0)$, with $M_\odot$ the Sun's mass and $GM_\odot$ the gravitational constant. The original orbit is that of the comet before it entered the planetary region and became subject to planetary perturbations and nongravitational forces. Positive values in $1/a_0$ indicate bound orbits, while negative values denote hyperbolic orbits. The few hyperbolic orbits shown in Figure 10.3, however, are almost certainly not from comets originating in interstellar space (Problems 10.1 and 10.2). The negative $1/a_0$ values are probably caused by errors in the calculation of the orbital elements, possibly induced by unaccounted for nongravitational forces (Section 10.2.3).

Based upon the large spike between 0 and 10^{-4} AU^{-1}, Oort postulated the existence of a vast spherical 'cloud' of comets surrounding our planetary system, at heliocentric distances $\gtrsim 10^4$ AU. The spike represents comets from the *Oort cloud* which enter the planetary region for the first time. These comets have semimajor axes of (1–5) $\times$ 10^4 AU and are randomly oriented on the celestial sphere, aside from a small signature of the galactic tidal field. The orbits of these comets are highly eccentric ellipses, which appear nearly parabolic when passing through the planetary region of the Solar System. In contrast, the orbits of short-period comets are less eccentric, usually prograde and appear to be concentrated near the ecliptic plane (inclination angles $i \lesssim 35°$).

The number of comets in the classical Oort cloud can be estimated from the observed flux of dynamically 'new' comets. Counting objects brighter than absolute magnitude H_{10} = 11 (for the definition of H_{10}, see Section 10.3.1.1), a total number of $\sim 10^{12}$ comets are hypothesized to populate the classical Oort cloud. Comets in the Oort cloud whose orbits are perturbed by nearby stars, close encounters with giant molecular clouds or by the tidal field of the galactic disk may enter the inner Solar System, and be observed as dynamically 'new' long-period comets.

The dynamical lifetime of comets in the classical Oort cloud to ejection by passing stars is about half the age of the Solar System. The importance of encounters with giant molecular clouds is much harder to estimate due to uncertainties in cloud parameters, but giant molecular clouds could be as much as several times as effective as stars at ejecting comets from the Solar System. This suggests that the Oort cloud needs to be replenished, either from the inside by, e.g., an unseen inner Oort cloud between 10^3 and 10^4 AU, or by capture from the interstellar medium. While, due to the rather high interstellar encounter velocities (20–30 km s^{-1}), capture of interstellar comets is exceedingly unlikely, the existence of a vast inner Oort cloud is, in fact, a natural consequence of the formation of our Solar System; dynamical models show that ejection of planetesimals from the planetary region results in an inner Oort cloud which is about 5–10 times more populated than the outer (classical) cloud. Under 'normal' circumstances, the inner Oort cloud contributes essentially nothing to the flux of new comets observable in the inner Solar System because it is virtually unperturbed. However, large perturbations caused by penetrating stellar encounters and giant molecular clouds could produce showers of comets lasting a few million years about once every 10^8 years, and also repopulate the outer Oort cloud.

On subsequent passes of comets through the planetary region of the Solar System, gravitational perturbations by the planets scatter comets in $1/a$. It may take a comet about 400 returns to the planetary region before its orbit is changed to that of a short-period comet. Dynamical calculations show that fewer than 0.1% of long-period comets evolve into short-period comets. While evolving inwards, comets usually preserve their orbital inclination. Most of the SP comets, however, have low orbital inclinations and a prograde sense of revolution. It was therefore suggested that most SP comets come from a flattened annulus of objects beyond the orbit of Neptune. Such a collection of bodies was postulated by K.E. Edgeworth in 1949 and by G.P. Kuiper in 1951 based upon a natural extension of the original solar nebula beyond the orbit of Neptune. The reservoir of SP comets is therefore referred to as either the *Kuiper belt* or the *Edgeworth–Kuiper belt*.

10.2.2 Kuiper Belt

From the observed number of SP comets, the total mass of bodies in the Kuiper belt is estimated to be 0.0026 $M_\oplus$, which translates roughly into 10^9–10^{10} objects. The first trans-neptunian, or Kuiper belt object (KBO) was discovered in 1992, which marked the onset of a flurry of search activities. As of September 2000, $\sim$350 KBOs have been detected. Most lie interior to 50 AU and have low inclinations ($\lesssim 30°$). The first small trans-neptunian object detected, 1992 QB_1, is very red, similar to 5145 Pholus, while others with known colors vary from gray to moderately red. Assuming albedos of $\sim$4%, the observed reflectivity of P/Halley, typical radii for the trans-neptunian objects are between 25 and 150 km. The brightest object (adjusted for heliocentric and geocentric distance), 2000 WR_{106}, has a radius of $\sim$450 km and albedo of $\sim$7%. Extrapolating the number of detected comets to the total number of KBOs with radii $R > 25$ km, there should be $\sim 10^5$ objects at $30 < r_{AU} < 50$, with a total mass of ~ 0.2 $M_\oplus$. There may be many more small KBOs, and many objects may be hidden at $r_{AU} > 50$ (Problem 10.5). An upper limit of $\sim$1 $M_\oplus$ worth of material contained in small bodies distributed in a flattened ring just beyond the orbit of Neptune has been deduced from the lack of detection of the gravitational perturbations of such a comet disk on P/Halley. Future tracking of the Voyager and Pioneer spacecraft is expected to yield improved bounds. Dynamical calculations show that the orbits of KBOs with $a > 45$ AU are stable over the age of the Solar System, while most bodies found with $a < 41$ AU appear to be in a 3:2 resonance orbit with Neptune, as is the planet Pluto; thus, these bodies are referred to as *plutinos*. Roughly 30–40% of the KBOs detected so far belong to this class of objects. Pluto itself may be thought of as an exceptionally large member of the Kuiper belt.

Although the eccentricities of most KBOs are relatively low, 3–4% of the known KBOs have been found on highly eccentric orbits. All of these highly eccentric KBOs have perihelia $\lesssim 40$ AU, and most were discovered near perihelion. The first of these discovered was 1996 TL_{66}, which has aphelion at 135 AU. Their eccentric orbits result from gravitational scattering by Neptune. This class of objects is referred to as *scattered KBOs* (SKBOs), in contrast to the *classical KBOs* (CKBOs, also known as *cubewanos*, after 1992 QB_1) on more circular orbits. *Resonant KBOs* (primarily plutinos) typically have intermediate values of eccentricity. The number of scattered KBOs discovered so far suggests the existence of a vast population of KBOs on highly eccentric orbits, with aphelia reaching beyond 200 AU. It has been estimated that the total number of SKBOs with radii > 50 km and semimajor axes $a > 50$ AU is similar to that of the CKBOs inside 50 AU.

Very few KBOs have been found in orbits with semimajor axes between 36 and 39 AU. It has been suggested that Neptune's orbit has evolved outward as a result of angular momentum exchange with planetesimals during the accretional phase of Solar System formation (Chapter 12), when the giant planets ejected many comets into the Oort cloud. While Neptune moved outwards, its mean motion resonances were pushed outwards as well. In this way Neptune may have swept up bodies into the currently stable 3:2 resonance, thereby cleaning the inner Kuiper belt. Dynamical

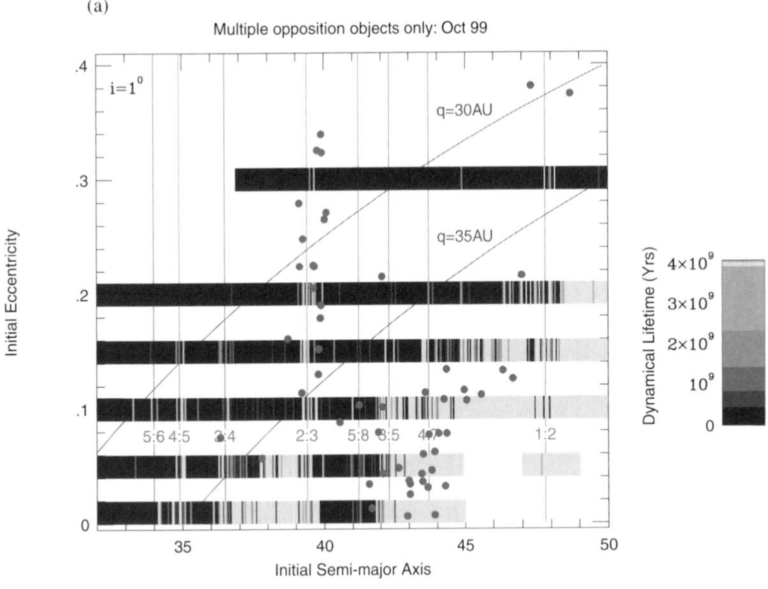

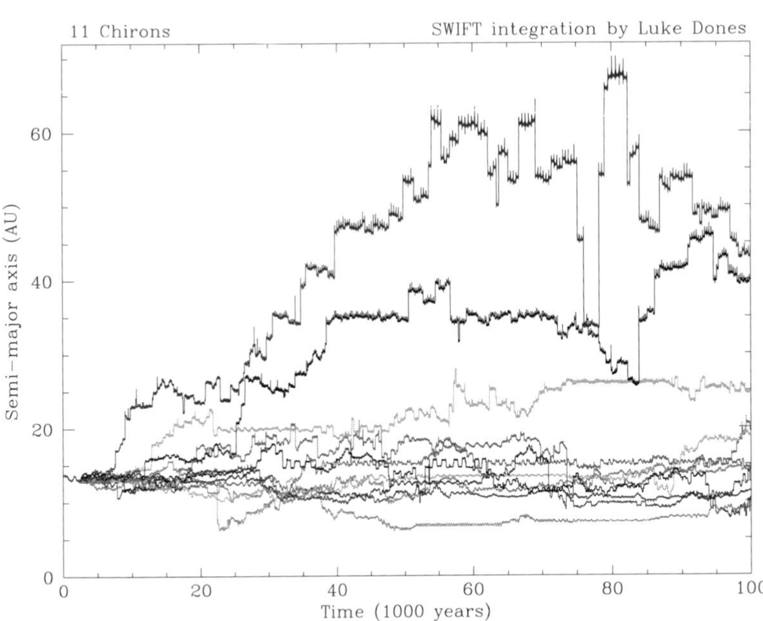

FIGURE 10.4 COLOR PLATE
(a) The dynamical lifetime for small particles in the Kuiper belt derived from 4 billion year integrations. Each particle is represented by a narrow vertical strip, the center of which is located at the particle's initial eccentricity and semimajor axis (initial orbital inclination for all objects was 1 degree). The color of each strip represents the dynamical lifetime of the particle. The yellow strips represent objects that survive for the length of the integration, 4×10^9 years. Dark regions are particularly unstable on these timescales. For reference, the locations of the important Neptune mean motion resonances are shown as blue vertical lines and two curves of constant perihelion distance, q, are indicated. The green dots show the orbits of the known Kuiper belt objects which have been observed at more than one opposition as of October 1999. (Duncan *et al.* 1995)
(b) Eleven simulations of the future orbital semimajor axis of P/Chiron, which currently crosses the orbits of both Saturn and Uranus. Initial orbital elements of the simulated bodies differed by about 1 part in 10^6, which is smaller than observational uncertainties. Note that the orbit is highly chaotic, with gross divergence of trajectories in less than 10^4 years. (Courtesy: L. Dones; adapted from Lissauer 1999a)

studies of the trans-neptunian region show that orbits with $a_{AU} < 35$ and $40 < a_{AU} < 42$ are unstable to gravitational perturbations by Neptune and Uranus (Fig. 10.4a). These studies further show that a small fraction of KBOs continues to move into these unstable zones, where their orbits will sooner or later suffer a major perturbation.

Some of the KBOs in the dynamically unstable zones ultimately evolve inwards and form a source of short-period comets. A few such bodies ($\sim$25 as of September 2000) probably have been detected. These bodies, including 2060 Chiron[1] and 5145 Pholus, are in highly unstable orbits that cross the orbits of Saturn, Uranus and Neptune (Fig. 10.4b). They typically have perihelia between 8 and 11 AU, and aphelia between 19 and 36 AU. Orbital eccentricities are $\sim$0.4–0.6. The objects are quite large, with radii $10 < R < 100$ km (smaller objects in

[1] After Chiron showed cometary activity, it also received a comet designation: 95P/Chiron.

similar orbits would have escaped detection). Collectively they are called *centaurs*. Typical dynamical lifetimes of the centaurs are 10^6–10^8 years. It is important to study these objects in detail, since they may represent the link between Kuiper belt objects (which are technically asteroids, as they do not exhibit comas) and short-period comets. Interestingly enough, new dynamical calculations of the centaur 5145 Pholus suggest that this body may have originated in the Oort cloud, rather than the Kuiper belt.

10.2.3 Nongravitational Forces

Although to lowest order comet trajectories are determined by the gravitational pull of the Sun and the planets[2], the observed orbits of most active comets deviate from these paths in small but significant ways. These variations can advance or retard the perihelion passage date of a comet by many days from one orbit to the next. The observation of these nongravitational effects was a major motivation for Whipple's *icy conglomerate* model of comet nuclei introduced in 1950 (Section 10.6). *Nongravitational forces* result from the momentum imparted to a comet's nucleus by the gas and dust which escape as the comet's ices sublime. The process is in certain respects analogous to rocket propulsion, but the magnitude of the effects is much smaller, because only a tiny fraction of the comet's mass is lost per orbit, mass escapes at a slower speed, and forces exerted at differing phases of the comet's orbit produce opposing effects. Nongravitational forces are in some respects analogous to the Yarkovski effect on small bodies, which is described in Section 2.7.3.

The standard model (also known as the *symmetric model*) for the nongravitational acceleration of comets is based on the hypothesis that ice evaporates from a rotating comet nucleus at a specified rate which depends only on heliocentric distance, and thus which is symmetric about perihelion. The equations of motion for a comet are written as:

$$\frac{d^2\mathbf{r}_\odot}{dt^2} = -\frac{GM_\odot}{r_\odot^2}\hat{\mathbf{r}}_\odot + \nabla\mathcal{R} + A_1\eta(r_\odot)\hat{\mathbf{r}}_\odot + A_2\eta(r_\odot)\hat{\mathbf{T}} + A_3\eta(r_\odot)\hat{\mathbf{n}} \qquad (10.1)$$

with

$$\eta(r_\odot) = \eta_1\left(\frac{r_\odot}{r_{\odot 0}}\right)^{-\eta_2}\left(1+\left(\frac{r_\odot}{r_{\odot 0}}\right)^{\eta_3}\right)^{-\eta_4}. \qquad (10.2)$$

The disturbing function, $\mathcal{R}$, accounts for planetary perturbations (eq. 2.23), A_1, A_2 and A_3 are the nongravitational acceleration coefficients, $\hat{\mathbf{r}}$, $\hat{\mathbf{T}}$ and $\hat{\mathbf{n}}$ are unit vectors pointing in the directions radially outwards from the Sun, perpendicular to $\hat{\mathbf{r}}$ in the plane of the comet's orbit (tangentially along the comet's orbit at perihelion), and normal to the comet's orbit, respectively. The variations of cometary activity with heliocentric distance are approximated by $\eta(r_\odot)$ in equation (10.2). The acceleration is given in AU day^{-2}. For water-ice, the values for the constants in equation (10.2) that are currently being used are: $r_{\odot 0} = 2.808$ AU, $\eta_1 = 0.111\,262$, $\eta_2 = 2.15$, $\eta_3 = 5.093$, $\eta_4 = 4.6142$.

The values of A_1, A_2 and A_3 are determined observationally for individual comets. The value of A_1 is generally much larger than A_2 and A_3, because most of the gas is released by the portion of a comet near the subsolar point, and escapes roughly normal to the surface. However, in the standard model, the radial portion of the nongravitational force is symmetric about perihelion, and this component of the force therefore returns to the comet's orbit after perihelion the same amount of orbital energy as it removes from the comet's orbit prior to perihelion. Forcing in the tangential direction is a consequence of the thermal lag between cometary noon and the time of maximum outgassing, and thus does not suffer from inbound/outbound cancellation. Instead, according to the standard theory, the tangential component of the nongravitational force produces a component of acceleration along the direction of the comet's motion if the comet rotates in the prograde direction (Problem 10.6). Such an acceleration increases the comet's orbital energy, and thus increases its orbital period. The situation is reversed for comets with retrograde rotation. Thus, even though A_2 is substantially smaller than A_1, the tangential term is the most important term in the standard theory. For most comets, forces normal to the orbit are small in magnitude, and as a result of symmetries they do not produce any secular effects, so the last term in equation (10.1) is generally ignored or found to be negligible in size.

The values of A_1 and A_2 are fit over both long and short intervals of time, and the solutions are not necessarily constant. Roughly half of the comets with well-studied orbits have values of A_2 that are nearly constant or vary slowly over many apparitions. However, the standard theory clearly breaks down for those comets which show more rapid variations in A_2. Nonuniform distribution of active areas on the surface of a comet can lead to pronounced

[2] Relativistic effects are significant for comets with small perihelion distances, and are included in most high-accuracy orbital calculations.

seasonal effects, as individual vents become more or less active. In such cases, not only is the standard theory inapplicable, but the radial component of the nongravitational force can become far more important than the tangential component (Problem 10.7).

The dynamics of comets is clearly intertwined with their evolution as physical objects. Cometary activity depends strongly on heliocentric distance. Outgassing, in turn, changes the orbits of comets, albeit in a less profound manner.

10.3 Gaseous Coma

In this and the following three sections, we describe the various parts of an active comet, and then we conclude this chapter with a discussion of cometary origins.

10.3.1 Solar Heating

It is generally accepted that cometary activity is triggered by solar heating, as exemplified by the fact that comets are usually inert when they are at large heliocentric distances, and only start to develop a coma when they get closer to the Sun. A cometary nucleus is covered with ice, which sublimates (evaporates directly from the solid state) when the comet approaches the Sun. When the sublimating gas *evolves off* the surface, dust is dragged along. The gas and dust form a comet's coma, and hide the nucleus from view. A comet is usually discovered after the coma has formed, when it is bright enough to be seen with relatively small telescopes. Many comets are still inert when they cross Jupiter's orbit, although some show activity at distances even beyond Uranus's orbit. At one extreme, P/Tempel 2, with an orbital period of 5.3 years and perihelion and aphelion distance of 1.4 AU and 4.7 AU respectively, is inert over a significant fraction of its orbit, while comets Shoemaker 1987o (C/1987 H1), Shoemaker 1984f (C/1984 K1), and Cernis (C/1983 O1) still had a dust coma/tail at distances over 20 AU. Comet P/Schwassmann–Wachmann 1, which orbits the Sun between 5.4 and 6.7 AU, is known to flare up substantially, sometimes by up to 8 magnitudes, and molecular gases (CO, HCN and CH_3N) were detected from C/Hale–Bopp (C/1995 O2) at a heliocentric distance $\gtrsim$ 6 AU.

Most comets brighten considerably when they get closer than 3 AU to the Sun. From the observed relationship between cometary brightness and heliocentric distance, it has been deduced that water-ice is the dominant volatile in most comets. Any activity at larger heliocentric distances, such as has been seen for example for C/Hale–Bopp (see Fig. 10.5 below), suggests the presence of small quantities of ices with lower vaporization temperatures. The activity, or 'outbursts', of some comets (especially those which are dynamically classified as new) at heliocentric distances well beyond the orbit of Jupiter, may result from the conversion of amorphous water-ice to crystalline ice, as discussed further in Section 10.6.4.

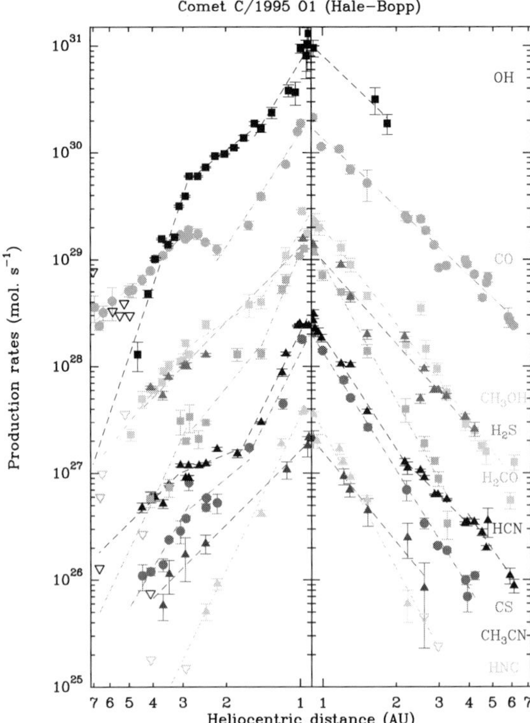

FIGURE 10.5 COLOR PLATE Time evolution of the observed production rates of C/Hale–Bopp, and fitted power laws (dashed lines) are plotted versus heliocentric distance. (Biver et al. 1999)

10.3.1.1 *Brightness*

The apparent brightness of a comet, B_ν, varies with heliocentric distance, $r_\odot$, and the distance to the observer, r_Δ, a behavior usually approximated by:

$$B_\nu \propto \frac{1}{r_\odot^\zeta r_\Delta^2}. \tag{10.3}$$

An inert object, such as an asteroid, has an index $\zeta = 2$, but comets usually show $\zeta > 2$, attributed to the fact that a comet's gas production rate increases with decreasing heliocentric distance. Although some comets indeed appear to follow a power law in $r_\odot$ over a large range of heliocentric distances, there are others which deviate significantly from a power law over many intervals in $r_\odot$. In Figure 10.5, we show the observed dependence of the gas production rate on heliocentric distance for a large number of molecules in C/Hale–Bopp. All molecules were observed at radio wavelengths. Power law fits to the production rates have been superposed. The pre-perihelion data have been split into three intervals. It is clear from the various curves that C/Hale–Bopp's brightness at $r_\odot > 3$ was dominated by the sublimation of gases more volatile than water. CO is the most volatile species, and showed a moderate increase in production rate at large heliocentric distances inbound, with $\zeta = 2.2$ at $r_\odot > 3$. This gas clearly must have been key to the observed brightness of the comet at these large heliocentric distances. Between 3 and 1.6 AU the production rate for CO actually decreased ($\zeta \approx 0$). It is possible that the CO in the upper layers of the nucleus had all been evaporated and that most of the solar energy went into the sublimation of water-ice, rather than heating deeper layers in the cometary crust. The CO production rate steepened considerably ($\zeta \approx 4.5$) at $r_\odot < 1.6$. Also the less volatile species CH_3OH, HCN, CH_3CN and H_2S were still overabundant compared to water (or OH) at $r_\odot > 3$. At a heliocentric distance of 3 AU, OH became the most abundant species, confirming that the composition of the comet is dominated by water-ice ($H_2O \rightarrow OH + H$). At $r_\odot < 1.5$ all species show a dramatic increase in production rates, with $\zeta \approx 4.5$ pre-perihelion, and $\zeta \approx 3.4$ post-perihelion. In addition we note that the ratios of H_2CO, CS, and HNC relative to HCN increased with decreasing heliocentric distance at $r_\odot > 1.5$, suggestive of an origin in the coma rather than the nucleus, through e.g., evaporation of grains and/or chemical reactions between species.

The brightness of a comet is generally expressed as the apparent magnitude at visual wavelengths, m_v:

$$m_v = -2.5 \log B_v = M_v + 2.5\zeta \log r_{AU} + 5 \log r_{\Delta,\mathrm{AU}}, \quad (10.4)$$

where $r_{\Delta,\mathrm{AU}}$ is the geocentric distance in AU, and M_v is the comet's absolute magnitude, which is equal to the apparent magnitude if the comet were at 1 AU from both the observer and the Sun. The maximum brightness of a comet is usually reached a few days after perihelion, and the brightness variation shows asymmetries between the branches before and after perihelion (the latter is clearly shown in Fig. 10.5). Pronounced differences have been observed in the brightening of old and new comets. Dynamically new comets often brighten slowly on their inbound journey, beginning at large heliocentric distances (at $r_\odot \gtrsim 5$ AU, with $\zeta \approx 2.5$). In contrast, most short-period comets do not brighten up much at large distances, but may 'flare up' when they get closer to perihelion (on average $\zeta \approx 5$). Generally, $\zeta = 4$ is adopted, and the correspondingly derived estimate of the comet's absolute magnitude is denoted as H_{10}.

10.3.1.2 *Gas Production Rate*

Assuming only solar heating, we can write the energy losses from the surface of the cometary nucleus as (Section 3.1.2):

$$(1 - A_b)\frac{\mathcal{F}_\odot e^{-\tau}}{r_{AU}^2}\pi R^2 = 4\pi R^2 \epsilon_{ir} \sigma T^4 + \frac{Q L_s}{N_A} + 4\pi R^2 K_T \frac{T}{z}. \quad (10.5)$$

The term on the left is the energy received from the Sun, with A_b the Bond albedo, $\mathcal{F}_\odot$ the solar constant, τ is the optical depth of the coma, r_{AU} the heliocentric distance in AU, and R the radius of the comet, which is assumed to be spherical. The first term on the right side represents losses caused by thermal infrared reradiation (cf. Section 3.1.2), and the last two terms represent losses due to sublimation of ices and to heat conduction into the nucleus, respectively. The infrared emissivity, ϵ_{ir}, is close to unity for most ices. The symbol L_s represents the latent heat of sublimation per mole, and N_A is Avogadro's number. Heat conduction into the surface is represented by the last term on the right, with K_T the thermal conductivity; this term is usually very small for comets and is therefore often ignored. Inside $r_\odot < 3$ AU the gas production rate, $\mathcal{Q}$ (molecules s^{-1}), for water molecules is usually adequately approximated by:

$$\mathcal{Q} \approx \frac{1.2 \times 10^{18} \pi R^2}{r_{AU}^2}. \quad (10.6)$$

This equation was derived under the assumption that $A_b = 0.1$. A better estimate for the mean variation in $\mathcal{Q}$ with heliocentric distance is, however, given by the form of equation (10.3), as exemplified by the power law fits superposed on Figure 10.5. For most comets, however, the scatter is so large that the simple form given by equation (10.6) is usually considered to be adequate.

Assuming a certain outflow velocity for the gas (e.g., thermal expansion, Section 10.3.2), both the temperature and coma density can be determined independently. We consider two extreme situations: (a) At large heliocentric distances the energy going into sublimation is negligible, and the temperature of the coma can be determined from a balance between insolation and reradiation (equilibrium temperature). (b) At small heliocentric distances, all energy is used for evaporation, and the evaporation rate varies as $r_\odot^{-2}$. The gas coming off the nucleus initially has the temperature of the surface. This is not the equilibrium blackbody temperature, since escaping gas carries away much of the heat as latent heat of sublimation. The temperature is determined by the gas that controls the evaporation. If outgassing is dominated by a single gas, and if the flux received on Earth is proportional to the number of (parent) molecules in the coma, the production rate and latent heat of the parent material can be deduced. If the sublimation is controlled by CO, the surface temperature of the nucleus is between 30 and 45 K at $0.2 < r_{AU} < 10$; for CO_2 it will be between 85 and 115 K; for H_2O it varies from 210 K at 0.2 AU, to 190 K at 1 AU, and to 90 K at 10 AU.

Since the side of the comet facing the Sun is hotter than the anti-sunward side, gas evolves predominantly from the sunward side. Images of P/Halley obtained with the Giotto spacecraft show clear jets coming off the nucleus in the direction of the Sun (Fig. 10.6a). HST images of C/Hyakutake (C/1996 B2) also clearly show most activity to happen at the sunward side (Fig. 10.6b). However, the subsolar/afternoon region does not always have to be brightest. Ices may appear as patchy spots on a cometary nucleus, and cometary activity can only occur above the icy spots. The asymmetric outgassing of a comet gives rise to nongravitational (jetting) forces, which distort a comet's orbit, as discussed in Section 10.2.3.

(a)

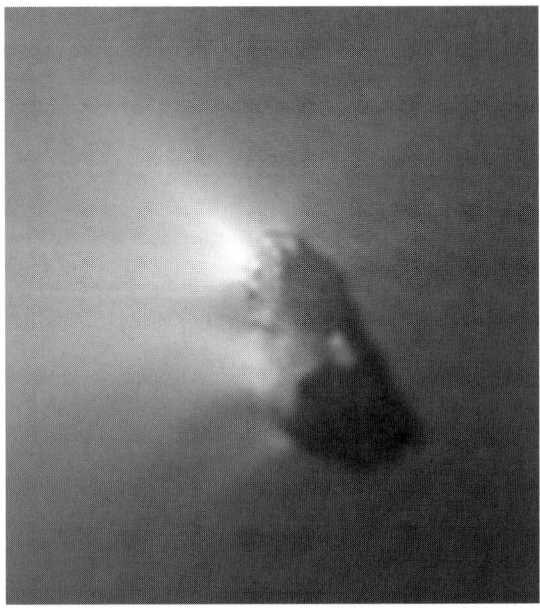

(b)

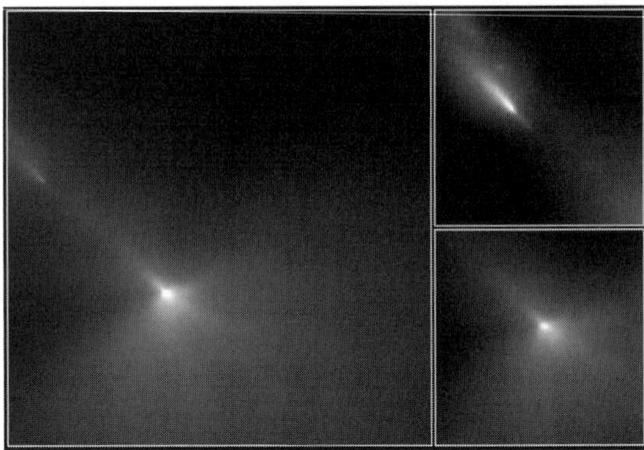

FIGURE 10.6 (a) The nucleus of P/Halley photographed by the Giotto spacecraft from a distance of $\sim$600 km. The resolution in this composite image varies from 800 m at the lower right to 80 m at the base of the jet at the upper left. The nuclear dimensions are $\sim 16 \times 8$ km. The jets flow in the sunward direction. (Courtesy: Halley Multicolor Camera Team; ESA) (b) Hubble Space Telescope images of the near-nucleus region of C/Hyakutake. The sunward and tailward directions are at approximately the 4 o'clock and 11 o'clock positions, respectively. A full-field view is shown on the left, an image that is 3340 km across, which shows that most of the dust is being produced on the sunward-facing hemisphere of the comet. Also at upper left are three small pieces which have broken off the comet and are forming their own tails. A close-up view of the nucleus region is shown in the bottom right, and of the comet fragments at the top right. (Courtesy: H.A. Weaver, HST Comet Hyakutake Observing Team, and NASA)

10.3.2 Outflow of Gas

The gas comes off the nucleus at the thermal expansion velocity, v_o, where:

$$\frac{1}{2}\mu_a m_{amu} v_o^2 = \frac{3}{2}kT, \tag{10.7}$$

with $\mu_a m_{amu}$ the molecular mass. Typical expansion velocities are ~ 0.5 km s^{-1} near $r_\odot \sim 1$ AU, which is much larger than the escape velocity from a comet (Problem 10.12). The coma is usually well developed at heliocentric distances $r_\odot \lesssim 3$ AU. Since the equilibrium temperature for a body at $r_\odot \approx 3$ AU is roughly equal to the sublimation temperature of water-ice, the appearance of a well-developed coma at smaller heliocentric distances has been taken as evidence that the volatile component of most comets consists primarily of water-ice.

The gas released from the nucleus expands into a near vacuum and rapidly reaches supersonic velocities. The *terminal velocity*, which is the velocity of the gas at large distances from the comet, is usually reached within $\sim$1000 km from the nucleus. The flow of gas can be calculated using the continuity equations for mass, momentum and energy (Section 7.1.3), including sources and sinks of these quantities. Numerical calculations show that terminal velocities of parent molecules are usually between 0.5 and 2 km s^{-1} at heliocentric distances near 1 AU, in agreement with observed velocities. Since the gas density in the coma decreases approximately as the square of the distance from the nucleus, the flow becomes a collisionless free molecular flow rather than a hydrodynamic flow at the collisional radius, $\mathcal{R}_c$, which in the kinematic sense can be defined as the location at which an outward-moving particle has a 50% chance of escaping to infinity without colliding with another particle:

$$\int_{\mathcal{R}_c}^{\infty} \frac{Q\sigma_x}{4\pi r^2 v_o} dr = 0.5, \tag{10.8}$$

where r is the cometocentric distance, and σ_x the collisional cross-section.

Radiative cooling processes and heating of abundant molecules through photolysis change the energetics in a comet's coma. The gas flow is usually not adiabatic (cf. Section 3.2.3.2), as both the gas temperature and outflow velocities are modified by collisions with particles (i.e., OH radicals and fast H atoms). Observations and numerical calculations show that the outflow velocity of heavy molecules starts to increase significantly at nuclear distances $\gtrsim$ 10^3–10^4 km. For a Halley-like comet at $r_{AU} \approx 1$, the outflow of gas can be described by hydrodynamical theory up to a distance of ~1000 km from the nucleus. Collisions are still important at large distances from the cometary nucleus; a free molecular flow is not reached until $\gtrsim 5 \times 10^4$ km from the nucleus (Problem 10.13). The presence of dust adds an extra complication to the models. Dust slows down the gas outflow in the first few tens of meters above the nucleus through friction on the outflowing gas and energy exchange between the gas and dust. There is increasing evidence (Section 10.3.5) that dust is a source of gas (e.g., CO, H_2CO, HCN, CN), and that the chemical composition of the gas may be changed by the presence of dust (e.g., recondensation of H_2O molecules onto dust grains). These effects have not yet been taken into account in coma models.

10.3.3 Photodissociation and Chemical Evolution

When a comet approaches the Sun, ice starts to sublimate and molecules evolve off the surface. The coma, except for the inner ~100 km, stays optically thin at most wavelengths. Hence, essentially all molecules in the coma are irradiated by sunlight at visible and UV wavelengths. The mean lifetimes of the molecules and radicals against dissociation and ionization, therefore, varies with $r_\odot^2$. The lifetime, t_ℓ, of a molecular species is thus given by:

$$\frac{1}{t_\ell} = \frac{1}{r_{AU}^2} \int_0^{\lambda_T} \sigma_x(\lambda)\mathcal{F}_\odot(\lambda)d\lambda, \tag{10.9}$$

where $\mathcal{F}_\odot(\lambda)$ is the solar flux at 1 AU in the wavelength range $(\lambda, \lambda + d\lambda)$, $\sigma_x(\lambda)$ is the photodestruction cross-section, and λ_T is the threshold wavelength for photodestruction (which is usually in the UV). The solar flux is well known, but the cross-section against photodestruction is not always known.

The primary constituent of a cometary nucleus is water-ice. The typical lifetime of water molecules at $r_\odot \approx 1$ AU is ~ 5–8 × 10^4 s. The main (~ 90%) initial photodissociation products of water are H and OH. The OH molecules receive an average excess speed of 1 km s^{-1} in the dissociation process, and the H atoms ~18 km s^{-1}. Other dissociation products are excited oxygen (roughly 5%), H_2O^+, OH^+, O^+ and H^+. A small number of H_2O molecules dissociate inside the collisional radius. The OH radicals produced in this region are quickly thermalized by numerous collisions with other molecules; the much lighter H atoms are not thermalized.

A significant fraction of OH is dissociated into O and H by solar Lyman α photons (1216 Å). The typical timescale for this process at $r_\odot$ = 1 AU is ~1.6–1.8 × 10^5 s, and the H atoms receive an excess velocity of 7 km s^{-1}. The H atoms form a large *hydrogen coma*, several ×10^7 km in extent. Hydrogen is lost either by photoionization or by charge exchange reactions with solar wind protons. Observations show that the latter is the more efficient of the two processes. Since charge exchange reactions depend on the solar wind properties, which vary with time (Section 7.1.1), lifetimes for H vary significantly, from ~ 3 × 10^5 to ~ 3 × 10^6 s.

To derive gas production rates from cometary observations, one needs models of the outflow of the gas and the time evolution (e.g., place of origin, dissociation) of the individual molecular species. The simplest and most widely used model is the so-called *Haser* model. This model assumes isotropic radial outflow and a finite lifetime of the molecules. The density distribution for *parent* molecules, which evolve directly off the nucleus, can be written:

$$N_p(r) = \frac{Q_p}{4\pi r^2 v_p} e^{-r/\mathcal{R}_p}, \tag{10.10a}$$

where N is the number density, and $\mathcal{R}_p = v_p t_{\ell p}$ is the scale length, with $t_{\ell p}$ the lifetime of the parent molecules. The subscript p stands for parent molecules. If the *daughter* molecules, i.e., the molecules produced upon dissociation (subscript d), continue to move radially outwards like their parents with $v_d = v_p$, then the distribution for daughter molecules is:

$$N_d(r) = \frac{Q_p}{4\pi r^2 v_d} \frac{\mathcal{R}_d}{\mathcal{R}_d - \mathcal{R}_p} \left(e^{-r/\mathcal{R}_d} - e^{-r/\mathcal{R}_p} \right) \tag{10.10b}$$

It is more likely, however, that dissociation products from parent molecules are ejected isotropically in the parent's frame of reference. As a result, not all molecules move radially outwards; some radicals actually move inwards, towards the nucleus. Obviously, the resulting velocity distribution of daughter molecules in such a model may differ substantially from the Haser model. The Haser model is generally a good approximation if a large fraction of the radicals is formed inside the collisional regime (short parent lifetime, high total outgassing). However, if a large fraction of the radicals is formed outside the collisional regime (low total outgassing, long parent lifetime), the velocity distribution of the daughter molecules will noticably influence the shape of 'observed' spectral lines. This situation can be better modeled by a random walk (Monte Carlo or vectorial) model. Figure 10.7 shows observed line profiles for a parent and a daughter molecule: HCN at a wavelength of 3 mm, and OH at a wavelength of 18 cm. On the OH line, profiles based upon a Haser and a random walk model are superposed. It is clear that the depression in OH emission predicted by the Haser outflow of gas is not observed; the observed emission typically increases towards the line center, as expected for a random walk model, rather than towards the edges of the line. In contrast, the HCN profile agrees well with a Haser model. Note that each hyperfine line is split; the stronger blueshifted component in each line suggests a higher gas production rate from the side facing the observer, which is also the side facing the Sun, and is thus hotter than the other side.

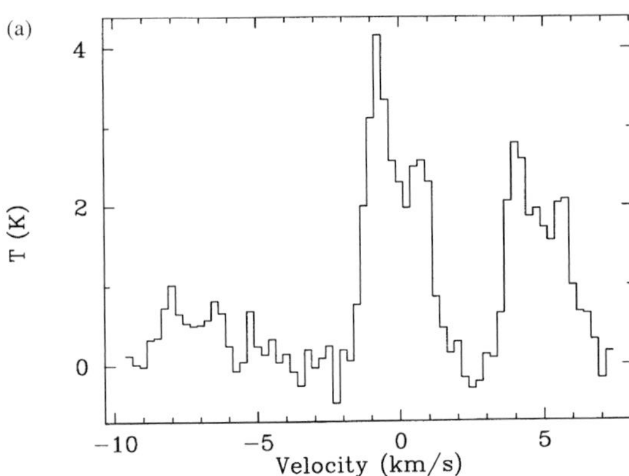

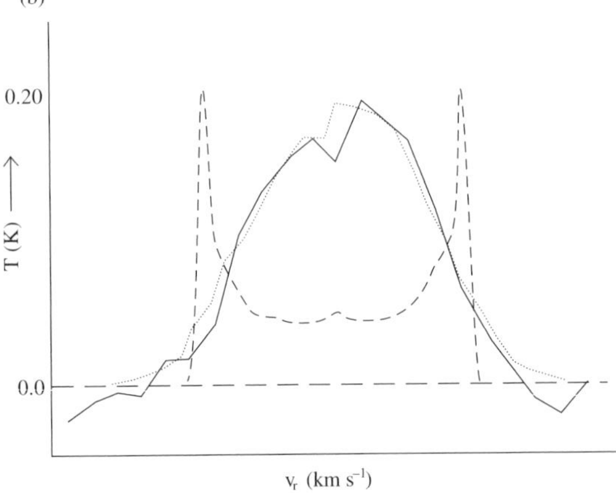

FIGURE 10.7 (a) HCN spectrum of C/Hale–Bopp, obtained with the BIMA array at a frequency of 89 GHz. The spectrum was taken in an $11'' \times 8''$ beam at the peak of the HCN emission on 3 April 1997. (Wright *et al.* 1998) (b) Observed OH line profile (18 cm) of C/Austin (1982g) (solid line), with superposed calculated profiles based upon the Haser model (dashed line) and the random walk (vectorial) model (dotted line) for cometary outflow. (Adapted from Bockelee-Morvan and Gerard 1984)

10.3.4 Excitation and Emission

The composition of a cometary coma can be determined from spectra (Fig. 10.8) and *in situ* mass spectrometer measurements. Most atoms and molecules in the coma are in the electronic ground state. Whenever an atom or molecule is excited, it usually falls back to a lower energy state through the emission of a photon. The main emission mechanism is *fluorescence*: absorption of a solar photon excites an atom or molecule, which is followed by spontaneous emission in a single- or multi-step decay process. Other emission mechanisms are collisional excitation (with neutrals, ions or electrons), and dissociation (radiative or collisional) of (parent) molecules leaving the observed radicals in an excited state.

10.3.4.1 *Fluorescence*

The strongest lines in cometary spectra are *resonance transitions*, i.e., transitions between the ground state and first excited level of an atom/molecule. The brightness of an emission line is determined by the number of molecules multiplied by the *g-factor* or emission rate per molecule. If most or all molecules are in the electronic ground state and excitation is caused by simple fluorescence, one can calculate the absorption rate of solar photons which excite the molecule, g_a, and multiply this by the appropriate branching ratio for the particular transition at wavelength

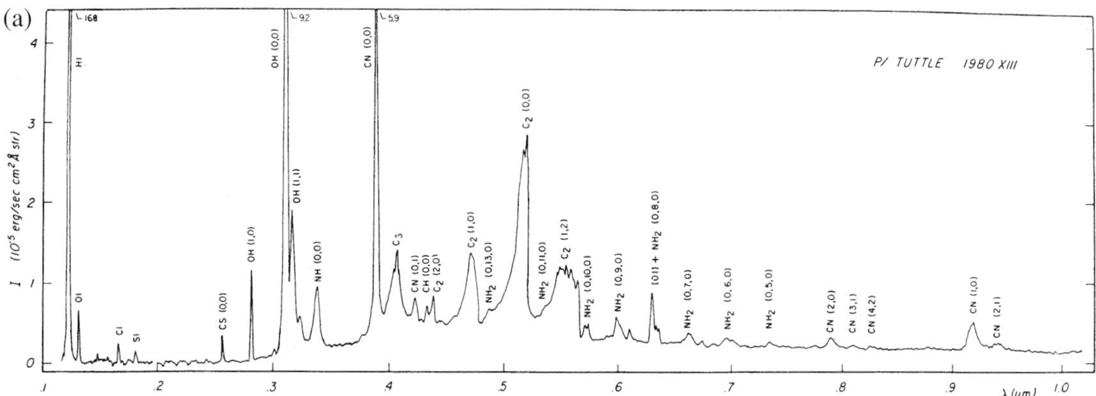

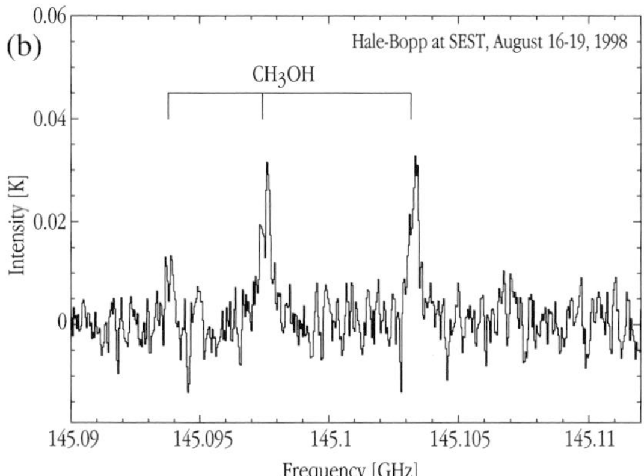

FIGURE 10.8 (a) A spectrum between 0.1 and 1.0 μm wavelength from P/Tuttle. Line identifications are indicated on the spectrum. (Courtesy: S. Larson) (b) Part of a radio spectrum with emission from CH_3OH molecules in the coma of C/Hale–Bopp, as observed with the 15-m SEST telescope at La Silla from 16 to 19 August 1998, when the comet's heliocentric distance was 6 AU . Three lines of this molecule were detected at 145.0938, 145.0974 and 145.1032 GHz. The total integration time is close to 12 hours. The intensity is indicated in units of antenna temperature. During the same observing run, emissions from CO at 230 GHz and HCN at 89 GHz were also detected. The CO production rate was 2.4×10^{28} mol s^{-1}; those of HCN and CH_3OH were about 200 and 20 times smaller, respectively. CO is clearly the primary driver of activity at this distance. The observations of these two organic species constitute the most distant detections ever made in any comet. It is possible that the detection of these two molecules was caused by an outburst (see also Fig. 10.5, and Section 10.3.5). (Courtesy: M. Gunnarsson, ESO, and Biver *et al.* 1999)

λ_{ul} in order to determine g:

$$g = g_a \frac{A_{ul}}{\sum_{j \le u} A_{jl}}, \qquad (10.11a)$$

$$g_a = \frac{B_{lu} \mathcal{F}_\odot(\lambda_{lu})}{r_\odot^2}, \qquad (10.11b)$$

where B_{lu} and A_{jl} are the Einstein coefficients (cf. Section 3.2.2.3) for transitions from the ground states and excited states, respectively, with u the upper, l the ground level and $l < j \le u$. The solar radiation density $\mathcal{F}_\odot(\lambda_{lu})$ is measured at Earth's orbit at the wavelength of absorption λ_{lu}. Note that in the case of pure resonance, $g = g_a$; in the case of resonance fluorescence, the de-excitation of the upper level may occur via more than one transition, and $g < g_a$. For many molecular species, several electronic levels are populated and calculation of the g-factors is quite complicated. Once the g-factor or emission rate for a given transition is known, the gas column density of the emitting molecules, N_c, can be obtained from the observed brightness:

$$N_c = \frac{B_\nu}{g} \frac{4\pi}{\Omega_s}, \qquad (10.12a)$$

with B_ν the brightness of the comet as measured by the observer in photons s^{-1} cm^{-2}, and Ω_s the solid angle over which radiation is received. If excitation is by simple resonance fluorescence, the gas production rate of the molecular species is simply related to the brightness:

$$Q = \frac{4\pi r_\Delta^2 B_\nu}{g t_\ell}, \qquad (10.12b)$$

with t_ℓ the lifetime of the emitting molecules. Note that the product $g t_\ell$ does not depend on the heliocentric distance. If the excitation is not simply resonance fluorescence, the situation is more complex since many energy levels may be populated. In this case every excitation/de-excitation process has to be included in the equations.

10.3.4.2 *Collisional Excitation*

Since the densities in the coma are usually very low, collisions between neutrals are generally not important, except very close to the nucleus. A typical cross-section, σ_x, for a neutral–neutral collision is proportional to a molecule's size, R: $\pi(R_1 + R_2)^2 \approx 10^{-15}$ cm^2. The timescale between collisions, t_c, is then given by:

$$t_c = \frac{1}{N\sigma_x v_o}, \tag{10.13}$$

with v_o the thermal velocity of the molecules and N the number density. Typical collision rates for molecules near the nucleus are of the order of 10 s^{-1} (Problem 10.16); collision rates are a factor of $\sim 10^4$ less at a cometocentric distance of 10^3 km. At 1 AU, a typical absorption rate of solar photons by water out of the electronic ground state is $\sim 10^{-3}$ s^{-1}. Hence, collisional excitation by neutrals is generally not important in a comet's coma, except very close to the nucleus. Collisions with ions or electrons are more significant, since the effective cross-sections are $\gtrsim 10^3$ larger than for neutrals, and the velocity of the electrons is much higher than that of neutrals or ions. Thus, collisional excitation in the inner coma, $\lesssim 10^3$ km from the nucleus, can be important. The energy available in a collision is the kinetic energy of the molecule. With a thermal velocity of 0.1 km s^{-1}, a water molecule may transfer 1.5×10^{-15} erg, or less than 0.001 eV. This amount of energy is not enough to excite vibrational or electronic transitions, but it can excite rotational transitions, which are observable at radio wavelengths.

10.3.4.3 *Radiation Pressure*

Radiation pressure from the Sun acts on molecules via absorption and re-emission of solar photons. The net effect is a tailward displacement of the coma. The acceleration of an atom/molecule caused by radiation pressure is:

$$\frac{dr_\odot^2}{dt^2} = \frac{h}{\mu_a m_{amu} r_\odot^2} \sum_i \frac{g_i}{\lambda_i}, \tag{10.14}$$

where h is Planck's constant, $\mu_a m_{amu}$ the molecular mass, g_i the g-factor for transition i and λ_i the wavelength of absorbed photons. For atomic hydrogen, Lyman α is the most important transition, which leads to a displacement of the hydrogen cloud by $\sim 10^6$ km. In the case of an isotropic uniform outflow of gas, a displacement of this magnitude should be visible in the data. However, comets generally display large anisotropies in their gas outflows, which makes it difficult to attribute any displacement to a particular process.

10.3.4.4 *Swings Effect*

Although the timescales involved in the various excitation processes strongly argue for fluorescence as the main emission mechanism, even stronger evidence for fluorescence is given by the *Swings effect.* As the comet orbits the Sun, its heliocentric velocity changes. The solar spectrum consists of countless *Fraunhofer absorption lines*, created in the cool photosphere overlying hotter parts of the Sun. In the comet's frame of reference, the solar spectrum is Doppler shifted, and the Fraunhofer absorption lines move in and out of the excitation frequencies. Hence, the g-factor (eq. 10.11) depends on the radial components of a comet's heliocentric velocity! There is an excellent correlation between the relative strength of cometary OH lines at UV wavelengths and the Fraunhofer absorption line spectrum, Doppler shifted to the reference frame of the comet. A cometary line is strong if at the appropriate Doppler shifted frequency the solar radiation is not weakened by absorption effects in the Sun's own photosphere. The cometary line is weak or absent if the solar intensity at the exciting frequency is weak or zero.

The atoms/molecules within a comet's coma have certain velocities with respect to the nucleus; hence their heliocentric radial velocity is slightly different from that of the nucleus. Thus, in the reference frame of each molecule the solar spectrum is Doppler shifted by slightly different amounts. This leads to a modification of the Swings effect, generally referred to as the *Greenstein effect.* The Greenstein effect may cause a certain side of a comet's coma to be brighter, which may manifest itself as an asymmetry in a spectral line profile (Problem 10.17).

10.3.5 Composition

The composition of a cometary coma can be determined from spectral line measurements of molecules and radicals. In addition, the composition of both the gas and dust grains has been measured *in situ* for P/Halley by ion mass spectrometers on board the Giotto and Vega spacecraft. However, identification in mass spectrometer data is somewhat hampered by the fact that masses of different species overlap, in particular in instruments of moderate resolution. A peak at $\mu_{amu} = 28$, for example, contains CO, N_2 and C_2H_4 as possible main constituents. It appears that, on average, the overall cometary composition (gas + dust) is, within a factor of two, similar to solar values, except for noble gases, hydrogen (deficient by a factor of ~ 700) and nitrogen (deficient by a factor of ~ 3). As all known meteorites are strongly deficient in all volatile (CHON) materials compared to solar values (Section 8.1),

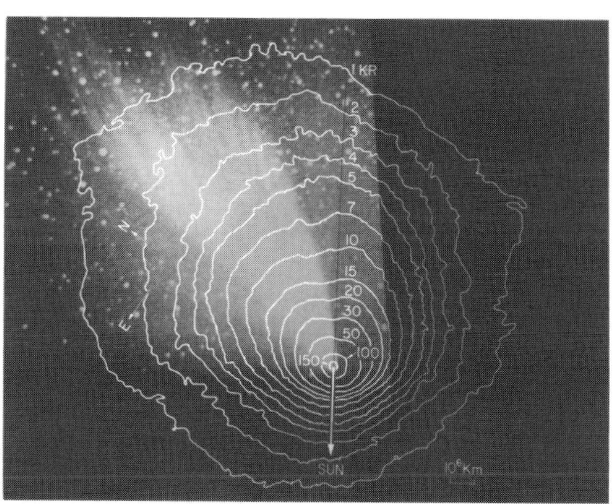

FIGURE 10.9 Isodensity contours of the hydrogen cloud (in Ly α) of C/West superposed on a photograph of the comet. The Lyman α emission was observed by Opal and Carruthers (1977), while the visible picture of the comet was taken by S. Koutchmy on the same day. (Courtesy: Koutchmy; Figure from Fernandez and Jockers 1983)

cometary dust can be considered as the most 'primitive' material ever sampled.

10.3.5.1 *Water Derived Atoms/Molecules*

Water. Although it was long surmised that H_2O is the dominant volatile species in a cometary nucleus, water itself was not detected until December 1985, when it was observed via infrared spectroscopy from the Kuiper Airborne Observatory (KAO) on P/Halley. Ten lines of the ν_3 band were detected in emission, in agreement with models for solar infrared fluorescence from rotationally relaxed cometary water. The water production rate was found to vary from day to day, in concert with visible lightcurves. Tentative detections have been reported of the 22 GHz water line in C/IRAS–Araki–Alcock (C/1983 VII) and C/Hale–Bopp, while several vibrational bands have been detected for C/Hyakutake and C/Hale–Bopp at infrared wavelengths using ground-based telescopes.

The ions H_2O^+ and H_3O^+ have been detected in many comets. It is interesting to point out that H_2O^+ emits in the red part of the spectrum, and sometimes reddish ion tails have been seen.

Hydrogen. Lyman α (at $\lambda = 1216$ Å) images of comets reveal the coma to be surrounded by a huge atomic hydrogen cloud. The H cloud produced by C/West (C/1976 VI) is shown in Figure 10.9. The Ly α is indicated by isodensity contours, superimposed on a photograph of the comet at visible wavelengths. The extent of the hydrogen cloud is roughly 10^7 km. Its asymmetric shape may be the result of solar radiation pressure. Models of the outflow velocity suggest the existence of two velocity distributions, one with $v \approx 7$ km/s, and one with $v \approx 20$ km s^{-1}, in agreement with excess velocities expected from the dissociation of OH and H_2O (Section 10.3.3). The extent of the cloud suggests that the lifetime of H atoms is determined by charge exchange reactions with solar wind protons.

Oxygen. Several transitions of atomic oxygen have been detected, including the red (6300 Å, 6364 Å), green (5577 Å) and UV (2973 Å) forbidden lines[3]. Most oxygen atoms originate from the dissociation of water or OH. About 5% of the H_2O (and OH) molecules produce oxygen atoms in an excited state, which radiate the forbidden red lines. Dissociation of CO_2 produces oxygen atoms in an excited state which radiate the forbidden green and UV lines. In contrast to fluorescence emission, this prompt emission occurs only once for each of these O atoms; thus, observations of these lines yield information on the atom's parent molecule. The forbidden lines are only observed in the inner coma, where dissociation of H_2O and CO_2 is expected.

Hydroxyl. The OH radical has been observed extensively at UV ($\sim$ 3000 Å) and radio (1665 and 1667 MHz) wavelengths. Solar photons excite the OH molecules, which upon de-excitation exhibit a number of UV lines around 3000 Å. The relative strengths of the various lines clearly exhibit the Swings and Greenstein effects. The rotational level of the ground state is split into two levels, each of which is split again by hyperfine structure effects. Four lines can be observed at radio wavelengths, near 18 cm, the most intense of which are the 1665 and 1667 MHz transitions. The intensity of the lines depends strongly on the heliocentric radial velocity of the comet (Swings and Greenstein effects). Usually these ground state levels would be populated in proportion to their statistical weights (Section 3.2.2). Since the transition between the lines is highly forbidden, a strong population inversion can build up (Fig. 10.10). The 3 K cosmic microwave background radiation can then induce a maser action, i.e., stimulate emission. The OH molecule is not excited when Fraunhofer absorption lines are Doppler shifted into the excitation frequency. In this case, the ground state levels of OH become anti-inverted, and the molecules absorb photons from the 3 K background radiation. Thus, depending on the heliocentric velocity, OH is seen either in emission (maser) or in absorption against the galactic

[3] Atoms may be excited to metastable states. Decay to lower energy levels from such states is said to be 'forbidden', because on Earth such atoms are collisionally de-excited. The spontaneous decay times for 'forbidden' transitions are very long: seconds to thousands of years. In the interplanetary medium where collisions are rare, the atoms eventually decay spontaneously, resulting in forbidden line emissions.

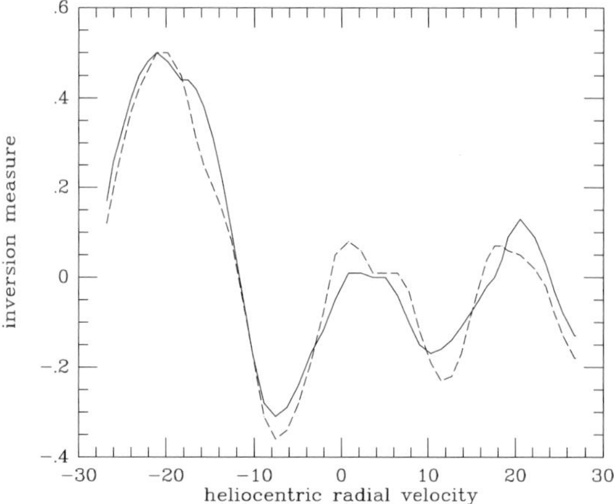

FIGURE 10.10 A graph of the inversion measure, $(n_u - n_l)/(n_u + n_l)$, for the OH transition at 1667 MHz, as a function of a comet's heliocentric velocity. When the inversion measure is positive, OH is seen in emission (maser); when the inversion measure is negative, OH is observed in absorption. The result for two pumping models are shown: ——, Despois *et al.* (1981) and – – – Schleicher (1983). (Adapted from de Pater *et al.* 1989)

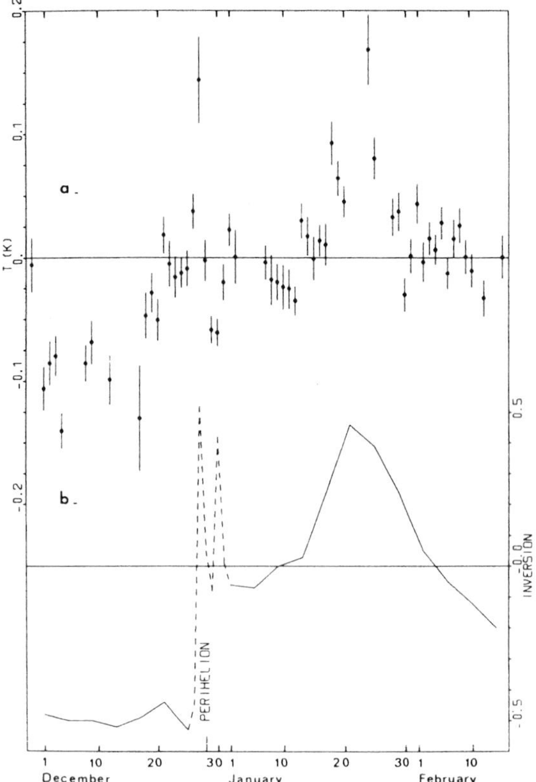

FIGURE 10.11 A comparison between radio OH data of C/Kohoutek in 1973–1974 (a) with the inversion measure predicted by the ultraviolet pumping model as a function of time (b). (Biraud *et al.* 1974)

background. A comparison of the population inversion expected from the solar Fraunhofer line spectrum with OH observations is shown in Figure 10.11. The agreement is striking, though not perfect. Deviations from the calculated values can be explained by insufficient knowledge of the Fraunhofer absorption spectrum, Greenstein effects, asymmetric outgassing of the comet and fluctuations in the gas production rate.

The population of the ground state energy levels is also sensitive to collisions. In the inner coma, inside the collisional radius, the transitions are 'quenched': the OH molecules are thermalized via collisions, and the maser activity ceases. This effect should show up as a 'hole' in the OH distribution around the coma. This hole may have been observed in high resolution radio images of P/Halley (Fig. 10.12).

10.3.5.2 *Carbon Compounds*

Observed emission lines from carbon species at visible and UV wavelengths are numerous (e.g., C, C_2, C_3, CH, CH_2, CN, CO and associated ions). Rather than giving an exhaustive list of carbon species detected in comets, we focus on the parent material which produces the various observed species. Likely candidates, in addition to dust grains, are carbon dioxide (CO_2), hydrogen cyanide (HCN), methane (CH_4) and more complex molecules such as formaldehyde (H_2CO) and methanol (CH_3OH), all of which have been detected in at least several comets. Spacecraft observations of P/Halley suggested the CO_2 production rate to be roughly 3% that of water. A CO_2 abundance of a few percent is also suggested by observations of prompt emission in CO bands at UV wavelengths (Cameron bands) in several comets. The total CO production rate is typically 15–20% that of water; less than half appears to originate from the nucleus itself, while the remaining molecules come from an 'extended' source region, which could be dust grains or complex molecules in the coma.

The lifetimes of CO and CO_2 are roughly an order of magnitude larger than for water, so most of the molecules enter the region of free molecular flow. A large fraction of CO and CO_2 becomes ionized; the ions are accelerated and swept away by the solar wind. Both CO^+ and CO_2^+ ions have been observed in many comets. Emission from the CO^+ ions shows up as the prominent bright ion tail, with a blue color since the energy of the transition between the two lowest electronic levels corresponds to photons in the blue part of the visible wavelength range.

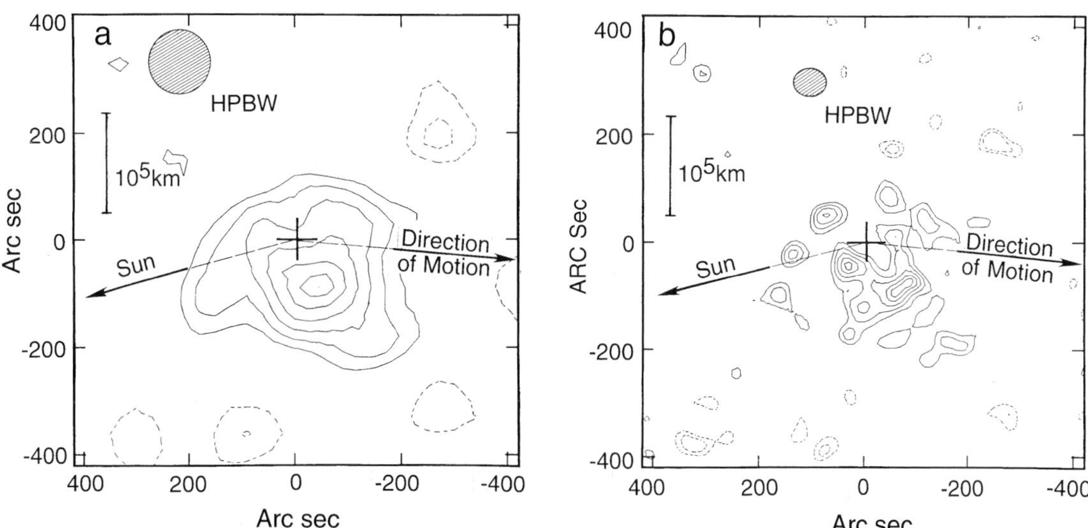

FIGURE 10.12 Radio image of the OH emission of P/Halley, taken at the peak flux density of the line (0.0 km s^{-1} in the reference frame of the comet). The cross in the center indicates the position of the nucleus. The shaded ellipse represents the resolution of the beam. Contour levels for the low-resolution image (a) are 4.9, 7.8, 10.8, 13.7, 16.7 and 18.6 mJy/beam; for the high resolution image (b) they are: 4.4, 6.0, 7.7, 9.3, and 10.4 mJy/beam. Dashed contours signify negative values (which are unphysical). (de Pater *et al.* 1986)

As discussed further in Section 10.7, the (relative) abundances of CH_4, CO, C_2H_6 (ethane), and that of formaldehyde (H_2CO) and methanol (CH_3OH) may provide information on the conditions under which cometary ices formed. Although some of these species have only been detected in a few comets to date (in particular C/Hyakutake and C/Hale–Bopp), there is growing evidence that cometary ices have similar characteristics to interstellar ice. The abundance ratio CH_4/CO in the condensed phase in the interstellar medium is typically around unity, with variations up to a factor of $\sim$10, while the abundance ratios in the gas phase are much smaller, typically $\sim$0.001–0.01. The ratio CH_4/CO has been directly detected in C/Wilson (C/1986 I), C/Hyakutake and C/Hale–Bopp, and was typically found to be of the order $\gtrsim 0.1$. Upper limits for P/Halley and C/Levy (C/1990 XX) yield values of $\lesssim$ 0.1–0.2. Ethane (C_2H_6) and acetylene (C_2H_2) were detected in C/Hyakutake, C/Hale–Bopp and C/Lee (C/1999H1) with abundances of $\sim$0.2–0.4% that of water, very comparable to the abundance of methane gas. C_2H_6 abundances similar to methane gas abundances are inconsistent with production mechanisms in thermochemical equilibrium in the primitive solar nebula or a giant planet subnebula, where one typically expects the ratio $C_2H_6/CH_4 \lesssim 10^{-3}$. C_2H_6 can be produced by photolysis of CH_4-rich ice or from H atom addition reactions on grain surfaces to acetylene, condensed from the gas phase.

In the interstellar medium methanol is typically more abundant than formaldehyde, by a factor of a few tens. Formaldehyde and methanol have been detected in a growing number of comets with abundances typically up to a few percent that of H_2O. The source of formaldehyde is clearly distributed throughout the coma, much like that of CO. It could originate from dust grains or polymerized formaldehyde (POM). Presence of the latter molecules had been postulated to explain the heavy ion mass spectra in P/Halley detected with the ion mass spectrometer on board the Giotto spacecraft. The abundance ratio of methanol-to-formaldehyde may contain information on its place of origin. Abundance ratios less than the interstellar value might be expected if comets formed in the primitive solar nebula, since methanol can be converted to formaldehyde by photoprocessing. Although the H_2CO/CH_3OH ratio does vary from comet to comet, it usually varies by a factor of a few rather than a few tens. Some comets (C/Austin 1990 V, C/Hale–Bopp) have a higher methanol abundance ($\sim$5% of H_2O) than other comets (P/Halley, C/Hyakutake, C/Levy have $CH_3OH \lesssim 1\%$ of H_2O), whereas the reverse may be true for formaldehyde. Whether there is a clear difference in the ratio for comets which originated in the Kuiper belt as opposed to those which originated in the Oort cloud is not known. If there is a difference, it might reflect a varying degree of photoprocessing on comets.

Some large carbon-containing molecules have also been detected. The ion mass spectrometer on board Giotto detected molecules with molecular weights up to $\mu_a = 120$, while ground-based telescopes sensitive to (sub-)millimeter wavelengths detected molecules such as HNCO, HCOOH, CH_3CN, NH_2CHO and CH_3OCHO in Comet Hale–Bopp.

10.3.5.3 *Nitrogen Compounds*

The major potential reservoirs for volatile nitrogen are N_2, HCN and NH_3. Molecular nitrogen (N_2) cannot be detected directly, since there are no allowed vibrational or rotational transitions in the observable spectral range. Its presence at an abundance of 0.2% has been inferred from the detection of N_2^+. Ammonia gas readily dissociates: $NH_3 \rightarrow NH_2 + H \rightarrow NH + H + H$. Both NH and NH_2 have been detected in a number of comets, and suggest NH_3 production rates 0.3–1% that of H_2O. Ammonia gas has been observed directly in C/IRAS–Araki–Alcock, C/Hyakutake and C/Hale–Bopp, with typical production rates (relative to water) varying from 0.3% for C/Hyakutake, to ~1.5% for C/Hale–Bopp and 6% for C/IRAS–Araki–Alcock. These numbers are consistent with the above values, and suggest that the NH_3 production rate may vary substantially from comet to comet. Hydrogen cyanide has been observed and imaged directly in several comets. It seems to originate both on the nucleus and as a distributed source, i.e., part of it may originate from dust grains. Its production rate is 0.3–0.6% that of water. HNC has been observed in C/Hyakutake and C/Hale–Bopp, at levels roughly 5–6 times less abundant than HCN, at similar ratios to those seen in warm molecular clouds. Based upon the relative increase in the ratio HNC/HCN in C/Hale–Bopp with decreasing heliocentric distance, it has been suggested that a large fraction of HNC is produced in the coma through dissociative electron recombination of the ion $HCNH^+$.

Nitrogen is also present in dust particles (CHON, see Section 10.4) and in the form of CN. Although a possible progenitor for CN is HCN, the production rate of HCN appears to be less than half that of CN. Comet-to-comet variations in the production rate of CN are strongly correlated with the dust-to-gas ratio in comets, and the CN emissions have revealed the presence of jets and shells in more than one comet. About 20–50% of the CN is therefore thought to come directly from the dust grains. When adding up all the available nitrogen, the N/C and N/O abundance ratios seem to be depleted relative to the solar ratios by a factor of 2–3. A similar N/C depletion may prevail in the giant planets (Section 4.3.3.4), although indirect results from the Galileo probe on Jupiter suggest N/C $\approx$ solar. A depletion in the N/C ratio suggests that N was more volatile than C and O in the outer solar nebula where the comets, and perhaps the giant planets, formed.

10.3.5.4 *Sulfur Compounds*

No firm measurements on the overall sulfur abundance in comets exists, although estimates for P/Halley and C/Hale–Bopp suggest that the overall S/O ratio is slightly higher than the solar value. While UV transitions of the radicals S and CS have been observed in many comets, much less data exist on possible parent molecules. Since water is the most abundant volatile in cometary comae, and S and O are chemically quite similar, one might expect that a relatively large fraction of cometary sulfur is present in the form of hydrogen sulfide (H_2S). This molecule has now been detected at millimeter wavelengths in four bright comets: C/Austin, C/Levy, C/Hyakutake and C/Hale–Bopp with an abundance ~0.2–1% that of water. Other sulfur-bearing molecules which have been detected in C/Hale–Bopp at radio wavelengths are SO, SO_2, OCS, H_2CS and NS. Several detections of S_2 have been reported. Because S_2 has a very short lifetime (~450 s), it has not been possible to detect this species, except during very close comet encounters and more recently using very sensitive equipment. It is interesting to search for S_2 since it is a very low temperature condensate, and therefore puts interesting constraints on comet formation theories (Section 10.7.1). The first detection of S_2 was obtained in 1983 during an outburst in Comet C/IRAS–Araki–Alcock, a comet which came very close to the Earth (0.03 AU). This detection was followed in 1996 by a detection in C/Hyakutake, another comet which came exceptionally close (~0.06 AU) to Earth. In 1999 S_2 was detected in C/Lee using a new infrared spectrograph on the 10-m Keck telescope. It is now believed that S_2 is present in all or most comets, and that we have not been able to see it because the instrumentation has not been sensitive enough. The Vega 1 and Giotto spacecraft detected sulfur in the grains of P/Halley, at a typical ratio S/O ~0.08.

10.3.5.5 *Dust, Alkalies and Metals*

Sodium emission has been seen in several long-period comets which came to within ~1.4 AU of the Sun. The emission is seen as coming from a tail of sodium atoms, which can be quite extensive, such as has been measured for C/Hale–Bopp (Fig. 10.13). Since the boiling point for sodium is ~1150 K, the sodium atoms detected might be bound in complex molecules rather than in rocky material. Models of C/Hale–Bopp imply that the emitting sodium atoms originate in the vicinity of the nucleus, and, under

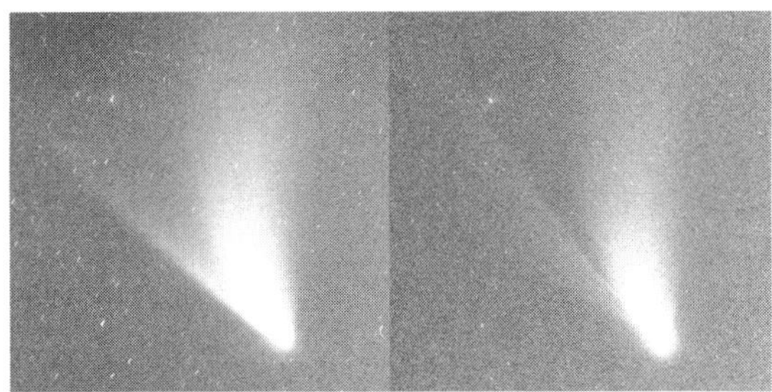

FIGURE 10.13 The thin straight sodium tail of C/Hale–Bopp stands out in the left image, which records the fluorescence (D-line) emission from sodium atoms. A traditional image of the plasma and dust tails is shown on the right. (Cremonese *et al.* 1997)

the assumption of cosmic abundances, about 0.1% of the atoms released have been observed in this 'sodium' tail.

Emission lines of calcium, potassium and metals have only been observed in *Sun-grazing comets*, which came to within 0.1 AU of the Sun. At $r_{AU} < 0.1$–0.2, potassium lines show up, and at even smaller heliocentric distances emission lines of Fe, Ni, Co, Mn, V, Cr, Cu, Si, Mg, Al, Ti, Ca and Ca^+ have been seen. These lines support the hypothesis that dust grains evaporate when close enough to the Sun. In one Sun-grazing comet, C/Ikeya–Seki (1965 VIII), the relative elemental abundances have been measured and found to be similar to those seen in carbonaceous chondritic meteorites.

The composition of dust grains has been measured *in situ* by impact-ionization mass analyzers on the Giotto and Vega spacecraft. Some dust grains appear to be dominated by ions from the elements H, C, N and O; in other grains the rock-forming elements Mg, Si and Fe are the most abundant ones. The dust is therefore referred to as either CHON or silicate particles.

It is interesting to note that, in contrast to bodies in the inner parts of our Solar System, numerous bodies in the cold outer parts are coated with organic material. Although life is only found in the inner Solar System where liquid water exists, the ingredients for life, i.e., the organic material, may originate in the outer Solar System and be transported inwards by comets. Cometary studies may therefore provide us with information on the origin of life.

10.3.5.6 *Isotope Ratios*

Isotopic abundance ratios yield information on fractionation effects from the time and the region where the comets or their constituent grains formed. In particular the D/H ratio as measured from HDO/H_2O yields important clues with regard to the origin of cometary water-ice. The big question is whether cometary ices condensed directly from gases within the primitive solar nebula, or whether they originated in the interstellar medium and were incorporated as such in the solar nebula. In both cases the deuterium fractionation must have occurred at relatively low temperatures. In the primitive solar nebula, deuterium fractionation occurs via reactions between neutrals, and depends upon temperature, pressure and time. At the low temperatures and pressures expected in the outer reaches of the solar system, deuterium fractionation goes to zero, and the D/H ratio should not be much larger than three times the protosolar value. Deuterium fractionation in cold interstellar clouds occurs through ion–molecule reactions and grain-surface chemistry. As observed in cold interstellar clouds, these processes apparently lead to a considerable (orders of magnitude) enhancement in the D/H ratio in some species compared to that in molecular hydrogen.

The D/H ratio in cometary water has been measured accurately for P/Halley from measurements of the ion and neutral mass spectrometers aboard Giotto, and was found to be equal to $(31 \pm 4) \times 10^{-5}$. Ground-based observations lead to D/H ratios in water outgassed from C/Hyakutake and C/Hale–Bopp of $(29 \pm 10) \times 10^{-5}$ and $(33 \pm 8) \times 10^{-5}$, respectively (Fig. 10.14). These ratios are roughly a factor of 10 higher than the protosolar value, and thus suggest an interstellar origin for cometary water-ice. The D/H ratio in hydrogen cyanide was determined for C/Hale–Bopp at a value of $(23 \pm 4) \times 10^{-4}$, which is roughly a factor of 10 higher than that found for cometary water. This is again suggestive of an interstellar origin. These values are discussed further in Section 10.7, where they are compared to D/H ratios for other Solar System bodies and terrestrial water.

Isotopic ratios $^{12}C/^{13}C$, $^{14}N/^{15}N$, $^{32}S/^{34}S$, and $^{16}O/^{18}O$ have been determined for P/Halley, C/Hale–Bopp, C/Ikeya (C/1963 A1), C/Kohoutek and C/Kobayashi–Berger–Milon (C/1975 N1), both from *in situ* measurements with the mass spectrometers on board Giotto (P/Halley), and from ground-based measurements of rare isotopes in HCN and CS. Results differ by a factor of $\lesssim 2$ from comet to comet, and are consistent with the terrestrial value. These measurements are thus consistent with theories that cometary ices originate in the outer parts of the primitive solar nebula.

10.3.6 X-Ray Emissions

Since the discovery of X-ray emission from C/Hyakutake by ROSAT and EUVE (Extreme Ultraviolet Explorer) (Fig. 10.15), such emissions have since been detected from six other comets. Hence it appears that X-ray emissions from comets are a general phenomenon. The emissions

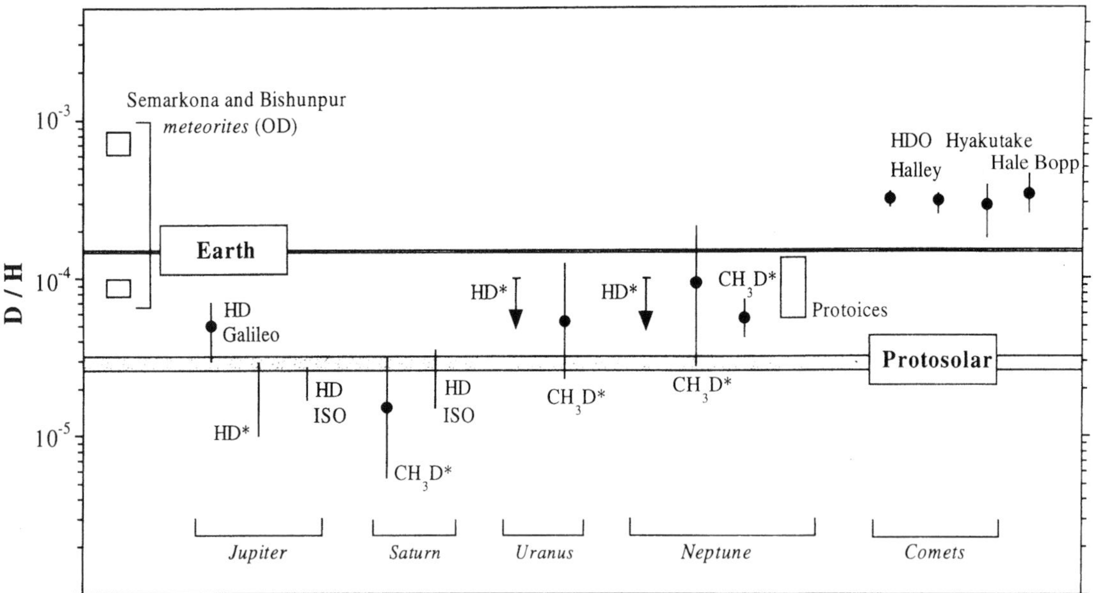

FIGURE 10.14 The D/H ratios of various bodies in our Solar System. The Earth and protosolar values are shown as horizontal lines. Asterisks (as HD*) indicate ground-based observations. (Adapted from Bockelee-Morvan *et al.* 1998)

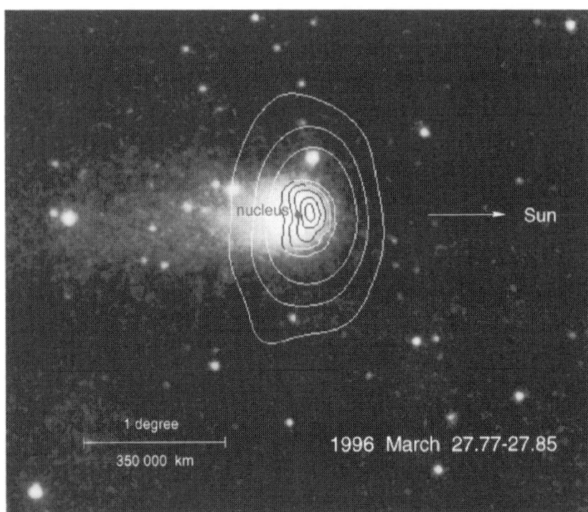

FIGURE 10.15 Image of the X-ray emission from C/Hyakutake, as observed with the ROSAT (Röntgen) satellite. The image shows the measured intensity of both the high (0.1–2.0 keV inner contours) and low (0.09–0.2 keV outer contours) energy bands as contour lines overlayed on an optical image taken with a common camera during the ROSAT observation. (Courtesy: Max-Planck-Institut für extraterrestrische Physik)

are typically displaced from the nucleus in the direction of the Sun by $\sim 50\,000$ km. To get a feeling for the strength of such emissions we note that sunlight scattered by comets at visible wavelengths is usually much weaker than that reflected by the Moon (a factor of 5 for C/Hyakutake, when it was at $r_\Delta \approx 0.12$ AU), while X-rays from comets exceed those from the Moon by many orders of magnitude (a factor of 600 for C/Hyakutake, at $r_\Delta \approx 0.12$ AU). X-ray emission from the Moon is observed from the sunlit crescent, and is simply caused by reflected solar X-rays. (The dark side of the Moon is actually darker than the X-ray background, because the Moon blocks out the X-ray background radiation from deep space.)

Several mechanisms have been suggested to produce the observed X-ray emission from comets. The measured X-ray brightness offsets, production rates, and maximum brightnesses in four comets favor charge transfer of solar wind heavy ions to cometary neutrals as a dominant process of X-ray excitation in comets. The soft X-rays from C/Hyakutake can also be attributed to collisions between cometary neutrals and dust with solar wind electrons (electron impact and recombination excitation). Electron bremsstrahlung or simply scattering of solar X-rays by real tiny dust particles may also lead to cometary X-rays. More exotic ideas include mini-flares in cometary atmospheres produced by the annihilation of magnetic field lines in current sheets within the solar wind.

10.3.7 Comet Taxonomy

Comets are a compositionally diverse group of bodies, as evidenced by large variations in dust-to-gas ratio, the relative abundances of various gaseous species and changes in brightness as a function of heliocentric distance. However, sorting comets into compositional classes is more difficult than grouping asteroids for several reasons. Nearly all observations of comets are directed at the coma and/or tail(s), and the physical processes involved in cometary outgassing and the chemical evolution of the coma are quite complicated and depend on the distance from the Sun. Relative abundances within individual comets vary with heliocentric distance, and dynamically new comets exhibit variations which are not symmetric about perihelion. The dust-to-gas ratio varies from comet to comet by over an order of magnitude, comets with small perihelion distances having the most gas relative to dust. In addition, most comets have relatively long orbital periods and are easily observable only near perihelion.

Strong evidence implies that observed differences in C_2 and C_3 abundances are attributable to bulk variations in composition between cometary nuclei. The abundances of these carbon chain molecules relative to CN vary by less than a factor of ~ 2 for the majority of comets studied. In contrast, some comets are depleted in carbon chain molecules by roughly an order of magnitude. Moreover, nearly all of the carbon chain depleted comets are members of the Jupiter family, and thus presumably originated in the Kuiper belt (Section 10.2.2), whereas 'normal' carbon chain abundances are found among all dynamical groupings of comets. This suggests that conditions were right for forming the parents of these carbon chain species, possibly large hydrocarbons like C_2H_2 and C_3H_4, in the region where Oort cloud comets formed, i.e., presumably in the giant planet region of the protoplanetary disk (cf. Section 10.7) and where some but not all of the Kuiper belt comets formed (presumably *in situ*). It is possible that the temperatures in the outermost sampled region of the Kuiper belt were too cold for the chemistry required to make these large hydrocarbons. Further details on cometary taxonomy can be found in the article by A'Hearn *et al.* (1995).

10.4 Dust

10.4.1 Morphology

When gases sublimate and evolve off the nucleus, dust is entrained in the gas and released from the cometary nucleus. The dust-to-mass ratio is usually between 0.1 and 10, and is strongly correlated with perihelion distance. Submicrometer-sized particles almost reach the speed of the escaping gases ($\sim$1 km s^{-1}), but larger particles barely attain the gravitational escape velocity of the nucleus ($\sim$1 m s^{-1}). At a distance of a few tens of nuclear radii, the dust decouples from the gas and becomes subject to solar radiation forces.

As discussed in Section 2.7, solar radiation pressure on (sub-)micrometer-sized dust grains 'blows' the particles outwards from the Sun. Particles of different sizes and with different release times become spatially separated in the dust tail. A schematic representation of tail formation is shown in Figure 10.16. The dashed lines in Figure 10.16a are the trajectories of the dust grains, if emitted at zero velocity. Usually, the grains have a non-zero initial velocity, and thus a slightly different trajectory than indicated. The exact path also depends on the ratio, β, between the solar radiation force and the gravitational force from the Sun. Thus, the dust orbits depend on size, shape/composition and initial velocity. The ensemble of particles released at different times together form a curved tail, as depicted by the shaded region.

Lines connecting particles of the same β at a given instant of time are called *syndynes* (tail in Fig. 10.16a, dashed lines in Fig. 10.16b, c). The width of the tail is equal to $2v_d t$, with v_d the velocity of the dust particles in the comet's frame of reference when released from the surface and t the time since release. One can also order the dust particles according to release time, regardless of their size or β. The locus of such points is called a *synchrone*; these are indicated by the solid lines in Figure 10.16b, c. The numbers indicate the time in days since the dust was released.

The simultaneous ejection of dust grains of different sizes gives rise to inhomogeneities in the dust tail, in particular the *dust jets*: highly collimated structures, which at larger distances become *streamers*: straight or slightly curved bands which converge at the nucleus. Observed infrequently are *striae*, which are parallel narrow bands at large distances from the nucleus that do not converge at the nucleus (Fig. 10.17). They usually intersect the comet–Sun line on the sunward side. Their origin has not yet been explained. It has been suggested that striae originate from the simultaneous destruction of large particles, or from the action of electromagnetic forces on charged dust grains.

Although dust tails always point away from the Sun, on some rare occasions the tail appears to point towards the Sun, as shown in an image of C/Hale–Bopp (Fig. 10.18) taken on 5 January 1998. Such phenomena, referred to as *anti-tails*, are caused by a particular viewing geometry between the Sun, the observer and the comet. The normal dust tail in this image is visible in the anti-solar direction, extending over a distance of more than 4° to the edge of the field. Since the photograph was taken near the time that the Earth was crossing the comet's orbital plane, the image also shows the 'neck-line structure', a narrow and very straight feature in the same direction which is caused

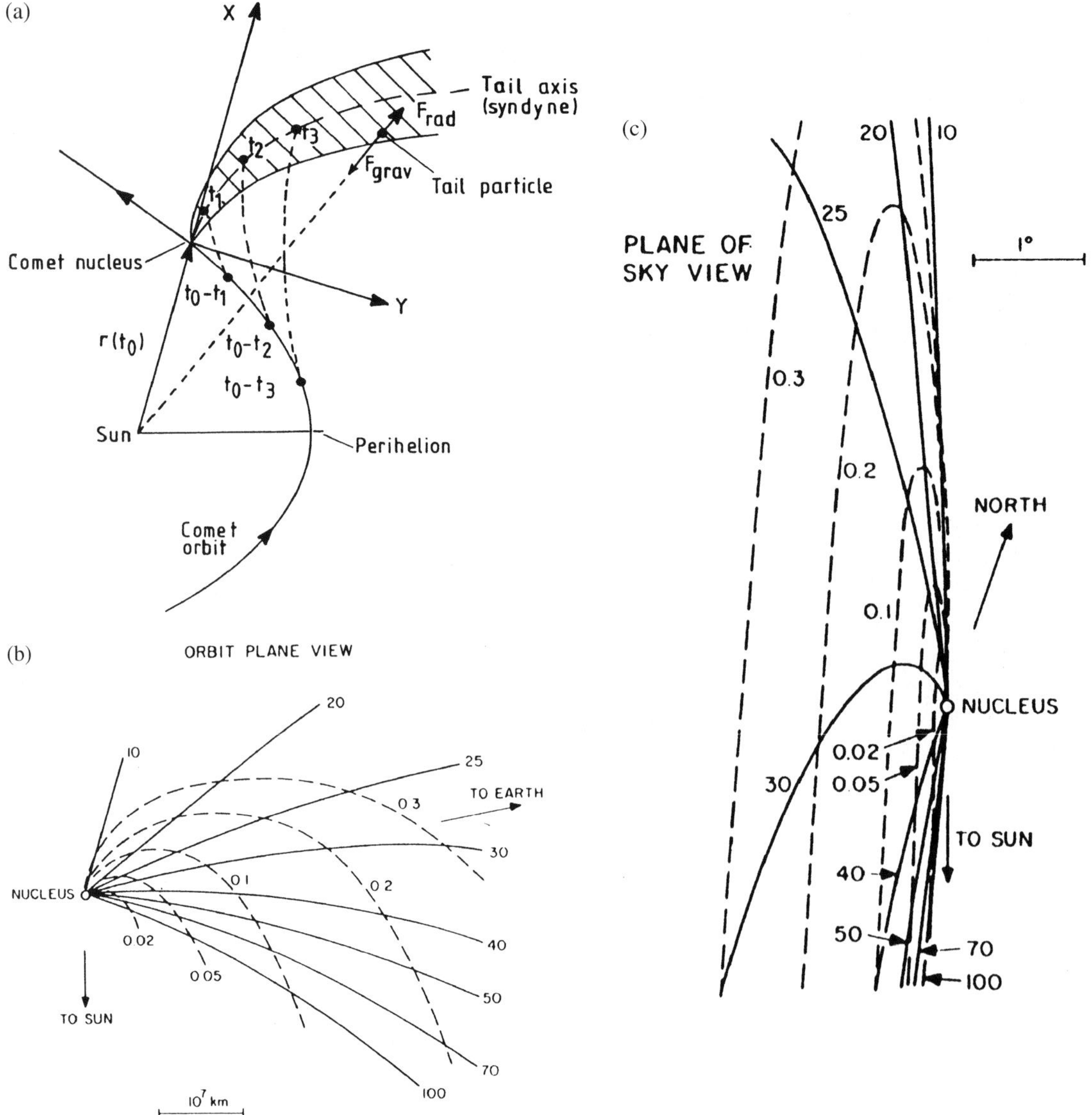

FIGURE 10.16 (a) A schematic presentation of the formation of a dust tail. Trajectories of dust particles that were released from the nucleus at different times t_1, t_2, and t_3 prior to the time of the observations at t_0 are shown as dashed curves. (Adapted from Finson and Probstein 1968) (b) Synchrones (solid lines with ages in days) and syndynes (dashed lines with β values) for C/Arend–Roland on 28 April 1957, in the orbit plane of view. (Sekanina 1976) (c) Synchrones (solid lines with ages in days) and syndynes (dashed lines with β values) for C/Arend–Roland on 28 April 1957, as viewed on the plane of the sky. This view explains the formation of anti-tails. (Sekanina 1976)

by sunlight reflected in the thin dust sheet in this plane. The true length of this feature depends on the exact geometry of the dust distribution, but may well be of the order of 1 AU or even more. The anti-tail is the narrow, sunward spike that may be seen at least 0.5° in the solar direction. This tail is composed of large (10–100 μm sized) dust grains (small β) that left the nucleus about 3 months before this photograph was taken. Because of the small β, these particles are still in the vicinity of the nucleus. Anti-tails have also been observed for C/Arend–Roland (C/1957 III) and C/Kohoutek; the synchrones and syndynes in Figure 10.16b (in the orbit plane) and 10.16c (projected onto the

FIGURE 10.17 C/West in March 1976 showing both streamers and striae. About a week after this picture was taken, the nucleus broke into four pieces. (Courtesy: Akira Fujii)

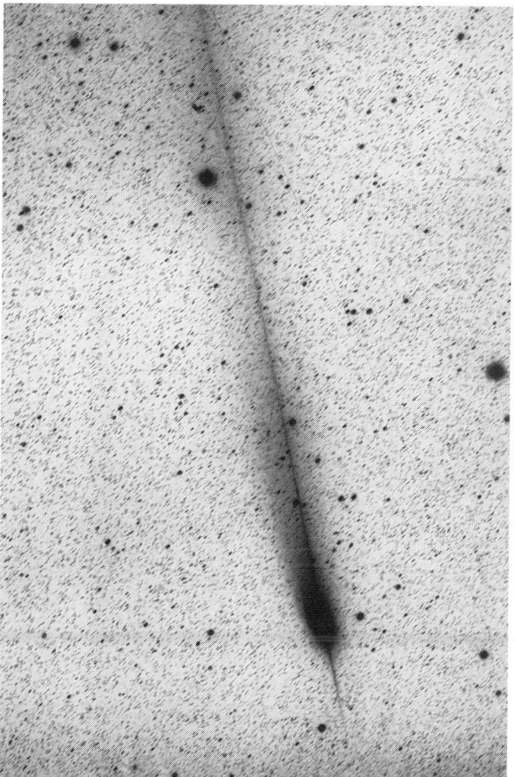

FIGURE 10.18 Negative print of C/Hale–Bopp as photographed with the ESO 1-m Schmidt Telescope on 5 January 1998. The exposure time was 1 hour. During this time, an artificial satellite crossed the field – its trail is seen as a very thin line to the right of the comet. (Courtesy: Guido Pizarro, ESO Press Photo 05a/98)

plane of the sky) were calculated for C/Arend–Roland at the time the anti-tail was visible.

10.4.2 Observed Dust Properties

10.4.2.1 *Optical and Infrared Properties*

Visible and infrared spectra of cometary dust tails have a number of features in common. Spectra of C/Bennett (C/1970 II) and C/Kohoutek are shown in Figure 10.19. Visible and near-infrared spectra typically resemble a solar spectrum, because sunlight is scattered off dust particles in the comet's coma and/or tail. The spectra are neutral to slightly reddish in color; there is no Rayleigh scattering (bluish color), which suggests that there are not many particles much smaller than the wavelength of visible light, i.e., a lower limit to the grain size is ~ 0.1 μm. Solar absorption lines are evident in a cometary spectrum, and the derived blackbody temperature is ~ 6000 K, the temperature of the Sun. At a heliocentric distance of 1 AU, thermal emission from the dust particles dominates at wavelengths longwards of 3 μm. At larger distances, the dust particles are colder, and the 'crossover' wavelength between reflected light and thermal emission shifts to longer wavelengths.

Assuming equilibrium between solar insolation and reradiation outwards (Section 3.1.2), the equilibrium temperature of a dark, rapidly rotating grain becomes (Problem 10.22):

$$T_{eq} = \frac{280}{\sqrt{r_{AU}}}\left(\frac{1-A_b}{\epsilon_\nu}\right)^{1/4} \approx \frac{280}{\sqrt{r_{AU}}}, \qquad (10.15)$$

where A_b is the Bond albedo of the grain and ϵ_ν is the emissivity. The term $((1 - A_b)/\epsilon_\nu)^{1/4}$ is usually close to unity, except for particles with sizes $2\pi R/\lambda < 1$, with R the radius of the particle and λ the radiating wavelength. Micrometer-sized grains absorb efficiently at UV and visible wavelengths, near the peak of the Sun's blackbody (Planck) curve, while the emission efficiency is low at infrared wavelengths (Section 3.1.2.2). For such particles $(1 - A_b)/\epsilon_{ir} > 1$, and the observed temperature is higher than the equilibrium temperature from equation

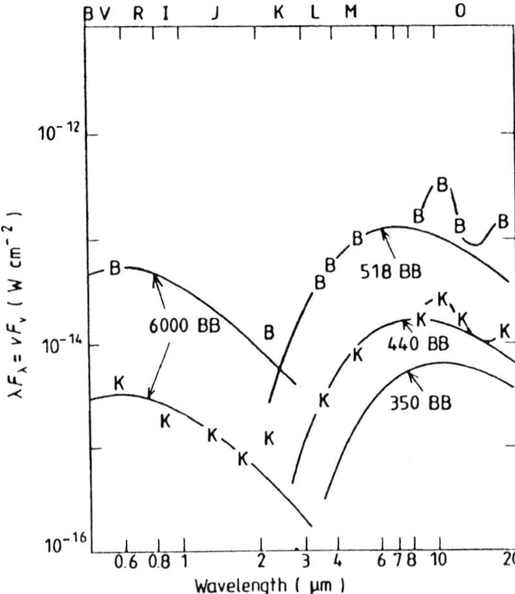

FIGURE 10.19 Broadband spectra of comets C/Bennett and C/Kohoutek at $r_\odot = 0.65$ AU. Blackbody curves were fitted to the data, indicated by solid lines BB, which are labeled by temperature. The value of 350 K corresponds to a rapidly rotating blackbody at a heliocentric distance of 0.65 AU. The 6000 K BB curves correspond to sunlight reflected by a gray body. The letters at the top denote the various observing bands. (Ney 1974)

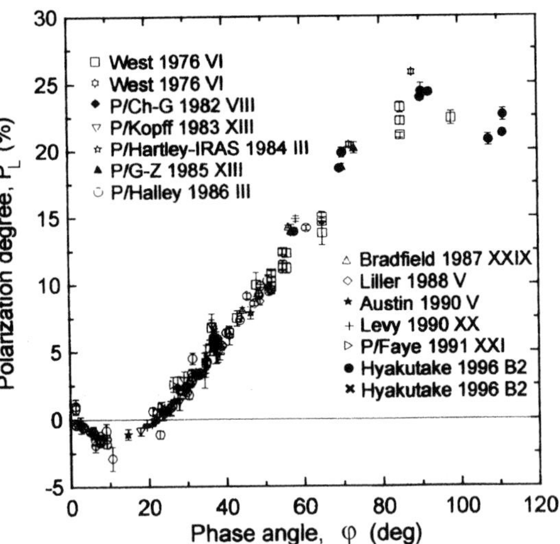

FIGURE 10.20 Phase dependence of the degree of linear polarization for various dusty comets measured by various people at wavelengths between 4850 Å and 6840 Å. (Kiselev and Velichko 1998)

(10.15). The observed thermal spectrum can thus be used to determine the size of the grains in a cometary tail. For C/Bennett and C/Kohoutek, the peak in thermal emission occurred around 7–8 μm, which suggests that the cometary grains must be less than a few micrometers in size. An independent constraint on grain sizes follows from emission features of silicates, which are near 10 and 18 μm. Such emission features can only be observed if the grains producing them are smaller than ~5 μm in radius. Thus, the dust grains in a comet's tail are likely between ~0.1 μm and a few μm in radius.

Simultaneous observations at visible and infrared wavelengths can be used to estimate the Bond albedo of the dust particles; the results depend, however, quite critically on the scattering phase function of the material. Also the polarization characteristics contain information on the nature of dust particles. As expected for scattered light, all comets exhibit linear polarization, the degree of which depends on the scattering phase (Sun–comet–observer) angle ϕ. We define the degree of polarization, P_L, as in equation (9.5), and show results for many dusty comets in Figure 10.20. P_L can be positive or negative, depending on the relative contributions of $I_\perp$ and $I_\parallel$. As shown, it is generally negative for backscattering, at phase angles $\phi \sim 0$–$22°$, with a minimum near -1.5% at $\phi \approx 10°$, increasing nearly linearly with a slope $h = 0.34\%$ per degree at blue, and 0.41% per degree at red wavelengths, to a maximum of ~24–26% at $\phi \approx 94°$. These polarization curves contain information on the composition and size of the grains. The curves can be reproduced if the grains consist of an equal mixture (in number) of silicate and graphite particles. The polarization measurements further rule out significant contributions by grains less than ~0.1 μm in size, in agreement with the observed lack of Rayleigh scattering.

10.4.2.2 *Radar Observations*

Several comets have been observed using radar techniques. In such experiments a monochromatic signal is transmitted to the comet and received after reflection off the object (cf. Section 9.3.6). The signal is Doppler broadened by the rotation of the comet; the echo bandwidth is determined by the rotation rate, the direction of the rotation axis and the effective size of the nucleus. C/IRAS–Araki–Alcock, P/Halley and C/Hyakutake each show a strong narrow-band radar echo from the nucleus, plus a broad-band much weaker echo, which is attributed to $\gtrsim$ centimeter-sized grains in a halo around the comet (Fig. 10.21). Thus, a comet may contain, in addition to (sub-)micrometer-sized dust in its coma and tail, a halo of centimeter-sized grains.

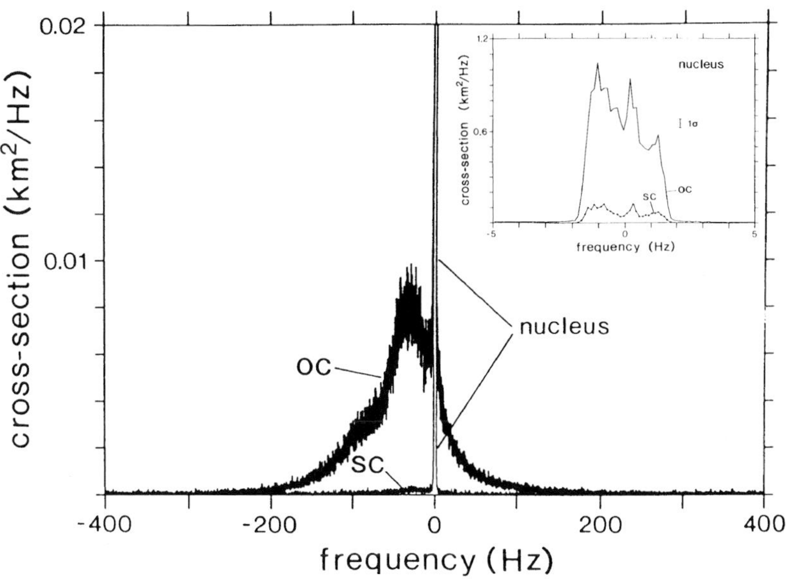

FIGURE 10.21 Radar echo power from C/IRAS–Araki–Alcock. The observed cross-section (km^2 Hz^{-1}) is plotted as a function of frequency. The strong spike is backscatter from the nucleus, while the 'skirt' is attributed to reflection off large icy grains in a coma surrounding the nucleus. The insert shows the radar echo from the nucleus (note scale on the y-axis). (Adapted from Harmon *et al.* 1989)

10.4.3 Escape of Dust

Dust grains entrained in the outflowing cometary gases vary in size and composition. Observations at optical and infrared wavelengths rule out significant populations of particles with sizes $\lesssim 0.1$ μm (see above). The largest size particle, $R_{d,max}$ (radius of the particle) which can be dragged off the surface of a spherical nucleus depends upon the gas production rate, $\mathcal{Q}$, the temperature of the surface, T, the radius of the nucleus, R, and the densities of the nucleus, ρ, and dust grain, ρ_d:

$$R_{d,max} = \frac{9\mu_a v_o \mathcal{Q} R}{4N_A G \rho \rho_d}, \qquad (10.16a)$$

where N_A is Avogadro's number, μ_a the molecular weight of the gas, G the gravitational constant, and v_o the outflow velocity (Problem 10.25). Equation (10.16a) is valid if the mean free path of the gas molecules is larger than the dust particle sizes. At heliocentric distances less than a few tenths of an AU, this may not be a valid assumption, and the drag force has to be replaced by Stokes law (eq. 2.55). The largest particle which can be dragged off the surface is then:

$$R_{d,max} = \left(\frac{27\nu_v \rho_g v_o}{8\pi R G \rho \rho_d}\right)^{1/2}, \qquad (10.16b)$$

where ρ_g is the density of the outflowing gas and the viscosity ν_v, between 200 and 300 K is approximately:

$$\nu_v \rho_g \approx 10^{-6} (\text{g cm}^{-1}\ \text{s}^{-1}). \qquad (10.16c)$$

If the nucleus consists predominantly of well-compacted water-ice with $\rho = \rho_d = 1$ g cm^{-3}, then the maximum size for icy particles dragged off the surface from a comet at 1 AU with a radius of 1 km is ~10 cm (Problem 10.26). Particle sizes may be larger during cometary outbursts, where part of the 'dust crust' is thrown into space. Icy particles are subjected to sublimation effects, similar to the cometary nucleus. The lifetime of an icy dust grain depends strongly on the composition or dielectric constant (absorptivity) of the grain; the absorptivity determines the temperature of the grain, which sets the sublimation rate. Depending on the absorptivity, the lifetime can vary by many orders of magnitude.

It must be clear that comets 'pollute' the environment significantly, not only via gases but also with dust. The gas will ionize and ultimately be swept away by the solar wind, but dust particles may stay around for a long time. In particular the larger particles, which are affected least by solar radiation pressure, may share the orbital properties of the comet for a long time. These particles are seen as *meteor streams*, or cometary *dust trails*, which are discussed in more detail in Section 9.2.4.

10.5 Magnetosphere

Although sublimating gases are neutral when released by a comet, ultimately all atoms and molecules get ionized. The dominant ionization processes are photoionization and charge exchange with solar wind protons. Cometary ions and electrons interact with the interplanetary magnetic field, and 'drape' the field lines around the comet, as depicted in Figure 10.22. This process induces a magnetic field, similar to the field around Venus. In the fol-

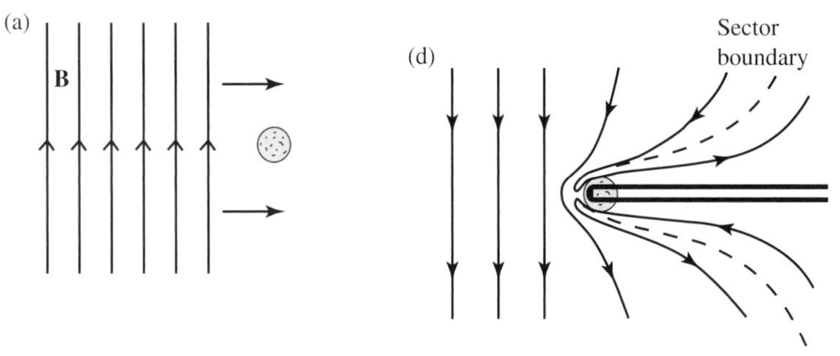

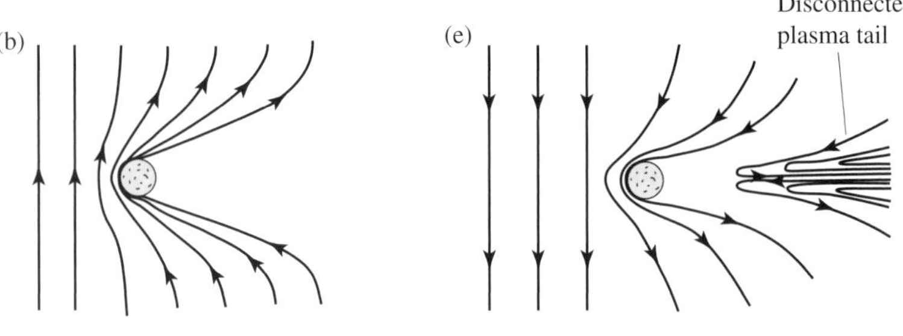

FIGURE 10.22 The field draping model of Alfvén, in which interplanetary magnetic field lines are deformed by a comet's ionosphere. The sequence from (a) to (c) shows the gradual draping of interplanetary magnetic field lines around the comet into a magnetic tail. When the comet encounters a sector boundary in the interplanetary magnetic field (where the magnetic field reverses direction; dashed line), the tail becomes disconnected, as depicted in the sequence from (c) to (e).

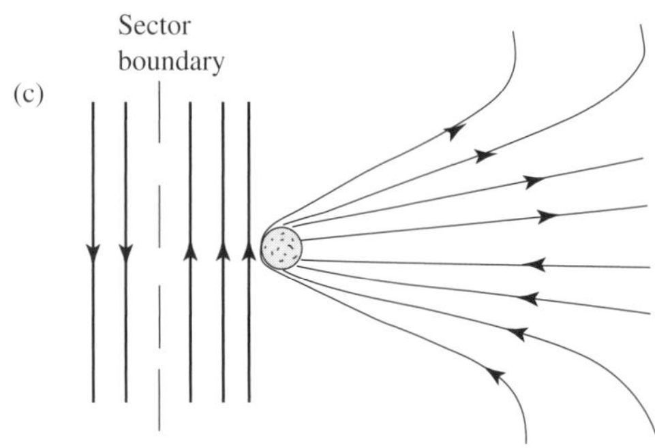

lowing, we describe the magnetic field morphology for a comet with a well-developed atmosphere; i.e., for a comet which is close to the Sun (within 1–2 AU). A rough sketch is shown in Figure 10.23.

10.5.1 Morphology

Since a comet's gravitational field is very weak, neutral gas expands supersonically outward from the nucleus (Section 10.3.2). When the neutrals get ionized, they are accelerated by the strong electric fields in the solar wind (Section 7.1.4):

$$\mathbf{E}_{sw} = \frac{\mathbf{v}_{sw} \times \mathbf{B}_{sw}}{c}. \tag{10.17}$$

The solar wind velocity, v_{sw}, is typically ~ 400 km s^{-1}, and the interplanetary magnetic field $B_{sw} \approx 5 \times 10^{-5}$ G at $r_\odot \approx 1$ AU. Cometary ions are accelerated to the solar wind speed, either directly (if $\mathbf{B}_{sw} \perp \mathbf{v}_{sw}$) or indirectly, via plasma instabilities (if $\mathbf{B}_{sw} \parallel \mathbf{v}_{sw}$). The ions gyrate around the local magnetic field lines. The pick-up of these heavy ions by the solar wind leads to a mass loading of the solar wind. Based upon conservation of momentum, the solar wind is (temporarily) slowed down. As a result of the mass loading process, when the solar wind has accumulated an admixture of $\sim 1\%$ (by number) cometary ions, a collisionless reverse shock forms: the *bow shock*. Note the difference between the cometary bow shock – caused by mass loading – and the bow shocks upstream from planets, which form to divert the solar wind flow around the impenetrable boundaries of the planets or their magnetospheres (Section 7.1). Downstream of the cometary bow shock, the solar wind is subsonic, as in a planetary magnetosheath, and continues to interact with the cometary atmosphere. The approximate location of the bow shock, $\mathcal{R}_{bs}$, depends on the gas production rate, $\mathcal{Q}$, and is inversely proportional to the solar wind momentum, $\rho_{sw} v_{sw}$. Using the equations for mass, momentum and energy for a plane-parallel supersonic flow in front of the bow shock and mass loading by the comet, the following relation has been derived for the standoff distance of a cometary bow shock:

$$\mathcal{R}_{bs} = \frac{\mathcal{Q} \mu_a m_{amu} \alpha (\gamma^2 - 1)}{4\pi v_o \rho_{sw} v_{sw}}, \tag{10.18}$$

with v_o the outflow velocity of the neutrals, α the ionization rate, γ the ratio of specific heats ($\gamma = 2$ for a magnetized flow), ρ_{sw} and v_{sw} the solar wind density and ve-

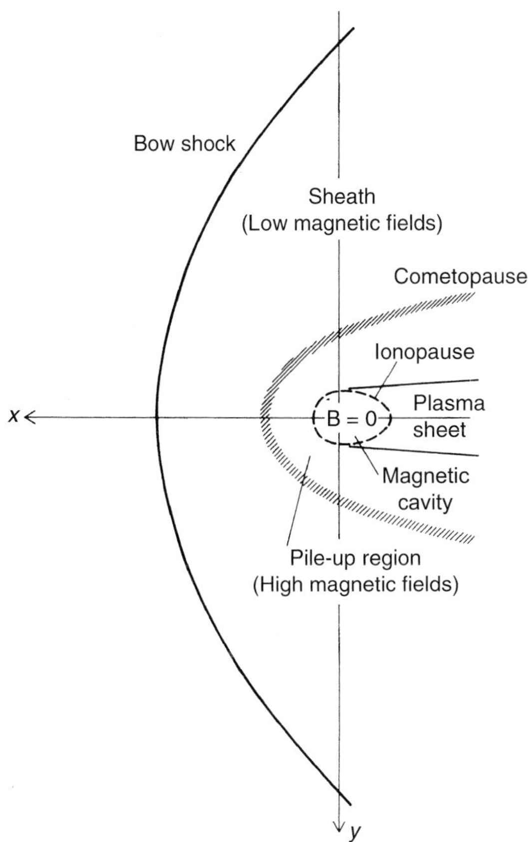

FIGURE 10.23 A cartoon sketch of the magnetic field morphology for a comet with a well-developed atmosphere. (Adapted from Neubauer 1991)

locity far from the comet. For P/Halley, with $\mathcal{Q} \approx 10^{30}$ molecules s^{-1} at $r_\odot \approx 0.9$ AU, $\mathcal{R}_{bs} \approx 10^6$ km (Problem 10.23).

The outflowing cometary neutrals collide with the incoming solar wind, which is consequently slowed down. The *cometopause* or *collisionopause* separates the collisionless solar wind plasma flow from the cometary gas, a flow in which collisions dominate. At the cometopause, significant momentum is transferred through collisions between the outflowing cometary neutrals and the solar wind ions. Although this boundary layer is not completely understood, there is general agreement that the region inside the cometopause is dominated by cometary ions (in particular H_2O^+ and H_3O^+) and compressed interplanetary magnetic field; in contrast, solar wind plasma, loaded with cometary ions created at large distances from the nucleus, dominates the region outside the cometopause. The cometopause was encountered by the Vega spacecraft $\sim 10^5$ km upstream from P/Halley, when the comet was at $r_\odot \approx 0.8$ AU. Its location can be approximated by the relation:

$$\mathcal{R}_{cp} = \frac{\sigma_x \mathcal{Q}}{4\pi v_o}, \tag{10.19}$$

with σ_x the collision cross-section for ions.

Inside the cometopause, the magnetic field is highly compressed, such that there is approximate pressure equilibrium between the magnetic pressure at the inside and the supersonic solar wind ram pressure at the outside:

$$\frac{B_c^2}{8\pi} = \rho_{sw} v_{sw}^2, \tag{10.20}$$

with B_c the magnetic field strength at the inside of the cometopause.

The magnetometer experiment on board the Giotto spacecraft discovered a well-defined boundary inside the cometopause, at a cometocentric distance of 4700 km, where the magnetic field suddenly dropped to zero. This boundary has been called the *ionopause*, a tangential contact discontinuity which separates the cometary plasma from the contaminated solar wind plasma. There is no magnetic field inside the ionopause: the nucleus is surrounded by a magnetic cavity. There is approximate pressure equilibrium between the magnetic field outside of the cavity and thermal pressure inside.

Interior to the ionopause, models predict the existence of an inner shock, which decelerates the supersonically outward flowing cometary ions and diverts them into the tail; this shock, however, has not been observed by any spacecraft. As in planetary magnetotails, a neutral sheet separates the two 'lobes' of the magneotail, formed by the folding of the interplanetary magnetic field lines. When the ICE spacecraft crossed the neutral sheet in the tail of P/Giacobini–Zinner, the polarity of the field reversed, as expected when crossing a neutral sheet in a planetary magnetosphere.

Figure 10.24 summarizes a model for P/Halley showing the calculated electron and neutral gas temperatures and velocities. Within 100 km of the cometary nucleus, the electron temperature is likely to be a few tens of degrees Kelvin, because water molecules are efficient coolants. Since electron–neutral collisions are frequent, electrons and neutrals should have the same temperature. When the collision rate drops, the electron temperature rises because newly produced electrons have a large excess kinetic energy. The electron temperature may increase to a few hundred degrees Kelvin at a cometocentric distance of 4000–5000 km.

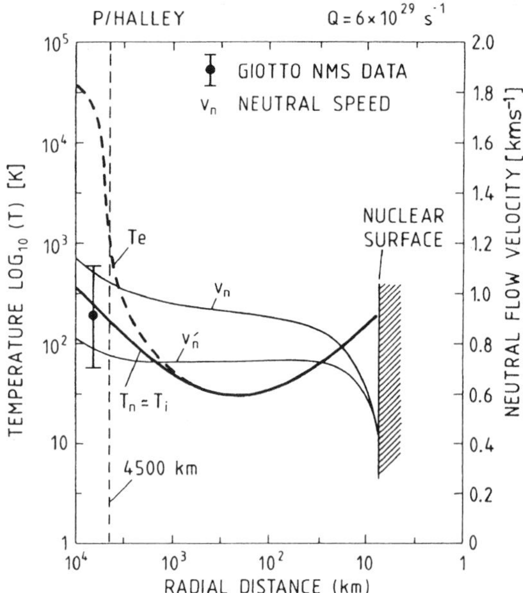

FIGURE 10.24 A model for various parameters in the inner coma of P/Halley: the electron and neutral gas temperatures, T_e and T_n, respectively, and the expansion velocity with (v_n') and without (v_n) IR radiative cooling by water molecules. (Ip and Axford 1990)

FIGURE 10.25 This image of C/Hyakutake was obtained with a 30 cm reflector, during a 60 second exposure on 8 April 1996 UT. The image was processed to bring out the ray structure and jets emanating from the nucleus. The raw image is shown in the insert. (Courtesy: Tim Puckett)

10.5.2 Plasma Tail

In the anti-solar direction, the cometary ions form an ion or plasma tail. The length of this tail can easily exceed 10^7 km, and the width of the main tail is roughly 10^5 km in diameter. The tail usually looks blue, as a result of fluorescent transitions of the abundant, long-lived CO^+ ions, though occasionally a reddish tail has been seen due to emissions by H_2O^+. The tail often consists of filaments, rays and bright knots, and its structure changes on timescales of minutes to hours. The knots are caused by enhanced densities, and their motion can be followed down the tail. Typical speeds are $\lesssim 100$ km s^{-1}; this is clearly much more than the cometary outflow speed, but less than the solar wind velocity. Closer to the nucleus, in the cometary head, Doppler shifts of the H_2O^+ ion indicate speeds of 20–40 km s^{-1}. The plasma is clearly accelerated down the tail. Solar wind speeds are reached at distances $> 10^7$ km from the nucleus. Density condensations in the main tail are often accompanied by condensations in adjacent filaments, at the same cometocentric distance.

In gas-rich comets, one can often see the tail rays from the plasma envelope. Figure 10.25 shows a processed image of C/Hyakutake which clearly shows the rays 'draped' around the head of the comet (compare the field line draping depicted in Fig. 10.22). Individual rays are also clearly seen in Fig. 10.26.

The precise structure of a cometary plasma tail depends on the interplanetary medium and its magnetic field. One can often see large disturbances in a plasma tail (Fig. 10.26), and sometimes the comet appears to lose its tail, and starts forming a new one. Such events have been attributed to 'disconnection' or magnetic reconnection events, caused by a sudden reversal of the interplanetary magnetic field. Such reversals take place when the comet meets an interplanetary sector boundary, where the large scale interplanetary magnetic field reverses (Fig. 10.22).

10.6 Nucleus

10.6.1 Size, Shape and Albedo

When a comet approaches the Sun, ices sublime and evolve off its surface, dragging dust particles along to form a bright coma, which prevents a clear view of the bare nucleus of a comet. It is very difficult to obtain data on cometary nuclei from Earth. Most observations are contaminated to some extent by the comet's coma (Problem 10.27). To subtract the coma from the data introduces an error of variable magnitude. Despite these problems, both size and albedo estimates have been made. Typical cometary diameters range from 1–20 km; however, this may be primarily an observational effect, with smaller bodies being more numerous, but difficult to detect, and larger bodies rarer, but still dominating the distribution by mass. The albedo of most comets is very low, between ~2 and 5%. Reflection spectra at visible–

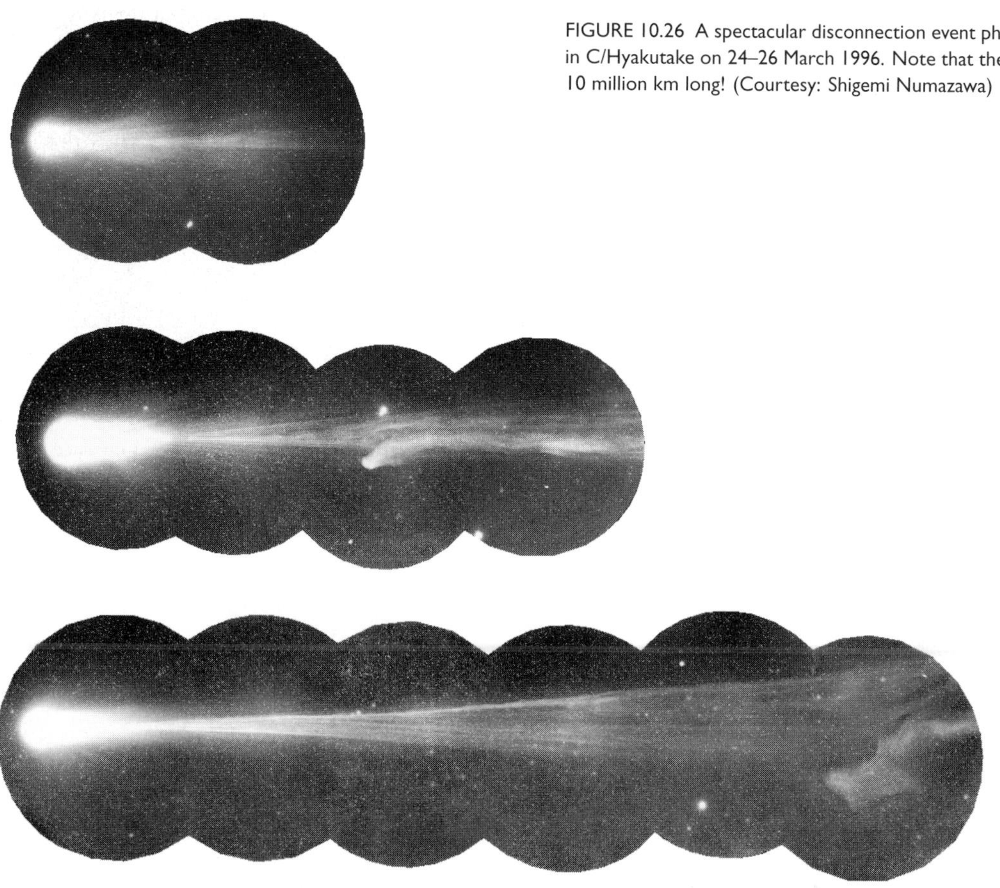

FIGURE 10.26 A spectacular disconnection event photographed in C/Hyakutake on 24–26 March 1996. Note that the tail is over 10 million km long! (Courtesy: Shigemi Numazawa)

IR wavelengths (3800–8500 Å) usually show smooth straight lines, slightly increasing towards longer wavelengths (Fig. 10.27).

Since comets are relatively small, they are most likely not spherical (Section 6.1), as confirmed by the Giotto spacecraft images of the nucleus of P/Halley (Fig. 10.6a). This body was measured to be $8 \times 8 \times 16$ km across, very irregular and dark (albedo $\sim 4\%$). Large (compared to the size of the object) craters are visible, and there are two clear jets in the sunward direction. Radar observations are indicative of a surface irregular on scales exceeding meters (Fig. 10.21; Section 9.3.6). It is interesting to note here that the two largest cometary nuclei for which sizes have been 'directly' measured (via radiometry), C/Schwassmann–Wachmann 1 and 2060 Chiron, have albedos which are much higher than for other comets, and they are also more spherical than smaller comets.

10.6.2 Rotation

The physics of cometary rotation is analogous to that of asteroid rotation (discussed in Section 9.2.3), with the added complication that outgassing of comets can result in torques which alter the rotational angular momentum vector. Comets which are (or were in the geologically recent past) active are thus more likely to exhibit complex (non-principal axis) rotation states than are asteroids. Moreover, outgassing-induced spin-up of cometary nuclei may cause some comets to split into two or more pieces.

Measuring the rotational properties of comets is non-trivial, because a comet's coma generally dominates the body's reflected light, except at large heliocentric distances where most comets are very faint and difficult to detect. As active regions of a comet are not uniformly distributed across its surface, rotation of these spots can produce periodic variations in coma brightness which can be used to estimate the rotation period. Radar measurements can probe through the coma, but the comet has to come very close to Earth to be observed. As of mid-1999, radar echoes of seven comets had been obtained. Such echoes can be used to deduce many properties of a cometary nucleus, including its spin state. Because it is very hard to obtain accurate rotation periods for comets, these have been estimated for only a small number of comets, and values range from a few hours to several days. These rotation periods are typ-

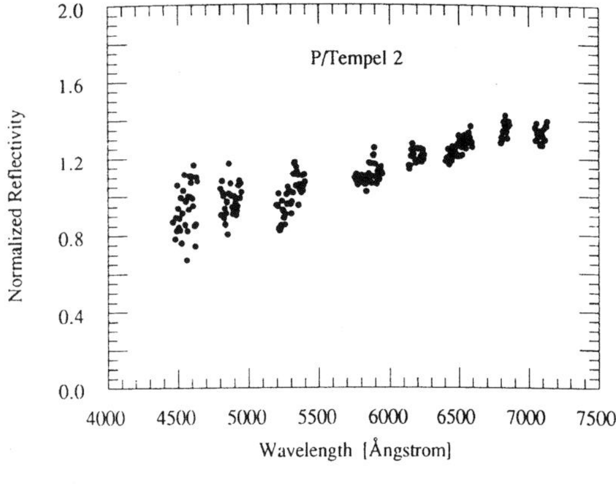

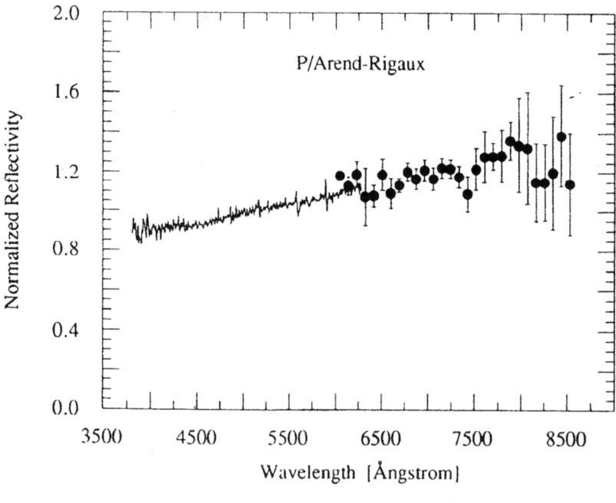

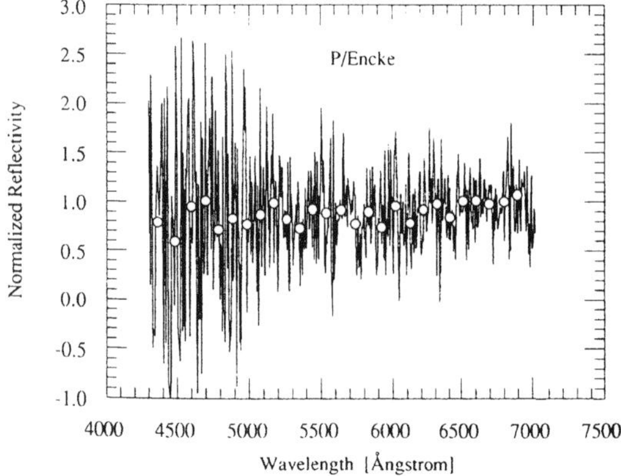

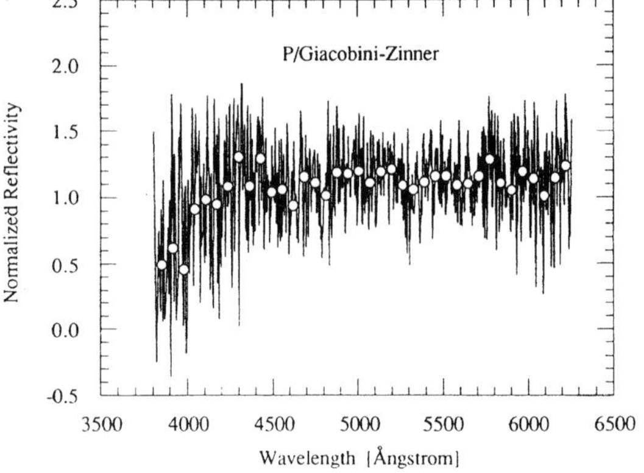

FIGURE 10.27 Reflection spectra of several cometary nuclei obtained with different telescopes. (Luu 1993)

ically somewhat longer than for main belt asteroids (Section 9.3.2).

Close-up images of the nucleus of P/Halley obtained by three spacecraft in 1986, together with studies of coma brightness variations observed from the ground, exclude pure rotation about either the short or the long principal axis. Thus, P/Halley is in a complex rotation state, with its instantaneous axis of rotation precessing about its rotational angular momentum vector. The particulars of P/Halley's spin are not fully constrained, but the most probable rotational state is one in which the long axis executes precessional motion about the angular momentum vector with a period ~3.7 days combined with rotation around the long axis at a period of ~7.3 days. Time series photometric measurements suggest that the nucleus of C/Schwassmann–Wachmann 1 may also be in a complex rotation state.

10.6.3 Splitting and Disruption

Comets sometimes appear to possess multiple nuclei which are clearly spatially separated. The gravitational fields of comets are too weak for these nuclei to be a bound binary or multiple system. Rather, they were presumably formed by a break-up (splitting) event, although no direct observations of this process exist. At least 20 comets are thought to have broken up. The first clear case was D/Biela in 1845/1846 (3D/1846 II). After the comet broke up, it was left with a large companion, which evolved as a separate comet. On its next return, 7 years later, D/Biela appeared as a double comet. Other comets which have split include C/West and C/Kohoutek. In many cases, only tiny fragments separate from the nucleus; such pieces last for at most a few weeks. Since one would expect the splitting of a comet to be accompanied by a 'flare-up' in the comet's brightness and a (temporary) increase in dust emission, a sudden increase in the observed brightness of a comet is often taken as evidence of a 'splitting' event. A clear example can be seen in visual lightcurves of C/West (Fig. 10.28), where a sudden increase in brightness before perihelion passage is attributed to the break-up of the primary nucleus. Near simultaneous images of C/West's dust tail show a substantial increase in the dust emission, in the form of broad streamers. Most comets show splitting events when they come close to the Sun. However,

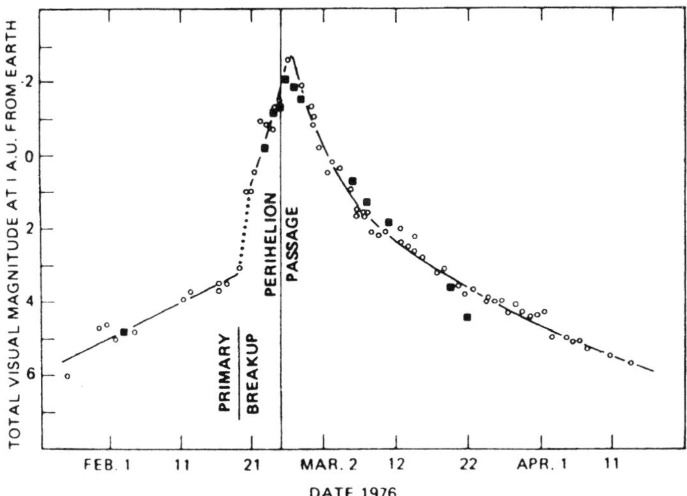

FIGURE 10.28 The lightcurve of C/West, which shows evidence of a splitting event (indicated in the figure). (Sekanina and Farrell 1978)

although comets usually are close to the Sun when they break up, some, like P/Schwassmann–Wachmann 1, show (frequent) outbursts even beyond Jupiter's orbit.

The best understood cause for the break-up of a cometary nucleus is tidal disruption during a close encounter with the Sun or a planet. Cometary nuclei may also split as a result of the centrifugal strain of rapid rotation, which can happen anywhere in the Solar System. Another possibility is a collision with interplanetary boulders. Although such events would occur most often in the asteroid belt, comets have not been observed to split while crossing this region.

Perhaps the most interesting case of a split comet was D/Shoemaker–Levy 9 (D/1993 F2), which was initially found to orbit and later to crash into Jupiter (July 1994; Section 5.4.5). Ground-based and HST images revealed the existence of over 20 cometary nuclei, strung out like pearls on a string (see Fig. 5.33). Orbital calculations show that the comet completed several orbits about Jupiter, and that it may have been captured by Jupiter around 1930. It was in a chaotic orbit, venturing close (1.3 $R_♃$) to the planet in July 1992. The tidal forces from the giant planet must have disrupted the comet just after its close approach. Simulations of the effect suggest the parent body to have had a low material strength, with a bulk density between 0.3 and 0.7 g cm^{-3}, which is less than that of water-ice. The individual kilometer-sized clumps appear to have consisted of loose agglomerates of material, although there is as of yet no general agreement on the structural composition of the cometary fragments.

Sun-grazing comets have their perihelion passage very close to the Sun. They sometimes break up near perihelion as a result of the Sun's strong tidal forces, they may evaporate completely or collide with the Sun. The SOHO (Solar and Heliospheric Observatory) spacecraft has detected $\sim$ 200 Sun-grazing comets over a period of less than 5 years, most, if not all, of which belong to the Kreutz Sun-grazing family of comets, named after H. Kreutz, who made the first extensive observations of Sun-grazing comets in the nineteenth century. These comets have similar orbital parameters, and are believed to be fragments of a single comet that broke up several thousand years ago. On 1 and 2 June 1998, SOHO discovered two Sun-grazing comets that plunged into the Sun (Fig. 10.29), an event which was followed the next day, although probably purely coincidently, by a large coronal mass ejection of hot gas (Section 7.1.1.2).

10.6.4 Surface Properties

The composition and make-up of a cometary nucleus has never been measured directly, not even by the Giotto spacecraft which made numerous *in situ* measurements of the immediate environment of P/Halley. The make-up (composition and structure) of a cometary nucleus must be determined from observations of its end-products: coma, plasma and dust tail, and from dynamical observations, in particular nongravitational forces and break-up of nuclei. Such observations led Whipple to propose his *dirty snowball* theory in 1950. A cometary nucleus in this model is a loosely bound agglomeration of weakly bonded, icy-conglomerate planetesimals, composed of frozen volatile material interspersed with meteoritic dust (Fig. 10.30). These conglomerates may be welded into a single nucleus by thermal processing and sintering. *Sintering* is a process by which weak chemical bonds form along grain contacts of a particulate material that is near but below its melting temperature. Based upon observations of the tidally dis-

FIGURE 10.29 The LASCO coronagraph on the Solar and Heliospheric Observatory (SOHO) spacecraft observed two comets plunging into the Sun's atmosphere in close succession, on 1 and 2 June 1998. (Courtesy: SOHO/LASCO consortium; ESA and NASA)

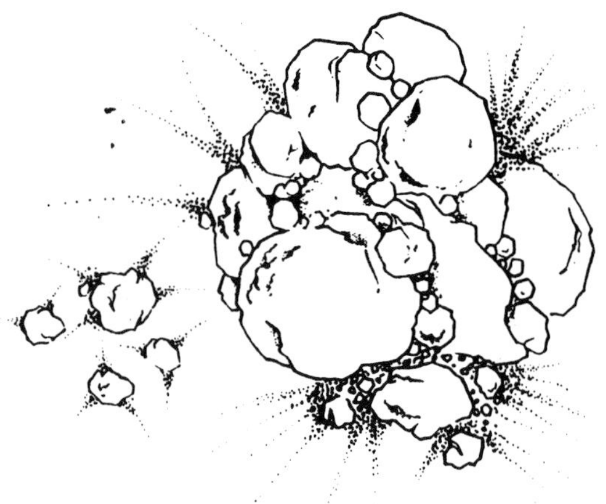

FIGURE 10.30 A schematic representation of a cometary nucleus according to the rubble-pile model, in which the individual fragments are lightly bonded by thermal processing or sintering. (Weissman 1986)

rupted D/Shoemaker–Levy 9 (see above), the comet nucleus retains its rubble-pile make-up, and does not weld into a coherent body. This theory is also consistent with radar observations of cometary nuclei.

Dynamically new comets, which have never entered the inner Solar System before, are likely to be covered by highly volatile ices. This volatile material sublimates and evolves off the surface as soon as its sublimation temperature is reached. This explains the relatively high activity in new comets at large heliocentric distances. In contrast, an old, periodic comet has lost much or all of its volatile surface material. Such a comet is covered by a *dust crust*. This crust is built up by dust grains too heavy to be dragged off the surface by the sublimating gases. Gas pressure from sublimating subsurface ice builds up to its highest value near perihelion. When the gas pressure is high enough, a portion of the dust crust is blown off, exposing fresh ice. This effect produces sudden increases in the activity of many periodic comets when they approach perihelion. During each perihelion passage a typical comet sublimates away a layer of ice only $\sim$1 m thick, which is small compared to the size of a comet (Problems 10.14, 10.15). However, periodic comets can develop thick dust crusts, and their activity can drop so low that no coma can be observed, even near perihelion. Such extinct comets are classified as asteroids.

One would expect the most volatile material of a comet to sublimate away long before perihelion is reached. Cometary activity at large heliocentric distances is often explained by the presence of such material, as was more recently confirmed by (radio) observations of the dependence of the production rate on heliocentric distance for a large number of (parent) molecules from C/Hale–Bopp (Fig. 10.5). Highly volatile material is, however, still very abundant, and even increases in abundance when a comet reaches perihelion. Hence, these highly volatile gases are prevented from sublimating away, even if the temperature is above their sublimation temperature. It has been suggested that some of these molecules are 'trapped' in the form of solid clathrate hydrates, where a guest molecule occupies a cage in the water-ice lattice. An alternative model is based upon the fact that water-ice in the cometary nucleus initially is likely amorphous rather than crystalline, since comets formed presumably at large heliocentric distances where the temperature is very low. Amorphous ice can easily trap large quantities of guest molecules. Timescales for crystallization are large, and depend exponentially on temperature. At a temperature of 140 K, amorphous ice will convert to a crystalline structure within about an hour, but at temperatures of $\sim$75 K this process takes about 40 billion years. The reaction is exothermic and irreversible. When a new comet enters the inner Solar System, one would expect amorphous ice to convert to a crystalline structure at $\sim$5 AU; because the reaction is exothermic it can be self-propagating within ice that is close to 140 K, and 10–15 m of ice may be converted at once. This transition may supply enough energy to 'blow off' the original primitive crust from a new comet, which could explain sudden brightenings at large heliocentric distances. The guest molecules trapped in the amorphous ice are set free. Those which have low sublimation temperatures diffuse through the crystalline ice and escape into space. Other molecules may recondense near the surface of the nucleus. These pockets of ice may sublimate away closer to the Sun, and cause *flare-ups* at smaller heliocentric distances.

10.6.5 Chemical Processing of Comets at Large Heliocentric Distances

Rather than a completely primitive crust, a new comet is thought to be covered by an *irradiation mantle*, caused by irradiation from cosmic rays and high energy photons at large heliocentric distances. Continuous bombardment of comets in the Oort cloud and Kuiper belt leaves the surface dark and red. Bombardment by energetic charged particles breaks up molecules on a comet's surface. The light hydrogen atoms may escape into interplanetary space, migrate through the ice matrix to form H_2, and/or initiate spin conversion in water and other symmetric molecules through exchange reactions. The heavier atoms/molecules stay on the surface, and may form new carbon-rich material (e.g., CHON particles, hydrocarbon chains), which is usually dark and red. The penetration depth of a charged particle depends upon its energy; low energy protons (1–300 keV) penetrate $\sim$10 μm into the crust, while GeV particles penetrate 1–2 m (if comet's density is 1 g cm^{-3}). The resulting refractory mantle is expected to be roughly a meter thick. The visible effect of irradiation by energetic photons (UV and X-ray) on a cometary surface is similar to that induced by charged particles (darkening and reddening of the material), but pertains only to the upper few μm of the surface. Most outer Solar System objects are dark, and some (e.g., Pholus, 1992 QB_1, 1993 HA_2) are very red.

Spectra of some objects in the Kuiper belt are neutral or slightly blue between 0.4 and 2.2 μm, resembling 2060 Chiron; while others show a strong reddening between 0.4 and 1.2 μm, being flat at longer wavelengths, which resembles the spectrum of 5145 Pholus.. Intermediate spectra have also been observed. This is thus in strong contrast to the theory of an irradiation mantle, discussed above, which predicts KBOs to be red. The data suggest that the reddest objects are also the faintest, so they presumably have the smallest diameters. This may suggest that the resurfacing model discussed above is best applicable to small bodies, which have the lowest gravitational pull and thus the largest loss of hydrogen atoms.

Since comets are rather small, they cannot have undergone much thermal evolution, and are regarded as the most pristine objects in our Solar System. Note, however, that only the outer layers of a comet are probed, and they may have undergone significant processing. We mentioned the radiation damage to cometary surfaces above. Other processes which could have altered a cometary surface are erosion and/or gardening by debris in the Oort cloud or by interstellar grains, and heating by passing stars or supernova explosions. There is a high probability that the Oort cloud has experienced heating events up to 30–50 K over the age of the Solar System. Such events would induce considerable effects/alterations in a comet's surface, such as changes in its porosity and sublimation of highly volatile material. These effects are discussed in detail by Mumma *et al.* (1993). Thus, although comets are still the most pristine objects in our Solar System, their surfaces may have undergone considerable processing while in the Oort cloud, an effect which complicates interpretation of cometary data considerably.

10.7 Comet Formation and Constraints on Theories of Solar System Formation

For centuries debates have been going on as to whether comets originated in the Solar System or come from interstellar space. The failure to observe comets on hyperbolic orbits (unless they have recently been perturbed onto escape orbits by the gravitational pull of the planets) is a convincing argument against an interstellar origin. The volatile nature of comets implies that they could not have formed near the Sun. Gravitational encounters with the giant planets scattered many small bodies out of the Solar System; however, a significant fraction could be perturbed by nearby stars or the galactic tidal field into the Oort cloud. An alternative viewpoint is that comets formed *in situ*, thousands of AU from the Sun; however, it is quite unlikely that the physical conditions necessary for these processes to operate existed.

10.7.1 Where did Comets and Cometary Ices Form?

Additional constraints on cometary origins come from the chemical composition of comets. All observed comets appear to be composed primarily of water-ice and more refractory dust, but more volatile species have also been observed. The total production rates of gas and dust, as well as the gas-to-dust mass ratios and the ratios of minor gases to that of water, vary considerably from one comet to the next, and even in time for any one comet. Thus there is growing evidence that comets are diverse, and that their nuclei are heterogeneous. The heterogeneity argues for a composition of chemically distinct bodies formed in different parts of the solar nebula.

The most volatile species detected in comets to date are S_2, N_2^+ and CO. The maximum ambient temperature in the region where the comets formed can be estimated from the condensation temperature of these gases, which is 20 K for S_2, 22 K for N_2 (the parent molecule of N_2^+), and 25 K for CO, if these gases condensed directly. The low spin temperature for water derived from observations of P/Halley suggests that water was last processed at a temperature of ~ 29 K. Thus, most comets must have formed in regions where the temperature was $\lesssim 20$–30 K, which corresponds to a heliocentric distance in the solar nebula $r_\odot \gtrsim 20$ AU, i.e., beyond the orbit of Uranus. As mentioned in Section 10.1, the formation of icy planetesimals at distances corresponding to the orbits of Uranus–Neptune and subsequent ejection into an inner Oort cloud is a natural consequence of current models of planet formation. Formation at distances in the solar nebula beyond the orbit of Neptune would also be consistent with a comet's chemical composition; however, the densities in the far regions of the solar nebula (e.g., the Oort cloud) were too low to explain accretion of objects. Finally, it is possible that these materials formed in the cold interstellar medium, and were incorporated into comets without substantial modifications.

The (relative) abundances of CH_4–CO–C_2H_6, and that of formaldehyde (H_2CO) and methanol (CH_3OH), as well as the D/H ratio in cometary water and other isotope ratios, may further provide information on the conditions under which cometary ices formed (Section 10.3.5). Although some of these species have only been detected in a few comets to date (in particular C/Hyakutake and C/Hale–Bopp), there is growing evidence that cometary ices have similar characteristics to interstellar ice. The abundance ratio CH_4/CO in comets is $\gtrsim 0.1$, similar to the ratio observed in the condensed phase of the interstellar medium. The abundances of ethane and acetylene are comparable to the methane abundance, which can be explained by grain surface chemistry in a cold environment. Since photoprocessing converts methanol into formaldehyde, one would not expect much methanol in comets if cometary ices condensed within the primitive solar nebula. The detection of comparable amounts of formaldehyde and methanol suggests an interstellar origin for these constituents. Isotopic abundance ratios yield information on fractionation effects from the time and the region where the comets formed. In particular, deuterium fractionation in water-ice can only be much larger than the protosolar value if fractionation occurred in cold interstellar clouds; thus the high D/H values reported for comets suggest an interstellar origin. Isotope ratios in C, N, S and O, on the other hand, are generally consistent with the terrestrial value. However, chemical fractionation in these heavier isotopes is inefficient compared to hydrogen, thus these studies do not necessarily distinguish between an origin in the primitive solar nebula or cold interstellar clouds. Hence, based upon the low condensation temperature of some materials (in particular S_2 and N_2), the high D/H ratio and the relative abundances of above mentioned hydrocarbons, it seems unavoidable to conclude that a significant fraction of the molecules contained in the icy grains from which the comets formed originated in cold interstellar clouds, and were incorporated in the forming comets without having been isotopically equilibrated within the solar nebula.

10.7.2 Dynamical Constraints

Objects in the Kuiper belt likely formed *in situ*, by accretion in the solar nebula beyond Neptune's orbit. Thus, models predict that Kuiper belt comets formed farther from the Sun than did the now more distant Oort cloud objects. Pluto may be viewed as the largest member of the inner Kuiper belt. Its moon Charon probably formed from a giant impact between Pluto and a planetesimal, similar to the scenario of our Moon forming from a huge impact with Earth. Triton, in many respects similar to Pluto, probably originated in either the Uranus/Neptune region, or the inner Kuiper belt. However, while Triton was captured by Neptune, Pluto survived by being in a 3:2 resonance with the large planet.

There is broad agreement that two distinct comet reservoirs exist: the Kuiper belt and Oort cloud, but many questions remain unanswered. The most distant detected KBOs are at a heliocentric distance of just over 50 AU; this is at least in part a selection effect of the direct detection method. As mentioned in Section 10.2.2, based

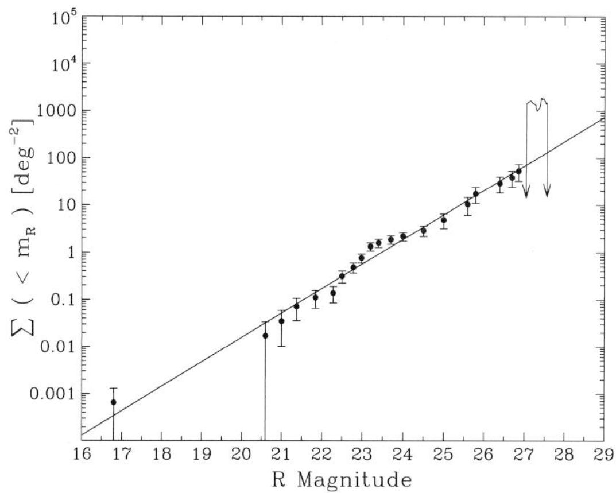

FIGURE 10.31 Cumulative sky density of Kuiper belt objects. The solid line is a power law (index $\zeta = 3.6$) fitted to all data, except for Tombaugh's (Pluto) point. (Chiang and Brown 1999)

upon the detection of the scattered KBOs there is a vast population of KBOs on highly eccentric orbits with semi-major axes well over 50 AU. In contrast, no KBOs with perihelia above 50 AU have been detected, although some should have been seen if the population of 100 km and larger objects were constant or increasing with heliocentric distance. It thus appears as if there is a 'cliff' at ~50 AU, beyond which the KBO population decreases substantially. Such a cliff might be caused by truncation of the planetary disk at the time of Solar System formation by perturbations from a passing star. Or, alternatively, is the region beyond 50 AU filled with only tiny KBOs? If so, the Kuiper belt might extend all the way out to the inner Oort cloud.

The size distribution of asteroids in the main belt is largely consistent with a power law size distribution expected for a population of collisionally interacting bodies ($\zeta = 3.5$ in eq. 9.2). Assuming uniform albedos, the radii of the KBOs fit a power law with $\zeta = 3.6 \pm 0.1$ (Fig. 10.31), i.e., similar to that obtained for asteroids. However, this number is highly uncertain, since there are no measurements of KBOs with radii smaller than 25 km (assuming albedos of 0.04). If there were a relative overabundance of small bodies, ζ would be larger than 3.6.

The Oort cloud and Kuiper belt have survived as comet reservoirs for over four billion years. Icy bodies in these reservoirs occasionally are perturbed into orbits which bring them into the planetary region. The galactic tide and passing stars provide the primary perturbations for bodies in the Oort cloud, whereas resonant perturbations of the outer planets (sometimes aided by orbit-alterating collisions) dominate for Kuiper belt objects. When these icy bodies approach the Sun, their most volatile constituents sublimate and evolve off the nucleus, taking along with them more refractory dust and producing comets which may be spectacular in their visual appearance. Most comets are quickly ejected from the Solar System as a result of gravitational perturbations by the planets. Some comets crash into the Sun, while others end their active lives releasing all of their volatiles or completely disintegrating. A small minority of comets collide with planets. The source regions of comets are gradually being depleted, but the average rate at which comets are being supplied to the planetary region will probably drop by at most a factor of a few between the present epoch and the end of the Sun's main sequence (hydrogen burning) lifetime six billion years hence.

10.8 Future

Several missions to comets have been planned over the next few years: Discovery mission 'Stardust', launched in February 1999, will fly to within 150 km of P/Wild-2 to capture cometary dust particles for analysis here on Earth. The dust particles will be captured in a material called 'aerogel', and the spacecraft is expected to bring this back to Earth in the year 2006. The mission 'Contour' (Comet Nucleus Tour) is scheduled for launch in the year 2002. It will image and take spectra of at least three comet nuclei (P/Encke, P/Schwassmann–Wachmann 3 and P/d'Arrest) and analyze the dust *in situ*. 'Deep Impact' will create and image a crater on the nucleus of P/Tempel 1 in the year 2005. The European comet mission, 'Rosetta', will be launched in the year 2003, to rendezvous and orbit P/Wirtanen. This spacecraft also has a lander to take *in situ* measurements on the comet's surface.

Several projects are being undertaken to search for small (~few km-sized) KBOs via stellar occultations. A small ground-based array of telescopes, TAOS (Taiwanese–American Occultation Survey), will search for KBOs by monitoring star clusters. This survey should be able to detect KBOs when they pass in front of a star and temporarily (for a fraction of a second) block out the star's light. This project will allow detection of small (~2 km-sized) KBOs at heliocentric distances $\lesssim$50 AU, as well as larger KBOs beyond 50 AU. A similar survey will be undertaken from space by the French satellite COROT. Because space-based observations are not influenced by the Earth's atmosphere, they should be able to achieve much higher photometric precision, and thus might be able to detect even smaller KBOs.

Further Reading

The following book contains a collection of review papers, written after the Giotto flyby of P/Halley:

Huebner, W.F., Ed., 1990. *Physics and Chemistry of Comets*. Springer-Verlag, Berlin.

The following two review papers are highly recommended:

Festou, M.C., H. Rickman, and R.M. West, 1993. Comets. 1 – Concepts and observations. *Astronomy and Astrophysics Reviews* **4**, no. 4, 363–447.

Festou, M.C., H. Rickman, and R.M. West, 1993. Comets. 2 – Models, evolution, origin and outlook. *Astronomy and Astrophysics Reviews* **5**, no.1,2, 37–163.

A paper describing and comparing the properties of 85 comets is:

A'Hearn, M.F., R.L. Millis, D.G. Schleicher, D.J. Osip, and P.V. Birch, 1995. The ensemble properties of comets: Results from narrowband photometry of 85 comets, 1976–1992. *Icarus* **118**, 223–270.

The most recent catalogue of cometary orbits is published by:

Marsden, B.G., and G.V. Williams, 1997. *Catalogue of Cometary Orbits*. The International Astronomical Union, Minor Planet Center and Smithsonian Astrophysical Observatory (Cambridge, MA).

Papers reviewing cometary properties and putting these into perspective with regard to the origin of the Solar System are given by:

Mumma, M.J., P.R. Weissman, and S.A. Stern, 1993. Comets and the origin of the Solar System: Reading the Rosetta Stone. *Protostars and Planets III*. E.H. Levy and J.I. Lunine, Eds. pp. 1177–1252. University of Arizona Press, Tucson.

Irvine, W.M., F.P. Schloerb, J. Crovisier, B. Fegley Jr., and M.J. Mumma, 2000. Comet: a link between interstellar and nebular chemistry. In *Protostars and Planets IV*. V. Manning, A.P. Boss, and S.S. Russell, Eds. pp. 1159–1200. University of Arizona Press, Tucson.

Papers summarizing our knowledge of the Kuiper belt can be found in:

Jewitt, D.C., and J.X. Luu, 2000. Physical nature of the Kuiper belt. In *Protostars and Planets IV*. V. Manning, A.P. Boss, and S.S. Russell, Eds. pp. 1201–1229. University of Arizona Press, Tucson.

Malhotra, R., M.J. Duncan, and H.F. Levison, 2000. Dynamics of the Kuiper belt. In *Protostars and Planets IV*. V. Manning, A.P. Boss, and S.S. Russell, Eds. pp. 1231–1254. University of Arizona Press, Tucson.

Farinella, P., D.R. Davis, and S.A. Stern, 2000. Formation and collisional evolution of the Edgeworth–Kuiper belt. In *Protostars and Planets IV*. V. Manning, A.P. Boss, and S.S. Russell, Eds. pp. 1255–1282. University of Arizona Press, Tucson.

Problems

10.1.E Estimate the largest negative value of $1/a_0$ that nongravitational forces are reasonably likely to produce for a comet that is initially in heliocentric orbit and is not subjected to significant planetary perturbations. You may assume that the comet's initial orbit is so eccentric that it may be taken as parabolic and that the comet outgasses 0.1% of its mass, with a 10% asymmetry and that this outgassing occurs in a single burst at perihelion, which is at 0.2 AU.

10.2.I Draw a theoretical histogram of $1/a_0$ for interstellar comets. You may assume that the velocity distribution (or, more precisely, the distribution of speed, since direction does not matter) of interstellar comets relative to the Sun at 'infinity' is comparable to that of neighboring stars, which can be approximated by a Maxwellian of mean velocity 30 km s^{-1}, i.e.,

$$N(v) \propto \frac{v}{v_*} e^{-(v/v_*)^2} \tag{10.21}$$

where $v_* = 30$ km s^{-1}.

10.3.E Draw a histogram of $1/a$ for short-period comets, using data in a recent edition of Marsden and Williams' comet catalogue, or on the web, via: http://cfa-www.harvard.edu/iau/ or http://pdssbn.astro.umd.edu/

10.4.I Estimate the frequency at which we would expect to observe interstellar comets. Assume that the Oort cloud contains 10^{13} comets that are large enough to be observable if they were to reach the inner Solar System, and that twice as many comets have been ejected from the Solar System over time. Postulate that the number of comets produced and released by a planetary system is proportional to the star's mass, and that our Solar System is average in this regard. Furthermore, assume that the density of stellar mass in the solar neighborhood is 0.065 $M_\odot$ parsec^{-3} (1 parsec = 3.1×10^{18} cm), the typical velocity of a comet relative to the Sun 'at infinity' is comparable to that of neighboring stars, 30 km s^{-1}, and a comet must come within 2 AU of the Sun to be visible.

(a) Neglecting gravitational focusing by the Sun, i.e., approximating the trajectories of the comets by straight lines.
(b) Including gravitational focusing by the Sun, i.e., considering the comets to be traveling on hyperbolic orbits about the Sun.
(c) Repeat parts (a) and (b) assuming that we can observe all comets that approach to within 5 AU of the Sun.

10.5.E The magnitude of a body is related to its flux density via the first part of equation (10.4). Calculate the apparent magnitude of a Kuiper belt object at a heliocentric distance of 40, 70 and 150 AU, assuming a visual albedo of 0.04 and a radius of 150 km. (Hint: What is the solar flux at these distances?)

10.6.I Use diagrams and/or equations to show that under the assumptions of the standard (symmetric) model of nongravitational forces, the tangential term of the acceleration resulting from cometary outgassing can produce a secular change in a comet's orbital period, but the radial term cannot.

10.7.E Show that if seasonal variations break the symmetry about perihelion of a comet's outgassing, then the radial component of the nongravitational force can secularly alter the comet's orbital period.

10.8.E Suppose a comet has a velocity of 40 km s^{-1} at perihelion. The perihelion distance is 1 AU. Calculate the aphelion distance, the velocity of the comet at aphelion and the orbital period of the comet.

10.9.E An asteroid and a comet have the same apparent brightness while both at $r_\Delta = 2$ AU, and $r_\odot = 3$ AU. At a later time they are both observed at $r_\Delta =$ 2 AU and $r_\odot = 2$ AU. Which object is brighter, and by approximately how much?

10.10.E The apparent visual magnitude of Comet X is $m_v = 21.0$ when it is discovered. At this time $r_\odot = r_\Delta = 10$ AU.

(a) Calculate the absolute magnitude of the comet, H_{10}.

(b) Estimate m_v when the comet reaches perihelion at $r_\odot = 0.3$ AU; at this time $r_\Delta = 1$ AU.

10.11.E Calculate the surface temperature of a comet with Bond albedo $A_b = 0.1$ at a heliocentric distance of 15 AU.

10.12.I At a distance $r_\odot = 1$ AU, outgassing is dominated by H_2O, and a comet's surface temperature is 190 K. The latent heat of sublimation $L_s \approx 5 \times 10^{11}$ ergs mole^{-1}. Assume the Bond albedo $A_b = 0.1$ and the infrared emissivity $\epsilon_{ir} = 0.9$.

(a) Derive equation (10.6), and use it to determine the gas production rate of this comet.

(b) Calculate the thermal expansion velocity of the gas, and compare this to the escape velocity of a typical comet (radius 1–10 km).

10.13.E (a) Perform the integration in equation (10.8) and solve for the collisional radius $\mathcal{R}_c$.

(b) Use your result to calculate the collision radius of Halley's comet at 1 AU from the Sun. Assume the comet outgasses $\sim 10^{30}$ molecules s^{-1}, the terminal expansion velocity $v = 1$ km s^{-1} and a typical molecular radius is 1.5 Å.

10.14.E A comet's perihelion distance is 1 AU, and its aphelion distance is 15 AU. In the following we make a very very crude calculation of the average rate of shrinkage of the comet.

(a) Calculate the comet's orbital period.

(b) Calculate how many meters of ice the comet will lose each time it orbits the Sun. (Hint: In order to simplify the calculations, you may assume that ice sublimates off the comet's surface during 1/10 its orbital period, that the average cometary distance over that period is 1.5 AU and that the density of the cometary ice is 1 g cm^{-3}.)

10.15.I Assume the comet from Problem 10.14 is active inside 3 AU and inert at large heliocentric distances. By making use of Kepler's second and third laws, together with the gas production rate as a function of heliocentric distance (eq. 10.6), calculate how many meters of ice the comet loses each orbit.

10.16.E A comet consists primarily of water-ice; when the ice sublimates, the water molecules flow off the surface at the thermal expansion velocity. Assume the comet to be at $r_\odot = 0.6$ AU, and the sublimation temperature of water to be 200 K.

(a) Calculate the thermal expansion velocity.

(b) The typical lifetime of H_2O molecules is 6×10^4 s, for OH it is 2×10^5 s, and for H it is 10^6 s. Assume the outflow velocity of OH to be equal to that of H_2O, and for H it is on average 12 km s^{-1}. Calculate the typical sizes of the H_2O, OH and H comae.

(c) Suppose the comet is perfectly spherical and homogeneous, with a radius of 10 km. Calculate the collisional radius, $\mathcal{R}_c$.

(d) When the particle density exceeds 10^6 molecules cm^{-3}, collisions with OH molecules are so numerous that these molecules are quickly thermalized. Calculate the radius inside which the coma density exceeds 10^6 molecules cm^{-3}.

(e) Calculate the time between collisions at a distance of 20 km (from the center of the comet), 100 km, 1000 km and 10^4 km.

(f) Compare your answers from (c) and (d); explain the significance of those numbers.

10.17.I A comet at a heliocentric distance of 1 AU shows a gas production rate $\mathcal{Q} = 10^{29}$ molecules s^{-1}. The outgassing is dominated by H_2O.

(a) Plot the H_2O number density as a function of distance from the comet. Assume an outflow velocity of 1 km s^{-1} and a lifetime for H_2O molecules of 6×10^4 s.

(b) Assume that all H_2O molecules dissociate into OH and H, and OH has a lifetime of 1.7×10^5 s. Plot, according to the Haser model, the OH number density as a function of distance from the comet.

(c) How, qualitatively, does $N_d(r)$ differ in the vectorial model from that given by the Haser model? Describe approximately using a graph and associated text.

10.18.I Assume the OH brightness of a comet is proportional to N_d, the number of OH molecules, and that the comet under consideration is not resolved by the telescope. We receive the OH emission at a frequency of 1667.0 MHz. There are 15 frequency channels, centered at 1667.0 MHz, which is the rest frequency of the comet (i.e., the comet's velocity is zero at 1667.0 MHz). Each channel is 1.11 kHz wide. Calculate the line profile (in relative numbers, as a function of frequency and gas flow velocity) you expect for the comet from Problem 10.16, assuming the Haser model represents the gas outflow accurately.

10.19.I Derive equation (10.10*b*) from equation (10.10*a*) using the assumptions of the Haser model. (Hint: Follow the total flux of molecules moving through concentric spheres centered on the comet.)

10.20.E Comet X is observed in the CO (1510 Å), C I (1657 Å) and OH (3090 Å) transitions, while at a geocentric distance of 0.8 AU and heliocentric distance of 0.4 AU. The g-factors at $r = 1$ AU for the three molecules are 2.2×10^{-7}, 2.5×10^{-5} and 1.2×10^{-3}, respectively. The lifetimes of the various molecules (at 1 AU) are approximately: 1×10^6, 2.5×10^5 and 1.6×10^5 seconds, respectively. The observed brightnesses are 30, 770 and 105 000 photons s^{-1} cm^{-2} for CO, C I and OH, respectively. Assume that the comet is unresolved in the telescope beam, and that it subtends an angle of 10 arcminutes on the sky.

(a) Calculate the column density for each of the three species.

(b) Assume that the comet is spherically symmetric and outgassing equally in all directions. Determine the gas production rate, $\mathcal{Q}$, for all three compounds. Compare the results and comment on the similarities/differences.

10.21.E Consider a comet in orbit around the Sun. The molecules evolve off the nucleus at 1 km s^{-1}.

(a) Sketch the OH line profiles at 1667 MHz that you would expect to observe when the radial component of the comet's heliocentric velocity $v_r = -20$ km s^{-1}. (Hint: Use the graph for the inversion measure shown in Fig. 10.10.)

(b) Repeat for $v_r = -8$ km s^{-1}.

(c) Repeat for $v_r = -14$ km s^{-1}.

10.22.E Derive equation (10.15). (Hint: Assume that the grain is spherical, and that it radiates uniformly in all directions.)

10.23.E Calculate approximately the standoff distance of the bow shock of Halley's comet at $r_\odot = 0.9$ AU. Assume an icy comet with a production rate of 10^{30} molecules s^{-1}, a solar wind density of 5 protons cm^{-3} and velocity of 400 km s^{-1}. Estimate roughly the ionization rate from molecular lifetimes.

10.24.I Comet X is on a nearly parabolic orbit with a perihelion at 0.5 AU. Calculate the (generalized) eccentricity of grains released near perihelion as a function of β. Give a formula for the separation between the grains and the nucleus as a function of β and t which is valid for the first few days following release.

10.25.I Derive equation (10.16*a*). (Hint: Set the upward force from gas pressure equal to the downward force from gravity and solve for $R_{d,max}$.)

10.26.E Calculate the size of the largest icy particle that can be released by gas drag from the surface of an icy comet with a radius of 1 km at heliocentric distances of 3, 1 and 0.3 AU. List and explain all of your assumptions.

10.27.E Make a very crude estimate of the relative brightness of a comet's coma and nucleus by assuming that the coma consists of 10 μm grains which emanate from a uniform 10 cm thick layer on the comet's surface, the composition of the comet is 50% ice and 50% dust and that the albedo of the grains is the same as that of the nucleus. Using your result, comment on the possibility of observing the nucleus of an active comet from the ground.

11 Planetary Rings

It (Saturn) is surrounded by a thin flat ring, nowhere touching, and inclined to the ecliptic.

Christiaan Huygens, published in Latin in anagram form in 1656

Each of the four giant planets in our Solar System is surrounded by flat, annular structures known as *planetary rings*. Planetary rings are composed of vast numbers of small satellites, which are unable to accrete into large moons because of their proximity to the planet.

When Galileo Galilei first observed Saturn's rings in 1610, he believed them to be two giant moons in orbit about the planet. However, these 'moons' appeared fixed in position, unlike the four satellites of Jupiter which he had previously observed. Moreover, Saturn's 'moons' had disappeared completely by the time Galileo resumed his observations of the planet in 1612. Many explanations were put forth to explain Saturn's 'strange appendages', which grew, shrank and disappeared every 15 years (Fig. 11.1). In 1656, Christiaan Huygens finally deduced the correct explanation, that Saturn's strange appendages are a flattened disk of material in Saturn's equatorial plane, which appear to vanish when the Earth passes through the plane of the disk (Fig. 11.2).

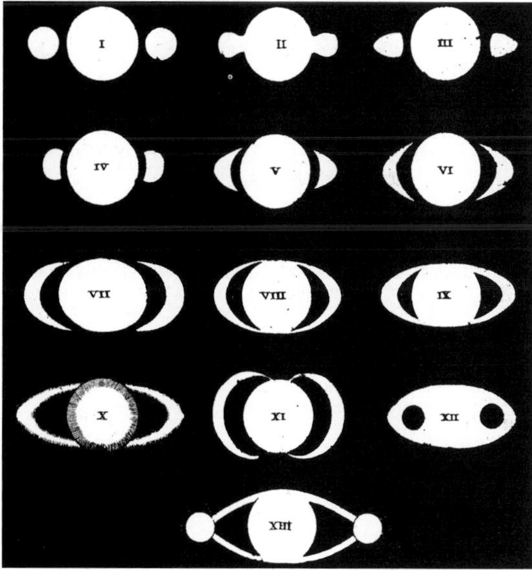

FIGURE 11.1 Seventeenth century drawings of Saturn and its rings.

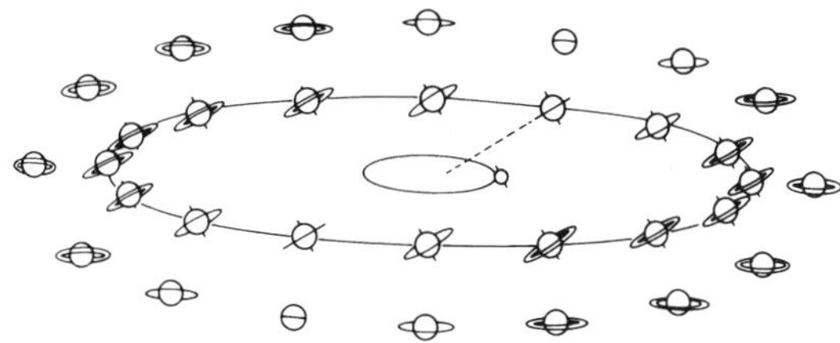

FIGURE 11.2 Cartoon views of Saturn and its rings over one saturnian orbit according to Huygens' model.

For more than three centuries, Saturn was the only planet known to possess rings. Although Saturn's rings are quite broad, little structure in the ring system was detected from Earth (Fig. 11.3). Observational and theoretical progress towards understanding the physics of planetary rings was slow. In March of 1977, an occultation of the star SAO 158687 revealed the narrow opaque rings of Uranus (Fig. 11.4) and launched a golden age of planetary ring exploration. The Voyager spacecraft first imaged and studied the broad but tenuous ring system of Jupiter in 1979 (Section 11.3.1). Pioneer 11 and the two Voyagers obtained close-up images of Saturn's spectacular ring system in 1979, 1980 and 1981 (Fig. 11.5; Section 11.3.2). Neptune's rings, whose most prominent features are azimuthally incomplete arcs, were discovered by stellar occultation in 1984. Voyager 2 obtained high-resolution images of the rings of Uranus in 1986 (Section 11.3.3) and the rings of Neptune in 1989 (Section 11.3.4). Technological advances allowed ground-based and Earth-orbiting studies of planetary rings over far more wavelengths and with much higher precision than previously attained. The Galileo spacecraft obtained high-resolution images of Jupiter's rings in the late 1990s. Finally, our theoretical understanding of rings advanced by leaps and bounds – despite these advances, we now have more out-

FIGURE 11.3 Ground-based photographs of Saturn and its rings over one-half of a saturnian orbit.

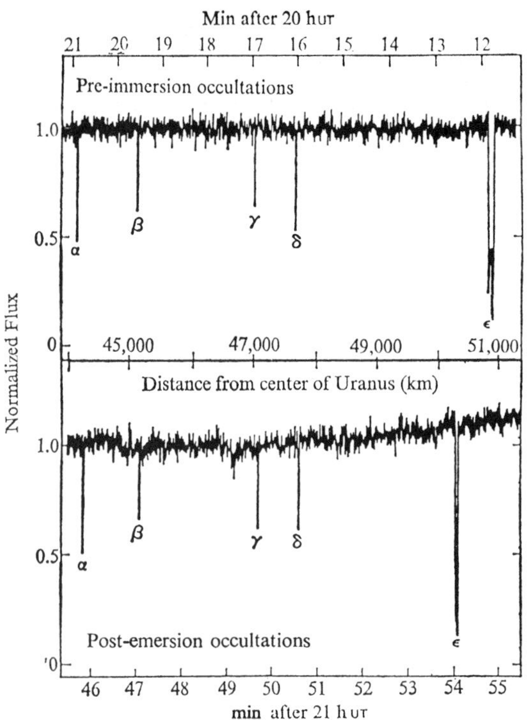

FIGURE 11.4 Lightcurve of the star SAO 158687 as it was observed to pass behind Uranus and its rings. Dips in the lightcurves corresponding to the occultation of the star by five rings are clearly seen both prior to immersion and following emersion of the star from behind the planet. Four of these pairs of features are symmetric about the planet, but the location, depth and duration of the outermost pair imply that the ϵ ring is both noncircular and nonuniform. (Adapted from Elliot *et al.* 1977)

standing questions concerning rings than researchers had in the mid-1970s!

In this chapter, we summarize our current observational and theoretical understanding of planetary rings. We begin with an explanation of why rings exist, and why they are generally located much closer to planets than are large moons (Fig. 11.6). A more detailed observational summary is then presented, followed by theoretical models for some of the features observed. We conclude with a discussion of the evolution of planetary ring systems and models of planetary ring formation.

11.1 Tidal Forces and Roche's Limit

The strong tidal forces close to a planet lead orbital debris to form a planetary ring rather than a moon. The closer a moon is to a planet, the stronger the tidal forces that it is subjected to. If it is too close, then the difference between the gravitational force exerted by the planet on the point of the moon nearest to (and farthest from) the planet and that exerted on the center of the moon is stronger than the moon's self-gravity. Under such circumstances, the moon is ripped apart, unless it is held together by mechanical strength, and a planetary ring results.

In order to understand tidal disruption more quantitatively, we make the following assumptions:

(1) The system consists of one large primary body (the planet) and one small secondary body (the moon).

(2) The orbit is circular, the rotational period of the moon is equal to its orbital period, and the moon's obliquity is nil. (These assumptions make the analysis far simpler, because the problem becomes stationary in a rotating frame.)

(3) The moon is spherical, and the planet can be treated as a point mass.

(4) The moon is held together by gravitational forces only.

The 'external' forces per unit mass on material in orbit about a planet of mass M_p are gravity:

$$\mathbf{g}_p = -\frac{GM_p}{r^2}\hat{\mathbf{r}}, \tag{11.1}$$

and centrifugal force:

$$\mathbf{g}_n = n^2 r\hat{\mathbf{r}}, \tag{11.2}$$

FIGURE 11.5 COLOR PLATE This approximately natural-color image shows Saturn, its rings, and four of its icy satellites. Three satellites (Tethys, Dione, and Rhea) are visible against the darkness of space, and another smaller satellite (Mimas) is visible against Saturn's cloud tops very near the left horizon and just below the rings. The dark shadows of Mimas and Tethys are also visible on Saturn's cloud tops, and the shadow of Saturn is seen across part of the rings. The pronounced concentric gap in the rings, the Cassini division, is a 3500 km wide region that is much less populated with ring particles than are the brighter B and A rings to either side. This image was synthesized from images taken in Voyager's blue and violet filters and was processed to recreate an approximately natural color and contrast. (NASA PIA00400)

where the origin is at the center of the planet and n is the angular velocity of the system. Steady state in the frame rotating with the system gives:

$$n^2 r\hat{\mathbf{r}} - \frac{GM_p\hat{\mathbf{r}}}{r^2} = 0, \tag{11.3a}$$

therefore:

$$n^2 = \frac{GM_p}{r^3}. \tag{11.3b}$$

Note that equation (11.3b) implies Kepler's third law for the case of circular orbits.

The sum of the gravitational force and the effect of the rotating frame of reference ('centrifugal force') is referred to as the *effective gravity*; the local effective gravity vector points normal to the *equipotential surface* in the rotating frame. The effective gravity, $\mathbf{g}_{eff}$, felt by an object that is at a distance r from the planet's center and traveling on a circular orbit at semimajor axis a is:

$$\mathbf{g}_{eff} = GM_p\left(\frac{r}{a^3} - \frac{1}{r^2}\right)\hat{\mathbf{r}}. \tag{11.4}$$

The (effective) tidal force upon such a body is:

$$\frac{d\mathbf{g}_{eff}}{dr} = GM_p\left(\frac{1}{a^3} + \frac{2}{r^3}\right)\hat{\mathbf{r}} = \frac{3GM_p}{a^3}\hat{\mathbf{r}}, \tag{11.5}$$

where the approximation $r^3 = a^3$ has been used in the last step. Note that equation (11.5) differs from equation (2.40) because it also includes a contribution from the centrifugal force. The moon's self-gravity just balances the tidal force at the surface of the moon when:

$$\frac{GM_s}{R_s^2} = \frac{3GM_pR_s}{a^3}, \tag{11.6}$$

where the subscript s refers to the satellite (moon). This occurs at a planetocentric distance of:

$$\frac{a}{R_p} = 3^{\frac{1}{3}}\left(\frac{\rho_p}{\rho_s}\right)^{\frac{1}{3}} = 1.44\left(\frac{\rho_p}{\rho_s}\right)^{\frac{1}{3}}. \tag{11.7}$$

In the above derivation, we made a number of simplifying assumptions. Let us now review the accuracy of these assumptions in order to assess the applicability of our calculations:

(1) We used the small moon/large planet approximation in order to neglect the influence of the moon on the planet and to neglect terms containing higher powers of the ratio of the radius of the moon to its orbital semimajor axis. This assumption is thus very accurate for bodies within our Solar System.

(2) All known interior moons have low eccentricities, so the approximation of circular orbits is very good. All moons near planets for which rotation rates have been measured are in synchronous rotation and have low obliquity. Young moons which have not had time to be tidally despun, and thus rotate rapidly, would be less stable.

(3) Although the giant planets are noticeably oblate, the departures of their gravitational potentials from those of point masses have only an $\mathcal{O}(1\%)$ effect on these tidal stability calculations. A much larger effect results from moons being stretched out along the planet–moon line due to the planet's gravitational tug (Fig. 2.9). This stretching brings the tips of the moon farther from its center, which both decreases the

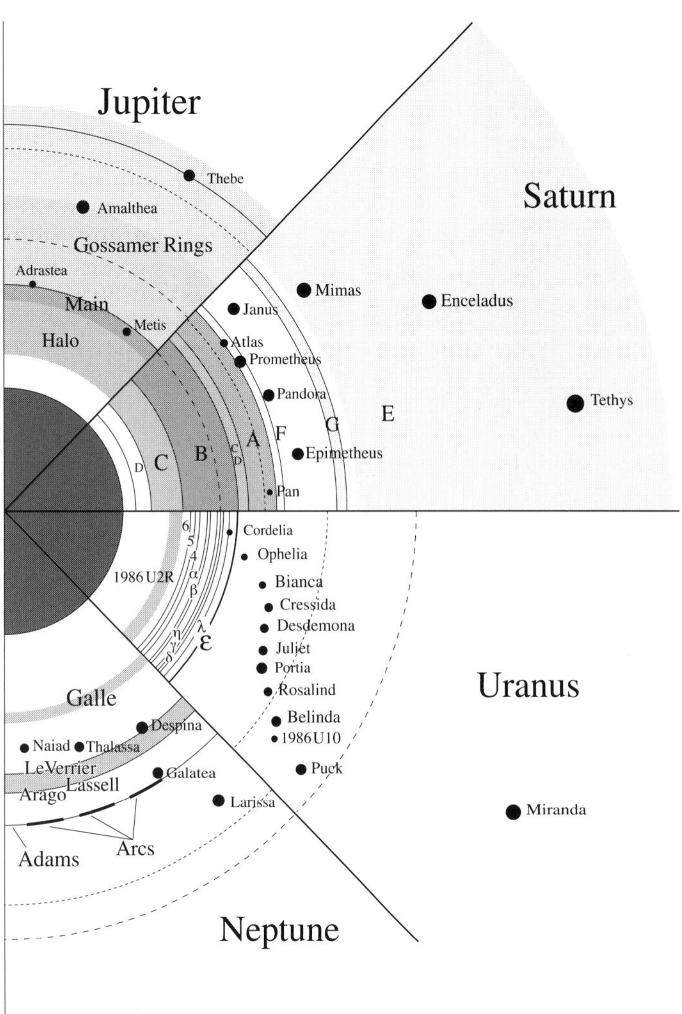

FIGURE 11.6 Diagram of the rings and inner moons of the four giant planets. The systems have been scaled according to planetary equatorial radius. The long-dashed curves denote the radius at which orbital motion is synchronous with planetary rotation. The short-dashed curves show the location of Roche's limit for particles of density 1 g cm^{-3}. (Courtesy: J.A. Burns, D. P. Hamilton, and M. R. Showalter)

magnitude of self-gravity and increases the tidal force. In 1847, Roche performed a self-consistent analysis for a liquid (fully deformable) moon and obtained:

$$\frac{a_R}{R_p} = 2.456\left(\frac{\rho_p}{\rho_s}\right)^{\frac{1}{3}}. \tag{11.8}$$

The location a_R is known as *Roche's limit* for tidal disruption.

(4) Most small bodies have significant internal strength, e.g., small moons are not always spherical. Physical coherence of small bodies allows moons smaller than ~100 km radius to be stable somewhat inside Roche's limit. Ring particles, which typically are so small that internal strength exceeds self-gravity by orders of magnitude provided the particles are not loose aggregates, can remain coherent well inside Roche's limit.

The concept of Roche's limit explains in a semi-quantitative manner why we observe rings near giant planets, small moons a bit farther away, and large moons only at greater distances (Fig. 11.6). However, the interspersing of some rings and moons implies that other factors are important in determining the precise configuration of a planet's satellite system. We will return to theories on the origin and evolution of ring/moon systems in Section 11.7.

11.2 Flattening and Spreading of Rings

A ring particle orbiting a planet passes through the planet's equatorial plane twice each orbit, unless its trajectory is diverted by a collision with another particle or by the ring's self-gravity. The average number of collisions that a particle experiences during each vertical oscillation is a few times as large as the optical depth of the rings, τ (Problem 11.6; collisions can be much more frequent in physically thin but optically thick self-gravitating rings). Typical orbital periods for particles in planetary rings are 6–15 hours. As τ is $\mathcal{O}(1)$ in most of Saturn's (and Uranus's) rings, collisions are very frequent. Collisions dissipate energy but conserve angular momentum. Thus, the particles settle into a thin disk on a timescale:

$$t_{flat} = \frac{\tau}{\mu}, \tag{11.9}$$

where μ is the frequency of the particles' vertical oscillations. An oblate planet exerts torques which alter the orbital angular momenta of orbiting particles. These torques cause inclined orbits about oblate planets to precess (cf. Section 2.5). When coupled with collisions among ring particles, a secular transfer of angular momentum between the planet and the ring can result. Only the component of the ring's angular momentum which lies along the planet's spin axis is conserved. Any net angular momentum of the disk parallel to the planet's equator is quickly dissipated. As collisions at high speeds damp relative motions rapidly, the ring settles into the planet's equatorial plane on a timescale of a few orbits (or $\sim \tau^{-1}$ orbits, if $\tau \ll 1$).

Several mechanisms act to maintain a nonzero thickness of the disk. Finite particle size implies that even particles on circular orbits in a planet's equatorial plane collide with a finite velocity, when an inner particle catches up with a particle farther out that moves less rapidly. Unless the collision is completely inelastic, i.e., unless the two particles stick, some of the energy involved goes into random particle motions. The ultimate consequence of these collisions is a spreading of the disk. Viewed in another way, spreading of the disk is the source of the energy required to maintain particle velocity dispersion in the presence of inelastic collisions. Gravitational scatterings between slowly moving particles also convert energy from ordered circular motions to random velocities, in this case without the losses resulting from the inelasticity of physical collisions.

Collective gravitational effects are important when the velocity dispersion of the particles, c_v, is so small that Toomre's stability parameter:

$$Q_T \equiv \frac{\kappa c_v}{\pi G \sigma_\rho}, \tag{11.10}$$

is less than unity. Here κ is the particles' epicyclic (radial) frequency (for orbits near the equatorial plane of an oblate planet, κ is slightly smaller than n, cf. eqs. 2.34 and 2.35), and σ_ρ is the surface mass density of the rings. The dispersion velocity of the ring particles, c_v, is related to the Gaussian scale height of the rings, H_v, via the formula:

$$c_v = H_v \mu. \tag{11.11}$$

When $Q_T < 1$, the disk is unstable to axisymmetric clumping of wavelength:

$$\lambda = \frac{4\pi G \sigma_\rho}{\kappa^2}. \tag{11.12}$$

For parameters typical of Saturn's rings, λ is on the order of 10–100 m. Clumping may be able to produce particles $\sim (2\pi)^{-1}$ times as large as this length scale. Such large bodies can then stir up random velocities of smaller bodies, and the observed size distribution (discussed in Section 11.3.2.4) may be maintained in this manner. External energy sources may also contribute to maintaining random velocities of particles against energy loss from inelastic collisions, especially near strong orbital resonances with moons (cf. Section 11.4).

As a result of continuing collisions, rings spread in the radial direction. The diffusion timescale is:

$$t_d = \frac{\ell^2}{\nu_v}, \tag{11.13}$$

where ℓ is the radial length scale (the width of the ring, or of a particular ringlet). The viscosity, ν_v, depends on particle collision velocities and the local optical depth approximately as:

$$\nu_v \approx \frac{c_v^2}{2\mu}\left(\frac{\tau}{1+\tau^2}\right) \approx \frac{c_v^2}{2n}\left(\frac{\tau}{1+\tau^2}\right). \tag{11.14}$$

Inserting $\ell = 6 \times 10^9$ cm (the approximate radial extent of Saturn's main rings) and a viscosity of $\nu_v = 100\ \text{cm}^2\ \text{s}^{-1}$, equation (11.14) yields a timescale comparable to the age of the Solar System. For smaller ℓ, timescales are much shorter (Problem 11.9). Even in regions of planetary rings where the viscosity is a few orders of magnitude lower than the value quoted above, viscous diffusion is expected to rapidly smooth out any fine scale density variations. Note that equation (11.14) is valid only if particles move several times their diameters relative to one another between collisions; the viscosity of dense, high τ, rings depends on other factors, including particle size.

Although under most circumstances viscosity acts to wipe out structure, it is possible that in some regions of planetary rings a *viscous instability* may occur. A ring is unstable to clumping in the radial direction if the viscous torque, $\nu_v \sigma_\rho$, is a decreasing function of surface density:

$$\frac{d}{d\sigma_\rho}\left(\nu_v \sigma_\rho\right) < 0 \rightarrow \text{instability}. \tag{11.15a}$$

If the surface density is proportional to optical depth, then equations (11.14) and (11.15a) may be combined to yield the stability condition:

$$\frac{\tau}{c_v}\frac{dc_v}{d\tau} + \frac{1}{\tau^2+1} < 0 \rightarrow \text{instability}. \tag{11.15b}$$

As c_v and τ are positive, equation (11.15b) implies that a ring is viscously unstable if the velocity dispersion of the ring particles decreases sufficiently rapidly as optical depth is increased. This would imply that more particles are able to diffuse from regions of lower optical depth into regions of higher optical depth than vice versa. A density perturbation is thus amplified as a result of diffusion and 'ringlets' may be formed. Note that as the second term in equation (11.15b) is always positive and decreases as τ increases, the instability is most likely to occur in regions of high optical depth, such as Saturn's B ring, where our expression for viscosity may not be valid. Furthermore, the velocity dispersion of ring particles depends on several factors, including the coefficient of restitution for inelastic collisions. Experimental results for low-velocity impacts of icy particles suggest that c_v does in fact decrease as τ increases, but probably not sufficiently rapidly to lead to viscous instability. Most structure in optically thick planetary ring systems must therefore be actively maintained, except on the largest length scales, where it may have resulted from 'initial' conditions.

11.3 Observations of Planetary Rings

Although the same basic physical processes govern the particles in all planetary ring systems, each system has its own distinctive character. Most rings lie inside or near the Roche limit. However, tenuous rings of ephemeral dust particles are observed in less tidally hostile environments. The differences in dynamical structure of the planetary ring systems are apparent in casual inspection of Figures 11.7–11.10, 11.14 and 11.15[1]. Saturn's ring particles have a high albedo, whereas particles in other ring systems are generally quite dark. Particle sizes range from submicrometer dust to bodies large enough to be considered moons. Indeed, there is no fundamental demarcation line between large ring particles and small moons. An operational definition useful at present is that bodies viewed as individual objects are defined to be moons and given names, whereas (smaller) bodies detected only as a collective ensemble are known as ring particles. This definition will become obsolete when extremely high-resolution images of rings revealing small bodies as distinct entities become available; at such time a new definition, probably based on the ability of a body to gravitationally clear a gap around itself (Section 11.4.3), will be needed.

Our knowledge of ring properties has been obtained almost exclusively from photons of various wavelengths that have been scattered, reflected, absorbed or emitted by ring particles, although a small amount of data is available from ring absorption of charged particles and impacts of microscopic particles on spacecraft passing through very tenuous regions of planetary ring systems. In this section, we summarize the properties of planetary ring particles and large scale structure in planetary rings. Theoretical explanations for some of this structure are discussed in Sections 11.4–11.6; the mechanisms responsible for creating/maintaining the particle size distribution are not well-understood theoretically, but a few general principles are discussed together with the observations.

11.3.1 Jupiter's Rings

The jovian ring system is extremely tenuous, and consists of three principal components, the main ring, the halo and the gossamer rings (Fig. 11.7 and Table 11.1). Jupiter's rings must contain a substantial fraction of tiny particles (in terms of surface area and optical depth, but not of mass), because they are observed to be brightest in forward scattering. The *main ring* is the most prominent component, especially in backscattered light, which suggests the presence of larger particles. The normal optical depth of the main ring $\tau \approx$ few $\times\ 10^{-6}$. The micrometer and submicrometer sized particles which present about half of the optical depth of the main ring and cover the bulk of the surface area in the other two components of the jovian ring system are ephemeral: several loss mechanisms, principally sputtering by energetic ions, limit the lifetimes of such particles to $\lesssim 10^3$ years; orbital evolution via processes discussed in Section 11.5.2 is also very rapid. An ongoing source of particles is thus required, unless we are observing the rings at a very special time. The primary formation mechanism for the dust grains in the main and gossamer rings is believed to be erosion (probably by micrometeorites) from the small moons bounding the rings, a theory based upon the morphology and vertical structure of the rings (Section 11.6).

[1] Note that the differences between ring systems are generally substantially larger than they appear to be in most processed images. Camera exposures and a variety of image processing techniques are usually selected to compensate for variations in overall brightness and are often used to stretch or filter these data in order to make structure more apparent to the eye. Thus, processed images usually display rings as fairly prominent features with internal brightness variations of order unity, even though many actual rings are tenuous and/or nearly uniform in brightness.

(a)

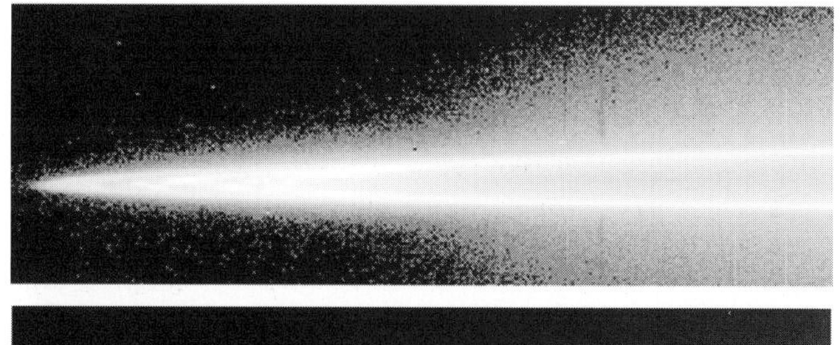

(b)

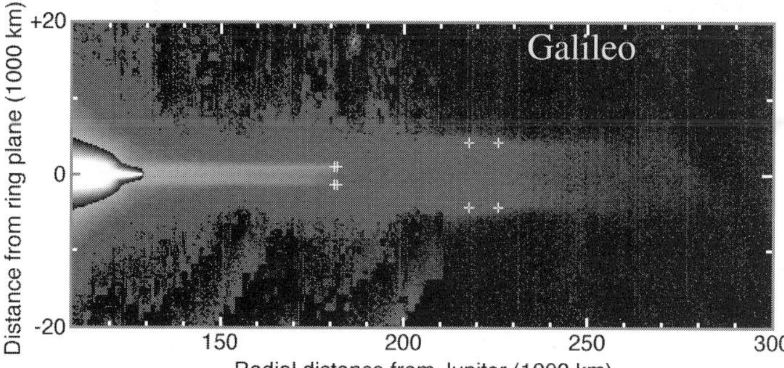

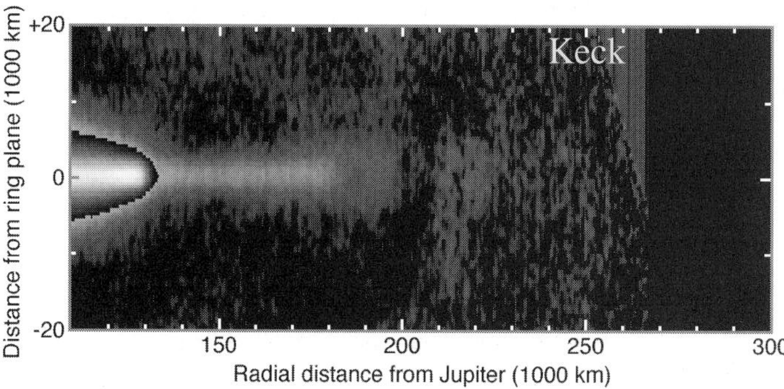

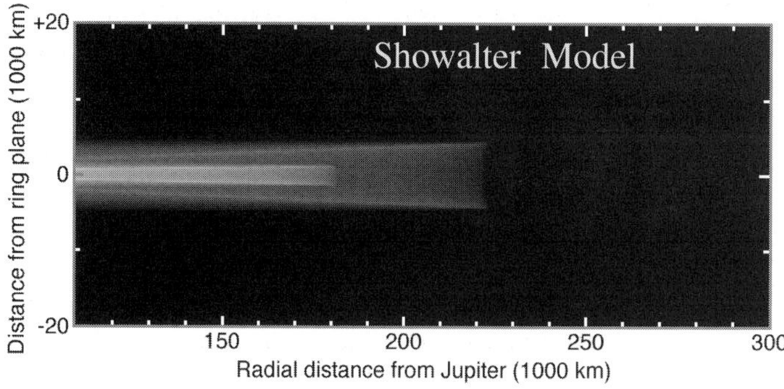

FIGURE 11.7 (a) Mosaics of Galileo images of Jupiter's ring system processed to emphasize the ring halo (upper panel) and the main ring (lower panel). The rings were viewed approximately edge-on. (NASA/Galileo, PIA01622) (b) Upper panel: A mosaic of Jupiter's gossamer rings made from four Galileo images. Images were obtained through the clear filter (central wavelength = 0.611 μm, passband = 0.440 μm) from within Jupiter's shadow (scattering angle = 1° to 3°) at an elevation of 0.15° (i.e., almost edge-on). The logarithm of the brightness is shown to reduce the dynamic range. Each gray-scale change represents roughly a 10-fold increase in brightness. The top and bottom edges of both gossamer rings are approximately twice as bright as their central cores, although this difference is subdued by the logarithmic scale. The two gossamer rings have crosses showing the four extremes of the eccentric and inclined motions of Amalthea and Thebe. Middle panel: Jupiter's ring at a backscattering phase angle of 1.1°, taken at 2.27 μm with the Keck 10-m telescope on 14 and 15 August 1997, when Earth's elevation above the ring plane was only 0.17°. The 0.6″ seeing (full width at half power, FWHP) corresponds to a resolution of 1800 km. In both the upper and middle panel part of the main ring and halo are visible at a jovicentric distance $\lesssim$ 130 000 km. The gossamer ring material in both images is visible to well beyond Thebe's orbit; in the Keck image one can discern it, albeit barely above the noise level, out to the frame's edge at ~257 000 km. Lower panel: A model of debris rings formed from Amalthea and Thebe ejecta. Each ring is composed of material created continually at its source moon and decaying inward at a uniform rate, retaining its initial inclination but having randomized nodes. (Burns et al. 1999)

TABLE 11.1 Properties of Jupiter's Ring System[a].

	Halo[b]	Main ring	Amalthea ring	Thebe ring
Radial location ($R_{♃}$)	1.3–1.72	1.72–1.806	1.8–2.55	1.8–3.15 (−3.8)
Vertical thickness	$\sim 5 \times 10^4$ km	$\lesssim 30$ km	~ 2300 km	~ 8500 km
Normal optical depth	$\sim 10^{-6}$–10^{-5}	$\sim 2 \times 10^{-6}$–10^{-5}	$\sim 10^{-7}$	$\sim 10^{-8}$ ($\sim 10^{-9}$)
Particle size	(sub-)micrometer	broad size distribution?	(sub-)micrometer	(sub-)micrometer

[a] Data from Ockert-Bell *et al.* (1999) and de Pater *et al.* (1999).

[b] Numbers quoted are based upon the Galileo data (visible light data, in forward scattered light). Relative to the main ring, the halo is much less bright and more spatially confined at longer wavelengths and in backscattered light.

(c)

FIGURE 11.7 Continued. (c) Galileo image C0416076420 of the night hemisphere of Europa, which was faintly illuminated by sunlight reflected off a nearly full Jupiter. This image was obtained when the Galileo spacecraft was eclipsed by Jupiter (phase angle 178.8°), in order to search for faint dust near Europa's limb. This long-exposure (96 ms) image fortuitously captured parts of Jupiter's ring system, which was illuminated by direct sunlight. The ring system's halo appears as an overexposed (in places saturated) smudge at the lower right edge of the image. The saturated elliptical arc jutting in from the halo is the main jovian ring, viewed from an elevation of 0.23° with a resolution of 67 km. The Amalthea gossamer ring stretches across the bottom left side of the image. The gossamer ring's brighter top and bottom edges are evident; this configuration is a consequence of the fact that particles on inclined orbits spend most of the time near the vertical extremes of their orbits. (Courtesy: Mark R. Showalter)

The mass of Jupiter's rings is poorly constrained. The dust component of the ring system is very low in mass. The macroscopic particles observed in backscattered radiation clearly provide a substantially larger contribution, but one that is uncertain by several orders of magnitude (Problem 11.11). At Jupiter's distance from the Sun, icy ring particles would evaporate rapidly, so Jupiter's ring particles are believed to be composed of more refractory materials. The red color of the jovian rings is very similar to that of the small moons Metis and Adrastea. This color is consistent with particles composed of silicates and/or carbonaceous compounds, with a possible contribution of sulfur from Io.

Very little small scale radial structure and no azimuthal features have been observed in Jupiter's ring system. The main ring extends radially inward about 0.1 $R_{♃}$ from the orbit of the tiny moon Adrastea, with a dip in brightness of 20–30% near Metis's orbit. The ring is $\lesssim 30$ km thick. Immediately interior to the main ring, at 1.72 $R_{♃}$, is the *halo*, which extends inwards to 1.3 $R_{♃}$. Its normal optical depth is very similar to that of the main ring, $\tau \approx$ few $\times 10^{-6}$. The observed full thickness of the halo is $\sim 12\,500$ km at half maximum, and it has a total thickness roughly four times as large. The particles within the halo probably had their inclinations increased by interactions with the planet's magnetic field at Jupiter's 3:2 Lorentz resonance (Section 11.5.2). The locations of the inner and outer boundaries of the halo coincide with *Lorentz resonances*.

The much fainter gossamer rings ($\tau \sim 10^{-7}$) consist of several parts (Fig. 11.7): immediately interior of Amalthea is the so-called *Amalthea ring*, which is uniform in brightness in the edge-on view in both forward and backscattered light. Interior of Thebe is the *Thebe ring*, fainter and thicker than the Amalthea ring. Exterior to Thebe one can, albeit barely, distinguish material out to ~ 3.7 $R_{♃}$ (Thebe is at 3.11 $R_{♃}$, Amalthea at 2.54 $R_{♃}$). In a very high resolution Galileo image (Fig. 11.7b), the upper and lower edges of the gossamer rings are much brighter than their central cores. The vertical location of the peak brightness of each of the gossamer rings, as well as the vertical extent seen in backscattered light, is proportional to distance from

FIGURE 11.8 (a) As the phase angle increases, the dusty F Ring and the spokes get brighter and brighter. (NASA/Voyager 1, FDS 34956.55) (b) This Voyager image shows the rings passing in front of the planet. As a result, you can see which rings are opaque (optically thick) and which are transparent (optically thin). Part of the middle B ring obscures the planet completely, while the planet easily shines through the fainter, inner C ring. (NASA/Voyager 1)

FIGURE 11.9 Close-up images of typical structures in Saturn's A, B and C rings are shown in panels (a)–(c) respectively. (d) Voyager 2 image of Saturn's faint inner D ring taken at phase angle 166°. The Sun's shadow cuts across the ring. (NASA PIA01388)

Jupiter's center. In Section 11.6.3 we show that this effect, as well as the thickness of the rings, is to be expected if the particles originate from the bounding satellites.

11.3.2 Saturn's Rings

Saturn's ring system is the most massive, the largest, the brightest and the most diverse in our Solar System (Figs. 11.5, 11.8 and 11.9); for these reasons it is also the best-studied. Most ring phenomena observed in other systems are present in Saturn's rings as well. The large scale structure and bulk properties of Saturn's rings are listed in Table 11.2. A schematic illustration of Saturn's rings and inner moons is shown in Figure 11.6.

(a)

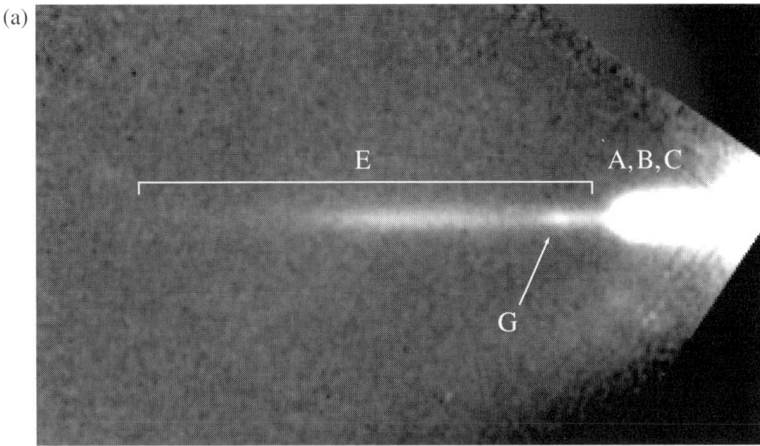

(b)

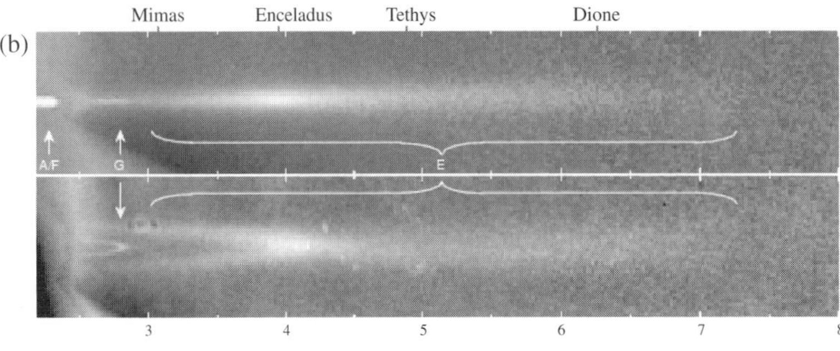

FIGURE 11.10 (a) Infrared photograph ($\lambda = 2.3$ μm) of Saturn's E and G rings, taken on 10 August 1995 from the Keck telescope. Even in this view of the dark face of the rings seen nearly edge-on, the main rings still appear substantially brighter than the E and G rings. Saturn is off to the right. (de Pater *et al.* 1996) (b) HST images of Saturn's G and E rings at visible wavelengths, seen edge-on in August 1995 (top panel) and open by 2.5° in November 1995 (bottom panel). The G ring is the relatively bright and narrow annulus whose ansa appears in the leftmost portion of the image. The E ring is much broader and more diffuse. Saturn is off to the left. (Courtesy: J.A. Burns, D.P. Hamilton, and M.R. Showalter)

TABLE 11.2 Properties of Saturn's Ring System[a].

		Main rings						
	D ring	C ring	B ring	Cassini division	A ring	F ring	G ring	E ring
Radial location (R_h)	1.09–1.24	1.24–1.53	1.53–1.95	1.95–2.03	2.03–2.27	2.32	2.75–2.87	3–8
Vertical thickness			<1 km	<1 km	<1 km			10^3–2×10^4 km (increases with radial location)
Normal optical depth	~10^{-5}–10^{-4}	0.05–0.2	1–3	0.1–0.15	0.4–1	1	10^{-5}–10^{-4}	10^{-7}–10^{-6}
Particle size	μm	mm–m	cm–10 m	cm–10 m	cm–10 m	μm–cm	μm–mm?	1 μm (very narrow distribution)

[a] Data for main rings primarily from Cuzzi *et al.* (1984); data for ethereal rings primarily from Burns *et al.* (1984); data for the D ring from Showalter (1996).

11.3.2.1 *Radial Structure of Saturn's Rings*

As seen through a small to moderate size telescope on Earth, Saturn appears to be surrounded by two rings (Fig. 11.3). The inner and brighter of the two is called the *B ring* (or Ring B) and the outer is known as the *A ring*. The dark region separating these two bright annuli is named the *Cassini division* after its discoverer; the Cassini division is not a true gap, rather it is a region in which the optical depth of the rings is only about 10% of that of the surrounding A and B rings. A larger telescope with good seeing can detect the faint *C ring*, which lies interior to the B ring. The *Encke gap*, a nearly empty annulus in the outer part of the A ring, can also be detected from the ground under good observing conditions. Rings A, B and C and the Cassini division are known collectively as *Saturn's main rings* or *Saturn's classical ring system*. Interior to the C ring lies the extremely tenuous *D ring*, which was imaged by the Voyager spacecraft but has not (yet) been detected from the ground. The narrow, multistranded, kinky *F ring* is 3000 km exterior to the outer edge of the A ring. Two tenuous dust rings lie well beyond Saturn's Roche limit (assuming a particle density equal to that of nonporous water-ice): the fairly narrow *G ring* and the extremely broad *E ring*.

The classically known components of Saturn's ring system are quite inhomogeneous upon close examination, with both radial and azimuthal variations being present (Figs. 11.9, 11.12, 11.18, 11.20, 11.23–11.25, 11.27, 11.29). The character of this structure is correlated with the overall optical depth of the region in which it exists. The A ring, with its moderate optical depth, $\tau \approx 1/2$, is relatively uniform in appearance (Fig. 11.9a). The observed features in the A ring are better understood than the majority of structure elsewhere in Saturn's ring system. Most of the A ring's structure results from resonant perturbations by external moons (cf. Section 11.4.2). The Encke gap is maintained by the embedded moonlet Pan, and an unseen moonlet is believed responsible for the narrow *Keeler gap* near the outer edge of the A ring; the theory of gap clearing by embedded moonlets is discussed in Section 11.4.3. The outer edges of the B and A rings are maintained by the Mimas 2:1 and Janus 7:6 resonances, which are the strongest resonances within the ring system (Section 11.4.1). The optically thick B ring (as well as a region of high optical depth in the inner portion of the A ring) contains irregular structure in the radial direction (Fig. 11.9b); the cause of this structure is still unknown. The optically thin C ring and Cassini division contain several gaps which may be produced by embedded moonlets. The causes of the large scale optical depth variations observed in the C ring and the Cassini division are not known.

Saturn's narrow, multistranded and clumpy F ring (Fig. 11.27) is confined in radius by the perturbations of the shepherding moons Prometheus and Pandora. The theory of this confinement mechanism is discussed in Section 11.4.3. The D ring is not well-studied, owing to its faintness; it is observed to be broken up into many ringlets. The G ring is also extremely faint; no internal structure of this ring has yet been identified. Although the E ring is also quite ethereal, it is so broad that it can readily be observed from Earth when the ring system appears almost edge-on (Fig. 11.10). The inner boundary of the E ring is fairly abrupt and lies around 3.0 $R_♄$, near the orbit of the moon Mimas. The peak density of this ring occurs at the orbit of Enceladus, near 4.0 $R_♄$. Enceladus, which has an anomalously bright and in many places smooth surface, is believed to be the primary source of the $\sim$1 μm size particles which make up the E ring. The density of the E ring drops gradually outside of this location, until it disappears into the sky background near 8.0 $R_♄$.

11.3.2.2 *Azimuthal Variations*

To a first approximation, Saturn's rings are uniform in longitude, i.e., the character of the rings varies much more substantially with distance from the planet than with longitude. This is, presumably, a consequence of the much shorter timescale for wiping out azimuthal structure via Kepler shear compared to radial diffusion times (Problem 11.7). However, various types of significant azimuthal structure have been observed in Saturn's rings. The most spectacular longitudinal structures seen in Saturn's rings are the nearly radial features known as *spokes* (Fig. 11.29), which are described in detail in Section 11.5.3.

Several narrow rings and ring edges are eccentric. Some of these features are well-modeled by keplerian ellipses which precess slowly as a result of the planet's quadrupole (and higher order) gravitational moments (cf. eq. 2.37). However, a few features, such as the outer edges of rings B (Fig. 11.11) and A and the edges of Encke's gap (Fig. 11.25), are multilobed patterns which are controlled by satellite resonances; the dynamical mechanisms responsible for such features are described in Section 11.4.3.

Saturn's F ring varies substantially with longitude in an irregular manner (Fig. 11.27). The literature often refers to some of these variations as 'braids', but there is no evidence that the various strands of the F ring are actually intertwined in a three-dimensional sense. The F ring's longitudinal variations have been attributed to the shepherding moons which radially confine the ring, as well as to the gravitational effects from large ring particles hypothesized to orbit within the F ring (cf. Section 11.4). The ringlet in Encke's gap, which orbits at the same radius as the moonlet Pan, shares some characteristics of longitudinal variability with the F ring.

The reflectivity of the A ring exhibits an intrinsic longitudinal variation known as the *azimuthal asymmetry*. The pattern is not symmetric about the *ansae* (the portions of the rings that appear farthest from the disk of the planet), rather the minimum brightness observed at low phase angles (e.g., from Earth) occurs 24° before each ansa, measured in the direction of the particle orbits. The amplitude of the azimuthal asymmetry is largest in the middle of the A ring, and greatest at low ring tilt angles, where the peak brightness is $\sim$ 40% larger than the brightness minima. The azimuthal asymmetry is believed to result from *density wakes*, which are elongated, temporary, optically thick groupings of ring particles that form near large parti-

FIGURE 11.11 The eccentric outer edge of Saturn's B ring as imaged at four different longitudes by Voyager 2. The upper portion of the images show the inner part of the Cassini division, and the lower part shows the outermost region of the B ring. The middle two slices were taken from high resolution (< 8 km pixel^{-1}) images of the east ansa and the outer two slices are from the west ansa. The width of the gap separating the B ring from the Cassini division varies by up to 140 km. These variations are caused by perturbations exerted by Mimas near its 2:1 inner Lindblad resonance. Also visible are variations in fine structure in the B ring and the eccentric Huygens ringlet within the variable width gap. All slices were taken within about 7 hours. (Smith *et al.* 1982)

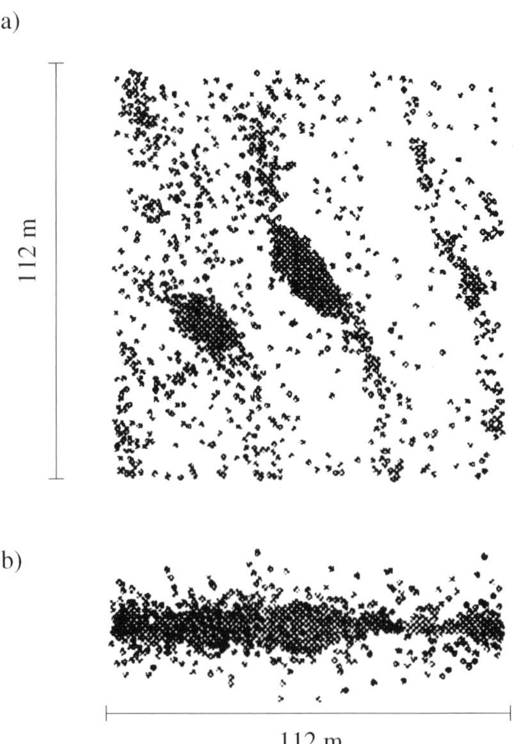

FIGURE 11.12 (a) Results of a numerical simulation of density wakes formed by particles orbiting in Saturn's A ring, 130 000 km from the center of the planet. The particles all have radii of 1 meter, density of 0.9 g cm^{-3} and coefficient of restitution equal to 0.5. The region shown is 112 m on each side and contains 1600 particles. (b) Vertical cross-section of the same group of ring particles. (Salo 1995)

cles or clusters of particles as a result of local gravitational forces (Fig. 11.12). Kepler shear causes such wakes to trail at an angle of 23° in planetary rings. At radio wavelengths, the west (dusk) ansa is brighter than the east (dawn) ansa by factors up to 2 for the C and B rings. The rings scatter radio waves emitted by Saturn, and, like at visible and near-infrared wavelengths for the A ring, the 'radio' asymmetry may also be due to multiple scattering in density wakes.

Azimuthal variations in brightness associated with the Sun/ring/spacecraft geometry have been observed at ring locations which are in vertical resonance with Saturn's moons; these variations are produced by spiral bending waves (cf. Section 11.4.2). Subtle azimuthal variations in resonantly excited spiral density waves have also been detected.

11.3.2.3 *Thickness*

Saturn's rings are extremely thin relative to their radial extent. Upper limits to the local thickness of the rings of ~150 m at several ring edges were obtained by the abruptness of some of the ring boundaries detected in stellar occultations by the rings observed from Voyager 2 and from diffraction patterns in the radio signal transmitted through the rings by Voyager 1. Estimates of ring thickness from the viscous damping of spiral bending waves and density waves and from the characteristics of the outer edge of the B ring yield values ranging from one meter to tens of meters.

The amount of light reflected from the rings when they appear edge-on from Earth is equivalent to a 1 km thick slab with reflectivity comparable to that of the lit face of the rings. Such a thickness is far too small to be resolvable from Earth-based telescopes. This equivalent thickness is probably dominated by the F ring, which is quite bright near ring plane crossing and is likely warped and/or inclined slightly to Saturn's equatorial plane. Additional contributions to the rings' edge-on brightness come from the actual thickness of the main rings, local corrugations of the ring plane in bending waves (Section 11.4.2), more gradual warping of the ring plane on larger scales by Saturn's moons, and the G and E rings.

11.3.2.4 *Particle Properties*

The particles in Saturn's main ring system are better characterized than are particles in other planetary rings. Spectra of infrared light reflected off Saturn's main rings are similar to that of water-ice, implying that water-ice is a major constituent. The high albedo of Saturn's rings suggests that impurities are few and/or not well mixed at the microscopic level. Small but significant color variations are seen from ring to ring. In general, particles in high optical depth rings appear brighter and redder than particles in regions of low optical depth. These differences may result from the more rapid contamination of particles in low τ regions by impacting micrometeoroids (particles in regions of high optical depth partially shield one another from such bombardment).

The frequent collisions among ring particles can lead to particle aggregation as well as erosion. Scale-independent processes of accretion and fragmentation lead to power law distributions of particle number vs. particle size. Such power law distributions are observed over broad ranges of radii in the asteroid belt and in most of those planetary rings for which adequate particle size information is available. Data on particle sizes are thus often fit to a distribution of the form:

$$N(R) = N_o \left(\frac{R}{R_o}\right)^{-\zeta} \qquad (R_{min} < R < R_{max}) \qquad (11.16)$$

and zero otherwise, where $N(R)dR$ is the number of particles with radii between R and $R + dR$ and N_o and R_o are normalization constants. The distribution is characterized by the values of its power law index, ζ, and the minimum and maximum particle sizes, R_{min} and R_{max}, respectively. A uniform power law over all radii implies infinite mass in either large or small radii particles (Problem 11.10); thus, such a power law must be truncated at large and/or small radius. Note that the value of upper size limit is not very important for the total mass or surface area of the system if the distribution is sufficiently steep (ζ significantly larger than 4) and that the lower limit is not a major factor provided the distribution is shallow enough (ζ significantly smaller than 3).

Radar signals, with wavelength several centimeters, have been bounced off Saturn's rings. The high radar reflectivity of the rings implies that a significant fraction of their surface area consists of particles with diameters of at least several centimeters. Radio signals sent through rings by Voyager 1 give information on particle sizes from a comparison of optical depths at two wavelengths and from diffraction patterns of the signal. The combination of these data implies a broad range of sizes from ~1 cm to 5–10 m, with approximately equal areas in equal logarithmic size intervals (i.e., $\zeta = 3$) and most of the mass being in the largest particles. The power law index in the size distribution is $2.8 < \zeta < 3.4$ for 1 cm $< R <$ 5 m and $\zeta > 5$ for $R > 10$ m. The particles in the C ring are somewhat smaller than those in the A ring, consistent with a tidal limit on particle size. Micrometer-sized dust particles are comparable in size to the wavelength of visible light, so they preferentially scatter in the forward direction. Micrometer-sized particles are most common in the E, G and F rings, but also dominate the spokes in the B ring and are apparent in the very outer part of the A ring, exterior to the Keeler gap. No images showing individual ring particles have yet been obtained, but imagined views from within Saturn's rings are shown in Figure 11.13.

It is likely that the micrometer-sized dust seen in spokes is rapidly reaccumulated by larger ring particles, causing the spoke to vanish. One model of the rings suggests that macroscopic particles agglomerate into loosely bound objects (called *dynamical ephemeral bodies*) of order ten meters in size which survive a few orbital periods before being collisionally disrupted. Attempts to derive characteristic particle sizes from first principles using Hertz's theory of elastic solids have not been successful.

The unusual narrow clustering of particle sizes in Saturn's E ring near 1 μm suggests that this population is not collisionally evolved. Models have been proposed in which the E ring particles are ice crystals formed in the plumes of hypothetical water-ice volcanoes or geysers on the moon Enceladus. Evidence for recent local resurfacing on Enceladus from crater counts provides credibility to arguments for volcanism on this small ($R \approx 250$ km) icy body. However, radiation pressure is probably much more efficient at dispersing ~ 1 μm grains over a broad annulus than it is at spreading out grains of larger or smaller size (Section 11.5.1). If a broad particle size distribution were created by erosion of dust off Enceladus, grains of other sizes would remain clustered near the orbit of their parent moon, and most would soon either reimpact the moon or be ground down via mutual collisions. Thus, the narrowness of the E ring's size distribution may be a consequence of survival advantages for 1 μm grains, rather than a monodisperse particle formation mechanism.

11.3.2.5 *Mass*

Saturn has by far the largest, brightest and most massive ring system within the Solar System. Nonetheless, the mass of Saturn's rings is too low to be measured by its gravitational effects on moons or spacecraft. Thus, it must be deduced from more circuitous theoretical arguments. Several different techniques have been used, and all give similar answers, lending confidence to the results.

(a)

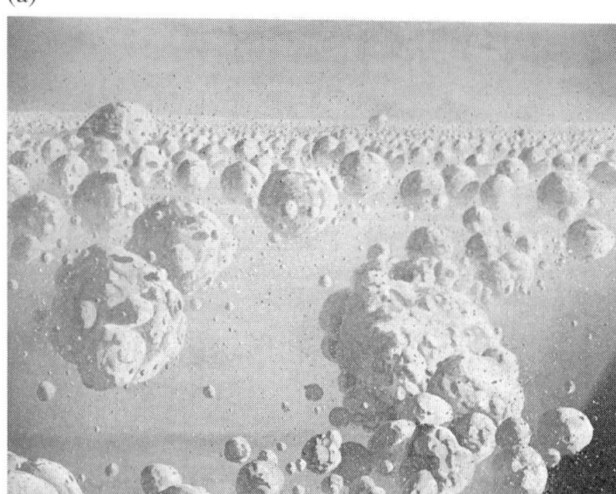

(b)

FIGURE 11.13 (a) Hypothetical view from within Saturn's rings, painted by W.K. Hartmann. (b) Computer-generated schematic of the particle size distribution in Saturn's A ring. (JPL/NASA)

The wavelengths of spiral density waves and spiral bending waves are proportional to the local surface mass density of the rings; therefore, we can deduce the mass where we see waves. The theory behind this analysis is presented in Section 11.4, and a sample exercise in this technique is given in Problem 11.14b. The surface density, σ_ρ, at the two wave locations observed in the B ring is $\sim$ 50–80 g cm^{-2}. Analysis of dozens of waves within the A ring reveals surface densities of $\sim$ 50 g cm^{-2} in the inner to middle A ring, dropping to $\lesssim$ 20 g cm^{-2} near this ring's outer edge. Measured values at two locations in the optically thin C ring are $\sim$1 g cm^{-2}, while an intermediate value of $\sim$10 g cm^{-2} has been estimated from one wavetrain in Cassini's division. Since observable spiral waves cover only a small fraction of the area of Saturn's rings, we assume that the *opacity*, σ_ρ/τ, is constant within a given region of the rings in order to estimate the mass of the ring system. This approximation is fairly good wherever there are several waves near one another, so that it can be checked, but it must be noted that the energy input by the moons into the waves may make regions in which strong waves propagate somewhat anomalous. The large amplitude of the eccentricity of the outer edge of the B ring is likely to be caused by a combination of resonant forcing by the moon Mimas and ring self-gravity; calculations based on this model suggest that the surface density near the (optically thick) outer edge of the B ring is $\sim$100 g cm^{-2}, consistent with the density wave results. The particle flux produced from cosmic ray interactions with the ring system suggests an average surface density of 100–200 g cm^{-2}, a few times as large as the average from density wave estimates.

The total mass of Saturn's ring system is estimated (using spiral waves) to be:

$$M_{rings} \sim 5 \times 10^{-8} \mathrm{M}_{♄} \sim M_{Mimas}, \tag{11.17}$$

where Mimas, the innermost and smallest of Saturn's classical, nearly spherical, moons, has a mean radius of 196 km. The particle size distribution derived from Voyager radio occultation measurements suggests a ring mass 40% less than the value in equation (11.17), assuming particle densities equal to that of water-ice. Uncertainties in these estimates are large, so the ring particles may either be nearly pure water-ice or ice/rock mixtures; fluffy, high porosity aggregates are unlikely to be common within the main rings of Saturn.

11.3.3 Uranus's Rings

Most of the material in the uranian ring system is confined to nine narrow annuli whose orbits lie between 41 000 and 52 000 km from the planet's center (Table 11.3, Figs. 11.4 and 11.14). These nine optically thick rings were discovered by observations of stars whose light was seen to diminish as they were occulted by Uranus's rings in the late 1970s. Most are 1–10 km wide and have eccentricities of order 10^{-3} and inclinations of 10^{-4}–10^{-3} radian. The outermost annulus, *Ring* ϵ, is the widest and most eccentric, with $e = 8 \times 10^{-3}$ and width ranging from 20 km at periapse to 96 km at apoapse. The majority of ring edges, including both the inner and outer boundaries of the ϵ ring, are quite sharp compared to ring width, but in a few cases a more gradual dropoff in optical depth is observed. The

TABLE 11.3 Properties of Uranus's Ring System[a].

	Rings 6, 5, 4, α, β, γ, η, δ	Ring λ	Ring ϵ	1986U2R + inter-ring dust
Radial location ($R_♅$)	~1.64–1.90	1.96	2.01	1.41–1.91
Radial width	1–10 km individual rings vary by up to factor of 2	~2 km	20–96 km varies with azimuth	~1000 km
Normal optical depth	~0.5–4	0.1–0.5	1–4	$\sim 10^{-6}$–10^{-3}
Particle size	$\gtrsim$ centimeter	submicrometer	meter	(sub-)micrometer

[a] Data from French *et al.* (1991) and Esposito *et al.* (1991).

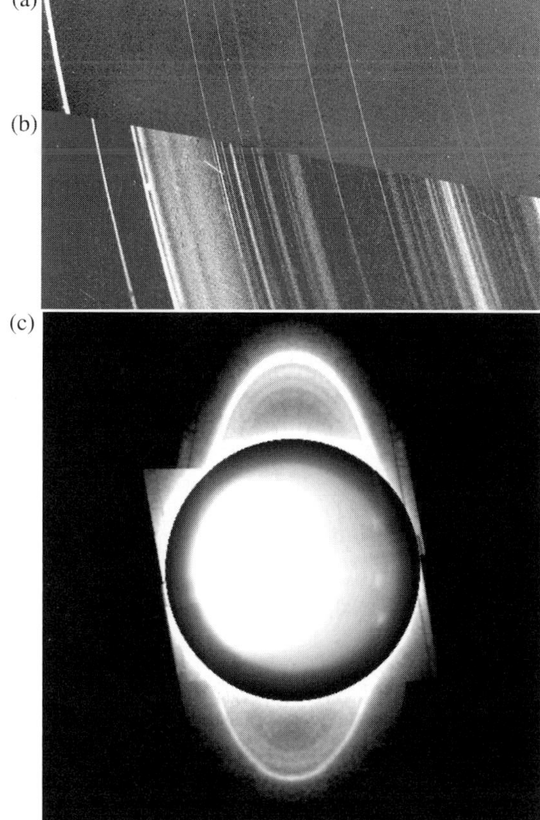

FIGURE 11.14 (a) Mosaic of two low phase angle (21°) high resolution (10 km/pixel) Voyager 2 images of the rings of Uranus. The planet's nine narrow optically thick rings are clearly visible, and the very narrow moderate optical depth λ ring is marginally detectable. (NASA/Voyager 2, PIA00035) (b) High phase angle (172°) Voyager 2 view of the uranian ring system. The forward scattering geometry dramatically enhances the visibility of the micrometer-sized dust particles. The streaks are trailed star images in this 96 second exposure. (NASA/Voyager 2, PIA00142) (c) An infrared image of Uranus and its rings taken with adaptive optics system at the Keck telescope. The planet is artificially darkened by a factor of ~ 15 (inside the dark circle). The ϵ ring is clearly visible (periapse is to the south), and the three rings interior to the ϵ ring are from the outside inwards: (1) the combined δ, γ, η rings, (2) combined β, α rings, and (3) combined 4, 5, 6 rings. In addition to the rings, the image shows the familiar circumpolar haze around Uranus's South pole, as well as the presence of three cloud features in the planet's northern hemisphere. (de Pater et. al. 2001)

uranian ring system also includes the narrow, moderate optical depth, dusty λ *ring* and wide, radially variable, low optical depth dust sheets (Fig. 11.14b).

In all of the uranian rings which possess measurable eccentricities, including the ϵ ring, e increases with distance from the planet, with the *eccentricity gradient*, $a\,de/da \sim 0.5$. Two of the uranian rings are not well modeled by precessing keplerian ellipses. The δ *ring* looks like an ellipse *centered* on Uranus, whereas the γ *ring* combines a standard eccentric ellipse pattern with temporally coherent radial motions of the particles (i.e., periapse times are correlated for all longitudes, so that the ring appears to 'breathe').

The particles in the nine uranian rings discovered from Earth are similar in size to those in Saturn's main ring system, ~ 1 cm to 10 m. These values are less well constrained than in the saturnian case, and there are indications that uranian ring particles may be somewhat larger. Uranus's ring particles are extremely dark, with a Bond albedo of ~0.03. They appear as dark as the darkest asteroids and carbonaceous chondrite meteorites. However, they probably consist of *radiation darkened ice*, which is a mixture of complex hydrocarbons embedded in ice produced as a result of the removal of H atoms via sputtering

TABLE 11.4 Properties of Neptune's Ring System[a].

	Galle, Lassell rings	Le Verrier ring	Adams ring
Radial location (R_{Ψ})	1.7; 2.2	2.1	2.5
Radial width (km)	2000; 4000	~100	15 (in arcs)
Normal optical depth	$\sim 10^{-4}$ (of dust)	~0.003	0.1 in arcs 0.003 elsewhere
Dust fraction	? (large particles not detected)	~50%	~50% in arcs ~30% elsewhere

[a] Data from Porco *et al.* (1995).

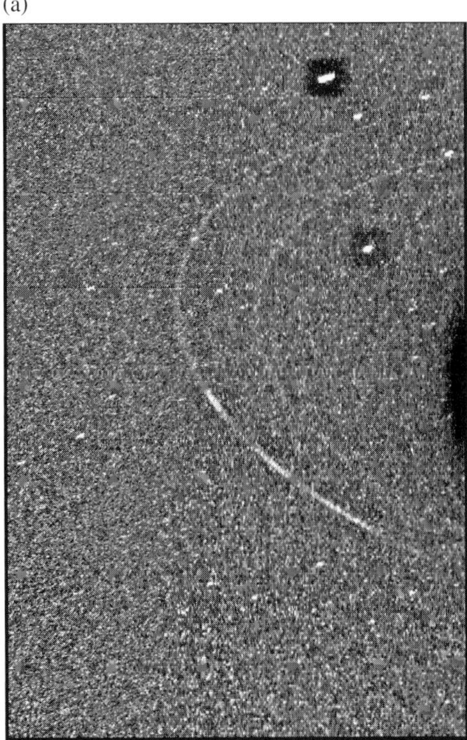

FIGURE 11.15 (a) Neptune's two most prominent rings, Adams (which includes higher optical depth arcs) and Le Verrier, as seen by Voyager 2. The rings appear faint in this image, FDS (FDS numbers refer to the Voyager flight data system timeline. Each image has a unique FDS number) 11350.23, taken in backscattered light (phase angle 15.5°) with a resolution of 19 km per pixel. The moon Larissa at the top of the image appears streaked as a result of its orbital motion. The other bright object in the field is a star. The image appears very noisy because a long exposure (111 s) and large stretch were needed to show the faintly illuminated, low optical depth dark rings of Neptune. (NASA/Voyager 2, PIA00053) (b) Image FDS 11412.51, another 111 second exposure with a resolution of 80 km per pixel. The rings are much brighter in forward scattered light (phase angle 134°), indicating that a substantial fraction of the ring optical depth consists of micrometer-size dust. (NASA/Voyager 2, PIA01493)

processes. Such radiation darkened ice may also account for the low albedos of cometary nuclei (cf. Section 10.6.1). Most of the surface area of the low τ rings discovered by Voyager is covered by submicrometer and micrometer-sized particles.

The mass of the ϵ ring estimated from the particle size distribution and an assumed particle density of 1 g cm^{-3} is 1–5 $\times$ 10^{19} g. Dynamical models for the maintenance of the ϵ ring's eccentricity by ring self-gravity yield a mass estimate of $\sim 5 \times 10^{18}$ g, but this estimate may not be very accurate as these models do not adequately reproduce certain aspects of ring structure. The combined mass of all of Uranus's other rings is probably a factor of a few lower than that of ring ϵ.

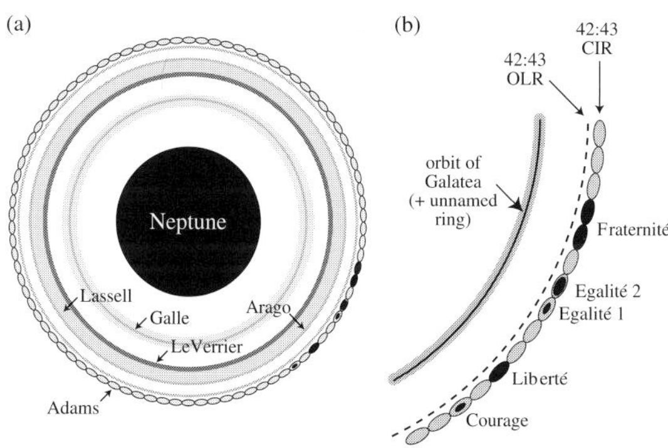

FIGURE 11.16 Cartoon sketch of Neptune's rings and associated moons as viewed from the south pole showing: (a) the location and names of the main rings; (b) a model for the arcs in the Adams ring as the filled centers of libration of the Galatea 42:43 resonance. (Murray and Dermott 1999)

11.3.4 Neptune's Rings

Neptune's ring system is quite diverse, showing structure in both radius and longitude (Fig. 11.15). Most of Neptune's rings have very low optical depth. The most prominent feature of the ring system is a set of five *arcs* of optical depth $\tau \approx 0.1$ within the *Adams ring*. The arcs vary in extent from $\sim 1°$ to $\sim 10°$, and are grouped in a $40°$ range in longitude; they are about 15 km wide. At other longitudes within the Adams ring, $\tau \approx 0.003$; a comparable optical depth has been measured for the *Le Verrier ring*. Neptune's other rings are even more tenuous. Several moderately large moons orbit within Neptune's rings; these satellites are believed to be responsible for much of the radial and longitudinal ring structure that has been observed. Figure 11.16 is a cartoon sketch of the known components of Neptune's ring system; ring locations and optical depths are listed in Table 11.4.

The particles in Neptune's rings are very dark. They probably are red, and they may be as dark as those in the uranian rings, but the properties of Neptune's ring particles are less well constrained by data currently available. The fraction of optical depth due to micrometer-sized dust is very high, $\sim 50\%$, and appears to vary from ring to ring. The limited data available are not sufficient to make even an order of magnitude estimate of the mass of Neptune's rings, although they suggest that the rings are significantly less massive than the rings of Uranus, unless they contain a substantial population of undetected large (~ 100 m) particles, which is unlikely given the paucity of smaller macroscopic ring particles.

11.4 Ring–Moon Interactions

Observations of planetary rings reveal a complex and diverse variety of structure, mostly in the radial direction, over a broad range of length scales, in contrast to naive theoretical expectations of smooth, structureless rings (Section 11.2.1). We now turn to theoretical explanations of the causes of some of these features. The processes responsible for some types of ring structure are well understood. Partial or speculative explanations are available for other features, but the causes of many structures remain elusive. The agreement between theory and observations is best for ring features believed to be produced by gravitational perturbations from known moons, and we examine such models first.

11.4.1 Resonances

A common process in many areas of physics is resonance excitation: When an oscillator is excited by a varying force whose period is very nearly equal to the oscillator's natural frequency, the response can be quite large even if the amplitude of the force is small (cf. eq. 2.26). In the planetary ring context, the perturbing force is the gravity of one of the planet's moons, which is generally much smaller than the gravitational force of the planet itself.

Resonances occur where the radial (or vertical) frequency of the ring particles is equal to the frequency of a component of a satellite's horizontal (or vertical) forcing, as sensed in the frame rotating at the frequency of the particle's orbit. In this case, the resonating particle is repeatedly near the same phase in its radial (vertical) oscillation when it experiences a particular phase of the satellite's forcing. This situation enables continued coherent 'kicks' from the satellite to build up the particle's radial (vertical) motion, and significant forced oscillations may thereby result. Particles nearest resonance have the largest eccentricities, as they receive the most coherent kicks; the forced eccentricity is inversely proportional to the distance from resonance for noninteracting particles in the linear regime. Collisions among ring particles and the self-gravity of the rings complicate the situation, and resonant forcing of planetary rings can produce a variety of features including gaps and spiral waves, which are discussed in detail in the following subsections.

The locations and strengths of resonances with any given moon can be calulated by decomposing the gravitational potential of the moon into its Fourier components. The *disturbance* (forcing) *frequency*, ω_f, can be written as the sum of integer multiples of the satellite's angular,

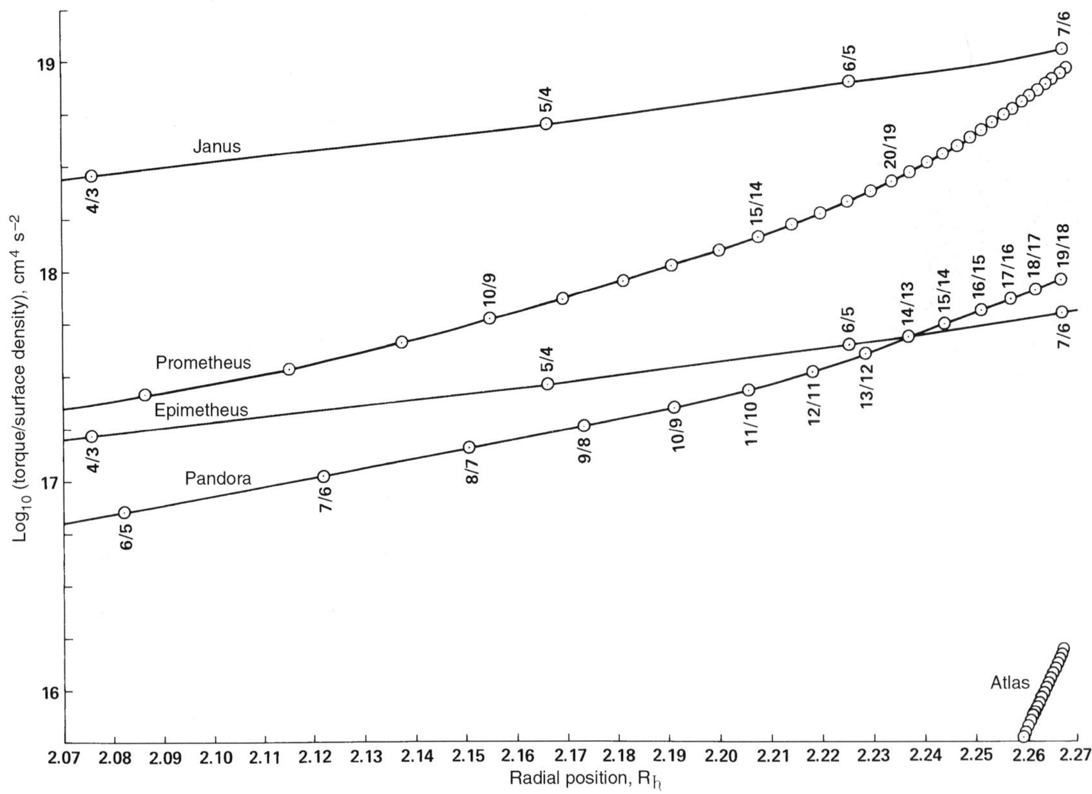

FIGURE 11.17 Locations and strengths of major Lindblad resonances in Saturn's A ring for Saturn's five closest 'ring moons' whose orbits lie interior to the main ring system: Janus, Epimetheus, Pandora, Prometheus and Atlas. The moons that orbit closer to the A ring have more closely spaced resonances with strength increasing outward more rapidly. (Adapted from Lissauer and Cuzzi 1982)

vertical and radial frequencies:

$$\omega_f = m_\theta n_s + m_z \mu_s + m_r \kappa_s, \tag{11.18a}$$

where the azimuthal symmetry number, m_θ, is a nonnegative integer, and m_z and m_r are integers, with m_z being even for horizontal forcing and odd for vertical forcing. The subscript s refers to the satellite (moon).

A particle placed at a distance $r = r_L$ from the planet is in horizontal (Lindblad) resonance if r_L satisfies:

$$\omega_f - m_\theta n(r_L) = \pm\kappa(r_L). \tag{11.18b}$$

Vertical resonance occurs if its radial position r_v satisfies:

$$\omega_f - m_\theta n(r_v) = \pm\mu(r_v). \tag{11.18c}$$

When equation (11.18b) is valid for the lower (upper) sign, we refer to r_L as the inner (outer) Lindblad or horizontal resonance, which are frequently abbreviated as ILR and OLR respectively. The radius r_v is called an inner (outer) vertical resonance, IVR (OVR), if equation (11.18c) is valid for the lower (upper) sign. Since all of Saturn's large satellites orbit the planet well outside the main ring system, the moon's angular frequency, n_s, is less than the angular frequency of the particle, and inner resonances are more important than outer ones. The differences between the orbital, radial and vertical[2] frequencies are at most a few percent within Saturn's rings. Thus, when $m_\theta \neq 1$, the approximation $\mu \approx n \approx \kappa$ may be used to obtain the ratio:

$$\frac{n(r_{L,v})}{n_s} \approx \frac{m_\theta + m_z + m_r}{m_\theta - 1}. \tag{11.19}$$

The notation $\frac{(m_\theta + m_z + m_r)}{(m_\theta - 1)}$ or $(m_\theta + m_z + m_r){:}(m_\theta - 1)$ is commonly used to identify a given resonance. If $n = \mu = \kappa$, the inner horizontal and vertical resonances would coincide: $r_L = r_v$. Since, due to Saturn's oblateness, $\mu > n > \kappa$, the positions r_L and r_v do not coincide: $r_v < r_L$.

The strength of the forcing by the satellite depends, to lowest order, on the satellite's mass, M_s, eccentricity, e,

[2] Disk self-gravity can significantly increase the vertical frequencies of particles in a thin ring. However, self-gravity does not alter the vertical frequency of the local disk midplane, which is the relevant quantity for vertical resonances.

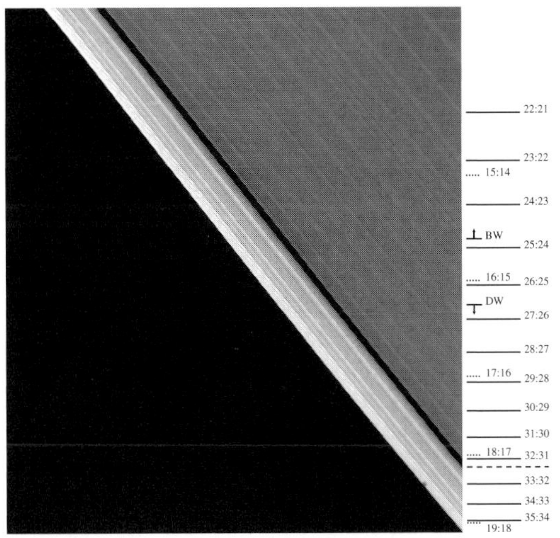

FIGURE 11.18 Voyager 2 image FDS 44005.04 of the outer portion of Saturn's A ring. Over a dozen density waves are evident in this image, although most are unresolved. The outer edge of the A ring is maintained by the 7:6 inner Lindblad resonance of the coorbital moons Janus and Epimetheus. (NASA/Voyager 2, PIA01953)

and inclination, i, as $M_s e^{|m_r|} \sin^{|m_z|} i$. The strongest horizontal resonances have $m_z = m_r = 0$, and are of the form $m_\theta : (m_\theta - 1)$. The strongest vertical resonances have $m_z = 1, m_r = 0$, and are of the form $(m_\theta + 1) : (m_\theta - 1)$. The location and strengths of such orbital resonances can be calculated from known satellite masses and orbital parameters and Saturn's gravity field. By far the lion's share of the strong resonances in Saturn's ring system lie within the outer A ring (Figs. 11.17 and 11.18), near the orbits of the moons that excite them.

Resonant forcing leads to a secular transfer of orbital angular momentum from Saturn's rings to its moons. These torques produce two classes of structure in Saturn's rings: gaps/ring boundaries and spiral density and bending waves. The outer edges of Saturn's two major rings are maintained by the two strongest resonances in the ring system. The outer edge of the B ring is located at Mimas's 2:1 ILR, and is shaped like a two-lobed oval *centered* on Saturn (Fig. 11.11). The A ring's outer edge is coincident with the 7:6 resonance of the coorbital moons Janus and Epimetheus, and appears to have a seven-lobed pattern, consistent with theoretical expectations, although the data are less conclusive in this case. In order for a resonance to clear a gap or maintain a sharp ring edge, it must exert enough torque to counterbalance the ring's viscous spreading. In the low optical depth C ring, resonances with moderate torques create gaps, but in the higher optical depth A and B rings, resonances with similar strengths excite spiral density waves.

Nearly empty gaps with embedded optically thick ringlets have been observed at strong resonances located in optically thin regions of the rings. Just exterior to the B ring lies a small gap with an opaque eccentric ringlet (Fig. 11.11). Similar ringlets are also observed in gaps at the strongest resonances in Saturn's C ring. Qualitatively similar features have been reproduced in numerical simulations of resonantly forced particulate rings. Note, however, that most empty gaps with embedded ringlets have also been observed at nonresonant locations.

11.4.2 Spiral Density Waves and Spiral Bending Waves

Spiral density and bending waves generated by gravitational perturbations of external moons have been observed at several dozen locations within Saturn's rings and have tentatively been detected within the rings of Uranus. They represent one of the best understood forms of structure within planetary rings, and have been very useful as diagnostics of ring properties such as surface mass density, σ_ρ, and local thickness. However, the angular momentum transfer associated with the excitation of density waves within Saturn's rings leads to characteristic orbital evolution timescales of Saturn's A ring and inner moons which are much shorter than the age of the Solar System, creating a major theoretical puzzle.

Spiral density waves are horizontal density oscillations that result from the bunching of streamlines of particles on eccentric orbits (Fig. 11.19a, b). Spiral bending waves, in contrast, are vertical corrugations of the ring plane resulting from the inclinations of particle orbits (Fig. 11.19c). In Saturn's rings, both types of spiral waves are excited at resonances with Saturn's moons and propagate as a result of the collective self-gravity of the particles within the ring disk (Fig. 11.20).

Ring particles move along paths that are very nearly keplerian ellipses with one focus at the center of Saturn. However, small perturbations caused by the wave force a coherent relationship between particle eccentricities/periapses (in the case of density waves) or inclinations/nodes (in the case of bending waves), which produces the observed spiral pattern. The theory of excitation and propagation of linear spiral waves within planetary waves has been reviewed by Shu in the book *Planetary Rings* edited by Greenberg and Brahic (1984). An abbreviated summary of aspects of the waves which may be observed and analyzed to determine ring properties such as surface density and thickness is presented below.

(a)

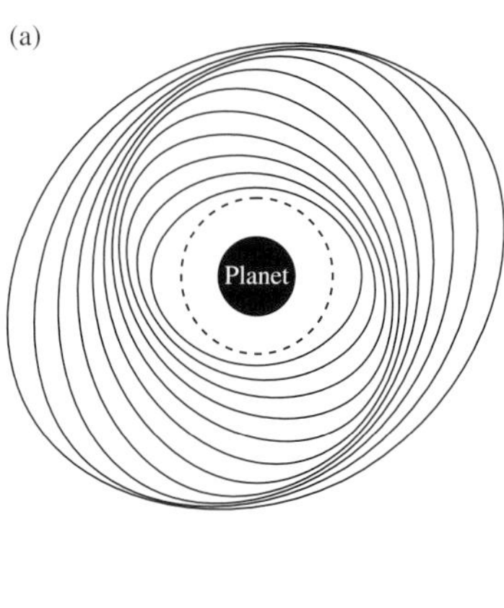

(b)

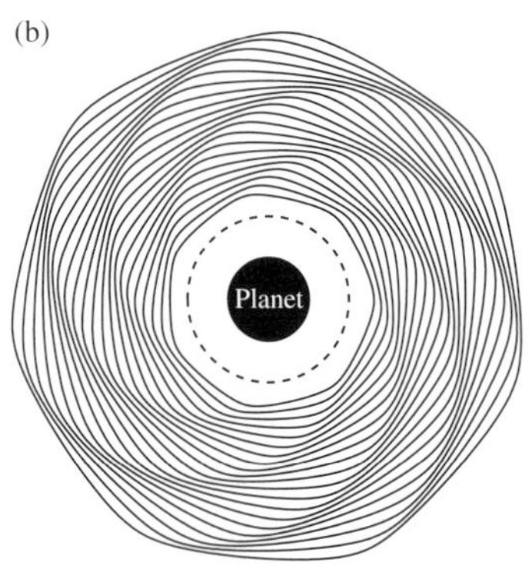

(c)

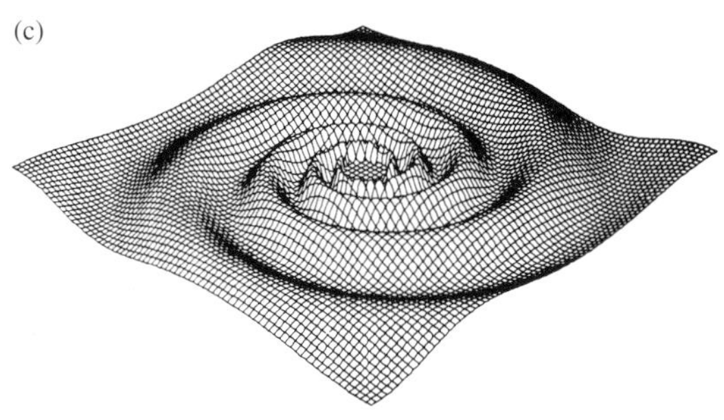

FIGURE 11.19 Schematic diagrams of the coplanar particle orbits that give rise to trailing spiral density waves near a resonance with an exterior satellite are shown in panels (a) and (b). (a) The two-armed spiral density wave associated with the 2:1 ($m = 2$) inner Lindblad resonance. (b) The seven-armed density wave associated with the 7:6 ($m = 7$) inner Lindblad resonances. The pattern rotates with the angular velocity of the satellite and propagates outwards from the exact resonance (denoted by a dashed circle). (Murray and Dermott 1999) (c) Schematic of a spiral bending wave showing variation of vertical displacement with angle and radius for a two-armed spiral. Spiral waves observed in Saturn's rings are much more tightly wound. (Shu *et al.* 1983)

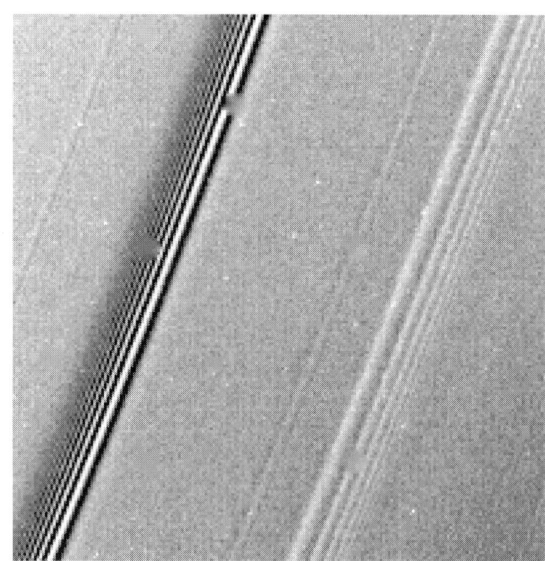

FIGURE 11.20 Voyager 2 image (FDS 43999.19) of a portion of the lit face of Saturn's A ring showing two prominent wave patterns. The feature on the left is the Mimas 5:3 bending wave; its contrast is high because the tilt of the local ring plane due to the wave was comparable to the solar elevation angle ($\sim 8^\circ$) when the image was taken. The Mimas 5:3 density wave is seen on the right. The separation between the locations of the two waves results from the nonclosure of orbits caused by Saturn's oblateness. The other linear features in the images are unresolved density waves excited by the moons Pandora and Prometheus. Saturn is off to the left. (NASA/Voyager 2)

11.4.2.1 *Theory of Bending Waves*

The vertical component of the gravitational force exerted by a satellite on an orbit inclined with respect to the plane of the rings excites motion of the ring particles in the direction perpendicular to the mean ring plane. The vertical excursions of the particles are generally quite small (up to ~ 400 m, resulting primarily from perturbations by Titan and the Sun) and vary coherently over scales of tens of thousands of kilometers, warping the rings like the brim of a hat. However, at vertical resonances in Saturn's rings, the natural vertical oscillation frequency of a particle, $\mu(r)$, is equal to the frequency at which one of Saturn's moons tugs the ring particles perpendicular to the ring midplane. Such coherent vertical perturbations can produce significant out-of-plane motions (cf. Section 2.3.2). Self-gravity of the ring disk supplies a restoring force that distributes the torque exerted by the moon upon the ring at the resonance to nearby (but not resonant) regions of the ring. This process enables bending waves to propagate away from resonance, creating a corrugated spiral pattern. The number of spiral arms is equal to the value of m_θ. Spiral density and bending waves with values of m_θ ranging from 1 to more than 30 have been detected in Saturn's rings. In contrast, most spiral galaxies have 2–4 spiral arms. For $m_\theta > 1$, bending waves propagate toward Saturn; *nodal bending waves* ($m_\theta = 1$) propagate away from the planet.

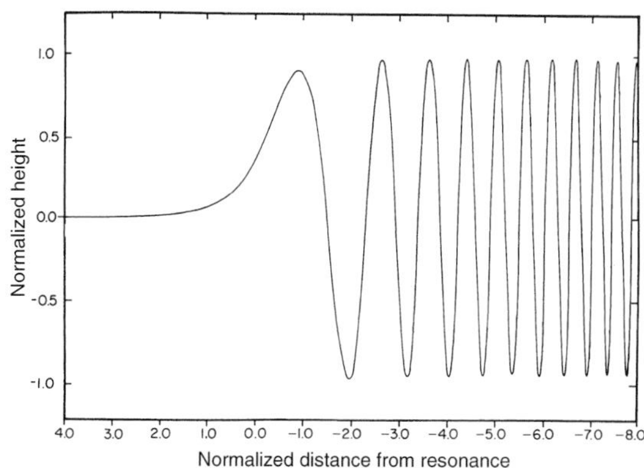

FIGURE 11.21 Theoretical height profile of an undamped linear spiral bending wave as a function of distance from resonance. The length and height scales are in arbitrary units. (Adapted from Shu *et al.* 1983)

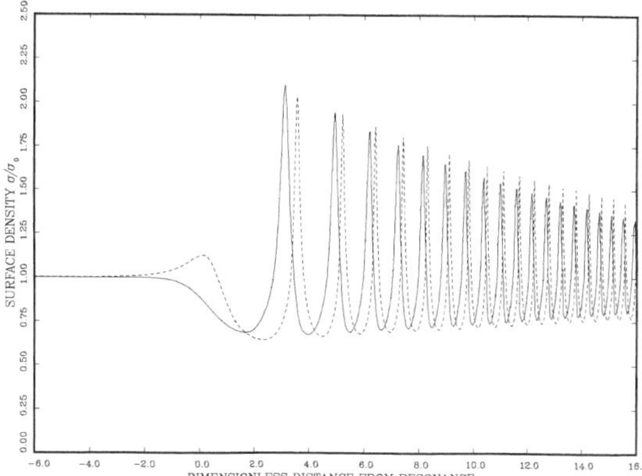

FIGURE 11.22 Theoretical surface density profile of a damped nonlinear spiral density wave. The solid and dashed lines represent two profiles of the same wave plotted at different azimuths. (Adapted from Shu *et al.* 1985)

In the inviscid linear theory, where viscous damping is ignored and the slope of the bent ring midplane is assumed to be small, the height of the local ring midplane relative to the Laplace (invariable) plane is given by a Fresnel integral, which is evaluated and plotted in Figure 11.21. In the asymptotic far-field approximation, the oscillations remote from resonance have a wavelength:

$$\lambda = \frac{4\pi^2 G\sigma_\rho}{m_\theta^2[\omega_f - n(r)] - \mu^2(r)}. \tag{11.20}$$

Equation (11.20) can be simplified by approximating the orbits of the ring particles as keplerian, $n(r) = (GM_p/r^3)^{\frac{1}{2}} = \mu(r)$, for the $m_\theta > 1$ case and approximating the departure from keplerian behavior to be due exclusively to the quadrupole term of Saturn's gravitational potential for the $m_\theta = 1$ case. The resulting formulae are:

$$\lambda(r) \approx 3.08\left(\frac{r_v}{R_\mathrm{h}}\right)^4 \frac{\sigma_\rho}{m_\theta - 1}\frac{1}{r_v - r} \quad (m_\theta > 1) \tag{11.21a}$$

$$\lambda(r) \approx 54.1\left(\frac{r_v}{R_\mathrm{h}}\right)^6 \sigma_\rho\frac{1}{r_v - r} \quad (m_\theta = 1), \tag{11.21b}$$

where λ, r and r_v are measured in kilometers, and σ_ρ is in g cm^{-2}. Equations (11.21) afford a means of deducing the surface density from measured wavelengths (cf. Problem 11.14b).

Inelastic collisions between ring particles act to damp bending waves. Larger velocities lead to more rapid damping. The damping rate of bending waves can be used to estimate the ring viscosity, which can be converted into an estimate of the ring thickness using equations (11.11) and (11.14).

11.4.2.2 *Theory of Density Waves*

The gravity of a moon on an arbitrary orbit about Saturn has a component which produces epicyclic (radial and azimuthal) motions of ring particles. However, as in the case of vertical excursions induced by moons on inclined orbits, the epicyclic excursions are generally extremely small. An exception occurs near Lindblad (horizontal) resonances (eq. 11.18*b*), where coherent perturbations are able to excite significant epicyclic motions. In a manner analogous to the situation at vertical resonances, self-gravity of the ring disk supplies a restoring force that enables density waves to propagate away from Lindblad resonance. All density waves identified within Saturn's rings are excited at inner Lindblad resonances and propagate outward, away from the planet.

The theory of spiral density waves is analogous to that of spiral bending waves, with the fractional perturbation in surface mass density, $\Delta\sigma_\rho/\sigma_\rho$, replacing the slope of the disk, dZ/dr. The relationship in the linear theory $(\Delta\sigma_\rho/\sigma_\rho \ll 1)$ which is analogous to equation (11.20) is:

$$\lambda = \frac{4\pi^2 G\sigma_\rho}{m_\theta^2[\omega_f - n(r)] - \kappa^2(r)}. \tag{11.22}$$

The approximations given by equations (11.21) are also valid for density waves, provided r_v is replaced by r_L.

Most density waves observed in Saturn's rings have $\Delta\sigma_\rho/\sigma_\rho \approx 1$. At such large amplitudes the linear theory breaks down, and a nonlinear model is required. The principal results of the nonlinear theory are as follows: (*i*) Nonlinear density waves depart from the smooth sinusoidal pattern predicted by the linear model, and become highly peaked (Fig. 11.22). (*ii*) The theoretical wave profiles have broad, shallow troughs with surface

(a)

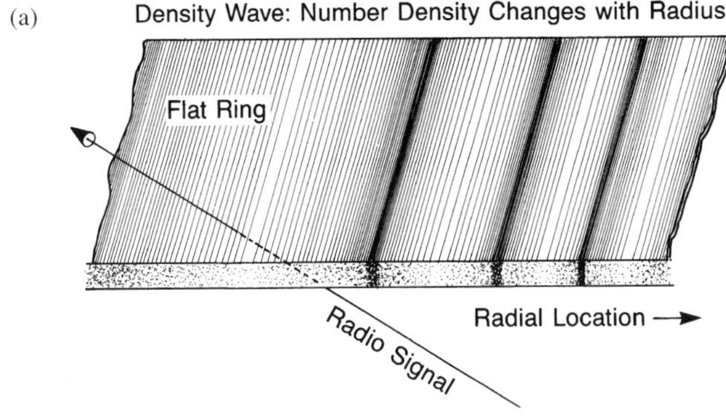

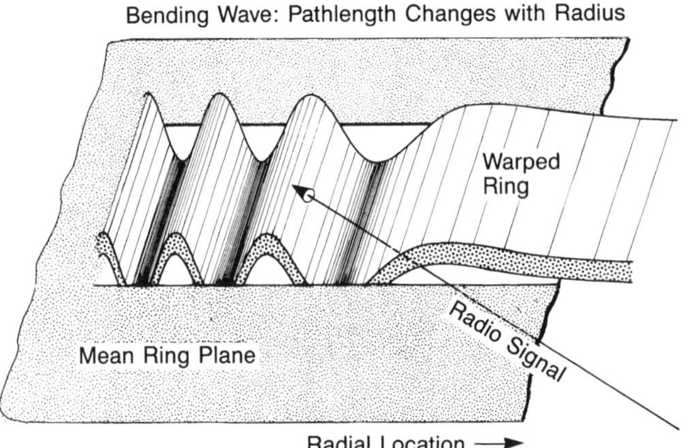

(b)

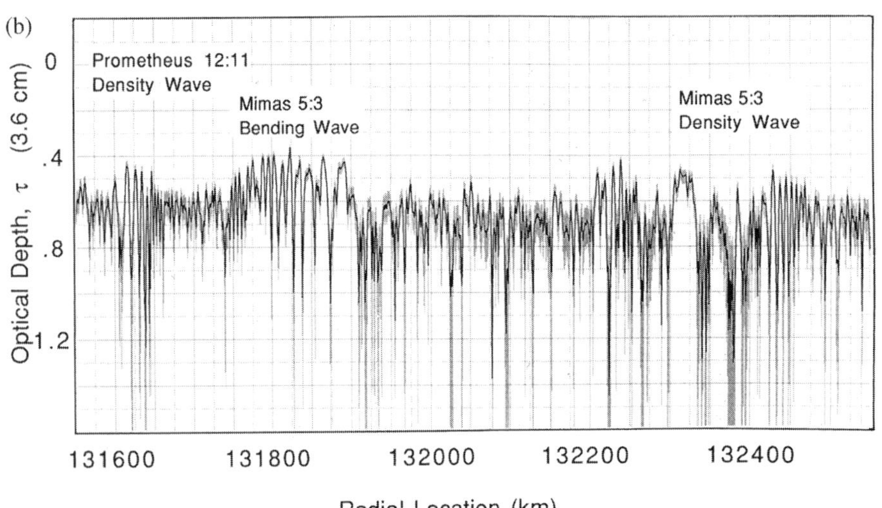

FIGURE 11.23 (a) Schematics of a radio occultation of a spiral density wave (top panel) and a bending wave (bottom panel). (Rosen *et al.* 1991) (b) Examples of wave features observed in the radio occultation data of Saturn's A ring. The solid curve is measured normal optical depth, $\tau(r)$, plotted to increase downward. The gray shaded region represents the 70% confidence bounds on the measurement. (Rosen 1989)

density never dropping below half of the ambient value, and are qualitatively similar to observed waves (compare Figs. 11.22 and 11.23b). (*iii*) The nonlinear torques exerted by Saturn's moons are similar to those calculated using linear theory.

11.4.2.3 *Observations of Spiral Waves in Saturn's Rings*

Spiral waves in planetary rings are extremely tightly wound, with typical *winding angles* (departures from circularity) being 10^{-5}–10^{-4} radian (compared to $\gtrsim 10^{-1}$ radian in most spiral galaxies). Such waves have very short

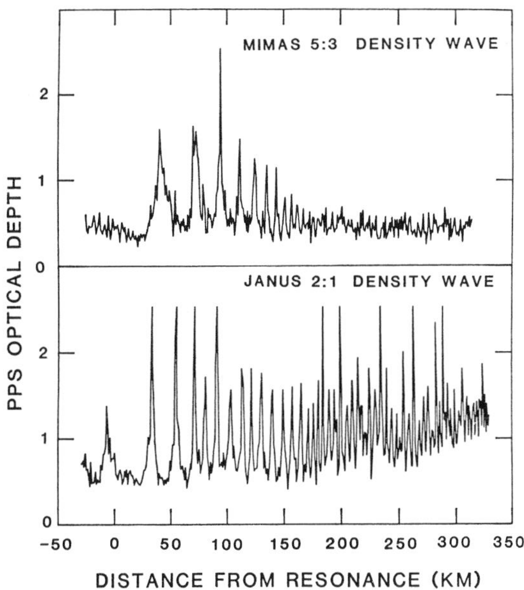

FIGURE 11.24 The Mimas 5:3 and Janus 2:1 density waves as viewed from the Voyager 2 PPS stellar occultation experiment, plotted so that $\tau(r)$ increases upwards. Note the sharp peaks and broad flat troughs caused by nonlinearities (cf. Fig. 11.21). (Esposito 1993)

wavelengths, of the order of 10 km. Spiral waves in Saturn's rings were detected by four instruments on the Voyager spacecraft. Brightness contrasts between crests and troughs of density waves, both in reflected light on the sunlit face of the rings (Fig. 11.20) and in diffuse transmission of sunlight on the dark side of the ring plane, can be seen in Voyager images. Bending waves are visible on Voyager images of the lit face of the rings (Fig. 11.20) as a result of the dependence of brightness on local solar elevation angle. Bending waves appear on images of the unlit face of the rings because the slant optical depth, through which sunlight diffused, depends on local ring slope.

Both density waves and bending waves were detected in Voyager radio occultation data (Fig. 11.23b). Density waves are observable by occultation experiments because bunching of particle streamlines increases the optical depth, τ, of crests. Bending waves, although they leave the optical depth normal to the ring plane unchanged, can be detected because the tilt of the ring plane causes oscillations in the observed slant optical depth (Fig. 11.23a). Similarly, the Voyager Photopolarimeter Subsystem (PPS) (Fig. 11.24) and Ultraviolet Spectrometer (UVS) observed the diminution of light as a star passed behind the rings, thereby observing changes in ring optical depth.

Five bending waves and several dozen density waves in Saturn's rings have thus far been identified with the resonances responsible for exciting them and have been analyzed to determine the local surface mass density of the rings. The surface density at most wave locations in the optically thick A and B rings is of order 50 g cm^{-2}. Measured values in the optically thin C ring are $\sim$1 g cm^{-2}; an intermediate value of $\sim$10 g cm^{-2} has been estimated for Cassini's division.

The damping behavior of three bending waves has also been analyzed to place upper bounds on the viscosity and local thickness of the rings. The A ring appears to have a local thickness of a few tens of meters; the thickness of the C ring is $\lesssim$ 5 m. Viscosity measurements from the damping of density waves are less reliable than those from bending waves, because most observed density waves are highly nonlinear in the regions where the bulk of the damping occurs. Even for bending waves, complications resulting from slight wave nonlinearities and nonuniformities in disk properties make such measurements less reliable than corresponding techniques used to estimate ring surface mass density.

In summary, resonantly excited density waves and bending waves are among the best understood features in planetary rings. Waves are seen to propagate from all of the strong satellite resonances, except those resonances that produce gaps. The locations of these waves agree with predicted values to within the observational uncertainties, which are often less than 1 part in 10^4. The wavelength behavior also agrees with the theory, and wavelength analysis has been used to obtain the best available estimates of the surface mass density of the rings; the total ring mass determined in this manner is roughly equivalent to that of an ice moon between 150 km and 200 km in radius. The damping behavior of spiral waves appears to be very complex, and theoretical studies suggest that it may be very sensitive to particle collision properties.

11.4.2.4 *Angular Momentum Transport by Density Waves*

Before Voyager arrived at Saturn, Peter Goldreich and Scott Tremaine predicted that torques from density waves excited by the moon Mimas at its 2:1 resonance had removed sufficient angular momentum from ring material to have cleared out Cassini's division, the 4000 km wide region of depressed surface mass density located between the broad high-density A and B rings. Although this hypothesis remains unverified, Voyager found a multitude of density waves excited by small newly discovered satellites orbiting near the rings. The torque at these individual resonances is less than that at Mimas's 2:1 resonance, but the sum of their torques is much greater. These waves are observed with amplitudes that agree with theoretical predictions to within a factor of order unity.

The back torque that the rings exert on the inner moons causes these moons to recede on a timescale short compared to the age of the Solar System; current estimates suggest that all of the small moons orbiting inside the orbit of Mimas should have been at the outer edge of the A ring within the past 10^8 years, with the journey of Prometheus, a relatively large moon located quite close to the A ring, occurring on a timescale of only a few million years. Resonance locking to outer, more massive, moons could slow the outward recession of the small inner moons; however, angular momentum removed from the ring particles should force the entire A ring into the B ring in a few times 10^8 yr. If the calculations of torques are correct, and if no currently unknown force counterbalances them, then the small inner moons and/or the rings must be 'new', i.e., much younger than the age of the Solar System. However, a 'recent' origin of Saturn's rings appears to be *a priori* highly unlikely. We return to this problem when we discuss the origins of planetary ring systems in Section 11.7.

11.4.3 Shepherding

Moons and rings repel one another through resonant transport of angular momentum via density waves. As is the case for viscous spreading of a ring, most of the angular momentum is transferred outwards and most of the mass inwards (in this case, the ring and moon are considered together as parts of the same total reservoir of energy and angular momentum). This is a general result for dissipative astrophysical disk systems, as a spread out disk of material on circular (nearly) keplerian orbits has a lower energy state for fixed total angular momentum than does more radially concentrated material. Analogously, resonant transfer of energy from an inner moon to a moon on an orbit farther out frees up energy for tidal heating (Problem 2.29), whereas transfer in the opposite direction is almost always unstable.

We now consider the process of *shepherding*, by which a moon repels ring material on nearby orbits. The essence of the interaction is that ring particles are gravitationally perturbed into eccentric orbits by a nearby moon. Collisions among ring particles damp these eccentricities; the net result is a secular repulsion between the ring and the moon. The details of the interactions depend on whether a single resonance dominates the angular momentum transport or the moon and ring are so close that individual resonances don't matter, either because the resonances overlap or because the synodic period between the ring and the moon is so long that collisions damp out perturbations between successive close approaches. However, many aspects of the basic qualitative picture are the same in both cases.

A moon exerts its strongest torques on a ring particle when they are near one another. Most of the interaction during a single encounter thus occurs near conjunction. If both the ring particles and the moon are initially on circular orbits, the perturbations received by each ring particle can only depend on its semimajor axis. By conservation of the Jacobi parameter (eq. 2.21), the amplitude of the eccentric excursions induced in the ring particles, ae, is much larger than the change in their semimajor axes. Moreover, for given semimajor axes of the moon and particle, any induced eccentricity can only depend on the longitude of the particle's orbit relative to where it passed the moon. Thus, if we may approximate the encounter by an impulse and the moon initially sends ring particles inwards, the particles will execute epicyclic motion, and be at periapse in $\frac{1}{4}$ orbit, at apoapse $\frac{3}{4}$ orbit after encounter, periapse at $\frac{5}{4}$ orbit after encounter etc. A plot of radius vs. time would be a sinusoid. In fact, the shape of an edge represents just such a plot. The motion of the particles relative to the moon is nearly constant, as long as the difference in their semimajor axes is large compared to the size of the induced epicyclic motion. Thus, distance along the edge is essentially a measure of time, with the normalization being the synodic velocity. The wavelength of the induced oscillation is the relative motion of the ring and moon during one epicyclic period:

$$\lambda_{edge} = 3\pi \, |\Delta a| \, \frac{n}{\kappa} \approx 3\pi \, |\Delta a|, \tag{11.23}$$

where $\Delta a \equiv a_r - a_s$ is the separation in the orbital semimajor axes of the ring and moon. Calculation of the amplitude of the oscillation is more complicated. Perturbations before and after encounter almost cancel out, except for a second order (in torque) term which results from the particles having been pulled slightly closer to the moon by the encounter, and thus the torque being slightly larger after closest approach than before. Particles initially on circular orbits are excited to eccentricities of:

$$e \approx 2.24 \frac{M_s}{M_p} \left(\frac{a}{\Delta a} \right)^2 . \tag{11.24}$$

The shape and amplitude differ if the ring or moon has an initial eccentricity, or if ae is not small compared to $|\Delta a|$, but the wavelength remains the same as long as the potential is keplerian and $|\Delta a| \ll a$. Wavy edges of Encke's gap excited by the moonlet Pan are visible in Voyager images (Fig. 11.25). This pattern satisfies the wavelength relationship given by equation (11.23), and the mass of Pan has been estimated from the observed wave amplitudes via

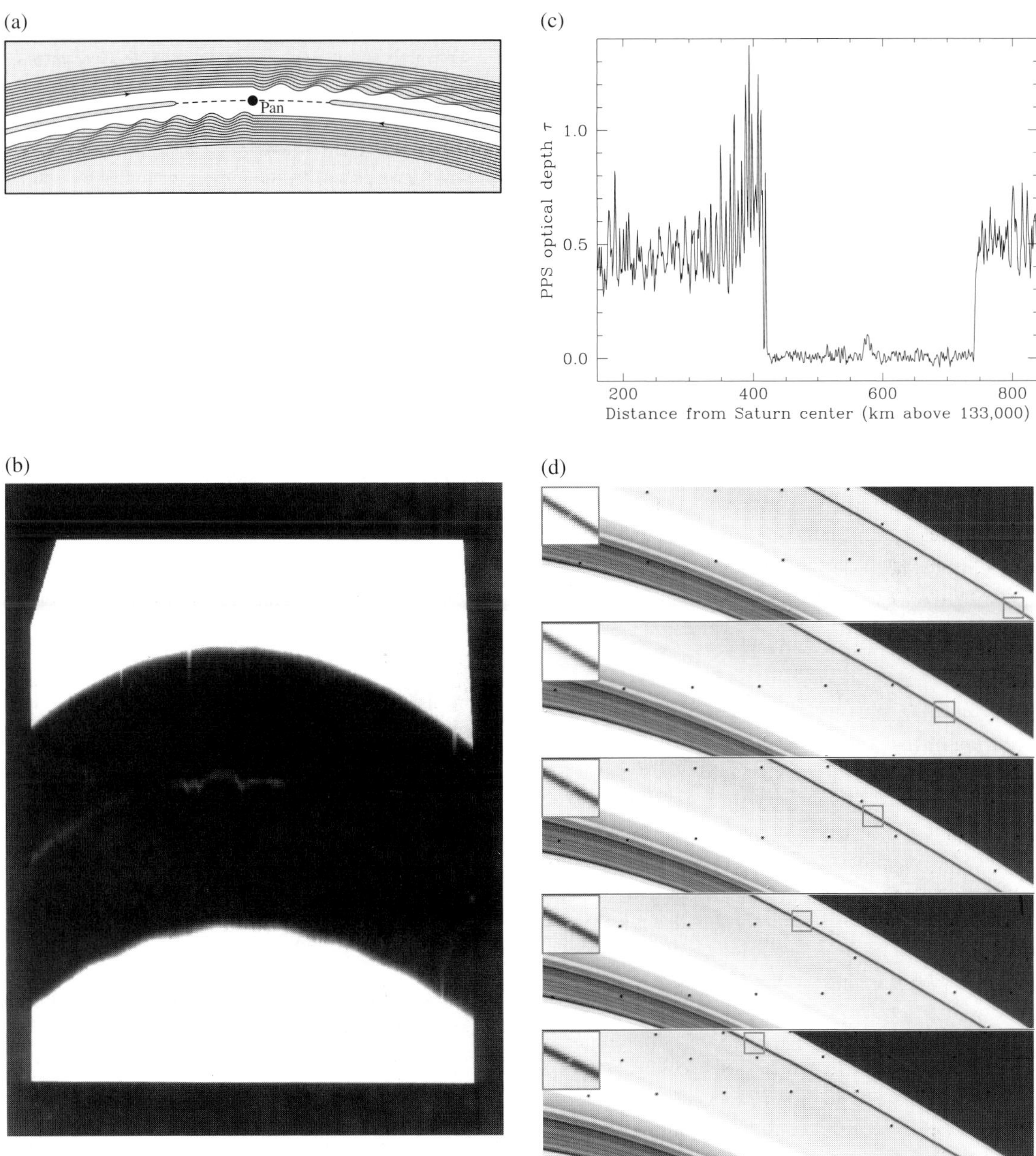

FIGURE 11.25 (a) Schematic illustration of the pattern that the moonlet Pan creates on and near the edges of the Encke gap in Saturn's rings. This pattern remains stationary in the frame rotating with the moonlet's orbital frequency. The radial scale is greatly exaggerated relative to the angular scale for clarity. (Murray and Dermott 1999) (b) Wavy edges are seen in this Voyager 2 image of Encke's gap, which also highlights the narrow kinky ringlet near the gap's center. The longitudinal scale has been compressed by a factor of 10 relative to the radial scale in order to increase visibility. (c) Optical depth obtained from the Voyager stellar occultation profile of the region of the Encke gap and environs. The regular pattern of oscillations in the region of the A ring immediately interior to the gap represents a cross-section of optical depth variations resulting from Pan's satellite wake. (Showalter 1991) (d) Sequence of five horizontal strips from Voyager 2 images FDS 43904.59 – 43905.23 showing Pan traversing the (nearly fixed) field of view. The time interval between successive frames is roughly 5 minutes. For each image a small box identifies the location of the moonlet, and the same region is shown enlarged at the upper left. The black dots in the images are reseau markings. (Within the optics of the Voyager cameras was a grid of black dots called 'reseau markings'. These appear superimposed upon each image sent back from Voyager, and are used to correct for geometric distortions in the camera. They are often suppressed in published images by replacing them by the local average brightness. Nevertheless, you can almost always find them if you look closely.) (Showalter 1991)

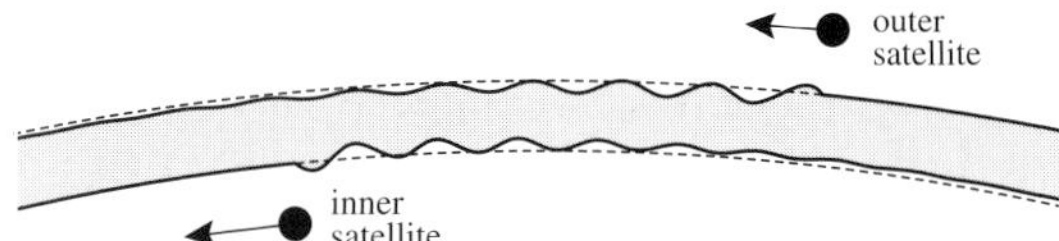

FIGURE 11.26 Schematic illustration of the shepherding of a planetary ring by a moon. Ring particles' eccentricities are excited as the particles pass by the moon. Interparticle collisions subsequently damp eccentricities and leave the particles orbiting farther from the moon than prior to the encounter. Particles orbiting closest to the moon are affected most strongly. (Murray and Dermott 1999)

equation (11.24). The characteristic length scales of longitudinal variations in Saturn's F ring are comparable to wavelengths predicted from forcing by the nearby moon Prometheus according to equation (11.24), but the situation is more cluttered and complicated than for the edges of Encke's gap.

The observable manifestations of Pan's interactions with Saturn's A ring extend beyond the wavy edges of Encke's gap. Pan also excites eccentricity of ring particles more distant than the gap edges, producing a pattern which decreases in amplitude and increases in wavelength with increasing distance from the moonlet. The variations in particle response with semimajor axis produce a *satellite wake* which is observable as radial variations in ring optical depth (Fig. 11.25c). The wake can be used to obtain an estimate of Pan's mass which is more accurate than that provided by available measurements of the amplitude of the wavy edges of Encke's gap.

In order to study the shepherding problem in more detail, let us move to the frame in which the moon's position is fixed (the frame rotating at the moon's orbital frequency). For our analysis, we assume that the moon's orbit is circular and exterior to that of the ring. We further assume that the separation in the orbital semimajor axes of the ring and moon is small, $|\Delta a| \ll a$, but that they are far enough apart that the moon's perturbations alter the radius of the ring's orbit by an amount $ae \ll |\Delta a|$. Interactions among the ring particles are ignored for the brief interval near conjunction with the moon, during which essentially all of the angular momentum transport occurs, but the orbits of the ring particles are assumed to be circularized between successive encounters with the moons as a result of interparticle collisions. Note that the eccentricity of the ring particles must be damped out by collisions, at least partially, before the next encounter with the moon, or the reverse angular momentum transfer may occur, and over long periods of time the net torque is zero. The system is sketched in Figure 11.26.

As the ring particles approach the moon, they are pulled in the forward direction by its gravity, thereby increasing the particles' energy (and angular momentum) and causing their orbits to move closer to that of the moon. The magnitude of the force is proportional to $M_s/\Delta a^2$. However, after the ring particle passes the moon, the tug is in the opposite direction, and removes energy and angular momentum. To the lowest order, the forces cancel, but a small asymmetry exists because the particle's orbit is slightly closer to that of the moon on the outbound journey as a result of perturbations prior to conjunction. The net torque is thus second order in the forcing. A more detailed analysis gives the magnitude of the torque to be:

$$T_g \approx 0.40 \frac{\Delta a}{|\Delta a|} \sigma_\rho \left[\frac{GM_s}{n(\Delta a)^2} \right]^2 . \tag{11.25}$$

A moon thus pushes material away from it (on both sides). Ring material between two moons is forced into a thin annulus between them. The annulus should be located nearer the smaller moon, so that the torques balance. Viscous diffusion maintains a finite width of this annulus. Saturn's F ring is confined between the moons Prometheus and Pandora (Fig. 11.27). The shepherding moons Cordelia and Ophelia confine the uranian ϵ ring, although in this case the ring edges are maintained by individual 'isolated' resonances of these moons. Uranus's other narrow rings are also believed to be confined by shepherding torques; a few ring edges may be maintained by resonances with Cordelia and Ophelia, but the small moons that are believed to hold most of Uranus's narrow rings in place have yet to be observed.

What about a small moon embedded in the middle of a ring plane? Material is cleared on both sides, so a gap is formed around it. Diffusion acts to fill in the gap and blur the edges, but optical depth gradients can occur on length scales comparable to typical particle collision distances, allowing sharp edges to be produced. If a moon is too small, then it can't clear a gap larger than its own size, so no gap is formed. As discussed above, Encke's gap is cleared by the moonlet Pan, and smaller moonlets probably maintain the Keeler gap as well as some of the other gaps in Saturn's rings.

11.4.4 Longitudinal Confinement

Neptune's Adams ring contains prominent arc-shaped features which orbit at the keplerian rate (Figs. 11.15 and 11.16). The arcs have been observed to persist for at least several years, which is more than their lifetime against Kepler shear (Problem 11.7). Thus, there must be some mechanism which confines the ring particles. A combination of

(a)

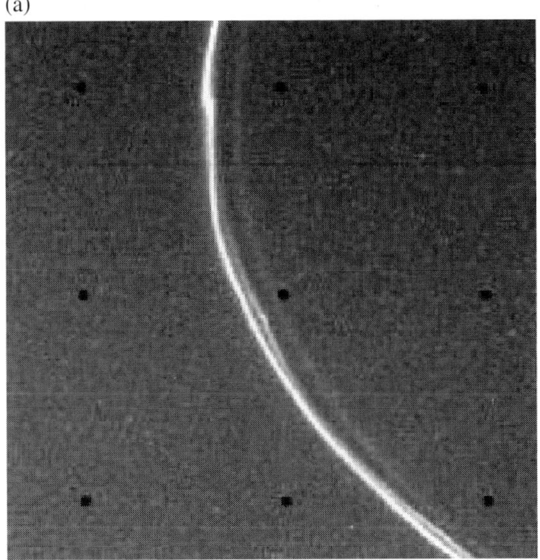

(b)

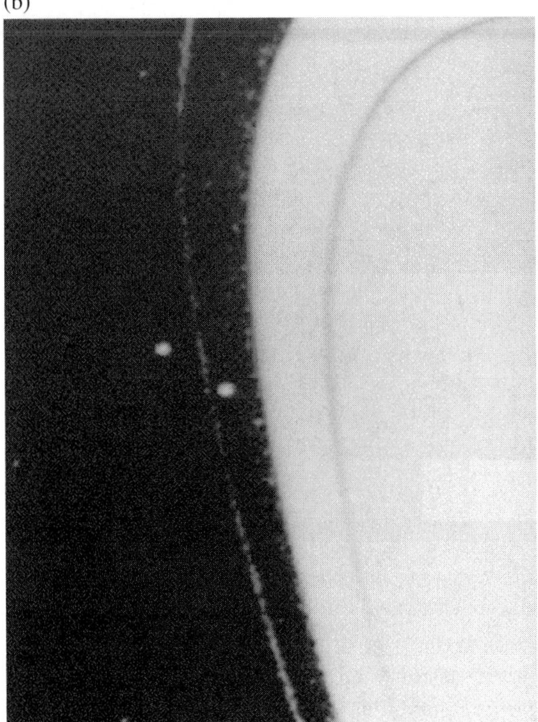

FIGURE 11.27 (a) Voyager 1 image (FDS 34930.48) of Saturn's F ring. The F ring is split into three components, the outer two of which appear to cross or at least approach one another to within the resolution of the image. Several bright clumps are visible in each of the outer two components. The regularly spaced black dots are reseau marks. (NASA/Voyager 1)
(b) Voyager 2 image of Saturn's F ring and its shepherding moons, Prometheus and Pandora. Prometheus, the inner shepherd, orbits faster and had just overtaken Pandora before this image was taken. (NASA/Voyager 2)

Lindblad and corotation resonances of a moon or moons is capable of confining rings in both radius and longitude.

The most prominent example of a corotation resonance is the 1:1 commensurability. Jupiter's 1:1 corotation resonance is responsible for the confinement of the Trojan asteroids (cf. Sections 2.3.2.2 and 9.1), which librate in 'tadpole' orbits about Jupiter's triangular Lagrangian points. A similar form of confinement is not stable for a dissipative collisional system such as a planetary ring. The triangular Lagrangian points, L_4 and L_5, are potential energy maxima, and thus are unstable to most forms of dissipation, including interparticle collisions. A ring would gradually spread in both radius and longitude as a result of such collisions. An arc ring could be confined about one of the triangular Lagrangian points of a moon if a second moon on a nearby or nearly resonant orbit exerted a shepherding torque on the ring at a Lindblad resonance.

Moons on eccentric or inclined orbits have corotation torques with various pattern speeds, and these torques can provide azimuthal confinement at orbital radii different from that of the moon. Although such corotation resonances are usually much weaker than the 1:1 resonance, in a nearly keplerian potential (such as those of all four ringed planets in the Solar System) these other corotation resonances are associated with a nearby Lindblad resonance, which may provide the torque required to counteract dissipation. Thus, an arc ring could be confined by the combination of a corotation resonance and a Lindblad resonance of a single moon. Indeed, dynamical models imply that the arcs in Neptune's Adams ring are confined in whole or in part by the nearby moon Galatea at its 42:43 corotation and Lindblad resonances. Large particles orbiting within the Adams ring may have a role in producing the detailed structure of the arcs.

11.5 Physics of Dust Rings

Thus far, we have analyzed the motions of ring particles by considering only the forces of gravity and physical collisions. While these are the dominant forces on ring particles that are $\gtrsim$ 1 mm in size, micrometer-size dust is significantly affected by electromagnetic forces. Radiation forces, primarily Poynting–Robertson drag (Section 2.7.2), affect all long-lived small particles in planetary rings. A planet's magnetic field can be an important influence on the motion of charged dust particles, with substantial effects occurring on both orbital and secular timescales.

11.5.1 Radiation Forces

A general introduction to the effects of radiation forces on the motion of Solar System particles was presented in Section 2.7. In Chapter 2, we concentrated on particles in heliocentric orbits. Here, we focus our discussion on those aspects of radiation forces pertinent to particles in planetocentric orbits. In most cases, radiation directly from the Sun dominates over radiation reflected from and emitted by the planet.

Radiation pressure has little effect on the orbits of most planetocentric grains. Even if $\beta \equiv F_r/F_g \gtrsim 1$, the planet's gravity is usually far greater than the Sun's, so the perturbation is small. The shapes of the orbits of small grains are altered slightly, and velocities of particles change when they enter and exit the shadows of planets. However, in most circumstances the net effects of these forces cancel out over time and they do not produce any secular evolution of particle orbits. A notable exception occurs for particles of radius $\sim 1 \pm 0.3$ μm in Saturn's E ring. The periapse precession rate induced by the Lorentz force (Section 11.5.2) on charged 1 μm grains at Enceladus's orbit is opposite in sign and approximately equal in magnitude to the precession resulting from Saturn's oblateness (Section 2.5.2). The very slow rate of periapse precession for particles within this narrow size interval which results from cancellation of precession rates allows perturbations to build up over many orbits, so that radiation pressure can alter almost circular orbits of micrometer-size dust into highly eccentric trajectories. Assuming the moon Enceladus, and possibly other saturnian moons as well, is the source of the particles in the E ring, such eccentricities would greatly reduce the chances that particles would recollide with their parent moons and be lost to the ring system. Thus, this model explains both the large radial extent of the E ring (produced by particles on eccentric orbits) and the unusual narrowness of the particle size distribution.

Poynting–Robertson drag, in contrast, leads to substantial evolution of tiny particles within planetary rings over a much broader range of parameter space. The secular rates of change of orbital semimajor axis and eccentricity are given by

$$\frac{da}{dt} = -\frac{a}{t_{pr}} \frac{5+\cos^2 i_*}{6} \tag{11.26}$$

and

$$\frac{de}{dt} = 0, \tag{11.27}$$

respectively. In equation (11.26), i_* represents the inclination of the particle's orbit *to the planet's orbital plane about the Sun*, and the characteristic decay time, t_{pr}, is given by

$$t_{pr} = \frac{1}{3\beta} \frac{r_\odot}{c} \frac{r_\odot}{GM/c^2} \approx 530 \frac{r_{AU}^2}{\beta} \quad \text{yr}, \tag{11.28}$$

where $r_\odot$ is the planet–Sun distance, and r_{AU} is this distance in astronomical units. The orbital eccentricity is thus constant, apart from minor short-period variations, and the particle's semimajor axis decreases in an exponential fashion. This contrasts to particles in heliocentric orbits, whose semimajor axes decrease more rapidly as they approach the Sun (cf. eq. 2.50). For heliocentric particles

$$\frac{da}{dt} = -\frac{2}{3t_{pr}} \frac{a_o^2}{a}, \tag{11.29}$$

where t_{pr} is given by equation (11.28) with a_o substituting for $r_\odot$. Poynting–Robertson decay times for microscopic grains given by equation (11.28) are short compared to the age of the Solar System, even for particles in orbit about Neptune. Jupiter's ring particles are thought to be ejecta from jovian moons, which spiral inwards through Poynting–Robertson drag, thereby forming very broad rings.

11.5.2 Interactions of Charged Grains with Planetary Magnetospheres

Planetary ring particles orbit close to their massive primaries, in environments characterized by high densities of energetic charged particles trapped by strong planetary magnetic fields. Uncharged dust grains are impacted by electrons more frequently than by ions because the thermal speeds of electrons are much larger. Grains thus acquire sufficient negative charge to achieve a balance between the rates at which they accumulate additional electrons and ions via electrostatic attraction and repulsion. For parameters typical of micrometer-size grains in planetary rings, equilibrium is achieved in less than an orbital period. In Jupiter's rings, grains reach equilibrium at a potential $\Phi_V \approx -10$ volts. The charge that a grain can accumulate depends on its proximity to other grains as well as the charged particle environment; if other grains are within the plasma's Debye shielding length (the characteristic distance beyond which the electric field of the grain is shielded by particles in the plasma having opposite charge), then they contribute to the repulsion of additional electrons, and a given potential can be maintained with less charge per grain. For an isolated grain of radius R, the potential, Φ_V, and charge, q, are related according to:

$$\Phi_V = \frac{q}{R}. \tag{11.30}$$

Other charging mechanisms, such as photoelectron currents, can perturb the equilibrium value of q. Stochastic variations in particle charge have minor effects on particle motions; however, systematic variations in charge which are experienced by grains on highly eccentric orbits can significantly affect particle trajectories.

The motion of charged grains within planetary magnetospheres is influenced by trapped plasma as well as by the planetary magnetic field itself. Electric and magnetic forces are most important for very small particles, because mass increases much faster than mean charge as particle radius grows. The collisions of charged particles with dust grains result in an exchange of angular momentum between the plasma and dust, a process known as *plasma drag*. The drag force depends upon the velocity of the grain relative to the plasma. The plasma rotates with the planet's magnetic field, so grains orbiting at the corotation radius, r_c, don't move relative to the plasma and thus feel no drag. Grains orbiting interior to r_c lose energy and angular momentum to the plasma and spiral inwards, whereas grains at $r > r_c$ gain energy and spiral outwards (unless energy losses from Poynting–Robertson drag exceed the energy gains from plasma drag, as is the case for Jupiter's rings). Jupiter's corotation radius is located at $r_c = 2.24\ R_{♃}$, within the gossamer rings. Although most of Jupiter's rings are formed from particle ejecta off Jupiter's moons brought inwards by Poynting–Robertson drag (Section 11.6.3), the feeble outward extension of the gossamer Thebe ring may have formed from plasma drag on some of the Thebe ejecta.

Charged dust grains are acted upon by the planet's magnetic field via the Lorentz force (eq. 7.43):

$$\mathbf{F}_L = \frac{q}{c}\mathbf{v} \times \mathbf{B}. \tag{11.31}$$

The Lorentz force couples charged dust grains to the magnetic field, and hence reinforces the plasma drag. The Lorentz force is especially important for Jupiter's rings, because they have many small particles, are located close to the planet, and Jupiter's magnetic field is very strong and inclined with respect to its rotation axis. The jovian Lorentz force is approximately 1% of Jupiter's gravitational force for typical ring particles (Problem 11.18). The tilt of the field with respect to the rotation axis implies that particles not orbiting at r_c experience a time-varying force as they move relative to the field (Fig. 11.28). At certain locations, the frequency at which a particle experiences variations in the Lorentz force is commensurate with the particle's epicyclic or vertical frequency, leading to *Lorentz resonances*. Lorentz resonances have many things in common with the gravitational resonances discussed in Section 11.4.1, but there are also several differences. The Lorentz force affects particle mean motions, so orbital, radial and vertical frequencies are functions of charge-to-mass ratio. Thus, the locations of Lorentz resonances vary (slightly) with particle size, in contrast to the sharply defined gravitational commensurabilities. The consequences of forcing at Lorentz resonances in Jupiter's rings also differ substantially from those at vertical and Lindblad resonances in Saturn's rings. Self-gravity and collisions are important in Saturn's rings, whereas self-gravity is negligible for Jupiter's ring system and collisions are very infrequent. Vertical forcing of particle motions by the strong Lorentz 3:2 resonance located at 1.71 $R_{♃}$ is responsible for the for-

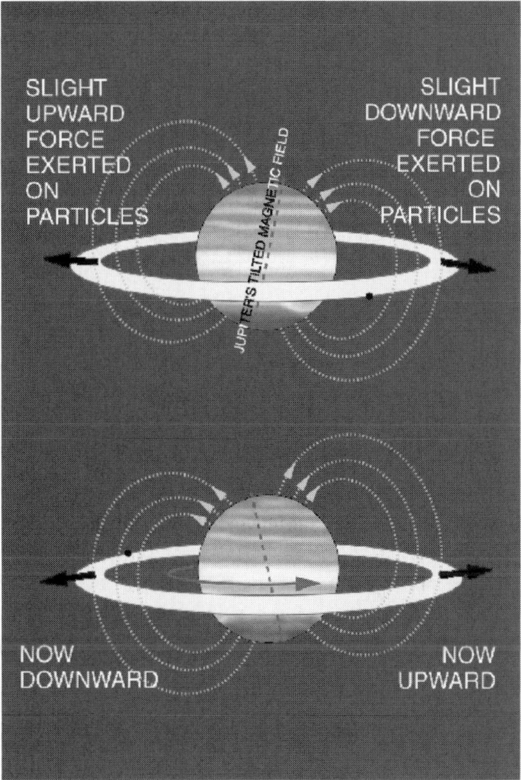

FIGURE 11.28 This graph illustrates the effect of the magnetic force (shown as arrows) on charged ring particles. Since Jupiter's dipolar magnetic field (shown dotted) is tilted about 10° from the planet's spin axis (vertical in these sketches), the direction of this out-of-plane force depends on where the particle is in the ring plane and on Jupiter's rotational orientation. With Jupiter's orientation as shown in the top panel, the magnetic force on a charged ring particle has a slightly upward (downward) force on particles at the left (right) of Jupiter. This situation is reversed in direction five hours later, after Jupiter has rotated 180° (into the orientation depicted in the bottom panel). Thus, every charged ring grain experiences an oscillating vertical force; the period of these forces depends on the orbital radius, so that at certain locations the periods become multiples of the particles' orbital periods, which leads to Lorentz resonances. (Courtesy: J.A. Burns)

mation of the halo: the orbits of (sub-)micrometer-sized inward-migrating particles are perturbed significantly and become much more inclined, thus forming the halo. At the location of the Lorentz 2:1 resonance, 1.40 $R_♃$, the particle orbits are perturbed again, this time into orbits which impact the planet, leading to a loss of halo particles. The normal optical depth of the halo is substantially reduced inwards of 1.40 $R_♃$.

11.5.3 Spokes in Saturn's Rings

Electric and/or magnetic effects are responsible for *spokes*, the only known planetary ring features which are predominantly radial in shape. They are centered within the B ring, where particles orbit synchronously with Saturn's magnetic field. Spokes appear darker than their surroundings in backscatter, but brighter in forward scatter (Fig. 11.29). The strongly forward-scattered appearance of spokes implies they contain a significant component of micrometer- and submicrometer-sized dust grains. Spokes form rapidly, on timescales of minutes to tens of minutes, and initially appear as linear features pointing towards the center of Saturn. One edge remains radial (and evolves with the period of Saturn's magnetic field) as long as new material is being added to the spoke. Dust in the spokes orbits at essentially the keplerian rate, so spokes get smeared out as they age. The spoke fades gradually as dust is reaccumulated by the larger ring particles, and disappears from view in about one-fourth to one-third of an orbital period. Spokes occur at all azimuths, but not with equal frequency. The formation of spokes occurs most frequently soon after the ring emerges from Saturn's shadow, is strongly correlated with magnetic field longitude, and is weakly correlated with the longitude of certain cloud features near Saturn's equator, where strong winds cause a $\sim 5\%$ reduction in rotation period relative to the rotation period of Saturn's magnetic field. No consensus exists concerning the physical mechanism responsible for spoke formation.

11.6 Meteoroid Bombardment of Planetary Rings

Planetary rings have a very large surface area to mass ratio, and thus are heavily bombarded with and significantly affected by the flux of small stray debris that is present throughout our Solar System. Such debris can change the orbits and composition of planetary rings, and may be responsible for the formation and destruction of planetary ring systems. The most direct evidence for erosive impacts on planetary rings is provided by observations of a ring 'atmosphere' of neutral hydrogen extending 0.5–1 $R_♄$ above Saturn's rings. Because collisions with ring particles and

(a)

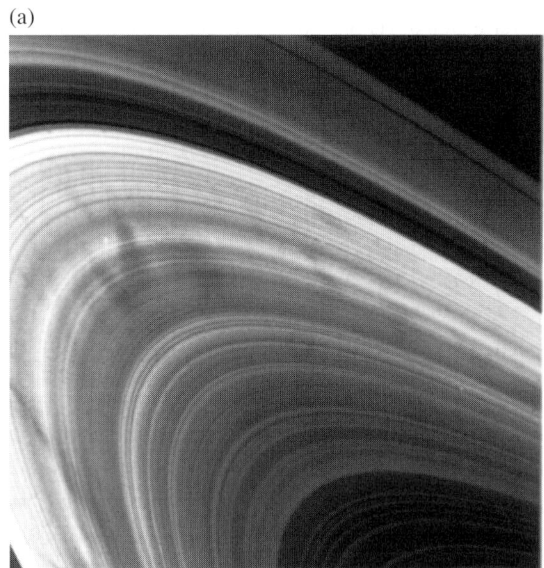

(b)

FIGURE 11.29 Two images of spokes in Saturn's B ring. The spokes appear dark in frame (a), which was taken in backscattered light, whereas they are brighter than the surrounding ring material in frame (b), which was imaged in forward scattered light. (NASA/Voyager)

the formation of H_2 rapidly deplete neutral H, the ring atmosphere must be replenished at a rate of $\sim 3 \times 10^{28}$ atoms per second, presumably by erosion of icy ring particles.

When hypervelocity dust strikes a ring particle, it can excavate $\sim 10^4$–10^5 times its mass in impact ejecta. Some of the debris from the impact is permanently lost to the rings, either by flying off on orbits which are unbound or intersect the planet, or by being vaporized/ionized and subsequently escaping. However, under most circumstances, the bulk of the ejecta is re-accreted by the ring system, although not necessarily at the same distance from the planet from which it originated. In this section, we first consider the changes in mass, composition and total ring angular momentum resulting from meteoroid impacts, and then discuss ballistic transport of ejecta and its possible role in creating ring structure.

11.6.1 Accretion of Interplanetary Debris

Interplanetary impactors add material to ring systems, but they also remove matter from the rings. The net effect is uncertain, either a gain or loss may result, depending on many factors including impact speed, proximity to the planet's atmosphere, depth within the planet's gravitational potential well, and the fragility and volatility of the materials involved. The impactor flux is also quite uncertain, but is consistent with Saturn's rings being bombarded with their own mass of material over the age of the Solar System, so material losses and gains from impacts may be a significant factor in the evolution of planetary ring systems.

Planetary rings possess substantial net orbital angular momentum, as all particles orbit in the same direction. Impacting debris from heliocentric orbits provides essentially no net angular momentum to an optically thick ring system. (Optically thin rings accrete a net negative angular momentum flux because bodies traveling in the retrograde direction have larger impact probabilities (Problem 11.20).) Ejecta lost from the ring system generally take with them 'positive' angular momentum. Thus, in most circumstances, the net effect of interplanetary impactors is to cause a ring system to lose (specific) angular momentum, and thus to decay slowly inwards.

Impacting debris can change the mineralogical composition of planetary rings in three ways: New material is brought to the ring system, which, on average, is roughly solar in composition (apart from a substantial depletion in volatiles). The more volatile and fragile components of the rings are preferentially lost from the ring system. High pressures and temperatures during the impacts produce chemical changes.

The most interesting observable chemical consequences of interplanetary debris on planetary rings may well be the 'pollution' of Saturn's rings. Observations suggest that Saturn's rings are almost pure water-ice. The largest fraction of other material is observed in regions of low optical depth, which should receive the largest fractional contamination by interplanetary debris. The age of Saturn's rings estimated by attributing all the observed darkening to interplanetary debris is on the order of 10^8 years. Although the uncertainties in the meteoroid flux and vaporization/loss fractions are quite high, it is interesting to note that this timescale is consistent with age estimates from density wave torques, and suggests that Saturn's ring system is not primordial.

11.6.2 Ballistic Transport

Most of the ejecta from hypervelocity impacts leave ring particles at speeds that are much smaller than the particles' orbital speed. This debris thus travels on orbits of low eccentrity and inclination, and unless the ring optical depth is very small, most ejecta reimpact ring particles within a few orbital periods. The rates at which a given region of a ring gains and loses material by this process depends on its optical depth and the optical depth of neighboring regions, within a 'throw distance' of the ejecta. Structure can be formed by this *ballistic transport*, especially near abrupt boundaries in ring optical depth. Numerical simulations of ballistic transport have successfully reproduced the ramp-like structure seen at the boundaries between the C ring and B ring and between the Cassini division and the A ring (Fig. 11.30), and provide the most plausible explanation for such features.

11.6.3 Mass Supply to Rings from Satellite Ejecta

Rings can gain mass, M_{rings}, due to impacts on a satellite:

$$\frac{dM_{rings}}{dt} = f_i Y_e Y_i \pi R^2, \tag{11.32}$$

where R is the radius of the satellite, f_i is the mass flux density of hypervelocity impactors, and Y_i is the impact yield or the ratio of ejected mass to projectile mass, which is typically of the order of $10v_i^2$, with the impact velocity v_i in km s^{-1}. Based upon empirical fits to hypervelocity cratering experiments, the fraction of ejecta that escapes the satellite, Y_e, can be approximated by:

$$Y_e \approx \left(\frac{v_{min}}{v_e}\right)^{9/4}, \tag{11.33}$$

with v_{min} the minimum speed at which ejecta are launched (typically 10–100 m s^{-1}) and v_e is the escape velocity.

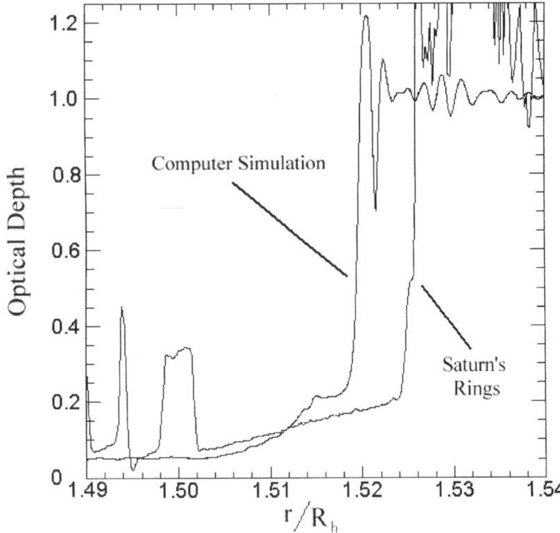

FIGURE 11.30 Comparison between observations of the transition region from Saturn's optically thin C ring to the optically thick B ring and a numerical simulation of the pattern produced by ballistic transport resulting from a few million years of meteoroid bombardment. The observed optical depth profile was obtained from Voyager 1 images of the ring taken against the planet's limb. The initial profile in the simulation was assumed to be sharp-edged, but the tendency for viscosity to spread such an edge was also included. Note the similarity between the sharpness of the inner B ring edge as well as the gradual ramp located just interior to it in both curves. (Courtesy Richard Durisen, for details about the simulations, see Durisen *et al.* 1996.)

For an isolated satellite, $v_e \propto R$, and is $\sim$ 10–100 m s^{-1} for spheres with sizes similar to Jupiter's small inner satellites. Using these approximations one can show that $Y_e \propto R^{-9/4}$, and therefore

$$\frac{dM_{rings}}{dt} \propto R^{-1/4} \quad (v_e > v_{min}), \tag{11.34a}$$

$$\frac{dM_{rings}}{dt} \propto R^{2} \quad (v_e < v_{min}). \tag{11.34b}$$

Thus smaller satellites, even though they have a smaller surface area, apparently provide much more material than larger ones because their gravitational force is much weaker. If the satellites reside within the Roche limit, the ejecta losses by impacts are even larger. The optimum source for ring material is a satellite for which $v_e \approx v_{min}$. If the satellite has a density of 2 g cm^{-3} and is covered by a soft regolith, it would be an optimum source for ring material if its radius $R \approx$ 5–10 km, roughly the size of Adrastea.

Ejecta from impacts on the jovian satellites Thebe, Amalthea, Metis and Adrastea migrate inwards via Poynting–Robertson drag, producing the entire morphology and mass of the jovian ring system. The ejecta start on orbits that resemble the orbit of the parent satellite. The orbits precess quickly (4 and 8 months at 2.5 and 3.1 $R_{♃}$, respectively), while the particles spiral slowly inward due to Poynting–Robertson drag; meanwhile the inclinations of the orbits are preserved. Within a relatively short period of time, the ensemble of particle orbits resembles a doughnut with a width (vertical extent) determined by the inclination of the parent satellite. The top and bottom edges are brightest, since a particle spends most of its time near its turning points. The inclination of the orbits of Thebe, Amalthea and Metis/Adrastea are 1.1°, 0.4° and 0.0°, respectively. The vertical extent of the Thebe ring is thus largest, as shown on the images (Fig. 11.7), while Jupiter's main ring lies in the equatorial plane. Although the Thebe and Amalthea gossamer rings extend inwards beyond Metis, they are outshone by the main ring.

11.7 Origins of Planetary Rings

The major outstanding issue in planetary rings is the origin and age of ring systems. Are ring systems primordial structures (dating from the epoch of planetary formation, $\sim 4.5 \times 10^9$ years ago) that are remnants of protosatellite accretion disks, or did they form more recently, as the result of the disruption of larger moons or interplanetary debris? Various evolutionary processes, such as angular momentum transport via resonantly excited density waves, occur on timescales far shorter than the age of the Solar System, suggesting that planetary rings must be geologically young. However, several arguments suggest that such a recent origin of Saturn's rings is extremely unlikely.

The strongest evidence for a geologically recent origin of planetary rings is the short timescales computed for evolutionary processes. The most convincing factors in the case of Saturn's rings are orbital evolution of rings and nearby moons resulting from resonant torques (responsible for density wave excitation) and ring pollution by accretion of interplanetary debris. Testing the validity of satellite torques is especially important, because models also suggest that angular momentum transport via resonant torques and density waves is a significant factor in the evolution of other, less well observed, astrophysical disk systems, such as protoplanetary disks and accretion disks in binary star systems. For example, density wave torques may have led to significant orbital evolution of Jupiter within the protoplanetary disk on a timescale of $\sim 10^4$–10^5 years.

Micrometer-size dust is removed from rings quite rapidly. Some loss mechanisms lead to permanent removal of grains from ring systems; examples include: sputtering, which is dominant for Jupiter's dust; gas drag, which is important for particles orbiting Uranus because of that planet's hot extended atmosphere; and Poynting–Robertson drag. Other dust removal mechanisms, such as re-accretion by large particles, which dominates for the dust comprising Saturn's spokes, allow for recycling. In all ring systems, the dust thus requires continual replenishment, which means macroscopic parent particles, but the dust mass is so small that in most cases a quasi-steady state could exist over geologic time.

Let us examine some specific origin scenarios. A stray body passing close to a planet may be tidally disrupted (e.g., Comet D/Shoemaker–Levy 9, cf. Section 10.6.3). However, under most circumstances, the vast majority of the pieces escape from or collide with the planet. Moreover, such an origin scenario does not explain why all four planetary ring systems orbit in the prograde direction. Ring particles are thus most likely (possibly second or later generation) products of circumplanetary disks.

Planetary rings may be the debris from the disruption of moons that got too close to their planets and were broken up by tidal stress, or that were destroyed by impacts and did not re-accrete because of tidal forces. There are two difficulties with the hypothesis that the rings that we observe are the products of recent disruptions of moons: Can moons form at or move inwards to the radii where planetary rings are seen? Why did the rings form recently, i.e., why is now special? It is more difficult for a body to accrete than to remain held together (Problem 11.2). Thus, a ring-parent moon must form beyond the planet's Roche limit and subsequently drift inwards. Tidal decay towards a planet only occurs for moons inside the orbit that is synchronous with the planet's spin period (unless the moon's orbit is retrograde), but Roche's limit is outside the synchronous orbit for Jupiter and Saturn. However, the presence of moons interspersed with ring particles in all four planetary ring systems (Problems 11.3 and 11.4) argues strongly for the viability of this model, especially for the rings of Jupiter, Uranus and Neptune, which are less massive than (or, in the case of Uranus possibly of comparable mass to) the nearby moons.

If rings don't last long, why do we observe them around all four giant planets? The flux of interplanetary debris has decreased substantially over geologic time, at least in the inner Solar System, where we have 'ground truth' from radioisotope dating of lunar craters (cf. Section 5.4.4.1). This question is most relevant to the case of the saturnian ring system, as it is far more massive than the rings of other planets. In the cases of Jupiter, Uranus and Neptune, it is easier to envisage the current ring/moon systems as remnants of eons of disruption, re-accretion (to the extent possible so near a planet) and gradual net losses of material. In the previous section we showed that Jupiter's ring system is formed from impact ejecta from small moons. That theory, in fact, suggests that all small moons orbiting near planets (including the martian satellites Phobos and Deimos) should cause the formation of a ring.

11.8 Summary

Jupiter's rings are broad but extremely tenuous. Most of the light observed from the jovian ring system has been scattered off micrometer-size dust of silicate composition. Saturn's main rings are broad and optically thick, with most of the area covered by centimeter–meter bodies composed primarily of water-ice; Saturn's broad outer rings are tenuous and consist of micrometer-size, (presumably) ice-rich particles. Most of the mass in the uranian rings is confined to narrow rings of particles similar in size to the bodies in Saturn's main ring system, but much darker in color. Neptune's brightest ring is narrow and highly variable in longitude; more tenuous, broader rings have also been observed around Neptune. The fraction of micrometer-size dust in Neptune's rings is larger than in Saturn's rings and Uranus's rings, but smaller than in Jupiter's. The particles in Neptune's rings have very low albedos.

The reason(s) why Saturn's main ring system is divided into four major components, and the causes of the major ring boundaries, remain for the most part unexplained, although there are two major exceptions. The classical identification of the outer edge of the B ring with the Mimas 2:1 inner Lindblad resonance was confirmed by Voyager, and the unexpected sharpness and noncircularity of this edge have been explained theoretically. The outer edge of the A ring has now been identified with the location of the 7:6 resonance of the coorbital moons Janus and Epimetheus. The two strongest resonances in the ring system are thus responsible for the two major ring outer boundaries. Attempts to explain inner boundaries of the rings have been less successful. In order to maintain an inner boundary, positive angular momentum must be transferred to the ring particles. If angular momentum is transferred inwards from an outer moon, excess energy, above that which is required for circular orbits, must be supplied. As this energy loss would rapidly damp the perturbing moon's eccentricity (or inclination), the mechanism is not effective over long periods of time, and some other process must be responsible for the maintenance of inner edges. This problem is absent for outer ring boundaries; the transfer of angular momentum outwards in this case leads to increase in noncircular orbital energy which can be dissipated by inter-particle collisions. (An analogous situation exists for the stability of orbital resonances between moons to tidal evolution. If the inner moon of a resonant pair is tidally forced outwards faster, excess energy is available, and can lead to tidal heating (cf. Problem 2.29), whereas if the outer moon is tidally receding from the planet more rapidly, the lock is unstable.) Ballistic transport resulting from cratering and disruptive impacts by hypervelocity micrometeoroids on ring particles has the potential of reproducing the morphology of the inner edges of rings A and B, but it requires special initial conditions, and does not explain why the edges are positioned as observed.

Moons produce rings (jovian ring system, Saturn's E ring), and moons cause much of the identified ring structure. Electromagnetic forces influence the motion of small grains. Some structure may be produced by internal instabilities within rings, and ballistic transport of ring material might also play a significant role. However, these processes do not explain all of the diversity observed in planetary ring systems, and the identification of other structure-producing mechanisms in planetary rings remains an active research area.

Further Reading

The following review book provides a comprehensive overview of our knowledge of planetary rings at that date:

Planetary Rings. R. Greenberg and A. Brahic, Eds. 1984, University of Arizona Press, Tucson. Particular attention should be given to the articles by Cuzzi *et al.* (Saturn's rings), Burns *et al.* (ethereal rings) and Shu (spiral waves).

More recent reviews are available for the jovian, uranian and neptunian rings:

Showalter, M.R., J.A. Burns, J.N. Cuzzi, and J.B. Pollack, 1987. Jupiter's ring system: new results on structure and particle properties, *Icarus* **69**, 458–498.

Burns, J.A., D.P. Hamilton, M.R. Showalter, P.D. Nicholson, I. de Pater, and P.C. Thomas, 1999. The formation of Jupiter's faint rings. *Science*, **284** 1146–1150.

French, R.G., P.D. Nicholson, C.C. Porco, and E.A. Marouf, 1991. Dynamics and structure of the uranian rings. *Uranus*, Eds. J.T. Bergstrahl, E.D. Miner, and M.S. Matthews. University of Arizona Press, Tucson, pp. 327–409.

Porco, C.C., P.D. Nicholson, J.N. Cuzzi, J.J. Lissauer, and L.W. Esposito, 1995. Neptune's ring system. In *Neptune*, Ed. D.P. Cruikshank. University of Arizona Press, Tucson, pp. 703–804.

Problems

11.1.E Derive equation (11.7) from equation (11.6). (Hint: Recall that the analysis assumes that the planet is spherical.)

11.2.I In Section 11.1, we estimated the limits for tidal stability of a spherical satellite orbiting near a planet by equating the self-gravity of the satellite to the tidal force of the planet at a point on the satellite's surface which lies along the line connecting the planet's center to the satellite's center. The results of this study are directly applicable to the ability of a spherical moon to accrete a much smaller particle which lands on the appropriate point on its surface. Perform a similar analysis for the mutual attraction of two spherical bodies of equal size and mass whose centers lie along a line which passes through the planet's center. The result is known as the *accretion radius*. Comment on the qualitative similarities and quantitative differences between the accretion radius, the tidal disruption radius estimated for a single spherical body (eq. 11.7), and Roche's tidal limit (eq. 11.8), which was calculated for a deformable body.

11.3.E Calculate the Roche limit of each of the giant planets as a function of satellite density. Compare your results with the observed positions and sizes of some of these planets' inner satellites and rings, and comment.

11.4.E Calculate the densities required for Neptune's six inner satellites assuming each is located just exterior to the Roche limit for its density. Are these densities realistic? What holds these moons together?

11.5.E Why are planetary rings so flat? Why do they always orbit in a planet's equatorial plane?

11.6.E If a ring has a normal optical depth τ, then using the geometrical optics approximation, a photon traveling perpendicular to the ring plane has a probability of $e^{-\tau}$ of passing through the rings without colliding with a ring particle. For this problem, you may assume that the ring particles are well-separated, and that the positions of their centers are uncorrelated.

(a) What fraction of light approaching the rings at an angle θ to the ring plane will pass through the rings without colliding with a particle?

(b) The cross-section of a ring particle to collisions with other ring particles is larger than its cross-section for photons. Assuming all ring particles are of equal size, what is the probability that a ring particle moving normal to the ring plane would pass through the rings without a collision?

(c) On average, how many ring particles does a line perpendicular to the ring plane pass through?

(d) On average, how many ring particles does a particle passing normal to the ring plane collide with? (As ring particles on keplerian orbits pass through the ring plane twice per orbit, two times your result is a good estimate for the number of collisions per orbit of a particle in a sparse ring. The in-plane component of the particles' random velocities increases the collision frequency by a small factor, the value of which depends upon the ratio of horizontal to vertical velocity dispersion. For a ring which is thin and massive enough, local self-gravity can increase the vertical frequency of particle orbits, and combined with the gravitational pull of individual particles, greatly increase particle collision frequencies.)

11.7.**E** (a) Calculate the time necessary for Kepler shear to spread rings with the following parameters over 360° in longitude:

(i) Width 1 km, orbit 80 000 km from Saturn.
(ii) Width 100 km, orbit 80 000 km from Saturn.
(iii) Width 1 km, orbit 120 000 km from Saturn.
(iv) Width 2 km, orbit 63 000 km from Neptune.

(b) Calculate the radial diffusion time for the doubling of the widths of the rings in part (a) assuming a viscosity $\nu_v = 100$ cm^2 s^{-1}. How do these times vary with viscosity (give the functional form)?

(c) Compare your results in parts (a) and (b), and comment on the observation that rings are generally observed to vary much more substantially with radius than with longitude.

11.8.**I** Consider the situation in which a small moon on an eccentric and inclined orbit near Roche's limit gets disrupted by a hypervelocity impact. The debris is initially quite localized, and has a velocity dispersion small compared to its orbital velocity. Using the results of the previous problem, describe the subsequent evolution of the debris swarm.

11.9.**I** Assume that an isolated ringlet of width $\Delta r =$ 20 km, full thickness $2H = 2c_v/\mu = 20$ m and optical depth $\tau = 1$ is in orbit 100 000 km from Saturn's center at time $t = 0$. The symbols c_v and μ represent the velocity dispersion and vertical frequency, respectively, and $\mu \approx n$. What is the approximate width of the ring at $t = 10^3$ years? At $t = 10^8$ years?
You should use the viscosity formula given by equation (11.14), and the diffusion relationship:

$$(\Delta r(t))^2 = (\Delta r(0))^2 + \nu_v t. \qquad (11.35)$$

You may make any plausible assumption about the time evolution of H and τ. (Choose something convenient and justify it.) Quote your answer in simple units.

11.10.**E** Assume that the differential size distribution of particles in a planetary ring is given by the power law $N(R) \sim R^{-\zeta}$ ($R_{min} < R < R_{max}$).

(a) Compute the critical values of ζ for which equal amounts of (i) mass and (ii) surface area are presented by particles in each factor of two interval in radius. For larger ζ (steeper distributions), most of the mass or surface area is contained in small particles, whereas for smaller ζ, most is in the largest bodies in the distribution. (Hint: This problem requires you to integrate over the size distribution.)

(b) Prove that it is impossible to have both $R_{min} = 0$ and $R_{max} = \infty$ for any value of ζ.

11.11.**E** Estimate (very crudely) the mass of Jupiter's ring system and associated inner moons. In parts (a)–(c) of this problem, you may assume that ring particles have a density of 1 g cm^{-3}.

(a) Of the dust in each of the three parts of the ring system, assuming particle radii of 0.5 μm in the halo and 1 μm in the other regions. Use the optical depths given in Table 11.1.

(b) Of the macroscopic particles in the main ring, assuming particle radii are

(i) all 5 cm;
(ii) all 5 m;
(iii) distributed as a power law with $N(R) \propto R^{-3}$ from 5 cm to 5 m;
(iv) distributed as a power law with $N(R) \propto R^{-2}$ from 5 cm to 5 m;
(v) distributed as a power law with $N(R) \propto R^{-3}$ from 1 cm to 500 m;
(vi) distributed as a power law with $N(R) \propto R^{-2}$ from 1 cm to 500 m.

(c) Of the moon Metis, using the size given in Table 1.5.

(d) Compare the uncertainties introduced by the assumed density to those resulting from uncertainties in the particle size distribution.

11.12.I The moonlet Pan orbits within the Encke gap in Saturn's rings.

(a) Calculate the location of Pan's 2:1 inner Lindblad resonance using the keplerian approximation for orbits.
(b) Calculate the location of Pan's 2:1 inner Lindblad resonance more precisely.
(c) Is this resonance within Saturn's rings? If yes, state which ring.
(d) What type of wave is such a resonance capable of exciting? How many spiral arms would it have?

11.13.E Saturn's equatorial radius $R_♄ = 60\,330$ km, and its gravitational moments are $J_2 = 1.63 \times 10^{-2}$, $J_4 = -9.17 \times 10^{-4}$.

(a) Calculate the location of the Enceladus 3:1 inner Lindblad (horizontal) resonance using the keplerian approximation for orbits.
(b) Is this resonance located within Saturn's ring system? If so, in which ring?

11.14.D (a) Calculate the location of the Enceladus 3:1 inner Lindblad resonance more precisely than you did in the previous problem. State whatever approximations you use.
(b) Calculate the location of the expected crests of the density wave excited by this resonance if the surface density of the ring is 50 g cm^{-2}. You may assume that a crest occurs at exact resonance. Is this a reasonable assumption? Comment.
(c) Calculate the location of the Enceladus 3:1 inner vertical resonance using the keplerian approximation for orbits.
(d) Calculate the location of the Enceladus 3:1 inner vertical resonance more precisely. State whatever approximations you use.

11.15.I The Cassini mission (scheduled to arrive at Saturn in 2004) discovers that the outer edge of Saturn's A ring is 'wavy', with a half-amplitude of 1 km and a wavelength of 1000 km.

(a) What is the semimajor axis of the orbit of the moon responsible for this pattern?
(b) Which, if any, other orbital elements of the moon can be determined from the information provided? What other observations of the ring edge and near-edge environment would be useful in determining additional orbital elements of the moon? Explain.
(c) What is the size of the moon in question (assume $\rho = 1$ g cm^{-3})?

11.16.E Imagine you were to construct a scale model of Saturn's rings 5 km in radius (about the size of San Francisco). How thick would this model be locally? What would be the height of the corrugations corresponding to the highest bending waves? How wide would the A ring be? What would be the width of the F ring?

11.17.E Briefly explain how shepherding of planetary rings works and why shepherding does not occur in the asteroid belt.

11.18.I Calculate the ratio of the Lorentz force to the gravitational force on an unshielded charged grain with a potential of −10 volts in Jupiter's ring as a function of grain size and density and of distance from Jupiter. Assume Jupiter's magnetic field can be approximated by a dipole with a surface magnetic field strength of 4 G.

11.19.E Assume that a spoke in Saturn's rings forms radially and stretches out from $r = 1.6\,R_♄$ to $r = 1.9\,R_♄$. Calculate the orbital period for particles at the two ends of the spoke. Sketch the spoke after the particles at $r = 1.6\,R_♄$ have completed one-fourth of an orbit.

11.20.I The probability that a photon traveling towards a planetary ring will interact with a ring particle depends upon the angle between the photon's path and the ring plane (Problem 11.6), but it does not depend (significantly) upon whether the in-plane component of the photon's velocity is in the direction of particle orbits or opposite to them, because photons move much more rapidly than do ring particles. However, the velocity of interplanetary debris is only slightly larger than that of ring particles, so particle motion during the passage of this debris through the ring disk cannot be neglected in determining collision probabilities. By decomposing the velocity of the debris into cylindrical polar coordinates in the planetocentric system, derive a formula for the impact probability as a function of speed (relative to the orbital velocity of the ring particles), direction, and ring normal optical depth, τ.

12 Planet Formation

From a consideration of the planetary motions, we are therefore brought to the conclusion, that in consequence of an excessive heat, the solar atmosphere originally extended beyond the orbits of all the planets, and that it has successively contracted itself within its present limits.

Pierre Simon de Laplace, *The System of the World*, 1796

The origin of the Solar System is one of the most fundamental problems of science. Together with the origin of the Universe, galaxy formation and the origin and evolution of life, it is a crucial piece in understanding where we, as a species and as individuals, come from. Because planets are so difficult to detect at astronomical distances, we have detailed knowledge of only one planetary system, the Solar System in which we live. Thus, at present, theoretical modeling provides the best means by which we can estimate the abundance and diversity of planetary systems in our galaxy, including those planets which may harbor conditions conducive to the formation and evolution of life (cf. Section 13.5). Models of planetary formation are developed using our single example of a planetary system, supplemented by astrophysical observations of star-forming regions and circumstellar disks. Data from other planetary systems around both main sequence stars and pulsars are now beginning to provide further constraints (see Chapter 13).

12.1 Observational Constraints

Any theory of the origin of our Solar System must explain the following observations:

Orbital Motions, Spacings and Planetary Rotation: The orbits of most planets and asteroids are nearly coplanar, and this plane is near that of the Sun's rotational equator. The planets orbit the Sun in a prograde direction (the same sense as the Sun rotates), and travel on nearly circular trajectories. Most planets rotate around their axis in the same direction in which they revolve around the Sun, and have obliquities of $< 30°$. Venus, Uranus and Pluto form an exception to this rule (see Tables 1.2 and 1.3). Major planets are confined to heliocentric distances $\lesssim 30$ AU, and the separation between orbits increases with distance from the Sun (Table 1.1). Pluto, the outermost ($a \approx 40$ AU) and smallest planet ($M = 0.0022\ M_{\oplus}$) in our Solar System, is the only planet which orbits the Sun in a highly eccentric ($e = 0.25$) and inclined ($i = 17.2°$) orbit. Some minor planets (asteroids) also move in highly eccentric and inclined orbits. Aside from the asteroid belt between 2.1 and 3.3 AU and the regions centered on Jupiter's stable triangular Lagrangian points, interplanetary space contains very little stray debris. However, most orbits within these empty regions are unstable to perturbations by the planets on timescales short compared to the age of the Solar System; bodies initially in orbits traversing these regions would likely collide with a planet or the Sun or be ejected from the Solar System. Thus, in a sense, the planets are about as closely spaced as they could possibly be.

Age: Radioisotope ^{207}Pb/^{206}Pb dating of chondritic meteorites, the oldest rocks known, yields an age of 4.56 ± 0.01 Gyr. Dating with other isotope systems yields similar ages. Rocks formed on the Moon and Earth are younger: lunar rocks are typically between 3 and 4.4 Gyr old, and terrestrial rocks are $\lesssim$4 Gyr old, although terrestrial mineral grains up to 4.4 Gyr old have been found.

Sizes and Densities of the Planets: The relatively small terrestrial planets and the asteroids, which are mainly composed of rocky material, lie closest to the Sun. The uncompressed (zero pressure) density increases significantly with decreasing heliocentric distance, which suggests a larger fraction of heavier elements, like metals and other refractory (high condensation temperature) material, in planets closer to the Sun. At larger distances we find the giants Jupiter and Saturn, and farther out the somewhat smaller Uranus and Neptune. The low densities of these planets imply lightweight material. Jupiter and Saturn are primarily composed of the two lightest elements, hydrogen and helium (Jupiter has $\sim$90% H and He by mass, Saturn $\sim$80%), while Uranus and Neptune contain relatively large amounts of ices and rock (they contain 5–20% H and He by mass).

Asteroid Belt: Between the orbits of Mars and Jupiter are countless minor planets. The total mass of this material is $\sim 1/20$ the mass of the Moon. Except for the largest asteroids ($R \gtrsim 100$ km), the size distribution of these objects is similar to that expected from a collisionally evolved population of bodies (Chapter 9).

Comets: There is a 'swarm' of ice-rich solid bodies orbiting the Sun at $\gtrsim 10^4$ AU, commonly referred to as the Oort cloud. There are roughly 10^{12}–10^{13} objects larger than one kilometer in this 'cloud'. The bodies are isotropically distributed around the Sun, aside for a slight flattening produced by galactic tidal forces. Closer to the Sun there is a second comet reservoir, the Kuiper belt, a flattened disk at heliocentric distances between 35 and ~ 500 AU.

Satellite Systems: Most planets, including all giant planets, have natural satellites. Most close-in satellites orbit in a prograde sense, in a plane closely aligned with the planet's rotational equator. They are locked in synchronous rotation, so their orbital periods are equal to their rotation periods. Some of the smaller, distant satellites (and Triton, Neptune's large and not so distant moon) orbit the planet in a retrograde sense, and/or on orbits with high eccentricity and inclination. All planetary satellites are primarily composed of a mixture of rock and ice in varying proportions. Jupiter's Galilean satellites imitate a miniature planetary system, with the density of the satellites decreasing with increasing distance to the main planet.

Meteorites: Meteorites display a large variety of spectral and mineralogical differences. The crystalline structure of many inclusions within primitive meteorites indicates rapid heating and cooling events. There is also evidence for (local) magnetic fields of the order of 1 G during the planet formation epoch.

Isotopic Composition: Although elemental abundances vary substantially among Solar System bodies, isotopic ratios are remarkably uniform. This is true even for bulk meteorite samples. Most isotopic variations that have been observed can be explained by mass fractionation, or as products of radioactive decay. The similarity of the isotopic ratios suggests a well mixed environment. However, small scale variations in the isotopic ratio of oxygen and a few trace elements in some primitive meteorites imply that the protoplanetary nebula was not completely mixed on the molecular level, i.e., that some pre-solar grains did not vaporize.

Differentiation and Melting: The interiors of all the major planets, many asteroids and most if not all large moons are differentiated, with most of the heavy material confined to their cores. This implies that all of these bodies were warm at some time in the past.

Composition of Planetary Atmospheres: The elements which make up the bulk of the atmospheres of both the terrestrial planets and planetary satellites can form compounds that are condensable at temperatures which prevail on Solar System bodies; hydrogen and noble gases are present in far less than solar abundances. Giant planet atmospheres consist primarily of H_2 and He, but appear to be enhanced in most if not all ice-forming elements; this enhancement increases from Jupiter to Saturn to Uranus/Neptune.

Surface Structure: Most planets and satellites show many impact craters, as well as past evidence of tectonic and/or volcanic activity. A few bodies show signs of volcanism at the present time. Other surfaces appear to be saturated with impact craters. At current impact rates, such a high density of craters cannot be produced over the age of the Solar System.

Angular Momentum Distribution: Although the planets contain $\lesssim 0.2\%$ of the Solar System's mass, over 98% of the angular momentum in the Solar System resides in the orbital motions of the giant planets. In contrast, the orbital angular momenta of the satellite systems of the giant planets are far less than the spin angular momenta of the planets themselves.

12.2 Nucleosynthesis: A Concise Summary

The nuclei of the atoms which compose stars, planets, life, etc. formed in a variety of astrophysical environments. Models of *nucleosynthesis*, together with observational data from meteorites and other bodies, yield clues about the history of the material which was eventually incorporated into our Solar System. The two most important environments for nucleosynthesis are the very early Universe and the interiors of stars. However, other environments are important for some isotopes; e.g., energetic cosmic rays can split nuclei with which they collide, and this *spallation* process is a major source of some rare odd-number light isotopes.

12.2.1 Primordial Nucleosynthesis

The Universe began in an extremely energetic *hot big bang* roughly 12 billion years ago. The very young Universe was filled with rapidly moving particles. Immense numbers of protons (ordinary hydrogen nuclei, ^{1}H) and neutrons, which are unstable with a half-life of 12 minutes and decay via the reaction:

$$^1\mathrm{n} \rightarrow {}^1\mathrm{p}^+ + \mathrm{e}^- + \overline{\nu}_\mathrm{e}, \tag{12.1}$$

were present. Protons and neutrons collided and sometimes fused together to form deuterium (^{2}H) nuclei, but particles were moving so rapidly that the latter nuclei were destroyed very soon after they formed. After about three minutes, the temperature cooled to the point that deuterium was stable for long enough to merge with protons, neutrons and other deuterium nuclei. Within the next few minutes, about one-fourth of the matter in the Universe agglomerated into alpha particles (^{4}He); most of the matter remained as protons, with small amounts forming deuterium, light helium (^{3}He) and tritium (^{3}H, which decays into ^{3}He with a half-life of 12 years), as well as very small but astrophysically significant amounts of the rare light elements lithium, beryllium and boron, and minute amounts of heavier elements. Big bang nucleosynthesis did not proceed much beyond helium because by the time the Universe was cool enough for nuclei to be stable, its density had dropped too low for fusion to continue to form heavier nuclei. After about 700 000 years, the Universe had cooled sufficiently for electrons to join the nuclei formed in the early minutes of the Universe, producing atoms.

12.2.2 Stellar Nucleosynthesis

Most nuclei heavier than helium, as well as a small but still significant fraction of the helium nuclei, were produced in stellar interiors. Main sequence stars, such as the Sun, convert matter into energy via nuclear reactions which ultimately transform hydrogen nuclei into alpha particles. In normal (nondegenerate) stars, thermal pressure acts to counter gravitational compression. Young stars contract as they radiate away their thermal energy, and this contraction leads to an increase in pressure and density in the stellar core. As energy loss from the surface continues, contraction does not stop until the core becomes hot enough to generate energy from *nuclear fusion*. The rates of fusion reactions increase steeply with temperature because only the tiny minority of nuclei in the high velocity tail of the Maxwell–Boltzmann distribution have enough kinetic energy to overcome Coulomb repulsion. If fusion is too rapid, the core expands and cools; if not enough energy is supplied by fusion, the core shrinks and heats up; in this fashion, equilibrium can be maintained. Deuterium fusion requires a lower temperature than fusion of ordinary hydrogen, so it occurs first and stars rapidly deplete their supply of deuterium, although a significant amount of deuterium can remain in the outer (cooler) portion of a star if it is not convectively mixed with the lower hot regions. The cores of very low mass objects get so dense that they are stopped from collapse by degenerate electron pressure before they reach a temperature large enough for fusion to occur at a significant rate. A *brown dwarf* is an object massive enough for substantial deuterium fusion ($\gtrsim 0.013$ $M_\odot$, depending on composition), but too low in mass for sustained energy generation via hydrogen fusion ($\lesssim 0.075$ $M_\odot$, with the precise value again depending on composition).

In main sequence stars of solar mass and smaller, the primary reaction sequence is the *pp-chain*. The principal branch of the pp-chain occurs as follows:

$$2(^1\mathrm{H} + {}^1\mathrm{H} \rightarrow {}^2\mathrm{H} + \mathrm{e}^+ + \nu_\mathrm{e}), \tag{12.2a}$$
$$2(^2\mathrm{H} + {}^1\mathrm{H} \rightarrow {}^3\mathrm{He} + \gamma), \tag{12.2b}$$
$$^3\mathrm{He} + {}^3\mathrm{He} \rightarrow {}^4\mathrm{He} + 2\,{}^1\mathrm{H} + 2\gamma. \tag{12.2c}$$

The reaction rate for the pp-chain becomes significant near $T_{nucl} = 3 \times 10^6$ K, and is roughly proportional to $(T/T_{nucl})^{10}$ near T_{nucl}. This steep temperature dependence implies that fusion acts as an effective thermostat. If the core gets too hot, it expands, cools, and energy production drops; if the core is too cold, it shrinks until adiabatic compression heats it enough for fusion rates to generate enough energy to balance the energy transported outwards. Moreover, the steep temperature dependence of fusion rates implies that more massive main sequence stars only require a slightly higher core temperature in order to generate a substantially higher luminosity than their smaller brethren. In main sequence stars more massive than the Sun, the core temperature is somewhat higher, and the even more temperature-sensitive catalytic *CNO cycle* predominates. The principal branch of the CNO cycle is:

$$^{12}\mathrm{C} + {}^1\mathrm{H} \rightarrow {}^{13}\mathrm{N} + \gamma, \tag{12.3a}$$
$$^{13}\mathrm{N} \rightarrow {}^{13}\mathrm{C} + \mathrm{e}^+ + \nu_\mathrm{e}, \tag{12.3b}$$
$$^{13}\mathrm{C} + {}^1\mathrm{H} \rightarrow {}^{14}\mathrm{N} + \gamma, \tag{12.3c}$$
$$^{14}\mathrm{N} + {}^1\mathrm{H} \rightarrow {}^{15}\mathrm{O} + \gamma, \tag{12.3d}$$
$$^{15}\mathrm{O} \rightarrow {}^{15}\mathrm{N} + \mathrm{e}^+ + \nu_\mathrm{e}, \tag{12.3e}$$
$$^{15}\mathrm{N} + {}^1\mathrm{H} \rightarrow {}^{12}\mathrm{C} + {}^4\mathrm{He}. \tag{12.3f}$$

The most stable nucleus is ^{56}Fe (Figure 12.1), so fusion up to this mass can release energy. However, fusion of heavier nuclei (up to $Z = 28$) requires higher temperatures in order to overcome the *Coulomb barrier* (electromagnetic repulsion between nuclei dominates the strong nuclear force unless the nuclei are very close, Figure 12.2). Moreover, no element with atomic mass 5 or 8 is stable, so to produce carbon from helium requires two fusions in immediate succession: first a pair of alpha particles combine to produce a (highly unstable) beryllium 8 nucleus, and then another alpha particle is added before this nucleus decays:

$$^4\mathrm{He} + {}^4\mathrm{He} \leftrightarrow {}^8\mathrm{Be} \tag{12.4a}$$

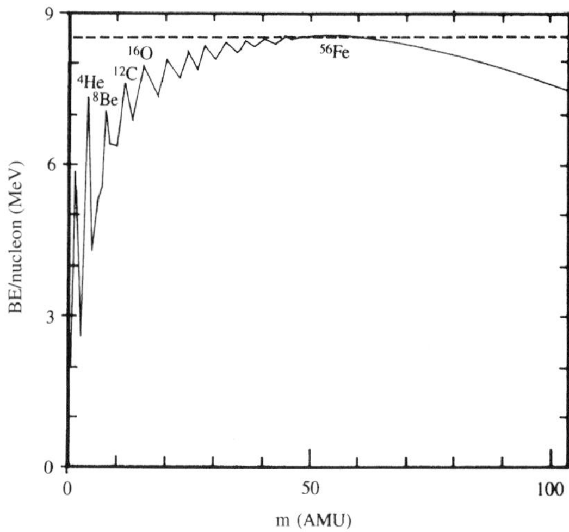

FIGURE 12.1 The nuclear binding energy (BE) per nucleon is shown as a function of atomic weight. Note the prominence of the alpha-particle multiples and the iron group. (Lewis 1995)

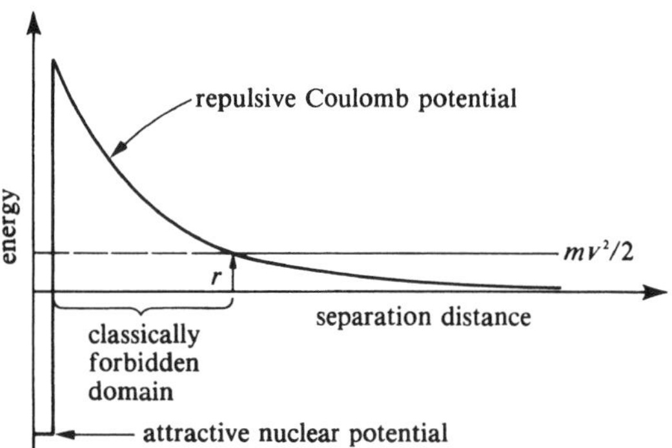

FIGURE 12.2 The electromagnetic repulsion between nuclei dominates the strong (but short-ranged) nuclear force unless the nuclei are very close. (Shu 1982)

followed immediately by:

$$^{8}\text{Be} + {}^{4}\text{He} \rightarrow {}^{12}\text{C} + \gamma. \tag{12.4b}$$

This *triple alpha process* requires much higher densities than do the pp-chain and CNO process described above. Helium fusion occurs when a sufficiently massive star ($\gtrsim$ 0.25 $M_\odot$) has exhausted the supply of hydrogen in its core, so the thermostat which maintained equilibrium during the star's main sequence phase is no longer active. Hydrogen fusion occurs in a shell surrounding the core, and total stellar energy production greatly exceeds that during the star's main sequence phase, so its outer layers expand and cool, and the star becomes a *red giant*.

Nuclear growth beyond carbon does not require two reactions in immediate succession (as does the triple alpha process), and thus could occur in a lower density environment, but the increased Coulomb barrier implies that even higher temperatures and thus larger stellar masses are required. Growth can proceed by successive addition of alpha particles:

$$^{12}\text{C} + {}^{4}\text{He} \rightarrow {}^{16}\text{O} + \gamma, \tag{12.5a}$$

$$^{16}\text{O} + {}^{4}\text{He} \rightarrow {}^{20}\text{Ne} + \gamma, \tag{12.5b}$$

$$^{20}\text{Ne} + {}^{4}\text{He} \rightarrow {}^{24}\text{Mg} + \gamma, \tag{12.5c}$$

or at somewhat higher temperatures by reactions such as:

$$^{12}\text{C} + {}^{12}\text{C} \rightarrow {}^{24}\text{Mg} + \gamma. \tag{12.5d}$$

Nuclei composed of 3–10 alpha particles are quite stable and easy to produce, so they are relatively abundant. Larger nuclei of this form are too proton-rich, and rapidly inverse β-decay into more neutron-rich nuclei.

Fairly large quantities of elements up to the iron binding energy peak can be produced by reactions of the type discussed above, but the Coulomb barrier (Fig. 12.2) is too great for significant quantities of substantially more massive elements such as lead and uranium to be generated in this manner. Such massive nuclei are produced primarily by the addition of free neutrons, which are uncharged and thus do not need to overcome electrical repulsion. Free neutrons are released by reactions such as:

$$^{4}\text{He} + {}^{13}\text{C} \rightarrow {}^{16}\text{O} + {}^{1}\text{n}, \tag{12.6a}$$

and

$$^{16}\text{O} + {}^{16}\text{O} \rightarrow {}^{31}\text{S} + {}^{1}\text{n}. \tag{12.6b}$$

Neutron addition does not produce a new element directly, but if enough neutrons are added, nuclei can become unstable and β-decay into elements of higher atomic number. The mix of elements and isotopes produced by neutron addition depends upon the flux of neutrons. When the time between successive neutron absorptions is long enough for most unstable nuclei to decay, the mixture of isotopes produced lies deep within the *valley of nuclear stability*, where the mixture of neutrons and protons leads to the greatest binding energy for a nucleus with a given total number of nucleons; this 'slow' type of heavy element nucleosynthesis is referred to as the *s-process*. Isotopes with atomic masses as large as 209 may be formed via the s-process. The rapid *r-process* chain of nuclear reactions occurs during explosive nucleosynthesis (such as supernovae), when there is a very high flux of neutrons, and produces a more neutron-rich distribution of elements. Uranium and other very heavy naturally occurring elements are produced via r-process nucleosynthesis. Rare proton-rich heavy nuclei are produced by *p-process nucleosynthesis*. The two leading models for p-process nucleosynthesis are removal of neutrons through partial nuclear photodissociation in a high temperature (10^9 K) environment and β-decay induced by a high neutrino flux. An alternative model is rapid proton capture in a hot, proton-rich stellar environment.

Note that most of the elements produced in stellar nucleosynthesis are never released from their parent stars; only material ejected by stellar winds and supernova explosions is available to enrich the interstellar medium and to form subsequent generations of stars and planets. The distributions of elements and isotopes found in individual interstellar grains and in the Solar System as a whole are indicative of the various environments in which stellar nucleosynthesis occurs, and the conditions under which material is released from stars.

12.3 Star Formation: A Brief Overview

In analogy with current theories on star formation, it is generally believed that our Solar System was 'born' in a dense molecular cloud, as the result of gravitational collapse. In the remainder of this chapter we review current ideas on star formation, the formation of a disk around a (proto)star, and, finally, the evolution of such a disk and the accretion (growth) of planets.

12.3.1 Molecular Cloud Cores

Our Milky Way galaxy contains a large number of cold, dense molecular clouds, varying in size from giant systems with masses of $\sim 10^5$–10^6 $M_\odot$ to small ~ 1 $M_\odot$ cores. The small cores are usually embedded in the larger complexes, and are observed at radio wavelengths in molecular line transitions such as NH_3, HCN, CS or H_2CO. Molecular clouds have typical temperatures of $\sim$10–30 K and densities of a few thousand molecules cm^{-3}. The dense cores may have densities 10–100 times larger than this. Molecular clouds consist mostly of H_2 and presumably He. Many other molecules are present, including CO, CN, CS, SiO, OH, H_2O, HCN, SO_2, H_2S, NH_3, H_2CO and numerous other combinations of H, C, N, and O, some containing more than a dozen atoms in a molecule. All of these more massive molecules combined, however, make up only a small fraction of the total mass of the cloud.

The typical interstellar cloud is stable against collapse. Its internal pressure (ordinary gas pressure augmented by magnetic fields, turbulent motions and rotation), is more than sufficient to balance the inward force of self-gravity. This excess pressure would cause the cloud to expand, were it not for the counterbalancing pressure of surrounding gas of higher temperature ($\sim 10^4$ K) and lower density (~ 0.1 atoms cm^{-3}).

In equilibrium systems where magnetic pressure and external pressure can be ignored, the *virial theorem* states that the gravitational potential energy, E_G, is equal to negative twice the kinetic energy, E_K (Problem 2.5). The kinetic energy of a gas cloud is primarily thermal energy, unless the cloud is highly turbulent or rapidly rotating. When $|E_G| > 2E_K$, the cloud may collapse under its own self-gravity. One can solve for the minimum mass of such a cloud, the *Jeans mass*, M_J (Problem 12.1):

$$M_J \approx \left(\frac{kT}{G\mu_a m_{amu}}\right)^{3/2} \frac{1}{\sqrt{\rho}}. \tag{12.7}$$

A cloud with $M > M_J$ will collapse if its only means of support is thermal pressure. Note that the critical mass, M_J, decreases if the density in the cloud increases. Low density but massive clouds may collapse into galaxies, less massive clouds with a higher density may collapse into clusters of stars or a single star. Observed cores within molecular clouds appear to be dense enough to gravitationally collapse into objects of stellar masses. However, the density in a small, cold (10 K) cloud would need to exceed $\sim 10^{-11}$ g cm^{-3} to form Jupiter-mass objects from gravitational collapse. This is much larger than the observed densities of interstellar clouds.

When a marginally stable molecular cloud passes through a spiral arm of a galaxy it is compressed; such compression may be sufficient to trigger collapse. Clouds pass through spiral arms since the clouds and stars in a galaxy orbit faster than the pattern of spiral density waves rotates. Other phenomena which may trigger gravitational collapse are (super)nova explosions, where shells of gas and dust are thrown into space, and stellar outflows or expanding HII (ionized hydrogen) regions. Collapse converts gravitational potential energy into kinetic energy of the collapsing material. If this energy is retained, either in ordered motion or random thermal motions, virial equilibrium may be achieved and the collapse ceases. However, if this energy is lost, for example via radiation, then the cloud becomes even more unstable. Once a gravitational collapse begins, densities increase, causing collapse to proceed faster.

12.3.2 Collapse of Molecular Cloud Cores and Star Formation

When the pressure in a cloud is small enough relative to the gravity that it can be ignored, the cloud collapses on a free-fall timescale, t_{ff} (Problem 12.2):

$$t_{ff} = \left(\frac{3\pi}{32G\rho}\right)^{1/2}. \tag{12.8}$$

Equation (12.8) implies that the denser clumps collapse more rapidly, which may cause separation and fragmentation to occur. As clumps are observed to be densest near their centers, the interiors cave in most quickly, leading to an inside out collapse. Rotation of the cloud may also lead to fragmentation. Rotation becomes a dominant effect when the centrifugal force balances the gravitational force. If the cloud rotates rapidly, it may break up into two or more subclouds, where the angular momentum is taken up by the individual fragments orbiting one another. Each of the subclouds may collapse into a star, forming a binary or multiple star system. The majority of stars are observed to be in such binary or multiple systems. Clumps with less angular momentum may form only one star. Since a clump has to contract by orders of magnitude to form a star, even an initially very slowly rotating clump contains much more angular momentum than the final star can take without breaking up. We expect, therefore, that virtually all single stars, and probably many binary/multiple systems, are surrounded by a flat disk of material at some stage during their formation. Although the star may contain most of the cloud's initial mass, most of the angular momentum is in the disk. Recall that in the Solar System today, 99.8% of the mass is in the Sun, and over 98% of the angular momentum in planetary orbits.

While a dense core is collapsing, its temperature rises as a result of the conversion of gravitational energy into kinetic energy. If the cloud is sufficiently transparent at infrared wavelengths, most of the thermal energy is radiated away, and the clump stays relatively cool. The increasing density eventually makes the cloud opaque, so thermal energy can no longer escape. Released gravitational energy then heats the protostar growing at the center of the core, thereby building up the internal pressure, until hydrostatic equilibrium (balance between gravity and pressure; eq. 3.65) is reached. When the temperature inside the protostar gets hot enough ($\sim 10^6$ K), nuclear reactions (the conversion of deuterium into helium) start. The energy generated by this process is sufficient to temporarily forestall further contraction. When the supply of deuterium becomes exhausted, the star shrinks and heats up until the central temperature reaches the 10^7 K value required for ^{1}H fusion.

During these accretion phases, the protostar is blocked from view by dust in the outer layers of the cloud. In the early phases of gravitational collapse, the dust stays relatively cool, ~ 30 K, and thus emits infrared radiation whose distribution peaks near 100 μm. When the protostar forms, the inner layers of dust heat up dramatically, as is observed at shorter infrared wavelengths.

12.3.3 Observations of Star Formation

Numerous young stars have been observed within several molecular clouds. The ages of these stars are estimated in several ways: kinematic ages of groups of stars (the size of the region divided by the relative velocities of the stars), the ages of individual stars on the *Hertzsprung–Russell* (H–R) *diagram* (Fig. 12.3), and the presence and intensity of Li absorption lines in the stellar spectra. (Lithium is convectively transported downwards to zones in which it can be destroyed by thermonuclear reactions in stars of mass $\lesssim 1\ M_\odot$.) These methods are in agreement that the youngest of the well-studied star forming regions are $\lesssim 10^7$ years old.

Young stars that are still contracting towards the main sequence are called *pre-main sequence stars*. Among these stars we find *T Tauri stars*, named after the first such star discovered, the variable star T in the constellation Taurus. T Tauri stars are usually found within dense patches of gas and dust. The luminosity of many T Tauri stars varies considerably in an irregular manner, over timescales as short as a few hours. The spectral energy distributions of T Tauri stars are much broader than blackbody spectra, are dominated by intense emission lines and show the presence of strong stellar winds. T Tauri stars typically have large starspots, which modulate their lightcurves and allow their rotation periods to be measured. Typical observed rotation periods for deeply embedded T Tauri stars are a few days, much shorter than the Sun's current 27 day period, yet several times as long as that required for rotational break-up (cf. Problem 9.11). T Tauri stars also emit substantially more X-rays than do older stars of similar mass; this implies that very energetic nonequilibrium processes occur during star formation.

The formation and early evolution of stars can be quite erratic and violent. *Bipolar outflows* of gas are ejected at ~ 100 km s^{-1} perpendicular to disks around accreting stars. Interaction of this gas with the surrounding interstellar medium produces shocks which are believed to be responsible for the bright emission nebulae known as *Herbig–Haro (HH) objects*. Outbursts, presumably related to enhancements in the stellar accretion rate by a factor $\gtrsim 100$ for tens of years, are seen in *FU Orionis stars*. Star and planet formation is discussed in great detail in the book *Protostars and Planets IV* (Mannings *et al.* 2000).

(a)

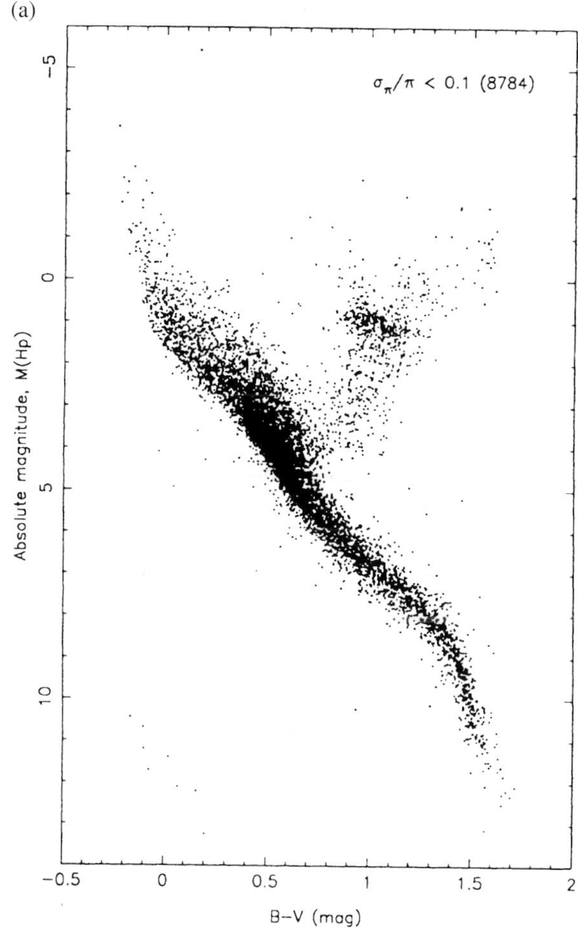

(b)

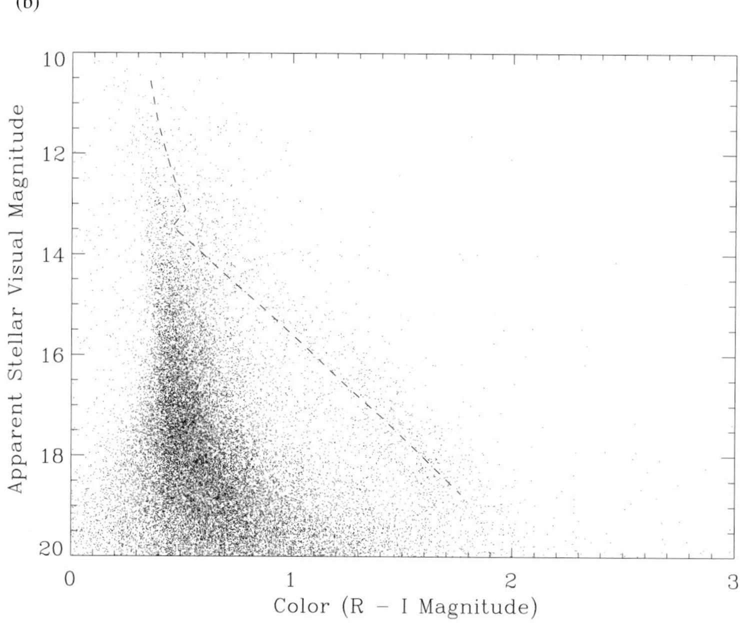

FIGURE 12.3 Field stars show a broad range of colors and magnitudes. A graph showing the absolute magnitude vs. color for many stars is called a *Hertzsprung–Russell diagram*. Most stars are observed to lie on the *main sequence*; stars which lie on the main sequence are generally supported against gravitational collapse by the thermal energy supplied from fusion of hydrogen. (a) Hertzsprung–Russell diagram for the single stars from the Hipparcos Catalogue with the most accurately known distances and magnitudes, i.e., nearby bright stars. (Perryman *et al.* 1995) (b) Color–magnitude (H–R) diagram for stars within an area of about 7 square degrees in the belt of Orion. This region contains a giant molecular cloud and many young stars in the Orion OB1b association. The bulk of the stars, lying near $R - I = 0.5$, are G and K stars at various distances in the galactic disk. The enhancement of the number of stars in a diagonal band above and to the right of the G and K field stars is due to low mass pre-main sequence stars in the Orion OB1b association. The age of the Orion OB1b association is about 1.7 million years. The dashed line represents a theoretical 1.7 million years isochrone for stars of mass between 2.0 and 0.1 $M_\odot$. The width of the locus of the pre-main sequence stars can be attributed to differences in reddening of light by interstellar dust, distance, and age of individual stars. (Courtesy Fred Walter and William Sherry)

12.3.4 Observations of Circumstellar Disks

Excess emission at infrared wavelengths, indicative of circumstellar material extending out to tens or hundreds of AU from the star, is observed in 25–50% of pre-main sequence solar mass stars, including T Tauri stars. The lack of near-infrared radiation from some disks suggests that gaps exist in the inner parts of these disks. This absence of near-infrared radiation is observed more frequently in the somewhat older objects. Direct evidence of circumstellar dust disks comes from millimeter and HST observations. Millimeter data have indicated disk-like structures with masses between $\sim$ 0.001 and 0.1 $M_\odot$ around several protostars. HST images have revealed disk-like structures of order 100 AU in radius around young stars; these structures are referred to as *proplyds* (Fig. 12.4). The disks around the stars HD 141569 and HR 4796A as imaged by HST (Fig. 12.5) appear to contain inner holes. The former disk, $\sim$750 AU wide, shows a dark band near the center, i.e., absence of material. This material may have accreted into a planet or been pushed away by a planet's shepherding torque. The ring of dust around HR 4796A, at a distance of $\sim$40 AU from the star, is less than 17 AU wide. Such a confinement might be explained via shepherding (cf. Section 11.4.3), and hence suggests the existence of unseen planets near the disk.

The Infrared Astronomical Satellite (IRAS) detected the presence of cool clouds of solid particles around the nearby star Vega and a few other main sequence stars (via photometry in the far IR). These circumstellar disks may extend up to a few hundred AU from the star, but may contain as little as one lunar mass of material. The first image of a circumstellar dust disk around a (main sequence) star, β Pictoris, was obtained with a CCD camera from the ground. HST images of this 1500 AU wide disk (Fig. 12.6) show that the inner part of the disk is warped, a feature which may be produced by the gravitational pull of one or more nearby planets or a brown dwarf.

(a)

(b)

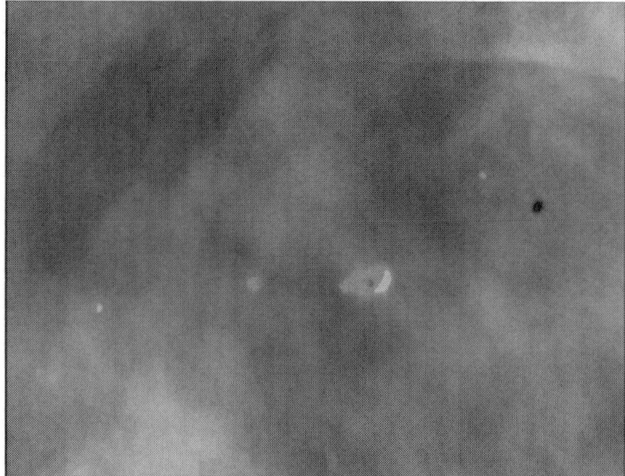

FIGURE 12.4 (a) This dark structure is a column of cool molecular hydrogen gas and dust that is an incubator for new stars. The stars are embedded inside finger-like protrusions extending from the top of the nebula. Each 'fingertip' is somewhat larger than our own Solar System. The pillar is slowly eroding away by the ultraviolet light from nearby hot stars, a process called *photoevaporation*. Small globules of especially dense gas buried within the cloud are uncovered by this erosion. Embryonic stars are forming inside at least some of the globules. The stars emerge as the gobules themselves succumb to photoevaporation. Such globules are found in the 'Eagle Nebula', a nearby star-forming region 7000 light-years away in the constellation Serpens. (Courtesy: Jeff Hester and Paul Scowen, and HST/NASA) (b) A Hubble Space Telescope view of a small portion of the Orion Nebula reveals five young stars. Four of the stars are surrounded by gas and dust. These proplyds may be protoplanetary disks. The proplyds which are close to hot stars are seen as bright objects, while the object far from the hottest stars appears dark. The field of view is only 0.14 light-years across. (Courtesy: C.R. O'Dell, HST/NASA)

12.4 Evolution of the Solar Nebula: The Protoplanetary Disk

In analogy with star formation in our galaxy at the present epoch, we assume that our Sun and planetary system formed in a molecular cloud. The growing Sun together with its surrounding disk are referred to as the *primitive solar nebula*; the planetary system formed from the *protoplanetary disk* within this nebula. A *minimum mass* of $\sim$0.02 $M_\odot$ for the protoplanetary disk can be derived from the present abundance of refractory elements in the planets and the assumption that the abundances of the elements

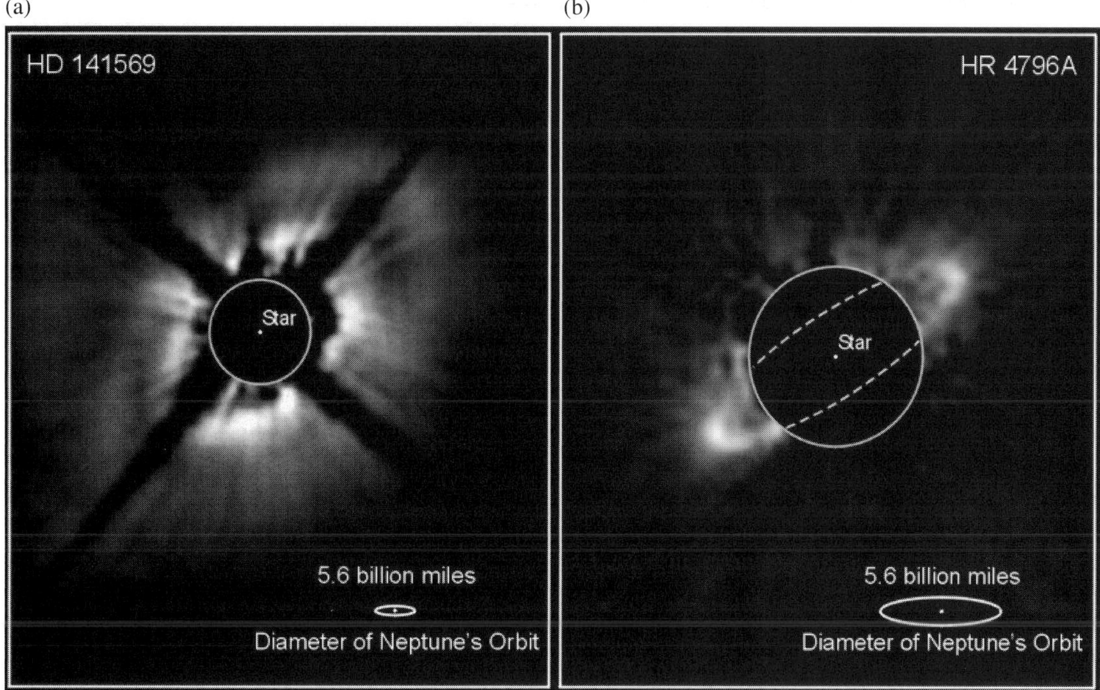

FIGURE 12.5 Coronographic images of disks around young stars. (a) A near-infrared image of a disk around the star HD 141569, located about 320 light-years away in the constellation Libra. A dark band separates a bright inner region of the ~ 750 AU wide disk from a fainter outer region. This band may be the result of the formation of a planet in the disk. (Courtesy: B. Smith and G. Schneider, HST/NASA) (b) A near-infrared image of a dust ring around the young ($\lesssim 10^7$ yr old) star HR 4796A. The ring, at a distance of ~ 40 AU from the central star, is less than 17 AU wide. The confinement of this ring suggests the presence of unseen planets in orbit around the star. In both cases a coronagraph was used to block off the light from the star. (Courtesy: E. Becklin and A. Weinberger, HST/NASA)

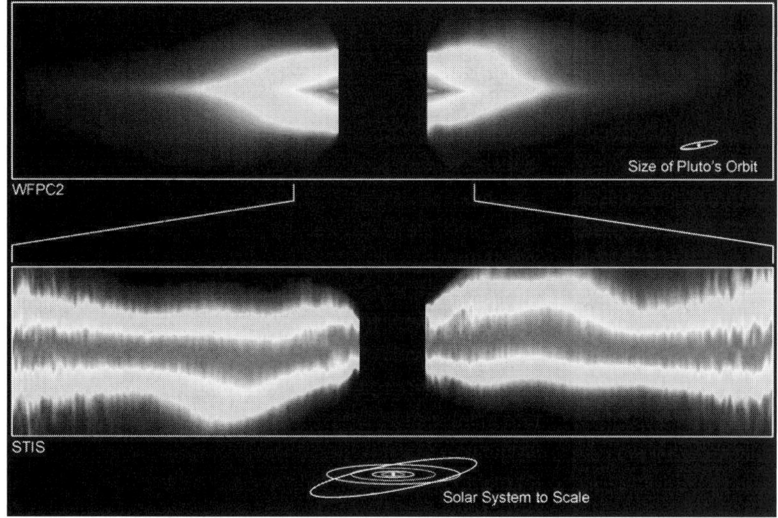

FIGURE 12.6 HST image of the inner portion of the dust disk around the star β Pictoris. The bright glare of the central star is blocked by a coronagraph. The warps in the disk might be caused by the gravitational pull of one or more unseen (planetary?) companions. (Courtesy: Al Schultz, HST/NASA)

throughout the nebula were solar. The actual mass was probably significantly larger, since the efficiency of the accretion of the refractory component of this mixture into planets is likely to have been much less than unity. The history of our solar nebula can be divided into three stages: infall, internal evolution and clearing.

12.4.1 Infall Stage

When a cloud core becomes dense enough that its self-gravity exceeds thermal, turbulent and magnetic support, it starts to collapse. Collapse proceeds from the inside out, and continues until the reservoir of cloud material is exhausted, or until a strong stellar wind reverses the flow. The duration for the infall stage is comparable to the free-fall collapse time of the core, $\sim 10^5$–10^6 yr.

Initially, gas and dust with low specific angular momentum falls to the core. Eventually, matter with high specific angular momentum falls towards the protostar, but cannot reach it due to centrifugal forces. Essentially, the material is on orbits which do not intersect the central, pressure-supported star. However, as the gas and dust mixture falls to the equatorial plane of the system, it is met by material falling from the other direction, and motions perpendicular to the plane cancel. The energy in this motion is dissipated as heat in the forming disk. Significant heating can occur, especially in the inner portion of the disk, where the material has fallen deep into the potential well.

Let us consider a parcel of gas which falls from infinity to a circular orbit at $r_\odot$. Half of the gravitational energy is converted to orbital kinetic energy; the other half:

$$\frac{GM_{protostar}}{2r_\odot} = \frac{v_c^2}{2} \tag{12.9}$$

per unit mass, is available for heat. At 1 AU, the circular velocity $v_c = 30$ km s^{-1} if $M_{protostar} = 1$ M$_\odot$. If no energy escapes the system, it follows that the temperature in a hydrogen gas would be $\sim 7 \times 10^4$ K (Problem 12.3). However, this very high temperature is never actually attained, as the timescale for radiative cooling is much shorter than the heating time.

Gas reaches supersonic velocities as it descends towards the midplane of the nebula. The gas slows abruptly when it passes through a *shock front* as it is accreted onto the disk. The highest temperature that is attained in the nebula depends upon the structure of the shock through which material passes. Typical post-shock temperatures for the protoplanetary disk are believed to be $\sim$1500 K at 1 AU, and $\sim$100 K at 10 AU. Equilibrium is reached when all forces balance, i.e., the gravitational force towards the center balances with the centripetal force outward and the gravitational force toward the midplane balances with the pressure gradient outwards. Provided the star's gravity dominates that of the disk, the pressure variation in the vertical direction is given by:

$$P_z = P_{z_0} e^{-z^2/H_z^2}, \tag{12.10}$$

where the Gaussian scale height, H_z, increases with heliocentric distance:

$$H_z = \sqrt{\frac{2kTr_\odot^3}{\mu_a m_{amu} G M_\odot}}. \tag{12.11}$$

12.4.2 Internal Dynamical Evolution of the Disk

Unless the collapsing cloud has negligible rotation, a significant amount of material lands within the disk. Redistribution of angular momentum within the disk can then provide additional mass to the star. The structure and evolution of the disk are primarily determined by the efficiency of the transport of angular momentum and heat. Angular momentum and mass can be transported in the following ways:

Magnetic Torques: If magnetic field lines from the star thread through the disk, then there is a tendency towards corotation, i.e., material orbiting more rapidly than the star's spin period loses angular momentum and that orbiting less rapidly gains angular momentum. The field lines couple the star to the disk if the gas in the disk is sufficiently ionized, which it tends to be in the innermost parts of the solar nebula, where temperatures are high. Thus, the spin rate of a rapidly rotating star is slowed by *magnetic braking* torques from the inner parts of the disk, where ionized gas couples to the stellar magnetic field but is slowed by frequent collisions with neutral gas orbiting at the keplerian rate.

Angular momentum is transferred outwards, from the star to the disk, but transfer to larger radii is inhibited by the lack of ionized gas in the disk and the weakness of stellar magnetic field lines at greater distances. Protoplanetary disks apparently do not extend all of the way to the star's surface. Rather, magnetic interactions between the star and the disk near the *corotation point* (where the keplerian orbital angular velocity in the disk equals the star's rotational angular velocity) funnel some of the disk's gas onto the star and expel other gas in a rapid centrifugally driven bipolar outflow which carries with it a substantial amount of angular momentum. (Although this 'bipolar' wind is concentrated near the star's poles, the gas does not move exactly parallel to the star's rotation axis, so it is able to carry away angular momentum.) This loss of angular momentum explains why protostars are observed to rotate substantially less rapidly than break-up speed (Section 12.3.3). The solid particles that are expelled along with the gas are subjected to brief but intense heating by bright starlight; this process may produce the chondrules and CAIs that are found in most primitive meteorites (Section 8.7.2).

Gravitational Torques: Local or global gravitational instabilities may lead to rapid transport of material within the protoplanetary disk. As discussed in Section 11.2, a thin rotating disk is unstable to local axisymmetric perturbations if Toomre's parameter, $Q_T < 1$ (eq. 11.10; note that for a gaseous disk, the sound speed replaces the vertical velocity). Nonaxisymmetric local instabilities also occur when $Q_T \lesssim 1$. These instabilities can produce spiral density waves, which transport mass and angular momentum on a dynamical timescale until a stable configuration is once again reached. This limits the disk mass to a value less than or comparable to the protostar's mass. Under most circumstances, a global one-armed spiral instability limits the disk mass to less than 1/3 that of the star.

Large protoplanets may clear annular gaps surrounding their orbits and excite density waves at resonant locations within the protoplanetary disk. These density waves transfer angular momentum outwards. Such processes have been observed in Saturn's rings (Section 11.4), although on a much smaller scale. If the protostar were to have enough angular momentum, its lowest energy rotation state would be triaxial, which would present an asymmetric and rotating gravitational potential to the disk, much like that due to a protoplanet. This would trigger resonances and density waves, which transport angular momentum from the protostar to the disk. However, observations imply that $\sim$1 $M_\odot$ protostars generally rotate far too slowly to become triaxial.

Viscous Torques: Since all molecules revolve around the protosun in roughly keplerian orbits, those closer to the center move faster than those farther away. Collisions speed up the outer molecules, hence driving them outwards, and slow down the inner molecules, which then fall towards the center. The net effect is that most of the matter diffuses inwards, angular momentum is transferred outwards, and the disk as a whole spreads.

The disk evolves on a diffusion timescale, given by equation (11.13), where the length scale, ℓ, is equal to the radius of the disk (or that portion of the disk under consideration). Viscous diffusion is thus more rapid in the inner portions of protoplanetary disks, provided the kinematic viscosity is roughly constant. Unfortunately, the magnitude of the viscosity in protoplanetary disks is uncertain by several orders of magnitude. Thermal motions of the gas in the disk produce a *molecular viscosity*, ν_v, of order:

$$\nu_v \sim \ell_{fp} c_s, \tag{12.12}$$

where ℓ_{fp} is the mean free path of the molecules and c_s is the sound speed. This molecular viscosity is far too small to produce significant viscous evolution over the lifetime of a protoplanetary disk (Problem 12.5). However, if turbulence, caused for example by convection in the nebula, plays a role, it may be able to transport substantial mass and angular momentum within the protoplanetary disk. The physics of turbulence is extremely complicated and poorly understood. Because turbulent velocities are unlikely to exceed the sound speed, and eddy sizes are unlikely to be larger than the scale height of the disk, the turbulent viscosity of accretion disks is often parameterized as:

$$\nu_v = \frac{2}{3} \alpha_v c_s H_z, \tag{12.13}$$

where the viscosity parameter $\alpha_v \lesssim 1$. Theoretical estimates based on convection in an optically thick disk yield values of α_v between 10^{-4} and 10^{-2}, which imply significant viscous evolution of (at least the inner portions of) protoplanetary disks (Problem 12.7).

The rate at which the disk evolves may thus depend on its optical depth perpendicular to the midplane. Very close to the protostar, well inside the distance of Mercury's orbit, the disk is too hot for grains to condense out, and interstellar grains all have evaporated. The dominant sources of opacity in this region are due to molecular transitions in, for example, H_2O and CO molecules and atomic hydrogen ionization processes. At larger distances from the star the temperature in the nebula is well below 2000 K, and micrometer-sized dust provides the dominant source of opacity. The magnitude of this *Rosseland mean* (energy-averaged) *opacity* varies with temperature roughly as T^2, aside from occasional sharp drops when abundant species evaporate (Fig. 12.7). The disk cools by radiating energy from its faces. Portions of the disk may become unstable against convection perpendicular to the disk's midplane, allowing heat to be convectively transported from the hot midplane of the disk to the faces, where it is radiated away into space. Turbulence, induced by this convection, may also mix material in the radial direction over a distance comparable to the size of the largest convective eddies. Under most circumstances, the temperature in the nebula decreases both with distance from the nebula midplane and with increasing distance from the Sun.

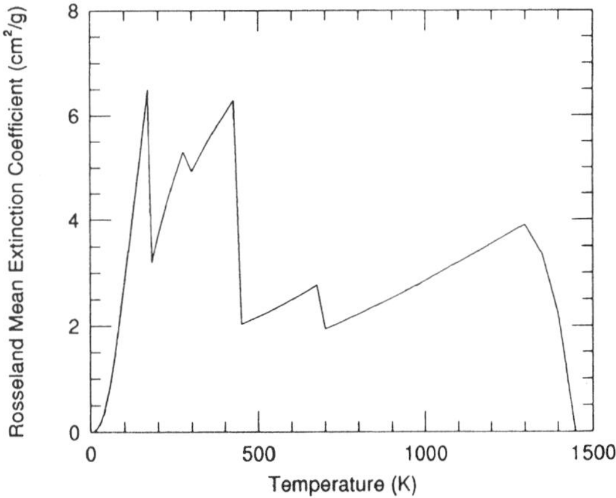

FIGURE 12.7 The Rosseland mean opacity of a solar composition mixture of gas and dust as a function of temperature. The size spectrum of the dust is assumed to resemble that of interstellar grains; the opacity is lower if the condensed material is contained in macroscopic bodies. The contribution of gas to the opacity is not included in these calculations, but at the low pressures occuring in protoplanetary disks, gas opacity is small compared to dust opacity at temperatures below 1400 K. (Adapted from Pollack *et al.* 1994)

12.4.3 Chemistry in the Disk

The chemical composition of protoplanetary disks is important because it determines what raw materials are available for planetesimal formation. Protoplanetary disks are regions in which interstellar matter is made into planets. Comets and chondritic meteorites are relics from the planetesimal-forming era of the Solar System's protoplanetary disk. Disks are dynamically evolving objects: the physical conditions within them, and thus their chemical composition, change over time as disk material accretes onto the central star, as planets form, and as the disk eventually disperses, leaving behind a planetary system such as the Solar System.

The initial chemical state of the disk depends upon the composition of the gas and dust in the interstellar medium and subsequent chemical processing during the collapse phase. The chemical composition varies both with time and with distance from the central (proto)star. Because the Sun formed from the same raw materials as its protoplanetary disk, the abundances of the elements in the Sun tell us what the original *elemental* composition of the disk was, but not the chemical compounds these elements formed.

The chemical evolution of interstellar matter as it is incorporated into planetesimals via the disk process is fundamental to our understanding of planet formation. Gas cools after passing through the shock front that it encounters while entering the protoplanetary disk, but can subsequently be heated as material is added above it and/or if dynamical evolution of the disk brings it closer to the protostar. The chemical composition can be calculated where nebular material has experienced temperatures high enough to completely evaporate and dissociate all incoming interstellar gas and dust (> 2000 K). At such high temperatures, the chemistry can be assumed to be in thermodynamic equilibrium, since the chemical reaction rates are rapid compared to the cooling rate of the disk. This situation is likely to occur near the protostar. When the nebula cools below a temperature at which the chemical reaction times become comparable to the timescale of cooling, the chemistry becomes more complicated. This *freeze-out temperature* is different for different species. For example, the CO/CH_4 and N_2/NH_3 ratios are sensitive functions of the temperature and pressure in the nebula. At the low pressures given by models of the solar nebula, carbon is thermodynamically most stable in the form of CO at $T \gtrsim$ 700 K, and in the form of CH_4 at lower temperatures. Nitrogen is most stable as N_2 at $T \gtrsim 300$ K, and as NH_3 at lower temperatures. Thus, were the protoplanetary disk in thermodynamic equilibrium, CO and N_2 would dominate in the warm inner nebula, while in the cold outer nebula, CH_4 and NH_3 would be the favored forms of C and N, respectively (Fig. 12.8a). The existence of N_2 and CO ices on Pluto and Triton, for example, indicates that the outer solar nebula did not have enough time to chemically equilibrate.

As a protoplanetary disk cools, elements condense out of the gas and undergo chemical reactions at different temperatures (Fig. 12.8b, c). Refractory minerals such as REE (Rare Earth Elements) and oxides of aluminum, calcium and titanium (e.g., corundum, Al_2O_3 and perovskite, $CaTiO_3$) condense at a temperature of ~1700 K. At $T \sim$ 1400 K, iron and nickel condense to form an alloy; at slightly lower temperatures magnesium silicates appear, including forsterite (Mg_2SiO_4) and enstatite ($MgSiO_3$). Upon further cooling, at $T \lesssim 1200$ K the first feldspars appear; initially the more refractory compounds such as plagioclase anorthite ($CaAl_2Si_2O_8$), and later ($T \sim 1100$ K) sodium and potassium feldspars ($(Na,K)AlSi_3O_8$). Note that, as essentially all of the aluminum condenses out of the gas at much higher temperatures, aluminum is available for inclusion in feldspars only if it is contained in small grains, which can reach equilibrium with the surrounding gas. If grain growth is rapid compared to cooling, different minerals will be formed. If chemical equilibrium is maintained, chemical reactions in the gas and with the dust take place as the temperature drops. Noteworthy are the reactions of iron with H_2S to form troilite (FeS) at ~700 K, and with water to form iron oxide ($Fe + H_2O \rightarrow FeO + H_2$) at

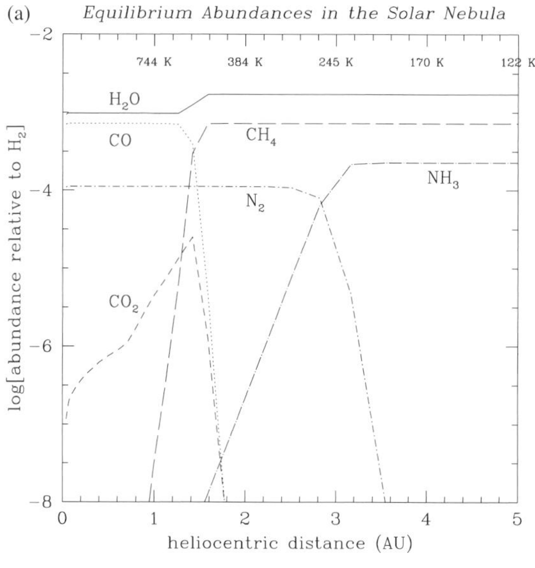

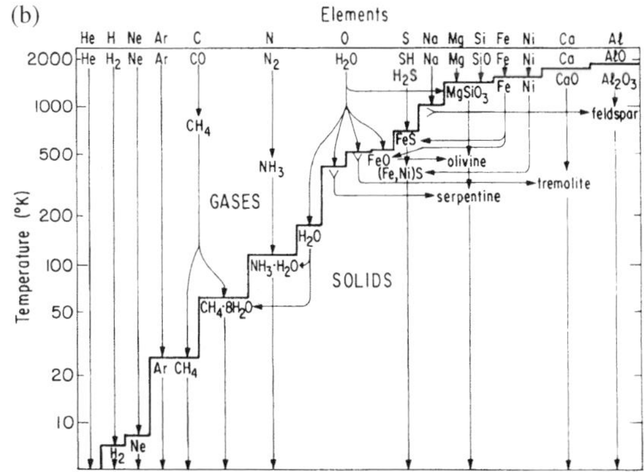

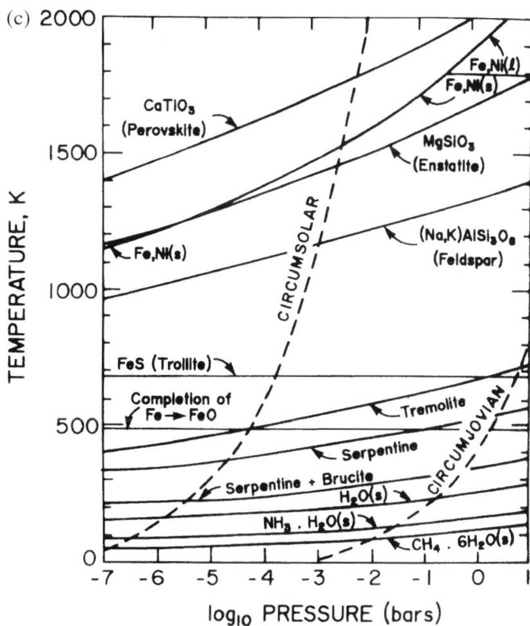

FIGURE 12.8 (a) Calculations of the thermodynamically stable forms of C, O and N at the time that icy condensates began to form at 5 AU. The mass accretion rate of the protoplanetary disk for this calculations is 10^{-7} $M_\odot$ per year. (Courtesy: Monika Kress) (b) Flow chart of major reactions during fully equilibrated cooling of solar nebula material from 2000 to 5 K. The 15 most abundant elements are listed across the top, and directly beneath are the dominant gas species of each element at 2000 K. The staircase curve separates gases from condensed phases. (Barshay and Lewis 1976) (c) Thermochemical equilibrium stability fields for condensed material in a solar composition medium. The species diagrammed are in primarily solid form below the lines and primarily gaseous at higher temperatures. The dashed lines are estimated temperature–pressure profiles for the circumsolar and circumjovian disks. (Prinn 1993)

$\sim$500 K. Reactions between FeO and compounds like enstatite and forsterite form olivines and pyroxenes of intermediate iron content (e.g., (Mg, Fe)$_2$SiO$_4$; (Mg, Fe)SiO$_3$).

Water plays an extremely important role below 500 K. The condensation sequence at low temperatures is quite uncertain, because reactions such as $CO + 3H_2 \rightarrow CH_4 + H_2O$, $N_2 + 3H_2 \rightarrow 2NH_3$ and the formation of hydrated silicates are thermodynamically favored at low temperatures, but they have high activation energies. Such reactions are *kinetically inhibited* because they take a very long time to reach equilibrium, longer than nebular evolution allows. These reactions are more likely to proceed to equilibrium at the higher densities believed to characterize circumplanetary nebulae. If equilibrium is maintained, water vapor reacts with olivines and pyroxenes to form hydrated silicates (e.g., serpentine $Mg_6Si_4O_{10}(OH)_8$, talc $Mg_3Si_4O_{10}(OH)_2$) and hydroxides (e.g., brucite $Mg(OH)_2$). Below 200 K, pure water-ice appears; at somewhat lower temperatures, ammonia and methane gas condense as hydrates and clathrates, respectively ($NH_3{\cdot}H_2O$, $CH_4{\cdot}6H_2O$). At temperatures of below $\sim$40 K, CH_4 and Ar ices form. The 'nonequilibrium' species CO and N_2 can form clathrates with water-ice below $\sim$60 K, and they can be physically 'trapped' in water-ice if it is cold enough. At $T \lesssim 25$ K, CO and N_2 will condense into ice.

In the cold outer regions of protoplanetary disks chemical equilibrium is not, however, achieved (or even closely approached), since the time required to reach equilibrium at these low gas densities ($\rho \sim 10^{-9}$ g cm^{-3}) and low temperatures is likely longer than characteristic cooling and condensation times, and may even exceed the lifetime of the disk. Thus, many chemical reactions may be kinetically inhibited from reaching equilibrium, and the chemistry in the outer nebula depends on the kinetics of the reactions involving interstellar medium constituents. We note that at the onset of gravitational collapse, roughly 40% of the carbon in the interstellar medium is in the form of dust and ~10% is contained in PAHs (polycyclic aromatic hydrocarbons), while most of the gas-phase C is in the form of CO molecules. The interstellar nitrogen is expected to be gaseous N_2, but a significant fraction of the nitrogen is also present as NH_3. It is feasible that CO and N_2 in the cold outer regions of the protoplanetary disk never converted to CH_4 and NH_3, and similarly it may be possible that interstellar grains never evaporated. Thus, the existence of NH_3 in cometary ices suggests that it may be of interstellar origin.

We have seen in Section 8.7 that the D/H and $^{15}N/^{14}N$ ratios in meteorites strongly suggest an origin in cold interstellar clouds. The relatively high D/H ratios observed in comets (Sections 10.3.5.6, 10.7.1) imply that ices and other volatile compounds did not equilibrate with H_2 in the protoplanetary disk. The high fractionation can be explained if the material formed in cold interstellar clouds. In addition, the (relative) abundances of CH_4:CO:C_2H_6 and those of H_2CO:CH_3OH, as well as the seemingly ubiquitous presence of S_2 and N_2 in comets also hint at compounds formed in cold interstellar clouds. These observations all suggest that at least a fraction of the molecules contained in the (icy) grains from which the planetesimals formed originated in cold interstellar clouds and/or in the cold outer reaches of our solar nebula, and were incorporated as such in the forming bodies. Thus, the role of 'disequilibrium' chemistry in the evolving protoplanetary disk should not be underestimated. The Galileo probe discovered near-equal enhancements (2.5–3 times solar) in the moderately volatile elements C, N, S, as well as the noble gases Ar, Kr, and Xe in Jupiter's atmosphere (at a pressure of ~10 bar). As these elements have a broad range of condensation temperatures, this similarly suggests that these elements were brought in by planetesimals which condensed at temperatures low enough for these elements to either be trapped within H_2O-ice or to be stable as solids.

Some models of protoplanetary disks take the evolution of molecular abundances into account while the matter accretes to the central star. In such models CO, N_2 and other gases as well as interstellar grains are brought in from the interstellar medium. When they are close enough to the star where the temperature is high, equilibrium chemistry becomes important, but at larger distances ($\gtrsim$ several AU from the Sun) disequilibrium chemistry is the norm. Of particular importance is cosmic ray ionization and the subsequent chemistry. The ions formed in the disk by cosmic rays influence the evolution of molecular species. For example, CO is transformed through such reactions into CO_2, H_2CO and CH_4, while N_2 is transformed to NH_3 and HCN. So over time, depending on the ionization rate, the CO and N_2 abundances decrease even though the equilibrium reactions ($CO \rightarrow CH_4$, $N_2 \rightarrow NH_3$) are kinetically inhibited. If the temperature is low enough for gases to freeze out, such gases may be adsorbed onto grains. The ices may sublimate again when the grains migrate inwards towards the star. Such more advanced models can thus explain the coexistence of certain ices (i.e., CH_4, CO and CO_2) in comets, and the survival of interstellar grains in the protoplanetary disk (a significant fraction of the water molecules in these grains are of interstellar origin). The chemistry is extremely complex, and detailed models of the formation of our Solar System are becoming increasingly more sophisticated.

Even though there still are many uncertainties in the chemical evolution of the cooling and evolving protoplanetary disk, it is clear that silicates and metal-rich condensates exist throughout almost the entire disk, but ices only in the outer parts. Because ice-forming elements are more abundant than refractory elements in the Sun and the interstellar medium (Table 8.1), the outer parts of the primitive solar nebula contained abundant ices. Well inside Mercury's orbit, the temperature was too high for solids to exist. However, the presence of complex disequilibrium compounds and both refractory and volatile grains of pre-solar origin (Sections 8.7.2, 10.3.5.6, 10.7.1) implies that the basic equilibrium condensation models are too simplistic.

12.4.4 Clearing Stage

Since there is no gas left between the planets, the gas must have been cleared away at some stage during the evolution process. One proposed explanation is that clearing occurred when the pre-main sequence Sun went through its T Tauri phase of stellar evolution, approximately 10^6–10^7 yr after the protostar formed. Models and observations of young stars suggest that the accreting Sun had a luminosity 20–30 times larger than the present solar luminosity. The gas in the disk may have been cleared out either by the strong solar wind associated with the Sun in its T Tauri phase of stellar evolution, or by photoevaporation from the faces of the disk induced either by the early Sun's high UV luminosity, or by UV photons from nearby massive stars (cf. Fig. 12.4). The timing of the gas loss is a crucial issue concerning the growth of giant planets, but is currently poorly constrained. Note that all four giant planets in our Solar System have roughly the same mass of rock and ice forming elements, but their H and He abundances vary by a factor of 100. This may be a consequence of the time at which the gas within the protoplanetary disk was dissipated.

12.5 Condensation and Growth of Solid Bodies

12.5.1 Timescales for Planetesimal Formation

Chondritic meteorites contain the oldest rocks known in our Solar System. As discussed in Chapter 8, the age of most chondrites (primitive meteorites) is 4.56 Gyr, and they formed within a period of $\lesssim 20$ Myr at the beginning of Solar System history. Evidence for live ^{26}Al ($t_{1/2} = 0.72$ Myr) in the protoplanetary disk suggests that the first solid planetary material formed only a few million years after the last injection of freshly nucleosynthesized matter, a timescale which is similar to that required for the collapse of a molecular cloud core. However, alternative models suggest that ^{26}Al was produced in the early Solar System by interactions between solar energetic particles and solid particles in the inner portion of the protoplanetary disk (Section 8.5.4). The presence of nearly pure ^{22}Ne in some meteorites (presumably from ^{22}Na decay, with $t_{1/2} = 2.6$ yr; Section 8.5.4.2), as well as the isotopic ratios measured in several different elements (i.e., D/H, ^{15}N/^{14}N; Section 8.7) implies the survival and inclusion in meteoritic material of some interstellar grains. As interstellar grains heat up considerably when falling to the midplane and passing through the accretion shock, most of the interstellar grains may evaporate completely before arrival in (at least the inner part of) the disk.

Evidence for extinct radionucleides in meteorites implies that the material which formed the Solar System contained an admixture of recently nucleosynthesized isotopes, which could have been produced in an asymptotic giant branch (AGB) star. It is possible that the gravitational collapse of the solar nebula was triggered by a strong wind from a nearby AGB star. Alternatively, some of these isotopes could have been produced by particle irradiation of the inner portions of the protoplanetary disk by the active protosun (Section 8.5.4).

12.5.2 Formation of Solid Planetesimals

As a disk of gaseous matter cools, various compounds condense into microscopic grains. For a disk of solar composition, the first substantial condensates are silicates and iron compounds. At lower temperatures, characteristic of the outer region of our planetary system, large quantities of water-ice and other ices can condense (cf. Section 12.4.3). In these regions there was also a significant fraction of pre-existing condensates from the interstellar medium and stellar atmospheres. Growth of solid particles then proceeds primarily by mutual collisions.

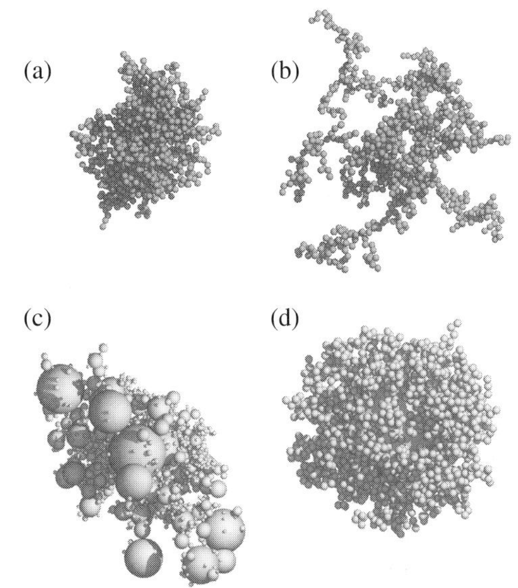

FIGURE 12.9 Examples of fractal aggregates produced by numerical simulations. In a ballistic particle-cluster agglomeration (BPCA) process, a seed particle grows by the accumulation of single particles that collide with random impact parameters and from random directions on linear trajectories (hit-and-stick process). A ballistic cluster-cluster agglomeration (BCCA) process proceeds through the coagulation of equal-mass aggregates (again on linear trajectories with random impact parameters and from random directions). (a) BPCA with 1024 monodisperse spherical particles; the simple BPCA process leads to aggregates with a fractal dimension of 3.0. (b) BCCA with 1024 monodisperse spherical particles; these aggregates have a fractal dimension of 1.9. (c) BPCA with 2001 spherical constituent particles, following a power law size distribution with an exponent of −3.15. (d) BPCA with 2000 monodisperse spherical particles, aggregated onto a large spherical core. (Blum *et al.* 1994)

The microphysics of the growth of subcentimeter-sized grains is quite different from the dynamical processes important to later stages of planetary accretion. The mechanical and chemical processes related to grain agglomeration are poorly understood. Data from smokestack studies and numerical models suggest that loosely packed fractal structures which are held together by van der Waals forces may be formed (Fig. 12.9). However, most primitive meteorites differ from the subject of these studies as they contain chondrules, which are small igneous inclusions ~1 mm in size. The large abundance of chondrules implies that a significant fraction of the hypothesized fluffy (very porous) aggregates were rapidly heated and cooled prior to being incorporated into larger bodies. Various models of chondrule formation exist, but no consensus has yet been reached (Section 8.7.2).

The motions of small grains in a protoplanetary disk are strongly coupled to the gas. For the parameter regime believed to have existed in the Solar System's protoplanetary disk, the coupling between the gas and solid particles smaller than 1 cm is well described by Epstein's drag law. When grains condense, the vertical component of the star's gravity causes the dust to sediment out towards the midplane of the disk. The acceleration of a grain is given by:

$$\frac{dv_z}{dt} = -\frac{\rho_g c_s}{R\rho} v_z - n^2 z, \tag{12.14}$$

where v_z is the grain's velocity in the z-direction (perpendicular to the midplane of the disk), ρ_g the gas density, ρ the grain's density, R is the radius of the grain, c_s the local speed of sound, which is equal to the thermal gas velocity, and the keplerian orbital angular velocity $n = \sqrt{GM_\odot / r_\odot^3}$. The equilibrium settling speed is:

$$v_z = \frac{n^2 z \rho R}{\rho_g c_s}. \tag{12.15}$$

Note that for particles of a given density, the settling rate is proportional to particle radius.

At a heliocentric distance of 1 AU, the temperature of the disk is approximately 500–800 K, and the gas density $\rho_g = 10^{-9}$ g cm^{-3}. The thermal velocity $c_s = 2.5 \times 10^5$ cm s^{-1} for an H_2 nebula. For 1 μm grains with a density of 1 g cm^{-3}, $v_z = 0.03(z/H_z)$ cm s^{-1}. With this sedimentation rate, it would take a 1 μm sized particle $\sim 10^6$ yr to fall half-way towards the disk midplane, or about 10^7 yr for 99.9% of the distance. Such a long settling time is inconsistent with timescales of grain condensation and growth into planetesimals based upon dating of meteorites and stellar evolution models, which predict the Sun to go through a T Tauri phase approximately 10^6–10^7 years after gravitational collapse started. There must therefore be additional processes at work.

Collisional growth of grains during their descent to the midplane of the disk shortens sedimentation times by several orders of magnitude, and differential settling velocities increase the collision rates between particles of differing sizes. For fluffy fractal aggregates, the settling rate increases much more gradually with increasing particle size, but the larger collision cross-sections of these low density agglomerates compensates for the slower settling rate. Current models suggest that (for the parameters believed appropriate to the terrestrial planet region of the solar nebula) the bulk of the solid material was able to agglomerate into bodies of macroscopic size within $\lesssim 10^4$ years at 1 AU. Most of these bodies were confined to a relatively thin region about the midplane of the disk in which the density of condensed material was comparable to, or exceeded, that of the gas.

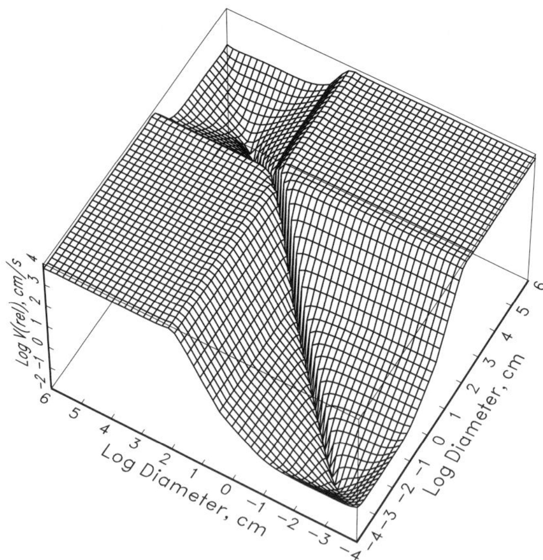

FIGURE 12.10 Contours of constant relative velocity (in cm s^{-1}) between pairs of particles of density 2 g cm^{-3} orbiting within a partially pressure-supported gaseous protoplanetary disk are displayed as a function of particle size. Sizes from 1 μm to 10 km are shown; relative velocities are due to thermal motions (dominant at sizes < 10 μm), as well as radial and transverse velocities induced by gas drag. Disk parameters are for the midplane at 1 AU in a nonturbulent minimum mass solar nebula: gas density $\rho_g = 3.4 \times 10^{-9}$ g cm^{-3}, $T = 320$ K, $\Delta v = 61.7$ m s^{-1}. The narrow 'valley' in the contour plot results from the fact that equal-sized bodies have identical velocities relative to the gas. (Courtesy: Stuart J. Weidenschilling)

At a larger scale, growth from centimeter-sized particles to kilometer-sized planetesimals depends primarily on the relative motions between the various bodies. The motions of (sub-)centimeter-sized material in the protoplanetary disk are strongly coupled to the gas (Fig. 12.10). The gas in the protoplanetary disk is partially supported against stellar gravity by a pressure gradient in the radial direction, so gas circles the star slightly less rapidly than the keplerian rate. The 'effective' gravity felt by the gas is (eq. 2.57):

$$g_{eff} = -\frac{GM_\odot}{r_\odot^2} - \frac{1}{\rho_g}\frac{dP}{dr_\odot}. \tag{12.16}$$

The second term on the right hand side of equation (12.16) is the acceleration produced by the pressure gradient. For circular orbits, the effective gravity must be balanced by centrifugal acceleration, $r_\odot n^2$. Since the pressure is much smaller than the gravity, we can approximate the angular velocity of the gas, n_{gas}, as:

$$n_{gas} \approx \sqrt{\frac{GM_\odot}{r_\odot^3}}(1-\eta), \tag{12.17}$$

where

$$\eta \equiv \frac{-r_\odot^2}{2GM_\odot \rho_g}\frac{dP}{dr_\odot} \approx 5 \times 10^{-3}. \tag{12.18}$$

For estimated protoplanetary disk parameters, the gas rotates $\sim 0.5\%$ slower than the keplerian speed.

Large particles moving at (nearly) the keplerian speed thus encounter a headwind which removes part of their orbital angular momentum and causes them to spiral inwards towards the star. Small grains drift less, as they are so strongly coupled to the gas that the headwind they encounter is very slow. Kilometer-sized planetesimals also drift inwards very slowly, because their surface area to mass ratio is small. Peak rates of inward drift occur for particles that collide with roughly their own mass of gas in one orbital period. Meter-sized bodies in the terrestrial planet region of the solar nebula drift inwards at the fastest rate (Problem 12.9), up to $\sim 10^6$ km yr^{-1}. Thus, a meter-sized body at 1 AU would spiral inwards approaching the Sun in ~100 years! As a consequence of the difference in (both radial and azimuthal) velocities, small (sub)centimeter grains can be swept up by the larger bodies, while gas drag on the meter-sized planetesimals may induce considerable radial motions. The radial migration can remove solids from the planetary region, or bring particles of various sizes together to enhance accretion rates. Thus, the material that survives to form planets must complete the transition from centimeter to kilometers size rather quickly, unless it is confined to a thin dust-dominated subdisk in which the gas is dragged along at essentially keplerian velocity. Bodies may collide with each other and stick together, forming larger agglomerates.

Two alternative hypotheses describe the growth through this size range. First, if the nebula is quiescent, the dust and small particles settle into a layer thin enough to be gravitationally unstable to clumping, and planetesimals presumably are formed as a result of this instability. The planetesimals produced by this mechanism have masses of order:

$$M_{planetesimal} \sim \frac{16\pi^2 G^2 \sigma_\rho^3}{n^4}, \tag{12.19}$$

where σ_ρ is the surface mass density of the particle layer at the time the instability occurs. Planetesimals formed in the inner regions of the solar nebula would have been ~1 km in radius, with larger planetesimals forming farther from the Sun (Problem 12.10). Second, in a turbulent nebula, growth continues via simple two-body collisions. Under these circumstances, there is no fine line between planetesimal formation and accretion from planetesimals to planets. However, the growth of solid bodies from millimeter size to kilometer size still presents particular problems. The physics of interparticle collisions in this size range is poorly understood. Furthermore, the high rate of orbital decay due to gas drag for meter-size particles implies that growth through this size range must occur very rapidly. One possibility is that a small fraction of the grains grow into solid planetesimals via fortuitous circumstances (e.g., being located at temporary nodes in the turbulent flow), and that these planetesimals subsequently sweep up many times their mass in small particles. Current models suggest that interactions between the gas and solids produced sufficient turbulence to prevent the particulate layer from becoming thin enough to be gravitationally unstable, at least in the terrestrial region of the solar nebula. Molecular forces can lead to ~1 km-sized planetesimals by coagulation, since the van der Waals (chemical) binding energies of $\sim 10^3$ erg g^{-1} are comparable to the gravitational binding energy of a 1 km body. When the planetesimals reach sizes of ~1 km, their mutual gravitational perturbations become important.

The large radial motions of planetesimals may provide an explanation for some of the anomalies observed in meteorite composition, where isotopically distinct components of a single meteorite must have condensed separately, at different heliocentric distances, and been brought together as solid bodies. However, note that current theories of planetesimal formation are clearly oversimplified. For example, planetesimal growth models do not account for chondrule formation and other violent and disruptive events which may occur in a turbulent nebula.

12.5.3 Growth from Planetesimals to Planetary Embryos

The primary factors controlling the growth of planetesimals into planets differ from those responsible for the accumulation of dust into planetesimals. Solid bodies larger than ~1 km in size face a headwind only slightly faster than that experienced by 10 m objects (for parameters thought to be representative of the terrestrial region of the solar nebula), and because of their much greater mass-to-surface-area ratio they suffer far less orbital decay from interactions with the gas in their path (Figure 12.11). The primary perturbations on the keplerian orbits of kilometer-sized and larger bodies in protoplanetary disks are mutual gravitational interactions and physical collisions. These interactions lead to accretion (and in some cases erosion and fragmentation) of planetesimals. Gravitational encounters are able to stir planetesimal random velocities up to the escape speed from the largest common planetesimals in the swarm. The most massive planetesimals have the largest gravitationally enhanced collision cross-sections, and accrete almost everything with which they collide. If the random velocities of most planetesimals remain much smaller than the escape speed from the largest bodies, then these large *planetary embryos* (also referred to as protoplanets) grow extremely rapidly. The size distribution of solid bodies becomes quite skewed, with a few large bodies growing much faster than the rest of the swarm in a process known as *runaway accretion*. Eventually planetary embryos accrete most of the (slowly moving) solids within their gravitational reach, and the runaway growth phase ends. We examine the growth process from kilometer-sized planetesimals to 10^3–10^4 kilometer-sized planetary embryos in detail below.

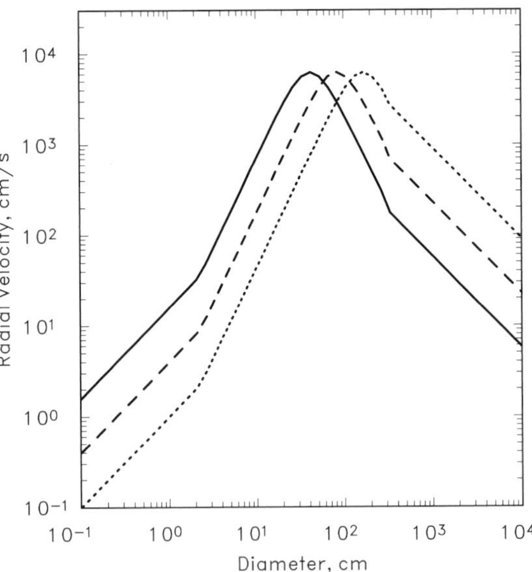

FIGURE 12.11 The inward radial drift rates of solid particles in a protoplanetary disk as a function of size for three values of density: 0.5 (solid line), 2.0 (dashed line), and 7.9 (dotted line) g cm^{-3}. Gas parameters are the same as for Figure 12.10. Small particles, with small mass/surface area ratios, are strongly coupled to the gas and compelled to move with (nearly) its angular velocity. As this is less than the keplerian orbital rate, they feel a residual component of the Sun's gravity, and settle inward at a terminal velocity at which gas drag balances this radial acceleration. Thus, larger and/or denser particles drift more rapidly in this regime. Bodies with large mass/surface area ratios travel in (nearly) keplerian orbits, moving faster than the gas. They experience a 'headwind' that causes their orbits to decay; larger and/or denser bodies are less affected by this drag, so the decay rate decreases with increasing particle radius. The radial velocity reaches a peak at the transition between these regimes, at sizes of about a meter. The abrupt changes in slope result from transitions between drag laws for different Knudsen and Reynolds numbers. (Courtesy: Stuart J. Weidenschilling)

12.5.3.1 *Planetesimal Velocities*

The distribution of planetesimal velocities is one of the key factors that controls the rate of planetary growth. Planetesimal velocities are modified by mutual gravitational interactions, physical collisions (which can be partially elastic, leading to rebound or fragmentation, or completely inelastic, leading to accretion) and gas drag. Gravitational scatterings and elastic collisions convert energy present in the ordered relative motions of orbiting particles (keplerian shear) into random motions and tend to reduce the velocities of the largest bodies in a swarm relative to those of smaller bodies. Inelastic collisions and gas drag damp eccentricities and inclinations, especially of small planetesimals.

The simplest analytic approach for calculating the evolution of planetesimal velocities uses a 'particle-in-a-box' approximation in which the evolution of the mean square planetesimal velocities is calculated via the methods of the kinetic theory of gases. In this approximation, one ignores the details of individual planetesimal orbits and uses a probability density to describe the distribution of orbital elements in the planetesimal population. During the final stages of planetesimal accumulation, the number of planetesimals eventually becomes small enough that direct N-body numerical integrations of individual planetesimal orbits is feasible.

12.5.3.2 *Collisions Between Planetesimals*

The size distibution of planetesimals evolves principally via physical collisions among its members. Physical collisions between solid bodies can lead to accretion, fragmentation, or inelastic rebound of relatively intact bodies; intermediate outcomes are possible as well. The outcome of a collision depends upon the internal strength of the planetesimals, the coefficient of restitution of the bodies, and most sensitively on the kinetic energy of the collision. The impact velocity at which two bodies of radii R_1 and R_2 and masses m_1 and m_2 collide is given by:

$$v_i = \sqrt{v^2 + v_e^2}, \tag{12.20}$$

where v is the velocity of m_2 relative to m_1 far from encounter, and v_e is the escape velocity from the point of contact:

$$v_e = \left(\frac{2G(m_1 + m_2)}{R_1 + R_2}\right)^{1/2}. \quad (12.21)$$

The impact velocity is thus at least as large the escape velocity, which for a rocky 10 km sized object is $\sim$6 m s^{-1}. The rebound velocity is equal to ϵv_i, where the coefficient of restitution $\epsilon \leq 1$. If $\epsilon v_i \leq v_e$, then the bodies remain bound gravitationally and soon recollide and accrete. Net disruption requires both fragmentation, which depends upon the internal strength of the bodies, and post-rebound velocities greater than the escape speed. Since relative velocities of planetesimals are generally less than the escape velocity from the largest common bodies in the swarm, the largest members of the swarm are likely to accrete the overwhelming bulk of the material with which they collide, unless ϵ is very close to unity. Fragmentation is most common for very small planetesimals. The largest bodies in the swarm accrete at a rate essentially identical to the collision rate. Subcentimeter-sized grains corotating with the gas may impact kilometer-sized planetesimals at speeds well above the escape velocities of these planetesimals. This process could lead to erosion of planetesimals via 'sandblasting'.

The simplest model for computing the collision rate of planetesimals ignores their motion around the Sun completely. A collision occurs when the separation between the centers of two particles equals the sum of their radii. The mean rate of growth of a planetary embryo's mass, M, is:

$$\frac{dM}{dt} = \rho_s v \pi R^2 \mathcal{F}_g, \quad (12.22)$$

where v is the average relative velocity between the large and small bodies, ρ_s the volume mass density of the swarm of planetesimals, and the planetary embryo's radius, R, is assumed to be much larger than the radii of the planetesimals. The last term in equation (12.22) is the gravitational enhancement factor, which in the 2 + 2-body approximation is given by:

$$\mathcal{F}_g = 1 + (v_e/v)^2. \quad (12.23)$$

The gravitational enhancement factor arises from the ratio of the distance of close approach to the asymptotic unperturbed impact parameter, b, in a two-body hyperbolic encounter, and can be derived using conservation of angular momentum and energy of the planetesimal relative to the planetary embryo (Problem 12.11). In the 2 + 2-body approximation, often abbreviated as the *two-body approximation*, one ignores the influence of planetesimals/planetary embryos on one another except during close encounters, and during such close encounters the influence of the Sun upon the bodies is neglected. Thus, the analysis reduces the problem to a pair of two-body calculations for the planetesimal.

It is often convenient to state the growth rate of the planet in terms of the surface density of the planetesimals in the disk rather than the volume density of the swarm. If the protosun's gravity is the dominant force in the vertical direction and if the relative velocity between planetesimals is isotropic, then the vertical Gaussian scale height H_z (eq. 12.11) of the planetesimal disk can be written as:

$$H_z = \frac{v}{\sqrt{3}n}. \quad (12.24)$$

The surface mass density, also referred to as the column density (g cm^{-2}) of solids in the disk, σ_ρ, can be written as:

$$\sigma_\rho = \sqrt{\pi}\rho_s H_z. \quad (12.25)$$

Equations (12.22)–(12.25) can be used to express the rate of growth in the planetary embryo's radius:

$$\frac{dR}{dt} = \sqrt{\frac{3}{\pi}}\frac{\sigma_\rho n}{4\rho_p}\mathcal{F}_g, \quad (12.26)$$

where ρ_p is the density of the planetary embryo. The planetary embryo's radius thus grows at a constant rate if $\mathcal{F}_g$ remains constant.

Random velocities of the planetesimals are determined by a balance between gravitational stirring and damping via inelastic collisions. If most of the mass is contained in the largest bodies, the equilibrium velocity dispersion is comparable to the escape speed of the largest bodies, implying $\mathcal{F}_g < 10$. Let us assume $\mathcal{F}_g = 7$ for the proto-Earth. In the minimum mass model, at 1 AU, the surface mass density $\sigma_\rho = 10$ g cm^{-2}, $n = 2 \times 10^{-7}$ s^{-1} and $\rho_p = 4.5$ g cm^{-3}, which implies a growth time for the Earth of 2×10^7 yr. More detailed calculations yield times closer to 10^8 years, since the accretion rate drops during the later stages of planetary growth due to a drop in overall density of the swarm of planetesimals when the Earth grew to nearly full size.

For the giant planets, however, growth times computed in this manner are much larger. For a minimum mass nebula, the surface density drops with heliocentric distance, approximately as $r^{-3/2}$, except for a jump by a factor of $\sim$3 at $\sim$4 AU due to the condensation of water-ice. At Jupiter's distance from the Sun, the surface mass density in a minimum mass nebula would be $\sigma_\rho \approx 3$ g cm^{-2}. Jupiter's heavy element mass is approximately 15–20 $M_\oplus$, which results in a growth time of over 10^8 yrs. For Neptune we get numbers many times the age of the Solar System

(Problem 12.13). Since at least Jupiter and Saturn must have formed within $\sim 10^7$ yrs, before the gas in the solar nebula was swept away, additional factors must be involved in the growth of giant planets.

12.5.3.3 *Runaway Growth of Planetary Embryos*

When the relative velocity between planetesimals is comparable to or larger than the escape velocity, $v \gtrsim v_e$, the growth rate is approximately proportional to R^2, and the evolutionary path of the planetesimals exhibits an orderly growth of the entire size distribution. When the relative velocity is small, $v \ll v_e$, one can show, by rewriting the escape velocity in terms of the protoplanet's radius, that the growth rate is proportional to R^4. In this situation, the planetary embryo rapidly grows larger than any other planetesimal, which can lead to *runaway growth* (see Fig. 12.12 and Problem 12.15). The runaway embryo can grow so much larger than the surrounding planetesimals that its $\mathcal{F}_g$ can exceed 1000; however, three-body stirring by the embryo prevents $\mathcal{F}_g$ from growing much larger than this. As embryos approach this maximum gravitational enhancement factor, larger embryos take longer to double in mass than do smaller ones, although embryos of all masses continue their runaway growth relative to surrounding planetesimals; this phase of rapid accretion of planetary embryos is known as *oligarchic growth*.

Runaway accretion requires low random velocities, and thus small radial excursions of planetesimals. The planetary embryo's feeding zone is therefore limited to the annulus of planetesimals which it can gravitationally perturb into intersecting orbits. Thus, rapid runaway growth ceases when a planetary embryo has consumed most of the planetesimals within its gravitational reach. Planetesimals within ~ 4 times the planetary embryo's Hill sphere eventually will come close enough to the planetary embryo during one of their orbits that they may be accreted (unless their semimajor axis is very similar to that of the embryo, in which case they may be locked in tadpole or horseshoe orbits that avoid close approaches, Section 2.2.2). The mass of a planetary embryo which has accreted all of the planetesimals within an annulus of width $2\Delta r_\odot$ is:

$$M = \int_{r_\odot - \Delta r_\odot}^{r_\odot + \Delta r_\odot} 2\pi r' \sigma_\rho(r') dr' \approx 4\pi r_\odot \Delta r_\odot \sigma_\rho(r_\odot). \tag{12.27}$$

Setting $\Delta r_\odot = 4 \; R_H$ (cf. eq. 2.28), we obtain the *isolation mass*, M_i (in grams), which is the largest mass to which a planetary embryo orbiting a 1 $M_\odot$ star can grow by runaway accretion:

$$M_i \approx 1.6 \times 10^{25} (r_{AU}^2 \sigma_\rho)^{3/2}, \tag{12.28}$$

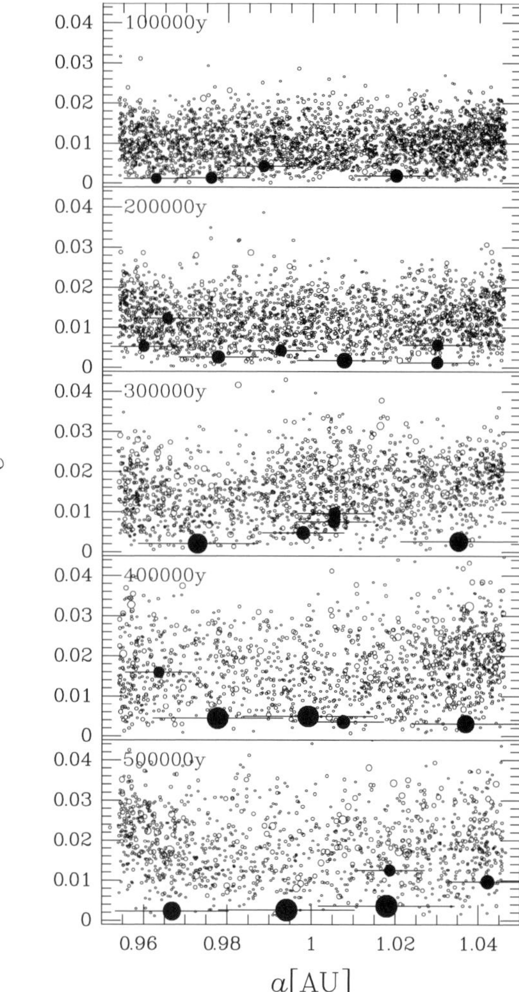

FIGURE 12.12 Snapshots of a planetesimal system on the a–e plane. The circles represent planetesimals and their radii are proportional to the radii of planetesimals. The system initially consists of 4000 planetesimals whose total mass is 1.3×10^{27} g. The initial mass distribution is a power with index $\zeta = -2.5$ over the mass range 2×10^{23} g $\leq m \leq 4 \times 10^{24}$ g. The system is followed using an N-body integrator, and physical collisions are assumed to always result in accretion. The numbers of planetesimals are 2712 (t = 100 000 yr), 2200 (t = 200 000 yr), 1784 (t = 300 000 yr), 1488 (t = 400 000 yr), and 1257 (t = 500 000 yr). The filled circles represent planetary embryos with mass larger than 2×10^{25} g, and lines from the center of each planetary embryo extend 5 R_H outwards and 5 R_H inwards. (Kokubo and Ida 1999)

where σ_ρ is in g cm^{-2}. For a minimum mass solar nebula, the mass at which runaway must have ceased in the Earth accretion zone would have been about 6 lunar masses, and in Jupiter's accretion zone roughly the mass of the Earth.

Runaway growth can persist beyond the isolation mass given by equation (12.28) only if additional mass can diffuse into the planet's accretion zone. Three plausible mechanisms for such diffusion are scattering between planetesimals, perturbations by planetary embryos in neighboring accretion zones and gas drag. Alternatively, radial motion of the planetary embryo may bring it into zones not depleted of planetesimals. Gravitational torques resulting from the excitation of spiral density waves in the gaseous component of the protoplanetary disk have the potential of inducing rapid radial migration of planets (Section 12.8). Gravitational focusing of gas could also vastly increase the rate of inward drift of planetary embryos.

The limits of runaway growth are less severe in the outer Solar System than in the terrestrial planet zone. If the surface density of condensed material at 5 AU was $\gtrsim 10$ g cm^{-2}, it is possible that runaway growth of Jupiter's core continued until it attained the mass necessary to rapidly capture its massive gas envelope. The 'excess' solid material in the outer Solar System could have been subsequently ejected to the Oort cloud or to interstellar space via gravitational scattering by the giant planets. In contrast, the small terrestrial planets, orbiting deep within the Sun's gravitational potential well, could not have ejected substantial amounts of material, so the total mass of solids in the terrestrial planet zone during the runaway accretion epoch was probably not substantially larger than the current mass of the terrestrial planets, implying that a high-velocity growth phase subsequent to runaway accretion was required to yield the present configuration of terrestrial planets.

12.6 Formation of the Terrestrial Planets

12.6.1 Dynamics of the Final Stages of Planetesimal Accumulation

The self-limiting nature of runaway growth implies that massive planetary embryos form at regular intervals in semimajor axis. The agglomeration of these embryos into a small number of widely spaced terrestrial planets necessarily requires a stage characterized by large orbital eccentricities, significant radial mixing, and giant impacts. At the end of the runaway phase, most of the original mass is contained in the large bodies, so their random velocities are no longer strongly damped by energy equipartition with the smaller planetesimals. Mutual gravitational scattering can pump up the relative velocities of the planetary embryos to values comparable to the surface escape velocity of the largest embryos, which is sufficient to ensure their mutual accumulation into planets. The large velocities imply small collision cross-sections and hence long accretion times.

Once the planetary embryos have perturbed one another into crossing orbits, their subsequent orbital evolution is governed by close gravitational encounters and violent, highly inelastic collisions. This process has been studied using N-body integrations of planetary embryo orbits, which include the gravitational effects of the giant planets, but neglect the population of numerous small bodies which must also have been present in the terrestrial zone; physical collisions are assumed to always lead to accretion (i.e., fragmentation is not considered). Few bodies initially in the terrestrial planet zone are lost; in contrast, most planetary embryos in the asteroid region are ejected from the system by a combination of jovian perturbations and mutual gravitational scatterings. As the simulations endeavor to reproduce our Solar System, they generally begin with about 2 $M_\oplus$ of material in the terrestrial planet zone, typically divided among several dozen or more protoplanets. The end result is the formation of 2–5 terrestrial planets on a timescale of about 10^8 years (Fig. 12.13). Some of these systems look quite similar to our Solar System, but most have fewer terrestrial planets which travel on more eccentric orbits. It is possible that the Solar System is by chance near the quiescent end of the distribution of terrestrial planets. Alternatively, processes such as fragmentation and gravitational interactions with a remaining population of small debris, thus far omitted from the calculations because of computational limitations, may lower the characteristic eccentricities and inclinations of the ensemble of terrestrial planets.

An important result of these N-body simulations is that planetary embryo orbits execute a random walk in semimajor axis as a consequence of successive close encounters. The resulting widespread mixing of material throughout the terrestrial planet region diminishes any chemical gradients that may have existed during the early stages of planetesimal formation, although some correlations between the final heliocentric distance of a planet and the region where most of its constituents originated are preserved in the simulations. Nonetheless, these dynamical studies imply that Mercury's high iron abundance is unlikely to have arisen from chemical fractionation in the solar nebula.

The mutual accumulation of numerous planetary embryos into a small number of planets must have entailed many collisions between protoplanets of comparable size. Mercury's silicate mantle may have been partially stripped off in such a giant impact, leaving behind an iron-rich core. Accretion simulations also lend support to the giant impact hypothesis for the origin of the Earth's Moon (Section 12.11); during the final stage of accumulation, an Earth-

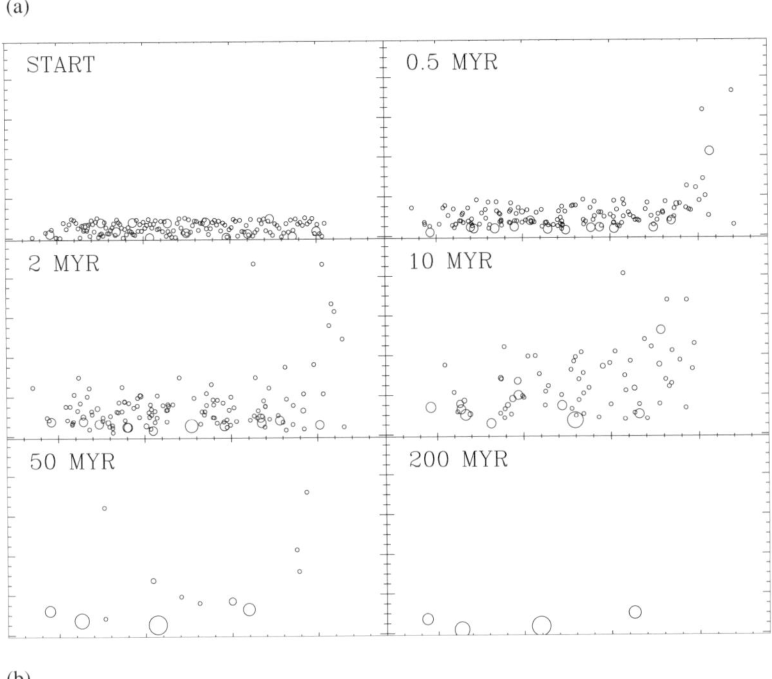

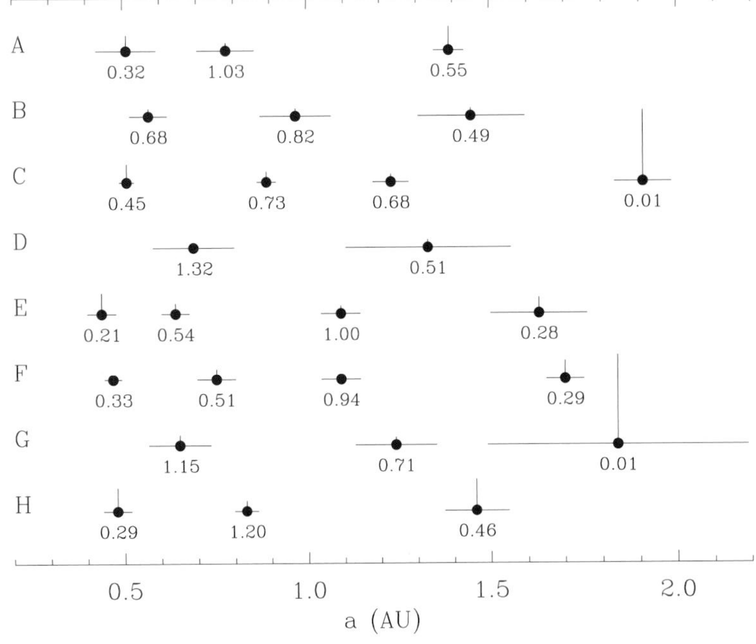

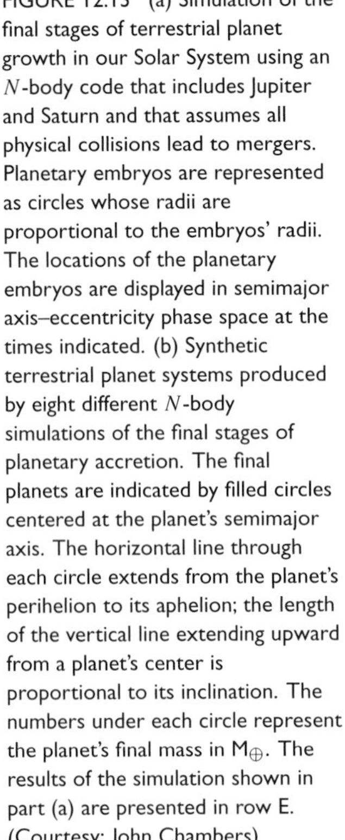
FIGURE 12.13 (a) Simulation of the final stages of terrestrial planet growth in our Solar System using an N-body code that includes Jupiter and Saturn and that assumes all physical collisions lead to mergers. Planetary embryos are represented as circles whose radii are proportional to the embryos' radii. The locations of the planetary embryos are displayed in semimajor axis–eccentricity phase space at the times indicated. (b) Synthetic terrestrial planet systems produced by eight different N-body simulations of the final stages of planetary accretion. The final planets are indicated by filled circles centered at the planet's semimajor axis. The horizontal line through each circle extends from the planet's perihelion to its aphelion; the length of the vertical line extending upward from a planet's center is proportional to its inclination. The numbers under each circle represent the planet's final mass in $M_\oplus$. The results of the simulation shown in part (a) are presented in row E. (Courtesy: John Chambers)

size planet is typically found to collide with several objects as large as the Moon and frequently one body as massive as Mars. The obliquities of the rotation axes of the giant planets provide independent evidence of the occurence of giant impacts during the accretionary epoch.

12.6.2 Accretional Heating and Planetary Differentiation

Impacting planetesimals provide a planet with energy as well as mass. This energy heats a growing planet. Under certain circumstances, a planet may become warm enough that portions melt, allowing denser material to sink and the planet to differentiate. The energy available to a growing planet is supplied by accreted planetesimals (which contribute both their kinetic energy 'at infinity' and the potential energy released as the planetesimal falls onto the planet's surface), gravitational potential energy released as the planet contracts (due to increased pressure) or differentiates, radioactive decay and exothermic chemical processes. The primary loss mechanism is radiation to space, although the planet may also cool via endothermic reactions or give up gravitational energy as a result of expansion if it heats up significantly or if water freezes. Energy may be transported within the planet via conduction, or if the planet is (partially or fully) molten, via convection. Energy transport is important to the global as well as local heat budgets, as radiative losses can only occur from the planet's surface or atmosphere.

Conduction is rather slow over planetary distances, and convection is rapid only in regions that are sufficiently molten to allow fluid motions to occur (Section 6.1.5.2). Thus, to a first approximation, a growing solid planet's temperature is given by a balance between accretion energy deposited at the planet's surface and radiative losses from the surface. The temperature of a given region changes slowly once it becomes buried deep below the surface (unless large scale melting and differentiation occur). For gradual accretion, the temperature at a given radius can thus be approximated by balancing the accretion energy source with radiative losses at the time when the material was accreted. For the 10^8 year accretion times estimated for the terrestrial planets, such an estimate implies far less heating than would be required to melt and differentiate the planet (cf. Problem 12.19).

However, modern theories of planetary growth imply that planets accumulate most of their mass in planetesimals of radius 10 km and larger. Impactors deposit $\sim 70\%$ of their kinetic energy as heat in the target rocks directly beneath the impact site, with the remaining $\sim 30\%$ being carried off with the ejecta. If an impactor is large, it may raise deeply buried heat to near the surface, where energy may be radiated away. A more important effect is that heat may become buried by deep ejecta blankets. The ejecta blankets produced by such large impactors are thick enough that most of the heat from the impacts remains buried. Planets can thus become quite warm, with temperature increasing rapidly with radius (Problem 12.19). Accretion energy can lead to the differentiation of planetary (but not asteroidal) sized bodies.

A planetary embryo can form a proto-atmosphere as it accretes solid bodies. When the mass of a growing planet reaches $\sim 0.01\ M_\oplus$, impacts are energetic enough for water to evaporate, while impact devolatilization of H_2O and CO_2 may occur a little sooner. Complete degassing of accreting planetesimals occurs when the radius of the protoplanet reaches about $0.3\ R_\oplus$. A massive proto-atmosphere increases the surface temperature of the protoplanet by a *blanketing effect*, which raises the temperature even more than the greenhouse effect. Solar radiation determines the temperature at the top of the atmosphere, and is scattered and absorbed at lower altitudes. The atmosphere provides a partially insulating blanket to the heat released from impacting planetesimals, so the surface becomes quite hot.

Calculations show that the proto-atmosphere's blanketing effect becomes important when the growing planet's mass exceeds $0.1\ M_\oplus$. When the planet's mass exceeds $0.2\ M_\oplus$ then the surface temperature exceeds ~ 1600 K, which is the melting temperature for most planetary materials. As a result, the surface melts and newly accreting planetesimals on the molten surface will also melt. Heavy material migrates downwards, while lighter elements float on top. This process of differentiation liberates a large amount of gravitational energy in the planet's interior; together with adiabatic compression due to the increase in the planet's mass, enough energy can be released to cause melting of a large fraction of the planet's interior, allowing the planet to differentiate throughout.

12.6.3 Accumulation (and Loss) of Atmospheric Volatiles

Atmospheric gases form a tenuous veneer surrounding many of the smaller planets and moons in the Solar System, amounting to far less than 1% of the mass of each body. These atmospheres consist primarily of high-Z ($Z \geq 3$) elements. The atmospheres of the terrestrial planets and other small bodies were probably outgassed from material accreted as solid planetesimals. The compositions of terrestrial planet atmospheres suggest that some but not all of these volatiles came from the outer Solar System. The problem of the origin of terrestrial planet atmospheres is not simply bringing the required volatiles to the planets, as losses were also important. Impacting planetesimals on a growing planet surrounded by a proto-atmosphere may lead to the following phenomena (Section 5.4.3):

(1) If the planetesimals are small enough to be stopped by atmospheric drag, all of their kinetic energy is deposited in the atmosphere. Most rocky objects with a size less than ~ 30 m are stopped in an atmosphere like that of Earth at the present time, and deposit all of their energy in the atmosphere.
(2) Ejecta excavated by larger impacting planetesimals

are slowed down by the atmosphere, and transfer kinetic energy to it. A detailed description of the interaction is, however, difficult. We note here that an atmosphere has a large compressibility, in contrast to a solid surface, and that the gas can be raised to very high temperatures and pressures. Additionally, the energy from atmospheric impacts is released over an extended area and over a period of tens of seconds.

(3) If the impactor is large, the energy transferred to the atmosphere may be sufficient to blow off part of the atmosphere via hydrodynamic escape (Section 4.8.3.1). If the size of the impactor is comparable to or larger than the atmospheric scale height, impact erosion blows off a large portion of the atmosphere, i.e., an atmospheric mass equal to the mass intercepted by the impactor (Section 4.8.3.2). At the same time the impactor also adds volatiles to an accreting planet. Whether this mass is more than that blown off from the atmosphere depends upon the size of the impactor, its volatile content and the density of the atmosphere. Impactors with radii of ~100 km and volatile content of 1% would yield a balance between impact erosion and accretion of volatiles for an atmospheric mass per unit area similar to that of the terrestrial ocean. A similarly sized impactor population with a volatile content of 0.01% would keep a present day Earth's atmosphere in equilibrium. Figure 12.14 graphs the mass per unit area of an atmosphere in equilibrium between impact erosion and addition of volatiles as a function of impactor radius and volatile content. Atmospheric blowoff is more likely to occur on smaller planets, like Mars. A growing planet may lose its atmosphere several times during the accretion period, since impacts with large planetesimals are quite common.

In addition to impact-related losses of atmospheric gases, the atmosphere may also disappear via Jeans escape (Section 4.8.1). In particular, light elements such as H and He easily escape from the top of a terrestrial atmosphere, while heavier gases may have escaped this way in the early hot proto-atmospheres. The present day terrestrial planet atmospheres were probably formed towards the end of the accretion epoch, by outgassing of the hot planet and impacts by small-sized planetesimals.

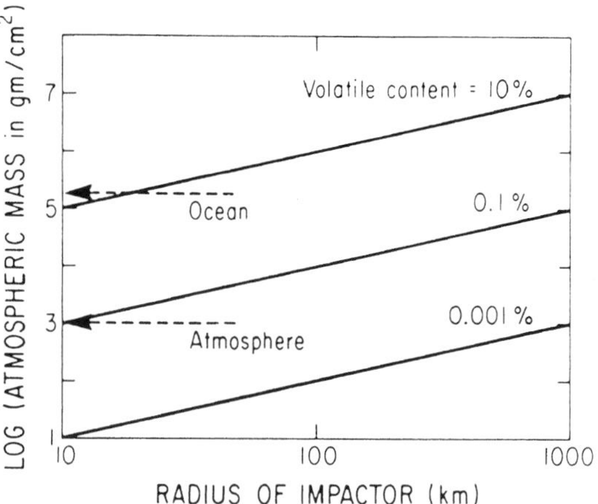

FIGURE 12.14 The mass per unit surface area of an atmosphere of a 1 $M_\oplus$ planet which is in equilibrium between the rate of addition of volatiles to the planet by accretion of material with the indicated volatile content and impact erosion of the atmosphere by impactors of the indicated radii. The arrows show the mass per unit surface area of the present terrestrial ocean and atmosphere. (Hunten *et al.* 1989)

12.7 Formation of the Giant Planets

The large amounts of H_2 and He contained in Jupiter and Saturn imply that these planets must have formed within $\sim 10^7$ yrs, before the gas in the protoplanetary disk was swept away. Any formation theory of the giant planets should account for these timescales. In addition, formation theories should explain the elemental and isotopic composition of these planets and variations therein from planet to planet, their presence and/or absence of internal heat fluxes, their axial tilts, and the orbital and compositional characteristics of their ring and satellite systems. In this section, we discuss the formation of the planets themselves; the formation of their satellites and ring systems is addressed in Section 12.11.

The heavy elements constitute less than 2% of the mass of a solar composition mixture. The giant planets, however, are enhanced in heavy elements relative to the solar value by roughly 5, 15, and 300 times for Jupiter, Saturn and Uranus/Neptune, respectively. Thus, all four giant planets accreted solid material much more effectively than gas from the surrounding nebula. Moreover, the total mass in heavy elements varies by only a factor of a few between the four planets, while the mass of H and He varies by about two orders of magnitude between Jupiter and Uranus/Neptune.

Table 4.5 shows the composition of the giant planet atmospheres. The enhancement in heavy elements increases from Jupiter to Neptune. The nitrogen mixing ratio derived from the NH_3 abundance, however, may form an exception to this rule. But if nitrogen is present as N_2 rather than NH_3, the abundance may scale similarly as the other heavy elements. We also note that sulfur, presumably present in the form of H_2S, has only been detected directly in Jupiter by the Galileo probe.

The D/H ratio in the giant planet atmospheres may provide important clues to the formation history of these planets. The D/H ratios in Jupiter and Saturn, measured from monodeuterated methane gas, are equal to the interstellar D/H ratio of 2×10^{-5}. Since 90% of Jupiter's and 75% of Saturn's mass consist of H and He, one would indeed expect the D/H ratios to agree with the interstellar value. Uranus and Neptune are only about 10% H and He by mass. The observed D/H values on Uranus and Neptune (Fig. 10.14) are higher than the interstellar value, which can be attributed to exchange of deuterium with an icy resevoir.

Two scenarios for the formation of the gaseous planets have been studied extensively:

- Gravitational collapse, or gas-instability hypothesis.
- Core-instability hypothesis, a combination of planetesimal accretion and gravitational accumulation of gas.

12.7.1 Gas-Instability Hypothesis (Giant Gaseous Protoplanets)

If the protoplanetary disk was extremely massive ($\sim$1 $M_\odot$), it would have been unstable to clumping. Under such circumstances, the giant planets (and, in some scenarios, the terrestrial planets too) could have formed by gravitational collapse just like the Sun itself, in subcondensations of the nebula. This theory was popular in the 1950s and 1960s, but fell out of favor for a variety of reasons (listed in decreasing order of importance):

(1) All of the planets are substantially enhanced in condensable material relative to solar composition. Although it is conceivable that some process could have removed large gaseous atmospheres from the terrestrial planets after the heavy material had settled to the core, such removal would have been more difficult in the cases of the much more massive planets Jupiter and Saturn. In addition, an initially solar composition for Uranus and Neptune is extremely hard to fathom. Uranus and Neptune are massive planets which are located a considerable distance from the solar energy source, and their H_2 and He abundances are $< 1\%$ those of hypothetical solar composition bodies with the same masses of heavy elements.

(2) In order for the planet-forming gravitational instabilities to have occurred, a disk mass comparable to that of the Sun would be required. Even then the planets could only have formed much farther from the Sun than their current locations, unless the surface density of the protoplanetary disk greatly exceeded that predicted by most models.

(3) The gas instability model does not account for asteroids, comets, moons and other small bodies, which are more easily understood within the planetesimal hypothesis.

(4) Even if the energetics of removing the outer regions of giant gaseous protoplanets were not a problem, these initially homogeneous bodies still had to differentiate into a heavy element enriched core and a heavy element depleted envelope prior to losing most of the mass in their envelopes. Calculations suggest heavy materials would not separate from lighter gases within massive giant gaseous protoplanets.

The above listed points do not preclude a modified version of giant gaseous protoplanets in which Jupiter (and possibly Saturn) formed via gravitational instability and subsequently accumulated their heavy element excesses by accreting planetesimals. However, the possibility of forming Jupiter and Saturn mass objects via disk instabilities has not been convincingly demonstrated, as gravitational instabilities might produce density waves which redistribute mass and stabilize the disk rather than bound planetary mass blobs. Moreover, the gradual progression of masses and composition discussed above argues for a single formation scenario for at least the four giant planets in our Solar System.

12.7.2 Core-Instability Hypothesis

In this theory the core of the giant planet forms first by accretion of planetesimals, while only a small amount of gas is accreted. Planetary embryos with $M \gtrsim 0.1$ $M_\oplus$ have surface escape velocities larger than the sound speed in the gaseous protoplanetary disk. Such a growing planetary core first attains a quasi-static atmosphere which undergoes Kelvin–Helmholtz contraction as the energy released by the planetesimal and gas accretion is radiated away at the photosphere. The contraction timescale is determined by the efficiency of radiative transfer, which is relatively low in some regions of the envelope. Spherically symmetric (1-D) models show that the minimum contraction timescale is a rapidly decreasing function of the core's mass. The gas accretion rate, which is initially very slow, accelerates with time and becomes comparable to the planetesimal bombardment rate after the core has acquired $\sim$10–20 $M_\oplus$. Once the gaseous component of the growing planet exceeds the solid component, gas accretion becomes very rapid, and leads to a runaway accretion of gas. This scenario gives a natural explanation

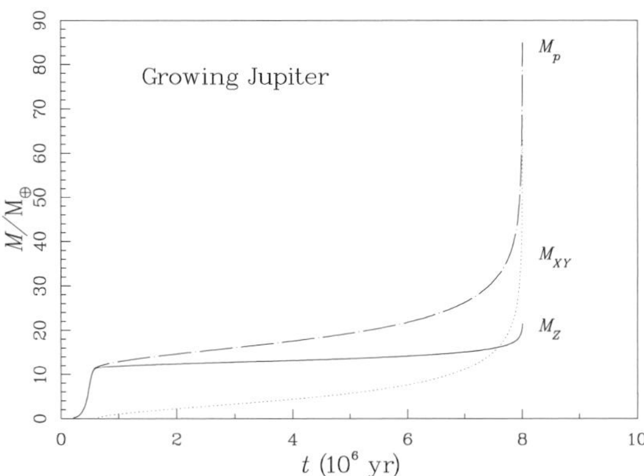

FIGURE 12.15 The mass of a growing giant planet as a function of time according to a model of Pollack *et al.* (1996). The planet's total mass is represented by the dot-dashed curve, the mass of the solid component is given by the solid curve, and the dotted curve represents the gas mass. The solid core grows rapidly by runaway accretion in the first million years. The rate of solid body accumulation decreases once the planet has accreted nearly all of the condensed material within its gravitational reach. The envelope accumulates gradually, with its settling rate determined by its ability to radiate away the energy of accretion. Eventually, the planet becomes sufficiently cool and massive that gas can be accreted rapidly. The substantial increase in the planet's total mass that results from this accretion of gas expands the planet's feeding zone into regions undepleted of solid planetesimals by previous accretion, causing an increase in the accumulation rate of solids. (Courtesy: Olenka Hubickyj)

for the similar masses of heavy material in the four planets. The fact that Uranus and Neptune contain less H_2 and He than Jupiter and Saturn suggests that the outer planets never quite reached runaway gas accretion conditions, possibly due to a slower accretion of planetesimals. As per equation (12.26), the rate at which accretion of solids takes place depends upon the surface density of condensates and the orbital frequency, both of which decrease with heliocentric distance.

The current composition of the atmospheres of the giant planets is largely determined by how much heavy material was mixed with the lightweight material in the planets' envelopes. Once the core mass exceeds about 0.01 $M_\oplus$, the temperature becomes high enough for water to evaporate into the protoplanet's envelope. While accretion continues, the envelope becomes more massive, and late accreting planetesimals have an increasing difficulty to penetrate through the growing envelope. These bodies sublimate in the envelopes of the giant planets, thereby enhancing the heavy element content of the mantles considerably.

During the runaway planetesimal accretion epoch, the protoplanet's mass increases rapidly (Fig. 12.15). The internal temperature and thermal pressure increase as well, preventing nebular gas from falling onto the protoplanet. When the feeding zone is depleted, the planetesimal accretion rate, and therefore the temperature and thermal pressure, decrease. This allows nebular gas to fall onto the protoplanet. It accumulates at a gradually increasing rate until the mass of gas contained in the protoplanet is comparable to the mass of solid material. The rate of gas accretion then accelerates more rapidly, and a gas runaway occurs. The gas runaway continues as long as there is gas in the region of the protoplanet's orbit. The protoplanet may cut off its own supply of gas by gravitationally clearing a gap within the disk, as the moonlet Pan does within Saturn's rings (Fig. 11.25), or it may accumulate all of the gas that remains in its region of the protoplanetary disk.

At the end of the runaway gas accretion phase, the protoplanet is roughly a few hundred times larger in radius than Jupiter is today, and it fills most of its Hill sphere. When accretion of gas ceases, the planet starts to contract. Initially, contraction takes place rapidly on a Kelvin–Helmholtz timescale, t_{KH}, which is the ratio of the planet's gravitational potential energy, E_G, to its luminosity, L:

$$t_{KH} \equiv \frac{E_G}{L} \sim \frac{GM^2}{RL}. \tag{12.29}$$

The temperatures in the envelope increase rapidly, so that the protoplanet's luminosity stays approximately constant, despite the planet's decrease in size. Vigorous convection mixes the envelope during this period, homogenizing the distribution of heavy elements. After $\sim 10^4$ yrs for Jupiter/Saturn and 2×10^5 yrs for Uranus/Neptune, contraction slowed down due to the increasing incompressibility of the fluid envelope, and the temperature and luminosity decreased with time. The slow cooling of the envelope is a major source of the excess thermal energy emitted into space by the giant planets.

12.8 Planetary Migration

Planetary orbits can *migrate* towards (or in some circumstances away from) their star as a consequence of angular momentum exchange between the protoplanetary disk and the planet. As is the case for moons near planetary rings (Section 11.4), protoplanets drift away from the disk material with which they interact. Calculations indicate that the torque exerted by the planet on the outer disk is usually stronger than that on the inner disk. Planets that are embedded within the gaseous disk migrate inwards on a timescale $\sim 10^5 (M_p/M_\oplus)^{-1}$ yr; this process is referred to as *Type 1 migration*. If Type 1 migration in protostellar disks is as efficient as these calculations suggest, protoplanets could migrate towards the stellar surface once their masses $M_p > 1\ M_\oplus$, because their growth timescale would then become longer than this migration timescale.

Orbital migration of a planet becomes unavoidable after it has acquired a sufficient mass to open up a gap in the disk. *Type 2 migration* is generally slower than Type 1 migration, and its speed does not vary with planetary mass. Observations of classical T Tauri stars indicate that gas in the inner region of protostellar disks is being continually depleted by accretion onto the young stellar objects. Without a mass supply, the surface density and the tidal angular momentum transfer rate interior to the planet's orbit decrease. In the outer disk, these quantities maintain their value as gas is prevented from viscously diffusing inwards by the protoplanet's tidal torque. The imbalance between the inner and outer disk leads to the inward orbital migration of the protoplanet. If the planet's mass is less than that of the disk, its orbital migration is coupled to the viscous evolution of the disk (Section 12.4.2), which is believed to occur on the timescale of $\sim 10^6$ yr.

These considerations have led to the speculation that tidal evolution may cause some first-born protogiant planets to migrate towards and eventually merge with their host stars. This 'infant mortality' would continue until the nebula mass is depleted to such an extent that the residual gas can no longer induce any significant evolution of the protogiant planets' orbits. According to this controversial scenario, Jupiter and Saturn may be the last survivors.

12.9 Small Bodies in Orbit About the Sun

12.9.1 Asteroids and Meteorites

Thousands of minor planets of radii > 10 km orbit between Mars and Jupiter (Chapter 9), yet the total mass of these bodies is $< 10^{-3}\ M_\oplus$. This is far less than would be expected for a planet accreting at ~ 3 AU within a smoothly varying protoplanetary disk. Why is there so little mass remaining in the asteroid region? Why is this mass spread among so many bodies? Why are the orbits of most asteroids more eccentric and inclined to the invariable plane of the Solar System than are those of the major planets? Why are the asteroids so diverse in composition, as indicated by their spectra and the wide variety of meteorites found on Earth?

Many small asteroids are differentiated (Chapter 9). Accretional heating and long-lived radionuclides could not have supplied sufficient energy to cause the melting required for differentiation. Proposed energy sources are electromagnetic induction heating and the decay of short-lived radionuclides, especially ^{26}Al. The observed segregation of asteroidal spectral types by semimajor axis places an upper limit on the amount of planetesimal mixing that could have occurred within the asteroid belt.

Proximity to Jupiter is most likely responsible for the mass depletion in the asteroid belt, as well as for the orbital properties of asteroids. Large planetary embryos scattered into the asteroid zone by Jupiter and/or direct resonant perturbations of Jupiter are capable of exciting eccentricities and inclinations of asteroid zone planetesimals. Much of the material once contained in planetesimals orbiting between Mars and Jupiter could thereby have been scattered into Jupiter-crossing orbits, from which it would have been ejected from the Solar System or accreted by Jupiter. Other planetesimals could have been ground to dust or even partially vaporized by high-velocity collisions. Alternatively/additionally, planetary embryos that formed within the present asteroid belt near resonances with Jupiter could have been resonantly pumped to high eccentricities and perturbed their nonresonant neighbors; orbital migration, as well as the dispersal of the gaseous component of the protoplanetary disk, could have enhanced these perturbations by sweeping resonance locations over a large portion of the asteroid region.

12.9.2 Comets

Current theories of Oort cloud formation imply that substantial quantities of small planetesimals, which formed between ~ 3 and 30 AU from the Sun, were ejected from the planetary region by gravitational perturbations from the giant planets. Accounting for the inefficiency in transporting bodies from the planetary region into bound Oort cloud orbits and for losses over the age of the Solar System, the mass of solid material ejected from the planetary region could have been 10–1000 $M_\oplus$. This implies that the minimum mass protoplanetary disk must be a moderate to severe underestimate of the actual mass of the disk out of which our planetary system formed. Clearing small bodies from the outer Solar System caused Jupiter to migrate inwards. In contrast, the other giant planets, which perturbed more planetesimals inwards to Jupiter-crossing orbits than directly outwards to the Oort cloud and beyond, should have migrated away from the Sun.

The Kuiper belt (of which Pluto may be considered an exceptionally large and close-in member) requires that planetesimals existed beyond the orbit of Neptune. Thus, the abrupt cutoff of observed massive planets beyond the orbit of Neptune (Pluto's mass is less than 10^{-4} times that of Neptune) cannot be explained solely by the lack of material in this region of the Solar System. Better observational

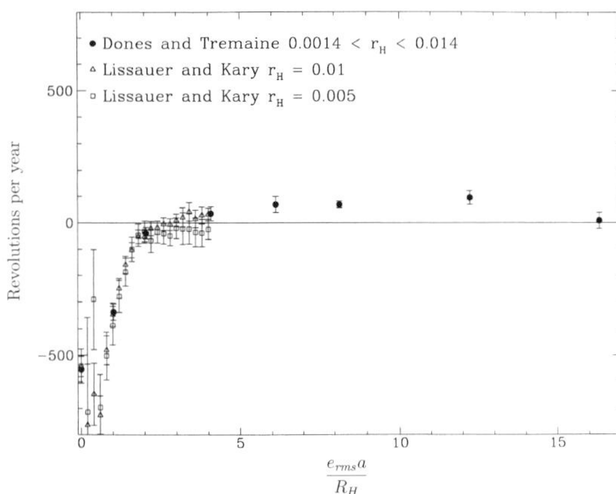

FIGURE 12.16 The number of rotations per orbit for an Earth-like planet which accreted within a uniform surface density 2-D disk of planetesimals is displayed as a function of the rms planetesimal eccentricity, e_{rms}, normalized by the size of the planet's Hill sphere. Negative values indicate rotation in the retrograde direction. The different symbols indicate different sets of numerical experiments, and the error bars result from statistical uncertainties. Note that the rapid prograde rotation observed for Earth and Mars cannot be produced by accretion of small planetesimals from a uniform disk regardless of the value of planetesimal rms eccentricity. (Courtesy Luke Dones)

estimates of the total mass, size distribution, and orbital characteristics of bodies in this region may provide helpful constraints on the dynamics of the accretionary process in the outer, loosely bound, regions of a protoplanetary disk.

12.10 Planetary Rotation

The origin of planetary rotation is one of the most fundamental questions of cosmogony. It has also proven to be one of the most difficult to answer. Planets accumulate rotational angular momentum from the relative motions of accreted material (Problem 12.18). The *obliquity* (or *axial tilt*) of a planet is the angle between its spin angular momentum and its orbital angular momentum. Planets with obliquity $< 90^\circ$ are said to have *prograde* rotation, whereas planets with obliquity $> 90^\circ$ have *retrograde* rotation. The stochastic nature of planetary accretion from planetesimals allows for a random component to the net spin angular momentum of a planet in any direction. As planets might accumulate a significant fraction of their mass and spin angular momentum from only a very few impacts, stochastic effects may be very important in determining planetary rotation. From the observed rotational properties of the planets, the size of the largest bodies to impact each planet during the accretionary epoch has been estimated to be 1–10% of the planet's final mass.

Very little net spin angular momentum is accumulated by a planet which accretes while on a circular orbit within a uniform surface density disk of small planetesimals (Fig. 12.16). A planet that partially clears a gap in the disk, and thus accretes a larger fraction of material from the edges of its accretion zone, may accumulate sufficient prograde angular momentum to explain the planetary rotation rates observed in our Solar System. Rapid prograde rotation can also result if planetesimal orbits decay slowly towards the protoplanet as a result of gas drag. Alternatively, stochastic impacts of large bodies may be the primary source of the rotational angular momentum of the terrestrial planets, with the observed preference of low obliquities being a chance occurrence.

Jupiter and Saturn are predominantly composed of hydrogen and helium, which they must have accreted hydrodynamically, in flows quite different from those which govern the dynamics of planetesimals. Such flows lead to prograde rotation. The nonzero obliquities of the giant planets were probably produced by giant impacts.

12.11 Origin of Planetary Satellites

The moons and rings of the giant planets are analogous to miniature planetary systems in many respects. In each case, multiple secondaries orbit their primary, with most traveling on nearly circular, coplanar prograde orbits having a certain regularity to their spacings. Satellites orbiting closest to giant planets (near the Roche limit) are generally small. Planetary rings dominate where tidal forces from the planet are sufficient to tear apart a moon held together solely by its own gravity. The outer regions of the satellite systems of all four giant planets contain small bodies on highly eccentric and inclined orbits. The diversity of planetary satellites suggests that they are formed by more than a single mechanism.

Planetary satellite systems consist of *regular* and *irregular* satellites. Regular satellites move on low eccentricity prograde orbits near the equatorial plane of their planet. They orbit close to the planet, well within the bounds of the planet's Hill sphere. These properties imply that regular satellites formed within a disk orbiting in the planet's equatorial plane. Irregular satellites generally travel on high eccentricity, high inclination orbits lying well exterior to a planet's regular satellite system; most irregular satellites are quite small. They are believed to be captured from heliocentric orbits. One possibility is a slow-down of nearby bodies due to gas drag in the protoplanet's envelope. Most of such planetesimals would then have ended up in the planet itself, but some survived and became captured satellites. Satellites may also have been captured via

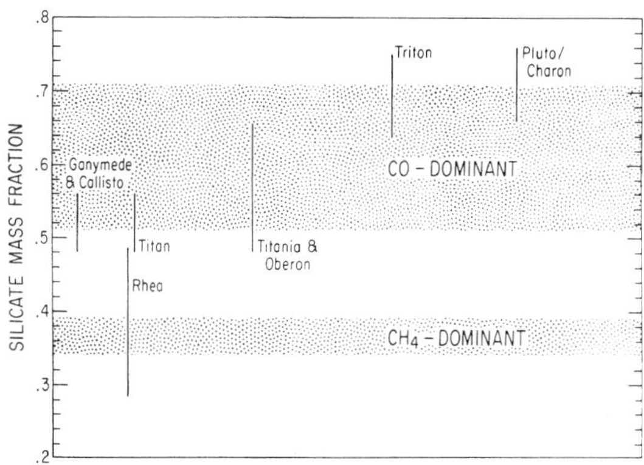

FIGURE 12.17 The mass ratio of rock (silicates, metals and other refractory compounds) to rock + ice in outer Solar System objects. The ranges for individual objects represent uncertainties in densities and assumptions regarding interior models. The shaded areas represent expected compositions for a solar composition mixture at outer Solar System temperatures, assuming that either carbon monoxide or methane is the dominant carbon-bearing species. (Lunine and Tittemore 1993)

collisions with regular satellites (possibly Triton, whose large mass would have allowed subsequent tidal circulation of its orbit, cf. Section 2.6).

The regular satellites of the giant planets likely formed by a solid body accretion process in a gas/dust disk surrounding the planet. Such disks may consist of material from the outer portions of the protoplanet's envelope, or directly captured from the protoplanetary disk. In this scenario one can explain the compositional differences between the regular and irregular satellites around Jupiter and Saturn. When youthful Jupiter's high luminosity is included, the model naturally accounts for the decrease in density of the Galilean satellites with increasing distance from Jupiter. The densities of the moons around Saturn and Uranus do not vary in such a systematic manner with distance from the planet; however, these lower-mass planets were never as luminous as was young Jupiter. Since tidal forces prevented material from accreting within a planet's Roche limit, rings were formed around the giant planets. Note, however, that most if not all of the ring systems that we see at present are not primordial (cf. Section 11.7).

The rock/(rock + ice) mass fraction of the icy satellites and Pluto yield clues to the place of formation of these bodies. Figure 12.17 shows the estimated rock fraction of various satellites and Pluto, together with the range in expected mass fractions for the protoplanetary disk, where CO was much more abundant than CH_4 (Section 12.4.3), and the circumplanetary disks, where CH_4 may have been dominant. The range in mass fractions is caused by the uncertainty in the solar C/O ratio, which lies between 0.43 and 0.60. The larger planetary satellites (Ganymede, Callisto, Titan) contain more rock than expected for a body accreted in a circumplanetary disk; however, vaporization and escape of volatile material during accretion could have depleted the ice. The uranian satellites also contain a substantial rock component. Io and Europa (not shown) are composed primarily of rock. Since these moons formed very close to Jupiter at high temperatures, there was less water-ice available for accretion. Triton and Pluto/Charon have very similar high rock mass fractions, consistent with that expected in a CO-rich solar nebula. One also would expect a large loss of volatile material from these small bodies if they were hit by sizable impactors subsequent to being differentiated.

Terrestrial planets presumably never possessed gas-rich circumplanetary disks; thus other explanations are required for the origins of the moons of Mars, Earth and Pluto. Mars's moons Phobos and Deimos are similar in composition to C-class asteroids (Section 9.5.2); these satellites are most likely (disruptively?) captured planetesimals.

The Earth's Moon is a very peculiar object whose origin has been hotly debated over the past century. The Moon/Earth mass ratio greatly exceeds that of any other satellite/planet aside from tiny Charon/Pluto, raising the question of how this much material was placed into orbit about Earth. The combination of low mean density and lack of volatiles implies that the Moon is not simply an amalgam of that solar composition material which is able to condense above a certain temperature; rather, the Moon's bulk composition appears to resemble Earth's mantle, albeit depleted in volatiles. The bulk composition of the lunar crust and mantle could be understood if the Moon equilibrated with a large iron core, but the Moon has no such core. Capture, coaccretion and fission models of lunar origin have all been studied in great detail, but none satisfies both the dynamical and chemical constraints in a straightforward manner. The favored hypothesis is the *giant impact model*, in which a collision between the Earth and a Mars-sized or larger planetary embryo ejects a lunar mass (or more) of material into Earth's orbit (Fig. 12.18). Assuming both bodies were differentiated prior to the impact, this model would explain the apparent similarities between the lunar composition and that of the Earth's mantle and at the same time the lack of volatile material on the Moon. Volatile material would have been vaporized completely by the impact, and most of the resulting gas would have escaped into interplanetary space. A range of impactor and impact parameters can place roughly one lunar mass of material primarily from the mantles of the im-

(a) Time = 0 seconds

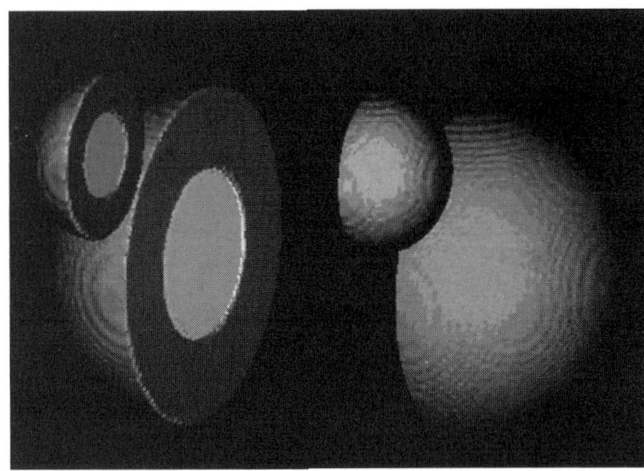

(b) Time = 804 seconds

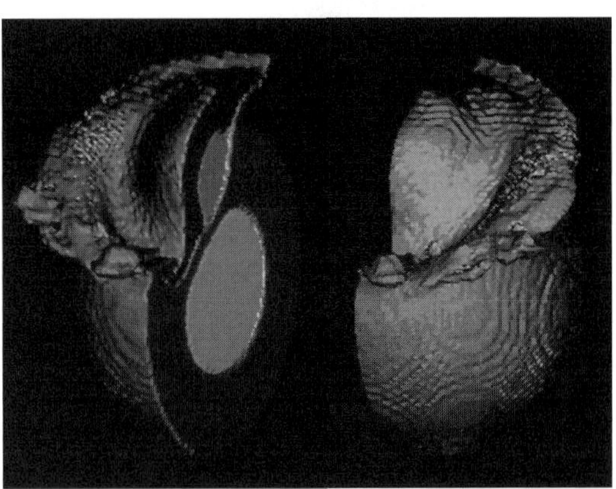

(c) Time = 1600 seconds

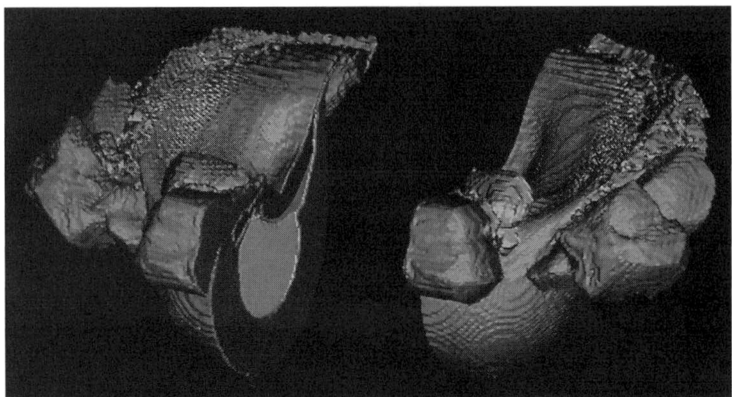

FIGURE 12.18 COLOR PLATE These computer-generated images illustrate the first 30 minutes after a Mars-size protoplanet and the proto-Earth collide with a velocity upon contact of 8 km s^{-1}. This contact velocity corresponds to a relative $v_\infty = 0$. The simulations were performed using the CTH three-dimensional hydrocode at Sandia National Laboratories. The metallic cores of the projectile and target are colored differently from their dunite mantles. (a) External and central cut views of the two bodies at the time of initial contact. (b) 804 seconds after initial contact. (c) 1600 seconds after initial contact. The vapor plume of mixed projectile and target mantle material is well developed. This plume eventually condenses into dust, some of which remains in orbit, available to accrete into the Moon. (Courtesy Jay Melosh)

pactor and/or Earth into terrestrial orbit. Once this material is cool enough to form condensed bodies, it can quickly accumulate into a single large moon (Fig. 12.19). A giant impact origin of the Pluto/Charon system has also been proposed, but no detailed model of such a collision has yet been published.

The wide variety of properties exhibited by the satellite systems of the four giant planets in our Solar System suggests that stochastic processes may be even more important for satellite formation than current models suggest they are in planetary growth. A possible explanation for this difference is that satellite systems are subjected to a very heavy bombardment of planetesimals on heliocentric orbits, which may fragment moons and also produce them. Deterministic models of satellite formation must thus be interpreted with caution.

12.12 Confronting Theory with Observations

The current theory of planetary growth from planetesimal accretion within a circumstellar disk provides excellent

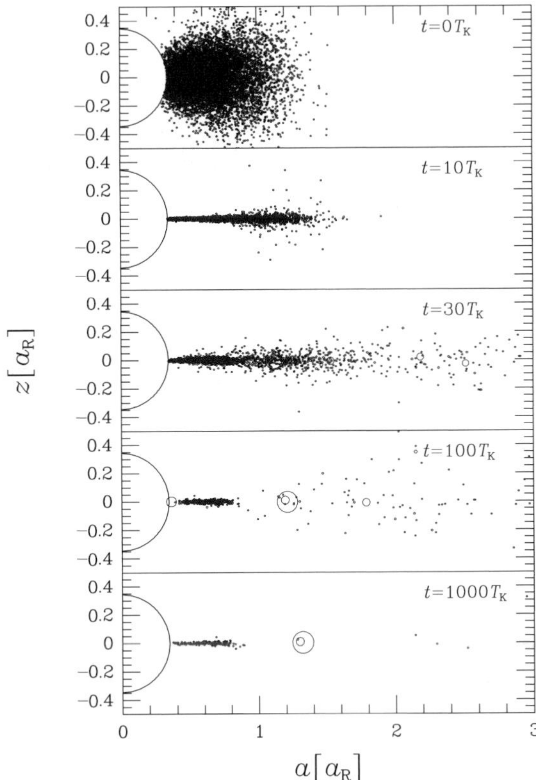

FIGURE 12.19 Snapshots of the protolunar disk in the r–z plane at $t = 0, 10, 30, 100, 1000\ T_K$, where T_K is the keplerian orbital period at the Roche limit. The initial number of disk particles is 10 000, and the disk mass is four times the present lunar mass. The semicircle centered at the coordinate origin stands for the Earth. Circles represent disk particles and their sizes are proportional to the physical sizes of the disk particles. The horizontal scale shows the semimajor axis of disc particles in units of the Roche limit radius, a_R (cf. eq. 11.8). (Kokubo *et al.* 2000)

explanations of the causes of many of the observed Solar System properties, but less complete or less satisfactory explanations for several others.

12.12.1 Dynamical State of the Solar System

Dynamical models of planetary accretion within a flattened disk of planetesimals produce moderately low eccentricity, almost coplanar orbits of planets, except at the outer fringes of the Solar System. The ultimate sizes and spacings of solid planets are determined by their ability to gravitationally perturb one another into crossing orbits. Such perturbations are often caused by weak resonant forcing and occur on timescales much longer than the bulk of planetesimal interactions discussed in Section 12.4.3. A more massive protoplanetary disk probably produces larger, but fewer planets. Stochastic processes are important in planetary accretion, so nearly identical initial conditions could lead to quite different outcomes; for example, the fact that there are four terrestrial planets in our Solar System as opposed to three or five is probably just the luck of the draw.

Jupiter probably played a major role in preventing the formation of a planet in the asteroid zone. Jovian resonances could have directly stirred planetesimals in the asteroid zone, or Jupiter could have scattered large failed planetary embryos inwards from 5 AU, yielding the same effect. The resulting stirring could have prevented further planetary growth and/or ejected an already formed planet from the Solar System.

The giant planets probably ejected a substantial mass of solid bodies from the planetary region. The majority of these planetesimals escaped from the Solar System, but $\gtrsim 10\%$ ended up in the Oort cloud. Perturbations from the galactic tide, passing stars, and giant molecular clouds have randomized the orbits of Oort cloud comets over the past 4.5×10^9 years. Aside from a small flattening caused by the tidal potential of the galaxy, the Oort cloud is nearly spherical with prograde as well as retrograde objects. The Kuiper belt likely formed *in situ*, from planetesimals orbiting exterior to Neptune's orbit. The dynamical structure of the Kuiper belt, especially the large number of objects on eccentric orbits in the 2:3 resonance with Neptune, suggests that Neptune migrated outwards by several AU during the final stages of planetary formation.

Stochastic impacts of large bodies provide sufficient angular momentum to produce the obliquities of the planets. Systematic prograde rotation of Jupiter and Saturn is a result of gas accretion, while the excess of prograde rotation among the other planets may have been produced in a systematic way via expansion of their accretion zones, or may just be a chance result. The gross features of the regular satellite systems of the giant planets can be understood if these planets had disks orbiting them; various models for the formation of such disks exist. The angular momentum distribution of the Solar System resulted from outward transport of angular momentum (via poorly characterized viscous, gravitational and/or magnetic torques) within the protosun/protoplanetary disk and a subsequent removal of most of the Sun's spin angular momentum by the solar wind.

The high cratering rate in the early Solar System and the lower rate at the current epoch are a consequence of the sweep-up of debris from planetary formation. The early high bombardment rate caused the planets, asteroids and satellites to be covered with craters. Huge impacts may have changed the spin orientation of Uranus, led to the formation of the Moon, and stripped off the outer layers of Mercury. Some large planetesimals may have been captured into planetocentric orbits, like Triton about Neptune.

12.12.2 Composition of Planetary Bodies

The masses and bulk compositions of the planets can be understood in a gross sense as resulting from planetary growth within a disk whose temperature and surface density decreased with distance from the growing Sun. The terrestrial planets are rocky because the more volatile elements could not condense (or survive in solid form) so close to the Sun, whereas comets and the moons of the giant planets retain ices because they grew in a colder environment. The condensation of water-ice beyond ~ 4 AU provided the outer planets with enough mass to gravitationally trap substantial amounts of H_2 and He from the solar nebula; longer accretion times at greater heliocentric distances may account for the decrease in the gas fractions of the giant planets with increasing semimajor axis.

Migration of some planetesimals over significant distances within the protoplanetary disk probably occurred, leading to a radial mixing of material which condensed in different regions of the solar nebula and/or had survived from the pre-solar era. This mixing helps explain the gross properties of meteorites; however, the causes of many detailed characteristics (especially the formation of chondrules and remnant magnetism) are poorly understood. The planetesimal hypothesis explains the similarity in ages among primitive meteorites and the fact that all Solar System rocks, whose components are believed to have at one point passed through a stage similar to primitive meteorites, are the same age or younger.

All planets were hot during the accretionary epoch. The present terrestrial planets show evidence of this early hot era in the form of extensive tectonic and/or volcanic activity. Jupiter, Saturn and Neptune have excess thermal emissions resulting from accretional and differentiation heating. Outgassing of the hot newly formed planets, combined with an influx of cometary volatiles, led to the formation of atmospheres on the terrestrial planets.

12.12.3 Conclusions

The planetesimal hypothesis provides a viable theory of the growth of the terrestrial planets, the cores of the giant planets and the smaller bodies present in the Solar System. The formation of solid bodies of planetary size should be a common event, at least around young stars which do not have binary companions orbiting at planetary distances. Planets could form by similar mechanisms within circumpulsar disks if such disks have adequate dimensions and masses. The formation of giant planets, which contain large quantities of H_2 and He, requires rapid growth of planetary cores, so that gravitational trapping of gas can occur prior to the dispersal of the gas from the protoplanetary region. According to the scenario outlined in this chapter, the largest bodies in any given zone are the most efficient accreters, in the sense that they double in mass the fastest. Such runaway accretion of a few large solid protoplanets can lead to giant planet core formation in $\sim 10^6$ years, provided disk masses are a few times as large as those given by 'minimum mass' models of the solar nebula. Thus, it appears possible that giant planets may also be common, although this conclusion must be regarded as somewhat more tentative.

Further Reading

The series of *Protostars and Planets* books contains review papers on formation processes; the latest volume in the series is:

Mannings, V., A.P. Boss, and S.S. Russel, Eds., 2000. *Protostars and Planets IV*. University of Arizona Press, Tucson, 1422pp.

Several papers related to different aspects of the formation of our Solar System are given by:

Lin, D.N.C., 1986. The nebular origin of the Solar System. In *The Solar System: Observations and Interpretations*. Ed. M.G. Kivelson. Rubey Vol. IV, pp. 28–87. Prentice Hall, Englewood Cliffs, NJ.

Lissauer, J.J., 1993. Planet formation. *Annu. Rev. Astron. Astrophys.* **31**, 129–174.

Lissauer, J.J., 1995. Urey Prize lecture: On the diversity of plausible planetary systems. *Icarus* **114**, 217–236.

Lissauer, J.J., J.B. Pollack, G.W. Wetherill, and D.J. Stevenson, 1995. Formation of the Neptune system. In *Neptune and Triton*. Ed. D.P. Cruikshank. University of Arizona Press, Tucson, pp. 37–108.

A good paper on the formation of terrestrial atmospheres is:

Ahrens, T.J., J.D. O'Keefe, and M.A. Lange, 1989. Formation of atmospheres during accretion of the terrestrial planets. In *Origin and Evolution of Planetary and Satellite Atmospheres*. Eds. S.K. Atreya, J.B. Pollack, and M.S. Matthews. University of Arizona Press, Tucson, pp. 328–385.

Equilibrium chemistry in the solar nebula is described in:

Prinn, R.G., and B. Fegley, Jr., 1989. Solar nebula chemistry: Origin of planetary, satellite and cometary volatiles. In *Origin and Evolution of Planetary and Satellite Atmospheres*. Eds. S.K. Atreya, J.B. Pollack, and M.S. Matthews. University of Arizona Press, Tucson, pp. 78–136.

A rather detailed chemical model of the evolution of our protoplanetary disk is written by:

Aikawa, Y., T. Umbebayashi, T. Nakano, and S.M. Miyama, 1999. Evolution of molecular abundances in protoplanetary disks with accretion flow. *Astrophys. J.* **519**, 705–725.

A (mostly still current) thorough review of nucleosynthesis within stars is given in:

Clayton, D.D., 1983. *Principles of Stellar Evolution and Nucleosynthesis*. University of Chicago Press.

An excellent popular account of big bang nucleosynthesis is provided by:

Weinberg, S., 1988. *The First Three Minutes*. Basic Books, New York.

Problems

12.1.E (a) Calculate the gravitational potential energy of a uniform spherical cloud of density ρ and radius R.

(b) Determine the Jeans mass, M_J, of an interstellar cloud of solar composition with density ρ and temperature T. (Hint: Set the gravitational potential energy equal to negative twice the cloud's kinetic energy and solve for the radius of the cloud.)

(c) Show that if the cloud collapses isothermally, it becomes more *unstable* as it shrinks.

(d) Show that if the cloud retains the gravitational energy of its collapse as heat, it becomes more *stable* as it shrinks.

12.2.E Derive the formula for the free-fall gravitational collapse time of a uniform spherical cloud of density ρ (eq. 12.8). (Hint: The trajectory of a gas parcel initially at rest at a distance r from the center of the cloud can be approximated as a very eccentric ellipse with semimajor axis $r/2$.)

12.3.E Consider an H_2 molecule which falls from ∞ to a circular orbit at 1 AU from a 1 $M_\odot$ star.

(a) Calculate the circular velocity at 1 AU, and determine the total mechanical (kinetic + potential) energy of a molecule on a circular orbit at 1 AU. Note that the total energy of the molecule at rest at infinity is zero.

(b) Calculate the temperature increase of the hydrogen gas assuming it has not suffered radiative losses.

12.4.I Show that the density of gas in a thin isothermal circumstellar disk varies in the direction perpendicular to the midplane of the disk as:

$$\rho = \rho_0 e^{-z^2/H_z^2}, \tag{12.30}$$

where the Gaussian scale height is given by equation (12.11). (Hint: Consider a balance between pressure and the component of the star's gravity perpendicular to the midplane of the disk.)

12.5.E (a) Compute the molecular viscosity in a protoplanetary disk at a radius of 10^{14} cm, where the mean free path is $\ell_{fp} = 10$ cm and the sound speed $c_s = 1$ km s^{-1}.

(b) What is the viscous accretion timescale of such a disk?

12.6.E Compute the value of α_v necessary for a turbulent disk with the parameters listed in Problem 12.5 to have a viscous evolution time of 10^6 years.

12.7.I (a) Diffusion inwards within a viscous circumstellar disk leads to accretion. The timescale for this accretion is equivalent to the diffusion timescale (cf. eq. 11.13) between the radius in question and the radius of the star. Derive a formula for the viscous accretion timescale (in years) of a protoplanetary disk with scale height $H_z = 0.1\, r$ and sound speed $c_s = 10^5\, r_{AU}^{-1/2}$ cm s^{-1} as a function of the viscosity parameter α_v and radius, r_{AU}.

(b) Evaluate your formula at 1 AU and 5 AU for $\alpha_v = 0.01$.

12.8.I As discussed in the text, the ice/rock ratio in a protoplanetary disk depends both on elemental composition and chemical state. In this problem, you will calculate the ice/rock ratio under a variety of assumptions.

(a) Assume anhydrous rock to consist of SiO, MgO, FeO and FeS, using up the entire inventory of all of these elements other than oxygen, which is the most abundant. Referring to the elemental abundances listed in Table 8.1, calculate the amount of oxygen available to combine with lighter elements, forming compounds such as CO and H_2O. Express your answer in atoms of available oxygen per 10^6 silicon atoms.

(b) Calculate the mass of rock in amu/silicon atom. Augment your result by 10% to approximately account for the less abundant rock-forming elements not included in your calculation.

(c) Assuming that all of the carbon is in CO, calculate the mass of H_2O ice (in amu/silicon atom), the H_2O ice/rock ratio, the $H_2O + CO$ ice/rock ratio and the $H_2O + CO + N_2$ ice/rock ratio.

(d) Assuming that all of the carbon is in CH_4, calculate the mass of H_2O ice, the H_2O ice/rock ratio, the $H_2O + CH_4$ ice/rock ratio and the $H_2O + CH_4 + NH_3$ ice/rock ratio.

(e) Repeat the above calculations assuming that the abundance of O is 10% greater than listed in Table 8.1. (This is within the uncertainty to which Solar System abundances are known, and probably significantly less than the differences between various protoplanetary disks.)

12.9.E (a) Calculate the amount of gas with which a particle R cm in radius orbiting at 1 AU from a 1 $M_\odot$ star collides during one year. You may assume that the density of the protoplanetary disk is 10^{-9} g cm^{-3} and $\eta = 5 \times 10^{-3}$.

(b) Assuming a particle density of 3 g cm^{-3}, calculate the radius of a particle which collides with its own mass of gas during one orbit.

12.10.E Estimate the masses and radii of planetesimals formed via gravitational instabilities in a quiescent (nonturbulent) protoplanetary disk orbiting a 1 $M_\odot$ star. Assume that the surface mass density of solid material in the disk varies as $\sigma_\rho = 10\, r_{AU}^{-1}$ g cm^{-2}. Perform your calculations at

(a) 1 AU.

(b) 5 AU.

(c) Repeat your calculation at 5 AU with a surface density twice as large (to account for the condensation of water-ice).

12.11.E Derive the two-body gravitational accretion cross-section of a planet of radius R and mass M for small planetesimals whose velocity at ∞ relative to the planet is v. (Hint: First determine the maximum unperturbed impact parameter of a planetesimal which collides with a planet. Note that the periapsis of this planetesimal's orbit is at a distance R from the planet's center. Use conservation of angular momentum and energy.)

12.12.E (a) Calculate the rate of growth, dR/dt, of a protoplanet of radius $R = 4000$ km and mass $M = 10^{27}$ g, in a planetesimal disk of surface density $\sigma_\rho = 10$ g cm^{-2}, temperature $T = 300$ K and velocity dispersion $v = 1$ km s^{-1}, at a distance of 2 AU from a star of mass 3 $M_\odot$. You may use the two-body approximation for planetesimal/protoplanet encounters.

(b) What will halt (or at least severely slow down) the accretion of such a planet? What will its mass be at this point? (Hint: See Problem 12.16.)

12.13.E Calculate the growth time for Neptune assuming ordered growth (i.e., not runaway accretion; use $\mathcal{F}_g = 10$) in a minimum mass nebula. (Hint: Determine the surface density by spreading Neptune's mass over an annulus from 25 to 35 AU.)

12.14.I Two asteroids, each of mass 10^{21} g, collide and accrete (i.e., the collision may be assumed to be completely inelastic). Their initial orbits were $a_1 = 2.75$ AU, $e_1 = 0.1$, $i_1 = 10°$; $a_2 = 3.0$ AU, $e_2 = 0$, $i_2 = 0°$.

(a) Calculate the orbital elements of the single body after accretion and the energy dissipated by the collision. (Hint: Convert to cartesian coordinates and use conservation of momentum.)

(b) In reality, is such a collision likely to result in accretion or disruption? Why?

12.15.I Consider a few relatively large planetesimals in a swarm of much smaller bodies. Assume that the densities of all bodies are the same and that the velocity dispersion is comparable to the escape speed of the *small* bodies.

(a) Show that the cross-sections (and the accretion rates) of the large planetesimals are proportional to the fourth power of their radii.

(b) Use this result to demonstrate that the largest planetesimal doubles in mass the fastest and thus 'runs away' from the rest of the distribution of bodies.

12.16.E Equation (12.28) gives the isolation mass for runaway growth of a planet around a 1 $M_\odot$ star as $M_i = 10^{24} \left(r_{AU}^2 \sigma_\rho\right)^{3/2}$ grams. Generalize this formula to stars of arbitrary mass.

12.17.I Compare a hypothetical planetary system which formed in a disk with the same size as the solar nebula but only half the surface mass density to our own Solar System. Assume that the star's mass is 1 $M_\odot$ and that it does not have any stellar companions. Concentrate on the final number, sizes and spacings of the planets. Explain your reasoning. Quote formulas and be quantitative where possible.

12.18.I A planet of mass M and radius R initially spins in the prograde direction with zero obliquity and rotation period P_{rot}. It is impacted nearly tangentially at its north pole by a body of mass m, whose velocity prior to encounter was small compared to the escape speed from the planet's surface.

(a) Derive an expression for the planet's spin period and obliquity after the impact. You may assume that the projectile was entirely absorbed.

(b) Numerically evaluate your result for $M = M_\oplus$, $R = R_\oplus$, $P_{rot} = 10^5$ seconds, $m = 0.02\ M_\oplus$. (Of course, a truly tangential impactor is likely to 'skip off' rather than being absorbed, but even for a trajectory only $\sim 10^\circ$ from the horizontal, most ejecta can be captured at the velocities considered here. The case of a polar impactor is also a singular extremum, but both of these effects together only add a factor of a few to the angular momentum provided by a given mass impacting with random geometry, and they make the algebra much simpler.)

12.19.E Consider a simple model for the accretion of the Earth. Suppose the radius of proto-Earth increases linearly with time from some starting point until accretion ends t_{acc} years later, i.e. $R(t) = R_\oplus(t/t_{acc})$. Ignore the insulating effects of large impacts (which bury hot ejecta) and any possible atmosphere (which can prevent the surface from radiating freely to space). For further simplicity neglect compressibility and heat conduction (this is *very* simple). Assume the surrounding nebula has a constant temperature $T_n = 300$ K, and that the solid particles have an average density $\rho = 4.5$ g cm^{-3} and heat capacity $C_P = 10^7$ erg g^{-1} K^{-1}. Then, neglecting internal heat sources and setting the emissivity $\epsilon = 1$, we have

$$\frac{GM(R)\rho}{R}\frac{dR}{dt} = \sigma(T^4 - T_n^4) + \rho C_P (T - T_n)\frac{dR}{dt}. \quad (12.31)$$

Equation (12.31) quantifies the energy balance (per unit area) at the surface of the growing Earth.

(a) Describe physically each of the three terms in equation (12.31).

(b) Find the approximate temperature of the planet as a function of radius assuming $dR/dt =$ constant, if $t_{acc} = 10^8$ yr and if $t_{acc} = 10^6$ yr.

(c) Find t_{acc} so that $T(R_\oplus) = 2000$ K at the end of accretion.

12.20.I A serious omission from equation (12.31) is the lack of a term involving heat transport within the planet. Convection is suppressed, as the outer layers are warmer than the inner ones in our simple model. (If we consider the possibility of variable accretion rates, such as the impact of a single large body allowing heat to be buried below the surface, then temperature may decrease sufficiently rapidly with radius to allow convection at some locations. However, rapid convection can only occur in a fluid, and if a planet melts, the thermal effects of differentiation must also be included. Solid-state convection can occur, albeit at a slow rate, in material slightly below its melting point, bringing heat out sufficiently rapidly to prevent melting, especially in icy satellites. In any case, we shall ignore convection here.) Radiation within a solid planet is negligible compared to conduction, as the mean free path of a photon is extremely small. Conduction can be included by the addition of a conductivity term to the right hand side of equation (12.31). However, this requires knowing the temperature as a function of both position and time, because conduction changes temperatures below the surface. A simpler approach is to examine the infinite conductivity limit instead. In this limit, the body is isothermal, so, as before, temperature is only a function of time. The last term in equation (12.31) is thus replaced by:

$$\frac{1}{4\pi R^2}\frac{d}{dt}\left(\frac{4\pi}{3}\rho R^3 C_P (T - T_n)\right). \quad (12.32)$$

(a) By differentiating this term, derive an equation identical to equation (12.31) except for one extra term on the right hand side.

(b) Describe, qualitatively, the effects of this new term on the planet's surface temperature.

(c) Find, approximately, the planet's temperature at the end of accretion if $t_{acc} = 10^8$ yr and if $t_{acc} = 10^6$ yr, assuming infinite conductivity.

12.21.**I** Find the initial temperature profile of the Earth, assuming that it was homogeneous and it accreted so fast (or that large impacts buried the heat so deep) that radiation losses were negligible. Do this both for the zero and infinite conductivity cases.

12.22.**E** Calculate the rise in temperature if the Earth differentiated from an initially homogeneous density distribution to a configuration in which one third of the planet's mass was contained in a core which had a density twice that of the surrounding mantle. You may assume infinite conductivity.

12.23.**I** Repeat the four previous problems for asteroids of radius 50 km and 500 km.

12.24.**I** Most of the heating by radioactive decay in our planetary system at the present epoch is due to decay of four isotopes, one of potassium, one of thorium and two of uranium. Chondritic elemental abundances are listed in Table 8.1. Isotopic fractions and decay properties are given in the *CRC Handbook*.

(a) What are these four major energy-producing isotopes? What energy is released per atom decayed? What energy is released per gram decayed? What is the *rate* of energy produced by one gram of the pure isotope? What is the rate released per gram of the element in its naturally occurring isotopic ratio? Note: Some of these isotopes decay into other isotopes with short half-lives (e.g., radon). The decay chain must be followed until a stable (or very long-lived) isotope is reached, adding the energy contribution of each decay along the path.

(b) What is the rate of heat production per gram of chondritic meteorite (or, equivalently, per gram of the Earth as a whole, neglecting the fact that the volatile element potassium is less abundant in the Earth than in CI chondrules) from each of these sources?

(c) What was the heat production rate from each of these sources 4.5×10^9 years ago?

(d) There are very many radioactive isotopes known. What characteristics do these isotopes share which make them by far the most important? (Hint: There are two very important characteristics shared by all four, and one other by three of the four.)

12.25.**E** (a) How long would it take radioactive decay at the early Solar System rate calculated in the previous problem to generate enough heat to melt a rock of chondritic composition, assuming an initial temperature of 300 K and no loss of energy from the system?

(b) How long would it take radioactive decay to generate as much energy as the gravitational potential energy obtained from accretion for an asteroid of radius 500 km? How long for an asteroid 50 km in radius?

12.26.**I** ^{26}Al is a radioactive isotope formed by a variety of nucleosynthetic processes, which has been observed in interstellar space. It has a half-life of 7.2×10^5 years, and thus none remains on Earth from the origin of the Solar System.

(a) How many *grams* of pure ^{26}Al would have had to have been present 4.5×10^9 years ago in order for one *atom* to likely be present today?

Evidence for extinct ^{26}Al exists in some primitive meteorites. As isotopes are chemically (almost) indistinguishable, isotopic ratios are nearly uniform throughout the Solar System. The major exceptions to this rule are deuterium, which, being twice as massive as hydrogen, tends to preferentially occupy positions within heavier molecules, and atmospheric gases, for which the lighter isotopes escape more easily. Other isotopic variations are caused by radioactive decay. The decay product of ^{26}Al is ^{26}Mg. Excess amounts of ^{26}Mg (compared to other Mg isotopes) have been found in some aluminum-rich meteoritic inclusions; these detections imply that the inclusions condensed with 'live' ^{26}Al.

(b) What ratio of ^{26}Al/^{27}Al would have been required for the heat produced by ^{26}Al decay to have equaled that produced by the four radioactive isotopes mentioned in Problem 12.24 4.5×10^9 years ago?

(c) What ratio of ^{26}Al/^{27}Al would have been required for the heat produced by ^{26}Al decay to have been enough to melt chondritic rock (assuming no loss of heat)? How long would it have taken for 90% of this energy to have been released?

12.27.**I** (a) Estimate the radius of the smallest icy satellite which could have melted as a result of accretional heating. For your calculation, you may assume (1) the moon is pure water-ice, (2) the specific heat of ice is 2×10^7 erg g^{-1} K^{-1} and the latent heat is 10^9 erg g^{-1} and (3) accretion was rapid (or large impacts buried the heat released) and random velocities of accreting bodies were small.

(b) How well do your results agree with the observed 'boundary' between spherical and nonspherical moons? What other heat sources may have been important? How could the energy requirement be substantially reduced?

12.28.**E** Describe and sketch the temperature profiles you would expect after the accretion of Mars due to:

(a) Accretion heating only.
(b) Radioactive heating only.
(c) Suppose Mars had accreted very slowly ($> 10^8$ years) from tiny planetesimals devoid of radioactive material. Would you expect Mars to be differentiated? Why or why not?

12.29.**I** From the results of the calculations described in the previous ten problems, what can you conclude regarding heating, melting and differentiation of asteroids and terrestrial planets?

13 Extrasolar Planets

Since one of the most wondrous and noble questions in Nature is whether there is one world or many, a question that the human mind desires to understand, it seems desirable for us to inquire about it.

Albertus Magnus, thirteenth century

The first twelve chapters of this book were concerned with general aspects of planetary physics/chemistry/atmospheres/geology and with specific objects within our Solar System. We now turn our attention to far more distant planets. What are the characteristics of planetary systems around stars other than the Sun? How many planets are typical? What are their masses and compositions? What are the orbital parameters of individual planets, and how are the paths of planets orbiting the same star related to one another? These questions are difficult to answer because planets are so faint that none have yet been directly observed over interstellar distances.

Many claims of the discovery of the first extrasolar planet made newspaper headlines from the 1940s to the 1990s, only to be quietly retracted when more data became available. Even the definition of what is a planet is controversial, since most of the objects claimed are more massive than Jupiter but not massive enough to sustain quasi-equilibrium fusion of protons in their core and become a star ($\gtrsim 0.075$ $M_\odot$). We adopt the definition that *planets* are not sufficiently massive for fusion to ever consume a majority of their deuterium ($\lesssim 0.013$ $M_\odot$), with *brown dwarfs* being intermediate objects that are large enough for deuterium fusion but not massive enough to sustain fusion of ordinary hydrogen. Gravitational contraction is a major source of the energy radiated by giant planets and brown dwarfs. These objects shrink and (after some initial warming) cool as they age (Fig. 13.1), so there is not a unique relationship between luminosity and mass.

Radial velocity surveys in the 1990s first demonstrated that planets with masses and orbits quite different from those within our own Solar System are present around main sequence stars in our region of the galaxy. All of the extrasolar planets thus far discovered orbiting main sequence stars induce variations in stellar reflex motion larger than would a planetary system like our own, and surveys accomplished to date are strongly biased against detecting low mass and long period planets. Our own Solar System may represent a biased sample of a different kind, because it contains a planet with conditions suitable for life to evolve to the point of being able to ask questions about other planetary systems.

13.1 Physics and Sizes of Giant Planets, Brown Dwarfs and Low-Mass Stars

Nuclear reactions maintain the temperature in the cores of low-mass stars close to $t_{nucl} \approx 3 \times 10^6$ K (Section 12.2.2). The virial theorem (eq. 2.61) can be used to show that the radii of such stars must be roughly proportional to mass: In equilibrium, the thermal energy and the gravitational potential energy are in balance:

$$\frac{GM_\star^2}{R_\star} \sim \frac{M_\star}{m_{amu} k T_{nucl}}. \tag{13.1}$$

Therefore

$$R_\star \propto M_\star, \tag{13.2a}$$

and the star's density

$$\rho_\star \propto M_\star^{-2}. \tag{13.2b}$$

At low densities, the hydrostatic structure of a star is determined primarily by a balance between gravity and thermal pressure. At sufficiently high densities, another source of pressure becomes significant. Electrons, because they have half-integer spins, must obey the *Pauli exclusion principle* and are accordingly forbidden from occupying identical quantum states. The electrons thus successively fill up the lowest available energy states. Those electrons that are forced into higher energy levels contribute to *degeneracy pressure*. The degeneracy pressure scales as $\rho^{5/3}$ and is important when it is comparable in magnitude to or larger than the ideal gas pressure (which scales as ρT, cf. eq. 3.66). Near T_{nucl}, the degeneracy pressure dominates when densities exceed a few hundred grams per cubic centimeter.

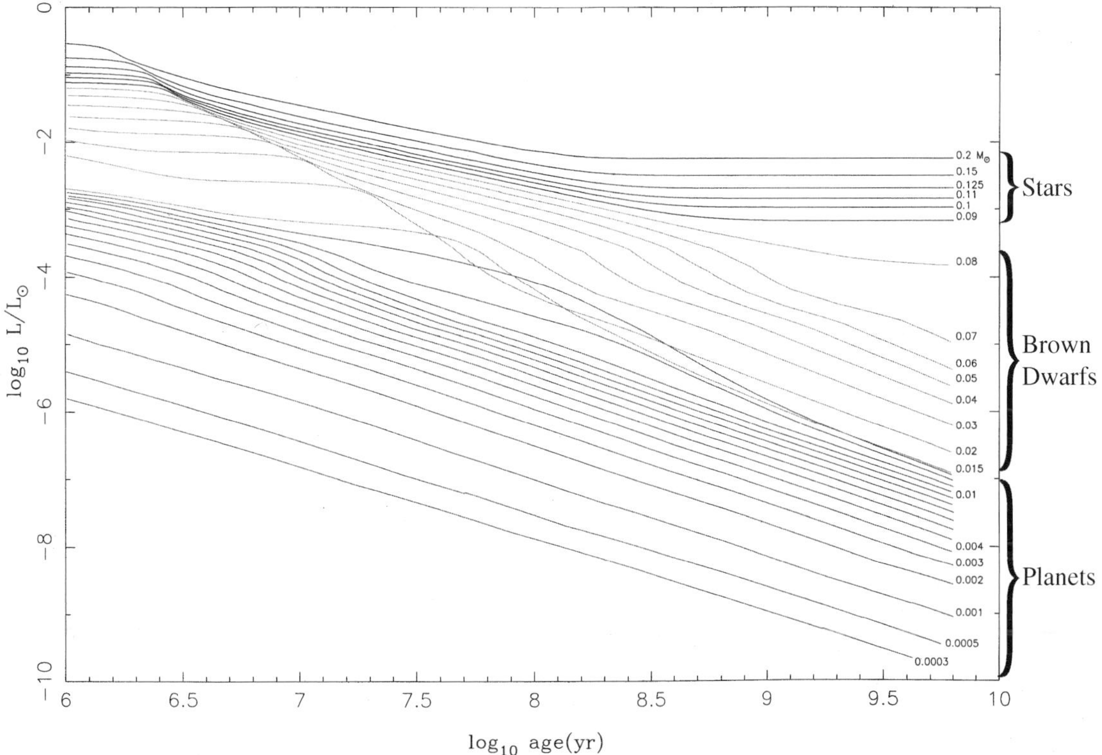

FIGURE 13.1 Evolution of the luminosity (in $L_\odot$) of isolated solar-metallicity M dwarf stars and substellar objects plotted as functions of time (in years) after formation. The stars, brown dwarfs and planets are shown as the upper, middle and lower sets of curves, respectively. The masses in $M_\odot$ label most of the curves, with the three lowest curves corresponding to the mass of Saturn, half the mass of Jupiter, and the mass of Jupiter. All of the objects become less luminous as they radiate away the energy released by gravitational contraction from large objects to sizes of order $R_{♃}$. Objects with $M \gtrsim 0.013\ M_\odot$ exhibit plateaus between 10^6 and 10^8 years as a result of deuterium burning (the initial deuterium mass fraction was assumed to be 2×10^{-5}). Deuterium burning occurs earliest and is quickest in the most massive objects. Stars ultimately level off in luminosity when they reach the hydrogen burning main sequence. In contrast, the luminosities of brown dwarfs and planets decline indefinitely. (Burrows *et al.* 1997)

Bodies supported primarily by degeneracy pressure are referred to as *compact objects*. In compact objects, the virial theorem implies that the energy of the degenerate particles (electrons in the case of brown dwarfs) is comparable to the gravitational potential energy:

$$\rho^{5/3} R^3 \sim \frac{GM^2}{R}. \tag{13.3}$$

Therefore, $R \propto M^{-1/3}$ (eq. 6.11; cf. Problem 6.4), and compact objects thus shrink if they are given more mass.

The most massive cool (old enough to be close to equilibrium) brown dwarfs are indeed expected to be slightly smaller than their lower mass brethren. However, as mass drops, *Coulomb pressure* begins to dominate over degeneracy. Since Coulomb pressure is characterized by constant density, $R \propto M^{1/3}$ (eq. 6.10). The net result is that all cool brown dwarfs and giant planets of solar composition, as well as the very lowest mass stars, are expected to have radii similar to that of Jupiter (Fig. 6.22).

Jovian mass planets orbiting very close to stars are subjected to intense stellar heating. This heating retards convection in the upper envelope of the planets. The excess entropy retained throughout such planets implies that their radii can be tens of percent larger than colder planets of the same mass, composition and age. This effect is greater for lower mass planets. Note that if a planet migrates close to a star after radiating away much of its initial accretion energy and shrinking to near 1 $R_{♃}$ (a process which requires tens of millions of years), its radius grows only slightly larger than when it was in colder environs, as the star's heating would only affect and expand the outer part of the planet's atmosphere which had cooled below the planets new T_{eff}.

13.2 Detecting Extrasolar Planets

Prior to the 1990s, our ability to understand how planets form was constrained because we had observed only one planetary system, our own Solar System. Dozens of extrasolar planets have been discovered within the past few years, and many more are likely to be found in the upcoming decades. Various methods for detecting planets around other stars are being used or studied for possible future use. As distant planets are extremely faint, most methods are indirect, in the sense that the planet is detected through its influence on the star that it orbits. The methods are sensitive to different classes of planets, and provide us with complementary information about the planets they do find, so most or all of them are likely to provide valuable contributions to our understanding of the diversity of planetary system characteristics. A brief review of detection techniques is presented in this section.

13.2.1 Pulsar Timing

The first confirmed detection of extrasolar planets was provided by *pulsar timing*. Pulsars, which are magnetized rotating neutron stars, emit radio waves that appear as periodic pulses to an observer on Earth. The pulse period can be determined very precisely, and the most stable pulsars rank among the best clocks known. The mean time of pulse arrival can be measured especially accurately for the rapidly rotating millisecond pulsars, whose frequent pulses provide an abundance of data. Even though pulses are emitted periodically, the times at which they reach the receiver are not equally spaced if the distance between the pulsar and the telescope varies in a nonlinear fashion. Earth's motion around the Sun and Earth's rotation cause such variations, which can be calculated and removed from the data. If periodic variations are present in this reduced data, they may indicate the presence of companions orbiting the pulsar.

As pulsar timing effectively measures variation in the position of the pulsar (relative to a trajectory with constant velocity with respect to the barycenter of our Solar System), it is most sensitive to massive planets with orbital periods comparable to the length of the interval over which timing measurements are available.

13.2.2 Radial Velocity Measurements

Radial velocity surveys have thus far been the most successful method for detecting planets around main sequence stars. By fitting the Doppler shift of a large number of features within a star's spectrum, the velocity at which the star is moving towards or away from an observer can be precisely measured. After removing the motion of the observer relative to the barycenter of the Solar System and other known motions, radial motions of the target star resulting from planets which are orbiting the star remain.

The semi-amplitude, K, of the radial velocity of a star of mass $M_\star$ that is induced by an orbiting planet of mass M_p is:

$$K = \left(\frac{2\pi G}{P_{orb}}\right)^{1/3} \frac{M_p \sin i}{(M_\star + M_p)^{2/3}} \frac{1}{\sqrt{1-e^2}}, \tag{13.4}$$

where P_{orb} is the orbital period, i is the angle between the normal to the orbital plane and the line of sight, and e is the orbit's eccentricity. As in the case of pulsar timing, radial velocity measurements yield the product of the planet's mass (divided by that of the star, whose mass can usually be estimated accurately from its spectral characteristics) and the sine of the angle between the orbital plane and the plane of the sky, as well as the period and the eccentricity of the orbit. This technique is most sensitive to massive planets and to planets in short-period orbits. Geoff Marcy's group is now achieving a precision of 3 m s^{-1} (representing a Doppler shift of one part in 10^8) on spectrally stable stars. With this precision, Jupiter-like planets orbiting Sun-like stars are detectable, although these detections will require a long timeline of observations (comparable to the planet's orbital period). Sub-Uranus mass planets orbiting very close to stars also could be detected; however, Earth-like planets orbiting at 1 AU are well beyond the capabilities currently envisioned for this technique. Precise radial velocity measurements require a large number of spectral lines, and thus cannot be achieved for the hottest stars (spectral types A, B and O), which have far fewer spectral features than do cooler stars like the Sun. Stellar rotation and intrinsic variability (including starspots) represent major sources of noise for radial velocity measurements.

13.2.3 Astrometry

Planets may be detected via the wobble that they induce in the motion of their stars projected onto the plane of the sky. This *astrometric* technique is most sensitive to massive planets orbiting about stars which are relatively close to Earth. The amplitude of the wobble, $\Delta\theta$, is given by the formula:

$$\Delta\theta(\text{arcsec}) = \frac{M_p}{M_\star}\frac{a_{AU}}{r_\odot}(\text{parsecs}), \tag{13.5}$$

where $r_\odot$ is the distance of the star from our Solar System. For example, a 1 $M_{♃}$ planet orbiting 5 AU from a 1 $M_\odot$ star located 10 parsecs (1 parsec = 3.26 light-years = 2.06

$\times$ 10^5 AU = 3.0857 $\times$ 10^{18} cm) from Earth would produce an astrometric wobble 0.5 milliarcseconds (mas) in amplitude. Because the star's motion is detectable in two dimensions, a better estimate of the planet's mass can be obtained astrometrically than using radial velocities, but it is somewhat more difficult to estimate the eccentricity of the orbit.

Planets on more distant orbits are ultimately easier to detect using astrometry because the amplitude of the star's motion is larger, but finding these planets requires a longer timeline of observations due to their greater orbital periods. Astrometric systems require considerable stability over long times to reduce the noise that can lead to false detections. The best long-term precision demonstrated by single ground-based telescopes is $\sim$1 mas. Interferometry from the ground should provide a precision of $\sim$20 microarcseconds (μas) in the first decade of the twenty-first century, and interferometers in space may reach $\sim$4 μas in the same time frame. No astrometric claim of detecting an extrasolar planet has yet been confirmed, but data obtained by the Hipparcos satellite in the early 1990s have shown that several candidate brown dwarfs observed in radial velocity surveys are actually low mass stellar companions whose orbits are viewed almost face-on.

13.2.4 Photometric Detection of Transits

If Earth lies in or near the orbital plane of an extrasolar planet, that planet passes in front of the disk of its star once each orbit as viewed from Earth. Precise *photometry* can reveal such transits, which can be distinguished from rotationally modulated starspots and intrinsic stellar variability by their periodicity, square-well shapes and relative spectral neutrality. Transit observations provide the size and orbital period of the detected planet.

For a transit to be observed, the orbit normal must be nearly 90° from the line of sight,

$$\cos i < \frac{R_\star + R_p}{a}, \tag{13.6}$$

where $R_\star$ and R_p are the stellar and planetary radii, respectively. Neglecting grazing transits (those with a duration less than half that of a central transit), the probability of detecting a planet orbiting 1 AU from a 1 $R_\odot$ star is 0.4%, whereas the probability of detecting a planet at 0.05 AU from the same star is 8%. Although geometrical considerations limit the fraction of planets detectable by this technique, thousands of stars can be surveyed within the field of view of one telescope, so transit photometry is quite efficient.

Scintillation in and variability of Earth's atmosphere limit photometric precision to roughly one-thousandth of a magnitude, allowing detection of transits by Jupiter-sized planets (but not by Earth-sized planets) from the ground. Far greater precision is achievable above the atmosphere, with planets as small as Earth likely to be detectable. One major advantage of the transit technique is that larger planets detected in this manner are likely to be observable via the radial velocity method as well, yielding a mass (as the inclination is known from the transits). Such a combined and complementary detection provides the density of the planet, an especially valuable datum for formation studies.

A space-based photometric telescope could also detect the sinusoidal phase modulation of light reflected by inner giant planets as they orbit their stars. The albedos of planets detected by both transit photometry and reflected light photometry could thus be measured.

13.2.5 Microlensing

Microlensing, which is produced by the general relativistic bending of the light from a distant star by a massive object (lens) passing between the source and the observer, is currently being used to investigate the distribution of faint stellar and substellar mass bodies within our galaxy. The lens amplifies the light from the source by a substantial factor when it passes closer to the line of sight than the radius of the *Einstein ring*, R_E, which is given by:

$$R_E = \sqrt{\frac{4GM_L r_{\oplus L}}{c^2}\left(1 - \frac{r_{\oplus L}}{r_{\oplus S}}\right)}, \tag{13.7}$$

where M_L is the mass of the lens, c is the speed of light and $r_{\oplus L}$ and $r_{\oplus S}$ are the distances from the Earth to the lens and the source respectively. The brightness of the source can increase several fold for a period of weeks during such an event, and the pattern of brightening can be used to determine (in a probabilistic manner) properties of the lens. If the lensing star has planetary companions, then these less massive bodies can produce characteristic blips on the observed lightcurve provided the line of sight passes within the planet's (much smaller) Einstein ring. Under favorable circumstances, planets as small as Earth can be detected. Microlensing provides information on the masses of detected planets and their (projected) distance from their star, but not on orbital eccentricities or inclinations. Careful monitoring of many microlensing events could provide a very useful data set on the distribution of planets within our galaxy. However, the properties of individual planets, and usually even of the stars that they orbit, could only be estimated in a statistical sense because of the many parameters which influence a microlensing lightcurve.

13.2.6 Imaging

Distant planets are very faint objects which are located near much brighter objects (the star or stars that they orbit), and thus they are extremely difficult to image. The reflected starlight from planets with orbits and sizes like those in our Solar System is roughly one-billionth as large as the stellar brightness, although the contrast decreases by ~ 3 orders of magnitude in the thermal infrared. Diffraction of light by telescope optics and atmospheric variability add to the difficulty of *direct detection* of extrasolar planets. However, a brown dwarf companion to the star Gl 229 has been imaged in the thermal infrared, and studied spectroscopically as well. Technological advances in interferometry and nulling (i.e., by choosing the right parameters the intensity of the star's image can be decreased to near-zero levels, cf. Section 9.3.5) should eventually allow for imaging and spectroscopic studies of planets orbiting nearby stars.

Many substellar objects which don't orbit stars have been imaged in the infrared. This newly discovered class of objects may have members less massive than the deuterium burning limit. The term *free-floating giant planets* has been used to describe these objects, even though they appear to be in many ways more akin to low-mass stars and brown dwarfs than to the planets within our Solar System.

13.2.7 Other Techniques

Several other methods can be used to detect and study extrasolar planets. Precise *timing of eclipses of eclipsing binary stars* has the potential of revealing the masses and orbits of unseen companions. Spectroscopy could be used to identify gases that would be stable in planetary atmospheres but not in stars, and Doppler variations of such signals could yield planetary orbital parameters. Planets transiting nearby stars and giant stars at larger distances could be detected as dark dots moving across *high resolution images of stellar disks* that will be obtainable using interferometry. Radio emissions similar to those detected from Jupiter (Section 7.5) could reveal the presence of extrasolar planets. Finally, *artificial signals* (cf. Section 13.6) from an alien civilization could betray the presence of the planets on which they live (and they might be willing to provide us with substantially more information!).

13.3 Observations of Extrasolar Planets

The first extrasolar planets were discovered in the early 1990s by Alexander Wolszczan and Dale Frail. They found periodicities in the pulse arrival time from pulsar PSR 1257+12 (which has a 6 millisecond rotational/pulse period) that remained after the motion of the telescope about the barycenter of the Solar System had been accounted for, and attributed these variations to two companions of the pulsar. One companion has an orbital period of 66.54 days and the product of its mass and orbital tilt to the plane of the sky is $M \sin i = 3.4\ \mathrm{M}_\oplus$; the other planet has a period of 98.22 days and $M \sin i = 2.8\ \mathrm{M}_\oplus$ (these masses assume that the pulsar is 1.4 times as massive as the Sun). Both planets have orbital eccentricities of around 0.02. Subsequent observations showed the effects of mutual perturbations of these two bodies on their orbits, thereby confirming the planet hypothesis and implying that $i \lesssim 60°$. Additionally, the data suggest that there is probably a lunar mass planet orbiting interior to the two near-resonant planets, and there are tentative indications of a planet with roughly Saturn's mass orbiting at around 40 AU from the pulsar. Pulsar timing has thus been demonstrated to be a very sensitive detector of planetary objects, but it only works for planets in orbit about a rare and distinctly nonsolar class of stellar remnants.

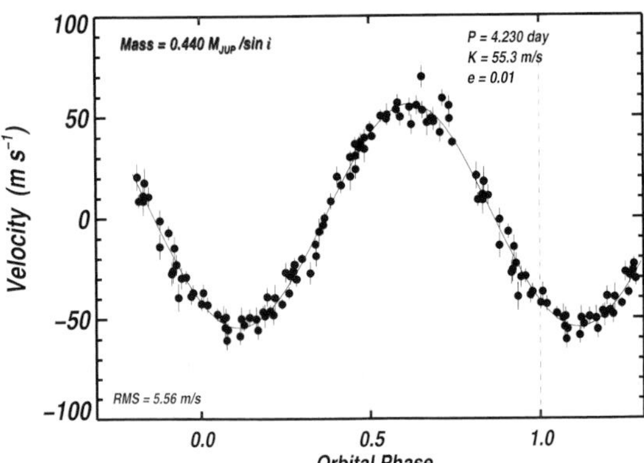

FIGURE 13.2 Radial velocity measurements of the star 51 Pegasi (points with error bars) as a function of phase of the orbital fit (solid line). (Courtesy: D. Fischer)

The first planet known to orbit a main sequence star other than the Sun is the $M \sin i = 0.47\ \mathrm{M}_{♃}$, $P_{orb} = 4.23$ days companion discovered to orbit the star 51 Pegasi by Michel Mayor and Didier Queloz in 1995 (Fig. 13.2). In the 1990s, radial velocity surveys identified 30 companions with $M \sin i < 13\ \mathrm{M}_{♃}$ orbiting main sequence stars other than the Sun. All planets thus far identified in radial velocity surveys share at least two of the following three characteristics, each of which acts to increase their detectability: Their masses exceed that of Saturn, their orbital semimajor axes are less than roughly 3 AU, and they dominate the radial velocity variations of their parent stars over a broad range of timescales (thus, the most massive

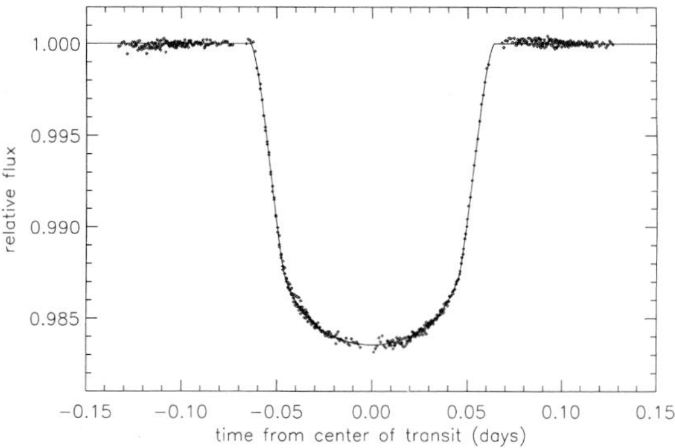

FIGURE 13.3 Superposed lightcurve of four transits of the planet of HD209458, observed with the HST from 25 April through 12 May 2000. Plotted points represent individual measurements, while the solid line is the fit to a model of a circular planet passing in front of a limb-darkened star. There is 1 time sample per minute, each with precision of about 0.01%. By fitting the shape of the curve measured with this precision, one can estimate the diameters of both the star and the planet, the inclination of the orbit, and one parameter describing the star's limb darkening. Differences between the observed and model lightcurves would allow detection of a planetary ring system similar in size to Saturn's, or of satellites as small as 1.5 $R_\oplus$, but no evidence for either rings or moons was found. (Courtesy: T. Brown)

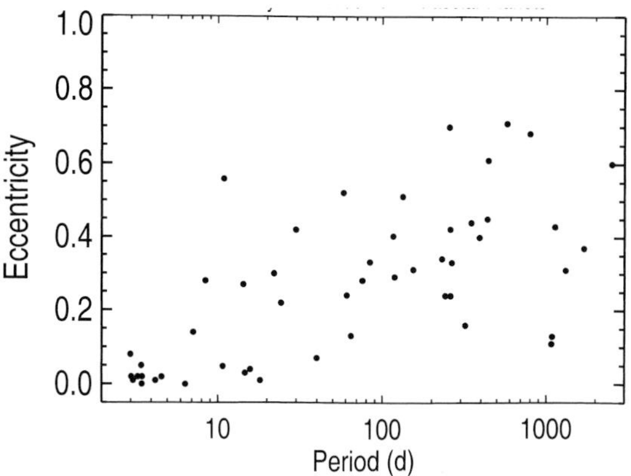

FIGURE 13.4 The eccentricities of extrasolar giant planets (known in August 2000) are plotted against orbital period. The eccentricities of planets with periods of less than a week are quite small (consistent with 0), presumably as a result of tidal damping, whereas the eccentricities of planets with longer orbital periods are generally much larger than those of the giant planets within our Solar System. (Courtesy: D. Fischer)

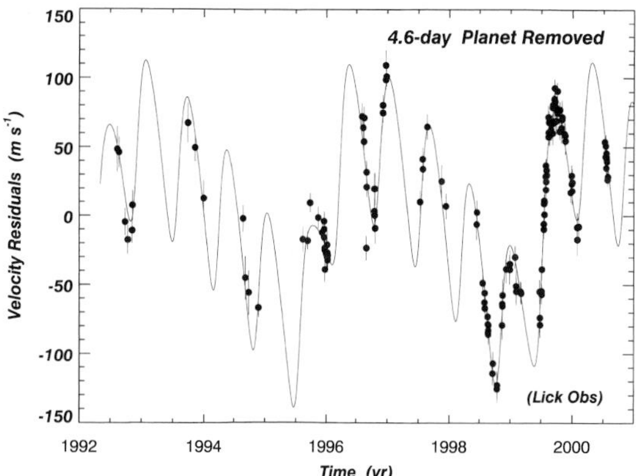

FIGURE 13.5 The variations in the radial velocity of the star υ Andromedae after subtracting off the star's motion due to its inner planet are shown as a function of time. Uncertainties of individual measurements are indicated. The solid curve represents the model response of the star to two additional planets. Estimated planetary parameters are shown in Table 13.1. (Courtesy: D. Fischer)

planet near these stars surpasses the second most massive planet by a factor larger than the ratio of the mass of Jupiter to that of Saturn).

Several of these planets, henceforth referred to as *vulcans*, after the hypothetical planet once believed to travel about the Sun within the orbit of Mercury, have periods less than one week. The orbits of the vulcans are all nearly circular; eccentric orbits this close to a star would be damped relatively rapidly by tidal forces (cf. Section 2.6). Vulcan planet companion to the Sun-like star HD209458 was discovered by the Doppler technique and subsequently observed to transit the star that it orbits (Fig. 13.3). This planet's orbital period is 3.525 days, its mass is 0.63 $M_♃$ and its radius is 1.35 ± 0.05 $R_♃$, implying that it is composed primarily of H_2 and He and that its envelope is bloated because the intense stellar radiation that it absorbs has prevented it from contracting. Some of the planets orbiting farther from their stars travel on quite eccentric paths (Fig. 13.4).

Three jovian mass planets have been detected in orbit about the star υ Andromedae (Fig. 13.6). The innermost of these objects is a vulcan planet; the other two planets are far more distant from the star and travel on eccentric orbit. (Table 13.1). Dynamical calculations imply that at least the outermost two planets have $\sin i > 1/5$, in order for the system to remain stable for the star's 2.5×10^9 years age.

Several stars are now known to possess two jovian-mass planets. In some cases, the planets have substantially different orbital periods, and are unlikely to perturb one another significantly. However, the stars GJ 876 and HD 82943 each has a pair of planets locked in a 2:1 orbital mean motion resonance with one another. These planets perturb one another substantially, and fits to the radial velocity data which account for these perturbations (Fig. 13.6) are far superior to those in which the stellar response to keplerian motions of two planets are superposed.

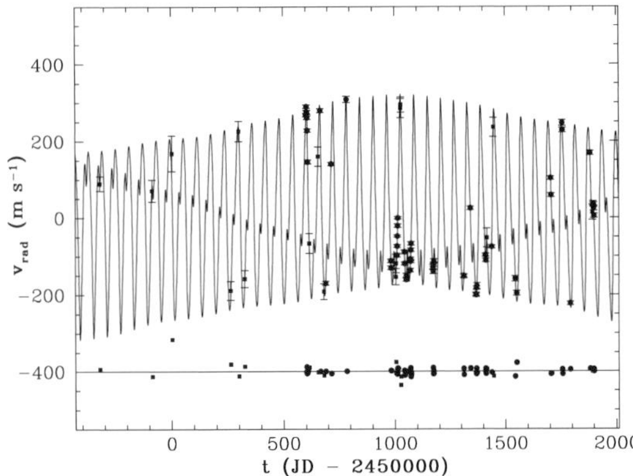

FIGURE 13.6 The variations in the radial velocity of the faint M dwarf star GJ 876 measured at the Lick and Keck observatories are shown as squares and circles respectively. The curve passing through most of the points was calculated by varying the parameters of two mutually interacting planets in orbit about the star to best fit the data. Error bars for the points represent intrinsic uncertainties of the measured velocities. The points below the main curve show the departures of the data from the best fit model. (Courtesy: Eugenio Rivera)

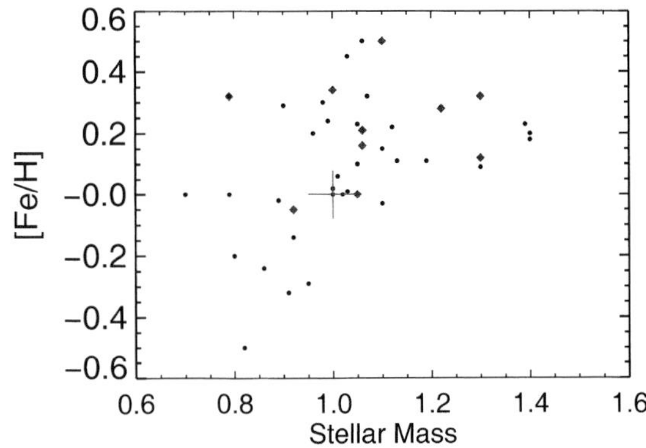

FIGURE 13.7 Metallicity of stars known to have planets (as of August 2000) is plotted against stellar mass. The heavy element abundance of the star is represented by the $\log_{10}$ of the iron to hydrogen ratio in its atmosphere normalized to the iron to hydrogen ratio in the Sun's atmosphere. Diamonds display the metallicity of stars having planets with $P_{orb} < 7$ days, the cross represents the Sun, and the dots show other stars known to possess planets. The Sun is more metal-rich than most stars in the solar neighborhood (the average value of [Fe/H] for stars near the Sun's mass in the solar neighborhood is -0.2), and the stars of 1 $M_\odot$ and larger which bear planets have substantially higher metallicity than does the sample of stars being searched for planets using the Doppler technique. (Courtesy: D. Fischer)

TABLE 13.1 Upsilon Andromedae's Planetary System.

Period (d)	a (AU)	e	ω	K (m s^{-1})	$M_p/\mathrm{M}_{♃}$
4.617	0.0592	0.014	33	71.1	0.693
240.8	0.827	0.25	250	57.3	2.03
1312.6	2.56	0.35	253	67.8	4.08

The majority of the planets identified in radial velocity surveys are the only known substellar companions of their star, and are much more massive ($M \sin i$) than any other companions that the star may possess which have periods of a few years or less. This is in sharp contrast to our Solar System, where planets of comparable size have orbital periods within a factor of two or three of their neighbors. These extrasolar planetary systems thus appear to be more overstable than is our own, which may indicate different mechanisms important in the formation process.

Most of the stars with detected planets are richer in elements heavier than helium (referred to by astronomers as 'metals') than is our Sun. The Sun itself has a higher metallicity than do most $\sim$1 $M_\odot$ stars in the solar neighborhood, and the vast majority of radial velocity surveys have not been biased towards metal-rich stars. Thus, stars with higher metal abundances appear to be more likely to harbor giant planets within a few AU than are metal-poor stars. This trend is stronger for stars with vulcan planets, and is more pronounced for higher mass stars (Fig. 13.7).

13.4 Models for the Formation of Planets Observed to Orbit Main Sequence Stars Other Than the Sun

The orbits of the extrasolar giant planets thus far observed are quite different from those of Jupiter, Saturn, Uranus and Neptune, and new models have been proposed to explain them. It has been suggested that the giant vulcans formed substantially farther from the star and subsequently migrated inwards to their current short-period orbits. Planetary orbital decay had been studied prior to the discovery of extrasolar planets, but no one predicted giant planets near stars because migration speeds were expected to increase as the planet approached the star, so the chance that a planet moved substantially inwards and was not subsequently lost was believed to be small (cf. Section 12.8). Two possible mechanisms have been proposed for stopping a planet less than one-tenth of an AU from the star: tidal torques from the star counteracting disk torques or a substantial reduction in disk torque once the planet was well within a nearly empty zone close to the star. However, the discovery of giant planets with orbital periods ranging

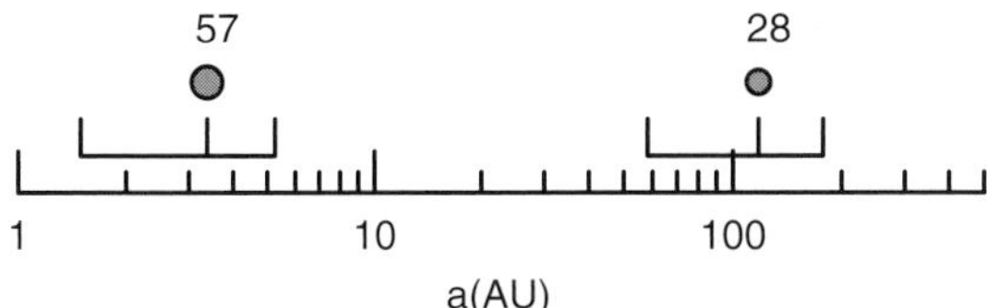

FIGURE 13.8 Orbits of the two remaining planets from a synthetic system of giant planets that was disrupted by chaotic scattering which sent most of the planets off into interstellar space. The planets' masses are given in units of $M_\oplus$ and the periastra, semimajor axes and apoastra are shown below. The inner planet in this system, which is twice as massive as the outer planet, would dominate the radial velocity signature of the star, especially for observations over a time span of a few decades. (Adapted from Levison *et al.* 1998)

from 15 days to 3 years, which feel negligible tidal torque from their star and are exterior to the region of the disk believed to be cleared by magnetic accretion onto the star (Section 12.4.2), cast doubt upon this model.

Some of the giant planets move on quite eccentric ($0.2 < e < 0.7$) orbits. These eccentric orbits may be the result of stochastic gravitational scatterings among massive planets (which have subsequently merged or been ejected to interstellar space, see Fig. 13.8), by perturbations of a binary companion or by past stellar companions (if the now single stars were once members of unstable multiple star systems). However, as neither scattering nor migration offer a good explanation for those planets with nearly circular orbits and periods from a few weeks to a few years, the possibility of giant planet formation quite close to stars should not be dismissed.

The total number of known extrasolar planets is still small, and the sample contains strong biases. Most solar-type stars could well have planetary systems which closely resemble our own. Nonetheless, if giant planets (even of relatively modest Uranus/Neptune masses) orbiting near or migrating through 1 AU are the norm, then terrestrial planets in habitable zones may be scarcer than they were previously believed to be. However, giant planets could have large moons which themselves might be habitable.

13.5 Planets and Life

One of the most basic questions that has been pondered by Natural Philosophers since antiquity concerns humanity's place in the Universe: Are we alone? This question has been approached from a wide variety of viewpoints, and similar reasoning has led to widely diverse answers. Aristotle believed that earth, the densest of the four elements, fell towards the center of the Universe, so no other worlds could possibly exist; in contrast, Democritus and other early atomists surmised that the ubiquity of physical laws implies innumerable Earth-like planets must exist in the heavens.

A major scientific debate concerning the possibility of life, advanced or otherwise, on Mars was ongoing at the beginning of the twentieth century. As we learned more about the current surface conditions on the planet whose climate is more Earth-like than any of our other neighbors, the chances of life seemed to be far more remote. However, recent theoretical and observational results suggest that early Mars may have been as hospitable to life as was early Earth (Section 4.9), and that the descendants of such life may survive deep under Mars's surface. Martian microbes may even have traveled to Earth within meteorites and are our very distant ancestors!

The conventional, very anthropocentric, picture is that to be habitable, a planet must have liquid water on its surface for a very long time. This single factor is unlikely to be either necessary or sufficient. Nonetheless, it provides a useful guide for habitability by life similar to that on Earth. The main sequence phase of low to moderate mass stars provides long-lived regions where planets may maintain liquid water on their surfaces; these regions are referred to as *continuously habitable zones*. The lifetime of continuously habitable zones is shorter for more massive stars; also, although the quantity of radiation received by planets in habitable zones is the same regardless of stellar type, the 'quality' differs, as more massive stars radiate a larger fraction of their energy at shorter wavelengths. Shorter-lived habitable zones occur at greater distances from post-main sequence stars that burn helium in their cores and shine brightly as red giants.

The greater flux of ultraviolet radiation could conceivably speed up biological evolution enough to compensate for a few $M_\odot$ star's shorter lifetime. At the other end of the spectrum, the smallest, faintest stars can live for trillions of years, but they emit almost all of their luminosity at infrared wavelengths and their luminosity varies because they emit large flares. Also, habitable zone planets would orbit so close to these faint stars that their rotation would be tidally synchronized; thus there would be no day–night cycle, and if their atmosphere was thin it would freeze out on the planet's cold, perpetually dark, hemisphere.

The minimum separation of $\sim 1\ M_\oplus$ planets on low eccentricity orbits required for the system to be stable for long periods of time is comparable to the width of a star's continuously habitable zone. Thus, orbital stability arguments support the possibility that most stars could have one or even two planets with liquid water on their surfaces, but unless greenhouse effects conspire to substantially compensate for increasing distance from the star, larger numbers of habitable planets around an individual star are unlikely (although it is conceivable that a giant planet orbiting at the appropriate distance from a star could possess several $\sim$1 $M_\oplus$ moons).

Because of the destruction that impacts may produce, impact frequency is an important factor in planetary habitability. The impact rate on the terrestrial planets of our Solar System was orders of magnitude larger 4 billion years ago than it is at present. In another planetary system, large impact fluxes could continue, making planets with Earth-like compositions and radiation fluxes hostile abodes for living organisms. Life on Earth has thrived thanks to billions of years of benign climate. Mars appears to have had a climate sufficiently mild for liquid water to have flowed on its surface when the Solar System was roughly one-tenth its current age (Section 4.9.2), but at the present epoch, the low atmospheric pressure and usually low temperature mean that liquid water would not be stable on the martian surface. Venus is too hot, with a massive carbon dioxide dominated atmosphere; we cannot say whether or not young Venus had a mild Earth-like climate. Indeed, as stellar evolution models predict that the young Sun was about 25% less luminous than at present, we don't understand why Earth, much less Mars, was warm enough to be covered by liquid oceans 4 billion years ago (Section 4.9.1).

Carbon dioxide on our planet cycles between the atmosphere, the oceans, life, fossil fuels and carbonate rocks on a wide range of timescales (Section 4.9). The carbonate rocks form the largest reservoir; they are produced by reactions involving water, in some cases living organisms act as catalysts, in other cases not. Carbon dioxide is recycled from carbonates back into the atmosphere as plates are subducted and heated within the Earth's mantle. Carbonates are not readily recycled on a geologically inactive planet such as Mars; in contrast they are not formed on planets like Venus, which lack surface water. Larger planets of a given composition remain geologically active for longer, as they have smaller surface area to mass ratios, and thus retain heat from accretion and radioactive decay longer. The number of variables involved in determining a planet's habitability precludes a complete discussion, but some of the major issues are summarized in Figure 13.9.

13.6 SETI

The Search for Extra-Terrestrial Intelligence (*SETI*) is an endeavor to detect signals from alien life forms. A clear detection of such a signal would likely change humanity's world view as much as any other scientific discovery in history. As our society is in its technological infancy, another civilization capable of communicating over interstellar distances is likely to be enormously advanced compared to our own – compare our technology to that of a mere millennium ago and then extrapolate millions or billions of years into the future! Thus, a dialog with extraterrestrials could alter our society in unimaginable ways.

The primary instrument used by SETI is the radiotelescope. Most radio waves propagate with little loss through the interstellar medium, and many wavelengths also easily pass through Earth's atmosphere. They are easy to generate and to detect. Radio thus appears to be an excellent means of interstellar communication, whether data is being exchanged between a community of civilizations around different stars or is being broadcast to the galaxy in order to reach unknown societies in their technological infancy. Signals used for local purposes, such as radar and TV on Earth, also escape and can be detected at great distances.

The first deliberate SETI radiotelescope observations were performed by Frank Drake in 1960. Since that time, improvements in receivers, data processing capabilities and radiotelescopes have doubled the capacity of SETI searches roughly every 8 months. Nonetheless, only a minuscule fraction of directions and frequencies have been searched, so one should not be discouraged at the lack of success thus far.

13.7 Conclusions

Prior to the discovery of extrasolar planets, models of planetary growth suggested that most single solar-type stars possess planetary systems that are grossly similar to our Solar System. Observations have subsequently demonstrated that nature is more creative than the human imagination. It was realized that stochastic factors are important in planetary growth, so that the number of terrestrial planets (as well as the presence or lack of an asteroid belt) would vary from star to star, even if their protoplanetary disks were initially very similar. The difficulty in accreting giant planet atmospheres prior to dispersal of circumstellar gas suggested that many systems might lack gas giants. The low eccentricities of the giant planets in our Solar System (especially Neptune) are difficult to account for, so systems with planets on highly eccentric orbits were viewed as possibilities, although researchers did not hazard to estimate the detailed characteristics of such systems. A maximum planetary mass similar to that of Jupiter was suggested as a possibility if Jupiter's mass was determined by a balance between a planet's gap clearing ability and viscous inflows, although it was noted that the value of the viscosity could well vary from disk to disk. Orbital mi-

FIGURE 13.9 COLOR PLATE Theoretical comparison of planets of different sizes with the same composition as Earth. (Left) A smaller planet would be less dense, because the pressure in the interior would be lower. Such a planet would have a larger ratio of surface area to mass, so its interior would cool faster. Its lower surface gravity and more rigid crust would allow for higher mountains and deeper valleys than are seen on Earth. Most important to life is that the mini-Earth would have a much smaller surface pressure as a result of four factors: larger surface area to mass, lower surface gravity, more volatiles sequestered in crust because there would be less crustal recycling and more atmospheric volatiles escaping to space. This would imply, among other things, lower surface temperature, resulting from less greenhouse gas in the atmosphere. Some remedial measures which could improve the habitability of such a mass-deprived planet are: (1) Move it closer to the star, so less greenhouse effect would be needed to keep surface temperatures comfortable. (2) Add extra atmospheric volatiles. (3) Include a larger fraction of long-lived radioactive nuclei than on Earth, to maintain crustal recycling. (Center) Earth; home sweet home. (Right) A larger planet made of the same material as Earth would be denser and have a hotter interior. Its higher surface gravity and more ductile crust would lead to muted topography. It would have a much greater atmospheric pressure, and, unless its greenhouse was strong enough to boil away the planet's water, much thicker oceans, probably covering the planet's entire surface. Some remedial measures which could improve the habitability of such a mass-gifted planet are: (1) Move it farther from the star. (2) Include a smaller fraction of atmospheric volatiles. It is not clear that more active crustal recycling would be a problem, within limits, but crustal activity would be lessened if the planet had a smaller inventory of radioactive isotopes. (Lissauer 1999b)

gration of some giant planets towards their parent star was also envisioned, but since migration rates increased as the planet approached the star, such planets were expected to be accreted by their star, and the existence of numerous giant planets with orbital periods ranging from a few days to several weeks was not predicted.

It must thus be admitted that theoretical models based upon observations within our Solar System failed to predict the types of planets thus far detected by radial velocity surveys. On the other hand, note that the surveys which prove that such planets exist also show that they are fairly rare, occurring in fewer than ten percent of the systems. The radial velocity surveys conducted thus far are quite biased in favor of detecting massive planets orbiting close to stars, and planets similar to those in our own Solar System would not yet have been detected. Thus, it is possible that the majority of single Sun-like stars possess planetary systems quite similar to our own. Alternatively, although theoretical considerations suggest that terrestrial planets are likely to grow around most Sun-like stars, they may typically be lost if most systems also contain giant planets which migrate into the central star.

We still do not know whether terrestrial planets on which liquid water flows are rare, are the norm for solar-type stars or have intermediate abundances. Nonetheless, even if planetary migration destroys some promising systems, planets qualifying as continuously habitable for long periods of time by the liquid water criterion are expected to be sufficiently common that if we are the only advanced life form in our sector of the galaxy, biological and/or local planetary factors are much more likely to be the principal limiting factor than are astronomical causes.

Further Reading

Good overviews of planet detection techniques and early results, concentrating on the radial velocity method, are given by:

Marcy, G.W. and R.P. Butler, 1998. Detection of extrasolar giant planets. *Annu. Rev. Astron. Astrophys.* **36**, 57–97.

Marcy, G.W., W.D. Cochran, and M. Mayor, 2000. Extrasolar planets around main-sequence stars. In *Protostars and Planets IV*. Eds. V. Mannings, A.P. Boss, and S.S. Russell. University of Arizona Press, Tucson, pp. 1285–1311.

Techniques which may be used in the future to detect and study extrasolar terrestrial planets are discussed in:

Woolf, N. and J.R. Angel, 1998. Astronomical searches for Earth-like planets and signs of life. *Annu. Rev. Astron. Astrophys.* **36**, 507–537.

A general overview of extrasolar planet research with frequently updated news on planetary discoveries is given by the Extrasolar Planet Encyclopedia:
http://www.usr.obspm.fr/departement/darc/planets/catalog.html

The web site of the most successful group at finding extrasolar planets (Geoff Marcy's radial velocity search team) is:
http://www.exoplanets.org

Problems

13.1.**E** A planet of mass $M_p = 2\ \mathrm{M}_{♃}$ travels on a circular orbit of radius 4 AU about a 1 $\mathrm{M}_\odot$ star. The Solar System lies in the plane of the orbit. Write the equation for the star's radial velocity variations caused by the planet, and sketch the resulting curve.

13.2.**I** A planet of mass $M_p = 2\ \mathrm{M}_{♃}$ travels on an orbit with a semimajor axis of 4 AU and an eccentricity of 0.5 about a 1 $\mathrm{M}_\odot$ star. The Solar System lies 60° from the plane of the orbit, and periapse occurs when the planet is closest to Earth. Write the equation for the star's radial velocity variations caused by the planet, and sketch the resulting curve.

13.3.**E** What is the amplitude of the astrometric wobble induced by a planet of mass $M_p = 2\ \mathrm{M}_{♃}$ that travels on a circular orbit of 4 AU radius about a 1 $\mathrm{M}_\odot$ star located 4 parsecs from the Sun?

13.4.**E** Calculate the probability of transits of the planets Venus and Jupiter being observable from another (randomly positioned) planetary system.

13.5.**E** (a) Calculate the ratio of the light reflected by Earth at 0.5 μm to that emitted by the Sun at the same wavelength.
(b) Calculate the ratio of the thermal radiation emitted by Earth at 20 μm to that emitted by the Sun at the same wavelength.
(c) Repeat the above calculations for Jupiter.

Appendix A: List of Symbols Used

a	semimajor axis of an orbit
a_{AU}	semimajor axis of planetary orbit in AU
A	surface area
$\mathcal{A}$	area enclosed by orbit
A_0	geometric albedo or head-on reflectance
A_b	Bond albedo
A_v	albedo at visual wavelengths
A_ν	albedo at frequency ν
$A_{0,\nu}$	geometric albedo at frequency ν
A_{rdr}	radar albedo
b	impact parameter
b_m	semiminor axis
$B, \mathbf{B}$	magnetic field strength and vector
B, B_ν	brightness, at frequency ν
B_e	magnetic field strength at the magnetic equator
B_o	surface magnetic field strength
B_{sw}	magnetic field strength in solar wind
c	speed of light in vacuum
c_s	speed of sound, velocity dispersion
c_A	Alfvén velocity
c_P	specific heat at constant pressure
c_V	specific heat at constant volume
C_D	drag coefficient
C_J	Jacobi's constant
C_P	thermal heat capacity (or molecular heat) at constant pressure
C_V	thermal heat capacity (or molecular heat) at constant volume
D	diameter
D_i	molecular diffusion coefficient, for species i
D_{st}	worldwide average of geomagnetic disturbances
$D_{\mathcal{LL}}$	radial diffusion coefficient
e	(generalized) eccentricity
E	total energy
$E, \mathbf{E}$	electric field strength and vector
E_G	gravitational potential energy
E_K	kinetic energy
E_{sw}	electric field due to $\mathbf{v}_{sw} \times \mathbf{B}_{sw}$ in solar wind
f	true anomaly (angle between planet's periapse and instantaneous position)
f_p	phase space density
f_C	Coriolis parameter
f_{osc}	oscillator strength
$\mathbf{F}$	force
$\mathbf{F}_c$	centripetal force
F_d	amplitude of the driving force
$\mathbf{F}_g$	force of gravity
$\mathbf{F}_{g,eff}$	effective gravitational force
$\mathbf{F}_D$	drag force
$\mathbf{F}_L$	Lorentz force
$\mathbf{F}_T$	tidal force
$\mathbf{F}_Y$	Yarkovski force
$\mathbf{F}_{rad}$	radiation force
$\mathcal{F}$	flux
$\mathcal{F}_e$	enhancement factor
$\mathcal{F}_g$	gravitational enhancement factor
$\mathcal{F}_\nu$	flux density at frequency ν
$\mathcal{F}_\odot$	solar constant
g, g_i	g-factor or emission rate
g_n	acceleration due to centripetal force
g_p	gravitational acceleration
g_r	radar backscatter gain
g_{eff}	effective gravitational acceleration
G	gravitational constant
h	Planck's constant
$\hbar$	normalized Planck's constant
h	vertical scale length
h_{cp}	height crater peak
H	scale height
H	enthalpy

H_{10}	absolute magnitude
H_v	Gaussian scale height
H_z	scale height in vertical direction
$\mathcal{H}$	heating rate
$\mathcal{H}_\nu$	Eddington flux
i	inclination angle
I_B	integral of $J_B/(2mv)$
I_ν	specific intensity
j	differential energy flux of particles
j_ν	mass emission coefficient
j_{ν_o}	mass emission coefficient at line center
$\mathbf{J}$	electric current
J, J_ν	mean intensity, at frequency ν
J_i	action variable
J_n	gravitational moments
$\mathbf{J}_u$	energy flux vector
J_B	second adiabatic invariant
k	Boltzmann constant
$\mathbf{k}$	wave vector
k_d	diffusivity
k_T	tidal Love number
k_{ri}	chemical reaction rate for reaction i
K_m	incompressibility modulus
K_T	thermal conductivity
$\mathcal{K}$	eddy diffusion coefficient
$\mathcal{K}_\nu$	the $\mathcal{K}$-integral
ℓ	characteristic length or depth scale
ℓ_{fp}	mean free path
$\mathbf{L}$	angular momentum
L	luminosity
L_1–L_5	Lagrangian points
L_e	electrical skin depth
L_s	latent heat of sublimation/condensation
L_T	thermal skin depth
$L_\odot$	solar luminosity
$\mathcal{L}$	McIlwain's parameter
m	mass
m_v	(visual) apparent magnitude
m_H	mass of hydrogen atom
m_{amu}	atomic mass unit
m_{gm}	mass of one gram-mole
M	mass
M_s	total mass of bodies
M_v	absolute magnitude (visual wavelengths)
M_J	Jeans mass
$M_\odot$	solar mass
$M_\oplus$	Earth mass
$\mathcal{M}_o$	magnetosonic Mach number
$\mathcal{M}_B$	magnetic dipole moment
n	energy level of an electron
n	index of refraction
n	mean angular velocity of body in orbit
N	number density of particles (cm^{-3})
N_c	column density (cm^{-2})
N_A	Avogadro's number
$\mathcal{N}_u$	Nusselt number
$\mathbf{p}$	momentum
p	as subscript: polarization, planet, particle
p_r	momentum due to radiation pressure
p_{sw}	momentum of solar wind motion
P	pressure
P_n	Legendre polynomials
P_L	linear polarization
P_{orb}	orbital period
P_{rot}	rotation period
P_{yr}	orbital period in years
$\mathcal{P}(\cos\phi)$	scattering phase function
q	electric charge
q	pericentric separation
q_r	rotation parameter
$q_{ph,\lambda}$	phase integral at wavelength λ
Q	amount of heat
Q_J	Joule heating
Q_T	Toomre's stability parameter
Q_{pr}	radiation pressure coefficient
$\mathcal{Q}$	gas production rate
r, $\mathbf{r}$	distance, separation
r_c	corotational radius
r_e	equatorial crossing distance of field line
r_g	guiding center
r_v	location of vertical resonance
r_L	location of horizontal (Lindblad) resonance
r_{AU}	heliocentric distance in AU
r_{Bohr}	Bohr radius
r_Δ	distance from observer
$r_\odot$	heliocentric distance
$r_\oplus$	geocentric distance
R	radius of object
R_0	Fresnel reflection coefficient
R_c	critical radius
R_c	field line curvature radius
R_d	radius dust grain
R_e	equatorial radius
R_p	polar radius
R_s	distance of closest approach
R_H	radius of Hill sphere
R_L	gyro, cyclotron, Larmor radius
R_{gas}	universal gas constant
R_{Sch}	Schwarzchild radius
$R_\parallel$	Fresnel reflection coefficient linearly polarized in the plane of incidence

$R_\perp$	Fresnel reflection coefficient linearly polarized normal to the plane of incidence
$\mathcal{R}$	disturbing function
$\mathcal{R}$	Rydberg's constant
$\mathcal{R}_c$	collisional radius
$\mathcal{R}_p, \mathcal{R}_d$	scale length of parent and daughter molecules, respectively
$\mathcal{R}_{bs}$	distance of bow shock from planet/comet
$\mathcal{R}_{cp}$	distance of cometopause from nucleus
$\mathcal{R}_{ip}$	distance of ionopause
$\mathcal{R}_{mp}$	distance of magnetopause
$\Re_a$	Rayleigh number
$\Re_e$	Reynolds number
$\Re_o$	Rossby number
S	entropy
S_ν	source function
SC/OC	ratio radar cross-sections in orthogonal directions
t	time
t_c	time between collisions
t_d	diffusion timescale
t_ℓ	lifetime
t_{damp}	damping timescale
t_{ff}	free-fall timescale
t_ϖ	time of periapse passage
$\tan\Delta$	loss tangent, ϵ_i/ϵ_r
T	temperature
T_a	atmospheric temperature
T_e	effective temperature
T_m	melting temperature
T_s	surface temperature
T_{cr}	critical temperature
T_{eq}	equilibrium temperature
T_{tr}	triple point
$\mathcal{T}$	potential temperature
u_a	annihilation rate
u_ν	radiation density
U	total energy
v, $\mathbf{v}$	velocity magnitude and vector
v'	velocity in rotating coordinate frame
v_c	circular orbit velocity
v_e	escape velocity
v_g	wave group velocity
v_i	impact velocity
v_o	thermal velocity
v_r	radial component of the velocity
v_P	P wave velocity
v_S	S wave velocity
v_θ	tangential component of the velocity
v_{ph}	wave phase velocity
v_{sw}	solar wind velocity
v_∞	terminal velocity (velocity at a large distance from the planet)
X_i	fractional concentration of constituent i
z_{ex}	altitude exobase
Z	atomic number
Z_p	partition function
x, y, z	Cartesian coordinate axes
r, θ, ϕ	spherical coordinate system
u, v, w	wind velocities along the x, y, z axes
α	pitch angle of a particle
α_e	particle's pitch angle at the magnetic equator
α_i	thermal diffusion parameter for constituent i
α_l	particle's loss cone
α_v	viscosity parameter
α_ν	mass extinction coefficient
β	ratio of radiation to gravitational force, F_{rad}/F_g
β_{cp}	ratio of corpuscular to radiation drag
γ	ratio of specific heats, C_P/C_V
γ_c	Lyapunov exponent (c for chaos)
γ_c^{-1}	Lyapunov timescale
γ_r	relativistic correction factor $\left(1-(v^2/c^2)\right)^{-1/2}$
γ_T	thermal inertia
Δ	thickness bow shock
Δh_g	geoid height anomaly
ϵ	flattening or geometric oblateness $((R_e - R_p)/R_e)$
ϵ, ϵ_ν	emissivity, at frequency ν
ϵ_i	imaginary part of the dielectric constant
ϵ_r	real part of the dielectric constant
ϵ_{ir}	infrared emissivity
ζ	exponent in power law distribution
$\eta(r), \eta_i$	nongravitational force parameters
$\eta_\nu(\phi)$	beaming factor at frequency ν, phase angle ϕ
θ	angle between line of sight and normal to surface
θ, θ'	(co-)latitude
θ_B	tilt angle magnetic axis
κ	epicyclic (radial) frequency
κ_ν	mass absorption coefficient
κ_{ν_o}	mass absorption coefficient at line center
λ	wavelength
λ_T	threshold wavelength for photodestruction
λ_{esc}	escape parameter

λ_{np}	longitude of magnetic north pole
Λ_2	response coefficient
μ	frequency of vertical oscillation
μ_a	molecular weight in amu
μ_b	magnetic moment μ_B/γ_r
μ_i	chemical potential of species i, or partial mole free energy
μ_r	reduced mass
μ_B	first adiabatic invariant
μ_θ	$\equiv \cos\theta$
μ_{rg}	rigidity modulus
ν	frequency
ν_c	critical frequency
ν_i, ν_e	collisional frequency for ions (i), electrons (e)
ν_v	kinematic viscosity
ν_v/ρ	absolute viscosity
ρ	density (g cm^{-3})
ρ_d	density of dust grain
ρ_g	gas density
ρ_p	density of (proto)planet
ρ_s	volume mass density of swarm planetesimals
σ	Stefan–Boltzmann constant
σ_c	Cowling conductivity
σ_h	Hall conductivity
σ_o	electrical conductivity
σ_p	Pederson conductivity
σ_x	(molecular) cross-section
σ_ν	mass scattering coefficient
σ_ρ	surface mass density
σ_{xr}	radar cross-section
τ	optical depth
ϕ	phase angle (scattering phase angle: Sun–target–observer angle, with Sun as source of light)
Φ_g	gravitational potential
Φ_i	upward particle flux in an atmosphere
Φ_ℓ	limiting flux
Φ_q	centrifugal potential
$\mathbf{\Phi}_B$	magnetic flux
Φ_J	Jeans escape rate
Φ_V	potential (esu)
Φ_ν	line shape
ψ	obliquity of a body (angle between rotation axis and orbit pole)
ω	argument of periapse
ω_d	forcing frequency
ω_o	frequency of oscillator, wave frequency
ω_p, ω_{pi}, ω_{pe}	plasma frequency (i for ions, e for electrons)
ω_B	gyro, cyclotron, Larmor frequency
ω_{Bi}, ω_{Be}	ion and electron cyclotron frequency, respectively
ω_{rot}	spin angular velocity
ω_{UHR}, ω_{LHR}	upper and lower hybrid resonance frequency
Ω	longitude of ascending node
Ω_s	solid angle
ϖ	longitude of periapse
ϖ_v	vorticity
ϖ_ν	single scattering albedo at frequency ν
ϖ_{pv}	potential vorticity
☉	Sun
☿	Mercury
♀	Venus
⊕	Earth
☾	Moon
♂	Mars
♃	Jupiter
♄	Saturn
⛢	Uranus
♆	Neptune
♇	Pluto

Appendix B: Units, Constants, Conversions

TABLE B.1 Prefixes.

Prefix	Value in SI units
p (pico-)	10^{-12}
n (nano-)	10^{-9}
μ (micro-)	10^{-6}
m (milli-)	10^{-3}
c (centi-)	10^{-2}
d (deci-)	10^{-1}
da (deca-)	10
h (hecto-)	10^{2}
k (kilo-)	10^{3}
M (mega- or million)	10^{6}
G (giga- or billion)	10^{9}
T (tera-)	10^{12}

TABLE B.2 Units.

Symbol	Value in cgs units
Å (angstrom)	10^{-8} cm
μm (micrometer)	10^{-6} cm
m (meter)	100 cm
km (kilometer)	10^{5} cm
kg (kilogram)	10^{3} g
t (tonne)	10^{6} g
J (joule)	10^{7} erg
eV (electronvolt)	1.602×10^{-12} erg
W (watt)	10^{7} erg s^{-1}
N (newton)	10^{5} dyne
atm (atmosphere)	1.013 25 bar
pascal	10 dyne cm^{-2}
bar	10^{6} dyne cm^{-2}
Hz (hertz)	1 cycle s^{-1}
Ω (ohm)	1.1126×10^{-12} esu
A (ampere)	2.998×10^{9} esu
γ (gamma)	10^{-5} gauss
T (tesla)	10^{4} gauss
Jy (jansky)	10^{-23} erg cm^{-2} Hz^{-1} s^{-1}

TABLE B.3 Physical Constants.

Symbol	Value in cgs units	Quantity
c	2.9979×10^{10} cm s^{-1}	Velocity of light
G	6.670×10^{-8} dyn cm^2 g^{-2}	Gravitational constant
h	6.626×10^{-27} erg s	Planck constant
k	1.3806×10^{-16} erg deg^{-1}	Boltzmann's constant
m_e	9.109×10^{-28} g	Electron mass
m_p	$1.672\,66 \times 10^{-24}$ g	Proton mass
m_{amu}	$1.660\,531 \times 10^{-24}$ g	Mass of unit atomic weight
n_o	2.686×10^{19} cm^{-3}	Loschmidt's number
N_A	6.022×10^{23} mole^{-1}	Avogadro's number
r_0	$0.529\,177 \times 10^{-8}$ cm	Bohr radius or atomic unit
R_{gas}	8.3143×10^7 erg deg^{-1} mole^{-1}	Universal gas constant
$\mathcal{R}$	1.0967×10^5 cm^{-1}	Rydberg constant
q	4.803×10^{-10} esu	Electron charge
ν_v	0.134 cm^2 s^{-1}	Kinematic viscosity of air at STP[a]
σ	5.6696×10^{-5} erg cm^{-2} deg^{-4} s^{-1}	Stefan–Boltzmann constant
c_P	1.0×10^7 erg g^{-1} K^{-1}	Isobaric specific heat capacity of air at STP
c_V	7.19×10^6 erg g^{-1} K^{-1}	Isochoric specific heat capacity of air at STP
c_P	1.2×10^7 erg g^{-1} K^{-1}	Typical value for the specific heat of rock
L_v	2.50×10^{10} erg g^{-1}	Specific latent heat of vaporization for water
L_s	2.83×10^{10} erg g^{-1}	Specific latent heat of sublimation for water-ice
ρ	1.293×10^{-3} g cm^{-3}	Density of air at STP

[a] STP: Standard temperature (273 K) and pressure (1 bar).

TABLE B.4 Astronomical Constants.

Symbol	Value in cgs units	Quantity
AU	1.496×10^{13} cm	Astronomical unit of distance
ly	9.4605×10^{17} cm	Light-year
pc	3.086×10^{18} cm	Parsec
$M_\odot$	1.989×10^{33} g	Solar mass
$R_\odot$	6.96×10^{10} cm	Solar radius
$L_\odot$	3.827×10^{33} erg s^{-1}	Solar luminosity
$\mathcal{F}_\odot$	1.37×10^6 erg cm^{-2} s^{-1}	Solar constant
$M_\oplus$	5.976×10^{27} g	Earth's mass
$R_\oplus$	6.378×10^8 cm	Earth's equatorial radius
g_p(eq)	978 cm s^{-2}	Equatorial gravity at sea level on Earth
g_p(pole)	983 cm s^{-2}	Polar gravity at sea level on Earth

Appendix C: Periodic Table of the Elements

Key to chart

Atomic number → 97 ← Oxidation states +3 +4
Bk ← Symbol
Atomic mass → (247) Berkelium ← Name

- Metals
- Non-metals
- Semi-metals
- Artificially prepared metals

1 H +1 −1 1.008 Hydrogen																	2 He 0 4.003 Helium
3 Li +1 6.94 Lithium	4 Be +2 9.01 Beryllium											5 B +3 10.81 Boron	6 C +2 +4 −4 12.01 Carbon	7 N +1 +2 +3 +4 +5 −1 −2 −3 14.00 Nitrogen	8 O −2 16.00 Oxygen	9 F −1 19.00 Fluorine	10 Ne 0 20.18 Neon
11 Na +1 22.99 Sodium	12 Mg +2 24.31 Magnesium											13 Al +3 26.98 Aluminum	14 Si +2 +4 −4 28.09 Silicon	15 P +3 +5 −3 30.97 Phosphorus	16 S +4 +6 −2 32.06 Sulfur	17 Cl +1 +5 +7 −1 35.45 Chlorine	18 Ar 0 39.95 Argon
19 K +1 39.10 Potassium	20 Ca +2 40.08 Calcium	21 Sc +3 44.96 Scandium	22 Ti +2 +3 +4 47.88 Titanium	23 V +2 +3 +4 +5 50.94 Vanadium	24 Cr +2 +3 +6 52.00 Chromium	25 Mn +2 +3 +4 +7 54.94 Manganese	26 Fe +2 +3 55.85 Iron	27 Co +2 +3 58.93 Cobalt	28 Ni +2 +3 58.69 Nickel	29 Cu +1 +2 63.55 Copper	30 Zn +2 63.39 Zinc	31 Ga +3 69.72 Gallium	32 Ge +2 +4 72.59 Germanium	33 As +3 +5 −3 74.92 Arsenic	34 Se +4 +6 −2 78.96 Selenium	35 Br +1 +5 −1 79.90 Bromine	36 Kr 0 83.80 Krypton
37 Rb +1 85.47 Rubidium	38 Sr +2 87.62 Strontium	39 Y +3 88.91 Yttrium	40 Zr +4 91.22 Zirconium	41 Nb +3 +5 92.91 Niobium	42 Mo +6 95.94 Molybdenum	43 Tc +4 +6 +7 (98) Technetium	44 Ru +3 101.07 Ruthenium	45 Rh +2 102.91 Rhodium	46 Pd +2 +4 106.42 Palladium	47 Ag +1 107.87 Silver	48 Cd +2 112.41 Cadmium	49 In +3 114.82 Indium	50 Sn +2 +4 118.71 Tin	51 Sb +3 +5 −3 121.75 Antimony	52 Te +4 +6 −2 127.60 Tellurium	53 I +1 +5 +7 −1 126.91 Iodine	54 Xe 0 131.29 Xenon
55 Cs +1 132.91 Caesium	56 Ba +2 137.33 Barium *	71 Lu +3 174.97 Lutetium	72 Hf +4 178.49 Hafnium	73 Ta +5 180.95 Tantalum	74 W +6 183.85 Tungsten	75 Re +4 +6 +7 186.21 Rhenium	76 Os +3 +4 190.2 Osmium	77 Ir +3 +4 192.22 Iridium	78 Pt +2 +4 195.97 Platinum	79 Au +1 +3 196.97 Gold	80 Hg +1 +2 200.59 Mercury	81 Tl +1 +3 204.38 Thallium	82 Pb +2 +4 207.2 Lead	83 Bi +3 +5 208.98 Bismuth	84 Po +2 +4 (209) Polonium	85 At (210) Astatine	86 Rn 0 (222) Radon
87 Fr +1 (223) Francium	88 Ra +2 226.03 Radium ◆	103 Lr +3 (260) Lawrencium	104 Rf +4 (261) Rutherfordium	105 Db (262) Dubnium	106 Sg (263) Seaborgium	107 Bh (262) Bohrium	108 Hs (265) Hassium	109 Mt (266) Meitnerium	110 (269)	111 (272)	112 (277)	113	114 (285)	115	116 (289)	117	118 (293)

* Lanthanide	57 La +3 138 Lanthanum	58 Ce +3 +4 140.12 Cerium	59 Pr +3 140.9 Praseodymium	60 Nd +3 144.24 Neodymium	61 Pm +3 (145) Promethium	62 Sm +2 +3 150.36 Samarium	63 Eu +2 +3 151.96 Europium	64 Gd +3 157.25 Gadolinium	65 Tb +3 158.92 Terbium	66 Dy +3 162.50 Dysprosium	67 Ho +3 164.93 Holmium	68 Er +3 167.26 Erbium	69 Tm +3 168.93 Thulium	70 Yb +2 +3 173.04 Ytterbium
◆ Actinide	89 Ac +3 227.02 Actinium	90 Th +4 232.04 Thorium	91 Pa +5 +4 231.04 Protactinium	92 U +3 +4 +5 +6 238.02 Uranium	93 Np +3 +4 +5 +6 237.05 Neptunium	94 Pu +3 +4 +5 +6 (244) Plutonium	95 Am +3 +4 +5 +6 (243) Americium	96 Cm +3 (247) Curium	97 Bk +3 +4 (247) Berkelium	98 Cf +3 (251) Californium	99 Es +3 (252) Einsteinium	100 Fm +3 (257) Fermium	101 Md +2 +3 (258) Mendelevium	102 No +2 +3 (259) Nobelium

Numbers in parentheses are mass numbers of most stable known isotope of radioactive elements that are rare or not found in nature.

References

Acuña, M.H., J.E.P. Connerney, N.F. Ness, R.P. Lin, D. Mitchell, C.W. Carlson, J. McFadden, K.A. Anderson, H. Rème, C. Mazelle, D. Vignes, P. Wasilewski, and P. Cloutier, 1999. Global distribution of crustal magnetization discovered by the Mars Global Surveyor MAG/ER experiment. *Science* **284**, 790–793.

A'Hearn, M.F., R.L. Millis, D.G. Schleicher, D.J. Osip, and P.V. Birch, 1995. The ensemble properties of comets: Results from narrowband photometry of 85 comets, 1976–1992. *Icarus* **118**, 223–270.

Ahrens, T.J., J.D. O'Keefe, and M.A. Lange, 1989. Formation of atmospheres during accretion of the terrestrial planets. In *Origin and Evolution of Planetary and Satellite Atmospheres*. Eds. S.K. Atreya, J.B. Pollack, and M.S. Matthews. University of Arizona Press, Tucson. pp. 328–385.

Aikawa, Y., T. Umbebayashi, T. Nakano, and S.M. Miyama, 1999. Evolution of molecular abundances in protoplanetary disks with acretion flow. *Astrophys. J.* **519**, 705–725.

Anders, E., and M. Ebihara, 1982. Solar-system abundances of the elements. *Geochim. Cosmochim. Acta* **46**, 2363–2380.

Anders, E., and N. Grevesse, 1989. Abundances of the elements: meteoritic and solar. *Geochim. Cosmochim. Acta* **53**, 197–214.

Anderson, J.D., E.L. Lau, W.L. Sjogren, G. Schubert, W.B. Moore, 1996b. Gravitational constraints on the internal structure of Ganymede. *Nature* **384**, 541–543.

Anderson, J.D., G. Schubert, R.A. Jacobson, E.L. Lau, W.B. Moore, W.L. Sjogren, 1998a. Europa's differentiated internal structure: inferences from four Galileo encounters. *Science* **281**, 2019–2022.

Anderson, J.D., G. Schubert, R.A. Jacobson, E.L. Lau, W.B. Moore, W.L. Sjogren, 1998b. Distribution of rock, metals, and ices in Callisto. *Science* **280**, 1573–1576.

Anderson, J.D., W.L. Sjogren, G. Schubert, 1996a. Galileo gravity results and the internal structure of Io. *Science* **272**, 709–712.

Atreya, S.K., 1986. *Atmospheres and Ionospheres of the Outer Planets and Their Satellites*. Springer-Verlag, Heidelberg.

Atreya, S.K., and T.M. Donhue, 1976. Model ionospheres of Jupiter. In *Jupiter*. Ed. T. Gehrels. University of Arizona Press, Tucson. pp. 304–318.

Atreya, S.K., M.H. Wong, T.C. Owen, P.R. Mahaffy, H.B. Niemann, I. de Pater, P. Drossart, and Th. Encrenaz, 1999. A comparison of the atmospheres of Jupiter and Saturn: deep atmospheric composition, cloud structure, vertical mixing, and origin. *Planet. Space Sci.* **47**, 1243–1262.

Bagenal, F., 1989. Torus–magnetosphere coupling. In *Time Variable Phenomena in the Jovian System*. Eds. M.J.S. Belton, R.A. West, and J. Rahe. Proceedings of a conference held in Flagstaff, Aug. 25–27, 1987, pp. 196–210.

Bagenal, F., 1992. Giant planet magnetospheres. *Annu. Rev. Earth Planet. Sci.* **22**, 289–328.

Bagenal, F., 1997. Plasma. In *Encyclopedia of Planetary Sciences*. Eds. J.H. Shirley and R.W. Fairbridge. Chapman and Hall, London. pp. 624–630.

Barnola, J.M., D. Raynaud, Y.S. Korotkevich, and C. Lorius, 1987. Vostok ice core provides 160,000-year record of atmospheric CO_2. *Nature* **329**, 408–414. (update via http://cdiac.esd.ornl.gov/trends/co2/vostok.htm)

Barshay, S.S., and J.S. Lewis, 1976. Chemistry of primitive solar material. *Annu. Rev. Astron. Astrophys.* **14**, 81–94.

Barth, C.A., A.I.F. Stewart, S.W. Bougher, D.M. Hunten, S.J. Bauer, A.F. Nagy, 1992. Aeronomy of the current Martian atmosphere. In *Mars*. Eds. H.H. Kieffer, B.M. Jakosky, C.W. Snyder, and M.S. Matthews. University of Arizona Press, Tucson. pp. 1054–1089.

Beatty, J.K., C.C. Peterson, and A. Chaikin, Eds., 1999. *The New Solar System*, 4th Edition. Sky Publishing Co., Cambridge, MA and Cambridge University Press, Cambridge, England.

Becker, G.E., and S.H. Autler, 1946 *Phys. Rev.* **70**, 300.

Belcher, J.W., 1987. The Jupiter–Io connection: an Alfvénic engine in space. *Science* **238**, 170–176.

Bell, J.F., D.R. Davis, W.K. Hartmann, and M.J. Gaffey, 1989. Asteroids: the big picture. In *Asteroids II*. Eds. R.P. Binzel, T. Gehrels, and M.S. Matthews. University of Arizona Press, Tucson. pp. 921–945.

Bernath, P.F., 1995. *Spectra of Atoms and Molecules*. Oxford University Press, Oxford.

Bezard, B., T. Encrenaz, E. Lellouch, and H. Feuchtgruber, 1999. A new look at the jovian planets. *Science* **283**, 800.

Bezard, B., D. Strobel, J.-P. Maillard, P. Drossart, and E. Lellouch, 1999. The origin of carbon monoxide in Jupiter. *BAAS* **31**, 1180.

Bida, T., T. Morgan, and R. Killen, 2000. Discovery of calcium in Mercury's atmosphere. *Nature* **404**, 159–161.

Binzel, R.P., 1989. An overview of the asteroids. In *Asteroids II*. Eds. R.P. Binzel, T. Gehrels, and M.S. Matthews. University of Arizona Press, Tucson. pp. 3–18.

Binzel, R.P., 1992. 1991 Urey prize lecture: Physical evolution in the Solar System – present observations as a key to the past. *Icarus* **100**, 274–287.

Binzel, R.P., P. Farinella, V. Zappala, and A. Cellino, 1989. Asteroid rotation rates: distributions and statistics. In *Asteroids II*. Eds. R.P. Binzel, T. Gehrels, and M.S. Matthews. University of Arizona Press, Tucson. pp. 416–441.

Biraud, F., G. Bourgois, J. Crovisier, R. Fillit, E. Gérard, and I. Kazès, 1974. OH observation of comet Kohoutek (1973f) at 18 cm wavelength. *Astron. Astrophys.* **34**, 163–166.

Bird, M.K., O. Funke, J. Neidhöfer, and I. de Pater, 1996. Multi-frequency radio observations of Jupiter at Effelsberg during the SL-9 impact. *Icarus* **121**, 450–456.

Bishop, J., S.K. Atreya, P.N. Romani, G.S. Orton, B.R. Sandel and R.V. Yelle, 1995. The middle and upper atmosphere of Neptune. In *Neptune*. Ed. D.P. Cruikshank. University of Arizona Press, Tucson. pp. 427–488.

Biver, N., A. Winnberg, D. Bockelée-Morvan, P. Colom, J. Crovisier, E. Gérard, B. Germain, F. Henry, E. Lellouch, F.T. Rantakyrö, J.K. Davies, M. Gunnarsson, H. Rickman, R. Moreno, G. Paubert, D. Despois, H. Rauer, 1999. Post-perihelion observations of the distant gaseous activity of Comet C/1995 O1 (Hale–Bopp) with the Swedish-ESO Submillimeter Telescope (SEST). *Asteroids, Comets and Meteors 1999.*

Blum, J., Th. Henning, V. Ossenkopf, R. Sablotny, R. Stognienko, E. Thamm, 1994. Fractal growth and optical behaviour of cosmic dust. In *Fractals in the Natural and Applied Sciences*. Ed. M.M. Novak. Elsevier Science B.V. (North-Holland). pp. 47–59.

Bockelee-Morvan, D., and E. Gerard, 1984. Radio observations of the hydroxyl radical in comets with high spectral resolution. Kinematics and asymmetries of the OH coma in C/Meier (1978XXI), C/Bradfield (1979X), and C/Austin (1982g). *Astron. Astrophys.* **131**, 111–122.

Bockelee-Morvan, D., D. Gautier, D.C. Lis, K. Young, J. Keene, T. Phillips, T. Owen, J. Crovisier, P.F. Goldsmith, E.A. Bergin, D. Despois, and A. Wootten, 1998. Deuterated water in comet C/1996 B2 (Hyakutake) and its implications for the origin of comets. *Icarus* **133**, 147–162.

Bouchez, A.H., M.E. Brown, and N.M. Schneider, 1999. Spectral identification of Io's visible eclipse emissions. *BAAS* **31**, # 78.02.

Bowell, E., B. Hapke, D. Domingue, K. Lumme, J. Peltoniemi, and A.W. Harris, 1989. Application of photometric models to asteroids. In *Asteroids II*. Eds. R.P. Binzel, T. Gehrels, and M.S. Matthews. University of Arizona Press, Tucson. pp. 524–556.

Brecht, S.H., I. de Pater, D.J. Larson, and M.E. Pesses, 2001. Modification of the jovian radiation belts by Shoemaker–Levy 9: An explanation of the data. *Icarus* **151**, 39–50.

Brouwer, D., and G.M. Clemence, 1961. *Methods of Celestial Mechanics*. Academic Press, New York, 598 pp.

Brown, G.C., and A.E. Mussett, 1981. *The Inaccessible Earth.* George Allen and Unwin, London.

Buie, M.W., D.J. Tholen, and L.H. Wasserman, 1997. Separate lightcurves of Pluto and Charon. *Icarus* **125**, 233–244.

Buratti, B.J., J.K. Hillier, and M. Wang, 1996. The lunar opposition surge: observations by Clementine. *Icarus* **124**, 490–499.

Burns, J.A., P.L. Lamy, and S. Soter, 1979. Radiation forces on small particles in the solar system. *Icarus* **40**, 1–48.

Burns, J.A., M.R. Showalter, and G.E. Morfill, 1984. The ethereal rings of Jupiter and Saturn. In *Planetary Rings*. Eds. R. Greenberg and A. Brahic, University of Arizona Press, Tucson. pp. 200–272.

Burns, J.A., D.P. Hamilton, M.R. Showalter, P.D. Nicholson, I. de Pater, and P.C. Thomas, 1999. The formation of Jupiter's faint rings. *Science* **284**, 1146–1150.

Burrows, A., M. Marley, W.B. Hubbard, J.I. Lunine, T. Guillot, D. Saumon, R. Freedman, D. Sudarsky, and C. Sharp, 1997. A non-gray theory of extrasolar giant planets and brown dwarfs. *Astrophys. J.* **491**, 856–875.

Carlson, B.E., A.A. Lacey, and W.B. Rossow, 1992. The abundance and distribution of water vapor in the Jovian troposphere as inferred from Voyager IRIS observations. *Astrophys. J.* **388**, 648–668.

Carlson, R.W., and the Galileo NIMS team, 1991. Galileo infrared imaging spectroscopy measurements at Venus. *Science* **253**, 1541–1548.

Carr, M.H., 1999. Mars: surface and interior. In *Encyclopedia of the Solar System*. Eds. P.R. Weissman, L. McFadden, and T.V. Johnson. Academic Press, Inc., New York. pp. 291–308.

Carr, T.D., M.D. Desch, and J.K. Alexander, 1983. Phenomenology of magnetospheric radio emissions. In *Physics of the Jovian Magnetosphere*. Ed. A.J. Dessler, Cambridge University Press, Cambridge. pp. 226–284.

Catermole, P., 1994. *Venus, The Geological Story*. Johns Hopkins University Press.

Chamberlain, J.W., and D.M. Hunten, 1987. *Theory of Planetary Atmospheres*, Academic Press, Inc., New York.

Chen, F.F., 1974. *Introduction to Plasma Physics*. Plenum Press, New York.

Chiang, E.I., and M.E. Brown, 1999. Keck pencil-beam survey for faint Kuiper belt objects. *Astron. J.* **118**, 1411–1422.

Clancy, R.T., B.J. Sandor, M.J. Wolff, P.R. Christensen, M.D. Smith, J.C. Pearl, B.J. Conrath, and R.J. Wilson, 2000. An intercomparison of ground-based millimeter, MGS TES, and Viking atmospheric temperature measurements: Seasonal and interannual variability of temperatures and dust loading in the global Mars atmosphere. *J. Geophys. Res.* **105**, 9553–9572.

Clark, B.E., J. Veverka, P. Helfenstein, P. Thomas, J.F. Bell III, J. Joseph, A. Harch, M.S. Robinson, S. Murchie, L. McFadden, and C. Chapman, 1999. NEAR photometry of asteroid 253 Mathilde. *Icarus* **140**, 53–65.

Clark, B.E., J. Veverka, P. Helfenstein, P. Thomas, J.F. Bell III, J. Joseph, A. Harch, B. Carcich, M.S. Robinson, S. Murchie, A. Cheng, N. Izenberg, L. McFadden, C. Chapman, W. Merline, and M. Malin, 1998. NEAR photometry of C-type asteroid 253 Mathilde. *LPSC* **29**, 1768–1769.

Clark, R.N., F.P. Fanale, and M.J. Gaffey, 1986. Surface composition of natural satellites. In *Satellites*. Eds. J.A. Burns, and M.S. Matthews. University of Arizona Press, Tucson. pp. 437–491

Clarke, J.T., S.A. Stern, and L.M. Trafton, 1992. Pluto's extended atmosphere: an escape model and initial observations. *Icarus* **95**, 173–179.

Clayton, D.D., 1983. *Principles of Stellar Evolution and Nucleosynthesis*, University of Chicago Press.

Conrath, B., and D. Gautier, 2000. Saturn helium abundance: A reanalysis of Voyager measurements. *Icarus* **144**, 124–134.

Conrath, B.J., F.M. Flasar, R. Hanel, V. Kunde, W. Maguire, J. Pearl, J. Pirraglia, R. Samuelson, P. Gierasch, A. Weir, B. Bezard, D. Gautier, D. Cruikshank, L. Horn, R. Springer, and W. Shaffer, 1989. *Science* **246**, 1454–1459.

Conrath, B.J., R.A. Hanel, and R.E. Samuelson, 1989. Thermal structure and heat balance of the outer planets. In *Origin and Evolution of Planetary and Satellite Atmospheres*. Eds. S.K. Atreya, J.B. Pollack, and M.S. Matthews. University of Arizona Press, Tucson. pp. 513–538.

Courtin, R., D. Gautier, A. Marten, B. Bezard, and R. Hanel, 1984. NH_3, PH_3, C_2H_2, C_2H_6, CH_3D, CH_4, and the Saturnian D/H isotopic ratio. *Astrophys. J.* **287**, 899–916.

Coustenis, A., and R.D. Lorenz, 1999. Titan. In *Encyclopedia of the Solar System*. Eds. P.R. Weissman, L. McFadden, and T.V. Johnson. Academic Press, Inc., New York. pp. 377–404.

Coustenis, A., A. Salama, E. Lellouch, Th. Encrenaz, G.L. Bjoraker, R.E. Samuelson, D. Gautier, Th. de Graauw, H. Feuchtgruber, M.F. Kessler, and G.S. Orton, 1998. Titan's atmosphere from ISO observations: temperature, composition and detection of water vapor. *BAAS* **30**, 1060.

Cowley, S.W.H., 1995. The Earth's magnetosphere: A brief beginner's guide. *EOS* **51**, 525–529.

Cremonese, G., H. Boehnhardt, J. Crovisier, H. Rauer, A. Fitzsimmons, M. Fulle, J. Licandro, D. Pollacco, G.P. Tozzi, R.M. West, 1997. Neutral sodium from comet Hale–Bopp: A third type of tail. *Astrophys. J. Lett.* **490**, L199–202.

Cuzzi, J.N., and J.D. Scargle, 1985. Wavy edges suggest moonlet in Encke's gap. *Astrophys. J.* **293**, 276–290.

Cuzzi, J.N., J.J. Lissauer, L.W. Esposito, J.B. Holberg, E.A. Marouf, G.L. Tyler, and A. Boishot, 1984. Saturn's rings: Properties and processes. In *Planetary Rings*. Eds. R. Greenberg, and A. Brahic, University of Arizona Press, Tucson. pp. 73–199.

Danby, J.M.A., 1988. *Fundamentals of Celestial Mechanics*, 2nd Edition. Willmann-Bell, Richmond, VA., 467 pp.

de Graauw, *et al.*, 1997. First results of ISO-SWS observations of Saturn: detection of CO_2, CH_3C_2H, C_4H_2 and tropospheric H_2O. *Astron. Astrophys.* **321**, L13–L16.

de Pater, I., 1980. 21 cm maps of Jupiter's radiation belts from all rotational aspects. *Astron. Astrophys.* **88**, 175–183.

de Pater, I., 1981. Radio maps of Jupiter's radiation belts and planetary disk at $\lambda = 6$ cm. *Astron. Astrophys.* **93**, 370–381.

de Pater, I., 1999. Solar System studies with the SKA. In *Perspectives in Radio Astronomy: Science with Large Millimeter Arrays*. Proceedings of a conference April 7–9, 1999. Ed. M.P. van Haarlem. (ASTRON/NFRA) pp. 327–338.

de Pater, I., and M.J. Klein, 1989. Time variability in Jupiter's synchrotron radiation. In *Time Variable Phenomena in the Jovian System*. Eds. M.J.S. Belton, R.A. West, and J. Rahe. Proceedings of a conference held in Flagstaff, Aug. 25–27, 1987, pp. 139–150.

de Pater, I., and D.L. Mitchell, 1993. Microwave observations of the planets: the importance of laboratory measurements. *J. Geophys. Res. Planets* **98**, 5471–5490.

de Pater, I., and R.J. Sault, 1998. An intercomparison of 3-D reconstruction techniques using data and models of Jupiter's synchrotron radiation. *J. Geophys. Res. Planets* **103**, 19,973–19,984.

de Pater, I., D. Dunn, K. Zahnle, and P.N. Romani, 2001. Comparison of Galileo probe data with ground-based radio measurements. *Icarus* **149**, 66–78.

de Pater, I., P. Palmer, S.J. Ostro, D.L. Mitchell, D.K. Yeomans, and L.E. Snyder, 1994. Radar aperture-synthesis observations of asteroids. *Icarus* **111**, 489–502.

de Pater, I., P. Palmer, and L.E. Snyder, 1986. The brightness distribution of OH around comet Halley. *Astrophys. J. Lett.* **304**, L33–L36.

de Pater, I., P. Palmer, and L.E. Snyder, 1991c. Review of interferometric imaging of comets. In *Comets in the Post-Halley Era*. Eds. R.L. Newburn, and J. Rahe. A book as a result from an international meeting on Comets in the Post-Halley Era, Bamberg, April 24–28, 1989, 175–207.

de Pater, I., P.N. Romani, and S.K. Atreya, 1989. Uranus' deep atmosphere revealed. *Icarus* **82**, 288–313.

de Pater, I., P.N. Romani, and S.K. Atreya, 1991b. Possible microwave absorption by H_2S gas in Uranus and Neptune's atmospheres. *Icarus* **91**, 220–233.

de Pater, I., F.P. Schloerb, and A. Rudolph, 1991a. CO on Venus imaged with the Hat Creek Radio Interferometer. *Icarus* **90**, 282–298.

de Pater, I., M.R. Showalter, J.A. Burns, P.D. Nicholson, M.C. Liu, D.P. Hamilton, and J.R. Graham, 1999. Keck infrared observations of Jupiter's ring system near Earth's 1997 ring plane crossing. *Icarus* **138**, 214–223.

de Pater, I., M.R. Showalter, J.J. Lissauer, and J.R. Graham, 1996. Keck infrared observations of Saturn's E and G rings during Earth's 1995 ring plane crossing. *Icarus* **121**, 195–198.

de Pater, I., F. van der Tak, R.G. Strom, and S.H. Brecht, 1997. The evolution of Jupiter's radiation belts after the impact of comet D/Shoemaker–Levy 9. *Icarus* **129**, 21–47. Erratum (Fig. reproduction): 1998, **131**, 231.

Dermott, S.F., and C.D. Murray, 1981. The dynamics of tadpole and horseshoe orbits I. Theory. *Icarus* **48**, 1–11.

Desch, M.D., M.L. Kaiser, P. Zarka, A. Lecacheux, Y. LeBlanc, M. Aubier, and A. Ortega-Molina, 1991. Uranus as a radio source. In *Uranus*. Eds. J.T. Bergstrahl, A.D. Miner, and M.S. Matthews. University of Arizona Press, Tucson. pp. 894–925.

Despois, D., E. Gerard, J. Crovisier, and I. Kazes, 1981. The OH radical in comets: observation and analysis of the hyperfine microwave transitions at 1667 MHz and 1665 MHz. *Astron. Astrophys.* **99**, 320–340.

Dollfus, A., M. Wolff, J.E. Geake, D.F. Lupishko, and L.M. Dougherty, 1989. Photopolarimetry of asteroids. In *Asteroids II*. Eds. R.P. Binzel, T. Gehrels, and M.S. Matthews. University of Arizona Press, Tucson. pp. 594–616.

Domingue, D.L., M. Robinson, B. Carcich, J. Joseph, P. Thomas, and B.E. Clark, 2001. Disk-integrated photometry of 433 Eros. *Icarus*, in press.

Dowling, T.E., 1999. Earth as a planet: atmosphere and oceans. In *Encyclopedia of the Solar System*. Eds. P.R. Weissman, L. McFadden, and T.V. Johnson. Academic Press, Inc., New York. pp. 191–208.

Drummond, J., A. Eckart, and E.K. Hege, 1988. Speckle interferometry of asteroids. IV. Reconstructed images of 4 Vesta. *Icarus* **73**, 1–14.

Duncan, M.J., and T. Quinn, 1993. The long-term dynamical evolution and stability of the Solar System. In *Protostars and Planets III*. Eds. E.H. Levy, and J.I. Lunine. University of Arizona Press, Tucson. pp. 1371–1394.

Duncan, M.J., H.F. Levison, and S.M. Budd, 1995. The dynamical structure of the Kuiper belt. *Astron. J.* **110**, 3073–3081.

Durisen, R.H., P.W. Bode, J.N. Cuzzi, S.E. Cederbloom, and B.W. Murphy, 1992. Ballistic transport in planetary ring systems due to particle erosion mechanisms. II. Theoretical models for Saturn's A- and B-ring inner edges. *Icarus* **100**, 364–393.

Durisen, R.H., P.W. Bode, S.G. Dyck, J.N. Cuzzi, J.D. Dull, and J.C. White II, 1996. Ballistic transport in planetary ring systems due to particle erosion mechanisms III. Torques and mass loading by meteoroid impacts. *Icarus* **124**, 220–236.

Elliot, J.L., E. Dunham, and D. Mink, 1977. The rings of Uranus. *Nature* **267**, 328–330.

Encrenaz, Th., B. Bezard, E. Lellouch, B. Schulz, P.N. Romani, and S.K. Atreya, 1999. Detection of C_2H_4 in Neptune from ISO/PHT-S observations. *BAAS* **31**, 1180.

Encrenaz, T., J.-P. Bibring, and M. Blanc, 1995. *The Solar System*, 2nd Edition. Springer-Verlag, Berlin.

Esposito, L.W., 1993. Understanding planetary rings. *Annu. Rev. Earth Planet. Sci.* **21**, 487–521.

Esposito, L.W., J-L. Bertaux, V. Krasnopolsky, V.I. Moroz, and L.V. Zasova, 1997. Chemistry of lower atmosphere and clouds. In *Venus II*. Eds. S.W. Bougher, D.M. Hunten, and R.J. Phillips. University of Arizona Press, Tucson. pp. 415–422.

Esposito, L.W., A. Brahic, J.A. Burns, and E.A. Marouf, 1991. Particle properties and processes in Uranus' rings. In *Uranus*. Eds. J.T. Bergstrahl, E.D. Miner, and M.S. Matthews. University of Arizona Press, Tucson. pp. 410–465.

Etheridge, D.M., L.P. Steele, R.L. Langenfelds, R.J. Francey, J.-M. Barnola, and V.I. Morgan, 1996. Natural and anthropogenic changes in atmospheric CO_2 over the last 1000 years from air in Antarctic ice and firn. *J. Geophys. Res.* **101**, 4115–4128.

Fahd, A.K., and P.G. Steffes, 1992. Laboratory measurements of the microwave and millimeter wave opacity of gaseous sulfur dioxide (SO_2) under simulated conditions for the Venus atmosphere. *Icarus* **97**, 200–210.

Farinella, P., D.R. Davis, and S.A. Stern, 2000. Formation and collisional evolution of the Edgeworth–Kuiper belt. In *Protostars and Planets IV*. Eds. V. Manning, A.P. Boss, and S.S. Russell. University of Arizona Press, Tucson. pp. 1255–1282.

Faust, J.A., *et al.*, 1997. Determination of the Charon/Pluto mass ratio from the center-of-light asymmetry. *Icarus* **126**, 362–372.

Fernandez, J.A., and K. Jockers, 1983. Nature and origin of comets. *Rep. Prog. Phys.* **46**, 665–772.

Festou, M.C., H. Rickman, and R.M. West, 1993a. Comets. 1 – Concepts and observations. *Astron. Astrophys. Rev.* **4**, 363–447.

Festou, M.C., H. Rickman, and R.M. West, 1993b. Comets. 2 – Models, evolution, origin and outlook. *Astron. Astrophys. Rev.* **5**, 37–163.

Finson, M.L., and R.F. Probstein, 1968. A theory of dust comets. 1. Model and equations. *Astrophys. J.* **154**, 327–380.

Fowler, C.M.R., 1990. *The Solid Earth: An Introduction to Global Geophysics*. Cambridge University Press, Cambridge.

Frankel, C., 1996. *Volcanoes of the Solar System*. Cambridge University Press, Cambridge.

French, R.G., P.D. Nicholson, C.C. Porco, and E.A. Marouf, 1991. Dynamics and structure of the uranian rings. In *Uranus*. Eds. J.T. Bergstrahl, E.D. Miner, and M.S. Matthews. University of Arizona Press, Tucson. pp. 327–409.

Gaffey, M.J., J.F. Bell, and D.P. Cruikshank, 1989. Reflectance spectroscopy and asteroid surface mineralogy. In *Asteroids II*. Eds. R.P. Binzel, T. Gehrels, and M.S. Matthews. University of Arizona Press, Tucson. pp. 98–127.

Gautier, D., and T. Owen, 1989. The composition of outer planet atmospheres. In *Origin and Evolution of Planetary and Satellite Atmospheres*. Eds. S.K. Atreya, J.B. Pollack, and M.S. Matthews. University of Arizona Press, Tucson. pp. 487–512.

Gautier, D., B.J. Conrath, T. Owen, I. de Pater, and S.K. Atreya, 1995. The troposphere of Neptune. In *Neptune and Triton*. Eds. D.P. Cruikshank, and M.S. Matthews, University of Arizona Press, Tucson. pp. 547–612.

Ghil, M., and S. Childress, 1987. *Topics in Geophysical Fluid Dynamics: Atmospheric Dynamics, Dynamo Theory, and Climate Dynamics*. Springer-Verlag, New York.

Gibbard, S.G., B. Macintosh, D. Gavel, C.E. Max, I. de Pater, A.M. Ghez, E.F. Young, and C.P. McKay, 1999. Titan: High resolution speckle images from the Keck telescope. *Icarus* **139**, 189–201.

Gibbard, S.G., B. Macintosh, D. Gavel, C.E. Max, I. de Pater, A.M. Ghez, E.F. Young, and C.P. McKay, 2001. Submitted.

Ginzburg, V., and S. Syrovatskii, 1965. Cosmic magnetobremsstrahlung (synchrotron radiation). *Annu. Rev. Astron. Astrophys.* **3**, 297–350.

Goguen, J.D., D.L. Blaney, G.J. Veeder, D.L. Matson, T.V. Johnson, P.D. Nicholson, T.L. Hayward, J.E. Van Cleve, D. Toomey, L. Bergknut, C. Kaminsky, W. Golisch, and D. Griep, 1997. IR photometry and spectra of Io occultations in 1997. *BAAS* **29**, 978.

Goldstein, H., 1980. *Classical Mechanics*, 2nd edition. Addison Wesley, MA.

Goody, R.M., and J.C.G. Walker, 1972. *Atmospheres*. Prentice Hall, Englewood-Cliffs, NJ.

Gosling, J.T., 1999. The solar wind. In *Encyclopedia of the Solar System*. Eds. P.R. Weissman, L. McFadden, and T.V. Johnson. Academic Press, Inc., New York. pp. 95–122.

Gradie, J.C., C.R. Chapman, and E.F. Tedesco, 1989. Distribution of taxonomic classes and the compositional structure of the asteroid belt. In *Asteroids II*. Eds. R.P. Binzel, T. Gehrels, and M.S. Matthews. University of Arizona Press, Tucson. pp. 316–335.

Graham, J.R., I. de Pater, J.G. Jernigan, M.C. Liu, and M.E. Brown, 1995. W.M. Keck telescope observations of the Comet P/Shoemaker–Levy 9 fragment R Jupiter collision. *Science* **267**, 1320–1323

Greeley, R., 1994. *Planetary Landscapes*. Chapman and Hall, New York, London.

Grevesse, N., D.L. Lambert, A.J. Sauval, E.F. van Dishoeck, C.B. Farmer, and R.H. Norton, 1991. Vibration rotation bands of CH in the solar infrared spectrum and the solar carbon abundance. *Astron. Astrophys. J.* **242**, 482–495.

Guillot, T., 1999. Interiors of giant planets inside and outside the solar system. *Science* **286**, 72–77.

Guillot, T., G. Chabrier, D. Gautier, and P. Morel, 1995. Effect of radiative transport on the evolution of Jupiter and Saturn. *Astrophys. J.* **450**, 463–472.

Gulkis, S., and I. de Pater, 1992. Radio astronomy, planetary. In *Encyclopedia of Physical Science and Technology*, **14**, Academic Press, Inc., New York. pp. 141–164.

Gurnett, D.A., W.S. Kurth, A. Roux, S.J. Bolton, and C.F. Kennel, 1996. Evidence of a magnetosphere at Ganymede from Galileo plasma wave observations. *Nature* **384**, 535–537.

Hamblin, W.K., and E.H. Christiansen, 1990. *Exploring the Planets*. Macmillan Publishing Company, New York.

Hamilton, D.P., 1993. Motion of dust in a planetary magnetosphere: orbit-averaged equations for oblateness, electromagnetic, and radiation forces with application to Saturn's E ring. *Icarus* **101**, 244–264.

Hanel, R.A., B.J. Conrath, D.E. Jennings, and R.E. Samuelson, 1992. *Exploration of the Solar System by Infrared Remote Sensing*. Cambridge University Press, Cambridge.

Hapke, B., R. Nelson, and W. Smythe, 1998. The opposition effect of the Moon: Coherent backscatter and shadow hiding. *Icarus* **133**, 89–97.

Harmon, J.K., D.B. Campbell, A.A. Hine, I.I Shapiro, and B.G. Marsden, 1989. Radar observations of comet IRAS–Araki–Alcock. *Astrophys. J.* **338**, 1071–1093.

Harmon, J.K., M.A. Slade, R.A. Velez, A. Crespo, M.J. Dryer, and J.M. Johnson, 1994. Radar mapping of Mercury's polar anomalies. *Nature* **369**, 213–215.

Hartmann, W.K., 1989. *Astronomy: The Cosmic Journey*. Wadsworth Publishing Company, Belmont, CA.

Hartmann, W.K., 1993. *Moons and Planets*, 3rd Edition. Wadsworth Publishing Company, Belmont, CA.

Hartmann, W.K., 1999. *Moons and Planets*, 4th Edition. Wadsworth Publishing Company, Belmont, CA.

Helfenstein, P., J. Veverka, and J. Hiller, 1997. The Lunar opposition effect: A test of alternative models. *Icarus* **128**, 2–14.

Helfenstein, P.J., *et al.*, 1994. Galileo photometry of asteroid 951 Gaspra. *Icarus* **107**, 37–60.

Herzberg, G., 1944. *Atomic Spectra and Atomic Structure*. Dover Publications, New York.

Hewins, R.H., R.H. Jones, and E.R.D. Scott, Eds. 1996. *Chondrules and the Protoplanetary Disk*. Cambridge University Press, Cambridge. 346 pp.

Hill, T.W., A.J. Dessler, and C.K. Goertz, 1983. Magnetospheric models. In *Physics of the Jovian Magnetosphere*. Ed. A.J. Dessler. Cambridge University Press, Cambridge.

Hofstadter, M.D., 1993. Microwave imaging of Neptune's troposphere. *Bull. Amer. Astron. Soc.* **25**, 1077.

Hood, L., and J. Jones, 1987. Geophysical constraints on lunar bulk composition and structure: A reassessment. Proc. 17th Lunar Planet. Sci. Conf., Part 2. *J. Geophys. Res.* **92**, E396–E410.

Hood, L., and M.T. Zuber, 1999. Recent refinements in geophysical constraints on lunar origin and evolution. In *Origin of the Earth and Moon*. Eds. R. Canup and K. Righter. University of Arizona Press, Tucson.

Howard, A.D., 1967. Drainage analysis in geological interpretation: a summation. *Am. Ass. Petrol. Geol. Bull.* **51**, 2246–2259.

Hubbard, W.B., 1984. *Planetary Interiors*. Van Nostrand Reinhold Company Inc., New York.

Hubbard, W.B., M. Podolak, and D.J. Stevenson, 1995. The interior of Neptune. In *Neptune and Triton*. Ed. D.P. Cruikshank. University of Arizona Press, Tucson. pp. 109–138.

Hudson, R.S., and S.J. Ostro, 1994. Shape of asteroid 4769 Castalia (1989 PB) from inversion of radar images. *Science* **263**, 940–943.

Huebner, W.F., Ed. 1990. *Physics and Chemistry of Comets*. Springer-Verlag, Berlin.

Hughes, W.J., 1995. The magnetopause, magnetotail, and magnetic reconnection. In *Introduction to Space Physics*. Eds. M.G. Kivelson, and C.T. Russell. Cambridge University Press, Cambridge. pp. 227–287.

Hundhausen, A.J., 1995. The solar wind. In *Introduction to Space Physics*. Eds. M.G. Kivelson, and C.T. Russell. Cambridge University Press, Cambridge. pp. 91–128.

Hunten, D.M., 1999. Venus: Atmosphere. In *Encyclopedia of the Solar System*. Eds. P.R. Weissman, L. McFadden, and T.V. Johnson. Academic Press, Inc., New York. 147–160.

Hunten, D.M., and J. Veverka, 1976. Stellar and spacecraft occultations by Jupiter: A critical review of derived temperature profiles. In *Jupiter*. Ed. T. Gehrels. University of Arizona Press, Tucson. pp. 247–283.

Hunten, D.M., T.M. Donahue, J.C.G. Walker, and J.F. Kasting, 1989. Escape of atmospheres and loss of water. In *Origin and Evolution of Planetary and Satellite Atmospheres*. Eds. S.K. Atreya, J.B. Pollack, and M.S. Matthews. University of Arizona Press, Tucson. pp. 386–422.

Hunten, D.M., T.H. Morgan, and D.E. Shemansky, 1988. The Mercury atmosphere. In *Mercury*. Eds. F. Vilas, C.R. Chapman, and M.S. Matthews. University of Arizona Press, Tucson. pp. 562–612.

Hunten, D.M., M.G. Tomasko, F.M. Flasar, R.E. Samuelson, D.F. Strobel, and D.J. Stevenson, 1984. Titan. In *Saturn*. Eds. T. Gehrels, M.S. Matthews. University of Arizona Press, Tucson. pp. 671–759.

Ingersoll, A.P., 1990. Atmospheres of the giant planets. In *The New Solar System*. Eds. J.K. Beatty, and A. Chaikin, 3rd Edition. Cambridge University Press and Sky Publishing Corporation. pp. 139–152.

Ingersoll, A.P., 1991. Atmospheric dynamics of the outer planets. *Science* **248**, 308–315.

Ingersoll, A.P., 1999. Atmospheres of the giant planets. In *The New Solar System*. Eds. J.K. Beatty, C.C. Petersen, and A. Chaikin, 4th Edition. Cambridge University Press and Sky Publishing Corporation. pp. 201–220.

Ip, W.-H., and W.I. Axford, 1990. The Plasma. In *Physics and Chemistry of Comets*. Ed. W.F. Huebner. Springer-Verlag, Berlin. pp. 177–233.

Irvine, W.M., F.P. Schloerb, J. Crovisier, B. Fegley Jr., and M.J. Mumma, 2000. Comets: A link between interstellar and nebular chemistry. In *Protostars and Planets IV*. Eds. V. Manning, A.P. Boss, and S.S. Russell. University of Arizona Press, Tucson. pp. 1159–1200.

Jacobs, J.A., 1987. *The Earth's Core*, 2nd Edition. Academic Press, New York.

Jacobson, M.Z., 1999. *Fundamentals of Atmospheric Modeling*. Cambridge University Press, New York. 656 pp.

Jeanloz, R., 1989. Physical chemistry at ultrahigh pressures and temperatures. *Annu. Rev. Phys. Chem.* **40**, 237–259.

Jewitt, D.C., and J.X. Luu, 2000. Physical nature of the Kuiper belt. In *Protostars and Planets IV*. Eds. V. Manning, A.P. Boss, and S.S. Russell. University of Arizona Press, Tucson. pp. 1201–1229.

Karkoschka, E., 1994. Spectrophotometry of the Jovian planets and Titan at 300 to 1000 nm wavelength: The methane spectrum. *Icarus* **111**, 174–192.

Kary, D.M., and L. Dones, 1996. Capture statistics of short-period comets: Implications for Comet D/Shoemaker–Levy 9. *Icarus* **121**, 207–224.

Keen, B.E., Ed. 1974. *Plasma Physics*. The Institute of Physics, London.

Kerridge, J.F., 1993. What can meteorites tell us about nebular conditions and processes during planetary accretion? *Icarus* **106**, 135–150.

Kerridge, J.F., M.S. Matthews, Eds., 1988. *Meteorites and the Early Solar System*. University of Arizona Press, Tucson. 1269 pp.

Kiselev, N.N., and F.P. Velichko, 1998. Polarimetry and photometry of comet C/1996 B2 Hyakutake. *Icarus* **133**, 286–292.

Kivelson, M.G., 1995. Pulsations and magnetohydrodynamic waves. In *Introduction to Space Physics*. Eds. M.G. Kivelson, and C.T. Russell. Cambridge University Press, Cambridge. pp. 330–355.

Kivelson, M.G., and F. Bagenal, 1999. Planetary magnetospheres. In *Encyclopedia of the Solar System*. Eds. P.R. Weissman, L. McFadden, and T.V. Johnson. Academic Press, Inc., New York. pp. 477–498.

Kivelson, M.G., and G. Schubert, 1997. Atmospheres of the terrestrial planets. In *The Solar System: Observations and Interpretations*. Ed. M.G. Kivelson. Rubey Vol. IV. Prentice Hall, Englewood Cliffs, NJ.

Klein, M.J., S. Gulkis, and S.J. Bolton, 1995. Changes in Jupiter's 13-cm synchrotron radio emission following the impacts of comet Shoemaker–Levy-9. *GRL* **22**, 1797–1800.

Klein, M.J., T.J. Thompson, and S.J. Bolton, 1989. Systematic observations and correlation studies of variations in the synchrotron radio emission from Jupiter. In *Time Variable Phenomena in the Jovian System*. Eds. M.J.S. Belton, R.A. West, and J. Rahe. Proceedings of a conference held in Flagstaff, Aug. 25–27, 1987. pp. 151–155.

Kliore, A.J., I.R. Patel, G.F. Lindal, D.N. Sweetnam, H.B. Hotz, J.H. Waite Jr., and T.R. McDonough, 1980. Structure of the ionosphere and atmosphere of Saturn from Pioneer 11, Saturn radio occultation. *J. Geophys. Res.* **85**, 5857–5870.

Kokubo, E., and S. Ida, 1999. Formation of protoplanets from planetesimals in the solar nebula. *Icarus* **143**, 15–27.

Kokubo, E., R.M. Canup, and S. Ida, 2000. Lunar accretion from an impact-generated disk. In *Origin of the Earth and Moon*. Eds. R.M. Canup, and K. Righter. University of Arizona Press, Tucson. pp. 145–163.

Konopliv, A.S., W.B. Banerdt, and W.L. Sjogren, 1999. Venus gravity: 180th degree and order model. *Icarus* **139**, 3–18.

Konopliv, A.S., A.B. Binder, L.L. Hood, A.B. Kucinskas, W.L. Sjogren, and J.G. Williams, 1998. Improved gravity field of the Moon from Lunar Prospector. *Science* **281**, 1476–1480.

Kowal, C.T., 1996. *Asteroids: Their Nature and Utilization*. 2nd Edition. Wiley, New York.

Kurth, W.S., 1997. Whistler. In *Encyclopedia of Planetary Sciences*. Eds. J.H. Shirley, and R.W. Fairbridge. Chapman and Hall, London. pp. 936–937.

Kurth, W.S., and D.A. Gurnett, 1991. Plasma waves in planetary magnetosphere. *J. Geophys. Res.* **96**, 18,977–18,991.

Lazzaro, D., M.A. Florczak, C.A. Angeli, J.M. Carvano, A.S. Betzler, A.A. Casati, M.A. Barucci, A. Doressoundiram, and M. Lazzarin, 1997. Photometric monitoring of 2060 Chiron's brightness at perihelion. *Planet. Space Sci.* **45**, 1606–1614.

Lellouch, E., M.J.S. Belton, I. de Pater, S. Gulkis, and T. Encrenaz, 1990. Io's atmosphere from microwave detection of SO_2. *Nature* **346**, 639–641.

Lellouch, E., M. Belton, I. de Pater, G. Paubert, S. Gulkis, and Th. Encrenaz, 1992. The structure, stability, and global distribution of Io's atmosphere. *Icarus* **98**, 271–295.

Lellouch, E., D.F. Strobel, and M. Belton, 1995. Detection of SO in the atmosphere of Io. *BAAS* **27**, 1155.

Levison, H.F., Lissauer, J.J., Duncan, M.J., 1998. Modeling the diversity of outer planetary systems. *Astron. J.* **116**, 1998–2014.

Levy, E.H., 1986. The generation of magnetic fields in planets. In *The Solar System: Observations and Interpretations*. Ed. M.G. Kivelson. Rubey Vol. IV. Prentice Hall, Englewood Cliffs, NJ. pp. 289–310.

Lewis, J.S., 1995. *Physics and Chemistry of the Solar System*. Academic Press, San Diego.

Li, X.D., and B. Romanowicz, 1996. Global mantle shear velocity model developed using nonlinear asymptotic coupling theory. *J. Geophys. Res.* **101**, No B10, pp. 22,245–22,272.

Lin, D.N.C., 1986. The nebular origin of the Solar System. In *The Solar System: Observations and Interpretations*, Rubey Vol. IV. Ed. M.G. Kivelson. Prentice Hall, Englewood Cliffs, NJ. pp. 28–87.

Lindal, G.F., 1992. The atmosphere of Neptune: An analysis of radio occultation data acquired with Voyager 2. *Astron. J.* **103**, 967–982.

Lindal, G.F., J.R. Lyons, D.N. Sweetnam, V.R. Eshelman, D.P. Hinson, and G.L. Tyler, 1987. The atmosphere of Uranus: Results of radio occultation measurements with Voyager 2. *J. Geophys. Res.* **92**, 14,987–15,002

Lipschutz, M.E., and L. Schultz, 1999. Meteorites. In *Encyclopedia of the Solar System*. Eds. P.R. Weissman, L. McFadden, and T.V. Johnson. Academic Press, Inc., New York. pp. 629–671.

Lissauer, J.J., 1993. Planet formation. *Annu. Rev. Astron. Astrophys.* **31**, 129–174.

Lissauer, J.J., 1995. Urey Prize lecture: On the diversity of plausible planetary systems. *Icarus* **114**, 217–236.

Lissauer, J.J., 1999a. Chaotic motion in the Solar System. *Rev. Mod. Phys.* **71**, 835–845.

Lissauer, J.J., 1999b. How common are habitable planets? *Nature* **402**, C11–C14

Lissauer, J.J., and J.N. Cuzzi, 1982. Resonances in Saturn's rings. *Astron. J.* **87**, 1051–1058.

Lissauer, J.J., and J.N. Cuzzi, 1985. Rings and moons: clues to understanding the solar nebula. In *Protostars and Planets II*. Eds. D. Black and M.S. Matthews. University of Arizona Press, Tucson. pp. 920–956.

Lissauer, J.J., J.B. Pollack, G.W. Wetherill, and D.J. Stevenson, 1995. Formation of the Neptune system. In *Neptune and Triton*. Ed. D.P. Cruikshank. University of Arizona Press, Tucson, pp. 37–108.

Lithgow-Bertelloni, C., and M.A. Richards, 1998. The dynamics of cenozoic and mesozoic plate motions. *Rev. Geophys.* **36**, 27–78.

Lodders, K., and B. Fegley, Jr., 1998. *The Planetary Scientist's Companion*. Oxford University Press, Oxford.

Luhmann, J.G., 1995. Plasma interactions with unmagnetized bodies. In *Introduction to Space Physics*. Eds. M.G. Kivelson and C.T. Russell. Cambridge University Press, Cambridge. pp. 203–226.

Luhmann, J.G., C.T. Russell, L.H. Brace, and O.L. Vaisberg, 1992. The intrinsic magnetic field and solar-wind interaction of Mars. In *Mars*. Eds. H.H. Kieffer, B.M. Jakosky, C.W. Snyder, and M.S. Matthews. University of Arizona Press, Tucson. pp. 1090–1134.

Lunine, J.I., and W.C. Tittemore, 1993. Origins of outer-planet satellites. In *Protostars and Planets III*. Eds. E.H. Levy and J.I. Lunine. University of Arizona Press, Tucson. pp. 1149–1176.

Luu, J.X., 1993a. Cometary activity in distant comets: Chiron. *PASP* **105**, 946–950.

Luu, J.X., 1993b. Spectral diversity among the nuclei of comets. *Icarus* **104**, 138–148.

Luu, J., D. Jewitt, and E.J. Cloutis, 1994. Near-infrared spectroscopy of primitive solar system objects. *Icarus* **109**, 133–144.

Lyons, L.R., and D.J. Williams, 1984. *Quantitative Aspects of Magnetospheric Physics*. Reidel Publishing Company, Dordrecht.

Macintosh, B., D. Gavel, S. Gibbard, C. Max, I de Pater, A. Ghez, and M. Eckart, 2001. Keck speckle imaging of volcanoes on Io. *Icarus*, submitted.

Magnusson, P., M.A. Barucci, J.D. Drummond, K. Lumme, S.J. Ostro, J. Surdej, R.C. Taylor, and V. Zappalà, 1989. Determination of pole orientations and shapes of asteroids. In *Asteroids II*. Eds. R.P. Binzel, T. Gehrels, and M.S. Matthews. University of Arizona Press, Tucson. pp. 66–97.

Malhotra, R., M.J. Duncan, and H.F. Levison, 2000. Dynamics of the Kuiper belt. In *Protostars and Planets IV*. Eds. V. Manning, A.P. Boss, and S.S. Russell. University of Arizona Press, Tucson. pp. 1231–1254.

Marcus, P.S., 1993. Jupiter's great red spot and other vortices. *Annu. Rev. Astron. Astrophys.* **31**, 523–573.

Marcy, G.W., and R.P. Butler, 1998. Detection of extrasolar giant planets. *Annu. Rev. Astron. Astrophys.* **36**, 57–97.

Marcy, G.W., W.D. Cochran, and M. Mayor, 2000. Extrasolar planets around main-sequence stars. In *Protostars and Planets IV*. Eds. V. Manning, A.P. Boss, and S.S. Russell. University of Arizona Press, Tucson. pp. 1285–1311.

Marcy, G.W., R.P. Butler, S.S. Vogt, D. Fischer, and J.J. Lissauer, 1998. A planetary companion to the M4 dwarf, Gliese 876. *Astrophys. J. Lett.* **505**, L147-L149.

Margot, J.-L, M.C. Nolan, L.A.M. Benner, S.J. Ostro, M.E. Brown, E.S. Howell, R.F. Jurgens, J.D. Giorgini, M.A. Slade, D.B. Campbell, I. de Pater, and H. Roe, 2001. Discovery and characterization of three binary asteroids, 2000 DP107, 2000 UG11, and 87 Sylvia. *Proc. of the Asteroids 2001 Conference*, June 11–16, 2001, Palermo, Italy.

Marley, M.S., 1999. Interiors of the giant planets. In *Encyclopedia of the Solar System*. Eds. P.R. Weissman, L. McFadden, and T.V. Johnson. Academic Press, Inc., pp. 339–356.

Marsden, B.G., and G.V. Williams, 1997. *Catalogue of Cometary Orbits*. The International Astronomical Union, Minor Planet Center and Smithsonian Astrophysical Observatory, Cambridge, MA.

Marten, A., D. Gautier, T. Owen, D. Sanders, H.E. Matthews, S.K. Atreya, R.J.P. Tilanus, and J. Deane, 1993. First observations of CO and HCN on Neptune and Uranus at millimeter wavelengths and their implications for atmospheric chemistry. *Astrophys. J.* **406**, 285–297.

Mayor, M., and D. Queloz, 1995. A Jupiter-mass companion to a solar-type star. *Nature* **378**, 355–359.

McKay, D.S., E.K. Gibson Jr., K.L. Thomas-Keprta, H. Vali, C.S. Romanek, S.J. Clemett, X.D.F. Chillier, C.R. Maechling, and R.N. Zare, 1996. Search for past life on Mars: possible relic biogenic activity in martian meteorite ALH84001. *Science* **273**, 924–930.

McKinnon, W.B., 1997. Galileo at Jupiter – meetings with remarkable moons. *Nature* **390**, 23–26.

McPherron, R.L., 1995. Magnetospheric dynamics. In *Introduction to Space Physics*. Eds. M.G. Kivelson, and C.T. Russell. Cambridge University Press, Cambridge. pp. 400–458.

McSween, H.Y., Jr., 1999. Meteorites. In *The New Solar System*. 4th Edition. Sky Publishing Co., Cambridge, MA and Cambridge University Press, Cambridge, England. pp. 351–364.

Megnin, C., and B. Romanowicz, 2000. The 3D shear velocity structure of the mantle from the inversion of body, surface, and higher mode waveforms. *Geophys. J. Int.* **143**, 709–728

Melosh, H.J., 1989. *Impact Cratering, A Geologic Process*. Oxford Monographs on Geology and Geophysics, no. 11. Oxford University Press, New York.

Merline, W.J., L.M. Close, C. Dumas, C.R. Chapman, F. Roddier, F. Ménard, D.C. Slater, G. Duvert, J.C. Shelton, and T. Morgan, 1999. Discovery of a moon orbiting the asteroid 45 Eugenia. *Nature* **401**, 565–568.

Merline, W.J., L.M. Close, C. Dumas, J.C. Shelton, F. Ménard, C.R. Chapman, and D.C. Slater, 2000. Discovery of companions to asteroids 762 Pulcova and 90 Antiope by direct imaging. *BAAS* **32**, # 13.06

Mewaldt, R.A., R.S. Selesnick, and J.R. Cummings, 1997. Anomalous cosmic rays: the principle source of high energy heavy ions in the radiation belts. In *Radiation Belts: Models and*

Standards. Geophysical Monograph **97**. Eds. J.F. Lemaire, D. Heyndericks, and D.N. Baker. American Geophysical Union, pp. 35–42.

Miller, R., and W.K. Hartmann, 1993. *The Grand Tour, A Traveler's Guide to the Solar System*. Workman Publishing, New York.

Millis, R.L., and J.L. Elliot, 1979. Direct determination of asteroid diameters from occultation observations. In *Asteroids*. Ed. T. Gehrels. University of Arizona Press, Tucson. pp. 98–118.

Millis, R.L., and 41 others, 1987. The size, shape, density and albedo of Ceres from its occultation of BD $+8^{\circ}471$. *Icarus* **72**, 507–518.

Mitchell, D.L., 1993. *Microwave Imaging of Mercury's Thermal Emission: Observations and Models*. Ph.D. Thesis, University of California, Berkeley.

Mitchell, D.L., and I. de Pater, 1994. Microwave imaging of Mercury's thermal emission: observations and models. *Icarus* **110**, 2–32.

Mitchell, D.L., J. Halekas, R.P. Lin, K.A. Anderson, S. Frey, L.L. Hood, M.H. Acuña, and A. Binder, 2001. The global distribution of lunar crustal magnetic fields: results from the Lunar Prospector magnetometer/electron reflectometer. *Science*, submitted.

Morgan, J., *et al.*, 1997. Size and morphology of the Chicsxulub impact crater. *Nature* **390**, 472–476.

Moroz, V.I., 1983. Stellar magnitude and albedo data of Venus. In *Venus*. Eds. D.M. Hunten, L. Colin, T.M. Donahue, and V.I. Moroz. University of Arizona Press, Tucson. pp. 27–68.

Morrison, D., and T. Owen, 1996. *The Planetary System*. Addison-Wesley Publishing Company, New York.

Mottola, S., and F. Lahulla, 2000. Mutual eclipse events in asteroidal binary system 1996 FG_3: Observations and a numerical model. *Icarus* **246**, 556–567.

Muhleman, D.O., G.S. Orton, and G.L. Berge, 1979. A model of the Venus atmosphere from radio, radar, and occultation observations. *Astrophys. J.* **234**, 733–745.

Mumma, M.J., P.R. Weissman, and S.A. Stern, 1993. Comets and the origin of the Solar System: Reading the Rosetta Stone. In *Protostars and Planets III*. Eds. E.H. Levy and J.I. Lunine. University of Arizona Press, Tucson. pp. 1177–1252.

Murray, C., and S. Dermott, 1999. *Solar System Dynamics*. Cambridge University Press, Cambridge.

Namouni, F., A.A. Christou, and C.D. Murray, 1999. Coorbital dynamics at large eccentricity and inclination. *Phys. Rev. Lett.* **83**, 2506–2509.

Ness, N.F., J.E.P. Connerney, R.P. Lepping, M. Schulz, and G.-H. Voigt, 1991. The magnetic field and magnetospheric configuration of Uranus. In *Uranus*. Eds. J.T. Bergstrahl, E.D. Miner, and M.S. Matthews. University of Arizona Press, Tucson. pp. 739–779.

Neubauer, F.M., 1991. The magnetic field structure of the cometary plasma environment. In *Comets in the Post-Halley Era*. Eds. R.L. Newburn, M. Neugebauer, and J. Rahe. A book as a result from an international meeting on Comets in the Post-Halley Era, Bamberg, April 24–28, 1989, pp. 1107–1124.

Ney, E.P., 1974. Multiband photometry of Comets Kohoutek, Bennett, Bradfield, and Encke. *Icarus* **23**, 551–560.

Nicholson, P.D., 1996. Earth-based observations of impact phenomena. In *The Collision of Comet Shoemaker–Levy 9 and Jupiter*. Eds. K.S. Noll, H.A. Weaver, and P.D. Feldman. Space Telescope Science Institute Symposium Series 9, IAU Colloquium 156, Cambridge University Press, Cambridge. pp. 81–110.

Niemann, H.B., S.K. Atreya, G.R. Carignan, T.M. Donahue, J.A. Haberman, D.N. Harpold, R.E. Harthe, D.M. Hunten, W.T. Kasprzak, P.R. Mahaffy, T.C. Owen, and S.H. Way, 1998. The composition of the Jovian atmosphere as determined by the Galileo probe mass spectrometer. *J. Geophys. Res.* **103**, 22,831–22,845.

Ockert-Bell, M.E., J.A. Burns, I.J. Dauber, P.C. Thomas, J. Veverka, M.J.S. Belton, and K.P. Klaasen, 1999. The structure of the jovian ring system as revealed by the Galileo imaging experiment. *Icarus* **138**, 188–213.

Opal, C.B., and G.R. Carruthers, 1977. Lyman-alpha observations of Comet West/1975n. *Icarus* **31**, 503–508.

Ostro, S.J., 1989. Radar observations of asteroids. In *Asteroids II*. Eds. R.P. Binzel, T. Gehrels, and M.S. Matthews. University of Arizona Press, Tucson. pp. 192–212.

Ostro, S.J., *et al.*, 1995. Radar images of asteroid 4179 Toutatis. *Science* **270**, 80–83.

Ostro, S.J., D.B. Campbell, J.F. Chandler, A.A. Hine, R.S. Hudson, K.D. Rosema, and I.I. Shapiro, 1991. Astroid 1986 DA: Radar evidence for a metallic composition. *Science* **252**, 1399–1404.

Ostro, S.J., R.S. Hudson, M.C. Nolan, J-L. Margot, D.J. Scheeres, D.B. Campbell, C. Magri, J.D. Giorgini, and D.K. Yeomans, 2000. Radar observations of asteroid 216 Kleopatra. *Science* **288**, 836–839.

Ostro, S.J., R.F. Jurgens, K.D. Rosema, R.S. Hudson, J.D. Giorgini, R. Winkler, D.K. Yeomans, D. Choate, R. Rose, M.A. Slade, S.D. Howard, D.J. Scheeres, and D.L. Mitchell, 1996. Radar observations of asteroid 1620 Geographos. *Icarus* **121**, 44–66.

Palme, H., and W.N. Boynton, 1993. Meteoritic constraints on conditions in the solar nebula. In *Protostars and Planets III*. Eds. E.H. Levy and J.I. Lunine. University of Arizona Press, Tucson. pp. 979–1004.

Parker, E.N., 1963. *Interplanetary Dynamical Processes*. Interscience, New York.

Pasachoff, J.M., and M.L. Kutner, 1978. *University Astronomy*. W.B. Saunders Company, Philadelphia.

Peale, S.J., 1976. Orbital resonances in the Solar System. *Annu. Rev. Astron. Astrophys.* **14**, 215–246.

Pedlovsky, J., 1979. *Geophysical Fluid Dynamics.* Springer-Verlag, New York.

Pieri, D.C., and A.M. Dziewonski, 1999. Earth as a planet: surface and interior. In *Encyclopedia of the Solar System.* Eds. P.R. Weissman, L. McFadden, and T.V. Johnson. Academic Press, Inc., New York. pp. 209–245.

Podosek, F.A., and P. Cassen, 1994. Theoretical, observational, and isotopic estimates of the lifetime of the solar nebula. *Meteoritics* **29**, 6–25.

Poirier, J.P., 1991. *Introduction to the Physics of the Earth's Interior.* Cambridge University Press, New York.

Pollack, J.B., D. Hollenbach, S. Beckwith, D.P. Simonelli, T. Roush, and W. Fong, 1994. Composition and radiative properties of grains in molecular clouds and accretion disks. *Astrophys. J.* **421**, 615–639.

Pollack, J.B., O. Hubickyj, P. Bodenheimer, J.J. Lissauer, M. Podolak, Y. Greenzweig, 1996. Formation of the giant planets by concurrent accretion of solids and gas. *Icarus* **124**, 62–85.

Porco, C.C., P.D. Nicholson, J.N. Cuzzi, J.J. Lissauer, and L.W. Esposito, 1995. Neptune's ring system. In *Neptune.* Ed. D.P. Cruikshank. University. of Arizona Press, Tucson. pp. 703–804.

Pravec, P., *et al.*, 2000. Two-period lightcurves of 1996 FG_3, 1998 PG, and (5407) 1992 AX: One probable and two possible asteroids. *Icarus* **146**, 190–203.

Press, F., and R. Siever, 1986. *Earth.* W.H. Freeman and Company, New York.

Press, F., and R. Siever, 1998. *Understanding Earth*, 2nd Edition. W.H. Freeman, New York.

Prinn, R.G., 1993. Chemistry and evolution of gaseous circumstellar disks. In *Protostars and Planets III.* Eds. E.H. Levy and J.I. Lunine. University of Arizona Press, Tucson. pp. 1014–1028.

Prinn, R.G., and B. Fegley, Jr., 1989. Solar nebula chemistry: Origin of planetary, satellite and cometary volatiles. In *Origin and Evolution of Planetary and Satellite Atmospheres.* Eds. S.K. Atreya, J.B. Pollack, and M.S. Matthews. University of Arizona Press, Tucson. pp. 78–136.

Pruppacher, H.R., and J.D. Klett, 1980. *Microphysics of Clouds and Precipitation.* D. Reidel Publishing Company, Dordrecht, Holland. 714pp.

Putnis, A., 1992. *Mineral Science.* Cambridge University Press, Cambridge.

Redman, R.O., P.A. Feldman, and H.E. Matthews, 1998. High-quality photometry of asteroids at millimeter and submillimeter wavelengths. *Astron. J.* **116**, 1478–1490.

Roederer, J.G., 1970. *Physics and Chemistry in Space 2: Dynamics of Geomagnetically Trapped Radiation.* Springer-Verlag, Berlin.

Roederer, J.G., 1972. Geomagnetic field distortions and their effects on radiation belt particles. *Rev. Geophys. Space Phys.*, **10**, 599–630.

Rosen, P.A., 1989. Ph.D. Thesis, Dept. of Electrical Engineering, Stanford University.

Rosen, P.A., G.L. Tyler, E.A. Marouf, and J.J. Lissauer, 1991. Resonance structures in Saturn's rings probed by radio occultation II: Results and interpretation. *Icarus* **93**, 25–44.

Rybicki, G.B., and A.P. Lightman, 1979. *Radiative Processes in Astrophysics.* John Wiley and Sons, New York.

Russell, C.T., 1995. A brief history of solar-terrestrial physics. In *Introduction to Space Physics.* Eds. M.G. Kivelson and C.T. Russell. Cambridge University Press, Cambridge. pp. 1–26.

Russell, C.T., D.N. Baker, and J.A. Slavin, 1988. The magnetosphere of Mercury. In *Mercury.* Eds. F. Vilas, C.R. Chapman, and M.S. Matthews. University of Arizona Press, Tucson. pp. 514–561.

Salby, M.L., 1996. *Fundamentals of Atmospheric Physics.* Academic Press, New York.

Salo, H., 1995. Numerical simulations of dense planetary rings. III. Self-gravitating identical particles. *Icarus* **117**, 287–312.

Samuelson, R.E., W.C. Maguire, R.A. Hanel, V.G. Kunde, D.E. Jennings, Y.L. Yung, and A.C. Aiken, 1983. CO_2 on Titan. *J. Geophys. Res.* **88**, 8709–8715.

Schleicher, D.G., 1983. Ph.D. Dissertation, University of Maryland.

Schloerb, F.P., 1985. Millimeter-wave spectroscopy of solar system objects: present and future, Proceedings of the ESO-IRAM-Onsala workshop on (sub) millimeter astronomy, Aspenas, Sweden, 17–20 June 1985. Eds. P.A. Shaver and K. Kjar. *ESO Conference and Workshop Proceedings* **22**, ESO, Garching bei Munich, pp. 603–616

Schneider, N.M., J.T. Trauger, 1995. The structure of the Io torus. *Astrophys. J.* **450**, 450–462.

Schneider, N.M., W.H. Smyth, and M.S. McGrath, 1987. Io's atmosphere and neutral clouds. In *Time Variable Phenomena in the Jovian System.* Eds. M.J.S. Belton, R.A. West, and J. Rahe. NASA SP-494, pp. 75–79.

Schneider, N.M., J.T. Trauger, J.K. Wilson, D.I. Brown, R.W. Evans, D.E. Shemansky, 1991. Molecular origin of Io's fast sodium. *Science* **253**, 1394–1397.

Scholl, H., L.D. Schmadel, and S. Roser, 1987. The mass of the asteroid 10 Hygieia derived from observations of 829 Academia. *Astron. Astrophys.* **179**, 311–316.

Schubert, G., W.B. Moore, J.D. Anderson, E.L. Lau, and W.L. Sjogren, 1996. Internal structure of Io and Ganymede. *EOS* **77**, F442, P21A-9.

Schulz, M., and L.J. Lanzerotti, 1974. *Physics and Chemistry in Space 7: Particle Diffusion in the Radiation Belts.* Springer-Verlag, Berlin.

Seiff, A., 1983. Thermal structure of the atmosphere of Venus. In *Venus*. Eds. D.M. Hunten, L. Colin, T.M. Donahue, and V.I. Moroz. University of Arizona Press, Tucson. pp. 215–279.

Seinfeld, J.H., and S.N. Pandis, 1998. *Atmospheric Chemistry and Physics: From Air Pollution to Climate Change*. John Wiley and Sons.

Sekanina, Z., 1976. Progress in our understanding of cometary dust tails. In *The Study of Comets*. Eds. B. Donn, M. Mumma, W. Jackson, M. A'Hearn, and R. Harrington. NASA SP–393, pp. 893–942.

Sekanina, Z., and J.A. Farrell, 1978. Comet West 1976. VI – Discrete bursts of dust, split nucleus, flare-ups, and particle evaporation. *Astron. J.* **83**, 1675–1680.

Shoemaker, E.M., 1960. Penetration mechanics of high velocity meteorites, illustrated by Meteor Crater, Arizona. *Rep. of the Int. Geol. Congress*, XXI Session, Norden, part XVIII, pp. 418–434, Copenhagen.

Showalter, M.R., 1991. Visual detection of 1981S13, Saturn's eighteenth satellite, and its role in the Encke gap. *Nature* **351**, 709–713.

Showalter, M.R., 1996. Saturn's D ring in the Voyager images. *Icarus* **124**, 677–689.

Showalter, M.R., J.A. Burns, J.N. Cuzzi, and J.B. Pollack, 1987. Jupiter's ring system: New results on structure and particle properties. *Icarus* **69**, 458–498.

Showalter, M.R., J.N. Cuzzi, E.A. Marouf, and L.W. Esposito, 1986. Satellite "wakes" and the orbit of the Encke gap moonlet. *Icarus* **66**, 297–323.

Showman, A.P., and R. Malhotra, 1999. The Galilean satellites. *Science* **286**, 77–84.

Shu, F.H., 1982. *The Physical Universe: An Introduction to Astronomy*. University Science Books, Berkeley, CA.

Shu, F.H., 1992a. *The Physics of Astrophysics, Vol. I: Radiation*. University Science Books, Mill Valley, CA.

Shu, F.H., 1992b. *The Physics of Astrophysics, Vol. II: Gas Dynamics*. University Science Books, Mill Valley, CA.

Shu, F.H., J.N. Cuzzi, and J.J. Lissauer, 1983. Bending waves in Saturn's rings. *Icarus* **53**, 185–206.

Shu, F.H., H. Shang, and T. Lee, 1996. Toward an astrophysical theory of chondrites. *Science* **271**, 1545–1552.

Shu, F.H., L. Dones, J.J. Lissauer, C. Yuan, and J.N. Cuzzi, 1985. Nonlinear spiral density waves: viscous damping. *Astrophys. J.* **299**, 542–573.

Shu, F.H., H. Shang, A.E. Glassgold, and T. Lee, 1997. X-rays and fluctuating X-winds from protostars. *Science* **277**, 1475–1479.

Simonelli, D., M. Wisz, A. Switala, D. Adolfini, J. Veverka, P.C. Thomas, and P. Helfenstein, 1998. Photometric properties of Phobos surface materials from Viking images. *Icarus* **131**, 52–77.

Smith, B.A., *et al.*, 1979. The Jupiter system through the eyes of Voyager 1. *Science* **204**, 951–971.

Smith, B.A., *et al.*, 1981. Encounter with Saturn: Voyager 1 imaging science results. *Science* **212**, 163–191.

Smith, B.A., *et al.*, 1982. A new look at the Saturn system: The Voyager 2 images. *Science* **215**, 504–537.

Smith, B.A., *et al.*, 1986. Voyager 2 in the Uranian system: Imaging science results. *Science* **233**, 43–64.

Smith, B.A., *et al.*, 1989. Voyager 2 at Neptune: Imaging science results. *Science* **246**, 1422–1449.

Smith, D.E., M.Y. Zuber, G.A. Neumann, and F.G. Lemoine, 1997. Topography of the Moon from the Clementine Lida. *J. Geophys. Res.* **102**, 1591.

Sprague, A.L., R.W.H. Kozlowski, D.M. Hunten, W.K. Wells, and F.A. Grosse, 1992. The sodium and potassium atmosphere of the Moon and its interaction with the surface. *Icarus* **96**, 27–42.

Stacey, F., 1992. *Physics of the Earth*. Brookfield Press. 513pp.

Stern, S.A., and R.V. Yelle, 1999. Pluto and Charon. In *Encyclopedia of the Solar System*. Eds. P.R. Weissman, L. McFadden, and T.V. Johnson. Academic Press, Inc., New York. pp. 499–518.

Stevenson, D.J., 1982. Interiors of the giant planets. *Annu. Rev. Earth Planet. Sci.* **10**, 257–295.

Stevenson, D.J., and E.E. Salpeter, 1976. Interior models of Jupiter. In *Jupiter*. Eds. T. Gehrels and M.S. Matthews. University of Arizona Press, Tucson. pp. 85–112.

Stix, M., 1987. In *Solar and Stellar Physics*, Lecture Notes Phys. **292**. Eds. E.H. Schröter and M. Schüssler. Springer, Berlin, Heidelberg. p. 15.

Stix, M., 1989. *The Sun*. Springer-Verlag, New York.

Stone, E.C., and E.D. Miner, 1989. The Voyager 2 encounter with the Neptunian system. *Science* **246**, 1417–1421.

Strom, R.G., 1999. Mercury. In *Encyclopedia of the Solar System*. Eds. P.R. Weissman, L. McFadden, and T.V. Johnson. Academic Press, Inc., New York. pp. 123–146.

Sykes, M.V., 1999. Infrared views of the Solar System from space. In *Encyclopedia of the Solar System*. Eds. P.R. Weissman, L. McFadden, and T.V. Johnson. Academic Press, Inc., New York. pp. 715–733.

Taylor, S.R., 1975. *Lunar Science: A Post-Apollo View*. Pergamon Press, New York.

Taylor, S.R., 1992. *Solar System Evolution: A New Perspective*. Cambridge University Press, Cambridge.

Taylor, S.R., 1999. The Moon. In *Encyclopedia of the Solar System*. Eds. P. Weissman, L. McFadden, and T.V. Johnson. Academic Press, Inc., New York. pp. 247–275.

Thieman, J.R., and M.L. Goldstein, 1981. Arcs in Saturn's radio spectra. *Nature* **292**, 728–730.

Tholen, D.J., and M.W. Buie, 1997. The orbit of Charon. I. New Hubble Space Telescope observations. *Icarus* **125**, 245–260.

Tholen, D.J., and E.F. Tedesco, 1994. Pluto's lightcurve: results from four oppositions. *Icarus* **108**, 200–208.

Thomas, P.C., 1999. Phobos and Deimos. In *Encyclopedia of the Solar System*. Eds. P.R. Weissman, L. McFadden, and T.V. Johnson. Academic Press, Inc., New York. pp. 309–314.

Thomas, P.C., D. Adinolfi, P. Helfenstein, D. Simonelli, and J. Veverka, 1996. The surface of Deimos: Contribution of materials and processes to its unique appearance. *Icarus* **123**, 536–556.

Thomas, P.C., M.J.S. Belton, B. Carcich, C.R. Chapman, M.E. Davies, R. Sullivan, and J. Veverka, 1996. The shape of Ida. *Icarus* **120**, 20–32.

Thomas, P.C., J.A. Burns, L. Rossier, D. Simonelli, J. Veverka, C.R. Chapman, K.P. Klaasen, and M.J.S. Belton, 1998. Small inner satellites of Jupiter. *Icarus*. **135**, 360–371.

Thomas, P.C., J. Veverka, D. Simonelli, P. Helfenstein, B. Carcich, M.J.S. Belton, M.E. Davies, and C. Chapman, 1994. The shape of Gaspra. *Icarus* **107**, 23–36.

Townes, C.H., and A.L. Schawlow, 1955. *Microwave Spectroscopy*. McGraw-Hill, New York.

Van der Tak, F., I. de Pater, A. Silva, and R. Millan, 1999. Variability of Saturn's brightness distribution. *Icarus* **142**, 125–147.

Veverka, J., P. Helfenstein, B. Hapke, and J.D. Goguen, 1988. Photometry and polarimetry of Mercury. In *Mercury*. Eds. F. Vilas, C.R. Chapman, and M.S. Matthews. University of Arizona Press, Tucson. pp. 37–58.

Veverka, J., P. Thomas, A. Harch, B. Clark, J.F. Bell III, B. Carcich, J. Joseph, C. Chapman, W. Merline, M. Robinson, M. Malin, L.A. McFadden, S. Murchie, S.E. Hawkins III, R. Farquhar, N. Izenberg, and A. Cheng, 1997. NEAR's flyby of 253 Mathilde: Images of a C asteroid. *Science* **278**, 2109.

Veverka, J., P.C. Thomas, P. Helfenstein, P. Lee, A. Harch, S. Calvo, C. Chapman, M.J.S. Belton, K. Klaasen, T.V. Johnson, and M. Davies, 1996. Dactyl: Galileo observations of Ida's satellite. *Icarus* **120**, 200–211.

Veverka, J., *et al.*, 2000. NEAR at Eros: imaging and spectral results. *Science* **289**, 2088–2097.

Wasson, J.T., 1985. *Meteorites: Their Record of Early Solar-System History*. W.H. Freeman, New York.

Weidenschilling, S.J., 1977. Aerodynamics of solid bodies in the solar nebula. *Mon. Not. R. Astron. Soc.* **180**, 57–70.

Weinberg, S., 1988. *The First Three Minutes*. Basic Books, New York.

Weissman, P.R., 1986. Are cometary nuclei primordial rubble piles? *Nature* **320**, 242–244.

West, R.A., 1999. Atmospheres of the giant planets. In *Encyclopedia of the Solar System*. Eds. P. Weissman, L. McFadden, and T.V. Johnson. Academic Press, Inc., New York. pp. 315–337.

Wilcox, J.M., and N.F. Ness, 1965. Quasi-stationary corotating structure in the interplanetary medium. *J. Geophys. Res.* **70**, 5793–5805.

Williams, J., 1992. *The Weather Book*, Vintage Books, New York.

Wisdom, J., 1983. Chaotic behavior and the origin of the 3/1 Kirkwood Gap. *Icarus* **56**, 51–74.

Wolf, R.A., 1995. Magnetospheric configuration. In *Introduction to Space Physics*. Eds. M.G. Kivelson, and C.T. Russell. Cambridge University Press, Cambridge. pp. 288–329.

Wong, M.H., I. de Pater, C. Heiles, R. Millan, R.J. Maddalena, M. Kesteven, R.M. Price, and M. Calabretta, 1996. Observations of Jupiter's 20-cm synchrotron emission during the impacts of comet P/Shoemaker–Levy 9, *Icarus* **121**, 457–468.

Wood, J.A., 1988. Chondritic meteorites and the solar nebula. *Annu. Rev. Earth Planet. Sci.* **16**, 53–72.

Woolf, N., and J.R. Angel, 1998. Astronomical searches for Earth-like planets and signs of life. *Annu. Rev. Astron. Astrophys.* **36**, 507–537.

Wright, M.C.H., I. de Pater, J.R. Forster, P. Palmer, L.E. Snyder, J.M. Veal, M.F. A'Hearn, L.M. Woodney, W.M. Jackson, Y.-J. Kuan, and A.J. Lovell, 1998. Mosaiced images and spectra of $J=1\rightarrow 0$ HCN and HCO^+ emission from Comet Hale–Bopp (1995 O1). *Astron. J.* **116**, 3018–3028.

Yelle, R.V., 1991. Non-LTE models of Titan's upper atmosphere. *Astrophys. J.* **383**, 380–400.

Yelle, R.V., D.F. Strobel, E. Lellouch, and D. Gautier, 1997. Engineering models for Titan's atmosphere. In *Huygens Science, Payload and Mission*, ESA SP-1177, pp. 243–256.

Yeomans, D., 2000. Small bodies of the Solar System. *Nature* **404**, 829–832.

Yeomans, D.K., *et al.*, 2000. Radio science results during the NEAR-Shoemaker spacecraft rendezvous with Eros. *Science* **289**, 2085–2088.

Yoder, C.F., 1995. Astrometric and geodetic properties of Earth and the Solar System. In *Global Earth Physics, A Handbook of Physical Constants*, AGU Reference Shelf 1, American Geophysical Union.

Zahnle, K., 1996. Dynamics and chemistry of SL9 plumes. In *The Collision of Comet Shoemaker–Levy 9 and Jupiter*. Eds. K.S. Noll, H.A. Weaver, and P.D. Feldman. Space Telescope Science Institute Symposium Series 9, IAU Colloquium 156, Cambridge University Press, Cambridge. pp. 183–212.

Zahnle, K.J., and N.H. Sleep, 1997. Impacts and the early evolution of life. In *Comets and the Origin and Evolution of Life*. Eds. P.J. Thomas, C.F. Chyba, and C.P. McKay. Springer, New York.

Zarka, P., 1992. The auroral radio emissions from planetary magnetospheres - What do we know, what don't we know, what do we learn from them? *Adv. Space Res.* **12**, (8)99–(8)115.

Zarka, P., B.M. Pederson, A. Lecacheux, M.L. Kaiser, M.D. Desch, W.M. Farrell, and W.S. Kurth, 1995. Radio emissions from Neptune. In *Neptune and Triton*. Eds. D.P. Cruikshank and M.S. Matthews. University of Arizona Press, Tucson. pp. 341–387.

Zharkov, V.N., and V.P. Trubitsyn, 1978. *Physics of Planetary Interiors*. Ed. W.B. Hubbard. Pachart Publishing House, Tucson.

Zinner, E., 1998. Stellar nucleosynthesis and the isotopic composition of presolar grains from primitive meteorites. *Annu. Rev. Earth Planet. Sci.* **26**, 147–188.

Index

"No. That's your business. But you're right. I'd like to know."

"I'll keep in touch with you, Father. Thanks for coming in."

It was a good time of day, McMahon thought, reaching the street. Next to dawn he loved it the best, the last hours of the sun when its heat was spent but a golden haze hung over the city. The youngsters were playing stickball, and great fat women leaned out their windows watching for their men to come home from work. There were flags in the windows of almost half the apartments. No college deferments here. Brogan was not a young man whose insight should be underestimated: he would not say to many people in this neighborhood that he joined the force to avoid the army.

Crossing Ninth Avenue he decided to walk downtown a few blocks to Ferguson and Kelly's funeral parlor. It was no new thing to him, trying to arbitrate the costs of a funeral: he generally did well until the family arrived to select the casket. This part of town, where the street markets commenced, was predominantly Italian. Sausages and cheeses hung in the windows over stacked canisters of olive oil, two-quart tins of tomatoes. The produce was all outdoors. The people were noisy and friendly and a priest was accepted as one of themselves, neither feared nor revered. It was a strange place for Ferguson and Kelly, but as he thought about it, he could not name an Italian in the undertaking business. That they left to the morbid Irish. But obviously in Italy Italians buried Italians. Could the circumstance here be the dominance of the Irish in the church? Since Muller was not a Catholic, or so he assumed, he would have to see Ferguson, a man he took to be of Scotch-Irish antecedence. He would rather have negotiated with Kelly. As he opened the door setting off the muted chimes, he wished he had telephoned. A typical McMahonism: taking the

should have a nice funeral. You know?" By the rubbing together of her fingers she suggested money. "Come to my house, Father. The people liked him. They will all give something." That, he felt, was genuine.

He asked for Brogan at the desk. The sergeant directed him to a room on the second floor. He went up by way of a staircase, the color and smell of which put him in mind of a cheap hotel. The windows were wire-meshed on the outside, sealing in the dirt of generations. He met Brogan and Lieutenant Traynor coming out of the room with Pedrito, a tall, skinny boy of eighteen, sallow and sullen, with a mop of black hair and a scraggle of beard.

The best he could do for him at the moment was to acknowledge an acquaintanceship. "Hello, Pedrito."

The boy nodded curtly.

"Keep your nose clean, young fellow. We'll be watching you," Traynor said.

"*Cochinos*," Pedrito snarled. Pigs. But by then he had reached the stairs.

"Makes you want to love them, doesn't it?" Traynor said. He went on down the hall.

Brogan led the priest into the interrogation room where an officer was removing the tape from a recorder. They waited until he had left the room.

"So you had to bring Carlos in anyway," McMahon said.

"*Si*," Brogan said. He searched a folder for the statements he wanted.

McMahon was not to be put off. "Why?"

Brogan shrugged. "The lieutenant didn't like it, not the way the kid told it to us. The doorknobs were what really put him in a flap."

"I don't get it."

"Well, Father, let me put it this way: he questioned the boy on whether Muller had molested him."

McMahon's temper snapped. "Balls."

"Exactly."

"Christ Jesus help us," McMahon said, but he already knew he was being unreasonable. The luring of a child to an abandoned building: it could be construed that way. Even the monsignor's first question was whether the man was a pervert.

Brogan half-sat on the desk. He indicated the chair to the priest. "What is it that bugs you, Father? You know yourself that a kid like Carlos, there's nothing he's going to learn from us he didn't know from the street already."

McMahon sat down and took in hand his own typed statement. What Brogan said was true: trying to shield the innocence of a child in Carlos' environment was almost as impossible as the restoration of virginity. He read the statement and signed it.

"But you're right," Brogan said. "That wasn't Muller's trouble."

"What was?"

Brogan shrugged. "Mrs. Phelan? Or vice versa. I have a notion she was hot for him. There's gossip in the building. Even we can get to it. She picked him up in a bar, nested him down in her back room. Like charity begins at home. Where was Phelan through all this? Where *is* Phelan?"

And what's his problem? McMahon kept the thought to himself, but he suspected Brogan was doing the same thing. He asked, "Is Pedrito in the clear?"

"As far as the homicide, he has to be. He works on a machine assembly line. Twenty witnesses to where he was from six A.M. to three this afternoon. And he wasn't a chum of the victim. That was Carlos' idea. To a kid, I guess, everybody over fifteen is the same age, especially if they come to his birthday party. They all drank wine that night and it was then Muller got the idea of building a house of doors for the youngster. Pedrito went with him. If he gets into no worse trouble than swiping doors, I'll settle."

McMahon said, "Why are you a cop, Brogan?"

The young detective colored. "To stay out of the draft. I'll take my law and order straight, Father."

The priest was not sure why, but he felt a kind of respect for Brogan saying it.

"Phelan has an assault record, by the way," Brogan added.

"Was he at the birthday party too?"

"No, but Mrs. Phelan was."

"It makes you wonder why there was gossip, if she's so popular with her tenants," McMahon said, "and they're not notoriously cooperative with the police, are they?"

"It's pretty simple, Father—it's not the infidelity, if that's what it is. Homicide is something you can get put away for a long time. They don't like Phelan."

That had to be it, McMahon realized. Priscilla Phelan had not calculated the relative values of her Spanish-speaking friends. "Do you want me to go over Carlos' story?"

"It won't be necessary, unless you want to see it. You can g over to the house if you want to—I'll fix it up—if you wan to see his things. There's not much there. He was travelin light, wherever he came from. A sign painter by his identific tion."

McMahon shook his head: he did not want to go near t Phelan apartment.

Brogan tapped his statement with a pencil. "I just thou by this you might be interested."

"I am," McMahon said. "He got to me and I'm not s why. Was it his courage? He was ready to die, but it wa though that was because he wanted to live, to live right u and over the threshold. And he said he would like to know That always sets a man up, doesn't it?"

"It sure does, Father."

"There was more to him than what he left in that ro feel pretty sure of that."

"Then have a look at his things."

step first and weighing the consequences only when he had no choice but to live up to them.

A half hour later he headed uptown again, a set of if-or-and figures in his pocket and a stiff Scotch whisky in his stomach. The whisky roused in him a feeling of kinship with every man on the street, and he went over in his mind the lines of his sermon. They were not so bad after all, he decided, with even a touch of poetry to them. Long ago the monsignor had said to him, "Remember, you're not addressing the sacred congregation in Rome. Simple truths are the most eloquent. Sincerity, that's the key." Which put him in mind of politicians and brought him round full circle to the banal again. Priests and politicians. He felt as restless as the birds scratching in the gutters. His spirits fell as low. He had not yet read his office of the day. That and music and his morning Mass were his refuge. All having little to do with the world around him. It was not that the world was too much with him, but that he was too much with the world. He wondered what Muller would have thought of that distinction.

The pawnbroker was closing the iron grill across his shop windows as McMahon approached. The grill gave the shop more distinction than the merchandise warranted, the grill and the three golden balls newly painted, and the sign, Gothic-lettered in fresh gold leaf—or so it appeared in the slanting rays of the sun: A. ROSENBERG.

McMahon thought at once of Muller, a sign painter, but he also thought of the curiosity of the Gothic lettering of the name. He was himself familiar with the type face from liturgical books, but ninety-nine out of a hundred would have to puzzle the letters to get the name. Ninth Avenue Gothic. He greeted the pawnbroker. They knew each other by sight, and there was a placard in the window announcing the girls' choir concert.

"I'm curious, Mr. Rosenberg." He pointed to the sign. "Why in those letters?"

Rosenberg looked up and shaded his eyes. "Beautiful, eh?"

McMahon agreed.

"Why not? It isn't the phone book. The whole neighborhood knows Rosenberg. They don't need the sign. Only Rosenberg needs it. I like it, that's why."

"Was it a man named Muller who painted it?"

"Would you believe me, Father, I don't know his last name? A beard and beautiful hands. I always notice hands."

"He's dead, you know," the priest said.

He had not expected the man's reaction, the little moan of a personal pain, and the mouth working under the gray mustache. Rosenberg put his hand to the grating and held onto it.

"I'm sorry," McMahon said, "I didn't realize he was a friend."

"When?"

"This morning." McMahon told him the circumstances.

"Come inside." He made a gesture to let the gate stand open. "Leave it. I don't have to keep union hours, thank God." He led the way through the shop to his desk in the back and lit the green-shaded lamp above it. He pulled out the swivel chair and made the priest take it. The desk, an ancient rolltop, was littered with papers, letters in foreign handwriting, some in Yiddish, or so McMahon presumed from a glance. "Coffee? Or a glass of cognac, Father."

"A little cognac, thank you."

"He liked the cognac, too, let me tell you. But he liked also the coffee. But most of all he liked to talk."

"I'd have thought that," the priest said.

Rosenberg got his glasses from the desk drawer and put them on. "Let me show you." He looked more like a scholar than a pawnbroker as he went to a shelf and looked up. His white hair fringed the collar of his coat. He started to take a large book from among several on the shelf, then changed his mind, and with a sweep of his hand, he abandoned them to McMahon's own scrutiny. "They are his, Bosch, Vermeer, the Italians. He did not like the Italians except for Botticelli."

McMahon took down the Bosch while Rosenberg went to a cabinet and brought out the bottle and two stemmed glasses. The book was so heavy McMahon had to lay it on the desk to open it. It revealed nothing but Bosch, but that was quite a lot, the contrasts of good and evil, and as in most things, he thought, the evil figures were by far the most interesting. There was no name, no mark at all in the front pages, only the smudges of use.

Rosenberg cleared a place on the desk, put the glasses down and poured the brandy, twice as much for the priest as he poured for himself. "Oh, yes. He liked to talk. I did not understand half of what he was talking about, but I liked to listen to him. I liked to listen," he repeated. "And now it is one more voice not to listen to." He lifted his glass. "To him, *shalom.*"

"Shalom," McMahon said. Peace.

He took a sip of the brandy and then put the book back on the shelf. He took down another volume and looked at the flyleaves: nothing. "How long have you had these, Mr. Rosenberg?" What he really wanted to ask at the moment was how much it would cost him to buy them if they were not redeemed.

"He brought in the Bosch maybe three weeks ago. The others in between. I have no claim on them, Father. It was a matter of some place to keep them for the time being. 'I don't like possessions,' he said. 'Why do people collect things? Because they cannot bear to be alone, alone with themselves.' Something like that. He was always saying things like that."

McMahon examined the books one by one while the pawnbroker was talking. Not one gave any clue to its ownership.

"He came in that first day and said, 'I am a one-man crusade to clean up the store fronts on Ninth Avenue. You clean the windows and I will paint.' 'How much?' I said. 'For a half day's work, what it costs me to live for a day.' He did not look as though he lived extravagantly. I paid him ten dollars and the paint, and it was to me a valuable investment to meet the man. He was a kind of salesman like that, you know? I was the first customer he worked it on, I found out afterwards. Next he went to the drugstore on the corner. That was a disgrace. He showed me as a model. A half dozen shops in all maybe. Until now. But I am thinking, where did he come from? Could it be there are little islands like this all over New York? Fresh paint and clean windows? How long would he have stayed? And why did he have to leave . . . that way." Rosenberg fell silent. He sipped the brandy, smacked his lips and nodded at his own thoughts.

McMahon returned to the chair. The brandy seared its way down his throat and seemed to grab at his stomach. He was reminded of the hour of the day. He had been late for lunch. Miss Lalor would be in a temper if he was late for dinner also. Then he thought how truly unimportant were Miss Lalor's tempers.

Rosenberg looked at him. "He was sitting here where I am one day, and I was there at the desk and he said to me: 'Rembrandt would have liked you. Right as you are, the light, the junk, the cupboards, everything. He would have made of you one of his famous Jews.' And I said to him, on the chance it would bring him out, you know: 'Rembrandt is already dead and I am not yet a famous Jew.' I looked him in the eye when I said it, but he only laughed and shook his head. All the same, Father, I think he was a painter of more than signs and store fronts."

"I think so too," McMahon said.

After a moment Rosenberg said, "You will tell the police you were here?"

"I shall have to."

"What can I tell them, a man I knew only by the name of Gus? Gust. He liked the 't' on the end of it. I do not like to think they will take his books away. I like to think he will come back for them."

"He won't."

"Or somebody then to talk to the way we talked."

"I will not tell them about the books. That is up to you," McMahon said. He would have to examine his own conscience on his moment of covetousness of them. Long ago he had wanted to be a painter even more than he had wanted to be a musician, but even less than he had wanted to be a priest. "Did he ever talk about himself?" He knew the answer before asking it. If he had, Rosenberg would have said so.

The pawnbroker shook his head.

"Not even where he lived, where he was bringing the books from? Or why?"

"No, not a word personal. Oblique: it was a word he liked, but he was talking about light, the light of a painter."

"If the police are told about them, they will want to check if they were stolen."

"He did not steal them," Rosenberg said vehemently.

"You are judging by our values, my friend," McMahon said. "And you and I both know now he was a very unconventional man."

"That is true, but even looking at it from a practical point of view, Father, it would be easier to steal a hippopotamus than Hieronymus Bosch."

McMahon laughed and finished his brandy. He held up his hand to stay the pawnbroker from pouring more. "I must go. If there is no family claim on his body, there will probably be a service at Ferguson and Kelly's."

"When?"

"That will depend on the police, I suppose. The autopsy. I will see that the newspapers get a notice."

"What kind of a service?"

"That's the question, isn't it? It's the people in the building where he lived who want it."

"Funerals are always for the living," Rosenberg said, getting to his feet with the priest. "I think he would have agreed to that."

"I will play some Bach on the organ," McMahon said.

"And Mahler. He liked Mahler."

McMahon said, "Gustave Muller, Gustav Mahler." The association had crossed his mind the first time he heard the complete name. He and Rosenberg looked at each other. "So it is possible we don't even know his real name."

"I am thinking, Father, what I am going to do: I am going to sit down and try to write the things I remember we talked about. It will not sound very much, the way I write it. I have tried to write before, and my mind it becomes a moth just trying to get at the light. But I will try and I will give it to you. Who knows? Maybe you and I can talk also."

"I would like that," the priest said.

At the door of the shop they shook hands. McMahon remembered that it was Friday. "Good *shabbos,*" he said.

"Gut shabbos." Rosenberg squinted up at the sky. "It will be a fine sunset."

5 ◆◆◆

It had been McMahon's intention to go directly home. But then that had been his intention when he had left the precinct station house well over an hour earlier. He found himself walking toward the sunset. Scotch and brandy and Gustave Muller. Benediction and rosary at eight, a sermon to be got into his head, Muller out. He paused at the parapet beneath which lay the railroad tracks and beyond which the West Side Highway arched against the sky. Every approaching car caught an instant of sunset in its windshield, passed, and seemed no more than a beetle on a rampart. He turned and walked back on the street where Muller had died. The rush of suburb-going traffic was over, the street again a silent wilderness, bulldozers and cranes the dinosaurs of the era. The one lone building stood, its walls raw brick where the walls it once met had been shorn away. At the very top, the windows shone like golden eyes.

He paused where the uniformed policeman stood by the basement grill and exchanged a few words with him. "Love Power" had been all but wiped out with the shuffle of many feet. At the top of the steps the double doorway was open. "Mind if I go up, officer?"

"I guess it'd be all right, Father. They're all through up there. Just stay away from the basement."

"Believe me, I will."

A house without doors, he noticed, climbing one flight of dusty stairs after the other. To have stolen the doorknobs Muller would have needed to be around for a while. And he was, of course. Carlos had said that, the man coming down when he called him. The turn-of-the-century gas fixtures were still in the hallways, and there were patches of a floral-patterned wallpaper where the paint had chipped away. On every landing he noticed a clutter of tinfoil and burnt-out photography bulbs. The police had gone over the building well. The roof hatch had been tilted to let in air. When he reached the fifth and final floor the room to the west was suffused with light, the blearing X's had been removed from the windows. There were spatters of paint on the floor, and squares of raw wood where bits of the surface had been cut out, he suspected for laboratory study. So the police too would now presume him to have been an easel painter as well as the painter of Mrs. Phelan's walls. Northern light was painters' light, and in the mornings here Muller would have had the best of it. Now, with the sun having gone down, the sky was changing fast, holding briefly the red and yellow tints, then almost palpably letting them go, yielding to the darker strokes of night. The room was utterly bare. Silence and peace: he could feel it. He found it himself only at the altar when he was no longer himself, at the moment of the transubstantiation. His conscience told him that he must go, but the wish to wait for night was very strong.

"Father McMahon?" The voice halloed up the stairwell.

He thought it would be Brogan and went to the top of the stairs.

"Stay there. I'm coming up."

McMahon went to the west window and waited. Torn wisps of cloud held the last pinks and lemon of the sunset.

"Some spot he found for himself, wouldn't you say, Father?"

"How did he find it?"

"We've been asking the same question. The building belongs to an old crank who wouldn't sell it to the developers. They went ahead without him, starting the wreckers next door. They wouldn't give him the ground to shore up his walls. The city condemned. The building's going but he still won't sell the land. It's in the courts and it's been in the papers, but he never heard of Gustave Muller."

"But the abandoned building could have been what attracted Muller to the neighborhood," McMahon said. "What did you find in this room?"

"An old army cot and three more doorknobs."

"Nothing else?"

"A few spatters of paint. He could have decorated the kid's doorknobs up here."

"He'd have needed a brush and paint," McMahon said.

"It wasn't here."

"And nothing like that where he lived?"

Brogan shook his head. "No Phelan yet either."

"And no weapon," McMahon said after a moment.

"It was a square-edged blade. Maybe a narrow chisel."

Or a palette knife, McMahon thought, but he did not say so.

"He cleaned up a few store fronts on Ninth Avenue," Brogan said. "A real eccentric, like they say."

McMahon felt relieved of having to tell him of his conversation with Rosenberg. But to compensate—his own conscience, he thought afterward—he reminded the detective of Muller's last words.

"I was going over your statement again, Father. That's some pretty fancy talk between you and him. What do you think he meant when he said he'd taken the knife from his killer?"

"I suppose I took it to mean that the man was not dangerous to anyone else."

"That's the way I read it, and that's pure crap, Father. Unless he killed himself and got somebody to get rid of the weapon for him."

"Who?"

Brogan shrugged. "And why? Nobody who had any sense would touch it. That leaves the kid."

"Carlos? I'm sure he ran all the way from here to the rectory."

"So am I. I think he told it the way it was."

McMahon could hardly read the dial hand on his watch. "I've got to get home."

"We've canvassed all the big art galleries, Father, on the

chance he painted something besides balls and walls. But maybe they wouldn't know him under that name. His Social Security number's a fake. He was on the run from something. We'll find out."

McMahon remembered his earlier mission that evening. "How long will you keep the body?"

"We've got the facilities. Till somebody claims it."

"The tenants at 987 would like a funeral service."

"A wake?"

"I suppose you could call it that. I've inquired about the costs at Ferguson and Kelly."

"So you need the mortal remains. I'll speak to Traynor. It's something the newspapers would pick up. The publicity might help us."

"I want to think about it first," McMahon said. "Hold off speaking to Traynor." The whole idea now became repugnant to him.

At the top of the stairs Brogan said, "You're right, Father. Somebody would be on our necks for it, some organization for the rights of corpses." A few steps down, he paused. "Hah! I remember a song my grandfather used to sing when he'd get a few drinks in him . . . 'If this wake goes on a minute, sure the corpse he must be in it. You'll have to get me drunk to keep me dead.' That's the end of it. I forget the beginning."

How fortunate, McMahon thought.

On the street Brogan asked: "Are you off duty now, Father?"

"No. I've taken French leave."

"What does that mean?"

McMahon rubbed the back of his neck. "I guess it means AWOL. It's Irish. I don't remember ever saying it before myself."

"I check out in a half hour. I was going to suggest, if you're free, have a meal and a couple of drinks with me."

"Where?"

"Downtown. The Village maybe."

"He wasn't the Village type," the priest said, although, God knew, he said it on shallow grounds.

"Maybe he wasn't, Father, but I was thinking about his killer. And I could use a good excuse for a few hours on the town. What do you say?"

"If you don't mind starting with Benediction and rosary. I'll be free after that."

"In mufti, Father."

"The best mufti I have," McMahon said.

6 ◆◆◆

It was only after McMahon had resisted the temptation to take the steak bone in his fingers that he remembered. "Holy God," he said, "it's Friday."

"Well, it's not a sin any more, is it?" Brogan wiped his fingers in his napkin. He had not been inhibited about taking the bone in hand.

"No, not for you, but a priest should hold to it."

"But tonight you're on French leave—was that it?—and if I know the French . . ." The young policeman rolled his eyes. His cheeks were flushed. They had had two stiff drinks before dinner.

McMahon brushed the crumbs from the lapel of his sport jacket. "I was trying to think where that term could have come from."

"World War I?"

"Much earlier, I think. From the time of Napoleon, I shouldn't be surprised, when the French fleet turned back from the west coast of Ireland and left Wolfe Tone in the lurch."

"That's the French for you," Brogan said solemnly.

"There was a MacMahon a general in the French army in those days."

"Was there now? Were you related?"

McMahon grinned. He was aware that after the drinks both of them were falling into a brogue of sorts. "Well, there were Wild Geese in the family, I'm told, Irish soldiers fighting in the French army."

"Ah, yes. We're a race that fights best when the cause is somebody else's. Wouldn't you say that, Joseph?"

McMahon flinched inwardly at the policeman's use of his

first name, the deferential young man of the afternoon. He laughed to cover his pulling-in in case it showed. But Brogan would not have noticed. McMahon would not be the first priest he had taken on the town. He said, "Well, we fight best for lost causes, and no man's our hero until we've made a martyr out of him." Nonsense, he thought. Poetic nonsense.

"Brian Boru and Kevin Barry?" Brogan suggested.

"I'm not sure about Brian Boru," McMahon said. "Shall we have coffee or another drink?"

"Irish coffee?"

"It's too early in the night for that," McMahon said.

"You're a man after my own heart." Brogan reached for his wallet. "Let's have a drink somewhere else."

"Down the middle," McMahon said of the check.

"Not tonight. Who knows? Before it's out we may turn up something that'll put the city in debt for our tab."

McMahon said nothing. He did not know which he liked the least: carousing on Brogan or on the taxpayer. But with the ten dollars he had borrowed from the monsignor on the way out and his own two, he would not pick up many tabs. Remember your prerogatives and not your pride, the old man had bidden him, not for the first time.

But McMahon enjoyed himself all the same. The streets were alive with youth and music, purveyors of flowers and chestnuts, carters of cameras and souvenirs, papier maché and art nouveau, sailors on leave and cops on vigil. He loved the young people, beards, beads and begonias, and if he had had his way, he and Brogan would have sat astride an old Morgan car parked near MacDougal Street, and he'd have conducted the singing himself. ". . . Now don't you know that's not the way to end the war . . ." a young troubadour sang to the off-key strum of his guitar.

"Beautiful!" McMahon shouted. "Sing it again."

A whole chorus of young people did.

A fire truck approached, its bell clanging. The youngsters

pushed back from the street, but coming abreast of them and making the wide turn even wider than necessary, the fireman gave a deafening blast with his bullhorn. A leprechaun of a boy cupped his hands around his mouth and shouted after the truck: "Yankee, go home!"

McMahon threw back his head and laughed as he had not laughed for a long time. A shaggy-haired girl came up to him and held out a string of beads. McMahon stooped and allowed her to put them around his neck. He offered her money but she would not take it. The Yankee-go-home boy came up behind her and said, "Excuse me, miss, but your skirt is showing."

Finally Brogan got McMahon away. He was looking for a particular bar. When he found it and they went in, he said, "My feet need a rest. Let's sit in a booth."

McMahon too was glad to sit down. He was trying to remember a line of Yeats. He got it the moment he stretched his legs under the table. "There midnight's all a-glimmer and noon a purple glow." So was the bar. There was this to be said for a priest's night out: it was so rare an occasion, the whole laughing world seemed to join him. With a few drinks McMahon became a democrat, as by the light of day assuredly he was not. He thought of Father Purdy, poor little Purdy, obsequious as a snail, pulling in, poking out on his way to the throne of God.

"I need a drink," he said.

"I'll go get them myself," Brogan said. The bar was crowded.

He drank more than he ought to, McMahon knew, and like Miniver Cheevy, he had reasons. Miniver Cheevy, child of scorn . . . something about the Medici. He would have sinned incessantly, could he have been one.

There was a small commotion at the bar before Brogan pushed his way through the men and returned with the drinks. His color was as high as the lights were low, and he was cursing under his breath. It was only then that McMahon realized they were in a homosexual hangout.

"What did you come in here for then?"

"On a hunch," Brogan said. He ruffled his shoulders and then settled back, seating himself so that he too could view the bar. "It takes all kinds," he said, "and sure the whisky comes out of the same keg."

"So do we all," McMahon said, aware of the sententiousness even as he said it. *"Slainte."* It was the one word of Gaelic he knew. He touched his glass to Brogan's.

"Did you ever know an Irish fag, Joseph?"

"Any number, but most of them in clergy's petticoats."

Brogan was shocked, for all his worldliness. The double standard had just quadrupled. "Is that a fact?" he murmured, but not believing it for a moment.

McMahon stared at the men at the bar, the tight little behinds in the snug narrow-legged pants. "Poor bastards," he said, and threw down his drink.

"You'll go for the next one yourself and see how you like it."

McMahon said, "I'll have the next one on the road home."

But Brogan sipped. He was in no hurry. He took a match packet from his pocket, and played with it, folding one match, then another into a fan. "Go on, get yourself a drink, Joseph. I dare you."

"Since you put it that way, I might."

McMahon approached the bar flanking it so that he came up last man where it curved to the wall. He glanced down the row of faces: young, aging, delicate, tired, gay . . . gay, gay, gay, but there wasn't a cruel face among them. But where that night would he have seen a face that he thought cruel? It was a second or two after his scanning of them all that he realized he knew one of the men. He turned abruptly from the bar, only to all but bump into Brogan who had come up behind him.

"Sit down," the detective said. "I'll fetch your drink."

"I don't want one."

"You'll need it," Brogan said. "Sit down."

McMahon did as he was told. The night had lost its glow and so had he. He watched Brogan fake amiability with the men, and then looked away. What had he expected from a cop? The taxpayers' money. He picked up the fanned match packet. Pierre's Unique. That's where they were, in Pierre's Unique. Eunuch, unique. Christ Jesus forgive me. He tore one match after another from the packet. Brogan returned.

"Which one is he?"

McMahon almost involuntarily put his hand to where at another time he would be wearing the collar.

"The high turtleneck sweater?"

"That's he."

"I'll be back in a minute," Brogan said, and then before leaving, "I picked those up when I was talking to his wife." He indicated the matches.

"Congratulations," McMahon said.

Brogan went to the phone booth between the doors to the rest rooms. McMahon watched him over his shoulder until he had dialed. Then he stared at the back of Daniel Phelan, the narrow back and the skinny hips and the rundown heels with the hole in one sock. The informer priest. Phelan, he knew, had not seen him. He was tempted to run. No, you'll stick it out, he told himself, and learn a little deeper how the troubled man confronts his trouble. He was, he knew, absolving Phelan of the murder rather more for reasons of his own self-disgust, and he was remembering, despite the wish not to, Priscilla Phelan's words to him that day, "Like a bull last night, Father, when I no longer wanted him . . ." He shuddered. With what? His own sexuality? He lifted the glass and studied the whisky. You keep it in a bottle, Joseph. Corked up tight. Up tight.

Brogan came from the phone and slipped easily into the

booth. His eyes were bright with excitement. "We've earned our night on the town, Joseph."

McMahon forced a tight-lipped smile.

The policeman leaned closer, touching the priest's glass with his. "Look, man. It's only for questioning. And it's better now than later for him. You know that."

McMahon grunted assent. He drank down his drink. He had not meant to. It had lost its savor.

"Take it easy on that, Joseph. We've a long ways to go when this bit of work is over."

"How long till they get here?"

Brogan shrugged.

"Do we have to wait?" He hated himself for saying it, but it had occurred to him that he might be expected take part in the arrest.

"Just to make sure," Brogan said. "Our work is all done—unless he tries a runout before they get here."

But Phelan made no move, scarcely even to raise his glass to his lips or to shift his weight one foot to the other. He might not even know Muller was dead, unless . . . What Brogan did not know was Phelan's performance with his wife the night before. Was that what the poor devil now was pondering? The fear to go home lest it be expected of him again? McMahon made a restless gesture, a sweep of his hand that upset his glass. The ice tinkled out.

"Easy, Joseph. You don't want to call attention."

They sat in silence, the hands of the Roman-numeraled clock above the back bar spanning the long slow minutes from eleven-three to eleven-eighteen.

When the two detectives walked in the men nearest the doorway, glancing round, stiffened a little, straightened a little, and then there was almost silence, talk and laughter cut mid-sound. There would have to have been some signal between Brogan and them, McMahon thought, but he did not see it. They seemed to know Phelan on sight. One on either side of him, they showed their identification. He went out with them without protest except for the motion to pay his check. The bartender waved him on. Had all this happened to him before? McMahon wondered. The resignation of the man was what troubled him. What did he know of Phelan except from Mrs. Phelan? And from the neighbors who didn't like him. He remembered the man's only police record, assault because a dog had lifted his leg on his shoe.

"Well, shall we go?" Brogan said. He could not restrain a little show of expansiveness. "There's a phone call I have to make."

"Make it." McMahon jerked his head toward the phone booth. He dreaded the crawl to the street.

Brogan grinned. "Not here. It would be a desecration."

He was about to get to his feet when the bartender came up to him. He was a big man, broad-shouldered. He jerked his thumb toward the door.

"Roger," Brogan said.

The bartender looked down at McMahon. "You, too, Padre. I can smell the cloth."

"Watch it," Brogan said. "Even a joint like this needs a license."

McMahon got up, again upsetting the glass. This time it broke. "I'm sorry," he said.

But Brogan with the back of his hand deliberately knocked over his own glass on the table.

McMahon got out as best he could with the burden of humiliation and anger. Brogan stood on the curb and stretched to his full height. He rubbed his belly with both hands and drew in several deep breaths. "Now for a telephone. There's a drugstore on the corner." He touched the priest's elbow to turn him in that direction. "You're game, aren't you, Joseph?"

"For what?"

"Aw, come off it, Joseph. They're nice girls, and they're clean. And they'll think you're a cop."

McMahon shook his head. "I've had too much to drink."

"You'll have coffee while I'm on the phone."

He went as far as the drugstore and had the coffee. It was as black as tar, as his own mood.

Brogan came from the phone and ordered coffee for himself. "Do you have any money, Joseph?"

"Twelve dollars."

"Buy a fifth of Scotch. I'll take care of the rest."

McMahon bought the whisky and set the bottle before Brogan on the counter. He was drunker than he had thought and yet he wanted more. He wanted what? Just whisky. Not a woman, not now. Just whisky and the last bitter dregs of the night.

Brogan drank his coffee. "They've a nice little place on Tenth Street. You'll be surprised. Or maybe you won't. They're nice girls."

"You said that."

"I know. But you won't believe me till you meet them." He picked up the bottle, crumpling the paper at the neck, and brandished it. "Tally-ho!"

McMahon did not even know if Brogan was married. He did not want to know. He followed him as though the bottle was a pipe and Brogan the piper. And then as they walked, his the careful walk of the man who knows he is drunk and has to take care of the drunkard in him, he began to say to himself, "Jesus help me, Jesus and Mary, help me. Jesus, Mary and Joseph. . . . Only Joseph can help you. He felt the sweat cold on his brow, on his back, beneath his armpits. As they turned the corner of Sixth Avenue, he heard the rumble of the subway. Students were going down the steps, books under their arms, and the gray of their faces telling of long days' work and nights spent in the classroom.

"Good night, Brogan." He heard a part of himself saying it,

and it was almost as though another part of him was surprised. "My apologies to the ladies."

Brogan stood—with the air of a man balancing himself on the top of the world, so that again a part of McMahon wanted to go on with him—and studied the priest for seriousness, for whether or not he wanted to be persuaded. Then he shrugged. "Okay. No hard feelings, Father?"

Father.

"No hard feelings," McMahon said, but he could not bring himself to a false thanksgiving.

7

The first thing McMahon thought of when he awakened in the morning was that walk to the subway entrance. Only Joseph can help you. He had never said that before. He sat on the side of the bed, his aching head in his hands, and castigated himself. There were times when he had called on a litany of saints to get him home safely, and once he had pictured himself—or dreamed it—carted home on the back of St. Christopher. On the floor at his feet was his breviary. He had managed his office, just the words, but he had managed it, and he had slept in his own bed. He would have to manage now, for he could not break his fast with an aspirin. He dreaded the first sip of wine at the altar. God forgive me. Lord, I am not worthy. . . . He showered and shaved and saw, whether he wanted to or not, the bloodshot eyes of a priest in the mirror. I will go unto the altar of God, to God who gives joy to my youth: the new liturgy had taken that from him, but he never entered the sanctuary that the words did not pass through his mind: to God who gives joy to my youth. Youth and joy. He was forty years old and the devil was hard on his tracks. And there was Muller again, with the dark spittle on his lips: *Do you believe in him, horns, tail and all?* I believe in evil. Deliver us from evil. Amen, amen, amen.

"Well, Joseph, you've made *The New York Times* as well as the *Daily News*." The monsignor was walking through the hall with a cup of coffee in his hand, the papers under his arm, when McMahon went down. The old man was an early riser. He had said the first Mass. The smell of coffee carried through the house, coffee and burnt bacon. The monsignor stopped at the office door and looked back at him. "Don't you want the papers?"

"Not till later, thank you."

"You spent the ten by the looks of you."

"Most of it."

"Did you have a good time at least?"

He thought of the youngsters and the fire truck, and the beads now lying on his bedside table like a rosary. "A fine time," he said. Then, remembering Phelan: "How did I make *The Times?*"

"Finding the victim."

"Ah, of course." He cleared his throat.

"You'll be in fine voice for the nuptial Mass," the old man said dryly.

McMahon had forgotten the wedding, and he had promised a final rehearsal after the eight o'clock Mass. At ten there was a funeral which Purdy would take. He was fonder of funerals than he was of weddings. But so was McMahon, to admit the truth. Or would have been that morning.

"They'll give us a good lunch at Costello's after the reception," the old man said. "But I hope to God they serve French champagne. The sweet stuff turns my stomach."

McMahon went out the side door and across the cement yard to the sacristy. He knelt on the prie-dieux near the sanctuary while Father Gonzales finished the seven-thirty Mass, which was in Spanish. Suppose it had been Gonzales whom Carlos had run into? Gonzales who knew nothing of Mrs. Phelan and her marital problems, whom Brogan would never have asked to go on the town with him. Or even going, Gonzales might not have been able to identify Phelan even if he wanted to. It was that which stuck in his craw. But there were more things than that in his craw. If it had not been Muller it would have been something else. What is it about, Lord? I shall try to be silent and hear.

He said his own Mass with no more than twenty or thirty people in the church, and most of them there for the wedding rehearsal. The lector read in Spanish and in English. When the priest raised his hand in the final blessing, he noticed the girl rise and leave the church by the side door. He noticed her because she left without genuflecting. He spoke to the wedding party from the altar to say he would be out in a few minutes.

He was removing the chasuble when the girl he had seen leave the church came into the sacristy. She was tall and quite thin, with heavy black hair down to her shoulders and large dark eyes.

"Excuse me, sir. Are you Father McMahon?"

"I am."

The eyes were not furtive, but she was uneasy. "Am I not allowed to come in here?" she said.

"Why not?" And trying to put her at ease, "We've no secrets."

A little smile. She wore no makeup. But she was not as young as he had thought at first.

He waited before removing the white alb. "Do you want to talk? I'm afraid I have a wedding party out there for a practice run, but it won't take long."

"I'll go," she said, and put out her hand as though to guide her turn back to the door. Then she shoved the hand into the pocket of her skirt. "Just tell me, what was he like, the man you found?"

"Are you Mim?"

Her head shot up, the lips parted and the eyes grew even wider than before. The face froze in his memory, for almost the instant he said the name she whirled around and was gone. He went to the door after her and called out. But she was running between the sunlight and shadow down the long passageway with all her might. He went out, vestments and all, but by the time he reached the street she was nowhere in sight. The restless groom was pacing the church steps. McMahon did not question him, and he conducted the rehearsal as he was, the white alb billowing out as he strode up and down the aisle.

The visitation haunted him all day, the face as vivid as a Rouault saint—if Rouault had ever painted saints. Between the funeral and wedding Masses, he went to see Mrs. Morales and explained to her and the other women on the stoop that the funeral arrangement was not possible.

The women talked among themselves in Spanish too rapid for his limping understanding of it. Mrs. Morales conducted the council with the hairbrush with which she had been grooming her older daughter's hair. Anita translated for him: "My mother will bake the cake and put out candles, Father. She wants to know, will you come tonight and say the prayers for the dead?"

"After nine," he said. "After confessions."

A woman in curlers—he remembered her from the stoop the day before—gave a toss of her head to the apartment windows alongside. "Father, he's back."

He knew she referred to Phelan, but why tell him? It was the same woman who had suggested to him that Carlos' mother was not home very often. A purveyor of mischief, Mrs. Vargas, no doubt.

"Good," he said and left quickly.

After the wedding Mass—he excused himself from the luncheon—he spent a half hour on the next month's calendar of parish activities, then an hour on music, feeling all the while that it should have been the other way around. Then having a few minutes before religious instruction, he took the musical score to *The Bells* into the choir loft and tried it on the pipe organ. The old church fairly vibrated. He pulled out stops that set free voices in the organ that might never have been sounded before in all its years of muted trebling beneath a spinster's hands. Then to the instructions: for baptized Protestants entering matrimony with Catholics, eager promises and runaway eyes. It was like stamping passports and letting the luggage go. On the subject of birth control, Father? No problem at all to celibates: the words went through his head even as he repeated the church doctrine as lately redefined by the Holy Father.

Miss Lalor made him a special tea and brought it up to his room on a tray, little sandwiches made up of the fish left from the supper he had not come home to the night before, and the pudding from lunch, but with fresh custard, all done daintily. The thing he was forever forgetting about Miss Lalor was that after her tempers, if you didn't appease them, she came round on a courtship of her own.

"I sent your other suit to the cleaners, Father. You got something muckety on it."

"Thank you."

"It struck me afterwards, if you'd wore it yesterday maybe you were saving it for the police?"

"No."

She lingered in the doorway, wanting to talk about the murder, but unsure of a safe way in. "Wasn't it nice, them mentioning in the papers that you're the director of the Girls' Choir?"

"Nice?" he said, scowling.

"I suppose you're right. There isn't anything they wouldn't turn to publicity nowadays. Eat something, Father. You're losing too much weight."

He said nothing, wanting her squat, corseted, lavender-scented presence removed from his doorway.

"Do you want the door closed, Father?"

"Please."

Alone, he conjured again the girl's face. He was trying, he told himself, to compare it with his memory of the dead man's, and there was not any comparison to be made except in his own sense of bereavement at losing both of them so soon. In all the city, and she had fled into the heart of Manhattan, where would you go to look for a dark-haired, dark-eyed girl named Mim? The Duminy Bar? He had thought of going there, but that was Priscilla Phelan's territory and he did not want to tread on that. Besides, if the girl had known him there, she would not have needed to come to McMahon to inquire what the dead man looked like. She had lost track of him and she had known him by another name, McMahon felt sure. Muller—Mahler: was that the association that had brought her? He felt no incumbence to go to the police. She had not identified the victim, only to herself. Of one thing he was sure: he would not again be used as Brogan had used him, and if he never saw Brogan again, so much the better.

A half hour later he went back to see Rosenberg. The pawnbroker was glad to see him, but he shook his head. "Ach, Father, for me writing is like trying to take fleas from a dog. As soon as I think I have one it disappears into another part of the anatomy. But I will keep trying."

"Perhaps we should talk," McMahon said.

"Nothing would give me more pleasure, but on Saturday afternoon I am busy like no other day in the week. You know how the old song goes. Nobody who can raise a buck wants to be broke on Saturday night."

Even as he spoke a well-dressed young man came into the shop, removing and winding his watch. Rosenberg asked the priest to wait. McMahon watched the transaction from the back of the shop, the gestures, the expressions. He did not hear the words, but the ceremony was as ancient as the charge of usury against the Jews. Another customer came in, this one in a Mexican serape. He reclaimed his guitar with a kind of shamefaced emotion, like someone getting his brother out of jail. Rosenberg, before handing it over to the boy, ran his own fingers over the strings. You see, he seemed to be saying, it has been in good hands.

He came back and entered both transactions in his ledger. He took off his glasses.

"Just one question today," McMahon said. "Did he ever speak of a girl, Mim, Min, something like that?"

"Many girls but not often by name. Nana Marie. I remember that one. I liked that name, Nana Marie."

Nim for short, McMahon thought. He wanted to be careful not to start the old man's thoughts in flight from whatever his association might be. It was a pleasant memory, whatever it was. Then very quietly the priest started: "Where were they together? What kind of neighborhood—or what were they doing?"

"Making love, I should think. Excuse me, Father."

"I'd think so too," he encouraged, "but where?"

"There would be a Greek church and it would be a poor neighborhood, for he loved poverty as much as honor. To him poverty was the only honor. No, that is not right. It is the climate of honor. God protect me! If only I could write it down and get it straight." He struck his temples with his fists.

"It will come, my friend. It will come. Perhaps we can help one another."

"I will not do this for the police, Father. Honor will not be confused with justice, not by Abel Rosenberg. Where in this world is justice, will you tell me that? And if there is a world in which there is justice, tell me why there is none in this? God is just, you will say, and I will say that is because man is not."

McMahon smiled. "I have said nothing, my friend. If I had, I'd have said, God is merciful, and that would have upset you even more."

"Pah! Mercy. Excuse me again. But it was the police who spoke of justice. They were here. I told them about the books. Now they will check the stores for their inventories. Let them. Nothing will be missing except the man himself."

"The Greek church," McMahon prompted gently.

"Very old—in a forgotten place, forgotten people: he said that. Beautiful old . . ." He made an elongated shape with his hands.

"Icons?" McMahon suggested.

He shook his head. "The delicate chains from the rafters to hold the candle bowls, beautiful in the dark of night. He was a janitor. There's an old-fashioned word for you, a janitor. He would not call himself a building superintendent, not my friend Gust.

"Not a man to pretend to greatness," McMahon said.

"It is so, and it was to not pretend to anything, that was how he wanted to live."

"You wonder what his talent was like," the priest mused. "I have a feeling it was valuable."

"Beware of feelings, Father. They are the biggest liars in us. They make truth what we want it to be." He looked to the front of the shop as the door opened again, and flung his hands in a gesture of hopelessness. "Look what this one is bringing me. An accordion. He will want a fortune for it, ivory and mother-of-pearl. And if he cannot redeem it, where will I find a street singer?"

McMahon said, "I must go. There will be a memorial tonight in the building where he lived. No funeral until the police are good and ready. I've promised to say a few prayers. You would be welcome, Mr. Rosenberg." He wrote the address on a card the pawnbroker gave him.

"Will you play Bach and Mahler?"

"If there's anything to play it on, I might."

"He would like it better than the prayers. And to tell you the truth, Father, I would too."

But Rosenberg did not go to the Morales apartment, and Father McMahon did not play Bach or Mahler. A visit to the stoop of 987 in the daytime was quite different from going up those steps at nine-thirty on a warm Saturday night. The sound of an electric guitar twanged through the building, and somewhere a Calypso singer was tuned in at top volume, and above it all, a cacophony of voices, one pitched higher than another.

A dim light shone in the windows of the Phelan apartment, but there was no light at all in the vestibule. He had to follow the voices. The ceiling bulbs in the hallway were caged in wire mesh. He began a slow, reluctant ascent. The smell of disinfectant was so strong it hurt his nostrils, yet it could not quite kill the undersmell of the communal bathroom on each floor. Behind a closed door a child was crying. He could just hear it through the raucous din, the loneliest of sounds, and one that angered him. A gang of teenagers thundered down the steps. He backed against the wall to let them pass, the girls rattling and sparkling with cheap jewelry, scented with heavy perfume, the boys with glossy hair and clattering, highly shined boots. He recognized Anita and called out after her.

"Upstairs, Father. Everybody."

On the second flight of stairs a fat grandmother was lumbering up, one painful step at a time, and at every pause she shouted a gutter invective, not for what was going on, but because it was going on without her.

A man leaned over the banister and baited her.

"Bastard!" She shook her fist at him.

He and another came down the steps and between them hauled her up. On the last step he groped her fat buttocks and goosed her. She shrieked and swung her arm around on him, almost tumbling them all down the stairs.

And among these people in so short a time, McMahon thought, Muller had found a welcome. He was far less confident of his own and he had been in the parish for eleven years.

But the women made way for him, pushing their men to the side, black men and brown and sallow-white, almost as varied as the colors of their shirts. Mrs. Morales shouted the two crowded rooms into silence. Someone turned off the record player, the last few notes wilting away. The guitarist was in another part of the building and someone closed the door on him. It helped a little. As Mrs. Morales led McMahon toward the inner room, he noticed Dan Phelan sitting in a corner, a glass in both hands, his eyes on the glass and his face as taut as the fingers around the glass. His wife sat on the arm of his chair. She was made up with her old flair, defiance in every feature.

A sideboard was spread with food in the second room, but it was not toward it that Carlos' mother drew him. It was toward a small round table where a candle burned alongside a shoebox. A chill ran down the priest's back when he saw what was in the box: a waxen colored doll laid out as in a coffin, a doll clothed as a man and made from child to man by the crude gluing on of a black beard. When his first shock was spent, McMahon realized that it was probably Anita who had given a swatch of her hair to the making of the beard. The unpliable hands of the doll were crossed over a bunch of violets.

Out of the corner of his eye the priest saw Mrs. Morales make the sign of the cross and he sensed rather than saw the others who had pressed into the room after him do the same. He lifted his eyes to the picture which hung on the wall behind the table, the placid, bearded, long-haired Christ with his forefinger touching the flaming heart. It was a picture familiar to him from childhood on, but in that instant, as alien as the shrouded doll.

He bent his head and prayed silently, but for himself.

The people were waiting to hear his words, but when, after a moment, he raised his head, they assumed his silent prayer appropriate and said their amens.

"You like it, Father, *si?*" Mrs. Morales said of the boxed figure, her gold teeth shining as she smiled.

"Beautiful," he said, and turned away determined not to look again at the picture of the Sacred Heart. Yet the words ran through his mind: Most Sacred Heart of Jesus, I place my trust in thee.

"Please, Father, take something to eat. The stuffed crabs I make myself."

"Have a drink, Father. We brought the whisky, and I'll have one with you." It was Phelan who spoke, having come up beside him, bottle in hand. He had a deep voice for so slight a man. McMahon tried to suppress the thought, but it came again, the wife's telling, 'Like a bull, Father.'

"Thank you," McMahon said, "I will have a drink."

Mrs. Morales gave him a glass, wiping it first with her apron.

When he held out his glass, two of the younger men present held theirs out to Phelan, too, and Pedrito Morales came up with his. Phelan poured without a word, generously, but with his lips clamped tight. He quarter-filled his own glass before setting the bottle on the table. Everyone held his glass, waiting. McMahon finally lifted his toward the effigy and said, "Peace be with him."

The men all drank. Pedrito coughed and wiped his mouth on his sleeve. "Irish piss," he said.

Phelan threw the contents of his glass in the boy's face. It all happened with flashing instancy: Pedrito flung his own glass over his shoulder and started for Phelan; Phelan, with one step backward, drew a knife from his pocket and switched the blade. McMahon was a few seconds reacting, for the last person in the room he expected to carry a knife was Phelan. He leaped between the men and ordered Phelan in the name of God to put it away. Phelan stood his ground and made jerky little stabs with the knife, trying to motion the priest out of his way. Behind McMahon, the men were derisive, their mockery the filth of two languages, and the women more contemptuous than the men.

Priscilla Phelan came up behind her husband and locked her arms around his neck, pulling his head back. McMahon caught his arm and twisted it until he let go the knife. The priest put his foot on the blade. It was not necessary. No one wanted it. Muller had been killed with a knife. McMahon picked it up, flicked the blade closed and put the knife in his own pocket.

Phelan had gone limp. He stood in a slouch, his eyes wild with hatred. His wife gave him a push toward the door. He pulled himself up straight then and walked with the controlled, exaggerated dignity of the drunk which McMahon knew well. At the door he spat and went out.

Priscilla Phelan tossed her red hair back over her shoulders. "This is my house, you bastards! You tell the police about this and out you go, every mother-selling one of you."

Mrs. Morales was scolding the instigators, the rilers among the crowd. The fat grandmother sat and rocked herself with pleasure. She clapped her hands. McMahon kept catching flashes of faces, of gestures, bare arms and laughing mouths, a girl draped over a chair, her legs fanning the air. And noise, noise, noise. Mrs. Phelan went out, her hips swaggering, and the men whistled and hooted, and one of them pranced a few steps as if to follow, stopped, and gave a roundhouse sweep of his arm, his thumb in the air. The Calypso music went on again. It was all over. And it wasn't that the party resumed, the explosion was part of the party, the language a kind of vernacular, and the noise was a way of life. No one apologized to the priest. Someone brought him his drink where he had put it down on the table when Phelan drew the knife. Mrs. Morales brought him a plate with two stuffed crabs and some pastries.

"Where is Carlos?" he asked her.

"In bed," she said. She indicated the door to a room off the parlor.

The boy could not have slept through that noise. "May I look in on him?"

She shrugged. It was up to him.

McMahon ate a few bites of the food and went to the bedroom door. He opened it, expecting to see the youngster wide-eyed and staring out at him as when he and Brogan had found him in the hut. Instead, he saw four bundled shapes beneath a blanket, children huddled together like puppies in a box, and all of them sound asleep.

The door to the Phelan apartment was open when he went down the stairs a few minutes later. She would be watching for him, and in any case, he wanted to get rid of the knife in his pocket. She called out to him to come in and then closed the door behind him.

Phelan stood, his back to them, and stared out the window. There was an Irish look to the apartment, which was merely to say McMahon felt a familiarity there not present for him in the rest of the building: it was the curtains, perhaps, just the curtains that made the difference, full, window-length, and white.

McMahon laid the knife on the side table on top of a copy of the *Daily News*.

"Dan's in real trouble now, Father. The police picked him up in a bar last night, and he can't even remember where he'd been in the morning."

He could remember, McMahon thought, but he was not telling. Protecting a man, perhaps. The police would have suspected that, picking him up where they had. He pointed to the knife. "Where did that come from?"

"It's been in the house for years," she said. "But why in the name of God he had to take it up there with him tonight, I don't know."

Phelan turned from the window. "Don't you, Priscilla? I think you do." He was about to sit down. "Would you like a drink, Father? There's another bottle, I think."

McMahon shook his head. "No, thanks."

Phelan slumped into the chair and shaded his eyes with his hand. A gentle hand, McMahon would have said.

"Do you know what I think, Father?" Mrs. Phelan said. "I

think he wants to be charged with the murder. Big shot! He wants to be a big shot to a houseful of freaks."

Phelan was shaking his head.

McMahon thought: it's being a freak in a houseful of big shots that's killing the man. He sat down on the couch near Phelan. "If I spoke to a doctor, Dan, would you go and see him?"

Phelan took his hand away from his eyes. A sad smile twisted at the corners of his mouth. "A psychiatrist?"

"Yes."

"He doesn't need a doctor. I told you that yesterday, Father. He's all right now."

Phelan looked from one to the other of them, and then at his own hands which he put palms together, the shape of prayer. No man ever showed more eloquently his sense of betrayal. McMahon got up. Twice, by inadvertence—or by some destiny that was tracking himself as well—McMahon had betrayed him. "Come and see me if you want to, Dan. Or call the rectory and we can meet somewhere else."

"In the jail maybe," his wife said, "if he keeps this up."

McMahon said nothing until she had followed him into the hallway. "Do you want to kill him or save him? I don't want the answer, but you'd better find it for yourself, Mrs. Phelan."

"Father . . ." She put her hands to her ears.

The electric guitar, the Calypso singer, and now someone on the drums.

"All I want is peace. Really, that's all I want."

"We could all say the same thing," McMahon said. "But where is it?"

"Thank you, Father," she said after him as he went down the steps.

For what? But he nodded and went on. A priest expected thanks, always thanks and only thanks, and a glass of whisky on the house. He was about to cross the street mid-block when he heard her scream, and he knew what it was the instant he heard it, and he felt that somewhere in his soul he had expected that also. After all, he had given the man back the knife. He ran the half block to the police call box where he had reported the death of Muller.

8 ◆◆◆

Phelan had botched his attempted suicide. He might die of the wound he had attempted to inflict on his own heart, but not if the surgical team at St. Jude's hospital could prevent it. He was in surgery for three hours, a nightmare of time McMahon spent with Mrs. Phelan and the police, one of whom called the priest's attention to the fact that for Muller to have lived long enough to talk with him, that job also had been botched. The consensus seemed to be—although no one said so in so many words—Phelan did not know much about anatomy. Then, to make matters worse, sitting in the small office provided by the night supervisor, Priscilla Phelan broke down and confessed to the police her affair with Muller.

Traynor said, "So you are unfaithful to him; you were unfaithful to a homosexual."

"He's not."

Traynor turned his cold gray eyes on McMahon. "Father?"

"I have nothing to say." Then: "Mrs. Phelan ought to know she has right of counsel before talking to you."

"I'm sure she was so informed," Traynor said with the quiet sarcasm that was always in his voice. He turned to the detective who had answered the first call. "Tonelli?"

"Yes, sir. I told her that."

Priscilla Phelan's eyes darted from one to the other of the detectives. "What are you trying to say to me?"

"You didn't know *that* about your husband?" Traynor said.

McMahon intervened. "Do you know it for a fact, lieutenant? Or do you know only that the man was picked up in a bar frequented by homosexuals?"

"You will make an excellent witness, Father. Or even counsel, you seem so well equipped."

Mrs. Phelan pounded her fists on her knees. "Give it to me straight, officer. What is it you're trying to tell me about Dan?"

"Father McMahon has pointed out the inadmissibility of hearsay—that's a nice word for it—hearsay on the street, in your home building, in the bar in question—hearsay. So all I can do for you, Mrs. Phelan, is ask a question. Without a lawyer, you don't need to answer it. You were unfaithful to your husband: did he attack *you* for it?"

"He screwed me!" she shouted. "Yes, you bastards, so don't try to give me that fag crap about Dan. Now get out of here and leave me alone."

"I wish we could do that, but we can't. We all have to wait—except Father, if he wishes to go—but we can pray the man lives to speak for himself." Traynor stretched his legs, put his hands behind his head and closed his eyes.

McMahon stayed. Three hours in hell. The detectives smoked, Priscilla Phelan chain-smoked. The detectives talked about their families, about baseball, about war and draft dodgers, about the kids today, and their own kids, most of them ashamed to say their father was a cop. My old man was a cop and I was proud of it. Dropouts and freakouts, but nobody mentioned fallouts. Then word came that Phelan had been taken to the intensive-care ward. There was a good chance that he would live. The detectives matched coins to see which of them would take the first four-hour shift at his bedside.

McMahon went as far as the ward door with Mrs. Phelan. She would be able to look in now and then on the trussed and tubed and taped figure of the man who had not been allowed to die, and she would take up the vigil meanwhile on the hard bench among the other watchers at the swinging door between life and death. Here was the documentary to the night's violence, and here among the watchers was a brotherhood of man that leapt all barriers, color, language, money. There was but one further reduction to the common denominator of humanity: the dead knew no prejudice.

After baptisms the next day, which was Sunday, Father McMahon commenced his search for the girl who had known Muller. Poverty and an Orthodox church, an old, forgotten church.

On Monday afternoon, having abandoned the classified phone book for an outdated map of the city, and Greek Orthodox for Russian or Ukranian, he walked down from Tompkins Square to Fifth Street and then east. Poverty was assuredly served by the street market on Avenue B. Used shoes hung on stands in clusters, there were bins of battered pots and patched-up toys, Mickey Mouse T-shirts and lingerie, chipped glassware and gaudy pottery. The vendors looked like gypsies but they spoke a voluble Yiddish, and Spanish if they had to. There were fruit stands and vegetables with Lexington Avenue prices. A woman might haggle the price of a washcloth, but she counted out twelve cents for a pound of potatoes and never said a word. McMahon walked on, his hands in his pockets, his eyes to the windows which always measured a neighborhood for him: the state of the curtains (that Irish core in him again); there were not many curtains at all now, but where there were, there were often window boxes as well, and he could see the red bloom of geraniums stretching their growth toward the sun. He looked in on a youth center: there were more boys on the street than inside, and children clamoring over old cars, spittling the dust and making faces on the windshields. And the flower children were here, bearded and beaded, and baited by the squares. He stopped a girl with an infant strapped on her back and inquired if she knew of a Greek or Russian church nearby.

"I know where there's a synagogue," she said, pointing, "and the mission house on the corner. We've started a nursery school there. It's cooperative and doesn't cost much—if you know of anyone who wants to join."

"I'll remember," he said, pleased, in slacks and sweater, to be taken for a native.

"He won't like it much." She gave a hoist to the child saddled on her back. "But he'll get used to it. He'll have to."

McMahon knew by her speech, the modulation of her voice, that the home she had come from was far from a city slum.

"He looks pretty sophisticated to me," McMahon said. And the child did, gazing solemnly around him for something worthy of his attention. That his nose leaked like a faucet seemed to disturb neither him nor his mother.

The girl, and she was only that, smiled broadly. "Man, he's so sophisticated he won't even talk."

McMahon thought about that as he walked on, the sophistication of not talking. The voice of silence. Something was happening to him in these hours of search he stole from the day's routine. He stole them chiefly from music, and therefore from himself. But walking the streets, he listened for another music, other sounds, and permitted unfamiliar thoughts to dangle in his mind: in a way, it was like listening for the voice of God instead of drowning it out with prayer. And save for the wine at the altar, he had not had a drink in two days.

St. Chrysostom Church stood out for him by its very inconspicuousness. A narrow building of gray stone, it hunched in the shadows of the soot-blackened fortress which was a public school built, according to the cornerstone, in 1896. McMahon tried the front door of the church. It was locked but he could hear voices within. He went to the side door. The smell of incense, to which he had thought himself inured, came at him like something alive rushing to get out the door.

There was no vestibule. The door opened directly into the church. Several older women and a few men arose from the benches along the walls just as he entered. The bearded Orthodox priest, fully robed, came from the sanctuary, censer in hand. He censed first the icons and then, one by one, each man and woman and McMahon too when he reached him. McMahon resisted the impulse to bow low in response as in the solemn high Mass of Rome. The priest moved on to those along the other wall. The church was lighted by small amber-glassed windows near the ceiling, and by the lamps hanging from the rafters on delicately wrought chains of varying lengths. Such had Rosenberg described from his conversations with Muller. The celebrant priest vanished from sight behind the gates of the sanctuary: ecumenism had not yet invaded this shrine of orthodoxy.

McMahon slipped out of the church. He wondered what the occasion was for the afternoon service. How little he knew of any calendar except his own. He had studied Greek in his seminary days, but the smattering of it left to him now was scarcely more than sufficed to tell him that the church inscription was in a different alphabet.

He began looking for "Nana Marie" on the hallway letter boxes of the building on the corner of Avenue A. A forlorn search, for while there were several names on most boxes, they were last names only. Mostly Spanish, some Slovak, one Irish name in the first two buildings. He stopped a square-faced sturdy policewoman, whose white belt and strap proclaimed her special duty at school crossings. She did not know anyone by the name of Nana Marie or Nim. By sight, maybe. She knew almost everyone by sight except the Puerto Ricans. Three quarters of her own people had moved away in the past ten years, respectable people with trades—bakers, upholsterers; her brother was a bricklayer who had moved to Staten Island. "Hunkies," she said with a kind of pride. "That's what they called us in the old days. We didn't like it, but we didn't beat up people for it, and the way this neighborhood used to be, you could eat off the street it was so clean."

While she spoke McMahon saw the girl and recognized her even in the distance, the wide skirt and a jacket open, and an overstuffed bag at her waist, the strap to which she shifted from one shoulder to the other as she came. He liked her walk, the jauntiness of it.

The policewoman, seeing that she had lost his interest, moved on. McMahon called out his thanks but he dared not take his eyes from the girl for fear she might vanish into a building on the way. But she came on, saying a word here and there to the children, to a boy polishing a car. She stopped, turning into the building where McMahon waited on the stoop, with the uncertain look on her face of having seen him somewhere before.

"I'm Joseph McMahon," he said. "Remember?"

She came up the steps. With a sidelong glance and a sly little smile, she said, "Hello, Joe."

She slung the strap from her shoulder, disentangling it from her hair, and gave him the woven bag, taking for granted that he would carry it and follow her into the building. She opened the mailbox marked "Lavery."

"I wondered who the Irishman was in this lot," he said. "Miss or Mrs.?"

"Nim," she said, and took two or three pieces of mail from the box.

"Nana Marie," McMahon said.

"Just Nim." She led him up one flight of dust-encrusted stairs after another. There were books in the bag and what he took to be a drawing pad.

"That's quite a climb," he said, waiting while she put her key in the lock of the front apartment on the top floor.

"Even for the cockroaches," she said.

She opened the door on a room that was as bright as the halls were dingy, one huge room, the walls broken through, but pillars had been left and painted like barber poles. An easel stood alongside the front windows. A floor to ceiling rack contained many canvases standing on their sides. Everything was pin neat, although to be sure, everything was not much.

She took her bag from him and hung it on the back of the door. "Would you care for a drink?"

"No. No thank you. Why did you run away when I asked if you were Nim?"

"Why have you come after me?"

"I wanted to know more," he said.

"About him?"

"Yes."

"Don't you think he'd have told you if he wanted you to know?"

"There wasn't time. He died too soon."

"Everybody does, except those who die too late and they're already dead."

"You're a very judgmental young lady. That's supposed to be my department."

"I guess I am," she said. "I don't have many chairs. That's the comfortable one. Please take it." She motioned him toward a rattan rocker and then took off her jacket and hung it over the bag on the door hook.

He took the moment to look around the room, at the posters, the paintings. But he wanted more to look at her.

"I like the floor myself," she said and sat on a rush mat, folded her legs and spread the wide skirt over them. "What did he tell you about me?"

"That once you said he would shake hands with the devil. Only that."

She leaned her elbows on her knees and her chin on her folded hands. There were marks of trouble or just possibly of dissipation that should not have lined the face of one as young as he thought her to be. The eyes filled with sadness. She tried to throw it off. "Do you think he did? Go like that, I mean: ready for anything—or nothing?"

"Let me tell you what those few minutes were like," McMahon said. "Sometimes I think I dreamt them, they're so vivid, surrealistic in the way things stand out in a dream, the hands, the shadows, the sounds. The smell of the place: all the incense in that church across the street this afternoon won't take the smell of that cellar from my memory. But it was what he said to me that makes connection, Miss Nim, that breaks through the dream. Or maybe strengthens it." He rocked back in the chair and wondered just for an instant why he felt so right in being in it, in being where he was. Within the compass of his gaze was a painting in grays, browns and ocher, a cubistic city with shapes like half-lidded eyes cornered in the windows. "You're a window-watcher, too," he said and pointed to the picture.

"You don't want to tell me what he said."

"I'm holding back, amn't I? I wonder why. It wasn't so much *what* he said either, but the sheer bravura of saying it at a time like that. He lived till he died. Do you know what I mean?"

She closed her eyes and nodded vehemently. It was what she had wanted to hear.

"You loved him, didn't you?"

"There hasn't been anyone since, and that's not like me." She gave a dry little laugh. "I wish I had his child, and that's not like me either."

"What is like you?"

She shook her head. "You were going to tell me . . ."

"This?" he said, indicating the room. "I've never gone into a house so clean, so neat."

"That's me on the outside," she said. "Please tell me!"

"I was writing a sermon," he said. "That's how it begins and I've got to tell it that way, because he asked me what I was doing when the youngster came for me."

"You were writing a sermon. What about? He'd want to know that too."

"Brotherhood," he said defensively, feeling that she also would have little affinity with sermons.

Her smile was flashing, brief. "I'm sorry."

"That's what religion is about," McMahon said, even more defensively.

"I'd like to think so." She drew a quick, deep breath. "Father McMahon was writing a sermon, and my poor runaway lover was dying in a stinking cellar."

He told her then, re-creating the scene down to the smallest detail he could remember, from the words on the sidewalk to the way the man had turned his head in on himself to die.

At that picture, she covered her face and wept, letting the sobs come out as they would. McMahon got up and went to the east windows, avoiding the easel where it stood by the windows overlooking the church, roughly north. He walked softly on the bare red-painted floor. He could see the river, a barge coming into view as it passed the new housing complex between Avenue A and the Drive. On the tarred rooftop two stories below him were old whisky bottles, beer cans, and things he recognized as contraceptives just before she spoke.

"Don't look down," she said. "In this part of the city, we always look up. That's where the sky is."

"A little bit of heaven," he murmured, turning back to her.

She blew her nose violently. "What?"

"That's one of the monsignor's favorite songs. A non sequitur, but it ran through my mind. I'll explain." He found himself then telling her in almost a stream of consciousness about his parish life, then about the tenement where Muller had lived and the Phelans, not about Mrs. Phelan's confidences of course, but of the police inquisition and then at the end, because it relieved him to tell it, of the way he had betrayed Phelan, identifying him in the Village bar.

"A homosexual bar?" Her intuition was quick. Or perhaps it was his own naïveté.

"Yes." He went on to describe the memorial in the Morales apartment and its aftermath.

She was sitting on the floor still, but now with one arm propped on the daybed so that he saw her face in profile, the long straight nose, the strong chin. "Do you know what I think, Joe—I can't call you 'Father.' You don't mind, do you?"

"Please. I like to hear the name once in a while. The monsignor calls me Joseph. Otherwise . . . please do. What is it that you think?"

She sighed and then said it: "My boy was sleeping with the Phelan woman."

McMahon did not say anything, and he looked quickly away when she surprised him with a flashing glance to his face.

"A man's got to sleep with someone . . . I guess. And he wasn't coming back." She got up and paced the room, almost boyish, her pacing, and yet she was very feminine.

"You're sure of that?"

"Oh, yes. That's one thing I've learned: no marriage band, they don't come back. Not that a marriage band brings them back either. I didn't try to look for him. I tried to forget him until I read in the paper about your finding a man named Gustave Muller."

"Mahler?" McMahon said.

"Exactly. *The Song of Earth*. I've got his records. You talk about the monsignor and his sentimental songs. Do you smoke?"

He shook his head.

She went to her jacket pocket and got a cigarette. He thought of taking the match from her and lighting the cigarette, but he didn't do it. She lit it herself and then inhaled deeply before going on. "Dark and velvet-toned: me and Kathleen Ferrier." She laughed self-consciously, having said it.

Again McMahon was tempted, this time to say that it was so. He said nothing.

"Gustave Muller. It sounds like a house painter, doesn't it? Do you know what he did for a buck here?"

"Janitor for the church. For how much? Ten dollars a week?"

"Twelve. We got along. I have a job from twelve to four. I'm a pretty good secretary. Forty dollars a week, take home twenty-five. Taxes, old-age security. Security—it's nice to think it can come out of a paycheck." She glanced at him. "But then you've got it made, haven't you? How did you know about the janitor work?" She spoke rapidly, nervously, as though she was afraid to stay on any one subject for very long.

"A man named Rosenberg: I got three things from him, janitor, the Orthodox Church and Nana Marie. He was a friend to whom Muller talked a great deal, a pawnbroker on Ninth Avenue. Nim, you said you didn't try to find him. I think somebody did. I don't think Phelan killed him. I feel as sure of that—a lot more sure than I am of salvation. My word for security."

"I like it better. I hate security."

Because you haven't got it, McMahon thought, no more than I salvation. "What I'm trying to say is, I think somebody he didn't expect to find him, did. He knew the man. That much I'm sure of."

"I know what you're saying. I don't care."

"That's remarkable, if you mean it. He did not care either. There's a philosophy at work here that I don't understand."

"If he had his reasons for not telling you, they're enough for me," she said. "I didn't know him by the name Muller. It was just the feeling, and pure accident I picked up *The Times* at the office that day. I didn't know his real name either. That was part of the way he chose to search himself—for himself. He wasn't afraid of much. I have a lot of fears, but he didn't. The only thing he was afraid of was becoming somebody that somebody else thought he ought to be. I've never seen his painting, do you know? And we lived together for over a year. But whatever's good in my work, and I'm beginning to believe in it myself, is what he taught me. There's a time to lie fallow, he would say. Come fallow with me. His puns were awful."

McMahon wished she would slow down for a moment. He

wanted to think about the phrase, becoming somebody that other people thought he ought to be. But she went on.

"Do you know how I met him? I was working in a gallery then on Tenth Street—secretary, messenger, a lot of things. Anyway, he came in out of the rain one night and I was there alone. It was a perfectly terrible night, the weather I mean. He crawled out from under what looked like a tarpaulin and dropped it by the door. He went from one painting to another—the artist is pretty bad, I think, but he's got a Madison Avenue gallery now—and he didn't say a word until he had looked at them all. The artist's name—well, I'll tell you, it's Kenyon, but he signs his things just with the letter K. Finally Stu said: 'K presumably knows K intimately. I know a painter who signs himself E—E for Ego, but for the opposite reason. Or is K for Kokomo?' " Nim reminded McMahon of the member of the monkey family written up for his art work.

" 'I'll tell Mr. K you recognized his style,' I said, something like that. He looked at me in that very intense way of his—like the person was a painting too. 'And who are you?' " She mimicked a slightly clipped speech. "I'd had it for the day, I'd almost had it for life at that period. 'I'm an aging hippie,' I said. 'Well, since I'm a dirty old man, let's close up this zoo and have coffee together.' We did and I talked a lot, mostly about me, and afterwards he walked me home here under that crazy tarpaulin of his and I remembered he did smell like a dirty old man, but I guess Sir Walter Raleigh did too, and I liked him quite a lot. I ought to say—maybe it doesn't matter—but before we left the gallery I asked him to sign the visitor's book, just to prove I'd been there after the boss went home. He thought about it for a couple of minutes before he signed. I think he was making up a name then, Stuart Robinson. I said I'd call him Stu, and he said it was appropriate since he was a reformed drunk, a member of AAA." She smiled wryly, remembering.

"I said, 'AA, you mean.' And he said, 'No, Alcoholics Anonymous Anonymous . . .' And another time, I asked him what his own painting was like. Another of his lousy puns: Anonymous Bosch." She pulled hard at her cigarette and sent the smoke to the ceiling. "Bosch and Tchelitchew. He said that was how nature hung him up too. He didn't like nature. The outdoors, I mean, trees and sky and water. They hurt too much."

McMahon thought of telling her about the art books, including Bosch, at Rosenberg's, but it would wait. He said: "When did he go away from here, Nim?"

"Sixty-three days before the day he died. How's that for remembering?"

"It's been easier," the priest said, "than it's going to be to forget."

She made a noise of ironic agreement. "Are you sure about that drink?"

"For now I am." He wanted to think about her and Muller.

She squashed the cigarette out in the tray at her feet, grinding the butt into shreds. "Why don't you ask me?"

"Why he left? I assume you would tell me if you wanted me to know."

"It's funny. I said something like that to you in the beginning. Didn't I? And if you'd asked me, I was going to say I thought it was because he was ready to start painting again, and he had to find that place for himself, where he could be the self he was looking for. But I'm not sure of that any more. He didn't say anything. He just went out one day with some of his books and he didn't come back. And that's what I told myself was the reason. But the real reason—I think now—I'm a very conventional person underneath. I wanted something conventional of him."

"Marriage?"

She shrugged. Daylight was beginning to fade. Her face was shadowed. "Security."

"I thought you said you hated security," McMahon said gently.

"I do. But that doesn't make me not want it."

He laughed. "I will have that drink you offered me."

"It's gin."

"I can take it."

She poured two glasses of gin over ice in the kitchen, McMahon watching from the doorway, and told of how Stu had taken out the walls and reinforced the beams with sewer-pipe rejects. McMahon rapped his knuckles against the striped pillars, and the sound was hollow. He told her of Carlos' hut and the doorknobs. They touched their glasses and drank to him.

"Whoever," Nim said.

On the second drink, McMahon proposed the toast, "To life."

"It's what we have, isn't it?" Nim said. "Will you have supper with me? I can make it stretch. You know, instead of hamburger, spaghetti."

He would have liked to stay, for he could feel the ache of her loneliness. Which must be his own, he thought. One does not feel another's pain that keenly. But he declined. "We have a tyrannous housekeeper, and I have work I must do tonight."

"Your father's business—isn't that what the boy Christ said when they found him in the temple?"

"Something like that," McMahon said.

"I don't think I've ever known a priest before, not to talk to like this."

"Not many people do," he said, which was his own truth at least. At the door, having thanked her, he said: "Nim, you could be wrong in why he left."

"I know. It's my way of beating myself, to think it."

"Then stop it, because I think you were right the first time. What we've got to do is find his painting."

Her eyes went moist as she thought about it. "It would almost be like finding his child—in me I mean, and that would be a miracle."

"I believe in them," McMahon said. "And I am convinced, he was not an anonymous man."

"Neither are you," she said.

Which meant, he supposed, going down the stairs, that to her all priests were just that. And if that had not been what she meant, he did not want to think about it.

9 ◆◆◆

"The signs are all go, Father," the desk nurse said sportingly. "Except that he can't talk. Or won't. The detectives, I suppose. I shouldn't say that. I don't really know."

"Do they question him?"

"Not that I know of. They just sit and take up space. So does he, Father. We need that bed for people who are dying in the wards because they can't get the care we could give them."

McMahon went into the ward, making his way around beds that could launch into space, so elaborate and delicate their equipment. On each of them lay miraculously live testimony to man's violence on the ground, especially when he used the automobile like a weapon.

Priscilla Phelan was standing at her husband's side, her hand on his taped wrist. A chair had been squeezed in the corner for the plainclothesman. A big man, he was doing his best to keep out of the way: the detail was not one of his own choice.

"It's Father McMahon," Mrs. Phelan said, bending close to her husband's ear.

He did not open his eyes.

The nurse had been right: he had better color than his wife, McMahon thought. It was a good face. There was even strength in the mouth, something the haunted eyes distracted from when you confronted him face on. Mrs. Phelan brushed the hair back from his forehead. She made a gesture of hopelessness to McMahon.

He spoke to the officer. "I'll be responsible while I'm here, detective."

"Okay, Father. I'll have a smoke."

"I will too," the wife said. "Danny, I'll be right outside. Or maybe I'll go home for a little while. I won't be long."

There was no response from Phelan.

McMahon waited, saying nothing, watching the face gradually, almost imperceptively relax when the silence told him his constant companions were gone. He opened his eyes, screwed them up against the shock of light and then looked at the priest.

"I wish she would never come back," he said.

"She will—until you make yourself better. You're like her child right now." Fleetingly he thought of Nim and her wish for Muller's child.

"That's what I've been for a long time, except . . ."

"I know."

"Do you? Yes, I remember now."

So did McMahon, the palmed hands, the portrait of betrayal. "Why don't you talk to the police and get them off your back?"

"You too, Father?"

"All right, Dan."

"I wish they'd go away."

"They're gone for now. Did you know him at all, the man Muller?"

"Just as the man in the back room. I didn't want to know him."

"Why?"

"I liked him, that's why. Christ almighty, leave me alone." Phelan tried to fling out the arm with the intravenal tube attached to it. The overhead bottle rocked in its cradle.

McMahon steadied it. "Easy, Dan, easy. I want to help you if I can, not torment you. You see, I don't believe for a minute you could use that knife on anybody but yourself. But let's get off the subject. Do you still want to die?"

"No. But I don't want to live much either."

"There's a difference, a marvelous difference. What would make you want to live?"

Phelan thought about it, his eyes almost closed. McMahon was afraid for a minute that he had turned off again. But the V deepened between his eyebrows. "If I could be like him."

"In what way?"

"Just in the way he had with people. And I don't mean my wife. The kids, you know . . . just the way he was with them. I can't explain it."

"I know what you mean. I felt that quality in him even in the last minutes of his life. We'll talk tomorrow if you want to, but only if you want to."

"I guess I do. This is kind of wild, Father, when you think about some things, but there was a . . . a holiness about him. Crazy?" He looked up at the priest: a little zeal came into the hitherto fugitive eyes.

McMahon said: "You're an idealist, Dan, and there aren't many left. I want you to live."

"So do the cops, but that's not what they call me."

McMahon leaned close to him. "Screw the cops," he said under his breath.

Phelan parted his teeth in a soundless, almost motionless laugh, and McMahon thanked God for it.

He patted Phelan's shoulder. "Get out of this ward. They need the bed for things you wouldn't believe are human, the way they come in."

"I know. I hear them all night long."

"Tomorrow, Dan."

In the hallway Brogan was waiting with the other detective and a third who was about to relieve his colleague at the bedside. It was the first time McMahon and Brogan had met since their night on the town.

Brogan offered his hand. "A lot of water under the bridge, eh, Father?"

"A lot."

"He talked to you, didn't he? I'm not asking what he said, Father. I mean he can talk if he wants to."

"He can talk."

"But will he? That's the question. Don't worry, Father. I'll wait him out. Got time for coffee?"

"I don't actually. I've got to make the rounds while I'm here." He had taken over from Father Gonzales in this, and Gonzales would take his church history class that afternoon. He had fallen behind in his work with the chorus and the recital was a week off.

Brogan walked down the hall with him. "I was going over your statement again, that crazy conversation with Muller? And you know what I think, Father? That part about taking the knife away from his killer—that could be pure Freud."

"I suppose it could," McMahon said with heavy solemnity.

"I made the mistake of trying it on Traynor. Oh, man. You know what they're calling me now at the station house?"

"Doctor Freud."

"I still think I've got something." He jerked his head in the direction of the ward. "Why won't he talk to us? You don't spend twelve hours sleeping off a drunk someplace you can't even find the next day. No, sir. When you wake up after a night out, you know where you are. Ask me."

McMahon had no such intention. "Brogan, I agree. The man who wants to go home can generally find his way there. And the man who doesn't want to go home needs help. That's where I'd call on Doctor Freud if I were you."

"Me?" Brogan said.

"I wasn't thinking of you. That's your connection." McMahon walked on down the corridor.

He had scheduled an extra hour with the girls' chorus after school that afternoon. He was now wishing that he had chosen a less ambitious program. *The Bells* just weren't swinging: because he wasn't, he knew that, and for that reason, he tried to be more patient than usual with the girls. It crossed his mind that Sister Justine could do as well with them as he was doing, and her programming would be more to the monsignor's liking, and more to the tastes of most of his audience. *His* audience. His audience was not the majority. His was an audience of one, himself. That bit of self-scrutiny out of the way, he went after the beat he wanted. He illustrated at the piano, going through the crescendo passage, accenting with his own baritone voice the preciseness, the mounting excitement he wanted.

"You've seen it in the movies, on television," he said, getting up from the piano. "The pioneers trapped in the stockade, the Indians creeping up on them, closer and closer, and then *Voom!* the soldiers, the United States cavalry racing to the rescue. Now, sopranos, you're the pioneers, you're scared, you feel it in the scalp of your head. Every note is a cry for help, urgent, more urgent. Mezzos, you're the Indians. You move in softly, carefully, but you keep coming on. These people have taken your land, your buffalo, your way of life. And nobody, but nobody has to tell the United States cavalry what to do." He motioned Sister Justine back to the piano and murmured, "May Rachmaninoff lie quiet in his grave."

After conducting the passage once from the podium, he went down the steps and up the raked aisle to hear what it sounded like from the back of the auditorium. There in the last row, flashing a broad smile, was Nim Lavery. After the surprise of seeing her, he was both pleased and angry. The anger was a throwback. He would have spent it in sarcasm on any other intruder.

She sensed this instantly. "I'm sorry. One of the sisters brought me here."

"The sisters," he said, rather enjoying Nim's discomfiture, "are not in charge of the choir." He turned back to the stage. "All right, take it from the beginning."

"I'll wait outside for you," Nim said, about to get up.

"No. It's too late to leave the stockade now." He sat down beside her, slouched in the seat and closed his eyes, trying to concentrate on the music. The girls, bless them, were good. They rose to an audience of more than one, and so did he, after all.

"That was just fine," he said, his voice ringing through the auditorium. "So let's quit while we're ahead. Choir dismissed."

"I'm sorry I came in here," Nim said.

"Are you?"

"For intruding on your privacy, yes. I didn't realize it until you came down the aisle."

"Did you like what you heard, at least?"

"Very much. If I closed my eyes it was like the Vienna Choir Boys."

"Better," he said. "These kids know what life's about. And that's where real singing comes from."

"Cowboys and Indians," Nim said.

He grunted, caught.

"Forgive me again," Nim said.

The auditorium was empty except for the girl who collected the music and Sister Justine who seemed uncertain of whether she should go or stay.

McMahon got up. "Come back to the rectory and maybe Miss Lalor will give us a cup of tea."

In the courtyard, Nim said, "Father McMahon . . ."

He looked at her sidewise.

"I can't call you Joe in that." She traced the shape of his collar, her hand at her own neck. "I wanted to tell you—I don't know if it will mean anything, but after you left yesterday, I was thinking about Stu and the things you and I had said. I remembered there was a showing of Tchelitchew drawings at the Burns Gallery. I went there at noon. Gustave Muller signed the visitors' book the day of the opening, a week ago Tuesday."

"So you see," McMahon said slowly, "he *had* begun a new phase of work." It did not necessarily follow, but he wanted her to believe it.

"But what you said about someone's finding him, that's where it could have happened, don't you think?"

"Yes. How many people signed the book?"

"Eighty or so. And not everybody who goes to an opening signs in. Especially when it's not new work."

"Still, I suppose we ought to tell this to the police."

"I was afraid you'd say that." She threw her hair back from her shoulders.

"What have you got against the police?"

"Prejudice."

"If you were in trouble, wouldn't you call them?"

"Yes."

And that seemed to be that until at the school gate she added: "Then maybe I'd be in more trouble."

On the rectory steps she hung back. "Are you sure it's all right, bringing me home to tea?"

McMahon laughed. "Miss Lalor is not my mother, though to be sure, she sometimes thinks so. There's nothing she likes better than to serve tea—unless it's to be asked to join the party."

"Please don't ask her." A smile fidgeted at the corners of her mouth. "I might call you Joe."

"I won't ask her," he said, and touched the bell as he opened the vestibule door. He took Nim into the study where he had been working when he saw Carlos.

"Ah, it's the young lady," Miss Lalor said, coming to the door. "I'm glad you found him, miss."

"Miss Lalor, this is Miss Lavery. She was a friend of the man who was murdered."

Miss Lalor gave Nim her most sincere look of sympathy. Her commiseration was rarely in words, only sounds and attitude. Like the priests she served, she sometimes tired of the tools, but never the materials. "Sit down, dear, and I'll bring you and Father a nice cup of tea." On her way out, she paused. "Lavery—that's a North of Ireland name, isn't it?"

"My great-grandfather came from Londonderry," Nim said.

"I've seen people from the North before with black eyes," the housekeeper said. "I've been told it's the Spanish, a long way back. Well, I'll get the tea."

When she was gone, Nim said: "My grandmother was Italian. My mother's people were Jewish."

"As long as there's a bit of Irish in there somewhere, to Miss Lalor you could be Greek, Gallic or Phoenician, and you'd still be Irish."

"Unless I were black. Am I right?"

"I'm afraid so. Except that the North are black Irish," McMahon said and grinned.

Nim studied the room with open curiosity, the crucifix, the framed blessing of the parish by Pius XII, the pictures of Popes Paul and John. "He's the one," she said of John.

"Ah, yes. As Miss Lalor would say, he's the one of them all. Mind, she's a Pius the Twelfth woman herself, but she's trying her best to catch up."

Nim, her hands behind her back, and with a childish sort of swagger continued to tour the small room. Again she stopped at the crucifix.

"It can't be all that strange," McMahon said, "Irish and Italian."

"My father was an agnostic, a physicist."

"Is he dead?"

"To me—almost from the day I was born."

"And your mother?"

"She was going to one of those rejuvenation farms the last I heard, and Dad had just been made a director in Dow Chemical."

"I see," McMahon said.

"You'd be pretty blind if you didn't." She found a straight chair, not that there was any in the room that wasn't, but she chose one without arms. She tugged at the short skirt, a hopeless gesture. "I didn't expect to be invited to tea," she said.

"I like them," McMahon said of the skirts girls were wearing now.

"For shame, Father!" Then she laughed. "There's a story, but I'll tell you another time. Something more important: it's been going through my mind all afternoon. It's Tchelitchew again. In a way he's passé now. Forgive me if I talk to you the way I'd expect you to treat me about music. I asked Mr. Burns why the exhibit. You know, so many good artists can't get a gallery, and he said it was because a collector, a friend of his, wanted it and was willing to offer some of his Tchelitchew drawings for sale. His name is Everett Wallenstein. The name is familiar but I can't place it. I'm sure Stu never mentioned it, but he didn't mention anybody, except maybe painters he thought important to me. I'd like to go and see this man, just to talk to him. But I don't want to do it alone. I don't think I could."

"All right," McMahon said. "Make the appointment and I'll go with you. Or have you done it already?"

Nim shook her head. "I wish you'd do it."

After tea they went into the office and McMahon looked up the name in the phone book. When he dialed the number he got an answering service from whom he elicited the information that Mr. Wallenstein would not be home until after six.

"Let's just go and camp on his doorstep, surprise him," Nim said.

"You're making an adventure of it."

"I don't know what I'm making. It's all instinctual. It's not like me to crash the gate. Or wasn't. I did that this afternoon too, didn't I? And me brought up on the nicest amenities."

McMahon glanced at the parish calendar for the day. He was free between six and eight o'clock if he was willing to take his supper cold after nine. "Why not?" he said. "Meet me at six-thirty in the Whelan drugstore on the corner of Eighth Street and Sixth Avenue."

10 ◆◆◆

The house on Charles Street had been beautifully restored, outside and presumably inside, the black shutters freshly painted, the brass knocker and the mail slot polished to a high gloss. McMahon lifted the latch on the gate. A hip-high fence of wrought iron bordered a garden of tulips and iris.

"I wish I'd worn my uniform," he said, in the sport jacket again.

"I almost wish you had too," Nim said. "I feel like we ought to be peddling *The Watch Tower.*"

"Not in my uniform."

The doorbell chimed deep within the house.

No one came. No sound from inside. Only the rumble of traffic on Greenwich Avenue.

"The bells," Nim said nervously, "the tintinnabulation of the bells, bells, bells."

"The Bells of St. Mary's, that's what Monsignor Casey thought we were doing. 'There's a tune to that, Joseph.' He mimicked the old man's accent.

"You're a snob," Nim said. "Did you see *La Plume de ma Tante?* You know, the monks ringing the bells and getting carried away. Literally, all hung up on the ropes. Wild. I was in college then. I met my father in New York and he took me to see it. That was one of our few good times together."

The door opened without their having heard the man approach. He was a tall young man, quite handsome and at the moment, sweating, as though he had been interrupted in the midst of some strenuous exercise. His hair was touseled from his having pulled on the velure sweatshirt.

"Forgive the intrusion, Mr. Wallenstein," McMahon said.

"We came on the chance that you might know an artist who was a friend of ours. I'm Joseph McMahon and this is Miss Lavery."

"And who is the artist?" the man asked coldly.

"That's the trouble. We're not sure of his name."

"Then how can you be sure he's an artist?" The man looked annoyed and McMahon did not blame him. But after a second or two of indecision, he said, "You may as well come in."

It wasn't camp, or what McMahon thought of as camp, but it was pure Victorian, the small, high-ceilinged parlor into which he led them. The lamp he turned on was the real Tiffany. It occurred to McMahon that he knew more about fashion and furnishings than he had been aware of knowing.

"I don't have a telephone," Nim said, an uneasy attempt at explaining why they had come without forewarning.

"That is understandable," their reluctant host said with a sudden turn of gallantry. His eyes reinforced his intention of compliment in a frank appraisal of her, head to toe. "Excuse me a moment while I get a towel." He touched his brow where the sweat was glistening. "I have a gymnasium of sorts in the basement."

When he was gone Nim said: "I really dig this place."

"I'm trying to figure out whether I do or not," McMahon said. He went closer to one of the paintings, a pastoral scene. The signature surprised him. He covered it with his hand. "Who would you say, Nim?"

She turned on another light and studied the painting for a moment. "It's way out, but I'd say . . ." She hesitated. "All right, I'll say it, early Kandinsky."

"Very good," Wallenstein said from the doorway.

"Am I right?" She was delighted with herself.

"My father bought that in 1912," Wallenstein said. He wiped his face and neck in the towel. His having combed his hair, the gray streaks in it showed up. Again McMahon had misjudged age. Wallenstein was in his forties. "Now about your friend."

McMahon told him of Muller's death. "I'm a priest, by the way." He had to add that, explaining why he had been taken to the dying man.

"Are you? No one turns out to be what he seems these days. I'm sorry, but I'm at a loss to know why you've come to me: I'm afraid I have not heard of Gustave Muller."

"Neither had I," Nim said, "but I lived with him for over a year."

Wallenstein did not say anything for a few seconds, but he looked at her in a way McMahon did not like, almost as though he was fantasying himself in that position. Then he said, "And whom did you think you were living with, Miss Lavery?"

"Stuart Robinson."

Wallenstein repeated the name. "That seems familiar. Perhaps I've seen his work. Why *did* you come to me?"

"He was at the opening of the Tchelitchew exhibit at the Burns Gallery."

"Ah, now I see. But my dear girl, so were a hundred or so other people."

"I wish I had gone," Nim said. "Maybe things would have turned out differently if I'd found him."

"Not if he hadn't wanted it," McMahon said.

"Poor fool, he," Wallenstein said, again with that look at the girl which made McMahon want to hit him. An irrational reaction, he knew. Was it the having of money that made the man arrogant in such a manner? Or the fact of Nim's having frankly admitted to living with a man?

"Did you admire him as an artist, Miss Lavery?"

"Yes," Nim said unhesitatingly, which seemed strange to McMahon, knowing that she had never seen Muller's work.

"I should like to see him," Wallenstein said. "But perhaps I have. How extraordinary that an artist should change his name. His technique, his medium, his philosophy, I can understand. I paint, myself, you see, and I am as jealous of my name as I am of my mistress."

McMahon said, almost before he knew he was going to say it: "Mr. Wallenstein, would you go to the city morgue with me in the morning? It's possible you would know him under yet another name."

"Yes, of course, if it's that important to you." No hesitation, and McMahon had expected it somehow. In fact, part of his intention was to discomfit the man. "Then we won't take any more of your time now. You've been very kind, sir." He got to his feet.

"Won't you have a drink? It is that time of day."

"It's past that time for me," McMahon said. "What hour may I call you in the morning?"

"After eight. Any time after eight will be fine."

When they reached the street, Nim said: "My God, the way you got me out of there, you'd have thought it was a house of prostitution."

"That was my very feeling."

Nim grinned. "He's an odd one, isn't he? Aren't you glad we came?"

"I don't know. Maybe it's just money. I spend half my life talking about it, trying to coax it out of penny banks and working people's pockets. I didn't like the man and that's a fact."

"But why?"

"I don't know why!" he exploded.

"Because he's decadent? I rather liked that. I like the filthy rich. It's the in-betweens that turn me off."

"Then go back and have a drink with him. He'd be delighted."

"Thank you very much. You've been very kind, sir. I won't take any more of your time." Having given him back his words to Wallenstein, she turned and ran for the bus that was pulling up at the corner of Greenwich Avenue. She boarded it without looking back.

McMahon walked to Eighth Avenue and then north, thinking at every tavern he passed that he needed a drink, and aware of that craven thing in him that made him watch for those with Irish names where a priest would never be allowed to put a cent on the bar. He realized then that he was in mufti. He needed a drink, but he needed more to remember that he was a priest.

11 ◆◆◆

That the mortal remains of a man should be pulled out on a tray like a slab of beef from a freezer chilled McMahon to his very bones. He felt Brogan's hand go tight on his arm to steady him. Wallenstein had gone paper pale too. McMahon stared at the tag, then the covering: the words "winding sheet" came to him, and he thought of Lazarus rising from the dead at the bidding of Christ. Thus he got through the self-imposed ordeal. Brogan had said he could wait outside, but he chose to accompany them. He had thought at first he did it to give Wallenstein moral support, but there was also a measure of self-mortification in the act.

When he had arranged with Brogan that Wallenstein should see if he could identify the victim, McMahon had told his first lie of commission: he told of the exhibit at the Burns Gallery where Muller signed his name to the visitors' book, but he attributed the discovery to himself, not Nim, whose name had not yet come into the investigation. "Remember the art books at the pawnshop? It was just a hunch."

"Sometimes they pay off. Bring him down, Father," Brogan had said.

"I thought at first I knew him," Wallenstein said when they left the morgue.

"Who did you think he was?" Brogan asked.

"A painter I studied with some years ago—at the Art Students League. But the nose—it wasn't the same, and I remember that chap's nose. I don't remember *his* name now. I would if I heard it of course."

They went into a small cell-like office within the building. A smell McMahon associated with embalming fluid stayed with him. He was glad to see Wallenstein light a cigarette.

"It would be in the school records," Brogan suggested.

"Yes, but I assure you, he is not the same man."

"As well as the names of others in the class," the detective went on doggedly. "How many?"

"Twenty or so."

"And how many people signed in at this gallery affair last week?"

"About a hundred. Ah, I see—cross-checking the names in case I'm mistaken. That is clever."

"That's how they train us, Mr. Wallenstein," Brogan said. He didn't like him either, McMahon thought, but Brogan's next question, put with the same aloofness, surprised the priest. It also told him Brogan's slant: "Have you ever come across a man named Phelan?"

"Phelan or Fallon? I knew a Steve Fallon at one time."

"What business was he in?"

"Interior design," Wallenstein said, his voice like ice. The tenor of the detective's questioning had come across to him too. He looked at his watch, a gesture Brogan ignored.

Brogan said: "Would you have any objection to telling us where you were last Friday morning from—say dawn till noon?"

"It is none of your damned business, if I may say so, sir. Three days a week, including Fridays, I rise at eight, my housekeeper brings me breakfast at eight-thirty and by nine o'clock I am in my studio on the top floor of the house I live in. Two days a week I go to Wall Street—Wallenstein and Warren. I have not varied that routine in five years."

"I've heard of Wallenstein and Warren," Brogan said with a sheepish attempt at a smile. "I've got my routine too, and the men at the top like me to stick to it." He got up and held out his hand. "No hard feelings, Mr. Wallenstein?"

Wallenstein shook the hand. McMahon noticed that afterwards Brogan flexed his fingers.

To McMahon, Brogan said: "Thanks, Father. Keep in touch."

The priest and Wallenstein walked to the parking lot in silence. Wallenstein had picked him up at the rectory and insisted on driving him back there. Before he turned the key in the ignition, Wallenstein sat a moment playing his fingers over the steering wheel, meticulously clean fingers, such as McMahon would not have expected in an artist. In a banker, yes, however. "A curious thing about the police," Wallenstein said, "they're human like the rest of us, but they don't mind our knowing it. They would have no place in a civilized society." He glanced at McMahon. "Tell me about this Lavery girl. She has the most striking face I've seen in a long while."

McMahon was nonplused at the directness of the man. Civilization, no doubt. He was caught in a civilized trap, and one to which his own vulnerability was hinge. Certainly he was not going to comment on Nim's beauty. "She came to me thinking that Muller might have been the man she knew."

"So you said. One might wonder why she did not go to the police."

"Very civilized," McMahon said curtly.

Wallenstein smiled. "Would you mind giving me her address, Father McMahon?"

"I don't think that's my place," he said, but saying it, and remembering Nim's and his last exchange, he realized that he might be assuming something that it was truly not his place to assume. Nim might want a liaison with Wallenstein. It was himself who did not want it for her.

"Perhaps then you might arrange a meeting among the three of us? You would go to dinner with me, let us say. I should like to see her painting. That's a twist, isn't it? But I am right in supposing her to be a painter, am I not?"

"Yes."

"It is very difficult for a woman to get a decent gallery.

That's gauche of me. And I've not seen her work. But her recognizing that Kandinsky on my wall—it's never been exhibited, you know."

"I didn't know," McMahon said almost sullenly. He forced himself to throw off his petulence. Then it occurred to him that he would need a fairly strong pretext for contacting Nim again himself, and he knew that he wanted to. "When?"

"I'm free tonight if it can be arranged," Wallenstein said.

"I'm not." He had to chair a meeting of the school funding committee at six, and he had not yet prepared the agenda. But how he longed to foreshorten those meetings. "At least not until after seven," he amended.

"That's fine with me. I assume you can contact Miss Lavery?"

"I'll try."

Wallenstein turned on the car motor. He gave McMahon his card. "Leave the message with whoever answers and I'll pick you up at seven-thirty."

McMahon sent Nim a telegram when he got back to the rectory. Then he put her, Wallenstein, Muller, the whole affair firmly out of his mind and concentrated on his parish duties. But in the afternoon, he managed his promised visit to Phelan.

He had been transferred to a semiprivate room. A detective McMahon had not seen before was on duty. The occupant of the other bed was a dark Puerto Rican, one leg in traction. Phelan's eyes were closed. So were the detective's. McMahon visited first with the patient in traction and his voluble family who were trying vainly to keep their voices lowered. The man had fallen down an elevator shaft. That much McMahon was able to understand. He promised to tell Father Gonzales to stop by on his next rounds.

The detective opened his eyes when he heard McMahon's voice. McMahon laid his hand on Phelan's wrist. "Dan, are you awake?"

Phelan opened his eyes and McMahon waited for the detective to leave them alone. Phelan followed the man's departure until he was out of sight. "You'd think they'd have to go to the bathroom once in a while, wouldn't you?"

"How do you feel?"

"Thoughtful. I guess that's the word. I've been lying here trying to figure out what it's all about, my marriage, my life that's been handed back to me in a glass tube."

McMahon pulled up the chair the detective had been sitting in. But he put his foot on it, the newspaper under his shoe. He wanted to be able to see the man while he talked.

"I wonder what else they could give me in a glass tube?" Phelan said with grim humor.

"A lot of your trouble's up here," McMahon said, pointing to his head. "That's where to work on it."

Phelan glanced at him and away. "Priscilla told you all about us. I keep forgetting that. I guess I want to."

"She told me the problem, yes. But she wants to find the solution to it. She loves you."

"Enough to give me up, do you think? To let me go?" This time Phelan looked at him.

"You'd rather do that than try to fix the problem?"

"It would be better for both of us. She needs somebody like . . . him. I'm thirty-one years old, Father. I got as far as two years at City College. I had a scholarship to St. Victor's Seminary in Pennsylvania. My mother wanted me to be a priest so badly she turned me off it."

"How badly did you want it?"

"Quite a lot. But I was scared—this thing, you know. I was scared of getting kicked out, I guess."

"And now you want your wife to kick you out."

"That's about it."

"What do you do for a living, Dan?"

"I'm a grip—a movie stagehand when I work. And when I don't work, I'm a stagehand at home. You know what I'd like,

Father? Another chance at St. Victor's. Look, if God gave me another crack at life, why not at the seminary? It wouldn't be the first time the church annulled a marriage on those grounds."

"That wouldn't solve your problem, Dan, and now you've added a history to it with this mess."

"Look, Father, this mess has castrated me. No problem—except Priscilla."

"When you're up and around again, you may feel differently. Mind now, I'm not saying you should stay married. That's something we don't have to meet for a while. I'll make a bargain with you: promise me you'll see a doctor I've got in mind, and I'll make inquiries about St. Victor's. If there's any chance, the doctor's word would go a long way in your favor."

"Let me think about it," Phelan said.

"I'll think about it too," McMahon said.

"Priscilla doesn't need me. She's got the house and some other real estate."

"She needs you."

"How?"

"Let me ask you a question, Dan: why did you marry her?"

Phelan stared at the chart at the bottom of his bed. "Mother love."

"All right. There's lots of reasons people marry. For her it could have been vice versa."

"But for her it's not enough. She wants another kid so she'd have two of us." He pounded his fist on the bed. "For Christ's sake, Father, get me out of it!"

What God has bound together, let no man put asunder. McMahon said, "Take it easy, Dan. I'll stop by tomorrow."

12 ◆◆◆

Nim was watching for them from the window high above. When McMahon stepped from the car she called out to him and waved. The children of the street gathered around Wallenstein's black Jaguar and examined it with awe: a horse would scarcely have given them more pleasure. Wallenstein kept a tight smile going, more teeth than heart, McMahon thought. One youngster, seeing the priest go up the steps, jerked his thumb at the car. *"Agente funerario."*

He waited at the top step. The hall door was propped open. Nim came swinging down the stairs, a blur at first of yellow and red. She slowed down and became a picture. Which was not her intention. She was shy of the man who waited there. "Good evening, Father McMahon." He had come dressed as a cleric, fortified in the armor of God.

"I'm sorry I was such a fool," he said of their last meeting.

"Comédie Humaine," she murmured. "Did he recognize Stu?"

"No."

"Then what are we celebrating?"

"Experience. I'll explain that later. He hopes to be invited to view your painting—after dinner."

Nim tossed back her hair as they went out the door. "That will depend entirely on the celebration."

Some of the boys who had gathered around the big car whistled, and one, a gamin of twelve or so, skipped to the door ahead of McMahon and opened it.

Nim stuck her tongue out at him as she swept in.

Wallenstein, almost elegantly casual in dress and manner, proved himself a host of similar bent. He took them to the Trattore Gatti on Fortieth Street. Over cocktails which, he said, he drank with practiced disapproval—as an aside, he expounded on the martini as being much less a duller of the palate than sherry—he drew them out on their tastes in food. Nim liked everything except liver.

"Especially paté." Wallenstein said. "It's like coating your tongue with velvet." They agreed to his ordering the dinner for all of them.

With the second martini they had oysters, with the pasta, oil and garlic sauce only, the poor man's spaghetti, McMahon thought, but with such a difference here—a Soave wine. When the *osso bucco* came, Wallenstein unbuttoned the cuffs of his British suit and turned them up while he disjointed and served the knuckles. The maître d' came to watch a craftsman at work. The waiter showed the claret bottle, but McMahon missed the label, leaning over to hear Nim whisper, "Dago red."

What had they talked about? McMahon wondered afterwards. He became a little drunk, as much with the food and the talk as the drinks, and he thought about the pleasure it would give him to recite the menu in every detail to the monsignor. He talked of Lili Boulenger and the trenchant music she had composed, so young, so ill. Wallenstein was delighted with his use of the word, trenchant, to describe music. "I knew this dinner would pay off," he said, but with such deliberate overexposure of self-interest that it seemed ingratiating. And Nim talked about her tutelage as a painter under Stuart Robinson. Trenchant, she felt, was a good word for his approach to art also. Then, in a kind of haze, McMahon heard her tell of her father's test as to whether she was an artist. He had taken her to see a Professor Broglio with whom he had studied as a boy.

"He wanted to know by what right I thought I had sufficient talent."

"Sufficient talent for what?" Wallenstein asked.

"To measure in dollars and cents," Nim said. "That's how

to calculate a woman's work. A man's can be a long-term investment." She was lucid if not faultless in her syllogism.

Wallenstein clicked his tongue at the cynicism. "And what did the professor say?"

"He said, 'Lavery, how do you know you cannot play the violin?' 'I just know it,' my father said. 'Any fool would know it the moment I picked one up.' 'Any fool but you,' Professor Broglio said. 'If you had it in you to play the violin, you would know it. It would only be a matter of learning how.' "

"Bully for him," Wallenstein said. "Father McMahon has told you that I would like to see your painting?" Suave as Soave, McMahon thought.

He pondered these fragments of their dinner talk as they drove downtown again. Both the beguilement and the booze were wearing off. The man was an enigma, and the whole experience of having sprung an evening such as this from a five-minute intrusion into his parlor seemed a kind of madness. Will you come into my parlor, said the spider. . . . He glanced at Nim who sat between them on the wide seat. She was holding herself, prim as a spinster, as Miss Lalor might, touching neither of them with arm, elbow or thigh.

"Contrast, the only true measure of enjoyment," Wallenstein said, turning into Fifth Street.

"I don't think my neighbors would appreciate the esthetic," Nim said. "But it suits me fine."

"Aren't you afraid, living here?"

"Sometimes. But not of my neighbors."

"Of what?"

She shrugged. "I guess of the people who come here to get away from respectability."

"Like me, for example."

"I hadn't thought of you that way."

"You are right. I am not that respectable," Wallenstein said, all quite as though McMahon was not in the car at all.

Nim, with the forthrightness McMahon admired in her,

said: "Do you really want to see my painting, Mr. Wallenstein?"

He took his eyes from the street for a moment and turned toward her. "Only if you want me to."

So, McMahon decided, whether he liked it or not, he had to give Wallenstein the benefit of the doubt.

Nim turned on all the lights and took her canvases from the rack, setting out three or four of them at a time, propping them against the beams, a chair she had brought from the kitchen, and the rattan rocker. McMahon and Wallenstein sat on the edge of the bed.

Wallenstein said nothing at first, but got up now and then to see a particular picture from another angle. He moved the lamps around to his own satisfaction. "Set that one aside," he said now and then so that when she had shown some thirty canvases in all and said herself that it was enough, he had picked out ten that he wanted to look at again.

McMahon found himself looking inward more than at the pictures. He simply could not relate. It distressed him, for he had wanted very much to see and like Nim's work. Why, God knew, but suddenly he was remembering the picture of the Sacred Heart that hung in the Morales house, and he wondered what they had done with the effigy of Muller.

"Have you shown at all, Miss Lavery?" Wallenstein could not have been more formal, McMahon thought, forcing himself back to the present.

"In a couple of group shows on Tenth Street," Nim said. "I've sold three paintings."

"From this period?" He indicated the pictures he had wanted set aside.

"Yes, as a matter of fact."

"Fortunate buyers. You ought to have a gallery, young lady."

Nim sat down in the rocker, smiling a little, a remote look

in her eyes. It was not so much remoteness, McMahon decided then, but more, as she glanced toward her work, a private involvement. Finally she said, "I was thinking, I'm not that young—as your young lady, and I was glad I'm not."

A strange thing happened: At a little sound from Wallenstein, McMahon looked at him. The man's eyes were watery. Wallenstein got up from the bed and went to Nim, and taking her hand, he lifted it briefly to his lips.

"I weep for myself, you know that, don't you?" He blew his nose. "Well. I am not an entrepreneur, but I do have friends. Thank you very much. He turned abruptly to McMahon. "Shall I drop you at the rectory, Father?"

"Please stay, Father McMahon," Nim said, "just for a little while."

McMahon said, "I'll walk home, thank you. I often walk the streets at night."

"That's where the message is," Wallenstein said, and put out his hand.

McMahon felt the almost hurtful clasp of a hand you would have thought would go limp in yours. He remembered Brogan flexing his fingers.

At the door Wallenstein paused. "May I ask, Miss Lavery, what does your father think of your painting now?"

"My father?" Nim said. "I've never shown it to him."

"Why?"

Nim shrugged and thought about it. "I guess it's my own kind of revenge."

"I like you," he said. "I wish we could be friends."

McMahon was not looking at them, only listening, but he could imagine Wallenstein's eyes on him, the guardian of honor.

When she closed the door after he had gone down the first two flights, Nim went to the window and looked down. He would have looked up, for presently she waved.

McMahon stared at the one painting facing him, blues and black and many greens: a fish? a raft afloat in the changing sea? a coffin?

Nim came and stood beside him. "What are you thinking?" She nodded at the picture.

"Just wondering . . ." He did not want to admit his literalness.

But she knew it anyway. "What do you want it to be?"

"Is that the criterion?"

"It's as good as any," she said. "I'm that way about music, poetry. Stu used to say that most artists are conservative about every art except their own. Please take off the collar. I want to call you Joe." She put away the canvases. "I suppose you've guessed by now, I mixed them up, but every painting he selected came out of the time Stu and I were together."

McMahon put his collar and the stud in his pocket. "What do you think that means?"

"One of two things," Nim said. "Either he knew Stu and isn't admitting it—or he knows painting and I'm good."

"I prefer that interpretation."

Nim smiled. "I'd suggest a drink, but gin would be sinful after a meal like that. Wasn't it the most?"

"It was a lot," McMahon said, "and for once I didn't hate myself, piling it in."

"Why do you hate yourself for that? Penance?"

"Pride. The handout. You know, nothing but the best for the priest, free. I'm afraid pride is my hangup on the road to —wherever I'm trying to go."

"Nothing but the best for the artist," Nim said with a defiant thrust of her chin, "and he damned well deserves it. Mr. Wallenstein got his money's worth out of both of us tonight, Joe. Maybe he cries easier than I think he does, but tears aren't anything money can turn on." She laughed. "It can't turn them off either."

"That sounds Talmudic," McMahon said.

"My Jewish grandparents."

McMahon did not say anything.

"You want to get out of here, don't you?" she said after a moment. "Let's walk. Let's turn another table: I'll walk *you* home."

"That's not a very good idea at this hour," he said.

"For the love of God, don't go fatherly on me. Let's just walk." She took up her red stole from the bed.

When they reached the street Nim paused on the stoop and breathed deeply. "I'm getting expert at holding my breath. But I don't mind the smell so much, not really. And my place is nice, don't you think?"

"I think it's beautiful."

"Then why wouldn't you tell me until I asked you?"

McMahon shrugged. "I suppose because I never think such things are expected of me."

"And don't you ever do things that aren't expected of you?"

"I try not to," he said truculently.

"Forgive me. I'm a curious sort of person. I ask too many questions."

They crossed the street. Nim paused at the gates of the Orthodox Church. "Are you willing? I'd love to show you. I sometimes come here alone at night."

"Isn't it locked?"

"I know where the key is. Stu was custodian, remember? And after he went away, I asked for the job. The old priest was shocked, but I think he'd have given it to me if it weren't for his wife."

"All right," McMahon said. Waiting at the top of the steps while she groped the frame of the side door for the cutout where the key was concealed, he subconsciously reverted to the self he was at the moment of entry into the holiest of places.

Inside, the door closed behind them, they stood in silence

in the presence of—what? Scented ghosts of ancient saints, their icons palely glittering in the light of glass-bound candles.

McMahon said the words aloud: "I will go unto the altar of God, to God who gives joy to my youth."

"That's lovely," Nim said softly. "What is it?"

"It's the opening of the old Mass. It's been changed now. It's not the same."

"Why did they change it?"

"I don't know. Nothing's the same—for me." He realized as he said it that it was true: he was as reactionary about the liturgy as the monsignor. And what was liturgy but form, and therefore was it the forms only that bound him in his faith? "I believe . . ." he said aloud but again speaking to himself. He held out his hands and turned them slowly in the flickering light; he shaped a crescent, with the consecrated fingers, the miracle-making hands in which the bread became the body of Christ. He gathered them into fists. "I do believe."

"In God, in joy, in youth," Nim said. "What else is there?"

"Love."

"That's what it's all about, isn't it?"

"I hope that is so," he said. He would have liked to pray, but for that he needed a church of his own, and that he knew to be the answer to the question he had asked himself about the forms. He had expected, he admitted now, the flesh to be his greatest temptation, but it was not. How long he stood there, his fists tight against himself, he did not know. "I need to go home," he said. "Wherever that is."

"I understand. And you want to go alone."

"Yes."

"Then go," she said quietly. "I'll stay here for a while. I only wish I could take you where I think it is."

"God bless you, Nim. You are a fine person."

"I am blessed and I am damned," she said. "Good-bye, Joe."

"Good night. That's all. Just good night." He groped his way along the back wall until he found the door for himself.

The monsignor's bedroom door was open at the top of the stairs, his light on. Everyone else had gone to bed. McMahon tapped at the door. It was expected of him, whether he had been to a wake or a concert, if the light was on. When he went in the old man was propped up in bed, his breviary in hand, his glasses halfway down his nose.

McMahon sat on the bedside chair and told him about the dinner which, in the perspective of what happened afterward, had lost its flavor. He was primed by his audience, however, and he took an almost wicked pleasure in coaxing the old man into the savoring of it.

"Did they open the oysters at the table? I've seen that done, you know." Then later, marking his place in the book with the faded red ribbon and setting it aside, the old man said, "Joseph . . ." and hesitated as though not sure of how to get into a delicate subject.

McMahon steeled himself. He had explained the dinner party as growing out of Nim's feeling that Wallenstein might have known Muller. Now he expected to be questioned on Nim and his involvement in the whole affair. On the long walk home he had made up his mind and he was prepared now to say that he did not intend to see her again.

But the monsignor said, "Wallenstein . . . it's an old name in New York finance. Is he part of the family?"

"I don't know," McMahon lied. He knew then what was coming, and he tried to stem the rise of his own anger.

"A dinner like that, Joseph?"

"There is money," he said.

"I would think so—and a patron of the arts. Did you talk about music?"

"Yes, Monsignor."

"Some of our greatest benefactors are Jewish, you know. It's their way of making up."

"For what?"

The old man looked at him over his glasses. "Joseph," he said in a tone that warned of his temper's rising.

But McMahon said, "I thought it was we who were trying to make up to them these days."

"That was not my meaning and you know it, Joseph. If you had to deal night and day with the support of a parish as I do —yes, I'll say it to you—if you lived up to the talents God gave you for directing a parish instead of diddling on that piano in there, if you left the highfalutin music to Carnegie Hall and taught the girls the songs of their own people that'd keep them singing at home and off the streets, then you'd be doing a priest's work."

"Yes, Monsignor," McMahon said tightly.

"How much did you get pledged at the meeting for the renovation of the school?"

"Not very much. I'll have the report on your desk in the morning."

"You've the time to write me reports, but not to tell me what happened," the old man said, revealing the true source of his wrath, McMahon's remoteness.

"There's a question on whether the school should be renovated, Monsignor."

"Would they have it fall in on the children's heads?"

"They recommended the referral of the matter to the archdiocese with the reminder that a new public school is about to be built within two blocks of the parish."

"They'd close the school?" the old man said in slow disbelief. Then: "Over my dead body! I've put twenty years of my life into this parish, and by the glory of God, if I've seen it integrated, I won't see it disintegrated." His face was an apoplectic red. "For shame, Father McMahon. Will you sign the report?"

"It's not my place, Monsignor."

"I'm to sign my own death warrant, am I? Isn't that what it amounts to?"

McMahon, his temper overcome by sympathy, said: "You'd be humiliated, Monsignor, having to go to the cardinal for three quarters of the money. There's no hope of raising more than a quarter of it in the parish."

"There's always a way. It's a matter of asking the right people when you know them."

McMahon could say nothing except, "Good night, Monsignor."

"Good night, good night . . ." But when McMahon reached the door, he said after him, "You don't want to beg from a Jew. Is that it, Joseph?"

"Maybe that's it," he said to get out. "Or maybe it's just that I don't want to beg any more." He closed the door after him.

In the hallway he saw Miss Lalor bumping down from her end of the house in her tent of a bathrobe, her hair as wild as a bunch of heather.

"It's all over," he said, "Good night, Miss Lalor."

But she had to say her piece too. "The two of you, shouting at the top of your voices. Isn't there enough of that on the streets, Father?"

"Mind your own business," McMahon said. "You are not my mother."

In his room he took his collar from his pocket and looked at it and then put it away for another wearing.

13 ◆◆◆

Phelan's eyes were open when McMahon visited him the next day. "Father, I'll see that doctor of yours, if you'll get rid of this bedside companion for me." He jerked his thumb at the detective who lumbered to his feet as the priest came in.

"Listen, Mac. If you think this is my idea of paradise, you're out of your ever-loving mind. You won't talk and they won't shut up." He threw a mean glance at the family reunion around the other patient.

"Have a smoke," McMahon said.

"I was off them for two weeks when I got this detail. Now I'm hooked again."

While the detective was on his way to the door, Phelan, baiting him, said after him: "That's right, man. You're really hooked." To McMahon he added, the man now out of hearing, "He's got six kids and his father-in-law lives with them."

"Which is probably why he doesn't have eight kids," McMahon said.

Phelan grinned.

"You're feeling better."

Phelan said: "Did you get in touch with St. Victor's?"

"Not yet. You told me you wanted to think about it. I did call a friend in the chancery council. There are other places besides St. Victor's, if you're serious, Dan."

"If I wasn't, I wouldn't be seeing this doctor you talk about. Is he a Catholic?"

"Yes."

"There aren't many of them, are there—Catholic head-shrinkers?"

"They don't shrink heads. They open you up so that you can look for yourself."

"That's even worse."

"What have you been doing these last few days lying here?"

"Trying to look. That's the truth," Phelan said after a moment.

"All he'll do is throw a little light your way. It's not like going to confession, Dan. It's not that at all. It may turn out he'll show you a lot of the things you're feeling guilty about are not a matter for confession at all."

Phelan looked up at him sharply. "Man, you're the new church, aren't you, Father?"

McMahon smiled. "Half and half." He did not want to think about himself. He had spent too much of the night doing that, half on his knees and half on his back, fighting fantasies he could not elude. Even the scriptures fed them: Nim-Naomi—even that. And the whole Orthodox church in the darkness when he had supposed his temptation to be of the spirit and not the flesh, turning to it as a countertemptation—fighting fire with fire—he had remembered how he came to the church in the first place, Rosenberg's saying that Muller and the girl had probably made love there: a joke on the pawnbroker's part, but the joker in McMahon's house of cards which he had built, thinking it the house of God. He turned away from the bed and the thoughtful man lying in it and saw the family who waited the moment he would come to them. A chorus of responses to the merest of his attentions. "Look, Father!" A child wheeled the pulley supporting his father's leg, sending it slowly up a few inches and then down again, proving to all the man's improvement.

"Congratulations," McMahon said. "I'll come over in a minute."

"Father," Phelan said, "tell the police I have nothing to give them. I did not kill him. I'd have killed my wife first. Say that to them and they'll understand it. I've got it all figured out, lying here—the way they think."

"Dan, I'll get Doctor Connelly here as soon as I can. That

will impress them, that and another suspect if we can find one."

"Do you think I should go to confession?"

"Do you think so yourself?" McMahon did not want to confess him: he wanted to hear no more now about the sexual problems of man or woman.

"Maybe I'd better wait and see what this doctor has to say."

"Good man," McMahon said, and brushed Phelan's cheek with his knuckles. "Meanwhile, I'll go round and speak to Traynor."

He spent a few minutes with the Puerto Rican family, basking, refreshing himself in the warmth of their welcome.

As he was about to leave the room, Phelan said: "You won't forget St. Victor's?"

"No."

"I want it a lot, Father."

"I want it for you too." Then he added: "If it's what you want." But he realized as he said it that he wanted it for himself: he was trying to provide the church with a substitute priest.

He used the hospital phone to call the psychiatrist, aware of the dime it saved him. Dr. Connelly, hearing as much of Phelan's history as the priest could give him, agreed to stop at St. Jude's sometime that afternoon. McMahon did not like himself for thinking it, but he suspected it was the police aspect of the case that made it more attractive to a man as busy as Connelly. But then McMahon did not like himself for anything just now.

He walked from the hospital to precinct headquarters. Brogan and Traynor were in a meeting, but when the sergeant phoned up that Father McMahon was at the desk, he was told to send him up.

"We've got a line on him, Father," Brogan said as soon as he walked into the office. "No matter what kind of a copout he was trying to pull, you can't fool the FBI."

McMahon felt a sinking sense of disappointment, disillusionment, and he thought at once of Nim.

Traynor, with his quick, appraising eye, saw the change of expression on the priest's face. "What Brogan is saying, Father—we've been able to trace the victim through his fingerprints, an operation coordinated by the Federal Bureau of Investigation."

"Oh," McMahon said, his relief in his voice.

Traynor was not a man to be gratuitously kind. "It's got through to you, hasn't it, Father?—my enemy the cop, the FBI, the Establishment. That's what we are now, the Establishment. Brogan, how does it feel to be part of the Establishment on eight thousand bucks a year?"

"I don't get it," Brogan said.

"I don't either," Traynor said, "a priest on the side of anarchy."

From their first meeting, McMahon thought, he and Traynor had rubbed one another the wrong way: there was nothing reasonable in their reaction to one another; it was almost chemical, and what the lieutenant had just said came out of that polarization. He said now, "Lieutenant, we've got to live by communication. I believe that, and I don't think that's anarchy. For my part, I'm trying to communicate on both levels." Then, hypocrite or pragmatist, he added: "Monsignor Casey sends you his warmest, by the way."

"Thank you. Fill him in, Brogan. I'll get in touch with Wisconsin. Maybe we'll have a vacancy in the morgue if nothing else before the day's over."

At another desk in the office Brogan showed McMahon the teletype transcript of the information turned up through the fingerprint search. It had come through Selective Service records. The man they knew as Muller had been born in Madison, Wisconsin, August 13, 1925, Thomas Stuart Chase. He had served as a lieutenant during the occupation of Germany following World War II. Honorably discharged, he had returned to Europe on the G.I. Bill to study painting. His next of kin was listed as an aunt, Muriel Chase of Madison.

"Where do you go from here?" McMahon asked.

"Aunt Muriel, if she's still alive," Brogan said. "But with his real name to work on, we have several directions—Social Security records for one, if he ever worked for a living."

McMahon refused himself the indulgence of any more arguments with the police. "What about Phelan?"

Brogan shrugged. "He still won't talk. Ask the boss."

A few minutes later McMahon did go over the Phelan situation with Traynor. Not his interest in the priesthood, but the fact that he would be under the care of Dr. Gerald Connelly. A premature statement, but he was himself satisfied in the likelihood of its happening.

"You don't think he'll make another attempt at suicide?"

"I do not."

Traynor was more flexible than he had anticipated. "I'll go along with your thinking, Father." To Brogan he said, "Cancel the detail. But when he gets out of the hospital, put a tail on him. Maybe his feet will talk if his tongue won't."

"Thank you, lieutenant," McMahon said.

"Thank *you,* Father."

When McMahon got back to the rectory after classes that afternoon he found a note tacked to the door of his room to call Detective Brogan. Whenever Miss Lalor was miffed with him, she pinned his messages to the door. Otherwise, she delivered them in person, deciphering them from amongst her shopping notes. He called Brogan.

"There's an Aunt Muriel all right," Brogan said. "She's on her way East now. I told her you wanted to arrange a funeral service and she said to go ahead. She'll take the ashes back to Wisconsin with her."

McMahon was stunned. The so-called memorial at the Morales house had been sufficient to the wishes of Muller's neighbors, and certainly to his own involvement. But Brogan had not known that. "I'll call Ferguson and Kelly," he said mechanically.

"Let me know when. Call me right back on it, Father. The boss wants a notice in the papers. It might turn up some interesting people."

And there he was, McMahon thought, being used again by the police. But that's what a priest was for, to be used by those who needed him. Even the police. Humility, Joseph. Your name saint was a humble man. Remember that . . . and he was used if ever a man was. He thought about that: he could not stem the flow of cynicism now. It was as though he had opened the floodgates. He tried to remember what it was exactly that Martin Luther had said at the end of his life about being powerless to close them; and even in the wake of this thought came further skepticism: Luther had not wanted to close them, but sentimental Catholic historians had perpetuated the legend.

He made the funeral arrangement for four o'clock the following day, Friday. And its being the first Friday of the month, he would have it announced at the morning Masses so that word would reach the tenement house without his having to go there. He called Brogan back.

McMahon sat on at the office desk, his head in his hands. Phelan was not the only one who needed help. Father Purdy looked in and asked if there was anything he could do. "Pray for me," McMahon said, and God must know it was an act of humility to ask the prayers of Father Purdy.

He composed a telegram to Nim and phoned it in: "Funeral Muller born Thomas Stuart Chase Friday 4 P.M. Ferguson and Kelly Parlor."

Where, telegraphing her about meeting Wallenstein, he had worried about the item on the parish phone bill, it scarcely entered his mind now.

After choral practice he went again to see Abel Rosenberg.

"I suppose I knew he was over there," Rosenberg said. "I have made some notes. Words on paper, that is all. A man's conversation, it is not the way it looks in books." He took a notebook from the middle drawer of the desk and opened it. "For example, I have said here: When you think of concentration camps, what do you see? He asked me that. And I said—and it was very hard for me—I said, I see my sister Ida and the children, and I do not look. So, he said, put yourself there and look. And he described to me the filth, the stench, the wire, the degradation. Oh, yes. He was there. And he made me transport myself there. All right, I said, I am there, I am there." Rosenberg's whole frail body quivered with the recollection of what another man had made him conjure. He did not look at the priest, he stared ahead, conjuring against the background of the desk with all its cubicles. "I remember saying, each one of these"—he put his finger to one after another of the cubicles—"is a bunk and there are so many of us our bodies touch, always touch and stick and stink. And he said, what are you doing? You are there, but what are you doing inside you? I am surviving. I am thinking of a red rose, a red rose I once gave my mother on her birthday. Describe it, he said. And I described it, the silken petals, the color of heart's blood, the leaves, the fragrance, the thorns . . . And he said, That is art and where it comes from. And so we had a cognac and I was an artist. Will you have a cognac, Father?"

McMahon was a moment realizing that the last words were addressed to him. "No, thank you, Mr. Rosenberg."

"What else?" He looked again at the notebook. "The picture Hitler wanted. He told me about a painter he knew who had won a prize in Austria and how one of the German collectors so admired it he ordered the gallery to send it to the Fuehrer for his private collection. The artist himself went into the gallery that night and slashed it into a hundred shreds. 'Why did he do that, do you think?' Gust asked me. 'It is obvious,' I said. He did not want Hitler to have it.' 'Or was it this, Rosenberg? If the picture was good enough for Hitler, it was no longer good enough for him? That is why I would have destroyed it.' Something like that, he said."

McMahon said: "Do you think that's what he did? Destroy his own painting?"

"The same thought has come into my head, Father. And I think it is possible that he did that."

"This changing of identity, dropping out and starting over."

"A pursuit of absolute beauty," Rosenberg said.

"That's enough to destroy a man in itself," McMahon said, and thought again of Muller's words, I took the knife away from him. Had he wanted to die? Had his killer done him a kindness in those terms? Was that the meaning of his silence, his refusal to name the man? "No," he said aloud, getting up with Rosenberg because someone had come into the shop. "I don't believe he wanted to die."

"It was the opposite, Father. He was destroying himself because he wanted to live."

"I found the girl, by the way—Nana Marie. You will like her. I think she'll come to the funeral."

"Then I will come also, and you will play Bach and Mahler after all."

That night, after confessions were over and the church emptied of people, he took the score for "Song of Earth" which he had borrowed from the library, and his own copy of Bach's "Jesu, Joy of Man's Desiring" to the choir loft and practiced them on the organ. When he came downstairs he found the monsignor sitting in the back pew. The old man got up, steadying himself on the bench.

"That was beautiful, Joseph. I am sorry for what I said last night about your diddling on the piano."

McMahon said, "Give me your blessing, Monsignor."

He knelt and the old man touched his forehead, crossing it with his thumb and said the words in the Latin beloved to both of them. McMahon kissed his hand.

14

Only a few people came to the funeral, Mrs. Morales with Carlos, the boy in his first communion suit, Mrs. Phelan came with two other women of the building. McMahon realized he had not seen her during his recent visits to the hospital. He said the obvious by way of greeting, "Dan is much better."

"You saved him, Father," she said, a little curl of irony at the corners of her mouth.

Rosenberg came, but Nim had not when it was time to begin the service. The newspapers were represented. A photographer snapped a picture of Traynor and Brogan when they entered the chapel with a woman McMahon assumed to be Muller-Chase's aunt. She was large—not stout, but big-boned—fairly on in her sixties and well-dressed in a way that fashion would not interfere with. Traynor introduced her to the priest whom she acknowledged as "Reverend." Neither cordial nor aloof, she was, McMahon felt, as cautious with her commitments to people as her nephew had been casual with his. Out of the numerous available chairs—the casket was discreetly out of sight behind curtains in the apse of the room—she chose to sit next to Rosenberg, and McMahon mused on what other circumstance under the sun would have brought two people of such disparate backgrounds together. Rosenberg got to his feet at her approach and then sat down again when she did.

McMahon read from the Psalms and then from the Book of Ruth, the passage including the words, "And thy people shall be my people." He thought of Nim, but that was not why he had selected it: it led him into what he knew of this man who was loved by the strangers whom he chose time and again as his people. He spoke briefly and then said that at the request of a friend he was going to play some music. Before he had reached the organ at the back of the chapel Brogan and Traynor left. As he started to play, Nim came in and sat down.

Miss Chase turned in her chair to look at the girl. McMahon was reminded of all the matron ladies who turned purposefully in church to show their disapproval of latecomers, and there flashed through his mind the picture of this woman in some Fundamentalist congregation where, for a lifetime, she would have sat in the same pew every Sunday and after every service complimented the minister at the door on the excellence of his sermon.

Mrs. Phelan, Mrs. Morales and the other women of the building fled the chapel as soon as McMahon got up from the organ—like Catholics at the *Ita, Missa est,* Go, the Mass is finished. And he remembered the impish glee with which, among all altar boys, he too had sung out the Thanks be to God. In the presence of the dead he almost always thought of childhood. There came back to his mind then that last picture of Muller, turning his head in on his wounded self.

He moved quickly to Nim and took her to meet Miss Chase who was introducing herself to Rosenberg as they went up.

"Only a friend," Rosenberg said of himself.

"Miss Chase," McMahon said, "may I present Miss Lavery, another friend of your nephew's?"

She appraised the girl with one swift glance, wise eyes; not unkind, just satisfying herself. She murmured the amenity and turned to McMahon. "It was a lovely service, Reverend McMahon, but I had thought my nephew to be an atheist."

"He may have been, but I don't know any prayers in that language."

Nim and Rosenberg were amused, but Miss Chase followed her own train of thought. "The Chases have been Congregationalists for generations. But Tom made his own choices from the cradle—if not to the grave." Her expression showed the turning off of futile memories. "There were very few I approved of and I see no reason to be hypocritical about it now."

"How long has it been since you saw him?" McMahon asked.

"Over twenty years. He came home briefly after the war. But he wanted to live in Paris. So, I let him go. Not that I could have stopped him. But it was a painful letting-go—the last of the Chase name." She stood a moment, making up her mind whether to go on. Another quick glance at Nim seemed to decide her. "I was his guardian and I consented to give him the money left in my trust. We had argued it at some length. I borrowed on the securities rather than sell them, and finally I was able to put the check for the full inheritance in his hands. Whereupon he tore it up and put it back in mine. And left with an army knapsack on his back, although he had been an officer. But I was thinking, Reverend McMahon, while you spoke: does he have an heir? Will one now show up, or several perhaps? Is there someone who will come some day to see me and call me Aunt Chase—as he did—never Aunt Muriel?"

"I don't know," McMahon said.

She looked at Nim for whom the question had been intended in the first place.

"I know of none, Miss Chase," Nim said, "but then I only knew him for a year or so."

"Did you really know him, child?"

"I knew . . . someone," Nim said.

"We all knew someone," Rosenberg said, "but who did we know?"

"A kind man in any case," McMahon said.

"You have a peculiar notion of kindness, Reverend McMahon, if I may say so." And reaching out the gloved hand, she put one finger under Nim's chin and lifted it. "Look at this girl's face and tell me if it was a kind man she knew."

"He was," Nim said although her eyes filled with tears. "It was just that he went away—and he wasn't coming back."

"I said that twenty years ago, but I didn't believe it either. Now it is so. I'll give you my address if you would like to write to me, Miss Lavery . . ."

Nim shook her head.

"I think I understand," the older woman said. "Now I suppose I must speak to these newspaper people. I said I would after the service." She drew a deep breath. "Tom was a painter—but what did he paint?"

"He *was* a painter," Nim said. "You must believe that, Miss Chase."

"I do believe it. One of his few communications in all these years was a note saying he thought I would approve his present situation: he was teaching in the art department at Columbia University. . . . Which is why I mistook you, Mr. Rosenberg . . ."

"Just a friend," Rosenberg said again.

"I have decided to offer a reward," she said, "after discussing it with the police. I would like very much to have something painted by Thomas Stuart Chase. I do not like publicity. But in this case, I like the lack of it even less." She shook hands with Nim and Rosenberg, and then taking off her glove, she gave her hand to McMahon. But first she slipped a folded bill out from the other glove and put it in his hand. "You will have many charities. I have few."

McMahon murmured his thanks and put the money in his coat pocket without looking at it. His mother had always tucked something into that pocket when they parted. He tried to think of that and not of the fives and tens pressed upon him by strangers who were so moved by his participation in an intimate moment of their lives. It always nicely separated participation from sharing.

McMahon went to the chapel door with her and then returned to where Rosenberg was telling Nim of the art books left in his keeping. McMahon realized they had had to introduce themselves.

"Rightly they belong to Miss Chase," Nim said.

"Rightly? What is rightly? Young lady, he would not have approved that kind of shilly-shally. I wish you to have them."

"Thank you," Nim said. "I want them very much."

McMahon both dreaded and looked forward to this moment, just the meeting of eyes with Nim, but it passed with a kind of glancing off. "So now we know something we didn't know before, Columbia University."

"Crazily, I know someone there," Nim said. "Or did. Remember my telling Mr. Wallenstein about my father taking me to see Professor Broglio? That was at Columbia. And if that old man isn't dead, he might still be there."

"I would suppose, Miss Nim," Rosenberg said, "you think people to be a lot older than they are. I know I did at your age. May I offer you both a cognac? My brother-in-law is in the shop, but it will be time to close soon. Then you can take some of the books home with you."

McMahon consulted his datebook on the rest of the day's schedule. He had been going to see Phelan, but he had nothing more to tell him and he was now in Dr. Connelly's care. McMahon hoped that it was so. In the parish hall it was bingo night, and that could go on very nicely without his blessing. "I think it's exactly what we need," he said.

There was something in walking into the back of that shop with its old-fashioned desk, the chairs with their seats hollowed by a thousand sittings, the brass-knobbed cupboards, the green lamps, the orchestra of musical instruments temporarily put by, something that sent him back in history, that in truth released him from the tensions of his own place and time.

When Rosenberg had brought the cognac bottle and the glasses and took up his place in the swivel chair, his white hair fairly shone beneath the glow of the lamp. McMahon said to Nim: "Our friend thought of Rembrandt, sitting here where we are. What was it he said, Mr. Rosenberg?"

"That Rembrandt would have painted me one of his famous Jews."

"I'm part Jewish," Nim said.

Rosenberg looked at her gravely. "So who would he have painted you? What was the name of the woman at the half door?"

Nim shook her head. "No half doors for me."

She would not look at McMahon but the color rose to her face.

Rosenberg said, "So we drink again to our strange and lovable friend." He sipped and smacked his lips. "Our friend who got faces all mixed up in the roots of things. I have made another note or two. Rooting among the roots. He would like that, to play with words. He did not like nature, and I would make a bet with you: the aunt is an outdoors woman. A grower of roses maybe."

"The faces mixed up in the roots of things," Nim said. "That's Tchelitchew again."

"Tchelitchew," Rosenberg repeated. "That's the name. You will laugh at me like he did. I have written down Charlie Chan. Rembrandt I know, but Tchelitchew is a foreigner— to Abel Rosenberg. He went to see this Tchelitchew and he said it was a terrible mistake. He said a man should not try to run from the devil. He should open his arms."

McMahon and Nim looked at each other. "So there it is," she said, "but what is it?"

"That's police work, Nim. They'll go over that list with a fine-tooth comb."

Nim nodded. "I don't know why, but I don't feel revengeful. I just want to know about him, Stu . . . Mr. Rosenberg is right on how he felt about nature. It troubled him. It hurt too much. Something." She put her glass down carefully and folded her hands beneath her chin. She almost always sat in that hunched position. "I'm going to try to say something the way he said it to me once. You see, when I knew him it was this whole thing about the young people in the East Village he was trying to dig, the flower children, the nonviolent, the dropout. He started this way: I am an artist because I am a violent man. Most artists are. But the violence is inside them. They go among their brothers, crying 'Peace! Peace!' because they don't know any peace themselves. They keep digging at their own souls. They draw the world's infections into themselves, and in the furnace of their genius they try to burn it out, to get at the essence of man."

Rosenberg said, "You got him down perfect. It is me in the concentration camp all over again."

Nim said, "There's one more line, the one about nature: Walden Pond! I would rather look in the toilet bowl."

"Oh, my, my, my," Rosenberg said and bowed his head.

Finally McMahon said: "If we could reach this Professor Broglio, Nim, would you want to talk to him—just to find out?"

"If he'd remember me," Nim said.

"Any man would remember you, Miss Nim," Rosenberg said. "Just keep this in mind, young lady: our friend remembered you or we would not be sitting here together now."

Nim smiled at him flashingly.

"May I use the phone?" McMahon said.

"Please." While McMahon drew the phone to him, Rosenberg said: "Wasn't it beautiful, the music Father played?"

"Very."

McMahon hesitated, his hand on the phone. "Something strange went through my mind while I was playing—I think Brogan and Traynor's walking out started it—but I thought: I should have been a painter, not a musician."

"A composer," Rosenberg suggested.

"No, a painter. And I wondered if it was possible for an artist to choose the wrong medium." Not until he had dialed the operator did it occur to him that in the moment of saying that of himself he had quite forgotten his priesthood. It upset him and he avoided Nim's eyes though he felt them. He asked Information for the university number, and then said to Nim: "Tell Mr. Rosenberg what the professor said to your father."

A few minutes later McMahon had the information that Professor Broglio's schedule included a Saturday class from ten until noon.

"So," Rosenberg said, "you can go and see him again."

Nim said, "I told that story to Stu once. Professor Broglio, he said, and I said, do you know him? Only the name, Nim. Only the name. Now I wonder. He had to know him."

"Tomorrow morning. Enough till then," McMahon said.

"Always to him life was a hall of mirrors, bump, bump, bump," Rosenberg said. "A little more cognac before we close up shop. How Gust loved his cognac."

"And gin," Nim said, "and Scotch, I suppose, and Irish, and all the wines of Paris."

"It is not right for you to be bitter," the pawnbroker said. Bitterness is judgment."

"Self-pity," Nim said. "Is that judgment? I suppose it is."

"So your father consented to your studying painting," Rosenberg gently turned the subject.

"Not bloody likely. I consented and I've paid my own way since. And I'm glad. For all the bump, bump, bumps, I like it this way. "

"That is much better," Rosenberg said.

Nim looked at McMahon. "But I was thinking—what you said about the possibility of choosing the wrong art? For years and years I've wanted to be able to play the piano."

"So you will teach one another," Rosenberg said, a born fixer.

McMahon shook his head. "I am an old dog," he said, and as though the devil put the words in his mouth, he added: "I learned my tricks too long ago." Realizing what he had said and how it could be construed as mockery of his priesthood, he drank the cognac down like whisky and got up from the chair. His anger with himself, his situation, was too strong to conceal. But where to hide the darkness? He went to the bookshelf for want of any place to go except out the door. That was where he should have gone, but his feet would not take him there. "You are right, Nim. Things are so crazy," he said, looking at the books as though he was actually seeing them, and trying to ease matters for the others as well as for himself. "I coveted these books the first time I saw them."

It would have been easy for her to say, Take them, have them, and thus set off a round of banal protest to drown out the troubled moment, but neither she nor Rosenberg spoke. The conflict was his own, and they would not intervene to ease or to aggravate it. It was a kind of test of character, the ability to endure oneself in nakedness, and on her part, and perhaps Rosenberg's, to endure the nakedness of another's spirit. Whereas on his part, it had always been a matter of clothing quickly. The cloth . . . I can smell the cloth: the bartender in the homosexual bar. Phelan: he thought of him seeking the cloth, and he, the naked priest, all but urging it upon the man that he might clothe himself. He stood there, his back to the others, his hands in his pockets, and let it happen to him, whatever it was that was happening, the surfacing awareness that he had said what he had wanted to say, that there was no devil, only his own contention between will and want.

Finally he heard the clink of glass on glass and Rosenberg brought him the cognac. "Drink. It is the last of the bottle. But there is always a new bottle if a man wants to go out and get it."

McMahon, taking his hand from his pocket, drew with it the folded money Miss Chase had given him. He looked at it and then held it closer to the light to be sure: it was a hundred-dollar bill.

"Rosenberg," he said, taking the glass from the old man's hand, "where could you buy a piano for a hundred dollars?"

15

Nim waited for him in the church while McMahon returned to the rectory, changed his clothes, and made his excuses to Miss Lalor. The monsignor had been invited out to dinner and she hated to cook for the other two. She would give them eggs and set aside for the next night that night's menu. No family ever lived under such a matriarchy.

He found Nim walking up and down the side aisle, trying to see the stations of the cross, but the light was poor which, considering the artistic merits of the sculptures, was fortunate. An old Irish parishioner, half blind, was making the stations, her bones creaking at every genuflection. "Is it Father McMahon?" she said as he passed. "I could tell your step."

"How are you, Mrs. Carroll?"

"I've aches and pains," she whispered, "but otherwise I'm fine, thank God."

McMahon and Nim went out the side door where he had tried to catch up with her the day she first came to see him. On the street he repeated Mrs. Carroll's "Aches and pains but otherwise I'm fine, thank God."

"You know," Nim said, "I dig everything about religion except the church."

"And God," McMahon suggested.

"Sometimes I even dig him. But that's when things are pretty bad."

"Try it sometime when things are good. You'll like him better."

"Who needs him then?" she said.

They took a bus uptown to the warehouse off Broadway in the eighties where Rosenberg had called a friend. "Those old pianos you're always trying to get rid of, Michael," the pawnbroker had said with a wink to Nim and McMahon. "I have some young friends who might take one of them off your hands." And it had been arranged that the night watchman show them the dozen or so relics at the back of the storeroom.

The click of Nim's heels echoed through the huge loft. "People buying pianos today won't get the likes of these even on the installment plan," the watchman said, making the most of the chance to play salesman. "Of course, you have to have a house to suit them. They like more room than most."

He excused himself to turn on another light. A big man, on in years, he walked with a limp, and McMahon supposed he might have been a mover and incurred an injury.

Nim said, "Do you think everything in this place is alive to him? Did you hear, they like more room?"

"I hope they have life," McMahon said, thinking of all the classroom pianos in all the parochial schools in the city.

The watchman returned and they went on to where a row of bruised and battered uprights, some with keyboards open and some closed, stood against the wall.

"Look at them," the watchman said, "smiling up at you."

"The trouble is," McMahon said, "I'm not a dentist." He put his hand to one and struck a chord, or what would have been a chord when the old strings could make it.

"Some has more tune," the watchman admitted.

"So does the Liberty Bell." McMahon lifted the keyboard lid on the next one and tried it. The improvement in sound was slight, but the promise in each separate key as he struck it was greater. He stood back and looked at it in line with the others: it had the added advantage of being smaller.

"Maybe we should choose by the pound," Nim said.

"Believe me," McMahon said. He left the keyboard open and tried the others. None was better and some were worse. He borrowed the watchman's flashlight and examined the strings and mallets. Nim stood on tiptoe and looked in too.

The wires were like ripples in a sea of dust. McMahon got down on the floor and tested the pedals by hand.

"You'd think it was a horse," the watchman said.

McMahon looked up at him. "Do you have a name for it too?"

"Dulcimer," the man said without hesitation. "It's a word I always liked. I wanted to call our daughter Dulcimer—you know, Dulcie for short? But the wife wouldn't have it. Sharon. It's a nice enough name."

"Now," McMahon said, getting up and dusting his hands and knees, "how do we get it from here to there?"

The watchman's face lengthened. "You've not arranged that?"

"There wasn't anything to arrange till now," McMahon said.

"True, true. What floor do you live on?"

"The fourth," Nim lied.

"The fifth," McMahon said.

"No elevator?"

"If I had an elevator I'd be buying on the installment plan," Nim said with asperity.

"Where do you live, Miss?"

She lifted her chin. "On Fifth Street near Avenue A."

"I was afraid it was San Francisco." The watchman took the weight off his feet, half-sitting on the closed board of the next piano. "We've got some black boys working for us who might moonlight it on Sunday if I can get hold of them. I'll phone from the office. But five floors. The young lady was right. I'll tell them four and you tell them the fifth."

"How much do you think it will cost?"

"The last time, fifteen dollars per man and twenty for the truck. Four men."

"Eighty dollars," McMahon said. Then: "Isn't there a bench or a stool goes with it?"

"No, sir. Them we can sell for money."

On the street again, the delivery arranged for Sunday morning, McMahon said, "Now we have twenty dollars to spend on dinner. Where shall we go?"

"The Brittany," Nim said. "It's cheaper than most and I like it better. Do you know where it is?"

"Yes," McMahon said. Nothing more: it was within a few blocks of the rectory.

"Or better. There's a fish house on Third Avenue in the forties," Nim said, trying hard not to show her realization of why he had been reluctant about the Brittany.

"And we can have a drink at Tim Ryan's," he said. "He's an old friend of mine."

"And broadminded," Nim said.

McMahon did not answer.

"I'm sorry. Don't priests ever take girls to dinner?"

"You know the answer to that as well as I do, Nim. An agreement between us—no games?"

"No games," she repeated, and took his arm as they crossed the street toward Central Park. "But it is a game all the same, and I'm playing it even though I know I'm going to get hurt."

He was aware of her hand on his arm, more than aware. "Nobody's going to get hurt," he said.

"Let's just say nobody's going to cry. Will you go with me to see Professor Broglio?"

"I can't," McMahon said. "I have dress rehearsal in the morning for Sunday's concert."

"But by noon couldn't you get to Columbia?"

"I can try."

"Could I come to the concert Sunday?"

"It's in the school auditorium at three," he said. "I'll write you a pass. Otherwise it's two dollars."

"Two dollars!" Nim said, and then, "Forgive me."

"Toward the building fund."

Before walking through the park, Nim took off her shoes. "I'm quite primitive, you know."

"Savage," McMahon said, and then walking hand in hand, finding one another's hands by some mutual impulse he certainly was not going to examine then, they managed to throw off the gray mood that had come on them with the choosing of a restaurant. But McMahon could not take her to Tim Ryan's —for her sake, not his own. He could see the Irishman's cold blue eyes measuring the woman who consorted with off-duty priests. Broadminded Tim.

They drank where they ate, and it was better, the settling in to be themselves and with a waiter who calculated his tip by every drink and instead of hurrying, urged them to take their time—a public-private place.

Nim said: "If tomorrow you could go any place in the world you wanted to, where would it be?"

"Somewhere by the sea—with a long shore and no people anywhere. The sand would be damp and hard so I could run on it. I love to run. I had a dream once—it was as though I was chasing myself along a beach, trying to catch up."

"Did you?"

He shook his head.

"You love the sea and Stu hated it, except for the things beneath it. Protoplasmic man."

McMahon thought of the painting of Nim's, one that Wallenstein had admired, the last she had put away after he left. The undersea quality was in it. "You knew him well, Nim. Maybe better than you know."

"How do you mean?"

"That painting of yours, the one you asked if I liked?"

"It was the last thing I did while he was with me."

"I was just wondering if it might have had to do with his going away from you."

"I was becoming too much him to him, is that it?" she said after a moment.

"Or he was becoming too much you."

"I'd never have thought of it that way."

"Only in isolation would he have had what he needed of himself. I don't think you'll agree with this, but whatever he called it, I think he was looking for God." The thought of Phelan skimmed through his mind: something of this had come through to him too.

"Or the devil. Maybe he wanted to kill the devil, and I said he would shake hands with him."

"The world's infections," McMahon mused, "into the cauldron of his genius."

"Now that's spooky," Nim said. "When I told you that, that's the word I should have said, cauldron. It's the word he used, but I couldn't think of it. I said furnace. There are times you're very like him, Joe."

"There are times I even feel like him, whatever that means."

"I know what it means to me," Nim said. Her face, when he looked up, had become bleakly sad.

"Not tonight, not tonight," McMahon said. "I think we can manage a cognac. Then I've got to go home."

16 ◆◆◆

"It may seem wrong to you, but I do not feel grief for him, the way you tell his dying." He sat, a small man, deep in his chair and closed his eyes for a moment. His head was as bald as Picasso's. "Even when I knew him he lived on the edge of life, daring the step that might take him over. In a way it was me who brought him into the department. A mistake. Or was it? Who knows what is a mistake? Add up all the mistakes of my life and I am a success. Add up my successes and I am a failure. A man should not be in one place for thirty years. And if there is anything we can be sure about Tom Chase, he did not make that mistake. His name was a natural gift."

Nim said, "Professor Broglio, please tell us everything you can about him."

The old man grunted and scratched the top of his head. "Everything. Even that has a beginning. It was the first and last time I was ever political and I would say the same for him, the Adlai Stevenson campaign. It was in the old Marcantonio district, First Avenue on the edge of Harlem. We worked in an empty store, artists and writers for Stevenson. It was quite insane, ladies in mink, painters in dungarees, a classless society. Ha! All that does not matter but that's where we met. It was afterwards I got to know him, his rage at the frightened intellectuals, that was the time. Everybody playing it safe—afraid to think anything new, much less to teach it on the campuses. So I said, you are not afraid: teach. Teach what? What you know, how to paint, how to draw. He was a fine craftsman. And I got him a class a week to teach drawing, here where everything is new, experimental." He spoke ironically, hardly lifting his heavy-lidded eyes. He got up from the desk then, moving slowly with difficulty in straightening up at first, and McMahon thought he would have spent many years in damp places to have become so arthritic. He moved a chair out of his way and took a picture down from the wall leaving a white space where it had hung. "I had forgotten this. It is his."

He brought it back and put it in Nim's hands. McMahon looked at it over her shoulder: a sketch of a market place such as he himself knew from the Italian section on Ninth Avenue—the stalls and baskets, and an aproned man putting something in the scale: a simple drawing of a complex scene, but the shape of the man told it all, the rhythm of his lifetime was in the line of his back.

"He did not sign it," Nim said.

"But it is his all the same."

"He must have had a body of painting," McMahon suggested, "for you to have been able to judge his work."

"I do not judge work, only how people go about it. Judgment I leave to my superiors. That is survival and he came to hate me for it. To answer your question, Father, there was a body of work. I took the head of the department to his place, a loft on Amsterdam Avenue, and the boss agreed. There was an opening that February. He was given the assignment on condition that he try to get a gallery and show within a year. Backwards, you say, but the boss had that kind of confidence in himself—and his contacts. Some painters are born, others are made."

"And was there a showing?" McMahon asked.

"Never to my knowledge. But his contract was renewed for a year. The serious students liked him. The others were in the majority, and before the year was out, he walked out on his contract. They were studying for credits, you see, and he did not approve of working in the arts for credit. He was hopeless, pursuing the absolute. Anarchy. And now you tell me what happened to him—and you, Miss Nim—what he was like when you knew him, and I repeat, hopeless. He was always running, even when he was standing still. But we did enjoy life together, wine . . . and women. We enjoyed, and that is the thing. Maybe it is the only thing." He reached out a gnarled hand and patted Nim's. "Do you know, Miss Nim, when your father brought you to see me, how long ago? . . ."

"Eight years," Nim said.

"I almost told you both about him then, what it is like to want everything so that you settle for nothing. But it was too complicated for me to try to explain in the presence of a scientist."

"A scientist," Nim said with a touch of derision.

"You are still at war with him," the old man said.

"No. It's an armed truce, if it's anything."

"But you are painting?" He turned her hand over and exposed the fingernails.

"Yes, thanks to him." She set the drawing on the desk.

"No. Thanks to yourself only. You may have the drawing."

"I do want it very much," she said.

"Then it is yours." He turned to McMahon. "Does she paint well, Father?"

"I think so. Do you know a man named Wallenstein?"

"I know the name very well, David Wallenstein is an important collector, mostly in the Impressionists."

"I think this man's name is Everett," McMahon said.

"That would be the son. A dilettante—so I have heard, I forget in what connection."

"He also collects—Tchelitchew among others."

"Ah, yes. There is a show, or was recently. Tchelitchew is not my dish."

"Mr. Wallenstein thinks I'm good," Nim said, with that characteristic thrust of the chin.

Broglio leaned back in his chair. "Who is to say a dilettante does not know? It is only himself he does not know."

McMahon asked: "Do you have the address where Chase had his loft, professor?"

While the old man got a ragged address book from the bottom drawer of his desk and groped its pages, Nim said: "Was there anyone else, professor—anyone who might have known him later?"

He wrote the address in a painfully neat, almost exquisite script, and writing it, seemed to ignore her question. He gave the card to McMahon. Then he said, "Is it that important to you, Miss Nim?"

"Yes."

He drew a deep breath and let it out in a sigh. "The truth brings its own kind of pain—to teller and listener. Yes, my dear, there was someone else, a woman by the name of Andrea Robinson, and I ought to tell you, her husband was a trustee of the university, a connection not unimportant to my protégé—and myself."

Nim took it face on, not so much as flinching at the name, Robinson, by which she had known the man. "Mrs. what Robinson, professor?"

"Mrs. Alexander Brewer Robinson, and I would think the address is still Park Avenue."

Afterwards they sat, Nim and McMahon, on a bench and looked down from Morningside Heights. Neither of them had anything to say for some time. "Well," Nim said finally, "I do like that touch, Stuart Robinson—where his mind went when he made up a name to give me. It gives things a kind of continuity. This one I shall take on myself, Joe. Andrea Robinson, Park Avenue." She looked at the sketch. "I've just realized: Miss Chase would probably pay five hundred dollars for this. Did you see *The Times* this morning?"

"No."

"I must say the story brightened up the obituary page."

McMahon took the card Broglio had given him from his pocket. "I've got to get back," he said, "but I'll go down by way of Amsterdam Avenue."

"Confessions, visits to the sick, things like that?"

"Yes."

"Then I'll see you tomorrow," Nim said, much too brightly for the way he knew she felt. "Good luck with the concert."

McMahon left her sitting on the stone bench. He looked back before he turned the corner, but she was not watching him. She was staring out over Harlem, her elbow on her knee, her chin in her hand, the framed drawing under her arm. He stood for a moment looking at her, and thought that this was how it was going to be, as though every parting might be the last.

The address on Amsterdam Avenue had long since become a number in a vast red brick housing complex.

17 ◆◆◆

Phelan, propped up on pillows, the bed slightly elevated, his hands folded on the white sheet looked serene as a saint who had banished his demons, a picture, McMahon realized, out of romantic, holy-card lore. Neither saint nor sinner was free of demons so long as he was mortal. All of El Greco's saints were tortured men, the fugue of the artist's violence on himself.

McMahon stood a moment, reluctant to waken the man, and thought of Broglio's comments on Muller-Chase. He was wrong about him, however right he had made it sound. He had thought at the time, watching the painfully exquisite hand in which the professor had written the Amsterdam address, this old man has exhausted himself in perfectionism, but he has perfected the wrong things: he is half-blind, having never looked into the sun. Muller might have blinded himself, but he had looked it in the face.

Phelan opened his eyes. "I wondered if you'd given me up to Dr. Connelly, Father."

"I've been busy," McMahon said. It had taken an act of will for him to have come that afternoon at all.

"Well, I'm going home tomorrow," Phelan said without enthusiasm.

"Good. What happened to your neighbor?" The other bed was empty, stripped, the fresh linens in a heap, yet to be spread.

"They went out of here like an Easter procession. Father, if I was given my choice, I'd like to be like them. They came and kissed me good-bye, all of them, the children, the women, even the uncle. A lifetime couldn't have made us closer. And you know, I started out hating every mortal one of them—noisy, busy, and so damned happy. About what? Life, I guess. And I said to myself, Dan, my boy, these are God's children—the way he made them, and if you want to be God's deputy, you'd better find out what they're about."

McMahon, troubled at the ease with which he had allowed himself to be drawn into Phelan's fantasy of the priestly life—using the situation to compensate the Lord for his own backsliding—wanted to break in on the reverie. But that, he felt, would also be a kind of wickedness. He listened him out.

"Like a miraculous conversion, I was with them. Like a hand lifted a veil from my eyes. The hand of God?" He thought about it for a second or two. "Or was it just the police finally going away? Or me seeing Dr. Connelly . . . I don't know whether I like him or not."

McMahon was glad to get on the subject of Dr. Connelly. "You've only seen him what—twice?"

Phelan nodded. "It's important that I do like him though, isn't it?"

"You have to get through to one another. It won't be as easy as it was with your neighbors." McMahon indicated the empty bed.

"My neighbors," Phelan repeated thoughtfully. "I'm going home a different man than I came. Another thing—Pedrito Morales came in to see me yesterday. He said it was his fault, what I did." Phelan laughed. "A Puerto Rican heart—pride, temper, honor. You might say I am now an honorary Puerto Rican. And Lord, how I hated those people until this fiasco. He apologized for what happened that night in his house. And I've been thinking, Father—what it takes to gain respect. A knife or a gun. It isn't right."

"We've all got violence in us, Dan. It depends on what we do with it." And he told Phelan what Muller had said of the artist's violence.

Phelan grinned. "That's great, you know. Really great. Remember one of the first things I said when you came to see me—I thought he was a holy man? I'm not as stupid as I thought I was. Or even as crooked inside." He didn't say anything for a minute. "They don't know yet who killed him?"

"No."

"Funny, young Morales coming here. Maybe they think I killed him—which, in a crazy way, is what I wanted them to think when I pulled out the knife upstairs. Priscilla knew it. How I've hated her for seeing through me all these years. No, I've hated myself for what she saw."

"For what you thought she saw," McMahon suggested.

"All right. For what *I* saw."

"Did you tell that to Dr. Connelly?"

"Something like it. Understandable, he said. To him, everything is understandable. That's what scares me. I'd rather go to confession."

"And tell as sins the things you don't understand?"

"That's what you're for, Father. Whose sin you shall forgive, they are forgiven. Whose sins you shall retain, they are retained."

"Dan, it's not that simple being a priest today. Three Our Fathers and three Hail Marys and pray to the Holy Spirit just won't do any more."

"I'm a Holy Ghost man myself," Phelan said, again with a smile.

"I'd like to be myself, sometimes. But that's a copout. The young people want to know what is sin and why, and the answer that the Church says so is not enough. It's not even enough for me in my own confession."

"It's enough for me," Phelan said.

"Then you'd better do some more thinking about this change you want to make in your life."

"What I'd like you to do for me, Father—it's asking a lot but I'm going to ask it: go and see Priscilla and tell her what we've been talking about, that you think maybe I have a vocation . . . something like that."

"I didn't say I thought you had a religious vocation. I don't know whether you do or not, and I don't think you know it yourself."

"Tell her something!" Phelan's agitation was sudden, and it revealed the instability, the fear beneath the dream of serenity. In a word, he was a desperate man.

"Are you afraid to go home, Dan?"

"Yes!"

"You're still a sick man. She won't make any demands on you."

"You know all about those demands, don't you, Father? You could give her advice on how to get me into bed. One more expert on how to fix a marriage. I don't want it fixed!"

"Easy, Dan . . . there's time, and there's Dr. Connelly. I'll tell her about him if you want me to."

"I'll get out of here," Phelan said, "and I'll run, believe me, I'll run, and the cops will bring me down then for sure. And there'll be one nice clean thing about it that way—you can give me a Christian burial."

McMahon, heavy with helplessness, said, "I'll go and see her tonight. What time do you go home tomorrow?"

"By eleven-thirty. That's check-out time. Or check-in time, whatever way you look at it."

"I'll be here if I can and go home with you."

"Morales said he'd borrow a car."

"I'll check up on the arrangements tonight, Dan. Just rest and make your plans the way you think you want them now. A few prayers may help."

"I know one thing I'll pray for—just that they don't make any goddamned fiesta over me coming home."

Miss Lalor had all her priests home to supper that night, amiable as a hen with chicks even if she had never laid an egg. And McMahon had been wrong about the menu: he had forgotten it was Saturday night. By way of her pattern of association, Irish with Boston, and Boston with custom, and the custom of baked beans on Saturday night, she had made it her own tradition. Even the monsignor could not change it, and no longer tried, for it was no longer a matter that troubled him. But the other priests: what was more painful than sitting cramped with gas in the confessional box? McMahon ate bread and butter. Food no longer interested him anyway.

Phelan had rightly anticipated the fiesta plans for his homecoming. The women were hanging balloons and pinning flowers of papier maché to the curtains. The amplified guitar was going strong again, vibrating through the building. Happiness was a loud noise. McMahon had a glass of beer with the women, and then managed with no great grace to tell Priscilla Phelan that he wanted to talk with her. He suggested the room she had given Muller.

"I rented it today," she said.

She did not want to talk with him, McMahon realized. She was living her own kind of fantasy. But he had promised Phelan. "I'll come back a little later." He asked Mrs. Morales if Pedrito was home. He was and so, a week to the day later, McMahon climbed the dim, ill-smelling stairway again.

Pedrito and his friends were at the kitchen table, also drinking beer. A deck of cards lay in an untidy heap where they had wearied of the game. One of the boys was on the phone trying to reach a girl named Felicita. McMahon declined a beer but took the chair that had been vacated by the boy on the telephone.

"So you're going to bring Mr. Phelan home tomorrow, Pedrito. What time?"

Pedrito shrugged. "I got the car all morning."

"Make it eleven and I'll come with you."

"If I was him I would not come home," Pedrito said.

"Why?"

"She's got a pig in the back room."

McMahon was a few seconds figuring out what he meant. "A policeman?"

"That's how we see him, Father."

"How does Mrs. Phelan see him?"

"Ask her." Pedrito kept his eyes down. He picked up one of the cards and flicked its edge with his thumb, a snapping sound. "We don't like pigs in the house."

The boy on the phone said, "Felicita! It's me, Marcelo." To the others he said, "I got her! Hey, I got her!"

"So. It ain't television, ask her about the other girls. How many?" This from the boy at the end of the table.

"Shut up," Pedrito said to him. But he got up from the table and jerked his head to the priest to follow him. They went into the next room where the effigy of Muller had been laid out. "See, Father, the poor bastard—the cops keep pushing him around, and *zook!* next time for real." Pedrito pantomimed the thrust of a knife into his own heart.

McMahon glanced at the picture of the Sacred Heart, the shiny drops of blood. "How do you know he's a cop?"

"It figures, that's all. He looks like one."

"Pedrito, maybe it's none of our business. If Dan doesn't have anything to hide, maybe he'd like the idea of someone living there. He didn't kick Muller out, did he?"

Pedrito thought about it. "I think I get what you say, Father. But what if he's got something to hide? Somebody? The cops know better, he don't use a knife on Muller, not him."

"That isn't our business either, Pedrito."

"My friend's business, Father, that's my business. I never liked him, but he got guts, and pride and honor. I don't like to see him hurt any more."

"Eleven at the hospital," McMahon said.

"And I don't like pigs no more than they like me," the boy said over the banister.

McMahon went down the stairs, his head throbbing with weariness, with too much confinement in too small places, the smell of breaths, of bodies, of waste, and with almost the taste in his mouth of the little lusts of man. Every year he had gone home for a few days after the concert, but this year he would not go home. His mother had died and the house was sold. But he longed to breathe clean air and find God in the skies, to push out the walls of the tabernacle where men called priests had boxed their Savior in like a butterfly. And stifled him? The God-is-dead school was also dead. But Fair House of Joy, where was it? He tried to track the association, and it went straight back to Nim and the songs of Kathleen Ferrier, and how Muller had linked them in his mind. And tomorrow *The Bells*. And the next day, and the next?

Mrs. Phelan left the women and went out on the stoop with him. She sat on the parapet. "Look, Father, you've been very kind. But I think I can handle matters fom here."

Another mind your own business, McMahon thought. "I wanted to tell you about Dr. Connelly."

"I know all about him."

"Dan ought to keep on seeing him, even if it's expensive."

"For what?"

"He can help him with a lot of his problems."

She lit a cigarette. "I thought you told me I could do that. All those sessions we had, Father?"

McMahon felt his temper rise under her sarcasm. "A priest can be wrong. I should have recommended a separation then, but I didn't."

"But now you do. Is that it, Father?"

"That's up to Dr. Connelly."

"No, Father. Dan doesn't want him and neither do I. Dan's my husband and I can take care of him."

"All right . . . your husband, Mrs. Phelan, but not your child. Dan's got to have a chance to stand on his own feet."

"Father, I've never said this to a priest before. Thanks for everything, but go to hell."

18 ◆◆◆

It was a clear and sparkling day, that Sunday in New York, when even windows that were not washed looked as though they had been. The streets looked cleaner, and somehow there was more sky. Father McMahon did not go to the hospital: he could not, for the monsignor was not feeling well and asked McMahon to sing the high Mass at eleven-thirty. McMahon welcomed it as an act of God, lifting him out of the slough of despair. Afterwards he walked to the river and back and then fixed his own breakfast in the kitchen. He was hungrier than he had thought he would be, and felt unburdened, almost as though his failure with the Phelans had humbled him in a way that was pleasing to God. It remained now for some enlightenment to come upon him in the freshness of his spirit. And for that he offered up the afternoon's concert.

The monsignor managed to rise from his sickbed and Miss Lalor wore her Easter hat. Half the parish turned out, and neighborhood people who, having been persuaded to buy tickets, used them. Considering the fact that he and his girls were playing opposite the first Met double-header of the season, McMahon accepted the number of men in the audience as a tribute—to the girls. He did not look for Nim; he avoided looking toward the reserved-seat section. The girls, dressed in white, convened in the back of the auditorium. He could hear the clack of Sister Justine's frog over the murmur of conversation and shuffling of feet and creaking of seats. At three o'clock sharp he went onstage from the wings, and one hand on the piano from which he would conduct, he bowed, a curt, formal nod, and sat down before the keyboard. Both his hands and his knees were trembling and he wished profoundly that the whole thing was over. But that was not so either: he merely wished to be lost in the music, the self submerged, sublimated, and therefore exalted. He marched the chorus in to "Pomp and Circumstance" which he disliked but which they loved, and it was they who were going to make the music of the day.

And make music they did. To be sure, there was a mistake here and there, but only the trained ear caught it, and the joy of young voices singing burst over everyone like the soaring of spring itself. The audience clapped and stamped, and Father McMahon crossed the stage and led the soloists to the front. He shook hands with each of them, and then to both his delight and embarrassment, little Marietta Hernandez stood on tiptoe and kissed his cheek.

He went back to the piano and without a signal to anyone began to play "Bell Bottom Rock." The girls caught on at once and went into their thing, that uninhibited rhythm, each her own, the jerk and the halt and the shoulder shakes, with the little buds of bosoms popping up like buttons. Over his shoulder, the back of his head to the audience, he said to Sister Justine where she stood in the wings, "Curtain!"

And on that wild improvisation, the curtains closed.

People flocked around him in the basketball court afterwards, so many people, good, warm people all. Only then did he allow himself to search the faces for Nim's. The flash of her smile when he saw her gave him a stab of pleasure. He worked his way unhurriedly toward her, greeting, accepting the praise of everyone who stopped him on the way. Nim wore a green suit with an orange scarf at the throat.

"It was great," she said, extending her hand, "really the most."

He shook her hand briefly. Even gloves, he noticed. "It was a lot." The words had become a kind of theme between them. He introduced her to the other people who came up, to Mrs. Morales who was so proud of her two girls.

"Where's Carlos?" McMahon asked.

"Home. He wiggle . . . like the girls, you know at the

end?" Mrs. Morales covered her gold teeth in self-conscious laughter.

"You liked that part, did you?"

Mrs. Morales rolled her eyes and sidled away from them.

"Now I know what it's all about," Nim said.

"Did the piano come?"

"It certainly did. Now that *was* a lot. Doors came off hinges. People came out I'd never seen before. Words came out I'd never heard before."

"Where did you put it?"

"In the icebox," Nim said, and he laughed, the ridiculous image somehow appropriate. "Against the wall to the kitchen," she said then. "In winter it's warmer there and not so damp."

People were leaving them alone now. "You ought to go," Nim said.

"We're merely talking."

"I know, but they want you too."

"What about the Robinson woman?"

"I called her. Very Park Avenue: 'Oh, yes, darling. I knew him well. Do come and see me.' Joe—Father," she quickly amended, "I can't go alone. I thought I could. She invited me for cocktails and I asked if I could bring a friend."

"When?"

"Today—after five."

"Let's go then," McMahon said. "Free drinks for artists and the clergy. Why not?"

"It was presumptuous of me," Nim said.

McMahon said, "No games. Remember?"

Nim nodded.

He looked at his watch. "Five-thirty in front of the Metropolitan Museum."

McMahon wore his sport jacket and slacks and his black sweater. He did not have much of a wardrobe altogether, and little need until now for more than he had. On impulse, he put on the beads the flower child had given him in the Village. He took them off again and put them in his pocket until he got out of the house.

"Have a good time," the monsignor said as he went through the hall. "You've earned it, Joseph." When he had almost reached the door, the old man called him back to the office. "Are you planning to go up home this spring?"

"I hadn't thought much about it, Monsignor."

"Well, let me know. There's other places you could go. Mind, I'm not trying to get rid of you, but sometimes I have the feeling, Joseph, you'd like to be rid of us for a while."

"It's the spring, Monsignor, and I'm a little tired. That's all."

"Suit yourself, suit yourself." The old man went back to the ledger open on the desk. It was auditing time and McMahon understood the pastor's clinging to his bed that morning. He wished the vacation had not been mentioned, the fantasies it started in his mind: a few days' freedom and three dollars and eighty cents in his pocket. But then the monsignor always gave him a generous gift out of the concert money.

Other guests were arriving at the Park Avenue address at the same time as Nim and McMahon, and within the building —what in the old days was called a town house—a very large party was in progress. The maid taking wraps suggested the elevator although a marble staircase ascended to where most of the people were: laughter and the cadence of many voices and the tinkling sound of expensive glass. Some of the guests wore evening clothes.

"What have you got us into?" McMahon said, steering Nim toward the staircase.

"I'll say this for Stu: no rags to riches for him. No, man. Riches to rags. I'm scared. I don't know why, but I'm scared."

"That's why I'm here," McMahon said. "Hansel and Gretel."

"Ugh," Nim said.

"No. Humperdinck."

Nim made a face, but they too were able to arrive smiling at the balcony beyond which was the great living room where, as Nim said, Mrs. Robinson was having a few friends in to cocktails. People were gathered in clusters in the soft, sparkling light of the chandeliers, all making the sounds of the very rich, McMahon thought. There was not a guffaw in the house. It was like walking onto a motion-picture set, not that he had ever done that either, but it was the unreality of the genuine thing, nature imitating art. A waiter came up to them with champagne.

Nim took a glass, but McMahon said, "I wonder if I might have Scotch instead?"

"Yes, sir."

"Savoir-faire," Nim murmured. "I wonder how we find our hostess."

"We may never find her," McMahon said, "and you know, it has just occurred to me, you could make a way of life out of this, just walking in on parties with the invited guests."

"It's been done," Nim said. "Here comes somebody. Stay with me, Joe."

The woman came, dark-haired and tall, with a kind of angular poise to her gait. She was sleek in black silk, shoulders bare and bare-V'd to the navel with bell-legged trousers that swished as she walked. She put her glass in the hand with the cigarette and offered Nim her free one. "I'm Andrea Robinson," she said.

"Nim Lavery. And this is my friend, Joseph McMahon."

"You are nice," the woman said, giving him her hand as well as her long-lashed eyes. "Are you a painter? Do I know you?"

"Miss Lavery is the painter," McMahon said.

"And you don't know me either," Nim said, almost belligerently. Then, in retreat: "I thought we could talk—when you said to come today. I didn't know . . ."

"But we can talk, my dear. When you called it seemed like a voice from the dead. Which in a way it is, isn't it? What I mean to say, I was always insisting that Tom come to things like this —my perversity, for he loathed them. I wanted the fact established that I had a world of my own, and he could not have cared less."

McMahon was disconcerted at the ease with which she slipped into intimacy. As though it were a negligée. He was glad to see the waiter return with a bottle of Chivas Regal, a glass with ice and a bottle of soda on the tray.

"Just the ice," McMahon said, and he let the man half-fill the glass with whisky.

"When was it that you and he were friends?" the Robinson woman asked Nim.

"Last year and the year before."

"Were they good years for him?"

"I think so," Nim said.

The woman smiled showing the lines in her face that were concealed most of the time. "I'm sure they were." She touched Nim's arm with her fingertips, and McMahon for the first time could see Muller-Chase's attraction to her. "And were they good years for you?"

"The best," Nim said.

"Then I'm glad for you—and I suppose for all his women, now that he is gone."

"There's only one question really, Mrs. Robinson," Nim said. "Do you know where his paintings are?"

"I would have asked you the same thing. It's eight years since I last saw him. His loft was as bare as the steppes of Asia, all of his canvases crated. I asked him where they were going. To the bottom of the Atlantic Ocean. You do that, I said, and I'll jump in after them. But you can jump in and out of things so easily, Andrea, he said. And I think I knew then that I would not see him any more."

"Did you?" McMahon asked.

"Not ever. But I understand now I must have barely missed him at a gallery opening two weeks ago. Or did I see him and not recognize the beard? I have the feeling I did. And I can't help wondering if seeing me, he didn't duck out. I don't suppose we'll ever know now and I'm glad. You can't go home again. Maybe I wasn't home to him, but I was something for a few years." She became flippant again. "It's like a French novel, isn't it, mistresses comparing notes?"

A tall man came up behind her, too young to be her husband, McMahon thought. Or too young to be a trustee of a university. He kissed the back of her neck. She whirled around, welcoming the touch even before she knew whose it was. "Chet, how nice that you could come." To Nim and McMahon she said, "Darlings, do go in. I'll come soon and introduce you."

She drifted from them, taking the young man whom she had not introduced with her by the hand to greet more people emerging from the elevator.

"Let's go," Nim said.

"No. Let's see it through," McMahon said.

"There isn't anything to see through. Not that I want to look at any more. I'm not as brave as I thought I was." She moved toward the stairs, but the newly arriving guests blocked her way.

"Nim, are you afraid we might meet Wallenstein again?"

"Yes."

"And you don't want him to think you're chasing after him. You don't want to have to remind him of his promise."

"It wasn't a promise."

"It's the only kind of promise these people make, you silly girl."

"It's all so . . . decadent," Nim said.

"Don't you remember—that's what you liked about his house when we were there."

"But this is so much more. And I was hurt, her talking that

way about Stu." She looked up at McMahon. "You like it here, don't you?"

"I like the booze," he said. His glass was almost empty.

"It isn't only that."

"And I want to know."

Nim finished her champagne and before she could find a place to put the glass a waiter came and replenished it. "I like this too," she said, "but one shouldn't get too fond of it."

"You sound like Lee at the Battle of Fredericksburg."

They found a corner for themselves in the living room, a corner lined with books, leather-bound classics, uncut, McMahon suspected, and then contemplated himself for a moment, this habit of supporting his own morale by criticizing the mores of the rich.

"You're always surprising me, the things you know," Nim said.

"Like what?"

"The Battle of Fredericksburg."

"I'm a kind of fraud," he said, "making the most of bits and pieces. That I know, for example, because Lee actually said it of the Battle of Marye's Heights where the Irish Brigade was decimated. Charge after charge—absolutely pointless. They had no hope of taking the Heights. Utter madness. That was when Lee is supposed to have said, 'It's fortunate that war's so terrible. We might become too fond of it.' "

McMahon had become aware while talking of a gray-bearded man edging his way toward them away from the group he had been with. "What Lee said was—if you'll forgive the intrusion—'It is well that war is so terrible, or we should grow too fond of it.' I'm Jacob Burke. I'm a Civil War buff and I overheard. You don't mind?"

"No." McMahon introduced himself and Nim.

"What do you think Meagher had in mind? I'm pretty sure it was Meagher that led the charge."

"Thomas Francis Meagher," McMahon said. "One of the Young Ireland emigrees of 1848. And I would think he had in mind proving that the Irish were not cowards. They'd had a rotten war record, as you know."

"Yes, yes, and they had reason—the inequity of the draft and all that. Not quite the same today, but something to the comparison. But you see, the question I ask: was it not simply that Meagher was a bad general? Pride, of course, but a bad field officer. And then I suggest, it's what you've done with history—McMahon, you say—your inverted Irish pride that takes satisfaction in that debacle. The lost cause mystique, hopeless heroism."

"You may have something," McMahon said. "The reason I happen to know the story: it's a legend the old soldiers tell whenever the Sixty-Ninth Division gets together."

"It's a great pity there are old soldiers left to perpetuate such legends," Burke said. "That's how they make young soldiers." And with that he nodded formally and went back to his own group.

"Now him I dig," Nim said.

"I'm sure there are others if we just charge in," McMahon said.

"Remember the Irish at Marye's Heights."

If he got a little drunk it would be very easy, McMahon thought. But he did not want to do that. He wanted to know more about their hostess because the fact had emerged that she was the first person they had come on from Muller-Chase's past who might have seen him within a week of his death. He thought about Rosenberg's note on the Tchelitchew affair, Muller's: a man should not run from the devil. He should open his arms.

"Nim, can you see your friend Stu in this room?"

"You're psychic—or intuitive maybe. Something. That's what I was trying to do right now. He'd be going around, his hands behind his back, making faces at all the portraits on the wall. Then he'd go off through the house looking for young people. There aren't any."

"We haven't looked through the house," McMahon said.

"I suppose we could," Nim said tentatively.

"Wait till I get a refill." There was a circular bar in the middle of the room.

When he returned, Nim said, "What are we looking for?"

"I don't know exactly."

"She could have seen him at the opening. Is that it?"

"That's it."

"Did she go to it to look for him? I might have, you know. Sometimes I think I wanted to, but I was too proud to look for him. She'd have known his hangup on Tchelitchew."

"Let's just look and listen," McMahon said.

"Stop, look and listen," Nim said. "Or maybe just look, listen and stop. Maybe that's best."

McMahon said, "When I was going to high school I used to race a train to the crossing every morning. If I didn't make it ahead of the train I was late for first class."

"If you hadn't made it, I should think you'd've been late for the last class also," she said.

McMahon laughed.

"Did you always want to be a priest?"

"No. Just most of the time."

"Where did you grow up?"

"Upstate—near Albany."

"Family?"

"Father and mother, both dead now. Two sisters—one in Boston married to a doctor. One a nun."

"What did your father do?"

"He worked on the New York Central, a conductor."

"Working class. I like that," Nim said.

"I think you've said that before—working class and the filthy rich."

"So today I have the best of two worlds."

On the steps they met Mrs. Robinson. "You're not going so soon? We haven't really talked at all."

What she meant, McMahon felt, was that she intended them to go now. He sensed a subtle imperative. He chanced then a gambit that implied an intimate knowledge of her circle. "I thought maybe Mr. Wallenstein might be here."

"I thought so too. Not even a RSVP, which isn't like him. But it is really. Everything bores him."

"Except Tchelitchew," McMahon said.

Did she tense a little at the name? He could not be sure. "He must bore him too now, for Wally to sell off the drawings."

"Did Wallenstein know Chase?"

This time her reaction was direct: "Are you a policeman, Mr. McMahon?"

"Good God, no. But I'd better watch that. I sound like one, don't I? I'm a musician." He had said too much and carried the whole thing off badly and he knew it.

"I don't mind the police," Mrs. Robinson said, "but I do like to see their identification." She gathered the folds of her silken trousers. "It's been so nice meeting you both." She swept past them.

"Let's go," Nim said.

"I guess we'd better after that."

At the foot of the stairs they met someone McMahon had not expected to meet, Brogan, and another detective, giving their hats to the maid. Brogan arched his eyebrows and looked from McMahon to Nim and then back at the priest again. "It's a small world."

"Isn't it?" McMahon murmured.

"Do you know Father McMahon? Detective Tomasino." Brogan contracted his introduction.

McMahon had no choice but to introduce Nim to them both.

"Miss Lavery," Brogan said. "Mrs. Robinson told me on the

phone that you'd be here. It's just Father McMahon I didn't expect."

It had had to come sooner or later, McMahon realized. So did Nim, for she said: "It was I who went to Father McMahon. I thought the man he found might have been someone I'd known."

"Father was a good one to go to," Brogan said easily, "if you weren't going to come to the police." He intercepted the maid and asked her if there was a place they could talk by themselves.

She led them to a room beneath the staircase, the library. "Will I bring you cocktails, sir?"

"That'd be grand," Brogan said, having caught the lilt in her voice. "Bourbon for me and my partner. Scotch for Father, and the young lady?"

Nim shook her head.

Brogan tried hard to play the comfortable host. "Sit down, sit down. That set the Irish lass back a bit, me calling you Father."

"If she's never set back more," McMahon said, "she'll bear up under the shock."

"Why didn't you come to the police, miss?"

"I had nothing to tell you."

"You could've told us his name."

"I didn't know his real name either. Not then."

Brogan turned to the priest. "Father, you don't mind me asking, what are you doing here?"

"We were trying to find out if Chase left any paintings. Just trying to find out."

"Did you?"

"It begins to look as though he destroyed them all."

"If there ever were any. I have a feeling we'd know it by now after that piece in yesterday's paper. I'm no great authority, but it seems to me if somebody paints a picture, he wants people to see it. For Christ's sake, when I was a kid I took home every chimney I ever made smoke come out of. My mother's still got one of them hanging in the kitchen."

Nim laughed. So did McMahon. Brogan was human again. To McMahon he said: "Is this the girl—the name you mentioned in your statement?"

"Nim," McMahon said.

"What I'd like you to do, miss, come into the station house tomorrow morning and give us a statement—where you met him, where you saw him last, just the facts. That way, you're checked out, and Tommy here and I have done our job. It's not like we spend all our time on this case, you know. We've had four more homicides in the precinct since. And I've still got the notion we know where our man is."

"Phelan?"

"I'm not saying a thing, Father."

"Sorry."

"No offense taken. But I ought to tell you, we know you were at Columbia University yesterday too. Just don't try to do our work for us. Now here come the drinks, so let's relax and enjoy them." Before the maid left he asked her to send Mrs. Robinson in.

Nim and McMahon exchanged glances. "I don't want to see her again," Nim said, not minding that the detectives heard it.

"It will take her a while with that crowd," Brogan said. "Let Father have his drink. You must have gone over that gallery list pretty close to wind up here."

"No. We came by way of the university. Miss Lavery knew a professor there who had known both Chase and Mrs. Robinson." McMahon was aware of what he was doing: trying to give their hostess her own back for having mentioned Nim to the police.

Brogan took out his notebook and pencil. "The professor's name?"

Nim spelled the name for him.

"You one-upped us there, Father. We came dead-end in a

housing development on Amsterdam Avenue. Any other leads you can give us?"

"No. We're not looking for leads," McMahon said.

"We're trying to get hold of your friend Wallenstein again, by the way. But he's gone off to some island in Maine. Drummond Island. Ever hear of it, Miss Lavery?"

Nim shook her head. "Could we go now?"

McMahon drank down most of the drink.

Brogan went to the door with them. "Keep in touch," he said to the priest, and then to Nim: "Don't forget tomorrow morning, Miss Lavery."

When they reached the street Nim said, "Why did you tell her you were a musician?"

McMahon buttoned his jacket. A wind was rising. And he did not want to give her the answer which he gave nonetheless: "I guess I had in mind to avoid scandal."

"That's how I figured it." She turned one way and then the other to get her bearings. "I think I'll take the Lexington subway. Thank you for coming here with me—for the concert—everything."

"We'll both take the Lexington subway," McMahon said. "I want to see that piano in its natural habitat."

Nim was able to smile again. "To hell with Mrs. Robinson. Right?"

"Right."

There was a feeling of exquisite pain to that subway ride: it was for McMahon like going forward and backward in time at once, fragments of memory and ploys into the unknown which yet was deeply known, felt, instinctual knowing . . . his mother's wake, his sister's vows, Nim at the sacristy door, that moment among the Orthodox icons, his own prostration, the anointing of his fingers, the first trembling elevation of the Host and the terror that he would drop it . . . as though Christ had not fallen thrice himself on the road to Calvary. Whatever else drink did for him, it clarified his images and made him tell himself the truth: conscience and longing, Joseph, the spirit and the flesh. He and Nim did not talk: just awareness, and to him, every face in that subway car was marked with the condition of mortality: choices made for peace or for the promise of peace or for the abandonment of peace in the abandonment of promise.

They came up from underground at Astor Place into a virtual star of possible directions. On Sunday there were not many people, mostly students from NYU, heading back to their digs, and a few stray drunks cast up from the Bowery. Following Nim's lead, they walked past the old Cooper Union. His grandfather had gone there to classes soon after he arrived from Ireland. He thought of his mother again: how when things turned out badly she would say, "Man proposes, God disposes." Which was wrong: it was the other way around.

"Do you like pizza?" Nim said.

"Yes." Pizza was a very good thing to think about at the moment.

"Pizza with sausage?"

"Yes."

"Or anchovy?"

"Sausage."

Actually he did not care much for either pizza or sausage.

But carrying the box up the long flight of stairs a few minutes later, he breathed deeply of the pie's fragrance. Nim turned on the lights. "I don't have window shades," she said. "No neighbors up this high, unless they use a telescope. Besides, I need all the morning light I can get. I get up at six, you know, and work till ten."

"I get up at six most mornings," McMahon said, "but I don't quite like it."

"You would if your work was going well. Put the pie in the kitchen. There's a light switch just inside the door."

They were strangers, McMahon thought. He was in an alien world. He snapped on the light and caught sight of a cockroach just before it disappeared beneath the stove. He put the pie down carefully in the center of the white-topped table.

Nim had brought a lamp and set it next to the piano. "Beautiful," he said. The old wood had been brought up to a high gloss, its scars stained out. He could smell the furniture polish.

"I'm afraid to get it tuned," she said. "This way I don't seem to play so badly."

McMahon lifted the lid and struck a chord, the notes going off in all directions. "This way I don't play so goodly," he said.

Nim's eyes caught his and held them. "I want to say this, Joe. It's all right, your being here. I don't expect anything of you—just talk and company."

He held his arms out to her and she rushed into them. "Talk and company," he said at her ear, holding her close against him. She would be standing on tiptoe.

"I said expect, I didn't say want."

Such lies we tell to prime the truth, he thought, and then thought no more, just holding her tightly, kissing her hair, then her forehead, her cheek, her lips. The moment of self-repossession came when, to balance herself, Nim reached out and by inadvertence touched the piano keys. They drew apart and after a moment, McMahon said, "We've imported a chaperon."

Nim looked down at the piano and struck one note, then another. She gathered her fingers into a fist and sprung them open again. "Of all the things I didn't need—was to get hung up on a priest."

"What about the priest?"

"Yes. What about him? I want to know. I have a thousand questions." She walked away from him, hugging herself. "Please, turn on the electric heater. It's cold in here and I want to be comfortable. I want to talk. I want you to talk. I'm twenty-nine years old, and I know quite a lot—about men, about a lot of things. Have you ever been with a woman?"

"Yes. But . . ." He couldn't say it.

She said it for him. "I know. The other women weren't like me. What's his name . . . Tim Ryan's place?"

McMahon went to the small space heater in the middle of the room, a string of extension cords linking it to the wall outlet. He snapped it on and, squatting, watched the gradual rise of the orange glow.

Nim changed her suit jacket for a pullover sweater. She came and stood beside him. "And afterwards, what?"

"Guilt. An agony of guilt, confession, penance, starting over."

"Starting over what?"

"Trying to live up to my vow. Prayer, work, more work, walking, walking, music . . . and trying to keep Christ alive in me."

"Whatever that means," Nim said. She lit a cigarette and sat down on the floor, cross-legged, near the heater.

"It means . . ." He went to the daybed and sat on the corner of it, groping his mind for something she would not put down out of hand. "It means trying to be His representative on earth, to teach, to help people live decent lives, to encourage them when they fail, to assure them of forgiveness . . ." It sounded so hollow to him, so thin against the monstrous clamor of his own heartbeat.

"It's playing Christ," she said, "it's forgiving sin and jangling the keys to heaven. I'm not making fun of it, Joe. I know what it's like to get rid of guilt, to hurt somebody and then make up for it. And I think I dig what it means to kneel in church and say, 'God, I need you.' And if a man comes down from the altar and says, 'Go in peace,' and I know he's a good man, I'll go in peace . . . I think. In other words, I'm trying to say I know what a priest is to people who believe in priests. But what is he to himself?"

It was a question he had not been able to ask himself. A

priest should be nothing to himself, God's mirror . . . bump, bump, bump. "A vessel, a vehicle . . . No. He's like a doctor in a way."

"But a doctor's a man first. And he doesn't deny his nature to become a doctor."

"Nim, there is only one answer to your logic: if a man has to deny his nature to become a priest, he shouldn't be one."

"That makes more sense than some other things you've said. Do you think celibacy makes you a better priest?"

"No, but once I believed it. Once I believed everything the Church taught. Immutable truth, which, it turns out now, is somehow being muted. And I sometimes wonder what will happen if the forms are taken away, one by one."

"They won't need priests any more—just man and God."

McMahon tried to smile. "You make me sound obsolete."

"Do you really believe you have the power to forgive sin?"

"In God's name, yes."

"And so of course you believe in sin."

"Don't you?"

"I don't think I do. Or maybe one sin—dishonesty."

"You just spoke of another," McMahon said, "hurting people."

Nim, moving away from the heater, put her cigarette out slowly, determinedly. "That's not sin—that's a condition of being human."

"Believe me, so is sin."

She laughed and got up. "Would you like a drink? I bought some Scotch."

"Thank you, I would." McMahon followed her to the kitchen door. "I suppose we ought to eat that pie before . . . soon."

"Before what?"

He caught at the least false excuse available. "I saw a cockroach."

"Only one? They generally come in tandem." She opened

the refrigerator door and then closed it again. "Joe, do you want to go to bed with me?"

"Very much."

"Then let's turn out the lights and see what happens."

"I love you very much, Nim. So let's leave the lights on."

"You do sometimes surprise me," she said, going toward him.

19

McMahon washed and washed again his hands in the sacristy before putting on the Mass vestments. Guilt was one thing, but guilt with joy, how manage that? His mouth should be filled with ashes, but it was the taste of apples on his tongue. I will go unto the altar of God: the Mass was the Mass if said by a priest, however guilty the man . . . to God who gives joy to my youth.

He did not intend to go again to Nim. That had been implicit in their parting. But neither could he yet confess those moments to be sin. God have mercy on us. Christ have mercy on us. God have mercy on us. And in the front pew sat Priscilla Phelan who had never come to weekday Mass before to his knowledge. What angel roused her from her bed, what demon? She kept staring at him as though to read the gospel of hypocrisy whenever he raised his eyes. Poor child of pain, he thought, who comes to steal vengeance from the Lord. What are we at all, frail creatures of flesh pretending spirit, and compensating our grief in the torture of one another? Our father . . . lead us not into temptation, but deliver us from evil . . . Tell us what is evil, or silence the voices that accuse.

But Priscilla Phelan had not come to accuse. She waited for him in the courtyard outside the sacristy, and when he came out, she said, "Father, I am sorry for what I said to you on the stoop the other night."

"It was said in anger, and I don't blame you. I tried to help and made a mess of it."

"I made the mess, not telling you the truth in the first place. Now I've done something else, Father."

McMahon drew a deep breath and tried to put down the feeling of nausea rising in his throat.

"I rented the back room to a policeman. Don't ask me why. I guess because I didn't want Dan going any place he'd get in trouble. But now I don't know how to tell him."

"I don't think I can advise you in this, Mrs. Phelan. Except . . . Dan is sick in several ways, and you can't just set about healing him yourself no matter how much you want to. Let him see Dr. Connelly, and me. I'm not much help, but he talks with me."

"He won't talk to me at all. He keeps watching the door, like I was a policeman too. I made myself come out this morning. I gave him a sleeping tablet. I told him it was a tranquilizer. And I locked the door."

"Then for God's sake go home and unlock it and leave it open. He tried to kill himself. That's the easiest way out for him, don't you see?"

She covered her face with her hands. "What have I done to the man? What have I done to him?"

"It was done long ago, the worst of it—and the rest you've done to one another. With my help."

"Come home with me, Father. I'm afraid."

Down the same street they walked quickly as he had walked with Carlos, but when they reached the Phelan apartment, the man was sleeping like a child. They stood and looked down at him in the bed, the lines in his face all but vanished and a gentle smile on his lips. His wife stood, her arms folded, and out of the corner of his eye, McMahon saw the little cradling motion of her body.

In the living room McMahon said, "Tell him I came and that I'm working on his project. That will give him a kind of peace."

"A kind of peace," she repeated. "And where do I go for mine?"

"I just don't know." He left her sobbing quietly to herself in a corner of the sofa in the darkest part of the room.

During the morning he worked out a letter of inquiry to the

Franciscan Brothers which might serve Phelan as a guide in writing one of his own. He tried not to think of Nim, of the hour she would arrive at the precinct headquarters, where she would give her statement and how she would sign it. Nana Marie Lavery, he supposed. He had never even seen her handwriting. He tried not to think of her, but his mind would not stay servant to his will. He was going to have to get out of the city somehow.

He heard the front doorbell ring and Miss Lalor thump down the hall from the kitchen. A couple of minutes later she knocked on the door and handed him in a special delivery letter. He dreaded to open it and yet the beat of his heart bade him make haste: the neat handwriting and no return address. Then he saw that it had been postmarked the day before so that it could not be from Nim. He tore open the envelope.

The invitation was engraved. It read:

"You are invited to an exhibition of paintings from the collection of David Wallenstein at Mr. Wallenstein's home, 1090 Fifth Avenue on Monday, the fifth of May, at 4 P.M."

Along with the invitation came the personal card of Everett Wallenstein on the back of which had been written: "Do come. It will be to the interest of our protégée. My father has invited the most important people of the art world." Wallenstein had initialed the note.

And Nim would have gotten the same invitation. McMahon cleared his calendar for four that afternoon.

He saw her the moment he stepped from the bus. She was walking up and down in front of the building; the green suit again, the lovely green suit with the orange scarf at the throat, and the dark hair shining in the sun.

"The fates," she said with a wry smile when he walked up to her. "What is it all about?"

"We'll know soon."

"I was afraid you wouldn't come. I don't think I'd have gone in if you hadn't."

"That's why I came."

"That's why I came too," she said, responding to her own notion of his real reason for coming. "One wants to know—in spite of everything."

People were going into the mansion, not many and mostly men, but people you would identify with painting; at least, knowing of the invitation, you would, McMahon thought.

She gazed out over Central Park for a moment. "You see, Joe, I don't think I believed for a minute that Mr. Wallenstein was really interested in my work."

"In you, yes," McMahon said.

She turned toward the building and they started in. "No. Only in Stu . . . that's what I think now anyway. We'll see."

The old gentleman opened the door to them himself, offering his hand to McMahon. "I am David Wallenstein." He had a cherubic look, the plump pink cheeks that would have rarely needed shaving, mild gray eyes with none of the son's hauteur in them. One could associate him with painting, with collecting, with money, McMahon thought, but not with the making of money. "You are the personal friends Everett invited. I have been curious, I must admit. A priest . . ." Then he looked at Nim. ". . . And a maiden." The sensuous mouth puckered just a little while he gazed at her. He continued to gaze unashamedly even as he suggested that they join the other guests. McMahon had the identical reaction to him he had had to the son, possibly stronger. Where the son had looked at her with an aloof appraisal, the father's eyes grew indulgent, as though by grace of association with young Wallenstein she became a household intimacy.

In the living room beyond the foyer a dozen people had gathered, only two of them women. They stood, highball glasses in hand, in easy camaraderie. When Wallenstein left

Nim and McMahon at the door they were accorded the cool glances of nonrecognition.

"I don't like him much," Nim said of the elder Wallenstein.

McMahon did not want to talk of his reaction. "What's yours is mine and what's mine is my own," he murmured.

"Exactly," Nim said.

The walls were hung with the Impressionists which had made the collection famous. It was, someone said, the first time to his knowledge the old boy had opened his doors to the public. He would lend a painting now and then, but never any number of them at a time. Someone else reminded the speaker that the present company could hardly be called the public. Nim and McMahon accepted drinks and went from Monet to Redon. They were waiting as no one else there seemed to be waiting. Two more people arrived, also well known to the others, and then the elder Wallenstein came to the wide doors and spoke from the foyer. "Finish your drinks, gentlemen—and ladies. Then I would ask that you leave your glasses here . . ."

"The son isn't here," Nim said uneasily.

"Where was it Brogan said yesterday that he'd gone? Drummond Island? You did go to the police this morning, didn't you?"

"Yes. It was mostly about him they questioned me. I think they've made some connection between him and Stu. It could be Drummond Island."

A servant collected the glasses.

Wallenstein raised his hand for attention again. "You will indulge me in a few moments of personal reminiscence. Then we shall look at some more pictures. My son and I have many years of alienation between us, and I am as aware as he is of the feelings of this distinguished group toward his . . . public image. For much of his apparent superficiality, shall I call it, I am to be blamed. Today I atone. I should have liked myself to be a painter, and from the time Everett was a child I wanted it for him. I surrounded him with the best my fortune could buy—not the worst of the famous but, as you will have observed, the best of the lesser famous as well, most of whom have proven my taste to be as excellent as their own.

"It occurred to me one day when Everett had become adamant about showing me anything he'd done, that I might have crippled him with such perfectionism. I am rewarded in a way I am not sure I deserved. We were not friends for a long time even though he took his place in the firm. And when he came into his inheritance from his mother, he bought his own house and lived separately—very separately. I was not invited, ever. But one day I invited myself. So much is prologue. Please come with me."

"My God, my God," McMahon said. Nim's fingers were digging into his arm.

They followed the others who followed Wallenstein up the wide hall staircase, the latter among them exchanging whispers of disbelief. They would not easily discard their image of young Wallenstein, the dilettante.

The elder Wallenstein opened the door to a sitting room and threw on the lights. There were perhaps twenty paintings hung in the room. At a glance McMahon would have said they were the work of several artists, the difference in styles, colors and textures, the range from objects to abstraction: but this he and Nim already knew to be likely of the work of Muller-Chase.

"Please, one by one," Wallenstein said, "but this one first."

They gathered round him beneath a nude, the back of a black woman bowed so that her head showed only where the hair tumbled down over her arm, the forlorn shape of an abandoned woman. "This," the old man said, "is what I saw first that day when I presumed to invade his studio."

Nim cleared her throat and pulled McMahon down to where she could whisper to him: "E for ego, self . . . remember?"

Wallenstein then said almost the same thing: "It may seem academic, gentlemen, but Everett insists that E does not stand for his name but for Ego. Perhaps the self which I tried to deny him?"

They went from one group of paintings to another, only three or four in each group. McMahon could not really say whether he liked the nonobjective work or not. The experts were noncommittal, silent.

"He has destroyed hundreds of canvases, preserving only those which tell the truth. His philosophy of art is interesting: he believes an artist is only himself for a little while in any environment. Then he becomes corroded—with praise or blame, with fashion, with what is expected of him by, forgive me, you gentlemen. An artist should preserve only what he achieved at the moment of greatest self-demand, self-recognition. The rest turns him into a barnacle. But perhaps you will listen to him differently from now on, eh, gentlemen?"

Something terrible came into those mild gray eyes again when he said that: greed? arrogance? Power. Maybe that was it, McMahon thought. They were moved along to a painting that McMahon and Nim both recognized. She squeezed his arm. The sketch Broglio had given her would have been the forerunner to this work: the lines of the grocer's back as he stood at the scale were even more poignant in the painting. Wallenstein said of it: "How many generations of heritage went into that, would you say? In Germany a hundred and fifty years ago, my great-grandfather sold coffee by the measure."

"What does it mean?" Nim whispered again. "What does it mean?"

They had come almost full circle of the room to where a silken drapery hung over another frame.

"Everett hung this collection himself in the last week," Wallenstein said. "And yesterday, before he left for Europe, he brought this latest canvas. I have obliged his request that even I not see it until today." He reached up and with the plump, well-manicured hand, drew away the drapery. An envelope was in the corner of the frame, but McMahon saw it only peripherally as the old man took it in his hand. It was the picture that shook both him and Nim so that it seemed like one shudder running through both of them. He put his arm around her and held on tightly.

Realism superimposed on impressionism: the background tone was the smoky amber of the Orthodox church with the suggestion of the brown beams arching, and the pendulous chains shone through the muskiness, the paint seeming not yet dry, but instead of the glass-bowled candles that hung in the church, in the painting hung the severed heads of men. Nim hid her face against McMahon's arm.

He kept staring at the painting, wanting to keep his own balance. The chains seemed to sway a little as though the heads were restless still. The others of the group were reacting now, a murmurous consternation, but the attention shifted to Wallenstein himself. When McMahon looked at him he saw that the cherubic pink of his face had turned to parchment pallor, the hand in which he held the note trembled more and more violently.

A man stepped forward to help him. Wallenstein crumpled the paper and thrust it into the man's hand, and with the same motion pushed the man out of his way, and then the others, his arm stiff before him, and walked from the room with the autonomic step of one who has measured the distance to his collapse.

The man into whose hands he had given the paper uncrumpled it and read what was written, passing it on then to his colleagues. Finally it came to McMahon and Nim.

> My dear father: The work you have shown is that of Thomas Stuart Chase, whom I killed with a knife when he

agreed that I should perpetrate this hoax. E is for Ego, but Chase's, not mine. Mine I now propose to find in the way he found his. E.

They walked until they found a coffee shop on Madison Avenue, a quiet place, it being well before the dinner hour. They sat opposite each other, not looking directly at one another; or, when their eyes met, it was not in concern or involvement with one another. McMahon leaned an elbow on the table and shaded his eyes with his hand, seeing still in his mind's eye, that last picture, then finally the others. When he looked up Nim was crying.

"What I'm doing, Nim—I'm thinking of the paintings themselves. They're safe. That's the important thing. I almost think it would be the important thing even if the fraud had been successful. I'm wondering if that isn't how Muller himself felt at the end. He could have told me . . ."

She was only half-listening to him, trying to keep from sobbing.

"Why are you crying?"

"Because for the first time in my life I know what it is I'm hating and it feels good."

"To hate—or to cry?"

"Both."

"What are you hating, Nim?"

"That man—from the first minute I saw him. The arrogance—no, the voraciousness—that's it. It's what I ran away from, Joe, that soft, mine-all-mine possessiveness. The smugness, smothering. . . . God, how I hate it, him."

"I see. You've got a lot of things mixed up, Nim."

"Maybe I do. He's like my mother, and in a crazy way, his son is like my father—you know—Wall Street—a physicist making Dow chemicals to please her. You're right. I am mixed up."

"Why can't you just think about the paintings?"

"Because they're there! In his house."

"You'd think it was he who killed your friend, and it wasn't."

"But it was. He'd kill all of us. He's Hitler, Joe, the way Stu told it to Mr. Rosenberg. And if Stu could see his pictures hanging there, he'd go in and destroy them."

"But he'd be wrong, Nim. And I don't think he would. Or the paintings wouldn't be there now. The devil has many disguises, but the Christ in us knows them every one. It's an old saw, but the devil quotes scripture. Hitler might have liked Mozart. But Mozart was still Mozart. All I'm saying, Nim—love the pictures. Love what you *can* love, even if you can't love what you want to love. That's where the mix-up is, I think."

Nim leaned her head back in the booth while the waitress served their coffee. When the girl had gone, she said: "I'm beginning to get things sorted out. I'm trying to protect something. What?"

"A couple of things," McMahon hedged.

"Me. I'm protecting me. I wanted him—I wanted his child—you—the pictures—I wanted to protect them. But as you say, they are protected. It's all right now, Joe. As long as I know what's going on in me. It's when I don't know. That's when I'm in trouble." She blew her nose. "Will you go to the police?"

"Someone there will." He realized and said: "I have the note in my pocket. I didn't know what to do with it. I want to think some things out first. I don't like to think about young Wallenstein either, but I'm going to. I don't like what I remember of him, the way he looked at you that first time in his house. He wanted everything too."

"Not me," Nim said. "He didn't want me. What he wanted was to see if I had anything of Stu's."

"When I think of the coldness," McMahon went on, following his own train of thought, "with which he looked at the face of the man in the morgue and then came out and talked to me about you—that dinner he gave us—the tears in your house: what did they mean, Nim?"

"He said it: they were for himself. Like mine just now. Not remorse, not grief, not even anger. He'd have seen Stu in that painting of mine and he'd have known of himself that he wasn't even as good as me. Remember? At the door he asked how my father felt about my painting now."

McMahon thought about it. "I wonder if that was the moment when he decided on this note that's in my pocket. Vengeance over all."

"Uber alles," Nim said. "He would have loved Stu once I think, wherever they met, however often. But it would have been a kind of hell for him to keep those paintings when he couldn't make his own. And that father to run away from."

"Maybe that's it and now he thinks he's free. Sometimes I've thought that Muller didn't live to finish what he was trying to say at the end, I took the knife away from him. Maybe if he'd finished it, he'd have said: and given him a brush."

"Maybe, maybe, maybe. It's all over as far as I'm concerned. I'll be able to think of the paintings now. Stu is dead, and here I am with you, ready for another kind of dying."

"I'm going away for a few days," McMahon said. "Some place cheap by the sea where there aren't any people this time of year." He looked up at her suddenly. He had not meant to, but he did.

Her eyes were waiting for his, expecting them, the question in them. She shook her head.

20 ◆◆◆

When he got back to the rectory there was a message for him: Phelan had been picked up again by the police. He went at once to the station house.

Priscilla Phelan was pleading with the officer at the charge desk to be allowed to see her husband. She was on the raw edge of profanity. McMahon knew too well her mercurial temper. He called out her name. She ran to meet him.

"He'll be all right," McMahon said. "Tell me what happened."

"He went for a walk and that bastard in the backroom followed him."

"Where?"

"To the building where Muller was killed."

"It was not far," McMahon said, as though that were the important thing. He asked the desk officer to phone and see if he might go up to the interrogation room. Permission was granted. "I'll come back as soon as I can," he told Mrs. Phelan. "Just sit down quietly and don't make any trouble."

"Stay with him, Father. Don't leave him alone with them."

The point on which they were interrogating Phelan when McMahon entered the room was where he had met Everett Wallenstein. Phelan, slumped in the straight chair, circled in light, denied having ever met the man at all. While Brogan questioned, Traynor strolled over to meet the priest. "Counselor," he said.

McMahon said, "You know that Wallenstein left a confession to the murder?"

Traynor nodded.

Brogan, aware then of the priest, said, "Just tell us what you were doing in that building today, Dan."

"I wanted to see something upstairs. That's all."

"The poor devil could hardly climb the stairs," Brogan said to McMahon, "but he still won't tell us what he was looking for. His fingerprints were all over that goddamned cot we found up there."

McMahon said, "The painting, Dan? Is that it? The severed heads?"

Phelan looked in his direction, trying to see through the wall of light.

"Let's sit and talk without all that light. Let's try it that way, lieutenant."

So, with the priest's help, Phelan told of how he had first followed Muller, and then discovered the painting he was working on at the top of the abandoned building. "I'd go there sometimes when I knew he wasn't there and just sit on the cot, just to be there."

"Sexual fantasies, Dan?"

Phelan nodded. "But religious too, Father. I got them all mixed up."

McMahon said to the detectives, "This is Dr. Connelly's work, not ours." Then to Phelan: "Dan, were you in the building when he was killed?"

"I'd been there most of the night. That's where I went when I left Priscilla. It's where I was before too. But when I heard the voices in the basement, you could hear them up the stairwell, I got out as fast as I could. I didn't want him to find me. I swear to God I didn't know what was going on. I didn't know. And later when I found out what happened, I just kept away."

"But why the hell didn't you go home?" Brogan said.

"Because that's where my wife was," Phelan shouted hoarsely.

"Okay," Traynor said. "He's in no shape to do any more talking now. You can take him home, Father." He got up from the table around which the four of them had been sitting.

Brogan said, "To think you brought Wallenstein in to us,

Father, and I let him go. Drummond Island—and all the time he was skipping the country."

"Where's Drummond Island?"

"It's away out off the coast of Maine. His old man owns it."

McMahon gave Traynor the crumpled note.

When the police were gone he sat down again with Phelan.

"I'd like to go to confession, Father." He looked around the small, bare room with its smoke-grayed walls. "I don't think I can make it to the church."

"It's a long ways in your condition," McMahon said.

Phelan closed his eyes for a moment. "It's funny, you saying that, Father. I was thinking this morning of my own father, a tough brute of an Irishman. He was always trying to squeeze a dollar out of my mother, and she'd say, 'It's a long ways to payday, lover.' Always lover. And he'd say, 'It'll be a longer one, Tess, if you don't give me the dollar now.' Why do the Irish drink so much, Father?"

"I'm not sure they all do," McMahon said, "but I drink too much myself."

"Why?"

"It's a long ways to payday."

Phelan managed a crack of a smile. "I can't get over that family in the hospital with me. Love, it's the only word for it."

"They're real," McMahon said.

"Oh, yes. They know what it's about. No dark corners. You know, they'd kill for one another, Father. They wouldn't say it was right. But they'd say it was love."

McMahon drew a long breath. "Dan, I want to tell you something that's on my conscience: I was with Detective Brogan in the bar on Eighth Street when they picked you up. I'd gone out for a few drinks with him. I didn't know why we'd wound up in that particular spot until I saw you. You were informed on by a priest, Dan."

"I wasn't running away. Just staying away—in a place I felt comfortable."

"I know."

After a minute Phelan said: "It's all so rotten, so crooked, getting sex out of pictures."

"Dan, it's a very common thing. Dr. Connelly will tell you that. That's why I want you to keep seeing him."

"He can't change me."

"He can help you live with yourself—whether it's with Priscilla or not."

"Does it come with a guarantee?"

"No. Only confession comes with a guarantee. And even that is on your own bond."

" 'Nothing is given to man': I like that song."

"He's given some choices if he's got the guts to make them."

Phelan put his hand to his breast, to where the knife wound was scarcely healed. "I chose—and I couldn't even do that right."

"I've written a draft letter that might be some help to you. I'd suggest you write the Franciscan Brothers if it turns out that's still what you want when you get things straightened out."

Phelan gave a dry little laugh. "It was a dream—lying in the nice white sheets of the hospital. I should've done it long ago if I was going to. I'm not going to give the Lord back this rotten lump of clay."

"He made it, Dan, and he'd have known what was going to happen to it—even as you and I don't know now. Do you remember the story of St. Ignatius? A soldier, a dissolute man in the days when even the pope was corrupt."

"I don't feel corrupt. I say it, but I don't feel it. Another sin?"

"You are the judge of that."

"The new church again. I want the old one!"

"Then I'm not your priest, Dan," McMahon said quietly.

"Take me home," Phelan said. "I don't think I could find it on my own."

"Home is in you. And you've got to find it yourself."

"All right, all right, Father. She'll be waiting for me, and maybe that's all right—if we can stand the dark corners."

"Dear man," McMahon said, "you are not the only one with the dark corners. They're in every mortal one of us. Which is strange, if in the beginning—if there was a beginning and God said it . . . Let there be light."

Phelan looked at him. "You don't want to hear my confession, do you, Father?"

"No, I don't, but I will if you want it."

Phelan shook his head and lifted himself carefully from the chair. "Let's go."

21 ◆◆◆

McMahon, with police permission to visit it, left the next day for Drummond Island. The Coast Guard ferried him and the local police officers out. The caretaker, a part-time lobsterman, agreed to put McMahon up for a few days in his own cottage. After they had sealed and padlocked the big house, the police returned to the mainland.

There were no long beaches on which to run, only sandy coves when the tide was out, sheltered half-around by cliffs where the scrub pine was bent by the gales of centuries. But there was the sea and the sky and the stillness except for the gulls and the wind and the wash of the waves which were in themselves a kind of stillness. And the sun on his naked body. It ought all to have healed him, but it only salted the wound. He saw Nim everywhere. Sometimes he would look for an instant into the sun, and the afterglow against his eyelids became a golden cross, then the monstrance at the moment of elevation. To him now sign without substance, ceremony without worship, man without God.

He walked miles around the island, groping through the tangle of fallen timber and new growth, toward and away from the sea. The wild flowers of spring, violets and Mayflowers, and those with names he had forgotten bloomed toward the sun, and jack-in-the-pulpits came out in the shade. Jack-in-the-pulpits: the name went round and round in his head. One night before the fire after the lobsterman had gone to bed, he got up and took an old encyclopedia down from among the musty books. "Jack-in-the-pulpit: . . . the part that looks like the preacher is the slender stalk . . . enclosed in a leaflike growth which resembles the pulpit . . . a sounding board that extends behind and over. . . . The plant grows from a root filled with a burning juice . . ." My God, my God, why hast thou forsaken me? But you've got it wrong, Joseph. My son, my son. . . .

And from here Muller-Chase, loaned this refuge by his friend, young Wallenstein, had fled, according to the lobsterman. "He couldn't take it somehow. He'd pick up a tangle of seaweed from the shore and say, 'This is more like it, the slime of the earth and the sea, and the little faces looking out, making mouths for help.' And you know, neighbor, I'd look and I'd see those faces, just for the minute, mind you. But I understood him and I knew why he always had to go back to the city. It was lonesome when he went, but I said to myself, I'd be more lonesome where he's going."

McMahon knew that for all his longing for the sea and the shore, for the salt air and the birds flying, forever flying, he too had to go back to the city, to the little faces looking out of windows and making mouths for help. On the plane back a week later he decided he would see Nim once more and tell her about the island and the lobsterman. And then he knew it would be better to write to her. And finally he knew that silence would be the best of all.

The monsignor embraced him, and Miss Lalor put on her hat and went out to buy a steak. Purdy and Gonzales between them bought a bottle of wine for dinner, and Purdy, on the monsignor's instructions, took it from the slanting shelf of the liquor store and carried it home like a baby. All this they laughed at at the table, and Miss Lalor grew as red in the face as the monsignor. She had had but a sip of the wine. One of the Irish who did not drink, McMahon thought, and he wondered about Phelan.

After dinner, while the other priests went about the parish work, the monsignor and McMahon went into the study. The old man brought out a bottle of cognac. "Napoleon, Joseph.

I've had it since Christmas. You look fine—as brown as one of themselves. Did you spend all the money?"

"Most of it on the fare," McMahon said.

"Good!"

"But I saved the ten dollars I owed you."

"Keep it till payday. You'll want plenty of work now, Joseph. You'll need it, hard work that will tire you out. I've got it for you. I'm not giving up on remodeling the school myself. I was thinking of a bazaar in the summer maybe . . ."

McMahon sipped the cognac.

"No toast, Joseph?"

"To peace," McMahon said. He could think of no other.

"Exactly. It'll come to you after a while. I know from experience. You wouldn't think it to look at me now. But the good Lord knows there's times he has to wait on the frailty of man. Sure, it's all he has to make priests out of, mere men. And he needs them."

"Why?"

"The people need them and he needs the people or there'd be none of us on earth at all."

"We need one another," McMahon said. "That's all."

"That's the way you feel now. When you've started your penance and with your morning Mass you'll feel different. You weren't asked to say Mass while . . . you were away?"

"No."

"Good."

"I was alone on an island, Monsignor—except for a lobsterman." He looked the old man straight in the face. He saw the change in expression, the little shadow of doubt in what he had believed of McMahon. Or was it that he doubted McMahon was telling the truth now?

It was the latter, for the old man said, "Well, I'm glad you're back."

"No scandal," McMahon said with deliberate provocation.

Again the old man was not sure. "What do you mean, Joseph?"

"I gave no scandal."

"You don't have to say it out to me." The old man, reassured, poured himself a drop more brandy and was about to add to McMahon's.

McMahon covered his glass.

"Why won't you drink, Joseph?"

McMahon pushed the glass a little farther from him on the table. "You know, Monsignor, it has just occurred to me, I may never drink again."

"It's better than some things. A priest needs a drop now and then."

McMahon could feel the crawling urge to move in his neck, in his spine. "I think I'll go into the church for a few minutes, Monsignor, if you'll forgive me."

"Go, by all means. As the kids say nowadays, that's where it is. I'm beginning to catch up. But at my age, Joseph, you never catch up. You just step aside. I'll be doing that soon. I'd do it now—if I knew who was taking my place. I suppose they're right, not letting us know or having a say in the matter. There's an instinct in all of us—father and son. Go. I wish you felt like talking. I've missed our talks, but I was missing them before you went away, a long time before, even before that artist's death and that girl. There, I've said it. It's out in the open and we'll both feel the better for it. Go. The Lord's waiting, and I'm waiting on the Lord."

McMahon got as far as the sacristy door, but he did not go into the church. He took the subway downtown and walked through the West Village to the East Village on Eighth Street, then down to Fifth. Though he got no answer when he rang Nim's bell, he went up the stairs where the light was as frail as that in Mrs. Phelan's tenement. He knew, reaching the top floor, that she was gone, goods, pictures and all, for in the corner outside her door was the one piece of furniture she could not take with her, the piano.

He stood and looked at it for a long time, and then out the window to the Orthodox church, its simple façade with its complex cross, in the great hulking shadow of the old school building, and he began in his mind to compose a letter: "I will find you, Nim, so you may as well help me. There will be a place. I can teach music, you know . . ."